TONYAN

COURSE 1

RM 1-041

3-8321

HANDBOOK OF COMPOSITES

HANDBOOK OF COMPOSITES

Edited by

George Lubin

CONSULTANT

Formerly—Chief Scientist
Grumman Aerospace Corporation

Sponsored by the Society of Plastics Engineers

VAN NOSTRAND REINHOLD COMPANY
New York

To BEATRICE
CARYN
GLENDA
SHIRLEY
TERRI

Library of Congress Catalog Card Number: 81-10341
ISBN: 0-442-24897-0

Manufactured in the United States of America

Van Nostrand Reinhold Company Inc.
115 Fifth Avenue
New York, New York 10003

Van Nostrand Reinhold Company Limited
Molly Millars Lane
Wokingham, Berkshire RG11 2PY, England

Van Nostrand Reinhold
480 La Trobe Street
Melbourne, Victoria 3000, Australia

Macmillan of Canada
Division of Canada Publishing Corporation
164 Commander Boulevard
Agincourt, Ontario M1S 3C7, Canada

15 14 13 12 11 10 9 8 7 6 5 4 3

Library of Congress Cataloging in Publication Data

Main entry under title:

Handbook of composites.

(A Society of Plastics Engineers technical monograph)
A follow on text to Handbook of fiberglass
and advanced plastics composites. 1969.
Includes index.
1. Composite materials—Handbooks, manuals, etc.
2. Fibrous composites—Handbooks, manuals, etc.
I. Lubin, George. II. Handbook of fiberglass and
advanced plastics composites. III. Series: Society
of Plastics Engineers technical monograph.
TA418.9.C6H33 1981 620.1'18 81-10341
ISBN 0-442-24897-0

FOREWORD

The development of advanced composites, spanning a brief period from inception to application of only 15 to 20 years, epitomizes the rapidity with which a generation's change in the state-of-the-art can take place. This is in marked contrast to past history, in which it has usually required 25 years or more of research before a new structural material was technologically ready.

In the mid-1950's the U.S. Air Force identified the promise for early application of a new class of materials—advanced composites—and established its feasibility by the fabrication of raw fiber with exceptional strength- and modulus-to-weight ratios. The practical fabrication of boron and graphite fibers was essential to the development of advanced composite materials. The National Aeronautics and Space Administration (NASA) and the U.S. Air Force then sponsored research and technology programs, which led first to flight demonstrations and then to practical applications of advanced composites in production aircraft and space vehicles (starting about 1970). Thus, we have spanned the period from a laboratory concept to a practical production material in a shorter period of time than ever before achieved. In contrast, the introduction of aluminum into aircraft took over 30 years, and 40 years later we are still learning how to design rationally with this metal.

The aerospace applications of advanced composites, though still in their infancy, encompass military and civil aircraft, missiles, launch vehicles, and spacecraft. I believe that the future holds a continued growth in the use of advanced composites by the aerospace industry. For example, aircraft applications, such as those in the Boeing 767, are just beginning in commercial transports after early experience in military aircraft. However, the largest future use is likely to be in new, lighter-weight automobiles to reduce fuel consumption. Forecasts indicate that the potential usage in automobiles in the early 1990's will amount to millions of pounds of advanced composites.

We find ourselves in a peculiar position. The hardware capability is progressing so rapidly that the knowledge and familiarity of the designer can hardly keep pace. We have an obligation now not just to mature this advanced technology and its applications, but also to communicate the state-of-the-art to the designer in a form in which it can be applied readily to practical structures. I believe that this book, *Handbook of Composites*, will clearly provide a portion of this missing link.

The obligation of the research community in a world whose technology changes so rapidly must transcend the traditional roles of the research scientist and engineer. It is no longer sufficient to bring the level of technology to the edge of the laboratory workbench and expect designers to recognize their potential in the clutter of experimental apparatus and research reports. The researcher must support those application-oriented programs that reduce the risk of applying new technology, so that it can be embodied in new products in a minimum amount of time. In the case of composites, the Government has supported not only the basic research in polymer chemistry leading to improved composite properties or environmental resistance, but also manufacturing research and application-verification programs, such as those contained in many U.S. Air Force and NASA studies of the last several years. However, even these research programs are insufficient.

It takes senior designers 15 to 20 years to acquire the skill to design with the materials of their industry. It follows that the transition from one structural material to another will be restricted by the number of designers and analysts who are proficient in applying the new material. In part, this transition can be achieved by applied research, which has as its objective the verification of new material concepts in flightworthy components.

Fundamental to the process of educating the designer is the existence of a data base (including textbooks and manuals), which provides the understanding required in design. Our educational process also requires textbooks and manuals to equip students with their first understanding of materials of construction. Unfortunately, the preparation of such texts is a difficult and time-consuming task whose reward is not occasioned by a quick demonstration in structural tests, but only by the slow acquisition of knowledge in the hands of the design community.

I commend the editor and the authors for their efforts to assist in the rapid dissemination of data on advanced composites to the user community. With sound basic research, applied research to verify application concepts, and rapid dissemination of data, we can assure the early and safe introduction of composite technology into new designs of both aerospace as well as a host of consumer products.

Dr. Alan M. Lovelace
Vice President Science & Engineering
General Dynamics
St. Louis, MO
Formerly—Deputy Administrator
NASA Headquarters
Washington, D.C.

PREFACE

The publication date of this book coincides with the 40th Anniversary of the Composites Industry, and this volume contains many of the industry's significant achievements, including new data about processes and materials utilized, test methodology, and design/analysis techniques. The composites industry is now mature, fast growing, and of prime importance for overall energy conservation.

This book modifies and updates the previous volume by the same editor, *Handbook of Fiberglass and Advanced Plastics Composites*. In this Handbook, advanced composites using organic matrices are highlighted since current and near-future structures will maximize the use of this material.

The major emphasis in this Handbook has been to present validated and useful data, to simplify understanding of composites, and to provide a text that can be used in educational programs.

The Handbook is organized into sequential chapters of materials, processes, design and analysis, typical applications, and includes an appendix containing tabular and graphical basis data. The section on resins now includes polybutadiene and vinyl esters, with an expanded chapter on polyimides.

Specialty features of the book include discussions of novel processes, recently developed advanced materials, environmental suitability, processing aids, practical solutions to manufacturing problems, and examples of composites products.

The fiber chapters include all of the widely used reinforcements. The chapter on whiskers has been deleted, a chapter on Kevlar added, and less emphasis was placed on boron which, unfortunately, is still expensive and limited in use.

The processing chapters are totally new in concept and intended to be more practical. A new chapter on parting agents has been added as well as chapters on environmental resistance.

The design chapter is primarily aimed at the high-performance composites, which require very sophisticated analysis techniques.

The application chapters cover marine, transportation, and aerospace industries and list many examples of the latest breakthroughs in the composites field. Additional tables of data are again included in the appendix.

ACKNOWLEDGMENTS

In the preparation of this Handbook the editor has been helped and encouraged by many friends and business associates and is particularly grateful to the following individuals: the editorial staff of Van Nostrand Reinhold Co., particularly Alberta Gordon, David Ziller, Susan Munger, Barry Levine, and Michelle Herman. To William Cruze, John Delmonte, and Ray Greene who supplied new material for the chronology. To Sam Dastin, Herman Erbacher, Peter Donohue, and Arnold London of Grumman Aerospace Corporation for reviewing many of the chapters, and to Noreen O'Neill for doing most of the typing. Particular thanks to Robert Forger of SPE for his advice and help during the organizational stages. And, finally, my greatest debt and gratitude to my wife Beatrice, who put up with the book for three years and who also helped during the many hours spent on the index.

George Lubin

CONTENTS

Section IV Applications

HANDBOOK OF COMPOSITES

CHAPTER 1
AN OVERVIEW OF COMPOSITES

Dominick V. Rosato
Plastics Seminars
University of Lowell
Waban, Massachusetts

1.1. INTRODUCTION

It has been estimated that in 1979 the composite industry used a sufficient quantity of resins, reinforcing agents, and fillers to produce 8 billion pounds of composites. Of this total, about 2 billion pounds employed fiberglass reinforcement, 2 billion pounds employed asbestos reinforcement, and 4 billion pounds employed such other reinforcements as cellulosic fiber, cotton, polyamide (nylon), and sisal. The total value of these composites to the industry was about 6 billion dollars.

In this complex age of the specialist, when the research and development (R&D) capabilities of humans appear to have exceeded their ability to grasp and control all of the knowledge available to them, it is not strange that current business orientations have a greater influence on overall performance (particularly profits) than they did in the past.[1,2] The dramatic and significant progress in the technology of composite materials (specifically, reinforced plastics—RP) has been proof of this fact since the early 1940's, at which time the high-structural-strength RP were developed.[3–14]

Although the basic elements of RP as composite materials were understood, if little appreciated, by science and technology even before 1940, it was the need for their high structural capabilities generated by World War II that provided the impetus for the ultimate development of RP as competitive materials. As early as 1941, government contracts were issued for the fabrication of semistructural parts using such base materials as combed and carded cotton fibers impregnated with phenolic resin cured at 2000 pounds per square inch (psi). In 1942, low-pressure-curing polyester-resin systems were developed. By the end of the war, fiberglass-reinforced plastics (FRP) had been used successfully in structural applications.

Following demobilization, RP based on a variety of reinforcing agents and resins gradually entered civilian life as a material used in the manufacture of boats, cars, appliance housings, trays, storage containers, and other items. The properties of the fiber reinforcement–resin matrix combination in a homogeneous composition have proved to be superior to those of traditional materials in many ways.

Although initially hindered by the relatively high cost of raw materials and slow, expensive processing methods for the more sophisticated parts, the RP industry has nonetheless exhibited such strong, steady progress that today growth is measured in the billions of dollars each year. At the present time, such significant advances are being made in the development of materials with increased structural strength- and modulus-to-weight ratios (see Fig. 1.1) that the possible future applications will require the implementation of faster and more efficient mass-production techniques.

1.2. DEFINITION OF COMPOSITES

A composite is a combined material created by the synthetic assembly of two or more components—a selected filler or reinforcing agent

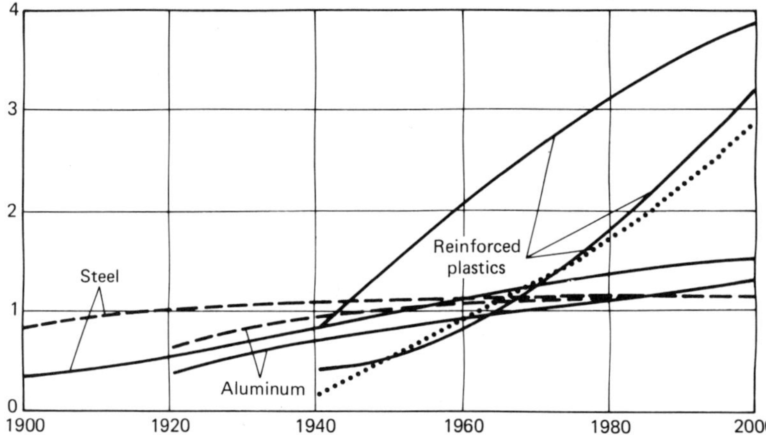

Fig 1.1. Past and forecasted growth in the unidirectional structural properties of RP and conventional materials. Ratio of tensile (———) or compressive (· · · ·) strength (psi) to density (lb/cu in.) × 10^6. Ratio of tensile modulus of elasticity (psi) to density (lb/cu in.) × 10^8 (----).

and a compatible matrix binder (i.e., a resin)—in order to obtain specific characteristics and properties. The components of a composite do not dissolve or otherwise merge completely into each other, but nevertheless do act in concert. The components as well as the interface between them can usually be physically identified, and it is the behavior and properties of the interface that generally control the properties of the composite. The properties of a composite cannot be achieved by any of the components acting alone.

Composites can be classified on the basis of the form of their structural components: fibrous (composed of fibers in a matrix), laminar (composed of layers of materials), and particulate (composed of particles in a matrix). The particulate class can be further subdivided into flake (flat flakes in a matrix) or skeletal (composed of a continuous skeletal matrix filled by a second material). In general, the reinforcing agent can be either fibrous, powdered, spherical, crystalline, or whiskered and either an organic, inorganic, metallic, or ceramic material. Typical resins include polyester, phenolic, epoxy, silicone, alkyd, melamine, polyimide, fluorocarbon, polycarbonate, acrylic, acetal, polypropylene, ABS (acrylonitrile–butadiene–styrene) copolymer, polyethylene, and polystyrene. Resins can be classified as thermoplastic (capable of being repeatedly hardened and softened by increases and decreases, respectively, in temperature) or thermoset (changing into a substantially infusible and insoluble material when cured by the application of heat or by chemical means). At present, the use of thermosetting resins (polyester, phenolic, polyimide, and epoxy) predominates.

Composites provide the designer, fabricator, equipment manufacturer, and consumer with sufficient flexibility to meet the demands presented by different environments (e.g., heat or high humidity) as well as any other special requirements. Thus, they totally eliminate the crippling necessity often faced by designers of restricting the performance requirements of designs to traditional experience. The goal in creating a composite is to combine similar or dissimilar materials in order to develop specific properties that are related to desired characteristics. Since composites can be designed to provide an almost unlimited selection of characteristics, they are employed in practically all industries. Composites are used to produce a variety of economical, efficient, and sophisticated items, ranging from toys and tennis rackets to reentry insulation shields and miniature printed circuits for spacecraft.

The use of reinforcing agents makes it possible for any thermoset- or thermoplastic-matrix property to be improved or changed to meet varying requirements. The composite industry employs many reinforcing agent–resin combinations to effect a diversity of performance and cost characteristics. These components are most frequently combined in two forms: the layered form (e.g., typical melamine–phenolic-impregnated paper sheets and polyester-impregnated fiberglass mat or fabric) or the molding-compound form (e.g., mineral- or cotton-filled phenolic or urea molding compounds). Many of the properties of the resulting composites are superior to those of their components. Although it is composed of several different materials, the composite itself behaves as a single product (e.g., the vinyl-coated fabric used in air mattresses and the laminated metal—bonded together with a plastic adhesive—used in helicopter blades).

The word "composite" evolved when industry required a more all-inclusive term to describe the final material resulting from the combination of many different reinforcing agents and matrices (resins). Composites are also often referred to as reinforced plastics (RP) or filled plastics. However, the term "filled plastic" is generally associated with low-cost materials in which no major changes in properties occur, although there are premium-priced molding compounds (e.g., melamine) that have fillers of glass or alpha cellulose. In the past decade, this term has also come to denote RP. It is sometimes difficult to differentiate between the basic reinforcement and filler, since their functions overlap.[15–28]

1.3. MATERIALS OF CONSTRUCTION

The primary reinforcing agents used in the production of composites at the present time are glass, paper (cellulosic fiber), cotton, polyamide and other natural fibers, asbestos, sisal, and jute. Specialty agents include carbon, graphite, boron, steel, and whiskers (a very short fiber form of reinforcement, usually of crystalline material), all of which are already of technological importance. These reinforcing agents are quite diverse with respect to cost, composition, and properties.

The many forms of fiberglass find wide use in the production of different commercial products (automobiles, appliances, etc.) as well as in the manufacture of parts for spacecraft, aircraft, and both surface-water and underwater vehicles. Fiberglass is the principal reinforcing agent for strengthening resins, since it provides rather significant advantages on a cost-to-performance basis.

Three types of paper reinforcement are in common use: kraft paper (high strength relative to other papers), alpha paper (electrical use), and rag paper (low moisture pickup with good machinability).

Cotton fabrics provide combinations of such different properties as strength, weather resistance, machinability, and toughness. The thickness and weight can be varied in order to control the extent to which these properties are manifested.

Polyamide (nylon) reinforcement is most often used in fabric form. It provides excellent electrical-grade laminates for conventional industrial use, has low water absorption, and exhibits good resistance to both abrasion and many chemicals.

Asbestos offers strength, heat and flame resistance, and heat-ablation and chemical resistance.[15] Sisal and jute are used as extenders for fiberglass or independently for such low-cost, low-strength structures as furniture backings and storage bins.

Of the specialty agents, carbon and graphite fibers provide a wide range of physical, mechanical, and chemical properties. Both have high tensile strength and modulus of elasticity (see Table 1.1) and are commercially available.

Thermosets (polyester, phenolic, polyimide, and epoxy) are the resins used most often in fiberglass composites. At present, however, the major concentration is being focused on the use of chopped glass fibers in thermoplastics (e.g., polycarbonates, ABS copolymer, acetal, and polystyrene).

Reinforcing fillers used in large quantities include alumina powder (aluminum oxide), asbestos, calcium carbonate, calcium silicate, cellulose flock, cotton (different forms), fiberglass, glass beads and spheres, granite, iron oxide powder, mica, quartz, sisal, silicon carbide, titanium oxide, and tungsten carbide. The choice of filler varies, depending to a great extent on the requirements of the end item and

Table 1.1. Properties of the Most Commonly Used Fiber Reinforcing Agents—Metallic and Nonmetallic.

					TENSILE STRENGTH				TENSILE MODULUS OF ELASTICITY			
FIBER/WIRE	DENSITY (lb/cu in.)	SPECIFIC GRAVITY	MELTING POINT (°F)	MELTING POINT (°C)	ULTIMATE (psi × 10^3)	ULTIMATE (MPa)	RATIO (in.) TO DENSITY (× 10^5)	RATIO (MPa·cclg) TO SPECIFIC GRAVITY (× 10^1)	MODULUS (psi · 10^6)	MODULUS (GPa)	RATIO (in.) TO DENSITY (× 10^7)	RATIO (GPa·cclg) TO SPECIFIC GRAVITY
Aluminum	0.097	2.70	1220	660	90	620	9	23	10.6	73	11	27
Aluminum oxide	0.144	3.97	3780	2082	100	689	7	17	76.0	323	53	132
Aluminum silica	0.140	3.90	3300	1816	600	4130	43	106	15.0	100	11	26
Asbestos	0.090	2.50	2770	1521	200	1380	22	55	25.0	172	28	60
Beryllium	0.067	1.84	2343	1284	190	1310	28	71	44.0	303	66	165
Beryllium carbide	0.088	2.44	3800	2093	150	1030	17	42	45.0	310	51	127
Beryllium oxide	0.109	3.03	4650	2566	75	517	7	17	51.0	352	47	116
Boron	0.091	2.30	3812	2100	500	3450	55	150	64.0	441	70	192
Carbon	0.051	1.76	6700	3700	400	2760	78	157	29.0	200	57	114
Glass												
E-glass	0.092	2.54	2400	1316	500	3450	54	136	10.5	72	11	28
S-glass	0.090	2.49	3000	1650	700	4820	78	194	12.4	85	14	34
R&D target	0.090	2.49	3000	1650	1000	6890	111	277	18.0	124	20	50
Graphite	0.054	1.50	6600	3650	400	2760	74	184	50.0	345	93	230
Molybdenum	0.367	10.20	4730	2610	200	1380	5	14	52.0	358	14	35
Polyamide	0.041	1.14	480	249	120	827	29	73	0.4	2.8	1	2.5
Polyester	0.050	1.40	480	249	100	689	20	49	0.6	4.1	1	2.9
Quartz (fused silica)	0.079	2.20	3500	1927					10.0	70	13	32
Steel	0.282	7.87	2920	1621	600	4130	21	53	29.0	200	10	25
Tantalum	0.598	16.60	5425	2996	90	620	2	3.7	28.0	193	5	12
Titanium	0.170	4.72	3035	1668	280	1930	16	41	16.7	115	10	24
Tungsten	0.695	19.30	6170	3410	620	4270	9	22	58.0	400	8	21
Tungsten monocarbide	0.565	15.70	5200	2871	106	730	2	4.6	104	717	18	46

Notes: 1. Boron fiber contains a tungsten boride core.
2. Whiskers that possess both an extremely high tensile strength and an extremely high tensile modulus of elasticity are also of interest:
Aluminum oxide — Tensile strength = 1.8×10^6 psi (12.4 GPa)
Graphite — Tensile strength = 3.0×10^6 psi (20.7 GPa); tensile modulus of elasticity = 10.0×10^{12} psi (6.9×10^2 GPa).
Iron — Tensile strength = 2.0×10^6 psi (13.8 GPa).

method of fabrication. Fillers are also used with long-fiber reinforcements.

Fillers offer a variety of benefits: increased strength and stiffness, heat resistance, heat conductivity, stability, wet strength, fabrication mobility, viscosity, abrasion resistance, and impact strength; reduced cost, shrinkage, exothermic heat, thermal-expansion coefficient, porosity, and crazing; and improved surface appearance. However, fillers also possess disadvantages: They may limit the method of fabrication, inhibit curing of certain resins, and shorten the pot life of the resin.

1.4. RESEARCH AND DEVELOPMENT

Major R&D efforts continue to be directed at RP and related composites for one basic reason. In composites, the potential exists for producing significant advances, as opposed to the relatively minor improvements possible in such wrought sheet-type metals as steel, aluminum, and titanium. Composites can be at least twice as efficient as any other current structural material. As a result of continued R&D with respect to both the materials themselves and their production techniques, composites have experienced greatly increased commercial availability and become competitive with other materials in the more lucrative big markets. There are numerous examples of how progressive thinking and action can result in profitable R&D programs.

1.5 FUTURE GROWTH OF THE COMPOSITE INDUSTRY

In this era of explosive advances in technology, business has to maintain a strong technological base to provide the critically necessary information that will allow both current products and, more importantly, new products to be profitable. The required level of this technology and to what extent it is necessary are, of course, directly related to the type of business and competitive situation of each individual company.[29,30]

The task of producing high-performance, profitable products in any industry (e.g., agriculture, construction, aerospace, metals, and plastics) is sufficiently complex to require a unique managerial organization if success is to be achieved. The broadest possible understanding of a variety of disciplines—including the physical sciences and technology—is required.

When reviewing modern trends in material technology, it becomes obvious that the overall plastics industry, including the RP (composites) industry, has had and will continue to have a rapid growth. The annual average growth rate for the overall industry as a whole has been 11%, whereas it has been 15% for the composites industry; this is in comparison with 6.5% for the chemical industry and 3.5% for the gross national product (GNP). Predictions for the future continue to envision the overall plastics industry and its composites group growing at these accelerated rates. In fact, the overall expansion is much greater than even the population explosion.

Greater demands for increased efficiency on a cost-to-performance basis will continue to grow as composite products inevitably move into larger-volume markets, which emphasize durability under different environmental conditions. Furthermore, as knowledge and confidence in the area of the durability of composite structures continue to expand, their use in both nonstructural and structural applications will gain even wider acceptance.

In recent years, the RP industry, similar to the plastics industry overall, has been involved in integration, mergers, and regrouping. At present, it appears that there will be much more of this type of activity in the future. For some markets, particularly the larger ones, the integration approach permits a company to progress from raw materials to end products more efficiently. Acquisitions have also been a real boon for many organizations, allowing them to expand in-house capability in a specific area. Companies that recognized the potential of RP in their infancy and prepared for expansion are still on the rise.

Unfortunately, there are many roadblocks associated with the growth of the RP industry, and these must be overcome before composites are as accepted and widely used as sheet steel. In general, the lack of total confidence on the part of the designer can rightfully be attributed to cost considerations or the reliability of design data. This reliability in primary structural applications is influenced by quality control evaluation procedures, particularly nondestructive testing.[31–47] To date, no industry comprehensive "design data book" on composites exists for use by the general public which would simplify the utilization of available and reliable design data. However, programs are now being undertaken to develop the "handbook" on "composite design data."

On the other hand, there are also designers who simply do not understand RP, probably due to the limited amount of time they have available to research the topic. However, since RP applications are continuing to expand, data will eventually be available in all handbooks, standards, and even college textbooks. In the meantime, government agencies, industry, societies, and associations are making continued efforts to update and develop new specifications, standards, and handbooks.[48]

Past and present performance, as well as the current areas of R&D, has laid the groundwork for the future growth of the composite industry. Effective exploitation of future opportunities is the key to the potential large-scale market penetration and consequent profitability of RP.

Composites should find expanding use in construction, public transportation, automobiles, watercraft, aircraft and missiles, electronics, storage tanks, pipelines, ordnance, appliances, furniture, and data processing equipment. There are also products to be used in the most rapidly growing U.S. industries—education, medicine, and recreation.

Although monumental technological breakthroughs are unlikely in the next decade, growth will continue to be manifested in steady, incremental advances, limited not by technology, but by economics. The real industrial breakthrough could occur as a result of the greater use of composites in primary structural applications. A fundamentally sound understanding of the mechanical and design properties of RP, matching that in metals, is developing and will soon provide increased opportunities for application. In the meantime, RP that utilize fibers of increased strength and higher modulus of elasticity in a suitable matrix offer promise of a structural material with mechanical properties substantially better than those of metals. In the usual pattern, the requirements of the military and the aerospace industry have provided and will continue to provide the impetus for R&D, thus creating new materials that eventually find application in commercial and industrial markets.

1.6 THE FIRST APPLICATION OF FIBERGLASS COMPOSITES

The first FRP for light airframe structures were conceived, developed, and designed by the Wright-Patterson Air Force Base, Structures and Materials Laboratory, Dayton, Ohio, in 1943. After analyzing test results on RP, theoretical calculations indicated that an efficient structure could be designed and fabricated using high-strength fiberglass–polyester resin laminate faces with a low-density core material. A survey of available military aircraft was performed in order to select a structural component that could adapt reasonably well to redesign in a sandwich structure. The aft fuselage of the two-place Vultee BT-15 basic trainer was selected. This component was completely redesigned and fabricated by the Wright-Patterson Air Force Base, Structures and Materials Laboratory. The first design was based on a balsa wood core with FRP skins. These skins were made of five plies of 3-mil-thick glass fabric impregnated with 42–45 wt % polyester resin. The lay-up was rubber-vacuum-bag molded in a sheet-metal female mold. Cellophane was used to separate the rubber blanket from the inner skin.

The static tests performed on the first fabricated fuselage demonstrated the very high structural efficiency that had been predicted. On the basis of the strength-to-weight ratio, the FRP sandwich structure was approximately 50% stronger than either the metal- or wood-type construction. In addition to its high structural efficiency, the sandwich section showed a remarkable absence of skin buckling under high torsional load. Whereas severe buckling of aluminum skin would occur at 100% of design load, there was almost no visual or measurable skin buckling with the FRP structure at 180% of design load. In order to eliminate the use of wood in this structure (the Air Force did not consider wood a desirable material to use, since the original poorly designed wooden gliders and airplanes were encountering many structural problems), three other fuselages were fabricated using a glass-fabric honeycomb core.

The properties of the RP used in this part were as follows: tensile strength of 40,000 psi (276 MPa), compressive strength of 34,000 psi (234 MPa), flexural strength of 57,000 psi (393 MPa), shear strength of 19,000 psi (131 MPa), modulus of elasticity in flexure of 2.75×10^6 psi (18.7 GPa), and a specific gravity of 1.8. The theoretical specific-strength-to-weight ratios were higher than those of the aluminum alloys and heat-treated steels then being used in structures. Unfortunately, the structural potential of the material was drastically reduced by its relatively low modulus of elasticity [e.g., magnesium alloys with approximately the same specific gravity have a modulus of 6.5×10^6 psi (44.2 GPa)]. The obvious solution that would permit utilization of the RP laminate in primary structures was to stabilize the material, since buckling would not occur until an appreciable portion of the compressive strength was developed. This stabilization was accomplished by the use of sandwich construction. Two high-strength outer faces were separated and supported by bonding them to a much thicker very light core. The core would require only the necessary strength in tension, compression, and shear to be able to adequately support the face materials up to a high stress level. At that time, cores with suitable physical properties ranged in density from 6 to 10 lb/cu ft. (17 to 28×10^4 kg/cu m).

References

1. E. W. Engstrom, Chairman, Executive Committee, RCA, speech to 1966 graduating class of Polytechnic Institute of Brooklyn.
2. R. L. Bisplinghoff, editorial, *AIAA J.* (June 1966).
3. "Reinforced Plastics: Where Will They Go from Here?," *Plastics World* (February 1966).
4. A. G. H. Dietz, "Composite Materials," Edgar Marburg Lecture, ASTM, June 16, 1965.
5. "Forging Military Aerospace Power," Air Force System Command Bulletin, 1965.
6. D. L. Grimes, "Why Develop New Composite Materials . . . Now?," *Research/Development*, 28–31 (September 1965).
7. D. V. Rosato and C. S. Grove, Jr., *Filament Winding: Its Development, Manufacture, Applications, and Design*, J. Wiley & Sons, Inc., New York, 1964.
8. *U.S. Plastics Industry Fact Sheet*, SPI Release from Basford, June 6, 1966.
9. "Building Construction: What's in It for Plastics?," *Plastics World* (December 1965/January 1966).
10. "Processing of Plastics: Structural Integrity of Filament Winding," Paper presented at Iowa State University Conference, April 29, 1966.
11. "Filament Wound $1 Million R&D Contract Release," North American Aviation, Inc., June 23, 1966.
12. *The Chemical Industry Fact Book*, Manufacturing Chemists' Association, 1962.
13. G. A. Rossi and J. H. Johnson, "Composite Sandwich for Small, Unmanned Deep-Submergence Vehicles," ASME 65-UNT-2, May 1965.
14. "Research in the Field of RP Sandwich Structure for Air Frame Use," University of Oklahoma, U.S. Army Fort Eustis Report 64–37, July 1964.
15. D. V. Rosato, *Asbestos: Its Industrial Applications*, Van Nostrand Reinhold, New York, 1959.
16. "Science and Technology and the U.S. Department of Commerce," Background Memorandum, U.S. Department of Commerce June 1966.
17. S. E. Tinkham, "Cost Estimating for Profit," *Plastics World* (May 1966).
18. J. E. Sayre, "Reinforced Polyesters—A Market Research Report," *Reinforced Plastics* (July–August 1965).
19. "Automated RP," *Plastics World* (September 1966).
20. "Sandwich Panel Design Criteria," Building Research Institute Publication 798, 1960.
21. "Filament Winding—Tool of the Space Age," *Reinforced Plastics* (July–August 1966).
22. T. A. Battaglini, "Current Trends and Future Needs for Plastics in Computers," Paper presented at SPI National Plastics Conference, June 1966.
23. G. R. Buck, "Ammunition Packaging Design," *Plastics World* (June 1965).
24. J. D. Matlack, "Plastics in Ammunition," Paper presented at SPI National Plastics Conference, June 1966.
25. D. V. Rosato and R. T. Schwartz, *Environmental Effects on Polymeric Materials*, J. Wiley & Sons, Inc., 1968.
26. "The All-RP Aircraft," *Reinforced Plastics* (March–April 1966).
27. "Composites Promise New Design Freedom," *Reinforced Plastics* (May–June 1965).
28. R. Reed, "Polybenzimidazole and Other Polyaromatics for High-Temperature Structural Laminates

and Adhesives," AFML TR 64-365, Part I, Vol. 1, November 1964.
29. E. C. Bursk, "A Rationale for Marketing Growth," *Industrial Marketing* (June 1966).
30. W. H. Bingham, "Roads to Growth: Independence or Merger," *Plastics World* (May 1966).
31. "Plastics for Aerospace Vehicles," Military Handbook, MIL-HDBK-17, Armed Forces Supply Support Center, Washington, D.C.
32. "Composite Construction," Military Handbook, MIL-HDBK-23, Armed Forces Supply Support Center, Washington, D.C.
33. C. C. Chambers, "The Changing Character of Professional Engineering," *Materials Research and Standards* (June 1966).
34. "Backing Industry Growth—Machinery," *Plastics World* (May 1965).
35. D. V. Rosato and G. Lubin, "The Application of Reinforced Plastics in Spacecraft." Fourth International RP Conference, London, England, British Plastics Federation, November 25-27, 1964.
36. D. V. Rosato, "Weighing Out the Aircraft Market—It's Not Pounds, but Profits and New Plastics That Count," AIAA Paper No. 68-320, Palm Springs, California, April 1-3, 1968.
37. D. V. Rosato, "Why Not Use Metal Wires in Filament Winding?," *Iron Age*, 102-103 (March 26, 1964).
38. D. V. Rosato, "RP Design in Antennas and Microwave Devices," SPE-RETEC, April 18, 1963.
39. D. V. Rosato, "Plastics in Missiles," *British Plastics*, 348-352 (August 1960).
40. "Asbestos-Reinforced Plastics," SPI-Asbestos Technical & Standards Committee of RP/C Division, 17th Annual RP Conference Preprint, February 6-8, 1962.
41. D. V. Rosato and L. J. Breindenbach, "Nonmetallic Composite Materials and Fabrication Techniques Applicable in Present and Future Solid Rocket Bodies," ARS Conference, Salt Lake City, Utah, February 1-3, 1961.
42. "Forecasts," *Fortune Magazine* (January 1967).
43. D. V. Rosato, W. K. Fallon, and Donald V. Rosato, *Markets for Plastics*, Van Nostrand Reinhold, New York, 1968.
44. "Composites Continue to Make Inroads," *Iron Age* (May 31, 1976).
45. "Materials Improve Helicopter Design," *Iron Age* (November 1, 1976).
46. D. N. Yates et al., "Designing with Plastics—The Need Is for Lighter Vehicles by 1985," Plastics Engineering (July 1977).
47. D. C. Hiler, "Carbon Fiber Composites," *Plastics World* (July 1977).
48. R. E. Chambers, F. J. Heger, A. G. H. Dietz, and D. V. Rosato, "Designing with Plastics," University of Lowell, Continuing Education Plastics Seminar.

Chronology of the Development of Reinforced-Plastic Composites and Their Application

1859 Butlerov described formaldehyde polymers.

1862 Baldwin's compression mold patented.

1867 First laminated boat (13 ft long) fabricated from manilla paper and glue. Four such boats were purchased by the U.S. Naval Academy in 1868.

1872 A. von Baeyer reported the reaction between phenols and aldehydes.

1899 Smith published his patent on phenolformaldehyde resin.

1907 Tech-Art Plastics Company (Loando Rubber Company) molded the first phenolic plastic compound.

1907 H. Lebach molded the first acid-resistant chemical tank from one-stage phenolic resin and asbestos.

1909 Dr. L. H. Baekeland granted "heat and pressure" patent on making phenolic resins usable (patent application in 1907).

1909 Westinghouse Electric and Manufacturing Company initiated phenolic laminating.

1909 Cold-molded bitumen, phenolic, and asbestos cement introduced.

1910 General Bakelite Company (Perth Amboy, New Jersey) formed.

1913 D. J. O'Conor's patent application for a plastic laminate sheet to replace moisture-absorbing vulcanized fiber assigned to Westinghouse Electric and Manufacturing Company (patent issued in 1918).

1913 H. Farber and D. J. O'Conor started Formica Products Company to produce RP electrical insulators.

1914 Redmanol Chemical Products Company founded to fabricate phenolic furniture.

1917 Formica Products Company made first contacts with the infant radio industry through the development of laminated components for U.S. Navy and Signal Corps communications apparatus.

1917 Bakelite-Micarta laminated propeller developed for the War Production Board.

1922 Bakelite Corporation formed by merger of General Bakelite Company and Redmanol Chemical Products Company; trade name Bakelite adopted.

1922 R. Kemp granted U.S. Patent 1,435,244 on producing an all-RP airplane.
1922 L. V. Redman engineered first laminate-production plant for Western Electric.
1924 Formica Products Company patented decorative laminates.
1926 Formica Products Company first to plasticize phenolic laminates.
1926 L. E. Shaw developed transfer molding; patent issued in 1928.
1926 Ciba patented aniline-formaldehyde resins.
1927 Benzoyl peroxide first offered in the United States by Lucidol.
1927 F. J. Stokes introduced both rotary- and single-punch preform presses for thermosets.
1928 Urea-formaldehyde introduced commercially.
1929 Synthane Corporation founded world's largest (at that time) laminated-plastic fabrication plant.
1930 Glass-fiber research initiated by Owens-Illinois and Corning Glass Works, supposedly after a molten-glass rod that was being used to apply lettering on a glass milk bottle resulted in the blowing of a fine fiber. This represented the start of the glass-wool-insulation business.
1931 Patent issued to Formica Products Company for a laminate consisting of a urea-formaldehyde surface on a phenolic-paper or similar core; this provided the major start for their decorative laminate business.
1931 Canvas-phenolic picker blocks and rayon buckets developed for textile machines.
1931 Toledo Scale Company's research project (at Mellon Institute) to replace the heavy and expensive porcelain in their weighing scales resulted in the development of urea resins. The demand for urea in scales required Toledo Scale Company to set up a special plastics organization—Plaskon Company.
1932 Dr. A. M. Howald of Toledo Synthetic Products Company molded first urea-formaldehyde Toledo scale housing.
1933 Ciba (now Ciba-Geigy Corporation) marketed "Cibanite," an aniline-formaldehyde molding material.
1935 Ciba (now Ciba-Geigy Corporation) patented melamine-formaldehyde resins.
1935 Owens-Corning Fiberglas Corporation was formed to manufacture glass fibers.
1936 C. Ellis patented unsaturated polyester resins.
1936 Ciba (now Ciba-Geigy Corporation) marketed melamine-formaldehyde molding compounds in Europe.
1937 A. Elmendorf's U.S. Patent 2,070,527 (assigned to Flexwood) on wood veneers reinforced with paper bonded in place issued.
1937 Lauroyl peroxide first offered in the United States by Lucidol.
1937 Automatic compression molding patented and introduced commercially by Stokes Machine Company.
1937 Urea-formaldehyde wood glues introduced by Aero Research Limited (Great Britain)—now Ciba-Geigy Corporation.
1937 The Society of the Plastics Industry (SPI) was incorporated.
1938 First epoxy resin patent granted to P. Castan (Swiss) and licensed to Ciba.
1939 Bakelite Corporation became part of Union Carbide and Carbon Corporation.
1939 U.S. Patent 2,161,533 on wood propellers reinforced with metal strips bonded in place granted to M. Scholz and A. Wagenitz.
1940 The RP industry came into existence with the use of glass fibers to reinforce plastics. The need for radomes (domes to protect aircraft radar antennas) required major R&D programs to study materials, manufacture, and design.
1940 Pittsburgh Plate Glass Company's "CR-39" and "CR-38"—the original low-pressure allyl-type polyester resins (used with cellulose pulp paper)—were commercially molded by King Plastics Company, Denver, Colorado.
1941 McDonnell Aircraft Company of St. Louis, Missouri, received a contract to fabricate a paper-phenolic structural wing box beam for the PT-19 airplane.
1941 General "Hap" Arnold sent a telegram to Wright-Patterson Air Force Base stating that a task force was to be set up so that plastics could be examined with the specific purpose of being used in aircraft wherever possible.
1941 King Plastics Company given Air Force contract to fabricate first plastic seats using combed and carded cotton fibers impregnated with urea and polyester.
1941 Society of Plastics Engineers (SPE) was founded.
1941 Henry Ford hit an RP automobile body with an axe to demonstrate its feasibility as a material in cars.
1941 to 1946 The RP industry made the following contributions to the war effort: cotton-phenolic ship bearings, asbestos cloth–phenolic high-strength switchgears, cotton–asbestos–phenolic brake linings, acetate–cotton bayonet scabbards, plywood–phenolic trainers, wings

and fuselage of British Mosquito bombers, and virtually thousands more.

1942 Patent granted to F. J. Stokes for automatic unscrewing of threaded closures and other threaded parts; this made possible for the first time a fully automatic machine that was available without a license.

1942 Polyester introduced commercially.

1942 British Patent 544,845 on cloth–phenolic laminates reinforced with metal granted to N. A. deBruyne; patent published in *British Plastics* (November 1942).

1942 Wright-Patterson Air Force Base personnel recognized the problem of curing RP at high pressure. One of the first contracts to develop a low-pressure-curing resin system was given to Dr. I. Muskat of Marco Chemical Company, Linden, New Jersey ($150,000 contract). Muskat had been previously employed by Pittsburgh Plate Glass Company, where he had worked on such low-pressure-curing resin systems.

1942 Wright-Patterson Air Force Base personnel had been visiting many different companies to accelerate activity in developing RP. On April 29, during a meeting of the SPI in Hot Springs, Virginia, D. L. Grimes of the U.S. Air Force made public an important government announcement. A program had been set up to cooperate with industry in collecting, at an accelerated rate, data on the use of plastics in aircraft structures (these data would later be issued in *ANC Bulletins on Design Criteria* and eventually become MIL HDBK-17 and MIL HDBK-23). At that time, there was no question as to the importance being attached to the program—alternative materials that could be used in place of such strategic metals as aluminum had to be found.

1942 First fiberglass boat molded by Basons Industries. When making the mold, no parting agent was used (which was logical at that time); hence, all attempts to release the mold failed, and the entire assembly was rolled into the Bronx River.

1942 Dow Corning Corporation made silicone commercially.

1942 The U.S. Navy replaced all of the electrical terminal boards on its vessels with new ones made of fiberglass–melamine and asbestos–melamine laminates.

1942 Owens-Corning Fiberglass Corporation received a $200,000 government contract to evaluate RP.

1942 By the end of the year, important RP parts for aircraft were being produced by different fabricators throughout the country—Uniroyal, Goodyear Tire and Rubber, Formica, Boeing, Douglas, Grumman, Westinghouse, General Motors, Swedlow, and others.

1942 Patent on "Redux"—a thermoplastic modified phenol-formaldehyde structural adhesive—was issued to Aero Research Limited (Great Britain)—now Ciba-Geigy corporation.

1942 Consolidated Papers, Incorporated, formed a special laminated-plastics division to supply the U.S. Government with industrial laminates for use in airplane parts, gliders (e.g., their floors), land mines, and ammunition boxes during World War II.

1942 First fiberglass laminates (ECC-11-148) with CR-38 and CR-39 low-pressure polyester resins were produced by Pittsburgh Plate Glass Company for aircraft, boat, and automobile parts. Cotton fabric—CR-39 laminates were produced by Goodyear Aerospace Corporation for use in aircraft fuel-cell backing-sheet materials.

1943 Studies continued examining all types of RP. McDonnell Aircraft made a paper product—called "Mitcherlich"—that was not affected by water; Wisconsin Consolidated Water Power and Paper Company, in conjunction with the U.S. Forest Products Laboratory, under a Wright-Patterson Air Force Base contract further developed this paper to reinforce resins.

1943 Wright-Patterson Air Force Base started in-house projects to build RP primary structural aircraft parts for the following reasons: (1) FP would provide an alternative to strategic metals, (2) the low weight of RP had the potential to result in the production of more efficient structures, and (3) the good electrical insulation properties of RP, as well as their electromagnetic transparency characteristics, made them desirable for use in radomes.

1943 The first FRP for light airframe structures were conceived, developed, and designed by the Wright-Patterson Air Force Base, Structures and Materials Laboratory. After analyzing test results on RP, theoretical calculations indicated that an efficient structure could be designed and fabricated using high-strength fiberglass-polyester resin laminate faces with a low-density core material. A survey of available military aircraft was performed in order to select a structural component that could adapt reasonably well to redesign in a sandwich structure. The aft fuselage of the two-lace Vultee BT-15 basic trainer was selected. The first design was based on a balsa wood core with FRP skins. The static tests performed on the first fabricated fuselage demonstrated the very high structural efficiency that had been predicted. On the basis of the strength-to-weight ratio, the FRP sandwich structure was approximately 50% stronger than either the metal-or wood-type construction. In addition to its high structural efficiency, the sandwich section showed a remarkable absence of skin buckling under high torsional load.

1943 L. Wittman of Republic Aviation Corporation developed RP tooling.

1943 Chrysler Corporation developed the "Cycleweld" bonding process, to join aircraft parts.

1943 Pregwood (resin-impregnated reinforced wood) used to construct airplane propellers.

1943 Laminates used in bomb tubes and bazook barrels.

1943 Copper-clad laminates for use in electronics introduced.

1943 Plaskon Company of the Toledo Scale Company purchased by Libby-Owens-Ford (LOF) Company.

1944 "Metlbond" adhesives developed by Dr. Haven of Convair and used at Consolidated Vultee Aircraft.

1944 Bazooka barrels (20 ft long and 5 in. in diameter) extruded from cloth-filled phenolic by Plastics Engineering Company.

1944 L. Wittman of Republic Aviation Corporation developed first low-pressure-thermosetting prepregs. These were later merchandised by Fabricon in 1945.

1944 Thermoset composites extruded at Plastics Engineering Company.

1944 U.S. Patent 2,513,268 on the use of a vinyl silicone coupling agent was issued.

1944 Resorcinol-formaldehyde adhesives introduced by Aero Research Limited (Great Britain)—now Ciba-Geigy Corporation.

1944 Cumene hydroperoxide was first produced by Hercules Company.

1944 Eagle wing radar antennas located below main wing of B-29 airplane eroded and were damaged during flights through Pacific rains. This resulted in an acceleration of the development of elastomeric, rubber-type rain-erosion coatings that could be applied over the surfaces of radomes and other RP structures.

1944 On March 24, the Vultee BT-15 airplane with the fuselage fabricated in an FRP sandwich structure was flown for the first time at Wright-Patterson Air Force Base. This was considered to be the first major FRP structural component of an airplane to be developed and flight-tested successfully.

1944 In April, FRP sandwich wings based on a cellular core of cellulose acetate wrapped in RP and with RP skins were designed for AT-6 aircraft. The first wing was actually fabricated in 1945 at the Wright-Patterson Air Force Base, Structures and Materials Laboratory. Static tests were carried out in 1946 and flight tests were made in 1953.

1945 L. S. Meyer of Western Products made the first production RP honeycomb by using soda straws to index corrugated sheets.

1945 First filament winding of RP hoops accomplished by G. Lubin and W. Greenberg of Bassons Industries.

1945 Benzoyl peroxide paste formulations were first offered by Lucidol.

1945 Fire-resistant plastic laminates (fiberglass-melamine) developed by Pacific Plastics.

1946 First automatic injection molding of RP carried out by Dr. A. M. Howald of Plaskon Company. (Toledo Scale Company)

1946 G. Lubin, F. Minikes, and M. Martin apply for first patent on a fiberglass spray technique (centrifugal spraying for molding RP pipes).

1946 "Araldite"—an epoxy resin—introduced by Ciba (now Ciba-Geigy Corporation) at Swiss Industries Fair.

1946 First FRP rocket-motor case developed by R. E. Young.

1946 The SPI organized its Low Pressure Laminates Industry Division.

1947 R. Bernard formed Glass Fibers, Incorporated.

1947 First TV tuner circuit using printed inductors built by RCA for Hallicrafters.

1947 Dr. I. Muskat of Marco Chemical Company developed first thixotropic resins.

1947 On a U.S. Navy Bureau of Ordnance contract, R. E. Young of M. W. Kellog designed a machine and began the first filament-winding production operation using fiberglass–polyester on rocket motors; he also manufactured filament-wound pipe for use on ships.

1947 Copper-clad laminates for commercial use developed by Synthane Corporation.

1947 Consoweld Corporation introduced 4 ft × 4 ft high-pressure laminated panels.

1947 Epoxy introduced commercially in the United States as an adhesive.

1947 R. Steele for Hexcel Corporation produced

first FRP honeycomb using bonded stack expansion process.

1948 On March 9, the U.S. Air Force awarded Douglas Aircraft Company a contract (first of its application type) to design and fabricate the outer wing panel for the C-54A flying laboratory. In a feasibility study of using an integral wing antenna, which would also be part of the outer wing panel, metal was replaced by RP.

1948 Winner Manufacturing Company produced the first RP (fiberglass–polyester) boats (28 ft long) for the U.S. Navy (by 1968, the Navy had over 2000 boats of this type in use).

1948 R. White and L. Seidel of Glastic Corporation made and molded first fiberglass–polyester premix.

1948 On March 11, the U.S. Air Force began a project to design, develop, and fabricate RP primary structural parts for use in supersonic aircraft and missiles.

1949 U.S. Plywood produced the first large machines for the continuous production of filament-wound tubes.

1949 Molded Resin Fiber Company received a contract for the mass production of trays using matched metal molds and hydraulic presses, the first time a contract had been awarded for this mass-production technique; 1000 trays per day were produced.

1949 "Lupersol DDM"—the first commercial methyl ethyl ketone peroxide—was offered by Lucidol.

1949 Crucible Steel Corporation patented "Formold," a steel used to produce dies for compression molding.

1950 Fiberglass–phenolic commercial molding compounds produced.

1950 Continuous RP-sheet production lines were operating to meet principally military demands for protective backing sheets for the self-sealing elastomeric fuel tanks of aircraft.

1950 Due to the unique fatigue characteristics of RP, the development of RP helicopter and aircraft propeller blades was emphasized by Bell Aircraft, Kaman Aircraft, Curtiss-Wright, Hamilton Standard, and others.

1950 Programs to develop RP armor plate for aircraft were conducted.

1950 Dropable RP aircraft fuel tanks developed and produced; these were extensively used in Korea.

1950 Lockheed Aircraft Corporation's Constellation airplane used 80-in.-long radome on the top side as well as 20-ft-long tub-shaped underbelly radomes.

1951 Ferro, Pittsburgh Plate Glass, Libby-Owens-Ford Glass, and Gustin-Bacon were all granted licenses by Owens-Corning Fiberglass to produce fiberglass.

1951 First chrome complex patent issued to R. K. Iler; this complex formed the basis of DuPont's "Volan" finishes.

1951 R. Steinman patented first allylsilane glass size, the predecessor of silane coupling agents.

1951 L. Meyer and A. Howell granted first pultrusion patent (for fishing rods).

1951 The International Standards Organization established a technical committee on plastics (ISO/TC61).

1952 First FRP landing mats developed by the U.S. Army (S. Goldfein) at Fort Belvoir.

1952 to 1960 Beginning of the era of the production of such large rocket motors as Matador.

1953 A 12,000-lb RP fairwater made for the submarine *Halfback*.

1953 Mobile Plastics Division of Carlyle Corporation produced the first preimpregnated roving.

1953 Production runs begin for Corvette automobiles with a fiberglass–polyester body.

1955 Winchester made the first fiberglass-epoxy shotgun barrels.

1955 Early ablation studies on reentry from outer space (Viking rocket flight tests) illustrated the outstanding characteristics of fiberglass–phenolic and asbestos–phenolic laminates.

1955 General Electric Company was first to use plastics as ablative materials

1955 Taylorcraft Model 20 airplane used RP in wings, engine cowling, doors, seats, fuel tanks, instrument panels, fuselage skin (from nose to trailing edge of fin), and elsewhere.

1955 Vertol H-21 helicopter produced at lower cost due to use of RP in fuselage.

1956 Cincinnati Developmental Laboratories made first FRP reentry nose cone using asbestos fiber and phenolic resin.

1956 First asbestos–phenolic reentry nose cone (monocoque construction) successfully used on Vanguard rocket flight.

1956 The SPI Low Pressure Laminates Industry Division changed its name to "Reinforced Plastics Division."

1956 First filament-wound rocket motor (Grand Central Rocket) used in third stage of Vanguard rocket flight.

1956 Fiberglass–epoxy laminates widely adopted for printed-circuit boards.
1957 First fiberglass–melamine reentry nose cone recovered from Redstone missile in the Pacific.
1957 First operational refrasil–phenolic ablative nose cone used in Jupiter missile flight (IRBM).
1958 First refrasil–phenolic reentry nose cone used on ICBM missile.
1958 LOF Glass Fibers, Incorporated, was sold to Johns-Manville Corporation.
1958 Cycloaliphatic epoxy resin patent issued to CIBA Corp.
1958 Graphite fibers commercially produced from rayon for the first time.
1958 Piper Aircraft Company began investigation of RP for use in the primary structures of an airplane; this plane flew for the first time in 1962.
1959 Texaco Incorporated reported the strength and stiffness of boron fibers.
1960 Boeing 727 jet airplanes each contained 5000 lb (2270 Kg) of FRP parts; the FRP parts were lower in cost and 33% lower in weight than the previously used metal parts.
1960 Douglas DC-8 jetliners each contained 2000 lb (908 Kg) of FRP parts; this included unique FRP structural parts of the spar and vertical tail section.
1960 Such high-strength and high-modulus fibers as S-glass (tensile strength $> 1 \times 10^6 = 6.9$ GPa) and boron tensile modulus of elasticity $> 50 \times 10^6$ psi = 345 GPa) developed.
1960 Dow Smith began production of filament-wound pressure pipe.
1960 Grumman Aircraft Corporation's Hawkeye E-2A airplane used rotating 24-ft-diameter radome located above wing.
1960 The Piper Aircraft Company's airplane with fiberglass-polyester skins and a paper honeycomb core flew for the first time.
1960 Mississippi State University flight-tested a Marvelette airplane with smooth RF surface structure to determine its effectiveness with respect to laminar boundary-layer control.
1960 The Dow Chemical Company started building the Windecker fiberglass–epoxy wing for a monocoque, low-wing airplane.
1960 Admiral L. Smith from the Special Project Office of the U.S. Navy handed down a decision to develop filament-wound fiberglass–epoxy motor cases for the Polaris missile; these were highly successful, being more efficient with respect to strength-to-weight ratio, reliability, and cost than the standard steel types.
1960 The Department of Defense established the Plastics Technical Evaluation Center (PLASTEC)—a centralized source of evaluated technical information on plastics, composites, and adhesives.
1960 Consoweld Corporation installed first of the world's largest (28 ft high and 300 tons in weight) laminating presses. It was the first to produce 180,000 sq ft of laminates daily.
1961 A. Shindo experimentally produces first high-modulus (24×10^6 psi = 165 GPa) graphite from polyacrylonitrile.
1961 The Spaulding Fiber Company began production of filament-wound high-voltage fuse tubes.
1961 Continuous graphite filament (modulus of elasticity = 6×10^6 psi = 41 GPa) commercially produced for the first time.
1961 First filament-winding symposium (SAMPE) held.
1962 First RP rudders for submarines produced by Republic Aviation Corporation.
1963 U.S. Navy submarine *Fairwater*, placed in service during 1952 and containing different primary parts made of FRP, successfully completed 11 years of duty.
1963 General B. A. Schriever initiated Project Forecast—the first program for the development of boron composites.
1964 First book on filament winding—*Filament Winding: Its Development, Manufacture, Applications, and Design* by D. V. Rosato and C. S. Grove, Jr.—published by John Wiley & Sons, Incorporated.
1964 Graphite fibers became available for RP research.
1964 Conical, filament-wound (S-glass) motor case developed for Sprint (two stage) missile.
1964 Liquid Nitrogen Processing (LNP) Corporation introduced the first fully and evenly dispersed fiberglass-reinforced thermoplastics (FRTP); nylon 66 was used.
1964 Commercial production of injection-molded FRTP for military weapon systems began.
1965 The Advanced Composite Section (under the direction of G. Peterson) created at Wright-Patterson Air Force Base, Structures and Materials Laboratory.
1965 Owens-Corning Fiberglass introduced shipable forming package (type 30) for filament winding, pultrusion, and weaving.

1965 Gibbs and Cox Company completed feasibility study for U.S. Navy on using FRP parts in minesweepers.

1965 Boron filaments became available to the public.

1965 Grumman/General Dynamics developed first application of FRP in high temperature resistant aircraft structure (fuselage of F-111) for service at 600° F.

1965 Owens-Corning Fiberglas began construction of RP underground gasoline tanks.

1966 Grumman developed first full-scale ballast tank for PX-15 submersible.

1967 Windecker Research Incorporated, Midland, Texas, flight-tested the first relatively all-RP single-wing airplane; it had been specifically designed (which took 7 years) to be made of fiberglass-epoxy.

1967 U.S. Navy began R&D with respect to using RP on minesweepers.

1967 Molded RP bathroom displayed at Canada's Expo 67.

1967 The SPI Reinforced Plastics Division changed its name to "Reinforced Plastics/Composites Division."

1967 The Dow Chemical "black box" became commercially available. The "black box" fed proportional amounts of chopped fiberglass and resin into an injection machine, where a screw provided the mixing and blending.

1968 Reed-Prentice obtained Dow Chemical's "black box" to develop and market the fiberglass-resin proportional mixer used with injection machines. It became known as the CMB (custom material blender).

1968 Boeing's SST design included a projected use of over 6000 lb of RP as well as over 6000 lb of unreinforced plastics.

1968 Aircraft R&D continued to concentrate on possible applications of practically all types of plastics; this also provided an important testing ground for determining how new plastics stood up to different environments. D. V. Rosato estimated that 50 million pounds of plastics (half of which was reinforced) worth 500 million dollars was used in U.S. aircraft in 1968.

1969 Boron-epoxy rudders installed (on an experimental basis) on F-4 aircraft by General Dynamics.

1970 The boron-epoxy horizontal stabilizer on F-14 aircraft (Grumman Aerospace) represented the first advanced composite part produced that was designed as a composite part and not as a substitute for metal.

1971 The SPI Reinforced Plastics/Composites Division changed its name to "Reinforced Plastics/Composites Institute.

1971 First commercially available Aramid Fibers–Kevlars®, introduced by DuPont.

1973 Owens-Corning Fiberglas mass-produced all-fiberglass-polyester bathroom units.

1975 In April, graphite-epoxy honeycomb sandwich antenna horns were used on an actual flight for the first time on the NATO III Communications Satellite. This represented a big step forward for U.S. Air Force Space Systems.

1975 Ford and General Motors found major applications for RP in the automotive industry.

1978 The U.S. Army experimented successfully with graphite–epoxy bridging members.

1980 Lubin and Donohue of Grumman published data on "real life testing of composites" which showed almost no degradation in mechanical strength for FRP laminates in service for up to 20 years.

Section I
Raw Materials

2
UNSATURATED POLYESTER RESINS

Ivor H. Updegraff
Retired, formerly with American Cyanamid Corporation Stamford, Connecticut

2.2. INTRODUCTION

The unsaturated polyester resins used in reinforced plastics (RP) are combinations of reactive polymers and reactive monomers. The idea for this combination was introduced by Carleton Ellis in the 1930's. Ellis discovered that unsaturated polyester resins made by reacting glycols with maleic anhydride could be cured to insoluble solids simply by adding a peroxide catalyst. He applied for a patent on this idea in 1936.[1]

Ellis later discovered that a more useful product could be made by combining the unsaturated polyester alkyd with such reactive monomers as vinyl acetate or styrene. This greatly reduced the viscosity, hence making it easy to add the catalyst and apply the resin; furthermore, the resulting cure was vigorous and complete. In fact, the polymerization reaction for the mixture was faster than that for either of the components taken separately. A patent on this process was applied for in 1937 and granted in 1941.[2] This new product met a definite need in the plastics industry. At the present time, some 40 years after Ellis applied for his patent, annual production of unsaturated polyester resins in the United States has reached approximately one billion pounds.[3]

Unsaturated polyester resins are very versatile materials. At room temperature, the liquid resins are stable for months or even years, but can be triggered to cure in a few minutes simply by adding a peroxide catalyst. Curing takes place by an addition reaction that involves the conversion of double bonds into single bonds; hence, no by-product is released. Styrene is by far the most commonly used diluent; it combines with the reactive double bonds of the polyester chains, linking them together to form a strong three-dimensional polymer network. The curing reaction is exothermic, often providing adequate heat for a satisfactory cure. It is estimated that a normal cure will convert 90% of the reactive double bonds into single bonds.

Polyester resins are used in the manufacture of a broad range of products, including boats, building panels, structural parts for automobiles, aircraft, and appliances, fishing rods, and golf club shafts. Approximately 80% of the polyester resin made in the United States is used with reinforcing agents—primarily fiberglass. Applications that do not involve reinforcing agents include buttons, furniture castings, cultured marble, and auto-body putty.

In contrast to most other plastics, which are based on a single prime ingredient, the polyester resins that are used in RP contain substantial amounts of several components (e.g., resin, catalyst, filler, and accelerator). These components can be combined in different ratios and, furthermore, an alternate material can be substituted for each component. Consequently, many different types of polyester resin are available. In each type, the formulator strives to emphasize the particular properties needed for a specific application.

Maleic anhydride is used to provide the reactive double bonds for most unsaturated polyester resins. It can be combined with glycols (usually propylene glycol) and esterified to form linear polyester chains having molecular weights of about 1000–3000. Ethylene glycol is less expensive than propylene glycol, but is used only in a few specialty resins because the resulting polyesters are not as compatible with the styrene monomer. As esterification proceeds, the "cis" configuration of the maleic anhydride is converted to the "trans" or fumarate structure. This is advantageous, since copolymerization of styrene with the trans double bond of a fumarate ester is much faster than copolymerization with the cis double bond of a maleate. Consequently, a high degree of isomerization to the trans structure is essential for producing a polyester resin of high reactivity.[4] Although the isomerization of maleic anhydride is usually 90% or better, to obtain maximum reactivity, some polyester resins are made with the more expensive fumaric acid.

Other dibasic acids or anhydrides, such as adipic acid, isophthalic acid, or phthalic anhydride, are often added to modify the final properties of the resin and adjust the concentration of the reactive double bonds. A typical polyester resin structure is shown below (R is the alkyl or aryl group of the modifying dibasic acid or anhydride):

$$\mathrm{H}\left[\mathrm{O{-}\overset{\displaystyle O}{\overset{\|}{C}}{-}R{-}\overset{\displaystyle O}{\overset{\|}{C}}{-}O{-}\overset{\displaystyle CH_3}{\overset{|}{C}H}{-}CH_2{-}O{-}\overset{\displaystyle O}{\overset{\|}{C}}{-}CH{=}CH{-}\overset{\displaystyle O}{\overset{\|}{C}}{-}O{-}\overset{\displaystyle CH_3}{\overset{|}{C}H}{-}CH_2}\right]_n\mathrm{OH}$$

As a result of their versatility and the low tooling costs associated with them, many companies fabricate products from polyester resins. These companies are often inexperienced with polyester resin chemistry and thus require continuing technical assistance. Fortunately, resin suppliers are ready and able to provide much detailed information on resin types, fabrication techniques, costs, and properties. Catalyst suppliers are also prepared to give detailed directions for using their products with various accelerators and inhibitors

2.2. TYPES OF UNSATURATED POLYESTER RESINS

The broad range of properties obtainable with polyester resins makes them suitable for a wide variety of applications. The following sections provide brief descriptions of seven specific types of unsaturated polyester resins.

2.2.1. General-Purpose Polyester Resins

This type of polyester resin is usually based on a blend of phthalic anhydride and maleic anhydride esterified with propylene glycol. The phthalic/maleic mol ratio may range from 2:1 to 1:2. The polyester alkyd is blended with styrene in a proportion of about two parts alkyd to one part styrene. As the name implies, resins of this type are used for a great variety of applications, such as trays, boats, shower stalls, swimming pools, and water tanks.

2.2.2. Flexible Polyester Resins

If a straight-chain dibasic acid (e.g., adipic or sebacic acid) is used instead of phthalic anhydride, the resulting unsaturated polyester resin is much softer and more flexible than the general-purpose type. The use of diethylene or dipropylene glycol in place of propylene glycol also provides flexibility. Such flexible polyester resins can be added to the rigid general-purpose type in order to overcome brittleness and make the cured product easier to machine. This is an advantage in the manufacture of cast polyester buttons. Flexible resins can also be made by replacing some of the phthalic anhydride with tall oil fatty acids. These monobasic fatty acids provide flexible groups at the ends of polymer chains. Such flexible resins are often used to make decorative furniture castings and picture frames. The flexible resin is combined with a cellulosic filler (such as pecan shell flour) and cast in silicone rubber molds. Excellent reproductions of wood carvings can be obtained by using silicone rubber molds made directly from the original carvings.

2.2.3. Resilient Polyester Resins

Polyester resins of this type fall between the rigid general-purpose and flexible types. They are intended for use where toughness is needed, such as in bowling balls, safety helmets, guards, gel coats, and aircraft and automotive parts. It is common practice to use isophthalic acid instead of phthalic anhydride in formulating resilient polyester resins. Resin preparation may be a stepwise reaction: First, the isophthalic acid and glycol are combined to give a polyester resin of low acid number; maleic anhydride is then added and the esterification continued. This will give polyester chains having the reactive unsaturation at the ends or between blocks of glycol–isophthalic polymer rather than randomly distributed along the polymer chain. Phthalic anhydride does not perform as efficiently as isophthalic acid in this type of esterification because the half-ester has a tendency to revert to the anhydride at the higher temperatures needed to form the high-molecular-weight polyester resin.

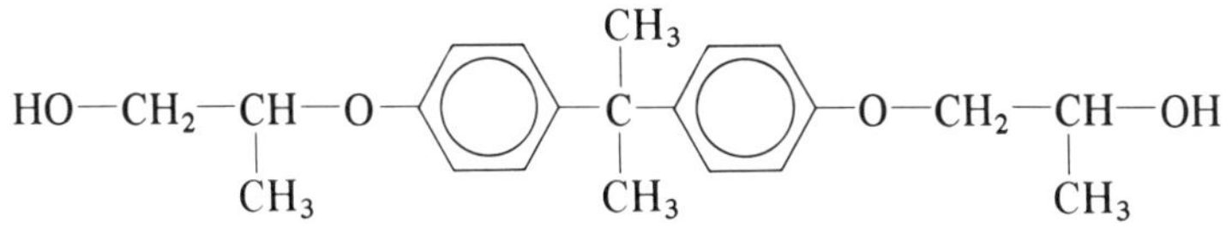

2.2.4. Low-Shrinkage (Low-Profile) Polyester Resins

In ordinary fiberglass-reinforced polyester moldings, the difference in shrinkage between the resin and fiberglass results in a pattern of sink marks on the molded surface. The use of low-shrinkage polyester resins can minimize this pattern so that moldings will not have to be sanded smooth before painting. This property is advantageous in automotive and appliance parts.

Low-shrinkage polyester resins include a thermoplastic component (such as polystyrene or polymethyl methacrylate) that is only partly soluble in the system. As curing proceeds, phase changes allow the formation of microvoids that compensate for the normal shrinkage of the polyester resin.

2.2.5. Weather-Resistant Polyester Resins

This type of polyester resin is formulated to have resistance to yellowing on exposure to sunlight by compounding with absorbers of ultraviolet (UV) light. Although part of the styrene can be replaced with methyl methacrylate, it is not practical to replace all of it, since the methyl methacrylate does not copolymerize very well with the fumarate double bonds of the unsaturated polyester resin. Resins of this type are used in gel coats, outdoor structural panels, and skylights.

2.2.6. Chemical-Resistant Polyester Resins

The lack of alkali resistance is the principal deficiency of polyester resins. The ester linkages are subject to hydrolysis in the presence of alkalies. Increasing the size of the glycol has the effect of reducing the concentration of ester linkages. Thus, a resin containing "bis glycol" (the reaction product of bisphenol A with propylene oxide),

$$HO-CH_2-\underset{CH_3}{\underset{|}{CH}}-O-C_6H_4-\overset{CH_3}{\overset{|}{\underset{CH_3}{\underset{|}{C}}}}-C_6H_4-O-CH_2-\underset{CH_3}{\underset{|}{CH}}-OH$$

or hydrogenated bisphenol A,

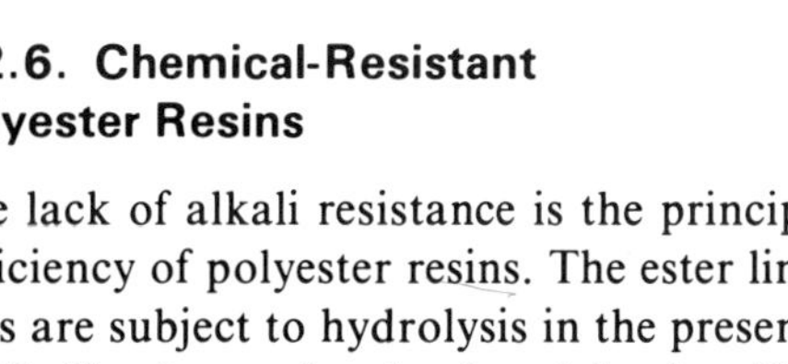

$$HO-C_6H_{10}-\overset{CH_3}{\overset{|}{\underset{CH_3}{\underset{|}{C}}}}-C_6H_{10}-OH$$

will contain far fewer ester linkages than a corresponding resin of the general-purpose type. These resins are used in the manufacture of such chemical-processing equipment as fume hoods, reaction vessels, tanks, and pipes.

2.2.7. Fire-Resistant Polyester Resins

Ordinary fiberglass-reinforced polyester moldings and laminates are combustible but have relatively low burning rates. Better resistance to ignition and burning can be achieved by using halogenated dibasic acids [e.g., tetrachlorophthalic, tetrabromophthalic, and chlorendic (the addition product of hexachlorocyclopentadiene with maleic anhydride), which is also known as "Het" acid] in place of phthalic anhydride. Dibromoneopentyl glycol has also been used. Further improvement in fire resistance can be achieved with various flame-retardant additives, such as phosphate esters and antimony oxide. Fire-resistant polyester resins are used in the manufacture of fume hoods, electrical equipment, and building panels, as well as for certain types of U.S. Navy boats.

The seven types of unsaturated polyester resin described in the preceding sections include most of those used in the RP industry. However, some specialty resins are manufactured to meet other specific requirements. For example, the use of triallylcyanurate in place of styrene can greatly improve heat resistance. Diallylphthalate is used in place of styrene in some molding compounds based on unsaturated polyester resins; being much less volatile than styrene, the molding compound can be handled as a putty or extruded rope without loss of monomer. Vinyl toluene is also used for this application. Special resins are also made for curing by UV light. They contain such light-sensitive catalytic agents as benzoin or benzoin ethers.

2.3. MANUFACTURE OF UNSATURATED POLYESTER RESINS

The unsaturated polyester resins used in RP are generally made by a batch process. The variety of products needed for a complete line of resins favors this process, since it offers quick and easy changeover. Continuous processes are used for large-volume general-purpose resins.

Stainless steel is the preferred material of construction for processing equipment, since it is resistant to polyester resin and other materials used in resin manufacture. Since the free-radical polymerization of polyester resins is inhibited by iron and copper ions, these metals are not used to build resin kettles or their subcomponents. Glass-lined reactors are preferred when halogenated starting materials are used.

In general, the glycol is charged to the reaction kettle, followed by the phthalic and maleic anhydrides. It is common practice to add 5–10% more glycol than the theoretical amount in order to compensate for losses through the condenser as well as by side reactions. Before stirring and heating, air is displaced by inert gas. The first stage of reaction—formation of the half-ester—takes place spontaneously at relatively low temperatures; the batch is then heated to complete the ester formation. The rate of flow of inert gas through the batch can be increased to remove water produced by the condensation reaction. A steam-heated partial condenser is often used to allow escape of water while returning glycol to the reaction kettle.

During the latter stages of esterification, the batch temperature is allowed to rise to 190–220° C (374–428° F). Higher temperature favors isomerization of maleate to fumarate, but it also induces side reactions that convert double bonds to other structures. There is an optimum temperature at which the amount of fumarate unsaturation will be at a maximum. For a general-purpose polyester resin, this occurs at 210° C (410° F).

Viscosity and acid number tests are performed to follow the progress of esterification; when these requirements are met, the batch is pumped to the cutting kettle. The required amount of styrene monomer may already be present in this kettle, so that the polyester alkyd dissolves as rapidly as it is pumped in. Additional inhibitor can be added at this point to avoid any polymerization as the hot alkyd contacts the styrene monomer. Cooling may also be needed to control the batch temperature. The properties of the batch are then determined and adjusted to meet product specifications. Complete production cycles range from 10 to 20 hr.

The method of polyester resin preparation described above is often referred to as the fusion process. The reactants are simply fused together and heated until the esterification reaction has progressed far enough to give the desired product. An alternative method is to use a small amount of xylene or toluene solvent to facilitate removal of the water by azeotropic distillation. The solvent represents only about 8% of the kettle charge; it is separated from the water by decantation and returned to the kettle. When esterification is complete, the condenser is set for straight distillation and the solvent is removed. A final vacuum strip can be applied for complete removal of any remaining solvent. Some side reactions may take place during esterification. For example, a branched-chain polymer may form by addition of a glycol hydroxyl group to the maleic or fumaric double bond. It has been estimated that side reactions consume as much as 10–15% of the reactive unsaturation.

The simplest continuous process for the manufacture of unsaturated polyester resins is by the reaction of a mixture of maleic and phthalic anhydrides with propylene oxide. Some glycol must be present to initiate the anhydride–epoxide chain reaction. Since the anhydride–epoxide polymerization reaction proceeds at a relatively low temperature, the maleate double bond is not isomerized to the more active trans configuration; therefore, a separate heat treatment must be applied after the polymer has been formed to bring about isomerization to fumarate and achieve the desired reactivity with styrene.

Continuous production from anhydrides and glycols is also possible by using a series of stirred reaction vessels, pumping partially staged resins from each heat cycle to the final cycle.

2.4. CURING UNSATURATED POLYESTER RESINS

Unsaturated polyester resins are cured by adding free-radical catalysts to start the chain reaction of polymerization. The free radicals can be derived from peroxides or from other unstable materials, such azo compounds, that can break into radical fragments when subjected to heat or irradiation with UV light or other high-energy radiation. Normally, the polyester resin contains an inhibitor, which is essentially a free-radical trap. When the catalyst is added, the inhibitor must first be overcome before the polymerization reaction can start. This induction period allows time for the catalyzed resin to be mechanically combined with a reinforcing agent and placed in position ready for cure before the polymerization reaction starts. Hydroquinone and related compounds, as well as several quaternary ammonium halides, are good inhibitors.

Most peroxide catalysts decompose rather slowly when added to the polyester resin. To get faster cure, accelerators (promoters) are used to speed up catalyst decomposition. Essentially, the accelerator is a catalyst for the catalyst, but it must be used with caution. The catalyst and accelerator are both reactive compounds and can interact violently, causing fire or explosion, when mixed together. These components should always be added to the resin separately, making sure that one is completely dissolved before adding the other. Many resins are supplied with the accelerator already added. The curing behavior of a polyester resin is a balance of the effects of inhibitor, catalyst, and accelerator.

Groups attached to the carbon atoms of the reactive double bond can influence reactivity in two important ways. Steric effects relate to the fact that bulky groups reduce reactivity by shielding the reactive double bond, making it less probable that a second reactive double bond will be in a favorable position to react. Polarity involves the tendency of the attached group to attract or donate electrons. Electron-donating groups, such as methyl, phenyl, and halogen, tend to make the double bond electronegative. This is also the case for the styrene, vinyl toluene, and chlorostyrene groups. Electron-withdrawing groups, such as nitrile and carbonyl, make the double bond (e.g., the fumarate double bonds in the polyester chain) electropositive. The opposite polarity of the styrene and fumarate double bonds of the alkyd accounts for the strong alternating tendency observed in the polymerization reaction of curing polyester resins. Styrene is more mobile than the unsaturated polyester chain and can also homopolymerize. Experience indicates that a 2:1 molar ratio of styrene to each alkyd double bond is about optimum.

Curing behavior is often described in terms of the SPI Gel Test.[5] A quantity of resin is catalyzed and placed in a test tube. A thermocouple probe is immersed in the resin and the assembly is then suspended in a constant-temperature water bath at 180°F (82°C). A recorder attached to the thermocouple draws a time–temperature curve—an exotherm curve—as the cure reaction proceeds. The temperature rises slowly as the sample approaches the bath temperature. During this time, the inhibitor is being consumed by the free radicals released by the catalyst. Polymer starts to form and the system quickly gels. The temperature then rises rapidly as polymerization accelerates due to the "gel effect," reaching a peak and then falling as the cured plastic cools back to the temperature of the bath. The curing behavior of the system can then be described in terms of the curve by noting the length of the induction period, the time required to reach the peak of the exotherm, and the maximum temperature of the exotherm. It is thus possible to make a detailed comparison of the influence of different amounts and kinds of catalysts, inhibitors, and accelerators. A typical exotherm curve is shown in Fig. 2.1.

2.4.1. Catalysts and Accelerators

A very wide range of catalyst–accelerator–inhibitor systems is available for use with polyester resins. For example, a general-purpose, hydroquinone-inhibited resin can be cured very rapidly by using an active peroxide catalyst, such as methyl ethyl ketone peroxide, in combination with an active accelerator, such as cobalt naphthenate or cobalt octoate. At the other extreme, the same polyester resin can be catalyzed with a much more stable peroxide catalyst, such as *t*-butyl perbenzoate, in order to make a polyester resin molding composition that includes calcium carbonate filler together with chopped fiberglass. Such a catalyzed premix molding compound might be stable in storage at room temperature for months yet would cure to a firm solid in 1 min. when pressed in a heated matched metal die at 140–160°C (284–320°F).

The selection of the proper catalyst and the amount to be used for any application depends on the resin, the curing temperature, the required working or pot life, and the gel time. Since none of the available catalysts can meet all of the requirements by itself, combinations of catalysts or of catalysts and accelerators must be used in order to obtain the best results. Table 2.1 presents SADT (self-accelerating decomposition temperature) data, storage temperature limits, labeling and shipping requirements, and disposal methods for typical organic peroxide catalysts.

When heat can be applied to the resin to effect full cure, the most commonly used catalyst is benzoyl peroxide (BPO), which is efficient, easy to handle, readily soluble in monomeric styrene, storable for long periods of time without loss of activity, and stable at room temperature, readily decomposing at elevated temperatures. Furthermore, BPO causes a high peak exothermic temperature, which aids in the full cure of the resin. The amount of BPO used in the resin varies from 0.5 to 2.0%, depending on the type of resin and monomer employed. In the paste form (in which BPO is usually compounded with 50% tricresyl phosphate), from 1.0 to 3.0% catalyst is used. The reactivity of a general-purpose polyester resin catalyzed with 1.0% BPO paste is shown in Fig. 2.2.

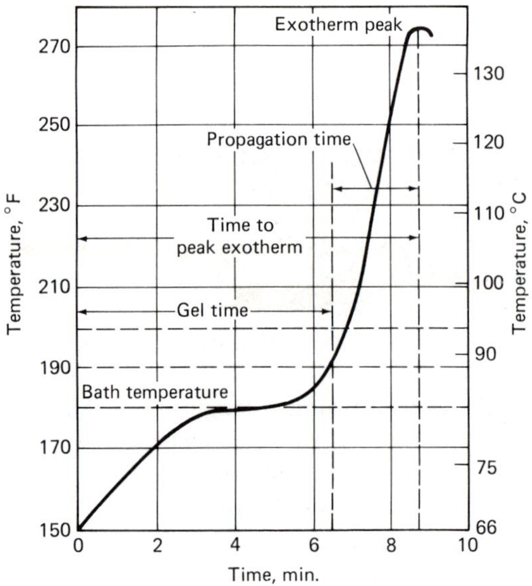

Figure 2.1. Typical exotherm curve for a polyester resin [10-g sample, 19-mm-diameter test tube, and 180°F (82°C) water bath].

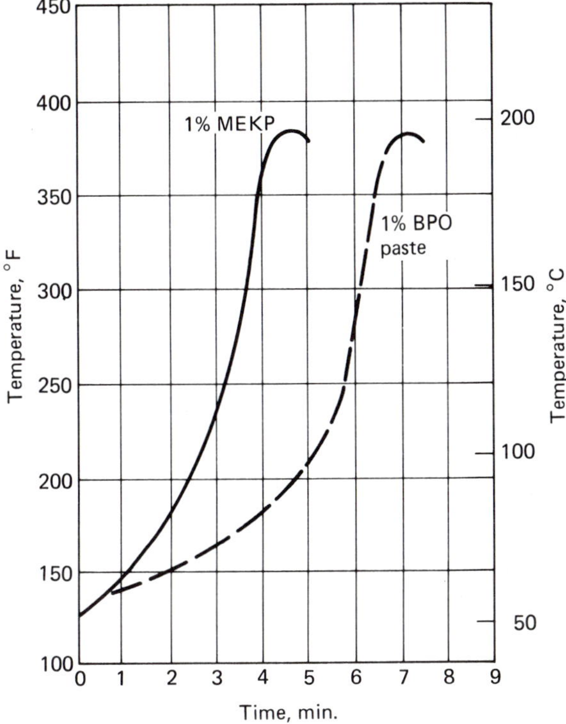

Figure 2.2 Reactivity of a general-purpose polyester resin catalyzed with 1.0% MEKP or 1.0% BPO paste [bath temperature = 180° F (82° C)].

It is occasionally desirable (and sometimes necessary) to start and even complete the cure at lower temperatures, so that the polymerization heat can be readily dissipated. This is of particular importance when using the wet hand lay-up method for large or complicated units, where it may not be possible to apply heat. In these cases, methyl ethyl ketone peroxide (MEKP) is generally used as the catalyst (see Fig. 2.2 for the reactivity of a general-purpose polyester resin catalyzed with 1.0% MEKP). Although it does not lead to full cure when used alone at ambient temperatures, with the addition of an accelerator (e.g., cobalt naphthenate), MEKP will cause gelation (see Fig. 2.3) and almost full cure within short periods of time, the specific time depending on the amounts of catalyst and accelerator actually used. From 0.5 to 2.0% MEKP and from 0.1 to 1.0% cobalt naphthenate can be used, depending on the required pot life of the resin.

Many catalysts and accelerators that will cause full cure with or without the presence of a heat source are currently available for use with polyester resins. The properties of the final product obtained can be varied through the use of different catalysts, as well as different combinations of catalysts and accelerators. These properties will also be affected to some extent by the operating conditions, such as type of tooling, mold temperature, and mold pressure.

Cobalt naphthenate is availabe as a solution containing 6% cobalt metal. It is used in conjunction with hydroperoxide catalysts to help initiate both gelation of the resin and propagation of full cure at room temperature. Figure 2.3 shows the effect of various concentrations of cobalt naphthenate on the gel time of a general-purpose polyester resin catalyzed with 0.5, 1.0, and 2.0% MEKP. An uncatalyzed resin containing cobalt naphthenate will have a storage life of several months. It should be noted that since cobalt naphthenate imparts a color to the resin, its use is not recommended if a clear or light-color final product is desired.

Table 2.1. Typical Organic Peroxide Catalysts (Courtesy of the Lucidol Division of Wallace and Tiernan, Incorporated).

					SADT DATA			STORAGE TEMP. (F)					
PRODUCT	FORM	TRADE NAME (1)	TYPICAL ASSAY %	ACTIVE O_2 CONTENT %	PACKAGE SIZE (Lbs.)	TEMP. (°F)	TYPE	MAX.	MIN.	RECOMMENDED	REQUIRED SHIPPING LABEL (2)	SHIPPING REQUIREMENTS	DISPOSAL METHOD
AROMATIC DIACYL PEROXIDES													
Benzoyl Peroxide	Solid	LUCIDOL	98.5	6.5	1	155	Sudden (Force and Smoke)	100	—	65–85	Yellow	Common Carrier	Hydrolysis
Benzoyl Peroxide, Wet	Solid	LUCIDOL-70	67–70	4.39–4.59				100	35	65–85	Yellow	Common Carrier	Hydrolysis
	Solid	LUCIDOL-78	75–80	4.95–5.28				100	35	65–85	Yellow	Common Carrier	Hydrolysis
Benzoyl Peroxide Compounded with Tricresyl Phosphate	Paste	LUPERCO ATC	50.0	3.3				100	—	65–85	Yellow	Common Carrier	Burning
Benzoyl Peroxide Compounded with Silicone Oil	Paste	LUPERCO AST	50.0	3.3				100	—	65–85	Yellow	Common Carrier	Burning
Benzoyl Peroxide Compounded with Plasticizer	Paste	LUPERCO ANS-50	50.0	3.3				100	—	65–85	Yellow	Common Carrier	Burning
	Paste	LUPERCO ANS-55	55.0	3.6				100	—	65–85	Yellow	Common Carrier	Burning
Benzoyl Peroxide Compounded with Fire Retardant	Paste	LUPERCO AFR	50.0	3.3	50	120	Mild Gassing	100	—	65–85	None	Common Carrier, Parcel Post	Burning
Plasticizer	Paste	LUPERCO AFR-55	55.0	3.6				100	—	65–85	None	Common Carrier, Parcel Post	Burning
2,4-Dichlorobenzoyl Peroxide Compounded with Dibutyl Phthalate	Paste	LUPERCO CDB	50.0	2.1				80	—	<80	Yellow	Common Carrier	Burning
2,4-Dichlorobenzoyl Peroxide Compounded with Silicone Oil	Paste	LUPERCO CST	50.0	2.1				80	—	<80	Yellow	Common Carrier	Burning
p-Chlorobenzoyl Peroxide Compounded with Dibutyl Phthalate	Paste	LUPERCO BDB	50.0	2.5				100	—	65–85	Yellow	Common Carrier	Burning

ALIPHATIC DIACYL PEROXIDES													
Lauroyl Peroxide	Solid	ALPEROX	96.5	3.87	65	120	Mild Gassing	80	—	<80	Yellow	Common Carrier	Burning, Hydrolysis
Decanoyl Peroxide	Solid	DECANOX	97.0	4.53	50	95	Mild Gassing	60	—	<60	Yellow	Refriger-ated	Burning, Hydrolysis
Acetyl Peroxide 25% Solution in Dimethyl Phthalate	Liquid		25.0	3.4	45	120	Sudden (Force and Smoke)	75	32	32–41	Yellow	Common Carrier	Burning, Hydrolysis
Propionyl Peroxide in High Boiling Hydrocarbon Solvent	Liquid	LUPERSOL 12	25.0	2.7	35	100	Mild	60	41	41–50	Yellow	Refriger-ated	Burning, Hydrolysis
Pelargonyl Peroxide	Liquid	LUPERSOL 9	97.0	4.9	35	75	Burning	50	—	<50	Yellow	Refriger-ated	Burning, Hydrolysis
DIBASIC ACID PEROXIDES													
Succinic Acid Peroxide	Solid		95.0	6.45	1	150	Burning	80	—	<80	Yellow	Common Carrier	Burning
KETONE PEROXIDES													
Bis (1-Hydroxycyclo-hexyl) Peroxide	Solid	LUPEROX 6	95.0	6.6				80	—	<80	Yellow	Common Carrier	Burning
Cyclohexanone Peroxides Compounded with	Solid	LUPERCO JDB-85	85.0	11.0				80	—	<80	Yellow	Common Carrier	Burning
Dibutyl Phthalate	Paste	LUPERCO JDB-50-T	45.0	5.8				80	—	<80	Yellow	Common Carrier	Burning
Methyl Ethyl Ketone Peroxides in	Liquid	LUPERSOL DDM	60.0	11.0	40	145	Mild Gassing	100	—	65–85	Yellow	Common Carrier	Burning, Hydrolysis
Dimethyl Phthalate	Liquid	LUPERSOL DELTA	60.0	11.0	40	145	Burning	100	—	65–85	Yellow	Common Carrier	Burning, Hydrolysis
	Liquid	LUPERSOL DELTA-X	60.0	11.0	40	140	Burning	100	—	65–85	Yellow	Common Carrier	Burning, Hydrolysis
Methyl Ethyl Ketone Peroxides in Dimethyl Phthalate and Diallyl Phthalate	Liquid	LUPERSOL DDA-30	30.0	5.5	40	140	Mild Gassing	100	—	65–85	Yellow	Common Carrier	Burning, Hydrolysis
Methyl Ethyl Ketone Peroxide	Liquid	LUPERSOL DNF	57.0	10.4	40	130	Mild Gassing	100	—	65–85	None	Common Carrier	Burning, Hydrolysis
	Liquid	LUPERSOL DSW		11.4	35	145	Mild Gassing	100	—	65–85	None	Common Carrier	Burning, Hydrolysis
3,5-Dimethyl-3,5-Dihydroxy-1,2-Peroxycyclopentane	Liquid	LUPERSOL 224		4.0	40	150	Mild Gassing	100	—	65–85	None	Common Carrier	Burning

Table 2.2. Continued

PRODUCT	FORM	TRADE NAME (1)	TYPICAL ASSAY %	ACTIVE O_2 CONTENT %	SADT DATA: PACKAGE SIZE (Lbs.)	SADT DATA: TEMP. (°F)	SADT DATA: TYPE	STORAGE TEMP. (F): MAX.	STORAGE TEMP. (F): MIN.	STORAGE TEMP. (F): RECOMMENDED	REQUIRED SHIPPING LABEL (2)	SHIPPING REQUIREMENTS	DISPOSAL METHOD
ALKYL PEROXYESTERS													
t-Butyl Peroxyisobutyrate in Mineral Spirits	Liquid	LUPERSOL 80	75.0	7.5	35	105	Sudden (Force and Smoke)	60	—	<60	Yellow	Refrigerated	Burning
Di-t-Butyl Diperoxyphthalate	Liquid	LUPERSOL KDB	50.0	5.1				100	41	65–85	Yellow	Common Carrier	Burning
t-Butyl Peroxybenzoate	Liquid		98.0	8.07	35	135	Burning	100	48	<75	Yellow	Common Carrier	Burning
2,5-Dimethyl-2,5-Bis (Benzoyl Peroxy) Hexane	Solid	LUPEROX 118	92.5	7.65	1	160	Burning	100	—	65–85	Yellow	Common Carrier	Burning
t-Butyl Peroxypivalate	Liquid	LUPERSOL 11	75.0	6.87	30	80	Rapid with Fire	30	0	0–30	Yellow	Refrigerated	Burning
t-Butyl Peroxyoctoate	Liquid		97.0	7.18	35	105	Burning	65	—	<65	Yellow	Refrigerated	Burning
5-Butyl Peroxyoctoate in Plasticizer	Liquid	LUPERSOL PDO	50.0	3.7	7	125	Mild	80	—	<65	Yellow	Common Carrier	Burning
t-Butyl Peroxymaleic Acid	Solid	LUPEROX PMA	98.0	8.3				100	—	65–85	Yellow	Common Carrier	Burning
2,5-Dimethyl-2,5-Bis (2-Ethyl Hexanoyl Peroxy) Hexane	Liquid	LUPERSOL 256	90.0	7.1	7	105	Mild Gassing	90	—	<60	Yellow	Refrigerated	Burning
t-Butyl Peroxyacetate in Benzene	Liquid	LUPERSOL 7	75.0	9.05	7	150	Violent	100	—	<80	Yellow	Common Carrier	Burning
t-Butyl Peroxyacetate in Mineral Spirits	Liquid	LUPERSOL 70	75.0	9.05	7	170	Violent	100	—	65–85	Yellow	Common Carrier	Burning
t-Butyl Peroxyisobutyrate in Benzene	Liquid	LUPERSOL 8	75.0	7.5	35	90	Mild Gassing	60	—	<60	Red	Refrigerated ated	Burning
ALKYL PEROXIDES													
Di-t-Butyl Peroxide	Liquid		99.0	10.8	30	175	Mild	85	—	<85	Red	Common Carrier	Burning
2,5-Dimethyl-2,5-Bis (t-Butyl Peroxy)	Liquid	LUPERSOL 101	90.0	9.92	30	180	Mild	100	46	65–85	Yellow	Common Carrier	Burning

Hexane	Solid	LUPERCO 101-XL	45.0	4.96				100	46	65–85	None	Common Carrier, Parcel Post	Burning
2,5-Dimethyl-2,5-Bis (t-Butyl Peroxy)	Liquid	LUPERSOL 130	90.0	10.05	35	195	Sudden	100	48	65–85	Yellow	Common Carrier	Burning
Hexyne-3	Solid	LUPERCO 130-XL	45.0	5.03				100	48	65–85	None	Common Carrier, Parcel Post	Burning
n-Butyl-4,4-Bis (t-Butyl) Peroxy) Valerate	Solid	LUPERCO 230-XL	50.0	5.6				100	—	65–85	None	Common Carrier, Parcel Post	Burning
ALKYL HYDROPEROXIDES													
t-Butyl Hydroperoxide	Liquid	t-Butyl Hydroperoxide-70	70.0	12.4	35	190	Mild with Burning	100	—	65–85	Yellow	Common Carrier	Burning
	Liquid	t-Butyl Hydroperoxide-90	90.0	16.0	35	180	Burning	100	—	65–85	Yellow	Common Carrier	Burning
2,5-Dimethyl-2,5-Bis (Hydroperoxy) Hexane	Solid	LUPEROX 2,5–2,5	70.0	12.6				100	40	40–100	Yellow	Common Carrier	Burning

(1) Registered Trademarks of the Lucidol Division, Wallace and Tiernan, Inc.
(2) As required by the U.S. Department of Transportation (formerly the Interstate Commerce Commission)

Both the pot life and the curing time of a catalyzed resin are affected by the type and amount of accelerator used. Other variables that affect the rate of cure of a resin are room temperature, age of resin, batch-to-batch variations in the resin, mass to be cured, amount of heat applied during the cure cycle, size and type of mold, and use of fillers.

The other accelerators that are used in conjunction with hydroperoxide catalysts help promote rapid gelation and initiate propagation of cure at room temperature. However, they require the application of heat in order for full cure of the resin to be accomplished. Dimethylaniline is the most commonly used accelerator of this type. It is generally used in concentrations ranging from 0.1 to 1.0%.

The effect of variations in catalyst or accelerator concentration on the gel time and peak exotherm temperature of a general-purpose polyester resin is illustrated by the data of Tables 2.2 and 2.3, respectively, for several catalyst–accelerator combinations. Table 2.4 gives the "kick-off" temperatures (i.e., the temperature at which a resin begins to polymerize) of the commonly used catalyst–accelerator combinations as determined from SPI exotherm curves. (For additional molding formulations see Chapters 13 and 15.) The following abbreviations for the accelerators are used in Tables 2.2–2.4:

Dimethylaniline	DMA
Diethylaniline	DEA
N,*N*-Dimethyl-*p*-toluidine	DMT
Lauryl mercaptan (*n*-dodecyl mercaptan)	LM
Manganese naphthenate (6%)	Mn^2
Cobalt naphthenate	Co^2

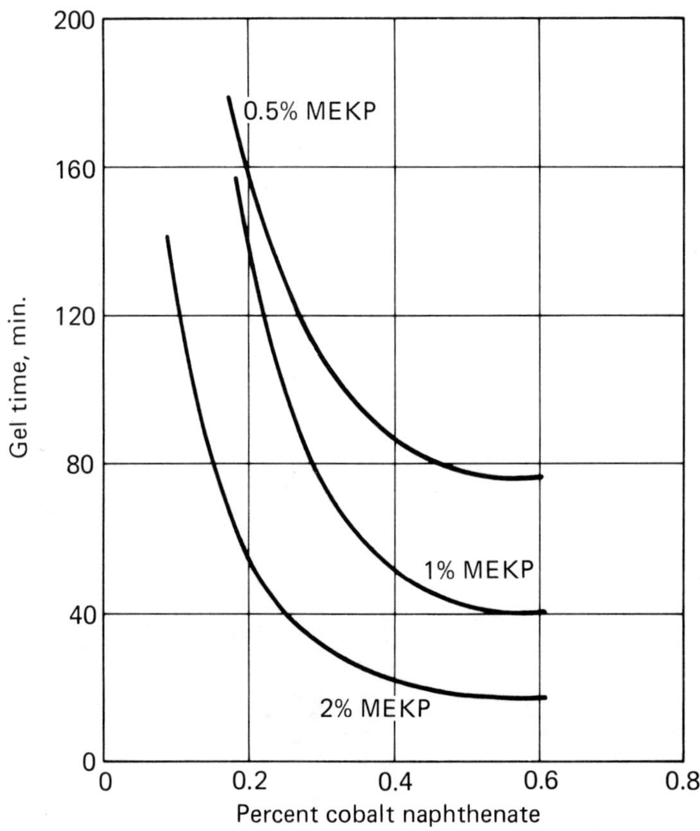

Figure 2.3. Effect of accelerator (cobalt naphthenate) concentration on the gel time of a general-purpose polyester resin catalyzed with MEKP.

Table 2.2. Effect of Variation in Catalyst Concentration on Gel Time and Peak Exotherm Temperature of a General-Purpose Polyester Resin for Several Catalyst–Accelerator Combinations.[5]

CATALYST	%	ACCELERATOR	GEL TIME (MIN.) AT 30°C (86°F)	PEAK EXOTHERM TEMP. AT ROOM TEMP. (75°F = 24°C) °F	°C
Benzoyl peroxide	2.0	0.2% DMA	4.5	330	166
	1.0		7.5	305	152
	0.5		12.0	275	135
	0.25		21.0	215	102
	0.20		25.5	200	93
	0.15		34.0	—	—
Benzoyl peroxide	2.0	0.1% DMT	3.0	340	171
	1.0		4.0	315	157
	0.5		6.5	285	141
	0.25		10.5	260	127
	0.20		12.0	230	110
	0.10		19.5	175	79
Acetyl peroxide	2.0	0.4% DMT	9.0	300	149
	1.0		13.0	270	132
	0.5		20.0	250	121
	0.2		35.0	—	—
Methyl ethyl ketone peroxide	2.0	0.4% Co[2]	9.0	305	152
	1.0		18.5	280	138
	0.75		25.0	210	99
	0.5		38.5	185	85
Bis(1-hydroxy-cyclohexyl) peroxide	2.0	0.1% Co[2]	6.5	285	141
	1.0		7.5	165	74
	0.5		14.0	115	46
	0.4		22.0	95	35
	0.3		>80	—	—

2.5 MOLDING AND LAMINATING PROCESSES FOR UNSATURATED POLYESTER RESINS

2.5.1. Hand Lay-up and Spray-up*

The methods of fabricating parts using polyester resins are more varied than those of any other plastic. The simplest technique, "hand lay-up," merely involves catalyzing the liquid resin and then applying it with layers of fiberglass cloth or mat to a form or mold using a brush, roller, or spray gun. The resin may contain a thixotropic agent to make it stay put on a vertical surface. In addition, the resin may contain a wax that is soluble in the liquid resin but insoluble in the cured plastic so that it exudes to the surface as the resin cures, protecting it from the air (which otherwise would inhibit cure and leave a tacky surface). Since the catalyzed resin has a limited pot life, no more should be catalyzed than can be conveniently applied within the pot life period. To avoid this inconvenience, it is possible to catalyze the resin as it is being applied. This technique is known as "spray-up" and is commonly used in the manufacture of boats. The boat mold is usually first sprayed with a pigmented gel coat, which becomes the exterior surface coating of the completed hull. Before the gel coat has cured completely, the

*Described in detail in Chapter 13.

Table 2.3. Effect of Variation in Accelerator Concentration on Gel Time and Peak Exotherm Temperature of a General-Purpose Polyester Resin for Several Catalyst–Accelerator Combinations.[6]

CATALYST	%	% DMA	% DMT	% CO[2]	GEL TIME (MIN.) AT 30°C (86°F)	PEAK EXOTHERM TEMP. AT ROOM TEMP. (75°F = 24°C) °F	°C
Benzoyl peroxide	1	0.4			4.0	305	152
		0.3			5.5	305	152
		0.2			8.0	305	152
		0.1			16.0	295	146
		0.05			35.0	250	121
Benzoyl peroxide	1		0.4		<2.0	245	118
			0.2		2.0	315	157
			0.1		4.0	315	157
			0.05		11.0	265	129
			0.03		31.0	No cure	
			0.02		>130.0	No cure	
Acetyl peroxide	1		0.4		13.0	270	132
			0.3		18.5	280	138
			0.2		30.0	290	143
			0.1		70.0	—	—
Methyl ethyl ketone peroxide				0.6	14.5	280	138
				0.5	16.0	285	141
				0.4	18.5	280	138
				0.3	22.5	265	129
				0.2	29.0	230	110
				0.1	47.0	170	77
Bis(1-hydroxy-cyclohexyl) peroxide	1			0.2	4.0	140	60
				0.1	7.0	165	74
				0.075	9.0	195	91
				0.05	13.5	225	107
				0.025	27.0	230	110

polyester resin and fiberglass reinforcement are sprayed up against it, thus making the gel coat an integral part of the finished molding. The spray-up gun not only sprays catalyzed resin, but also feeds fiberglass roving into a chopper. The pieces of cut roving are then carried with the resin spray to the mold surface. Considerable operator skill is needed to deposit a uniform coating in some areas while building up the thickness at points of stress. The resin-glass mixture must be rolled to remove trapped air before curing. Two types of catalyst–accelerator systems are in common use. The simplest makes use of the accelerator cobalt naphthenate (or cobalt octoate) in the resin and injects the catalyst, MEKP, at the spray gun. The other requires two separate resin containers and is known as the "two-pot system." One container is catalyzed with BPO, a fairly stable catalyst that gives a pot life of many hours. The other resin container is accelerated with an amine (e.g., dimethylaniline); this solution is also quite stable at room temperature. The gun mixes two equal streams to produce a catalyzed and promoted resin that will cure at room temperature. However, it is still necessary to roll the resin–glass mix in order to remove trapped air and distribute the resin where needed. The hand lay-up and spray-up techniques impose no limitation on the size of object that can be molded.

2.5.2. Preform Molding*

It is not uncommon to form the fiberglass reinforcement into the shape of the product to be molded. A small amount of binder resin can be used to hold the preform together. The preform is then laid in a matched metal mold and the resin poured over it. Closing the mold forces the resin into the fiberglass mat, eliminating air and excess resin. By using pressure, it is possible to achieve a higher proportion of fiberglass reinforcement and hence greater strength. Furthermore, the glass fibers are mostly oriented parallel to the surface of the molding. A modification of this technique uses a half-mold (or half-form) closed with a flexible sheet of polyvinyl alcohol film. Pressure is applied either by placing the whole assembly in an autoclave or by drawing a vacuum on the flexible covering (the so-called "bag").

2.5.3. Centrifugal Casting

Centrifugal casting makes use of centrifugal force to deposit and hold the resin and fiberglass reinforcement against the inside walls of a spinning mold. The assembly can be placed inside an oven so that the resin will be heated and cured while it is being held against the inside of the spinning mold.

*Described in detail in Chapter 15.

2.5.4. Pultrusion and Filament Winding†

Pultrusion and filament winding make use of continuous strands of reinforcement that are first drawn through a bath of catalyzed resin and then through either a heated die or wound on a form. It is possible to achieve very high strength by such techniques. Filament winding can be used to produce tanks, pipes, and pressure vessels, whereas continuous pultrusion can be used to produce structural shapes, golf club shafts, fishing rods, and so on.

2.5.5. Matched Metal Die Molding‡

High-volume production items, such as automotive and appliance parts, are commonly made in matched metal molds. The polyester

†Described in detail in Chapters 16 and 17.
‡Described in detail in Chapter 15.

Table 2.4. "Kick-off" Temperatures of Commonly Used Catalyst-Accelerator Combinations as Determined from SPI Exotherm Curves.

	CONTROL		DMA		DEA		DMT		LM		HQ		MN[2]		CO[2]	
CATALYST	°F	°C	°F	°C	°F	°C	°F	°C	°F	°C	°F	°C	°F	°C	°F	°C
Diacyl Peroxides																
Benzoyl peroxide	200	93	85	29	140	60	80	27	195	91	245	118	205	96	200	93
Lauroyl peroxide	180	82	160	71	190	88	130	54	180	82	190	88	190	88	180	82
Acetyl peroxide	195	91	150	66	180	82	140	60	190	88	210	99	200	93	200	93
Ketone Peroxides																
Methyl ethyl ketone peroxide	210	99	200	93	210	99	205	96	190	88	275	135	215	102	100	38
Cyclohexanone peroxide	230	110	210	99	—	—	—	—	—	—	—	—	—	—	100	38
Hydroperoxides																
t-Butyl hydroperoxide	270	132	240	116	250	121	240	116	240	116	295	146	260	127	250	121
Cumene hydroperoxide	250	121	230	110	—	—	—	—	—	—	—	—	245	118	245	118
Alkyl Peresters																
t-Butyl perbenzoate	245	118	230	110	235	113	230	110	165	74	280	138	250	121	215	102
Alkyl Peroxides																
Di-*t*-butyl peroxide	280	138	250	121	270	132	260	127	280	138	320	160	285	141	290	143

Notes. 1. Exotherm curves were obtained using a general-purpose polyester resin with 0.2% accelerator and a catalyst concentration equivalent to 1.0% benzoyl peroxide.
2. "Kick-off" temperature values are accurate to ±10°F (±5.6°C).

resin molding compound is supplied as a premixed blend of resin, filler, and chopped fiberglass. This premixed molding compound is known as bulk molding compound (BMC). It is already catalyzed and inhibited against premature curing in storage. The user merely places a weighed quantity of the dough-like molding compound or extruded rope into the hot mold cavity, closes the press, and holds pressure for a minute or two, the cured molding then being ready for removal.

A modification of this technique makes use of a similar type of molding compound in the form of flat sheets. The soft premix dough of polyester resin and filler is spread on sheets of fiberglass mat. This sheet molding compound (SMC) can then be sandwiched between sheets of polyethylene film and stored until ready for use. Thickening may take place during the first few days of storage, so that when the SMC is unrolled it is a nontacky sheet that can easily be cut to the desired shape and compression molded into finished parts. The thickening that takes place on storage is not due to free-radical-induced polymerization, but rather is the result of interaction between a thickening agent (such as calcium or magnesium oxide) and the polyester alkyd. The polyester resin is usually of the low-shrinkage (low-profile) type, so that the molded parts will not have to be sanded smooth before painting.

2.5.6. Automatic Injection Molding

The automatic injection molding of polyester resin molding compounds has been discussed and may now be in use to produce small parts. The process is similar to the injection molding of thermoplastics, except that instead of being plasticized in a hot cylinder and then injected into a much cooler mold cavity, the polyester resin is plasticized in a warm cylinder and injected into a hot mold cavity where the cure reaction hardens the product to a thermoset solid. The great advantage of this process is its high production rate and reduced labor cost. The charge is automatically placed in the mold and the finished part is automatically removed.

2.5.7. Evaluation of Catalyst-Accelerator-Inhibitor Systems

Each of the above mentioned fabrication methods requires specific curing conditions. For example, hand lay-up requires about 0.5 hr to apply the resin, wet the glass reinforcement, and free the film of air bubbles. The catalyst, accelerator, and inhibitor requirements are best established by a series of trials using small quantities of the resin. The test temperature should be the same as that of the fabrication method in question. The two most widely used catalyst–accelerator systems are MEKP–cobalt naphthenate (or cobalt octoate) and BPO–dimethylaniline. The MEKP is supplied as a 60% solution of the peroxide in dimethyl phthalate, whereas BPO is supplied either as a 100% active powder or a 50% active paste. The paste is recommended because it is easier to dissolve in the resin.

A starting concentration for gel-time tests of most resins might be 1% MEKP solution (60% active) with 0.2% cobalt naphthenate solution (6% cobalt) or 1.5% BPO with 0.4% dimethyl aniline. After the gel times have been determined for several concentrations, it will be possible to estimate the concentrations required to give the desired rate of cure. To assure adequate cure, the catalyst concentration should not be less than 1.0% BPO or 0.5% MEKP. A little extra inhibitor (e.g., hydroquinone or 2,5-di-*t*-butyl hydroquinone) can be added to extend the gel time if necessary.

2.5.8. Surfacing Materials

Surfacing veils and mats are used to improve the appearance of molded laminates as well as their resistance to weathering, corrosion, and abrasion. When a gel coat of resin is not used during the molding operation, the reinforcing material near the outer surface of the laminate is only covered by a thin layer of resin. This resin layer may be worn away in the course of the normal use of the product, thus possibly exposing the fibers. This is undesirable, since the exposed fibers can absorb moisture, wick, pick up dirt more readily, and, in general, cause a reduction in the physical and mechanical properties of the laminate.

This reduction in properties can be eliminated and the service life of a laminate increased by the use of surfacing veils and mats. The most common surfacing materials for RP are glass mat, dacron, orlon, dynel, and nylon. The surfacing veil or mat keeps the reinforcing fibers of the laminate away from the mold surface and helps provide a more resin-rich surface on the part. This surface is more resistant to weathering, corrosion, and abrasion and improves the appearance of the molded laminate by its tendency to mask any fiber pattern. It is also smoother, since the veil or mat is made of relatively fine-textured materials.

Dacron is composed of linear polyester fibers. It is made into nonwoven fabric or mat from untwisted, continuous yarn. Woven and knitted fabrics made from either continuous or staple yarns are also used.

Orlon is a synthetic fiber material consisting primarily of acrylonitrile. It is available for use in a woven fabric of continuous, spun fibers.

Dynel is manufactured by extruding an acetate solution of acrylonitrile–vinyl copolymer. It has a low softening point of 250°F (121°C) and is classified as self-extinguishing. Light- to medium-denier spun dynel is used for producing woven fabrics.

Nylon is a synthetic fiber consisting of continuous, spun filaments and is used as tow or roving and as a woven fabric.

Other miscellaneous surfacing materials that are used in various forms include cotton, wool, and paper.

2.6. PROPERTIES OF CURED UNSATURATED POLYESTER RESINS

Unsaturated polyester resins can be formulated to produce cured plastics ranging from hard and brittle to soft and flexible. Table 2.5 presents representative properties of typical unreinforced resin castings.[7] A fairly broad range of values is shown for most properties in order to cover the results expected from different formulations.

Polyester resins are usually combined with reinforcing agents (primarily fiberglass), and this can bring about dramatic changes in properties. For example, impact strength can be increased by as much as 50 times by using fiberglass reinforcement. Table 2.6 presents the properties of three types of fiberglass-reinforced polyester resins.[7] Major advantages of the reinforced polyester resin systems are as follows:

1. *High Strength with Light Weight.* Strength-to-weight ratios are high compared to those of most metals.
2. *Low Tooling Costs.* Simple forms or molds operating at low pressure are adequate to produce parts.
3. *Parts Consolidation.* A single molded part, such as a one-piece hood–fender assembly for trucks, can replace a number of metal parts and allow great design flexibility. Thus, reinforced polyester resins can offer cost–performance benefits considerably beyond those available with many other materials.

Some limitations must be considered in designing parts to be made from fiberglass-reinforced unsaturated polyester resins. Strength uniformity is very sensitive to the amount and orientation of the glass fibers. Fatigue resistance of the fiberglass-reinforced polyester resin may be inadequate for some highly stressed parts. Abrasion resistance and resistance to solvents and chemicals must be considered in some applications. The wicking of moisture along bundles of glass fibers can be troublesome, especially if the surface has been abraded to expose the fiberglass reinforcement.

2.6.1. Effects of Additives

2.6.1.1. Fillers

In general, fillers are used with polyester resins to help reduce costs—they are much less expensive than the resin itself as well as most reinforcing agents. However, there are other advantages associated with the use of fillers.

Table 2.5. Properties of Cast Polyester Resins.

PROPERTY	ASTM TEST METHOD	RIGID	FLEXIBLE
Mechanical Properties			
Specific gravity	D792	1.10–1.46	1.01–1.20
Flexural yield strength, psi (MPa)	D790	8500–23,000 (60–160)	—
Compressive strength, psi (MPa)	D695	13,000–30,000 (90–200)	—
Tensile strength, psi (MPa)	D638	6000–13,000 (40–90)	500–3000 (3.4–21)
Elongation, %	D638	<5.0	40–310
Tensile modulus of elasticity, psi $\times 10^5$ (GPa)	D638	3.0–6.4 (20–44)	—
Impact strength [$\frac{1}{2} \times \frac{1}{2}$ in. (13 × 13 mm) notched bar, Izod test], ft-lb/in. of notch (J/m)	D256	0.2–0.4 (16–32)	>7.0 (>370)
Rockwell hardness	D785	M70–M115 (50–75 Barcol)	84–94 (Shore D)
Refractive index	D542	1.523–1.570	1.537–1.550
Electrical Properties			
Volume resistivity (50% RH, 23°C = 73°F), Ω-cm	D257	10^{15}	—
Dielectric strength [1/8 in. (3.2 mm) thickness]			
Short-time test, V/mil	D149	380–500	250–400
Step-by-step test, V/mil	D149	280–400	277
Dielectric constant			—
60 Hz	D150	3.00–4.36	4.4–8.1
10^3 Hz	D150	2.8–5.2	4.5–7.1
10^6 Hz	D150	2.8–4.1	4.1–5.9
Dissipation (power) factor			—
60 Hz	D150	0.003–0.028	0.026–0.310
10^3 Hz	D150	0.005–0.025	0.016–0.050
10^6 Hz	D150	0.006–0.026	0.023–0.060
Arc resistance, sec	D495	125	135
Resistance Characteristics			
Heat resistance (continuous), °F (°C)	—	250 (121)	250 (121)
Water absorption [24 hr, 1/8 in. (3.1 mm) thickness], %	D570	0.15–0.60	0.5–2.5
Effect of sunlight	—	Yellows	—

For example, their use permits higher curing temperatures by reducing the concentration of reactive materials, which, in turn, tends to reduce the peak exotherm temperature. Furthermore, the use of fillers reduces the tendency of the resin to craze or crack during cure, reduces the degree of shrinkage, and helps produce smoother molded surfaces.

Clays and carbonates are the most commonly used fillers. In general, clay fillers produce a lower peak exotherm temperature than carbonate fillers, but they tend to darken when cured at high temperatures. The amount of filler used can vary from 10 to 70 wt %, depending on the filler, resin, and intended use; 20–30 wt % is normally used. Fillers should be pretested in the resin because some types tend to shorten both the storage life and pot life of a resin.

Since both clay and carbonate fillers yield a more rigid product, the flexural modulus of elasticity of a resin increases with the use of either type. The electrical properties of a resin are also changed by the addition of fillers. Table 2.7 shows the effect of fillers on the properties of a typical general-purpose resin.

In selecting a filler, its absorbent qualities and particle size must be considered, since both can affect the properties of the molded product. Fillers that are highly absorbent can cause washing of the fiberglass reinforcement as a result of poor resin flow (i.e., the resin mix may develop a strong hydraulic surge and incomplete resin flow, which, in turn, could lead to a large void content in the molded part). Fillers of large particle size may tend to be filtered out as the resin flows through the reinforcing material during cure; this would result in an uneven distribution of the filler.

Fillers must be well dispersed in the resin

mixture. It is good practice to initially blend the filler with as little resin as possible until the mix forms a heavy paste. This will produce a greater shearing action, which tends to cause proper breakdown of the filler particles.

2.6.1.2. Dyes and Pigments

Dyes and pigments can be readily added to polyester resins. Since most dyes and pigments are used in small amounts, they do not appreciably affect the properties of the cured resin. However, they do often inhibit or accelerate cure. In addition, some dye and pigment colors tend to impair weathering properties.

In general, dyes are added to the resin by first dissolving them in monomeric styrene and then mixing the resulting solution into the

Table 2.6. Properties of Fiberglass-Reinforced Polyester Resins.

PROPERTY	ASTM TEST METHOD	WOVEN CLOTH	CHOPPED ROVING	SHEET MOLDING COMPOUND
Mechanical Properties				
Specific gravity	D792	1.5–2.1	1.35–2.30	1.65–2.60
Flexural yield strength, psi (MPa)	D790	40,000–80,000 (276–550)	10,000–40,000 (69–276)	10,000–36,000 (69–248)
Flexural modulus of elasticity, psi × 10^5 (GPa)	D790	—	10–30 (69–207)	10–22 (69–150)
Compressive strength, psi (MPa)	D695	25,000–50,000 (172–344)	15,000–30,000 (103–206)	15,000–30,000 (103–206)
Tensile strength, psi (MPa)	D638	30,000–50,000 (206–344)	15,000–30,000 (103–206)	8000–20,000 (55–138)
Elongation, %	D638	0.5–2.0	0.5–5.0	—
Tensile modulus of elasticity, psi × 10^5 (GPa)	D638	15.0–45.0 (103–310)	8.0–20.0 (55–138)	—
Impact strength [½ × ½ in. (13 × 13 mm) notched bar, Izod test], ft-lb/in. of notch (J/m)	D256	5.0–30.0 (267–1600)	2.0–20.0 (107–1070)	7.0–22.0 (374–1175)
Barcol hardness	—	60–80	50–80	50–70
Electrical Properties				
Volume resistivity (50% RH, 23°C = 73°F), Ω-cm	D257	10^{14}	10^{14}	10^{14}–10^{15}
Dielectric strength [1/8 in. (3.2 mm) thickness]				
Short-time test, V/mil	D149	350–500	350–500	380–450
Step-by-step test, V/mil	D149	300–400	300–450	350–400
Dielectric constant				
60 Hz	D150	4.1–5.5	3.8–6.0	4.4–6.3
10^3 Hz	D150	4.2–6.0	4.0–6.0	4.4–6.1
10^6 Hz	D150	4.0–5.5	3.5–5.5	4.2–5.8
Dissipation (power) factor				
60 Hz	D150	0.01–0.04	0.01–0.04	0.007–0.021
10^3 Hz	D150	0.01–0.06	0.01–0.05	0.007–0.015
10^6 Hz	D150	0.01–0.03	0.01–0.03	0.016–0.024
Arc resistance, sec	D495	60–120	120–180	120–200
Resistance Characteristics				
Heat resistance (continuous), °F (°C)	—	300–350 (149–177)	300–350 (149–177)	300–400 (149–204)
Water absorption [24 hr, 1/8 in. (3.1 mm) thickness], %	D570	0.05–0.50	0.1–1.0	0.10–0.15
Effect of sunlight	—	Slight	Slight	Slight

Table 2.7. Effect of Fillers on the Properties of a Typical General-Purpose Resin.

PROPERTY	UNFILLED	30 PARTS CALCIUM CARBONATE FILLER	30 PARTS CLAY FILLER (ALUMINUM SILICATE)
Mechanical Properties			
Specific gravity (25°C = 77°F)			
Uncured	1.20	1.34	1.34
Cured	1.30	1.48	1.47
Heat distortion point, °F (°C)	174 (79)	181 (83)	181 (83)
Ultimate flexural strength, psi (MPa)	17,600 (121)	9000 (62)	10,550 (73)
Flexural modulus of elasticity, psi $\times 10^6$ (GPa)	0.63 (4.3)	1.03 (7.1)	0.94 (6.5)
Electrical Properties			
Dielectric strength (100-mil cast specimen, 25°C = 77°F), V/mil	440	400	435
Dielectric constant			
60 Hz	3.4	4.1	4.1
10^3 Hz	3.3	4.0	4.0
10^6 Hz	3.2	3.9	3.7
Dissipation (power) factor			
60 Hz	0.004	0.010	0.038
10^3 Hz	0.006	0.009	0.024
10^6 Hz	0.019	0.016	0.017
Dielectric loss factor			
60 Hz	0.013	0.040	0.015
10^3 Hz	0.045	0.034	0.094
10^6 Hz	0.058	0.067	0.063
Arc resistance (25°C = 77°F), sec	100	85	125

resin. This permits filtering of the solution in order to obtain good clarity. Dyes are usually employed in concentrations of 0.05–0.1% (based on the monomer and resin).

Pigments are added to the resin by one of two methods. On the one hand, they can be thoroughly ground into a small amount of the resin or a grinding medium before being added to the bulk resin. On the other hand, they can be ground into a carrying vehicle that is compatible with the resin. From 1 to 5% of the dispersion is usually employed, the exact amount depending on the reactivity of the resin, as well as the depth of color and degree of opacity desired in the finished product.

2.6.1.3. Flame-Retardant Substances

Flame-retardant substances are sometimes added to the resin in order to impart certain desired burning-rate properties. These substances can either be combined with the resin physically or a reactive modifier that unites them chemically can be used. Most organic phosphate esters and halogenated hydrocarbons (e.g., chlorinated waxes), as well as antimony trioxide, are combined with the resin physically. On the other hand, chlorendic acid and its derivatives are combined chemically. The addition of flame-retardant substances can affect the color, tensile strength, electrical properties, and molding characteristics of the resin.

2.6.1.4. Ultraviolet Absorbers

Ultraviolet absorbers are added to resin mixes that are subject to exposure to UV rays (from natural sunlight or from fluorescent light). These substances must have good initial color and be compatible with the catalysts and

Table 2.8. Cost of Unsaturated Polyester Resins (1980).

RESIN	COST
General purpose (orthophthalic)	51¢/lb
Resilient (isophthalic)	54¢/lb
Low shrinkage	60¢/lb
Corrosion (chemical) resistant	82–87¢/lb
Fire resistant	75–82¢/lb

accelerators used. They usually have a minimal effect on the polymerization reaction.

2.7. COST OF UNSATURATED POLYESTER RESINS

Unsaturated polyester resins are relatively inexpensive, and this accounts for their wide usage in the plastics industry. Epoxy resins, although somewhat more expensive, are preferred in some applications because of their low shrinkage, good adhesion, and resistance to attack by chemicals. Table 2.8 shows the approximate cost per pound of the different types of unsaturated polyester resin.

References

1. C. Ellis, U.S. Patent 2,195,362, March 26, 1940; appl. May 21, 1936 (Ellis-Foster).
2. C. Ellis, U.S. Patent 2,255,313, September 9, 1941; appl. August 6, 1937 (Ellis-Foster).
3. *Modern Plastics* **55** (1), 50 (1978).
4. E. E. Parker, *Ind. Eng. Chem.* **58** (4), 53 (1966).
5. A. L. Smith, Sixth Annual Technical Session, Reinforced Plastics Division, SPI, 1951.
6. Technical Bulletins, Lucidol Division, Wallace and Tiernan, Inc.
7. *Modern Plastics Encyclopedia*, Vol. 52 (10A), 1975–1976, p. 482.

Supplementary References

G. Lubin (ed.), *Handbook of Fiberglass and Advanced Plastics Composites*, Van Nostrand Reinhold, New York, 1969, Chapter 2.

J. Bjorksten, *Polyesters and Their Applications*, Van Nostrand Reinhold, New York, 1956.

J. R. Lawrence, *Polyester Resins*, Van Nostrand Reinhold, New York, 1960.

H. V. Boenig, *Unsaturated Polyesters: Structure and Properties*, Elsevier Publishing Co., New York, 1964.

R. E. Park and E. E. Parker, in: *Kirk-Othmer Encyclopedia of Chemical Technology*, 2nd supplement, Wiley-Interscience, New York, 1960, p. 902.

R. M. Nowak and L. C. Rubins, in: *Kirk-Othmer Encyclopedia of Chemical Technology*, Vol. 20, Wiley-Interscience, New York, 1969, p. 791.

H. V. Boenig, in: *Encyclopedia of Polymer Science and Technology*, Vol. 11, Wiley-Interscience, New York, 1969, p. 129.

E. E. Parker and J. R. Peffer, in: *Polymerization Processes* (edited by C. E. Schildknecht and I. Skeist), Wiley-Interscience, New York, 1977, p. 68.

3
VINYL ESTER RESINS

Matthew B. Launikitis
Shell Development Company
Houston, Texas

3.1. INTRODUCTION

3.1.1. Chemical Definition

Vinyl ester resins are thermosetting resins that consist of a polymer backbone with an acrylate (R = H) or methacrylate (R = CH_3) termination $R\text{—}[\text{—}O\text{—}\overset{\overset{\displaystyle O}{\|}}{C}\text{—}\overset{\overset{\displaystyle R}{|}}{C}\text{=}C]$. Although vinyl ester resins have sometimes been classified as polyesters, they are typically diesters that (depending on the polymer backbone) contain recurring ether linkages.

The backbone component of vinyl ester resins can be derived from an epoxide resin, polyester resin, urethane resin, and so on, but those based on epoxide resins are of particular commercial significance.

3.1.2. Characteristics

Vinyl esters resins can be used in a neat form (e.g., no diluent) or they can contain either a vinyl-type reactive comonomer (e.g., styrene, vinyl toluene, and trimethylol propane triacrylate) or a nonreactive diluent (e.g., methyl ethyl ketone and toluene). Typically, the methacrylate vinyl ester resins contain styrene and are used in chemical-resistant fiberglass-reinforced plastics (FRP), whereas the acrylate vinyl ester resins are supplied undiluted, with appropriate coreactants added during the formulation of ultraviolet (UV)-cure coatings and inks.

The physical and handling properties of vinyl ester resins depend on the source of vinyl termination (methacrylate or acrylate), the amount and type of coreactant, and the type and molecular weight of the resin backbone.[1]

Upon cure, the styrenated methacrylate vinyl ester resins exhibit excellent resistance to acids, bases, and solvents. The acrylate vinyl ester resins are more susceptible to hydrolysis than the methacrylate vinyl ester resins and thus are not generally used in applications requiring premium chemical resistance.[1] Conversely, acrylate vinyl ester resins are preferred in radiation-cure systems because of their greater reactivity.

Undiluted vinyl ester resins vary from semisolid to solid and therefore utilize both reactive and nonreactive diluents to provide workable viscosity levels and enhanced reactivity.[2]

Epoxide backbones of various molecular weights are used in vinyl ester resins. Higher molecular weights produce greater toughness and resiliency, lower solvent resistance, and lower heat resistance.

Unlike polyesters, vinyl ester resins possess a low ester content and a low vinyl functionality, which result in a greater resistance to hydrolysis as well as lower peak exotherm temperatures and less shrinkage upon cure. Like polyesters, vinyl ester resins exhibit a finite storage life that is controlled by the addition of polymerization inhibitors (free-radical traps) during manufacture.

3.1.3. History

Although various types of vinyl ester resin were prepared in laboratory-scale quantities during the late 1950's, it was not until 1965 that they were commercially introduced by Shell Chemical Company under the trade name of EPOCRYL® Resins.[3] These resins were identified as epoxy (meth)acrylates and

BISPHENOL A EPOXY-BASED (METHACRYLATE) VINYL ESTER RESIN (n = 0–3)

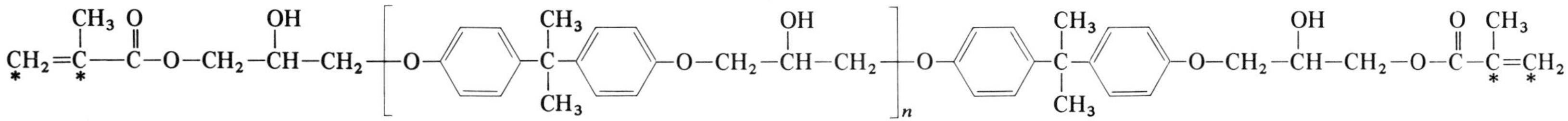

BISPHENOL A EPOXY-BASED (ACRYLATE) VINYL ESTER RESIN (n = 0)

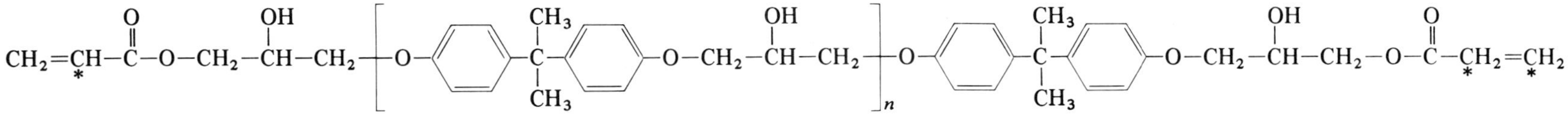

PHENOLIC-NOVOLAC EPOXY-BASED VINYL ESTER RESIN (n = 0-1)

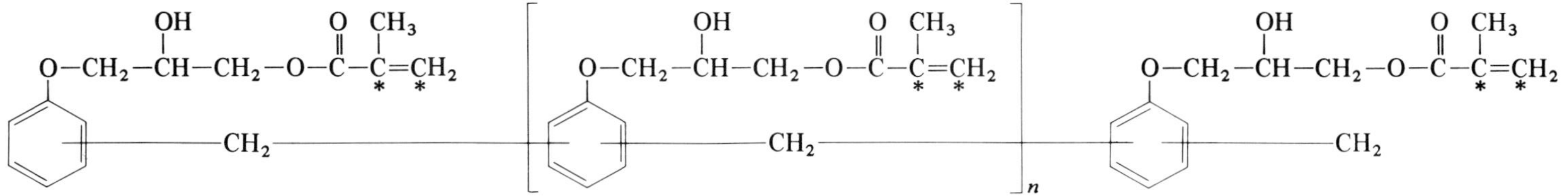

TETRABROMO BISPHENOL A EPOXY-BASED VINYL ESTER RESIN (n = 0–3)

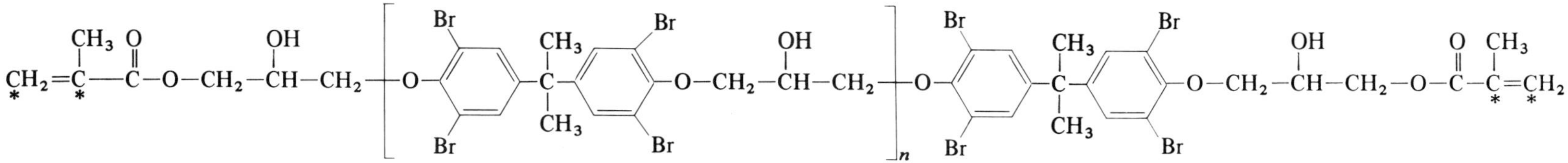

Figure 3.1. Major types of epoxy-based vinyl ester resin. An asterisk (*) denotes a reactive site.

Table 3.1. Commercially Available Vinyl Ester Resins and Their End Uses.

RESIN	DESCRIPTION	CHARACTERISTICS AND END USES
EPOCRYL® Resin 12[a]	A neat dimethacrylate ester of a low-molecular-weight bisphenol A epoxy resin	Nominal 1,000,000 cp (1 kPa·sec) viscosity designed for molding, adhesives, and electrical prepreg
EPOCRYL® Resin 370[a]	A neat diacrylate ester of a low-molecular-weight bisphenol A epoxy resin	Base resin for formulation of UV-cure inks and coatings
CoRezyn VE-8100[b] Derakane® 411-C-50[c] EPOCRYL® Resin 321[a]	A dimethacrylate ester of an intermediate-molecular-weight bisphenol A epoxy resin containing 50 wt % styrene	Nominal 100 cP (1 dPa·sec) viscosity designed for chemical-resistant FRP applications: casting, filament winding, flooring, and pultrusion.
Corrolite 31-345[d] CoRezyn VE-8300[b] Derakane® 411-45[c] EPOCRYL® Resin 322[a]	A dimethacrylate ester of a high-molecular-weight bisphenol A epoxy resin containing 45 wt % styrene	Nominal 500 cP (5 dPa·sec) viscosity designed for chemical-resistant FRP applications: hand lay-up and filament winding
Derakane® 470-36[c]	A methacrylate ester of a phenolic-novolac epoxy resin containing 36 wt % styrene	Nominal 200 cP (2 dPa·sec) viscosity designed for solvent resistance and high-service-temperature FRP applications
EPOCRYL® Resin 480[a]	A methacrylate ester of a low-molecular-weight bisphenol A epoxy resin containing 40 wt % styrene	Nominal 500 cP (5 dPa·sec) viscosity designed for solvent resistance and high-service-temperature FRP applications. Exhibits high tensile strengths and elongation with high modulus. Designed for filament winding and hand lay-up applications
Derakane® 510A40[c]	A dimethacrylate ester of a brominated bisphenol A epoxy resin containing 40 wt % styrene	Nominal 350 cP (3.5 dPa·sec) viscosity designed to impart fire-retardancy for chemical-resistant FRP applications

[a]Shell Chemical Company.
[b]Interplastic Corporation.
[c]Dow Chemical Company.
[d]Reichhold Chemical Company.

were shown to provide chemical resistance superior to that of the premium (chemical-resistant) polyester resins of that period. In 1966 Dow Chemical Company coined the Derakane® vinyl ester resin designation with the introduction of a similar series of resins for molding applications.[4] In 1977, Interplastic Corporation and Reichhold Chemical Company introduced vinyl ester resins under the CoRezyn and Corrolite product designations, respectively. Figure 3.1 shows the chemical structures of the major types of epoxy-based vinyl ester resin, whereas Tables 3.1 lists the various types of commercially available vinyl ester resin and their end uses.

3.1.4. Manufacture

Vinyl ester resins are produced by the addition of ethylenically unsaturated carboxylic acids (methacrylic or acrylic acid) to an epoxide resin (usually of the bisphenol A–epichlorohydrin type):

$$\overbrace{\mathrm{C-C}}^{\mathrm{O}}\mathrm{-C-R-C-}\overbrace{\mathrm{C-C}}^{\mathrm{O}} + 2\,\mathrm{HO-}\overset{\overset{\mathrm{O}}{\|}}{\mathrm{C}}\mathrm{-}\overset{\overset{\mathrm{R'}}{|}}{\mathrm{C}}\mathrm{=C} \xrightarrow{\text{Heat catalyst}}$$

$$\mathrm{C=}\overset{\overset{\mathrm{R'}}{|}}{\mathrm{C}}\mathrm{-}\overset{\overset{\mathrm{O}}{\|}}{\mathrm{C}}\mathrm{-O-C-}\overset{\overset{\mathrm{OH}}{|}}{\mathrm{C}}\mathrm{-C-R-C-}\overset{\overset{\mathrm{OH}}{|}}{\mathrm{C}}\mathrm{-C-O-}\overset{\overset{\mathrm{O}}{\|}}{\mathrm{C}}\mathrm{-}\overset{\overset{\mathrm{R'}}{|}}{\mathrm{C}}\mathrm{=C} + \text{coreactant} + \text{inhibitors}$$

$$\text{where } \mathrm{R} = -\mathrm{O}-\mathrm{C_6H_4}-\underset{\underset{\mathrm{C}}{|}}{\overset{\overset{\mathrm{C}}{|}}{\mathrm{C}}}-\mathrm{C_6H_4}-\mathrm{O}- \text{ and } \mathrm{R'} = \mathrm{H} \text{ or } -\mathrm{CH_3}$$

The reaction of acid addition to the epoxide ring (esterification) is exothermic and produces a hydroxyl group without the formation of by-products (e.g., as in polyesterifications where water is produced). Appropriate diluents and polymerization inhibitors are added during or after esterification.

Epoxide resins that have been used to produce vinyl ester resins include the bisphenol A types (general-purpose and heat-resistant vinyl esters), the phenolic-novolac types (heat-resistant vinyl esters), and the tetrabromo bisphenol A types (fire-retardant vinyl esters). For acrylate vinyl ester resins, the low-molecular-weight bisphenol A epoxies are generally used as the backbone component (see Fig. 3.1).

3.1.5. Cure

Vinyl ester resins, like unsaturated polyester resins, contain double bonds that react and crosslink in the presence of free radicals produced by chemical, thermal, or radiation sources (see Fig. 3.2). Cure proceeds by a free-radical mechanism comprising initiation (induction period), propagation, and termination. The rate-determining step is initiation, in which the catalyst "swamps-out" the polymerization inhibitors subsequent to reaction with the double bonds of the vinyl ester (and its coreactant).

3.2. PERFORMANCE PROPERTIES

3.2.1. Resin Properties

The resin properties of vinyl esters are typically characterized by viscosity, diluent content, and reactivity. The viscosity of vinyl ester resins as a function of molecular weight, diluent content, temperature, and coreactant content and type is shown in Fig. 3.3. The time–temperature reactivity as a function of catalyst type and concentration for a commercial vinyl ester resin is shown in Fig. 3.4. Other typical vinyl ester resin solution properties are listed in Table 3.2.

3.2.2. Mechanical-Strength Properties

Except for the thermal distortion temperature, the mechanical-strength (tensile and flexural) properties of cured vinyl ester resin castings

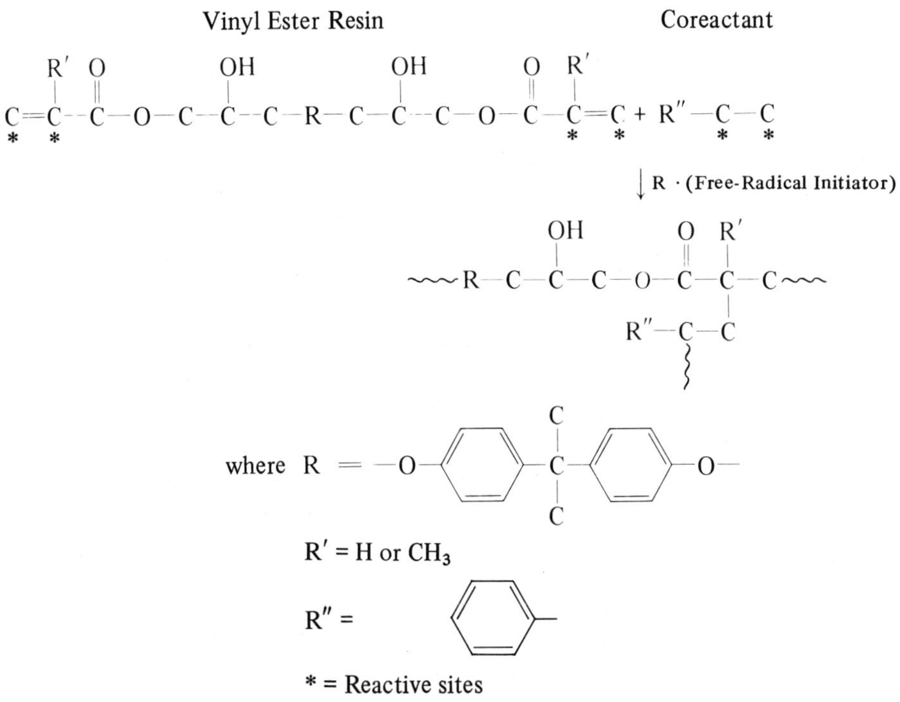

Vinyl Ester Resin

Coreactant

R · (Free-Radical Initiator)

where R =

$R' = H$ or CH_3

$R'' =$

* = Reactive sites

Figure 3.2. Model cure of vinyl ester resins.

vary only slightly as a function of the molecular weight and type of epoxide backbone; variations in coreactant type and amount also produce moderate changes in the mechanical-strength properties (see Table 3.3). As with other polymers, fibrous or particulate fillers generally enhance the mechanical-strength properties of these resins (see Table 3.4).

3.2.3. Electrical Properties

Vinyl ester resins have not been widely used in electrical applications even though they possess good dielectric strength and resistivity characteristics as shown in Table 3.5 and Fig. 3.5.

3.2.4. Chemical Resistance

At present, the largest commercial use of vinyl ester resins (methacrylates) is in chemical-resistant FRP equipment: pipe, ducts, scrubbers, flue stacks, and storage tanks. They exhibit excellent acid, base, and solvent resistance over a wide temperature range in both resin castings and glass laminates. Figure 3.6 shows typical chemical resistance data for a vinyl ester resin used in these applications.

The excellent acid–base resistance of epoxy-based methacrylate vinyl ester resins is attributable to their low ester content, the protection of the ester linkages by the adjacent methyl shield, and the steric hindrance of the coreactant,[1] usually styrene.

3.2.5. Shrinkage

Upon cure, the volumetric shrinkage of vinyl ester resins is generally lower than that of unsaturated polyester resins but higher than that of their parent epoxy resins. The volumetric shrinkage increases with increased backbone double bond and coreactant content (see Table 3.6), but decreases with increased filler loading or fiber content.

3.2.6. Adhesion

Because of their high shrinkage values (compared to epoxy resins), vinyl ester resins

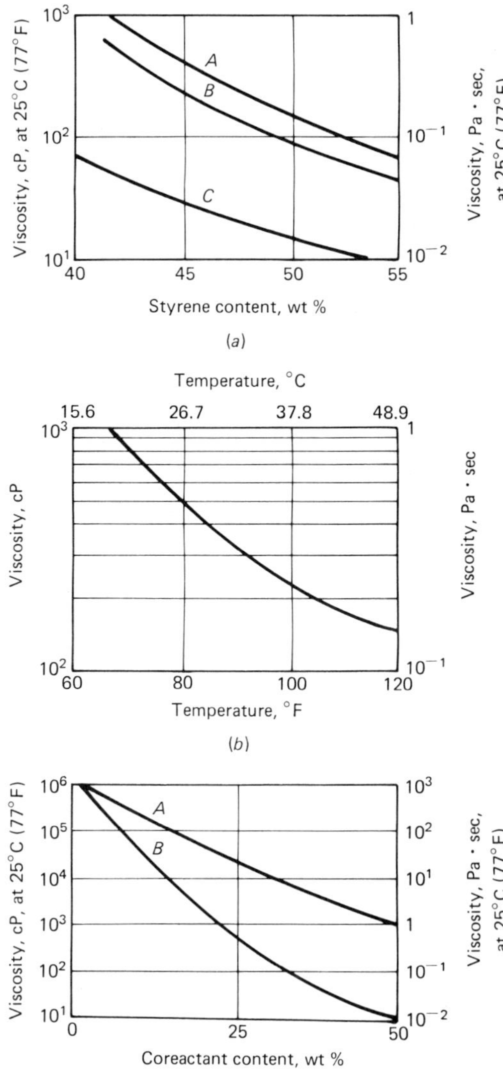

Figure 3.3. Viscosity characterization of vinyl ester resins. (*a*) Viscosity versus styrene content and resin molecular weight (methacrylate vinyl esters in styrene): curve *A*, molecular weight = 1200 (EPOCRYL® Resin 322); curve *B*, molecular weight = 900 (EPOCRYL® 321); curve *C*, molecular weight = 500 (EPOCRYL® Resin 12). (*b*) Viscosity versus temperature [resin molecular weight = 1200 (EPOCRYL® Resin 322), styrene content = 45 wt %]. (*c*) Viscosity versus coreactant content and type: curve *A*, acrylate ester-trimethylol propane triacrylate (EPOCRYL®; Resin 370) curve *B*, methacrylate vinyl ester-styrene (EPOCRYL® Resin 12).

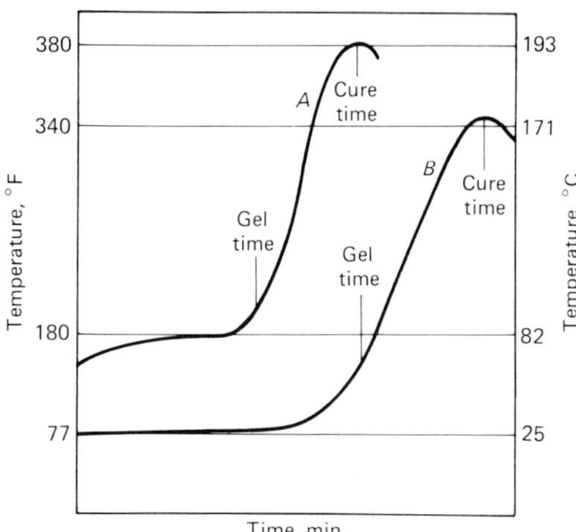

Figure 3.4. Reactivity of vinyl ester resins: time-temperature cure response. Curve *A*: 100 parts by weight EPOCRYL® Resin 322 (Shell Chemical Company), 1.0 part by weight benzoyl peroxide (BPO), cure temperature = 180° F (82° C), gel time = 12 min., cure time = 15 min., peak exotherm temperature = 380° F (193° C). Curve *B*: 100 parts by weight EPOCRYL® Resin 322, 2.0 parts by weight methyl ethyl ketone peroxide (60% MEKP), 0.4 part by weight cobalt naphthenate (6% cobalt), cure temperature = 77° F (25° C), gel time = 15 min., cure time = 25 min., peak exotherm temperature = 340° F (171° C).

exhibit only moderate adhesive strengths (see Table 3.7). To reduce shrinkage and obtain maximum strength, adhesive formulations are based on vinyl ester resins with low backbone double bond content, low coreactant level, and high filler content. The addition of various rubber compounds to reduce shrinkage[5] also improves the peel strength of vinyl ester adhesives.

Table 3.2. Typical Solution Properties of Vinyl Ester Resins

PROPERTY	RANGE
Gardner color	2–8
Appearance	Clear, straw yellow to amber
Specific gravity	1.03–1.22
Flash point, TOC (tag open cup) °C	32–93
Water content, wt %	0.01–0.20
Acidity, eq/100 g	0.001–0.050
Epoxide content, eq/100	0.001–0.020

Table 3.3. Mechanical-Strength Properties of Unfilled Vinyl Ester Resin Castings.

	Effect of Resin Molecular Weight (Bisphenol A Epoxy-Based Vinyl Ester Resins Containing 50 wt % Styrene)		
Resin	EPOCRYL® Resin 12[a]	EPOCRYL® Resin 321[a]	EPOCRYL® Resin 322[a]
Vinyl ester molecular weight	500	900	1200
Property			
Thermal distortion temperature, °F (°C)	240 (116)	210 (99)	200 (93)
Tensile properties at 77°F (25°C)			
Strength, psi (MPa)	11,700 (80.7)	11,600 (79.9)	11,400 (78.6)
Modulus of elasticity, psi × 10^6 (GPa)	0.48 (3.3)	0.46 (3.2)	0.44 (3.0)
Elongation, %	4.5	5.0	5.5
Flexural properties at 77°F (25°C)			
Strength, psi (MPa)	19,600 (135.1)	19,000 (131.0)	18,600 (128.2)
Modulus of elasticity, psi × 10^6 (GPa)	0.50 (3.4)	0.46 (3.2)	0.44 (3.0)
	Effect of Epoxy Resin Backbone Type (Vinyl Ester Resins Containing 40 wt % Styrene)		
Resin	EPOCRYL® Resin 480[a]	Derakane® 470[b]	Derakane® 510A[b]
Backbone epoxy resin type	Bisphenol A	Phenolic novolac	Tetrabromo bisphenol A
Property			
Thermal distortion temperature, °F (°C)	250 (121)	255 (124)	210 (99)
Tensile properties at 77°F (25°C)			
Strength, psi (MPa)	12,200 (84.1)	11,000 (75.8)	10,600 (73.1)
Modulus of elasticity, psi × 10^6 (GPa)	0.50 (3.4)	0.50 (3.4)	—
Elongation, %	5.0	3.5	4.0
Flexural properties at 77°F (25°C)			
Strength, psi (MPa)	18,000 (124.1)	18,000 (124.1)	18,000 (124.1)
Modulus of elasticity, psi × 10^6 (GPa)	0.50 (3.4)	0.50 (3.4)	0.50 (3.4)
	Effect of Styrene Content (Bisphenol A Epoxy-Based Vinyl Ester Resin EPOCRYL® Resin 12[a])		
Styrene content, wt %	0	15	50
Property			
Thermal distortion temperature, °F (°C)	300 (149)	280 (138)	240 (116)
Tensile properties at 77°F (25°C)			
Strength, psi (MPa)	—	11,800 (81.3)	11,700 (80.7)
Modulus of elasticity, psi × 10^6 (GPa)	0.59 (4.1)	0.53 (3.7)	0.48 (3.3)
Elongation, %	2.0	2.3	4.5
Flexural properties at 77°F (25°C)			
Strength, psi (MPa)	18,400 (126.8)	19,600 (135.1)	18,200 (125.5)
Modulus of elasticity, psi × 10^6 (GPa)	0.58 (4.0)	0.55 (3.8)	0.50 (3.4)

[a]Shell Chemical Company.
[b]Dow Chemical Company.

Table 3.4. Mechanical-Strength Properties of Glass-Reinforced Vinyl Ester Resins

Resin	EPOCRYL® Resin 322[a]			
Glass content, wt %	0	25	40	75
Glass type	—[b]	Mat[c]	Woven roving[d]	Roving[e]
Tensile properties at 77° F (25° C)				
Strength, psi (MPa)	11,400 (78.6)	14,000 (97)	28,000 (193)	>55,000 (>379)
Modulus of elasticity, psi × 10^6 (GPa)	0.44 (3.0)	0.90 (6.2)	1.55 (10.7)	6.3 (43.4)
Elongation, %	5.5	2.0	1.4	0.8
Flexural properties at 77° F (25° C)				
Strength, psi (MPa)	18,600 (128.2)	19,000 (131.0)	32,000 (220.5)	111,000 (764.7)
Modulus of elasticity, psi × 10^6 (GPa)	0.44 (3.0)	0.70 (4.8)	1.25 (8.6)	3.8 (26.2)

[a]Shell Chemical Company.
[b]125-mil-thick unfilled casting.
[c]120-mil-thick laminate: 2 layers chopped mat, 2 layers C-veil.
[d]250-mil-thick laminate: 2 layers woven roving, 4 layers chopped mat, 1 layer C-veil.
[e]500-mil-thick laminate: roving.

3.3. APPLICATIONS

3.3.1. Radiation Cure

Although both the methacrylate and the acrylate vinyl ester resins have been employed in radiation-cure applications, use of the latter is preferred in UV-cure coating and ink formulations (see Table 3.8) because they cure faster. In many UV applications, cure is complete within 0.1 sec. Electron beams and other radiation sources have also been used to cure vinyl ester resins. The major advantages of radiation cure as compared to thermal cure include greater energy efficiency, reduction or elimination of volatile emissions, faster line speeds, ambient temperature cures, low equipment costs, and low floor space requirements.

3.3.2. Fiberglass-Reinforced Plastics

Vinyl ester resins can be fabricated into FRP composites via filament winding, hand lay-up, centrifugal casting, bag molding, or pultrusion. Vinyl ester resin-based FRP equipment (e.g., pipes, ducts, scrubbers, flue stacks, and

Table 3.5. Electrical Properties of Unfilled Vinyl Ester Resin Castings.

PROPERTY	EPOCRYL® RESIN 12[a] (25 wt % Styrene)	EPOCRYL® RESIN 12[a] (50 wt % Styrene)
Dielectric constant 1MHz/1kHz 25°C (77°F)	3.16/3.5	3.15/3.33
Dissipation factor 1MHz/1kHz 25°C (77°F)	0.020/0.009	0.014/0.008
Insulation resistance, Ω [35 days at 60°C (140°F), 95% RH]	$> 10^{12}$	$> 10^{12}$
Volume resistivity, Ω-cm	$> 10^{16}$	1.3×10^{16}
Surface resistivity, Ω	$> 10^{16}$	4.5×10^{16}

[a]Shell Chemical Company.

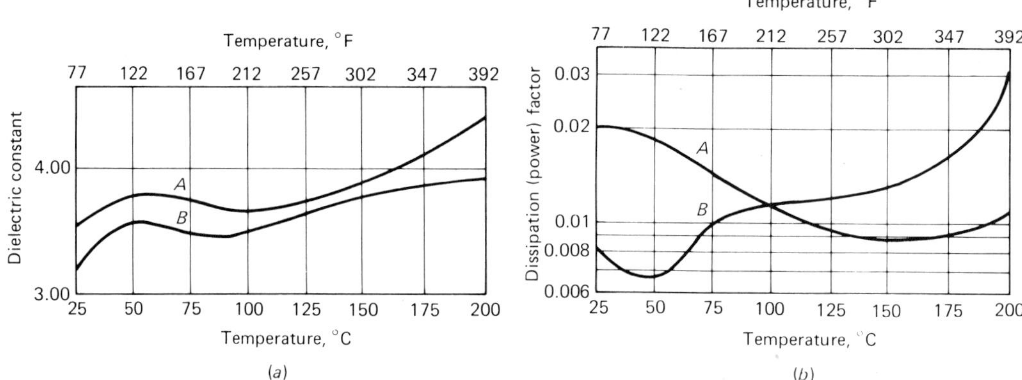

Figure 3.5. Electrical properties of unfilled vinyl ester resin castings. (a) Dielectric constant versus temperature for EPOCRYL® Resin 12 (Shell Chemical Company) plus 25 wt % styrene: curve *A*, 1 kHz; curve *B*, 1 MHz. (*b*) Dissipation factor versus temperature for EPOCRYL® Resin 12 plus 25 wt % styrene: curve *A*, 1 kHz; curve *B*, 1 MHz.

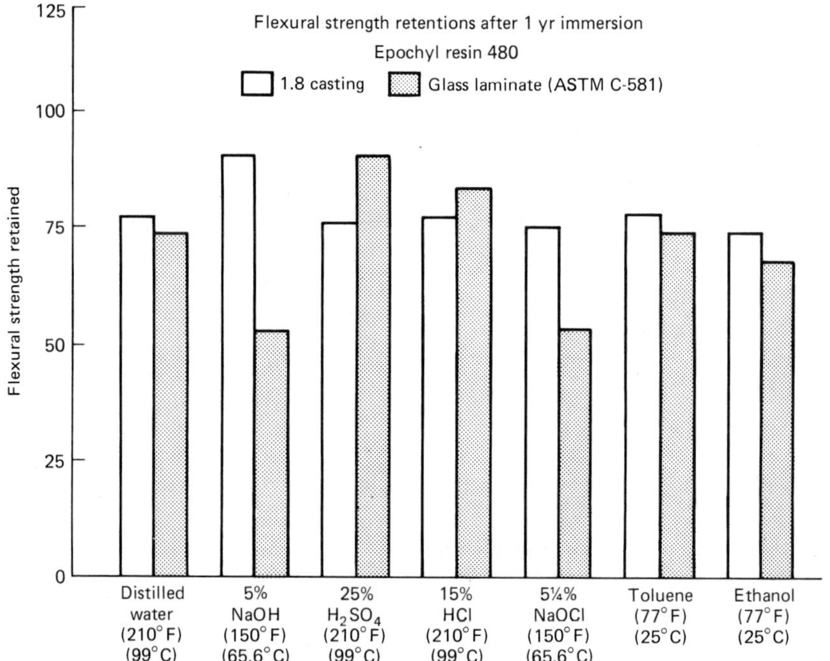

Figure 3.6. Chemical resistance of vinyl ester resins.

Table 3.6. Volumetric Shrinkage of Vinyl Ester Resins.

RESIN MOLECULAR WEIGHT	STYRENE CONTENT, WT %	DOUBLE BOND CONTENT, eq/100 g	VOLUMETRIC SHRINKAGE, %: FROM GEL POINT TO FULL CURE	VOLUMETRIC SHRINKAGE, %: FROM LIQUID RESIN TO FULL CURE
500[a]	0	0.40	4.3	5.4
500[a]	25	0.55	6.8	7.6
500[a]	50	0.70	9.1	10.3
900[b]	50	0.61	—	7.0
1200[c]	50	0.58	—	5.4

[a]EPOCRYL® Resin 12.
[b]EPOCRYL® Resin 321.
[c]EPOCRYL® Resin 322.

Table 3.7. Typical Vinyl Ester Adhesive Formulation.

FORMULATION	PARTS BY WEIGHT
EPOCRYL® Resin 12[a]	95
Styrene	5
Asbestos floats	30
Luperco ATC[b]	1

ADHESIVE STRENGTH
[Aluminum Clad 2024 T-3; Cure: 2 hr at 176° F (80° C)]

TEMPERATURE, °F	(°C)	LAP SHEAR, psi	(MPa)
77	(25)	1360	(9.38)
180	(82)	1620	(11.17)
240	(116)	1360	(9.38)
300	(149)	1200	(8.27)

[a]Shell Chemical Company.
[b]Lucidol Division, Wallace and Tiernan, Inc.

Table 3.8. Selected Vinyl Ester Resin UV-Cure Formulations and Typical Properties.

MEDIUM-VISCOSITY CLEAR COATING		
EPOCRYL® Resin 370,[a] parts by weight	60	60
HDODA,[b] parts by weight	40	40
Viscosity (77° F = 25° C), P (Pa · sec)	8 (0.8)	8 (0.8)
DEAP[c] or Vicure 10,[d] parts by weight	2	2
Cure	UV only	UV plus heat[e]
Cure speed, ft/min./lamp[f]	125	125
Double methyl ethyl ketone rubs	>100	>100
Pencil hardness[g]	H	6H
Weight retained, %[g]	99	91
Scotch® tape adhesion, %[g]	10	100
BLACK-PIGMENTED INK BASE		
EPOCRYL® Resin 370,[a] parts by weight	75	
TMPTA,[h] parts by weight	25	
IGEPAL CO 430,[i] parts by weight	0.4	
Viscosity (77° F = 25° C), P (Pa · sec)	380 (3.8)	
Furnace black,[j] parts by weight	10	
Benzophenone, parts by weight	7	
Michler's ketone,[k] parts by weight	3.5	
Cure speed, ft/min./lamp[l]	50–100	
Double methyl ethyl ketone rubs	5–20	

[a]Shell Chemical Company.
[b]1,6-Hexanediol diacrylate—Celanese Chemical Company.
[c]Diethoxy acetophenone—Union Carbide Corporation.
[d]Stauffer Chemical Company.
[e]5 min. at 399° F (204° C).
[f]200-W/in. mercury lamp with drawdowns at 0.6 mil thickness on Bonderized 100 steel panels.
[g]Bonderized 100 steel panel.
[h]Trimethylol propane triacrylate—Celanese Chemical Company.
[i]GAF Corporation.
[j]Mineral Pigments Company—M8405.
[k]Biddle Sawyer Corporation.
[l]200-W/in. mercury lamp using "Quickpeek" ink proofer at 0.1–0.2 mil thickness on paper and metal substrates. Cure speed varies with degree of through-cure desired.

Table 3.9. Vinyl Ester Resin Molding Compound Formulations.

COMPOSITION A	PARTS BY WEIGHT	COMPOSITION B[a]	PARTS BY WEIGHT
EPOCRYL® Resin 12[b]	40	XD-7608.02[c]	25
Styrene	30	XD-7609.00[c]	15
Polyvinyl acetate	10	Calcium carbonate	60
Calcium carbonate	150	Zinc stearate	1.6
ASP400 clay	17	*t*-Butyl Perbenzoate	0.04
Zinc stearate	4	Lupersol 256[d]	0.04
t-Butyl perbenzoate	0.5	Magnesium oxide	1
		Derakane 470-45[e]	1.5

[a]"Dow Experimental Vinyl Ester Resin SMC"—Dow Chemical Company.
[b]Shell Chemical Company.
[c]"Dow Experimental Vinyl Ester Resin"—Dow Chemical Company.
[d]Lucidol Division, Wallace and Tiernan, Inc.
[e]Dow Chemical Company.

Table 3.10. Mechanical-Strength Properties of Vinyl Ester Resin Molding Compounds.

COMPOSITION A[a]	
Glass content, wt %	29
Cure	2.5 min. at 302°F (150°C)
Flexural properties at 77°F (25°C)	
Strength, psi (MPa)	28,000 (193)
Modulus of elasticity, psi × 10^6 (GPa)	2.3 (15.8)
Tensile properties at 77°F (25°C)	
Strength, psi (MPa)	13,000 (89.6)
COMPOSITION B[a]	
Glass content, wt %	28
Cure	1.5 min. at 302°F (150°C)
Flexural properties at 77°F (25°C)	
Strength, psi (MPa)	28,000 (193)
Modulus of elasticity, psi × 10^6 (GPa)	1.6 (11.0)
Tensile properties at 77°F (25°C)	
Strength, psi (MPa)	13,300 (91.7)
Modulus of elasticity, psi × 10^6 (GPa)	1.58 (10.9)

[a]From Table 3.9.

storage tanks) is used by the chemical processing, petroleum, agricultural, mining, and food industries (as well as utilities) because it exhibits excellent chemical resistance[6,7] low maintenance requirements, design flexibility, and ease of installation. Vinyl ester resin-based composites have demonstrated economy and better price-to-performance characteristics than steel and its alloys in many corrosive industrial environments. Vinyl ester resins can also be formulated into chemical-resistant linings, grouts, flooring compounds, and mortars which can be applied by spraying or trowelling.

3.3.3. Molding

Vinyl ester resin-based bulk molding compounds (BMC) or sheet molding compounds (SMC) are used in compression-molded pipe fittings, appliance housings, pump impellers, and automotive components. These compounds typically contain approximately equal proportions by weight of resin, chopped fiberglass, and fillers. Latent catalysts, pigments, mold release agents, and thickening agents are also added to these formulations. Typical formulations and strength properties are shown in Tables 3.9 and 3.10, respectively.

REFERENCES

1. M. B. Launikitis, "A New Heat Resistant Vinyl Ester Resin," 31st Annual Technical Conference, SPI, 1976.
2. H. E. De La Mare et al., "A Review of Epoxy-Derived Acrylates: Synthesis, Chemistry, and Photocure," Paper FC76-494, presented 1976, The Association for Finishing Processes of the SME.
3. C. A. May and H. A. Newey, "Epoxy-Acrylic Resins for FRP Structures," 20th Reinforced Plastics Technical and Management Conference, SPI, 1965.
4. W. H. Linlow, "Derakane Resins—A Valuable Addition to Thermosetting Resin Technology," 21st Annual Technical Conference, SPI, 1966.
5. W. J. McCarthy and H. H. Bowerman, "Vinyl Terminated Liquid Polymer for Polyester Modification," 29th Annual Technical Conference, SPI, 1974.
6. Shell Chemical Company, Bulletin SC: 161-77, "EPOCRYL Resins—Corrosion Control Guide."
7. Dow Chemical Company, "Chemical Resistance Guide—Derakane Vinyl Ester Resins."

4

POLYBUTADIENE RESINS

Maxwell Stander
Naval Air Systems Command
Washington, D.C.

4.1. INTRODUCTION

Polybutadiene resins are high-hydrocarbon-content, thermosetting resins whose excellent electrical properties, excellent chemical resistance, high thermal distortion temperatures, low water absorption, and ready cure with peroxide catalysts make them of interest for many applications. On the basis of these properties, polybutadiene resins are formulated to be used in compression, transfer, and injection molding, in the preparation of prepregs, and in wet lay-up laminating. Various polybutadiene derivatives extend their use as resin modifiers, coatings, adhesives, and potting compounds.

4.2. HISTORY

The polybutadiene resins originated around 1955 with the introduction of the Buton® molding compounds by Enjay Laboratories.[1,2] The resin used in these molding compounds was based on a high proportion of liquid 1,2-polybutadiene, some butadiene–styrene copolymers, and an adduct of the two. Similar resins have since been manufactured by Richardson Company and Lithium Corporation. A high-vinyl polybutadiene with small amounts of isocyanate chain extender and some peroxide catalysts became commercially available in 1968 under the trade name HYSTYL®. This resin was originally marketed by Hystyl Development Company, but it is now marketed by Dynachem Corporation and by Nippon Soda Company, Limited, under the trade name NISSO-PB®. NISSO-PB® is identified as a liquid atactic polybutadiene having a molecular weight of 1000–4000 with over 90% of its total unsaturation as 1, 2 vinyl. It is available in three types: the B series having no terminal functional group, the G series with a primary hydroxyl group on both ends, and the C series with carboxyl groups on both ends. Other polybutadiene resins were formerly marketed under the trade name Dienite® (Firestone Synthetic Rubber and Latex Company), whereas some are now available under the RICON® (Colorado Chemical Specialties, Incorporated) trade name. Dienite® resins were mixtures of 1, 2 and 1,4 polymers (Dienite® PD-702, PD-703) or blended with coreactive monomers, such as vinyl toluene (PM-520, PM-503) or styrene–butadiene (PD-701, PD-753).

4.3. CHEMISTRY

The structures and typical properties of the basic polybutadiene resins are shown in Table 4.1.

Commercially available polybutadiene resins usually consist of a mixture of low-molecular-weight 1,2-polybutadiene and 1,4-polybutadiene resins. These isomers differ with respect to the reactive sites involved in polymerization. Hence, 1,2-polybutadiene monomers have a high reactivity because of the pendant vinyl group, whereas 1,4-polybutadiene monomers have a lower reactivity because of the presence of the double bond in the main polymer chain. For this reason, resins that contain a higher

Table 4.1. Structures and Typical Properties of the Basic Polybutadiene Resins.

RESIN TYPE	STRUCTURE	MOLECULAR WEIGHT	SPECIFIC GRAVITY	VISCOSITY AT 45°C (113°F), P
1,2-Polybutadiene	$\left[H - CH(CH{=}CH_2) - CH_2 - \right]$	1000–4000	0.86	5–350
1,4-Polybutadiene	$[H_2C - CH{=}CH - CH_2 -] -$	1000–4000	0.86	5–350
Hydroxy-terminated polybutadiene	$\left[HO - CH(CH{=}CH_2) - CH_2 - \right]$	1000–3000	0.88	25–550
Carboxy-terminated polybutadiene	$\left[HOOC - CH(CH{=}CH_2) - CH_2 - \right]$	1000–2500	0.90	50–800

content of the 1,2-monomer are more rapidly and easily cured, and that those containing a relatively high proportion of the 1,4-monomer are most commonly used in the preparation of elastomeric materials.

To render the 1,2-polybutadiene resin more suitable for composite processing, a high-molecular-weight, narrow-molecular-weight distribution has been formulated. The chemical reactivity of the basic resin has been made more versatile by the addition of chemically functional terminal groups (e.g., hydroxyl, carboxyl, and isocyanate groups), as well as by the preparation of blends that incorporate such reactive monomers as styrene and vinyl toluene. The terminal hydroxyl groups open up the potential for polyurethane reactions, whereas the terminal carboxyl groups permit the employment of epoxy chemistry. Isocyanate-terminated polybutadienes are used mainly for potting compounds.

The high vinyl content of the polybutadiene resins (>85%) allows them to be cured effectively by peroxide catalysts. The chemically functional terminal groups permit an increase in molecular weight prior to curing, so that a workable pot life can be achieved (molecular weight increase will decrease flow before crosslinking takes place, with resultant gelation and introduction of rigid polymer chains). The chain-extension step can be controlled in order to obtain polymers ranging from high-viscosity liquids to rigid, high-molecular-weight solids. This capability to chain extend is the basis for the wide application of the polybutadiene resins in molding compounds, coatings, adhesives, potting compounds, and thermosetting laminates. The polybutadiene (pBD) derivatives listed below have potential applications as modifiers for other resins as well as in the preparation of special laminates:

Polybutadiene Derivative	*Average Molecular Weight*
α,ω-pBD glycol	2000
pBD dicarboxylic acid	1400
Epoxidized pBD	1100
Maleated pBD	—
Acrylated pBD	2600
Styrene pBD	2200

Polybutadiene resins are usually supplied as blends with reactive monomers or as solutions in heptane.

4.4. RESIN-CURE CHARACTERISTICS

The resemblance of the curing process of polybutadiene resins to that of the well-known polyester resins by application of such free-radical catalysts as the peroxides renders them of greatest interest in composite technology. Resin-cure characteristics fall into three categories:[3] low-temperature gel, high-temperature cure, and thermal cyclization. At lower temperatures, molecular weight extension and viscosity increases occur which can lead to gelation and B staging. High-temperature cures begin at about 121°C (250°F), where it is postulated that vinyl activity predominates; hard resins are produced. Thermal cyclization begins above 232°C (450°F), where the remaining unsaturation reacts to form a highly crosslinked structure.

Due to the high peak exotherm temperature, catalyst concentration must be carefully controlled in accordance with the usual mass considerations. Such commonly employed peroxide catalysts as Lupersol® 101 (Lucidol Division, Wallace and Tiernan, Incorporated) can be used in amounts of 1–2%. Combinations of such catalysts as Lupersol® 101 (1 part per hundred—phr) and benzoyl peroxide (1 phr) can be used to give a total cure cycle of about 2 hr. Typical data for the processing of a prepreg are given below:[4]

Molding temperature	149–204°C (300–400°F)
Molding pressure	0.07–6.9 MPa (10–1000 psi)
Cure cycle	10 min. at 177°C (351°F) for a 3.2-mm (1/8-in.)-thick laminate
Post cure	None

4.5. CHEMICAL STRUCTURE AND PROPERTIES

Examination of the chemical structure of the polybutadiene resins provides a clear explanation of their excellent electrical properties and chemical resistance. The high hydrocarbon content and minimum aromatic content are responsible for their low dielectric constant, low dissipation (power) factor, and excellent chemical resistance. The low aromaticity accounts for their good arc and track resistance, resulting from the resistance to the formation of carbon when pyrolyzed under a high-voltage arc similar to the behavior of polyethylene. The absence of ester groups, which make the polyester resins vulnerable to chemical attack by acids and bases, explains the hydrophobicity of the polybutadiene resins as well as their resistance to attack by acids and bases.

4.6. COMPOSITE APPLICATION

Due to their unique combination of excellent electrical properties and chemical resistance, polybutadiene-based composites have become primary candidates for use in radome construction. The trend toward the use of operating frequencies above the K-band (10.90–36.00 GHz) has rendered the conventional epoxy–E-glass composites structurally inadequate for this purpose as a result of their high dielectric constants (4.5–5.0). This can be understood if we consider that the radome wall thickness is a function of both the dielectric constant and the operating wavelength as defined by the following equation:[5]

$$d = \frac{n\lambda_0}{2(E - \sin^2\theta)^{\frac{1}{2}}}$$

where

d = radome wall thickness
E = dielectric constant
θ = angle of incidence
λ_0 = free-space wavelength
n = integer $\geqslant 0$ ($n = 0$ for a thin wall; $n = 1$ for a half-wave wall).

Since the radome wall thickness is directly related to the operating wavelength, but inversely related to the dielectric constant, the combination of the increase in operating frequency together with composites having a high dielectric constant create the problem of inadequate wall thickness at higher-wavelength applications. It is clear that both shorter wavelengths and higher dielectric constants will result in a requirement for thinner radome walls. However, thin walls present the structural problem of impact damage, which is particularly evident in the serious rain erosion of thin-faced sandwich constructions. Another problem associated with the higher

Table 4.2. Electrical Properties of Candidate Resins at 24 GHz.[6]

RESIN	CURE TEMPERATURE, °C (°F)	CURE TIME, hr	DIELECTRIC CONSTANT	DISSIPATION (POWER) FACTOR
Dienite® PD-702[a]	177 (351)	36	2.34	0.0026
	232[d] (450)	22	2.34	0.0019
Dienite® PM-502[a]	177 (351)	36	2.37	0.0031
	232[d] (450)	22	2.37	0.0025
Dienite® X-545[a]	177 (351)	36	2.36	0.0030
	232[d] (450)	22	2.37	0.0020
Dienite® PD-701[a]	232[d] (450)	92	2.37	0.0024
RICON® 100[b]	177 (351)	23	2.40	0.0036
	232[d] (450)	24	2.40	0.0014
PPQ 401[c]	371 (700)	—	3.18	0.0059

[a]Firestone Synthetic Rubber and Latex Company.
[b]Colorado Chemical Specialties Company.
[c]Whittaker Research and Development Corporation.
[d]Post cure.

dielectric materials is the reduction in allowable physical thickness variations in the radome wall, which leads to costly machining or addition of material to provide the correct electrical thickness. Furthermore, operation in both aircraft and marine environments has additionally required that the composites used in radome construction be stable over a wide temperature range and after exposure to high humidity. The stringent requirements that result from higher-frequency operation and environmental considerations are not easily met by the conventional composite systems. However, as indicated in the following tables, polybutadiene-based composites can more readily satisfy these requirements.

Table 4.2[6] illustrates the very low dielectric constant and dissipation (power) factor values of several polybutadiene resins and a polyphenylquioxaline (PPQ) resin for purposes of comparison. Table 4.3[6] illustrates the low weight gain experienced by five of the resins listed in Table 4.2 after an accelerated temperature–humidity exposure. Table 4.4[7] illustrates the electrical stability of unfilled HYSTYL® resin castings over a wide frequency range, whereas the properties of laminates of this resin reinforced with style 181 E-glass fiber are shown in Table 4.5.[4] The electrical stability of polybutadiene and other

Table 4.3. Weight Gain for Polybutadiene Resins after Exposure to 99% RH at 49°C (120°F).[6]

RESIN	MAXIMUM CURE TEMPERATURE, °C (°F)	WEIGHT GAIN (%) EXPOSURE TIME (hr) 22	45	212	474	905
Dienite® PD-702[a]	177 (351)	0.02	0.07	0.09	0.17	0.17
	232[c] (450)	0.03	0.12	0.33	0.66	0.69
Dienite® PM-502[a]	177 (351)	0.017	0.03	0.03	0.06	0.06
	232[c] (450)	0.026	0.04	0.04	0.05	0.05
Dienite® X-545[a]	177 (351)	0.02	0.05	0.06	0.11	0.11
	232[c] (450)	0.02	0.04	0.04	0.08	0.08
Dienite® PD-701[a]	177 (351)	0.03	0.04	0.08	0.22	0.44
RICON® 100[b]	177 (351)	0.00	0.01	-[d]	0.05	0.06
	232[c] (450)	0.03	0.06	0.09	0.22	0.33

[a]Firestone Synthetic Rubber and Latex Company.
[b]Colorado Chemical Specialties Company.
[c]Post cure.
[d]No data available.

Table 4.4. Electrical Properties of Unfilled HYSTYL®[a] Resin Castings at 23°C (73°F).

FREQUENCY	DIELECTRIC CONSTANT	DISSIPATION (POWER) FACTOR
60 Hz	2.56	0.0062
100 Hz	2.54	0.0062
500 Hz	2.52	0.0056
1 kHz	2.51	0.0049
5 kHz	2.50	0.0042
10 kHz	2.49	0.0040
50 kHz	2.48	0.0037
100 kHz	2.46	0.0029

[a]Dynachem Corporation.

laminates after exposure to accelerated humidity aging is illustrated in Table 4.6.[6] It can be seen that the condensation polyimide PPQ 401–astroquartz laminate shows appreciable changes in electrical properties, whereas the PD 753–astroquartz laminate is virtually unaffected. Polybutadiene laminates, such as those listed in Tables 4.5 and 4.6, are made in standard commercial prepreg treaters and dried in conventional towers at about 121°C (250°F) for 4 hr. Since there is practically no solvent left in the prepreg, the laminates can be molded at low pressure.

The electrical properties of one of the humidity-conditioned specimens of Table 4.6 were also determined over a temperature range 22–261°C (72–502°F). The results of these tests are presented in Table 4.7.[6]

4.7. PROBLEM AREAS

Although composites based on the polybutadiene resins possess excellent electrical properties and chemical resistance, they are inferior to those based on epoxy resins with respect to mechanical properties at room temperature. In addition, their elevated-temperature mechanical properties are second to those of composites based on the condensation- and addition-type polyimide resins. For this reason, studies have been initiated to upgrade and optimize the mechanical properties of polybutadiene-based composites by means of post cure and resin modifications. Post cure was effective in enhancing the elevated-temperature mechanical properties without changing those at room temperature. A promising approach appears to be the use of reactive monomers. Table 4.8[6] compares the flexural strengths and flexural moduli of elasticity of RICON® 431 and RICON® 431

Table 4.5. Properties of HYSTYL®[a] Resin Laminates.[4b]

Prepreg resin content	23–30 nominal wt %
Volatile content of prepreg	1.5 wt. %
Resin flow at 1.4 Pa (200 psi)	14 nominal %
Tack	Medium
Drape	Good
Molding pressure	1.4 Pa (200 psi)
Cure temperature	177°C (351°F)
Cure time	10 min.
Specific gravity	1.7–1.9
Barcol hardness	60–80
Ultimate flexural strength	310–503 MPa [(45–73) × 10^3 psi]
Flexural modulus of elasticity	17.1–27.5 GPa [(2.5–4) × 10^6 psi]
Ultimate tensile strength	268–413 MPa [(39–60) × 10^3 psi]
Tensile modulus of elasticity	16.9–26.8 GPa [(2.45–3.9) × 10^6 psi]
Ultimate compressive strength	283–467 MPa [(41–68) × 10^3 psi]
Dielectric constant (maximum)	
at 1 MHz	3.6
at 10,000 MHz	3.1
Dissipation (power) factor (maximum)	
at 1 MHz	0.001
at 10,000 MHz	0.005

[a]Dynachem Corporation.
[b]Reinforced with style 181 E-glass fabric.

Table 4.6. Electrical Properties of Laminates at 24 GHz and Room Temperature (23°C = 74°F) before and after Humidity Aging.[6]

LAMINATE	BEFORE AGING		AFTER 30 DAYS AT 49°C (120°F) AND 99% RH	
	DIELECTRIC CONSTANT	DISSIPATION (POWER) FACTOR	DIELECTRIC CONSTANT	DISSIPATION (POWER) FACTOR
PD 753–astroquartz	3.08	0.006	3.09	0.007
PPQ 401–astroquartz	3.19	0.004	3.53	0.045
BPI 373–astroquartz[a]	3.35	0.004	3.85	0.092
PD 753–Kevlar (dried at 121°C = 250°F)	3.52	0.003	4.06	0.138
PD 753–Kevlar (treated 10 min. at 427°C = 800°F and sized with Kerimid 601 finish)	3.33	0.006	3.47	0.036

[a] Polyimide resin—Brunswick Corporation.

modified with 15 phr of triallyl cyanurate (TAC) at both room temperature and elevated temperatures.

Other formulations also hold interest for use in laminate processing. In particular, attention is currently being focused on the development of the special low-flow, low-pressure-cure resins known as high-vinyl-content, modified epoxy (HME) resins. These resins are essentially the reaction product of a carboxy-terminated polybutadiene (carboxyl equivalent weight = 855) and a solid multi-functional epoxy (epoxy equivalent weight = 140.2). These are reacted in solution, in a ratio of approximately 100 (parts by weight) pBD to 14 (parts by weight) epoxy, to form a prepolymer, hot melt solid useful for prepregging. Cure is effected by a peroxide initiator. Although the processability of the HME-resin-based composite is excellent and cure can be accomplished in one stage at 177°C (351°F) for 2 hr, low transverse properties severely limit its structural applications. This deficiency may be associated with high cross-link density leading to brittleness and low adhesion to carbon fibers.

Various fiber reinforcing agents, such as E-glass, S-glass, quartz, and aramid (Kevlar 49), have been used to prepare polybutadiene-based structural laminates. Table 4.9[6] presents the typical mechanical properties of the Kevlar 49-reinforced composites that are being considered for radome applications. The adhesion and wetting of the Kevlar 49 fibers require enhancement in order to improve certain mechanical properties, in particular, interlaminar shear and transverse tensile strength. Low moisture absorption of the Kevlar 49 fibers is an additional requirement for radome applications.

Table 4.7. Electrical Properties of PD 753/Astroquartz at 24 GHz versus Temperature after 30 Days at 49°C (120°F) and 99% RH.[6]

TEMPERATURE, °C (°F)	DIELECTRIC CONSTANT	DISSIPATION (POWER) FACTOR
22 (72)	3.09	0.006
73 (163)	3.08	0.005
121 (250)	3.04	0.005
177 (351)	2.99	0.005
220 (428)	2.92	0.004
261 (502)	2.90	0.003
26 (79)	3.07	0.003

4.8. STORAGE AND HANDLING

No problems arise in the storage and handling of the polybutadiene resin systems other than the usual ones generally associated with the handling of such volatile, flammable, organic solvents as heptane or toluene. Storage at temperatures of 0°C (32°F), 20°C (68°F), and 35°C (95°F) for 10 weeks has produced no noticeable changes in viscosity or laminate properties. Longer storage periods at temperatures over 35°C (95°F) are to be avoided because of the tendency of the solution to form gels.

Table 4.8. Flexural Properties of Glass-Reinforced RICON® 431 and RICON® 431 + 15 phr TAC versus Temperature.[6]

RESIN	PROPERTY	TEMPERATURE, °C (°F) 23 (73)	121 (250)	149 (300)	177 (351)
RICON® 431	Flexural strength, MPa (psi × 10^3)	293 (42)	186 (27.0)	139 (20.2)	115 (16.7)
	Flexural modulus of elasticity, GPa (psi × 10^6)	23.0 (3.34)	17.8 (2.58)	16.5 (2.39)	15.4 (2.23)
RICON® 431 + 15 phr TAC	Flexural strength, MPa (psi × 10^3)	502 (72.8)	304 (44.1)	236 (34.2)	192 (27.8)
	Flexural modulus of elasticity, GPa (psi × 10^6)	27.5 (4.0)	21.8 (3.16)	19.8 (2.87)	20.3 (2.94)

Notes. 1. Reinforcement: 7781 E glass.
2. Post cure: 24 hr at 288°C (550°F).

Table 4.9. Mechanical Properties of Unidirectional Kevlar 49-Reinforced Composites[6a]

PROPERTY	EPOXY[b]	DIENITE® PM-502
Flexural strength, MPa (psi × 10^3)	516–757 (75–110)	351 (51)
Flexural modulus of elasticity, GPa (psi × 10^6)	62–83 (9–12)	56 (8.1)
Interlaminar shear, MPa (psi × 10^3)	58–83 (8–12)	13.9 (2)
Transverse flexural strength, MPa (psi × 10^3)	—	13.9 (2)
Transverse flexural modulus of elasticity, MPa (psi × 10^3)	—	13.9 (2)

[a]Fiber, V/O = 60%.
[b]Du Pont data.

References

1. H. Clark and B. M. Vanderbilt, "A Hydrocarbon Thermoset Resin for Reinforced Plastics," 14th Annual Meeting, Reinforced Plastics Division, SPI, February 1959.
2. J. J. Jaruzelski, B. M. Vanderbilt, and T. Lewis, "Buton Hydrocarbon Resin Molding Compounds," 19th Annual Meeting, Reinforced Plastics Division, SPI, February 1964.
3. Colorado Chemical Specialties Company, Technical Bulletin CCS-102, RICON Laminating Resins.
4. HYSTYL Polybutadiene Resins Technical Bulletin.
5. J. B. E. Walton, "Techniques for Airborne Radome Design," Technical Report AFAL-TR-66-391, Vol. II, December 1966.
6. V. A. Chase, "Investigation of Reinforced Plastics for Naval Aircraft Electromagnetic (EM) Windows," January 1975; prepared for the Naval Air Systems Command, Contract N00019-74-C-0055.
7. M. Bianchi, and E. N. Dorman, "New, High-Performance, Versatile Polybutadiene Systems for Electrical and RP Applications," 26th Annual Technical Conference, SPI, 1971.

5
EPOXY RESINS

Lynn S. Penn
CIBA-Geigy Corp.
Ardsley, New York
and
T. T. Chiao
Lawrence Livermore Laboratory
University of California
Livermore, California

5.1. INTRODUCTION

Epoxy resins are among the best matrix materials for many fiber composites. This is true for several reasons:

- Epoxy resins adhere well to a wide variety of fillers, reinforcing agents, and substrates.
- The wide variety of available epoxy resins and curing agents can be formulated to give a broad range of properties after cure and to meet a diverse spectrum of processing requirements.
- The chemical reaction between epoxy resins and a curing agent does not release any volatiles or water. Hence, the shrinkage after cure is usually lower than that for phenolic or polyester resins.
- Cured epoxy resins are not only resistant to chemicals, but they also provide good electrical insulation.

We have written this chapter specifically for engineers and chemists who possess little background in the area of epoxy resins. The applications of epoxy resins, the currently available epoxy resins and curing agents, some typical curing reactions, and the formulation of epoxy resin systems are discussed. We also describe procedures for optimizing and monitoring the cure in order to enable the reader to obtain the best possible performance from a chosen system. Typical properties of several cured epoxy resin systems are given and test methods for characterizing the system, from separate components to cured resin, are presented. We also furnish an extensive bibliography, including textbooks,[1-4] journal articles, and laboratory reports.

5.2. APPLICATIONS

Epoxy resins are used in various composites and in many structural parts; they are also used as potting and encapsulating compounds, tooling compounds, molding powders, and adhesives.

5.2.1. Potting and Encapsulation

Epoxy resins are used for potting and encapsulation because they are very resistant to alkali, acid, and humidity. They can attain a high heat distortion temperature and have low shrinkage and high volume resistivity. In addition to providing environmental protection, the epoxy resin also can be used to consolidate multiple parts for convenient assembly. In the electronics industry, for example, epoxy resins are used for encapsulating welded modules, transformers, and motors, as well as for potting electrical cable splices.

5.2.2. Tooling

Epoxy resins have been used since World War II in tooling (e.g., as dies for stamping sheet metal or as models for production parts). Particulate or fibrous reinforcing agents are easily incorporated to reduce cost and improve dimensional stability. Epoxy resins are used instead of metals for two principal reasons: They are more economical to make and they can be modified quickly and inexpensively. Furthermore, epoxy resins have high dimensional stability, good mechanical properties, and low shrinkage, which enable the production of replicate parts to within extremely close tolerances.

5.2.3. Molding

Epoxy molding compounds (powdered, partially cured mixtures of resin and curing agent which become fluid when heated) are used for the production of all kinds of structural parts. Fillers and reinforcing agents are easily incorporated into epoxy resins to make molding compounds. The epoxy resins provide low shrinkage, good bonds to fillers and reinforcing agents, chemical stability, and good flow properties.

5.2.4. Adhesive Bonding

Epoxy resins provide the highest adhesive strength of any known polymeric material. They can thoroughly wet a wide variety of substrates with minimal shrinkage. Thus, epoxy resins can be used to join many dissimilar materials. They also can be formulated to cure at many different temperatures and times, an important consideration when producing adhesives commercially.

5.2.5. Filament Winding and Laminating of Composites

One of the most important uses of epoxy resins as matrix materials is in laminated and filament-wound composites for structural parts. These composite structural parts have been used for many aircraft, space, military, and industrial applications. Laminates are also used for printed circuit boards in the electrical industry. In addition, composite storage tanks and pipes are used extensively in the chemical and petroleum industries. Epoxy resins are adaptable and can be formulated for different processes: wet filament winding or wet lay-up and dry winding or lay-up with preimpregnated fiber strands, fabrics, or tapes (prepregs). In general, epoxies are more expensive than most other resins, but their superior performance often makes them more economical in the long run.

5.3. UNCURED EPOXY RESINS

By definition, any molecule containing the epoxy group, $-\overset{\overset{O}{/\ \backslash}}{\underset{|}{C}-\underset{|}{C}}-$, is called an epoxy. Table 5.1 lists the chemical structures, epoxy equivalent weights, and viscosities of the commonly used aromatic epoxy resins. The epoxy equivalent weight is defined as the number of grams of resin containing one chemical equivalent of epoxy group. The degree of viscosity is important for processing; the same resin is often available in different viscosities.

Table 5.2 lists the structures and characteristics of epoxies that are both commonly used and commercially available as reactive diluents to make the uncured resin mixtures more fluid (for easier processing). Small amounts of low-molecular-weight diluents, particularly the difunctional ones, do not seriously degrade the properties of the cured resin.

Table 5.3 lists the structures and characteristics of some higher-molecular-weight epoxy resins with relatively flexible aliphatic carbon backbones. These epoxy resins, which are known as flexibilizers, are added to aromatic epoxy formulations when greater elongation and reduced inherent brittleness of the cured resin (unfortunately, at the expense of tensile strength) are desired. It should be noted that although most flexibilizers do not have low viscosities, the few that do can also act as diluents.

Table 5.1. Structures and Characteristics of Commonly Used Aromatic Epoxy Resins.

EPOXY EQUIVALENT WEIGHT, g/eq	VISCOSITY AT 25°C (77°F), Pa · sec (cP)	COMMENTS
	Diglycidyl Ether of Bisphenol A (DGEBA)	
	$\overset{O}{CH_2-CH}-CH_2-O-C_6H_4-C(CH_3)_2-C_6H_4-O-CH_2-\overset{O}{CH-CH_2}$	
171–177	3.5–5.5 (3500–5500)	May crystallize on storage. Example: DER 332 (Dow Chemical Company).
180–188	6.5–9.5 (6500–9500)	Contains small amount of higher polymer to prevent crystallization. Examples: DER 330 (Dow Chemical Company), Epon 828 (Shell Chemical Company).
185–200	10.0–19.0 (10,000–19,000)	Contains small amount of higher polymer to prevent crystallization. Example: Epi-Rez 510 (Celanese Plastics Materials Company) and DER 331 (Dow Chemical Company).
	Diglycidyl Ether of Bisphenol A	
	$\overset{O}{CH_2-CH}-CH_2-O\left[-C_6H_4-C(CH_3)_2-C_6H_4-O-CH_2-CH(OH)-CH_2-\right]_n-O-C_6H_4-C(CH_3)_2-C_6H_4-O-CH_2-\overset{O}{CH-CH_2}$	
450–550	Melting Point 65–75°C (149–167°F)	$n \approx 2$; used for adhesives, coatings, and prepregs. Examples: Epon 1001 (Shell Chemical Company) and DER 661 (Dow Chemical Company).
	Diglycidyl Ether of Bisphenol F (DGEBF)	
	$\overset{O}{CH_2-CH}-CH_2-O-C_6H_4-CH_2-C_6H_4-O-CH_2-\overset{O}{CH-CH_2}$	
158–165	5.0–8.0 (5000–8000 cP)	Isomeric mixture that will not crystallize on storage. Example: XD 7818 (Dow Chemical Company).
	Polyglycidyl Ether of Phenol-Formaldehyde Novolac	
	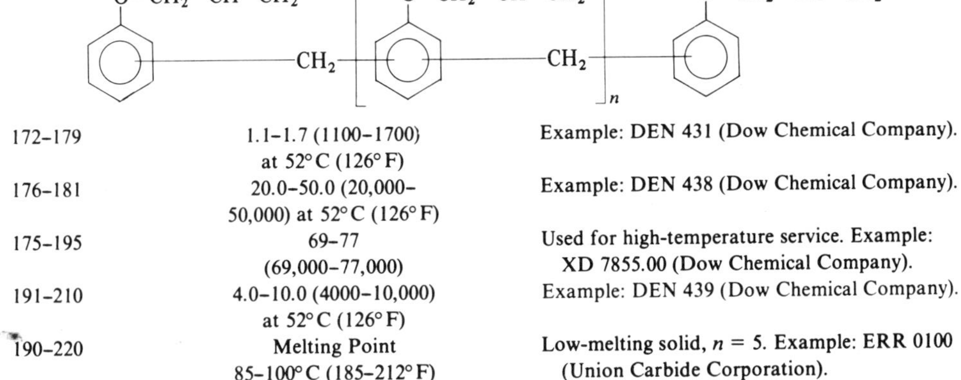	
172–179	1.1–1.7 (1100–1700) at 52°C (126°F)	Example: DEN 431 (Dow Chemical Company).
176–181	20.0–50.0 (20,000–50,000) at 52°C (126°F)	Example: DEN 438 (Dow Chemical Company).
175–195	69–77 (69,000–77,000)	Used for high-temperature service. Example: XD 7855.00 (Dow Chemical Company).
191–210	4.0–10.0 (4000–10,000) at 52°C (126°F)	Example: DEN 439 (Dow Chemical Company).
190–220	Melting Point 85–100°C (185–212°F)	Low-melting solid, $n = 5$. Example: ERR 0100 (Union Carbide Corporation).

Table 5.1. Continued

EPOXY EQUIVALENT WEIGHT, g/eq	VISCOSITY AT 25°C, (77° F), Pa · sec (cP)	COMMENTS
	Polyglycidyl Ether of *o*-Cresol-Formaldehyde Novolac	
	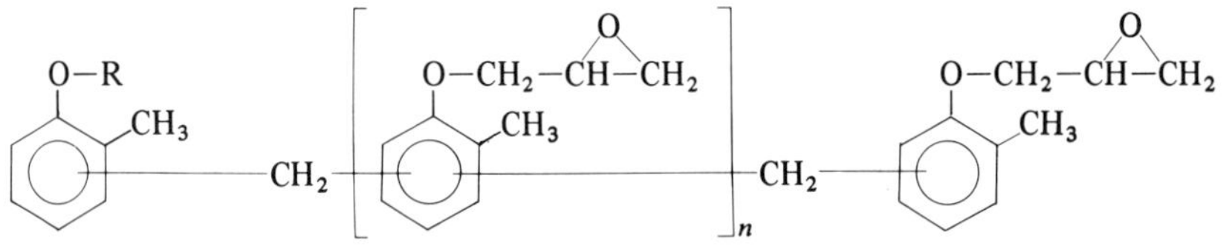	
200	Melting Point 35°C (95°F)	Used for high-temperature service; R represents chlorohydrins, glycols, and/or polymeric ethers. Molecular weight = 540. Example: ECN 1235 (Ciba-Geigy Corporation).
225	Melting Point 73°C (163°F)	Same as above, but molecular weight = 1080. Example: ECN 1273 (Ciba-Geigy Corporation).
230	Melting Point 80°C (176°F)	Same as above, but molecular weight = 1170. Example: ECN 1280 (Ciba-Geigy Corporation).
235	Melting Point 99°C (210°F)	Same as above, but molecular weight = 1270. Example: ECN 1299 (Ciba-Geigy Corporation).
	N,N,N′,N′-Tetraglycidyl Methylenedianiline	
	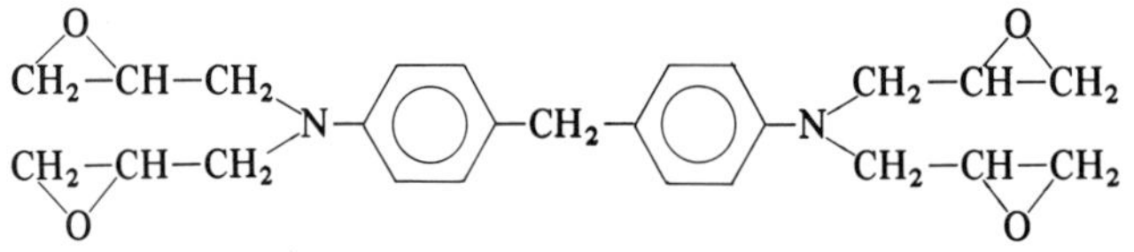	
117–133	10.0–15.0 at 50°C (122°F) (10,000–15,000)	Used for prepregs. Example: MY 720 (Ciba-Geigy Corporation).
	Triglycidyl *p*-Aminophenol	
	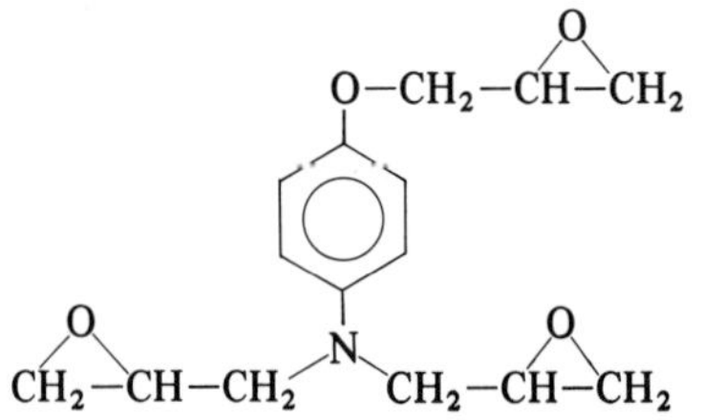	
95–107	0.55–0.85 (550–850)	Used extensively for prepregs and adhesives. Example: ERL-0510 (Ciba-Geigy Corporation).

5.4. CURING AGENTS AND CURING MECHANISMS

The epoxy group can bond chemically with other molecules, forming a large three-dimensional network. This process, called curing, changes a liquid resin into a solid. A major area of epoxy technology is devoted to the study of curing-agent properties.

5.4.1. Amine Curing Agents

The chemical structures, amine hydrogen equivalent weights, and viscosities of the commonly

Table 5.2. Structures and Characteristics of Commonly Used, Commercially Available Epoxy Reactive Diluents.

EPOXY EQUIVALENT WEIGHT, g/eq	VISCOSITY AT 25°C (77°F), Pa · sec (cP)	COMMENTS
	Butyl Glycidyl Ether	
	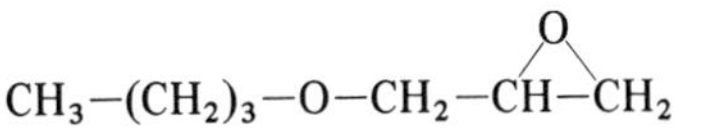 $CH_3-(CH_2)_3-O-CH_2-CH(O)CH_2$ (epoxide ring)	
130–149	0.002–0.003 (2–3)	Example: Epotuf 37-149 (Reichhold Chemicals, Incorporated).
	Heptyl Glycidyl Ether	
	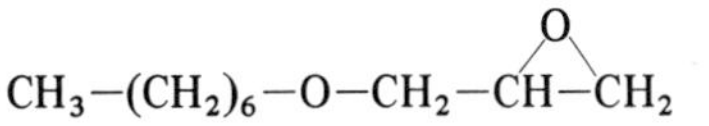 $CH_3-(CH_2)_6-O-CH_2-CH(O)CH_2$ (epoxide ring)	
227	0.004 (4)	Example: Epotuf 37-147 (Reichhold Chemicals, Incorporated).
	Octyl Glycidyl Ether	
	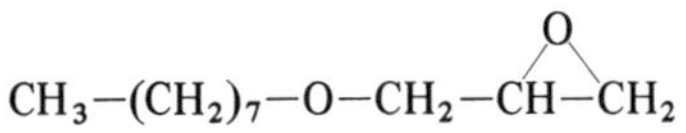 $CH_3-(CH_2)_7-O-CH_2-CH(O)CH_2$ (epoxide ring)	
286	0.010 (10)	Example: Epotuf 37-146 (Reichhold Chemicals, Incorporated).
	Allyl Glycidyl Ether	
	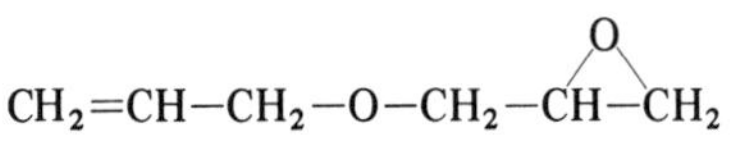 $CH_2=CH-CH_2-O-CH_2-CH(O)CH_2$ (epoxide ring)	
114	0.001 (1)	Available from Shell Chemical Company.
	p-t-Butyl Phenyl Glycidyl Ether	
	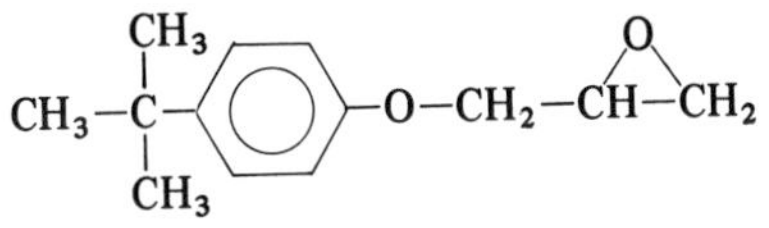 $(CH_3)_3C-C_6H_4-O-CH_2-CH(O)CH_2$ (epoxide ring)	
220–245	0.015–0.030 (15–30)	Available from Fisher Scientific Company.
	Phenyl Glycidyl Ether (PGE)	
	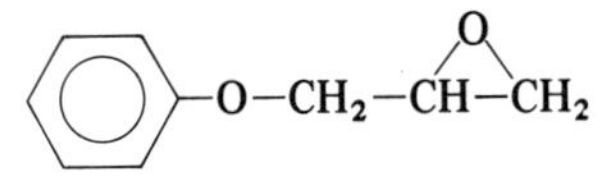 $C_6H_5-O-CH_2-CH(O)CH_2$ (epoxide ring)	
150	0.006 (6)	Available from Shell Chemical Company.
	Cresyl Glycidyl Ether	
	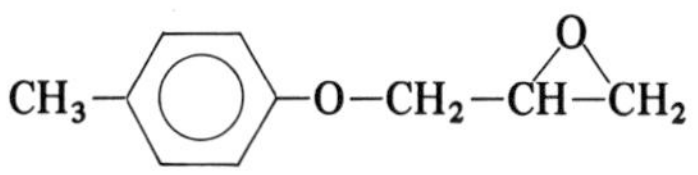 $CH_3-C_6H_4-O-CH_2-CH(O)CH_2$ (epoxide ring)	
170–190	0.005–0.050	Example: EpiRez 5011 (Celanese Corporation).

Table 5.2. Continued

EPOXY EQUIVALENT WEIGHT, g/eq	VISCOSITY AT 25°C (77°F), Pa · sec (cP)	COMMENTS
Diglycidyl Ether of 1,4-Butanediol (BDE)		
$CH_2(O)CH-CH_2-O-CH_2-CH_2-CH_2-CH_2-O-CH_2-CH(O)CH_2$ (epoxide rings at both ends)		
120–140	0.010–0.025 (10–25)	Example: RD-2 (Ciba-Geigy Corporation); not a pure compound.
Diglycidyl Ether of Neopentyl Glycol		
$CH_2(O)CH-CH_2-O-CH_2-C(CH_3)_2-CH_2-O-CH_2-CH(O)CH_2$ (epoxide rings at both ends)		
145–160	0.012–0.030 (12–30)	E.g., Dow XD 7114; not a pure compound
Diglycidyl Ether of Polypropylene Glycol		
$CH_2(O)CH-CH_2-O-[CH(CH_3)-CH_2-O]_n-CH_2-CH(O)CH_2$ (epoxide rings at both ends)		
175–205	0.030–0.060 (30–60)	$n = 4$. Example: DER 736 (Dow Chemical Company).
305–335	0.055–0.100 (55–100)	$n = 9$. Example: DER 732 Dow Chemical Company).
Vinyl Cyclohexene Dioxide		
Cyclohexane ring (S) with a ring-fused epoxide O, bearing $-CH(O)CH_2$ (epoxide)		
76	0.020 (20)	Example: ERL 4206 (Union Carbide Corporation).
Diglycidyl Ether of Resorcinol		
Benzene ring bearing $-O-CH_2-CH(O)CH_2$ and $-O-CH_2-CH(O)CH_2$ (epoxide rings)		
128	0.300–0.500 (300–500)	Example: ERE 1359 (Ciba-Geigy Corporation).

Table 5.3. Structures and Characteristics of Flexible Epoxy Resins (Flexibilizers).

EPOXY EQUIVALENT WEIGHT, g/eq	VISCOSITY AT 25°C (77°F), Pa · sec (cP)	COMMENTS
	Diglycidyl Ether of Polypropylene Glycol	
	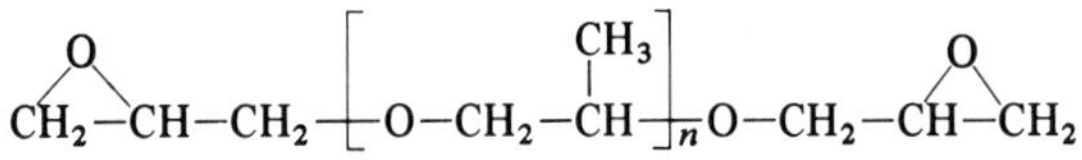 $\overset{O}{\overbrace{CH_2-CH}}-CH_2-\left[O-CH_2-\overset{CH_3}{\overset{\mid}{CH}}\right]_n O-CH_2-\overset{O}{\overbrace{CH-CH_2}}$	
175–205	0.030–0.060 (30–60)	$n = 4$; has a long flexible backbone; can also act as a diluent because of its low viscosity. Example: DER 736 (Dow Chemical Company).
305–335	0.055–0.100 (55–100)	$n = 9$; has a long flexible backbone. Example: DER 732 (Dow Chemical Company).
	Diglycidyl Ester of Linoleic Acid Dimer	
	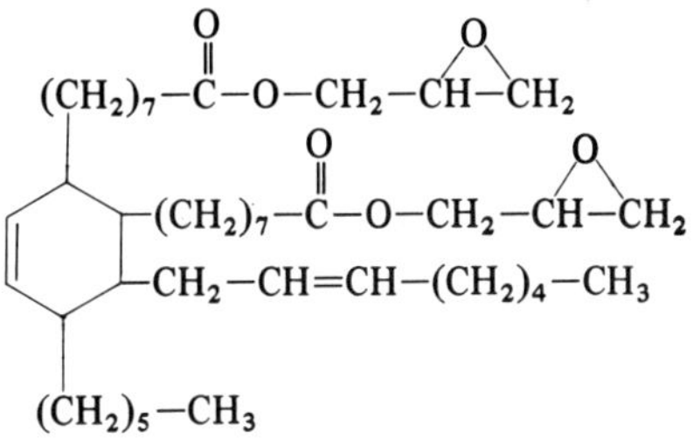 $(CH_2)_7-\overset{O}{\overset{\parallel}{C}}-O-CH_2-\overset{O}{\overbrace{CH-CH_2}}$ (cyclohexene ring bearing:) $-(CH_2)_7-\overset{O}{\overset{\parallel}{C}}-O-CH_2-\overset{O}{\overbrace{CH-CH_2}}$ $-CH_2-CH=CH-(CH_2)_4-CH_3$ $(CH_2)_5-CH_3$	
390–470	0.400–0.900 (400–900)	Example: Epon 871 (Shell Chemical Company).

Table 5.4. Structures and Characteristics of Commonly Used Amine Curing Agents.

AMINE HYDROGEN EQUIVALENT WEIGHT, g/eq	VISCOSITY AT 25°C (77°F), Pa · sec (cP)	COMMENTS
	Diethylenetriamine (DETA)	
	$H_2N-CH_2-CH_2-NH-CH_2-CH_2-NH_2$	
20	0.0055–0.0085 (5.5–8.5)	Available from Dow Chemical Company as DEH 20 and also from Union Carbide Corporation.
	Triethylenetetramine (TETA)	
	$H_2N-\left[(CH_2)_2-NH\right]_2 CH_2-CH_2-NH_2$	
24	0.020–0.023 (20–23)	Available from Dow Chemical Company as DEH 24 and also from Union Carbide Corporation.
	Diethylaminepropylamine (DEAPA)	
	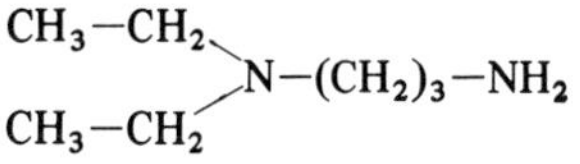 $(CH_3-CH_2)(CH_3-CH_2)N-(CH_2)_3-NH_2$	
65	<5.0	Available from Union Carbide Corporation.

Table 5.4 Continued

AMINE HYDROGEN EQUIVALENT WEIGHT, g/eq	VISCOSITY AT 25°C (77°F), Pa · sec (cP)	COMMENTS
	Tetraethylenepentamine $H_2N{-}[(CH_2)_2{-}NH]_3{-}CH_2{-}CH_2{-}NH_2$	
26–27	Boiling Point 152°C (306°F)	Available from Dow Chemical Company as DEH 26 and also from Union Carbide Corporation.
	Aliphatic Polyether Triamine (APTA) $CH_3{-}CH_2{-}C(CH_2{-}[O{-}CH_2{-}CH{-}(CH_3)]_x{-}NH_2)(CH_2{-}[O{-}CH_2{-}CH{-}(CH_3)]_y{-}NH_2)(CH_2{-}[O{-}CH_2{-}CH{-}(CH_3)]_z{-}NH_2)$	
77–82	0.072–0.080 (72–80)	$x + y + z \approx 5.3$. Available from Jefferson Chemical Company.
	Dicyandiamide (DICY) $H_2N{-}C(NH_2){=}N{-}C{\equiv}N$	
28	Melting Point 207–209°C (405–408°F)	Acts as latent curing agent. Used for adhesives, prepregs, and powder coatings. Available from Pacific Anchor Chemical Corporation.
	4,4′-Methylenedianiline (MDA) $H_2N{-}C_6H_4{-}CH_2{-}C_6H_4{-}NH_2$	
50	Melting Point 89°C (192°F)	Available from Allied Chemical Corporation.
	m-Phenylenediamine (MPDA) $C_6H_4(NH_2)_2$	
27	Melting Point 60°C (140°F)	Available from E. I. du Pont de Nemours & Company, Incorporated.
	4,4′-Diaminodiphenylsulfone (DDS) $H_2N{-}C_6H_4{-}SO_2{-}C_6H_4{-}NH_2$	
62	Melting Point 170–180°C (338–356°F)	Used mainly in prepregs; yields good shelf life and high-temperature properties. Example: Eporal (Ciba-Geigy Corporation).

Table 5.4 Continued

AMINE HYDROGEN EQUIVALENT WEIGHT, g/eq	VISCOSITY AT 25°C (77°F), Pa · sec (cP)	COMMENTS
	2,6-Diaminopyridine (DAP) 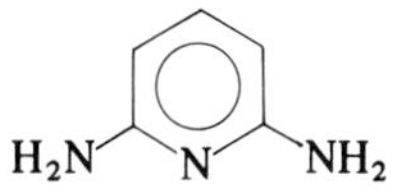	
27	Melting Point ~121°C (250°F)	Gives good pot life, about 10 times that of MPDA. Available from Reilly Tar & Chemical.
	33.3% MPDA-33.3% MDA-33.3% Isopropyl MPDA 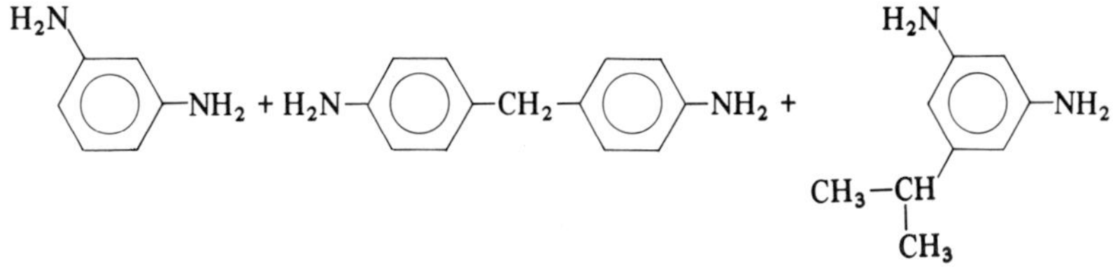	
36	5 (5000)	Eutectic mixture. Available from Hysol Corporation.
	40% MDA-60% Diethyl MDA 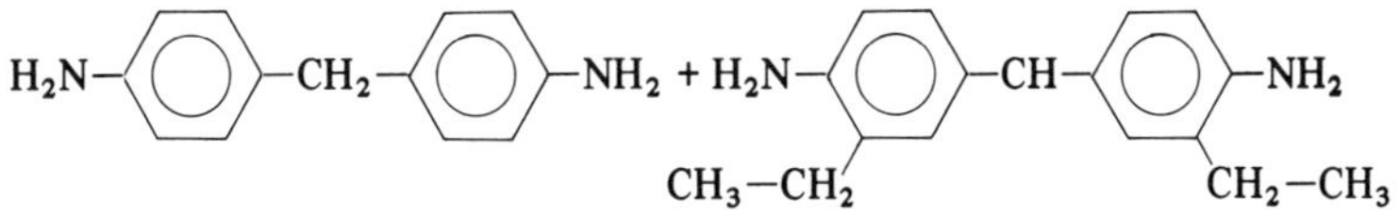	
57	2-5 (2000-5000)	Available from Ciba-Geigy Corporation as HY 932.
	40% MPDA-60% MDA 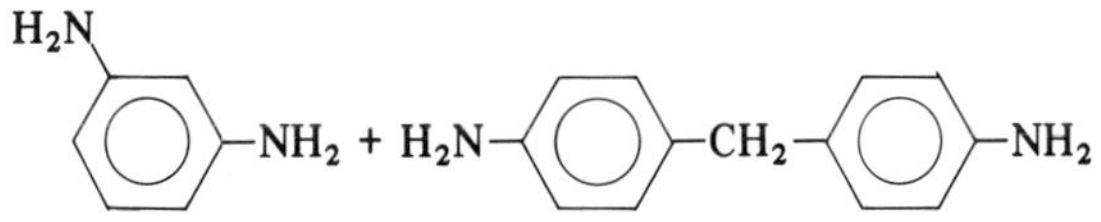	
38	1.50 (1500)	Eutectic mixture. Available from UniRoyal, Incorporated as Tonox 60-40.
	Aminopolyamide 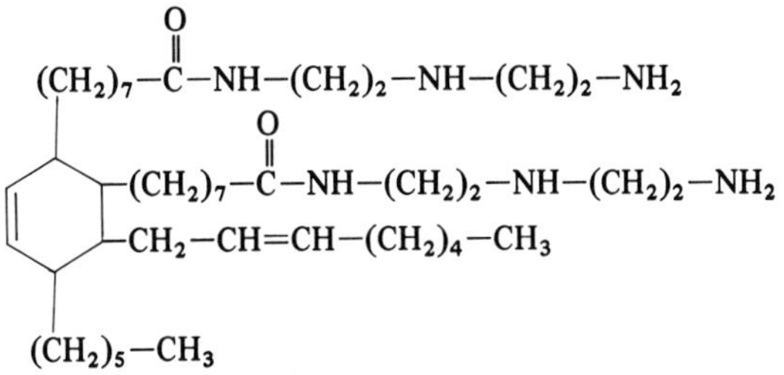	

Table 5.4 Continued

AMINE HYDROGEN EQUIVALENT WEIGHT, g/eq	VISCOSITY AT 25°C (77°F), Pa · sec (cP)	COMMENTS
180	50 (50,000)	Flexibilizes cured system. Available from General Mills, Incorporated as Versamid 125.
250	Solid at 23°C (73°F) 80–120 at 40°C (104°F) (80,000–120,000)	Different ration of DETA to dimer acid than used in above preparation; gives higher molecular weight. Available from General Mills, Incorporated as Versamid 115.
620	Melting Point 43–53°C (109–127°F)	Different ratio of DETA to dimer acid than used in above preparation; gives higher molecular weight. Available from General Mills, Incorporated as Versamid.
2,4,6-Tris(dimethylaminomethyl)phenol		
—	0.3 (300)	Lewis base used as the sole curing agent in small nonstoichiometric amount to provide room-temperature cure. Available from Rohm & Haas company as DMP-30.
2-Ethyl-4-methylimidazole (EMI)		
110	4–8 (4000–8000)	Lewis base used as the sole curing agent in small nonstoichiometric amount to provide long pot life and good elevated-temperature properties. Available from Houdry Process & Chemical Company as EMI-24.

used aromatic and aliphatic amine curing agents are listed in Table 5.4. It can be seen that many of the structures given in Table 5.4 include reactive groups on each end; these permit the formation of a crosslink between epoxy molecules. For example, an amine end-group with two hydrogens on the nitrogen (a *primary* amine) reacts with an epoxy molecule as follows:[5,6]

When another amine hydrogen combines with a second epoxy molecule, a crosslink is formed (page 67). The curing agents in Table 5.4 that contain *secondary* amine end-groups (one hydrogen on the nitrogen) react in much the same way.

For thorough crosslinking, the hydrogens of the primary and secondary amines should

$$H_2N\sim\sim\sim NH_2 + H_2C\overset{O}{\frown}CH\sim\sim\sim \longrightarrow H_2N\sim\sim\sim NH-CH_2-\overset{OH}{\overset{|}{C}H}\sim\sim\sim$$

$$H_2N\sim\sim N(NH)-CH_2-CH(OH)\sim\sim + H_2C\overset{O}{-}CH\sim\sim \rightarrow H_2N\sim\sim N(-CH_2-CH(OH)\sim)-CH_2-CH(OH)\sim\sim$$

be matched 1:1 with the epoxy groups. The amounts of curing agent and epoxy resin needed in order to obtain the 1:1 stoichiometric quantity are calculated as follows:

$$\frac{\left(\dfrac{\text{Molecular weight of amine}}{\text{Number of available hydrogens per molecule}}\right)}{\text{Epoxy equivalent weight}} \times 100 = \begin{array}{l}\text{Parts by weight of amine to be}\\ \text{used with 100 parts by weight of}\\ \text{epoxy resin}\end{array}$$

The reaction between aliphatic amines and epoxy groups will usually proceed at room temperature; however, heat is required when rigid aromatic amines are used. The carbon-nitrogen bond formed in crosslinking is stable against most inorganic acids and alkalis. However, it is less stable against organic acids than linkages formed by other curing agents. In addition, the electrical-insulating capabilities of amine-cured epoxy resins are not as high as those obtained with other curing agents, probably because of the polarities formed by the hydroxyl groups during the cure.

Tertiary amines (no hydrogen on the nitrogen) are Lewis bases that cure epoxy resins in an entirely differently manner than the aromatic and aliphatic amines. They are added to an epoxy resin in small nonstoichiometric amounts that have been empirically determined to give the best properties. The curing agent operates as a true catalyst by initiating a self-perpetuating anionic polymerization:

This homopolymerization of epoxy molecule to epoxy molecule results in a polyether. The ether linkages (C–O–C) are fairly stable against most acids (both organic and inorganic) and alkalis. Furthermore, like ester linkages, they are more thermally stable than the carbon-nitrogen linkages formed by an amine cure. Of the two Lewis bases listed in Table 5.4, EMI is the most efficient, that is, it produces the highest degree of crosslinking and the highest heat distortion temperature.[7]

There are two curing agents in Table 5.4 that deserve special comment: aminopolyamide and dicyandiamide. The versatile, fatty amino polyamide is a flexible polymer that contains primary and secondary amine functional groups as well as amide functional groups ($-\overset{O}{\overset{\|}{C}}-NH-$). Although referred to in the trade as an amide, this curing agent, which also acts as a flexibilizer, cures epoxy resins

$$R_3N + CH_2\overset{O}{-}CH-CH_2R \rightarrow R_3\overset{+}{N}-CH_2-CH(O^-)-CH_2R$$

$$R_3\overset{+}{N}-CH_2-CH(O^-)-CH_2R + CH_2\overset{O}{-}CH-CH_2R' \rightarrow$$

$$R_3\overset{+}{N}-CH_2-CH(-O-CH_2-CH(O^-)-CH_2-R')-CH_2R \xrightarrow{CH_2\overset{O}{-}CH-CH_2-R''} \text{and so on.}$$

mainly through its amine groups.[8,9] Amino-polyamide has a more adjustable mixing ratio (without loss of good properties) than other curing agents. It produces increased flexibility and impact strength in cured products, and it has less skin-irritation potential than the regular amine curing agents. However, an amino-polyamide-cured system is less resistant to chemicals and solvents than are the other epoxy resin systems.

The second curing agent of note is the popular dicyandiamide (DICY). The reaction mechanisms of DICY are a complex mixture of addition polymerization and homopolymerization of the epoxy. It is usually employed with solid epoxy resins in prepreg laminates because it is a latent curing agent (i.e., little or no reaction with the epoxy resin occurs at room temperature, but full cure is rapidly initiated at elevated temperatures). This characteristic is due to the fact that DICY decomposes at 145–154°C (293–309°F) to give ammonia and several amine-like species, all of which react with the epoxy resin to produce a cured product. Dicyandiamide is especially valuable as a curing agent for epoxy resins that must provide good adhesive strength and resistance to aging at elevated temperatures.

In general, amine curing agents present no storage problems. However, they may cause adverse skin reactions in some people and thus should be handled with care.

5.4.2. Anhydride Curing Agents

Table 5.5 lists the chemical structures, anhydride equivalent weights, and melting points of the commonly used anhydride curing agents. Anhydride curing agents usually require careful storage so as to prevent degradation as a result of moisture absorption. They also usually require the application of heat in order to initiate full cure. In fact, the epoxy-anhydride reaction can be so sluggish that often a small amount of accelerator is added to speed up the curing process. However, some anhydride-cured resin systems are quite heat-resistant up to a temperature of 200°C (392°F).[10]

Anhydride curing agents react with epoxy resins to form ester.[11,12] For the curing reaction to occur, the anhydride ring must be opened. Small amounts of proton-containing compounds (e.g., acids, alcohols, phenols, and water) or Lewis bases will open the ring.[13,14] The carboxyl groups thus formed react with the epoxy groups:

R—CH—C(=O)—O—C(=O)—CH—R (cyclic anhydride) + R_3N: (Lewis base accelerator) → R—CH(—C(=O)···$\overset{+}{N}R_3$)—CH(R)—C(=O)—O^- $\xrightarrow{R\ CH_2—CH—CH_2\ (epoxide)}$

R—CH(—C(=O)···$\overset{+}{N}R_3$)—CH(R)—C(=O)—O—CH_2—CH(O^-)—CH_2—R → and so on.

Lewis base accelerator

Theoretically, one anhydride group reacts with one epoxy group. The amounts of resin and curing agent that contain identical quantities of the two functional groups (i.e., 1 : 1 stoichiometry) can be determined as follows:

$$\frac{\left(\dfrac{\text{Molecular weight of anhydride}}{\text{Number of anhydride groups}}\right)}{\text{Epoxy equivalent weight}} \times 100 = \begin{array}{l}\text{Parts by weight of anhydride to be used}\\ \text{with 100 parts by weight of epoxy resin}\end{array}$$

Table 5.5. Structures and Characteristics of Commonly Used Anhydride Curing Agents.

ANHYDRIDE EQUIVALENT WEIGHT, g/eq	MELTING POINT, °C (°F)	COMMENTS
	Phthalic Anhydride (PA)	
148	130 (266)	Available from Allied Chemical Corporation.
	Hexahydrophthalic Anhydride (HHPA) 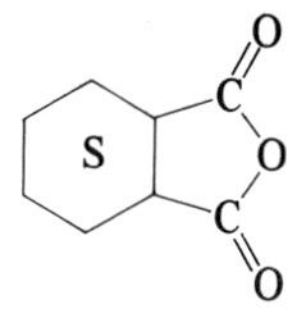	
154	40 (104)	Available from Allied Chemical Corporation.
	Nadic Methyl Anhydride (NMA) Maleic Anhydride Adduct of Methyl Cyclopentadiene	
180	Liquid at 25°C (77°F) (0.200 Pa · sec) 200 cP	Widely used for prepregs. Available from Allied Chemical Corporation.
	Dodecenyl Succinic Anhydride (DDSA)	
270	Liquid at 25°C (77°F) (0.200 Pa · sec) 200 cP	Available from Allied Chemical Corporation.
	Chlorendic Anhydride (CA)	
371	231–235 (448–455)	Needs no cure accelerator, but is high-melting and hard to handle; good fire retardant. Available from Velsicol Chemical Corporation.

Table 5.5. Continued

ANHYDRIDE EQUIVALENT WEIGHT, g/eq	MELTING POINT, °C (°F)	COMMENTS
	Trimellitic Anhydride (TMA)	
	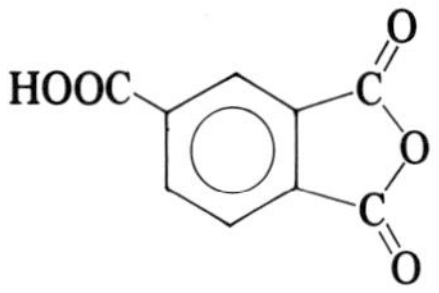	
193	161–164 (322–327)	Available from Allied Chemical Corporation.
	Maleic Anhydride (MA)	
	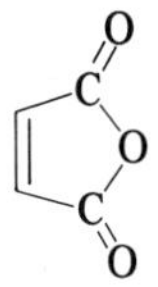	
100	53 (127)	Available from Allied Chemical Corporation.
	Succinic Anhydride (SA)	
	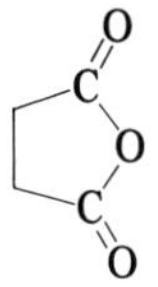	
100	120 (248)	Available from Allied Chemical Corporation.
	Methyltetrahydrophthalic Anhydride	
	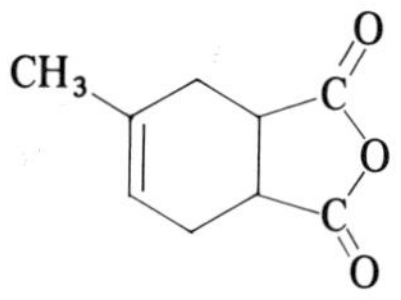	
166	4 (39) (0.06 Pa · sec, 60 cP) at 25° C = 77° F	Available from Allied Chemical Corporation.
	3,3′,4,4′-Benzophenone-tetracarboxylic Dianhydride (BTDA)	
	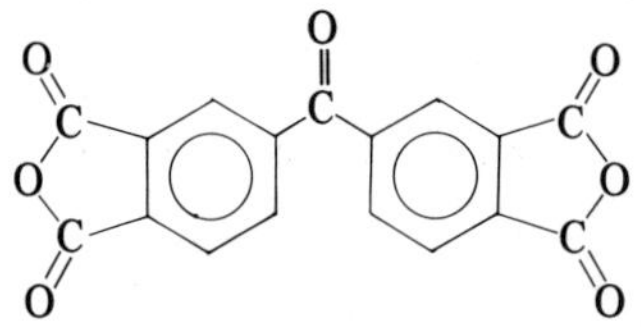	
161	221 (430)	Used mainly in powder coatings; when used as minor component in fiber composite matrix, it improves high-temperature properties. Available from Gulf Oil Chemical Company.

Anhydride curing agents differ in the extent to which they directly react with the epoxy group rather than catalyze the etherification between epoxy groups. To obtain optimum performance, that is, to maximize the extent of the anhydride–epoxy reaction, the hydroxyl content in the uncured system must be controlled and high-temperature cures must be used.

The ester linkage formed by the anhydride–epoxy reaction is stable in organic acids and in some inorganic acids, but decomposes in alkalis. It is more thermally stable and provides better electrical insulation than the linkage formed in an amine cure.

5.4.3. Lewis Acid Catalytic Curing Agents

Only one Lewis acid—boron trifluoride—has achieved great popularity as a curing agent for epoxy resins in composites. Added in small amounts to the epoxy resin alone, it functions as a catalyst by cationically homopolymerizing the epoxy molecules into a polyether. Boron trifluoride causes very rapid (occurring in minutes) and very exothermic polymerization of the epoxy resin, and blocking techniques to halt the room-temperature reaction must be used when other than very small amounts of resin are being cured. When blocked with monoethylamine to form the complex BF_3MEA (see Table 5.6), boron trifluoride is a latent curing agent at room temperature, but becomes active above 90°C (194°F) causing a rapid cure of the epoxy resin with a controllable release of heat. For prepregs, which are often stored for weeks before fabrication into a component, the use of latent curing agent is an absolute necessity. Epoxy resin systems containing BF_3MEA are popular for use in potting, tooling, laminating, and filament winding. It has been found, however, that prepregs and cured systems containing BF_3MEA generally have poor resistance to humidity.

5.4.4. Accelerators

Accelerators are added to an epoxy resin–curing agent mixture to speed up a sluggish reaction. They are added in small nonstoichiometric amounts that have been empirically determined to give the best properties.

Some of the tertiary amine catalytic curing agents for epoxy resins described in Section 5.4.1 can also act as accelerators when used with an epoxy resin–curing agent mixture. They are used most often to speed up an epoxy–anhydride cure.

Stannous octoate is a Lewis acid accelerator that promotes the curing reaction of epoxy-anhydride systems used in filament winding applications and as powder coatings. It permits a room temperature cure for some systems.

Table 5.6 lists the chemical structures and characteristics of several commonly used accelerators. There are numerous accelerators that are not described in this section. The patent literature and journals have reported a great variety of systems that appear promising.[15–20]

5.5. CURED EPOXY RESIN SYSTEMS

Some generalizations can be made concerning the relationship between the chemical structure and properties of a cured epoxy resin:

- The greater the number of aromatic rings a cured epoxy resin contains, the greater its thermal stability and chemical resistance.
- Epoxy resins cured with aromatic curing agents are likely to be more rigid and often make a stronger cured product than those cured with aliphatic curing agents. However, these epoxy resin systems require higher cure temperatures because their very rigidity reduces the molecular mobility needed to properly position two reactive end groups for reaction.
- A lower crosslink density can improve toughness (if strength is not significantly lowered) by permitting greater elongation before breakage.
- A lower crosslink density can also result in reduced shrinkage during cure.
- A higher crosslink density yields an improved resistance to chemical attack.
- A higher crosslink density also leads to an

Table 5.6. Structures and Characteristics of Commonly Used Accelerators.

AMINE HYDROGEN EQUIVALENT WEIGHT, g/eq	MELTING POINT, °C (°F)	COMMENTS
	Benzyldimethylamine (BDMA)	
	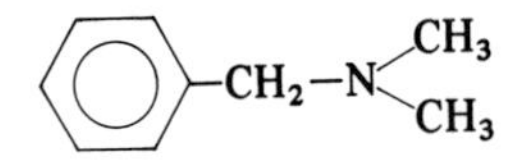 $C_6H_5-CH_2-N(CH_3)_2$	
—	Liquid at 25°C (77°F) (0.1 Pa · sec, 100 cP)	Lewis base used as an accelerator mainly for 1 cell epoxy–anhydride mixtures.
	2,4,6-Tris(dimethylaminomethyl)phenol	
	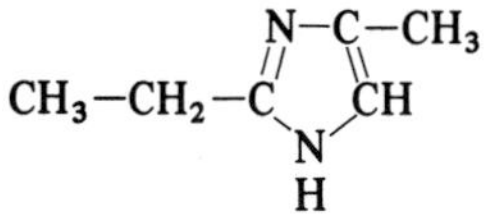 CH_3-CH_2-C ring: $N-C(-CH_3)=CH-N(H)-$ (imidazole)	
—	Liquid at 25°C (77°F) (0.3 Pa · sec, 300 cP)	Lewis base used as an accelerator for epoxy–anhydride mixtures to provide room-temperature cure. Available from Rohm & Haas as DMP-30.
	2-Ethyl-4-Methylimidazole (EMI)	
	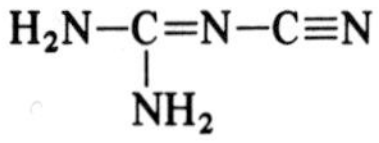 $H_2N-C(NH_2)=N-C\equiv N$	
110	Liquid at 25°C (77°F) (4–8 Pa · sec, 4000–8000 cP)	Lewis base used as an accelerator for epoxy–anhydride mixtures to provide long pot life and good elevated-temperature properties. Available from Houdry Process & Chemical Company as EMI-24.
	Dicyandiamide (DICY)	
	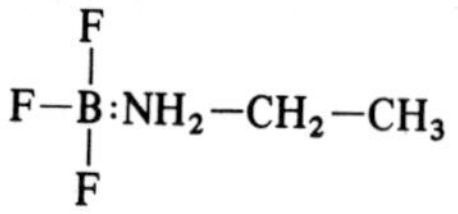 $F_3B:NH_2-CH_2-CH_3$	
28	207–209 (405–408)	Used for adhesives, prepregs, and powder coatings. Available from Pacific Anchor Chemical Corporation.
	Boron Trifluoride–Monoethylene Amine (BF_3MEA)	
	 $Sn(O-C(=O)-CH_2CH_2CH_2CH_2CH_2CH_2-CH_3)_2$	
—	85–90 (185–194)	Blocked Lewis acid; used as an accelerator for epoxy–DDS (see Table 5.4) systems in high temperature service; provides latency. Available from Allied Chemical Corporation.

Table 5.6. Continued

AMINE HYDROGEN EQUIVALENT WEIGHT, g/eq	MELTING POINT, °C (°F)	COMMENTS
Stannous Octoate		
$(H_3C)_2N-CH_2-$ / $-CH_2-N(CH_3)_2$ / OH / CH_2 / N / H_3C CH_3	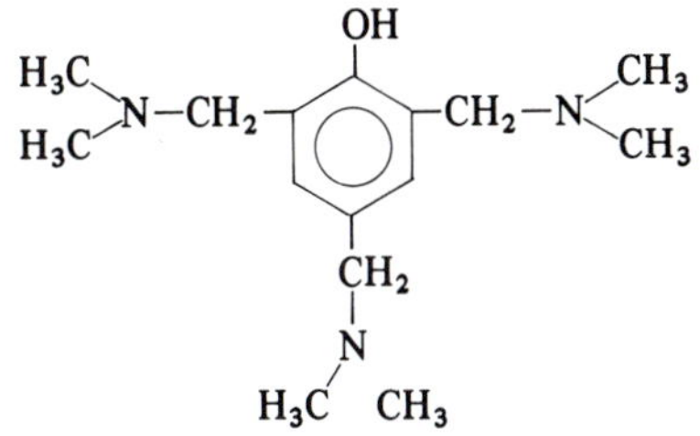	
—	—	Lewis acid; used as an accelerator for epoxy–anhydride mixtures. Available from Union Carbide Corporation.

increase in the heat distortion temperature (and glass transition temperature), but too high a crosslink density lowers the strain-to-failure (increased brittleness).

- The replacement of aromatic molecular sections by aliphatic or cycloaliphatic sections imparts greater flexibility and extensibility to the cured resin, provided that the crosslink density is not increased.
- The performance of anhydride-cured systems is better in an acidic medium than in a basic medium.

Table 5.7 illustrates the range of property values that can be achieved with different unfilled cast epoxy resin systems. Since epoxy resins are viscoelastic materials, their property values are both temperature and time (rate, frequency) dependent.

Figure 5.1 compares the chemical stabilities (in terms of flexural modulus of elasticity retention) of some typical unfilled cast epoxy resin systems in hostile chemical environments as a function of the type of curing agent. The amine-cured systems are quite resistant, the aromatic amine systems being even more stable than the aliphatic amine systems. The anhydride-cured system breaks down completely in the strong base and organic solvent, but withstands the acid, weak base, and water. The BF_3MEA-catalyzed system is seriously affected by the organic solvent and long exposure to moisture.

Table 5.7. Property Values of Unfilled Cast Epoxy Resin Systems.

PROPERTY	TYPICAL VALUE AT 23°C (73°F)
Specific gravity	1.2–1.3
Rockwell hardness	M100–M110
Impact strength [notched bar, Izod test.] (ft-lb/in) J/m of notch	0.1–1.0 (5.3–53)
Coefficient of thermal conductivity, W/m · °K (cal/cm · sec · °C)	0.17–0.21 [(4–5) × 10^{-4}]
Coefficient of thermal linear expansion, m/m/°C (in/in/°F)	(5–8) × 10^{-5} [(2.8–4.4) × 10^{-5}]
Specific heat, J/kg · °K (Btu/lb m · °F)	(1.25–1.80) × 10^3 (0.30–0.43)
Volume resistivity, Ω-m	(1–50) × 10^{17}
Dielectric constant (at 60 Hz)	2.5–4.5
Tensile strength, MPa (psi)	55–130 [(8.0–18.9) × 10^3]
Tensile of modulus of elasticity, MPa (psi)	2800–4200 [(406–609) × 10^3]
Poisson's ratio	0.20–0.33
Flexural strength, MPa (psi)	125 (18.1 × 10^3)

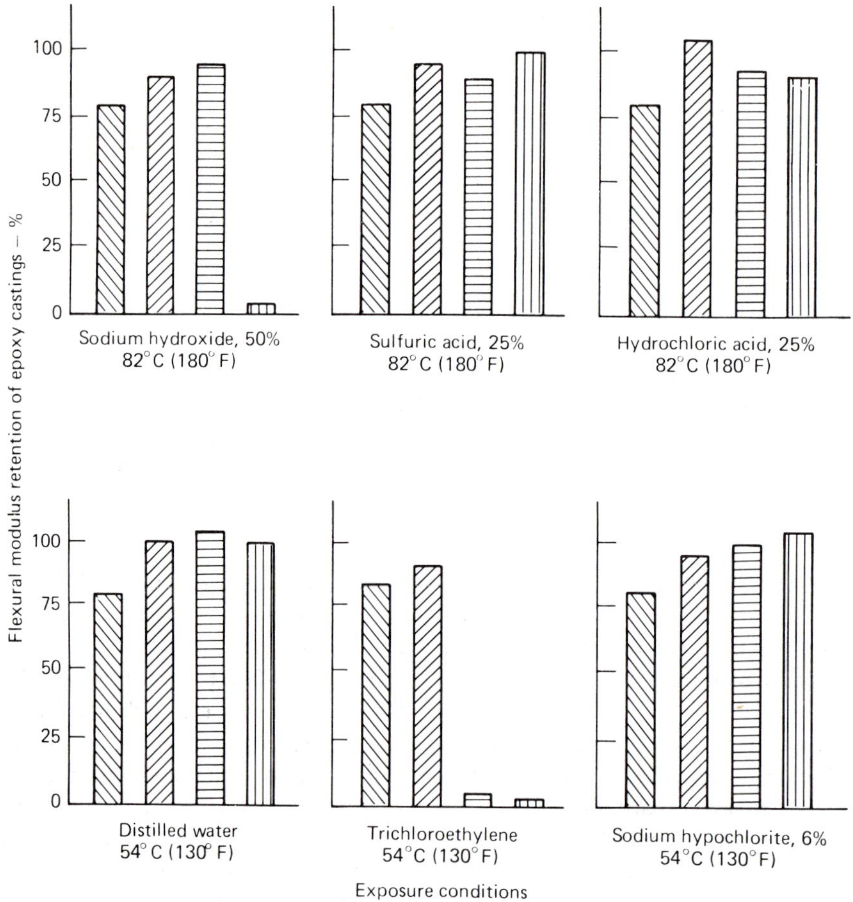

Figure 5.1. Chemical stability (in terms of flexural modulus of elasticity retention) of typical unfilled cast epoxy resin systems in hostile chemical environments as a function of the types of curing agent.[10] The curing agents used are TETA (aliphatic amine—see Table 5.4) ▧, MPDA (aromatic amine—see Table 5.4) ▨, BF_3MEA (catalyst for Homopolymerization—see Table 5.6) ▤, and PA (anhydride—see Table 5.5 ▥.

5.5.1. Properties of Specific Cured Epoxy Resin Systems

Epoxy resin systems can be formulated to accommodate most of the limitations presented by specific curing requirements. For example, although the use of high-cure-temperature systems in large components, which are inconvenient to heat, and thick-walled parts, which require that thermal stresses be kept to a minimum (i.e., room-temperature cure or only mild heating), would be inappropriate, low-cure-temperature systems are available for use in these applications.[21] Aliphatic amine-cured epoxy resins represent such systems, since they offer fairly good properties with a room-temperature cure (however, mild heating usually improves the properties).[22] Of course, these systems cannot be used at high temperatures. Table 5.8 presents the complete properties of an epoxy–aliphatic amine system suitable for advanced fiber composites.

For higher-temperature applications, the epoxy resin system must have a glass transition temperature (T_g) well above the temperature at which the composite will be used (the heat distortion point is a good estimate of T_g[23]), since the resin properties change drastically at and above T_g. Epoxy resins cured by aromatic amines, acid anhydrides, BF_3MEA (see Table 5.6), and DICY (see Table 5.4) can attain a high T_g; however, they also must be cured at temperatures above 120°C (248°F).

Table 5.8. An Aliphatic Amine-Cured Epoxy Resin System (Room-Temperature Curable).[22]

Resin system components		Parts by weight
Epoxy: DGEBA (see Table 5.1), e.g., DER 332 (Dow Chemical Company)		100
Curing agent: APTA (see Table 5.4), e.g., Jeffamine T-403 (Jefferson Chemical Company)		45 (36)[a]
Cure cycle: 16 hr at 60°C (140°F) for improved properties over room-temperature cure		

Property	SI	(English)
Viscosity at 25°C (77°F): Pa·sec (cP)	0.8	(800)
Density at 25°C (77°F) g/cm^3		
Cured	1.16[b]	
Volumetric shrinkage,		
After gelation	4.4[b]	
After cure	4.4[b]	
Water absorption, wt % gain		
After 2 hr in boiling water (ASTM D570-63)	0.75[b]	
Impact strength [notched bar, Izod test], J/m of notch	11.0	
Shear properties		
Failure stress, MPa (Ksi)	61	(8.85)
Modulus of elasticity, GPa (Ksi)	1.27	(184)
Tensile properties		
Modulus of elasticity, GPa (Ksi)	3.24	(470)
Heat distortion temperature at 1820 kPa (264 psi), °C (°F)	62	(144)[b]
Coefficient of thermal linear expansion from 298–374°K, °C^{-1}	6×10^{-5}	
Average specific heat from 286 to 367°K (ASTM C351-61), J/kg·°K (Btu/16 m. °F)	1.75×10^{3}[b]	(0.42)
Coefficient of thermal conductivity, W/m·°K		
At 298°K	0.133[b]	
At 318°K	0.174[b]	
At 336°K	0.210[b]	
Compressive properties		
Modulus of elasticity, GPa (Ksi)	3.48	(504)

[a]Earlier lots of curing agent required 36 parts per hundred for 1 : 1 stoichiometry, whereas later lots required 45 parts per hundred; this illustrates the importance of chemically analyzing all incoming lots of resin components.
[b]Cured for 24 hr at 60°C (140°F) + 24 hr at 77°C (171°F).

Table 5.9 presents the properties of an aromatic amine-cured epoxy resin system suitable for wet filament winding. Table 5.10 describes a popular anhydride-cured epoxy resin system with properties similar to those of the aromatic amine-cured system of Table 5.9.

Only a few epoxy resins and curing agents have been accepted by the aircraft industry for use in prepregs to make composites that withstand very high temperatures of about 177°C (351°F). The accepted epoxy resins are the polyglycidyl ethers of the novolacs, triglycidyl *p*-aminophenol, and tetraglycidyl diaminodiphenyl methane. Suitable curing agents are (see Table 5.4) DICY, DDS, MDA, MPDA, and (see Table 5.6) BF_3MEA. Table 5.11 gives the heat distortion temperatures of four accepted systems that have actually been used in graphite composites. Table 5.12 presents data on a currently popular epoxy resin system for aerospace applications involving high temperature and moisture. At the present

Table 5.9. An Aromatic Amine-Cured Epoxy Resin System.[24]

Resin system components	Parts by weight
Epoxy: DGEBA (see Table 5.1), e.g., Epon 826 (Shell Chemical Company)	100
Diluent: BDE (see Table 5.2), e.g., RD-2 (Ciba-Geigy Corporation)	25
Curing agent: MDA–MPDA eutectic (see Table 5.4), e.g., Tonox 60-40 (UniRoyal)	29

Cure cycle: 3 hr at 60°C (140°F) + 2 hr at 120°C (248°F)	
Viscosity at 25°C (77°F): Pa·sec (cP)	1.20 (1200)
Time to reach 2.0 Pa·sec, hr	6
Gel time for a 30-g mass at 25°C (77°F), hr	23
Density at 25°C (77°F), Mg/m^3	
Uncured	1.15
Cured	1.21
Volumetric shrinkage, %	
After gelation	3.7
After cure	5.4
Tensile properties	
Modulus of elasticity, GPa (Ksi)	2.68 (389)

Water absorption, wt % gain		
After 6 hr in boiling water (ASTM D570-63)	0.93	
Glass transition temperature by thermal mechanical analysis at 10°C (18°F)/min., °C (°F)	126–133 (259–271)	
Heat distortion temperature 1820 kPa (264 psi), °C (°F)	121	(250)
Coefficient of thermal linear expansion at 298–375°K, $°C^{-1}$	6.81×10^{-5}	
Specific heat by ice calorimetry, J/kg·°K		
At 363°K	1.54×10^3	
At 424°K	1.71×10^3	
Coefficient of thermal conductivity, W/m·°K		
At 325°K	0.243	
At 356°K	0.244	
At 389°K	0.256	
Shear properties		
Failure stress, MPa (Ksi)	52	(7.54)
Compressive properties		
Maximum stress, MPa (Ksi)	111	(16.1)
Strain at maximum stress, %	8.0	
Modulus of elasticity, GPa (Ksi)	2.9	(420)

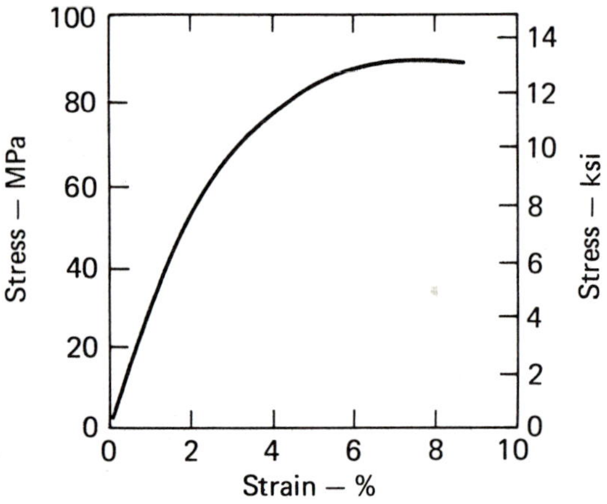

Table 5.10. An Anhydride-Cured Epoxy Resin System.[2,10]

Resin system components	Parts by weight
Epoxy: DGEBF (see Table 5.1), e.g., Epon 828 (Shell Chemical Company)	100
Curing agent: NMA (see Table 5.5)	90
Accelerator: BDMA (see Table 5.6)	1

Cure cycle: 3 hr at 120°C (248°F) + 24 hr at 150°C (302°F)	
Viscosity at 27°C (81°F): Pa·sec (cP)	1.78 (1780)
Time to reach 100 Pa·sec (1000 cP), days	5–6
Pot life of a 500-g mass at 23°C (73°F), days	4–6

Heat distortion temperature 1820 kPa (264 psi), °C (°F)	128	(262)
Solvent absorption, wt % gain		
After 24 hr in boiling water (ASTM D570-63)	0.67	
After 3 hr in boiling acetone	1.9	

Tensile properties	23°C (73°F)	100°C (212°F)
Maximum stress, MPa (Ksi)	72.4 (10.5)	46.2 (6.70)
Strain at maximum stress, %	2.7	7.2
Modulus of elasticity, GPa (Ksi)	3.45 (500)	1.38 (200)

Table 5.11. Epoxy Resin System Formulations Suitable for High-Temperature Composite Prepregs.

RESIN SYSTEM	MATERIAL RATIO (BY WEIGHT)	CURE CYCLE, hr/°C (°F)	HEAT DISTORTION TEMPERATURE, °C (°F)
Epoxy–aromatic amine			
DGEBA[a]–DDS[b]	100:30	1/125 (257) + 1/200 (392)	175 (347)
Epoxy–Lewis acid			
DGEBA[a]–BF_3MEA[c]	100:3	3/120 (248) + 1/200 (392)	174 (345)
Epoxy–anhydride			
Tetraglycidyl methylene-dianiline[a]–NMA[d]	100:110	2/100 (212) + 3/180 (356) + 24/200 (392)	250 (482)
Epoxy–aromatic amine			
Epoxy novolac[a]–MDA[b]	100:28	16/55 (131) + 2/175 (347) + 4/200 (392)	200 (392)

[a]See Table 5.1.
[b]See Table 5.4.
[c]See Table 5.6.
[d]See Table 5.5.

time, a limited amount of data has been published on the neat resin properties of such systems.

For some applications, a more flexible matrix is desirable. However, flexibility is obtained at the expense of strength. Figure 5.2 demonstrates the dramatic difference in tensile strength between a typical epoxy resin system with a flexibilizer and the same system without one. A flexibilizer such as the diglycidyl ether of linoleic acid dimer increases elongation by about eight times but decreases the strength to one-half its original value. Table 5.13 gives the property data on the flexible system depicted in Fig. 5.2.

Recently, there has been great interest in the rubberized epoxy resin systems because they are believed to have potential for improving composite performance.[27–29] Table 5.14 presents data on a rubberized epoxy resin system that was used in filament-wound pressure vessels and gave burst pressure results slightly above those of many other high-performance formulations.[30] Apparently, the resin mixture

Table 5.12. A Popular Epoxy Resin System for High-Temperature Composite Prepregs.[25]

Resin system components		Parts by weight
Epoxy: High-molecular-weight DGEBA (see Table 5.1), e.g., Araldite 6005 (Ciba-Geigy Corporation)		100
Curing agent: DDS (see Table 5.4)		36
Accelerator: BF_3MEA (see Table 5.6)		0.5
Cure cycle: 24 hr at 120°C (248°F) + 4 hr at 175°C (347°F)		
Gel time for a 30-g mass, hr	100°C (212°F)	120°C (248°F)
Without accelerator	3	2.17
With accelerator	3	1.93

Tensile properties at 23°C (73°F)		Water absorption, wt % gain	
Maximum stress, MPa (Ksi)	58.9 (8.54)	After 2 hr in boiling water (ASTM D570-63)	0.6
Strain at failure, %	3.3[a]		
Modulus of elasticity, GPa (Ksi)	2.34 (339)	Heat distortion temperature 1820 kPa (264 psi), °C (°F)	190 (374)

[a]No value available for strain at maximum stress.

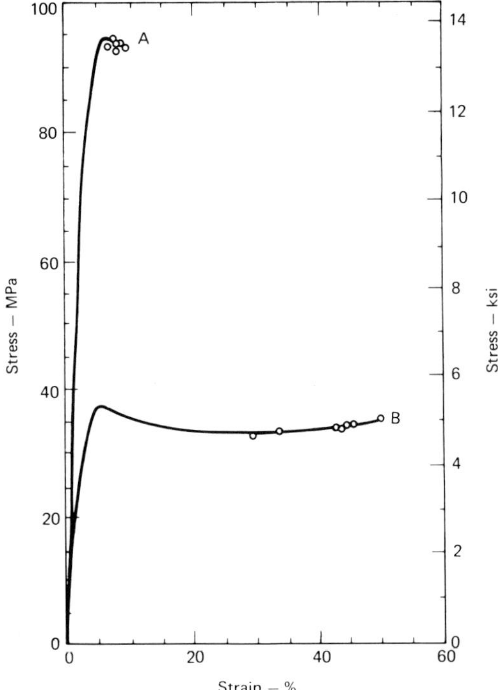

Figure 5.2 Effect of a flexibilizer on the tensile strength of an epoxy resin system:[26] (A)DGEBF (see Table 5.1)/diglycidyl ether of neopentyl glycol (see Table 5.2)/MDA–MPDA eutectic (see Table 5.4) in a ratio (Parts by weight) of 100:30:32; (B) diglycidyl ether of linoleic acid dimer (see Table 5.3)/DGEBF/diglycidyl ether of neopentyl glycol/MDA–MPDA eutectic in a ratio (parts by weight) of 50:30:20:16.4.

Table 5.13. A Flexible Epoxy Resin System.[26]

Resin system components	Parts by weight
Epoxy: DGEBF (see Table 5.1), e.g., XD 7818 (Dow Chemical Company)	30
Flexible epoxy: Diglycidyl ester of linoleic acid dimer (see Table 5.3), e.g., Epon 871 (Shell Chemical Company)	50
Diluent: Diglycidyl ether of neopentyl glycol (see Table 5.2), e.g., XD 7114 (Dow Chemical Company)	20
Curing agent: MDA–MPDA eutectic (see Table 5.4), e.g., Tonox 60-40 (UniRoyal)	16.4

Cure cycle: 3 hr at 70°C (158°F) + 5 hr at 130°C (266°F)	
Viscosity at 25°C (77°F), Pa·sec (cP)	0.54 (540)
Gel time for a 30-g mass at 25°C (77°F), hr	29.2
Water absorption, wt % gain	
After 6 hr in boiling water (ASTM D570-63)	2.4
Glass transition temperature by thermal mechanical analysis at 10°C (18°F)/min., °C (°F)	66 (151)
Tensile properties	
Modulus of elasticity, GPa (Ksi)	1.45 (210)

Stress – MPa: 40, 30, 20, 10, 0; Stress – ksi: 4, 2, 0; Strain—%: 0, 20, 40, 60

Table 5.14. A Rubberized Epoxy Resin System for Filament Winding.[29]

Resin system components	Parts by weight
Epoxy: DGEBF (see Table 5.1), e.g., XD 7818 (Dow Chemical Company)	50
Rubberized epoxy: DGEBF + 10% CTBN (see Table 5.1), e.g., XD 7575.02 (Dow Chemical Company)	50
Diluent: Vinyl cyclohexene dioxide (see Table 5.2), e.g., ERL 4206 (Union Carbide Corporation)	30
Curing agent: MDA–MPDA eutectic (see Table 5.4), e.g., Tonox 60-40 (UniRoyal)	38

Cure cycle: 5.5 hr at 60°C (140°F) + 3 hr at 120°C (248°F)	
Viscosity at 25°C (77°F), Pa·sec (cP)	0.7 (700)
Gel time for a 30-g mass at 25°C (77°F), hr	40
Density at 25°C (77°F), Mg/m^3	
Cured	1.22
Volumetric shrinkage, %	
After gelation	4.69
After cure	5.32
Tensile properties	
Modulus of elasticity, GPa (Ksi)	4.12 (538)
Water absorption, wt % gain	
After 6 hr in boiling water (ASTM D570-63)	1.23
Glass transition temperature by thermal mechanical analysis at 10°C (18°F)/min., °C (°F)	105–125 (221–257)
Heat distortion point at 1820 kPa (264 psi), °C (°F)	110 (230)
Coefficient of thermal linear expansion at 213–408°K, °C^{-1}	4.79×10^{-5}
Specific heat by differential scanning calorimetry, J/kg·°K	
At 313°K	1.352×10^3
At 343°K	1.558×10^3
At 373°K	1.771×10^3
Shear properties	
Failure stress, MPa (Ksi)	84 (12.2)
Modulus of elasticity, GPa (Ksi)	2.18 (316)
Compressive properties	
Maximum stress, MPa (Ksi)	134 (19.4)
Strain at maximum stress, %	5.68
Modulus of elasticity, GPa (Ksi)	3.51 (509)

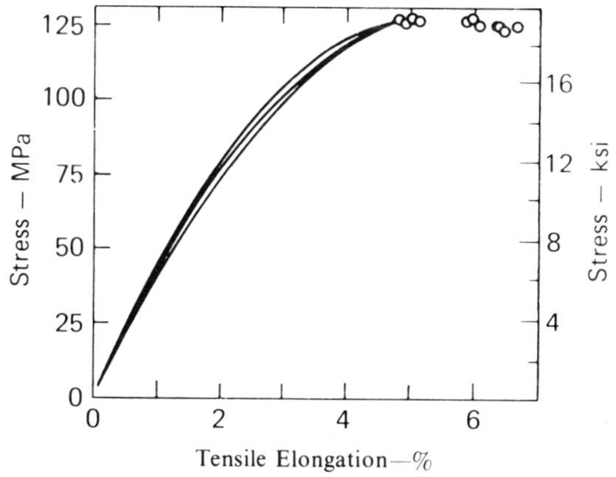

Stress — MPa / Stress — ksi vs. Tensile Elongation—%

cures to form a two-phase system containing small rubber particles that improve the fracture toughness of the neat resin.[31, 32]

5.6. PROCESSING CONSIDERATIONS

5.6.1. Quality Control

The quality control of base resins and prepregs involves checking the identities and amounts of the ingredients as well as measuring the quantities of volatiles and impurities. The molecular weight and molecular weight distribution, because of their effect on polymer properties, should be checked for consistency from lot to lot. For the unmixed epoxy resin components, this is easily done by means of chromatography or viscosity measurements. For prepregs, where the resin components are mixed and the cured network has already started to form, the state of cure advancement must be evaluated and controlled. The resin phase can be dissolved off for chromatography or, where this is impossible, the methods described in Section 5.6.3 can be used on the intact prepreg. The test techniques available for the comprehensive characterization of starting materials are listed in Table 5.15. Typical specifications for the quality control

Table 5.15. Test Techniques for Quality Control of Starting Materials.

TEST	REFERENCES	ASTM STANDARD
Quantitative Chemical Analysis for Functional Groups and Impurities		
Epoxy equivalent weight	33–35	D1652
Amine equivalent weight	36	D2073
Primary amine	36–38	—
Secondary amine	37, 38	—
Tertiary amine	37, 38	—
Hydrolyzable chlorine	—	D1726
Total chlorine	—	D1847
Water content	—	D2072, D1364
Anhydride equivalent weight	39	—
Sulfur content	40	—
Hydroxyl equivalent weight	41, 42	—
Acid content	39	D465, D1930
Total nitrogen		
Number and Relative Amounts of Components, Limited Component Identification, Molecular Weight (Size), and Molecular Weight (Size) Distribution		
Gas chromatography	43	—
Liquid chromatography	44	—
Gel permeation chromatography	44–47	D3016
Size exclusion chromatography	48	—
Qualitative Identification of Molecular Species		
Infrared curve	—	D3168
Indirect Measure of Overall Molecular Weight (Size) and Molecular Weight (Size) Distribution		
Viscosity	—	D2393

of a reactive epoxy diluent—diglycidyl ether of neopentyl glycol (see Table 5.2)—are shown in Table 5.16.

5.6.2. Processing Parameters

Any resin system must be formulated to meet processing requirements.

Wet lay-up and wet filament winding methods require a low viscosity in the uncured resin system to permit good wetting of the fiber phase and good resin distribution in the composite. Reactive diluents (see Table 5.2) are often added to reduce viscosity. The pot life (working life) also must be long enough to allow fabrication of the desired component without the complications of a rapidly advancing cure.

Aliphatic amine curing agents (see Table 5.4) react faster with epoxy resins than do aromatic amines. The former have pot lives ranging from minutes to a few hours, whereas the latter have longer pot lives—often 24 hr or more.[24, 49, 50] Anhydride curing agents (see Table 5.5) can produce long pot lives (e.g., 2 months for NMA). However, when an accelerator is used, the pot lives of epoxy–anhydride systems can be as short as a few hours, depending on the amount and type of accelerator used.

Prepregs are normally stored at low temperatures to inhibit further reaction of the partially cured (B-staged) resin until desired. A controllable and reproducible cure advancement that allows both sufficient pot life as well as shaping and arrangement of the prepreg is critically important. Table 5.17 lists the test techniques of processing-factor measurement.

Table 5.16. Typical Specifications for the Quality Control of a Reactive Epoxy Diluent—Diglycidyl Ether of Neopentyl Glycol. (See Table 5.2)

Structure

The diluent is mainly the diglycidyl ether of neopentyl glycol (see Table 5.2):

$$\overset{\displaystyle O}{\overbrace{CH_2-CH}}-CH_2-O-CH_2-\underset{\displaystyle CH_3}{\overset{\displaystyle CH_3}{\overset{|}{\underset{|}{C}}}}-CH_2-O-CH_2-\overset{\displaystyle O}{\overbrace{CH-CH_2}}$$

Requirements

The materials must conform to the following specifications:

- The infrared absorption curve must match the one given below, i.e., the frequency and order of intensity of the peaks must be the same.

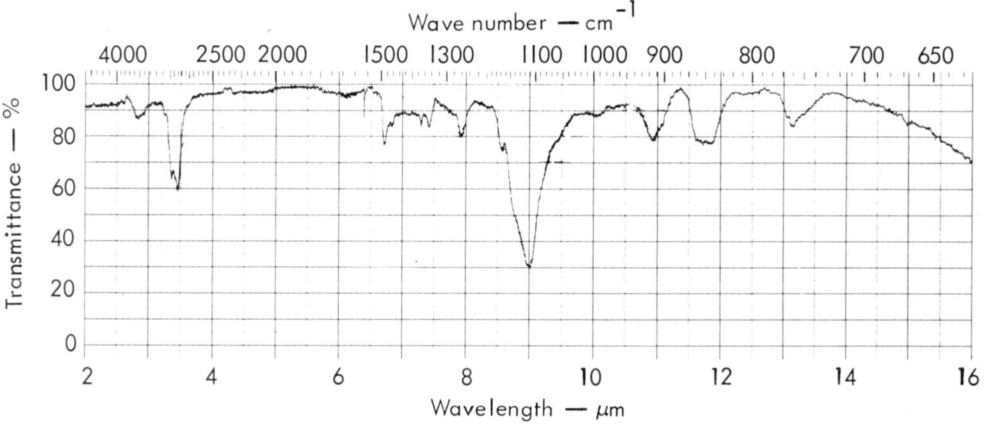

- The epoxy equivalent weight must be between 145 and 160 g/eq.
- The viscosity must be between 0.012 and 0.030 Pa·sec (12 and 30 cP) at 25°C (77°F).
- The hydrolyzable chlorine content must not exceed 0.50 wt %.
- The gas chromatogram should match the one given below. It is especially important that the peak area should show at least one 45% component.

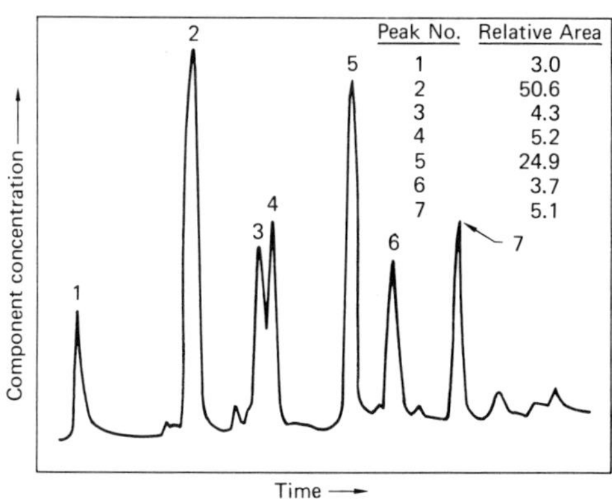

Methods

- Infrared spectrum. Record on any spectrophotometer and compare with a known spectrum.
- Epoxy equivalent weight. Determine by accepted acid titration method, e.g., ASTM D1652-73 or Ref. 34.
- Viscosity. Determine on a rotating spindle (Brookfield Model LVT) viscometer at 25°C (77°F) with No. 1 spindle and 60-rpm spindle speed (see ASTM D2393-68).
- Hydrolyzable chlorine content. Determine by mild treatment with alcoholic KOH followed by acid titration (see ASTM D1726-73). Report the average of at least two determinations.

Table 5.16. Continued

- Gas chromatogram. Inject material into gas chromatograph column (1 m by 3.2 mm) containing 10% SP 2100[a] on 100/120-mesh Supelcoport. A temperature schedule of 100–200°C (212–392°F) at 10°C (18°F)/min. and then 200–325°C (392–617°F) at 15°C (27°F)/min. gives good results. Detect the gas-eluted components by flame ionization and plot the amount of eluted solute against time. Later determinations should be done at the identical instrument settings to facilitate comparisons.

[a]SP 2100 is a silicone-phase, high-temperature partitioning fluid.

5.6.3. Optimum Cure and Monitoring of Cure

To determine the best cure cycle within the constraints of short processing time, good properties, and low thermal stresses, a variety of techniques is available. The degree of cure can be defined in terms of any of the properties that change during the curing reaction and reach a constant value at the end of the cure. Physical and mechanical measurements can be correlated to chemical changes so that the choice of a cure cycle can be based both on gross resin properties and on molecular structure.[6]

The critical crosslinking reactions occur in the resin system. Thus, many composite cure-monitoring methods have been developed for the resin alone. The total composite curing process is assumed to be similar enough to the cure process of the pure resin that the results for the resin are transferable to the composite. Table 5.18 lists the common cure-monitoring techniques for epoxy resins.

Wet chemical or physical analysis methods (e.g., solvent swell or titration) are often used to directly measure the chemical reaction during cure. In addition, infrared spectroscopy[6, 87] has been widely employed with success. Figures 5.3 and 5.4 demonstrate the use of near-infrared spectroscopy for monitoring the disappearance of epoxy groups with time. Although nuclear magnetic resonance (NMR) can also be used to track functional group disappearance,[63] it has not yet achieved widespread popularity.

The thermal properties (such as thermal conductivity, specific heat, and heat content) of an epoxy resin system depend on the extent of cure and hence can be used to assess cure advancement. Heat content measured by differential scanning calorimetry (DSC) has been used successfully to this end. Figure 5.5 compares the DSC traces of the same temperature-resistant epoxy resin system in two different stages of cure advancement; the difference is vivid.

Dielectric measurement has become an increasingly popular cure-monitoring technique. The mobility of molecular segments and therefore the extent of cure is inferred by the response of molecular dipoles to an oscillating electric field. An example of the instrumental output, that is, the dielectric response [dissipation (power) factor versus time] of a curing resin, is shown in Fig. 5.6. The equipment is easy to use, the method is quite sensitive, and the output is relatively simple to interpret. Fortunately, both DSC and dielectric analysis can be readily used to measure the advancement of cure in prepregs.[45, 78, 79, 82, 88]

Monitoring of shrinkage during cure can enable the selection of the cure cycle that minimizes the residual stresses in the final

Table 5.17. Test Techniques for Measuring Processing Factors.

TEST	REFERENCES	ASTM STANDARD
Viscosity	51, 52	D2393
Pot life	51–53	—
Gel time	—	D2471
Peak exotherm temperature	54–56	D2471
B-stage	See Table 5.19	

Table 5.18. Cure-Monitoring Techniques for Epoxy Resins.

TECHNIQUE	REFERENCES	ASTM STANDARD
Chemical and Spectroscopic Methods to Detect Changes in Molecular Structure		
Titration of functional groups	6, 35, 57, 58	—
Infrared spectroscopy	6, 58–62	—
Nuclear magnetic resonance	63	—
Physical Methods to Detect Growth in the Crosslink Network		
Solvent swell	59	—
Shrinkage during cure	64–70	D2566
Torsional braid and dynamic mechanical analysis	71–73	—
Mechanical properties	See Table 5.20	
Thermal Methods to Detect the Extent of Cure		
Differential scanning calorimetry	45, 60, 62, 74–79	—
Thermal conductivity	80	C177
Specific heat	74, 80	—
Measurement of Electrical Properties to Detect Molecular Segment Mobility		
Dielectrometer	78, 79, 81–85	—
Ion graphing	86	—

product. Because the gel temperature, maximum cure temperature, and exotherm influence the total shrinkage, different pathways to the same final cure temperature may produce different shrinkages.

Finally, the degree of cure can be estimated from the mechanical properties of the resin.[89–92] The available methods range from simple hardness tests to more complex static mechanical tests or sensitive dynamic mechanical tests (see Table 5.19). Once a cure cycle is selected, extensive monitoring procedures need not be used again—provided that adequate quality control of the starting material and processing procedures is exercised during production.

5.7. TESTS ON CURED RESIN SYSTEMS

It is advisable to follow standardized procedures for the testing of cured resins whenever possible. This allows the valid comparison of test results obtained by different experimen-

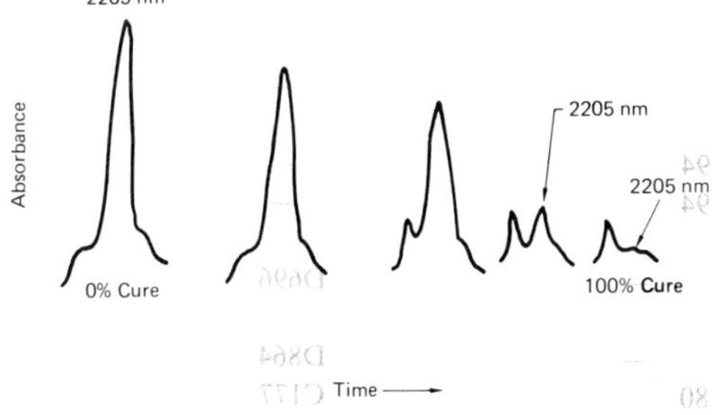

Figure 5.3. Cure monitoring by near-infrared spectroscopy: The above partial apectra show the disappearance of the epoxy ring C–H vibration at 2205 nm with time at 100° C (212° F) for the epoxy resin system DGEBF (see Table 5.1)/diglycidyl ether of neopentyl glycol (see Table 5.2)/MDA–MPDA eutectic (see Table 5.4)/DAP (see Table 5.4) in a ratio (parts by weight) of 100:30:13.2:13.2.

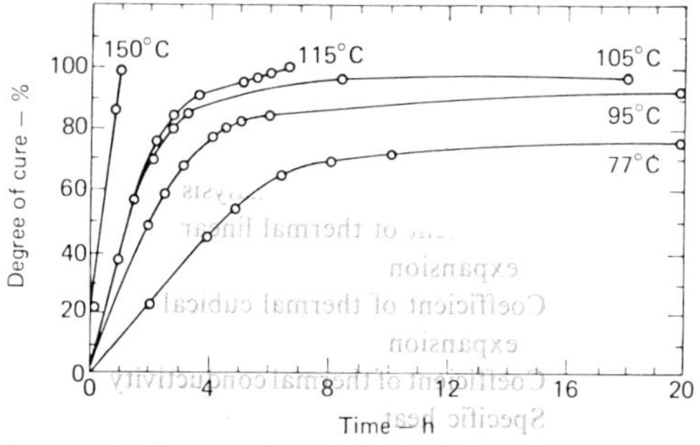

Figure 5.4. Degree of cure by near-infrared spectroscopy as a function of time and temperature for the epoxy resin system DGEBF (see Table 5.1)/diglycidyl ether of neopentyl glycol (see Table 5.2)/MDA–MPDA eutectic (see Table 5.4)/DAP (see Table 5.4) in a ratio (parts by weight) of 100:30:13.2:13.2.

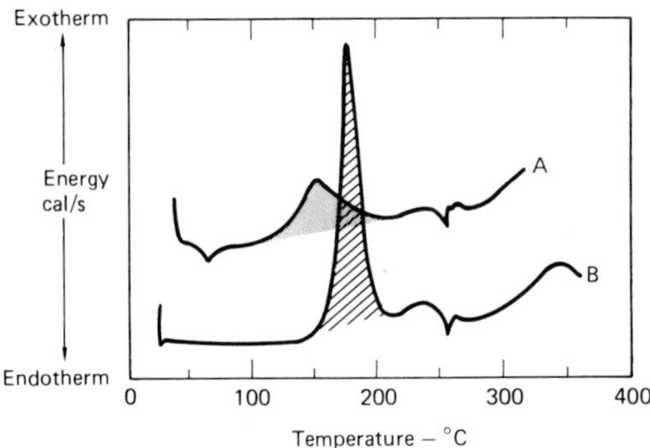

Figure 5.5 Comparison of DSC traces at 5° C (9° F)/min. of a temperature-resistant epoxy resin system in two different stages of cure advancement. Stage A is displaced upward for easy viewing; shaded areas are the hearts of reaction for the cure.

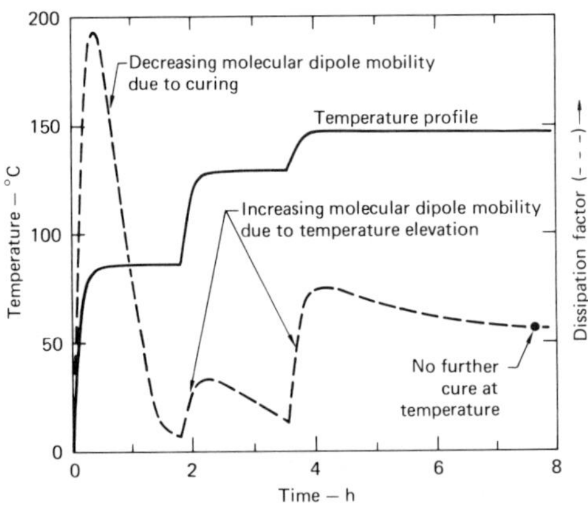

Figure 5.6. Monitoring of cure by dielectric response: dissipation (power) factor (dashed curve, 2 V full scale, at 500 Hz) versus time for the epoxy resin system DGEBF (see Table 5.1)/aliphatic amine. The applied temperature is given by the solid curve.

Table 5.19. Test Techniques for Cured or Partially Cured Epoxy Resin Systems.

TEST	REFERENCES	ASTM STANDARD
Chemical and Physical Tests		
Density by displacement	—	D792
Chemical resistance	—	D543
Water absorption	—	D570
Light and water exposure	—	D1499
Electrical Tests		
Volume resistivity	—	D257
Surface resistivity	—	D257
Dielectric strength	—	D149
Dielectric breakdown voltage	—	D149
Permittivity, dielectric constant	—	D150
Dielectric ac loss characteristics, dissipation (power) factor	—	D150
Thermal Tests		
Heat distortion point (deflection temperature)	23	D648
Glass transition temperature	23, 93, 94	D696
Thermal mechanical analysis	95	—
Thermal gravimetric analysis	96	—
Differential scanning calorimetry	94	—
Differential thermal analysis	94	—
Coefficient of thermal linear expansion	—	D696
Coefficient of thermal cubical expansion	—	D864
Coefficient of thermal conductivity	80	C177
Specific heat	80, 97, 98	—
Mechanical Tests		
Tension	99, 100	D638
Compression	—	D695
Flexure	—	D790
Impact resistance	101, 102	D256
Fracture toughness	103–105	—

Table 5.19. Continued

TEST	REFERENCES	ASTM STANDARD
Hardness: Rockwell	—	D785
Barcol	—	D2583
Shore	—	D1706
Dynamic properties by the torsion pendulum method	106	D2236
Flexural fatigue	—	D671

ters or in different laboratories. Table 5.19 lists the applicable ASTM methods. Other possible sources include Military Handbook MIL-HDBK-23 and Federal Test Method Standard No. 406. Additional references are listed in Table 5.19.

Acknowledgments

This work was performed under the auspices of the U.S. Department of Energy by the Lawrence Livermore Laboratory under contract No. W-7405-Eng-48. The authors would like to thank H. A. Newey, J. K. Lepper, and M. T. Lewis for their contributions to this work. In addition, the authors thank P. L. Lien for the editing.

References

1. H. Lee and K. Neville, *Handbook of Epoxy Resins*, McGraw-Hill, New York, 1967.
2. W. G. Potter, *Epoxide Resins* (published for the Plastics Institute), Iliffe Books, London, 1970.
3. C. A. May and Y. Tanaka (editors), *Epoxy Resins: Chemistry and Technology*, Marcel Dekker, Inc., New York, 1973.
4. P. F. Bruins, *Epoxy Resin Technology*, Wiley-Interscience, New York, 1968.
5. Kh. Arutyunyan, S. Davtyan, B. Rozenberg, and N. Yenikolopyan, "Kinetics of Hardening of an ED-5 Epoxide Oligomer by Reaction with *m*-Phenylenediamine under Adiabatic Conditions," *Polym. Sci. USSR* **16**, 2452 (1975).
6. M. Ochi, Y. Tanaka, and M. Shimbo, "Curing Mechanism of Epoxy Resin," *Nippon Kagaku Kaishi* **9**, 1600 (1975).
7. K. Okahashi, O. Hayashi, and K. Shibayama, "Kinetics of Curing Reaction of Epoxy Resins with Imidazole," *Mitsubishi Denki Giho* **47** (5), 547 (1973).
8. R. B. Prime, "Kinetics of Epoxy Cure: 2. The System Bisphenol-A Diglycidyl Ether/Polyamide," *Polymer* **13**, 455 (1972).
9. G. Piloyan, I. Ryabchikov, and O. Novikova, "Determination of Activation Energies of Chemical Reactions by Differential Thermal Analysis," *Nature* **212**, 1229 (1966).
10. Shell Chemical Company, *Epon Resins for Casting*, New York, 1967.
11. W. Fisch and W. Hofmann, "Über den Härtungsmechanismus der Äthoxylinharze," *J. Polym. Sci.* **12**, 497 (1954).
12. W. Fisch, W. Hofmann, and J. Koskikallio, "The Curing Mechanism of Epoxy Resins," *J. Appl. Chem.* **6**, 429 (1956).
13. S. L. Kaplan, A. Katzakian, and E. L. Mitch, "Fast Curing Acid/Epoxy, Anhydride/Epoxy Resins," 30th Annual Conference, Reinforced Plastics/Composites Institute, SPI, Washington, D.C., February 4–7, 1975, Section 8-C.
14. K. M. Hollands and I. L. Kalnin, "The Kinetics of Gelation of Some Accelerated Acid Anhydride Cured Epoxy Resins," paper presented at Symposium on Epoxy Resins," sponsored by Division of Organic Coatings and Plastics Chemistry at the 155th Meeting of the ACS, San Francisco, California, April 3–4, 1968, p. 60.
15. N. B. Godfrey, "Polyamine and Hydrocarbon Sulfonic Acid Accelerator Combination for Epoxy Curing," U.S. Patent 3,666,721, May 30, 1972.
16. N. B. Godfrey, "Polyamine and Hydrocarbon Sulfonic Acid Accelerator Combination for Epoxy Curing," U.S. Patent 3,785,997, January 15, 1974.
17. A. Lombardi and T. Smeal, "Catalyst Systems for Lowering Epoxy Resin Cure Temperatures," U.S. Patent 3,903,048, September 3, 1975.
18. H. Waddill, "Piperazine and Alkanolamine Accelerator for Epoxy Curing," U.S. Patent 3,875,072, April 1, 1975.
19. H. Waddill, "Method of Accelerating Epoxy Curing," U.S. Patent 3,943,104, March 9, 1976.
20. A. M. Partansky, "A Study of Accelerators for Epoxy–Amine Condensation Reaction," paper presented at Symposium on Epoxy Resins, sponsored by Division of Organic Coatings and Plastics Chemistry at the 155th Meeting of the ACS, San Francisco, California, April 3–4, 1968, p. 29.
21. T. T. Chiao, E. S. Jessop, and H. A. Newey, "A Moderate-Temperature-Curable Epoxy for Advanced Composites," *SAMPE Quart.* **6** (3), 38 (1975).

22. T. T. Chiao and R. L. Moore, "A Room-Temperature-Curable Epoxy for Advanced Fiber Composites," 29th Annual Conference, Reinforced Plastics/Composites Institute, SPI, Washington, D.C., February 5–8, 1974, Section 16-B.
23. T. Kamon, K. Saito, Y. Miwa, and K. Saeki, "Relation between Epoxy Resin Glass Transition Temperature and Heat Distortion Temperature," *Kobunshi Ronbunshu* **31** (11), 665 (1974).
24. T. T. Chiao, E. S. Jessop, and H. A. Newey, "An Epoxy System for Filament Winding," *SAMPE Quart.* **6** (1), 28 (1974).
25. Ciba-Geigy Corporation technical literature.
26. J. Rinde, E. T. Mones, and H. A. Newey, "Flexible Epoxies for Wet Filament Winding,"32nd Annual Conference, Reinforced Plastics/Composites Institute, SPI, Washington, D.C.,February 8–11, 1977, Section 11-D.
27. J. M. Scott and D. C. Phillips, "Carbon Fiber Composites with Rubber Toughened Matrices," UK Atomic Energy Authority, Harwell, Rept. AERE-R7793 (1974).
28. J. V. Larsen, "Fracture Energy of CTBN/Epoxy-Carbon Fiber Composites," 26th Annual Conference, Reinforced Plastics/Composites Institute, SPI, Washington, D.C., February 9–12, 1971, Section 10-D.
29. L. S. Penn, B. S. Morra, and E. T. Mones, "A Rubberized Epoxy System for Wet Filament Winding," *Composites* **8**, 31 (1977).
30. M. A. Hamstad and E. S. Jessop, "Performance of Filament-Wound Vessels from an Organic Fiber in Several Epoxy Matrices," in: *Proceedings of the 7th National SAMPE Technical Conference, Albuquerque, New Mexico, October 14–16, 1975*, Vol. 7, (1975), p. 202.
31. R. Drake and A. Siebert, "Elastomer-Modified Epoxy Resins for Structural Applications," *SAMPE Quart.* **6**, 11 (1975).
32. F. J. McGarry, "Building Design with Fibre Reinforced Materials," *Proc. Roy. Soc. London* **A319**, 59 (1970).
33. H. Lee and K. Neville, *Handbook of Epoxy Resins*, McGraw-Hill, New York, 1967, p. *4*-14.
34. R. R. Jay, "Direct Titration of Epoxy Compounds and Aziridines," *Anal. Chem.* **36**, 667 (1964).
35. H. Jahn and P. Goetzky, "Analysis of Epoxides and Epoxy Resins," in: *Epoxy Resins:* Chemistry and Technology (edited by C. A. May and Y. Tanaka), Marcel Dekker, Inc., New York, 1973, p. 653.
36. J. S. Fritz, "Titration of Bases in Nonaqueous Solvents," *Anal. Chem.* **22**, 1028 (1950).
37. S. Siggia, J. G. Hanna, and I. R. Kervenski, "Quantitative Analysis of Mixtures of Primary, Secondary, and Tertiary Aromatic Amines," *Anal. Chem.* **22**, 1295 (1950).
38. C. D. Wagner, R. H. Prown, and E. D. Peters, "The Analysis of Aliphatic Amine Mixtures; Determination of Tertiary Aliphatic Amines in the Presence of Primary and Secondary Amines and Ammonia," *J. Amer. Chem. Soc.* **69**, 2609 (1947).
39. H. Jahn and P. Goetzky, "Analysis of Epoxides and Epoxy Resins," in: *Epoxy Resins:* Chemistry and Technology (edited by C. A. May and Y. Tanaka), Marcel Dekker, Inc., New York, 1973, p. 683.
40. W. Selig, "Micro and Semimicro Determination of Sulfur in Organic Compounds by Potentiometric Titration with Lead Perchlorate," *Microchimica Acta (Wien)*, 168 (1970).
41. "Determination of Hydroxyl Group in Organic Compounds," Shell Chemical Company, Shell Method Series 237/55 (1960).
42. D. H. Reed, F. E. Critchfield, and D. K. Elder, "Phenyl Isoyanate Method for Determination of Hydroxyl Equivalent Weights of Polyoxyalkylene Compounds," *Anal. Chem.* **35**, 571 (1963).
43. O. E. Schupp III, "Gas Chromatography," in: *Technique of Organic Chemistry* (edited by E. S. Perry and A. Weissberger), Vol. 13, (Wiley-Interscience, New York, 1968.
44. D. W. Hadad, "Chemical Quality Assurance of Epoxy "Resin Formulations by Gel Permeation, Liquid, and Thin Layer Chromatography," in: *Proceedings of the 22nd National SAMPE Symposium and Exhibition, San Diego, California, April 26–28, 1977*, Vol. 22 (1977), p. 301.
45. I. L. Kalnin, M. Meisters, and H. J. Notarius, "Characterization of Epoxy Resin Advancement in Fiber Reinforced Composite Prepregs," 26th Annual Conference, Reinforced Plastics/Composites Institute, SPI, Washington, D.C., February 9–12, 1971, Section 14-A.
46. H. Batzer and S. Zahir, "Studies in the Molecular Weight Distribution of Epoxide Resins. I. Gel Permeation Chromatography of Epoxide Resins," *J. Appl. Polym. Sci.* **19**, 585 (1975).
47. B. Miller, "Why 'Good' Resin Makes Bad Parts—And What You Can Do about It," *Plastics World* (February 16, 1976).
48. S. D. Abbott, "High Performance Size Exclusion Chromatography Using Porous Silica Packings, paper presented at 27th Pittsburgh Conference, Cleveland, Ohio, March 15, 1976.
49. L. S. Penn and T. T. Chiao, "A Long Pot-Life Epoxy System for Filament Winding," in: *Proceedings of the 7th National SAMPE Technical Conference, Albuquerque, New Mexico, October 14–16, 1975*, Vol. 7 (1975), p. 177.
50. T. T. Chiao, E. S. Jessop, and L. S. Penn, "Screening of Epoxy Systems for High Performance Filament Winding Applications," in: *Proceedings of the 7th National SAMPE Technical Conference, Albuquerque, New Mexico, October 14–16, 1975*, Vol. 7 (1975), p. 167.
51. T. Kakurai and T. Noguchi, "Viscometric Studies on the Gelation of Epoxy Resins with Amines," *Kobunshi Kagaku* **19**, 542 (1962); "Gelation in the Reaction

of Epoxy Resin with Amines," *Kobunshi Kagaku* **20**, 17 (1963).

52. L. J. Gough and I. T. Smith, "A Gel Point Method for the Estimation of Overall Apparent Activation Energies of Polymerization," *J. Appl. Polym. Sci.* **3**, 362 (1960).
53. B. A. Hills, "Gelation Timing," *J. Oil Colour Chem. Assoc.* **45**, 251 (1962).
54. H. G. Manfield, "The Measurement of Exotherms of Casting and Laminating Resins," *Brit. Plastics* **26**, 230 (1953).
55. G. Mensching, "Eigenschaftsbildende Einflusse bei der Giessharztechnologie, Speziell in der Elektrotechnik," *Plaste Kautschuk* **8**, 179 (1961).
56. D. D. Smith, "Computerized Exotherm Calculations for Large Epoxy Castings," Bendix Corporation, *Bendix Technical Journal* (Spring 1969), p. 79.
57. W. Selig, "Estimation of the Curing Rate of Epoxy Resins: Improvement in Methods," *Z. Anal. Chem.* **255**, 130 (1971).
58. W. Selig and G. L. Crossman, "Estimation of the Cure Rate of Epichlorohydrin/Bisphenol A Type Epoxy Resins," *Z. Anal. Chem.* **253**, 279 (1971).
59. H. Dannenberg and W. R. Harp, Jr., "Determination of Cure and Analysis of Cured Epoxy Resins," *Anal. Chem.* **28**, 86 (1956).
60. J. F. Carpenter, "Assessment of Composite Starting Materials: Physiochemical Quality Control of Prepregs," paper presented at AIAA/ASME Symposium on Aircraft Composites: The Emerging Methodology for Structural Assurance, San Diego, California, March 24–25, 1977; McDonnell Aircraft Company Rept. MCAIR No. 77-001 (1977).
61. H. Dannenberg, "Determination of Functional Groups in Epoxy Resins by Near Infrared Spectroscopy," *SPE Trans.* **3**, 78 (1963).
62. M. A. Acitelli, R. B. Prime, and E. Sacher, "Kinetics of Epoxy Cure: 1. The System Bisphenol-A Diglycidyl Ether/*m*-Phenylene Diamine," *Polymer* **12**, 335 (1971).
63. W. B. Moniz, C. F. Poranski, Jr., and S. A. Sojka, "Carbon-13 Fourier Transform NMR—An Important New Analysis Tool," Naval Research Laboratory, Washington, D.C., Report of NRL Progress (August 1975).
64. O. D. Lascoe, "Shrink Tests Developed for Tool Plastic," *Tool Engineering* **39** (11), 117 (1957).
65. W. Fisch and W. Hofmann, "Reaction Mechanisms, Chemical Structure, and Changes in Properties during the Curing of Epoxy Resins," *Plastics Tech.* **7** (8), 28 (1961).
66. T. Shimazaki and I. Motoki, "Occurrence of Stress which Accompanies Shrinkage during Epoxy Curing," *Kobunshi Kagaku* **28** (319), 884 (1971).
67. H. L. Parry and H. H. Mackay, "Cure Shrinkage of Epoxy Systems," *SPE J.* **14** (7), 22 (1958).
68. V. V. Bolotkin and K. S. Bolotina, "Shrinkage of Epoxy Cements during Hardening," *Mekh. Polim.* **1**, 178 (1972) [p. 165 of English translation].
69. H. Lee and K. Neville, *Handbook of Epoxy Resins*, McGraw-Hill, New York, 1967, pp. *6*-29 and *17*-11.
70. B. Rosen and A. Fornof, *Absolute and Differential Dilatometry for Measurement of Unrestrained Shrinkage on Resin Curing*, ASTM, Philadelphia, Pennsylvania, ASTM Special Technical Publication No. 327, 1962, p. 40.
71. A. Lewis, "Dynamic Mechanical Behavior during the Thermosetting Curing Process," *SPE Trans.* **3**, 201 (1963).
72. J. K. Gillham, J. A. Benci, and A. Noshay, "Isothermal Transitions of a Thermosetting System," Technical Report No. 18 prepared for Office of Naval Research (August 1973); available from National Technical Information Service, Springfield, Virginia, Publication No. AD-765 765.
73. J. K. Gillham and P. G. Babayevsky, "Epoxy Thermosetting Systems: Dynamic Mechanical Analysis of Aromatic Diamines with the Diglycidyl Ether of Bisphenol A," Technical Report No. 17 prepared for Office of Naval Research (April 1973); available from National Technical Information Service, Springfield, Virginia, Publication No. AD-759 995.
74. W. W. Wendtland, *Thermal Methods of Analysis*, 2nd ed., J. Wiley and Sons, New York, 1974.
75. R. Fava, "Differential Scanning Calorimetry of Epoxy Resins," *Polymer* **9**, 137 (1968).
76. O. R. Abolafia, "Application of Differential Scanning Calorimetry to Epoxy Curing Studies," paper presented at 27th Annual Technical Conference, Society of Plastics Engineers, Chicago, Illinois, May 5–8, 1969, Technical Papers **15**, 610 (1969).
77. R. B. Prime, "Dynamic Cure Analysis of Thermosetting Polymers," *Anal. Calorimetry* **2**, 201 (1970); *Proceedings of the Symposium on Analytical Calorimetry* at the meeting of the American Chemical Society, Chicago, Illinois, September 13–18, 1970 (edited by R. S. Porter and J. F. Johnson), Plenum Press, 1970.
78. C. A. May, D. K. Whearty, and J. S. Fritzen, "Composite Cure Studies by Dielectric and Calorimetric Analyses," in: *Proceedings of the 21st National SAMPE Symposium, Los Angeles, California, April 6–8, 1976*, Vol. 21 (1976), p. 803.
79. J. F. Carpenter and T. T. Bartels, "Characterization and Control of Composite Prepregs and Adhesives," in: *Proceedings of the 7th National SAMPE Technical Confidence, Albuquerque, New Mexico, October 14–16, 1975*, Vol. 7 (1975), p. 43.
80. R. P. Kreahling and D. E. Kline, "Thermal Conductivity, Specific Heat, and Dynamic Mechanical Behavior of Diglycidyl Ether of Bisphenol A Cured with *m*-Phenylene Diamine," *J. Appl. Polym. Sci.* **13**, 2411 (1969).
81. S. Yalof and W. Wrasidlo, "Crosschecking between Dielectric Measurements, DTA, and Other Methods of Thermal Analysis in Research and Production," *J. Appl. Polym. Sci.* **16**, 2159 (1972).
82. M. J. Yokota, "In-Process Controlled Curing of

Resin Matrix Composites," in: *Proceedings of the 22nd National SAMPE Symposium and Exhibition, San Diego, California, April 26–28, 1977*, Vol. 22 (1977), p. 416.

83. J. Delmonte, "Electric Properties of Epoxy Resins during Polymerization," *J. Appl. Polym. Sci.* **2**, 108 (1959).
84. S. A. Yalof, "The Relationship between Mechanical and Dielectric Properties of Polymers," in: *Proceedings of the 4th National SAMPE Technical Conference, Palo Alto, California, October 17–19, 1972*, Vol. 4 (1972), p. 391.
85. S. Yalof, "Tracking Adhesive Behavior with Dynamic Dielectric Spectroscopy," *Adhesives Age*, 23 (April 1975).
86. D. J. Crabtree, "Ion Graphing as an In-Process Cure Monitoring Procedure for Composite and Adhesively Bonded Structures," in: *Proceedings of the 22nd National SAMPE Symposium and Exhibition, San Diego, California, April 26–28, 1977*, Vol. 22 (1977), p. 636.
87. P. V. Sidyakin, "An Infrared Spectroscopic Study of the Curing of Epoxide Resins with Amines," *Vysokomol. Soedin.* **A14**, 979 (1972); English translation: *Polym. Sci. USSR* **14**, 1087 (1973).
88. C. A. May, T. E. Helminiak, and H. A. Newey, "Chemical Characterization Plan for Advanced Composite Prepregs," in: *Proceedings of the 8th National SAMPE Technical Conference, Seattle, Washington, October 12–14, 1976*, Vol. 8 (1976), p. 274.
89. R. Arridge and J. Speake, "Mechanical Relaxation Studies of the Cure of Epoxy Resins: 1. Measurement of Cure," *Polymer* **13**, 443 (1972).
90. D. E. Kline, "Dynamic Mechanical Properties of Epoxy Resins during Polymerization," *J. Appl. Polym. Sci.* **4**, 123 (1960).
91. P. Babayevsky and J. Gillham, "Epoxy Thermosetting Systems: Dynamic Mechanical Analysis of the Reactions of Aromatic Diamines with the Diglycidyl Ether of Bisphenol A," *J. Appl. Polym. Sci.* **17**, 2067 (1973).
92. P. Eyerer and S. Wintergerst, "Hardening of Thin Layers of Epoxy Resins," *Adhäsion*, 106 (1971).
93. D. H. Kaelble, "The Dynamic Mechanical Properties of Epoxy Resins," *SPE J.* **15**, 1071 (1959).
94. W. W. Wendtland, *Thermal Methods of Analysis*, 2nd ed., J. Wiley and Sons, New York, 1974, p. 251.
95. W. W. Wendtland, *Thermal Methods of Analysis*, 2nd ed., J. Wiley and Sons, New York, 1974, p. 428.
96. W. W. Wendtland, *Thermal Methods of Analysis*, 2nd ed., J. Wiley and Sons, New York, 1974, p. 6.
97. D. C. Ginnings and R. J. Corruccini, "Enthalpy, Specific Heat and Entropy of Aluminum Oxide from 0° to 900°C," Research Paper RP 1797, *J. Res. Natl. Bureau Standards* **38** (June 1947).
98. A. P. Gray, "Simple Generalized Theory for the Analysis of Dynamic Thermal Measurement," *Perkin-Elmer Instrument News* **16** (2) (1970).
99. D. H. Kaelble, "Dynamic and Tensile Properties of Epoxy Resins," *J. Appl. Polym. Sci.* **9**, 1213 (1965).
100. T. T. Chiao, A. D. Cummins, and R. L. Moore, "Fabrication and Testing of Epoxy Tensile Specimens," *Composites* **3**, 10 (1972).
101. W. B. Hillig, *Impact Response Characteristics of Polymeric Matrices*, General Electric Company, Technical Information Series Rept. No. 74CRD213 (September 1974).
102. W. B. Hillig, *Impact Response Characteristics of Polymeric Matrices*, General Electric Company, Technical Information Series Rept. No. 75CRD198 (August 1975).
103. K. Selby and L. Miller, "Fracture Toughness and Mechanical Behaviour of an Epoxy Resin," *J. Mater. Sci.* **10**, 12 (1975).
104. Y. W. Mai and A. G. Atkins, "On the Velocity-Dependent Fracture Toughness of Epoxy Resins," *J. Mater. Sci.* **10**, 2000 (1975).
105. W. F. Brown, Jr., and J. E. Srawley, *Plane Strain Crack Toughness Testing of High Strength Metallic Materials*, ASTM, Philadelphia, Pennsylvania, ASTM Special Technical Publication No. 410 (1966).
106. R. D. Ezell, *A. Torsional Pendulum for Measuring Dynamic Mechanical Properties of Polymers*, Naval Ordnance Laboratory (Naval Surface Weapons Laboratory), Silver Spring, Maryland, Rept. NOLTR 73-183 (November 1973).

6
HIGH-TEMPERATURE RESINS

Tito T. Serafini
National Aeronautics and Space Administration
Lewis Research Center
Cleveland, Ohio

6.1. INTRODUCTION

The composites discussed in this chapter employ high-temperature resins as matrix materials. High-temperature resins are linear or crosslinked, aromatic/heterocyclic polymers that have a high glass transition temperature (T_g) and can withstand continuous exposure in air at temperatures above 600° F (316° C) without exhibiting a significant loss of structural integrity. Although these polymers do undergo thermo-oxidative degradation during elevated-temperature exposure in air, they degrade at relatively slow rates. Furthermore, it is speculated that these polymers degrade into stable residues, thus increasing their service life at elevated temperatures.

The key to achieving high-temperature resins is the synthesis of polymers that contain a multiplicity of aromatic/heterocyclic molecular structural units. These structural units, which contain a minimum number of oxidizable hydrogen atoms, are able to absorb thermal energy. Unfortunately, the structural units that are responsible for the thermal and oxidative (thermo-oxidative) stability of these polymers are also responsible for their intractable nature, which makes it extremely difficult or (most often) impossible to process them into useful articles.

During the 1960's, polymer chemists synthesized scores of aromatic/heterocyclic polymers that, on the basis of thermogravimetric analysis (TGA), appeared to possess more-than-adequate thermo-oxidative stability at elevated temperatures. However, attempts to utilize these polymers as matrix resins for advanced composites either failed or were not considered to represent economically feasible methodology. Therefore, in the early 1970's, the future of high-temperature-polymer matrix composites looked rather bleak and uncertain. This potentially useful class of materials appeared destined to remain a "laboratory curiosity." However, developments within the class of high-temperature polymers known as *polyimides* during 1972–1974 not only rekindled interest and work in the area of high-temperature-polymer matrix composites, but also made it possible to realize much of their potential. At the present time, polyimide–advanced fiber composites are being used or considered for use in a variety of 600° F (316° C) structural applications. The polyimide matrix materials being most widely used by prepreg manufacturers are either the NR-150B series of condensation-type polyimides (E.I. du Pont de Nemours & Company, Incorporated) or the monomeric reactant, addition-type polyimide PMR-15 developed at NASA's Lewis Research Center. Therefore, the data presented in this chapter will primarily be for the NR-150B and PMR-15 polyimides.

6.2. CONDENSATION POLYIMIDES

As shown in Fig. 6.1, condensation-type aromatic polyimides can be prepared by reacting aromatic diamines with aromatic dianhydrides, with aromatic tetracarboxylic acids, or with dialkyl esters of aromatic tetracarboxylic acids. The diamine–dianhydride reaction is preferred for preparing polyimide films and coatings, whereas the other two combinations of reactants are preferred for polyimide matrix

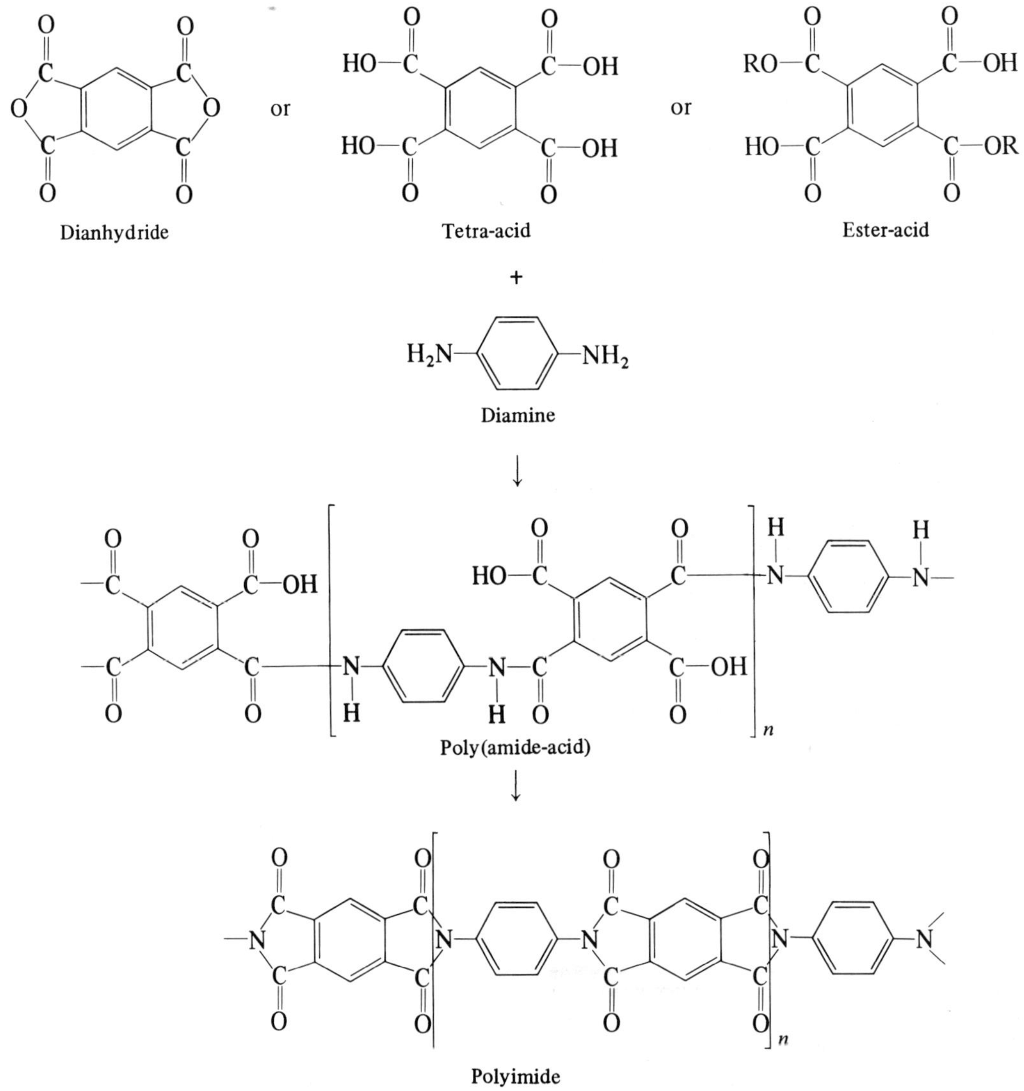

Dianhydride or Tetra-acid or Ester-acid

+

Diamine

↓

Poly(amide-acid)

↓

Polyimide

Figure 6.1. General reactions for the preparation of aromatic polyimides.

resins. The idealized reaction sequence, as presented in Fig. 6.1, involves the formation of a linear poly(amide-acid) followed by cyclization (imidization) to the polyimide, which is accompanied by the release of water, an alkyl alcohol, or both. Optimum thermo-oxidative stability results from the complete conversion of the amide-acid groups to the stable heterocyclic imide rings.

Polyimide prepreg, or precursor, solutions are prepared by simply dissolving the appropriate reactants in a high-boiling point aprotic solvent, (e.g., *N*-methylpyrrolidone, NMP) or in a solvent mixture containing an aprotic solvent. However, the high boiling points of aprotic solvents and their tendency to enter into complex formation make it difficult to volatilize them during the processing of prepreg lay-ups. Moreover, complex formation can interfere with imidization, thus resulting in lower-than-optimum polymer thermo-oxidative stability. The difficulties encountered in fabricating high-quality, void-free composites (in particular, thick sections

Table 6.1. Commercially Available Condensation-Type Polyimide Precursor Solutions.

DESIGNATION	PRODUCER
Skybond Resins	Monsanto Plastics & Resins Company
Pyralin	E. I. du Pont de Nemours & Company, Inc.
NR-150 (A and B)	E. I. du Pont de Nemours & Company, Inc.

and complex shapes) are traceable to the use of the high-boiling-point aprotic solvents as well as to the condensation reactions that release water, alcohol, or both. During the heating of the prepreg in order to effect chain extension and solvent removal, appreciable imidization also occurs, hence converting the resin to a more intractable state. Lower resin flow together with continued volatilization of solvent and evolution of condensation reaction by-products can result in composites that have high void contents (>5 vol %), which adversely affects their mechanical and thermo-oxidative properties.

6.2.1. Commercially Available Condensation-Type Polyimide Precursor Solutions

The condensation-type polyimide precursor solutions currently being produced are listed in Table 6.1. The Skybond Resins® and Pyralin® polyimide precursor solutions were first offered commercially in the early 1960's, whereas the NR-150B series was introduced in 1972. Compared to prepregs based on the Skybond Resins or Pyralin polyimide precursor solutions, prepregs made with the NR-150B series are somewhat easier to process, and composites with low void contents (< 1 vol %) can be fabricated. At the present time, fabricators are focusing considerable attention on the NR-150B polyimide prepreg materials.

A detailed discussion of the chemistry of the Skybond Resins, Pyralin, NR-150B polyimide precursor solutions is beyond the scope of this chapter. However, some discussion of their chemistry will be helpful in understanding their different processing characteristics.

All of the "Skybond Resins" listed in Table 6.2[1] are based on 3,3′,4,4′-benzophenonetetracarboxylic dianhydride (BTDA) (see Fig. 6.2). Skybonds 700, 703, 709, and 710 consist of an aromatic diamine and a bis(alkylester-acid) that is prepared by reacting BTDA with an alkyl alcohol (e.g., ethanol) (see Fig. 6.3). In Skybond 703, the aromatic diamine is thought to be 4,4′-methylenedianiline (MDA). Skybond 705 is a poly(amide-acid) and is prepared by reacting BTDA with an aromatic diamine. It is very likely that the Pyralin polyimide precursor solutions are also based on BTDA. The important point that needs to be made is that prior to complete conversion of the intermediate amide-acid groups to imide rings it is very likely that crosslinking occurs through the carbonyl and amino functional groups of the BTDA and amine, respectively, further contributing to the intractability of the resin.

Table 6.2. Skybond Resins.[a,1]

RESIN NUMBER	SOLIDS, %	BROOKFIELD VISCOSITY, CPS	SPECIFIC GRAVITY	pH	SOLVENT
700	60–64	2500–7000	1.15–1.18	4.0–4.7	NMP[b]
703	63–67	3000–7000	1.15–1.18	4.0–4.7	NMP
705	16.5–21	300–2000	1.05–1.08	—	NMP and xylene
709[c]	53–57	2000–9000	1.10–1.15	4.4–5.2	NMP, xylene, and ethanol
710[c]	53–57	1000–7000	1.09–1.14	4.0–4.8	NMP, xylene, and ethanol

[a]Monsanto Plastics & Resins Company.
[b]NMP = *N*-methyl-2-pyrrolidone.
[c]Specialty product.

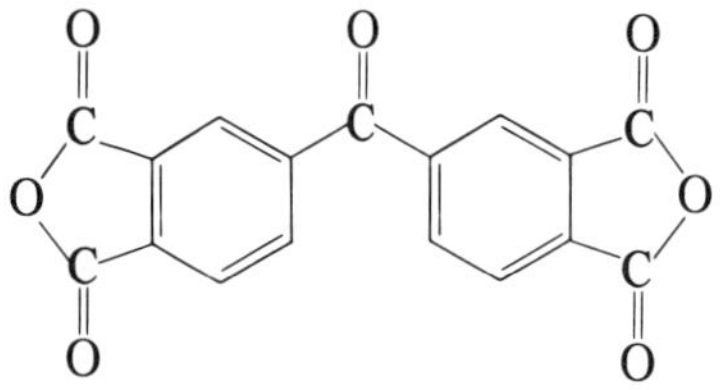

3,3′,4,4′-Benzophenonetetracarboxylic dianhydride

Figure 6.2. A major raw material used in the preparation of Skybond polyimide precursor solutions.

The NR-150 polyimide resins are based on 2,2-bis(3′,4′-dicarboxyphenyl)hexafluoropropane dianhydride (6F) (see Fig. 6.4a). The tetra-acid form of 6F (6FTA) is generally used in formulating NR-150 polyimide precursor solutions (see Fig. 6.4b). Table 6.3 describes a number of different NR-150 polyimide precursor solutions.[2] The NR-150B polyimide precursor solutions were developed for matrix applications, whereas the NR-150A series was developed for adhesive applications.

Comparing Figs. 6.2 and 6.4a, it can be seen that the major difference between BTDA and 6F rests in the nature of the group joining the two anhydride portions of the molecules. The hexafluoroisopropylidene group contained in 6F is not reactive with amines; hence, the crosslinking that occurs in polyimide resins employing BTDA is not present in NR-150 polyimide resins.[3] The NR-150 precursor solutions form linear, amorphous polyimide resins that remain melt-fusible above their T_g. Consequently, voids formed as a result of the evolution of volatile materials can be eliminated from NR-150 polyimide resins by applying pressure while heating them above their T_g.

6.2.2. Unreinforced NR-150 Polyimide Resins

The completely heterocyclic composition of NR-150 polyimide resins confers these materials with outstanding thermal and thermo-oxidative stabilities. The linear structure of the cured resins results in unexpectedly good mechanical properties, particularly as regards ultimate elongation. Studies reported by Gibbs[3] on the properties of two unreinforced NR-150 polyimide resins are summarized in Table 6.4. The 6F/PPD/MPD (95:5) resin exhibited excellent retention of properties even after 100 hr of exposure in nitrogen or air at 700° F (371° C).

Studies have also shown that NR-150A polyimide resins do not exhibit any loss of room-temperature tensile strength and elongation after a 2-week exposure in boiling water. The NR-150A system also displayed low dielectric constant and dissipation (power) factor values over a temperature range of 72–424° F (22°–218° C) (see Table 6.5).[4] Other NR-150 polyimide resins can be expected to exhibit similar hydrolytic stability and electrical properties.

6.2.3. Condensation-Type Polyimide Prepreg Materials

The major suppliers of prepreg materials prepared from condensation-type polyimide precursor resins are listed in Table 6.6. Skybond 703 is probably the most widely used precursor in the Skybond series. Pyralin and Skybond precursor resins are limited to wet impregnation (solution) prepregging; accord-

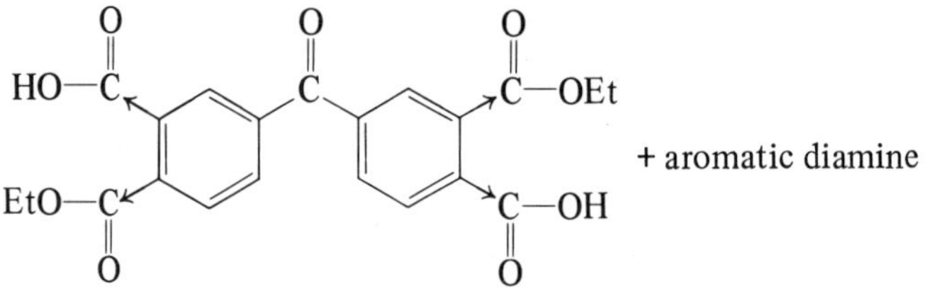

+ aromatic diamine

Figure 6.3. Constituents of Skybonds 700, 703, 709, and 710.

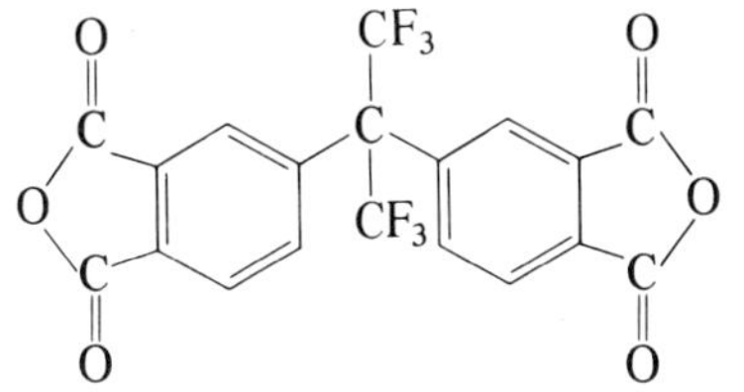

(*a*) 2,2-Bis(3′,4′-dicarboxyphenyl)hexafluoropropane dianhydride (6F)

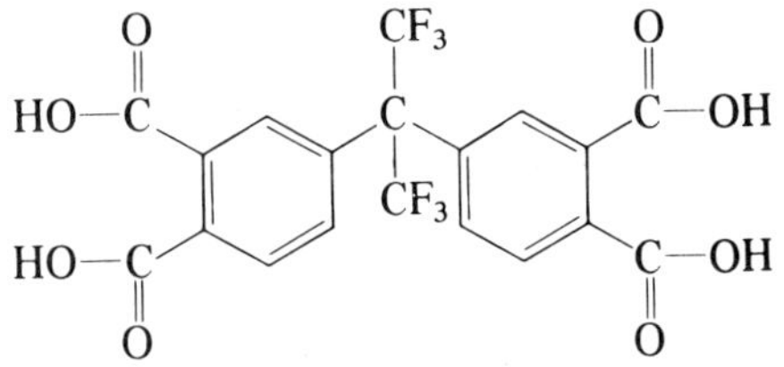

(*b*) Tetra-acid of 6F (6FTA)

Figure 6.4. Starting material and tetra-acid used in the preparation of NR-150 polyimides.

ingly, they are generally used to impregnate woven fabrics made from such materials as fiberglass, silica, quartz, and graphite. In contrast, NR-150 resins are amenable to both solution and hot-melt prepregging.

6.2.4. Condensation-Type Polyimide Composites

6.2.4.1. Glass-Fabric Composites

Table 6.7[5] summarizes the mechanical properties of Skybond 700–style 181 E glass fabric (A1100, soft finish) 12-ply laminates. The elevated-temperature properties were obtained from tests conducted at the aging temperature.

Table 6.3. NR-150[a] Polyimide Precursor Solutions.[2]

DESIGNATION	SOLVENT SYSTEM, WT %	MONOMER SYSTEM, MOLAR RATIO	YIELD WT % CRS[b]	TYPICAL VISCOSITY AT 23° C (73° F), POISES
NR-150A2	0.25 EtOH, 0.75 NMP	6FTA/ODA,[c] 1:1	48	25
NR-150B2	0.75 EtOH, 0.25 NMP	6FTA/PPD[d]/MPD,[e] 1:0.95:0.05	54	25
NR-150A2G	NMP	6FTA/ODA, 1:1	50	1000
NR-150B2G	NMP	6FTA/PPD/MPD, 1:0.95:0.05	48	1000
NR-150A2 S5X	NMP	6FTA/ODA, 1:1	60	Solid
NR-150B2 S2X	EtOH	6FTA/PPD/MPD, 1:0.95:0.05	60	150
NR-150B2 S3X	0.55 EtOH, 0.45 NMP	6FTA/PPD/MPD, 1:0.95:0.05	65	Solid
NR-150B2 S5X	NMP	6FTA/PPD/MPD, 1:0.95:0.05	60	Solid
NR-150B2 S6X	Diglyme	6FTA/PPD/MPD, 1:0.95:0.05	48	80
NR-050X	EtOH	6FDD[f]/PPD, 1.67:2.67	—	2
NR-055X	NMP	PA[g] from 6F/PPD/MPD, 1.05:0.95:0.05	25	490
NR-056X	Diglyme	6FTA/PPD/ODA, 1:0.75:0.25	48	60
NR-058X	Diglyme	PA from 6F/PPD/ODA, 1:0.75:0.25	10	3

[a]E. I. du Pont de Nemours & Company, Incorporated
[b]CRS = cured resin solids
[c]ODA = 4,4′-oxydianiline
[d]PPD = *p*-phenylenediamine
[e]MPD = *m*-phenylenediamine
[f]DD = diethylester diacid
[g]PA = polyamide acid

Table 6.4. Aging of Unreinforced NP-150 Polyimides[3]

	6F/ODA		6F/PPD/MPD (95:5)		
	CONTROL	10,000 HOURS AT 500° F (260° C) IN AIR	CONTROL	100 HOURS AT 700° F (371° C) IN NITROGEN	100 HOURS AT 700° F (371° C) IN AIR
Tg, °F (°C)	536–572 (280–300)	570 (299)	662–700 (350–371)	736 (391)	707 (375)
Tensile Strength psi (MPa)					
73° F (23° C)	16,400 (113)	—	16,000 (110)	—	—
500° F (260° C)	4,300 (30)	4,900 (34)	—	—	—
600° F (316° C)	—	—	4,500 (31)	4,300 (30)	4,000 (28)
Tensile Modulus, psi (MPa)					
73° F (23° C)	520,000 (3580)	—	580,000 (4000)	—	—
500° F (260° C)	200,000 (1380)	180,000 (1240)	—	—	—
600° F (316° C)	—	—	150,000 (1030)	171,000 (1180)	122,000 (840)
Elongation, %					
(23° C)	8	—	6	—	—
500° F (260° C)	60	4	—	—	—
600° F (316° C)	—	—	65	8	1
Flexural Strength, psi (MPa) 23° C (73° F)	14,100 (97)	—	—	—	—
Flexural Modulus, psi (MPa) 23° C (73° F)	578,000 (3980)	—	—	—	—
Notched Izod, ft.lb./in. (joules/m) 23° C (73° F)	0.7 (37)	—	0.8 (43)	—	—
Density, g/cc	1.42	—	1.46	—	—
% Weight Loss	—	4.8	—	0.3	1.5

Prepreg and laminate processing conditions are listed in Table 6.8.[5]

Figure 6.5 shows the thermo-oxidative stability of Skybond 700–style 181 E glass fabric (A1100, soft finish) laminates at various temperatures in the 550–800° F (288–427° C) range.[6] These data show that if a flexural strength of 20 Ksi (138 MPa) were considered to be the lower limit of acceptability, then the service life of the Skybond 700–style 181 E glass fabric laminates would be about 10 hr at 800° F (427° C), 100 hr at 700° F (371° C), 1000 hr at 600° F (316° C), and 10,000 hours at 550° F (288° C). Based on an elevated-temperature flexural strength of 50 Ksi (345 MPa), the value of 20 Ksi (138 MPa) represents a retention level of approximately 40%, which is good considering that the unreported void content is probably in the range of 5–10 vol %.

Table 6.5. Electrical Properties of 6F/ODA[a] Measured at 9.375 GHz.[b]

TEMPERATURE, °C	°F	DIELECTRIC CONSTANT	LOSS TANGENT
22	72	2.914	0.0016
107	225	2.921	0.0025
190	374	2.909	0.0042
218	424	2.878	0.0066

[a]Polymer derived from thermally curing DMF-based binder solution.
[b]Data supplied by Hitco, Gardena, Calif.

Figure 6.6 compares the thermo-oxidative stabilities of Skybond 700–glass fabric laminates made with either a soft or a hard finish.[6] The soft-finish fabric is more readily wetted by the precursor solution, thus resulting in

Table 6.6. Suppliers of Condensation-Type Polyimide Prepreg Materials.

SUPPLIER	PRECURSOR RESIN NR-150	PYRALIN	SKYBOND
E. I. du Pont de Nemours & Company, Inc. Fabrics and Finishes Department Saugus, California	F[a]	F	—
Ferro Corporation Composites Division Culver City, California	FT[b]	—	F
Fiberite Corporation Winona, Minnesota	FT	—	F
Hexcel Corporation Dublin, California	FT	—	FT
U.S. Polymeric Santa Ana, California	FT	—	FT

[a]F = fabric prepreg.
[b]FT = fabric and unidirectional-fiber tape.

laminates that have higher mechanical and thermo-oxidative stabilities.

Data for condensation-type polyimide–7781 E glass fabric laminates from a commercial source are presented in Table 6.9, and the processing conditions are listed in Table 6.10. Comparison of the data given in Tables 6.9 and 6.7 suggests that both systems employ the same polyimide precursor solution. More importantly, the data presented in Table 6.9 confirm the superior thermo-oxidative stability of condensation-type polyimides. Even though the void content is in the range of 12–22 vol %, laminates exposed in air at 550° F (288° C) for 1000 hr retained approximately 70% of their 500° F (260° C), 0.5-hr flexural strength.

The electrical properties in Skybond 700–style 181 E glass fabric (A1100, soft finish) 12-ply laminates are presented in Table 6.11. As will be discussed in Section 6.4, the Skybond and Pyralin glass-fabric prepreg materials are extensively used for the production of radomes.

Although most of the current work with NR-150B polyimides is concerned with their use and performance as matrix resins for graphite-fiber composites, some of the early studies also investigated the properties of NR-150B–glass fabric laminates. Table 6.12 summarizes the results from aging of autoclave-molded S-2 glass fabric laminates in air at 288° C. Laminate processing conditions are summarized in Table 6.13. The composition of the NR-150B matrix employed corresponds to the composition of NR-150B2 (see Table 6.3); the solvent, however, was reported to be either NMP or DMF (dimethyl formamide). The data presented in Table 6.12 illustrate the excellent thermo-oxidative stability of the NR-150B2 system. The low-void-content laminates (<1 vol% voids) exhibited a weight loss of only 1.5% after 1500 hr exposure at 550° F (288° C). It should be noted that the low-void laminates were obtained using a cure cycle that included the application of pressure (200 psi = 1.4 MPa) at 752° F (400° C), a temperature well above their apparent T_g. Comparison of the mechanical property data for these laminates (with either low or high void contents) with the data in Table 6.9 (recall that the F174 system uses a BTDA-derived precursor resin) for autoclave-cured laminates clearly shows the superiority of the NR-150B2 polyimide resins. The flexural strength of the NR-150B2–S-2 glass fabric laminates increased after aging in air at 550° F (288° C). Excellent property retention was also observed for the NR-150B2–S-2 glass fabric laminates exposed at either 600° F (316° C) or 700° F (317° C) (see Table 6.14).

Table 6.7. Mechanical Properties of Skybond® 700 Laminates.[5]

	High Temperature-High Pressure	Vacuum Bag
	Flexural strength flatwise, psi (MPa)	
Standard conditions 75° F (24° C)	75–85,000 (520–590)	76–83,500 (520–580)
One-half hour at 700° F (371° C)	45–60,000 (310–410)	22–32,000 (150–220)
100 hours at 700° F (371° C)	20–35,000 (140–240)	20–24,000 (140–170)
Weight loss 100 hours at 700° F (371° C)	3.0%	<5%
	Modulus of elasticity Msi (GPa)	
Standard conditions 75° F (24° C)	3.12 (21.5)	2.8 (19.3)
335 hours at 570° F (299° C)	3.12 (21.5)	—
100 hours at 700° F (371° C)		1.8 (12.4)
	Ultimate tensile strength, psi (MPa)	
Standard conditions 75° F (24° C)	57,000 (390)	50,300 (347)
335 hours at 570° F (299° C)	42,000 (290)	—
100 hours at 482° F (250° C) (at R.T.)		48,800 (336)
100 hours at 572° F (300° C) (at R.T.)		48,200 (332)
	Elongation	
Standard conditions 75° F (24° C) per cent	1.90%	2.0%
Tested at 75° F (24° C), 335 hours' aging at 570° F (299° C)	1.40%	—
100 hours at 482° (250° C) (at R.T.)		1.7%
100 hours at 572° F (300° C) (at R.T.)		2.0%

600° F (316° C) Long-Term Aging Study (High Temperature High Pressure Laminates)

Property	R.T.	500 hrs.	860 hrs.	1850 hrs.
Flexural strength psi (MPa)	75,000 (520)	29,000 (200)	20,000 (140)	10,950 (75.5)
Flexural modulus psi × 10^6 (MPa)		2.61 (18,000)	2.59 (17,900)	2.08 (14,300)
Weight loss (%)		2.2	3.4	7.9

550° F (288° C) Long-Term Aging Data (High Temperature High Pressure Laminates)

	R.T.	2300 hrs.	4500 hrs.	9000 hrs.
Flexural strength psi (MPa)	83,000 (570)	41,200 (284)	32,000 (220)	15,000 (130)
Flexural modulus Msi × (GPa)	3.02 (20.8)	2.63 (18.1)	2.95 (20.3)	2.00 (13.8)
Weight loss (%)		3.6	5.0	12.0

6.2.4.2. Graphite-Fiber Composites

In an attempt to realize the potential benefits of using graphite-fiber-reinforced polyimide composites for various aerospace structural applications, many studies employing Skybond and Pyralin graphite-fiber prepreg materials were conducted in the early to mid 1970's. The major objective of these studies was to develop processing methodology that would provide low-void-content composites. Although much excellent processing technology was developed, the efforts to achieve low-void-content composites were unsuccessful. For example, because Skybond 710 was con-

Table 6.8. Typical Processing Conditions for Skybond 700/E Glass Fabric Laminates.[5]

Prepreg

Stage at 250° (121°C) to 350°F (177°C) to achieve a volatile content of 8.0–8.5% for press laminates and 12–16% for vacuum bag laminates.

High-pressure laminates

Insert lay-up into preheated press at 600°F (316°C) and hold for 1.5 to 2.5 minutes at 10–25 psi (69–172 MPa).

Increase pressure to 250 psi (1.7 MPa) and hold for 30 minutes. Turn off heaters and release pressure when press cools to 140°F (60°C).

Vacuum-bag laminates

Apply full vacuum to bagged lay-up and heat to 350° F (177°C) at a heating rate of at least 5°F/min. (3°C/min). Hold for 5 minutes and increase pressure to 100 psi (0.7 MPa) and hold for an additional 25 minutes. Cool to 150°F (66°C) under full vacuum and pressure.

Post-cure

Time[1] (hrs.)	Temperature, °F (°C)
2	392 (200)
2	437 (225)
2	482 (250)
2	572 (300)
2	617 (325)
2	662 (350)
4	700 (371)

[1] In-air, unrestrained.

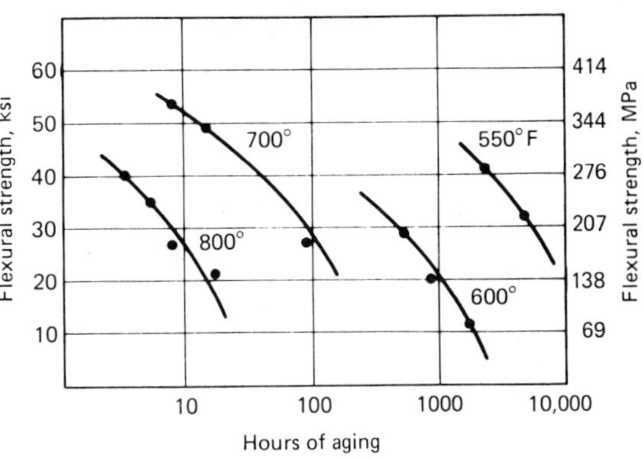

Figure 6.5. Thermo-oxidative stability of Skybond 700/glass fabric laminates at various temperatures.[6]

sidered to be more processable than the other polyimide resins that were then available, it was selected for use in a program[7] to develop design data for graphite-fiber-reinforced polyimide composites. It was concluded that the lowest void contents that could be achieved ranged between 5 and 10 vol %. This level of voids cannot be tolerated in highly loaded composite structures, and, therefore, in recent years there has been little interest in using the Skybond and/or Pyralin polyimide precursor solutions as matrix resins with graphite fibers.

Results obtained by Gibbs[3] for NR-150B2 reinforced with Hercules Magnamite® HMS graphite fibers are summarized in Table 6.15. The laminates were compression molded in matched metal dies using a final molding pressure of 2500 psi (17.2 MPa). The data show that the laminates exhibited outstanding elevated-temperature property-retention characteristics. Results obtained for NR-150B2 reinforced with Hercules Magnamite® AS graphite fibers are presented in Table 6.16. The higher rate of weight loss observed for the AS-reinforced laminates compared to the HMS-reinforced laminates reflects the lower elevated-temperature oxidative stability of the AS fibers. The effects of graphite-fiber type on composite stability will be discussed further in section 6.3.1.

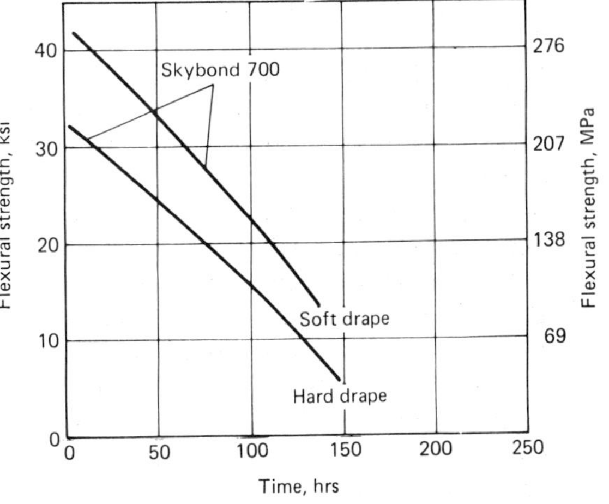

Figure 6.6. Thermo-oxidative stability of Skybond 700/glass fabric laminates.[6]

Table 6.9. Mechanical Properties of F 174[a]/7781 Glass Fabric Laminates.[6]

	VACUUM CURE CYCLE	AUTOCLAVE CURE CYCLE
Flexural Strength, psi × 10^3/ Flexural Modulus, psi × 10^6 (MPa/MPa × 10^3)		
Room Temp. Dry	75/3.4 (520/23)	78/3.5 (540/24)
Room Temp. Wet	63/3.4 (430/23)	66/3.5 (460/24)
500° F (260° C) − ½ hour	55/2.8 (380/19)	57/2.9 (390/20)
600° F (316° C) − ½ hour	40/2.7 (280/19)	43/2.8 (300/19)
500° F (260° C) − 100 hours	68/2.6 (470/18)	72/2.9 (500/20)
600° F (316° C) − 100 hours	44/2.4 (300/17)	48/2.5 (330/17)
700° F (371° C) − 100 hours	16/2.1 (110/14)	17/2.2 (120/15)
550° F (288° C) − 1000 hours	38/2.6 (260/18)	40/2.8 (280/19)
Compressive Strength, psi × 10^3/ Compressive Modulus, psi × 10^6 (MPa/MPa × 10^3)		
Room Temp. Dry	56/4.0 (390/28)	58/4.1 (400/28)
Room Temp. Wet	52/3.6 (360/25)	46/3.8 (320/26)
500° F (260° C) − ½ hour	50/3.3 (340/23)	53/3.5 (370/24)
600° F (316° C) − ½ hour	48/2.6 (330/18)	51/2.8 (350/19)
550° F (288° C) − 1000 hours	28 (190)	30 (210)
Tensile Strength, psi × 10^3/ Tensile Modulus, psi × 10^6 (MPa/MPa × 10^3)		
Room Temp. Dry	54/3.6 (370/25)	58/4.0 (400/28)
Room Temp. Wet	53/3.4 (370/23)	55/3.7 (380/26)
500° F (260° C) − ½ hour	48/3.2 (330/22)	52/3.5 (360/24)
600° F (316° C) − ½ hour	46/3.1 (320/21)	48/3.3 (330/23)
500° F (288° C) − 1000 hours	24 (170)	25 (170)
Interlaminar Shear, psi (MPa)	3600 (25)	3800 (26)
Laminate Resin Content %	22–30	22–30
Specific Gravity	1.80	1.85
Barcol Hardness	70	75
Void Content %	12–22	12–22

[a] Hexcel Corporation designation.

6.3. ADDITION POLYIMIDES

In 1968, Rhone-Poulenc (France) and TRW Incorporated (United States) each introduced a series of polyimides for which the final cure occurred by means of an addition reaction that did not result in the release of void-producing volatile materials.

The polyimide series developed by the French[8] was based on the reaction of nonstoichiometric quantities of bismaleimides and aromatic diamines (see Fig. 6.7). Two types of reaction are possible: (1) addition of the diamine on the double bond and (2) radical polymerization of the double bond.

The polyimide series developed at TRW Incorporated was the result of the serendipitous discovery that norbornenyl-terminated imide oligomers underwent a reverse Diels-Alder reaction at elevated temperatures 527–662° F (275–350° C), which resulted in the formation of macromolecules.[9] This discovery formed the basis for the P13N polyimides (see Fig. 6.8), which were the object of considerable interest for a short period of time. The postulated reaction mechanism for the reverse Diels–Alder polymerization of the norbornenyl chain-end is shown in Fig. 6.9.

Although the final curing step for the P13N and bismaleimide-based polyimides occurs by means of an addition reaction, both must first proceed through an intermediate condensation reaction that initially forms the thermally stable imide ring structures. For the bismaleimides, the dehydrocyclization step is conducted chemically, and the desired product

Table 6.10. Typical Processing Conditions for F 174/Glass Fabric Laminates.[6]

Oven and Autoclave Cure:

Edge bleed and surface bleed. Sufficient plies of bleeder fabric to remove volatiles and excess resin.

Cure Cycle: Apply vacuum pressure. Minimum of 28″ Hg. (95.2×10^{-3} MPa). Raise temp. to 240° F (116° C) at 2–4° F/min., (1–2° C/min.) raise temp. to 270° F (132° C) at 1–2° F/min. (0.5–1° C/min.) hold 30 min. If autoclave, apply 50 psi, (0.3 MPa) raise temp. to 250° F (177° C) at 1–2° F/min. (0.5–1° C/min). Cure 1 hour at 350° F (177° C).

Post Cure Cycle: ½ hr. @ 200° F (93° C),
½ hr. @ 250° F (121° C),
½ hr. @ 300° F (149° C),
½ hr. @ 350° F (177° C),
2 hrs. @ 400° F (204° C),
3 hrs. @500° F (260° C),
4 hrs. @550° F (288° C),
4 hrs. @ 600° F (316° C).

(bismaleimide) is redissolved in a high-boiling-point solvent (e.g., NMP or DMF). For P13N polyimides, the formation of the norbornenyl-terminated prepolymer proceeds by the thermal dehydrocyclization of norbornenyl-terminated amide-acid prepolymers. The prepolymer approach embodied in the P13N methodology also mandated the use of the above-mentioned high-boiling-point solvents. The difficulty in removing these solvents from prepreg lay-ups prior to final cure together with the "boardy" nature of the devolatilized prepreg materials were two of the major factors that led to the abandonment of the P13N polyimides and the earliest bismaleimides for practical applications.

6.3.1. PMR Polyimide Resins

Studies conducted at the NASA Lewis Research Center (LeRC) led to the development of a novel class of addition polyimides known as PMR (for *in situ* polymerization of monomer reactants) polyimides.[10–14] In the PMR approach, the reinforcing fibers are impreg-

Table 6.11. Electrical Properties of Skybond 700/Glass Fabric Laminates.[5]

(High temperature—high pressure laminates)

Property	As is	D 24/23	D 48/50	C 96/35/90
Dielectric strength				
— Short time parallel to laminate (volts)	55,000	—	32,000	—
— Step-by-step parallel to laminate (volts)	38,000	—	16,000	—
— Short time (volts/mil)	179	—	—	—
— Stepwise (volts/mil)	140	—	—	—
Dielectric constant (1MC)	4.10	4.30	4.81	
Dissipation factor (1MC)	.00445	.00639	.01650	—
Insulation resistance (megohms)	1.9×10^7	—	—	1.4×10^2
Volume resistivity (ohm-cms)	2.47×10^{15}	—	—	1.16×10^{11}
Surface resistivity (ohms)	3.35×10^{14}	—	—	2.90×10^{10}

X-BAND DATA (8.5 KMC)

Temperature	Dielectric constant	Dissipation factor
Room temperature	3.74	0.016
50° C	3.74	0.015
100° C	3.74	0.014
150° C	3.74	0.018
200° C	3.74	0.013
250° C	3.74	0.010
300° C	3.70	0.015

Table 6.12. Air Aging of S-2 Glass Fabric /6F/PPD/MPD (95/5) Laminates at 550° (288°C) Effect of Void Content.[3]

	LESS THAN 1% VOIDS[a]		7–10% VOIDS[b]	
	CONTROL	3000 HOURS EXPOSURE	CONTROL	1500 HOURS EXPOSURE
Weight % Resin	33		33	
Volume % Fibers	56		50	
Tg, °F (°C)	590 (310)	630 (332)	597 (314)	610 (321)
Flexural Strength, psi (MPa)				
73°F (23°C)	90,900 (627)	60,900 (420)	72,700 (501)	60,600 (418)
550°F (288°C)	51,800 (357)	55,400 (382)	46,000 (317)	52,000 (359)
Flexural Modulus, psi (MPa)				
73°F (23°C)	4.5×10^6 (31×10^3)	4.2×10^6 (29×10^3)	3.8×10^6 (26×10^3)	3.4×10^6 (23×10^3)
550°F (288°C)	4.1×10^6 (28×10^3)	4.1×10^6 (28×10^3)	3.8×10^6 (26×10^3)	3.4×10^6 (23×10^3)
Short Beam Shear Strength, psi (MPa)				
73°F (23°C)	11,400 (79)	8,900 (61)	10,400 (72)	
550°F (288°C)	5,800 (40)	5,100 (35)	4,400 (3)	4,600 (32)
% Weight Loss		1.5		1.5

[a]Final molding pressure of 200 psi (1.4 MPa).
[b]Vacuum bag molded.

Table 6.13. Typical Processing Conditions for NR-150B2[a]-S2 Glass Fabric Laminates.[3]

Initial staging

Apply 1-2 psi (7×10^{-3}–14×10^{-3} MPa) vacuum to bagged lay-up and heat at a rate of 9–14°F/min. (5–8°C/min.) to 392°F (200°C). Hold at 392°F (200°C) for 1 hour, and increase vacuum to 12–15 psi (83×10^{-3}–103×10^{-3} MPa) and hold for an additional 45 minutes.

Final molding

Full vacuum at 752°F (400°C) and 200 psi (1.4 MPa) for 1 hour.

[a] E. I. du Pont de Nemours & Company, Incorporated.

nated with a solution that consists of a mixture of monomers dissolved in a low-boiling-point alkyl alcohol solvent. The monomers are essentially unreactive at room temperature, but react *in situ* at elevated temperatures to form a thermo-oxidatively stable polyimide resin. Although the PMR approach utilizes the reverse Diels–Alder polymerization of norbornenyl groups, it eliminates the necessity of prepolymer synthesis and circumvents most of the shortcomings associated with the P13N prepolymer approach, such as variable solution shelf-life, need for DMF, and less-than-desirable thermo-oxidative stability at 600°F (316°C).

The PMR approach utilizes a dialkyl ester of an aromatic tetracarboxylic acid, an aromatic diamine, and a monoalkyl ester of 5-norbornene-2,3-dicarboxylic acid (NE). These monomers are dissolved in an alkyl alcohol (e.g., methanol or ethanol), and the solution is used to impregnate the fibers. If the solids content of the solution is maintained below 75 wt %, then wet-winding procedures can be employed to prepare the prepreg. At solids contents of 80–90 wt %, the PMR solution can be used to prepare the prepreg by means of a hot-melt technique.

Monomer reactant combinations for two PMR polyimide resins that differ in chemical composition have been identified. The earliest ("first-generation") PMR resin was designated PMR-15, and a subsequent "second-generation" resin was designated PMR II.[15]

For either PMR resin, the number of moles of each monomer is governed by the following ratio:

$$n : n + 1 : 2$$

where n, $n + 1$, and 2 are the number of moles of the dialkyl ester of the aromatic tetracarboxylic acid, the aromatic diamine, and NE, respectively.

In the initial study[10] that established the feasibility of the PMR approach, it was noted that composites made from monomer solu-

Table 6.14. Air Aging of S-2 Glass Fabric /6F/PPD/MPD (95/5) Laminates.[3]

	600°F (316°C)		700° (371°C)	
	CONTROL	1300 HOURS EXPOSURE	CONTROL	100 HOURS EXPOSURE
% Void Content	7		2	
Weight % Resin	32		32	
Volume % Fibers	50		54	
Tg, °F (°C)	662 (350)	662 (350)	666 (352)	745 (396)
Flexural Strength, psi (MPa)				
73° (23°C)	87,000 (600)		86,000 (590)	
600°F (316°C)	43,000 (300)	47,000 (320)	46,000 (320)	36,000 (250)
Flexural Modulus, Msi (GPa)				
73°F (23°C)	4.1 (280)		4.3 (30.0)	3.4 (23.0)
600°F (316°C)	3.2 (22.0)	4.4 (30.0)	3.0 (21.0)	3.8 (26.0)
Short Beam Shear Strength, psi (MPa)				
73°F (23°C)	9,000 (62)		9,800 (68)	5,000 (34)
600°F (316°C)	3,700 (26)	3,400 (23)	4,100 (28)	3,800 (26)
% Weight Loss		1.1		1.6

Table 6.15. Air Aging of Magnamite® HMS Graphite Fiber/6F/PPD/MPD (95/5) Unidirectional Laminates.[3]

	CONTROL	8600 HOURS EXPOSURE AT 500° F (260° C)	3000 HOURS EXPOSURE AT 600° F (316° C)	500 HOURS EXPOSURE AT 649° (343° C)
% Void Content	<2			
Weight % Resin	38			
Volume % Fibers	52			
Tg, ° F (° C)	662 (350)		680 (360)	718 (381)
Flexural Strength, psi (MPa)				
73° F (23° C)	126,000 (870)			
500° F (260° C)	94,000 (650)	80,000 (550)		
600° F (316° C)	99,000 (680)		81,700 (563)	
649° F (343° C)	72,000 (500)			74,700 (515)
Flexural Modulus, Msi (GPa)				
73° F (23° C)	21 (145)			
500° F (260° C)	21 (145)	15 (103)		
600° F (316° C)	20 (138)		19 (131)	
649° F (343° C)	17 (117)			17 (117)
Short Beam Shear Strength, psi (MPa)				
73° F (23° C)	7,400 (51)			
500° F (260° C)	4,400 (30)	5,200 (36)		
600° F (316° C)	4,600 (32)		5,900 (41)	
649° F (343° C)	4,600 (32)			5,800 (40)
% Weight Loss		0.8	2.5	1.5

tions containing the dimethyl ester of 3,3′,4,4′-benzophenonetetracarboxylic acid (BTDE), 4,4′-methylenedianiline (MDA), and NE exhibited a higher level of thermo-oxidative stability than did composites prepared from a monomer solution consisting of the dimethyl ester of pyromellitic acid, MDA, and NE. This unexpected observation was confirmed in a subsequent study,[14] and the optimum number of moles of BTDE (n) which provided the best overall balance of processing characteristics and thermo-oxidative stability was found to

Table 6.16. Air Aging of Magnamite AS Graphite Fiber/6F/PPD/MPD (95/5) Unidirectional Laminates.[3]

	6F/PPD/MPD (95/5)	
	CONTROL	500 HOURS EXPOSURE AT 600° F (316° C)
% Void Content	<2	
Weight % Resin	40	
Volume % Fibers	55	
Tg, ° F (° C)	676 (358)	653 (345)
Flexural Strength, psi (MPa)		
73° F (23° C)		
500° F (260° F)		
600° F (316° C)	97,000 (670)	
Short Beam Shear Strength, psi (MPa)		
73° F (23° C)		
500° F (260° C)		
600° F (316° C)	6,400 (44)	4,200 (29)
% Weight Loss		6.0

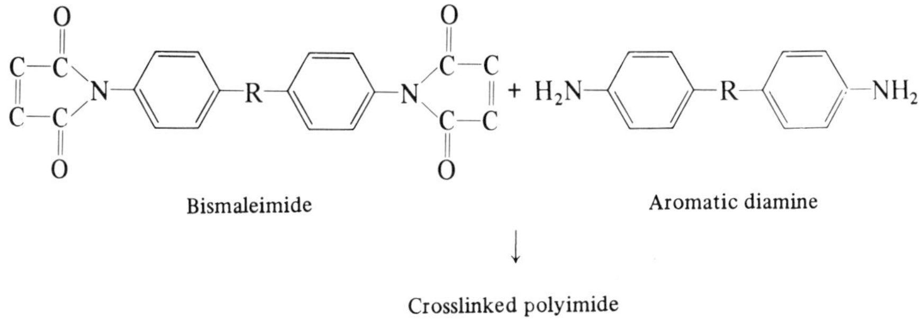

Bismaleimide + Aromatic diamine

↓

Crosslinked polyimide

Figure 6.7. Generalized reaction scheme for preparation of bismaleimide-based polyimides.

be 2.087; this corresponds to a PMR polyimide resin with a formulated molecular weight (FMW) of 1500. The FMW is considered to be the average molecular weight of imidized prepolymer that could have been formed if an amide-acid prepolymer had been synthesized. The equation for the FMW of a PMR polyimide resin prepared from n moles of BTDE, $n + 1$ moles of MDA, and 2 moles of NE is

$$FMW = n MW_{BTDE} + (n + 1) MW_{MDA} + 2MW_{NE} - 2(n + 1)[MW_{H_2O} + MW_{CH_3OH}]$$

where MW_{BTDE}, MW_{MDA}, and so on are the molecular weights of the monomer reactants and by-products. It is now common practice to denote the stoichiometry of a PMR resin by dividing the FMW by 100. The PMR matrices that employ BTDE are referred to as "first-generation" materials. The first-generation PMR matrix prepared from BTDE, MDA, and NE which has an FMW of 1500 is widely known as PMR-15. Prepreg materials based on PMR-15 are commercially available from the major suppliers. The structures of the monomers used in PMR-15 are shown in Table 6.17.

These early studies[10, 14] also clearly demonstrated the efficacy and versatility of the PMR approach. By varying the chemical nature of either the dialkyl ester acid or aromatic diamine (or both) and the monomer reactant stoichiometry, PMR resins having a broad range of processing characteristics and properties could easily be synthesized. A modified PMR-15—LARC-160—has been developed by substituting an aromatic polyamine for MDA.[16] Other studies[17, 18] have shown that the approach has excellent potential for "tailor-making" matrix resins with specific properties. Figure 6.10 shows the effect of FMW on resin flow for PMR polyimide–HTS graphite fiber composites. It can be seen that significantly higher resin flow is achieved by reducing the FMW. However, as shown in Figure 6.11, the PMR compositions that exhibit increased resin flow are less thermo-oxidatively stable at 550° F (288° C). The lower resin flow and increased thermo-oxidative stability in going from PMR-10 to PMR-15 clearly show the sensitivity of these properties to imide ring or alicyclic content. The reduction in resin flow with increased FMW also serves to quantitatively account for the intractable nature of linear, high-molecular-weight condensation polyimides. Tables 6.18 and 6.19 summarize the properties of PMR-15 neat resin and PMR-15 reinforced with HTS graphite fiber.[19]

Replacement of BTDE with the dimethyl ester of 4,4′-(hexafluoroisopropylidene)-bis-(phthalic acid) (HFDE) significantly improved the thermo-oxidative stability of "first-generation" PMR resins.[18] However, the initial 600° F (316° C) mechanical properties of HFDE/MDA/NE PMR polyimide composites were considerably poorer than those of BTDE/MDA/NE PMR polyimide composites. On the other hand, graphite-fiber-reinforced PMR polyimide composites prepared from a monomer solution consisting of HFDE, p-phenylenediamine (PPDA), and NE at an FMW of 1267 ($n = 1.67$) were found to exhibit improved thermo-oxidative stability and retention of mechanical properties at 600° F (316° C) compared to

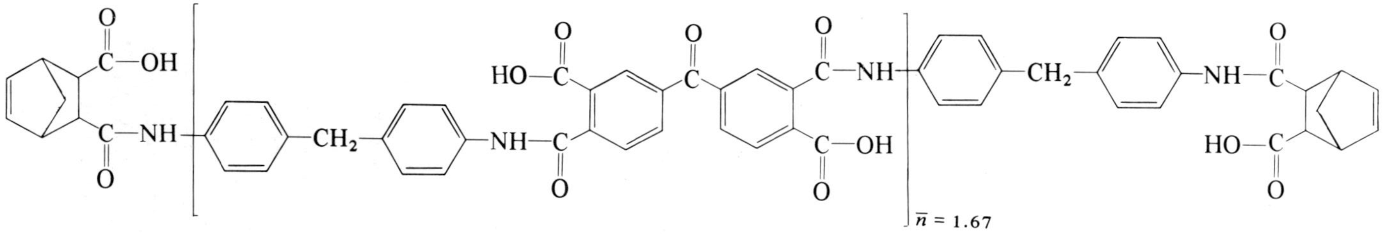

$\bar{n} = 1.67$

Amide-acid prepolymer

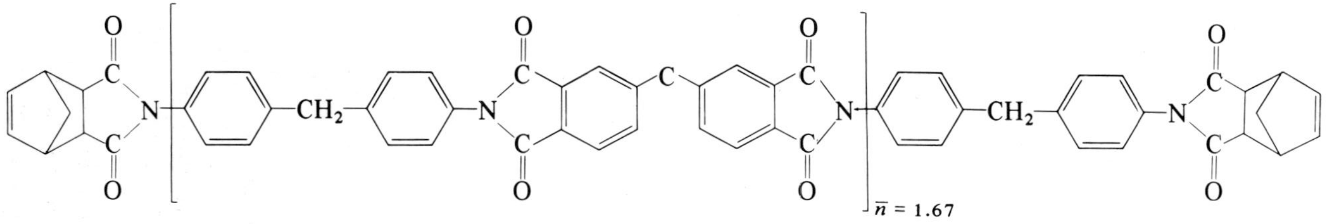

$\bar{n} = 1.67$

Imidized prepolymer

Figure 6.8. P13N prepolymers.

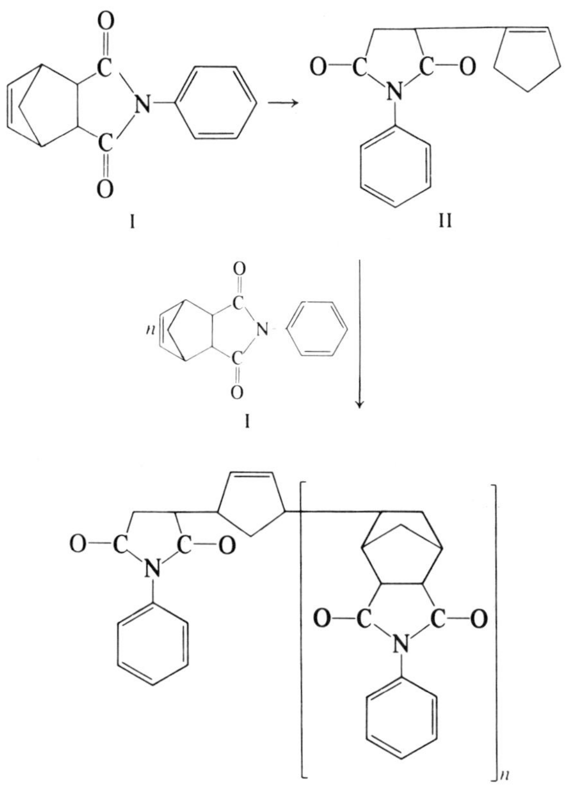

Figure 6.9. Pyrolytic polymerization reaction mechanism.

Table 6.17. Monomers Used For PMR-15 Polyimide

STRUCTURE	NAME	ABBREVIATION
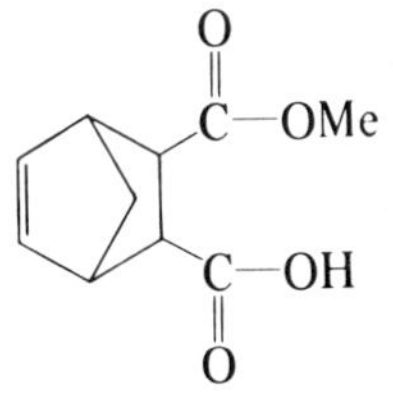	Monomethyl ester of 5-Norbornene-2,3-Dicarboxylic Acid	NE
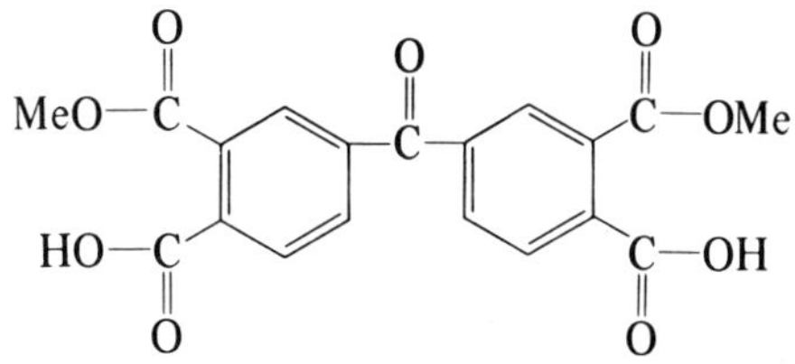	Dimethyl Ester of 3,3',4,4'-Benzophenonetetracarboxylic Acid	BTDE
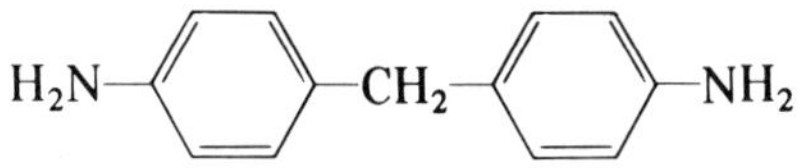 H_2N–C_6H_4–CH_2–C_6H_4–NH_2	4,4' Methylenedianiline	MDA

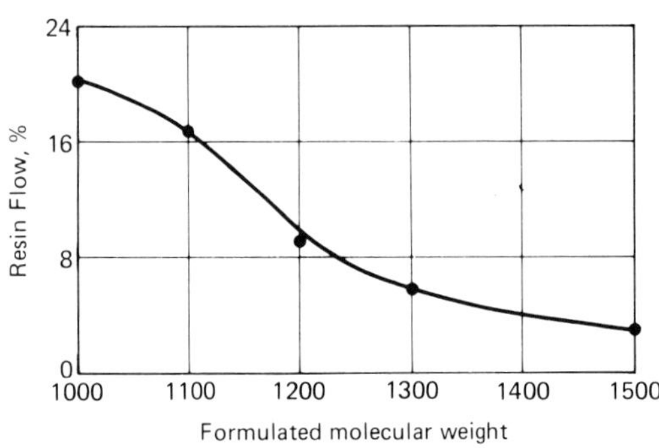

Figure 6.10. Percent resin flow for PMR PI/MTS graphite fiber composites.[17]

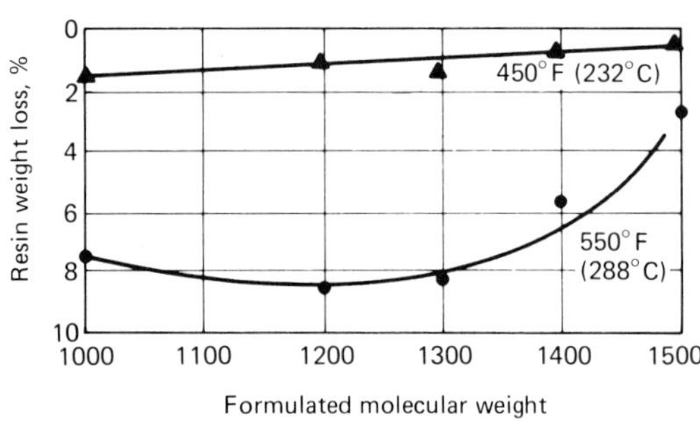

Figure 6.11. Percent resin weight loss for PMR PI/MTS graphite fiber composites after 600-hour exposure in air at 232° C (450° F) and 288° C (550° F).[17]

PMR-15 polyimide composites.[15] The HFDE-based PMR compositions are referred to as "second-generation" materials to differentiate them from the "first-generation" BTDE-based PMR materials. The "second-generation" resin consisting of HFDE, PPDA, and NE with n = 1.67 is known as PMR II. The structures of the monomers used in second-generation polyimide resins are shown in Table 6.20. The interlaminar shear strengths of composites based on PMR II (n = 1.67) and PMR-13 (a first-generation composition with $n = 1.67$) reinforced with HTS-1 graphite fiber are compared in Fig. 6.12. It can be seen that the PMR II composites have a 600° F (316° C) service life that is at least twice that of the earlier PMR composites. The PMR II and PMR-13 polyimide resins were selected for the comparison because they contain the same number of imide rings.

Further improvement in PMR polyimide composite performance at elevated temperatures has been made possible by the recent development of such new, high-strength, intermediate-modulus graphite fibers as Hercules HTS-2 and Celion 6000 (Celanese Corporation). The thermo-oxidative stability of these fibers is considerably superior to that of such older high-strength fiber types as HTS-1. The improved thermo-oxidative stability of HTS-2 and Celion 6000 compared to HTS-1 at 600° F (316° C) in air can be seen in Fig. 6.13. The interlaminar shear strength retention of PMR-15 composites fabricated from HTS-1, HTS-2, and Celion 6000 after isothermal exposure in air at 600° F (316° C) is shown in Fig. 6.14.[19] It can be seen that both the HTS-2 and Celion 6000 composites exhibit 100% retention of their initial interlaminar shear strength values during the first 1000 hr of exposure. The use of the improved fibers increases the service life of PMR-15 composites at 600° F (316° C) in air from about 400 hr to more than 1000 hr. This level of stability is outstanding and quite unexpected, particularly in view of the aliphatic nature of the PMR-15 crosslink structure.

6.3.1.1. PMR-Composite Processing

High-pressure (compression) and low-pressure (autoclave) molding cycles have been developed for the fabrication of PMR composites. Although the thermally induced crosslinking addition cure reaction of the norborne-

Table 6.18. Room Temperature Properties of PMR-15 Neat Resin

Tensile strength, N/cm^2, psi	5580; 8100
Tensile modulus, N/cm^2; psi	0.32×10^6; 0.47×10^6
Compressive yield strength, N/cm^2; psi	11 400; 16 500
Compressive strength, N/cm^2; psi	18 700; 27 200
Thermal coefficient of expansion, cm/cm/°C; in./in./°F	50.4×10^{-6}; 28.0×10^{-6}

Table 6.19. Properties of HT-S/PMR-15 Composites
[Fiber volume, 55 vol %; tested along 0° ply direction.]

Laminate	Tensile strength		Tensile modulus		Compressive strength		Flexural strength	
	N/cm^2	psi	N/cm^2	psi	N/cm^2	psi	N/cm^2	psi
6[0]	124 000	180 000	15.0×10^6	21.7×10^6	93 000	135 000	141 900	206 000
6[90]	6 710	9 740	.79	1.15	23 430	34 000	11 270	16 350
4[±10, ∓10]	103 400	150 000	14.5	21.1	------	-------	158 500	230 000
8[0, +10, 0, -10, -10, 0, +10, 0]	131 500	191 000	13.1	19.0	------	-------	135 700	197 000
13[±40, 9(0), ∓40]	85 400	124 000	9.6	14.0	------	-------	99 900	145 000

Laminate	Flexural modulus		Short-beam interlaminar shear		Miniature Izod impact energy		Coefficient of thermal expansion	
			N/cm^2	psi	cm-N	in.-lb	$\frac{cm/cm}{^\circ C}$	$\frac{in./in.}{^\circ F}$
6[0]	12.1×10^6	17.6×10^6	11 020	16 000	171.8	15.2	0	0
6[90]	.74	1.07	------	------	20.3	1.8	26.1×10^{-6}	14.5×10^{-6}
4[±10, ∓10]	11.6	16.8	------	------	-----	----	---------	---------
8[0, +10, 0, -10, -10, 0, +10, 0]	13.1	19.0	------	------	-----	----	---------	---------
13[±40, 9(0), ∓40]	6.4	9.3	------	------	-----	----	---------	---------

Table 6.20. Monomers Used for Second Generation PMR Polyimides

STRUCTURE	NAME	ABBREVIATION
	Monomethyl Ester of 5-Norbornene-2, 3-Dicarboxylic Acid	NE
	Dimethyl Ester of 4,4′-(Hexafluoroisopropylidene)-Bis(Phthalic Acid)	HFDE
	p-Phenylenediamine	PPDA

nyl group occurs at temperatures in the 527–662° F (275–350° C) range, nearly all of the processes developed use a maximum cure temperature of 600° F (316° C). Cure times of 1–2 hr followed by a free-standing post-cure in air at 600° F (316° C) for 4–16 hr are also normally employed. Compression-molding cycles generally employ high rates of heating [9–18° F (5 to 10° C)/min.] and pressures in the range of 3.45–6.90 MPa (500–1000 psi). Vacuum-bag autoclave processes at low heating rates 3.6–7.2° F (2–4° C)/min.] and pressures of 1.38 MPa (200 psi) or less have been successfully used to fabricate void-free composites. The successful application of autoclave processing methodology to PMR polyimide resins is due to the presence of a thermal transition—"melt-flow"—that occurs over a fairly broad temperature range.[21] The lower limit of the melt-flow temperature range depends on a number of factors, including the chemical nature and stoichiometry of the

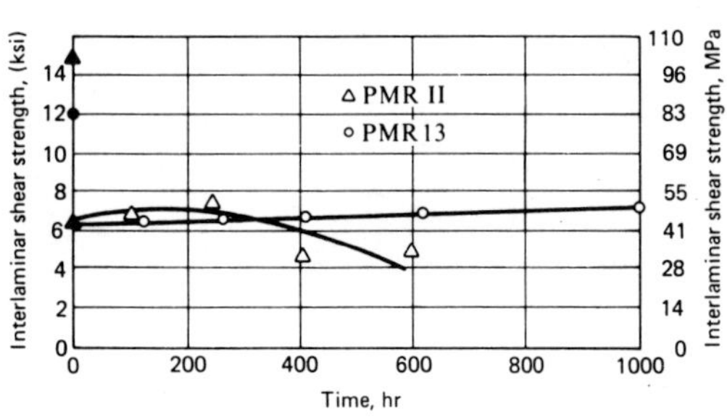

Figure 6.12. Interlaminar shear strength of PMR polyimide (n = 1.67)/ HTS graphite fiber composites exposed and tested at 600° F.[15]

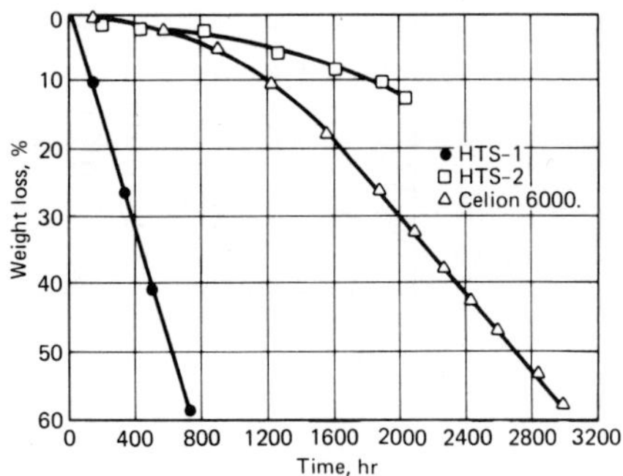

Figure 6.13. Weight loss of graphite fibers exposed in air at 600° F.[19]

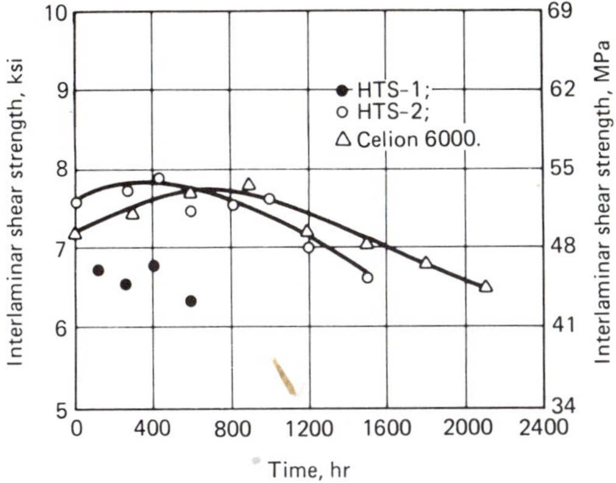

Figure 6.14. Interlaminar shear strength of PMR 15/graphite fiber composites exposed and tested in air at 600° F.[20]

monomer reactant mixture as well as the prior thermal history of the PMR prepreg. Differential scanning calorimetry studies have shown the presence of four thermal transitions that occur during the overall cure of a PMR polyimide resin.[22] The first, second, and third transitions are endothermic and related to the following: (1) melting of the monomer reactant mixture below 212° F (100° C), (2) *in situ* reaction of the monomers at 284° F (140° C), and (3) melting of the norbornenyl-terminated prepolymers in the 347–482° F (175–250° C) range (the melt-flow temperature range). The fourth transition, centered near 644° F (340° C), is exothermic and related to the addition crosslinking reaction. To a large extent, the excellent processing characteristics of PMR polyimide resins can be attributed to the presence of these widely separated and chemically distinct thermal transitions.

6.3.2. Addition-Type Polyimide Prepreg Materials

The major suppliers of the bismaleimide, PMR-15, and LARC-160 polyimide prepreg materials are listed in Table 6.21. Ferro Corporation will also supply imidized, unreinforced molding powder.

6.4. POLYIMIDE COMPOSITE APPLICATIONS

6.4.1. Condensation Polyimides

The high void contents (typically 10 vol % or more) of the Pyralin and Skybond polyimide composites have not deterred the extensive use of these relatively inexpensive materials in such secondary structural applications as sound suppressant panels in turbofan engines and radomes (see Figs. 6.15–6.17).

The major application for the NR-150B2 polyimide composites is in nozzle flaps for the Pratt & Whitney F100 engine. The nozzle flaps are fabricated by Hamilton Standard using a proprietary compression-molding process. A full engine set (such as the one shown

Table 6.21. Suppliers of Addition-Type Polyimide Prepreg Materials.

		RESIN TYPE	
SUPPLIER	BISMALEIMIDE	PMR-15	LARC-160
Ferro Corporation Composites Division Culver City, California	FT[a]	FT	FT
Fiberite Corporation Winona, Minnesota	GMC[b]	FT	FT
Hexcel Corportion Dublin, California	FT	FT	FT
U.S. Polymeric Santa Ana, California	FT	FT	FT

[a] FT = fabric and unidirectional-fiber tape.
[b] GMC = glass molding compound.

Figure 6.15. EA-6B track and jammer pod radome.

in Fig. 6.17) has successfully undergone many hours of flight testing.

6.4.2. Addition Polyimides

The PMR-15 polyimide composites have been or are being used in a number of diverse structural components. Some of these components are listed in Table 6.22 (Fig. 6.20). Figure 6.18 shows the inner cowl for the experimental QCSEE (Quiet Clean Short Haul Experimental Engine) turbofan engine developed for NASA by General Electric.[23] The cowl has a maximum diameter of about 90 cm and is of graphite fabric honeycomb sandwich construction; the cowl has successfully withstood over

Figure 6.16. EF-111A weapons bay radome.

Table 6.22. Applications of PMR Technology.

COMPONENT	AGENCY	CONTRACTOR
JT8D reverser stang fairing	NASA-LeRc	McDonnell-Douglas
F404 duct	NACA-LeRC/Navy	General Electric
Ion engine beam shield	NASA-LeRC	Hughes
Utility Missle	DOD	Boeing
AEDC supersonic wind tunnel compressor blades	Air Force	Hamilton Standard
QCSEE inner cowl	NASA-LeRC	General Electric
Shuttle orbiter aft body flap	NASA-LaRC	Boeing
Pratt & Whitney F-100 augmentor duct	Air Force	Composites Horizons
Tokamak dielectric insulation	DOE	Grumman

100 hr of ground engine testing without displaying any degradation.

Figure 6.19 shows a prototype of the shuttle orbiter aft body flap fabricated by Boeing Company. Boeing routinely fabricates 4 ft × 8 ft. (1.2 m × 2.4 m) PMR-15–graphite fiber panels that must pass the severe NDE inspection requirements established by NASA-Langley Research Center.

Figure 6.20 is a photograph of a DC-9 airplane. The inserts schematically show the design of the currently used metal fairing and the redesigned composite fairing now being developed by the Douglas Aircraft Company under the NASA-LeRC ECI (Engine Component Improvement) program.[24] Studies have shown that a redesigned fairing provides an opportunity to reduce both baseline drag

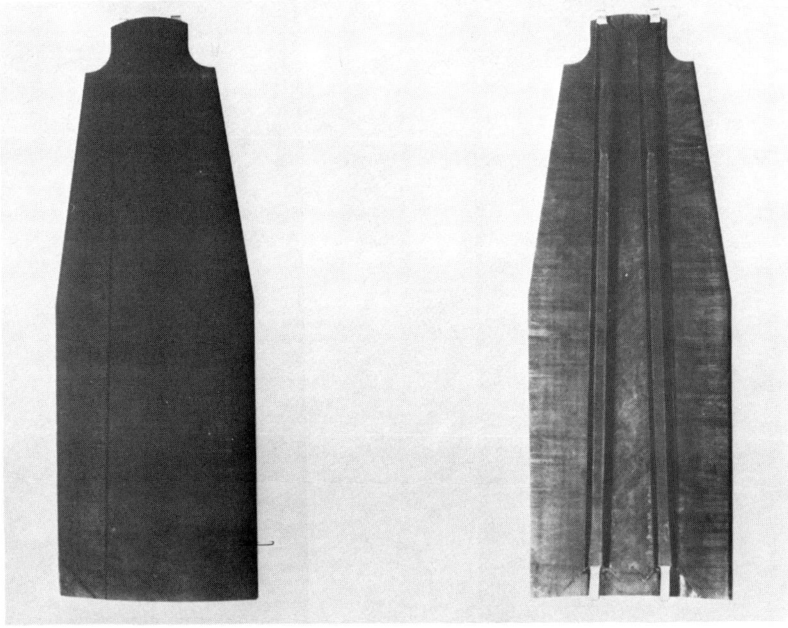

Figure 6.17. Graphite fiber-NR-150B2 external nozzle flap for F-100 engine.

Figure 6.18. PMR-15-graphite fabric inner cowl for QCSEE engine.[23]

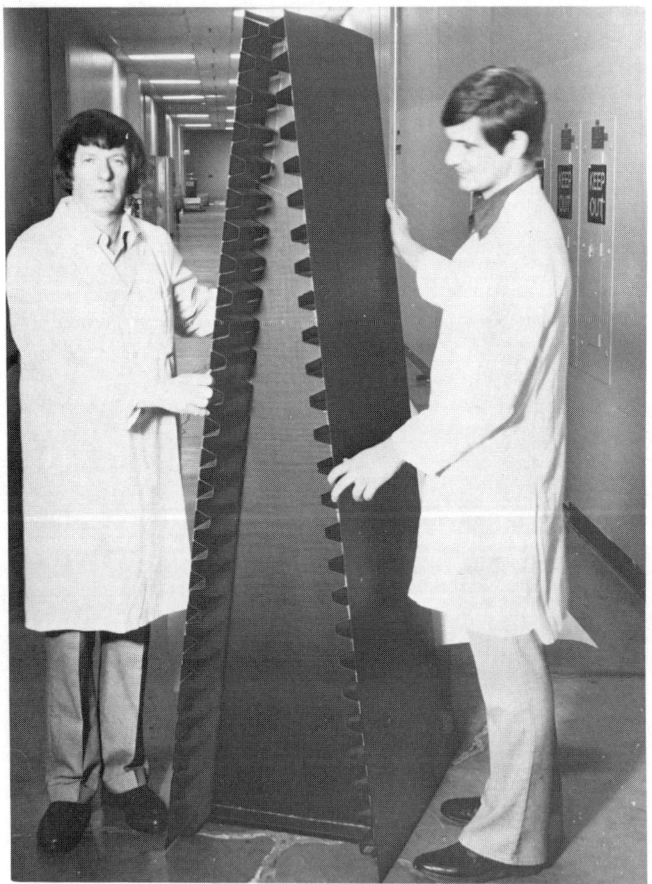

Figure 6.19. PMR-15-graphite fiber prototype of shuttle orbiter aft body flap.

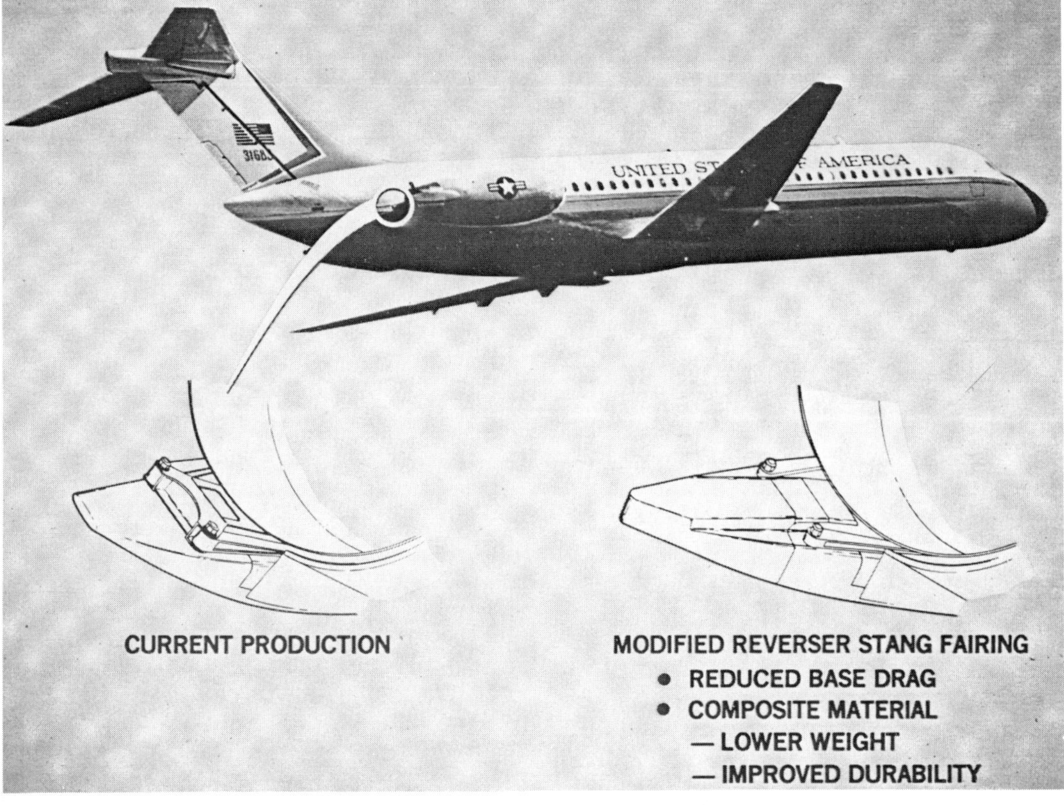

Figure 6.20. PMR-15–Kevlar reverser stang fairing for JT8D nacelle.

and fuel consumption. A composite fairing made from PMR-15 and Kevlar fabric has been autoclave fabricated and flight tested.

REFERENCES

1. G. P. Harris, Personal communication, Monsanto Plastics & Resins Company.
2. J. E. Goode, Personal communication, E. I. du Pont de Nemours & Company, Inc.
3. H. H. Gibbs, J. Appl. Polym. Sci., Applied Polymer Symposia, **35**, pp. 207–222, (1979).
4. H. H. Gibbs and C. V. Breder, *Advances in Chemistry Series, No. 142, Copolymers, Polyblends, and Composites*, American Chemical Society, 1975, p. 442.
5. Monsanto Technical Bulletin 5042D, Monsanto Company.
6. NASA CR-132685, *Industry Technology Assessment of Graphite–Polyimide Composite Materials: Monsanto Polyimides*, p. 301, June, 1975.
7. NASA CR-120413, *Development of Design Data for Graphite-Reinforced Epoxy and Polyimide Composites.*
8. French Patent 1,455,514.
9. E. A. Burns, H. R. Lubowitz, and J. F. Jones, NASA CR-72460 (1968).
10. T. T. Serafini, P. Delvigs, and G. R. Lightsey, *J. Appl. Polym. Sci.* **16**, 905 (1972).
11. T. T. Serafini, P. Delvigs, and G. R. Lightsey, U. S. Patent 3,745,149 (July 10, 1973).
12. T. T. Serafini and P. Delvigs, *Appl. Polym. Sympos.*, No. 22, 89 (1973).
13. T. T. Serafini, in: *Proceedings of the 1975 International Conference on Composite Materials* (edited by E. Scala), Vol. 1, AIME, New York (1976), p. 202.
14. P. Delvigs, T. T. Serafini, and G. R. Lightsey, NASA TN D-6877 (1972).
15. T. T. Serafini, R. D. Vannucci, and W. B. Alston, NASA TM-71894 in *Proceedings of the 21st SAMPE National Symposium* (April 1976).
16. T. L. St. Clair and R. A. Jewell, in: *Proceedings of the 23rd SAMPE National Symposium* (May 1978).
17. T. T. Serafini and R. D. Vannucci, NASA TM-71616 in *Proceedings of the 30th Annual Conference of the Reinforced Plastics/Composites Institute of the Society of the Plastics Industry*, (February, 1975).

18. R. D. Vannucci and W. B. Alston, NASA TM-71816 in *Proceedings of the 31st Annual Conference of the Reinforced Plastics/Composites Institute of the Society of the Plastics Industry*, (February 1976).
19. M. P. Hanson and C. C. Chamis, NASA TND-7698 in *Proceedings of the 29th Annual Conference of the Reinforced Plastics/Composites Institute of the Society of the Plastics Industry*, (February 1974).
20. P. Delvigs, W. B. Alston, and R. D. Vannucci, NASA TM-79062 in *Proceedings of the 24th SAMPE National Symposium* (May 1979).
21. R. D. Vannucci, NASA TM-73701; in: *Proceedings of the Ninth SAMPE Technical Conference* (October 1977).
22. R. W. Lauver, J. Polym. Sci., **17,** p. 2529 (1979).
23. C. L. Ruggles, NASA CR-135279 (1978).
24. *JT8D Nacelle Drag Reduction in DC-9,* NASA Contract No. NAS 3-21763.

7

GLASS-FILLED THERMOPLASTICS

Allen M. Shibley
Consultant, formerly at
Plastics Technical Evaluation Center
Picatinny Arsenal
Dover, New Jersey

7.1. INTRODUCTION

The reinforced plastics discussed in this chapter are conveniently described as short-fiber molding compounds (in contrast to other composites that contain continuous reinforcing agents or particulate fillers). In the simplest terms, they are defined as composites that contain randomly distributed, discontinuous fibers held together in a thermoplastic matrix. Other ingredients such as colorants, flame retardants, and lubricants can be dispersed within the matrix. Nearly all thermoplastic molding resins [e.g., polypropylene, nylon, ABS (acrylonitrile-butadiene-styrene), polycarbonate, acetal, SAN (styrene–acrylonitrile), polystyrene, high-density polyethylene and polysulfone] have been modified by fiber addition. Fiberglass is the primary reinforcing agent and consequently, will be emphasized in this chapter. Carbon/graphite, aramid, and asbestos fibers are also of interest as reinforcing agents.

The fibers are incorporated into the resin by one of several extrusion methods that delivers the compound in a pelletized form. Parts are usually fabricated by injection molding the pellets. Conventional molds and plunger or reciprocating screw-type machines are used for this purpose. Finished molding compounds can be purchased directly from material suppliers or, alternatively, molders can make their own by mixing the fibers or fiber concentrates with the base resin. A list of domestic suppliers of glass-filled thermoplastics (GFTP) and the resin types that they employ is presented in Table 7.1.[1] A list of foreign suppliers can be found in Ref. 2.

Although this chapter is mainly concerned with GFTP, and the associated injection-molding process, the significance of other fillers, material forms, and fabrication techniques is not overlooked. The GFTP have dominated the market, and their superior mechanical properties are well known. However, other fillers are now being used more frequently in order to allow the development of specific electrical, chemical, heat-resistant, and wear-resistant material grades. In some formulations, fillers are added to reduce the overall cost or to facilitate the processing. Mica, glass beads, talc, wollastonite, calcium carbonates, silica, and silicates are just a few of the fillers being used for the above purposes.[3,4]

Several recent innovations have expanded the field of GFTP and provided alternative fabrication techniques:

- Reinforced thermoplastic sheet for stamping or thermoforming. This method is applicable to relatively large parts and is currently being adapted for the automotive industry. The glass fibers contained in these materials are longer than those contained in pelletized compounds, being comparable to glass mat. In some

Table 7.1. Domestic Suppliers of GFTP.[1]

BASE RESIN	SUPPLIERS
ABS	Fiberfil Division, Dart Industries, Inc.; Liquid Nitrogen Processing Corporation; Thermofil, Inc.
Acetal	E. I. du Pont de Nemours & Company, Inc.; Fiberfil Division, Dart Industries, Inc.; Liquid Nitrogen Processing Corporation; Thermofil, Inc.
Fluorocarbon	Liquid Nitrogen Processing Corporation
Nylon 6	Adell Plastics, Inc.; Allied Chemical Corporation; Fiberfil Division, Dart Industries, Inc.; Liquid Nitrogen Processing Corporation; Thermofil, Inc.
Nylon 66	Adell Plastics, Inc.; E. I. du Pont de Nemours & Company, Inc.; Fiberfil Division, Dart Industries, Inc.; Liquid Nitrogen Processing Corporation; Polymer Corporation; Thermofil, Inc.; Wellman, Inc.
Nylon 610	Liquid Nitrogen Processing Corporation
Nylon 612	E. I. du Pont de Nemours & Company, Inc.; Fiberfil Division, Dart Industries, Inc.; Liquid Nitrogen Processing Corporation; Polymer Corporation
Nylon 11	Liquid Nitrogen Processing Corporation; Rilsan Corporation
Nylon 12	Emser Industries, Inc.; Liquid Nitrogen Processing Corporation
Phenylene oxide	Liquid Nitrogen Processing Corporation; Thermofil, Inc.
Polyarylether	Thermofil, Inc.; Fiberfil Division, Dart Industries, Inc.
Polycarbonate	Adell Plastics, Inc.; Fiberfil Division, Dart Industries, Inc.; General Electric Company; Liquid Nitrogen Processing Corporation; Mobay Chemical Corporation; Thermofil, Inc.
Polyester	Eastman Chemical Products, Inc.; GAF Corporation; General Electric Company; Liquid Nitrogen Processing Corporation
Polyethersulfone	Liquid Nitrogen Processing Corporation
Polyethylene (high density)	Fiberfil Division, Dart Industries, Inc.; Liquid Nitrogen Processing Corporation; Thermofil, Inc.
Polyphenylene sulfide	Liquid Nitrogen Processing Corporation; Thermofil, Inc.
Polypropylene	Adell Plastics, Inc.; Fiberfil, Inc.; Hercules, Inc.; Liquid Nitrogen Processing Corporation; Thermofil, Inc.
Polystyrene (general purpose)	Thermofil, Inc.
Polysulfone	Fiberfil Division, Dart Industries, Inc.; Liquid Nitrogen Processing Corporation; Thermofil, Inc.
SAN	Fiberfil Division, Dart Industries, Inc.; Liquid Nitrogen Processing Corporation; Thermofil, Inc.
Urethane elastomer	Fiberfil Division, Dart Industries, Inc.; Liquid Nitrogen Processing Corporation

cases the reinforcement is in the form of continuous roving.

- Reinforced foams. These structural foams are also suited for large molded parts. Processing methods based on direct nitrogen injection into the resin require extensive modification of the injection-molding equipment. Materials containing the blowing agent can be injection molded with only minor modifications.
- Investigations are being conducted to determine the feasibility of a pultrusion process with a thermoplastic binder and continuous strands of fiberglass or graphite.
- A filament-winding process based on thermoplastic-coated rovings is currently under development.

7.2. MANUFACTURE OF GFTP

7.2.1. Strand Coating

The first GFTP in the United States were made by a process closely resembling wire coating.[5] Glass roving was drawn through the crosshead or offset die of an extruder and removed as a continuous resin-coated strand. The strand was then cooled and chopped into discrete lengths to form the molding compound. This method produces pellets in which the glass is contained as a concentrated core surrounded by the resin. Break up of the strand and dispersion of the fibers occur during the injection-molding cycle. When necessary, glass finishes can be modified to facilitate fiber dispersion.

7.2.2. Extruder Compounding

In this process, the glass and resin are fed directly into the extruder for mixing. The numerous variations in the processing are related to (1) extruder type (single screw or twin screw), (2) screw design, (3) method of feeding, and (4) method of pelletizing. The glass feed is normally introduced in the form of chopped fiber. Certain twin-screw extruders are designed so that continuous rovings enter at a downstream port. At this point, the resin is in a molten state, and compounding takes place with minimal fiber damage. In other instances, a powder resin is preblended with the chopped fibers prior to extrusion. In general, the effects of these process variables on the compound are indicated by the uniformity of fiber distribution, fiber length, and fiber alignment.

7.2.3. In-Plant Fabrication

Although molders can produce GFTP by either of the above two methods, they can also purchase compounds with high glass concentrations for blending in the injection-molding machine. It is also possible to injection mold dry blends of some resins. In another process, the glass is chopped at the feed hopper of the injection-molding machine, where it is mixed with the resin as it enters the feed port. Reciprocating screw-type, preplasticizing machines are easier to adapt to these operations than the plunger types.

7.2.4. Fiberglass Feed

All commercial GFTP in the United States are made with E glass. Although G-diameter fibers (0.35–0.40 mil = 0.009–0.01 mm) are still being used, the general trend is toward K-diameter (0.50–0.55 mil = 0.013–0.014 mm) or larger. The fibers can be purchased already chopped into lengths of 1/8 or 1/4 in. (3 or 6 mm), and similar lengths are the rule when the strands are cut in-plant. However, occasionally, the length is increased to 1/2 in. (12 mm). The 1/4- and 1/2-in. (6- and 12-mm) fibers are commercially designated as "long glass," whereas the 1/8-in. (3-mm) fiber is correspondingly known as "short glass." A milled glass, nominally 1/32 in. (0.75 mm) in length, is sometimes utilized. All fibers are treated with sizes that are compatible with specific resins. Some sizes are said to be thermally stable and/or capable of "coupling" the fiber to the resin.

7.2.5. Resin Feed

For the most part, resins of injection-molding grade are used in GFTP. The majority of resins are available over a range of molecular weights. Typically, higher molecular weights are associated with superior mechanical properties. On the other hand, the higher-molecular-weight resins are more difficult to process. The general solution is to utilize the highest molecular weight practical and employ other means to improve processing. The addition of lubricants and the use of higher process temperatures are steps in this direction.

7.3. INJECTION MOLDING OF GFTP

As might be expected, the addition of fiberglass to a thermoplastic resin increases the melt viscosity of the mix and alters the flow characteristics during processing. The viscosity increases with increased fiber content and length. Under restricted flow conditions in either the injection machine or the mold, the fibers tend to become oriented in one direction, and a separation of the resin from the glass may occur. The mold shrinkage of a filled resin is substantially lower than that of the unfilled resin, and this decrease in shrinkage can be highly anisotropic. Filled resins also cool faster than unfilled resins. All of the above factors must be taken into account in establishing the molding conditions for a particular compound. Normally, conventional injection-molding systems can process

GFTP with only minor adjustments in operating conditions. However, mold and part design require detailed consideration.

The superior performance of in-line reciprocating screw-type machines has been widely acknowledged. Closer control of the shot weight and melt temperature, faster plasticizing, and more uniform mixing are cited as major advantages of this class of machine. Plunger-type machines have a less severe shearing action and thus produce moldings with somewhat higher mechanical strengths. However, product uniformity and surface appearance are inferior. The machine requirements for processing GFTP are as follows:

1. The machine should be capable of supplying injection pressures of at least 20,000 psi (138 MPa).
2. Speed control of the injection stroke is recommended.
3. Clamping pressures should be on the order of 5–10 tons/sq in. (69–138 MPa) of projected mold area.
4. The cylinder should have a minimum of three heating zones; the nozzle requires separate heating.
5. The nozzle should be short with a straight bore, its diameter being as large as possible for a given sprue bushing; a reverse taper or positive cut-off is helpful in preventing drooling, which may occur with some materials.

The following mold-design features are applicable to GFTP:

1. Sprues should be well tapered, highly polished, and as short as possible; cold-slug wells are desirable.
2. Runners should be of large, circular cross section, short, and well polished; trapezoidal sections are a second choice; runner extensions for trapping cold slugs and overflow wells to prevent the formation of welds within the part are recommended; hot runners (or no runners) are preferable when using molds made from such softer metals as kirksite.
3. The location and size of the gate are of major importance; gating should be balanced to avoid warping; multiple gating has been used for the same reason; multiple gating, however, can create weld line problems; normally, gates are positioned to promote random flow; round or rectangular shapes are preferred; gates should be located at or close to the thickest section; gate size is dictated by the material to be molded; as a rule of thumb, the size should range from two-thirds to full section thickness; gates should be back-tapered to the runner.
4. Additional venting is needed to facilitate rapid filling of the cavity; the size of the vents can be somewhat larger than is normal for the base resin.
5. To ensure ease of part removal, a sufficient draft angle should be allowed in the design, deep undercuts avoided, and additional knockout pins added to the mold; where possible, the pins should be located at ribs, bosses, or similar strategic positions.
6. Close temperature control of both cores and cavities is indicated.

7.3.1. Operating Conditions

Filled thermoplastics require greater injection pressures and higher processing temperatures than the corresponding unfilled base resins (pressure increases of up to 75% are not uncommon). In conjunction with the higher pressures, the maximum shot weight is reduced below the nominal machine rating. Charge weights of from 50 to 75% of machine capacity are typical (in some cases, these are as low as 25%). The processing temperatures of several filled and unfilled thermoplastics are compared in Table 7.2.[1] Injection speeds are usually at machine maximum to ensure fast mold filling, since rapid fill improves weld lines, reduces the tendency toward fiber alignment, and gives a better surface finish to the part. Screw speeds and back pressure of the machine are balanced to yield a homogeneous plastification without excessive working of the compound and subsequent fiber degradation.

The customary practice of adding ground

Table 7.2. Processing Temperatures of Several Filled and Unfilled Thermoplastics.[1]

	PROCESSING TEMPERATURE, °C (°F)[a]	
BASE RESIN	UNFILLED	FIBERGLASS FILLED
ABS	200–260 (392–500)	260–280 (500–536)
Acetal	175–200 (347–392)	200–230 (392–446)
Nylon 66	250–290 (482–554)	260–305 (500–581)
Nylon 6	215–260 (419–500)	215–280 (419–536)
Nylon 612	230–290 (446–554)	280–305 (536–581)
Nylon 12	175–180 (347–356)	175–180 (347–356)
Polyphenylene oxide	275–305 (527–581)	300–325 (572–617)
Polycarbonate	260–315 (500–599)	280–345 (536–653)
Polyethersulfone	—[b]	360 (680)
Polyethylene (high density)	215–230 (419–446)	230–250 (446–482)
Polypropylene	175–230 (347–446)	240–260 (464–500)
Polystyrene (GP)	200–235 (392–455)	240–280 (464–536)
SAN	—	260–280 (500–536)

[a]Temperatures vary with material grade and manufacturer.
[b]Data for comparable unfilled resin not available.

scrap to the feed has a variable effect that depends on the specific resin involved, the fiber length in the ground compound, and the specific property level to be realized. As a general rule, 25 wt % scrap can be readily tolerated. However, such high-impact materials as long-glass-reinforced nylon are exceptions. In these cases, the impact strength may decrease rapidly as the amount of regrind is increased.

7.4. PROPERTIES OF GFTP

7.4.1. General

The physical and mechanical properties of GFTP listed in Table 7.3[1] are useful in characterizing specific compounds and for comparative purposes. The data presented, which were obtained from material suppliers, were determined by appropriate ASTM (American Society for Testing and Materials) test methods. Representative examples for each base resin were selected, but other composites are available in most resin categories; however, these composites may contain different fiberglass contents or such additional modifiers as flame retardants, lubricants, molybdenum disulfide, tetrafluoroethylene (TFE) resin, talc, carbon, or mineral fillers.

The effects of fiber addition on the properties of thermoplastic resins are discussed in greater detail in the following sections. Although emphasis is placed on the mechanical properties, secondary effects on the thermal, physical, and electrical properties are briefly noted.

7.4.2. Effect of Fiberglass Reinforcement

7.4.2.1. Fiber Diameter

Optimum mechanical strengths of GFTP have been attained with the use of G-diameter fibers. However, the use of K-diameter fibers provides almost identical values, the impact strength being slightly higher.[6] The G- and K-diameter fibers have been extensively used in commercial GFTP, but the extent to which larger-diameter fibers have been employed is not known. Although slightly lower mechanical strength values can be expected to result from the use of larger-diameter fibers, improved coupling agents and sizes can possibly compensate for this effect.

7.4.2.2. Length

A comparison of fiber lengths showed that optimum mechanical properties were attained when 1/4-in. (6-mm) fibers were used. Other

Table 7.3. Physical and Mechanical Properties of GFTP,[1]

BASE RESIN	ABS			
SUPPLIER	BORG WARNER	FIBERFIL		THERMOFIL
Fiberglass content, wt %	0	20 (S)[a]	20 (L)[a]	10
Specific gravity	1.05	1.23	1.23	1.10
Tensile properties				
Yield strength, psi × 10^3 (MPa)	6.9 (48)	12.0 (83)	13.0 (90)	9.5 (66)
Modulus of elasticity, psi × 10^5 (GPa)	3.8 (2.6)	9.0 (6.2)	8.0 (5.5)	6.7 (4.6)
Yield elongation, %	—	1.5	2.0	2.0–3.0
Flexural properties				
Yield strength, psi × 10^3 (MPa)	12.3 (85)	17 (117)	20 (138)	14.8 (102)
Modulus of elasticity, psi × 10^5 (GPa)	4.0 (2.8)	8.5 (5.9)	8 (5.5)	6.5 (4.5)
Compressive properties				
Yield strength, psi × 10^5 (MPa)	—	14 (97)	14 (97)	12 (83)
Impact strength (notched bar, Izod test), ft-lb/in. of notch (J/m)				
$^1/_8$ in. (3 mm)	4.5 (240)	—	—	1.2 (64)
$^1/_4$ in. (6 mm)	—	1 (53)	2 (107)	—
Water absorption, %				
After 24 hr	—	0.3	0.2	0.3
At saturation	—	0.7	0.7	—
Thermal distortion temperature, °C (°F)				
At 66 psi (0.5 MPa)	101 (214)	107 (225)	116 (240)	102 (215)
At 264 psi (1.8 MPa)	93 (199)	102 (215)	110 (230)	98 (208)
Mold shrinkage				
E-03 in./in.	6	2–5	1–3	3

BASE RESIN	ABS	ACETAL				
SUPPLIER	LIQUID NITROGEN PROCESSING CORPORATION	E. I. DU PONT DE NEMOURS & COMPANY, INC.		FIBERFIL	LIQUID NITROGEN PROCESSING CORPORATION	THERMOFIL
Fiberglass content, wt %	30	0	20	20 (S)	10	30
Specific gravity	1.28	1.42	1.56	1.55	1.47	1.63
Tensile properties						
Yield strength, psi × 10^3 (MPa)	14.5 (100)	10.0 (69)	8.5 (59)	9.0 (62)	12.5 (86)	12.0 (83)
Modulus of elasticity, psi × 10^5 (GPa)	—	4.5 (3.1)	9.0 (6.2)	10.0 (69)	—	11.2 (77)
Yield elongation, %	3.0–4.0	12.0	7.0	2.2	6.0	2.0
Flexural properties						
Yield strength, psi × 10^3 (MPa)	18.5 (128)	14.1 (97)	10.7 (74)	11.5 (79)	18.5 (128)	16.5 (114)
Modulus of elasticity, psi × 10^5 (GPa)	11 (7.6)	4.1 (2.8)	7.3 (5.0)	7.0 (48)	5.5 (38)	10.5 (72)
Compressive properties						
Yield strength, psi × 10^5 (MPa)	14.5 (100)	5.2 (36)	5.2 (36)	—	—	11.8 (81)
Impact strength (notched bar, Izod test), ft-lb/in. of notch (J/m)						
$^1/_8$ in. (3 mm)	—	1.4 (75)	0.8 (43)	—	—	0.8 (43)
$^1/_4$ in. (6 mm)	1.4 (75)	—	—	0.8 (43)	1.4 (75)	—
Water absorption, %						
After 24 hr	0.14	0.25	0.25	0.27	0.24	0.2
At saturation	0.60	0.9	1.0	—	—	—
Thermal distortion temperature, °C (°F)						
At 66 psi (0.5 MPa)	110 (230)	170 (338)	174 (345)	163 (325)	—	158 (316)
At 264 psi (1.8 MPa)	104 (220)	124 (255)	157 (315)	160 (320)	149 (300)	—
Mold shrinkage						
E-03 in./in.	1	20	13–21	5–8	8	5

Table 7.3. Continued

BASE RESIN	NYLON 6			
SUPPLIER	ALLIED CHEMICAL CORPORATION			FIBERFIL
Fiberglass content, wt %	0	15	30	30 (L)
Specific gravity	1.13	1.22	1.37	1.40
Tensile properties				
Yield strength, psi × 10^3 (MPa)	11.8 (81)	15.0 (103)	23.5 (162)	24.0 (165)
Modulus of elasticity, psi × 10^5 (GPa)	—	—	—	10.0 (6.9)
Yield elongation, %	—	—	—	2.0
Flexural properties				
Yield strength, psi × 10^3 (MPa)	16.4 (113)	23.5 (162)	35.0 (241)	28.0 (193)
Modulus of elasticity, psi × 10^5 (GPa)	4.0 (2.8)	7.5 (5.2)	13.5 (9.3)	10 (69)
Compressive properties				
Yield strength, psi × 10^5 (MPa)	—	—	—	24.0 (165)
Impact strength (notched bar, Izod test), ft-lb/in. of notch (J/m)				
1/8 in. (3 mm)	—	—	—	—
1/4 in. (6 mm)	1.1 (59)	0.9 (48)	2.1 (112)	3.0 (160)
Water absorption, %				
After 24 hr	1.6	1.4	1.1	1.3
At saturation	—	—	—	6.4
Thermal distortion temperature, °C (°F)				
At 66 psi (0.5 MPa)	185 (365)	216 (420)	216 (420)	218 (425)
At 264 psi (1.8 MPa)	66 (150)	199 (390)	216 (420)	216 (420)
Mold shrinkage				
E-03 in./in.	13	6	4	3–5

BASE RESIN	NYLON 6		NYLON 66		NYLON 610	
SUPPLIER	LIQUID NITROGEN PROCESSING CORPORATION	THERMOFIL	ADELL PLASTICS, INC.	LIQUID NITROGEN PROCESSING CORPORATION	LIQUID NITROGEN PROCESSING CORPORATION	
Fiberglass content, wt %	40	15	30	30	20	50
Specific gravity	1.46	1.20	1.29	1.37	1.22	1.50
Tensile properties						
Yield strength, psi × 10^3 (MPa)	26.0 (179)	15.0 (103)	27.0 (186)	26.0 (179)	18.0 (124)	29.0 (200)
Modulus of elasticity, psi × 10^5 (GPa)	—	7.5 (5.2)	12.0 (8.3)	—	—	—
Yield elongation, %	2.0–3.0	2.0	3.0–4.0	3.0–4.0	3.0–4.0	2.0–3.0
Flexural properties						
Yield strength, psi × 10^3 (MPa)	36.0 (248)	22.0 (152)	38.5 (265)	38.0 (262)	26.0 (179)	43.0 (296)
Modulus of elasticity, psi × 10^5 (GPa)	15 (10.3)	8.0 (5.5)	13.0 (9.0)	13.0 (9.0)	9.0 (6.2)	19.0 (13.1)
Compressive properties						
Yield strength, psi × 10^5 (MPa)	23.0 (159)	17.0 (117)	35.5 (245)	24.0 (165)	17.0 (117)	25.0 (172)
Impact strength (notched bar, Izod test), ft-lb/in. of notch (J/m)						
1/8 in. (3 mm)	—	1.0 (53)	2.0 (107)	—	—	—
1/4 in. (6 mm)	3.0 (160)	—	—	2.0 (107)	1.1 (59)	4.5 (240)
Water absorption, %						
After 24 hr	1.46	1.0	0.8	0.9	0.22	0.16
At saturation	—	—	—	3.8	2.0	1.4
Thermal distortion temperature, °C (°F)						
At 66 psi (0.5 MPa)	218 (425)	—	—	260 (500)	216 (420)	224 (435)
At 264 psi (1.8 MPa)	216 (420)	227 (440)	249 (480)	254 (490)	210 (410)	218 (425)
Mold shrinkage						
E-03 in./in.	3	5	2–5	4	4	2.5

Table 7.3. Continued

BASE RESIN	NYLON 612	NYLON 11	NYLON 12	PHENYLENE OXIDE
SUPPLIER	E. I. DU PONT DE NEMOURS & COMPANY, INC.	RILSAN	EMSER	LIQUID NITROGEN PROCESSING CORPORATION
Fiberglass content, wt %	33	30	30	30
Specific gravity	1.32	1.26	1.22	1.28
Tensile properties				
Yield strength, psi × 10^3 (MPa)	24.0 (165)	13.0 (90)	18.0 (124)	18.5 (128)
Modulus of elasticity, psi × 10^5 (GPa)	—	—	9.6 (6.6)	—
Yield elongation, %	4	4.0–6.0	5	3.0–4.0
Flexural properties				
Yield strength, psi × 10^3 (MPa)	—	21 (145)	26.0 (179)	23.0 (159)
Modulus of elasticity, psi × 10^5 (GPa)	12.9 (8.9)	4.6 (3.2)	9.8 (6.8)	11.5 (7.9)
Compressive properties				
Yield strength, psi × 10^5 (MPa)	23.0 (159)	13.0 (90)	13.0 (90)	—
Impact strength (notched bar, Izod test), ft-lb/in. of notch (J/m)				
$^1/_8$ in. (3 mm)	2.6 (139)	1.7 (91)	2.6 (139)	—
$^1/_4$ in. (6 mm)	2.4 (128)	—	—	1.7 (90)
Water absorption, %				
After 24 hr	0.2	0.2	0.15	0.06
At saturation	2.0	—	1.1	0.11
Thermal distortion temperature, °C (°F)				
At 66 psi (0.5 MPa)	—	180 (356)	166 (330)	160 (320)
At 264 psi (1.8 MPa)	210 (410)	173 (343)	120 (248)	154 (310)
Mold shrinkage				
E-03 in./in.	3–9	—	5	1

BASE RESIN	POLYARYLETHER	POLYESTER (PBT)	POLYETHERSULFONE	POLYPHENYLENE SULFIDE	SAN
SUPPLIER	THERMOFIL	EASTMAN	LIQUID NITROGEN PROCESSING CORPORATION	THERMOFIL	FIBERFIL, INC.
Fiberglass content, wt %	20	30	40	40	20 (L)
Specific gravity	1.21	1.53	1.68	1.64	1.22
Tensile properties					
Yield strength, psi × 10^3 (MPa)	14.5 (100)	18.5 (128)	22.0 (152)	22.0 (152)	13.0 (90)
Modulus of elasticity, psi × 10^5 (GPa)	9.3 (6.4)	—	—	20.5 (14.1)	12.0 (8.3)
Yield elongation, %	5.0	5.0	3.0–4.0	3.0	—
Flexural properties					
Yield strength, psi × 10^3 (MPa)	18.5 (128)	28.0 (193)	30.0 (207)	37.0 (255)	20.0 (138)
Modulus of elasticity, psi × 10^5 (GPa)	7.5 (5.2)	12.0 (8.3)	16.0 (11.0)	19.0 (13.1)	11.0 (7.6)
Compressive properties					
Yield strength, psi × 10^5 (MPa)	17.8 (123)	—	—	21.0 (145)	—
Impact strength (notched bar, Izod test), ft-lb/in. of notch (J/m)					
$^1/_8$ in. (3 mm)	1.8 (96)	1.6 (85)	1.6 (85)	1.5 (80)	—
$^1/_4$ in. (6 mm)	—	—	—	—	2.5 (133)
Water absorption, %					
After 24 hr	0.06	0.07	0.31	0.01	0.22
At saturation	—	—	—	—	—
Thermal distortion temperature, °C (°F)					
At 66 psi (0.5 MPa)	—	207 (405)	—	266 (510)	—
At 264 psi (1.8 MPa)	143 (290)	—	216 (420)	—	104 (220)
Mold shrinkage					
E-03 in./in.	3	—	1.5	2	1–3

Table 7.3. Continued

BASE RESIN	POLYCARBONATE		POLYPROPYLENE	
SUPPLIER	GENERAL ELECTRIC		THERMOFIL	LIQUID NITROGEN PROCESSING CORPORATION
Fiberglass content, wt %	10	40	10	40
Specific gravity	1.25	1.52	0.98	1.22
Tensile properties				
Yield strength, psi $\times 10^3$ (MPa)	9.6 (66)	23.0 (159)	6.3 (43)	15.0 (103)
Modulus of elasticity, psi $\times 10^5$ (GPa)	4.5 (3.1)	16.8 (11.6)	3.6 (2.5)	—
Yield elongation, %	6.0–8.0	—	4.0	4.0–5.0
Flexural properties				
Yield strength, psi $\times 10^3$ (MPa)	15.0 (103)	27.0 (186)	7.8 (54)	22.0 (152)
Modulus of elasticity, psi $\times 10^5$ (GPa)	5.0 (3.4)	14.0 (9.7)	3.5 (2.4)	10.0 (69)
Compressive properties				
Yield strength, psi $\times 10^5$ (MPa)	14.0 (97)	21.0 (145)	6.0 (41)	9.8 (6.8)
Impact strength (notched bar, Izod test), ft-lb/in. of notch (J/m)				
1/8 in. (3 mm)	5.0 (267)	2.5 (133)	0.8 (43)	—
1/4 in. (6 mm)	2.0 (107)	2.0 (107)	—	2.0 (107)
Water absorption, %				
After 24 hr	0.12	0.12	0.05	0.05
At saturation	0.31	0.23	—	0.09
Thermal distortion temperature, °C (°F)				
At 66 psi (0.5 MPa)	146 (295)	154 (310)	—	166 (330)
At 264 psi (1.8 MPa)	142 (288)	146 (295)	127 (260)	154 (310)
Mold shrinkage				
E-03 in./in.	2–4	1–2	7	3

[a]S = short glass; L = long glass.

lengths tested varied from 1/8 to 3/4 in. (3 to 19 mm).[6] A similar investigation indicated only marginal differences between 1/8 and 1/4 in. (3 and 6 mm) lengths and that strand integrity was a significant factor.[7] The results will be equivalent when the glass is a continuous fiber, since pellet lengths of 1/8 or 1/4 in. (3 or 6 mm) are normal. The actual fiber length and distribution of lengths in the molding or pellets are not the same as in the feed, but are functions of the overall processing (fiber screening, preblending, compounding, and molding). Estimates reveal that the average fiber length can be as low as 25–30 mils (0.63–0.75 mm) in the compound and 20 mils (0.5 mm) in the finished molding, regardless of whether 1/8- or 1/4-in. (3- or 6-mm) fiber was used.[8,9] The fiber-length distributions for nylon 6 and polypropylene are presented in Fig. 7.1.[9] From these and other results it can be inferred that the minimum fiber aspect ratios (length/diameter) will be on the order of 50–125, and a considerable portion of the glass will be at this distribution level. With milled glass, aspect ratios of 8 have been reported.[10]

7.4.2.3. Surface Finish on Fiber

The coupling agent and size on the fiber are important factors in determining the performance of GFTP. The silanes have been found to be effective coupling agents for thermoplastics, the specific one used depending on the resin (in particular, whether it is amorphous or crystalline). The primary function of the coupling agent is to promote adhesion of the resin to the fiber, thus improving the resistance to moisture and minimizing the effects of moisture on the physical and mechanical properties of the composite. The

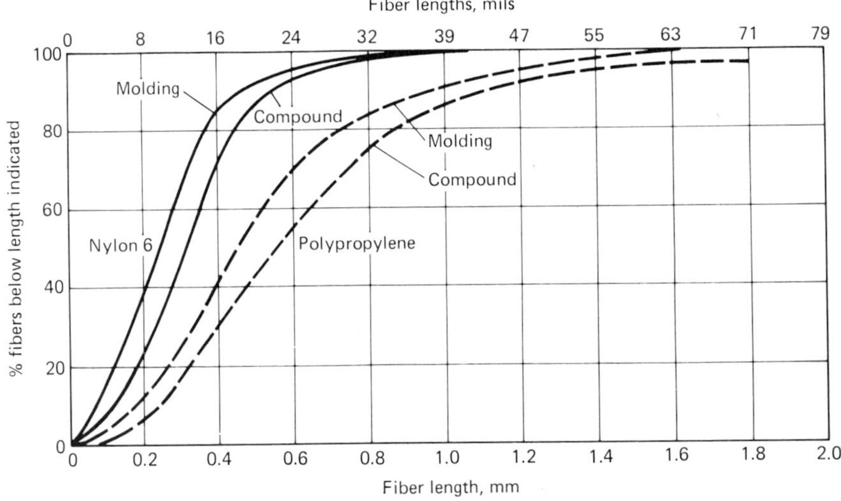

Figure 7.1. Fiber-length distributions for nylon 6 and polypropylene.[9]

coupling agent also affects fiber wet-out (air displacement), fiber dispersion, viscosity during processing, and finished-surface appearance. The size protects the fiber bundle during processing and also has an effect on viscosity and dispersion.

The effect of the interfacial bond on the mechanical properties and failure modes of short-fiber composites is a complex subject that will not be discussed here. Specific investigations in this area can be found in the literature.[10–13]

7.4.2.4 Fiber Alignment and Dispersion

The fiber alignment and dispersion are functions of the molding process, mold design, and flow properties of the mix. Several studies have been undertaken to quantitatively characterize fiber alignment and to determine the effects of viscosity, fill rate, mold temperature, and gate geometry with respect to controlling it. The complexity of the problem has limited these investigations to simple mold shapes. Results indicate the possibility of attaining increased mechanical strength and stiffness by

Table 7.4. Tensile Properties of Random Planar and Unidirectional Filled Thermoplastics.[a13]

BASE RESIN	FIBER TYPE AND ALIGNMENT[b]		TENSILE STRENGTH psi × 10^3	TENSILE STRENGTH MPa	TENSILE MODULUS OF ELASTICITY psi × 10^5	TENSILE MODULUS OF ELASTICITY GPa	ELONGATION, %
High-density polyethylene	Glass	R	6.1	42.1	4.9	3.4	1.9
		U	20.7	142.7	13.6	9.4	1.9
	Kevlar 49	R	10.7	73.8	7.5	5.2	2.3
		U	19.1	131.7	15.4	10.6	2.3
	Graphite	R	8.5	58.6	9.7	6.7	1.0
		U	16.0	110.3	28.0	19.3	0.7
Polycarbonate	Glass	R	8.9	61.4	5.7	3.9	2.0
		U	16.6	114.4	10.9	7.5	1.8
	Kevlar 49	R	16.8	115.8	7.4	5.1	3.0
		U	23.7	163.4	14.2	9.8	2.1
	Graphite	R	10.0	69.0	12.5	8.6	0.8
		U	25.6	176.5	18.3	12.6	1.7

[a]Fiber content = 20 vol %.
[b]R = random planar; U = unidirectional.

Table 7.5. Effect of Fiberglass Content on the Physical and Mechanical Properties of Nylon 66.[16]

Glass content, wt %	0	10	20	30	40	50	60
Specific gravity	1.14	1.21	1.28	1.37	1.46	1.57	1.70
Tensile properties							
Yield strength							
psi × 10^3 (MPa)	12 (83)	13 (90)	19 (131)	25 (172)	31 (214)	32 (221)	33 (228)
Elongation, %	60	3.5	3.5	3.0	2.5	2.5	1.5
Flexural properties							
Yield strength							
psi × 10^3 (MPa)	15 (103)	20 (138)	29 (200)	34 (234)	42 (290)	46 (317)	50 (345)
Modulus of elasticity,							
psi × 10^5 (GPa)	4 (2.8)	6 (4.1)	9 (6.2)	13 (9.0)	16 (11.0)	22 (15.2)	28 (19.3)
Compressive properties							
Yield strength							
psi × 10^3 (MPa)	4.9 (34)	13 (90)	23 (159)	27 (186)	28 (193)	29 (200)	30 (207)
Thermal distortion temperature at 264 psi (1.82 MPa), °C (°F)	66 (150)	243 (470)	246 (475)	252 (485)	260 (500)	260 (500)	260 (500)
Coefficient of linear thermal expansion E-05 in./in./°F (m/m/°C)	4.5 (8.1)	1.6 (2.9)	1.4 (2.5)	1.3 (2.3)	1.2 (2.2)	1.0 (1.8)	0.9 (1.6)
Water absorption after 24 hr, %	1.6	1.1	0.9	0.9	0.6	0.5	0.4
Mold shrinkage E-03 in./in.	15	6.5	5.0	4.0	3.5	3.0	2.5

controlled fiber alignment.[14,15] The more conventional approach is to obtain random fiber distribution and homogeneous properties rather than anisotropy. However, complete random distribution is difficult to achieve, and most parts will exhibit some degree of anisotropy. Table 7.4 compares the tensile properties of random planar and unidirectional GFTP.[13]

7.4.2.5. Fiberglass Content

Glass contents of up to 60 wt % have been reached with some resins, the equivalent glass volume being approximately 50%. The glass contents of most commercial GFTP range from 20 to 40 wt %. The optimum content is based on a trade-off between desired strength levels, strength increments with added glass, overall cost, and ease of processing.

Table 7.5 illustrates the variation in mechanical properties of nylon 66 as a function of increased glass content.[16] In this particular case, the glass content ranges from 0 to 60 wt %.

The effect of fiberglass content on the tensile strengths and moduli of elasticity of nylon 6 and polycarbonate is illustrated in Figs. 7.2 and 7.3, respectively.[17] Other thermoplastics will exhibit similar trends. The initial increase in tensile strength is usually linear, but tends to level off at higher glass contents. The tensile modulus of elasticity follows a similar pattern, except that the change in slope is less pronounced. The variations in flexural strength and modulus of elasticity as a function of glass content are almost identical to those in the corresponding tensile properties. However, in almost every case, the increase in flexural strength is greater than the corresponding increase in tensile strength. In contrast, compressive strength is modified to a lesser extent than either flexural or tensile strength. The reduced fiber efficiency at the higher loadings, particularly evident in tension, can be attributed to fiber conglomeration, poorer dispersion within the matrix resin, and excessive shearing action.

The results of notched-bar Izod impact-strength tests are varied. Some thermoplastics show a linear increase in strength with increased fiberglass content or peak at an

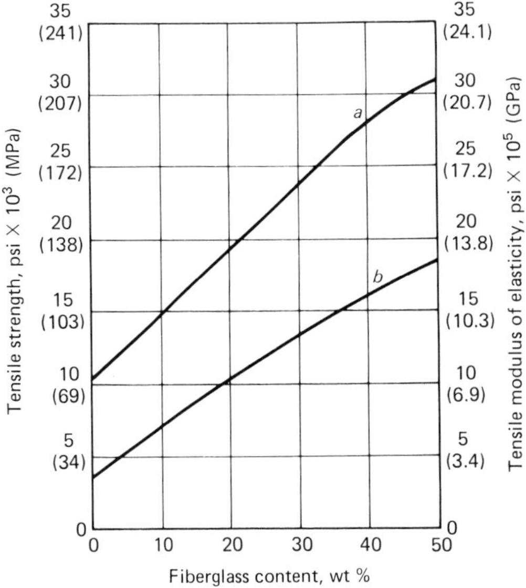

Figure 7.2. Tensile strength and modulus of elasticity of nylon 6 as a function of fiberglass content: (*a*) tensile strength; (*b*) tensile modulus of elasticity.

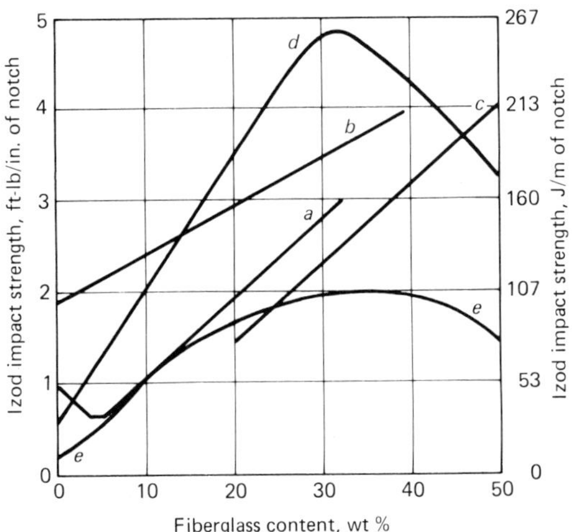

Figure 7.4. Izod impact strength as a function of fiberglass content: (*a*) nylon 66—Nyafil G-1;[17] (*b*) polycarbonate—Polycarbfil;[17] (*c*) nylon 6—LNP;[1] (*d*) polypropylene, direct mold;[6] (*e*) SAN, direct mold.[6]

intermediate value. For other resins the effect is reversed, that is, impact strength decreases. Based on this test method, some composites are tougher than the base resin whereas others are more brittle. Tests with unnotched specimens may be more indicative of actual material toughness. Tensile impact-strength tests, for which data are limited, reveal the strong sensitivity of some resins to strain rate. This is

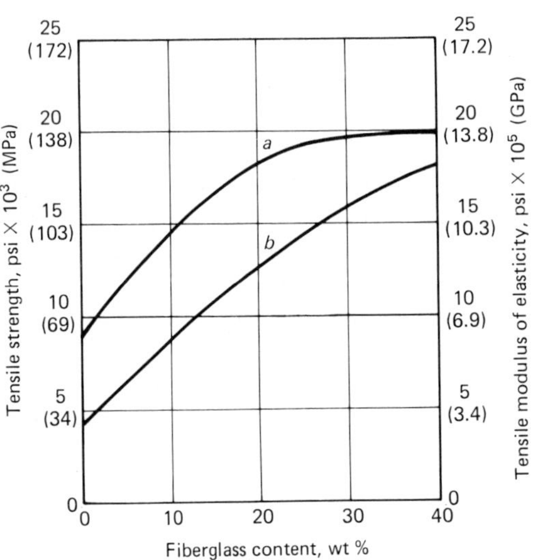

Figure 7.3. Tensile strength and modulus of elasticity of polycarbonate as a function of fiberglass content: (*a*) tensile strength; (*b*) tensile modulus of elasticity.[17]

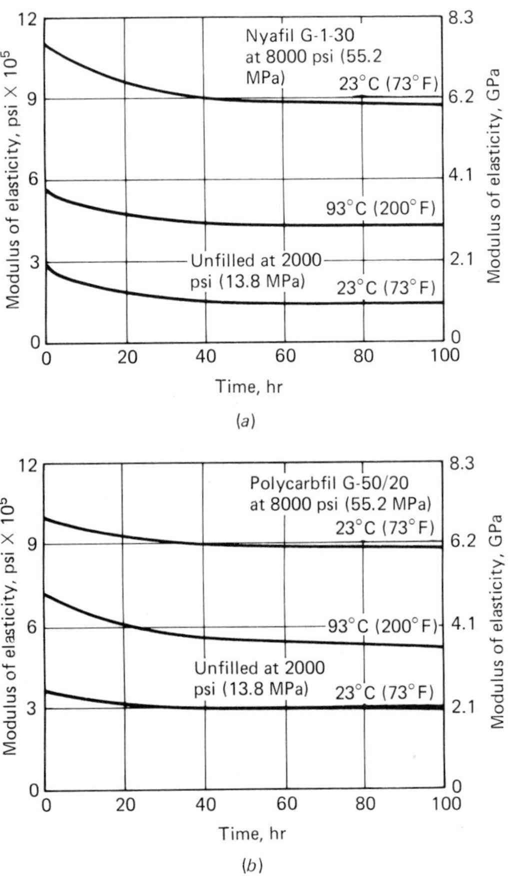

Figure 7.5. Flexural creep—apparent modulus of elasticity versus time: (*a*) nylon 66; (*b*) polycarbonate.[17]

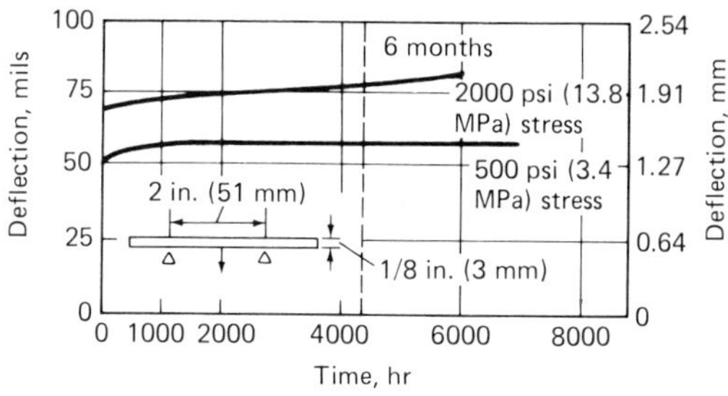

Figure 7.6. Flexural creep—deflection versus time for a glass-filled polyester at 150° C (302° F).[18]

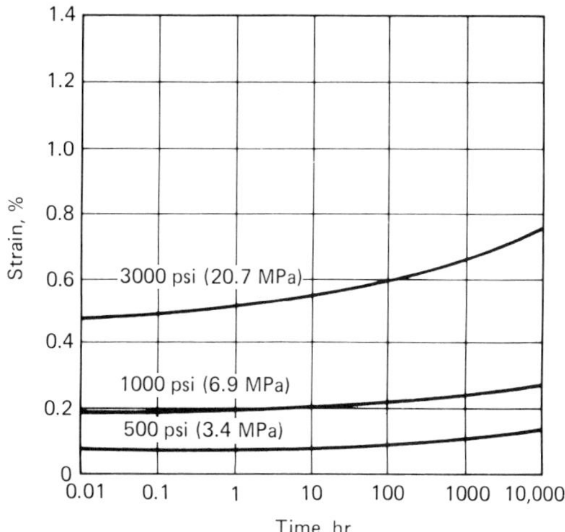

Figure 7.7. Flexural creep—strain versus time for Zytel (E. I. du Pont de Nemours & Company, Inc.) 70 G-33-HS1-L (nylon 66–33 wt % fiberglass) at 125° C (257° F), dry.[19]

particularly apparent with the tougher resins and composites. Fig. 7.4 shows the relationship between the Izod impact strength and fiberglass content for several materials, two of which were molded from blends rather than compounded pellets.

The creep resistance of thermoplastics subjected to various loading conditions is increased by the addition of glass. Fig. 7.5 compares the apparent flexural moduli of elasticity of glass-filled nylon 66 and polycarbonate with those of the unfilled resins.[17] Fig. 7.6 depicts the deflection as a function of time for a glass-filled polyester under a flexural load.[18] The flexural strain as a function of time for glass-filled (33 wt %) nylon 66 tested at 125°C (257°F) is shown in Fig. 7.7.[19] The tensile stress relaxation of several heat-resistant resins (filled and unfilled) at 23°C (73°F) and 149°C (300°F) is presented in Table 7.6.[20] For most unfilled thermoplastics, the apparent modulus of elasticity in creep (stress/creep strain) depends on the applied load. For 20-wt % glass-filled ABS, SAN, and polystyrene as well as 15-wt % glass-filled nylon 6, the

Table 7.6. Tensile Stress Relaxation of Filled and Unfilled Heat-Resistant Resins under an Applied Load of 2500 psi (17.2 MPa).[20]

BASE RESIN	FIBERGLASS CONTENT, WT %	DECREASE AT 23° C (73° F), %			DECREASE AT 149° C (300° F), %		
		1 hr	5 hr	15 hr	1 hr	5 hr	15 hr
ETFE[a]	20	12	17	21	30	32	38
FEP[b]	20	20	31	38	—[c]	—[c]	—[c]
Polyphenylene sulfide	40	2	5	9	28	28	28
Polyethersulfone	40	7	12	16	52	63	70
Nylon 66	50	2	5	8	17	22	24
Polyester	40	9	20	28	54	58	74
Polysulfone	40	5	9	14	—[c]	—[c]	—[c]
Polyimide	30	7	14	19	38	40	43
Polyamide-imide	0	0.5	1	2	3	4	8
Polyarylsulfone	0	5	7	11	7	10	16
Poly-*p*-oxybenzoate	0	8	11	16	40	42	48

[a]Ethylene-tetrafluoro-ethylene.
[b]Fluorinated ethylene-propylene.
[c]Could not sustain load.

Table 7.7. Tensile Strengths of Filled and Unfilled Heat-Resistant Resins at Elevated Temperatures.[20]

		TENSILE STRENGTH, psi × 10^3 (MPa)					
BASE RESIN	FIBERGLASS CONTENT, WT %	23° C (73° F)	93° C (200° F)	149° C (300° F)	177° C (350° F)	204° C (400° F)	232° C (450° F)
ETFE	20	11.3 (77.9)	6.85 (47.2)	4.0 (27.6)	2.0 (13.8)	—[a]	—[a]
FEP	20	5.0 (34.5)	4.2 (29.0)	2.3 (15.9)	1.2 (8.3)	—[a]	—[a]
Polyphenylene sulfide	40	23.2 (160)	11.2 (77.2)	8.1 (55.8)	4.8 (33.1)	1.1 (7.6)	—[a]
Polyethersulfone	40	22.7 (157)	19.4 (134)	13.1 (90.3)	4.9 (33.8)	3.1 (21.4)	—[a]
Nylon 66	50	31.2 (215)	16.0 (110)	12.4 (85.5)	7.3 (50.3)	2.2 (15.2)	—[a]
Polyester	40	19.4 (134)	7.4 (51.0)	4.1 (28.3)	0.6 (4.1)	—[a]	—[a]
Polysulfone	40	17.3 (119)	14.9 (103)	2.3 (15.9)	1.1 (7.6)	—[a]	—[a]
Polyimide	30	13.0 (89.6)	6.2 (42.7)	4.8 (33.1)	3.1 (21.4)	2.3 (15.9)	1.8 (12.4)
Polyamide-imide	0	27.4 (189)[b]	19.9 (137)	16.2 (112)	11.4 (78.6)	8.2 (56.5)	6.9 (47.6)
Polyarylsulfone	0	11.1 (76.5)	10.4 (71.7)	8.7 (60.0)	7.4 (51.0)	5.7 (39.3)	3.2 (22.1)
Poly-*p*-oxybenzoate	0	13.9 (95.8)	11.2 (77.2)	9.3 (64.1)	7.8 (53.8)	6.4 (44.1)	3.9 (26.9)

[a]Could not sustain load.
[b]Annealed prior to test.

apparent moduli of elasticity are independent of load up to a stress level of 5000 psi (34.5 MPa).[21] At a 40-wt % glass level and 5000 psi (34.5 MPa), polystyrene, rigid polyvinyl chloride, polysulfone, and polycarbonate show negligible creep rates after 1000 hr.[22]

The elevated-temperature tensile strengths of the heat-resistant filled and unfilled resins of Table 7.6 are listed in Table 7.7, whereas the effect of thermal aging at 260°C (500°F) is shown in Table 7.8.[20] Similar data for nylon 66 composites are presented in Figs. 7.8 and 7.9, respectively.[19] Stress-strain relations for two nylon 66 composites at several temperatures are plotted in Fig. 7.10.[19] Fig. 7.11 compares the tensile strengths of filled and unfilled polyphenylene sulfide as a function of temperature.[18]

The effect of temperature on the shear strengths of three nylon composites is presented in Fig. 7.12.[19]

The tensile strengths of nylon 66 at two fiberglass loadings and of nylon 612 at one loading as a function of humidity are compared in Fig. 7.13.[19]

Typical data for the fatigue resistance (S–N curve) of two nylon composites are presented in Fig. 7.14.[19]

The effect of weathering on the tensile strengths of several nylon composites for a period of up to 3 years is shown in Fig. 7.15.[19]

7.4.3. Thermal Properties

The thermal deflection temperature (TDT) as measured by ASTM D648 is increased by the coefficient of linear thermal expansion. In each case the decrease is approximately linear.

The thermal deflection temperature (TDT) as measured by ASTM D648 is increased by the addition of fiberglass; however, the increase is

Table 7.8. Tensile Strengths of Filled and Unfilled Heat-Resistant Resins after Thermal Aging at 260°C (500°F).[a20]

BASE RESIN	FIBERGLASS CONTENT, wt %	TENSILE STRENGTH, psi × 10^3 (MPa) 0 hr	100 hr	250 hr	500 hr	750 hr	1000 hr	1500 hr
ETFE	20	11.3 (77.9)	11.5 (79.3)	10.0 (69.0)	7.0 (48.3)	5.0 (34.5)	3.8 (26.2)	2.3 (15.9)
FEP	20	5.0 (34.5)	5.1 (35.2)	4.8 (33.1)	4.7 (32.4)	4.7 (32.4)	4.6 (31.7)	4.5 (31.0)
Polyphenylene sulfide	40	23.2 (160)	16.4 (113)	16.0 (110)	15.5 (107)	15.0 (103)	14.5 (100)	13.8 (95.2)
Polyethersulfone	40	22.7 (157)	15.6 (108)	14.8 (102)	14.3 (98.6)	13.7 (94.5)	12.2 (84.1)	10.5 (72.4)
Polyimide	30	13.0 (89.6)	15.0 (103)	14.3 (98.6)	13.4 (92.4)	12.8 (88.3)	12.0 (82.7)	11.2 (77.2)
Polyamide-imide	0	27.4 (189)	27.2 (188)	26.6 (183)	26.0 (179)	24.5 (169)	23.5 (162)	22.0 (152)
Polyarylsulfone	0	13.1 (90.3)	11.5 (79.3)	10.5 (72.4)	10.0 (69.0)	9.5 (65.5)	8.4 (57.9)	7.6 (52.4)
Poly-p-oxy benzoate	0	23.0 (159)	18.0 (124)	16.3 (112)	16.0 (110)	15.5 (107)	15.1 (104)	13.0 (89.6)
Nylon 66	50	31.0 (214)	17.6 (121)	10.3 (11.0)	9.4 (64.8)	—	—	—
Polyester	40	22.1 (152)	Melted	—	—	—	—	—
Polysulfone	40	20.3 (140)	Melted	—	—	—	—	—

[a]Tested at 23°C (73°F).

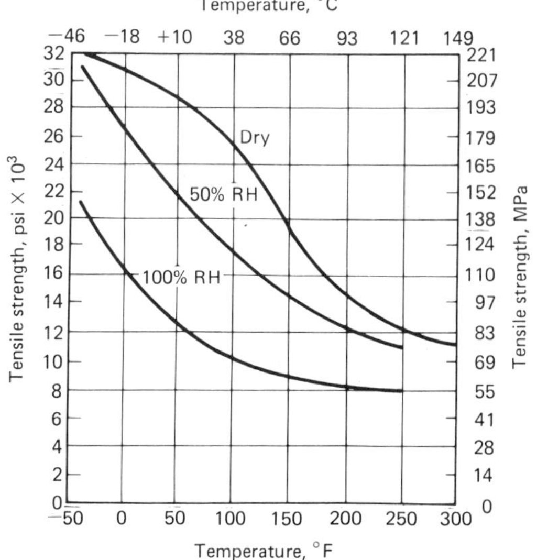

Figure 7.8. Tensile strength of Zytel 70 G-33 (nylon 66–33 wt % fiberglass) as a function of temperature (RH = relative humidity).[19]

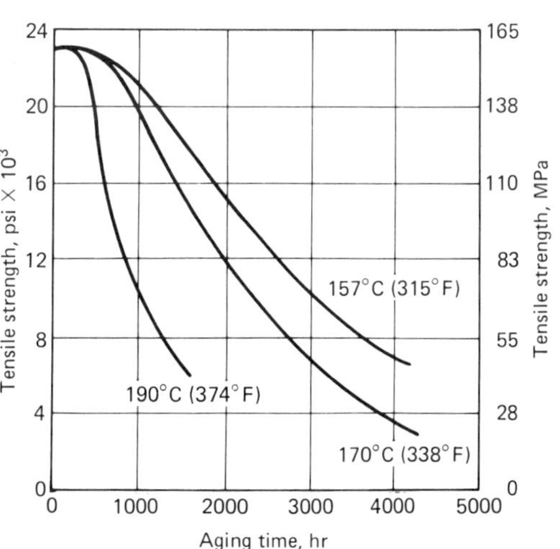

Figure 7.9. Effect of oven aging on the tensile strength of Zytel 71 G-33 (nylon 66–33 wt % fiberglass).[19]

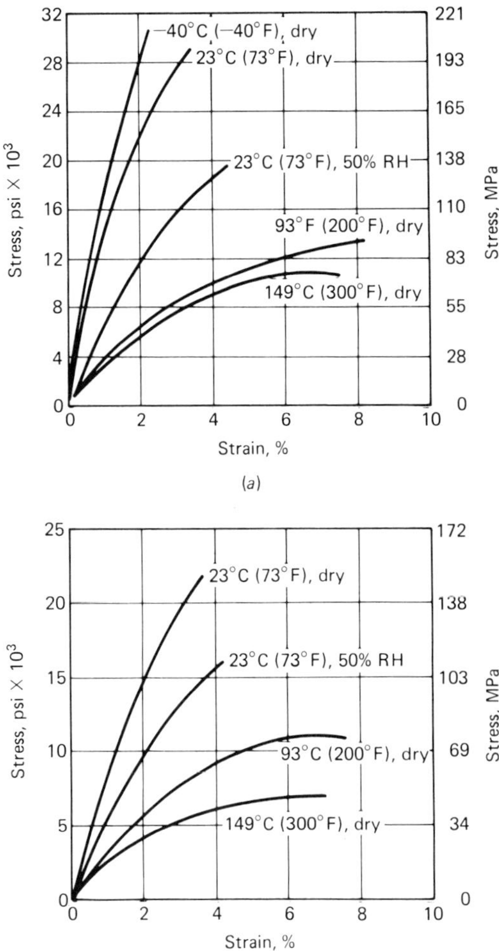

Figure 7.10. Stress–strain relations for two nylon 66 composites under various conditions: (*a*) Zytel 70 G-33 (nylon 66–33 wt % fiberglass); (*b*) Zytel 71 G-33 (nylon 66–33 wt % fiberglass).[19]

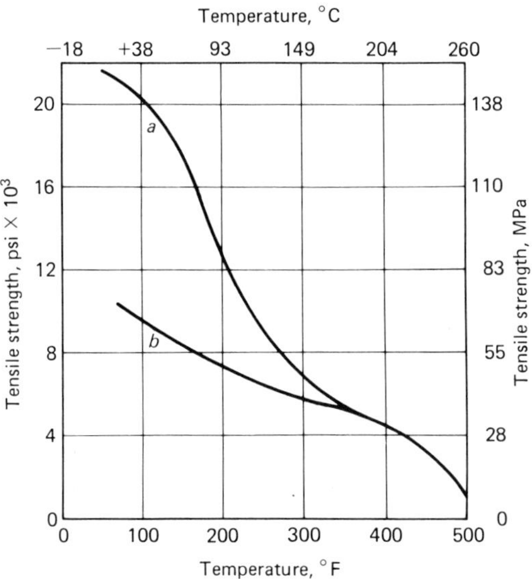

Figure 7.11. Tensile strength of polyphenylene sulfide as a function of temperature: (*a*) filled with 40 wt % fiberglass; (*b*) unfilled.[18]

less marked at higher fiberglass contents. The TDT is also dependent on the glass transition temperature or melting point of the resin. It does not correlate directly with the maximum service temperature of a material, but is intended primarily for comparative purposes. As a general rule, the TDT should not be exceeded if the molded part is load bearing or self-supporting. Resin systems that are considered chemically coupled, such as nylon, polyphenylene sulfide, alkane-modified polyimide, polyester, and polysulfone, appear capable of sustaining loads up to the transition temperature. The TDTs of systems not optimally coupled, including ETFE and FEP, do not approach the transition temperature as closely. The tensile strength and stress relaxation at elevated temperatures are criteria that should be investigated when appraising material performance.[20]

Direct comparisons of maximum service temperatures for filled and unfilled thermo-

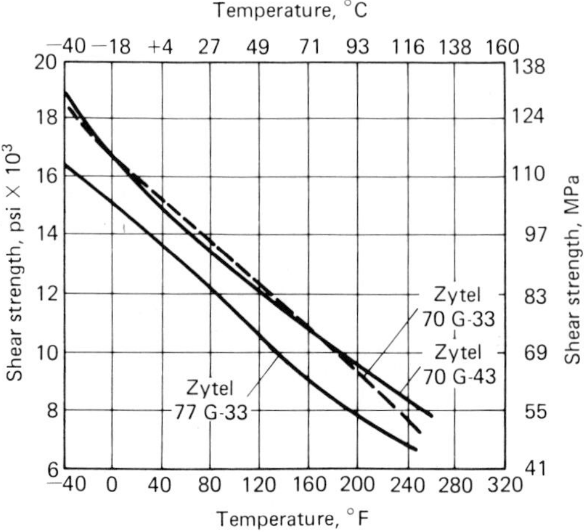

Figure 7.12. Shear strength of Zytel 70 G-33 (nylon 66–33 wt % fiberglass), Zytel 70 G-43 (nylon 66–43 wt % fiberglass), and Zytel 77 G-33 (nylon 612–33 wt % fiberglass) as a function of temperature.[19]

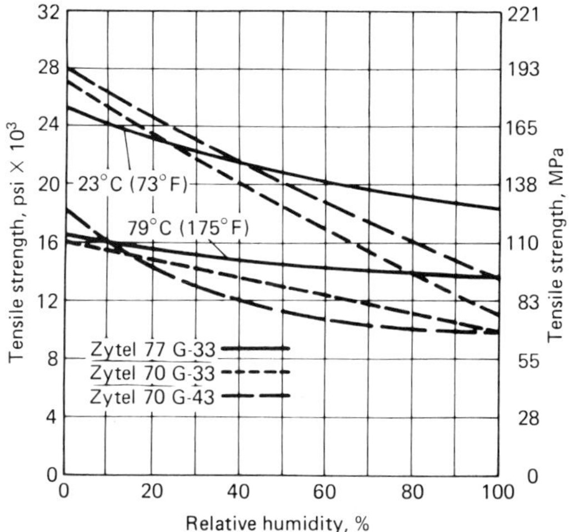

Figure 7.13. Tensile strength of Zytel 70 G-33 (nylon 66–33 wt % fiberglass), Zytel 70 G-43 (nylon 66–43 wt % fiberglass), and Zytel 77 G-33 (nylon 612–33 wt % fiberglass) as a function of humidity.[19]

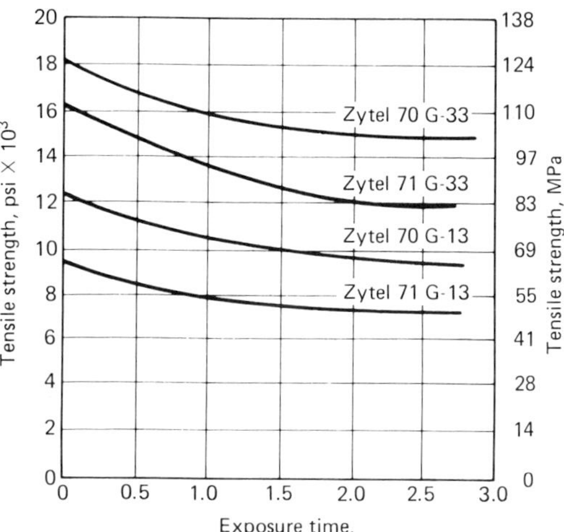

Figure 7.15. Effect of weathering (in Florida) on the tensile strength of Zytel 70G-33 (nylon 66–33 wt % of fiberglass), Zytel 71 G-33 (nylon -33 wt % fiberglass), Zytel 70 G-13 (nylon 66–33 wt % fiberglass), and Zytel 71 G-13 (nylon 66–13 wt % fiberglass).[19]

plastic resins are difficult to make. Considerable differences exist between the base resins in the filled and unfilled compounds. The latter may have a higher average molecular weight and therefore be more heat resistant. Performance is influenced by heat stabilizers and additives in the system. In this connection, it should be noted that flame retardants generally reduce the maximum service temperature. It can be concluded that although fiberglass addition only raises the useful service temperature to a slight extent, the filled resins have the additional advantage of higher initial strengths. The addition of fiberglass also has a slight effect on the softening point of amorphous thermoplastic resins, but no effect on crystalline melting points.

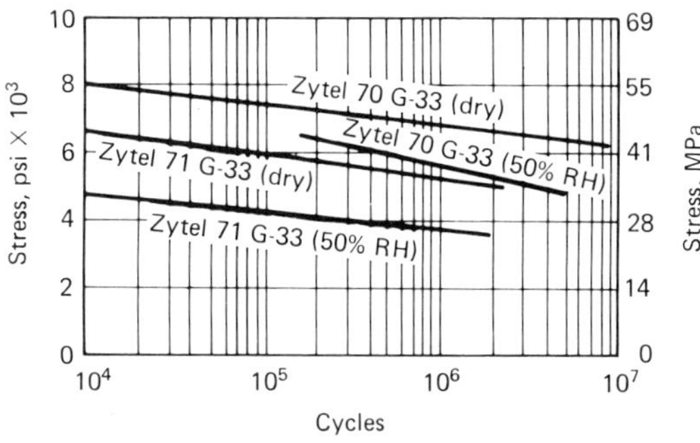

Figure 7.14. Fatigue resistance (tension–compression) of Zytel 70 G-33 (nylon 66–33 wt % fiberglass) and Zytel 71 G-33 (nylon 66–33 wt % fiberglass) at 1800 cycles/min.[19]

7.5. THEORETICAL CONSIDERATIONS

The analytical methods developed for continuous-fiber-filled materials have become indispensable tools in composite technology. Analogous procedures for discontinuous fibers are currently being investigated, but are not yet fully established. Solutions are complicated by such factors as (1) variation in fiber lengths and distribution, (2) spatial arrangements of the fibers, (3) stress concentrations at fiber ends, (4) interfacial reactions, and (5) the nonlinear response of the matrix. Several models have been proposed to determine the strength and stiffness of short-fiber composites. The method summarized here[23] has been reported in several publications by Halpin and others.

7.5.1. Quasi-Isotropic Laminate Analogy

It has been demonstrated that short-fiber systems are sufficiently close to a two-dimensional random array to be modeled as a quasi-isotropic laminate. To be more specific, the in-plane stiffness of a random short-fiber material of known fiber volume and aspect ratio is equivalent to that of a quasi-isotropic laminate of the same fiber volume and aspect ratio. The assumption is made that a linear strain field exists through the thickness, which is compatible with classical laminated plate theory.

The stress–strain response of an orthotropic (unidirectional) ply is characterized by a high modulus of elasticity, strength, and elongation in the principal direction (parallel to the fibers), whereas the corresponding values in the transverse direction are relatively low. When a number of plies are laminated at several orientations, the stress–strain relation will be intermediate to the longitudinal and transverse relations. As the number of oriented plies is increased, the isotropic strength is approached asymptotically. Four ply directions are sufficient, and a $(0°/90°/\pm45°)_s$ laminate can be selected for isotropic simulation.

A maximum strain theory developed by Petit and Waddoups[24] is modified to predict the strengths for the case of a random short-fiber composite. The first step is to calculate the engineering moduli of elasticity for an aligned short-fiber ply from the fiber and matrix properties:

Longitudinal modulus of elasticity:

$$E_{11} = E_f V_f + E_m V_m \tag{7.1}$$

Traverse modulus of elasticity:

$$E_{22} = \frac{E_m[1 + \zeta V_f(E_f/E_m - 1)(E_f/E_m + \zeta)]}{1 - V_f(E_f/E_m - 1)(E_f/E_m + \zeta)} \tag{7.2}$$

Shear modulus of elasticity:

$$G_{12} = \frac{G_m[1 + \zeta V_f(G_f/G_m - 1)(G_f/G_m + \zeta)]}{1 - V_f(G_f/G_m - 1)(G_f/G_m + \zeta)} \tag{7.3}$$

Major Poisson's ratio:

$$\nu_{12} = \nu_f V_f + \nu_m V_m \tag{7.4}$$

where V = volume fraction
f = fiber
m = matrix
ζ = a measure of reinforcement efficiency

The longitudinal modulus of elasticity and Poisson's ratio are approximated by the "rule of mixtures" which is not the case for the transverse and shear moduli of elasticity. The fiber modulus of elasticity, which provides the major portion of the composite modulus, is a function of the aspect ratio. The short-fiber stiffness approaches that of continuous fibers at large aspect ratios. Strength is also a function of the aspect ratio and approaches a limit as the latter is increased. However, it does not attain the continuous-fiber value.

The Halpin–Kardos equations[23] have been proposed for estimating a strength reduction factor (SRF). The SRF is used to calculate allowable strains (longitudinal, transverse, and shear) in the layered composite. This procedure is in compliance with the observed fact that there is a critical fiber aspect ratio. When this ratio is equalled or exceeded, the fiber will break at its intrinsic strength, but at a reduced composite strength. Below the critical value, the composite will fail in the matrix or interfacial region. The critical ratio is expressed as

$$(l/d)_c = 1/2\,(\sigma_f/\tau_m)$$

where σ_f = fiber strength
τ_m = matrix shear strength.

The simulation is continued by loading the laminate in increments and examining each ply for failure. A ply fails when the permissible strain is exceeded. As each ply fails, it is deleted, composite moduli of elasticity are recalculated, and the procedure is continued until all plies have failed. The results of such an analysis are shown in Fig. 7.16.[25] Similar predictions for a quasi-isotropic laminate with continuous fibers are presented in Fig. 7.17.[23] Figure 7.18 presents a comparison of the theoretical and experimental tensile strengths of a randomly oriented short-fiber laminate as a function of fiber volume.[26]

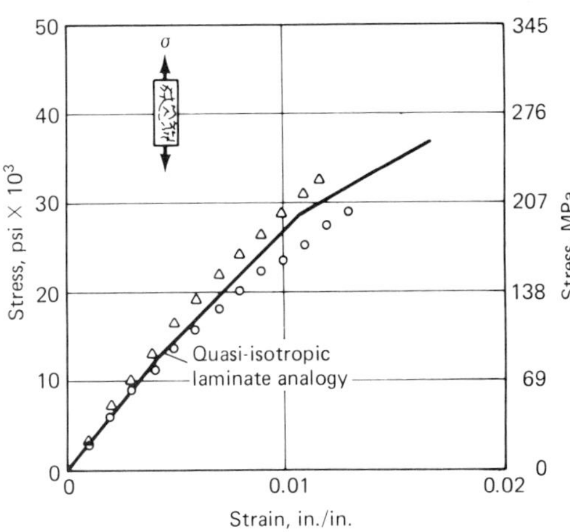

Figure 7.16. Stress–strain for a randomly oriented short-fiber—epoxy laminate with the maximum strain theory prediction.[25]

The quasi-isotropic laminate analogy is not only applicable to completely random composites, but can be used equally well when some degree of anisotropy exists and in the more realistic case of variable fiber aspect ratios. It has been noted previously that most GFTP moldings range between random and anisotropic and that fiber lengths vary significantly. It can be expected that some fibers will fall below the critical aspect ratio. The latter condition can be accommodated in the quasi-isotropic analysis. The fact that the above analyses have been conducted with thermoset resins should not discourage their use with thermoplastics. Prospective modifications of the theory may account for the nonlinear resin behavior and provide more accurate appraisals of the stress concentrations at the fiber ends.

7.6. APPLICATIONS OF GFTP

The first use of GFTP in the United States was for an Army land mine injection molded from glass-filled polystyrene. The development of the material for this part marked the beginning of the glass-reinforced thermoplastics division of the plastics industry. Government interest in this and similar materials has continued, although at a relatively modest level. The military accounts for perhaps 5% of total consumption in such diverse items as cartridge cases, rotating bands and sabots for artillery shells, fuse housings, sleeve bearings, gears, and vehicular components. Current government projects are concerned with the continuous extrusion or pultrusion of graphite

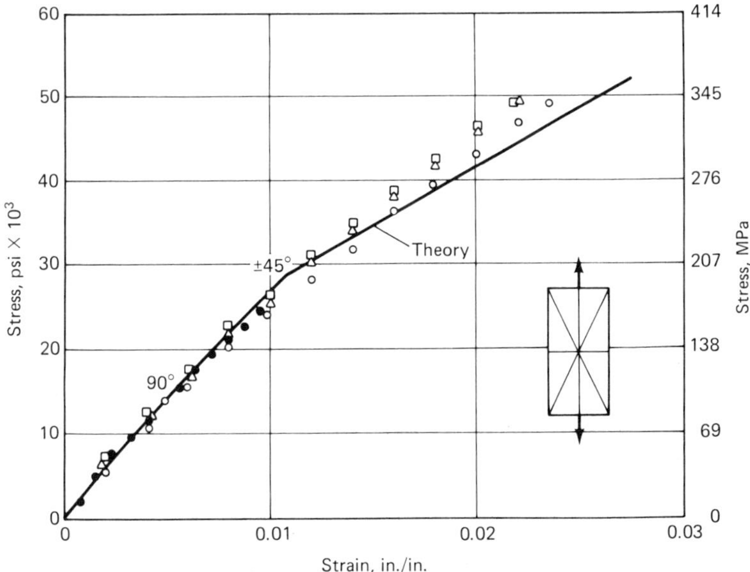

Figure 7.17. Stress–strain relationship for a continuous-fiber quasi-isotropic glass–epoxy laminate with the maximum strain theory prediction.[23]

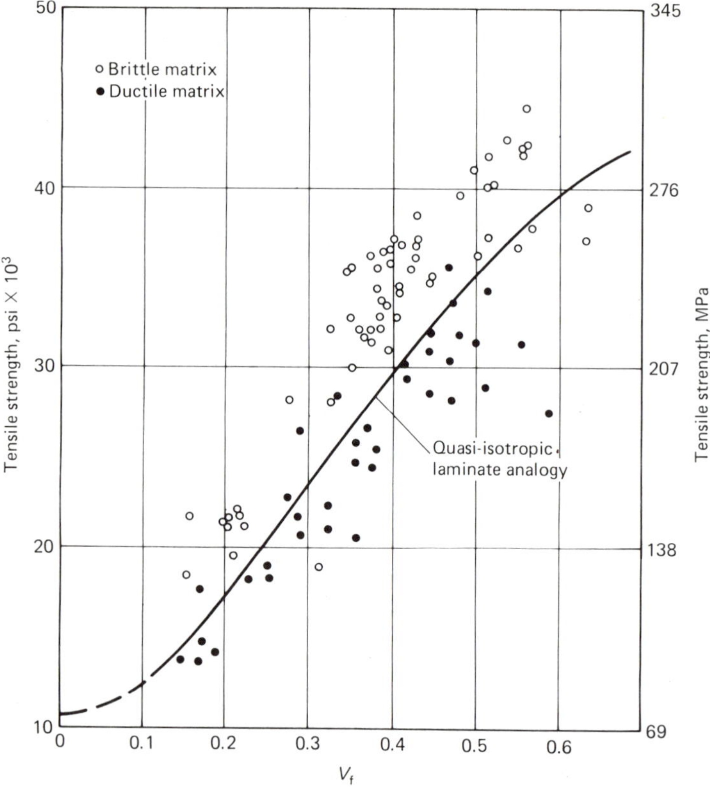

Figure 7.18. Effect of fiber volume fraction (V_f) on the tensile strength of a randomly oriented short-fiber ($l/d \approx 500$) glass–epoxy laminate as predicted by the maximum strain theory: ○) brittle matrix; ●) ductile matrix.[26]

strands with a thermoplastic resin and the filament winding of fiberglass or other reinforcements with such binders as the polysulfones.

The major outlet for filled thermoplastics is the automotive industry, which absorbs more than 50% of the production. The appliance and business machine fields, which absorb about 30%, are the second largest users. From a material viewpoint, polypropylene comprises approximately 30% of the market, the polyamides at 20% and the polystyrenes at 15% being the next most frequently used materials. The dominance of polypropylene is easy to understand, since—with the exception of polystyrene—it is the least expensive member of this class of materials. Comparative material costs, based on the price per cubic inch of molding, are listed in Table 7.9 for several filled and unfilled thermoplastic resins.

Impact resistance or toughness, rigidity, dimensional stability, and resistance to creep are the important properties that have led to the increased acceptance of GFTP. Development of the "engineered plastics," with an emphasis on improved properties at higher temperatures, has been instrumental in opening new market areas. Secondary features, such as surface finishes amenable to electroplating, low friction and wear, low moisture pick-up, and a combination of good electrical properties with flame retardance, have also contributed to a growing market.

Parts molded from GFTP have been relatively small in size compared to SMC (sheet molding compounds), BMC (bulk molding compounds), or other thermoset moldings. Typical examples are automotive instrument panels, pump bodies, appliance housings, washing machine agitators, plumbing fixtures,

Table 7.9. Comparative Material Costs.

BASE RESIN TYPE	FIBERGLASS CONTENT, wt %	COST, CENTS/ cu in.
Acetal (Delrin)	0	5.13
	20	6.54
Nylon 66 (Zytel)	0	4.78
	13	5.47
	33	6.18
	43	6.77
Polypropylene (GP)	0	0.98
	20	2.74
	30	3.10
SAN	0	1.74
	20	3.22
	30	3.55
ABS	0	1.82
	20	3.29
	30	3.56
Polyphenylene oxide (Mod)	0	4.47
	20	5.57
	30	6.08
Polycarbonate	0	4.90
	20	8.39
	30	8.63
Polyester	0	4.64
	20	5.66
	30	5.77
Polysulfone	0	13.44
	20	16.25
	30	15.92
Polystyrene (GP)	0	1.13
	20	2.60
	30	2.91
Aluminum SAE-309 (360)—ingot	—	5.72–5.77

Source: E. I. du Pont de Nemours & Company, Incorporated, October 1978.

and various gears and bearings. The advent of the stamping materials bodes well for the extension of GFTP applications to larger parts and components.

REFERENCES

1. *Plast. Technol.* **24**, No. 6 (May 1978).
2. W. V. Titow and B. J. Lanham, *Reinforced Thermoplastics* Applied Science Publishers Ltd., London, 1975.
3. J. Milewski, *Plast. Eng.* (November 1978).
4. T. Ferrigno, *Plast. Eng.* (November 1978).
5. R. Bradt, U.S. Patent 2,877,501 (assigned to Fiberfil, Inc.), March 1959.
6. J. Englehardt et al., 22nd Annual Conference, Reinforced Plastics Division, SPI, 1967, Section 10-E.
7. R. Richards and D. Sims, *Composites* **2** (December 1971).
8. A. Bernardo, *SPE J.* **26** (1970).
9. L. Goettler, Monsanto Research Corporation, HPC 72-149, December 1972.
10. P. Mallick and L. Broutman, 29th Annual Conference, Reinforced Plastics/Composites Institute, SPI, 1974, Section 13-B.
11. E. Plueddemann, and G. Stark, 32nd Annual Conference, Reinforced Plastics/Composites Institute, SPI, 1977, Section 4-C.
12. E. Plueddemann, 28th Annual Conference, Reinforced Plastics/Composites Institute, SPI, 1973, Section 21-E.
13. B. Blumentritt, B. Vu, and S. Cooper, *Composites* **6** (May 1975).
14. L. Goettler, Monsanto Research Corporation, HPC 69-91, December 1969.
15. B. Blumentritt, *Polym. Eng. Sci.* **14** (1974).
16. *Plastics World*, Rosato Plastics Seminar.
17. Fiberfil Division, Dart Industries, Inc., Catalog No. 475-7854.
18. J. Titus, Plastec Special Report 9, May 1971.
19. *Zytel Design Handbook A*-78648, E. I. du Pont de Nemours & Company, Inc.
20. J. Theberge et al., 30th Annual Conference, Reinforced Plastics/Composites Institute, SPI, 1975, Section 17-E.
21. J. Theberge, B. Arkles, and P. Cloud, 27th Annual Conference, Reinforced Plastics/Composites Institute, SPI, 1972, Section 14-C.
22. J. Theberge, *Mod. Plast.* (June 1968).
23. J. Halpin and J. Kardos, *Polym. Eng. Sci.* **18**, No. 6 (mid-May 1978).
24. P. Petit and M. E. Waddoups, *J. Composite Materials* **3** (January 1969).
25. K. Jerina and J. Halpin, AFML-TR-72-148, October 1972.
26. R. Lavengood, *Polym. Eng. Sci.* **12**, No. 1 (January 1972).

8
FIBERGLASS REINFORCEMENT

Charles E. Knox
Technical Director
Uniglass Industries
New York, New York

8.1. INTRODUCTION

For over 3500 years, mankind has been aware of the fact that molten glass could be drawn into fine lengths (which were originally used for both making and decorating ornamental glass objects). Late in the 19th century, it was theorized that glass drawn into very fine fibers would be suitable for use in various textile applications. Although experimental glass fibers blended with silk fibers were woven into novel dresses and gowns in France and the United States, commercial fiberglass did not become a reality until 1939 with the formation of Owens-Corning Fiberglass Corporation (an outgrowth of research efforts by Owens-Illinois and Corning Glass Works).

From these humble beginnings, textile fiberglass has grown into a multimillion dollar industry. However, textile fiberglass only accounts for approximately 27% of all fiberglass production, the balance consisting of fiberglass insulation products.

8.2. FIBERGLASS PRODUCTION

Two forms of fiberglass can be produced—continuous fiber and staple (discontinuous) fiber. Both forms are made by the same manufacturing process up until fiber drawing.

Silica sand, limestone, boric acid, and other ingredients (e.g., clay, coal, and fluorospar) are dry mixed and melted in a high-refractory furnace. The temperature of this melt varies for each glass composition, but is generally about 2300°F (1260°C). The molten glass flows directly to the fiber-drawing furnace in the *direct-melt process*, whereas it flows into a marble-making machine in the *marble process* (the marbles are subsequently remelted and drawn into fibers). Most fiberglass is currently produced by the direct-melt process, which is illustrated schematically in Fig. 8.1.

Continuous fibers are produced by introducing molten glass into the fiber-drawing furnace (a platinum alloy tank called a *bushing*), where it is gravity fed through a multiplicity of tiny orifices in the base of the bushing. The droplets of molten glass extruding from each orifice (see Figs. 8.2 and 8.3) are gathered together, mechanically attenuated to the proper dimensions, passed through a light water spray (quench), and passed over a belt that applies a protective, lubricating binder or size to the individual fibers (*filament*). The filaments are then gathered together in a suitably shaped shoe to form a bundle of filaments called a *strand*. The fiberglass strand is then wound onto a tube at speeds up to 2 miles (3.2 km)/min. This "cake" is then conditioned or dried prior to further processing into sellable products.

Staple fibers are produced by passing a jet of air across the orifices in the base of the bushing, thus pulling individual fibers 8–15 in. (20.3–38.1 cm) long from the molten glass extruding from each orifice. These filaments are collected on a rotating vacuum drum, sprayed with a binder, and gathered into a strand. Again, conditioning or drying is re-

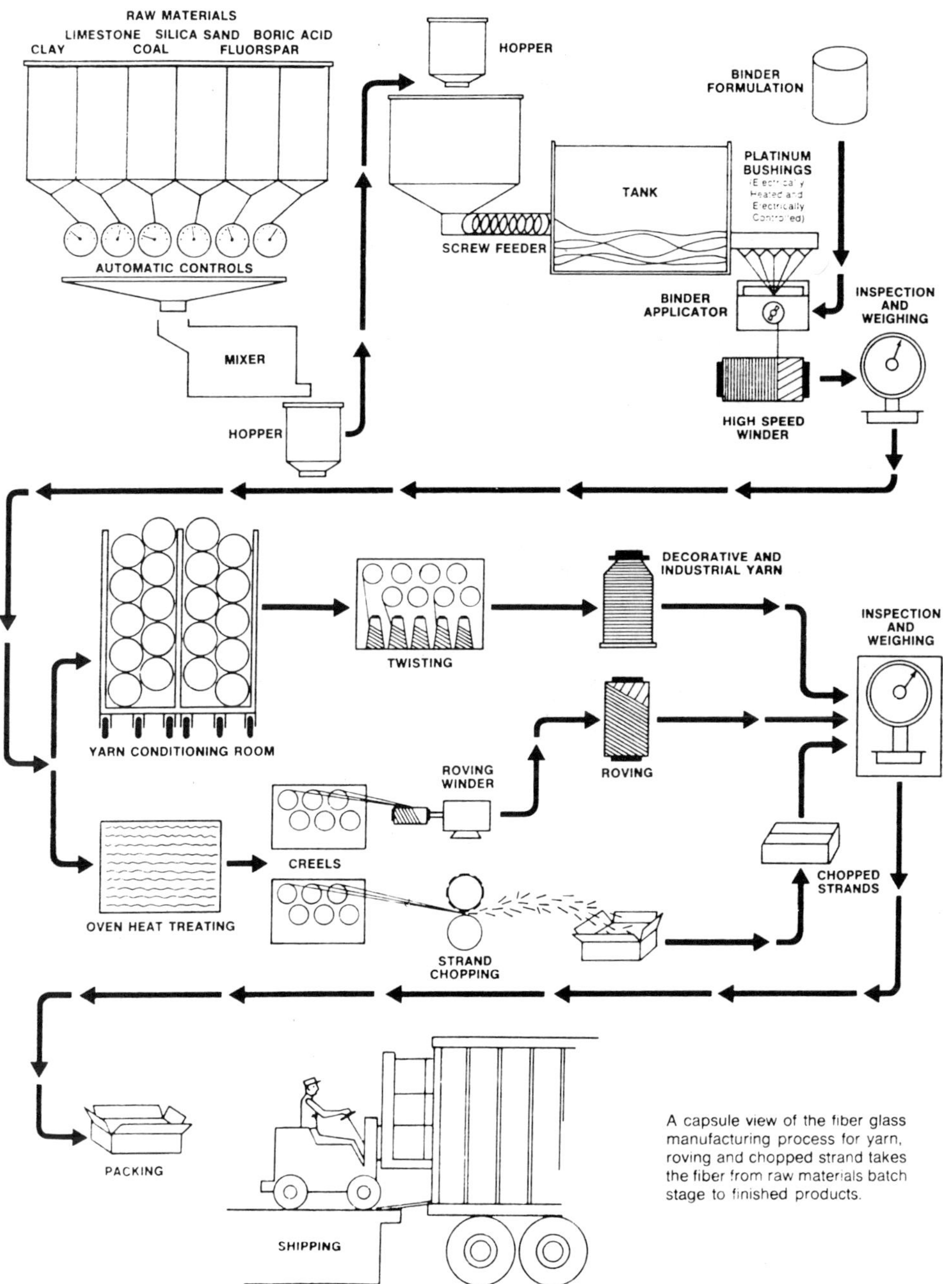

Figure 8.1. Direct-melt fiberglass-manufacturing process.

quired prior to subsequent processing into a sellable product.

Each fiber drawn from a bushing orifice must be controlled in order to obtain reproducible filament and strand dimensions and properties. This control is achieved by regulating the viscosity and temperature of the molten glass as well as the drawing speed (whether mechanical or air blown). Therefore, it is possible to produce a great number of filament diameters by varying the number of orifices in the bushing and the drawing conditions.

Over the years, the fiberglass industry has

Figure 8.2. Molten-glass flows from tiny orifices in platinum bushings. *Courtesy of PPG Industries.*

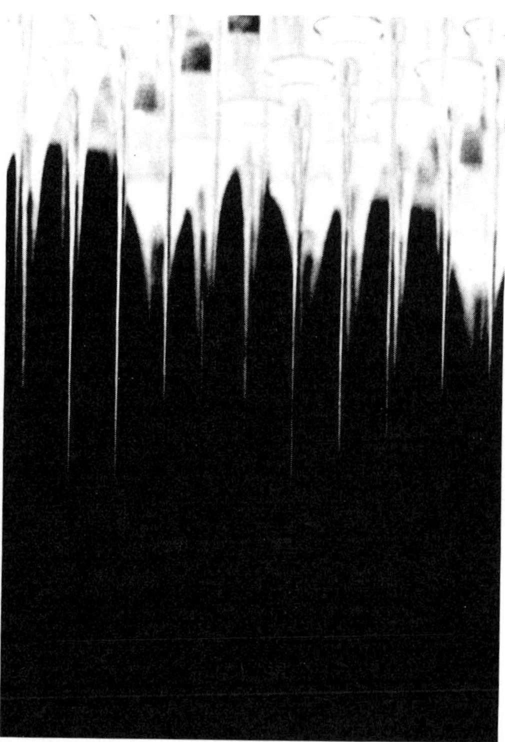

Figure 8.3. Molten-glass flows from tiny orifices in platinum bushings. *Courtesy of Owens-Corning Fiberglass Corporation.*

established a number of standard filament diameters; these are listed and defined in Table 8.1. The micron (μm) equivalents have been rounded off for proposed use in the International System of Units (SI).

8.3. GLASS COMPOSITION

Glass is an amorphous material that is neither solid nor liquid; it does not possess either the crystalline structure of solids or the flow characteristics of liquids. Chemically, glass is comprised primarily of a silica (SiO_2) backbone in the form of an $(-SiO_4-)_n$ polymer. However, silica by itself—that is, quartz—requires an extremely high temperature for liquefaction and drawing. Therefore, modifiers are needed to reduce temperatures to workable levels as well as to obtain molten-glass viscosities suitable for drawing. These modifiers are selected for their contribution to both glass properties and manufacturing capability.

Table 8.1. Fiberglass Filament Designations.

FILAMENT DESIGNATION	FILAMENT DIAMETER in. × 10^{-4}	FILAMENT DIAMETER μm
B	1.5	3.8
C	1.8	4.5
D	2.1	5
DE	2.5	6
E	2.9	7
G	3.6	9
H	4.2	10
K	5.1	13

High-alkali glass (otherwise known as soda or bottle glass) is the most common composition; it is generally used in the manufacture of containers and plate glass. A high-alkali composition (soda-lime silica) known as *A-glass* is drawn into fibers for applications where good chemical resistance is advantageous.

On the other hand, a high alkali content in the glass is detrimental to electrical properties;

hence, a low-alkali composition (alumino-borosilicate) known as *E-glass* which exhibits excellent electrical insulation properties was developed. At present, E-glass constitutes the majority of textile fiberglass production.

For special applications in which neither A-glass nor E-glass fibers are suitable, compositions have been tailored to meet the desired performance characteristics. Where extremely good chemical resistance is desired, *C-glass* (sodium borosilicate) fibers can be used. For applications requiring high tensile strength (such as in structural composites for the aircraft/aerospace industry), *S-glass* (magnesium aluminosilicate) fibers have been produced. The tensile strenth (in single-fiber form) of S-glass is approximately 40% higher than that of E-glass, thus resulting in higher-strength composites. Furthermore, at elevated temperatures, S-glass will retain significantly higher tensile strengths than E-glass (see Table 8.3). The S-glass fibers are available in a high-quality-performance grade and in a moderate-cost/performance grade (S2-glass).

Typical compositions for the above glass types are given in Table 8.2.

Several specialty glass compositions have been developed to take advantage of specific properties; however, they have not become commercial fiberglass products.

The *M-glass* composition was formulated to yield fibers having a high modulus of elasticity (16.4×10^6 psi = 113 GPa). Unfortunately, the presence of beryllia (beryllium oxide) in the glass prevented its commercialization.

The low-dielectric *D-glass* composition was developed for high-performance electronic applications. Its low dielectric constant (3.8) relative to that of E-glass (5.9) would be advantageous in such applications as radomes.

The L-glass (lead glass) composition has the advantage of radiation protection; hence, it is suitable for use in x-ray technologists' gowns and as a tracer yarn in composites where nondestructive x-ray examination can verify fiber alignment.

8.4. FIBERGLASS PROPERTIES

Composition is the primary determinant of fiberglass properties; however, various properties are also governed by the thermal history of the glass. The successful growth of fiberglass in many applications can be directly attributed to its inherent properties (see Table 8.3):

High Tensile Strength. Fiberglass has a very high tensile strength that exceeds those of other textile fibers. Its strength-to-weight ratio makes it stronger than steel wire in some applications.

Heat and Fire Resistance. As a result of its inorganic nature, fiberglass neither burns nor supports combustion. The high liquidus temperature of fiberglass means that it is able to withstand elevated-temperature environments.

Table 8.2. Fiberglass Compositions (wt. %).

	GRADE OF GLASS			
COMPONENTS	A (HIGH ALKALI)	C (CHEMICAL)	E (ELECTRICAL)	S (HIGH STRENGTH)
Silicon oxide	72.0	64.6	54.3	64.2
Aluminum oxide	0.6	4.1	15.2	24.8
Ferrous oxide	—	—	—	0.21
Calcium oxide	10.0	13.2	17.2	0.01
Magnesium oxide	2.5	3.3	4.7	10.27
Sodium oxide	14.2	7.7	0.6	0.27
Potassium oxide	—	1.7	—	—
Boron oxide	—	4.7	8.0	0.01
Barium oxide	—	0.9	—	0.2
Miscellaneous	0.7			

Table 8.3. Properties of Fiberglass.

PROPERTY	GRADE OF GLASS A	C	E	S
Physical Properties				
Specific gravity	2.50	2.49	2.54	2.48
Mohs hardness	—	6.5	6.5	6.5
Mechanical Properties				
Tensile strength, psi × 10^3 (MPa)				
At 72°F (22°C)	440 (3033)	440 (3033)	500 (3448)	665 (4585)
At 700°F (371°C)	—	—	380 (2620)	545 (3758)
At 1000°F (538°C)	—	—	250 (1724)	350 (2413)
Tensile modulus of elasticity at 72°F (22°C), psi × 10^6 (GPa)	—	10.0 (69.0)	10.5 (72.4)	12.4 (85.5)
Yield elongation, %	—	4.8	4.8	5.7
Elastic recovery, %	—	100	100	100
Thermal Properties				
Coefficient of thermal linear expansion, in/in/°F × 10^{-6} (m/m/°C)	4.8 (8.6)	4.0 (7.2)	2.8 (5.0)	3.1 (5.6)
Coefficient of thermal conductivity, Btu-in./hr/sq ft/°F (watt/motor °K)	—	—	72 (10.4)	—
Specific heat at 72°F (22°C)	—	0.212	0.197	0.176
Softening point, °F (°C)	1340 (727)	1380 (749)	1545 (841)	—
Electrical Properties				
Dielectric strength, V/mil	—	—	498	—
Dielectric constant at 72°F (22°C)				
At 60 Hz	—	—	5.9–6.4	5.0–5.4
At 10^6 Hz	6.9	7.0	6.3	5.1
Dissipation (power) factor at 72°F (22°C)				
At 60 Hz	—	—	0.005	0.003
At 10^6 Hz	—	—	0.002	0.003
Volume resistivity at 72°F (22°C) and 500 V DC, Ω-cm	—	—	10^{15}	10^{16}
Surface resistivity at 72°F (22°C) and 500 V DC, Ω-cm	—	—	10^{13}	10^{14}
Optical Properties				
Index of refraction	—	—	1.547	1.523
Acoustical Properties				
Velocity of sound, ft/sec (m/sec)	—	—	17,500 (5330)	19,200 (5850)

Chemical Resistance. Fiberglass is not attacked or degraded by most chemicals. It is also not affected by fungus, bacteria, or insect attack.

Moisture Resistance. Fiberglass does not absorb moisture; therefore, it does not swell, stretch, or disintegrate. Fiberglass resists rot and retains its maximum mechanical strength in humid environments.

Thermal Properties. Fiberglass has a low coefficient of thermal linear expansion and a high coefficient of thermal conductivity; hence, excellent performance in thermal environments (especially where rapid heat dissipation is desired) is ensured.

Electrial Properties. Since it is nonconductive, fiberglass is ideal for use in electrical insulation, where advantage can be taken of its high dielectric strength and low dielectric constant.

The physical, mechanical, thermal, and electrical properties of A-glass, C-glass, E-glass, and S-glass are presented in Table 8.3. Individual applications usually take advantage of one or more of these properties. For example, the aircraft/aerospace composites employed in radomes efficiently utilize the high-strength contribution of fiberglass reinforcement as well as its electrical properties. On the other hand, printed-circuit boards combine the excellent electrical properties and superior dimensional stability offered by the presence of fiberglass to ensure continuity of the printed wiring under adverse environmental and operating conditions.

Many applications that use fiberglass as a resin-matrix reinforcement require maximum performance and property retention in high-moisture environments. Hence, E-glass fibers are given prime consideration because of their excellent resistance to attack by water (E-glass fibers show a weight loss of 1.7% when exposed to boiling water for 1 hr as compared to weight losses of 0.13% for C-glass and 11.1% for A-glass. Although C-glass shows lower weight loss to 1 hr exposure than E-glass, weight loss with time is greater for C-glass than E-glass; this results in greater degradation of C-glass properties. Since composite performance is time dependent, E-glass, showing less degradation with time, is the preferred reinforcement. The high tensile strength and low dielectric constant of E-glass are other important factors).

On the other hand, both E-glass and A-glass are leached by acids and alkalies, whereas C-glass has excellent resistance to such attack. Thus, C-glass fibers are the material of choice in such environments as battery separators.

8.5 FIBERGLASS FORMS

8.5.1. Fiberglass Roving

Fiberglass roving is a collection of parallel, continuous strands or filaments. Conventional rovings are produced by winding together the number of single strands necessary to achieve the required yield (i.e., number of yards of roving weighing 1 lb or 0.454 kg). Single-strand roving, as the name implies, consists of a single strand of fiberglass filaments. These filaments are drawn from a bushing that contains the proper number of orifices to result in the single strand having the required yield.

Rovings are generally made with either G or K filaments (see Table 8.1), although coarser diameters are also available. Roving yields vary from 1800 to 225 yd/lb (3600 to 450 m/kg) (276 to 2222 Tex*).

8.5.2 Woven Roving

Many rovings are woven into a heavy, coarse-weave fabric for applications that require rapid thickness build-up over large areas. This characteristic is especially useful in the manufacture of fiberglass boats, various marine products, and many types of tooling.

Woven rovings are available in weights of from 12 to 40 oz/sq yd (407 to 1356 g/sq m) and thicknesses of from 0.02 to 0.04 in. (0.51 to 1.02 mm). The readily available woven rovings are listed in Table 8.4.

Fiberglass woven rovings can be combined with promoted thermosetting polyester resins by means of the hand lay-up fabrication technique typically employed in the manufacture of boats. Laminates prepared in this manner have low glass-to-resin ratios and somewhat limited cross-linking of the resin during polymerization; hence, the mechanical properties of these laminates are not fully developed.

Improved mechanical properties can be achieved by reducing the resin content and using a heat-curable polyester resin system. In general, fire-retardant polyester resin systems yield lower mechanical strengths than general-purpose polyester resin systems. Typical woven roving–polyester resin laminate properties are given in Table 8.4.

8.5.3. Fiberglass Mat

There are three basic forms of fiber glass mat: chopped-strand mat, continuous-strand mat, and surfacing mat or veil.

Chopped-strand mat is a nonwoven ma-

*Tex = a unit of fiber fineness assessed by the weight in grams of 1000 m of yarn; the lower the number, the finer the yarn.

Table 8.4. Woven Roving and Polyester Laminate Data.

WOVEN ROVING CONSTRUCTIONS AND PROPERTIES

COUNT per in. (per cm)	WEIGHT oz./sq. yd. (g/m^2)	THICKNESS in. (mm)	WEAVE
5 × 8 (2 × 3.2)	18.0 (610)	0.031 (0.787)	Plain
5 × 8 (2 × 3.2)	24.0 (814)	0.038 (0.965)	Plain
5 × 6 (2 × 2.4)	30.0 (1020)	0.049 (1.24)	Plain
5 × 8 (2 × 3.2)	36.0 (1220)	0.052 (1.32)	Plain

LAMINATE PROPERTIES—24 OUNCE (814 g) WOVEN ROVING/POLYESTER RESIN

Resin Type[a]	FR	FR	GP
Resin Content, wt. %	49.0	45.6	42.5
Thickness, in. (mm)	0.299 (7.59)	0.265 (6.73)	0.263 (6.68)
Cure[b]	RT	P + T	P + T
Flexural Strength, psi (MPa)			
Cond. A[c]	34,200 (235.8)	49,800 (343.4)	73,100 (504.0)
Cond. D2/100[d]	38,800 (267.5)	44,400 (306.1)	59,600 (410.9)
Flexural Modulus, psi (GPa)			
Cond. A	2.06×10^6 (14.2)	2.41×10^6 (16.6)	2.92×10^6 (20.0)
Cond. D2/100	2.12×10^6 (14.6)	2.31×10^6 (15.9)	2.83×10^6 (19.5)
Compressive Strength, psi (MPa)			
Cond. A	29,700 (204.8)	25,700 (177.2)	47,300 (326.1)
Cond. D2/100	27,300 (188.2)	25,200 (173.8)	40,100 (276.5)
Tensile Strength, psi (MPa)			
Cond. A	42,000 (289.6)	48,900 (337.2)	51,400 (354.4)
Cond. D2/100	40,600 (279.9)	47,400 (326.8)	50,400 (347.5)

[a]FR = Fire Retardant; GP = General Purpose.
[b]RT = Room Temperature; P + T = Pressure and Elevated Temperature.
[c]Cond. A = as-received condition.
[d]Cond. D2/100 immersed for 2 hr in water at 100°C.

terial in which the fiberglass strands from cake or roving are chopped into 1–2-in. (25.4–50.8-mm) lengths, evenly distributed at random onto a horizontal plane, and bound together with an appropriate chemical binder. These mats weigh from 0.75 to 3 oz/sq ft (229 to 916 g/sq m) and are available in widths of from 2 to 76 in. (50.8 to 1930.4 mm).

Continuous-strand mat consists of unchopped continuous strands of fiberglass deposited and interlocked in a spiral fashion. This mat is open and springy, but due to its mechanical interlocking does not require much binder for adequate handling strength.

Surfacing mat or veil is a very thin mat of single, continuous filaments used as a decorative surface-reinforcing layer in hand lay-up or press-molding processes to minimize telegraphing of the primary reinforcement through to the finished surface of a component.

8.5.4. Textile Fiberglass Yarn

A yarn is an assemblage of fibers or strands which can suitably be woven into textile materials. The continuous, individual strand as it comes from the bushing represents the simplest form of textile fiberglass yarn and is referred to as a "singles" yarn. In order for this yarn to be properly and efficiently utilized in a weaving operation, additional strand integrity is introduced by twisting it slightly, that is, usually less than 1 turn/in. (40 turns/m).

However, many woven fabrics require heavier yarns than can be drawn from the bushing. These can be produced by combining single strands via twisting and plying operations. Typically, this simply involves twisting two or more single strands together and subsequently plying (i.e., twisting two or more of the twisted strands together).

A yarn or strand has what is known as an "S" twist if, when held in a vertical position, the spirals conform in slope to the central position of the letter S; it has a "Z" twist if the spirals conform in slope to the central portion of the letter Z. Strands that are simply twisted (greater than 1 turn/in. = 40 turns/m) will kink, corkscrew, and unravel because of their twist in one direction only. The plying operation normally eliminates this problem by countering the twist in the twisted "singles" yarn with an opposite twist in the plied yarn. For example, "singles" yarns having a "Z" twist are plied with an "S" twist, thus resulting in a "balanced" yarn. The twisting and plying operations permit the yarn strength, diameter, and flexibility to be varied, and are important steps in producing the variety of fabrics which composite fabricators require.

8.5.5. Textured Yarns

Textile fiberglass yarns ("singles" or plied) can be subjected to a jet of air impinging on their surface, which causes random but controlled breakage of surface filaments and a general fluffing of the yarn surface. This is referred to as *texturizing* or *bulking* of the yarn, and the degree to which it occurs is controlled by the air pressure and yarn feed rate. Although the mechanical damage to the surface filaments weakens the yarn, its bulkiness allows greater resin absorption than is the case with continuous-filament yarn. The presence of texturized yarn in woven fabrics is advantageous when low glass-to-resin ratios and maximum fiber strength contributions are desired.

8.5.6. Yarn Nomenclature

The wide variety of fiberglass yarns that can be produced makes it necessary to have an exact system for yarn identification. Fiberglass yarn nomenclature is based on both alphabetical and numerical designations.

The first letter of the alphabetical designation identifies the glass composition (see Table 8.2), the second letter specifies the filament type (C = continuous, S = staple, and T = texturized), and the third (and fourth) letter identifies the filament diameter (see Table 8.1).

The first series of numbers in the numerical designation represents 1/100th of the basic strand yield, whereas the second series, which resembles a fraction, specifies the number of single strands twisted together (the "numerator") and the number of the twisted yarns plied together (the "denominator"). The total number of basic strands in a plied yarn is determined by multiplying these two digits (0 being multiplied as 1), whereas its yield is obtained by dividing the basic strand yield by the total number of strands in the yarn. A third number combined with either an "S" or "Z" termination will sometimes be included as part of the numerical designation. This specifies the final number of turns per inch and the direction of twist in the yarn.

For example, a yarn designated as ECG 150 4/2 3.8S contains E-glass continuous filaments of G diameter with a basic strand yield of 15,000 yd/lb (33 Tex). Four basic strands of ECG 150 1/0 are twisted together ("Z" twist) to form ECG 150 4/04.0Z. Plying two of these strand together (using the "S" twist to create balance) results in an ECG 150 4/2 3.8S yarn. Thus, this yarn contains 8 (4 × 2) basic 150 strands with a bare glass yield of 1875 (15,000 ÷ 8) yd/bl (264 Tex).

The nomenclature, (including designations in the Tex system) and typical properties (including SI values) of commercially available weaving yarns are listed in Table 8.5.

8.5.7 Fiberglass Fabric

The properties and contribution to product performance of fiberglass fabric are dependent on the fabric construction, that is, fabric count, warp-yarn and filling-yarn construction, and weave pattern.

Fabric count consists of the number of warp

Table 8.5. Commercially Available Fiberglass Weaving Yarns.

YARN DESIGNATION		YIELD		MINIMUM BREAKING STRENGTH	
GLASS SYSTEM	TEX SYSTEM	yd/lb	TEX	lb	N
E-Glass Yarns					
ECD 1800 1/0	EC5 2.75 1 × 0	180,000	2.75	0.25	1.11
ECD 1800 1/2	EC5 2.75 1 × 2	90,000	5.5	0.5	2.22
ECD 900 1/0	EC5 5.5 1 × 0	90,000	5.5	0.5	2.22
ECD 900 1/2	EC5 5.5 1 × 2	45,000	11	1.1	4.89
ECD 450 1/0	EC5 11 1 × 0	45,000	11	1.1	4.89
ECD 450 1/2	EC5 11 1 × 2	22,500	22	2.2	9.79
ECD 450 1/3	EC5 11 1 × 3	15,000	33	3.3	14.7
ECD 450 2/2	EC5 11 2 × 2	11,250	44	4.4	19.6
ECD 450 3/2	EC5 11 3 × 2	7,500	66	6.6	29.4
ECD 225 1/0	EC5 22 1 × 0	22,500	22	2.2	9.78
ECD 225 1/2	EC5 22 1 × 2	11,250	44	4.4	19.6
ECD 225 1/3	EC5 22 1 × 3	7,500	66	6.6	29.4
ECD 225 2/2	EC5 22 2 × 2	5,625	88	8.8	39.1
ECD 225 2/3	EC5 22 2 × 3	3,750	132	14.4	64.1
ECDE 150 1/0	EC6 33 1 × 0	15,000	33	3.5	15.6
ECDE 150 1/2	EC6 33 1 × 2	7,500	66	7.0	31.1
ECG 150 1/0	EC9 33 1 × 0	15,000	33	3.5	15.6
ECG 150 1/2	EC9 33 1 × 2	7,500	66	7.0	31.1
ECG 150 1/3	EC9 33 1 × 3	5,000	99	9.0	40.0
ECG 150 2/2	EC9 33 2 × 2	3,750	132	12.0	53.4
ECG 150 2/3	EC9 33 2 × 3	2,500	198	18.0	80.1
ECG 150 2/4	EC9 33 2 × 4	1,875	264	24.0	107
ECG 150 3/3	EC9 33 3 × 3	1,667	297	27.0	120
ECG 150 4/4	EC9 33 4 × 4	938	528	48.0	214
ECDE 75 1/0	EC6 66 1 × 0	7,500	66	5.7	25.4
ECG 75 1/0	EC9 66 1 × 0	7,500	66	5.7	25.4
ECG 75 1/2	EC9 66 1 × 2	3,750	132	11.4	50.7
ECG 75 1/3	EC9 66 1 × 3	2,500	198	17.1	76.1
ECG 75 2/2	EC9 66 2 × 2	1,875	264	22.8	101
ECG 75 2/3	EC9 66 2 × 3	1,250	396	34.2	152
ECG 75 2/4	EC9 66 2 × 4	938	528	45.6	203
ECH 55 1/0	EC10 90 1 × 0	5,500	90	9.5	42.3
ECDE 37 1/0	EC6 134 1 × 0	3,700	134	11.2	49.8
ECG 37 1/0	EC9 134 1 × 0	3,700	134	11.2	49.8
ECG 37 1/2	EC9 134 1 × 2	1,850	268	22.8	101
ECG 37 1/3	EC9 134 1 × 3	1,230	403	34.2	152
ECH 25 1/0	EC10 198 1 × 0	2,500	198	17.0	75.6
ECK 18 1/0	EC13 275 1 × 0	1,800	275	23.0	102
S-Glass Yarns					
SCD 450 1/2	SC5 11 1 × 2	22,500	22	—	—
SCG 150 1/2	SC9 33 1 × 2	7,500	66	8.2	36.5
SCG 150 2/2	SC9 33 2 × 2	3,750	132	16.4	72.9
Textured Yarns					
ETDE 150 1/0	ET6 33 1 × 0	14,200	35	1.24	5.52
ETDE 75 1/0	ET6 66 1 × 0	7,100	70	2.2	9.79
ETDE 37 1/0	ET6 134 1 × 0	3,450	144	4.7	20.9

yarns ("ends") per inch (centimeter) of fabric width and the number of filling yarns ("picks") per inch (centimeter) in the lengthwise direction. The warp yarn ("end") is the yarn lying in the lengthwise (machine) direction of the fabric, whereas the filling yarn ("pick") is the yarn lying in the crosswise direction of the fabric (i.e., at right angles to the warp yarn). Therefore, fabric weight, thickness, and breaking strength are proportional to the number and types of yarn used in weaving. This can be realized by examination of the fabric constructions in Table 8.6.

There are a variety of weave patterns that can be used to interlace the warp and filling yarns so as to form a stable fabric. The weave pattern controls the handling characteristics of a fabric and (to some degree) the properties of the product using it as reinforcement. Some applications require that all fabric-construction variables be specifically designed into the fabric so that the desired performance criteria can be met.

The *plain weave*, in which one warp yarn interlaces over and under one filling yarn (and vice versa), demonstrates the greatest degree of stability with respect to yarn slippage and fabric distortion; however, this stability is also a function of fabric count and yarn content.

The *basket weave* has two or more warp yarns interlacing over and under two or more filling yarns. Although the basket weave is less stable than the plain weave, it is more pliable and will conform more readily to simple contours.

The *twill weave* interlaces one or more warp yarns over and under two or more filling yarns in a regular pattern. This produces either a straight or a broken diagonal line in the fabric, which, consequently, has greater pliability and better drapability than both plain-weave and basket-weave fabrics.

A *crowfoot satin weave* has one warp yarn interlacing over three and under one filling yarn in an irregular pattern. This results in a pliable fabric that is capable of conforming to complex or compound contours.

The *8-shaft satin weave* has one warp yarn interlacing over seven and under one filling yarn in an irregular pattern. This yields a very pliable fabric that will readily conform to compound contours. Since this weave pattern will allow comparatively high fabric counts, it contributes maximum strength to composites in all directions.

Fabrics woven with heavy warp yarns and fine filling yarns in either the crowfoot or long-shaft satin weave patterns are called *unidirectional fabrics*. These fabrics are characterized by a high strength contribution to composites in the heavy-yarn direction.

Nonwoven unidirectional fabrics can be produced by chemically bonding the "warp" and "filling" yarns rather than interlacing them. Although the chemical bonding contributes to the stability of these nonwoven products, they tend to be somewhat firm and do not readily conform to complex or compound contours.

8.5.8. Other Woven Forms

Fiberglass yarns are also woven into tapes, contoured fabrics, fluted-core fabrics, and three-dimensional fabrics.

Tapes are narrow (less than 12 in. or 30.5 cm wide) woven fabrics which may contain a woven *feathered* edge (i.e., the filling yarns protrude beyond the exterior warp yarns.

Contoured fabrics are woven in a given geometrical shape that often resembles the shape of the component for which they are to serve as reinforcement. These fabrics are woven on looms specially designed for this purpose.

Fluted-core fabrics comprise two parallel layers of fabric tied together by stringers of woven fabric such that the cross-sectional configuration is triangular or rectangular.

Three-dimensional fabrics are not, in the true sense, three-dimensional; rather, they are planar fabrics woven with yarns in three distinct directions within the fabric plane; that is, yarns are interwoven in (1) the machine direction, (2) +45° from the machine direction, and (3) −45° from the machine direction.

8.5.9. Milled Fibers

Continuous fiberglass strands can be hammer-milled into very short fiber lengths (generally

Table 8.6. Glass-Fabric Construction and Properties.

STYLE	COUNT/in. (cm)	WARP YARN, GLASS SYSTEM (TEX SYSTEM)	FILLING YARN GLASS SYSTEM (TEX SYSTEM)	WEAVE	WEIGHT, oz/sq yd (g/sq m)	THICKNESS, in. (mm)	BREAKING STRENGTH, lb./in. (N)
104	60 × 52 (23.6 × 20.5)	ECD 900 1/0 (EC5 5.5 1 × 0)	ECD 1800 1/0 (EC5 2.75 1 × 0)	Plain	0.58 (19.7)	0.0012 (0.030)	40 × 15 (350 × 130)
106	56 × 56 (22.0 × 22.0)	ECD 900 1/0 (EC5 5.5 1 × 0)	ECD 900 1/0 (EC5 5.5 1 × 0)	Plain	0.73 (24.7)	0.0015 (0.038)	45 × 40 (394 × 350)
107	60 × 35 (23.6 × 13.8)	ECD 900 1/2 (EC5 5.5 1 × 2)	ECD 900 1/0 (EC5 5.5 1 × 0)	Plain	1.05 (35.6)	0.0017 (0.043)	70 × 20 (613 × 175)
108	60 × 47 (23.6 × 18.5)	ECD 900 1/2 (EC5 5.5 1 × 2)	ECD 900 1/2 (EC5 5.5 1 × 2)	Plain	1.44 (48.8)	0.0020 (0.051)	70 × 40 (613 × 350)
112	40 × 39 (15.7 × 15.4)	ECD 450 1/2 (EC5 11 1 × 2)	ECD 450 1/2 (EC5 11 1 × 2)	Plain	2.12 (71.9)	0.003 (0.076)	82 × 80 (718 × 700)
113	60 × 64 (23.6 × 25.2)	ECD 450 1/2 (EC5 11 1 × 2)	ECD 900 1/2 (EC5 5.5 1 × 2)	Plain	2.47 (83.7)	0.003 (0.076)	123 × 60 (1077 × 525)
116	60 × 58 (23.6 × 22.8)	ECD 450 1/2 (EC5 11 1 × 2)	ECD 450 1/2 (EC5 11 1 × 2)	Plain	3.16 (107)	0.004 (0.102)	125 × 120 (1095 × 1050)
120	60 × 58 (23.6 × 22.8)	ECD 450 1/2 (EC5 11 1 × 2)	ECD 450 1/2 (EC5 11 1 × 2)	Crowfoot	3.16 (107)	0.004 (0.102)	125 × 120 (1095 × 1050)
128	42 × 32 (16.6 × 12.6)	ECD 225 1/3 (EC5 22 1 × 3)	ECD 225 1/3 (EC5 22 1 × 3)	Plain	6.00 (203)	0.007 (0.178)	250 × 200 (2185 × 1750)
143	49 × 30 (19.3 × 11.8)	ECD 225 3/2 (EC5 22 3 × 2)	ECD 450 1/2 (EC5 11 1 × 2)	Crowfoot	8.78 (298)	0.009 (0.229)	600 × 60 (5250 × 525)
162	28 × 16 (11.0 × 3.2)	ECD 225 2/5 (EC5 22 2 × 5)	ECD 225 2/5 (EC5 22 2 × 5)	Plain	12.2 (414)	0.015 (0.381)	450 × 350 (3940 × 3065)
164	20 × 18 (7.9 × 7.1)	ECD 225 4/3 (EC5 22 4 × 3)	ECD 225 4/3 (EC5 22 4 × 3)	Plain	12.6 (427)	0.016 (0.406)	500 × 450 (4375 × 3940)
181	57 × 54 (22.4 × 21.3)	ECD 225 1/3 (EC5 22 1 × 3)	ECD 225 1/3 (EC5 22 1 × 3)	Satin	8.90 (302)	0.009 (0.229)	350 × 340 (3065 × 2975)
341	30 × 49 (11.8 × 19.3)	ECD 450 1/2 (EC5 11 1 × 2)	ECD 225 3/2 (EC5 22 3 × 2)	Crowfoot	8.78 (298)	0.009 (0.229)	50 × 600 (438 × 5250)
1070	60 × 35 (23.6 × 13.8)	ECD 450 1/0 (EC5 11 1 × 0)	ECD 900 1/0 (EC5 5.5 1 × 0)	Plain	1.05 (35.6)	0.0017 (0.043)	70 × 20 (613 × 175)
1080	60 × 47 (23.6 × 18.5)	ECD 450 1/0 (EC5 11 1 × 0)	ECD 450 1/0 (EC5 11 1 × 0)	Plain	1.44 (48.8)	0.002 (0.051)	70 × 40 (613 × 350)
1125	40 × 39 (15.7 × 15.4)	ECD 450 1/2 (EC5 11 1 × 2)	ECG 150 1/0 (EC5 33 1 × 0)	Plain	2.60 (88.1)	0.0035 (0.089)	90 × 130 (788 × 1135)
1165	60 × 52 (23.6 × 20.5)	ECD 450 1/2 (EC5 11 1 × 2)	ECG 150 1/0 (EC9 331 × 0)	Plain	3.66 (124)	0.0043 (0.109)	150 × 135 (1310 × 1180)

Notes. 1. All fabrics listed in this table which contain ECD 225 (EC5 22) yarn can also be woven with the original ECE 225 (EC7 22) yarn equivalents. 2. Breaking strength of fabrics in SI units (N) was determined by testing a specimen of 5 cm width.

Table 8.6. Continued.

STYLE	COUNT/in. (cm)	WARP YARN, GLASS SYSTEM (TEX SYSTEM)	FILLING YARN, GLASS SYSTEM (TEX SYSTEM)	WEAVE	WEIGHT, oz/sq yd (g/sq m)	THICKNESS, in. (mm)	BREAKING STRENGTH, lb/in. (N)
1522	24 × 22 (9.4 × 8.7)	ECG 150 1/2 (EC9 33 1 × 2)	ECG 150 1/2 (EC9 33 1 × 2)	Plain	3.70 (125)	0.0055 (0.140)	160 × 135 (1400 × 1180)
1523	28 × 20 (11.0 × 7.9)	ECG 150 3/2 (EC9 33 3 × 2)	ECG 150 3/2 (EC9 33 3 × 2)	Plain	11.9 (403)	0.014 (0.356)	525 × 400 (4595 × 3500)
1526	34 × 32 (13.4 × 12.6)	ECG 150 1/2 (EC9 33 1 × 2)	ECG 150 1/2 (EC9 33 1 × 2)	Plain	5.45 (185)	0.0065 (0.165)	225 × 195 (1970 × 1705)
1527	17 × 17 (6.7 × 6.7)	ECG 150 3/3 (EC9 33 3 × 3)	ECG 150 3/3 (EC9 33 3 × 3)	Plain	12.9 (437)	0.015 (0.381)	500 × 485 (4375 × 4245)
1528	44 × 32 (17.3 × 12.6)	ECG 150 1/2 (EC9 33 1 × 2)	ECG 150 1/2 (EC9 33 1 × 2)	Plain	6.00 (203)	0.007 (0.178)	250 × 200 (2185 × 1750)
1543	49 × 30 (19.3 × 11.8)	ECG 150 2/2 (EC9 33 2 × 2)	ECD 450 1/2 (EC5 11 1 × 2)	Crowfoot	8.78 (298)	0.009 (0.229)	600 × 60 (5250 × 525)
1557	57 × 30 (22.4 × 11.8)	ECG 150 1/2 (EC9 33 1 × 2)	ECD 225 1/0 (EC5 22 1 × 0)	Crowfoot	5.42 (184)	0.0055 (0.140)	370 × 60 (3240 × 525)
1564	20 × 18 (7.9 × 7.1)	ECG 150 4/2 (EC9 33 4 × 2)	ECG 150 4/2 (EC9 33 4 × 2)	Plain	12.7 (431)	0.016 (0.406)	500 × 450 (4375 × 3940)
1581	57 × 54 (22.4 × 21.3)	ECG 150 1/2 (EC9 33 1 × 2)	ECG 150 1/2 (EC9 33 1 × 2)	Satin	8.92 (302)	0.0085 (0.216)	340 × 330 (2975 × 2885)
1582	60 × 56 (23.6 × 22.0)	ECG 150 1/3 (EC9 33 1 × 3)	ECG 150 1/3 (EC9 33 1 × 3)	Satin	13.4 (454)	0.014 (0.356)	525 × 500 (4595 × 4375)
1583	54 × 48 (21.3 × 18.9)	ECG 150 2/2 (EC9 33 2 × 2)	ECG 150 2/2 (EC9 33 2 × 2)	Satin	16.8 (570)	0.018 (0.457)	650 × 590 (5690 × 5165)
1584	44 × 35 (17.3 × 13.8)	ECG 150 4/2 (EC9 33 4 × 2)	ECG 150 4/2 (EC9 33 4 × 2)	Satin	25.4 (861)	0.027 (0.686)	950 × 800 (8315 × 7005)
1588	42 × 36 (16.5 × 14.2)	ECG 150 4/4 (EC9 33 4 × 4)	ECG 150 4/4 (EC9 33 4 × 4)	Satin	51.8 (1756)	0.050 (1.27)	1900 × 1350 (16,635 × 11,820)
1674	40 × 32 (15.7 × 12.6)	ECG 150 1/0 (EC9 33 1 × 0)	ECG 150 1/0 (EC9 33 1 × 0)	Plain	2.82 (95.6)	0.0042 (0.107)	140 × 95 (1225 × 832)
1675	40 × 32 (15.7 × 12.6)	ECDE 150 1/0 (EC6 33 1 × 0)	ECDE 150 1/0 (EC6 33 1 × 0)	Plain	2.82 (95.6)	0.0042 (0.107)	140 × 95 (1225 × 832)
1676	56 × 48 (22.0 × 18.9)	ECDE 150 1/0 (EC6 33 1 × 0)	ECDE 150 1/0 (EC6 33 1 × 0)	Plain	4.10 (139)	0.0048 (0.122)	195 × 150 (1705 × 1310)
1677	40 × 40 (15.7 × 15.7)	ECDE 150 1/0 (EC6 33 1 × 0)	ECDE 150 1/0 (EC6 33 1 × 0)	Plain	3.20 (108)	0.0045 (0.114)	140 × 130 (1225 × 1135)
1680	72 × 70 (28.3 × 27.6)	ECDE 150 1/0 (EC6 33 1 × 0)	ECDE 150 1/0 (EC6 33 1 × 0)	Satin	5.70 (193)	0.006 (0.152)	275 × 225 (2405 × 1970)
1681	56 × 36 (22.0 × 14.2)	ECDE 150 1/0 (EC6 33 1 × 0)	ECDE 150 1/0 (EC6 33 1 × 0)	Plain	3.60 (122)	0.005 (0.127)	195 × 110 (1705 × 963)

Table 8.6. Continued.

STYLE	COUNT/in. (cm)	WARP YARN, GLASS SYSTEM (TEX SYSTEM)	FILLING YARN, GLASS SYSTEM (TEX SYSTEM)	WEAVE	WEIGHT, oz/sq yd (g/sq m)	THICKNESS, in. (mm)	BREAKING STRENGTH, lb/in. (N)
1800	16 × 14 (6.3 × 5.5)	ECK 18 1/0 (EC13 275 1 × 0)	ECK 18 1/0 (EC13 275 1 × 0)	Plain	9.66 (327)	0.014 (0.356)	450 × 410 (3940 × 3590)
2112	40 × 39 (15.7 × 15.4)	ECD 225 1/0 (EC5 22 1 × 0)	ECD 225 1/0 (EC5 22 1 × 0)	Plain	2.10 (71.2)	0.0034 (0.086)	90 × 80 (788 × 700)
2113	60 × 56 (23.6 × 22.0)	ECD 225 1/0 (EC5 22 1 × 0)	ECD 450 1/0 (EC5 11 1 × 0)	Plain	2.38 (80.7)	0.0032 (0.081)	140 × 60 (1225 × 525)
2116	60 × 58 (23.6 × 22.8)	ECD 225 1/0 (EC5 22 1 × 0)	ECD 225 1/0 (EC5 22 1 × 0)	Plain	3.16 (107)	0.004 (0.102)	125 × 120 (1095 × 1050)
2120	60 × 58 (23.6 × 22.8)	ECD 225 1/0 (EC5 22 1 × 0)	ECD 225 1/0 (EC5 22 1 × 0)	Crowfoot	3.16 (107)	0.004 (0.102)	125 × 120 (1095 × 1050)
2125	40 × 39 (15.7 × 15.4)	ECD 225 1/0 (EC5 22 1 × 0)	ECG 150 1/0 (EC9 33 1 × 0)	Plain	2.62 (88.8)	0.0037 (0.094)	90 × 130 (788 × 1135)
2165	60 × 52 (23.6 × 20.5)	ECD 225 1/0 (EC5 22 1 × 0)	ECG 150 1/0 (EC9 33 1 × 0)	Plain	3.70 (125)	0.0045 (0.114)	125 × 140 (1095 × 1220)
2523	28 × 20 (11.0 × 7.9)	ECH 25 1/0 (EC10 198 1 × 0)	ECH 25 1/0 (EC10 198 1 × 0)	Plain	11.9 (403)	0.013 (0.330)	580 × 385 (5075 × 3370)
2532	16 × 14 (6.3 × 5.5)	ECH 25 1/0 (EC10 198 1 × 0)	ECH 25 1/0 (EC10 198 1 × 0)	Plain	7.25 (246)	0.010 (0.254)	300 × 280 (2625 × 2450)
3732	48 × 32 (18.9 × 12.6)	ECG 37 1/0 (EC9 134 1 × 0)	ECG 37 1/0 (EC9 134 1 × 0)	Crowfoot	12.7 (431)	0.0134 (0.340)	571 × 423 (5000 × 3700)
3733	18 × 18 (7.1 × 7.1)	ECG 37 1/0 (EC9 134 1 × 0)	ECG 37 1/0 (EC9 134 1 × 0)	Plain	5.80 (197)	0.008 (0.203)	250 × 200 (2185 × 1750)
3743	49 × 30 (19.3 × 11.8)	ECG 37 1/0 (EC9 134 1 × 0)	ECD 225 1/0 (EC5 22 1 × 0)	Crowfoot	8.45 (286)	0.008 (0.203)	600 × 60 (5250 × 525)
7500	16 × 14 (6.3 × 5.5)	ECG 75 2/2 (EC9 66 2 × 2)	ECG 75 2/2 (EC9 66 2 × 2)	Plain	9.66 (327)	0.014 (0.356)	450 × 410 (3940 × 3590)
7532	16 × 14 6.3 × 5.5	ECG 75 1/3 (EC9 66 1 × 3)	ECG 75 1/3 (EC9 66 1 × 3)	Plain	7.50 (254)	0.010 (0.254)	335 × 316 (2930 × 2765)
7533	18 × 18 (7.1 × 7.1)	ECG 75 1/2 (EC9 66 1 × 2)	ECG 75 1/2 (EC9 66 1 × 2)	Plain	5.70 (193)	0.008 (0.203)	230 × 220 (2010 × 1925)
7544	28 × 14 (11.0 × 5.5)	ECG 75 2/2 (EC9 66 2 × 2)	ECG 75 2/2 (EC9 66 2 × 2)	2 × 1 BSK.	18.00 (610)	0.022 (0.559)	745 × 830 (6520 × 7265)
7587	42 × 21 (16.5 × 8.3)	ECG 75 2/2 (EC9 66 2 × 2)	ECG 75 2/2 (EC9 66 2 × 2)	Mock leno	20.8 (705)	0.028 (0.711)	965 × 525 (8450 × 4600)
7626	34 × 32 (13.4 × 12.6)	ECG 75 1/0 (EC9 66 1 × 0)	ECG 75 1/0 (EC9 66 1 × 0)	Plain	5.40 (183)	0.0066 (0.168)	225 × 200 (1970 × 1750)

Table 8.6. Continued.

STYLE	COUNT/in. (cm)	WARP YARN, GLASS SYSTEM (TEX SYSTEM)	FILLING YARN, GLASS SYSTEM (TEX SYSTEM)	WEAVE	WEIGHT, oz/sq yd (g/sq m)	THICKNESS, in. (mm)	BREAKING STRENGTH, lb/in. (N)
7628	44 × 32 (17.3 × 12.6)	ECG 75 1/0 (EC9 66 1 × 0)	ECG 75 1/0 (EC9 66 1 × 0)	Plain	6.00 (203)	0.007 (0.178)	250 × 200 (2190 × 1750)
7637	44 × 22 (17.3 × 8.7)	ECG 75 1/0 (EC9 66 1 × 0)	ECG 37 1/0 (EC9 134 1 × 0)	Plain	7.09 (240)	0.009 (0.229)	250 × 250 (2190 × 2190)
7743	120 × 20 (47.2 × 7.9)	ECDE 75 1/0 (EC6 66 1 × 0)	ECG 150 1/0 (EC9 33 1 × 0)	Satin	10.22 (346)	0.011 (0.279)	800 × 60 (7000 × 525)
7781	60 × 54 (23.6 × 21.3)	ECDE 75 1/0 (EC6 66 1 × 0)	ECDE 75 1/0 (EC6 66 1 × 0)	Satin	8.92 (302)	0.0085 (0.216)	340 × 330 (2980 × 2890)
76281	44 × 32 (17.3 × 12.6)	ECG 75 1/0 (EC9 66 1 × 0)	ECG 75 1/0 (EC9 66 1 × 0)	Crowfoot	6.00 (203)	0.0068 (0.173)	250 × 200 (2190 × 1750)

1/64–1/4 in. (0.40–6.35 mm). The actual lengths are determined by the diameter of the screen openings through which the fibers pass during milling.

Milled fibers are used as inert fillers for a variety of thermoplastic and thermosetting resins.

8.6 SURFACE TREATMENTS

8.6.1. Rovings

In order for fiberglass rovings to be compatible with processing methods and materials, a chemical size is applied during the basic fiber-forming operation. This size must be carefully formulated so as to hold individual filaments together, lubricate the roving for contact with processing equipment, and allow the glass filaments to be thoroughly wetted when combined with other materials.

The vast majority of fiberglass roving sizes employ polyvinyl acetate as a film former and are modified with methacrylic chromic chloride complex and/or organosilane coupling agents, antistatic agents, and various lubricants in order to impart desired strand characteristics (e.g., choppability, abrasion resistance, and resin wetout).

The film former acts to adhere the filaments together while imparting strand integrity so as to prevent filamentary abrasion during fiber drawing, strand conversion into rovings, and use of rovings by fabricators.

The coupling agent contains one or more groups that can react chemically with the hydroxyl groups on the glass surface as well as one or more groups that can react with other materials, especially thermoplastic and thermosetting resins. Thus, the coupling agent creates a chemical bridge between the glass surface and the resin matrix, which leads to stress transfer from the comparatively weak matrix (lower Young's modulus) to the stronger and stiffer reinforcement.

The antistatic agents and lubricants are selected to impart such desired handling characteristics as hardness (softness) and choppability to the roving.

Selection of a roving–size combination must be directed toward the intended processing method, resin compatibility, and end-use performance.

A majority of continuous rovings are used in processes where reinforcement must be chopped into relatively short lengths (0.5–2 in. = 12.7–50.8 mm) in order to develop uniform strength patterns. This form is used in spray-up, preform molding, continuous laminating, and sheet or bulk molding compounds. The roving must be capable of being chopped and dispersed with minimum ribbonization of the strand into individual filaments.

Rovings for chopping are available in varying degrees of softness (which is an indication of strand integrity). The softer rovings may be more difficult to chop, but are recommended for applications with intricate shapes and extremely sharp or difficult radii. Harder rovings are generally recommended for components with simple radii and in those cases where chopping performance is of prime importance. A single strand within the conventional roving can be dyed and hence act as a tracer in order to allow the monitoring of the chopped strands for uniformity of distribution.

Rovings used in filament winding or pultrusion must have good strand integrity, a controlled level of broken filaments, good wetting by the resin, and uniform processability under constant applied tension. In these applications, single-strand rovings offer some advantage over the more conventional ones.

The mechanical properties of filament-wound composites are determined by a test procedure developed at the Naval Ordnance Laboratory (NOL) when filament-wound NOL rings are used for the test. The typical mechanical properties of parallel wound NOL rings include flexural strengths of approximately 175,000 psi (1207 MPa) and flexural moduli of elasticity of 8.0×10^6 psi (55.2 GPa).[1] The S-glass rovings have found widespread use in applications that require high composite tensile strengths, such as pressure vessels made by filament winding. A special hollow fiberglass roving that enables weight reduction of fiberglass-reinforced composites is also available for filament winding.

8.6.2. Textile Yarns

Continuous fiberglass strands intended for weaving are treated at the bushing with a starch–oil binder. A typical starch–oil binder consists of partially or fully dextrinized starch or amylose, hydrogenated vegetable oil, cationic wetting agent, emulsifying agent, and water. Small amounts of a film former (e.g., gelatin or polyvinyl alcohol), a small amount of fungicide (to prevent fungus growth), and a trace of disinfectant (e.g., pine oil) may also be added.

These binders protect the fibers from damage by lubrication during their formation and such subsequent textile operations as twisting, plying, and weaving.

8.6.3. Fiberglass Fabrics

Loomstate fabrics containing the original yarn binders applied at the bushing are not compatible with thermosetting resins; hence, they are used primarily in combination with thermoplastic resins and elastomer coatings as well as with various varnishes for flexible electrical insulation materials.

When used in conjunction with thermosetting resins, the starch–oil binder acts as a barrier between the glass surface and the resin boundary. The hydrophilic character of the binder allows moisture to penetrate the glass–resin interface, which leads to degradation of composite properties in wet and humid environments. Therefore, the binder must be removed prior to combination with other materials. This is accomplished by exposing the fabric to carefully controlled time-temperature cycles, which allows efficient and complete removal of all organic matter. Unfortunately, fabrics in this heat-cleaned state manifest the weakness of incomplete interfacial bonding between the glass surface and the resin, which is especially important with respect to resistance to degradation in wet or humid environments. However, this weakness can be overcome by treating the fabric with such chemical coupling agents as the aforementioned methacrylic chromic chloride complex or organosilanes.

These coupling agents function by reacting at one end of the molecule with the hydroxyl groups on the glass surface and at the other end with the resin. In this way, a chemical bridge is formed between the glass surface and the resin matrix, thus imparting to the composite an improved resistance to moisture attack and degradation. Chemical functionality of the coupling agent determines the resistance to other environments (e.g., chemical and thermal). Table 8.7 lists and describes the commonly used coupling agents, whereas Table 8.8 shows the effect of coupling agents on the mechanical-strength properties of typical laminates (this effect is especially noticeable with respect to wet strength retention).

Table 8.7. Fiberglass-Fabric Finishes

1. Heat-Cleaned Finishes

112	Total removal of all organics by exposure to carefully controlled time–temperature cycles. Fabric is normally white in color.
111, 210	Partial removal of organics under carefully controlled conditions of oxidation. Fabric retains a high percentage of its original loomstate tensile strength and is normally tan in color.

2. Polyester-Resin-Compatible Finishes

Volan A	A methacrylic chromic chloride complex applied to heat-cleaned (112) fabric. It produces a light-green laminate and is also compatible with both epoxy and phenolic resins.
550	A blend of Volan A and silane coupling agents which gives a light-green, transparent laminate with improved physical properties (especially wet strength) as compared to Volan A. It is also compatible with epoxy resins.
A174, Z6030	A γ-methacryloxypropyltrimethoxysilane applied to heat-cleaned fabric. It produces clear laminates having superior mechanical properties and excellent wet strength retention.
A172, Garan, Z6075	Vinylsilanes applied to heat-cleaned fabric. They produce clear laminates having high mechanical strengths and good wet strength retention.

3. Epoxy-Resin-Compatible Finishes

Volan A	See description under *Polyester-Resin-Compatible Finishes.*
550	See description under *Polyester-Resin-Compatible Finishes.*
A1100	A γ-aminopropyltriethoxysilane applied to heat-cleaned fabric which is available in either a firm or a soft hand for improved drapabiity. A 1100 may exhibit shelf-life problems. It is also used with phenolic and polyimide resins.
A1120, Z6020	An N-(β-aminoethyl)-γ-aminopropyltrimethoxysilane that exhibits a slower reactivity with resins than A1100. It is also used with phenolic and polyimide resins.
A187, Z6040	A γ-glycidoxypropyltrimethoxysilane that produces clear laminates having high mechanical strengths and good resistance to moisture degradation. It is widely used in rigid electrical laminates for printed-circuit boards.
Z6032	A cationic styrene-functional amine hydrochloride silane that exhibits good resin wetout and resin-to-glass bonding. Its laminates exhibit superior mechanical properties, excellent wet strength retention, and good electrical and thermal properties.

4. Phenolic-Resin-Compatible Finishes

A1100	See description under *Epoxy-Resin-Compatible Finishes.*
A1120, Z6020	See description under *Epoxy-Resin-Compatible Finishes.*
Z6032	See description under *Epoxy-Resin-Compatible Finishes.*

5. Polyimide-Resin-Compatible Finishes

A1100	See description under *Epoxy-Resin-Compatible Finishes.*
A1120, Z6020	See description under *Epoxy-Resin-Compatible Finishes.*

6. S Glass Finishes

Table 8.7. Continued.

The S-glass rovings and textile yarns are available from the manufacturer (Owens-Corning Fiberglas Corporation) with epoxy-resin-compatible treatments applied as the strands are drawn from the bushing.

The 901 treatment is an epoxy-resin-compatible direct size that has a limited shelf life; hence, these materials must be transported and stored under controlled temperature–humidity conditions.

The shelf-life limitations of the 901 treatment are not evident in the 904 treatment, which is also an epoxy-resin-compatible direct size.

Fabrics woven with S-glass yarns are also available with silane finishes similar to those listed in this table.

The coupling agents listed in Sections 1-5 of this table are commercially available—either individually or in combinations—from fiberglass-fabric weavers under their proprietary finish numbering system as outlined below:

	COMMERCIAL FIBERGLASS-FABRIC WEAVER			
RESIN COMPATIBILITY	BI[a]	CS[b]	JPS[c]	UI[d]
Polyester + chrome	Volan A	Volan A	Volan A	Volan A
+ chrome/silane	I-550	CS-550	S-550	UM-550
+ silane	I-593	CS-271	S-910	UM-686
		CS-272A	Garan	
			A172	
Epoxy + chrome	Volan A	Volan A	Volan A	Volan A
+ chrome/silane	I-550	CS-550	S-550	UM-550
+ silane	A1100	A1100	A1100	A1100
	I-588	Z6040	S-920	UM-675
	I-589	CS-272A	S-935	UM-702
	I-399	CS-307		UM-718
		CS-344		UM-724
Phenolic + silane	A1100	A1100	A1100	A1100
	I-516		S-935	UM-696
	I-589			
Melamine + silane	A1100	CS-616	615	
	I-516		616	UM-616
	I-608			
Silicone + heat cleaned	112	112		npH[e]
	I-514 (npH)	npH		
+ silane	I-594			
Polyimide + silane	A1100	CS-290		A1100
	I-589	CS-309		UM-696

[a]BI = Burlington Industries, Burlington Glass Fabrics Division.
[b]CS = Clark Schwebel Fiber Glass Corporation.
[c]JPS = J. P. Stevens & Company, Incorporated.
[d]UI = Uniglass Industries, Division of United Merchants & Manufacturers, Incorporated.
[e]npH = neutral pH.

8.7 DESIGN CONSIDERATIONS

Since fiberglass-reinforced thermosetting and thermoplastic resins are engineering materials, the contribution of each component and the process by which they are combined determine the various property and performance criteria of the end product.

Until recently, the engineering and design considerations inherent to the combination of highly dissimilar materials into a composite engineering material were poorly understood.

Table 8.8. Effect of Coupling Agents on the Mechanical-Strength Properties of Typical Laminates.

STYLE 1581 GLASS FABRIC–POLYESTER RESIN					
Finish	Loomstate	112	Volan	A172	A174, Z6030
Flexural strength, psi × 10^3 (MPa)					
Cond. A	40.4 (278.6)	32.3 (222.7)	73.2 (504.7)	68.8 (474.4)	75.5 (520.6)
Cond. D2/100	25.3 (174.4)	31.8 (219.3)	59.0 (406.8)	66.1 (455.8)	77.3 (533.0)
Compressive strength, psi × 10^3 (MPa)					
Cond. A	23.8 (164.1)	40.4 (278.6)	48.0 (331.0)	47.1 (324.8)	66.2 (456.4)
Cond. D2/100	12.5 (86.2)	12.1 (83.4)	38.9 (268.2)	44.9 (309.6)	58.8 (405.4)
Tensile strength, psi × 10^3 (MPa)					
Cond. A	48.1 (331.6)	20.6 (142.0)	49.9 (344.1)	50.0 (344.8)	52.2 (359.9)
Cond D2/100			48.8 (336.5)	51.3 (353.7)	50.7 (349.6)
STYLE 1581 GLASS FABRIC–EPOXY RESIN					
Flexural strength, psi × 10^3 (MPa)					
Cond. A	78.2 (539.2)	76.0 (524.0)	80.9 (557.8)	71.3 (491.6)	85.7 (590.9)
Cond. D2/100	76.7 (528.8)	72.4 (499.2)	83.1 (573.0)	68.9 (475.1)	82.4 (568.1)
Compressive strength, psi × 10^3 (MPa)					
Cond. A	64.3 (443.3)	63.4 (437.1)	70.7 (487.5)	63.0 (434.4)	67.5 (465.4)
Cond. D2/100	58.5 (403.3)	54.1 (373.0)	64.7 (446.1)	62.9 (433.7)	59.5 (410.3)
Tensile strength, psi × 10^3 (MPa)					
Cond. A	56.9 (392.3)	48.8 (336.5)	57.8 (398.5)	50.9 (351.0)	58.2 (401.3)
Cond. D2/100	55.8 (384.7)	47.2 (325.4)	56.1 (386.8)	49.4 (340.6)	52.2 (359.9)

Fortunately, the wide use of fiberglass-reinforced plastics (FRP) as an engineering material during the past 20 years was not hampered by this limitation. However, recent investigations of "advanced composites" (i.e., resins reinforced with fiber forms other than fiberglass) have been translated into a greater understanding of the engineering and design capabilities of all forms of FRP.

However, there are basic considerations inherent to the use of fiberglass reinforcing agents in composites. The type, form, amount, and alignment of the fibers have a strong bearing on the physical and mechanical properties of a composite. Although the thermal and flammability properties of a composite are primarily determined by the resin matrix (e.g., brominated resins yield better fire resistance than nonbrominated resins), the chemical and/or electrical properties may be influenced by the resin matrix and/or the selected reinforcing agent.

8.7.1. Glass Composition

As can be seen from Tables 8.2 and 8.3, glass composition has a direct influence on fiber properties, which, in turn—based on the rule of mixtures—are translated into composite performance. Data on epoxy laminates reinforced with style 1581 glass fabric woven with various glass compositions (E-glass, S-glass, and an experimental[4] YM31A high-modulus-of-elasticity glass) are given in Table 8.9.[2]

It has been reported[2,3] that E-glass fiber contributes 210,000–225,000 psi (1450–1550 MPa) to the tensile strength of a composite, which is considerably lower than the 500,000-psi (3450-MPa) tensile strength of the virgin filaments (see Table 8.3). On the other hand, S-glass fiber contributes approximately 340,000 psi (2340 MPa) to the composite tensile strength, which reflects the higher tensile strength of the virgin filaments.

Similarly, an experimental 16×10^6-psi (110-GPa) modulus-of-elasticity (YM31A or M-glass) composition yields composites with a relatively higher modulus of elasticity (stiffness) than E-glass composites (see Table 8.9).

8.7.2. Filament Diameter

A variety of E-glass filament diameters have been evaluated with respect to their influence

Table 8.9. Fabric Construction and Epoxy-Laminate Mechanical Properties for Various Glass Compositions.[2]

FABRIC CONSTRUCTION			
Fabric no.	E1581	S1581	YM31A[a]
Count/in. (cm)	57 × 54 (22.4 × 21.3)	57 × 54 (22.4 × 21.3)	57 × 54 (22.4 × 21.3)
Warp yarn			
Glass system	ECG 150 1/2	SCG 150 1/2	MCG 150 1/2
Tex system	EC9 33 1 × 2	SC9 33 1 × 2	MC9 33 1 × 2
Filling yarn			
Glass system	ECG 150 1/2	SCG 150 1/2	MCG 150 1/2
Tex system	EC9 33 1 × 2	SC9 33 1 × 2	MC9 33 1 × 2
Weave	Long-shaft satin	Long-shaft satin	Long-shaft satin
LAMINATE MECHANICAL PROPERTIES			
Flexural strength, psi × 10^3 (MPa)			
Cond. A	88.6 (610.9)	87.1 (600.6)	89.8 (619.2)
Flexural modulus of elasticity, psi × 10^6 (GPa)			
Cond. A	3.78 (26.1)	3.79 (26.1)	5.00 (34.5)
Compressive strength, psi × 10^3 (MPa)			
Cond. A	64.8 (446.8)	56.4 (388.9)	61.0 (420.6)
Tensile strength, psi × 10^3 (MPa)			
Cond. A	60.0 (413.7)	61.4 (423.4)	64.5 (444.7)

[a]Data from Ref. 4.

on composite properties. The fabrics studied were woven with the same count and weave pattern as style 1581, using both plied and singles yarn containing different filament diameters as described in Table 8.10[2].

Filament diameter does not appear to play an important role in determining the mechanical properties of such plied-yarn fabrics as style 1581 [woven with ECG 150 1/2 (EC9 33 1 × 2) yarn] and style 1281 [woven with ECB 150 1/2 (EC3.8 33 1 × 2) yarn]. On the other hand, in comparison with commercially available singles-yarn style 7781 woven with ECDE 75 1/0 (EC6 66 1 × 0) yarn, singles-yarn style B7781 woven with finer filaments [ECB 75 1/0 (EC3.8 66 1 × 0) yarn] exhibits a higher flexural strength, flexural modulus of elasticity, and tensile strength, whereas singles-yarn style G7681 woven with coarser filaments [ECG 75 1/0 (EC9 66 1 × 0) yarn] exhibits a higher compressive strength.

Generally speaking, fabrics woven with singles yarn yield higher composite strengths than fabrics woven with plied yarns. The plied yarn versus singles yarn phenomenon also exists in S-glass fabrics (see Table 8.11[2]), the singles-yarn fabric again manifesting markedly higher composite strengths. This is attributed to the parallel alignment of filaments in singles yarn, which exposes more surface area for glass-to-resin bonding.

8.7.3. Weave Pattern

The handling characteristics of a fabric are determined by the amount of yarn and the weave pattern holding the yarns together. If the weave pattern is too tight, the fabric will not conform to various contours and will not accept resin, thus resulting in a weak composite. On the other hand, if the weave pattern is too open or loose, the fabric will not contain sufficient fiber to develop its maximum possible strength and will easily distort, hence resulting in an inability to align the fibers with preferred strength axes.

In general, the longer the float (i.e, the portion of a warp or filling yarn that extends unbound over two or more yarns lying 90° to it), the higher the composite strength. Thus, as shown in Table 8.12, extending the length of

Table 8.10. Composite Mechanical Properties as a Function of Glass-Filament Diameter.[2]

PLIED-YARN FABRICS			
Fabric style	181	1581	1281
Count/in. (cm)	57 × 54 (22.4 × 21.3)	57 × 54 (22.4 × 21.3)	57 × 54 (22.4 × 21.3)
Warp yarn			
Glass system	ECE 225 1/3	ECG 150 1/2	ECB 150 1/2
Tex system	EC7 22 1 × 3	EC9 33 1 × 2	EC3.8 33 1 × 2
Filling yarn			
Glass system	ECE 225 1/3	ECG 150 1/2	ECB 150 1/2
Tex system	EC7 22 1 × 3	EC9 33 1 × 2	EC3.8 33 1 × 2
Weave	Long-shaft satin	Long-shaft satin	Long-shaft satin
Flexural strength, psi × 10^3 (MPa)			
Cond. A	93.0 (641.2)	88.6 (610.9)	92.4 (637.1)
Flexural modulus of elasticity, psi × 10^6 (GPa)			
Cond. A	3.88 (26.8)	3.78 (26.1)	4.06 (28.0)
Compressive strength, psi × 10^3 (MPa)			
Cond. A	58.1 (400.6)	64.8 (446.8)	64.9 (447.5)
Tensile strength, psi × 10^3 (MPa)			
Cond. A	60.0 (413.7)	60.0 (413.7)	64.8 (446.8)
SINGLES-YARN FABRICS			
Fabric style	G7681	DE7781	B7781
Count/in. (cm)	60 × 54 (23.6 × 21.3)	60 × 54 (23.6 × 21.3)	60 × 54 (23.6 × 21.3)
Warm yarn			
Glass system	ECG 75 1/0	ECDE 75 1/0	ECB 75 1/0
Tex system	EC9 66 1 × 0	EC6 661 × 0	EC3.8 66 1 × 0
Filling yarn			
Glass system	ECG 75 1/0	ECDE 75 1/0	ECB 75 1/0
Tex system	EC9 66 1 × 0	EC6 66 1 × 0	EC3.8 66 1 × 0
Weave	Long-shaft satin	Long-shaft satin	Long-shaft satin
Flexural strength, psi × 10^3 (MPa)			
Cond. A	92.6 (638.5)	94.9 (654.3)	101.0 (696.4)
Flexural modulus of elasticity, psi × 10^6 (GPa)			
Cond. A	3.78 (26.1)	3.90 (26.9)	4.02 (27.7)
Compressive strength, psi × 10^3 (MPa)			
Cond. A	76.7 (528.8)	67.7 (466.8)	69.9 (482.0)
Tensile strength, psi × 10^3 (MPa)			
Cond. A	64.2 (442.7)	66.3 (457.1)	69.4 (478.5)

the float, which hence reduces the interlacing frequency, will increase the composite strength. Furthermore, the transition from a plain weave to a crowfoot weave also results in improved composite mechanical properties (see Table 8.12).

8.7.4. Glass-to-Resin Ratio

Since FRP tend to follow the rule of mixtures, it would be expected that composite properties are proportional to the quantity of each component present. In actual practice, this tendency does exist to the extent that the higher the glass-to-resin ratio, the greater the mechanical strength of the composite.

In filament winding, parallel strands of fiberglass are wound around a cylindrical mandrel; the consequent round filament cross section allows close packing which results in a very low resin-content capability. Therefore,

Table 8.11. Mechanical Properties of S-Glass Fabrics-Epoxy Composites as a Function of Plied versus Unplied Yarn.[2]

Fabric style	S1581	S7681	S7681
Finish	Volan A	Volan A	UM670
Flexural strength, psi × 10^3 (MPa)			
Cond. A	84.8 (584.7)	120.0 (827.4)	116.0 (799.8)
Flexural modulus of elasticity, psi × 10^6 (GPa)			
Cond. A	3.10 (21.4)	4.18 (28.8)	5.21 (35.9)
Compressive strength, psi × 10^3 (MPa)			
Cond. A	61.3 (422.7)	55.8 (384.7)	73.9 (509.5)
Tensile strength, psi × 10^3 (MPa)			
Cond. A	70.6 (486.8)	87.5 (603.3)	86.4 (595.7)

fiberglass loadings in the neighborhood of 80 wt % (i.e., resin contents of 20 wt % lead to high mechanical strengths, since the effective contribution of the fiberglass is not drastically diluted by the low contribution of the resin matrix.

However, any deviation from the parallel alignment of the glass fibers in filament winding reduces the degree of packing and hence (proportionally) the optimum glass-to-resin ratios achievable. Thus, the filament winding of fiberglass strands in a helical pattern results in higher practical resin contents than is the case for parallel-wound fibers, which hence slightly reduces the mechanical-strength level.

Woven fiberglass fabrics, by virtue of the weave interlacing, reduce the effective glass-to-resin ratios even further. The optimum me-

Table 8.12. Effect of Weave Pattern on Composite Mechanical Properties.

Fabric style	7628[a]	76281[b]	16-149[c]	7781[d]
Finish	Volan A	Volan A	Volan A	Volan A
Resin	Polyester	Polyester	Polyester	Polyester
Plies	18	18	12	12
Resin content, wt. %	37.1	36.7	36.5	37.6
Thickness, in. (mm)	0.124 (3.15)	0.121 (3.07)	0.120 (3.05)	0.120 (3.05)
Flexural strength, psi × 10^3 (MPa)				
Cond. A	53.8 (371.0)	84.7 (584.0)	63.2 (435.8)	87.0 (600.0)
Flexural modulus of elasticity, psi × 10^6 (GPa)				
Cond. A	3.89 (26.8)	3.41 (23.5)	3.81 (26.3)	3.24 (22.3)
Compressive strength, psi × 10^3 (MPa)				
Cond. A	25.7 (177.2)	57.0 (393.0)	48.0 (331.0)	64.3 (443.3)
Tensile strength, psi × 10^3 (MPa)				
Cond. A	45.9 (316.5)	59.2 (408.2)	58.7 (404.7)	60.0 (413.7)

[a]Plain weave; data from Ref. 3.
[b]Crowfoot satin weave; data from Ref. 3.
[c]A 5-shaft satin weave version of style 7781.
[d]An 8-shaft satin weave.

Table 8.13. Laminate Mechanical Properties as a Function of Glass-to-Resin Ratio (Style 181 Glass Fabric-Epoxy Resin Composites).

Plies	7	9	11	13	14	16	18	20
Resin content, wt. %	55.0	49.2	44.0	35.6	31.2	28.0	22.5	22.2
Thickness, in. (mm)	0.117 (2.97)	0.126(3.20)	0.127 (3.23)	0.126 (3.20)	0.125 (3.18)	0.131 (3.33)	0.132 (3.35)	0.147 (3.73)
Flexural strength, psi $\times 10^3$ (MPa)								
Cond. A	45.6 (314.4)	58.8 (405.4)	64.8 (446.8)	83.0 (572.3)	81.2 (559.9)	92.1 (635.0)	87.4 (602.6)	91.9 (633.7)
Flexural modulus of elasticity, psi $\times 10^6$ (GPa)								
Cond. A	2.26 (15.6)	2.62 (18.1)	2.92 (20.1)	3.33 (23.0)	3.82 (26.3)	4.04 (27.9)	4.37 (30.1)	4.64 (32.0)
Compressive strength, psi $\times 10^3$ (MPa)								
Cond. A	45.2 (311.7)	47.9 (330.3)	44.7 (308.2)	53.2 (366.8)	56.9 (392.3)	52.4 (361.3)	54.3 (374.4)	67.8 (467.5)
Tensile strength, psi $\times 10^3$ (MPa)								
Cond. A	30.8 (212.4)	35.7 (246.2)	40.5 (279.2)	51.3 (353.7)	53.7 (370.3)	61.0 (420.6)	64.7 (446.1)	65.1 (448.9)
Tensile load/ply, lb (N)	514 (2286)	494 (2197)	468 (2082)	498 (2215)	480 (2135)	500 (2224)	476 (2117)	480 (2135)

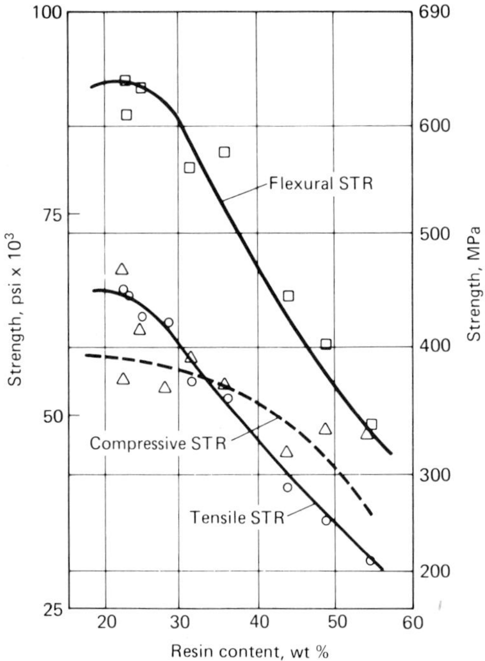

Figure 8.4. Laminate mechanical strength as a Function of resin content: ○) tensile strength; △) compressive strength; □) flexural strength.

chanical strengths of glass-fabric-reinforced composites are achieved at resin contents of 30–32 wt %, although a 36–38 wt % resin content is normally used.

The correlation between glass-to-resin ratio (expressed as weight percent of resin present) and laminate strength has been investigated[5] and it was shown that the mechanical strengths of laminates increased as resin content decreased from 55 to 22 wt % (see Table 8.13 and Fig. 8.4).

8.7.5. Fiber Distribution

Composites reinforced with glass strands aligned parallel to each other have their maximum mechanical strength and stiffness in the direction of strand alignment. Such parallel alignment is found in pultrusion and certain filament-winding applications.

When the reinforcement is aligned at right angles to itself, that is, half of the strands are laid at right angles to the other half (0° and 90°), the mechanical strength at either angle is less than that of the parallel alignment. As the distribution of the strands varies between the 0 and 90° angles, the mechanical strength varies proportionally.

Since woven fabrics are bidirectional in the 0 and 90° angles, they contribute to the mechanical strength of composites accordingly.

The rotation of alternate layers of reinforcement from the 0°, 90° alignment to a +45°, −45° alignment further reduces the mechanical strength in the primary direction, but increases it in the +45 and −45° directions as a result of the presence of reinforcing members in those directions.

However, the yarn distribution between the 0 and 90° directions can be varied as part of

Table 8.14. Effect of Fabric Yarn Distribution on Laminate Mechanical Properties.

Fabric type (style)	Bidirectional (1581)		Unidirectional (7743)	
Test direction	Warp	Filling	Warp	Filling
Yarn content, %	52	48	90	10
Flexural strength, psi × 10^3 (MPa)				
Cond. A	84.2 (580.6)	78.2 (539.2)	95.0 (655.0)	23.3 (160.6)
Cond. D2/100	75.5 (520.5)	68.5 (472.3)	87.4 (602.6)	24.1 (166.2)
Flexural modulus of elasticity, psi × 10^6 (GPa)				
Cond. A	4.02 (27.7)	3.65 (25.2)	4.95 (34.1)	2.47 (17.0)
Cond. D2/100	3.87 (26.7)	3.41 (23.5)	4.80 (33.1)	1.46 (10.1)
Compressive strength, psi × 10^3 (MPa)				
Cond. A	62.7 (432.3)	63.4 (437.1)	64.5 (444.7)	28.9 (199.3)
cond. D2/100	51.8 (357.2)	54.5 (375.8)	51.1 (352.3)	23.6 (162.7)
Tensile strength, psi × 10^3 (MPa)				
Cond. A	57.9 (399.2)	54.9 (378.5)	94.0 (648.1)	12.1 (83.4)
Cond. D2/100	55.7 (384.0)	50.2 (346.1)	91.6 (631.6)	11.7 (80.7)

the fabric design. A "balanced" fabric with equal yarn distribution in the warp and filling directions will have comparable (but not necessarily equal) laminate properties in those directions.

Weaving fiberglass yarn primarily in the warp direction and incorporating the minimum amount of filling yarn needed to give fabric stability results in a unidirectional fabric. Composites reinforced with unidirectional fabrics have their maximum mechanical strength in the direction corresponding to the greatest concentration of yarn.

The differences in laminate strength between the warp and filling directions in both bidirectional and unidirectional fabrics are illustrated by the data in Table 8.14.

Fiberglass reinforcement aligned in a random manner within the matrix (such as is found with chopped-strand mat) yields composites that exhibit relatively uniform mechanical strength in all directions (approaching isotropic characteristics) rather than a strength concentration in one or two directions. However, this leads to composites of relatively low mechanical strength in all directions, as illustrated by the data in Table 8.15.

Table 8.15. Chopped-Strand Mat-Polyester Resin Laminate Data

Resin content, wt. %	69.8
Thickness, in. (mm)	0.238 (6.05)
Flexural strength, psi (MPa)	
Cond. A	26,400 (182.0)
Cond. D2/100	30,400 (209.6)
Flexural modulus of elasticity, psi (GPa)	
Cond. A	0.99×10^6 (6.83)
Cond. D2/100	0.94×10^6 (6.48)
Compressive strength, psi (MPa)	
Cond. A	27,600 (190.3)
Cond. D2/100	23,200 (160.0)
Tensile strength, psi (MPa)	
Cond. A	14,100 (97.2)
Cond. D2/100	13,800 (95.1)

REFERENCES

1. C. F. Pitt and J. Harvey, 20th Anniversary Technical Conference, SPI Reinforced Plastics Division, 1965, Section 9-C.
2. C. E. Knox, *Non-Metallic Materials (SAMPE)* **4**, 127 (1972).
3. R. C. Horton and R. G. Adams, 21st Annual Conference, SPI Reinforced Plastics Division, 1966, Section 10-A.
4. C. E. Knox, *New Horizons in Materials and Processing (SAMPE)* **18**, 527 (1973).
5. G. P. Peterson, "Properties of High Modulus Reinforced Plastics," *SPE J.*, 57 (January 1961).

9
HIGH SILICA AND QUARTZ

Hugh Shulock and Richard R. Saffadi
Alpha Associates, Incorporated
Woodbridge, New Jersey

9.1. DEFINITIONS

9.1.1. High Silica

The term *high silica* can be used to describe any high-purity glass. However, for use in reinforced plastics, we shall define high silica as a high-purity glass of 95%-plus purity SiO_2 produced by a leaching process. High-silica fibers and fabrics are flexible materials that are similar in appearance to—and produced from—fiberglass. The fiberglass, with a silica content of 65%, is subjected to a hot-acid treatment that removes virtually all of the impurities while leaving the silica intact. This is commonly called the *leaching process*.

Most textile forms of glass (e.g., chopped fiber, mat, yarn, and fabric) are available in silica. Fabrics are available in various weights, thicknesses, and purities (from 95 to 99.4% SiO_2).

9.1.2. Quartz

The word *quartz* can also denote any high-purity glass, but we shall define quartz fibers as those fibers produced from high-purity (99.95% SiO_2) natural quartz crystals. The crystals are formed into rods from which filaments having one-fifth the diameter of human hair are drawn. Up to 240 filaments are combined to form a flexible, high-strength fiber that can be made into yarn and then woven into a fabric. All textile forms (including chopped fiber, mat, rovings, cordage, sleevings, tapes, and fabrics) are available in quartz grades. Figure 9.1 presents a flowchart for the production of quartz rods, yarn, fabric, roving, and mat.

Quartz fibers with a purity of 99.95% SiO_2 (exclusive of binder) retain virtually all of the characteristics and properties of solid quartz; however, they also have many of the properties of fiberglass and are extremely flexible.

9.2. HISTORY

9.2.1. High Silica

High silica was first developed by Owens-Corning Fiberglas Corporation, which held the original patents on the product. It was first commercially produced by HITCO Corporation (H. I. Thompson Corporation) under the trade name Refrasil.® High-silica fabrics are also produced under the trade names Sil Temp® (Haveg Corporation), Astrosil® (J. P. Stevens and Company, Incorporated), and Alphasil® (Alpha Associates, Incorporated).

9.2.2. Quartz

There are many fiber products that utilize the word quartz in their trade names. Since the word quartz can denote any high-silica glass product, these products are accurately named. However, they are not pure fused quartz.

Quartz fibers, like glass fibers, have been available for thousands of years. However, it was a French company, Quartz and Silice and Cie., that made the first commercially practical quartz fibers. Quartz fibers are now available in the United States from J. P. Stevens and

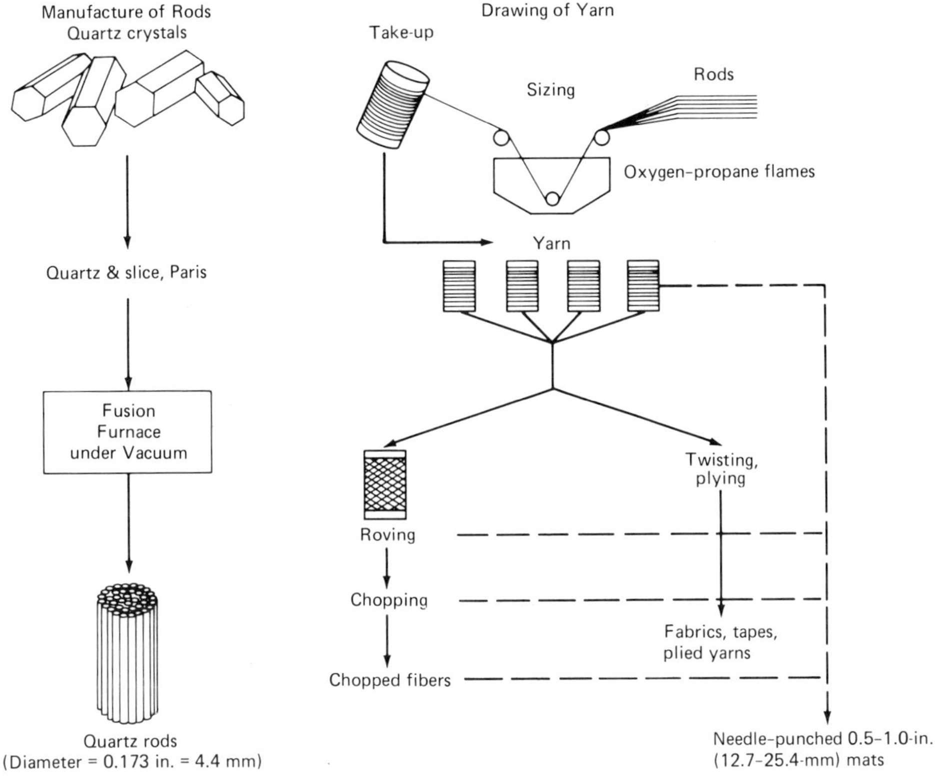

Figure 9.1. Flowchart for the production of quartz rods, yarn, roving, fabric, and mat.

Company, Incorporated, under the trade name Astroquartz,® General Electric Company, Quartz Lamp Department, and Alpha Associates, Incorporated, under the trade name Alphaquartz.®

Natural quartz crystals are found abundantly throughout the world. However, the high-purity crystals required for quartz fibers are relatively rare, being principally mined in Brazil.

9.3. FORMS

High-silica and quartz fibers are available in virtually all textile forms (see Table 9.1). However, as a result of their greater strength and flexibility, quartz fibers are more readily processed into a wide variety of textile types, weights, and thicknesses. In addition, a diversity of binders is available for use with quartz fibers; these binders are designed to aid in processing and are compatible with end-use requirements.

9.4. USES

High silica and quartz are both used in a wide variety of similar products. The selection of what type of raw material to use is generally dictated by a combination of performance requirements, manufacturing needs, and cost (see Sec. 9.5).

Table 9.2 lists the most widely known end uses and product recommendations, but should not be considered a complete listing. New products and end uses are being developed continually, and manufacturers should be contacted for the most current information.

9.5. COSTS AND PRICING

Sand is the principal raw material used to produce the fiberglass from which high silica is made; it is abundantly available and costs approximately 15 cents/lb (33 cents/Kg). Quartz fibers are produced from natural, pure fused-quartz crystals that cost approximately

Table 9.1. Available High-Silica and Quartz Forms and Their Sources.

FORM	SOURCE	
	HIGH SILICA	QUARTZ
Woven fabrics	Alpha Associates, Inc. Haveg Corporation Hitco Corporation J. P. Stevens and Company, Inc.	Alpha Associates, Inc. J. P. Stevens and Company, Inc.
Woven tapes	Hitco Corporation	Alpha Associates, Inc.
Yarn	Hitco Corporation	Alpha Associates, Inc. General Electric Company J. P. Stevens and Company, Inc.
Chopped fibers	Hitco Corporation	Alpha Associates, Inc. J. P. Stevens and Company, Inc.
Roving	—	General Electric Company J. P. Stevens and Company, Inc.
Cordage	Hitco Corporation	Alpha Associates, Inc.
Sewing thread	Hitco Corporation	Alpha Associates, Inc.
Mat	Haveg Corporation Hitco Corporation Johns-Manville Company	Alpha Associates, Inc. General Electric Company J. P. Stevens and Company, Inc. Johns-Manville Company
Wool	—	Alpha Associates, Inc. General Electric Company J. P. Stevens and Company, Inc.
Braided sleeving	Hitco Corporation	Alpha Associates, Inc.
Knit fabrics	—	Alpha Associates, Inc.
Knit tapes	—	Alpha Associates, Inc.

\$5.00/lb (\$11.00/Kg). The great disparity in the raw-material costs of high silica and quartz is not always reflected in the respective final product, since there are great differences in the manufacturing costs. High silica is sold in a price range of \$4.00–50.00/lb (\$8.80–110.00/Kg), depending on the form and quantity purchased. Quartz products are priced from \$45.00 to \$150.00/lb (\$99.00 to \$330.00/Kg), again depending on the quantity and form. It is sometimes less expensive in the final product to use the more expensive but stronger, more flexible, and more easily processed quartz than the less expensive high silica: The great difference in strength may make it possible to use considerably less of a quartz reinforcement as compared to a high-silica reinforcement. However, if thermal protection is all that is required, then high silica is the obvious choice.

It is suggested that product manufacturers be contacted for their product recommendations and current prices. Comparative prices of the most widely used products from the largest volume categories are shown in Table 9.3.

9.6. PHYSICAL AND MECHANICAL PROPERTIES

High-silica and quartz materials have higher strength-to-weight ratios than most other high-temperature materials; however, quartz has approximately five times the tensile strength of high silica. High-silica and quartz yarns are perfectly elastic, and their elongation at break is approximately 1%.

Quartz fibers retain virtually all of the characteristics and properties of solid quartz.

Table 9.2. End Uses and Product Recommendations for High Silica and Quartz.

END USE	PRODUCT RECOMMENDATION	
	HIGH SILICA	QUARTZ
Ablative		
Reentry heat shields		
Tape wrapped	Fabric—18.5 oz/sq yd (0.63 Kg/sq m)	Fabric—19.5 oz/sq yd (0.66 Kg/sq m)
Filament wound		Roving—20 end
Woven—three dimensional		Yarn—to order
Nose cones		
Compression molded	Fabric—18.5 oz/sq yd (0.63 Kg/sq m) —10.3 oz/sq yd (0.35 Kg/sq m)	Fabric—19.5 oz/sq yd (0.66 Kg/sq m) —8.4 oz/sq yd (0.28 Kg/sq m)
Filament wound		Roving—20 end
Woven—three dimensional		Yarn—to order
Nozzles and exit cones		
Tape wrapped	Fabric—18.5 oz/sq yd (0.63 Kg/sq m)	Fabric—19.5 oz/sq yd (0.66 Kg/sq m)
Filament wound	Cordage—⅛ in. (3.175 mm) diameter	Roving—20 end Cordage—⅛ in. (3.175 mm) diameter
Compression molded	Fabric—18.5 oz/sq yd (0.63 Kg/sq m) —10.3 oz/sq yd (0.35 Kg/sq m) Chopped squares—½-in. (12.7-mm) squares Chopped fiber—¼ in. (6.35 mm) length	Fabric—18.5 oz/sq yd (0.63 Kg/sq m) —8.4 oz/sq yd (0.28 Kg/sq m) Chopped squares—½-in. (12.7-mm) squares Chopped fiber—¼ in. (6.4 mm) length
Structural		
Fins and struts		
Lay-up (vacuum or autoclave molded)		Fabric—8.4 oz/sq yd (0.28 Kg/sq m)
Filament wound		Roving—20 end
Compression molded		Fabric—8.4 oz/sq yd (0.28 Kg/sq m)
Electrical		
High-pressure laminates		Fabric—5.6 oz/sq yd (0.19 Kg/sq m)
Antennas		
Lay-up (vacuum or autoclave molded)		Fabric—8.4 oz/sq yd (0.28 Kg/sq m)
Filament wound		Roving—20 end
Radomes		
Lay-up		Fabric—8.4 oz/sq yd (0.28 Kg/sq m)
Filament wound		Roving—20 end

Table 9.2. Continued.

END USE	PRODUCT RECOMMENDATION	
	HIGH SILICA	QUARTZ
Thermal		
Blast shields		
Compression molded	Fabric—18.5 oz/sq yd (0.63 Kg/sq m)	Fabric—19.5 oz/sq yd (0.66 Kg/sq m)
	Mat	Mat
	Chopped fiber—¼ in. (6.35 mm) length	Chopped fiber—¼ in. (6.35 mm) length
	Chopped squares—½-in. (12.7-mm) squares	Chopped squares—½-in. (12.7-mm) squares
Separators		
Compression molded	Fabric—18.5 oz/sq yd (0.63 Kg/sq m)	

Table 9.3. Comparative Prices of High-Silica and Quartz Products.

HIGH SILICA		QUARTZ	
PRODUCT DESCRIPTION	PRICE	PRODUCT DESCRIPTION	PRICE
Silica fabric—heavy	$8.00/sq yd ($9.57/sq m)	Quartz fabric—heavy	$75.00/sq yd ($89.70/sq m)
Weight—18.5 oz/sq yd (0.63 Kg/sq m)		Weight—19.5 oz/sq yd (0.66 Kg/sq m)	
Thickness—0.26 in. (6.6 mm)		Thickness—0.27 in. (6.9 mm)	
Strength—50 lb/in. (8800 N/m)		Strength—480 lb/in. (84,000 N/m)	
Silica fabric—light	$6.00/sq yd ($7.18/sq m)	Quartz fabric—light	$45.00/sq yd ($53.82/sq m)
Weight—10.3 oz/sq yd (0.35 Kg/sq m)		Weight—8.4 oz/sq yd (0.28 Kg/sq m)	
Thickness—0.013 in. (0.33 mm)		Thickness—0.11 in. (2.8 mm)	
Strength—30 lb/in. (5250 N/m)		Strength—185 lb/in. (32,375 N/m)	
Silica yard	$25.00/lb ($29.90/sq m)	Quartz yarn	$60.00/lb ($71.76/sq m)
Teflon binder		Teflon binder	
Yield—3000 yd/lb (6050 m/Kg)		Yield—7500 yd/lb (15,120 m/Kg)	
Average breaking strength—2.5 lb (11.1 N)		Average breaking strength—3.0 lb (13.3 N)	

Table 9.4. Physical and Mechanical Properties of High-Silica and Quartz Fibers.

PROPERTY	HIGH SILICA	QUARTZ
Filament diameter in. (mm)		
Yarn and fabric	0.0004 (0.010)	0.0004 (0.010)
Mat	0.00005–0.0004 (0.0013–0.010)	0.00005–0.0006 (0.0013–0.015)
Filament tensile strength, psi × 10^3 (MPa)		
At room temperature	—	130 (896)
At 400° F (204° C)	—	99 (682)
Specific gravity	1.74	2.2
Hardness—Moh's scale		5–6
Young's modulus, psi × 10^6 (GPa)	—	10 (6.89)

Table 9.5. Physical and Mechanical Properties of High-Silica and Quartz Yarns.

PROPERTY	HIGH SILICA		QUARTZ	
Type number	#100	#300-2/2	#300-2/4	#300-4/4
Nominal diameter, in. (mm)	0.020 (0.51)	0.005 (0.13)	0.010 (0.25)	0.014 (0.36)
Approximate yield, yd/lb (m/Kg)	3050 (6150)	7500 (15,100)	3750 (7560)	1875 (3780)
Minimum breaking strength, lb (N)	2.5 (11.1)	3.0 (13.3)	5.0 (22.2)	11.0 (48.9)
Linear shrinkage, %	12	1	1	1

Table 9.6. Physical and Mechanical Properties of High-Silica and Quartz Fabrics.

PROPERTY	HIGH SILICA		QUARTZ	
Type number	#82	#84	#581	#570
Weight, oz/sq yd (Kg/sq m)	10.3 (0.35)	18.5 (0.63)	8.4 (0.28)	19.5 (0.66)
Thickness, in. (mm)	0.013 (0.33)	0.026 (0.66)	0.011 (0.28)	0.027 (0.69)
Thread count/in. (cm)				
Warp	—	50 (20)	57 (22)	38 (15)
Filling	—	40 (16)	54 (21)	24 (9)
Breaking strength, lb (N)				
Warp	30 (133)	70 (311)	185 (823)	480 (2135)
Filling	25 (111)	45 (200)	170 (756)	400 (1779)
Weave	8H satin	8H satin	8H satin	8H satin
Maximum areal shrinkage, %	5	5	1	1
Moisture content, %	5	5	1	1
pH	4.0–7.0	4.0–7.0	6.0–8.0	6.0–8.0
Silica content, %	98.0–99.2	98.0–99.2	99.9+	99.9+

Tables 9.4–9.6 present the physical and mechanical properties of high-silica and quartz fibers, yarns, and fabrics. Table 9.7 provides a typical chemical analysis of high silica and quartz.

9.7. CHEMICAL PROPERTIES

High silica and quartz have similar chemical properties. They are not affected by halogens or common acids in the liquid or gaseous state with the exception of hydrofluoric and hot phosphoric acids. Silica products are not recommended for use with either hot or cold bases, although weak alkali solutions can be used for certain applications. Silica is not soluble in water or organic solvents.

9.8. THERMAL PROPERTIES

Since high silica and quartz are so similar in nature, they also have similar thermal characteristics. The major difference is the higher melt viscosity of quartz as a result of its higher silica content (see Fig. 9.2). High silica and quartz do not melt or vaporize until the temperature exceeds 3000° F (1649° C). At continuous temperatures in excess of 1800° F (982° C), both forms will begin to devitrify into a crystallized form known as cristobalite. This conversion tends to stiffen the material, but causes no change in its physical form or insulating properties.

High silica and quartz manifest excellent resistance to thermal shock. Silica products can be heated to 2000° F (1093° C) and rapidly quenched in water without any apparent change.

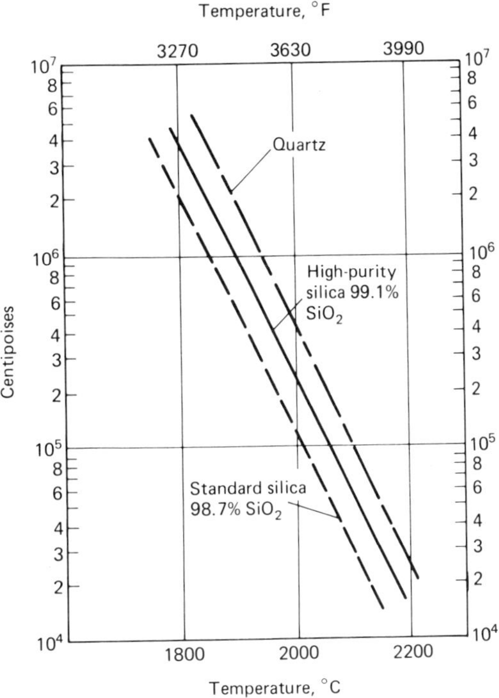

Figure 9.2. Melt viscosity versus temperature for silica (standard and high) and quartz.

Table 9.7. Typical Chemical Analysis of High Silica and Quartz.[a]

COMPONENT (ELEMENT)	CONTENT[b] HIGH SILICA (HIGH PURITY)	QUARTZ
SiO_2 (exclusive of binders)	99.23	99.95
Phosphorous	0	3
Sodium	4	9
Potassium	3	5
Lithium	1	1
Boron	205	10
Calcium	5	23
Magnesium	6	2
Strontium	0	0
Iron	16	3
Titanium	2800	12
Aluminum	900	100
Manganese	1	2
Copper	2	1
Cadmium		0.5
Antimony		0.5
Chromium	150	0

[a]From J. P. Stevens and Company, Incorporated, Central Research Laboratory, Garfield, New Jersey.
[b]The SiO_2 content is given in percent; all element contents are given in parts per million.

9.9. COMPOSITE PROPERTIES

Tables 9.8–9.10 provide data on several aspects of the properties of high-silica and quartz laminates. High silica and quartz can be used with most resin systems and are easily impregnated on most conventional coating equipment (either vertical or horizontal). Preimpregnated

Table 9.8. Typical High-Silica and Quartz Laminate Properties (Phenolic Resin System).[a]

PROPERTY	HEAVYWEIGHT FABRIC		LIGHTWEIGHT FABRIC	
	HIGH SILICA	QUARTZ	HIGH SILICA	QUARTZ
Fabric Description				
Weight, oz/sq yd (Kg/sq m)	18.5 (0.63)	19.5 (0.66)	10.3 (0.35)	8.4 (0.28)
Finish	None	A1100	None	A1100
Weave	8H satin	5H satin	8H satin	8H satin
Laminate Preparation				
Phenolic resin	V-204[b]	91-LD[c]	SC-1008[d]	V-204[a]
Cure	60 min. at 325° F (163° C)		60 min. at 325° F (163° C)	
Post-cure	180 min. at 325° F (163° C)		180 min. at 325° F (163° C)	
Molding pressure	250 psi (1723 Pa)		250 psi (1720 Pa)	
Laminate Properties				
Flexural strength, psi × 10^3 (MPa)				
Average at 75 ± 5° F (24 ± 2.8° C)	33.3 (229)		36.6 (252)	95.4 (657)
Average at 500 ± 10° F (260 ± 5.6° C) for 0.5 hr	24.8 (171)	35.5 (245)	21.1 (145)	60.0 (413)
Tensile strength, psi × 10^3 (MPa)				
Average at 75 ± 5° F (24 ± 2.8° C)	23.1 (159)	56.8 (391)	22.7 (156)	72.1 (497)
Average at 500 ± 10° F (260 ± 5.6° C) for 0.5 hr	19.1 (132)	50.0 (345)	15.3 (105)	47.0 (324)
Specific gravity	1.72	1.73	1.62	1.80
Resin content, wt %	30.3	34.0	37.0	30.7
Thickness, in. (mm)	0.124 (3.15)	0.127 (3.23)	0.152 (3.86)	0.118 (3.00)
Number of plies	6	6	12	12

[a]From J. P. Stevens and Company, Incorporated, Central Research Laboratory, Garfield, New Jersey.
[b]Barrett Division, Allied Chemical Corporation.
[c]Cincinnati Testing Labs, Division of Studebaker-Packard.
[d]Monsanto Chemical Company.

Table 9.9. Comparison of Quartz-Fabric-Laminate Mechanical and Electrical Properties for Various Resin Binders.

PROPERTY	RESIN BINDER				
	PHENOLIC V-204[a]	SILICONE[b]	EPOXY EPON 828/CL[c]	TEFLON[d]	PBI AF-R-100
Fabric Description					
Weight, oz/sq yd (Kg/sq m)	8.4 (0.28)	8.4 (0.28)	8.4 (0.28)	8.4 (0.28)	8.4 (0.28)
Weave	8H satin	8H satin	8H satin	8H satin	8H satin
Finish	A1100	None	A1100	Teflon	A1100
Mechanical Properties					
Flexural strength, psi $\times 10^3$ (MPa)					
Dry at 75° ± 5° F (24 ± 2.8° C)	95.4 (657)	34.0 (234)	98.6 (679)	—	—
Tensile strength, psi $\times 10^3$ (MPa)					
Dry at 75 ± 5° F (24 ± 2.8° C)	72.1 (497)	34.0 (234)	79.2 (546)	—	—
Electrical Properties					
Dielectric constant					
X-band, dry, room temperature, 9375 MHz	3.86	2.93	3.470	2.47	3.360
Loss tangent					
X-band, dry, room temperature	0.040	0.00098	0.0092	0.0007	0.0034

[a] U.S. Polymeric, Incorporated, data.
[b] Coast Manufacturing and Supply Company data.
[c] Ferro Corporation and Grumman Aircraft Engineering Corporation data.
[d] Custom Materials, Incorporated, data.

Table 9.10. Comparison of Mechanical and Electrical Properties of Silicone-Resin Laminates—Quartz, D-Glass, and E-Glass Fabrics.

PROPERTY	E-GLASS[a]	D-GLASS[b]	QUARTZ[c]
Fabric Description			
Style number	181	181	581
Weight, oz/sq yd (Kg/sq m)	9.0 (0.31)	9.0 (0.31)	8.4 (0.28)
Finish	None	None	None
Mechanical Properties			
Tensile strength, psi × 10^3 (MPa)			
Dry, room temperature	39.5 (272)	25.0 (172)	34.0 (234)
Compressive strength, psi × 10^3 (MPa)			
Dry, room temperature	27.3 (188)	18.3 (126)	24.0 (165)
Flexural strength, psi × 10^3 (MPa)			
Dry, room temperature	—	41.8 (288)	36.0 (248)
Flexural modulus of elasticity, psi × 10^6 (GPa)			
Dry, room temperature	—	2.90 (20)	2.80 (19)
Electrical Properties			
Dielectric constant			
X-band, dry, room temperature	3.946	3.60	2.93
Dissipation (power) factor			
X-band, dry, room temperature	0.0082	0.002	0.00098

[a]Summary Card, "Mechanical Properties of Aerospace Fiberglas Fabric Laminates," Owens-Corning Fiberglas Corporation, Office of Aerospace and Defense, April 1966.
[b]F. Trevarno, "130 Silicone Data Bulletin #54b," Coast Manufacturing and Supply Company, January 1965.
[c]F. Trevarno, "Structural Materials Data Bulletin #109," Coast Manufacturing and Supply Company.

fabrics are easily cut (either straight or on a bias) for use in tape winding or in lay-up techniques. The lighter-weight fabrics, which are usually more flexible, are normally used on highly contoured parts. Fabricated parts can also be made from chopped squares or with molding compounds made from shredded fabric or chopped fibers. High-silica materials are usually employed without a finish or binder. However, many types of resin-compatible finishes and modifying additives can be applied to silica fabrics.

All test data were obtained using panels made from unfinished silica fabrics. Quartz fabrics are generally supplied with chemical finishes that are compatible with the resin system to be used. Resin-compatible finishes substantially improve laminate physical properties and reduce moisture absorption. Most of the finishes, binders, and additives that are available for fiberglass fabrics can also be applied to quartz fabrics. Standard commercial binders are available for phenolic, epoxy-, polyimide-, silicone-, polyester-, and fluorocarbon-resin systems.

All of the property data presented in Tables 9.8–9.10 were obtained under laboratory conditions using flat panels prepared in accordance with MIL R-9300. These values are for comparative reference purposes only, and they should not be used for design criteria or specification writing.

10
BORON AND OTHER HIGH-STRENGTH, HIGH-MODULUS, LOW-DENSITY FILAMENTARY REINFORCING AGENTS

Harold E. DeBolt
Specialty Materials Division
Avco Corporation
Lowell, Massachusetts

10.1. INTRODUCTION

The development of high-strength, high-modulus, low-density filaments and composites, as well as their application to the primary load-carrying structures of various present-day aircraft systems, stemmed mainly from a continuing drive by the Air Force Materials Laboratory (AFML) to develop improved structural materials that would reduce structural weight and increase systems performance. Although fiberglass-reinforced composites, which offer low density and high strength, had previously been applied to radomes, fairings, and other secondary aerospace structures, they could not be considered for use in primary load-carrying structures due to their inherent lack of stiffness. C. P. Talley of Texaco Experiment Incorporated reported in 1959[1] that high-strength and high-stiffness boron filaments could be produced via chemical vapor deposition. Subsequently, he and his co-workers demonstrated that (1) continuous filaments that possessed good mechanical properties could be formed by the chemical vapor deposition process and (2) when combined with an organic matrix, these filaments resulted in a composite having both the desired strength and the desired stiffness.[2-4] Triggered by this development, and charged with the overall objective of providing improved materials for structural applications, the Air Force initiated an intensive research and development program on boron filaments and their composites. This effort further developed and refined the boron deposition process, established a usable prepreg material form, and explored the potential applications and systems payoffs of this new structural material. In the late 1960's, boron emerged as a viable engineering material and was immediately applied to selected structures of emerging advanced aircraft systems. In addition to its early application in aircraft structures, boron-reinforced composites were the forerunner of all high-performance composites in that they established the viability and unlimited potential of the latter as a new and unique class of structural materials.

During the 1960's, while boron-filament development was proceeding, filaments made from silicon carbide (SiC), titanium boride (TiB_2), and boron carbide (B_4C) were also being studied. Early work on filaments other than those made of boron at Texaco Experiment Incorporated,[5] General Technologies Corporation,[6] United Aircraft Research Lab,[7,8], and Dow Corning Corporation[9] indicated that these alternative candidate materials—particularly Sic—had promise. In 1972–1973, AFML[10] sponsored a program to advance the SiC-filament technology because it offered the promise of lower cost in plastic-matrix applications as well as compatibility with metal-matrix materials. More recently, additional efforts directed toward advancing the technology of SiC in metal-matrix applications have been sponsored by AFML[11] and the Office of Naval Research (ONR)[12].

Boron filaments have been modified by coating them with SiC and B_4C to improve their performance in aluminum- and titanium-matrix composites. Boron coated with SiC—Borsic® (United Aircraft)—has been commercially available in the United States for several years, whereas boron coated with B_4C was developed by Société Nationale des Poudres et Explosifs (SNPE)[13] in France and is now commercially available in the United States from the Specialty Materials Division of Avco Corporation. These materials are intended primarily for use in metal-matrix applications.

In the following sections, the production methods, properties, structure/morphology, and prepreg methods for each of the filaments commercially available for use in composites—boron, boron coated with SiC, boron coated with B_4C, and SiC—will be described. Several earlier works[4, 14, 15] have described in detail most facets of the production and characterization of boron, and only a summary of these works will be included here.

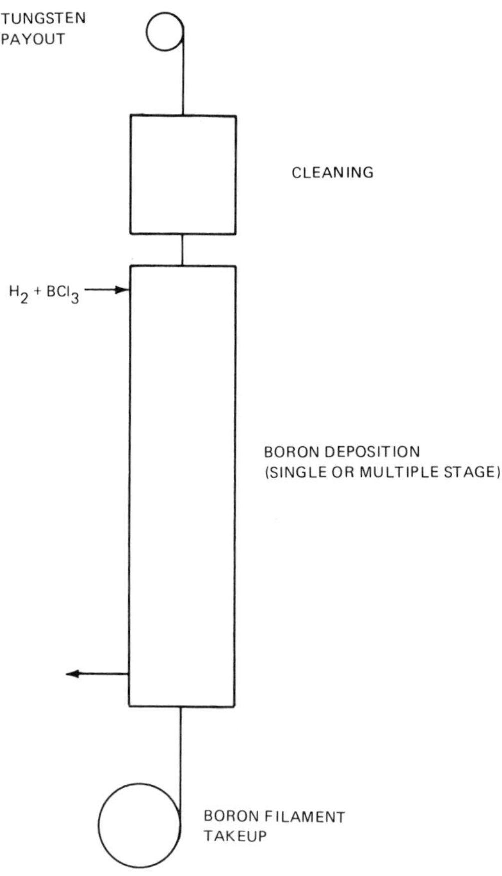

Figure 10.1. Schematic diagram of boron–tungsten deposition reactor.

10.2. BORON-FILAMENT TECHNOLOGY

10.2.1. Boron–Tungsten Substrate Filaments

Boron filaments are currently produced by chemical vapor deposition from a gaseous mixture of hydrogen (H_2) and boron trichloride (BCl_3) on (primarily) an electrically heated tungsten substrate of 0.5 mil (12.5 μm) diameter. The final filament diameter is either 4 mils (100 μm), 5.6 mils (140 μm), or 8 mils (200 μm), in descending order of production quantity; however, both larger and smaller diameters have been produced in experimental quantities. The U. S. production capacity of boron filaments is on the order of 50,000 lb (22,700 kg)/year.

Boron–tungsten filaments are produced in the reactor shown schematically in Fig. 10.1. The tungsten substrate is oriented vertically, housed in a glass tube having an inner diameter of approximately 0.4 in. (1 cm). Mercury-filled assemblies are used at both ends to provide electrical contact with the substrate and to isolate the reactive gases from the atmosphere (the reactive gases are at atmospheric pressure). In the first portion of the reactor, the substrate is heated in a nonreactive gas to outgas and clean it; in the second portion, it is heated in approximately an equimolar mixture of H_2 and BCl_3 to a peak temperature of approximately 1350°C (2460°F), at which point the boron is deposited on the tungsten substrate. The second portion of the reactor can be either single or multiple stage. Electrical power is used to heat the filament [either entirely DC or a combination of DC and ultra-high-frequency (UHF) power are employed, the latter being used to control the temperature profile in the reactor when the larger-diameter filaments (8 mils ≃ 200 μm and up) are produced]. The overall length of a boron–tungsten deposition reactor is about 2 m (6.6 ft), and the filament-production rate is in the range of 2 lb (907 g)/week.

The chemical reaction that produces elemental boron is as follows:

$$2BCl_3 + 3H_2 \rightleftharpoons 2B + 6HCl \quad (10.1)$$

Chemical equilibrium prevents reaction (10.1) from going to completion; in fact, only about 2% of the ingoing BCl_3 can be reacted to form boron. The unreacted BCl_3 is reclaimed in a condenser that operates at a temperature of approximately −80° C (−112° F). The HCl is removed from the exhaust H_2, and the unused H_2 is either ejected into the atmosphere or recycled.

The operating conditions in the reactor must be properly maintained in order for highest quality filament to be produced. If the deposition kinetics are slightly higher than optimum, crystalline boron spots are formed at widely spaced intervals; these spots result in weak points that make the material difficult to handle. If the specific deposition rate is entirely too high, fully crystalline boron, which has a tensile strength below 200 ksi (1379 MPa), is deposited. On the other hand, if the specific deposition rate is too low, the tensile strength decreases as well. Therefore, it is desirable to maintain near-optimal conditions in order to maximize the production rate and retain a high filament strength.

It should be noted that the temperature profile in any given stage of a boron–tungsten deposition reactor is not constant as a result of variations in the filament resistance along the length of the stage. In general, the resistance, temperature, and deposition rate decrease as the outlet end of the reactor or stage is approached. The temperature profile can be made to approximate a constant more closely by subdividing the reactor into two or more electrical stages and either controlling the power to each stage separately or adding additional UHF power (e.g., in the 30–100-MHz range) in the latter portion. The production rate can be increased by flattening the temperature profile; however, the rate increase obtainable by this method is limited by the fact that, for reasons not completely understood at this time, filament strength is degraded when excessive rates are obtained.

Unfortunately, boron–tungsten filaments are a relatively expensive material to produce. The tungsten substrate and boron trichloride alone contribute about \$50 and \$40, respectively, to the cost per pound of boron filaments. The high cost of tungsten wire has led to studies of a lower-cost carbon substrate.

10.2.2. Boron–Carbon Substrate Filaments

In 1969, AFML initiated a series of programs[16–19] aimed at reducing the cost of boron filaments by reducing the cost of the substrate. A carbon monofilament of 1.3 mil (33 μm) diameter made by melt-spinning, oxidizing, and carbonizing coal-tar pitch via a process first utilized by Great Lakes Carbon Research Corporation showed promise as a much less expensive substrate. However, the process of boron deposition on a carbon monofilament substrate was complicated by one of the properties of amorphous boron—its elongation during deposition. Thus, before discussing the process of boron deposition on carbon, it is appropriate to describe more fully the phenomenon of boron elongation.

10.2.2.1. Boron Elongation

During the very early stages of boron-filament development, researchers at Texaco Experiment Incorporated[4, 14] observed that boron filaments exhibited a residual stress pattern that was attributable, in part, to growth strains. Later, Mehalso and Diefendorf[16, 20] and Wawner and DeBolt[21] measured these growth strains for both tungsten and carbon substrates. The axial component of these growth strains when depositing boron on a carbon substrate is illustrated in Fig. 10.2. Note that not only is the magnitude of the growth strain substantial, but, in fact, it is also sufficient to exceed the strain-to-failure of the carbon substrate. When boron was deposited

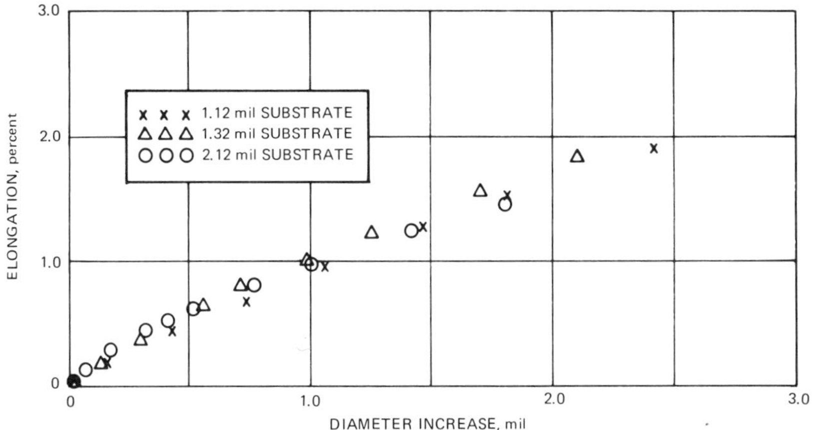

Figure 10.2. Boron elongation as a function of the thickness of boron deposited on a carbon substrate.

directly on the carbon monofilament substrate, breakage of the latter occurred when the deposition thickness reached approximately 0.85 mil (22 μm). Figure 10.3 shows a cross section of a boron–carbon filament that had broken by this mechanism. The substrate breakage occurred at a large number of points within the reactor just a few hundred microseconds after an initial break had occurred at the weakest point; the subsequent breaks were triggered by the shock wave originating from the first break point. The substrate break points were separated by distances varying from a few to many hundred substrate diameters. Since the substrate contributes little to the final properties of a filament, the breaks themselves were not catastrophic. However, at one or more points of breakage, the inner surface of the boron was damaged, hence resulting in closely spaced stress-raised points in the boron; this was catastrophic. As a result, strong boron filaments greater than approximately 2.95 mils (75 μm) in diameter could not be produced using the carbon substrate because of the catastrophic weakening of the filament at the substrate break points. When the breakage occurred within the reactor, it was visually observed as a series

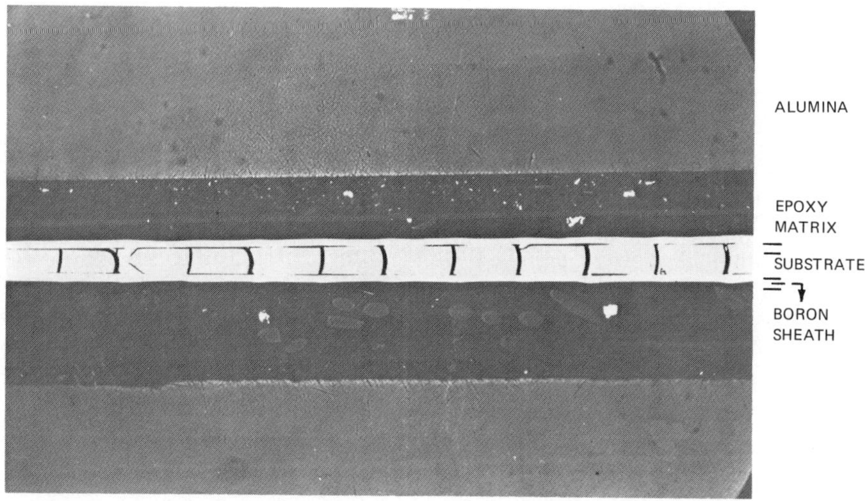

Figure 10.3. Photomicrograph of an axially sectioned portion of a boron–carbon filament that had exhibited the "light-bulb" effect during deposition.

of closely spaced hot spots that resembled a string of lights, which led to the name "light-bulb flaw."

10.2.2.2. Boron-on-Carbon Deposition Process

It was found possible to avoid the catastrophic strength degradation resulting from substrate breakage by coating the carbon substrate with a layer of pyrolytic graphite before applying the boron. The layer of pyrolytic graphite not only made possible the deposition of a thicker layer of boron before substrate breakage occurred,[22] but also prevented damage to the inner surface of the boron when substrate breakage did occur during the production of filaments greater than approximately 4.7 mils (120 μm) in diameter. It was learned that the filament strength was critically dependent on the deposition conditions for the layer of pyrolytic graphite.[17,23,24]

The deposition reactor used for producing boron–carbon filaments is different from its tungsten-substrate counterpart as a result of the need for the pyrolytic graphite layer. Figure 10.4 shows a schematic diagram of the reactor used to produce boron–carbon filaments. The pyrolytic graphite is applied on-line immediately preceding the deposition of the boron layer in order to avoid the damage to and contamination of the substrate which may occur if it is applied in a separate step. The temperature profile in the boron-deposition stage as well as that of a tungsten-

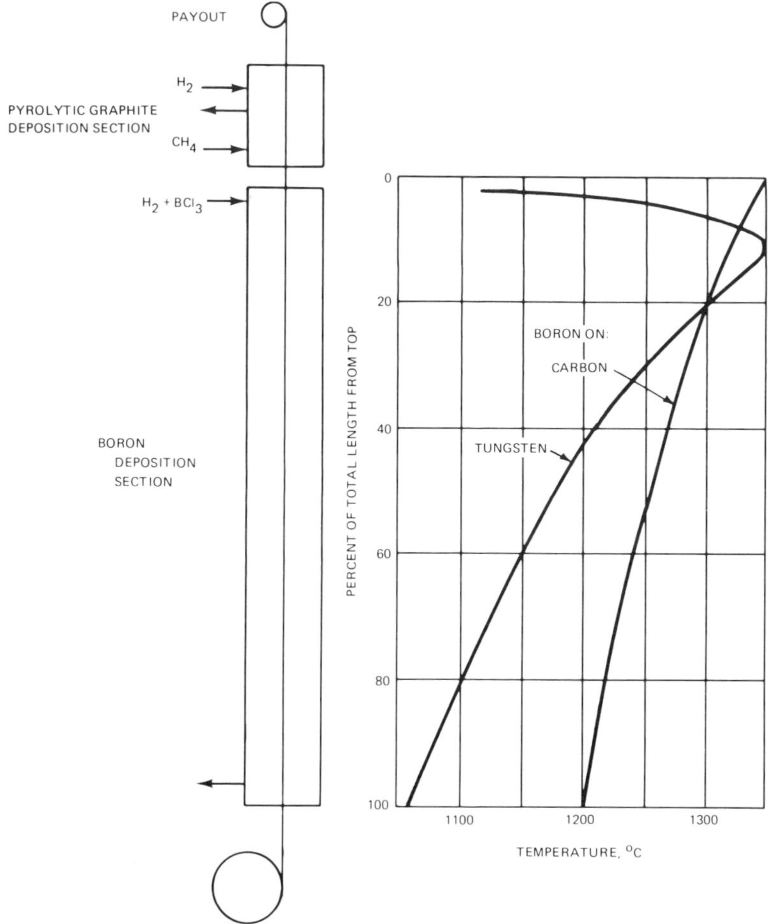

Figure 10.4. Schematic diagram of a boron–carbon deposition reactor and temperature profiles.

substrate reactor are also shown in Fig. 10.4. In the carbon-substrate reactor, the temperature rises quickly to its maximum value, the time required being limited by the thermal response of the substrate. The temperature then decreases monotonically due to the decreased resistance of the filament caused by the conductive boron layer. The carbon substrate with its pyrolytic graphite layer has a negative temperature coefficient of resistance rather than the large positive coefficient that is the case with tungsten and its borides; accordingly, there is no large amplification of the temperature decrease with thicker boron deposits as is the case with a tungsten substrate. The higher average temperature in the reactor results in approximately a 40% increase in the production rate and drawing speed of the boron–carbon deposition reactor (as compared to the equivalent boron–tungsten deposition reactor) for 4-mil (100-μm) filament production in spite of the additional reactor length used for the deposition of the pyrolytic graphite layer. This increase in production rate provides an additional economic advantage of boron–carbon filaments over their tungsten counterpart besides the savings in substrate cost.

10.2.2.3. Application of Boron–Carbon Filaments

Most of the boron filaments currently produced are used to make the empennage structure for the F-14 and F-15 aircraft. This application has been qualified using boron–tungsten filaments of 4 mil (100 μm) diameter. The tensile modulus of elasticity of 4-mil (100-μm) boron–tungsten filaments is 58 Msi (400 GPa), whereas that of 4-mil (100-μ) boron–carbon filaments is 52 Msi (358 GPa). This greater than 10% difference is due to the fact that the carbon substrate, which has a tensile modulus of elasticity of only about 5 Msi·(34 GPa) comprises approximately 10% of the volume of the filaments. Therefore, if the boron–carbon filaments were directly substituted for the boron–tungsten filaments, then the tensile modulus of elasticity of the composite would be unacceptably lower. However, the desired composite modulus could be maintained by increasing the volume fraction of boron, and tests were carried out using this approach. The boron–carbon filament diameter was increased to 4.2 mils (107 μm), the resin fraction was reduced, and the distance between filament centerlines was left unchanged, so that the tensile modulus of elasticity of the composite and the thickness per ply of boron were the same as for boron–tungsten composites of 4 mil (100 μm) diameter. Although the tensile modulus of elasticity of composites made in this way was acceptable, other properties—particularly the transverse strength and strain-to-failure—were adversely affected. This change was attributed to a combination of factors, including the fact that the surface texture of boron–carbon filaments was somewhat smoother than that of boron–tungsten filaments (see Fig. 10.5), the filament separation was reduced in both transverse directions, and the resin fraction was reduced. Note that the above discussion has concentrated on the problems of incorporating a new filament form into an existing, fixed design. The real potential of boron–carbon filaments, however, resides in new applications where the design would be tailored to the properties of the filament rather than back-fit in an attempt to meet pre-existing criteria and properties.

10.2.3. Properties of Boron–Tungsten Filaments

References 4, 14, and 15 (which are all excellent summary works) reviewed the properties of boron–tungsten filaments up to their date of publication. In this section, the data reported earlier will be briefly reviewed, and new, recently obtained data will be described in detail.

The tensile strength of boron–tungsten filaments has improved steadily over the past decade from an average of under 400 ksi (2756 MPa) to over 500 ksi (3445 MPa) at the time of this writing. Average tensile strength refers to the average over all production (e.g., as one would find for material put into a large composite structure). Figure 10.6 shows a histogram of tensile-strength data for boron–tungsten filaments of 4 mil (100 μm) diameter

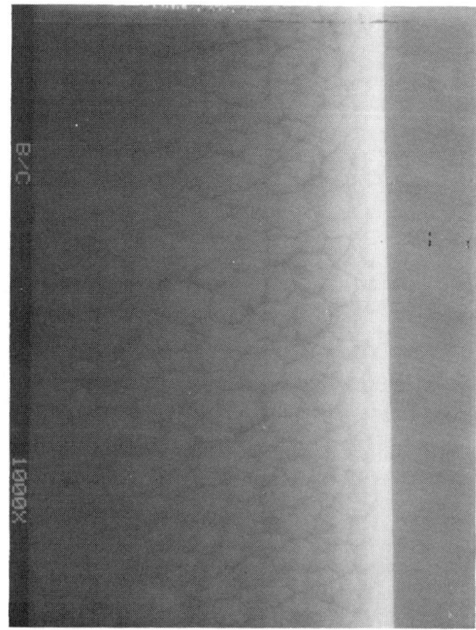

Figure 10.5. Surface morphology of boron–tungsten (*a*) and boron–carbon (*b*) filaments.

as they are made at the time of this writing. The average tensile strength is about 500 ksi (3445 MPa) with a standard deviation of 76 ksi (524 MPa) or 15%. The tensile strength of boron–tungsten filaments of 5.6 mil (140 μm) diameter is slightly higher—approximately 520 ksi (3583 MPa). A tensile-strength histogram of B_4C-coated boron–tungsten filaments of 5.6 mil (140 μm) diameter is also shown in Fig. 10.6; these filaments will be discussed further in Sec. 10.4. Wawner[14] reported that the intrinsic tensile strength of boron–tungsten filaments, as measured by bending them, exceeds 1 Msi (6890 MPa). Segments of split filaments frequently exhibited 1 Msi (6890 MPa) tensile strength after etching the split filament to remove the core and the stress-raising flaws at the boron–core interface. Recently, under the sponsorship of the National Aeronautics and Space Administration (NASA), Lewis Research Center,[25,26] this work has been extended and has shown that average tensile-strength levels over 1 Msi (6890 MPa) can be obtained. The method used to make 1-Msi (6890-MPa) boron–tungsten filaments is to produce a filament of approximately 5 mil (125 μm) diameter on a tungsten substrate of 1 mil (25 μm) diameter using a peak temperature about 100° C (180° F) lower than that normally used in the boron–tungsten deposition reactor. A filament produced in this manner readily splits after production into three equal segments. The segments are etched first in H_2O_2 solution to remove the core and then lightly a second time in hot nitric acid solution to remove the stress-raising flaws and sharp edges. Average tensile strengths of greater than 1 Msi (6890 MPa) were obtained from these filaments. At the time of this writing, studies were underway to determine (1) how much of this filament strength is realized in a composite and (2) how this affects the composite impact strength.

The tensile strength of boron–tungsten filaments can be increased by etching away part of its outer portion.[27] This improvement in tensile strength can be attributed to the decrease in residual tensile stresses at the inner surface of the boron and the increase in compressive stresses in the core due to the removal of the outer region of the filament, which contains a compressive residual stress. The tensile-strength-limiting flaws of the boron–tungsten filaments currently being pro-

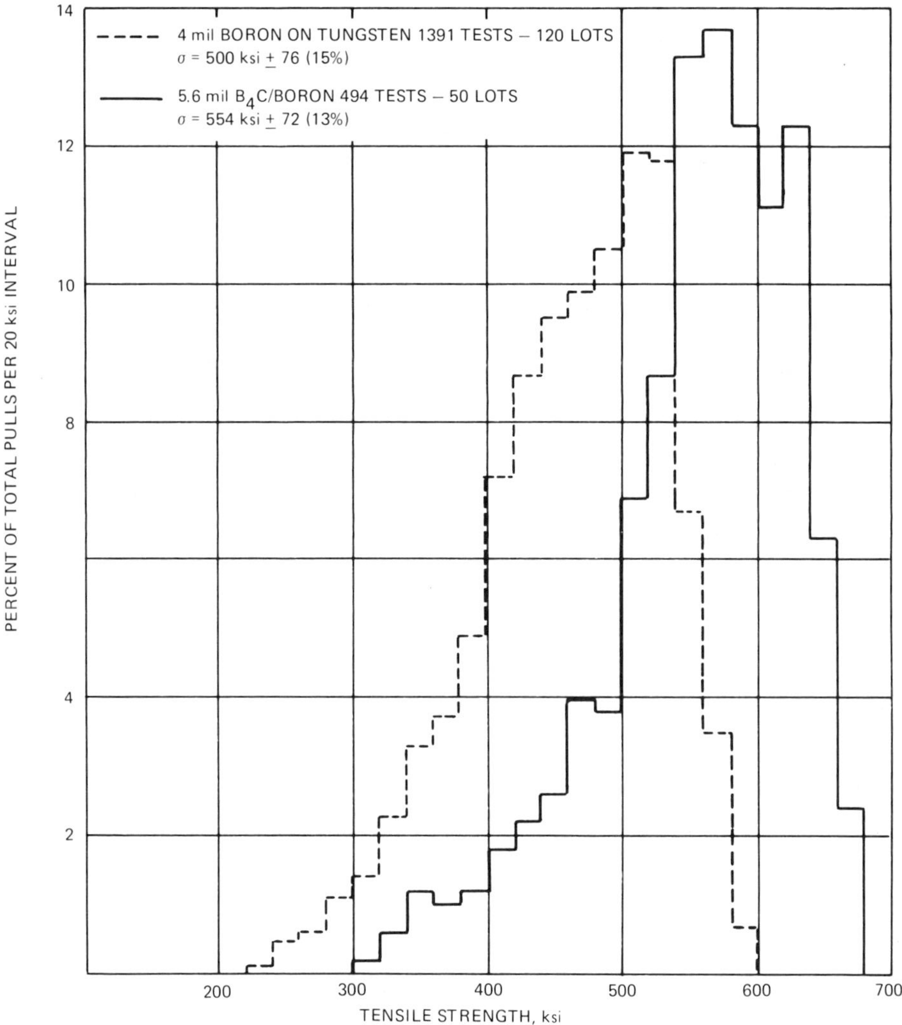

Figure 10.6. Tensile-strength histograms of boron–tungsten and B_4C-coated boron–tungsten filaments (1979).

duced are either in the core or at the boron–core interface; therefore, an improvement in tensile strength can be obtained by shifting the residual stresses toward lower tensile or higher compressive stresses in the core region.

10.2.4. Structure and Morphology of Boron–Tungsten Filaments

The reader is referred to summary works 4, 14, and 15 for a detailed description of the structure and morphology of boron–tungsten filaments. However, a brief summary is included here for completeness.

Although the boron deposited by chemical vapor deposition from a mixture of BCl_3 and H_2 is of the polycrystalline β-rhombohedral form, it has such a small crystallite size (on the order of 20 Å) that it is referred to as amorphous. However, in a deposition-temperature range of 1300–1400° C (2372–2552° F), localized spots of polycrystalline β-rhombohedral boron having a much larger crystallite size are formed. As the temperature is increased, these polycrystalline spots occur with greater frequency, and, above approximately 1400° C (2552° F), all of the boron is deposited in a polycrystalline form with a crystallite size on the order of several thousand angstrom units. Since this polycrystalline form is weak, it is avoided in the production of boron filaments.

A number of localized flaws that are stress raisers and hence degrade the strength of the filament have been identified during the production of boron filaments. The production of high-strength boron filaments is simply a matter of minimizing the degrading effects of these flaws. The polycrystalline boron spots mentioned above are avoided by keeping the peak deposition temperature below the point at which they form. Inclusion of such foreign material as carbonaceous debris, tungsten oxide, and mercurous chloride causes stress-raising flaws that are observed as enlarged bumps or beads on the external surface. The frequency of occurrence of these flaws is maintained low by keeping the mercury and the hole through which the filament passes clean, particularly at the intermediate electrodes within the boron-deposition region. Residual debris on the substrate causes similar flaws; to avoid this, the substrate is cleaned by heating it in a nonreactive gas (or gas mixture) such as pure hydrogen prior to depositing the boron. The filament can also pick up debris by rubbing or slapping against the reactor wall in response to electrostatic forces. This problem is avoided by keeping the filament centered within the reactor tube and by operating at an adequate tension, which is controlled at the substrate payout. Cracks in the tungsten substrate result in a filament with low tensile strength. This problem is avoided by proper tungsten-drawing methods. A flaw originating at the tip of a radial crack—a crack-tip flaw—causes tensile-strength degradation; this type of flaw, which arises when the peak deposition temperature is too low or when the production rate is pushed too high by multiple-stage or UHF boosting methods, is avoided by keeping the temperature profile within the proper limits. Improvements in the average tensile strength of boron–tungsten filaments over the past decade have come about as a result of operating the deposition process in such a way as to minimize the effects of the above-mentioned flaws.

The surface of a boron filament has a kernel structure that resembles an ear of corn (see Fig. 10.5). These kernels originate at nucleation areas on the substrate, which is not perfectly smooth. In Fig. 10.5, it can be noted that they are larger, more pronounced, and more coherently aligned on a boron–tungsten filament than on a boron–carbon filament; this difference is the result of the relatively coarser surface, including die grooves, of the tungsten substrate. The smaller kernels on boron–carbon filaments originate on fine-scale surface kernels on the pyrolytic graphite layer of the carbon substrate. As was mentioned in Sec. 10.2.2.3, the finer-scale surface texture of boron–carbon filaments is one of the properties considered responsible for the lower transverse strength of epoxy composites reinforced with boron–carbon filaments.

10.3. SILICON CARBIDE FILAMENTS

In the early 1960's, materials with high specific strengths and moduli of elasticity were being studied aggressively as reinforcing agents for plastic- and metal-matrix composites. Under AFML sponsorship, Texaco Research Incorporated did research[1] on a broad range of similar materials, including boron, silicon carbide, and boron carbide, but concentrated the majority of their experimental investigations on boron. Concurrently, AFML also sponsored work at General Technologies Corporation (GTC)[6] on a diverse group of materials, but here the experimental research concentrated most heavily on SiC. By 1966, GTC had produced SiC filaments with occasional average tensile strengths in the 400–500-ksi (2756–3445-MPa) range,[28] and overall mean tensile strengths in the 300–400-ksi (2067–3445-MPa) range. During that same period, White and Davis[29] of Texaco Experiment Incorporated developed a process to make SiC filaments with average tensile strengths in the mid 400-ksi (low 3000-MPa) range using a blend of two alkylsilanes: methyltrichlorosilane and methyl hydrogen dichlorosilane. Galasso et al.[7] at United Aircraft (now United Technologies) Research Laboratories also reported the good tensile-strength properties of SiC filaments. During 1966–1968 (the latter part under Naval Air Systems sponsorship), Michalik and Weinstein[30] at Marquardt Corporation studied SiC filaments modified by the addition of boron[31] or a combination of boron and titanium,[32] which also showed good tensile-strength potential. Avco Corporation[10, 33, 34] under AFML sponsorship, developed a process for making tungsten-substrate SiC filaments economically which involved recycling of the silanes used and which produced average tensile strengths in the high 400-ksi (high 3300-MPa) range. Later,[35] it was found that average tensile strengths greater than 500 ksi (3445 MPa) could be obtained using a carbon-monofilament substrate, with occasional tensile-test pulls approaching 1 Msi (6890 MPa).

10.3.1. Economics of SiC Filaments

It is primarily for economic reasons that SiC filaments are of current interest as a reinforcing agent. However, they also offer the potential of performing in metal-matrix applications with simpler higher-temperature consolidation where boron is degraded by rapid chemical attack by molten aluminum. The major reasons for the lower cost of SiC filaments are as follows:

1. Only about 8 lb (kg) of silane is required per pound (kilogram) of SiC as compared to about 15 lb (kg) of BCl_3 per pound (kilogram) of boron.
2. The cost of silane is currently about 75 cents/lb ($1.65/kg) as compared to about $2.50/lb ($5.50/kg) for BCl_3.
3. At present, SiC filaments incorporate a carbon substrate rather than a tungsten substrate. Thus, as in the case of boron filaments, both the economic advantage of a lower-cost substrate and the concomitant increase in reactor operating speed are realized.
4. The production rate in pounds (kilograms)/hour of an SiC deposition reactor is greater than that of a boron deposition reactor by more than a factor of two, hence resulting in less capital cost per unit of production.
5. The silanes are less volatile than BCl_3, and therefore a lower-cost refrigeration facility is required for the production of SiC filaments.

The first cost-reduction factor noted above is primarily the result of the fact that the ratio of the combined silicon + carbon weight in the silane molecule to the total weight is higher than the weight fraction of boron in BCl_3. The theoretical weight ratio for SiC from silane is about 3.5 (varying slightly as a function of the silane compound actually used), whereas the theoretical weight ratio for boron from BCl_3 is 10.84. However, the quantity of silane needed per unit of SiC is higher than the theoretical value because (1) silane polymers are synthesized and rejected in the process and (2) some silanes are lost through the recycling condensers. Similarly, the quantity of BCl_3 required to make boron is higher than the theoretical value because of (1) condensation losses and (2) synthesis of small amounts of diborane in the process, some of which is lost through the recycling condensation step.

Silanes are cheaper than BCl_3 primarily because the production rate of the former, which are used as precursors for a large silicone market, is much higher than the production rate of the latter, which is used almost exclusively for boron-filament production. The facility cost factors (items 3 and 4 above) represent a substantial fraction of the filament cost structure at any level of production of either boron or SiC.

10.3.2. Production Process of SiC Filaments

The production process for making SiC filaments is very similar to that for making boron–carbon filaments (see Fig. 10.4). A blend of alkyl silanes (mixed with hydrogen) is used in place of BCl_3.

One difference between the SiC- and boron filament production processes is that the recovery–recycle system of the latter includes a distillation step that separates out and rejects the silane polymers produced. These polymers are primarily alkylchlorodisilanes and disilmethylenes, that is, compounds having two silicon atoms, an Si–C–Si structure, and methyl groups as well as chlorine atoms bonded to the silicon atoms. The amount of silane lost in producing these rejected polymers is approximately equal to the amount of silane consumed in the formation of SiC.

10.3.3. Properties of SiC Filaments

As discussed above, the average tensile strength of SiC filaments has improved since the earliest days of its development to the 500-ksi (3445-MPa) range currently being attained. Figure 10.7 presents a tensile-strength histogram of SiC–carbon filaments as they are now produced. The tensile-strength histogram of SiC–tungsten filaments is similar, but shifted downward, the average tensile strength being near 475 ksi (3273 MPa).

The tensile strength of SiC filaments, just like that of boron filaments, is limited by localized flaws. The strength-limiting flaws in the case of SiC–carbon filaments are normally such substrate surface anomalies as surface specks, dirt particles, scratches, and adhesion points. The surface specks are bumps at the outer surface of an otherwise smooth melt-spun filament which are attributed to a localized high-viscosity region of the pitch when the filament was spun, residual particulate material in the pitch, or both. Adhesion points refer to marks at the surface caused by filaments sticking to each other (although not firmly enough to cause breakage during rewinding) in the package during the oxidation and carbonization steps of carbon-monofilament manufacture. These surface imperfections result in a stress-raising region at the inner surface of the SiC deposit even after the deposition of a layer of pyrolytic graphite, which determines the localized strength. Since the SiC deposit is already polycrystalline with a crystallite size greater than that of boron, a flaw that is observed in boron filaments is not present in SiC filaments—the localized crystal spot. However, weak points at the surface caused by either abrasion during handling or damage induced by the electric current passing through the SiC layer at the outgoing electrode are sometimes observed. The electrical damage is prevented by doping the SiC deposit with oxygen (approximately 100 ppm of oxygen is added to the H_2–silane mixture) to improve its conductivity.

The modulus of elasticity of SiC is 65 Msi (448 GPa) as compared with 58 Msi (400 GPa) for boron. However, the overall modulus of SiC filaments having a 5.6 mil (140 μm) diameter with a carbon substrate of 1.3 mil

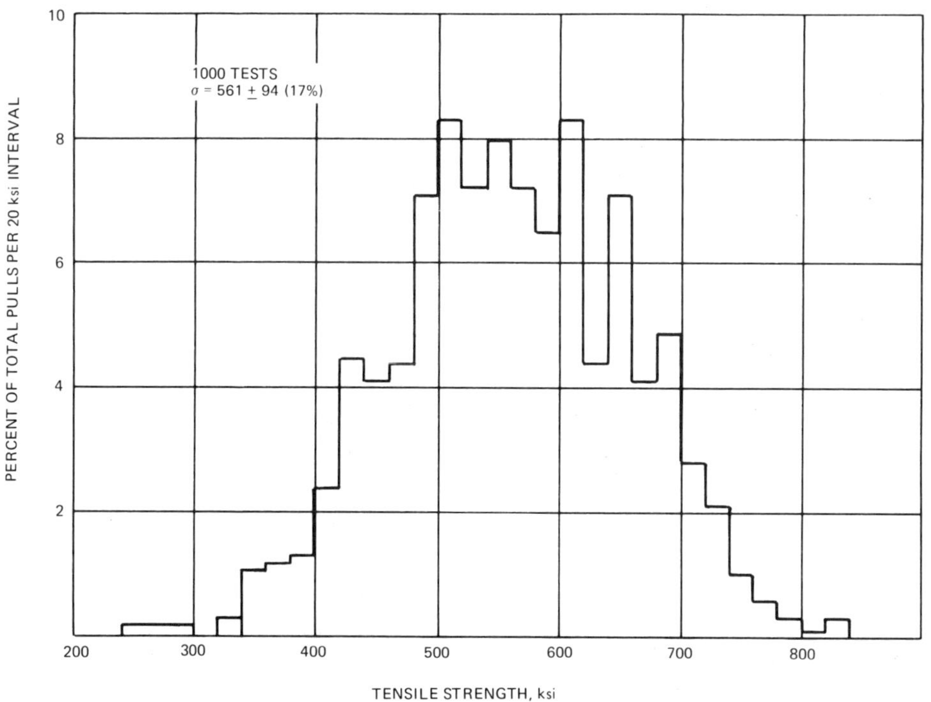

Figure 10.7. Tensile-strength historgram of SiC–carbon filaments (1979).

Table 10.1. Density and Length per Pound for Current Reinforcing Filament Types.

FILAMENT	SIZE mil (μm)	FILAMENT DENSITY lb/in^3 (g/cm^3)	FEET PER POUND
Boron–tungsten	4.0 (100)	0.0940 (2.59)	70,800
	5.6 (140)	0.0890 (2.46)	38,025
	8.0 (200)	0.0865 (2.40)	19,160
Boron–carbon	4.0 (100)	0.0802 (2.22)	82,600
	4.2 (107)	0.0806 (2.23)	74,600
	5.6 (140)	0.0822 (2.27)	40,175
Silicone carbide–carbon	4.0 (100)	0.1076 (2.980)	61,600
	5.6 (140)	0.1111 (3.075)	30,450
	8.0 (200)	0.1130 (3.128)	14,670

(33 μm) diameter is approximately 62 Msi (427 GPa), since the carbon substrate takes up 5% of the filament volume but contributes little to the modulus. These modulus of elasticity values are all very high compared with the values of 30 Msi (207 GPa) for steel and 10 Msi (69 GPa) for aluminum.

The density of SiC as deposited by chemical vapor deposition is 3.18. However, the overall density of SiC filaments having a 5.6 mil (140 μm) diameter with a carbon substrate of 1.3 mil (33 μm) diameter is 3.08 g/cm^3, the decreased value resulting from the fact that the carbon substrate has a lower density than SiC. For comparison, the density of boron as deposited by chemical vapor deposition is 2.33, whereas the overall density of 5.6-mil (140-μm) boron filaments with a tungsten substrate of 0.5 mil (12.5 μm) diameter is 2.46 g/cm^3 and that of 5.6-mil (140-μm) boron filaments with a carbon substrate of 1.3 mil (33 μm) diameter is 2.27 g/cm^3. These values, together with the lengths of the various filaments per pound are listed in Table 10.1.

10.3.4. Structure and Morphology of SiC Filaments

The SiC deposited on either a tungsten or carbon substrate is of the polycrystalline β-form with a crystallite size considerably larger than that reported for boron. Randon et al.[36] (from x-ray analyses) reported a crystallite size of approximately 275 Å which increased with a 2-hr heat treatment above 1200° C (2372° F). Measurements of SiC–carbon filaments[37] indicate a crystallite size that is at least as large as 1000 Å. These crystallite sizes are in sharp contrast with the 20–30 Å reported for amorphous boron.

The surface texture of SiC deposited on either a tungsten or carbon substrate is very smooth. Figure 10.8 shows the surface texture of an SiC filament as currently produced on a carbon substrate. An SiC–tungsten filament with a rough surface can be strong, but this nonsmooth surface texture is usually associated with deposition conditions that result in a low tensile strength.

An additional advantage of SiC–carbon

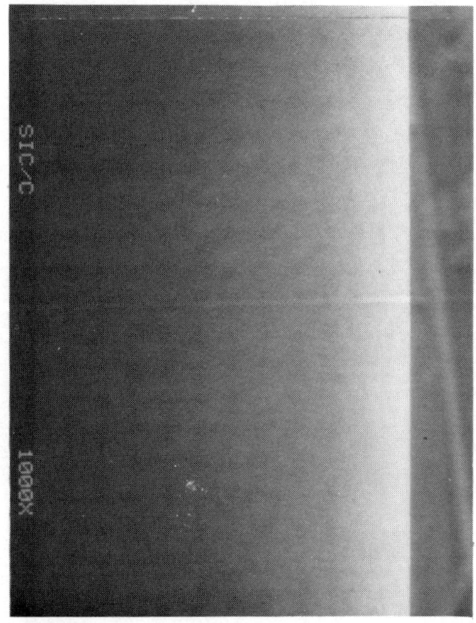

Figure 10.8. Surface morphology of SiC–carbon filaments.

(rather than SiC–tungsten) filaments is related to their properties under conditions of high-temperature exposure. Inasmuch as the carbon-monofilament substrate is chemically inert and hence does not react with the SiC mantle, the tensile-strength properties of the filament are retained considerably longer under conditions of high-temperature exposure. This is particularly important for metal-matrix-composite applications in which, typically, the SiC filament may be subjected to contact with molten aluminum for varying periods of time. In the case of SiC–tungsten filaments, core–mantle interactions at high temperatures lead to the formation of such diffusion compounds as tungsten carbides, which results in substantial losses of tensile strength over a period of time.

10.4. BORON FILAMENTS COATED WITH DIFFUSION BARRIERS

Boron filaments react with titanium and aluminum at an unacceptably rapid rate[38] at the temperatures used for some composite work. The reaction with titanium degrades the boron-filament tensile strength at 899–954°C (1650°F–1750°F)—the diffusion-bonding temperature range for titanium—within the time normally required to complete the diffusion-bonding cycle. Although degradation is negligible in aluminum at its diffusion-bonding temperature (approximately 496°C = 925°F), it becomes rapid and catastrophic if the temperature rises above the melting point of one of the aluminum alloy components present. For both titanium work and braze-bonding processing in aluminum, it is desirable to have a diffusion-barrier layer on the boron filament so as to improve the filament properties in titanium and permit the use of higher temperatures in aluminum processing.

Two diffusion barrier layers for use with boron filaments have been developed to the point of commercial availability: silicon carbide (SiC) and boron carbide (B_4C). The SiC layer was developed in the late 1960's,[39] and boron filaments with an SiC coating were marketed under the trade name Borsic®. In the early 1970's, SNPE (France) developed a process for applying a B_4C layer of 0.28 mil (7 μm) thickness to boron filaments, and evaluated these filaments in titanium.[40]

In 1978, the Specialty Materials Division of Avco Corporation concluded a licensing agreement with SNPE to produce and sell this B_4C-coated boron filament, and it is today available from Avco Corporation in production quantities.

The work by Naslain et al.[40] showed that the B_4C as deposited in the SNPE process was a better diffusion barrier in titanium than the SiC layer of 0.05 mil (1.25 μm) thickness found in Borsic®. The performance of the B_4C diffusion barrier is a sensitive function of the deposition conditions used to apply it. This is considered the reason for the lack of agreement between Naslain et al.'s study and earlier work,[41] which indicated that a B_4C barrier did not perform as well as an SiC barrier. In fact, it is still not known whether the deposition conditions for an SiC layer can be improved to the extent that its performance will equal that of B_4C deposited under optimum conditions.

The application of the B_4C layer on boron filaments results in an increased average tensile strength. Figure 10.6 shows tensile-strength histograms of B_4C-coated and uncoated boron–tungsten filaments.

Boron nitride (BN) has also been applied to boron filaments as a diffusion barrier.[15,42,43] The BN is formed on the outside of a boron filament by a two-step process: First of all, a layer of boron oxide (B_2O_3) is formed by holding the filament in air at approximately 1000°C (1832°F) for about 30 sec; this oxide-coated filament is then held in NH_3 at approximately 1100°C (2012°F) also for about 30 sec. The BN layer is at most 0.02 mil (0.5 μm) thick under the optimum conditions of formation. As a result of the exposure to the conditions required to form the BN layer,[15] the tensile strength of the filament is increased; the BN layer also renders the filament strongly resistant to attack by molten aluminum. However, poor bonding between the aluminum and the BN layer was a problem, the result being a transverse strength significantly below those expected for well-bonded reinforcing agents.

Researchers[44–47] at Watervliet Arsenal have

applied tungsten carbide (WC) and tantalum carbide (TaC) coatings to SiC filaments as diffusion barriers in superalloy composites. Without these barriers, the superalloy reacted with the SiC filament at an unacceptably high rate. With the barrier, a practicable fraction of the filament strength was retained for hundreds of hours at temperatures in the 1000–1100° C (1832–2012° F) range. Hafnium carbide (HfC) and titanium carbide (TiC) have also been tested as diffusion barriers for SiC filaments in superalloy composites. Cornie and Hakim[48] of Westinghouse Electric Corporation have studied these with favorable results.

10.5. PREPREG TECHNOLOGY

Since the vast majority of all of the boron filament sold is in the form of continuous boron–epoxy preimpregnated tapes or broadgoods, it is worthwhile to discuss prepreg construction and some methods of assembly. Continuous boron–epoxy prepreg tapes or broadgoods are typically sold in widths of ¼, 3, 6, or 48 in. (0.64, 7.6, 15.2, or 121.9 cm). The boron filaments are unidirectionally aligned and usually occupy about 50% of the composite volume. Typical loadings are upwards of 200 filaments/in. of width, 78.7 filaments/cm) in order to obtain the desired volume fraction. The resin content is typically 30–35 wt %. The tape is backed on one side with a fine layer of style 104 glass scrim cloth (thickness = 0.0012 in. = 0.003 cm). The scrim cloth not only serves to give the tape some lateral integrity during handling and lay-up, but, more importantly, maintains the filament spacing and collimation during the bonding process.

The tape or broadgood rolls are shipped in continuous lengths of up to 1000 ft (305 m) under dry ice or refrigerated conditions to prevent premature tack loss.

The epoxy resins employed generally cure at either 177° C (350° F), which meets most aerospace specifications, or 121° C (250° F), which is suitable for lower-temperature applications. The curing agent is part of any specific resin formulation, and thus the resin will slowly cure even at room-temperature conditions, albeit over a time span of many weeks. Boron–epoxy prepregs have been stored at −18° C (0° F) for several years without any loss of properties.

The assembly of the prepreg tapes begins with a creel of boron filaments. In the Avco Corporation process, a narrow ribbon of ⅛ or ¼ in. (0.32 or 0.64 cm) width is first formed and wound up. Subsequently, tape of 3 or 6 in. (7.6 or 15.2 cm) width is formed by collimating the narrow ribbons on a separate piece of equipment.[49] Occasionally, the final tape is directly formed in a one-step process from the roughly 600 or 1200 filament spools.

Two techniques have been employed to accomplish the resin coating of the filaments:

1. Direct coating of the filaments on a coating roller using either a hot-melt resin directly or a solvent system at reduced temperature. In the latter case, the solvent must be evaporated.
2. Transfer of a resin film from a backing paper onto the collimated filaments. This system requires the prior or on-line preparation of a resin film.

Boron–epoxy broadgoods are produced by the collimation of eight boron tapes of 6 in. (15.2 cm) width. Since the precursor tapes contain the scrim cloth, the material of 48 in. (121.9 cm) width has a discontinuous scrim. A discussion of broadgoods production can be found in Ref. 50.

10.6. RESIN-MATRIX-COMPOSITE PROPERTIES

10.6.1. Boron Filament–Resin Matrix Composites

The vast majority of all of the boron filament sold is in the form of continuous boron–epoxy preimpregnated tapes and broadgoods that range in width from ¼ to 48 in. (0.64 to 121.9 cm). The epoxy-resin systems in large-scale use in boron prepregs are Avco 5505, Avco 5506, 3M SP290, and 3M SP292. These resin systems are formulated to cure at 177° C (350° F); however, new systems that cure at 121° C (250° F) have also been developed.

Molded panels and shapes are produced using conventional vacuum bag/bleeder cloth technology and curing the part in an autoclave at 50–85 psi (344.5–585.7 Pa). The parts exhibit high strengths and stiffnesses consistent with both the intrinsic filament properties and the orientation of the filament layers. Although various boron–epoxy composite lay-up orientations have been used, the most popular appears to be the (0°, ±45°, 90°) construction.

The following mechanical-test and design data are presented as an introduction to the properties of boron–epoxy laminates. The mechanical-test data were obtained in laboratory testing, and the design data are those used for the design of the (0°, ±45°, 90°) family of laminates. The source of these data is the *Advanced Composites Design Guide* prepared under United States Air Force funding by the Rockwell International Corporation.[51]

The boron–epoxy laminate design data presented in this section are based on a generic 50 v/o boron–epoxy composite that consists of boron–tungsten filaments of 4 mil (100μm) diameter imbedded in an epoxy-resin matrix qualified for continuous 177° C (350° F) service (e.g., Avco 5505/4 and 3 M SP-292) and that cures to a nominal lamina thickness of 0.0052 in. (0.132 mm). The data for a generic boron–epoxy composite that utilizes filaments of 5.6 mil (140 μm) diameter are assumed to be the same unless specifically noted. The 5.6-mil

Table 10.2. Key Basic Properties of a Unidirectional (0°) Boron–Epoxy Laminate (V_f = 0.50)

PROPERTY	UNIT	SYMBOL	VALUE	
			ROOM TEMPERATURE	1700° C (350° F)
Design Strengths ("A" Basis)				
Ultimate longitudinal tensile strength	ksi (MPa)	F_L^{tu}	192.0 (1323)	157.0 (1082)
Ultimate transverse tensile strength	ksi (MPa)	F_T^{tu}	10.4 (72)	6.0 (41)
Ultimate longitudinal compressive strength	ksi (MPa)	F_L^{cu}	353.0 (2432)	116.0 (799)
Ultimate transverse compressive strength	ksi (MPa)	F_T^{cu}	40.0 (276)	11.0 (76)
Ultimate in-plane shear strength	ksi (MPa)	F_{LT}^{su}	15.3 (105)	5.5 (38)
Ultimate interlaminar shear strength	ksi (MPa)	F^{isu}	13.0 (90)	7.0 (48)
Ultimate longitudinal strain	μin./in.	ϵ_L^{tu}	65500.0	7600.0
Ultimate transverse strain	μin./in.	ϵ_T^{tu}	4000.0	7600.0
Elastic Properties (Typical)				
Longitudinal tensile modulus	Msi (GPa)	E_L^t	30.0 (207)	29.9 (206)
Transverse tensile modulus	Msi (GPa)	E_T^t	2.7 (19)	1.13 (7.8)
Longitudinal compressive modulus	Msi (GPa)	E_L^c	30.0 (207)	29.9 (206)
Transverse compressive modulus	Msi (GPa)	E_T^c	2.7 (19)	1.13 (7.8)
In-plane shear modulus	Msi (GPa)	G_{LT}	0.7 (4.8)	0.32 (2.2)
Longitudinal Poisson's ratio	—	ν_{LT}	0.21	0.21
Transverse Poisson's ratio	—	ν_{TL}	0.019	0.008
Physical Constants (Typical)				
Density	$lb/in.^3$ (g/cm^3)	ρ	0.0725 (2.007)	0.0725 (2.007)
Longitudinal coefficient of thermal linear expansion	μin./in./° F	α_L	2.3	3.0
Transverse coefficient of thermal linear expansion	μin./in./° F	α_T	10.6	19.6

Table 10.3. Key Basic Properties of a ±45° Boron–Epoxy Laminate (V_f = 0.50)

PROPERTY	UNIT	SYMBOL	VALUE	
			ROOM TEMPERATURE	177° C (350° F)
Design Strengths ("A" Basis)				
Ultimate longitudinal tensile strength	Ksi (MPa)	F_x^{tu}	29.0 (200)	11.0 (76)
Ultimate transverse tensile strength	Ksi (MPa)	F_y^{tu}	29.0 (200)	11.0 (76)
Ultimate longitudinal compressive strength	Ksi (MPa)	F_x^{cu}	30.0 (207)	11.0 (76)
Ultimate transverse compressive strength	Ksi (MPa)	F_y^{cu}	30.0 (207)	11.0 (76)
Ultimate in-plane shear strength	Ksi (MPa)	F_{xy}^{su}	77.0 (531)	59.7 (411)
Ultimate longitudinal strain	μin./in.	ϵ_x^{tu}	26,000.0	50,000.0
Ultimate transverse strain	μin./in.	ϵ_y^{tu}	26,000.0	50,000.0
Elastic Properties (Typical)				
Longitudinal tensile modulus	Msi (GPa)	E_x^t	2.59 (18)	1.16 (8.0)
Transverse tensile modulus	Msi (GPa)	E_y^t	2.59 (18)	1.16 (8.0)
Longitudinal compressive modulus	Msi (GPa)	E_x^c	2.59 (18)	1.16 (8.0)
Transverse compressive modulus	Msi (GPa)	E_y^c	2.59 (18)	1.16 (8.0)
In-plane shear modulus	Msi (GPa)	G_{xy}	7.92 (55)	7.65 (53)
Longitudinal Poisson's ratio	—	ν_{xy}	0.848	0.927
Transverse Poisson's ratio	—	ν_{yz}	0.848	0.927
Physical Constants (Typical)				
Density	lb/in.^{3} (g/cm^3)		0.0725 (2.007)	0.0725 (2.007)
Longitudinal coefficient of thermal linear expansion	μin./in.°F	α_x	3.10	3.78
Transverse coefficient of thermal linear expansion	μin./in.°F	α_y	3.10	3.78

(140-μm) composites cure to a nominal 0.0070 in. (0.178 mm) lamina thickness.

Key room-temperature and 177°C (350°F) design strengths, elastic properties, and physical constants are tabulated in Table 10.2 for a unidirectional (0°) laminate and in Table 10.3 for a ±45° orientation. Tensile strength, tensile modulus of elasticity, and shear data for the design of (0°, ±45°, 90°) laminates are given in Figs. 10.9–10.11.

The "A" basis unidirectional (0°) stress–strain curves in longitudinal and transverse tension and compression are presented in Figs. 10.12–10.15. The "A" basis in-plane shear stress–strain curves are presented in Fig. 10.16.

Typical ±45° stress–strain curves are presented in Figs. 10.17 and 10.18 for longitudinal tension and in-plane shear, respectively.

The effect of temperature variations on unidirectional (0°) boron–epoxy laminate properties is shown in Figs. 10.19–10.22 by means of percent of room-temperature property versus temperature curves. The properties at −54°C (−65°F) are considered to be the same as at room temperature.

The effect of temperature variation on the coefficient of thermal linear expansion is shown in Fig. 10.24. Fatigue data for boron–epoxy laminates are shown in Fig. 10.24.

The data that have been presented here constitute a representative sampling of the voluminous information assembled in the *Advanced Composites Design Guide*, which

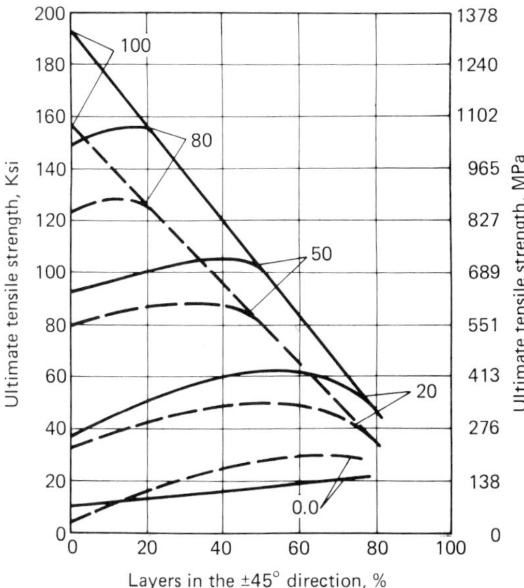

Figure 10.9. Tensile strength design data for (0°, ±45°, 90°) boron–epoxy laminate (V_f = 0.50): ———) room-temperature values; ----) values at 177° C (350° F). The numbers adjacent to the curves specify the percentage of layers in the 0° direction.

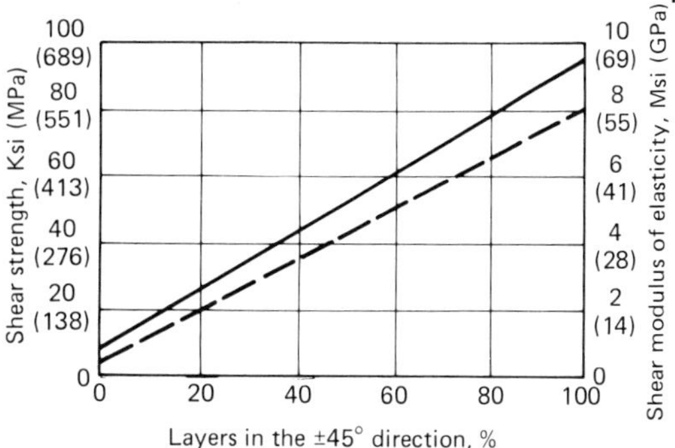

Figure 10.11. Shear-property design data for (0°, ±45°, 90°) boron–epoxy laminate (V_f = 0.50): ———) room-temperatures values; ----) values at 177° C (350° F).

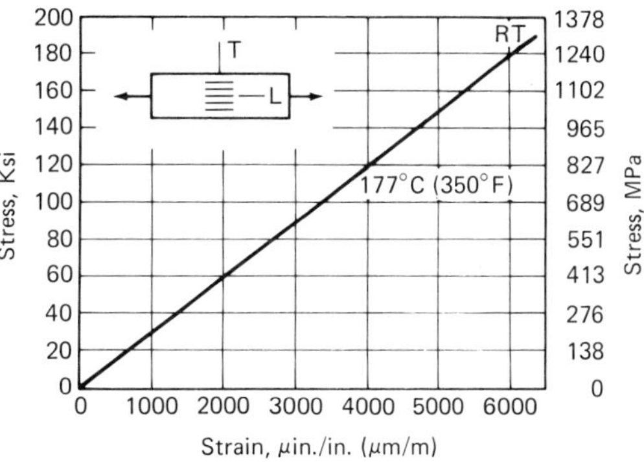

Figure 10.12. "A" basis stress–strain curves for unidirectional (0°) boron–epoxy laminate in longitudinal tension at room temperature (RT) and 177° C (350° F).

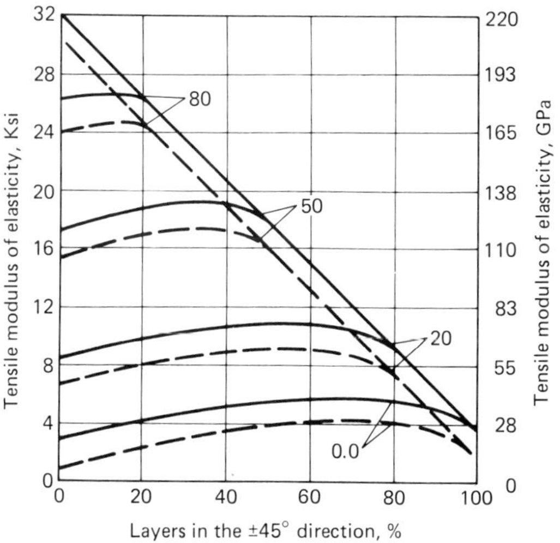

Figure 10.10. Tensile modulus of elasticity design data for (0°, ±45°, 90°) boron–epoxy laminate (V_f = 0.50): ———) room-temperature values; ----) values at 177° C (350° F). The numbers adjacent to the curves specify the percentage of layers in the 0° direction.

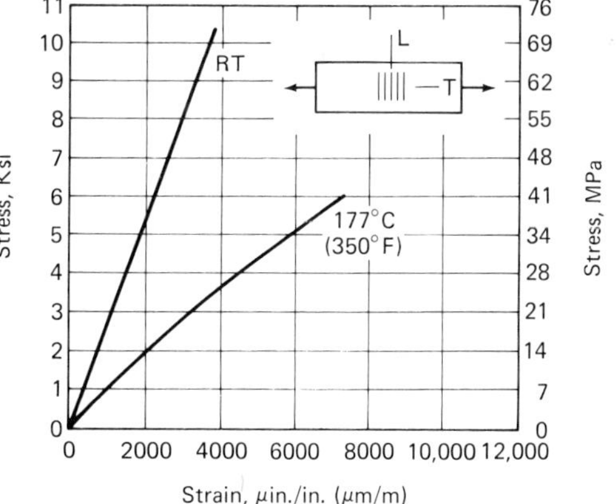

Figure 10.13. "A" basis stress–strain curves for unidirectional (0°) boron–epoxy laminate in transverse tension at room temperature (RT) and 177° C (350° F).

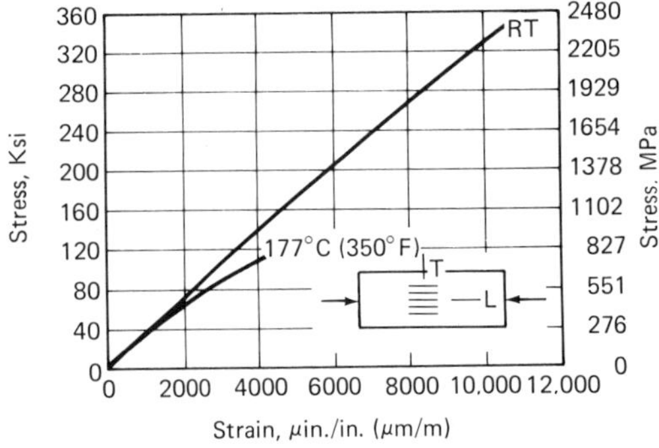

Figure 10.14. "A" basis stress–strain curves for unidirectional (0°) boron–epoxy laminate in longitudinal compression at room temperature (RT) and 177° C (350° F).

should be consulted for a more complete compilation.[51]

10.6.2. Silicon Carbide Filament–Resin Matrix Composites

The resin-matrix-composite data that have been developed for SiC–epoxy laminates are far less extensive than the data available for boron–epoxy laminates as a result of the newness of the material form. (It should be noted that the diameter of SiC filaments is exclusively 5.6 mils = 140 μm.) Table 10.4 presents the typical properties of an SiC–epoxy laminate which have been demonstrated in tests conducted up to the time of this writing with the use of a resin system that cures at 177° C (350° F).

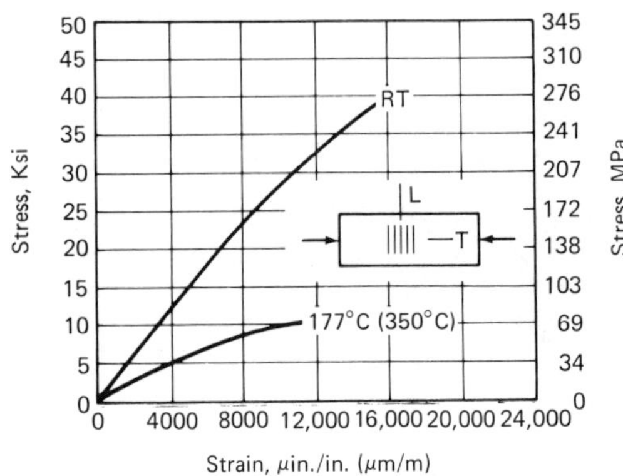

Figure 10.15. "A" basis stress–strain curves for unidirectional (0°) boron–epoxy laminate in transverse compression at room temperature (RT) and 177° C (350° F).

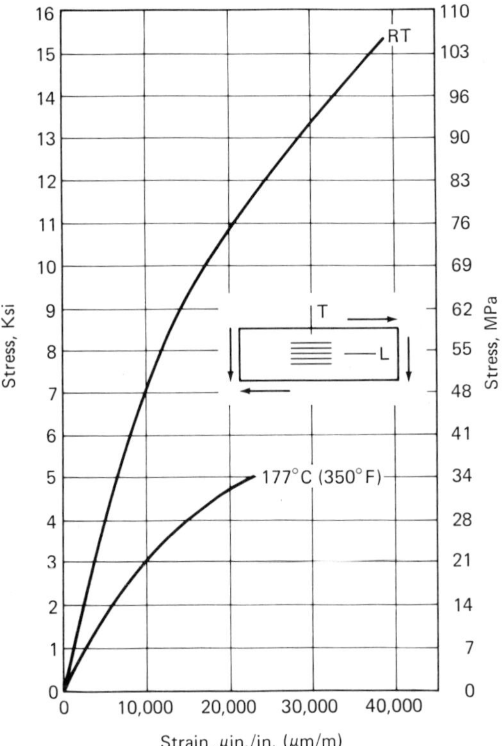

Figure 10.16. "A" basis stress–strain curves for a unidirectional (0°) boron–epoxy laminate in in-plane shear at room temperature (RT) and 177° C (350° F).

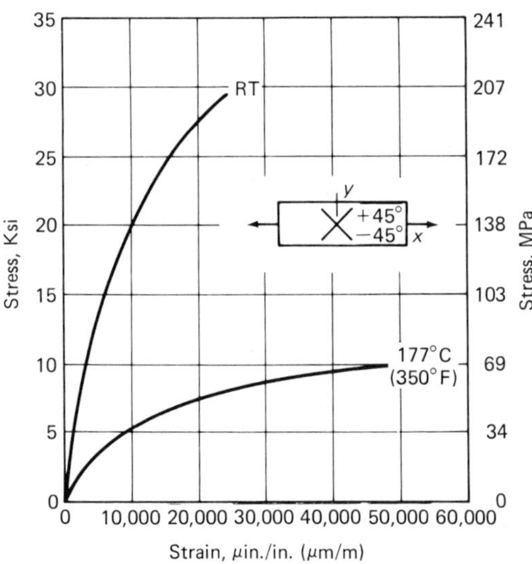

Figure 10.17. Typical stress–strain curves for a ±45° boron–epoxy laminate in longitudinal tension at room temperature (RT) and 177° C (350° F).

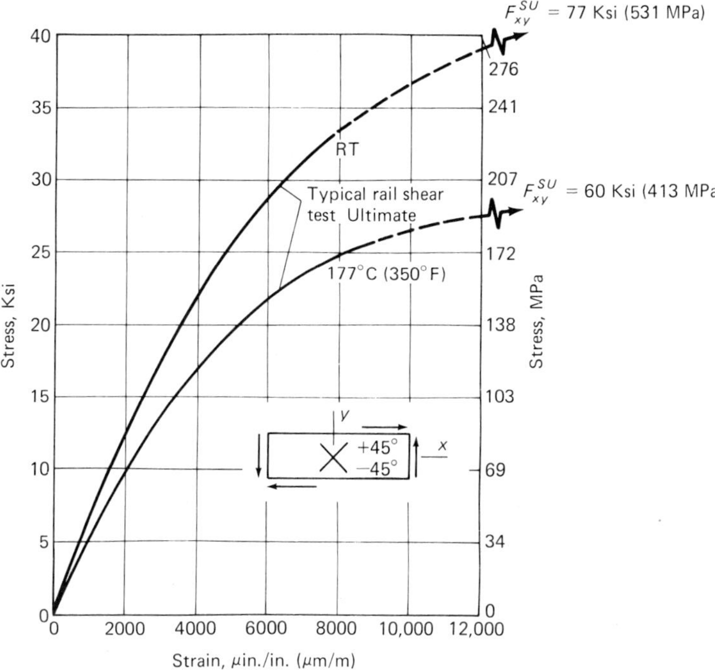

Figure 10.18. Typical stress–strain for a ±45° boron–epoxy laminate in in-plane shear at room temperature (RT) and 177° C (350° F).

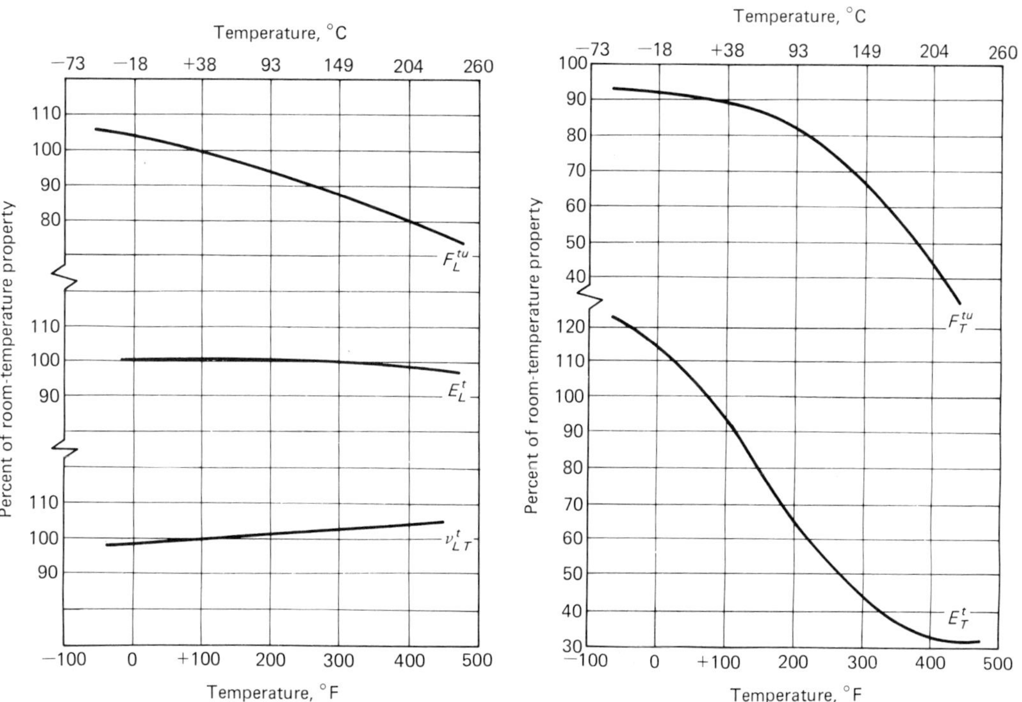

Figure 10.19. Properties (ultimate tensile strength, tensile modulus of elasticity, and Poisson's ratio) of a unidirectional (0°) boron–epoxy laminate in longitudinal tension as a function of temperature.

Figure 10.20. Properties (ultimate tensile strength and tensile modulus of elasticity) of a unidirectional (0°) boron–epoxy laminate in transverse tension as a function of temperature.

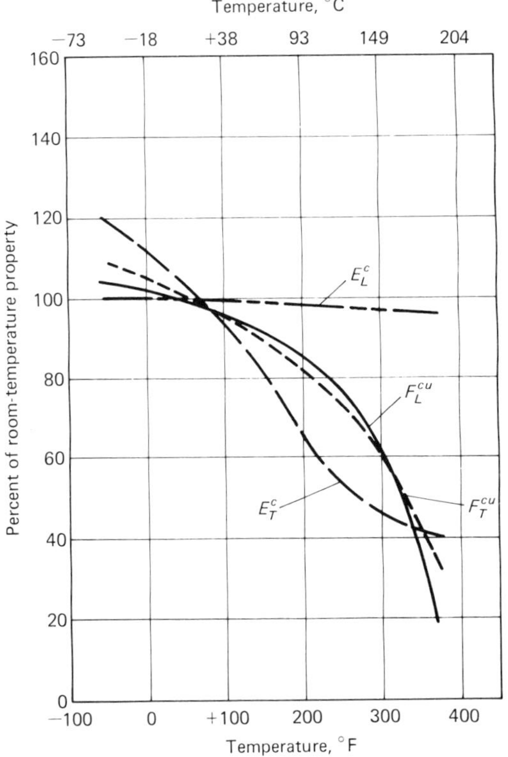

Figure 10.21. Properties (ultimate compressive strength and compressive modulus of elasticity) of a unidirectional (0°) boron–epoxy laminate in longitudinal and transverse compression as a function of temperature.

Inasmuch as the properties of SiC filaments are very similar or even superior to those of boron filaments, it is anticipated that the generation of more complete data will show the properties of SiC composites to also correlate well with those of their boron-composite counterparts.

It should be again noted that one of the main features of attraction of SiC filament-resin matrix composites is the potential of significantly lower cost as compared to boron composites. Recent cost projections for SiC filament indicate that sale prices as low as $40/lb ($88/kg) are realistically achievable in reasonably large volumes (approximately 100,000 lb = 45,359 kg). Since it is difficult to project boron-filament prices dropping much below $100/lb ($220/kg)—except at huge volumes—the advantages of a fiber-composite material that possesses properties that are equivalent or superior to those of boron, but is available at a much lower price, may prove to be quite important in the future.

10.7. APPLICATION OF BORON FILAMENTS

The superior mechanical properties of vapor-deposited boron filaments have proven attractive to many designers, and, as a result, boron filaments have been specified for many applications. These applications range from the aerospace industry's use of boron–epoxy and boron–aluminum composites to reduce the weight of various components of military and civil aircraft, on the one hand, to the leisure-product area, where such items as golf clubs and tennis rackets are reinforced with boron-epoxy composites to improve their strength, stiffness, and sensitivity, on the other hand. Table 10.5 summarizes all of the known

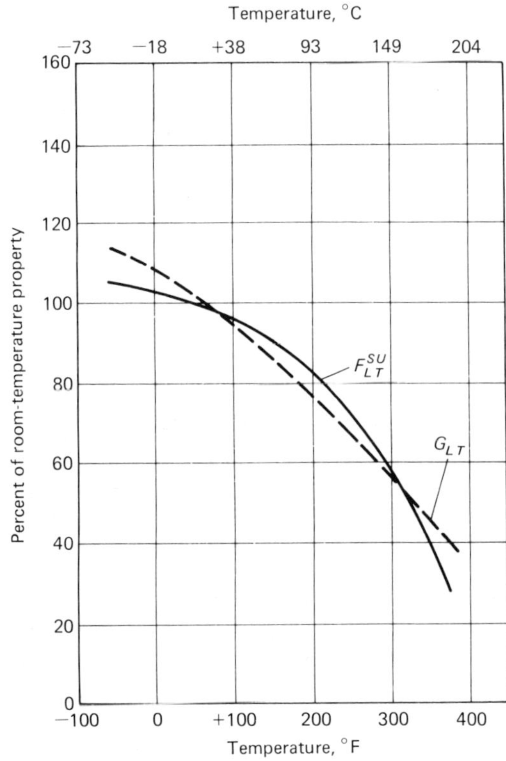

Figure 10.22. Properties (ultimate shear strength and shear modulus of elasticity) of a unidirectional (0°) boron-epoxy laminate in in-plane shear as a function of temperature.

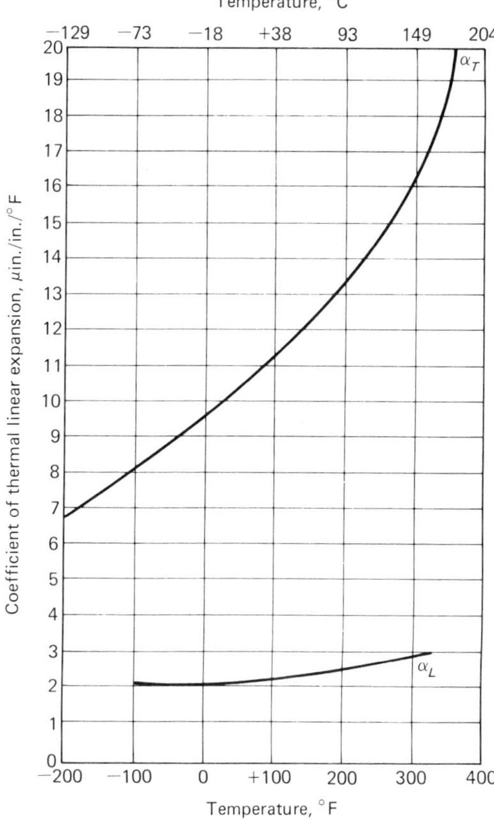

Figure 10.23. Instantaneous longitudinal and transverse coefficients of thermal linear expansion of a unidirectional (0°) boron–epoxy laminate.

applications of boron and SiC filaments and composites; it not only covers those applications that have resulted in large-volume orders, but also the very many successful research and development programs conducted by various private and government agencies. The primary and most successful applications are, of course, the F-14 (Grumman and Aerospace Corporation) F-15 (McDonnell-Douglas Corporation) fighter aircraft, where 145 and 215 lb (66 and 98 kg) of boron filament, respectively are used in every production vehicle. In each case, the boron–epoxy laminate is first bonded to titanium edge members and then bonded and bolted to the empennage substructure to form the skin surface of the horizontal stabilizer of the F-14 or the horizontal and vertical stabilizers as well as rudder of the F-15. Other boron–epoxy aircraft structures are similar in design, graphite fibers sometimes being added in the lower stressed areas in order to reduce cost. The Mirage 2000 rudder is an example of such a boron–graphite hybrid structure; it uses the higher strength/stiffness boron filaments along the critical +45° axis, whereas the graphite fibers are employed in the span and chordwise directions. The horizontal and vertical stabilizers for the B-1 bomber are a second example; although a boron/graphite–epoxy structure was called for in the design of this aircraft, production was halted by government decision. The above represent excellent examples of the optimum use of hybrid systems: The high-performance boron filaments are employed to carry the heavy wing-bending loads (and also the very heavy load-transfer stresses in the root joint area), whereas the graphite fibers are used to carry the torsion and chordwise loads.

A very innovative application of boron–epoxy laminates is their use as a means of increasing the fatigue life of metal components. In this application, instead of riveting it onto a metal patch, which sometimes does more harm than good by adding more fatigue-critical bolt holes, the boron–epoxy laminate is bonded directly to the metal surface; since boron–epoxy laminates are materials of very high modulus of elasticity and high strength, this procedure effectively (and with minimum weight) reduces the stress level in the base-metal components.

The F-111 and Air Machi (Australian Air Force) aircraft that are in service today use boron-epoxy repair patches to increase the service life of various critical structural components. Of the many attempts to incorporate boron–aluminum composites into aircraft, only the tubes of this material for the fuselage of the space shuttle have provided a successful application to date. These extremely high-performance, diffusion-bonded tubes are made by a unique process that incorporates the titanium "end-load" transition sleeves into a "one-shot" high-isostatic-pressure tool. These tubes range from 2 to 4 in. (5.08 to 10.16 cm) in diameter and form a truss structure that supports the lower section of the shuttle fuselage. The future of boron-aluminum com-

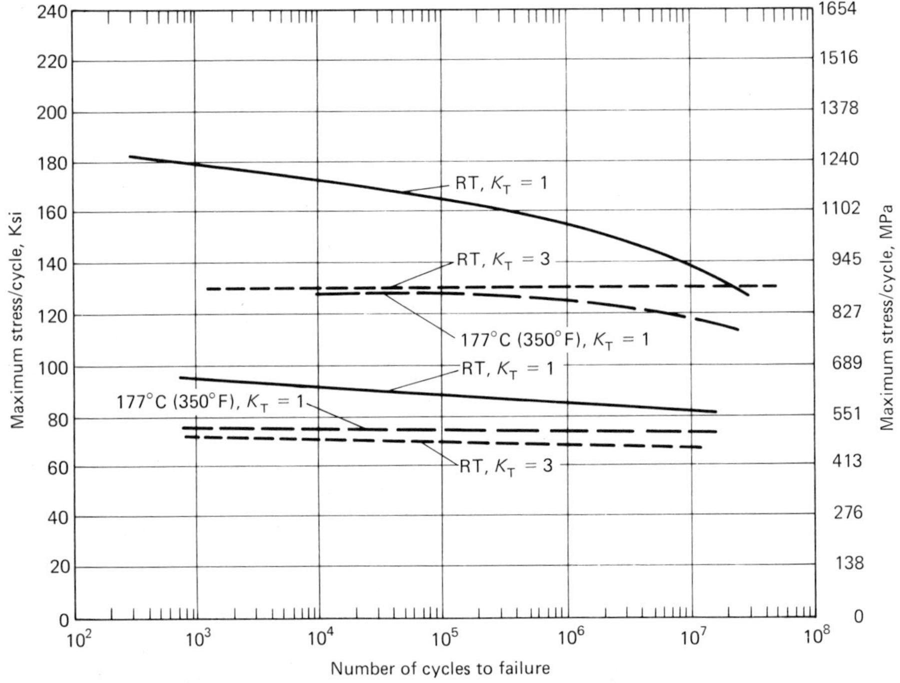

Figure 10.24. Constant-amplitude-fatigue design data for boron–epoxy laminates at room temperature (RT) and 177° C (350° F) (R = 0.1), 2000–3000 cpm; K_T denotes notch dimensions, not stress-concentration factor.

Table 10.4. Typical Properties of SiC–Epoxy Laminates (Nominal V_f = 0.50)

LAMINATE PROPERTIES	ROOM TEMPERATURE	177° C (350° F)
Longitudinal flexural strength, ksi (MPa)	330 (2274)	213 (1468)
Longitudinal flexural modulus of elasticity, Mpsi (GPa)	30.6 (211)	25.5 (176)
Transverse flexural strength, ksi (MPa)	18.3 (126)	11.3 (77.9)
Transverse modulus of elasticity, Mpsi (GPa)	3.1 (21.4)	1.2 (8.3)
Horizontal shear strength, ksi (MPa)	17.2 (119)	9.1 (62.7)
Transverse tensile strength, ksi (MPa)	11.0 (75.8)	—
Transverse tensile modulus of elasticity, Mpsi (GPa)	3.1 (21.4)	—
Transverse tensile strain, %	0.48	—
Density, lb/in.^{3} (g/cm^3)	0.80 (22.1)	—
PREPREG CHARACTERISTICS		
Resin content	26 wt %	
Volatile content	<1 wt %	
Tack (23° C = 73° F)	Good	
Filament count	140/in. (55/cm)	

Table 10.5. Applications of Boron and SiC Filaments and Composites

PRODUCT	COMPONENT AND/OR PRODUCER	COMPOSITE
	PRODUCTION	
F-14	Horizontal stabilizer	Boron–epoxy
F-15	Tail section, cabin floor, and stabilator	Boron–epoxy
Uttas helicopter	Structural beam reinforcement	Boron–epoxy
F-111	Wing pivot doubler	Boron–epoxy
Mirage 2000	Rudder	Boron/graphite–epoxy
Space shuttle	Fuselage	Boron–aluminum tubes
Fishing rods	Browning, Shakespeare, Rodon, and Phoenix	—
Tennis rackets	Head—Spalding, DuraFiber, Snauwaert & Depla, and Browning	—
Golf shafts	Aldilla	—
	RESEARCH AND DEVELOPMENT	
F-14	Overwing and fairing	Boron + other filaments/fibers—epoxy
A-7	Outer wing	Boron/graphite–epoxy
C-130	Wing box	Boron–epoxy-reinforced aluminum
F-4	Rudder	Boron–epoxy
707	Foreflap	Boron–epoxy
F-100	Engine fanblades	Boron–aluminum
C-5A	Wing slat	Boron–epoxy
CH-54	Fuselage string and tail skid	Boron–epoxy
Avco B210	Bicycle tubes	Boron–aluminum
B-1	Horizontal and vertical stabilizers and wing slat	Boron/graphite–epoxy
F-111	Horizontal stabilizer	Boron–epoxy
CH-47	Rotor blade	Boron–epoxy
Compressor blades	—	SiC–titanium
Advanced structures	—	SiC–titanium

posites in the aerospace industry probably lies in the fanblade market, where substantial performance benefits are possible if the resistance to damage by foreign objects can be proven. The development of B_4C–boron- and SiC-reinforced aluminum castings and moldings represents a further possibility.

On the commercial front, although boron–aluminum composites have not been at all successful due to their high processing costs, with the advent of the B_4C and SiC filaments as well as higher allowable processing temperatures, lower cost metal-matrix products are possible. For the last two years, Avco Corporation has been experimenting with a boron–aluminum bicycle, and they have succeeded in reducing the weight of a high-performance frame by over 40%. Lower cost processing could make this frame an attractive product.

In contrast, boron–epoxy composites have been very successful in the commercial market. Many fishing rod and tennis racket designs now utilize (in selected areas) a relatively small amount of boron–epoxy laminate to enhance local strength or increase sensitivity. The dynamic response of boron–epoxy fishing rods is particularly exciting. Tennis rackets that have a facing of boron–epoxy laminate (wood and glass rackets) provide a stiffer feeling, but do so without the "elbow-breaking" problems normally associated with the conventionally stiff rackets.

The future market for laminates based on

the large vapor-deposited filaments (e.g., boron or low-cost SiC) and epoxy resins will probably follow the previous examples of selective reinforcement (i.e., where a small quantity of the high-modulus/high-strength filaments can be cost effective) of both metals and graphite composite structures. In the metal-matrix area, however, a large potential market exists in both medium- and high-temperature aerospace applications. The manufacturer of aerojet fanblades and compression blades from SiC-titanium laminates offers the potential for increased performance and high impact resistance (foreign object damage). As a result of their exceptional properties at 316–482° C (600–900° F), SiC filaments also offer significant weight savings in the selectively reinforced structures that are now being developed by the U.S. Air Force for advanced aircraft and missile structures. Both B_4C–aluminum and SiC–aluminum composites are also potentially attractive for use in hot-molded or cast heavily loaded aircraft structures operating in the 204–260° C (400–500° F) temperature range.

REFERENCES

1. C. P. Talley, "Mechanical Properties of Glassy Boron," *J. Appl. Phys.* **30**, 1114 (1959).
2. C. P. Talley, "Boron Reinforcements for Structural Composites," ASD-TDR-62-257, April 1963.
3. C. P. Talley, "Boron Filament Process Development," AFML-TR-67-120, May 1967.
4. L. E. Line, Jr., and U. V. Henderson, Jr., "Boron Filament and Other Reinforcements Produced by Chemical Vapor Plating," in: *Handbook of Fiberglass and Advanced Plastic Composites* (edited by G. Lubin), Van Nostrand Reinhold, New York, 1971, Chapter 10.
5. C. P. Talley, "High Modulus, High Strength Reinforcements for Structural Composites," ML-TDR-64-88, November 1963–August 1965.
6. L. C. McCandless et al., "High Strength, High Modulus, Low Density Continuous Filaments for Use as Reinforcements in Composites," AFML-TR-65-265, Parts I–IV, 1965–1969 (Contract AF 33615-67-C-1646).
7. F. Galasso, M. Basche, and D. Kuehl, "Preparation, Structure, and Properties of Continuous Silicon Carbide Filaments," *Appl. Phys. Lett.* **9**, No. 1, 37 (July 1966).
8. F. Galasso, *High Modulus Fibers and Composites*, Gordon & Breach, New York, 1970.
9. W. B. Robbins, "Deposition Chamber for Manufacture of Refractory Coated Filaments," U.S. Patent 3,367,304, February 6, 1968.
10. H. E. DeBolt and V. J. Krukonis, "Improvement of Manufacturing Methods for the Production of Low Cost Silicon Carbide Filament," AFML-TR-73-140, 1973.
11. J. A. Cornie, J. Henshaw, P. Fuce, and A. Hauze, "Development of Silicon Carbide Aluminum Material," AFML-TR-78-167, November 1978.
12. "Surface Enhancement of Silicon Carbide Filament for Metal Matrix Composites," Contract N00014-79-C-0349, 1979.
13. M. Leclercq et al., "Boron Filaments with a Boron Carbide Antidiffusion Coating, and Metal Matrix Made Therefrom," U.S. Patent 3,864,224, November 5, 1974.
14. F. E. Wawner, "Boron Filaments," in *Modern Composite Materials* (edited by L. J. Broutman and R. H. Krock), Addison-Wesley, Reading, Massachusetts, 1967, Chapter 10.
15. V. J. Krukonis, "Chemical Vapor Deposition of Boron Filament," in: *Boron and Refractory Borides* (edited by V. I. Matkovich), Springer-Verlag, New York, 1977, Section D. I.
16. H. E. DeBolt, R. J. Diefendorf, P. E. Gruber, L. A. Joó, V. J. Krukonis, J. A. McKee, R. M. Mehalso, "Lower Cost High Strength Boron Filament," AFML-TR-70-287, June 1971.
17. H. E. DeBolt, V. Krukonis, J. McKee, "Development and Demonstration of a Low Cost Boron Filament Formation Process," AFML-TR-72-271, May 1972.
18. J. A. McKee, H. E. DeBolt, V. J. Krukonis, R. Prescott, G. E. Sharpe, "Production Process for Large Diameter Carbon Filament Substrate," AFML-TR-74-141, July 1974.
19. H. E. DeBolt, R. Prescott, J. McKee, G. Sharpe, "Carbon Monofilament Production," AFML-TR-76-231, December 1976.
20. R. M. Mehalso and R. J. Diefendorf, "High Strength, High Modulus Boron Vapor Deposited on a Carbon Monofilament Substrate," in: *Proceedings of a Symposium on Composites and Carbon, April 26–28, 1971,* American Ceramic Society, Columbus, Ohio, 1972, pp. 51–58.
21. F. E. Wawner, Jr., and H. E. DeBolt, "Investigation of Elongation and Its Relationship to Residual Stresses in Boron Filaments," Contract N00014-76-C-0694, A056528, 1977.
22. M. Basche, R. Fanti, F. Galasso, V. Kuntz, R. Schile, "Method for Producing Boron-Carbon Fillers," U.S. Patent 3,679,475, July 25, 1972; R. E. 28, 312, January 21, 1975; "Node Free Boron Composite Filament," U.S. Patent 3,861,953
23. H. E. DeBolt, "Carbon Filament Coated with Boron and Method of Making Same," U.S. Patent 4,142,008, February 27, 1949.
24. H. E. DeBolt, V. Krukonis, R. M. Neff, and F. E.

Wawner, Jr., "Chemical Vapor Deposition of Boron on a Carbon Monofilament Substrate," Paper Presented at the 17th National SAMPE Symposium, Los Angeles, California, 1972.

25. "High Strength Boron," Contract NAS3-20577, NASA Lewis Research Center, 1976.
26. D. Behrendt, "Some Properties of an Advanced Boron Fiber," NASA TM 79065, Paper Presented at Meeting of the American Ceramic Society, January 22–24, 1979.
27. R. J. Smith, "Changes in Boron Fiber Strength Due to Surface Removal by Chemical Etching," NASA TM D-8219, April 1976.
28. J. C. Withers, L. C. McCandless, and R. T. Schwartz, "Continuous Silicon Carbide Filaments," Paper Presented at the 10th National SAMPE Symposium, San Diego, California, November, 1966.
29. R. C. White and H. R. Davis, "Silicón Carbide Structure," U.S. Patent 3,508,954, April 28, 1970.
30. S. J. Michalik and M. Weinstein, "Silicon Carbide Filaments," Contract N00019-68-C-0177, Naval Air Systems, 505276, February 1969.
31. P. E. Elkins, G. M. Mallen, and H. Shimizer, "Modified Silicon Carbide Continuous Filaments," Paper Presented at the 10th National SAMPE Symposium, November 1966.
32. G. M. Mallen and H. Shimizer, "Research on Modified Silicon Carbide Filaments for High Temperature Applications," Contract N00019-67-C-0465, Naval Air Systems, 50186, July 1968.
33. H. E. DeBolt, V. J. Krukonis, and F. E. Wawner, Jr., "High Strength, High Modulus Silicon Carbide Filament via Chemical Vapor Deposition," in: *Silicon Carbide—1973, Proceedings of the 3rd International Conference on Silicon Carbide, September 17–20, 1973*, University of South Carolina Press.
34. H. E. DeBolt, V. J. Krukonis, and F. E. Wawner, Jr., "Low Cost, High Strength CVD Silicon Carbide Filament for High Temperature Reinforcement Applications," Paper Presented at the 19th National SAMPE Symposium, April 1974.
35. J. Marsden, Unpublished Work, Union Carbide Corporation, Tarrytown, New York.
36. J. L. Randon, G. Slama, and A. Vignes, "Silicon Carbide Fibers with Improved Mechanical Properties: Study of Thermal Stability," in: *Silicon Carbide—1973, Proceedings of the 3rd International Conference on Silicon Carbide, September 17–20,* 1973, University of South Carolina Press, Los Angeles, California, 1973.
37. J. Cornie, A. Unpublished Work.
38. A. G. Metcalfe, "Physical-Chemical Aspects of the Interface," in: *Modern Composite Materials* (edited by L. J. Broutman and R. H. Krock), Addison-Wesley, Reading, Massachusetts, 1974, Volume 1, (Interfaces in Metal Matrix Composites—edited by A. G. Metcalf) Chapter 3.
39. M. Basche, R. Fanti, and F. Galasso, *Fiber Sci. Technol.* **1**, 19 (1968).
40. R. Naslain, J. Thebault, and R. Pailler, "Chemical Compatibility in Metal–Boron Composites," in: *Proceedings of the 1975 International Conference on Composite Materials*, The Metallurgical Society of AIME, 1975, Volume 1, pp. 116–136.
41. A. G. Metcalfe, "Fiber Reinforced Titanium Alloys," in: *Modern Composite Materials* (edited by L. J. Broutman and R. H. Krock), Academic Press, Reading, Massachusetts, 19, Volume 4 (*Metal Matrix Composites*—edited by K. Krieder), Chapter 6.
42. J. L. Camahort, "Protective Coating by Surface Nitriding of Boron Filament," *J. Comp. Mater.* **2**, 104–111 (1968).
43. V. J. Krukonis and J. L. Camahort, "Exploratory Development and Evaluation of Low Cost Boron–Aluminum Composites," AFML-TR-75-197, AD-A027069, November 1975 (Contract F33615-74-C-5082).
44. W. J. Heffernan, R. W. Haskell, and I. Ahmed, "A Continuous CVD Process for Coating Filaments with Tantalum Carbide," in: *Proceedings of the Fourth International Conference of the Electrochemical Society*, The Metallurgical Society of AIME, 1979 p. 489.
45. I. Ahmed, D. N. Hill, and W. J. Heffernan, "Effect of CVD Coatings on the Mechanical Behavior of SiC (C) Filaments," (to be published).
46. I. Ahmed et al., "Development of High Strength Filament Reinforced Superalloy Composites," WVT-TR-74005, Ad-776346.
47. I. Ahmed, D. N. Hill, and W. J. Heffernan, "Silicon Carbide Filaments as Reinforcements for High Temperature Alloy Matrices," in: *Proceedings of the 1975 International Conference on Composite Materials, April, 1975,* 1975, Volume 1, pp. 85–102.
48. J. A. Cornie and J. Hakim, in: *Proceedings of the 5th International Conference on Chemical Vapor Deposition,* Electrochemical Society, Inc. 1975, p. 634.
49. T. Schoenberg et al., "Establishment of a Manufacturing Process for the Production of Boron Epoxy Tape," AFML-TR-73-185, July 1973.
50. T. Schoenberg et al., "Boron Broadgoods Manufacturing Process," AFML-TR-78-34, April 1978.
51. *Advanced Composites Design Guide*, Third Edition, Second Revision, Air Force Flight Dynamics Laboratory (FBC), Wright-Patterson Air Force Base, Dayton, Ohio, 1976.

11
GRAPHITE FIBERS AND COMPOSITES

Dennis M. Riggs*
Exxon Enterprises, Materials Division,
Fountain Inn, South Carolina

Richard J. Shuford
Composites Development Division
Army Materials and Mechanics Research Center
Watertown, Massachusetts

Robert W. Lewis
Composites Development Division
Army Materials and Mechanics Research Center
Watertown, Massachusetts

11.1. INTRODUCTION

The technological demand for materials that exhibit improved strength and stiffness characteristics has led to considerable research and development in the field of fiber-reinforced resin-matrix composites. The high mechanical strengths, high moduli of elasticity, and low densities of fibers made from such substances as carbon, boron, and glass, when combined with the toughness of various epoxy, polyester, and polyimide resins, produce a class of materials possessing specific tensile properties that can match or exceed (depending on the fiber—matrix combination) those of the best metal alloys currently in production. This is important because materials that are strong, stiff, and, in addition, *lightweight* are necessary for many applications. By using composite technology, it is now possible to tailor-make structural materials for specific applications. (See Appendix B.)

*The authors wish to acknowledge the contribution of Alan M. Litman, Rebecca M. Jurta, and John J. Deluca in the preparation of this chapter, Cheryl A. Burns for her patience and skill in typing the manuscript, and especially William W. Houghton without whose tireless editing efforts the task would have been much more difficult.

Carbon fibers are by far the predominant high-strength, high-modulus reinforcing agent currently used in the fabrication of high-performance resin-matrix composites. The unique properties obtainable in graphite fiber–resin matrix composites are directly attributable to the properties of the graphite fibers themselves, which, in turn, are directly attributable to the highly anisotropic nature of the graphite crystal. The crystallographic structure of a perfect single crystal of graphite is shown in part (*a*) of Fig. 11.1.[1] As can be seen, the graphite crystal is composed of many sheet-like layers of carbon atoms which are stacked one on top of the other and separated by a distance of 3.35 Å. In the plane of the sheets (i.e., the basal plane), the carbon atoms are linked together by very strong covalent bonds. As a result, the theoretical tensile modulus of elasticity and ultimate tensile strength of the crystal in a direction parallel to the basal planes are very high—on the order of 146 Msi (1000 GPa)[2] and 15 Msi (100 GPa),[3,4] respectively. On the other hand, relatively weak Van der Waals bonds hold the sheet-like layers of carbon atoms together in a direction normal to the basal plane; consequently, the mechan-

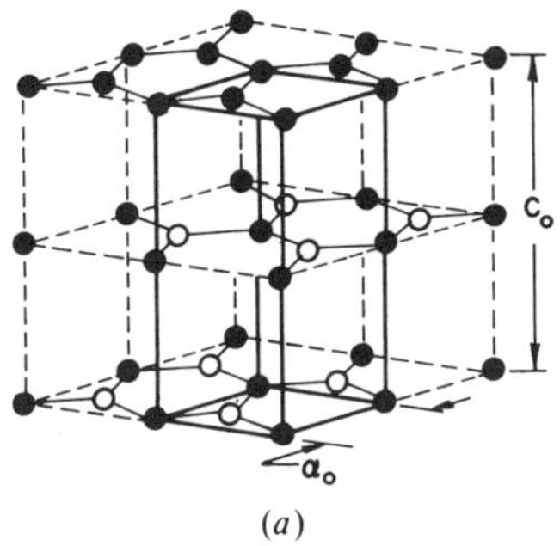

(a)

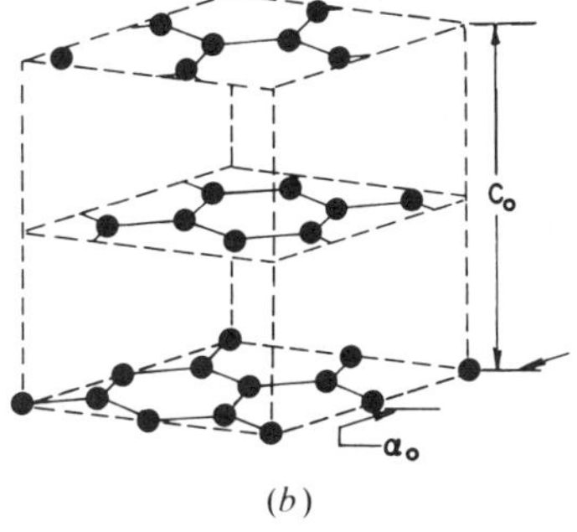

(b)

Figure 11.1. Crystallographic structure of (a) graphite and (b) turbostratic graphite.[1]

ical properties of the crystal in this direction are much poorer. For example, the theoretical tensile modulus of elasticity normal to the basal plane is only 5 Msi (35 GPa).

In a graphite fiber, the structure of the crystallites is not that of the perfect single crystal shown in part (a) of Fig. 11.1. Instead, the stacking arrangement of the various carbon sheets is slightly displaced, thus forming a "turbostratic" graphite [see part (b) of Fig. 11.1]. The interlayer "d" spacing of turbostratic graphite is on the order of 3.40–3.45 Å (as compared to the 3.35 Å separation of "perfect" graphite); however, the orientation of the graphite sheets is still more or less parallel to the fiber axis. Thus, even though the crystal structure is turbostratic, the properties attributable to the strong covalent bonding parallel to the basal planes should nevertheless be evident in the longitudinal properties of the graphite fibers, especially as the degree of alignment of the basal planes parallel to the fiber axis increases. This is indeed what is observed, as evidenced by the fact that the higher-modulus fibers possess higher degrees of preferred orientation.

In this chapter, the term "graphite fiber" will refer to those fibers that (1) have been heat-treated at temperatures in excess of 1700° C (3092° F), (2) possess high degrees of preferred orientation, and (3) have high tensile moduli of elasticity on the order of 50 Msi (345 GPa). The term "carbon fiber" will refer to fibers that have a tensile modulus of elasticity of up to approximately 50 MSi (345 GPa) and lower degrees of preferred orientation; "carbon fibers" are generally processed at temperatures lower than 1700° C (3092° F).

11.2. HISTORICAL REVIEW OF THE DEVELOPMENT OF CARBON FIBER

The current technology for producing carbon fibers generally centers on the thermal decomposition of various organic precursors. Rayon, polyacrylonitrile (PAN), pitch, polyesters,[5] polyamides,[6] polyvinyl alcohol,[7] polyvinylidene chloride,[8] poly-p-phenylene,[9] and phenolic resins[10,11] have all been considered and investigated as potential precursor materials for producing carbon fibers. Of these, rayon, polyacrylonitrile, and pitch have been found to offer the greatest potential in terms of carbon yield, cost, and so on. Hence, probably all current, commercially available carbon fibers are made using one of these three precursor materials.

Carbon fibers based on rayon (or cellulose) were first investigated as long ago as 1880. It was during this year that Thomas Edison first patented his incandescent lamp.[12] The filaments that he used in the lamp were made of carbon and had been produced by pyrolyzing natural or regenerated cellulose fibers. The carbon filaments themselves were very fragile; they contained numerous pores and were susceptible both to forming "hot spots" and to high-temperature oxidation. In an effort to overcome these difficulties and extend the life expectancy of the filaments, methods for applying pyrolytic coatings to the filaments were developed. Over the next 30 years, numerous process patents[13,14] were granted for the pyrolytic coatings, the last going to Whitney[15] in 1909. At about this time, however, lamps with flexible tungsten filaments initially became available, and thus interest in carbon filaments dropped significantly. It was not until the mid-1950's that interest in fibrous carbon was again revived. At this time, Soltes[16] and Abbott[17] developed processes for converting both natural cellulose and rayon into fibrous carbon materials. Essentially, the carbon fibers were produced by heat treating the precursors to temperatures on the order of 1000° C (1832° F) in inert atmospheres. It was found that the tensile strengths of these fibers were as high as 40 Ksi (275 MPa).[18] The driving force behind this renewed interest in carbon fibers was the search for superrefractory reinforcing agents to be used in ablative composites for rocket and missile components. It was also recognized that carbon fibers could be used for a variety of other applications, such as resistance heaters, filters for hot, corrosive gases and liquids, reinforced plastics, thermal insulation, and nuclear moderators. In 1959, Union Carbide Corporation began production of carbon cloths, felts, yarns, and battings using rayon fibers as the precursor material. [19,20] The carbon fibers were "batch processed" by heat treating the rayon precursors in inert atmospheres at about 900° C (1652° F) subsequently carbonizing the fibers at temperatures generally greater than 2500° C (4532° F). The fibers produced by this method possessed tensile strengths between 48 and 130 Ksi (330 and 900 MPa).

In 1960, the potential elastic properties obtainable from highly ordered graphite crystals were demonstrated when Bacon[21] reported on the growth, structure, and properties of graphite whiskers grown in a DC arc at about 3600° C (6512° F) under a pressure of 92 atm. He showed that when the graphite basal plane (0002) is oriented parallel to the whisker axis, tensile strengths in excess of 3 Msi (20 GPa) and tensile modulus of elasticity values over 100 Msi (690 GPa) were possible. The theoretical maximum tensile modulus of elasticity for graphite parallel to the basal planes is 146 Msi (1000 GPa).[2] Methods for obtaining the high degree of preferred orientation in the fibers necessary for the high tensile strength and modulus values were subsequently investigated by a number of researchers. In 1967, a continuous process for making rayon-base carbon fibers was patented by Bacon et al.[22] This development resulted in commercial-scale production of continuous lengths of carbon fiber having tensile strengths between 100 and 150 Ksi (690 and 1030 MPa) and tensile modulus of elasticity values on the order of 6 Msi (40 GPa). The properties of these fibers, however, were still considerably inferior to those of the graphite whiskers. In 1964, Tang and Bacon,[23,24] based on infrared, x-ray, and transmission electron microscopy, postulated a mechanism that described the thermal degradation of cellulose molecules to carbon and graphite. From this, it was shown that the fiber structure in the graphitized filament was directly related to the molecular fiber structure in the starting material. It was found that the tensile modulus of elasticity increased by a factor of two or three when the precursor filaments were subjected to a 150% stretch prior to carbonization.[25] In addition, in 1965 it was found[26,27] that when stress is applied to the continuous fibers at temperatures in excess of 2500° C (4532° F), plastic deformation occurs within the fiber, which thereby tends to orient the graphite basal planes parallel to the fiber axis (i.e., the direction of applied stress) and hence significantly increases the tensile modulus of elasticity of the fibers. This development led to the commercial production of high-strength, high-modulus Thornel 25 by Union Carbide Corporation. The average tensile modulus of elasticity of this fiber was 25 Msi (170 GPa), whereas the tensile strength averaged 184 Ksi (1270 MPa). By 1970, it was possible, using this same general process, to obtain commercially available rayon-base carbon fibers (Thornel 75) with a tensile modulus of elasticity of 75 Msi (520 GPa) and a tensile strength of 385 Ksi (2650 MPa).

The use of polyacrylonitrile (PAN) as a precursor material for carbon fibers was first reported in a patent application by Tsunoda[28] in 1960. In his patent, Tsunoda revealed that carbon fibers could be processed from PAN precursors by first crosslinking—or "stabilizing"— the polymer in an oxidizing atmosphere at 220° C (428° F) and then heat treating the stabilized fibers at 1000° C

(1832° F) in an inert atmosphere. Although Tsunoda's PAN-base carbon fibers exhibited rather poor properties, work continued on this system. In 1961, Shindo[29] published the results of his work on carbonized PAN fibers. He showed the variation in the tensile strength and modulus of elasticity of the fibers with heat-treatment temperature and also discussed the electrical properties. Shindo reported values of tensile strength on the order of 80–100 Ksi (550–690 MPa) and tensile moduli of elasticity of 25 Mpsi (170 GPa). In 1964, Watt and Johnson[30] of the Royal Aircraft Establishment filed a patent for high-strength, high-modulus graphitized fiber based on a PAN precursor. Watt and Johnson had independently developed their process without the knowledge that the work had been done previously by Shindo et al.[31] In 1966, carbon fibers of very high tensile modulus of elasticity—70 Msi (480 GPa) and tensile strength—300 Ksi (2070 MPa)—were reportedly made using a PAN precursor by Prescott and Stendege[32] as well as Watt et al.[30,33,34] and Clark and Bailey[39]. The high tensile modulus of elasticity was due to a tension that was applied to the fibers during an early stage of their preparation. Since 1966, numerous other patents[35–37] have been granted for carbon fibers based on PAN precursors. Basically, all of these processes involve first stretching the PAN precursor to obtain a high degree of molecular orientation of the polymer molecules and then subjecting it to a stabilizing treatment in an oxidizing atmosphere while the fibers are held under tension. Subsequently, the fibers are carbonized at temperatures ranging from 1000 to 3500° C (1832 to 6332° F). The mechanism of stabilization[38–40] as well as the development of the structure and properties[41–46] of carbon fibers have been studied in detail by a number of investigators. Currently, it is possible to obtain commercially produced PAN-base carbon fibers with tensile modulus of elasticity values ranging from 30 to 116 Msi (210 to 800 GPa) and tensile strengths of 235–475 Ksi (1620–3275 MPa).

The use of pitch as a precursor material for carbon fibers was reported by a number of investigators,[47–49] starting with Otani[50] in 1965. The fibers were generally made by first melt spinning a low-cost, isotropic molten pitch or similar petroleum product and then oxidizing the as-spun filaments. This treatment was followed by carbonization at about 1000° C (1832° F) in an inert atmosphere. The properties of the isotropic, carbonized fibers were generally poor, tensile modulus of elasticity values being approximately 5–10 Msi (35–70 GPa).[47] Improvements in this process were reported by Otani et al.[51] in 1969 and Hawthorne et al.[52,53] in 1970 and 1971. It was found that by stretching the fibers either during the initial stages of carbonization or at temperatures above 2500° C (4532° F), a high degree of basal-plane preferred orientation could be induced, and thus tensile strengths as high as 375 Ksi (2585 MPa) and tensile moduli of elasticity in excess of 70 Msi (480 GPa) could be attained.

Another method for producing carbon fibers from pitch was developed and patented in 1973.[54,55] In this method, the pitch was transformed into a liquid-crystalline (mesophase) state prior to spinning into filamentary form. The liquid crystals were formed by extended heat treatment at about 400–450° C (752–842° F) in an inert gas atmosphere.[56] Once the pitch had been at least partially converted to the liquid-crystalline state, the material was spun, oxidized, and subsequently carbonized at 1000–3000° C (1832–5432° F). The fibers obtained in this manner possessed a high degree of preferred orientation due to the formation of the liquid-crystalline state prior to spinning. Depending on the final heat-treatment temperature, the tensile modulus of elasticity values for these fibers could be as high as 100 Msi (690 GPa).[57] A similar process was described in 1975 by Fujimaki et al.[58] The mechanism of mesophase formation was discussed in detail by Riggs[59] in 1976.

11.3. CARBON-FIBER PRECURSORS

As discussed in the previous section, a number of different polymeric fibrous materials[60–64] have been tested and evaluated as potential precursors for the production of carbon fibers. The pyrolysis of organic precursor fibers into carbon fibers has been the preferred method of production because of the inherent advantages this technique offers compared to other graphite-filament formation processes, such as the growing of graphite whiskers in an arc[21] or the chemical vapor deposition (CVD) of pyrolytic carbon on appropriate substrates. Specifically,[65] the pyrolysis of organic precursors into carbon fibers is a practical method for producing large quantities of graphite fibers that have reproducible and desirable properties. In addition, the pyrolysis technique tends to be much less expensive than both the whisker-growth and CVD processes because it is faster and meticulous control of the system atmosphere and pressure is not so critical. Furthermore, compared to the whisker-growth and CVD processes, the pyrolysis technique makes it much easier to obtain very long, fine-diameter, flexible carbon fibers. This is important because the fine, flexible fibers can be fabricated into composites using techniques developed for the production of fiberglass composites.

Candidate organic materials for pyrolysis into carbon fibers having good properties should satisfy four criteria.[65] First of all, the precursor should possess the appropriate strength and handling characteristics needed "to hold the fibers together" during all stages of the conversion process to carbon. Secondly, the precursor should not "melt" during any stage of the conversion process. This can be accomplished by either selecting infusible precursor materials or by thermosetting thermoplastic precursors prior to the conversion process. Thirdly, the precursor material must not completely volatilize during the pyrolysis process, that is, the carbon yield of the precursor fiber after pyrolysis should be appreciable enough to justify its use on an economic basis. Furthermore, in order to obtain the optimal possible properties, the carbon atoms should tend to array themselves in aligned graphitic structures during pyrolysis: In general, the more highly graphitic and oriented the fibers, the better the mechanical properties. Finally, the precursor material should be as inexpensive as possible. This is necessary, since the cost of the finished carbon fiber will partially reflect the cost of the precursor. In addition, the more inexpensive the carbon fiber becomes, the more widely it will tend to be used.

As noted in the preceding section, the organic precursor materials that have best satisfied these four criteria to date are rayon, polyacrylonitrile (PAN), and pitch.

11.3.1 Carbon Fibers from Polyacrylonitrile (PAN)

11.3.1.1. The PAN Conversion Process

The process by which PAN is converted to carbon fibers involves five steps:

1. Spinning the PAN precursor.
2. Stretching the precursor.
3. Stabilization at 220° C (428° F) in air under tension.
4. Carbonization at ~1500° C (2732° F) in inert atmospheres.
5. Graphitization at ~3000° C (5432° F) in inert atmospheres.

Each of these steps will be discussed in the following sections.

11.3.1.2. Characteristics and Copolymers of PAN

Polyacrylonitrile (PAN) is an atactic, linear polymer consisting of a carbon–hydrogen backbone with polar carbon–nitrogen (nitrile) pendant groups attached. The structure of an ideal PAN molecule is shown in Fig. 11.2.[59]

The polarity of the nitrile groups significantly affects the physical properties of the polymer. The second-order transition (glass-transition) temperature of PAN is quite high—on the order of 120° C (248° F)—primarily due to the relatively strong intermolecular bonding forces that result from the polarity of the

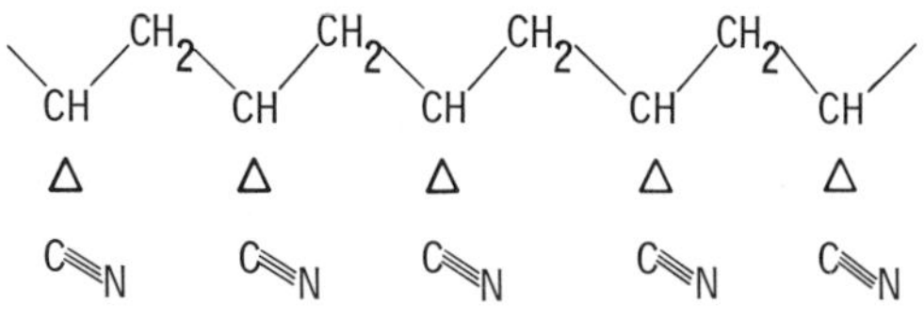

Figure 11.2. Structure of the ideal PAN molecule.[59]

nitrile groups. It is for this reason that, in making fibers, PAN is frequently copolymerized with other monomers, such as methyl acrylate (Courtelle 94% PAN–6% methyl acrylate)[66] or vinyl acetate. These additions lower the glass-transition temperature sufficiently enough to allow textile manufacturers to stretch the as-spun fibers in boiling water.[66,67] The strong polarity of the nitrile group also significantly affects the solution characteristics of the polymer. This is reflected in the high value of the solubility parameter[68] [15.4 $(cal/cc)^{1/2}$], which is a measure of the cohesive energy density of the polymer. Thus, only highly polar solvents can be used to solubilize PAN. The strong intermolecular bonding caused primarily by the polarity of the nitrile groups is also important when producing carbon fibers from PAN or its copolymers because the fiber will tend to decompose before it melts. This has very important implications, as will be seen in the next section.

Both PAN and its copolymers are generally produced by a wet-spinning process in which first a solution of the polymer and an appropriate solvent[69–72] are spun into a coagulating bath and then the fibers are washed, stretched, and dried. The molecular structure and properties of the as-spun fibers will be significantly influenced by the spinning-process variables. Wet spinning usually results in fibers that have a round cross section. However, the effect of the spinning-process variables on the properties of carbonized fibers is debatable.

Wide-angle x-ray diffraction patterns obtained from PAN fibers show a major reflection that corresponds to a chain-to-chain separation distance of 6 Å. A typical diffaction pattern that was obtained from a stretched PAN fiber is shown in Fig. 11.3.[59] There appears to be no order in the direction of the fiber axis, and the maximum degree of crystallinity is only about 50%. A preferred orientation in the precursor is necessary in order to obtain high-performance properties in the carbonized fibers.

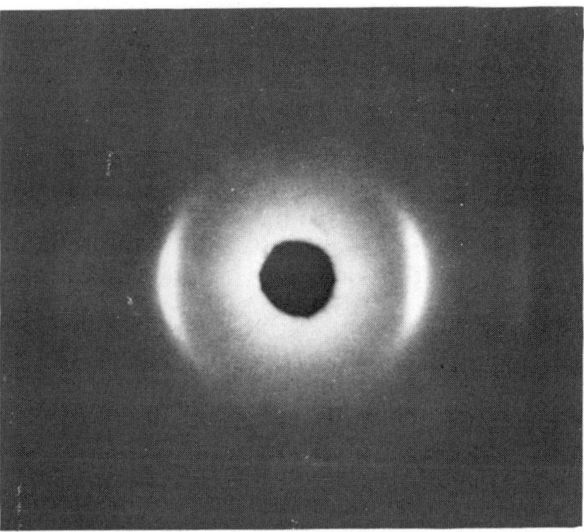

Figure 11.3. X-ray diffraction pattern obtained from a stretched PAN fiber.[59]

Electron micrographs obtained from the as-spun fibers show that the PAN molecules group themselves together to form fibrils, which, in turn, then join together to form a three-dimensional fibrillar network.[67,69,73] These fibrils have also been observed in swollen fibers.[74] The fibrillar network is probably the precursor of the graphite lamellar network that develops as a result of heat treatment. The length of the precursor fibrils between network junctions is markedly affected by the coagulation temperature, lower bath temperatures giving denser (tighter) network structures.[67] The composition of the coagulation bath, the extrusion rate, and the take-up rate also influence the network structure.[70] The conformation of the PAN molecules in the fibrils that compose the three-dimensional network is that of an irregular helix.[67,75,76] The irregularity of the helix is most likely due to the atacticity of the PAN itself.[77,78]

The orientation of the fibrils in the three-dimensional network can be accomplished by stretching the fibers while they are still in the coagulating bath or by stretching the copolymer fibers in boiling water. These stretching operations improve the mechanical properties of the fiber, but this enhancement is not due to an increase in crystallinity or molecular order; rather, it is attributed to a preferred orientation of the fibrils. This is shown schematically in Fig. 11.4.[73]

Fibers based on PAN can also be made by

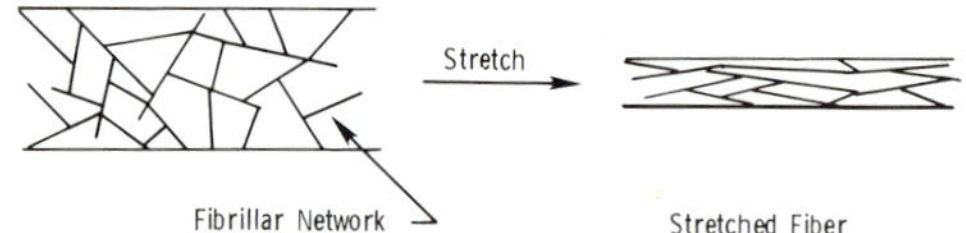

Figure 11.4. Effect of stretching on the Fibrillar network of a PAN fiber.[73]

means of a dry-spinning technique. In this method, the polymer solution is extruded through the spinneret vertically downward into an ascending current of hot air, which evaporates most of the solvent before the fibers are collected. This results in a fiber with a "dog-bone" cross section.

11.3.1.3. Stabilization of PAN

In order to produce good, high-strength, high-modulus carbon fibers from PAN or its copolymers, it is first necessary to induce a preferred molecular orientation parallel to the fiber axis and then stabilize the fiber against the relaxation phenomena and chain scission reactions[79, 80] that would tend to occur during the carbonization process. Since the glass-transition temperature of PAN-base fibers is quite low compared to the carbonization temperatures, a stabilization process must either increase the inherent stiffness of the PAN molecules or "tie" the molecules together in order to limit—or totally eliminate—the relaxation and chain scission tendencies.

Before PAN became a popular precursor for carbon fibers, it was noticed that the fibers would change color when they were heat treated at low temperatures in inert or oxidizing atmospheres. Howtz[81] explained this thermal coloration as being a result of the PAN cyclizing to form a "ladder" polymer. Schematically, the reaction would ideally proceed as shown in Fig. 11.5.[59]

The formation of the ladder polymer significantly enhances the thermal stability of the material due to the inherent stiffness of the "ladder" as well as the strength of the carbon–carbon double bonds. Consequently, cyclization of PAN molecules into ladder polymers became a viable technique for stabilizing PAN fibers prior to carbonization.

Since Howtz's early paper,[81] many other works[34, 80, 82–90] have dealt with the low-temperature heat treatment of PAN. Although most of these papers agree that a ladder polymer does form, considerable disagreement exists on its chemical structure; the extent of these differences as regards the proposed chemical structures is particularly dependent on whether the fiber was heat treated in an inert or oxidizing atmosphere.

The majority of the data reported on the cyclization of PAN in inert atmospheres were collected under dynamic heating conditions. It was found (using a variety of analytical techniques) that a large[84, 91–96] exotherm—corresponding to rapid cyclization of the nitrile groups—occurs at a temperature of approximately 270° C (518° F). The evolution of large amounts of hydrogen cyanide (HCN) and ammonia (NH_3) is associated with this exotherm.[91] The hydrogen cyanide is probably the result of the formation of the carbon double bonds, whereas the ammonia seems to be the product of termination reactions. Under isothermal conditions at temperatures less than 270° C (518° F), much less hydrogen cyanide is produced.[97] Cyclization proceeds (as evidenced by the color changes occurring in the fiber), but much less carbon–carbon unsaturation (as compared to that exhibited by fibers heat treated under dynamic conditions to temperatures greater than 270° C = 518° F) develops.

The implications of these findings are important. In an inert atmosphere, the fibers must be cyclized either isothermally or under a very slow rate of temperature increase. If the temperature approaches 270° C (518° F) without substantial cyclization having occurred, very rapid cyclization will take place, and the

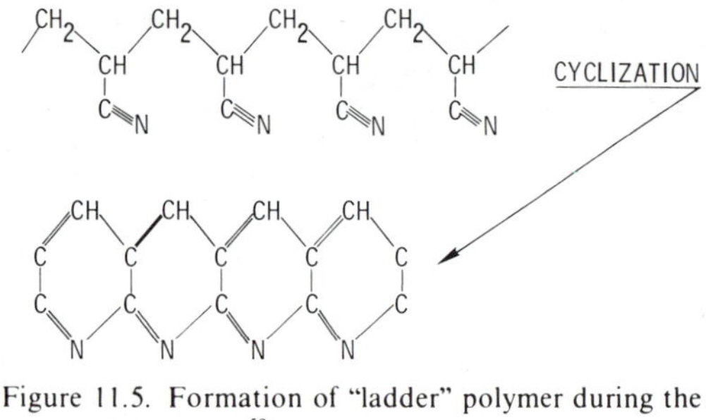

Figure 11.5. Formation of "ladder" polymer during the cyclization of PAN.[59]

corresponding large exotherm will be sufficient enough to cause partial "melting" of—and hence loss of orientation in—the fibers. This will have very deleterious effects on the properties of the subsequently processed carbon fibers, particularly with regard to the tensile modulus of elasticity and strength. In addition, the loss of the CN groups from the PAN backbone (as evidenced by the large amounts of HCN evolved) will disrupt the cyclization sequences. This will also result in somewhat poorer carbon-fiber properties. Under isothermal conditions, on the other hand, the cyclization sequence will tend to be longer, and the orientation within the fiber will be maintained (as long as the fibers are held in tension), but the PAN backbone itself will be saturated and thus prone to scission reactions during carbonization.[98,99] Schematically, the molecules will appear as shown in Fig. 11.6.[98]

When PAN is stabilized in an oxidizing atmosphere, not only cyclization, but also reactions between the polymer and the oxygen occur. The exact nature of the oxidized polymer has been subject to considerable debate.[69] Originally, the three structures shown in Fig. 11.7[69] were proposed for the oxidized polymer. Structure I was based on infrared and elemental analysis of the oxidized material. As can be seen, the oxygen is supposedly present as bridging ether groups.[34,82] Structure II shows oxygen present only in the form of carbonyl groups. This structure was based on evidence obtained from model compounds, elemental analysis, infrared spectroscopy, mechanical properties, and analysis of gases evolved during carbonization.[69] Structure III shows each nitrogen atom donating its lone pair of electrons to an oxygen atom. This structure was proposed after a study of model compounds.[85]

More recent studies on the oxidation of PAN, however, have given clear evidence that the oxygen present in the oxidized polymer is in the form of both hydroxyl and carbonyl groups.[79] In addition, infrared spectroscopy has provided evidence that carbon double bonds are readily produced upon oxidation;[80,100] thus, oxidized PAN tends to be more thermally stable than PAN[69] cyclized in inert atmospheres. A proposed chemical structure for oxidized PAN is shown in Fig. 11.8.[69,79]

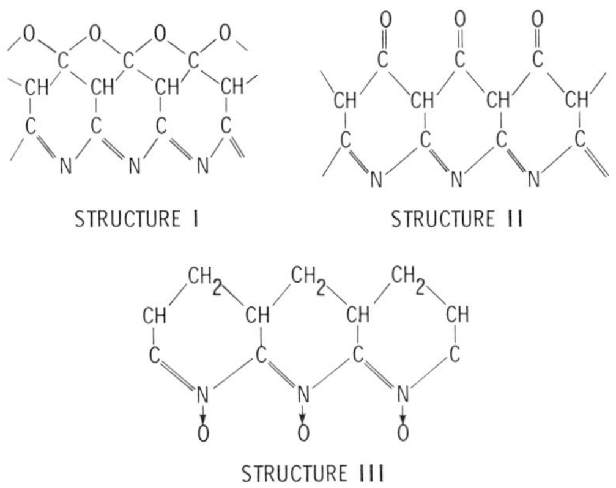

The rate of oxidation of PAN fibers appears to be controlled by the rate of diffusion of oxygen to the center of the fiber.[90,101-103]

At present, the oxidation process is the preferred method of stabilizing PAN fibers for subsequent conversion to carbon because it imparts a greater thermal stability to the fiber than does the inert-gas stabilization technique. Specifically, the fibers, in general, are first stretched (in order to align the fibrillar networks within each fiber parallel to the fiber axis) and then oxidized in air at temperatures of approximately 200–220° C (392–428° F) for varying lengths of time. The fibers are held in tension during this treatment in order to

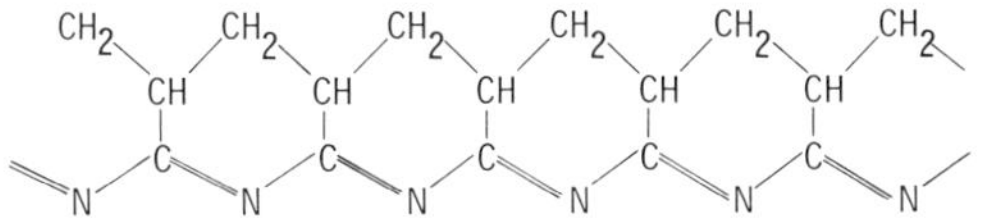

STRUCTURE I

STRUCTURE II

STRUCTURE III

Figure 11.7. Proposed structures for oxidized PAN.[69]

Figure 11.6. Ideal structure of PAN cyclized in an inert atmosphere.[98]

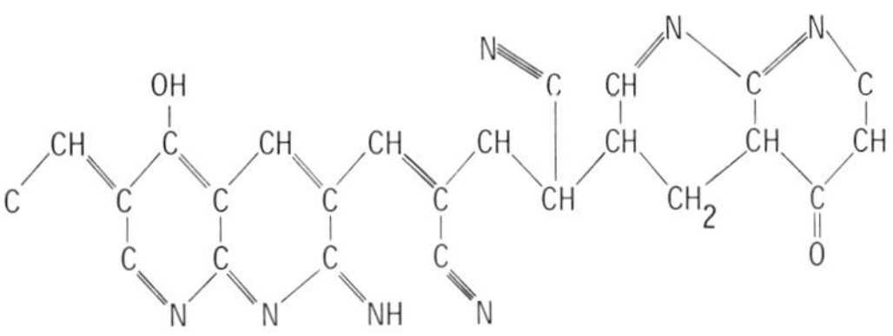

Figure 11.8. Realistic molecular-structure model for oxidized PAN.[69,79]

maintain a degree of orientation. Presumably, the fused heterocyclic structures formed during stabilization are also oriented more or less parallel to the fiber axis.

11.3.1.4. Carbonization and Graphitization

Carbonization is the process of pyrolyzing stabilized PAN-base fibers until they are essentially transformed into carbon fibers. It is during this stage that the high mechanical-property levels found in most commercially available carbon fibers are developed. The development of these properties is directly related to the formation and orientation of turbostratic graphite-like fibrils or ribbons within each individual fiber.

Carbonization heat treatments are generally carried out in an inert atmosphere at temperatures ranging from 1000 to 1500°C (1832 to 2732°F). It is within this temperature range that most (if not all) noncarbon elements are driven from the precursor fiber.[104] At temperatures less than 1000°C (1832°F), a considerable amount of gaseous products, such as methane, hydrogen cyanide, water, carbon dioxide, carbon monoxide, hydrogen, ammonia, and various other hydrocarbons,[97,105-107] are evolved from the precursor fiber. The rapid evolution of these gases necessitates a slow heating rate [21°C (37.8°F)/min.] for the precursor up to 1000°C (1832°F). The chemical composition of the fiber at 1000°C (1832°F) is approximately 94% carbon and 6% nitrogen.[105,108] Nitrogen evolution from the precursor generally starts at about 600°C (1112°F), the maximum rate occurring at approximately 900°C (1652°F). At 1300°C (2372°F) the nitrogen content of the fibers is approximately 0.3%.[105,108] As will be shown in section 11.3.1.7, the nitrogen content of the fiber plays an important role in determining its electrical properties.

The mass loss associated with the PAN carbonization process is shown as a function of temperature in Fig. 11.9.[46] As can be seen, approximately 55–60 wt % of the original PAN-base fiber is lost as a result of carbonization at 1600°C (2912°F). Conversely, the curves indicate a carbon yield for PAN of about 40–45%, which is quite satisfactory (compared, e.g., to rayon); hence, this is partly the reason why PAN represents such an attractive precursor for carbon fibers.

In general, graphitization heat treatments are specifically carried out at temperatures in excess of 1800°C (3272°F) in order to improve the tensile modulus of elasticity of the fiber by improving the crystallite structure and preferred orientation of the graphite-like crystallites within each individual fiber. At this point, it should be mentioned that the term "graphiti-

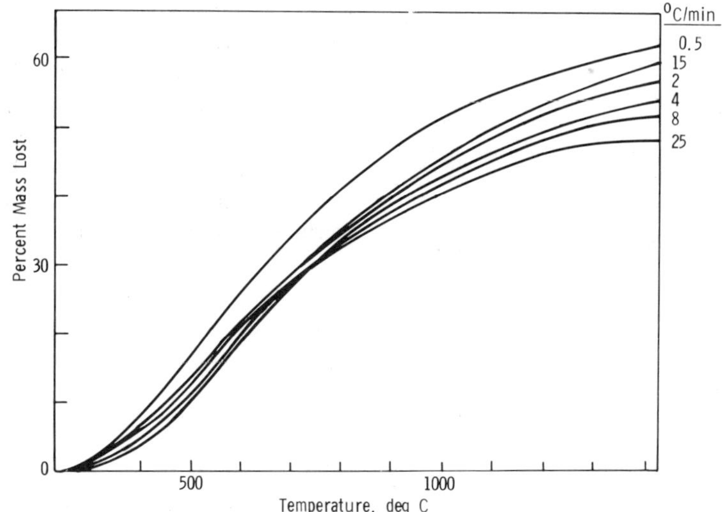

Figure 11.9. Percent mass lost versus temperature for various heating rates of courtelle fiber.[46]

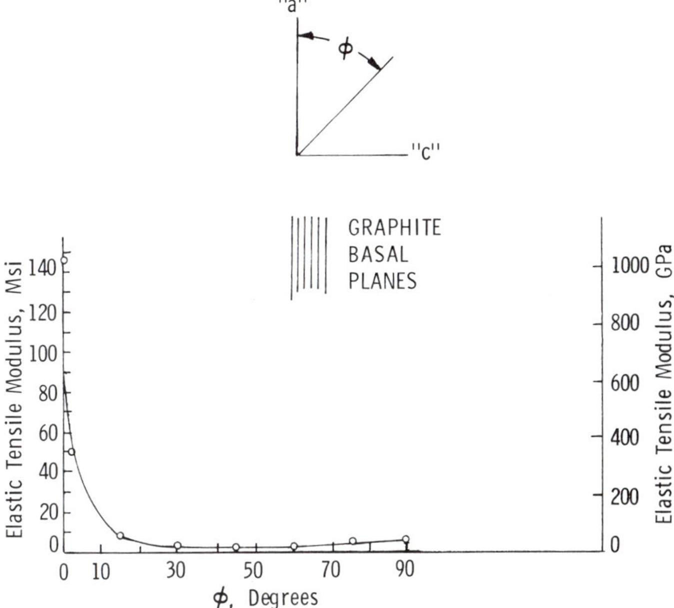

Figure 11.10. Calculated tensile moduli of elasticity E of a single crystal of graphite as a function of angular displacement ϕ from the "a" axis.[45]

zation" is strictly a misnomer, since the process does not result in the formation of the true graphite crystal structure in the fibers. The crystal structure of a graphite fiber is that of a turbostratic graphite in which the layer planes or carbon atoms are slightly displaced from their ideal positions. The interlayer "d" spacings for turbastratic crystals range from approximately 3.37 to 3.45 Å as compared to 3.35 Å for the true graphite structure. In addition, the term "graphite fiber" is used (more or less) to describe fibers that have a carbon content in excess of 99%, whereas the term "carbon fiber" describes fibers that have a carbon content of 80–95%. The carbon content, of course, is a function of the heat-treatment temperature.

The density of carbonized and graphitized fibers generally ranges from approximately 1.7 to 2.1 g/cm^3 (0.061 to 0.076 lb/in.^{3}) as compared to a value of 2.26 g/cm^3 (0.082 lb/in.^{3}) for true graphite and 1.2 g/cm^3 (0.043 lb/in.^{3}) for the precursor PAN fibers. In addition, the diameter of the PAN precursor generally decreases by approximately a factor of two as a result of the heat treatments. The final diameter of the fibers generally averages about 7–10 μm (0.28–0.39 mil).

11.3.1.5. Structure of PAN-Base Carbon Fibers

The high tensile strengths and moduli of elasticity obtainable in carbon fibers are a direct result of the anisotropies found in the graphite crystals. In order to achieve these high-performance properties, it is necessary to first form graphite crystallites in the fiber and to then align the layer planes of the graphite crystals more or less parallel to the fiber axis. Consider, for example, a perfect single crystal of graphite. The tensile modulus of elasticity of the crystal will vary with angular displacement from the direction of the basal plane as shown in Fig. 11.10.[45] Parallel to the basal plane, the tensile modulus of elasticity of the crystal is an outstanding 146 Msi (1000 GPa). However, a small off-axis angular displacement of only 15° will decrease the modulus of the crystal to 10 Msi (70 GPa). Thus, high-modulus carbon fibers must necessarily possess a very high degree of axial preferred orientation of the graphite basal planes parallel to the fiber axis.

The development of graphitic basal-plane orientation can apparently be traced back to the as-spun PAN fiber. As was shown previously, a fibrillar network is known to exist in the as-spun fibers. For wet-spun fibers, the nature of this network can be significantly influenced by various spinning parameters, such as the temperature and composition of the coagulation bath, the stretch rate of the fibers, and the drying temperature. It is probably reasonable to assume that this network can be carried over into the carbonized filaments. Recall that since adjacent molecules comprising each fibril are spaced fairly close together, cyclization and carbonization reactions between them should tend to preserve the overall skeletal network. Evidence that this might occur can be seen in an analogous case: It has been shown that the fibrous structure of wood is preserved in coal (even anthracite).[103]

The development of graphite-like sheet structures in pyrolyzed PAN begins to occur at temperatures as low as 400–600°C (752–1112°F)[108-110] by a dehydrogenation mechanism. The dehydrogenation of adjacent cyclized PAN molecules conceivably results in the formation of graphite-like structures consisting of three hexagons in the lateral direction bounded by nitrogen atoms on the periphery.[108] At temperatures in excess of 600°C (1112°F), it is known that the nitrogen in the heterocyclic rings becomes unstable,[105,107,111] and denitrogenation reactions occur. The effect of the removal of the nitrogen from the heterocyclic rings is that extensive lateral molecular growth of graphite sheets becomes possible, especially among closely spaced adjacent species. The growth of the graphite-like sheets tends to result in the formation of long, undulating ribbons of turbostratic graphite oriented in the general direction of the fiber axis. Whether or not these ribbons form on the skeletal network of the fibrils from the as-spun fibers is not resolved. The mechanism for the formation of graphite ribbons in PAN fibers described above is represented schematically in Fig. 11.11.[108]

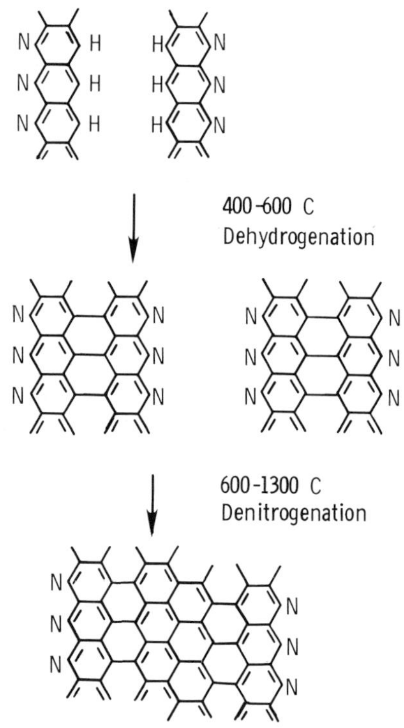

400-600 C
Dehydrogenation

600-1300 C
Denitrogenation

Figure 11.11. Proposed mechanism for the formation of graphite ribbons in PAN fibers.[108]

Ruland et al.[112-115] as well as Johnson and Tyson[116] have studied the microstructure and texture of carbonized fibers via x-ray diffraction and electron microscopy; they conclude that the ribbon-like structures found in the carbon fibers would appear as shown in Fig. 11.12.[112] As can be seen from this model, a large amount of microporosity can exist in the fibers depending on the wavelength and amplitude of the undulation. As the fibers are heat treated to higher temperatures, the wavelength of the undulations increases, the amplitude decreases, and the microporosity tends to close.[45,117] This model also illustrates the fact that the ribbon structure can be characterized by the parameters L_c and L_a, which can be obtained from x-ray data[63,112,116] (L_c represents the stack height of the ribbon, whereas L_a is a measure of the mean length of "straight" basal planes in the ribbon). Both L_a and L_c increase with increasing heat-treatment temperature. In addition, the degree of preferred orientation of the ribbons parallel to the fiber axis and, consequently, the tensile modulus of elasticity of the fibers increase as the heat-treatment temperature increases. To illustrate these changes, consider the work of Diefendorf and Tokarsky,[44] who found that a carbonized fiber that has a tensile modulus of

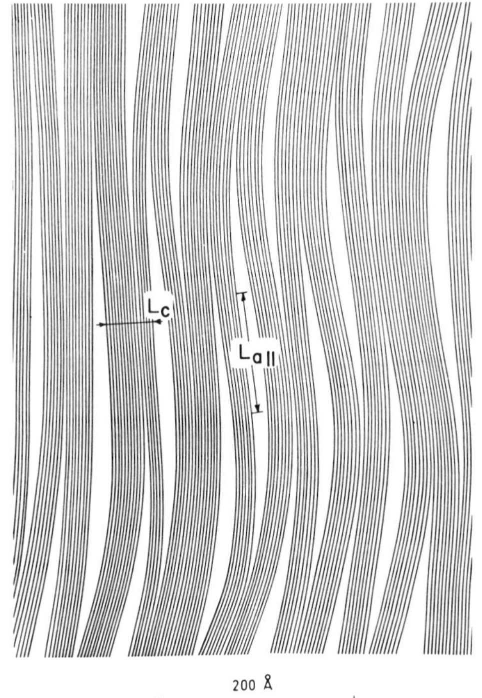

Figure 11.12. Model of carbon-fiber ribbon structure as proposed by Ruland et al.[112]

elasticity of 6 Msi (40 GPa) also has an almost random orientation of the ribbons with respect to the fiber axis. The thickness of each ribbon (L_c) is about 5 or 6 layer planes of turbostratic graphite, and its width is approximately 20 Å. On the other hand, in fibers that have a tensile modulus of elasticity of 40 Msi (275 GPa), the ribbons are typically about 13 layer planes thick and 40 Å wide. In these fibers, the amplitude of the undulation is greater than the wavelength, and approximately three-quarters of the basal planes are oriented within 12° of the fiber axis. Finally, fibers that have a tensile modulus of elasticity of 100 Msi (690 GPa) exhibit ribbons approximately 30 layer planes thick and 90 Å wide. The amplitude of the undulations is almost zero, and the ribbons are essentially parallel to the fiber axis. Diefendorf and Tokarsky have also shown that the amplitude and wavelength of the ribbons vary from the center of the fibers to the surface, the more highly oriented ribbons being closest to the fiber surface. This is significant because it means that most of the load-carrying capacity of the fibers comes from the "skin." The variation in ribbon

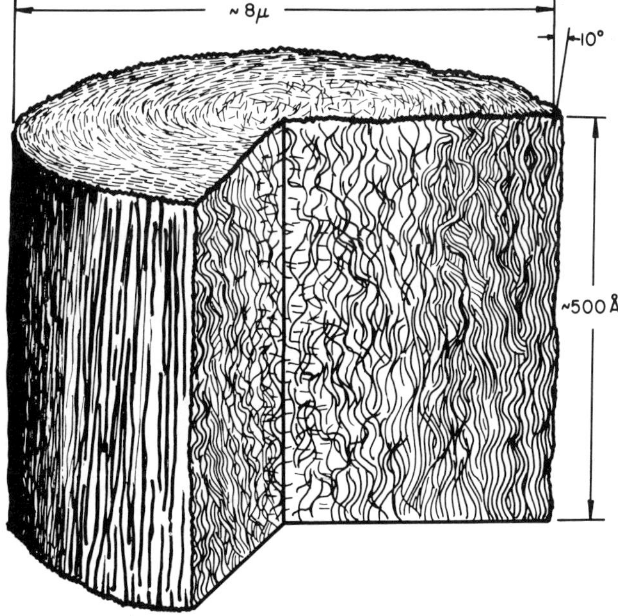

Figure 11.13a. Three-dimensional structural model of a PAN-base carbon fiber (round cross section) with a tensile modulus of elasticity of 41 Msi (280 GPa).[45]

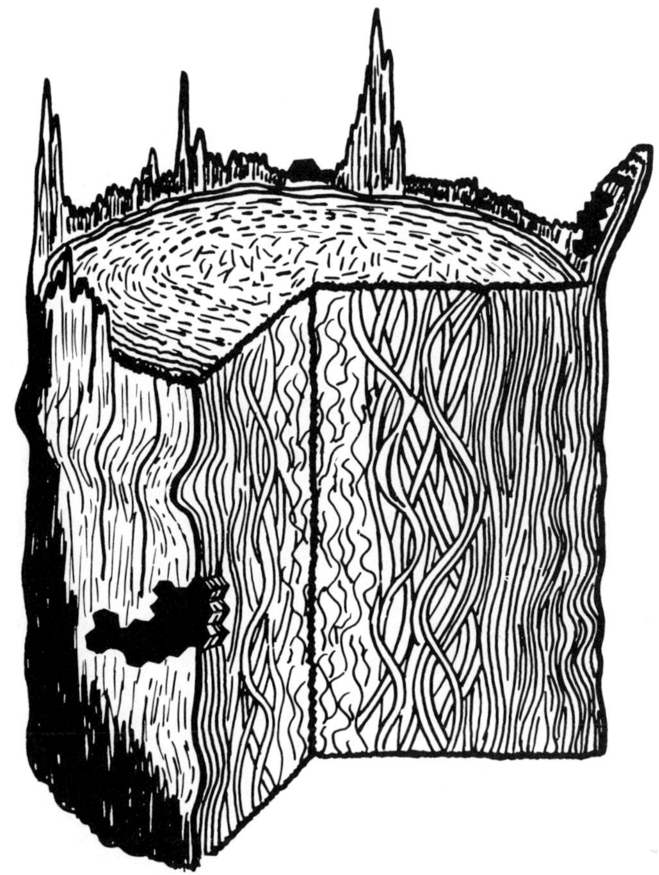

Figure 11.13b. Three-dimensional structural model of a Courtelle-type PAN-base carbon fiber (round cross section) with a tensile modulus of elasticity of 60 Msi (415 GPa).[45]

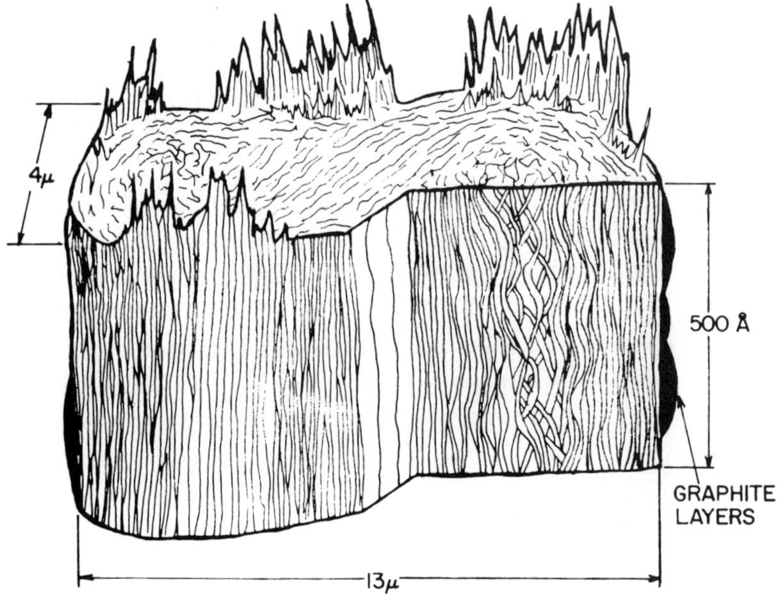

Figure 11.13c. Three-dimensional structural model of a PAN-base carbon fiber (dog-bone cross section) with a tensile modulus of elasticity of 116 Msi (800 GPa).[45]

orientation is illustrated in Fig. 11.13.[45] The variation in preferred orientation from surface to core is found to exist not only in the PAN-base carbon fibers with a round cross section (Fig. 11.13a and b; precursor produced by a wet-spinning process), but also in the PAN-base carbon fibers with a "dogbone" cross section (Fig. 11.13c; dry-spun precursors).

Recently, a number of investigators[112, 118–122] have used high-resolution electron microscopy to study the graphitic layering in ribbon structures of carbonized fibers as well as the orientations and crystal structures present in other carbonized materials. These investigations have shown that pyrolysis of the fibers to only 320°C (608°F) can result in the formation of small crystallites with a thickness of approximately 4 layer planes.[108, 120]

11.3.1.6. Mechanical Properties of PAN-Base Carbon Fibers

The tensile modulus of elasticity of PAN-base carbon fibers is intimately related to the heat-treatment temperatures that they have experienced. This is understandable, since, as was shown in the previous section, the tensile modulus of elasticity is determined by the structure and degree of orientation of the graphite ribbons comprising each fiber, which, in turn, are functions of the pyrolysis temperature.[44] Figure 11.14 plots the tensile modulus of elasticity of PAN-base carbon fibers as a function of heat-treatment temperature.[123] As can be seen, the tensile modulus of elasticity of the fibers begins to develop at very low temperatures and continues to rise as the pyrolysis temperature increases. For example, fibers with a tensile modulus of elasticity of 30 Msi (210 GPa) can be produced by pyrolyzing the PAN precursor at 1000–1100°C (1832–2012°F), whereas those with a tensile modulus of elasticity of 50 Msi (345 GPa) must be processed at 2000°C (3632°F). However, other factors besides temperature can influence the development of the tensile modulus of elasticity. Watt and Johnson,[124] for instance, have shown conclusively that the tensile modulus of elasticity of the carbonized fibers will be significantly greater for any given heat-treatment temperature the more parallel the PAN fibrillar network is to the fiber axis during the stabilization step. This necessitates stretching the PAN prior to stabi-

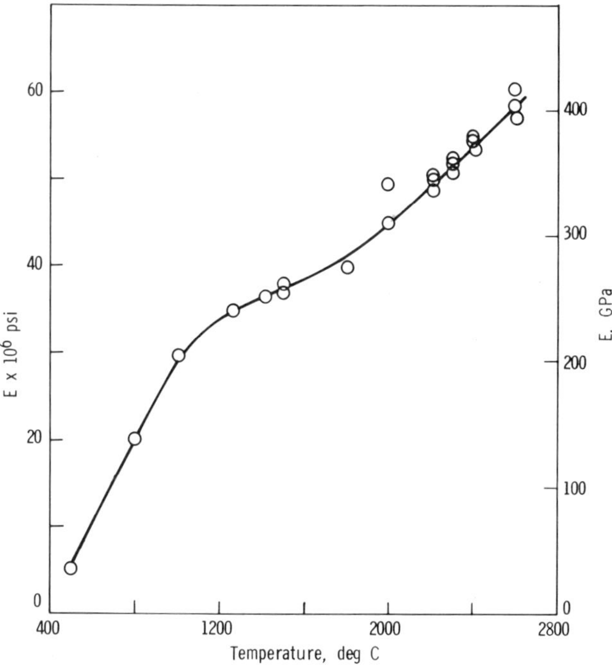

Figure 11.14. Tensile modulus of elasticity (E) of a PAN-base carbon fiber versus heat-treatment temperature.[123]

lization (in order to acquire the orientation) and then maintaining tension on the fibers during stabilization (in order to obviate relaxation).

The factors governing the tensile strength of PAN-base carbon fibers are much more difficult to understand than those governing the tensile modulus of elasticity. Figure 11.15 presents a typical tensile strength versus heat-treatment temperature curve for PAN-base carbon fibers.[123] As can be seen, the tensile strength does not continuously rise as the heat-treatment temperature increases. Rather, a maximum tensile strength of about 450 ksi (3100 MPa) is observed for fibers processed at approximately 1200–1400°C (2192–2552°F). For fibers processed at higher heat-treatment temperatures, the tensile strength drops off significantly.[125]

The tensile strength of the fibers is probably controlled by the presence of discrete flaws both within the volume of the material and on the surface.[126] A summary of strength-limiting factors for PAN-base carbon fibers has been published by Barnet and Norr.[127] Many of the volume flaws originate from defects that can be classified into four types: inorganic inclusions, organic inclusions, irregular voids from rapid coagulation, and cylindrical voids precipitated by dissolved gases. During heat treatment, these various defects are transformed into the diverse imperfections found in the final carbon fiber.[128–132] Basal-plane (Mrozowski) cracks are probably one of the more important volume flaws that tend to limit the tensile strength of PAN-base carbon fibers.[133,134] Mrozowski cracks occur as a result of anisotropic thermal contractions within the ribbon structure upon cool-down from heat treatment. Although these cracks, which are generally aligned along the fiber axis, do affect the tensile strength of the fiber, they do not affect its tensile modulus of elasticity. In general, Mrozowski cracks form only in fibers that have been heat treated at temperatures above 1500°C (2732°F); thus, they could partially account for the drop in the tensile strength of the fibers.

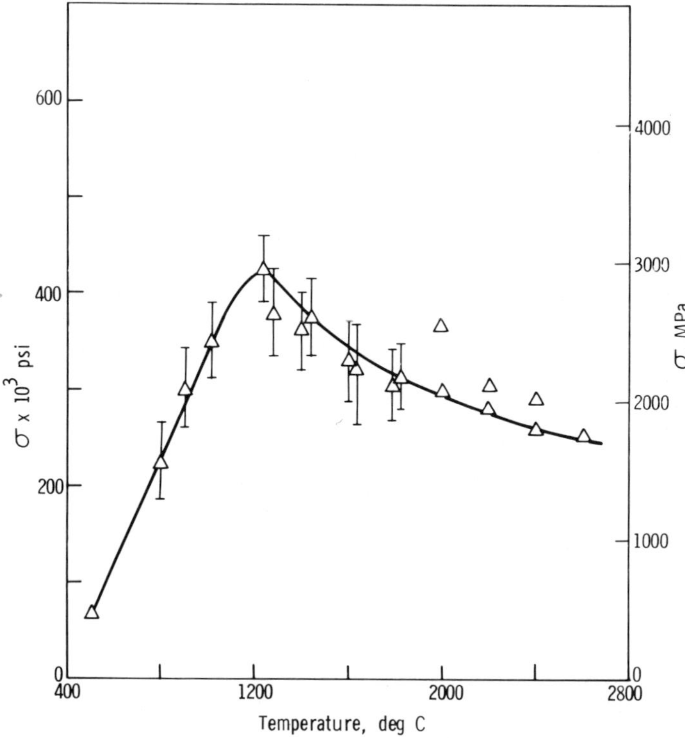

Figure 11.15. Tensile strength (σ) of a PAN-base carbon fiber versus heat-treatment temperature.[123]

Surface flaws also tend to limit the tensile strength of the carbonized fibers. Murphy and Jones[135] have suggested that the surface irregularities introduced immediately after the polymer fiber is spun are retained upon carbonization and thus result in local areas of misaligned graphite basal planes at the fiber surface. Oxidation treatments,[123] which presumably either remove the surface defects or reduce their effect, have been shown to increase the tensile-strength levels that can be attained.

Finally, Tokarsky and Diefendorf[136] feel that the tensile strength of the fibers will be a function not only of the surface and volume flaws, but also of the axial and radial textures of the fibers. The decrease in tensile strength is believed to be caused by the development of a radial preferred orientation that causes microcracking to occur on cool-down from processing temperatures.

Removal of volume flaws by careful processing[137] and of surface flaws by oxidation treatments,[123] has led to significant increases in the tensile strength of the fibers. In addition, LeMaistre[41] has shown that PAN-base carbon fibers with a dog-bone cross section (i.e., dry-spun fibers) can exhibit increasing levels of tensile strength up to a heat-treatment temperature of 2800°C (5072°F). He has proposed that differences between the residual-stress pattern of dog-bone PAN and that of circular fibers of PAN are responsible for the increasing tensile strength.

The typical strain-to-failure for a lower-modulus (30 Msi = 210 GPa), high-strength (475 ksi = 3275 MPa), circular-cross-section fiber is approximately 1.5–1.6%, whereas for a higher-modulus fiber (65 Msi = 450 GPa) with a correspondingly lower tensile strength (235 ksi = 1620 MPa), it is approximately 0.36%.[45] Dog-bone PAN-base carbon fibers, on the other hand, exhibit a different mode of behavior: For a fiber with a tensile modulus of elasticity of 30 Msi (210 GPa), the tensile strength is only 350 ksi (2410 MPa), and the strain-to-failure is 0.92%; a fiber with a tensile modulus of elasticity of 1 Msi (6.9 GPa) exhibits a tensile strength of 420 ksi (2895 MPa) and a strain-to-failure of 0.72%.

11.3.1.7. Electrical Properties of PAN-Base Carbon Fibers

The electrical properties of PAN-base carbon fibers have been studied in detail by Robson et al.[138-141] Their studies included investigations of the g-anisotropy factor (obtained from electron spin resonance measurements), resistivity, thermoelectric power, and magneto-resistance properties of the fibers as a function of heat-treatment temperature. Figures 11.16–11.18[138, 139] present curves of g-value anisotropy, thermoelectric power, and transverse magneto-resistance, respectively, as a function of heat-treatment temperature. As can be seen, sharp inflections or discontinuities occur for all of these parameters at an approximate heat-treatment temperature of 1750° C (3182° F). These results are significant because they indicate that major changes are occurring

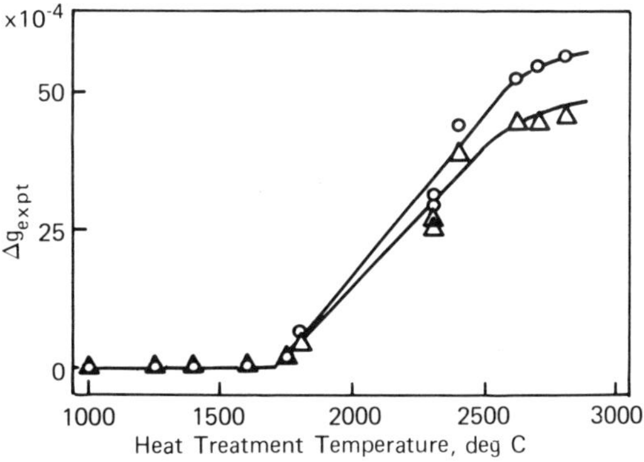

Figure 11.16. Variation in g-value anisotropy of PAN-base carbon fibers as a function of heat-treatment temperature (HTT):) 300° K;)77° K.[138]

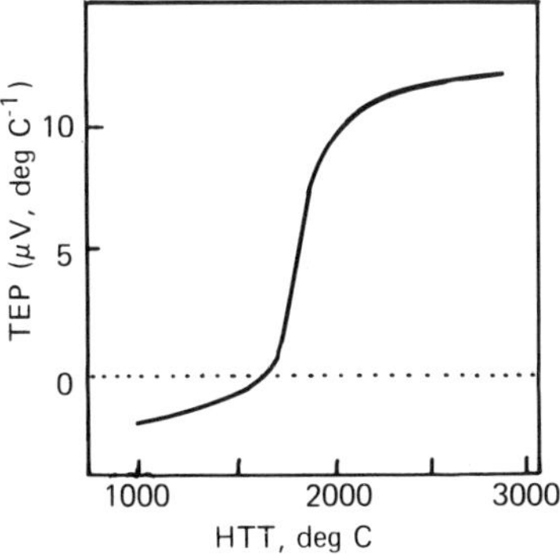

Figure 11.17. Variation in the thermoelectric power (TEP) of PAN-base carbon fibers as a function of heat-treatment temperature (HTT).[139]

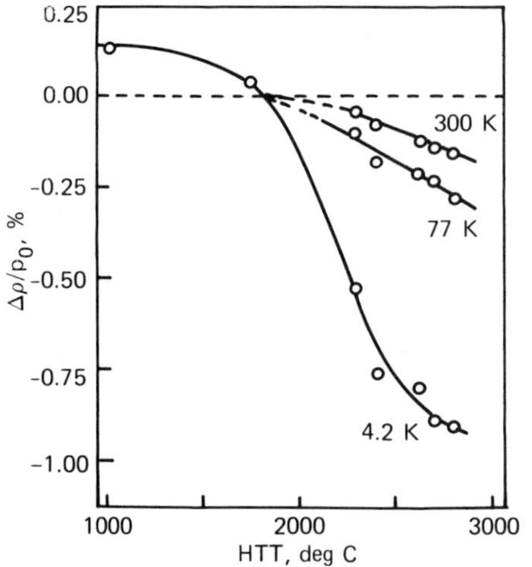

Figure 11.18. Variation of transverse magnetoresistance of PAN-base carbon fibers as a function of heat-treatment temperature (HTT)—magnetic field = 14.1 KG; measurements taken at 300, 77, and 4.2° K.[139]

in the electronic structure of the fibers at this heat-treatment temperature. Moreover, these changes in electronic structure correlate nicely with the observation that the last traces of nitrogen are eliminated from the fibers at about the same heat-treatment temperature. Marchand and Zanchetta,[142] by means of diamagnetic susceptibility studies, have shown that nitrogen has a definite inhibitive effect on graphitization; thus, when the nitrogen is eliminated from the system, the degree of graphitization increases and hence the electronic structure of the fibers abruptly changes. Apparently, the g-value anisotropy, thermoelectric power, and transverse magnetoresistance are all strongly controlled by the degree of graphitization. This result suggests that it may be possible to control the electrical properties of the fibers by maintaining nitrogen within the hexagonal network of carbon atoms.

The electrical-resistivity properties of PAN-base carbon fibers are also quite interesting. Conversion of the PAN to carbon results in a drop in resistivity from a value of 10^6 Ω-cm for the precursor to about 10^{-3}Ω-cm for fibers heat treated at 3000° C (5432° F). Interestingly enough, the major portion of this decrease occurs at processing temperatures between 500 and 1000° C (932 and 1832° F). The maximum rate of resistivity decrease apparently occurs at approximately 600–700° C (1112–1292° F).[143] Once again, denitrogenation appears to be the cause. Recall from Sec. 11.2.1.4 that graphite-like sheet structures begin to develop by a dehydrogenation mechanism at approximately 400–600° (752–1112° F). However, extensive lateral growth of the sheet-like structures does not occur until the nitrogen in the heterocyclic rings becomes unstable (at about 600° C = 1112° F). Thus, as denitrogenation occurs, the electronic structure of the fibers continuously changes (the nitrogen acts as electron traps), and hence the resistivity changes dramatically. These results suggest that if the nitrogen could be retained in the heterocyclic rings while the mechanical properties develop, then a high-performance, high-resistance fiber would be possible.

At heat-treatment temperatures above 1000° C (1832° F), the resistivity slowly decreases as the processing temperature increases. Even when processed at 1750° C (3152° F), where an apparent electronic structural change occurs, the electrical resistivity appears to be relatively insensitive and varies smoothly through this region (see Fig. 11.19[139]). As can be seen from Fig. 11.19, the longitudinal resistivity drops from about 4000 μΩ-cm at a heat-treatment temperature of 1000° C (1832° F) to approximately 1000 μΩ-cm for fibers processed at 3000° C (5432° F). The continued decrease in the resistivity of the carbon fibers is a result of both the loss of the remaining small amount of nitrogen in the fibers and the increasing preferential alignment of the graphitic crystallites parallel to the

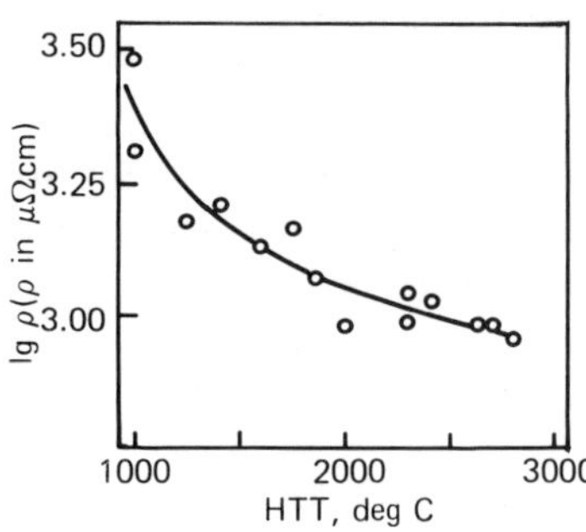

Figure 11.19. Variation in longitudinal resistivity (ρ) of PAN-base cargon fibers as a function of heat-treatment temperature (HTT).[139]

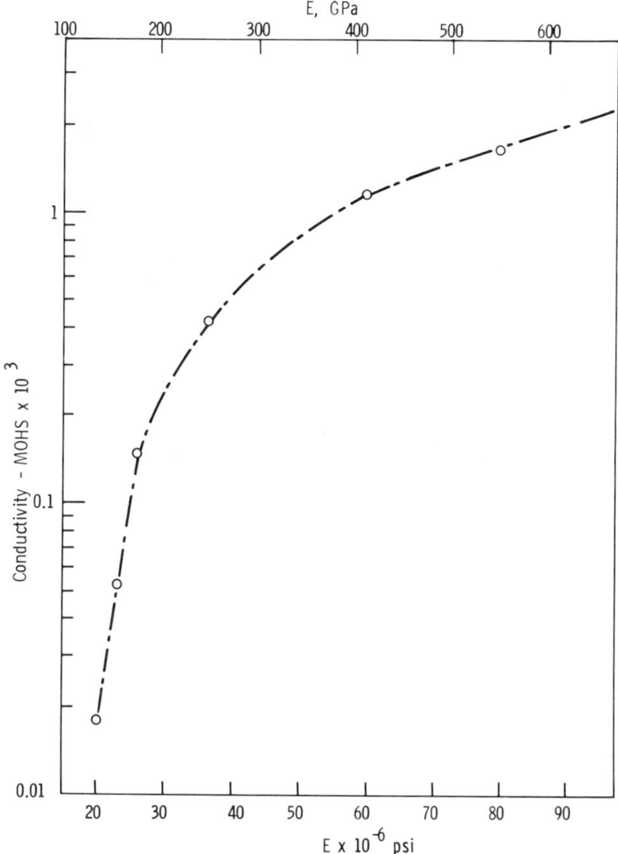

Figure 11.20. Conductivity of PAN-base carbon fibers as a function of tensile modulus of elasticity. (*Courtesy of* R.Gray, NSWC, Dahlgren, Virginia.)

fiber axis. Recall that the preferred orientation of the crystallites will increase as the heat-treatment temperature increases. The high basal-plane conductivity provides a path of low resistivity parallel to the fiber axis. As the degree of preferred orientation of the crystallites increases, so does the conductivity of the fibers. This is readily apparent from the curve of conductivity versus tensile modulus of elasticity presented in Fig. 11.20. Recall that the tensile modulus of elasticity is also highly dependent on preferred orientation.

11.3.2. Carbon Fibers from Pitch

11.3.2.1. Pitch Conversion Processes

Pitch-base carbon fibers are produced by two processes. The first of these processes results in low-modulus fibers unless stress-graphitization at extremely high temperatures is employed.[47-53] The precursor for this process is a low-softening-point isotropic pitch. The processing scheme is as follows:

1. Melt spin isotropic pitch.
2. Thermoset at relatively low temperatures for long periods of time.
3. Carbonize in an inert atmosphere.
4. Stress-graphitize at extremely high temperatures.

The high-performance fibers produced in this manner tend to be relatively expensive due to both the very long thermosetting times required and the need for a high-temperature stretch procedure. Tensile strengths as high as 375 ksi (2585 MPa) and tensile moduli of elasticity in excess of 70 Msi (480 GPa) can be attained in fibers produced by this technique.

On the other hand, the non-stress-graphitized fibers produced by this process tend to have tensile modulus of elasticity values in the range of 5–10 Msi (35–70 GPa).[47] These fibers are usually employed as substrates for the chemical vapor deposition of boron in order to make boron filaments. Since the isotropic-pitch process is not really commercially significant, the remainder of this section will emphasize only the mesophase-pitch process.

The processing scheme for making mesophase-pitch-base carbon fibers is as follows:

1. Heat treat at 400–450° C (752–842° F) in an inert atmosphere for an extended period of time in order to transform it into a liquid-crystalline (mesophase) state.
2. Spin the mesophase pitch into fibers.
3. Thermoset the fibers.
4. Carbonize the fibers.
5. Graphitize the fibers.

Since long thermosetting times and stress-graphitization treatments are not required, the high-performance carbon fibers produced by this process are low in cost.

11.3.2.2. Characteristics of Pitch

Pitch and other similar materials are generally formed as by-products of the destructive distillation of coal, crude oils, natural asphalts, and various synthetic compounds (e.g., polyvinylchloride—PVC) in thermal- or catalytic-conversion processes.[144, 145] Petroleum-refining companies generally consider these materials to be waste products mainly because they can only be further processed (cracked) into a more useful form (e.g., gasoline, jet fuel, or lubricating oils) with great difficulty.[146] This difficulty arises because these materials have fairly high average molecular weights and, moreover, are generally aromatic in character; thus, they tend to rapidly react to form coke and/or carbon particles during the various conversion processes.[147] This is detrimental because the coke and carbon particles settle out of the flow stream onto the catalytic beds in the converters, which results in poisoning of the catalyst and significant drops in conversion yields. Thus, it is economically unfeasible to process pitch into lighter hydrocarbons; consequently, pitch is a very cheap and readily available source of carbon for use as a carbon-fiber precursor.

Depending on the criteria used to differentiate the molecular structures, pitch is generally considered to be composed (more or less) of four generic fractions.[148, 149] These four generic fractions, which have been termed saturates, naphthene-aromatics, polar-aromatics, and asphaltenes, significantly differ from each other in molecular weight and degree of aromaticity. For instance, the saturates are that part of a pitch which is essentially aliphatic in nature and also possesses a low molecular weight; the saturates are quite similar to waxes. The naphthene-aromatic part of a pitch, on the other hand, consists of both very-low-molecular-weight aromatic molecules and various saturated ring structures. In contrast to the naphthene-aromatics, the polar-aromatics tend to be higher in molecular weight and consist of more heterocyclic-type molecules. Finally, the asphaltenes are not only the highest molecular-weight fraction of a pitch, but also have the highest degree of aromaticity. These are large, alkylated, plate-like molecules of condensed aromatic rings. This fraction is the most thermally stable part of the pitch,[150–154] and it is these molecules that result in the formation of liquid crystals and, eventually, turbostratic graphite in pitch-base carbon fibers.

The properties and thermal stability of carbonaceous materials (ranging from oils through tars and pitches) significantly depend on the relative proportions of each of the four generic fractions present in the material. For example, Fig. 11.21[154] shows the relative compositions of various carbonaceous materials. As can be seen, oils predominately contain naphthene-aromatics and saturates, whereas pitches (e.g., Ashland 260 petroleum pitch, Ashland Oil Company are predominantly composed of the more aromatic asphaltenes. In addition, the greater the percentage of asphaltenes, the higher the softening point of the material. Thus, Ashland 260, which contains 80% asphaltenes, has a much higher softening point (177° C = 260° F) than DAU Bottoms

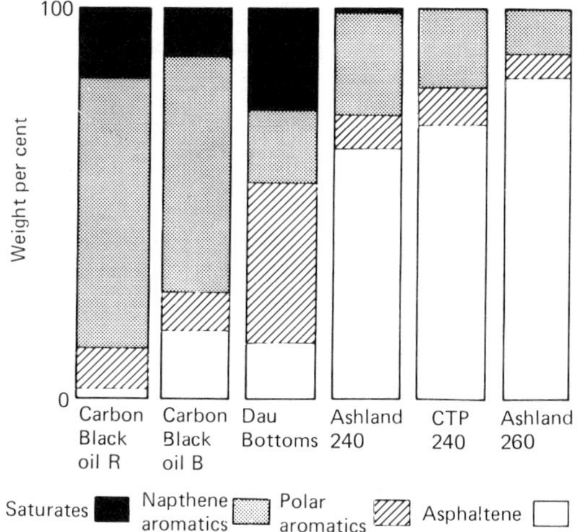

Figure 11.21. Relative compositions of various carbonaceous materials.

(a refinery sludge made by Exxon with a softening point slightly above room temperature), which has only 18% asphaltenes. The asphaltene content is also significant with regard to the pyrolysis yield of carbonaceous materials. Figure 11.22 plots the percent carbon yield of a number of oils and pitches as a function of temperature.[59] As can be seen, the carbon yield significantly increases as the relative fraction of asphaltenes present in the original material increases; in fact, the carbon yield of these materials can exceed 60%. Moreover, the carbon yield of the asphaltenes themselves is also quite high; however, it does depend on various molecular structural features, which is an important consideration in the production of carbon fibers.

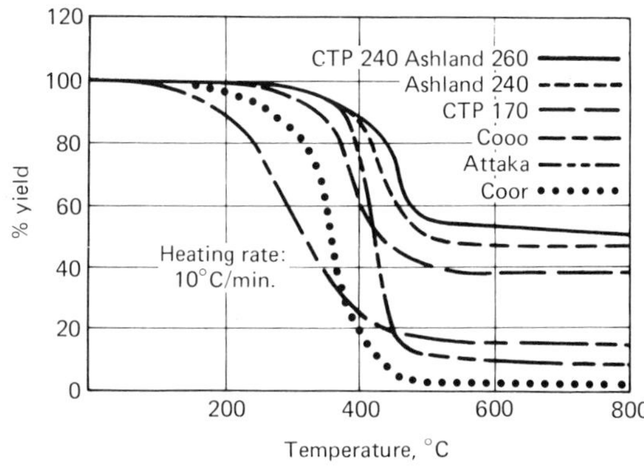

Figure 11.22. Percent carbon yields of various oils and pitches as a function of temperature (heating rate—10° C = 18° F/min.).[59]

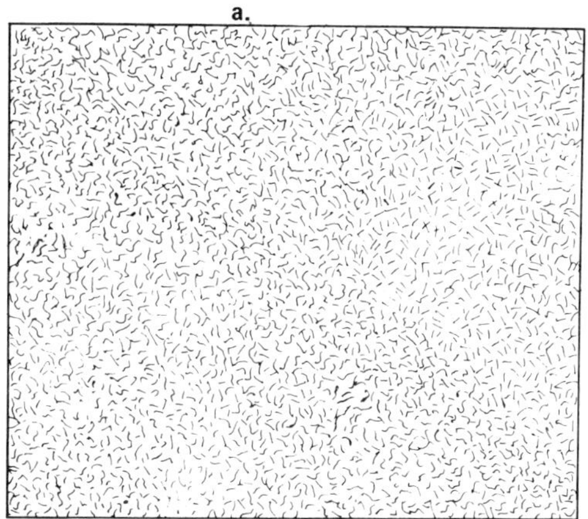

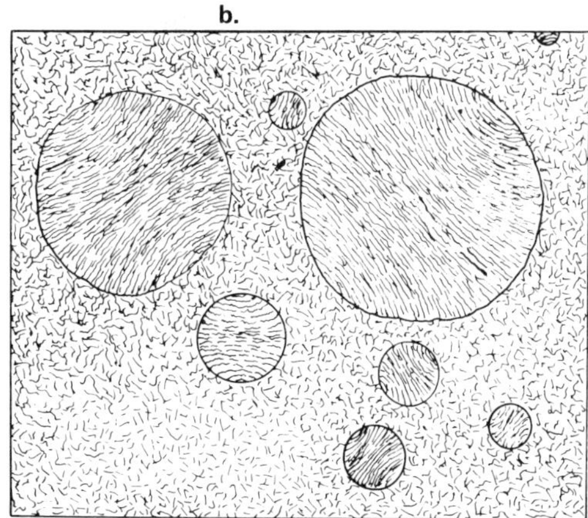

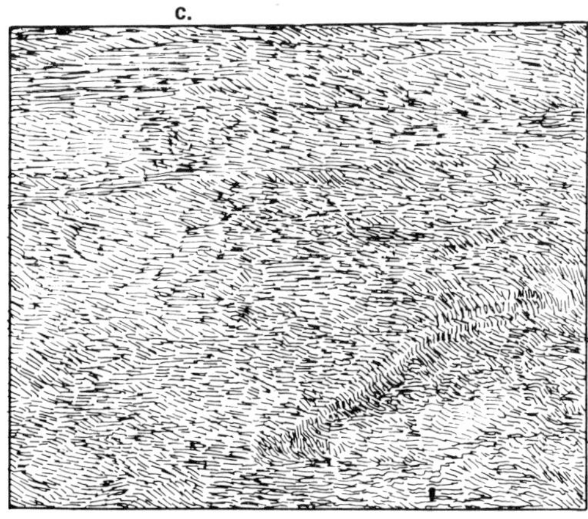

Figure 11.23. Development of liquid crystallinity in pitch as a result of heat treatment: (a) isotropic; (b) nucleation; (c) coalesced.[59]

In addition to its high carbon yield, pitch is also known to undergo a liquid-crystalline (mesophase) transformation when heat treated under appropriate conditions.[155-164] This is important with respect to fiber production because it allows one to spin a highly ordered as-spun filament from a thermodynamically stable, ordered fluid. This ordering is a result of the size, shape, structural features, and reactivities associated with the molecules[165-168]—especially the asphaltenes—comprising the pitch.[59] Essentially, by building up the aromaticity and molecular weight through distillation and condensation reactions, it becomes thermodynamically favorable for the molecules to stack up one on top of the other through a process similar to nucleation and growth. Even though the molecules are preferentially ordered, they are still in a fluid state. This ordering phenomenon is depicted in Fig. 11.23.[59] As can be seen, after a period of time at a given temperature, the isotropic fluid pitch nucleates small spheres of liquid-crystalline material, which then grow until they impinge upon one another and coalesce. This material can then be placed in a spinneret and, with proper processing technology, made into highly ordered as-spun fibers (Fig. 11.24[59]). These fibers can then be converted into carbon fibers by various heat treatments.

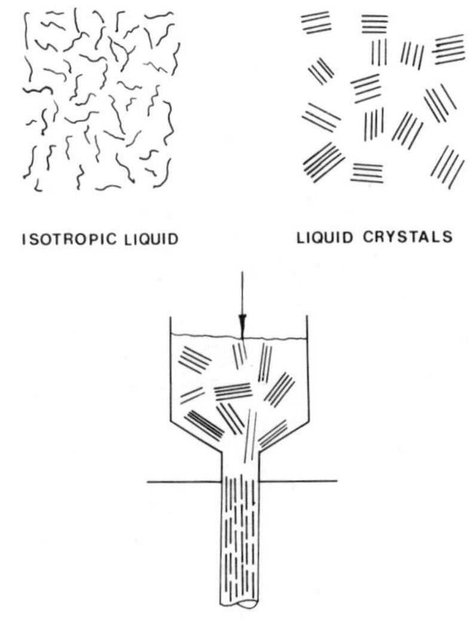

Figure 11.24. Processing of liquid crystals of pitch into a fiber.[59]

The development and properties of carbon fibers made from mesophase pitch will be discussed in the following section.

11.3.2.3. Spinning and Thermosetting Mesophase-Pitch Fibers

Mesophase-pitch can be spun into filamentary form by means of a variety of techniques commonly used for producing conventional polymer fibers.[169-176] By far, melt spinning is the most important technique adopted for this purpose. The material to be spun (generally about 50–90% mesophase) is loaded into either a monofilament or a multifilament die, heated to the proper temperature range, and pushed through the die with pressurized inert gas. The fiber is generally formed at rates of about 500 ft/(127 m)/min. and experiences a draw ratio of approximately 1000:1. The final diameter, which usually ranges from 10 to 15 μm (0.39 to 0.59 mil), can be varied by varying the take-up rate of the fibers.

The draw ratio plays an important role not only in determining the fiber diameter, but also in terms of controlling the degree of induced molecular orientation in the fibers.[177] Those fibers that experience a small draw ratio (i.e., spinneret cross-sectional area/fiber cross-sectional area) exhibit a relatively small amount of molecular orientation. The highest degree of orientation in this case occurs at the fiber skin as a result of the shear field set up by the interaction of the flowing pitch with the spinneret walls. However, this orientation tends to be very nonuniform and drops off rapidly toward the core of the fiber.

Fibers that experience a high draw ratio possess a much higher degree of molecular orientation, and the orientation itself tends to be more uniform.[178] As shown by Fig. 11.25, the molecular orientation in these fibers can be readily detected under polarized light.[59] There exist three distinct types of ordering, the one that is manifested depending on the ratio between the shear-induced orientation from the spinneret walls and the tensile-induced orientation from the fiber take-up wheel. These orientations, which are readily visible in the carbonized fibers, are known as radial, onion-skin, and random textures. These three

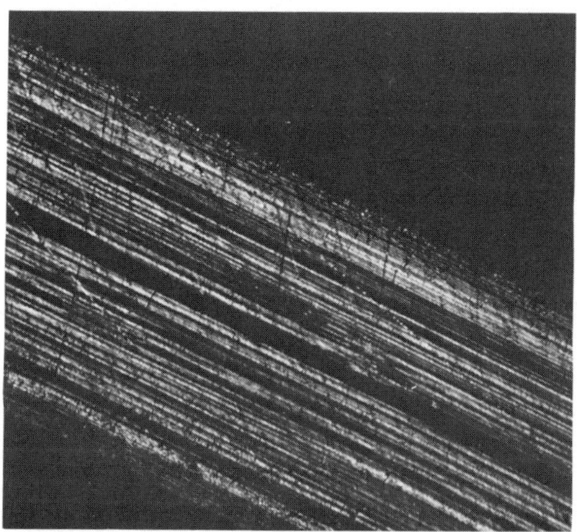

Figure 11.25. Polarized-light micrograph of an as-spun mesophase-pitch fiber.[59]

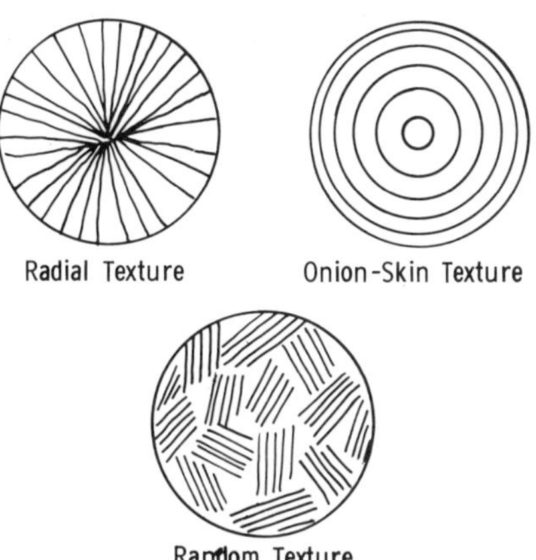

Figure 11.26. Textures observed in pitch fibers.[59]

textures are depicted schematically in Fig. 11.26.[59]

When examining the cross section of a carbonized fiber, the radial texture exhibits crystallites that emanate from the center similar to the spokes of a wheel. The onion-skin texture, on the other hand, is a wrap-around texture in which the appearance of the crystallites resembles the ends of a scroll or a sliced onion. Finally, the random texture has no distinctive ordering of the crystallites; however, the graphite-like layers are oriented parallel to the fiber axis.

Since mesophase pitches tend to be somewhat thermoplastic, it is necessary to thermoset the fibers prior to carbonizing treatments in order to obviate relaxation tendencies at higher temperatures. The fibers can be thermoset either by heat treating them at approximately 300° C (572° F) for a short period of time (2.5 hr) in an oxygen-containing atmosphere or by immersing them in strong, oxidizing liquids. During these treatments, the large plate-like molecules that form the mesophase are linked together via oxidative polymerization reactions; as a result, the fibers are stabilized against melting, which, in turn, allows them to be carbonized. Special care must be taken to ensure that the thermosetting temperature is low enough to minimize relaxation of the orientation in the as-spun fibers—too high a temperature will result in a loss of

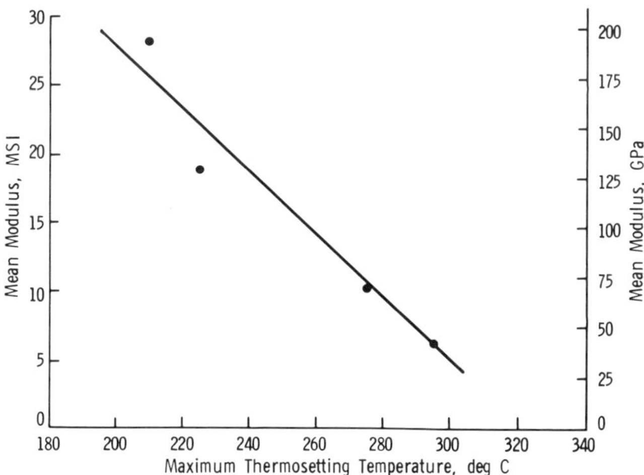

Figure 11.27. Effect of thermosetting temperature on the tensile modulus of elasticity of carbonized pitch (asphaltene) fibers at 1000° C (1832° F).[59]

orientation and hence a sharp degradation of the mechanical properties of the subsequently carbonized fibers.[178] This effect with respect to the tensile modulus of elasticity can be seen very vividly in Fig. 11.27.[59]

11.3.2.4. Carbonization and Graphitization

Subsequent to thermosetting the as-spun fibers, carbonization and graphitization heat treatments at temperatures approaching 3000°C (5432°F) are carried out.[179–189] These heat treatments result in the conversion of the mesophase pitch to coke, then carbon, and, ultimately, graphite.

During the processing, the fibers are (generally) first pulled through a thermosetting furnace, then a precarbonizing furnace, and, finally, a graphite-tube resistance furnace. The precarbonizing furnaces are used primarily to heat the fibers up to about 1000°C (1832°F) in two stages. The fibers are brought to a temperature of about 950°C (1742°F) in order to reduce the rate of evolution of the residual gases that are produced during the conversion of the thermoset pitch fibers to carbon.[182] The rapid evolution of gases is detrimental in that it tends to produce such structural flaws as bubbles and cracks. Rapid heating in the precarbonizing stages results in graphitized fibers that have poor physical properties.

After precarbonization, the fibers are heat treated at temperatures ranging from 1200 to 3000°C (2192 to 5432°F). The final heat-treatment temperature will significantly affect the properties of the carbonized fibers.

11.3.2.5. Structure and Mechanical Properties of Pitch-Base Carbon Fibers

As discussed previously, three distinct types of transverse textures are observed in mesophase-pitch base carbon fibers: radial, onion-skin, and random.

When pitch-base carbon fibers first became available, they tended to exhibit the radial texture exclusively. This texture provided carbonized fibers that had a "missing wedge" as a result of longitudinal cracks in the as-spun fibers. During carbonization, high tangential tensile stresses tended to open up the crack, and hence the fiber cross section appeared as though a "piece of pie" was missing. More recently, however, the filaments in the yarn generally show a mixture of textures, primarily the random and radial. Onion-skin textures occur only rarely and, in addition, are found only in monofilament materials.[169]

Quantitative measurements of preferred orientation relative to the fiber axis and of apparent crystalline stack height (L_c) were obtained on carbonized pitch fibers by Barr et al.[169] A preferred orientation parameter FWHM (full width at half maximum) of approximately 30° was obtained on as-spun fibers from the 002 reflection. Thermosetting treatments apparently had little effect on this parameter. As the fibers were carbonized, the

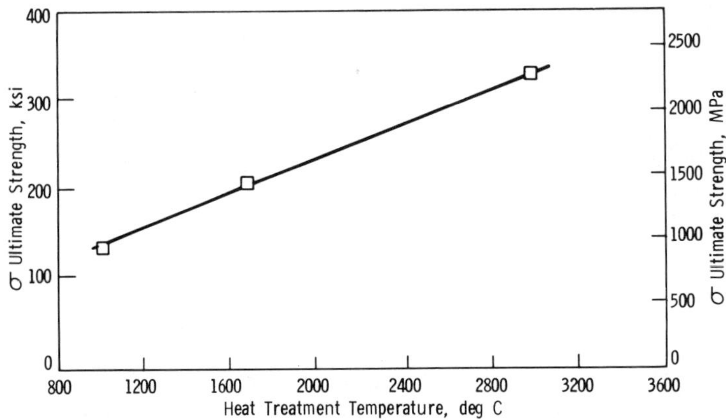

Figure 11.28. Ultimate tensile strength (σ) of pitch-base carbon fibers versus heat-treatment temperature (HTT) from data presented in Refs. 190 and 191.

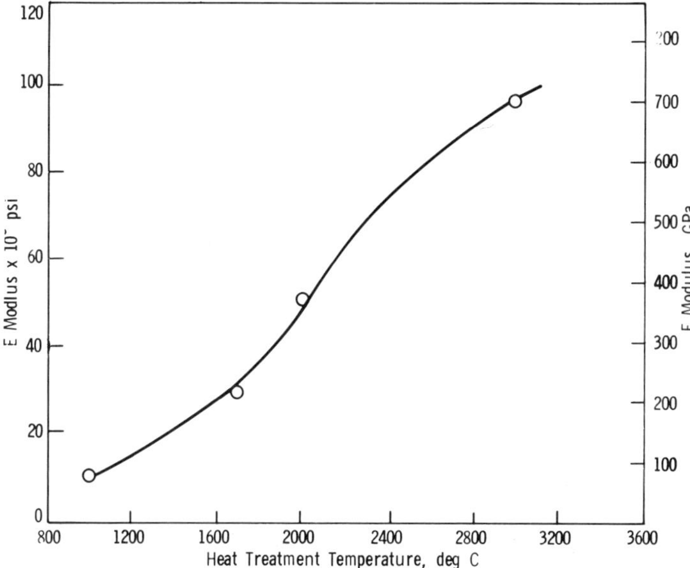

Figure 11.29. Tensile modulus of elasticity (E) versus final heat-treatment temperature of pitch-base carbon fibers (HHT).[191]

degree of preferred orientation increased markedly. Heat treatment at 3000° C (5432° F) results in an FWHM of approximately 5°, which is similar to the preferred orientation of graphite whiskers. Fibers with this high degree of preferred orientation exhibited a tensile modulus of elasticity of 128 Msi (880 GPa).

Figure 11.28 plots the variation in ultimate tensile strength of pitch-base carbon fibers as a function of the heat-treatment temperature.[190, 191] As can be seen, the ultimate tensile strength apparently increases linearly with processing temperature, ranging from an average of 200 ksi (1380 MPa) for fibers heat treated at 1700° C (3092° F) to 320 ksi (2205 MPa) for those processed at 3000° C (5432° F). These ultimate tensile strength values are relatively low, and, as a result, the strain-to-failure for the fibers is also low. This could be a major limitation with respect to their use in advanced composites. The primary source of ultimate failure appears to be micro- and macroporosity. As further developmental work progresses in this area, enhanced mechanical properties can be expected.

The variation in the tensile modulus of elasticity of pitch-base carbon fibers as a function of the final heat-treatment temperature is shown in Fig. 11.29.[191] As can be seen, the tensile modulus of elasticity increases rapidly as the processing temperature is increased. This is a result of the higher degree of preferred orientation imparted to the fibers by the higher heat-treatment temperature. Fibers heat treated at 1700° C (3092° F) exhibit tensile modulus of elasticity values of approximately 30 Msi (210 GPa).

11.3.2.6. Electrical Properties of Pitch-Base Carbon Fibers

The magnetoresistance, electrical resistivity, and electron spin resonance properties of pitch-base carbon fibers were discussed in detail by Bright and Singer.[186] Their objective was to correlate the electronic properties and structure of the fibers in order to demonstrate that the former are a sensitive indicator of the degree of graphitic order. They found that the ultimate degree of graphitization attainable by these fibers was comparable to that of similarly heat-treated pyrolytic carbons. This is in contrast to PAN-base carbon fibers, which are ultimately capable only of properties comparable to pitch-base fibers heat treated in the 1700–2300° C (3092–4172° F) range.

Figure 11.30 is a plot of the g-value anisotropy of two series of fibers as a function of the

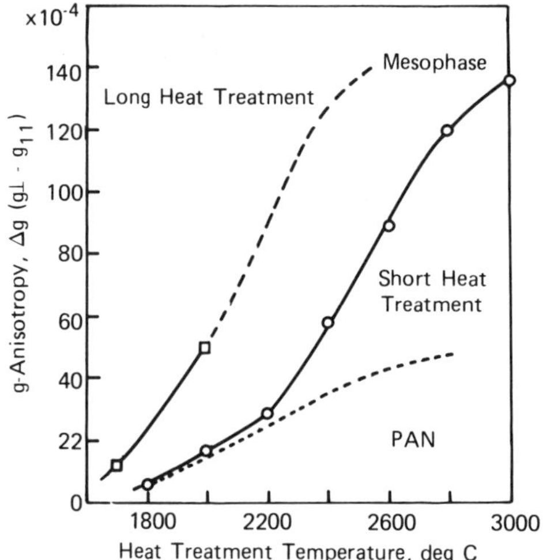

Figure 11.30. Variation in g-value anisotropy of pitch-base[191] and PAN-base[138] carbon fibers as a function of heat-treatment temperature (HTT).[191]

heat-treatment temperature.[191] As can be seen, the pitch-base carbon fibers apparently have much larger g-value anisotropies than the PAN-base carbon fibers discussed previously. Bright and Singer[186] claim that this is a direct consequence of the unique graphitizable character of mesophase-pitch-base carbon fibers. The g-value anisotropy of PAN-base carbon fibers saturates at high heat-treatment temperatures to the same extent as pitch-base carbon fibers heat treated in the 2000–2300°C (3932–4172°F) range. These results strongly indicate the highly graphitizable nature of pitch-base carbon fibers (compared to the more non-graphitizable nature of PAN-base carbon fibers). Interestingly enough, the g-value anisotropy of pitch-carbon base fibers changes markedly above 1700°C (3092°F) in a manner similar to that discussed previously for PAN-based carbon fibers. This change could also be related to the loss of the last traces of impurity atoms in the pitch; however, this is not confirmed.

The electrical resistivity of pitch-base carbon fibers as a function of the heat-treatment temperature is shown in Fig. 11.31.[191] As can be seen, pitch-base carbon fibers are excellent conductors. The resistivity drops from about 10^{-3} Ω-cm for fibers heat treated at 1700°C (3092°F) to about 2×10^{-4} Ω-cm for fibers heat treated at 3000°C (5432°F). For comparison, a plot of the conductivity of pitch-base, PAN-base, and rayon-base carbon fibers as a function of the tensile modulus of elasticity is shown in Fig. 11.32. Since pitch-base carbon fibers are highly graphitizable (compared to PAN-base carbon fibers), their conductivity is significantly better—particularly at lower tensile modulus of elasticity values.

11.3.3. Carbon Fibers from Rayon

11.3.3.1. The Rayon Conversion Process

The process by which rayon is converted to carbon fibers involves four steps:

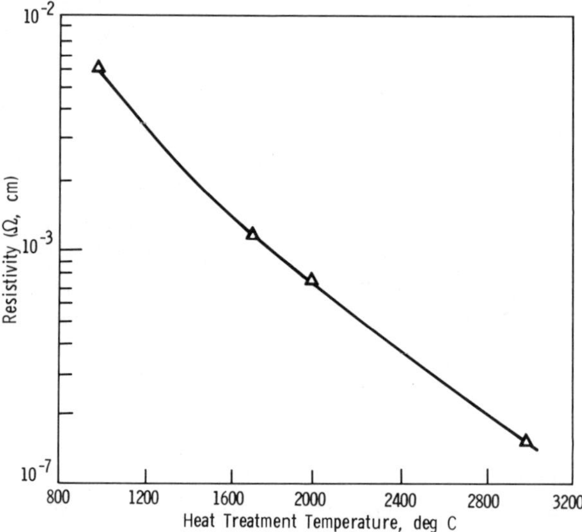

Figure 11.31. Electrical resistivity of pitch-base carbon fibers as a function of heat-treatment temperature (HTT).[191]

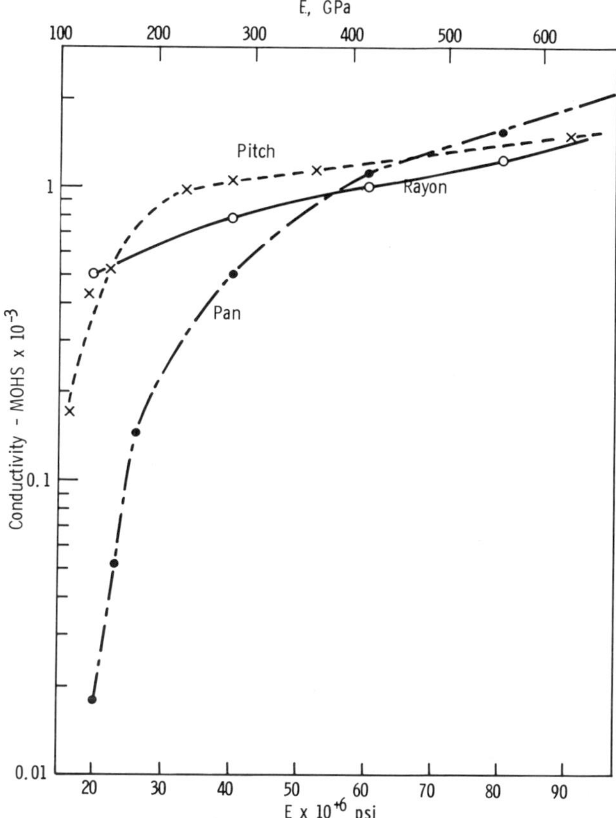

Figure 11.32. Conductivity of pitch-base, PAN-base, and rayon-base carbon fibers as a function of the tensile modulus of elasticity. (*Courtesy of* R. Gray, NSWC, Dahlgren, Virginia.)

1. Fiber spinning.
2. Stabilization.
3. Carbonization.
4. Stretch-graphitization.

Each of these steps will be discussed in the following sections.

11.3.3.2. Characteristics of Rayon

Rayon is a cellulosic material produced by wet spinning an extraction product of wood pulp. These fibers have been particularly popular for a number of years in the production of textiles for clothing and tire cord. However, it is only since 1959[20] that these fibers have been employed extensively in the production of high-strength, high-modulus carbon fibers[187, 188] for use in composite materials.[22, 26, 27]

Currently, a variety of rayon-fiber types are on the market. Saponified acetate rayon, tire-cord rayon, and cuprammonium rayon[18] have all been investigated as precursors for carbon fibers. Unfortunately, the carbon fibers produced from these materials are rather weak due to the high internal void contents and modifiers present in the precursors.[18, 189] On the other hand, polynosic rayons and viscose rayons[18, 187, 188] with a high degree of polymerization produce carbon fibers that have good properties.

The molecular structure of rayon is shown in Fig. 11.33.[192] As can be seen, a large percentage of the cellulosic structure is composed of hydrogen and oxygen. Consequently, the theoretical char yield after carbonization is only ~ 55%.[69] In actuality, however, the carbon yield generally ranges from 10 to 30%,[69, 193] which is considered to be quite low. This is one of the important limitations associated with the use of rayon as a carbon-fiber precursor. An extensive amount of work has been done on developing processes for increasing the pyrolyzation yield of these materials.[194]

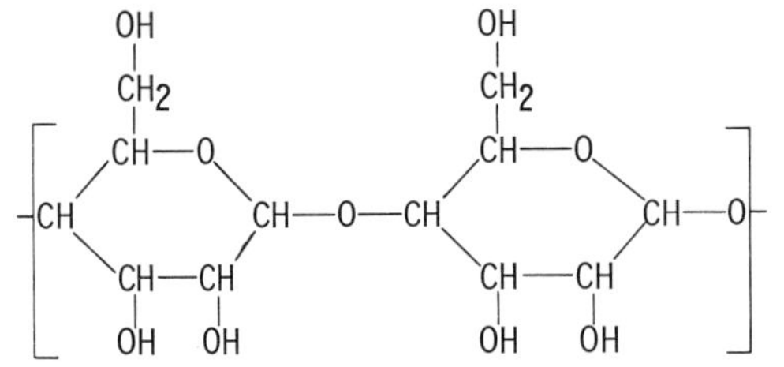

MOLECULAR STRUCTURE OF CELLULOSE

Figure 11.33. Molecular structure of rayon (cellulose).[192]

Shindo et al.[195,196] developed a process for increasing the carbon yield by heat treating the rayon fiber in flowing HCl vapor; char yields of 37% were reported. Similarly, low-temperature heat treatment of the fibers in other reactive atmospheres (e.g., air[194,197-200] oxygen,[18] and chlorine[201,202] also resulted in enhanced carbon yields.

An effective method for enhancing the carbon yield and increasing the processing rate is impregnating the rayon with carbonization promoters and flame retardants.[18] Bacon[18] has stated that these materials promote the dehydration of the cellulose structure and stabilize the molecules against excessive formation of volatile tars. Nitrogenous salts of strong acids,[203] acids or acidic salts,[195,204-206] metal halides,[207] various derivatives of phosphoric acid,[208-209] chlorosilanes,[210-213] and various other substances[18,190,214-218] have been shown to be effective impregnators.

11.3.3.3. Rayon-Precursor Processing Steps

The processing of rayon fiber precursors into carbon fibers requires three distinct processing steps: a low-temperature stabilizing heat treatment, a carbonization treatment at 1300° C (2372° F), and a stretch-graphitization treatment at 2800–3000° C (5072–5432° F).

Depending on the atmosphere, the low-temperature heat treatment is generally carried out at temperatures up to 400° C (752° F). This low-temperature heat treatment (stabilization treatment) is designed to stabilize the fibers for subsequent carbonization by converting the rayon to a char.

When the stabilization is conducted in an inert atmosphere, long processing times are required. Generally, the rayon is heat treated to 400° C (752° F) at an approximate rate of 10° C (18° F)/hr.[18] A number of reactions occur during this heat treatment. First of all, physically adsorbed water is desorbed. This is followed by the formation of more water by the reaction of hydroxyl groups that have split off the main cellulose molecule. Depolymerization and the accompanying evolution of H_2O, CO, and CO_2 occur next. Finally, aromatization and the subsequent development of graphite-like layers occur.[23] As a result of the depolymerization step, prestretching the fibers and stabilizing them under tension (as is done with PAN fibers) are ineffective. Bacon[18] has concluded that no amount of tension or stretching is capable of maintaining or increasing the preferred molecular orientation during the stabilizing heat treatment. The depolymerization step is also responsible for the formation of tarry substances—in particular, levoglucosan. These substances tend to be volatile and thus result in low carbon yields.[213] Control of these reactions is essential for the preparation of high-quality carbon fibers.

Stabilization in reactive atmospheres (e.g., those discussed in the previous section) result in significantly improved processing rates and carbon yields. The reactive atmospheres tend to inhibit the formation of tars.[18,193,195,214] Processing times ranging from 20 min. to 10 hr are generally used when stabilizing rayon in a reactive atmosphere.[18]

Chemical preimpregnation is again a very effective method for controlling the carbon yield and processing time. Processing times of only a few minutes are required. Tar formation and infusibility are hindered by the crosslinking reactions catalyzed by the impregnant.

The second step involved in processing carbon fibers from rayon is carbonization. Rayon is generally carbonized at temperatures ranging from 1000 to 1500° C (1832 to 2732° F) in inert atmospheres. Processing times are fairly rapid. Carbonizing under tension is an effective method for increasing the preferred orientation and hence enhancing the mechanical properties of the graphitized filaments.[219] Stretching the carbonizing filaments is also effective–particularly when the graphitic structure is developing and, therefore, quite plastic.

Graphitization of the filaments is carried out at temperatures greater than 2800° C (5072° F). Very short residence times are required at this temperature. Since the filaments tend to be quite plastic in this temperature range, they can be stretched as much as 100%, which significantly improves both the degree of preferred orientation and hence the mechanical properties.[220–222] Tensile modulus of elasticity values approaching 100 million psi are possible for fibers stretched 100% at 2800° C (5072° F),[18] whereas nonstretched fibers exhibit tensile moduli of elasticity of approximately 10 million psi.

The excessive cost of the processing is an inherent disadvantage of stress-graphitization. This expense is reflected in the cost of rayon-base graphite fibers.

11.3.3.4. Structure of Rayon-Base Carbon Fibers

Precursor rayon fibers possess a fairly high degree of three-dimensional crystalline order. Depending on the dividing line used to separate crystalline areas from "aligned amorphous"[69] regions, the degree of crystallinity generally ranges from 30 to 50%.[192] Microfibrils[223–226] similar to those observed in PAN precursors can also be observed in the case of rayon fibers.

During heat treatment, the crystalline structure and hence the preferred orientation of the cellulose break down at approximately 240–280° C (464–536° F). Above 1000° C (1832° F), a reorientation[114] occurs; however,this reorientation is not strong enough to allow for high-performance properties—only through stress-graphitization at very high temperatures can high-strength, high-modulus properties be obtained. The influence of the precursor microfibrillar network on the development of the graphitic layers during heat treatment has not been determined.[18]

Low-angle x-ray diffraction studies have shown that the void structure of the rayon precursors remains in the fibers even after heat treatment at relatively high temperatures.[69, 227] The precursor voids are converted to needle-shaped pores the diameter of which is dependent on the heat-treatment temperature.[69, 228] However, the diameter of these voids is generally less than 100 Å, and they tend to be oriented in the same direction as the graphite ribbon structure[229] comprising the microstructure of the fibers. As was discussed with respect to PAN-base carbon fibers, the preferred orientation of the graphite ribbon structure will determine the properties of the graphitized filaments; this is also true for rayon-base carbon fibers. The x-ray orientation halfwidth angle for non-stress-graphitized filaments that have a tensile modulus of elasticity of 5 Msi (35 GPa) is ~25°, whereas that for stress-graphitized fibers that have a tensile modulus of elasticity of 80 Msi (550 GPa) is 5°.[229]

The density of rayon-base carbon fibers ranges from 1.3 g/cm^3 (0.047 $lb/in.^3$) for the low-modulus variety to over 1.9 g/cm^3 (0.069 $lb/in.^3$) for those that have a very high modulus.[229]

11.3.3.5. Mechanical Properties of Rayon-Base Carbon Fibers

The tensile modulus of elasticity of rayon-base carbon fibers is highly dependent on a number of processing variables. Thus, such factors as the final heat-treatment temperature, the time spent at the heat-treatment temperature, the amount the fiber is stretched during "low"-temperature carbonizing heat treatments and, most importantly, the amount the fiber is stretched during the "high"-temperature (>2800° C = 5072° F) heat treatment will determine the ultimate tensile modulus of elasticity of the fiber. Rayon-base carbon fibers with tensile modulus of elasticity values greater than 100 Msi (690 MPa) can be produced.

The tensile strength of rayon-base carbon fibers can also be quite high. The tensile strength increases[229] from 100 ksi (690 MPa) to over 500 ksi (3445 MPa) as the tensile modulus of elasticity of the fibers increases from 10 to 110 Msi (690 to 760 GPa). This relationship can be seen in Fig. 11.34.[229] Since these fibers behave elastically through failure, the strain-to-failure will decrease from about

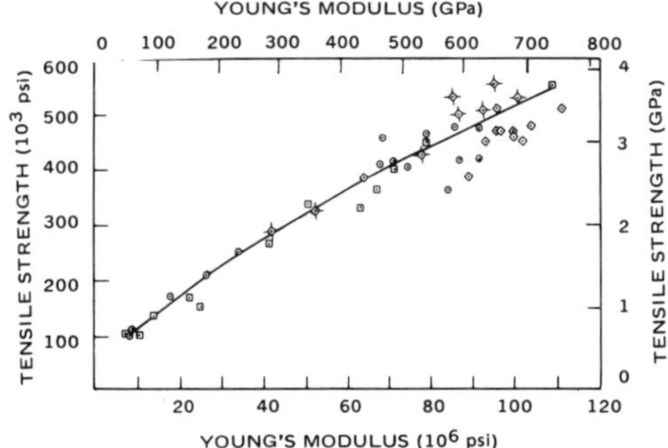

Figure 11.34. Ultimate tensile strength of rayon-base carbon fibers as a function of tensile modulus of elasticity.[229]

1% for the lower-modulus fibers to about 0.5% for the higher-modulus variety.[229]

11.3.3.6. Electrical Conductivity of Rayon-Base Carbon Fibers

The electrical conductivity of rayon-base carbon fibers is strongly correlated with their tensile modulus of elasticity. As was pointed out in the previous section, the tensile modulus of elasticity is dependent on a number of factors; therefore, it is more convenient to directly correlate electrical conductivity with the tensile modulus of elasticity rather than express it as a function of the processing variables. Figure 11.35[229] plots the electrical conductivity of rayon-base carbon fibers as a function of the tensile modulus of elasticity. As can be seen, the electrical conductivity ranges from approximately 400 Ω–cm^{-1} for a fiber that has a tensile modulus of elasticity of 10 Msi (70 GPa) to roughly 1900 Ω–cm^{-1} for a fiber that has a tensile modulus of elasticity of 110 Msi (760 GPa).

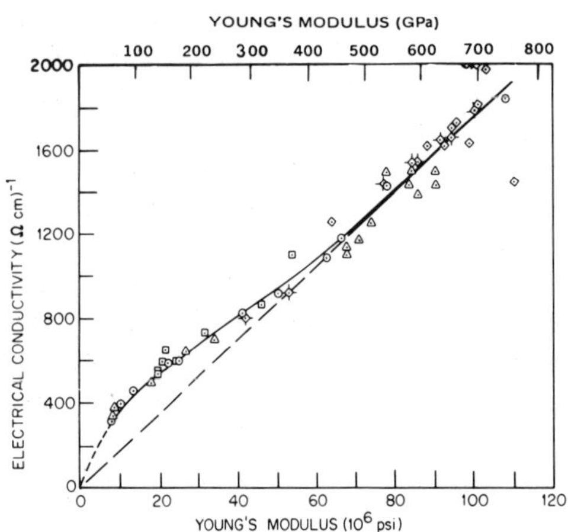

Figure 11.35. Electrical conductivity of rayon-base carbon fibers as a function of tensile modulus of elasticity.[229]

11.3.4. Summary of Carbon-Fiber Production Processes

The preceding sections have described the three basic processes currently, being utilized to produce carbon fibers. Each process offers distinct advantages and has drawbacks in terms of both fiber cost and properties. The PAN process results in carbon and graphite fibers with a relatively low cost and good properties: Fibers with a tensile modulus of elasticity of ~ 30 Msi (210 GPa) are currently selling for about \$30/lb (\$66/Kg), whereas higher-modulus varieties sell for approximately \$145/lb (\$320/Kg). This price is predominantly a reflection of the cost the PAN precursor and the high processing temperatures required to attain the property levels.

Pitch-base carbon fibers are currently the lowest-cost fiber on the market: A fiber with a tensile modulus of elasticity of ~30 Msi (210 GPa) currently sells for \$20/lb (\$44/Kg). Since the pitch precursors are so inexpensive, the cost of the carbon fibers is mainly a function of the heat-treatment conditions. The fibers themselves also offer the greatest cost-property relationship and are predicted to become the most widely used type of carbon fiber in the future.

Rayon-base carbon fibers are very expensive (>\$300/lb = \$660/Kg) due to the extremely high temperatures required for their "stretch-graphitization." The fact that high-performance fibers cannot be obtained without stretch-graphitization is partially the reason why rayon-base carbon fibers are becoming less and less available.

11.4. AVAILABLE GRAPHITE-FIBER FORMS

Table 11.1 presents selected properties of commercially available graphite fibers from different organic precursors. Graphite fibers are available in various forms: continuous, chopped, woven fabric, or mat. Tows, yarns,

Table 11.1. Selected Properties of Commercially Available Graphite Fibers.

FIBER NAME	MANUFACTURER	PRECURSOR	ULTIMATE TENSILE STRENGTH, Ksi (GPa)	TENSILE MODULUS OF ELASTICITY, Msi (GPa)	DENSITY, g/cm^3 ($kg/m^3 \times 10^{-3}$, $lb/in.^3$)	ELECTRICAL CONDUCTIVITY, (Ω-cm^{-1}) [Ω-$m^{-1} \times 10^{-4}$]	AXIAL THERMAL CONDUCTIVITY, W/cm·°C (W/m·°C, Btu/ft-lb·°F)	AXIAL THERMAL EXPANSION, 10^{-6}/°C (10^{-8}/°F)	FORMS AVAILABLE
Fortafil® 3(0)[a]	Great Lakes Carbon Corporation	PAN	360 (2.48)	27 (186)	1.73 (1730, 0.063)	570 (5.7)	0.20 (20, 11.6)	−0.11 (−6.11)	Chopped Supplied as 40,000 or 160,000 continuous tow—no twist Non-surface-treated
Fortafil® 3[a]	Great Lakes Carbon Corporation	PAN	360 (2.48)	30 (207)	1.71 (1710, 0.062)	570 (5.7)	0.20 (20, 11.6)	−0.11 (−6.11)	Surface treated Supplied as 160,000 + 40,000 end continuous two—no twist Chopped
Fortafil® 5[a]	Great Lakes Carbon Corporation	PAN	400 (2.76)	48 (331)	1.80 (1800, 0.065)	1050 (10.5)	1.44 (144, 83.2)	−0.5 (−27.8)	40,000 end continuous tow Chopped
Hi-Tex® 12000[b]	HITCO Corporation	PAN	395 (2.72)	34 (234)	1.80 (1800, 0.065)	—	—	—	12,000 end tow
Hi-Tex® 6000[b]	HITCO Corporation	PAN	420 (2.89)	34 (234)	1.80 (1800, 0.065)	—	—	—	6000 end tow
Hi-Tex® 3000[b]	HITCO Corporation	PAN	440 (3.03)	35 (241)	1.80 (1800, 0.065)	—	—	—	3000 end tow
Hi-Tex® 1500[b]	HITCO Corporation	PAN	460 (3.17)	36 (248)	1.80 (1800, 0.065)	—	—	—	1500 end tow
Panex® 30[c]	Stackpole Fibers Company	PAN	400 (2.76)	32 (220)	1.74 (1740, 0.063)	—	—	—	160,000 end tow
Panex® 1/4CF-30[c]	Stackpole Fibers Company	PAN	325 (2.24)	30 (207)	1.73 (1730, 0.063)	—	—	—	Chopped
Panex® 30R[c]	Stackpole Fibers Company	PAN	225 (1.55)	38 (262)	1.75 (1750, 0.063)	—	—	—	Spun yarns and rovings—1.2 turns/in. (0.5 turn/cm)
Panex® 30Y/800d[c]	Stackpole Fibers Company	PAN	225 (1.55)	38 (262)	1.75 (1750, 0.063)	—	—	—	Spun yarns and rovings—2 plies/yarn and 4 turns/in. (1.6 turns/cm)
Panex® 30Y/300d[c]	Stackpole Fibers Company	PAN	225 (1.55)	38 (262)	1.75 (1750, 0.063)	—	—	—	Spun yarns and rovings—2 plies/yarn and 4.5 turns/in. (1.8 turns/cm)
Celion® GY-70[d]	Celanese Corporation	PAN	270 (1.86)	75 (517)	1.96 (1960, 0.071)	1538 (15.38)	—	—	384 filaments/yarn—0.5 turn/in. (0.2 turn/cm)
Celion® 6000[d]	Celanese Corporation	PAN	400 (2.76)	34 (234)	1.76 (1760, 0.064)	667 (6.67)	—	—	6000 filaments/yarn or tow
Celion® 3000[d]	Celanese Corporation	PAN	400 (2.76)	34 (234)	1.76 (1760, 0.064)	667 (6.67)	—	—	3000 filaments/yarn or tow
Celion® 1000[d]	Celanese Corporation	PAN	360 (2.48)	34 (234)	1.76 (1760, 0.064)	667 (6.67)	—	—	1000 filaments/yarn—0.35 turn/in. (0.14 turn/cm)
AS[e]	Hercules Incorporated	PAN	450 (3.10)	32 (220)	1.77 (1770, 0.064)	—	—	—	10,000 filaments/tow
NTS[e]	Hercules Incorporated	PAN	400 (2.76)	36 (248)	1.80 (1800, 0.065)	—	—	—	10,000 filaments/tow
NMS[e]	Hercules Incorporated	PAN	340 (2.34)	50 (344)	1.86 (1860, 0.067)	—	—	—	10,000 filaments/tow
	Polycarbon Incorporated[i]		150–180 (1.03–1.24)	5–8 (34–55)	1.45 (1450, 0.052)	—	—	—	Yarn—720 filaments/ply, 1–20 plies, 2 twists/in. (0.8 twist/cm)

Table 11.1. Continued.

FIBER NAME	MANUFACTURER	PRECURSOR	ULTIMATE TENSILE STRENGTH, Ksi (GPa)	TENSILE MODULUS OF ELASTICITY, Msi (GPa)	DENSITY, g/cm^3 ($(kg/m^3 \times 10^{-3}$, $lb/in.^3$)	ELECTRICAL CONDUCTIVITY, (Ω-cm^{-1}) [Ω-$m^{-1} \times 10^{-4}$]	AXIAL THERMAL CONDUCTIVITY, W/cm·°C W/m·°C, Btu/ft-lb·°F)	AXIAL THERMAL EXPANSION, 10^{-6}/°C (10^{-8}/°F)	FORMS AVAILABLE
	Carborundum Company[j]		100–180 (0.69–1.24)	3–5 (21–34)	1.50 (1500, 0.054)	— —	— —	— —	Spun yarn—5–20 plies/yarn, 2.6 twists/in. ply (1.0 twist/cm ply), and 1.1–1.7 twists/in. yarn (0.43–0.67 twist/cm yarn) Chopped fiber
Thornel® 50[f]	Union Carbide Corporation	Rayon	320(2.20)	57 (393)	1.67 (1670, 0.060)	—	—	—	Yarn—720 filaments/ply and 2 plies/yarn
Thornel® 300[g] WYP 90-1/0	Union Carbide Corporation	PAN	385 (2.65)	33 (227)	1.75 (1750, 0.063)	—	(0.2051, 0.1186)	—	Yarn—1000 filaments/ply, 1 ply/yarn, and 4 twists/in. (1.6 twists/cm)
Thornel® 300[f] WYP 30-1/0	Union Carbide Corporation	PAN	360 (2.48)	34 (234)	1.76 (1760, 0.064)	—	(0.2051, 0.1186)	—	Yarn—3000 filaments/ply, 1 ply/yarn, and 4 twists/in. (1.6 twists/cm)
Thornel® 75[h]	Union Carbide Corporation	Rayon	385 (2.65)	76 (524)	1.82 (1820, 0.066)	—	—	—	Yarn—720 filaments/ply and 2 plies/yarn
P55 BS	Union Carbide Corporation	Pitch	300 (2.07)	55 (379)	—	—	—	—	2000 filaments
P75	Union Carbide Corporation	Pitch	300 (2.07)	75 (517)	—	—	—	—	Yarn—diameter - 0.017
P100	Union Carbide Corporation	Pitch	300 (2.07)	100 (689)	—	—	—	—	Slight twist

[a]Technical Data Sheet, Great Lakes Carbon Corporation, June 1977.
[b]Technical Data Sheet 78SCF103, HITCO Corporation.
[c]Technical Data Sheet, Stackpole Fibers Company, March 1977.
[d]Technical Data Sheet, Celanese Corporation, April 1977.
[e]Technical Information Bulletin, Hercules Incorporated, 1977.
[f]Bulletin #465-205, Union Carbide Corporation.
[g]Bulletin #465-223 and Bulletin #465-220, Union Carbide Corporation, 1976.
[h]Bulletin #465-220, Union Carbide Corporation.
[i]Product Data Sheet CF671, Polycarbon Incorporated.
[j]Graphite Products Booklet, Carborundum Company, Form A7002, January 1973.

rovings, and tape are the most common forms of continuous graphite fibers sold today. These fibers can be purchased with or without a sizing agent or surface treatment (see section 11.5). A tow consists of numerous filaments in a straight-laid bundle and is specified by their number. Depending on the organic precursor and manufacturer, typical filament counts vary from 400 to 10,000 or as high as 160,000. A yarn is a twisted tow, whereas a roving is a number of ends or strands collected in a parallel bundle with little or no twist and is specified by the number of ends. Finally, a tape consists of numerous (e.g., 300) tows or yarns side-by-side on a backing or stitched together.

Graphite fibers can be woven into a fabric by interlacing yarns to form different patterns or weaves. The woven fabric is assigned a style number that specifies the type of weave. Discontinuous graphite fibers are available in short lengths (e.g., 3 or 6 mm = 0.12 or 0.24 in.) and are finding wide use in the automotive and manufacturing industries. In addition, graphite fibers can be supplied in the form of felt or mat by a random spinning process.

The type and form of graphite fiber used in the fabrication of a composite would depend on the particular application (e.g., aerospace, automotive, medical, or industrial) and method of manufacture (e.g., tape lay-up, injection molding, or pultrusion). Graphite-fiber composites will be discussed in detail in subsequent sections of this chapter.

11.5. GRAPHITE-FIBER TREATMENTS

Graphite fibers are subjected to post treatments (including surface treatments and/or application of organic sizings) in order to improve their compatibility with the resin matrix and/or their "handleability." The mechanical properties—including flexural strength, interlaminar shear strength (ILSS), and mode of fracture—of a graphite-fiber-reinforced composite depend on the nature of the resin–fiber bond.[106, 230]

Organic coatings (sizings) in an amount of 0.5–7 wt % are typically applied by the manufacturer by passing the heated fibers through a sizing bath. The sizing agents most commonly employed are polyvinyl alcohol, epoxy, polyimide, and water. These coatings are applied to both untreated and surface-treated fibers. They not only improve the handleability and abrasion resistance of the fibers, but also affect their adhesion to the matrix.

The inverse relationship between the ILSS of a unidirectional composite and the modulus of elasticity of the reinforcing fibers was first reported by Simon et al. (see Fig. 11.36)[231] and confirmed by Goan and Porsen.[232]

This effect has been attributed to the poor interfacial bond between the high-modulus fiber and the resin.[106, 191, 233–236] Surface treatments that improve the ILSS of graphite-fiber composites by a factor of 2–3 are available; these include wet and dry oxidation, applica-

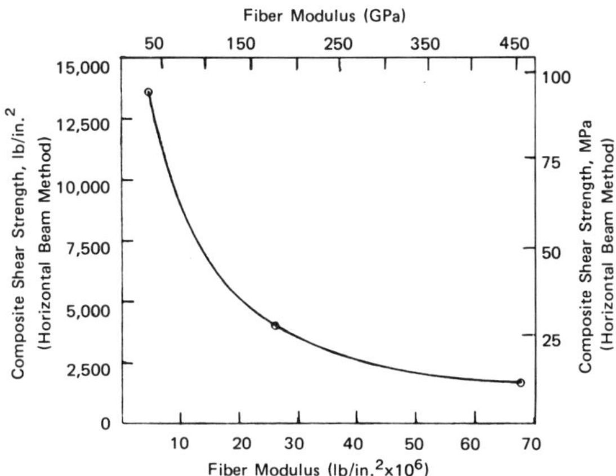

Figure 11.36. Shear strength of an amine-cured carbon fiber–epoxy resin composite as a function of the modulus of elasticity of the reinforcing fibers.[231]

tion of organic or inorganic coatings, whiskering, and irradiation. (The interested reader is referred to two reviews: McKee and Mimeault[234] and Larsen et al.[237]) Table 11.2 lists the various forms of these techniques, gives pertinent references, and describes the effect of the treatment on the mechanical properties of graphite fibers and composites.

As pointed out by McKee and Mimeault[234] as well as Rand and Robinson,[238] the mechanism(s) that account for the changes in the physical and chemical properties of the fiber

Table 11.2. Effect of Surface Treatments on the Mechanical Properties of Graphite Fibers and Composites.

TECHNIQUE	REFERENCES	EFFECT ON MECHANICAL PROPERTIES
Oxidative (Wet)		
HNO_3	232–234, 237, 238, 245, 247, 250, 251, 256	A 25–200% increase in ILSS Decrease in tensile strength Effect varies with degree of graphitization and precursor
$KMNO_4 \cdot H_2SO_4$	106, 247, 251	A 100–200% increase in ILSS Slight decrease in tensile and flexural strengths
Sodium hypochlorite	232, 233, 237, 256, 257	A 30–100% increase in ILSS
Chromic acid	248, 256	Decrease in tensile strength Excess degradation of fiber
Electrolytic NaOH	106, 237	A 70–120% increase in ILSS Slight decrease in tensile strength
Oxidative (Dry)		
Vacuum desorption	234, 247	A 20% increase in ILSS
Air	230, 232, 233, 237, 244, 247, 256, 257	A 10–200% increase in ILSS Difficult to control
Oxygen or ozone	232, 251	A 20–40% increase in ILSS Difficult to control
Catalytic oxidation	234	A 50–100% increase in ILSS
Coatings		
HNO_3 + polymer coatings	232, 241	Slight increase in ILSS (over HNO_3)
Air + alternating and block copolymers	246	A 50–100% increase in ILSS
Vapor-Phase Deposition		
Pyrolytic carbon	234, 242, 243	A 25–60% increase in ILSS that varies with fiber type
Silica/silicon	232, 247	Slight increase in ILSS Increase in oxidative resistance
Metals	234	Increase in oxidative resistance
Whiskering		
SiC	232	A 200–400% increase in ILSS that varies with fiber type
Irradiation	262	Slight increase in ILSS Slight increase in flexural strength

Table 11.3. Recent Patents on Surface Treatments for Graphite Fibers.

PATENT	COMPANY	DATE	TECHNIQUE	EFFECT ON MECHANICAL PROPERTIES
U.S. 3,791,840	Union Carbide Corporation	February 12, 1974	Aqueous hypochlorous acid	Improved ILSS
U.S. 3,801,350	U.S. Air Force	April 2, 1974	$NaIO_4$, $SnCl_4$	80% increase in shear strength
U.S. 3,816,598	Lockheed Aircraft	June 11, 1974	Heat treatment at 475–600° C (887–1112° F) in atmosphere containing 2–10% O_2	Improved ILSS
U.S. 3,821,013	Celanese Corporation	June 28, 1974	Coating of amorphorus carbon and Ti metal	200% increase in ILSS
U.S. 3,832,297	Hercules Incorporated	August 27, 1974	Electrolized in NH_4OH	Improved adhesion to resin matrix
U.S. 3,833,402	U.S. Navy	September 3, 1974	2% ferrocene in benzene, 800° C (1472° F) Fe coats graphite	Increased shear strength
U.S. 3,837,904	Great Lakes Carbon Corporation	September 24, 1974	Acrylonitrile–butadiene polymer, *m*-phenylenediamine	Increased handleability, good flexibility
U.S. 3,853,600	Celanese Corporation	December 10, 1974	Polyphenylene coating	Improved ILSS
U.S. 3,859,187	Celanese Corporation	January 7, 1975	Aqueous NaOCl at moderate temperature	Increased ILSS
U.S. 3,876,444	General Electric Company	April 8, 1975	Solution of formate, acetate, and nitrate salts of Cu, Pb, Co, Cd, and V_2O_5	Roughens surface, increased resin–fiber bonding
U.S. 3,931,329	U.S. Navy	January 6, 1976	Bromine	70% increase in tenacity
Germany 2,606,623	Asahi Chemical Industries	September 2, 1976	Epoxide containing vinyl polymers	Improved adhesion to resin matrix
Britain 1,455,331	U.K. Atomic Energy	November 10, 1976	Carbon coated and air oxidized	Improved ILSS
Japan 77 34,095	Showa Denko K.K.	March 15, 1977	$NaClO_3$ or $NaNO_3$ and heat treating at 1200° C (2192° F)	40% increase in ILSS

Table 11.3. Continued.

PATENT	COMPANY	DATE	TECHNIQUE	EFFECT ON MECHANICAL PROPERTIES
Japan 77 53,092	Nippon Carbon Company	April 28, 1977	Oxidized at > 700° C (1292° F) in atmosphere containing Cl or HCl	Improved flexural and tensile strengths
Japan 77 74,655	Asahi Chemical Industries	June 22, 1977	Aqueous solution of $K_2Cr_2O_7$ impregnated with bis(4-(diglycidyl-amino)phenyl)alkane	Improved ILSS
Japan 77 148,227	Mitsubishi Rayon	December 9, 1977	Blending siloxane with acrylic polymer	Carbon fibers with 50% increase in tenacity and tensile modulus of elasticity
Japan 78 09,272	Toray Industries	April 4, 1978	Coated with metal carbide (TiC)	Improved adhesion to resin matrix

(and subsequent mechanical properties of the composite) after oxidation are not well understood despite the numerous analytical techniques that have been employed to characterize the carbon-fiber surface: scanning electron microscopy (SEM),[106, 230, 232–234, 237–248] BET (Brunaver, Emmet, & Teller) surface-area measurements,[230, 232, 234, 237, 238, 244, 247, 249] laser Raman spectroscopy,[237, 251] infrared spectroscopy,[252] wicking rates,[236] contact angles,[232, 234, 253], adsorption isotherms,[238] electron spectroscopy for chemical analysis (ESCA),[106, 254–256] auger electron spectroscopy,[255] mass spectroscopy,[238] optical microscopy,[237, 257] thermogravimetric analysis,[234] and chemical titration of surface groups.[234, 245, 247] Researchers[234, 238, 247, 249] have used these techniques to correlate changes in the interfacial bond strength with changes in the area, functionality, and roughness of the fiber surface. Tuinstra and Koenig[251] showed a relationship between the Raman spectrum of the fiber surface (and hence the "amount of crystal boundary") and the shear strength of the composite. Others have suggested that the effect of postoxidation treatment on PAN precursor fibers was to remove the surface defects.[234] In general, the effect of oxidation on the shear properties of the composite varies with the degree of graphitization (i.e., heat-treatment temperature) and type of organic precursor.[232, 234] In addition, recent studies have shown that variations in the processing conditions (e.g., dwell time in the mold, viscosity of resin, and void content) can have a deleterious effect on the mechanical properties of the composite.[257–259]

Novak,[230] Brewis et al.[256] and Herrick[260] observed that the impact strength and mode of failure (when loaded in shear) of a graphite fiber–epoxy resin composite were related to (inversely proportional in the case of impact strength) its ILSS. Thus, graphite-fiber-reinforced epoxy-resin composites with a low ILSS (untreated fibers) tended to fail in shear (fiber pull-out) when loaded in shear. As the ILSS increased as a result of various surface treatments, the mode of failure changed to a combination of shear and tension. Hence, as the ILSS increased, failure became brittle, and hence the impact strength of the composite decreased. Recent work has shown that thermoplastic coatings increase the strength and fracture toughness of graphite fiber–epoxy resin composites.[239, 261]

The ideal surface treatment should increase the shear strength of the composite, have little effect on the tensile strength, and offer short processing times, good control, and low cost. As can be seen from Table 11.2, wet-oxidation treatments are widely used, whereas some dry-oxidation treatments are difficult to control. Although whiskering has many advantages, it may be too costly for commercial application. New surface treatments are continuously being developed by carbon-fiber manufacturers and other researchers. Table 11.3 lists some of the recent patents and their effect on the mechanical properties of graphite-fiber composites.

11.6. GRAPHITE-FIBER TEST METHODS AND METHODOLOGY

Manufacturers have developed techniques for determining certain physical and mechanical properties of random samples from each batch of graphite fibers they produce.[263] These tests can be performed on single filaments, dry strands, or resin-impregnated strands.[264] These data are invaluable to the manufacturer for quality control purposes and to the user as a material evaluation/selection guide.[265, 266] Table 11.4 lists some of the standard tests applicable to graphite fibers.

ASTM-D3379-75 describes the standard test method for determining the tensile strength and modulus of elasticity of graphite-fiber single filaments.[267] The technique involves mounting individual filaments (selected at random, on slotted cardboard tabs (see Fig. 11.37[267]), cutting or burning the center portion of the tabs, and then stressing the sample at a constant cross-head speed until failure. ASTM-D3544-76 serves as a guide for reporting test methods and results, including number of tests, selection technique, standard deviation, and coefficient of variation.[268] Figure 11.38

Table 11.4. Graphite-Fiber Test Methods.

PROPERTY MEASURED	TECHNIQUES/COMMENTS	REFERENCE
Tensile strength and modulus of elasticity	Single filaments	264, 267, 269–271, 283, 284
	Strands (dry)	264, 269–271
	Strands (resin impregnated)	263, 269, 272–274
	Effect of strain rate	274
	Effect of temperature	271, 274
	Electrical resistivity	269
	Effect of gage length/diameter	270, 275–278, 282, 284, 285
	Reporting test methods/results	264–268
	Sampling techniques	265, 269
Shear modulus of elasticity	Torsion pendulum	271
	Vibrating piezoelectric crystal	271
	Effect of temperature	271
Denier	Vibroscope	269, 282
	Microscopic area	269
	Weighed yarn	263, 269
Density	Density gradient column	279, 282, 286
	Displacement	280
Specific gravity	Displacement	272, 280
Cross-sectional area	Optical	267, 283
	Calculated (density × denier)	263, 267
Diameter	Optical	264, 284
Thermal conductivity	Unidirectional composite	281
	Strands (dry)	281

shows how one manufacturer uses statistical methods in determining confidence levels and material acceptance/rejection criteria.[265]

Single-filament tests have certain disadvantages: inaccurate denier measurements, operator variability and selectivity, and overall slowness of the method.[269] Figure 11.39 shows the effect of operator variability on the tensile strength of graphite-fiber single filaments.[269] As a result of these potential problems, techniques for measuring tensile properties on dry and resin-impregnated strands have been developed.[263, 264, 269–274]

In general, the ultimate tensile strength and modulus of elasticity of single filaments are higher than those measured on dry strands.[264] Tests on resin-impregnated strands showed good agreement between the theoretical bundle strength (based on average single-filament strength) and the strength distribution of the

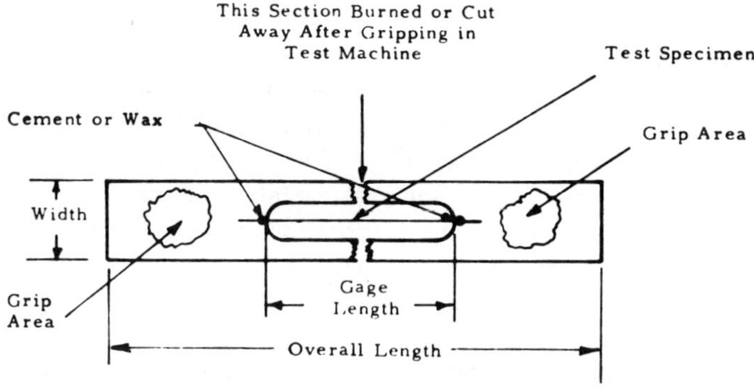

Figure 11.37. Tab showing typical specimen mounting method.[267]

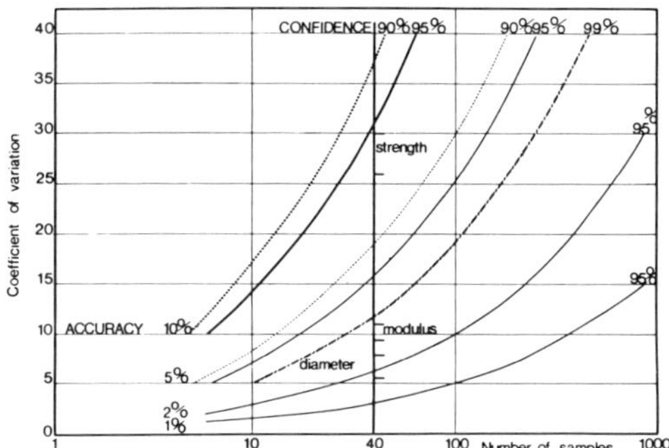

Figure 11.38. Accuracy/confidence graph.[265]

impregnated strand.[272] Table 11.5 gives the average tensile strength and modulus of elasticity for high-modulus graphite fibers as determined by single-filament, dry-strand, and resin-impregnated-strand tests.[269] The average modulus values recorded for each method are in close agreement, whereas the strength values from the impregnated strand are lower than those obtained by the other two methods. Recent work on testing epoxy-impregnated graphite-fiber strands at liquid-nitrogen temperature and various strain rates showed that these variables had only a minimal effect on the tensile strength and modulus of elasticity.[269] It is well known that the tensile properties of graphite fibers are dependent on the initial processing parameters (heat-treatment temperature, time, and elongation), type of organic precursor, surface treatment, and presence of flaws or broken filaments.[265,270,275] In addition, other researchers have found a dependence of graphite-fiber tensile properties on fiber diameter and gage length.[270,275–278]

The shear modulus of graphite fibers was determined by two methods: a torsion pendulum and a dynamic process using vibrating piezoelectric crystals. Using the latter technique, only a small change in shear modulus was noted at test temperatures from ambient to 800°C (1472°F).[271]

Various methods have been developed to measure the denier of graphite fibers.[269] (Denier is defined as the weight in grams of a 9000-m length of fiber.) Table 11.6 gives a comparison of three different denier measurements as determined by microscopic area, vibroscope, and weighed yarn.[269] The latter technique is the most commonly used in industry. Density or specific gravity is determined in accordance with ASTM specification D1505-68 or D792-66, respectively.[279,280] The cross-sectional area of the fibers can be measured optically or calculated from values of density and denier.[263,267,269]

The axial thermal conductivity of various commercially available graphite fibers was given in ASTM STP580 on the basis of two techniques.[281] One technique involved thermal-conductivity measurements on unidirectional-composite samples fabricated by impregnating a fiber strand with epoxy resin. This technique was found to be accurate to ±10%. The other method involved making a direct thermal-conductivity measurement on a dry-fiber strand. In this case, the thermal

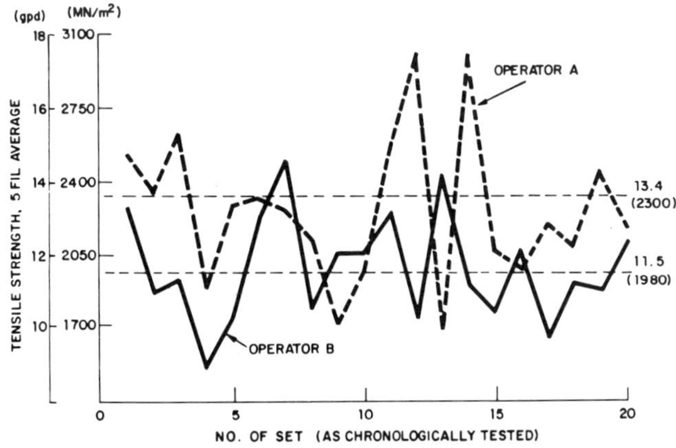

Figure 11.39. within-operator and operator-to-operator variations in tenacity on sets of five filaments each selected from a graphite-yarn control specimen.[269]

Table 11.5. Average Tensile Strength and Modulus of Elasticity of High-Modulus Graphite Fiber as Determined by Various Methods.[a,b,269]

	SINGLE FILAMENT				DRY STRAND				IMPREGNATED STRAND			
	STRENGTH,		MODULUS,		STRENGTH,		MODULUS,		STRENGTH,		MODULUS,	
SPECIMEN	gpd (MN/m^2)	ksi (MPa)	gpd (GN/m^2)	Msi (GPa)	gpd (MN/m^2)	ksi (MPa)	gpd (GN/m^2)	Msi (GPa)	gpd (MN/m^2)	ksi (MPa)	gpd (GN/m^2)	Msi (GPa)
1	10.0 (1720)	250 (1720)	2950 (507)	428 (2950)	9.3 (1600)	232 (1600)	2990 (514)	434 (2990)	7.4 (1270)	184 (1270)	2960 (509)	430 (2960)
2	9.6 (1650)	240 (1650)	2980 (513)	432 (2980)	10.3 (1770)	257 (1770)	3120 (537)	453 (3120)	7.9 (1360)	197 (1360)	3030 (521)	440 (3030)
3	9.7 (1670)	242 (1670)	2960 (509)	430 (2960)	9.7 (1670)	242 (1670)	3080 (530)	447 (3080)	8.0 (1380)	200 (1380)	2990 (514)	434 (2990)
4	10.2 (1750)	254 (1750)	3050 (525)	442 (3050)	9.5 (1630)	236 (1630)	2990 (514)	434 (2990)	8.6 (1480)	215 (1480)	2930 (504)	425 (2930)
5	9.1 (1570)	228 (1570)	3170 (545)	460 (3170)	9.9 (1700)	247 (1700)	2980 (513)	433 (2980)	8.6 (1480)	215 (1480)	3010 (518)	437 (3010)
Overall average	9.7 (1670)	242 (1670)	3020 (519)	438 (3020)	9.7 (1670)	242 (1670)	3030 (521)	440 (3030)	8.1 (1390)	202 (1390)	2980 (513)	433 (2980)

[a]Five ends were taken from each sample and all tests were performed on the same ends in accordance with the following:
1. 5 filaments/end or 25-filament specimen tested as single filaments.
2. 5 ends/specimen for dry strand; modulus and three breaking energies from each.
3. 5 ends/specimen for impregnated strand.

[b]Single-filament data are based on weighed-yarn dpf's.

Table 11.6. Denier/Filament as Determined by Three Independent Methods on the Same Filaments, Ends, or Both.[269]

SPECIMEN	MICROSCOPIC AREA	VIBROSCOPE[a]	WEIGHED YARN[b]
1	1.06	1.04	0.98
2	1.02	1.07	0.99
3	1.00	1.08	1.02
4	1.13	1.06	1.02
5	0.96	0.95	0.96
Overall average	1.03	1.04	0.99

[a] Average for 25 filaments—same 5 filaments from each of 5 ends, measurements unconnected.
[b] Average for 5 ends—the same 5 ends as above.

conductivity of the fiber was determined by extrapolation to a lower temperature differential (ΔT) or shorter gage length. The thermal conductivity was found to vary linearly with increasing electrical conductivity or tensile modulus of elasticity above a certain threshold level (see Fig. 11.40).[281]

11.7. GRAPHITE FIBER-ORGANIC MATRIX COMPOSITE MATERIALS

To realize the many advantages of graphite fibers discussed in the preceding section, it is necessary to place them in some sort of matrix in order to distribute the load over a wide area; in other words, it is necessary to form a composite. Although the matrix can be metallic, ceramic, or organic, only the latter will be considered in this chapter.

In an organic matrix–graphite fiber composite, the fibers contribute their superior tensile properties for rigidity and strength, whereas the matrix adds such properties as corrosion resistance and impact strength. Epoxy resins are by far the most commonly used today, although polyester, polysulfone, polyimide, and various thermoplastic resins are also used for certain applications. This section of the chapter will cover the physical and mechanical properties of graphite composites and their performance under environmental and fatigue conditions.

The main advantage of graphite-fiber composites is their very high specific tensile strength and stiffness, that is, modulus of elasticity (see Fig. 11.41).[287] It is primarily for this reason that graphite-fiber-reinforced

Figure 11.40. Thermal conductivity of carbon fibers as a function of tensile modulus of elasticity.[281]

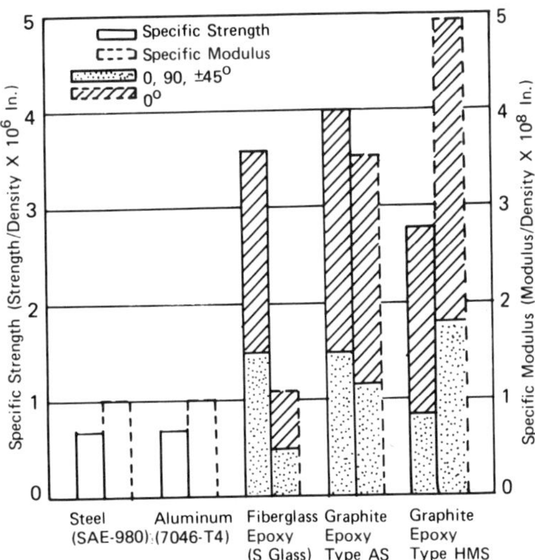

Figure 11.41. Comparison of specific tensile strength and modulus of elasticity of commonly used materials.[287]

Table 11.7. Typical Mechanical Properties of Commercially Available Graphite Fiber-Epoxy Resin Composites.[a]

PROPERTY	SYMBOL	ASTM TEST METHOD	UNITS	CELANESE CORPORATION[b] CELION 6000 (62%[e])	GY 70 (62%[e])
Unidirectional Laminate					
Longitudinal (0°) Properties					
Tensile strength	σ^{11}	D-3039	ksi (MPa)	239 (1647)	113.6 (783)
Tensile modulus of elasticity	E^{11}	D-3039	Msi (GPa)	21.9 (151)	42–47 (289–324)
Ultimate tensile strain	ϵ^{11}	D-3039	%	1.1	0.2
Compressive strength	σ^{11}	D-3410	ksi (MPa)	213 (1468)	90–102 (620–703)
Compressive modulus of elasticity	E^{c}_{11}	D-3410	Msi (GPa)	20.4 (141)	44.7 (308)
Ultimate compressive strain	ϵ_{11}	D-3410	%	1.7	—
Flexural strength (4 point)	σ^{f}_{11}	D-790	ksi (MPa)	254 (1750)	115 (792)
Flexural modulus of elasticity	E^{f}_{11}	D-790	Msi (GPa)	19.6 (135)	37 (255)
Interlaminar shear strength (short beam)	σ_{ILS}	D-2344	ksi (MPa)	18.1 (125)	8.6 (59)
Transverse (90°) Properties (π/zckd)					
Strength	σ_{22}	D-3039	ksi (MPa)	7.9 (54.4)	3.5 (24)
Modulus of elasticity	E_{22}	D-3039	Msi (GPa)	1.62 (11.2)	1 (7)
Ultimate strain	ϵ_{22}	D-3039	%	0.49	—

[a]This table is intended for comparison use only; manufacturers' data sheets should be consulted for design data.
[b]Product Data, Material Properties of Composites, Celanese Corporation.
[c]Hercules Incorporated, Product Data Sheet #842.
[d]*Plastic Compounding*, March–April 1979.
[e]Volume percent fiber.

composites are replacing metal in applications where weight savings are important (e.g., the aerospace and transportation industries). Table 11.7 lists the typical mechanical properties of commercially available medium- and high-modulus graphite fiber–epoxy resin composites. The directional strength and stiffness of these composites depend on the fiber orientation of the laminates.[288] Figures 11.42 and 11.43 show typical computer-generated design curves for relating the tensile modulus of elasticity and strength of Celion graphite fiber–epoxy resin composites with ply stacking arrangements.[288]

11.7.1. Environmental Stability

The mechanical properties of graphite fiber-organic matrix composites are known to vary as a function of such environmental conditions as surrounding temperature and relative humidity.[289,290] An excellent state-of-the-art

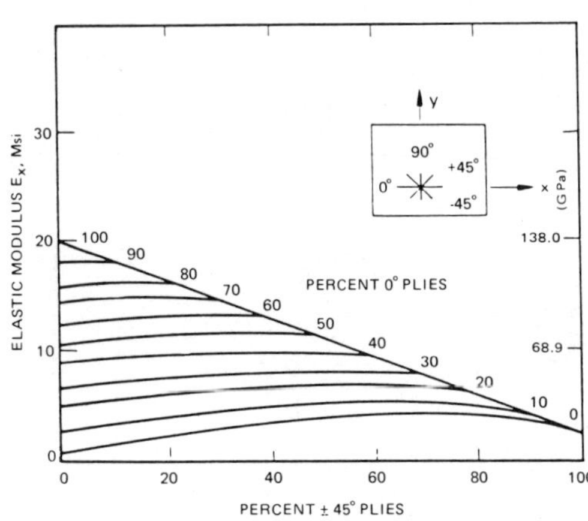

Figure 11.42. Tensile modulus of elasticity, E_x, of high-strength Celion–epoxy laminates at room temperature.[288]

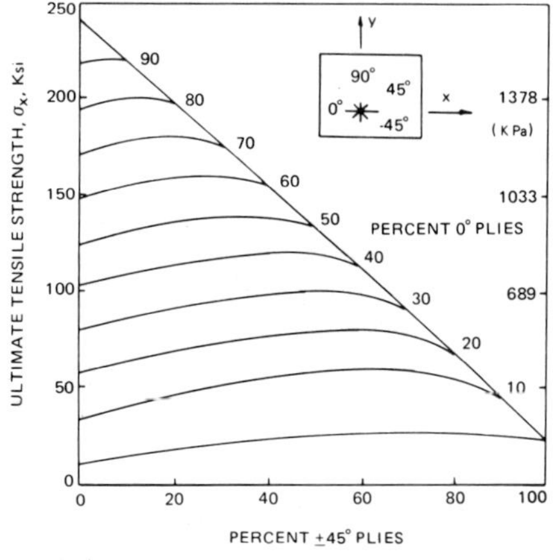

Figure 11.43. Ultimate tensile strength, σ_x, of Celion–epoxy laminates at room temperature.[288]

Table 11.7. Continued.

HERCULES INCORPORATED[c]			GREAT LAKES CARBON CORPORATION[d]		
AS (62%[e])	HMS (62%[e])	UNION CARBIDE CORPORATION[d] THORNEL 300 (65%[e])	FORTAFIL 3 (60%[e])	FORTAFIL 5 (60%[e])	STACKPOLE FIBERS COMPANY PANEX 30 (60%[e])
270 (1860)	165 (1137)	220 (1516)	190 (1309)	160 (1102)	225 (1550)
21 (145)	30 (207)	20 (138)	—	—	18 (124)
1.2	0.55	—	—	—	1.2
260 (1791)	55 (379)	230 (1585)	190 (1309)	160 (1102)	180 (1240)
20 (138)	15.5 (107)	20 (138)	—	—	18 (124)
—	0.42	—	—	—	—
260 (1791)	140 (965)	—	215 (1481)	180 (1240)	250 (1723)
17.5 (121)	24.5 (169)	—	18 (124)	23.5 (162)	18 (124)
18 (124)	8 (55)	18 (124)	12 (83)	11 (76)	16.5 (114)
9.4 (65)	4.5 (31)	—	—	—	12–13 (83–90)
1.36(9.4)	1.1 (8)	—	—	—	1.45 (10)
0.70	0.44	—	—	—	—

review of the effect of temperature and moisture on the behavior of these advanced composites was recently compiled by ASTM.[289]

11.7.1.1. Thermal Effects

The matrix-dominated mechanical properties of state-of-the-art graphite–epoxy composites may be severely degraded at temperatures between 150 and 200° C (302 and 392° F).[291, 292] The glass-transition temperature (T_g) of epoxy resins, which depends on the particular resin system used and the cure temperature, is typically between 125 and 175° C (257 and 347° F). For applications approaching the T_g, other high-temperature resin matrices must be employed.[291, 293, 294] Figure 11.44 demonstrates the higher elevated-temperature flexural-strength properties of graphite-fiber-reinforced phenolic and polyimide composites compared to epoxies.[291]

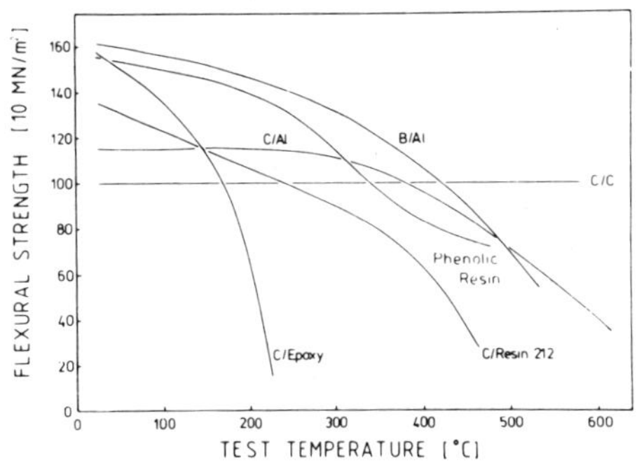

Figure 11.44. Flexural strength of unidirectionally fiber-reinforced composites as a function of test temperature.[291]

The effect of thermal cycling at temperatures between −54 and 149° C (−65 and +300° F) on the mechanical properties of HTS–Epon 828 composites was found to be dependent on the fiber content, rate of thermal cycling, and residual stress in the composite.[295] It has also been shown that at test temperatures from −40 to +120° C (−40 to 248° F), the fatigue properties of types I and II graphite fiber–epoxy composites was controlled by fiber type and orientation, method of fabrication, and dynamic-stress levels.[296] Camahort et al. showed that the degree of microcracking in graphite–epoxy composites resulting from thermal cycling between −195 and 100° C (−320 and 212° F) varied with the resin system and cure temperature.[297]

At these elevated temperatures, the thermo-oxidative stability of the graphite fibers (as well as the resin) may influence the mechanical properties of the composite.[298, 299] McMahon reported a correlation between the amount of trace metals present in graphite fibers (determined by emission spectroscopy) and their thermal stability in air.[298] Most commercially available medium-modulus PAN-base carbon

Table 11.8. Significant Temperatures as Determined by Dynamic Thermogravimetric Analysis.[300]

SUPPLIER	MATERIAL FIBER/RESIN	TEMPERATURE, °C (°F) INITIAL WEIGHT LOSS	RESIN DECOMPOSITION	FIBER DECOMPOSITION
General Dynamics	Celion 6000/5208	270 (518)	345 (653)	720 (1328)
↓	T6300/5208	165 (329)	345 (653)	645 (1193)
	T300/5208	210 (410)	360 (680)	642 (1188)
	T300/934	175 (347)	345 (653)	635 (1175)
	AS/3501	—	324 (615)	591 (1096)
	T300/F178	115 (239)	386 (727)	619 (1146)
Lockheed/Sunnyvale	UC PITCH/934	Ambient	292 (558)	770 (1418)
↓	T300/5208	110 (230)	364 (687)	610 (1130)
	T300/F178	140 (284)	387 (729)	617 (1143)
	T300/PMR-15	—	559 (1038)	559 (1038)
	AS/2080	100 (212)	603 (1117)	603 (1117)
Lockheed/Georgia	T300/5208	270 (518)	343 (649)	632 (1170)
↓	AS/3501	185 (365)	342 (648)	602 (1116)
	UMA/976	80 (176)	310 (590)	715 (1319)
	GY70/7534	175 (347)	327 (621)	876 (1609)
Lockheed/Burbank	T300/5208	130 (266)	347 (657)	621 (1150)
↓	T300/934	135 (275)	342 (648)	635 (1175)
McDonnell Douglas	T300/5208	135 (275)	358 (676)	602 (1116)
↓	AS/3501	200 (392)	354 (669)	593 (1099)
Northrop	AS/3501	—	340 (644)	592 (1098)
Grumman	AS/3501	190 (374)	345 (653)	600 (1212)
MIT	MOD 11/PPQ	520 (968)	623 (1153)	687 (1269)
NASA/Lockheed	HTS/PMR-15	370 (698)	571 (1060)	800 (1472)

fibers (AS and T300), which contain large amounts of alkali metals, are apparently less stable than Celion (Celanese Corporation) fibers, higher-modulus PAN-base carbon fibers, and pitch-base carbon fibers. These latter fibers have excellent thermo-oxidative stability at a temperature of 482° C (900° F) for times up to 1000 hr. The thermo-oxidative stability of several commercially available medium- and high-modulus graphite-fiber composites containing various organic matrices was recently compiled.[300] The dynamic and isothermal aging data for these composites are given in Tables 11.8 and 11.9. The thermal stability of the composites decreased in the following order: polyphenylquinoxaline > polyimides > epoxies. Recent studies indicate that certain thermoplastic resins (e.g., polyethersulfone) have thermo-oxidative stabilities greater than or equal to polyimides and polyphenylquinoxalines.[301]

11.7.1.2. Moisture and Temperature Effects

Graphite fibers are inert to moisture and most organic solvents.[302] However, all of the organic resins used in advanced composites absorb moisture to some extent. Figure 11.45 shows absorption curves for different types of neat epoxy resins.[303] Resin systems that are chemically different have their own unique moisture-absorption properties. Similar absorption curves are obtained for comparable graphite–epoxy composites. Figure 11.46 shows moisture uptake for AS-3501-5 composites at 71°C (160°F) and 75 and 95% relative humidity.[304] Moisture absorption in a fiber-reinforced composite can occur by three mech-

Table 11.9. Fractional Lives as Determined by Isothermal Aging[300]

SUPPLIER	MATERIAL FIBER/RESIN	AGING TEMPERATURE °C (°F) 400 (752) t_{20}, HR	400 (752) t_{50}, HR	500 (932) t_{20}, HR	500 (932) t_{50}, HR
General Dynamics	Celion 6000/5208	3.5	—[a]	0.3	12.5
	T6300/5208	2.3	35.0	0.2	3.6
	T300/5208	4.0	44.0	0.4	3.7
	T300/5208	3.7	35.7	0.3	3.5
	AS/3501	2.6	30.5	0.3	1.3
↓	T300/F178	10.0	42.5	0.5	2.0
Lockheed/Sunnyvale	UC PITCH/934	1.8	—[b]	0.3	—[b]
	T300/5208	2.6	25.5	0.3	0.9
	T300/F178	6.0	—[c]	0.5	2.7
	T300/PME-15	9.5	38.0	1.5	4.7
↓	AS/2080	6.7	18.5	0.7	1.2
Lockheed/Georgia	T300/5208	3.2	46.0	0.3	2.7
	AS/3501	1.5	38.5	0.2	1.2
	UMA/976	1.3	—[a]	0.2	25.0
↓	GY70/7534	3.0	—[a]	0.3	—[a]
Lockheed/Burbank	T300/5208	1.7	36.0	0.2	2.2
↓	T300/934	1.3	23.0	0.2	2.7
McDonnell/Douglas	T300/5208	5.0	38.5	0.4	1.2
↓	AS/3501	0.3	7.2	0.3	1.2
Northrop	AS/3501	2.5	23.5	0.2	1.2
Grumman	AS/3501	3.7	43.5	0.3	1.5
MIT	MOD 11/PPQ	50.0	—[a]	2.4	12.5
NASA/Lockheed	HTS/PMR-15	10.5	39.0	0.6	1.2

[a]Greater than 50% after 72 hr.
[b]Terminated—had plateaued at >50%.
[c]Terminated at <50% for convenience.

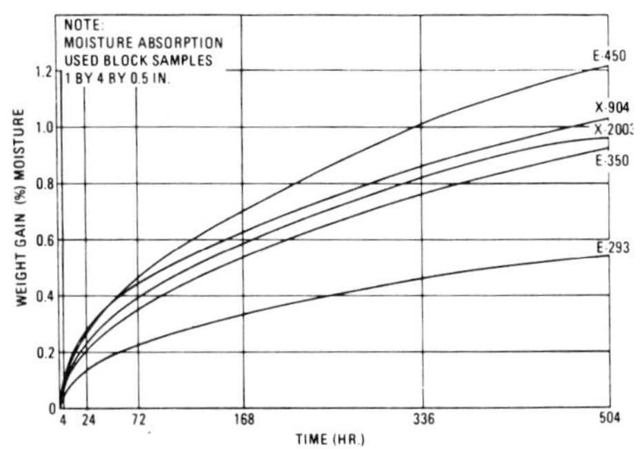

Figure 11.45. Moisture absorption of several neat epoxy resins—weight gain versus time (block samples 1 in. × 4 in. × 0.5 in. (2.5 cm × 12 cm × 1.3 cm).[303]

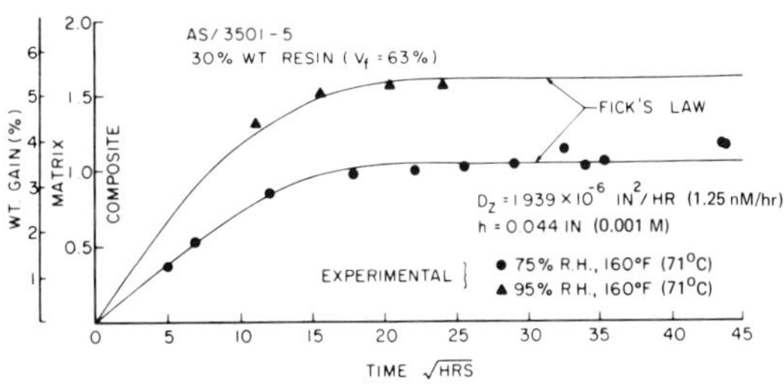

Figure 11.46. Moisture gain of AS-3501-5 composites (30 wt % resin, V_f = 63%) as a function of time CO_z = 1.939 × 10^{-6} in.^{2}/hr (1.25 nm/hr) and h = 0.044 in. (0.001m)]: experimental—) 75% RH, 71°C (160°F);) 95% RH, 71°C (160°F).[304]

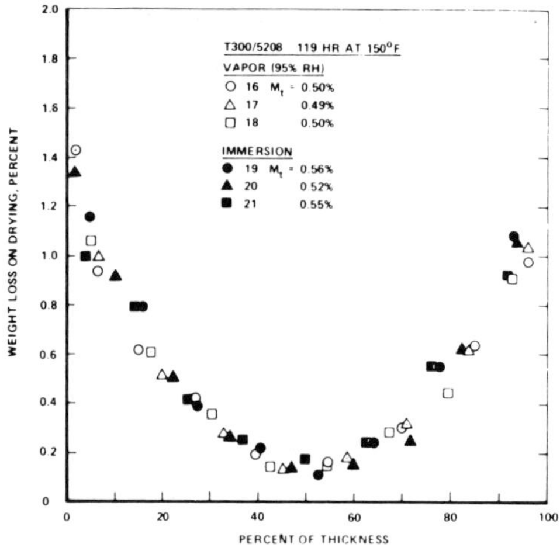

Figure 11.47. Comparison of diffusion effects in water-vapor and water-immersion environments: 16-ply uni directional T300-5208 laminate dried and then conditioned 119 hr at 66° C (150° F)—specimens 16, 17 and 18 at 98% RH, average weight gain = 0.54%; specimens 19, 20, and 21 immersed in water, gain = 0.59% average weight (specimens from panel 1SV1136, cured 10/14/77, not postcured; resin content = 29.3 wt %, density = 1.579).[310]

anisms: through the resin (diffusion), along the fiber–matrix interface, and through the cracks and voids.[305,306]

In general, fiber-reinforced composites absorb moisture by a controlled diffusion process that is dependent on temperature and relative humidity and is predicted by Fick's second law.[289,305] The most commonly used method of determining the rate and amount of moisture absorption and desorption is to measure the weight gain or loss of the composite. Within the last several years, however, other promising techniques have been reported. Carter and Kibler[307] developed an incremental grinding method for determining the equilibrium moisture content and diffusion coefficient of graphite–epoxy and graphite–polyimide composites exposed to moisture. Using this faster method, they were able to distinguish between different resin compositions and degree of cure (postcure). Recently, Belani and Broutman showed a linear relationship between the electrical resistivity and square of the moisture content for unidirectional graphite–epoxy composites.[308] Moisture causes the resin to swell, transverse to the fiber direction, which effectively decreases the fiber-to-fiber contact and hence increases its electrical resistance. Browning used near-infrared spectroscopy to determine the degree of cure of epoxy resins and the amount of absorbed moisture.[309] Absorbed water was determined by monitoring the growth of the OH peak at 1.91μ. In addition, a technique for quantitatively determining the moisture profile through the thickness of a graphite–epoxy composite has recently been developed.[310] This technique involves microtoming thin, uniform layers and weighting the aged sections immediately after cutting and drying to constant weight. Figure 11.47 shows the moisture profile obtained for a T300–5208 composite.[310] Identical moisture profiles were obtained by either immersion in water at 66° C (150° F) or exposure to 95% relative humidity at 66° C (150° F) (see Fig. 11.47). Other researchers have found a correlation between the results of mechanical tests on graphite–epoxy composites after boiling-water exposures and those obtained from a high-humidity environment of equivalent moisture content.[311] The advantage of the boiling-water tests is the shorter exposure time required. Kaelble et al.,[312,313] using ultrasonics, demonstrated a correlation between the amount of moisture absorbed by a graphite–epoxy composite and the decrease in its interlaminar shear strength as well as the corresponding decrease in the ultrasonic attenuation through it. The interested reader is referred to a recent review of nondestructive methods for detecting moisture degradation of fiber-reinforced composites.[314]

Absorbed moisture swells and plasticizes the resin (presumably by breaking epoxy–epoxy hydrogen bonds), lowers the T_g of the resin, and (as a result) degrades the matrix-dominated mechanical properties—especially at elevated temperatures.[289,309,311,315,316]

The equilibrium moisture content and degree of physical-property degradation depend on the following variables: specific resin system,[289,303,317,318] fiber content and orientation,[311,317,319–322] fiber surface treatment,[312,313] degree of cure,[294,307] specimen geometry,[317] temperature, humidity, and exposure time,[289,309,322] loading history,[323] quality of composite,[306,324] and mechanical test.[311,322]

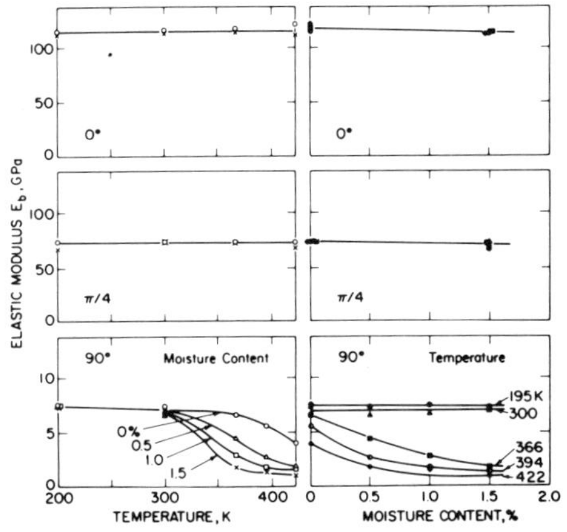

Figure 11.48. Buckling modulus of Thornel 300–Fiberite 1034 as a function of temperature and moisture content.[319]

Figures 11.48[319] and 11.49[318] show typical data. An excellent review of the effect of moisture and temperature on the tensile strength and modulus of elasticity of graphite–epoxy composites is given by Shen and Springer.[319, 320.]

In general, graphite–phenolic and graphite–polyimide composites are more resistant to physical-property degradation by moisture than current state-of-the-art graphite–epoxy composites.[291] The effect of fiber surface treatments on the moisture absorption and ILSS of a graphite–epoxy composite is shown in Fig. 11.50[312] Composites made with graphite fibers that have been surface treated and then reduced in an inert atmosphere were more resistant to moisture than those made with fibers that had been sensitized to water by treatment with Gantrez solution. Gantrez solution (General Aniline & Film Corporation) as a high molecular weight copolymer of maleic anhydride and methyl vinyl ether which is highly hydrophilic and water soluble.[312] In a comparison between graphite fibers treated with nitric acid, silanes, and bromine, Harris et al.[325] found that the acid-etched fiber had the highest weight gain when exposed to steam at 100° C (212° F). The rate of moisture diffusion onto a composite can be reduced by the addition of such protective coatings as polyurethane[322, 323] and glass scrim cloth.[292]

Due to the uneven diffusion of moisture through the material, internal residual stresses are produced in graphite–epoxy composites.[303, 309] In addition, the mode of failure in tensile tests of unidirectional graphite–epoxy composites changes from brittle to ductile as a result of moisture absorption.[304] Figures 11.51[309] and 11.52[304] show the effect of moisture on the stress–strain curve of an amine-cured TGMDA epoxy resin and AS-3501-5 uni-

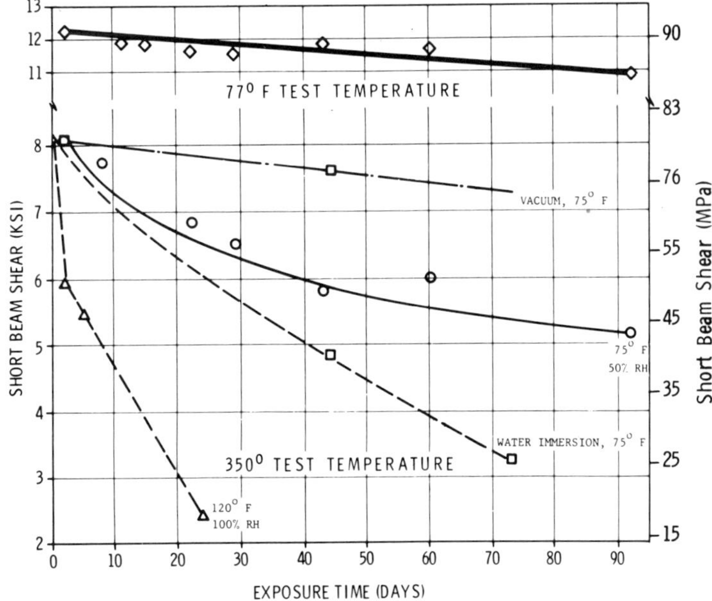

Figure 11.49. Short-beam shear strength of graphite–epoxy composites as a function of environment and exposure time.[318]

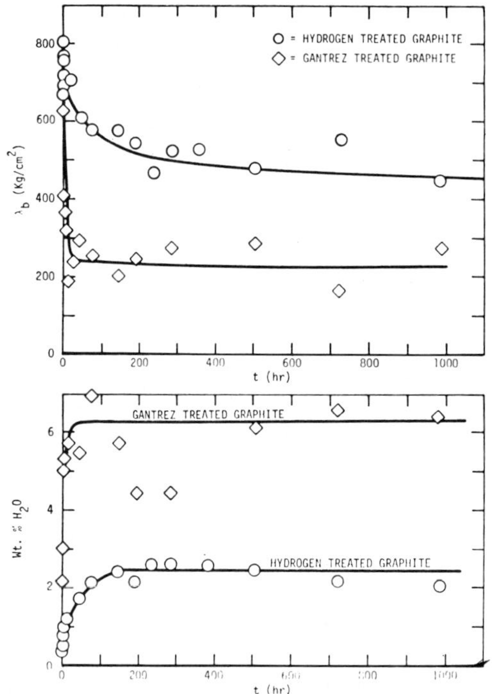

Figure 11.50. Dependence of interlaminar shear strength/ λ (upper view) and moisture content as Wt.%H_2O (lower view) of graphite-epoxy composite on hydrothermal aging time *t* at 100° C in H_2O.[312]

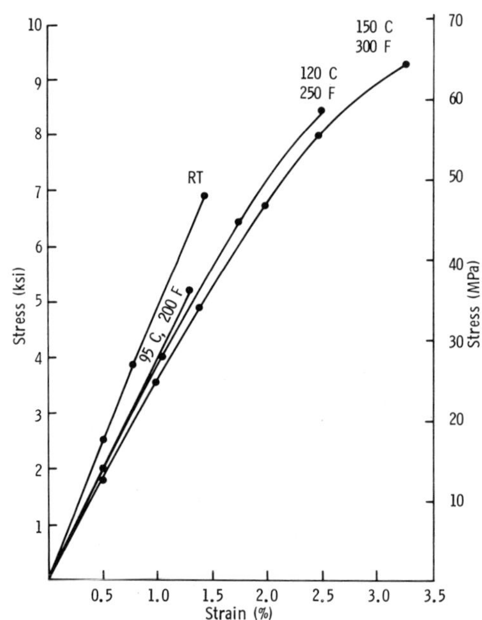

Figure 11.51a. Stress–strain behavior of dry, control amine-cured TGMDA epoxy-resin specimens.[309]

directional composite, respectively. In most cases, the effect is reversible, that is, the matrix-dominated mechanical properties are regained after drying.[311, 316] When moisture absorption was accompanied by rapid thermal spikes, however, there was a permanent change in the moisture-absorption characteristics of the material.[309, 326] This is thought to be due to the development of stress-induced cracks, which facilitates moisture absorption. A recent review paper by Loos and Springer was concerned with the effect of thermal spiking on graphite–epoxy composites.[290] They conclude that both the degree of mechanical-property degradation and the moisture-absorption characteristics due to thermal spiking depend on the composite: T300–5208 seems to be particularly sensitive to thermal spikes, whereas T300–1034 is not.

Similarly, Kaelble et al.[312, 313] found a permanent reduction in the ILSS of a graphite–epoxy composite (HTS–BP907) that had been immersed in water for 1000 hr at 100° C (212° F) and then desiccated (see Fig. 11.53). This

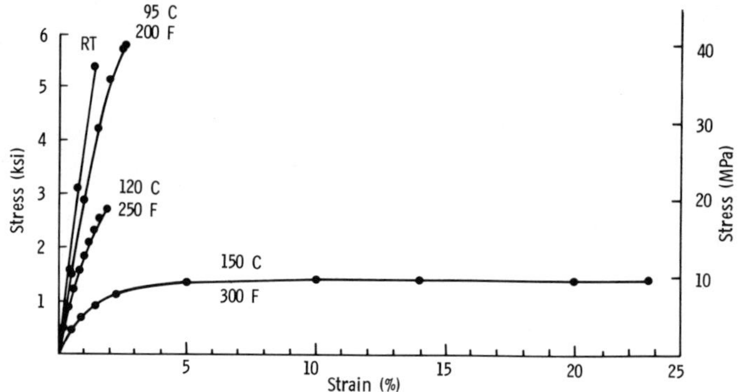

Figure 11.51b. Stress–strain behavior of wet (equilibrium weight gain) amine-cured TGMDA epoxy-resin specimens.[309]

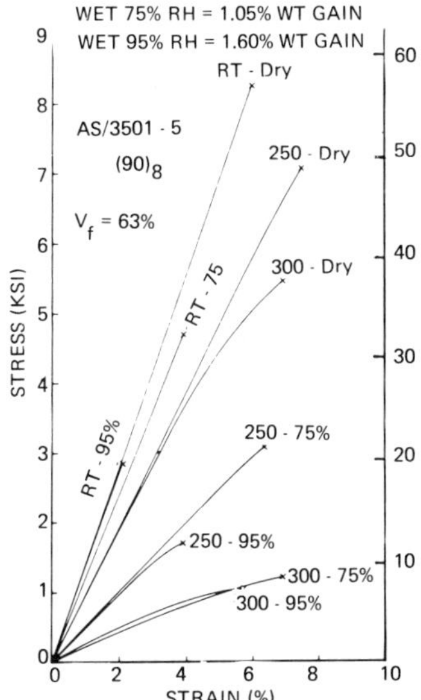

Figure 11.52. Typical transverse stress–strain curves for AS-3501-5 unidirectional composites as a function of temperature and humidity.[304]

indicates a degradation in the resin–fiber interface.[312, 325] In addition, the fracture toughness of the composite increased as a result of moisture absorption.[313] A combination of moisture and temperature had a degrading effect on the fatigue properties of graphite-epoxy composites. A correlation between debonding time and degradation of fatigue properties has been shown.[323] Lubin et al.[388] found that the moisture absorbtion effects on graphite are almost completely reversible. Removal of moisture by natural or forced drying essentially restores the original mechanical strengths.

Also, it was found[389] that the only effective moisture protection for graphite composites is obtained by bonding or co-curing a thin metal (Al) foil on the outside of the laminate.

11.8. THERMOPLASTIC VERSUS THERMOSET RESINS

11.8.1. Introduction

Certain structural applications require that advanced reinforced composites maintain their physical and mechanical properties at elevated temperatures (>177° C = 350° F) and high humidities.[327] These elevated temperatures are above the maximum service temperatures of most state-of-the-art thermosetting epoxy resins. In addition (as previously discussed), the mechanical properties of epoxy resins may be severely degraded by a combination of moisture and temperature. New, modified (135 or 177° C = 275 or 350° F) epoxy-resin systems are currently being developed under U.S. Air Force sponsorship at TRW Incorporated and Hercules Incorporated.[328] These high-vinyl modified epoxy resins (HME) have been shown to possess improved moisture resistance, better processibility, and lower cost when compared with conventional epoxy resins.[328] Figures 11.54 and 11.55 show the

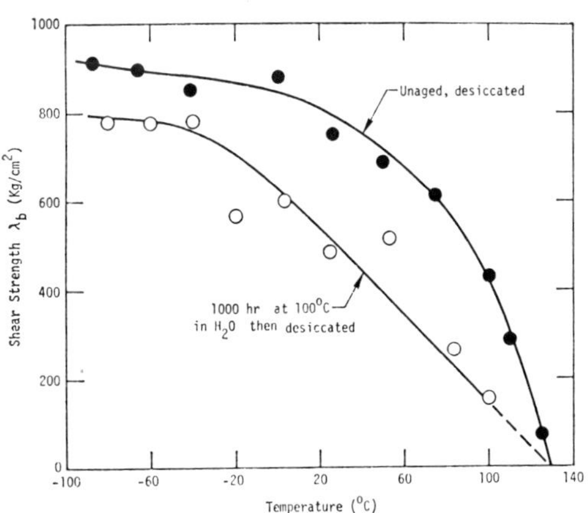

Figure 11.53. Temperature dependence of interlaminar shear strength λ_b for unaged (t = 0) and hydrothermally aged (t = 1000 hr) samples of composite SC-2-3.[312]

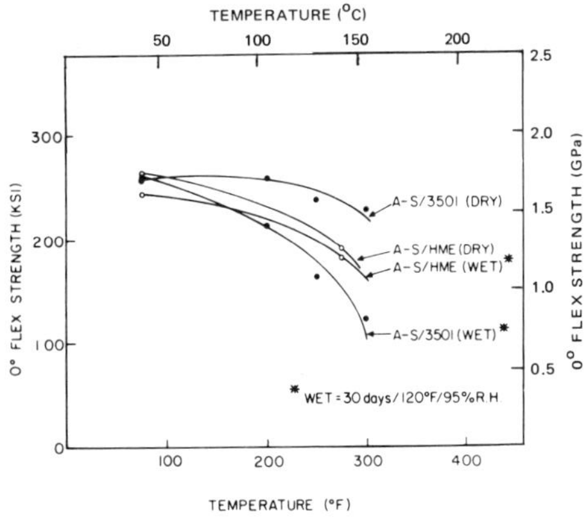

Figure 11.54. Effect of moisture and temperature on the longitudinal (0°) flexural strength of 135° C (275° F) AS-HME composites compared to a current graphite–epoxy material—wet samples held 30 days at 49° C (120° F) and 95% RG.[328]

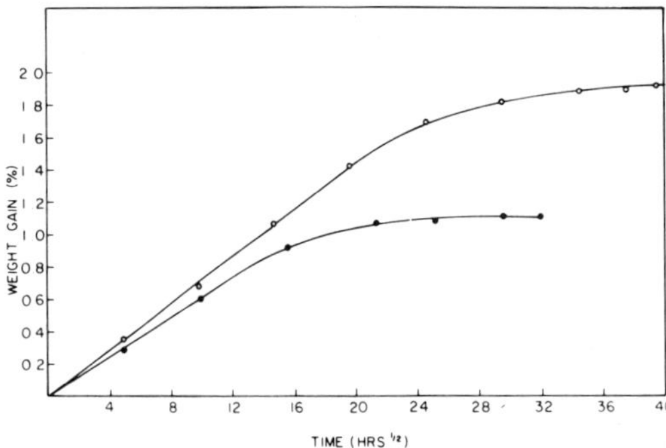

Figure 11.55. Moisture weight gain [71°C (160°F) and 95–100% RH] versus time for celion 6000–HME 350 (○) compared to current graphite-epoxy (AS-3501-5A) material (●).[328]

effect of temperature and humidity on the flexural strength and moisture content of HME composites.[328] French researchers have recently reported the synthesis of a new, 250° C- (482° F-) cure, low-cost thermosetting resin (PSP) obtained by the condensation of aromatic aldehydes with methylated derivatives of pyridine.[329] Figure 11.56 shows the change in flexural strength of HTS -PSP and other high-performance composites at 150° C (302° F) after water immersion.[329]

Thermosetting polyimide resins were developed in the 1960's to meet the high-performance and high-temperature requirements of aerospace materials.[330] These high-molecular-weight insoluble polyimide resins were typically prepared by the condensation polymerization of such soluble precursors as polyamic acids, which are formed from diamines and anhydrides. Good quality, low-void-content laminates were difficult to achieve due to the evolution of volatile condensation products (i.e., water or alcohols) or high-boiling solvents (i.e., dimethylformamide, dimethylacetamide, or n-methyl pyrrolidone). Later, addition-polymerization processes that led to the commercialization of P13N and P105A polyimide laminating resins were developed.[331, 332] Within the last few years, more processible polyimide resins—PMR,[327, 333, 334] NR-105,[335, 336] and LaRC-160[337, 338]—as well as higher-temperature acetylene-terminated ones—for example, Thermid 600,[339–341] the polyphenylquinoxalines,[342, 343] the polybenzimidazoles,[344] and the polyimidequinoxalines[344]—have been developed.

Linear polyimides (unless crosslinked) behave like thermoplastic resins and as such maintain their mechanical properties up to temperatures just below their glass-transition temperature.[331, 332] Due to the relatively high cost of precursors, the toxicity of solvents, and the inherent processing problems associated with the earlier polyimide resins, their use has been limited to aerospace applications. Numerous thermoplastic resins, having a wide range of thermal and mechanical properties, are commercially available as a matrix for continuous or discontinuous (chopped) graphite fibers.[335, 345–357] Table 11.10–11.12 review some of the recent data that have been published on continuous and thermoplastic composites as well as listing the advantages and disadvantages of continuous-fiber thermoplastic versus continuous-fiber thermoset composites. Tables 11.11 and 11.12 present the typical mechanical properties of continuous-fiber-reinforced thermoplastic and thermoset composites. Figures 11.57[355] and 11.58[354] show

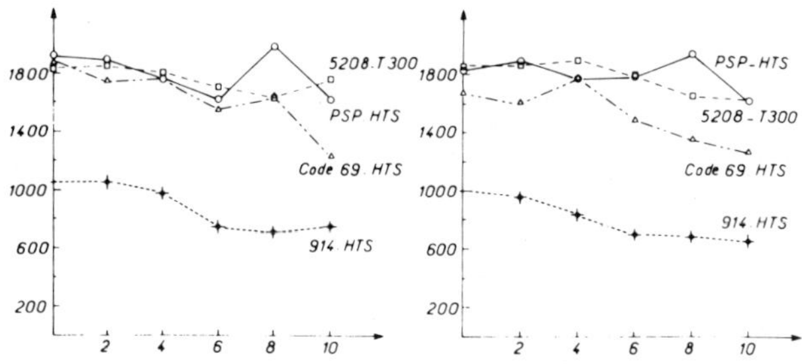

Figure 11.56. Flexural strength at 150° C (302° F) of high-performance laminates after water-immersion cyclings at room temperture: (*a*) without temperature spike; (*b*) with one 10-mn/150° C (302° F) spike per cycle.[329]

Table 11.10. Advantages and Disadvantages of Continuous-Fiber-Reinforced Thermoplastic Composites as Compared to Continuous-Fiber-Reinforced Thermoset Composites.

ADVANTAGES

1. Infinite storage capability (shelf life).
2. Variety of thermoplastic materials available to meet such specific requirements as abrasion resistance, UV resistance, good electrical properties, and weathering resistance.
3. Shorter "cure" or fabrication time.
4. Post-formable.
5. Ease of handling—no "tack" problems.
6. Ease of repair—welding, solvent bonding, adhesive bonding.
7. Reprocessing capability.
8. Lower cost.

DISADVANTAGES

1. Creep.
2. Limited temperature capability.
3. Limited processing technology available.
4. Limited chemical resistance of materials used to date.

the effect of humidity aging and boiling-water immersion, respectively, on the ILSS of various unidirectional graphite-fiber-reinforced thermoplastic and thermoset composites. Additional data on thermoplastic/graphite composites can be found in Chap. 28.

11.9. THERMOPLASTIC-COMPOSITE FABRICATION TECHNIQUES

11.9.1. Continuous-Fiber Composites

In general, there are four techniques available for the production of continuous-fiber-reinforced thermoplastic (CFRTP) "prepregs" for the fabrication of CFRTP composites: (1) solvent coating, (2) film lamination, (3) melt calendering, and (4) fiber weaving. To date, the solvent-coating method is by far the most widely used.

In the solvent-coating process, the thermoplastic resin is dissolved in a solvent(s), the solution is applied to the continuous fibers, and then the solvent is evaporated. Both unidirectional tapes and woven-fabric broadgoods are easily "prepregged" in this manner. The resultant "prepreg" materials are either crossplied (if unidirectional tape), depending on the desired mechanical properties, and/or layed up in sheets and consolidated under heat and pressure. The advantages of the solvent-coating prepregging technique are that the material can be processed similar to epoxy prepregs and that reasonable production rates are attainable. The disadvantages include the following: Polymer selection is limited to those polymers that are easily dissolved in room-temperature solvents; any residual solvent remaining in the prepreg material may produce voids in the finished laminated composite; a variable resin content; and the toxicity, flammability, and cost of the solvents.

Film lamination involves the lamination of thermoplastic films on continuous-fiber materials. The disadvantages associated with this process are the restriction of press platen size and the availability of specific thermoplastic film materials. Long cycle times (due to the heating and cooling of the platens) as well as the tendency of the fibers to misalign during the lamination process represent additional limitations. Advantages include a possible lower void content as well as a more consistent resin content.

The continuous calendering process developed by Union Carbide Corporation involves the use of a two- or three-roll calender. Melted thermoplastic material exits from a heated extruder die directly between the "nips" of the calender rolls. At the same time, continuous-fabric material is also fed between the nip rolls, thus allowing the fabric to be completely wet out by the resin. The formed composite is then fed through a series of cooling rolls and

Table 11.11. Typical Mechanical Properties of Continuous Graphite-Fiber-Reinforced Thermoplastic and Thermoset Composites

PROPERTY	SYMBOL	UNITS	T300–520g[a] GRAPHITE–EPOXY		T300–PR313[b] GRAPHITE–EPOXY RC[c] = 29–33% VC[d] = 1.7–2.4%	
			22° C (72° F)	277° C (350° F)	22° C (72° F)	177° C (350° F)
Unidirectional Laminate						
Longitudinal (0°) Properties						
Tensile strength	F_x^{tu}	Ksi (MPa)	218 (1502)	208 (1433)	197.7 (1362)	141.7 (976)
Tensile modulus of elasticity	E_x^t	Msi (GPa)	26.3 (181)	28.5 (196)	20.3 (140)	19.3 (133)
Compressive strength	F_x^{cu}	Ksi (MPa)	218 (1502)	206 (1419)	157.4 (1084)	148.1 (1020)
Compressive modulus of elasticity	E_x^c	Msi (GPa)	23.0 (158)	22.5 (155)	19.8 (136)	25.0 (172)
Flexural strength	F_x^{fu}	Ksi (MPa)	247 (1702)	196 (1350)	200.7 (1383)	96.4 (664)
Flexural modulus of elasticity	E_x^f	Msi (GPa)	—	—	17.77 (122)	16.29 (112)
Interlaminar shear strength (short beam)	F^{isu}	Ksi (MPa)	15.9 (110)	8.9 (61)	12.69 (87.4)	7.18 (49.5)
Transverse (90°) Properties						
Tensile strength	F_y^{tu}	Ksi (MPa)	3.85 (26.5)	2.89 (19.9)	4.9 (33.8)	3.7 (25.5)
Tensile modulus of elasticity	E_y^t	Msi (GPa)	1.50 (10.3)	1.78 (12.3)	1.3 (9.0)	1.05 (7.2)

[a]Source: Ref. 322, pp. 95–96.
[b]Source: Ref. 355.
[c]RC = resin content by weight.
[d]VC = void content by volume.

cut to desired lengths. The resin content is controlled by the distance between the nip rolls.

The continuous calendering process offers great potential and has several advantages:

1. Resin content easily controlled.
2. No possibility of voids.
3. Fast and continuous.
4. Economical and low cost.

Another important advantage is that the processing technology available for the fabrication of CFRTP composites is not limited to materials that are easily dissolvable (e.g., polysulfone); thus, such materials as nylon and polyphenylene sulfide may now be explored. A disadvantage of this process is that it is currently limited to woven materials.

A novel approach to the fabrication of thermoplastic prepregs and composites may involve the use of commercially available thermoplastic fibers. Thus, nylon, polyester, and polypropylene fibers can be woven-in during the weaving of graphite-fabric materials, heat and pressure being applied to consolidate them. The potential advantages to this process include the following: less energy required (due to the elimination of the heat necessary for prepregging), accurate resin control, improved chemical resistance, and ease of handling. A disadvantage is that it would be limited to woven materials.

The final net fabrication methods that use CFRTP prepreg materials are currently limited to either autoclave or laminating techniques. Other processing techniques, such a filament winding and pultrusion of CFRTP materials, are being investigated by both Army Materials and Mechanics Research Center (AMMRC) and Boeing-Seattle.

11.9.2. Discontinuous-Fiber Composites

The production and processing techniques for discontinuous (chopped) graphite-fiber-reinforced thermoplastic composites are the same as those for fiberglass-reinforced thermoplastic materials. Single- and twin-screw extruders are used in the production of initial feedstock material by combining ¼–½in. [(6.4–12.7) × 10^{-3} m] pellets.

The graphite-fiber thermoplastic feedstock is processed into its final net shape by conventional plastic processing techniques. Injection

Table 11.11. Continued.

T300-F178[b] GRAPHITE-POLYIMIDE RC = 35% VC = 0%		AS-4397[b] GRAPHITE-POLYIMIDE RC = 27.5-31% VC = 0%		AS-3004[b] GRAPHITE-POLYSULFONE RC = 33-34% VC = 0-1.9%	
22° C (72° F)	177° C (350° F)	22° C (72° F)	177° C (350° F)	22° C (72° F)	177° C (350° F)
156.7 (1080)	152.2 (1049)	203.3 (1401)	187.4 (1291)	187.9 (1295)	179.1 (1234)
20.25 (140)	19.56 (135)	18.3 (126)	18.9 (130)	16.3 (112)	17.5 (121)
180.0 (1240)	120.0 (827)	206.1 (1420)	164.6 (1134)	102.1 (703)	90.2 (621)
18.2 (125)	20.5 (141)	18.7 (129)	19.1 (132)	17.3 (119)	18.5 (127)
204.0 (1406)	179.1 (1234)	224.4 (1546)	178.8 (1232)	191.5 (1319)	135.2 (932)
16.89 (116)	18.43 (127)	18.4 (127)	17.3 (119)	17.8 (123)	20.0 (138)
14.8 (102)	10.2 (70.3)	13.62 (93.8)	9.79 (67.5)	11.6 (79.9)	8.4 (57.9)
2.82 (19.4)	2.97 (20.5)	5.37 (37)	3.81 (26.3)	5.02 (34.6)	5.39 (39.1)
1.50 (10.3)	1.15 (7.9)	1.39 (9.6)	1.09 (7.5)	1.15 (7.9)	1.07 (7.37)

molding followed by extrusion and compression molding is by far the most widely used technique. Table 11.13 lists typical reported properties of several discontinuous graphite-fiber-reinforced thermoplastic materials. (Figure 11.75 illustrates typical injection-molded graphite-fiber thermoplastic applications.)

Dibenedetto and Salee[358] found that the static tensile strengths and fatigue levels of discontinuous graphite fiber–nylon 66 composites could be characterized by a two- or three-parameter Weibull function. In addition, the fatigue-failure mechanism was influenced by the fabrication technique: Compression-molded composites exhibited a brittle failure, whereas injection-molded samples showed considerable ductile flow prior to ultimate failure.

11.9.3. Summary

Continuous and discontinuous graphite-fiber-reinforced composites are currently finding more and more applications in both the aerospace and commercial industries. With the introduction of high-performance thermoplastic resins that exhibit the advantages of thermoplastics (e.g., reusable scrap, complex shapes easily molded, and short cycle times), new applications are being developed—especially in replacing conventional epoxy resins. In addition, the wide selection of thermoplastic resins available allows the engineer to select a specific resin for a specific application.

11.10. FATIGUE BEHAVIOR OF GRAPHITE-FIBER COMPOSITES

One of the major advantages of graphite-fiber composites, second only to their high specific strength and stiffness (i.e., modulus of elasticity), is their outstanding fatigue resistance. Although the word "fatigue" had apparently not been coined at the time, engineers recognized the superiority of composite materials with respect to fatigue resistance as early as 1854. Speaking of the Wheeling, Virginia, "iron" suspension bridge that collapsed unexpectedly, writers at the time submitted that[359] "by frequent changes of pressure or strain in iron, a certain disturbance of its particles takes place, the metal deteriorates, and suddenly, when not expected, the very same strain or weight which it had sometimes supported—will break it to pieces." We now

Table 11.12. Typical Mechanical Properties of Continuous Graphite-Fiber-Reinforced Thermoplastic Composites.

PROPERTY	SYMBOL	UNITS	HMS–P1700[a] GRAPHITE-POLYSULFONE 22° C (72° F)	HMS–P1700[a] GRAPHITE-POLYSULFONE 163° C (325° F)	HMS–300P[a] GRAPHITE-POLYETHERSULFONE 22° C (72° F)	HMS–300P[a] GRAPHITE-POLYETHERSULFONE 163° C (325° F)
Unidirectional Laminate						
Longitudinal Properties (0°)						
Flexural strength	F_x^{FY}	ksi (GPa)	155–168 (1.07–1.16)	11–18 (0.076–0.124)	175–202 (1.21–1.39)	103–112 (0.710–0.772)
Flexural modulus of elasticity	F_x^{F}	Msi (GPa)	26–28 (179–193)	2.7–4.2 (18.6–28.9)	27.1–27.5 (187–189)	20.9–23.7 (144–163)
Interlaminar shear strength (short beam)	F^{isu}	ksi (MPa)	9.3–10.2 (64.1–70.3)	1.6–2.3 (11.0–15.8)	9.9–11.2 68.2–77.2)	6.2–6.7 (42.7–46.2)

[a]Source: Ref. 351, pp. 30–32.

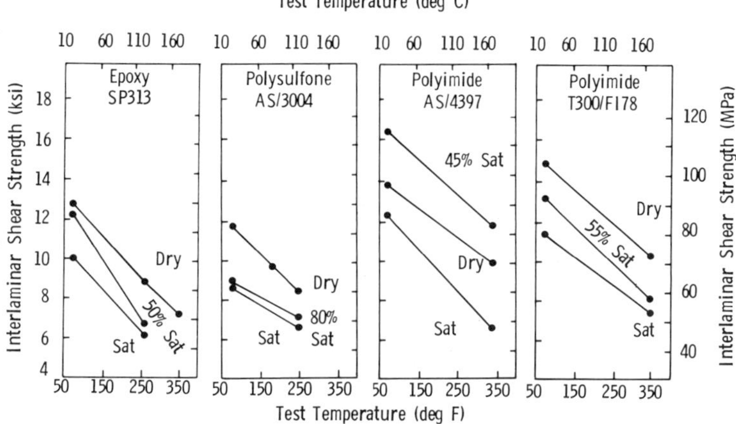

Figure 11.57. Comparative interlaminar shear strength retention of various continuous-fiber-reinforced thermoset and thermoplastic composites after humidity aging.[355]

call that disturbance of the particles fatigue. They further suggested that "it would seem more prudent to build—with stone piers and wooden superstructure." Wood, of course, is a natural composite with excellent properties, including good fatigue life.

The fatigue characteristics of unidirectional composite materials and metals are compared in Fig. 11.59[360] It is evident that in fatigue-critical structures, composites (especially graphite-fiber composites) offer significant advantages over the more common structural materials. It is for this reason that graphite composites are ideal for helicopters (which are actually flying fatigue machines) or such automotive applications as leaf springs.

Typical fatigue data for graphite composites are presented in Figs. 11.60–11.63.[360] It is emphasized that these data are typical and do not represent an insight into the mechanisms of fatigue or the origins of failure under fatigue loading. These subjects are beyond the scope of this chapter. However, several researchers have investigated the mechanisms of fatigue in graphite composites, and the interested reader is referred to the several publications in that area.[361–368]

The effect of fiber orientation on the fatigue behavior of graphite-epoxy composites is shown in Fig. 11.60[360] and 11.64.[355, 368, 369] It is evident that when the matrix dominates, as with the ±45 and 90° data, significantly lower

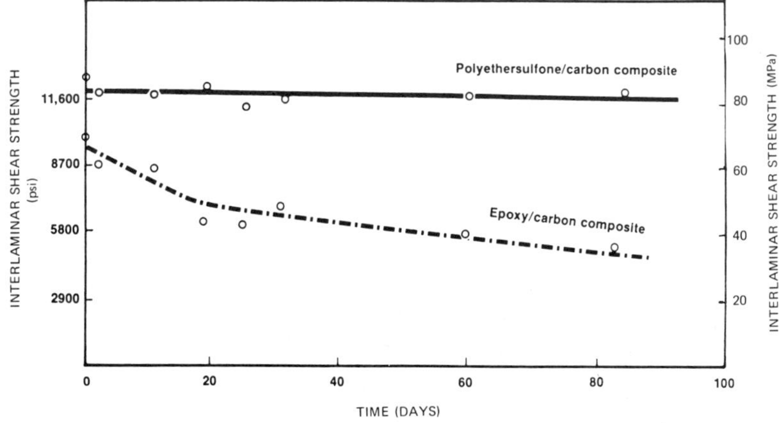

Figure 11.58. Variation of interlaminar shear strength of a thermoset and a thermoplastic composite as a function of immersion time in boiling water.[354]

Table 11.13. Typical Properties of Discontinuous Graphite-Fiber-Reinforced Thermoplastic Composites.

	NYLON 66 30% C (LNP)[a]	NYLON 66 30% C (U.C.)[b]	POLYSULFONE 30% C (LNP)[a]	POLYESTER 30% C (LNP)[a]	POLY-PHENYLENE SULFIDE 30% C (LNP)[a]	ETFE[c] 30% C (LNP)[a]	VF_2–TFE[d] 20% C (LNP)[a]	POLY-CARBONATE 30% C (U.C.)[b]	POLY-PROPYLENE 30% C (U.C.)[b]
Specific gravity	1.28	1.30	1.32	1.47	1.45	1.73	1.77	1.36	1.06
Water absorption (24 hr), %	0.5	—	0.15	0.04	0.04	0.015	0.03	—	—
Equilibrium, %[e]	2.4	—	0.38	0.23	0.10	0.24	—	—	—
Mold shrinkage, %[f]	1.5–2.5	3	2–3	1–2	1	1.5–2.5	2.5–3.5	2	5
Tensile strength, ksi (MPa)	35 (241)	15.8 (109)	19 (131)	20 (138)	27 (186)	15 (103)	12.3 (84.7)	11.5 (79.2)	5.4 (37. 2)
Tensile elongation, %	3–4	2.7	2–3	2–3	2–3	2–3	3–4	2.1	2.2
Flexural strength, ksi (MPa)	51 (351)	25.3 (174)	25.5 (176)	29 (200)	34 (234)	20 (138)	17.2 (119)	17.1 (118)	6.7 (46.2)
Flexural modulus of elasticity, Msi (GPa)	2.9 (20)	1.49 (10.3)	2.05 (14.1)	2.0 (13.8)	2.45 (16.9)	1.65 (11.4)	1.10 (7.58)	1.08 (7.44)	0.60 (4.1)
Shear strength, Ksi (MPa)	13 (89.6)	—	7 (48.2)	—	—	7 (48.2)	7.5 (51.7)	—	—
Izod impact strength[g] ft-lb/in (Joules/m)	1.5 (80.1)	1.6 (85.4)	1.1 (58.7)	1.2 (64.1)	1.1 (58.7)	4–5 (213–267)	2.6 (138.8)	3.0 (160.2)	0.7 (37.4)
Izod impact strength[h] ft-lb/in (Joules/m)	12.0 (641)	13.0 (694)	4–5 (213–267)	4–5 (213–267)	5–6 (267–320)	10.0 (534)	7–8 (374–427)	5.1 (272)	3.2 (171)
Thermal deflection temperature at 264 psi (1.82 MPa), °F (°C)	495 (257)	464 (240)	365 (185)	430 (221)	500 (260)	465 (241)	295 (146)	293 (145)	248 (120)
Coefficient of linear thermal expansion, in/in./°F $\times 10^{-5}$ (m/m/°C $\times 10^{-5}$)	1.05 (1.89)	1.5 (2.7)	0.7 (1.26)	0.5 (0.9)	0.6 (1.08)	0.8 (1.44)	3.0 (5.4)	1.6 (2.8)	1.6 (2.8)
Thermal conductivity, BTU-in/hr ft^2 °F (W/m °C)	7.0 (12.1)	—	5.5 (9.5)	6.5 (11.2)	5.2 (9.0)	5.6 (9.7)	3.0 (5.7)	—	—
Surface resistivity, ohms/sq	3–5	—	1–3	2–4	1–3	3–5	3–5	—	—

[a]LNP = Liquid Nitrogen Processing Corporation; source: Ref. 352.
[b]U.C. = Union Carbide Corporation; source: Technical Information Bulletin No. 465–236, Union Carbide Corporation.
[c]ETFE = ethylene tetrafluoroethylene.
[d]VF = vinylidene fluoride.
[e]equilibrium after continuous immersion.
[f]⅛-inch section.
[g]¼-inch notched.
[h]¼-inch unnotched.

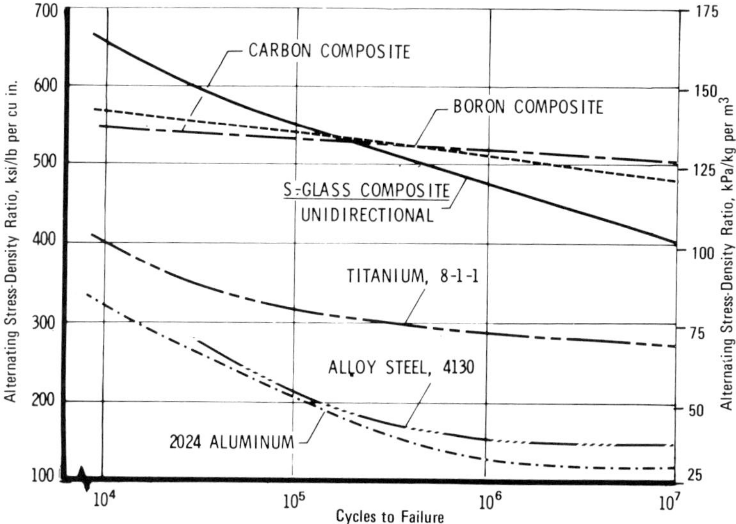

Figure 11.59. Typical tension–tension fatigue data (flat coupons, rotor-blade spar materials, zero stress ratio).[360]

fatigue lives are obtained. However, the fatigue values for unidirectional composites are significantly greater than those for aluminum and steel (see Fig. 11.64); moreover, even quasi-isotropic laminates exhibit values (at 10^7 cycles) two to four times those of steel and aluminum.

The effect of temperature on the fatigue behavior of a graphite–epoxy composite is shown in Fig. 11.61.[360] As temperature increases, the curves shift to lower stress values. Similar behavior is obtained on other matrices: polysulfone (see Fig. 11.62[360]) and polyimide (see Fig. 11.63[360]); in the latter case, however, the matrix is apparently exhibiting increased ductility at the elevated temperature.

The ratio of the maximum load a structure can withstand after cycling to the ultimate tensile strength is of interest to the designer of graphite-composite structures that will be utilized in a fatigue environment. Table 11.14

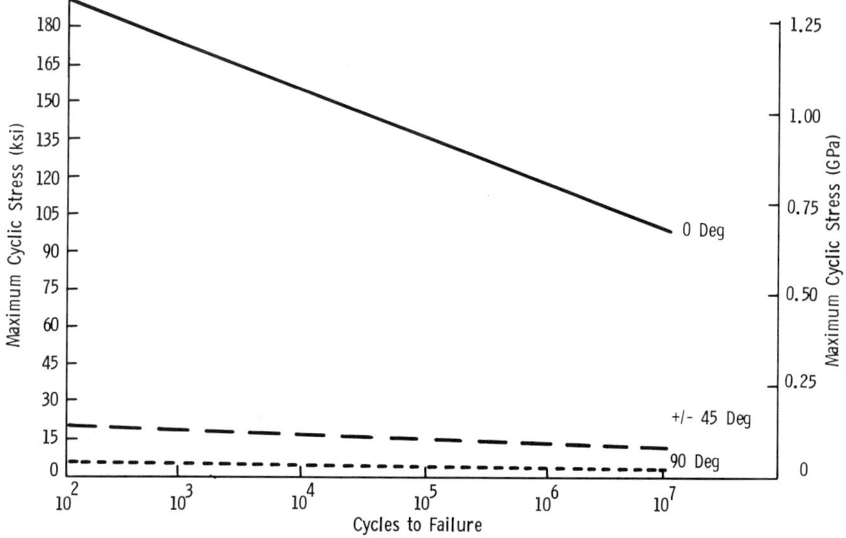

Figure 11.60. Effect of fiber orientation on room-temperature fatigue properties of an SP313 graphite–epoxy composite (fiber content = 61.4 vol. %; thickness: 0°—6 plies, 90°—15 plies, ±45°—8 plies; $R = 0.1$, 30 Hz).[360]

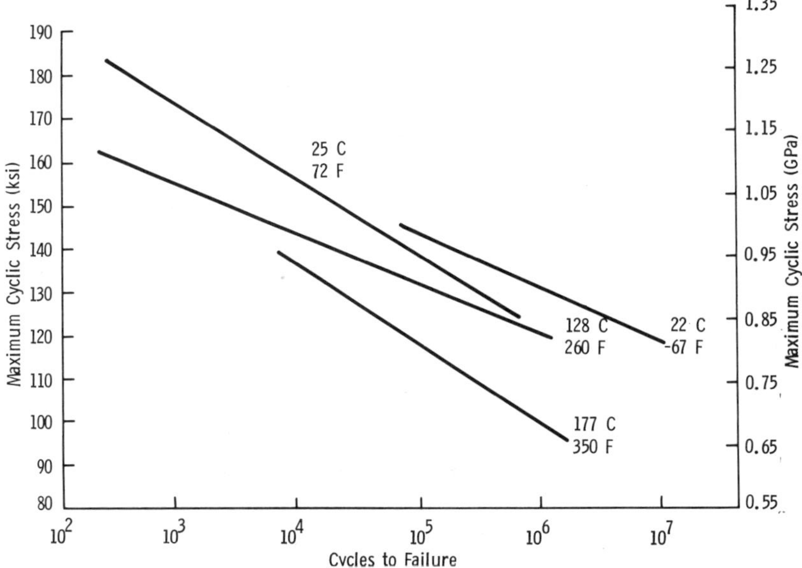

Figure 11.61. Effect of temperature on fatigue of an SP313 graphite–epoxy composite (Fiber content = 61.4 vol. %; thickness: 0°—6 plies, R = 0.1, 30 Hz).[360]

shows this ratio for a number of materials under a variety of conditions. Except for the graphite–polyimide value, all of the graphite-composite materials in tensile fatigue have ratios of 0.53–0.58 irrespective of temperature, configuration, or matrix. If tension-compression measurements are made, however, substantially lower values can be obtained, depending on the load applied and test fixtures used[368] (see Table 11.14). In designing for compression-critical structures, this lower fatigue life must be considered, the graphite perhaps being hybridized with boron (see Sec. 11.12).

11.11. CREEP BEHAVIOR OF GRAPHITE-FIBER COMPOSITES

Graphite-composite systems show outstanding resistance to long-term creep. Chiao et al.[370] have shown that after 10^5 hr, the 10% failure probability of graphite–epoxy strands

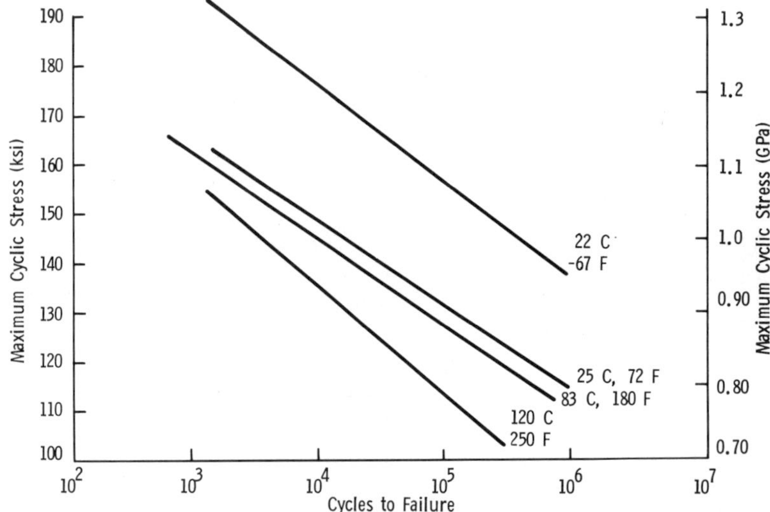

Figure 11.62. Effect of temperature on fatigue of an AS-3004 graphite–polysulfone composite (0° Fiber orientation, thickness = 6 plies, fiber content = 57.2 vol. %, R = 0.1, 30 Hz).[360]

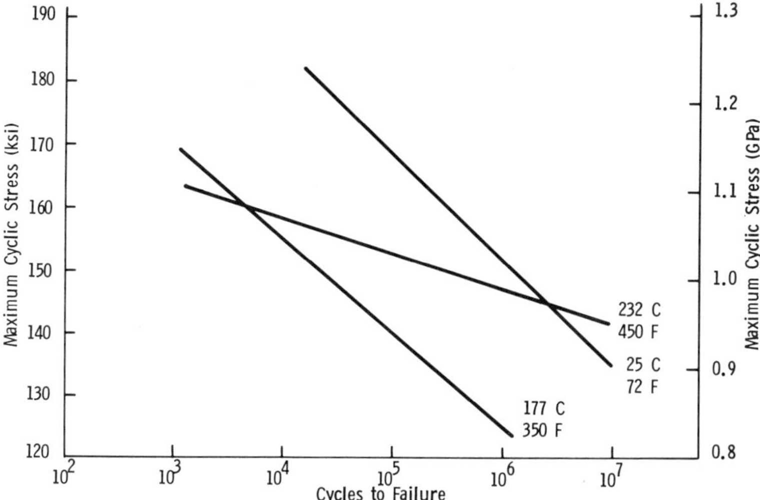

Figure 11.63. Effect of temperature on fatigue of an AS–4397 graphite–polyimide composite (0° fiber orientation, fiber content = 63.6 vol. %, $R = 0.1$, 30 Hz).[360]

is still about 80% of the ultimate tensile strength. Hofer and Bennett[371] have investigated the effects of temperature, humidity, and applied load on the creep behavior of graphite and graphite–glass hybrids (typical data are shown in Fig. 11.65[360]). They found that creep strains are quite low even when substantial glass content is present. Thus, the resistance of graphite composites to long-term creep is one of their major advantages[369] —at least under tensile loads; however, as in the case of fatigue behavior, caution should probably be exercised when compressive loads are applied.

11.12. GRAPHITE-HYBRID COMPOSITES

Hybrid composites are materials that combine two or more fibers in the same laminate,

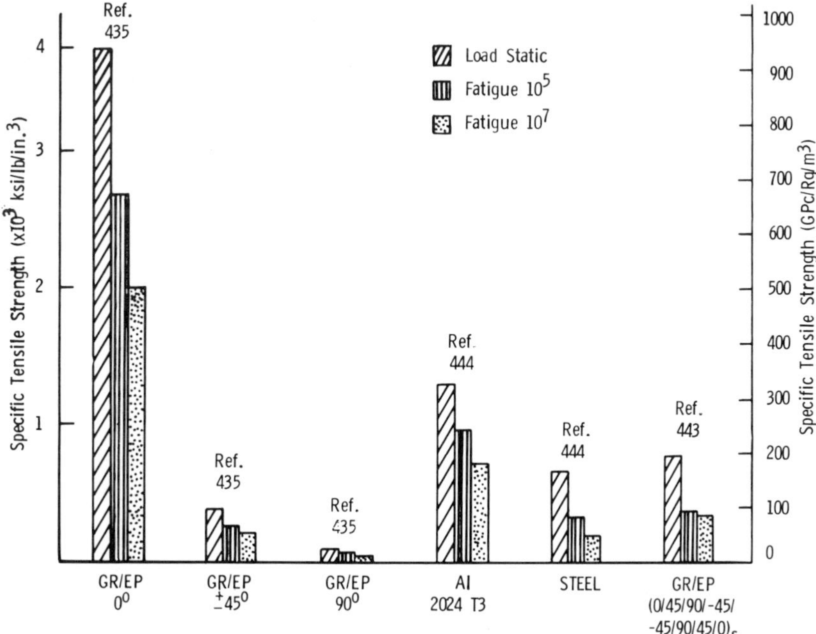

Figure 11.64. Specific tensile strengths of static-and fatigue-loaded graphite–epoxy composite (GR/EP), aluminum (Al 2024 T3), and steel.[355, 368, 369]

Table 11.14. Fatigue of Graphite Composites: Ratio of Failure Stress at 10^7 Cycles to the Ultimate Tensile Strength.[a]

MATERIAL	CONFIGURATION	REFERENCE	FREQUENCY (Hz)	RATIO
Graphite-epoxy[b]	0°	355	30	0.53
Graphite-epoxy[b]	±45°	355	30	0.56
Graphite-epoxy[b]	90°	355	30	0.54
2024 T3 aluminum	—	369	—	0.28
4130 steel	—	369	—	0.44
Graphite-epoxy[c]	0°	322	30	0.56
Graphite-epoxy[d,e]	Q.I.[j]	368	10	0.55
Graphite-epoxy[d,f]	Q.I.[j]	368	10	0.38
Graphite-epoxy[d,g]	Q.I.[j]	368	10	0.26
Graphite-epoxy[b] (127°C = 260°F)	0°	355	30	0.58
Graphite-epoxy[b] (177°C = 350°F)	0°	355	30	0.54
Graphite-polyimide[h]	0°	355	30	0.67
Graphite-polysulfone[i]	0°	355	30	0.56

[a] Values are for tension fatigue ($R = 0.1$) at room temperature (22°C = 72°F) unless otherwise stated.
[b] T300-PR313 epoxy (fiber content = 61 vol. %).
[c] T300-5208 epoxy (fiber content = 60-70 vol. %).
[d] T300-934 epoxy (fiber content = 60-66 vol. %).
[e] $R = 0$.
[f] Tension-compression fatigue (applied compressive load = 0.15 UTS).
[g] Tension-compression fatigue (applied compressive load = 0.23 UTS).
[h] AS-4397 polyimide (fiber content = 64 vol. %).
[i] AS-P1700 polysulfone (fiber content = 57 vol. %).
[j] Q.I. = quasi-isotropic: $(0/45/90/\text{-}45/\text{-}45/90/45/0)_s$.

generally either to enhance a property considered deficient in one fiber composite alone (e.g., the impact strength of graphite composites or the stiffness of glass composites) or to lower the cost (e.g., by substituting glass fibers for graphite fibers in non-stiffness-critical directions or portions of the structure). In particular, graphite fibers will probably be used selectively in automotive applications, that is, only in those areas where their stiffness is required, in order to keep the overall material cost down.

The number of possible material combinations and the diversity of properties obtained from hybrid composites are far too numerous to mention here. Instead, the discussion will be limited to hybrid types and their typical properties and rationale for use. For more

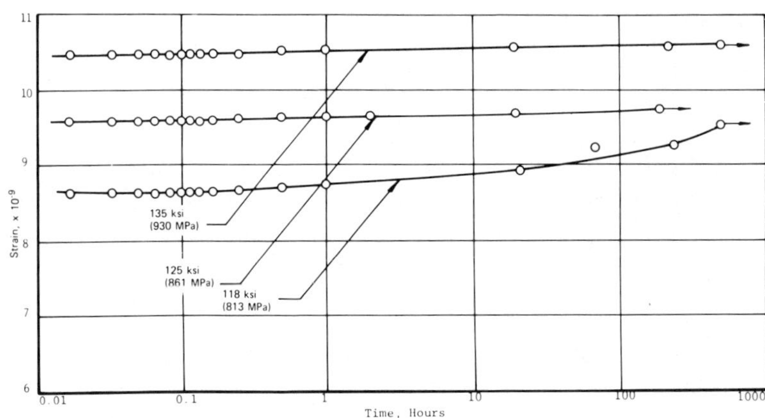

Figure 11.65. Creep strain versus time curves for +OL/OL/OR/OR/OL/OL− T300 graphite-1014-S glass-narmco 5208 hybrid composites tested dry at t = 21°C (70°F): L = Glass, R = Graphite.[360]

detailed information, the reader can refer to several recent reviews of hybrid composites.[372–376]

There are five major types of hybrid composite:

1. "Random"—the fibers are randomly mixed throughout the resin and composite with no preferential concentration of either fiber.
2. "Intraply"—the fibers are combined in a regular fashion in each ply of the composite, either through woven cloth or hybrid tapes; however, each ply can be different.
3. "Interply"—the composite consists of discrete layers of one fiber only (e.g., a graphite tube with outer layers of glass–epoxy).
4. Selective reinforcement (e.g., bonded-on stiffeners).
5. The so-called "super hybrids" that consist of resin-composite plies, metal-composite plies, and metal foils stacked in a specified sequence.[372]

The most common hybrids employed—and the ones for which most data exist—are the intraply and interply types.

Graphite-hybrid composites can be fabricated using conventional techniques and combined with boron, glass, or aramid fibers as well as with metals in a laminated structure. The intraply and interply hybrids usually have the same matrix and are fabricated by the cocuring process. Chamis and Lark[372] have given a state-of-the-art review, including 53 references, on the analysis, design, application, and fabrication of hybrids. They concluded that considerable data have been generated on the tensile and impact properties of hybrids, but that only limited data exist on the thermal properties, moisture effects, and residual-strain effects. In particular, to minimize thermal effects, the hybrid composite should be fabricated with the fiber sequences symmetrical with respect to the neutral axis[373] in order to prevent undesirable warping. The stress and structural analysis methods, design procedures, fabrication methods, and quality-assurance techniques that are used for conventional composites are also suitable for hybrids.[372]

There appear to be three underlying philosophies or rationales used in the hybridization of graphite composites: 1) the addition of another fiber to a predominantly graphite composite in order to overcome the disadvantages of graphite; 2) the addition of graphite fibers to a predominantly nongraphite composite or structure in order to take advantage of the benefits of graphite; and 3) to produce a lower cost structure.

The first category is primarily based on impact strength. It is generally agreed that the impact resistance of graphite-fiber composites can be improved by the addition of high-strength fibers with a greater strain-to-failure than graphite, and several energy-absorbing mechanisms have been proposed.[373] Typical data[377] are shown in Table 11.15, where greater-than-rule-of-mixture results are obtained. Other data on interply Kevlar–graphite hybrids are given by Helfinstine,[378] who showed that the stacking sequence affects the behavior more significantly than does the hybrid fiber ratios; Summerscales and Short[373] give 14 other references on the impact resistance of graphite-hybrid composites.

The compressive strength of Kevlar composites—their primary disadvantage—can be markedly improved by hybridization with graphite.[376,379] As shown in Table 11.16,[379] it is more than doubled when a 50:50 ratio of graphite to Kevlar is used. The effect of hybridization on the compressive modulus of

Table 11.15. Impact Resistance of Hybrid Composites.[377]

COMPOSITE (EPOXY MATRIX) WT %	UNNOTCHED IZOD IMPACT STRENGTH	
	ft-lb/in.	J/m
100% graphite	28	1495
75% graphite–25% Kevlar	34	1815
50% graphite–50% Kevlar	44	2349
100% Kevlar	48	2562
100% graphite	28	1495
75% graphite–25% glass	44	2349
50% graphite–50% glass	56	2989
100% glass	72	3843

Table 11.16. Properties of Unidirectional Thornel 300, Kevlar 49, and Hybrid Composites (Nominal Fiber Content = 60 vol. %).[379]

FIBER CONTENT, WT %			TENSION				COMPRESSION				FLEXURE							
			MODULUS OF ELASTICITY		ULTIMATE STRESS		STRESS AT 0.02% OFFSET		ULTIMATE STRESS		STRESS AT 0.02% OFFSET		ULTIMATE STRESS		SHORT-BEAM SHEAR STRESS		PREPREG COST	
THORNEL 300	KEVLAR 49	SPECIFIC GRAVITY	Msi	GPa	ksi	MPa	ksi	MPa	ksi	MPa	ksi	MPa	ksi	MPa	ksi	MPa	$/lb	$/kg
100	0	1.60	21.1	145	227	1564	98.4	678	146	1006	233	1605	233	1605	13.2	91	60	132
75	25	1.56	17.4	120	186	1282	68.8	474	136	937	181	1247	197	1357	11.0	76	48	106
50	50	1.51	15.7	108	176	1213	59.9	413	99.8	688	120	827	160	1102	8.1	56	35	77
0	100	1.35	11.2	77	183	1261	26.4	182	41.5	286	49.2	339	91.9	633	7.1	49	10	22

[a]Data supplied by Fiberite Corporation.

Table 11.17. The 0° Properties of 0°, ±45°, 90° 11-Ply Laminates (Room-Temperature Design Allowables—"B" Basis).*

MATERIAL	CONFIGURATION	TENSILE STRENGTH ksi	TENSILE STRENGTH MPa	COMPRESSIVE STRENGTH ksi	COMPRESSIVE STRENGTH MPa	COMPRESSIVE MODULUS OF ELASTICITY Msi	COMPRESSIVE MODULUS OF ELASTICITY GPa
Boron–epoxy[a]	$(0°_3, \pm45, 90)_s$	120	827	220	1516	19	131
High-strength graphite–epoxy[a]	$(0°_3, \pm45, 90)_s$	112	772	110	758	13	90
Hybrid: Boron–high-strength graphite–epoxy[b]	$(0°_{3b}, \pm \text{Gr}, \overline{90}\ \text{Gr})_s$	100	689	214	1474	18	124

**Advanced Composites Design Guide.* 3rd Ed. (2nd rev.) Structures Division, Air Force Flight Dynamics Laboratory, Wright-Patterson Air Force Base, Dayton, Ohio, September 1976.

[a] Section 1.2.2.

[b] Section 1.2.8.

elasticity can be even more impressive: It has been found[380] to double upon addition of approximately 5 wt % graphite fibers to a 50 wt % glass composite. Similarly, if graphite–epoxy is to be used in a compression-critical structure (e.g., aircraft horizontal stabilizers, wing slats, and fairings), then hybridization with boron can provide significant improvements in compressive properties. As shown in Table 11.17, a 50:50 mixture of boron and graphite can provide composites with 90% of the compressive strength and modulus of elasticity of an all-boron structure.

The tensile properties of hybrids have been reported in the recent literature.[372–376, 379, 383] Although rule-of-mixture values are sometimes obtained (see tensile modulus of elasticity values in Table 11.16 and Fig. 11.66[375]), the tensile modulus of elasticity of some hybrids has been found to exceed these values (see Fig. 11.67[383]). The data in Fig. 11.66 were obtained on woven-cloth intraply hybrids, whereas those in Fig. 11.67 were obtained on interply hybrids. Chamis and Lark[372] attribute the apparent "synergistic effect" shown in Fig. 11.67 to the stacking of the stiffer graphite plies further away from the neutral plane. This is certainly true for the flexural properties of hybrids. Kalnin[383] has shown that the flexural modulus of elasticity obeys the rule of mixtures if the graphite and glass plies are laid up alternately, whereas changing the stacking sequence of a 60 wt % graphite–40% glass hybrid causes the modulus to vary between 12 and 40 Msi (85 and 275 GPa) when the graphite layers are innermost and outermost, respectively (see Fig. 11.68[383]). Hence, it is quite apparent, that the properties of hybrids are dependent on the type of hybrid and the stacking sequence.

The fatigue behavior of fiberglass compos-

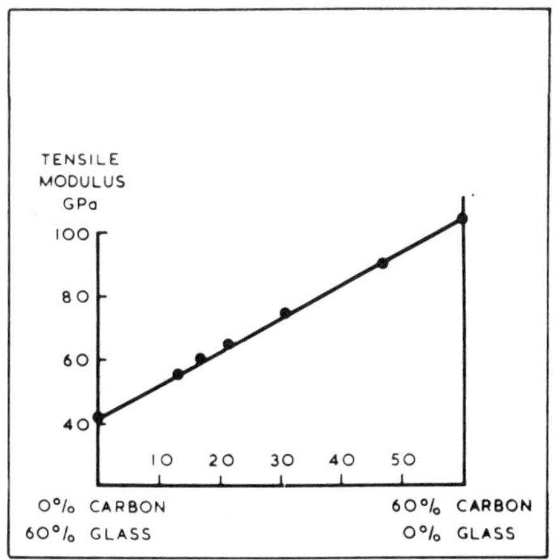

Figure 11.66. Tensile modulus of elasticity of a carbon-glass hybrid as a function of composition.[375]

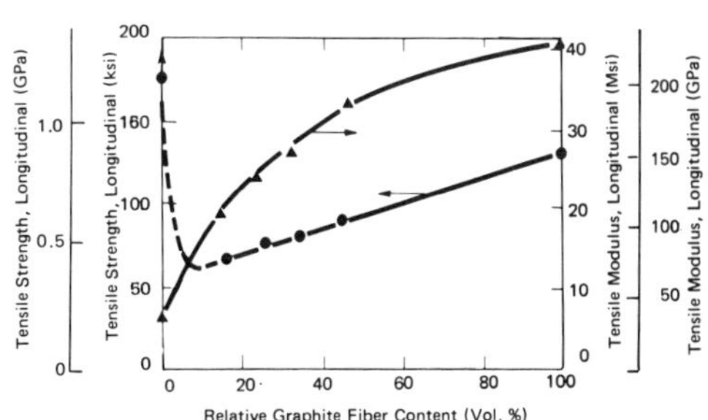

Figure 11.67. Longitudinal tensile strength and modulus of elasticity of hybrids.[383]

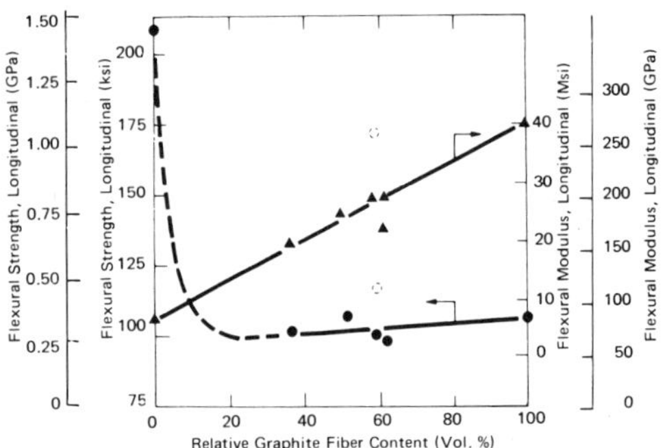

Figure 11.68. Longitudinal flexural strength and modulus of elasticity of fiberglass–graphite hybrids.[383]

ites can be markedly enhanced through hybridization with graphite.[384] Figure 11.69[384] shows the fatigue behavior of quasi-isotropic composites (balanced laminates of 0°, ±45°, 90° oriented plies) of glass and graphite. It is evident that the fatigue life of the hybrids is equivalent to that of the all-graphite composite and substantially greater than that of the all-glass composite. Although the quantitative reasons for this phenomenon have not yet been determined, it is qualitatively consistent with the fact that the graphite fibers prevent the glass fibers in the hybrid from being strained to the same extent as in an all-glass composite under identical applied loads. Consequently, they are not as greatly fatigued. As pointed out by Hofer et al.,[385] however, this effect depends on the type of hybrid and lay up. Thus, intraply hybrids gave better mechanical performance than interply hybrids.

Other mechanical properties—most notably torsional shear (see Fig. 11.70[383] and Poisson's ratio—have also been reviewed.[373] Further data can be found in the *Advanced Composites Design Guide.*[381]

In addition to forming hybrid composites for the purpose of mechanical-property enhancement, graphite fibers can be added to enhance the thermal, electrical, and abrasion characteristics of composites. Thus, several percent of carbon fiber added to asbestos-phenolic brake linings can increase their wear life 25–60% and decrease rotor wear.[369] A thin veil mat of carbon fibers on the surface of molded sheet molding compound automotive parts can make the composites electrically conductive, thereby suppressing radio-frequency interference generated by the engine. In addition, it is easier to electrostatically spray paint the conductive composites.[369]

11.13. APPLICATIONS OF GRAPHITE COMPOSITES

The major markets for advanced graphite-fiber composites, at least in the near future, are the aircraft and aerospace industries, marine products, and wheeled vehicles, all of which are discussed in other chapters. However, there are other areas—principally the recreation and industrail equipment markets—where graphite-fiber composites can be used to advantage.

Golf club shafts represented the first high-volume commercial application of graphite-fiber composites when they appeared in the marketplace during 1973. The higher specific

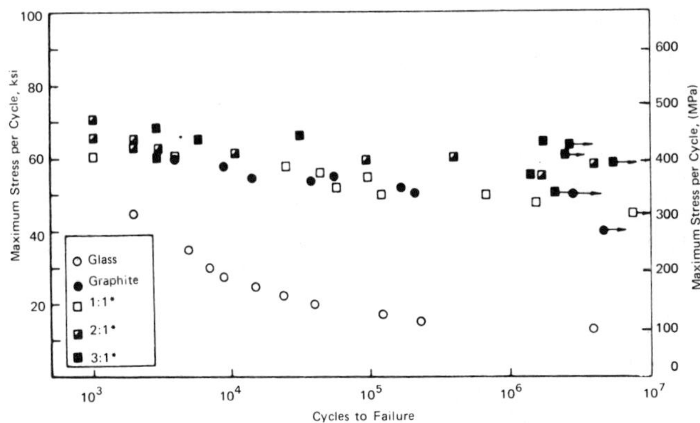

Figure 11.69. Comparison of the fatigue stress–strain behavior of quasi-isotropic T300 graphite–s-glass–NARMCO 5208 hybride composites: ○)glass; ●)graphite; graphite-to-glass ratios—1:1 (□), 2:1 (◪), and 3:1 (■).[384]

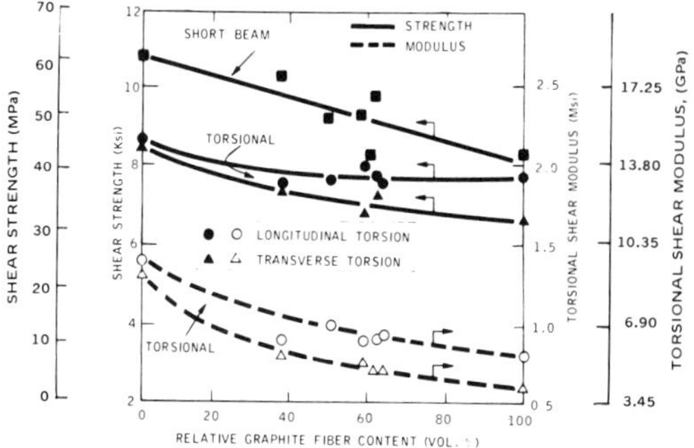

Figure 11.70. Shear strength and modulus of elasticity of hybrids.[383]

stiffness of the graphite enabled a lighter shaft to be built, thus allowing more weight in the head of the club. Initially capturing about 3% of the market, graphite golf club shafts has leveled off to ~1% of the market—~200,000 shafts/year (1978).[386] Since then, the use of graphite-fiber composites has continued to expand in the leisure market to include tennis rackets, fishing rods (see Fig. 11.71), skis, ski poles, bicycle frames (see Fig. 11.72), archery equipment, and other products.

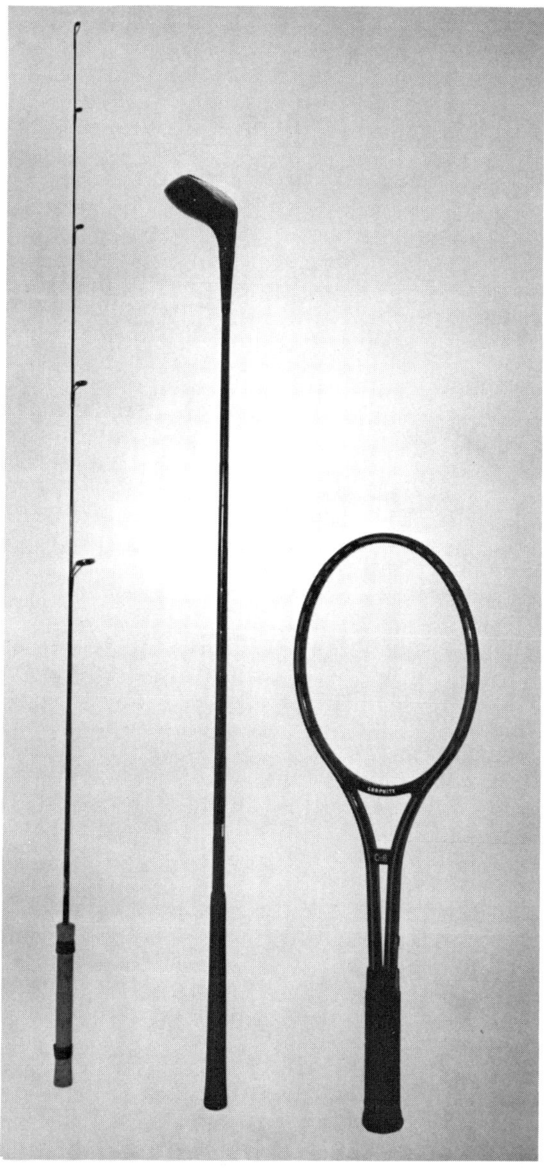

Figure 11.71. Example of athletic equipment that use graphite-fiber reinforcement.

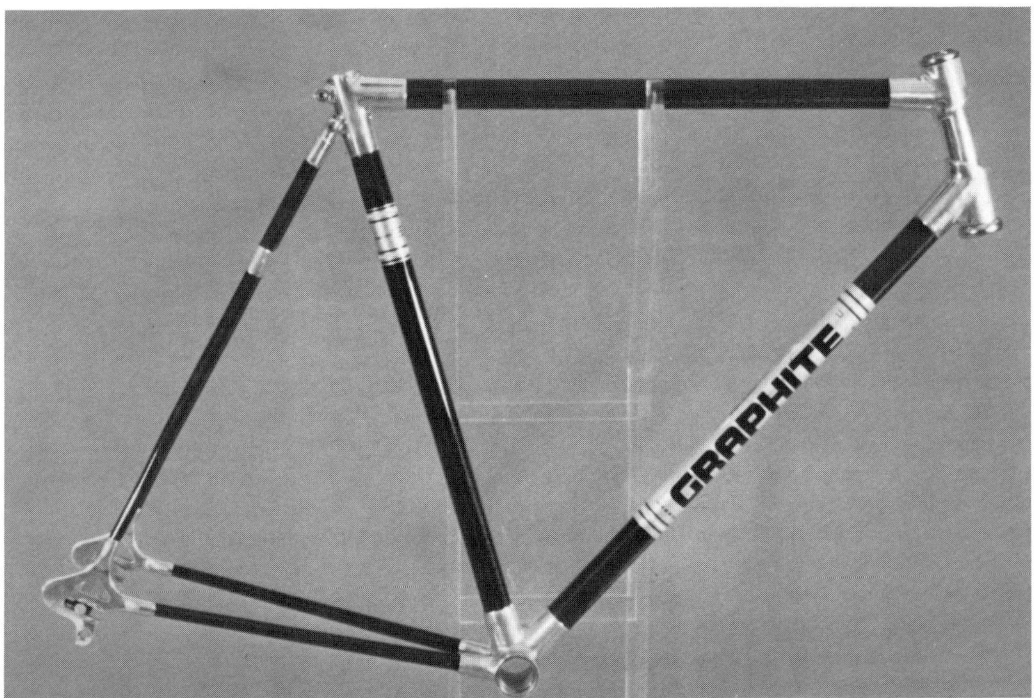

Figure 11.72. Graphite-fiber-reinforced bicycle frame (*Courtesy of* Union Carbide Corporation.)

Figure 11.73. Graphite-composite picking stick on weaving loom.[387]

Figure 11.74. Distributor tube molded from graphite–nylon 66 composite.[387]

The industrial market is beginning to become significant[386, 387] with applications occurring in agricultural machinery, food-processing equipment, material-handling equipment, industrial machinery, medical and radiological applications, musical instruments, and the like. Indeed, many manufacturers feel[387] that ultimately the bulk of the market will go to industrial applications. Any application that is inertial-mass limited would be ideal for graphite composites. Thus, picking sticks, which activate the reciprocating motion of a

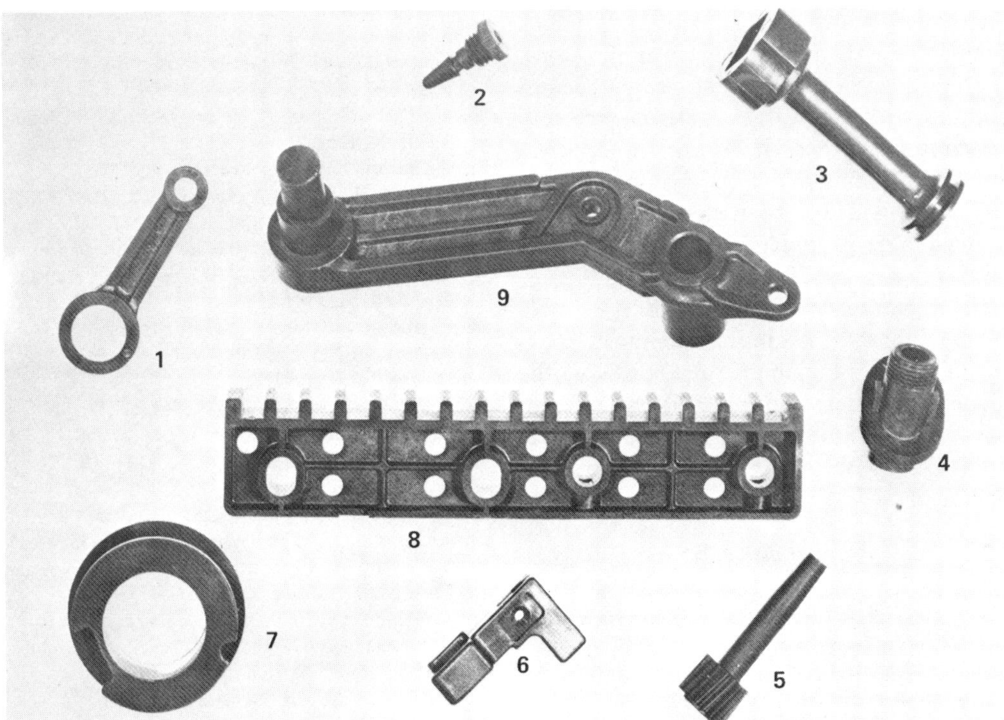

Figure 11.75. Typical small injection-molded parts incorporating Thornel carbon fiber in the resin system: 1) connecting rod, power tool; 2) electrode, vending machine; 3)piston, electric power tool; 4)wiper pivot, automobile; 5) gear, automobile speedometer; 6) middle hook, heddle frame; 7) governor, gasoline engine; 8) guide rack, heddle frame; 9) chain-tensioning device. (*Courtesy of* Union Carbide Corporation.)

shuttle in a weaving loom, are now being made from graphite pultrusions (see Fig. 11.73),[387] despite a cost that is 20–30 times that of the wooden stick they replace. The advantages are a 40% reduction in inertia, more efficient picking action, 12–15-fold increase in service life, two-thirds reduction in machine downtime, and significant noise reduction.[382] Other textile machinery applications include[386] heddle frames, lay bars, flyer arms, needles, sinkers, guide bars, and faller bars.

Chopped graphite-fiber-reinforced thermoplastics are also being used in industrial applications. A distributor tube for a cigarette manufacturing machine (see Fig. 11.74[387]) is being molded from graphite fiber reinforced Nylon 6,6. The part is in contact with hardened steel surfaces in the rotor module, and the excellent wear properties of the composite against steel extends the component lifetime.[387] Other advantages are conduction away of static electricity and cost (~1/3 the cost of an aluminum counterpart). Other components made from injection-molded, chopped-graphite thermoplastics are shown in Fig. 11.75, which demonstrates their wide range of applications.

The many advantages of graphite composites—strength, stiffness, fatigue and creep resistance, low thermal expansion, electrical and thermal conductivity, abrasion and wear resistance, and low coefficient of friction—portend a large potential for them in the military, aerospace, industrial, and commercial markets. Coupled with the lower energy utilization of graphite composites, occurring both in manufacture and use, the ultimate potential of graphite fibers and composites will be limited only by its cost-effectiveness. Otherwise, the potential is virtually unlimited.

REFERENCES

1. C. L. Mantell (editor), *Carbon and Graphite Handbook*, Wiley-Interscience, New York, 1968.
2. O. L. Blakslee et al., "Elastic Constants of Compression-Annealed Pyrolytic Graphite," *J. Appl. Phys.* **41**, 3373 (1970).
3. W. S. Williams et al., "Bending Behavior and Tensile Strength of Carbon Fibers," *J. Appl. Phys.* **41**, 4893 (1970).
4. G. B. Spence, in: *Proceedings of the Fifth Conference on Carbon*, Vol. 11, Pergamon Press, New York, 1962, p. 53.
5. Courtaulds Limited, U.S. Patent 3,533,741, 1970.
6. H. M. Ezekiel, "Graphite Fibers from an Aromatic Polyamide Yarn," *Appl. Polym. Symp.* No. 9, 315 (1969).
7. A. Shindo et al. "Highly Crystallite-Oriented Carbon Fibers from Polymeric Fibers," Appl. Polym. Symp. No. 9, *Appl. Polym. Symp.* No. 9, 305 (1969).
8. E. A. Boucher et al., "Preparation and Structure of Saran-Carbon Fibers," *Carbon* **8**, 597 (1970).
9. North American Aviation Inc., French Patent 1,535,800, 1968.
10. J. Economy, and R. Y. Lin, "Carbonization and Hot Stretching of a Phenolic Fibre," *J. Mater. Sci.* **6**, 1151 (1971).
11. K. Kawamura, and G. M. Jenkins, "A New Glassy Carbon Fibre," *J. Mater. Sci.* **5**, 262 (1970).
12. T. Edison, U.S. Patent 223,898, 1880.
13. H. Maxim, U.S. Patent 230,309, 1880.
14. T. Edison, U.S. Patent 248,416, 1881.
15. W. Whitney, U.S. Patent 916,905, 1909.
16. W. Soltes, U.S. Patent 3,011,981, 1961.
17. W. Abbott, U.S. Patent 3,053,775, 1962.
18. R. Bacon, "Carbon Fibers from Rayon Precursors," in: *Chemistry and Physics of Carbon*, (edited by P. L. Walder, and P. A. Thrower), volume 9, Marcel-Dekker, Inc., New York, 1973, p. 2.
19. G. E. Cranch, "Unique Properties of Flexible Carbon Fibers," in: *Proceedings of the Fifth Conference on Carbon*, Vol. 11, Pergamon Press, New York, 1962, p. 589.
20. C. E. Ford and C. V. Mitchell, U.S. Patent 3,107,152, 1963.
21. R. Bacon, "Growth, Structure, and Properties of Graphite Whiskers," *J. Appl. Phys.*, **31**, 283 (1960).
22. R. Bacon et al., U. S. Patent 3,305,315, 1967.
23. M. M. Tang and R. Bacon, "Carbonization of Cellulose Fibers. I. Low Temperature Pyrolysis," *Carbon* **2**, 211 (1964).
24. R. Bacon and M. M. Tang, "Carbonization of Cellulose Fibers. II. Physical Property Study," *Carbon* **2**, 221 (1964).
25. R. Bacon and M. M. Tang, unpublished data
26. R. Bacon and W. H. Smith, *Soc. Chem. Ind. London Chem. Eng. Group*, 203–213 (1966).
27. W. J. Spry, British Patent 1,093,084, 1967.
28. Y. Tsunoda, U.S. Patent 3,285,969, 1966.
29. A. Shindo, *J. Ceram. Assoc. Japan* **69**, C195 (1961).
30. W. Watt and W. Johnson, "New Materials Make Their Mark," *Nature* **220**, 835 (1968).
31. A. Shindo et al., Japanese Patent 4405, 1962.
32. R. Prescott and A. Standege, "High Elastic Modulus Carbon Fibre," *Nature* **211**, 169 (1966).
33. W. Watt et al., *The Engineer* **221**, (1966).
34. W. Watt et al., U.S. Patent 3,412,062, 1968.
35. W. Johnston et al., British Patent 1,148,874, 1969.
36. A. Standege and R. Prescott, Belgian Patent 690,072, 1966.
37. J. W. Johnston, French Patent 1,551,282, 1968.

38. D. M. Riggs, Masters' Thesis, Rensselaer Polytechnic Institute, Troy, New York, 1975.
39. A. J. Clark and J. E. Bailey, *Carbon '72 Preprints*, 296, 1972.
40. E. Fitzer and A. K. Fiedler, *Carbon '72 Preprints*, 299, 1972.
41. C. W. LeMaistre, Doctoral Thesis, Rensselaer Polytechnic Institute, Troy, New York, 1971.
42. E. W. Tokarsky, Doctoral Thesis, Rensselaer Polytechnic Institute, Troy, New York, 1973.
43. R. J. Diefendorf and E. W. Tokarsky, "The Relationships of Structure to Properties in Graphite Fibers, Part I," AFML-TR-72-133, 1971.
44. R. J. Diefendorf and E. W. Tokarsky, "The Relationships of Structure to Properties in Graphite Fibers, Part II," AFML-TR-72-133, 1973.
45. R. J. Diefendorf and E. W. Tokarsky, "The Relationships of Structure to Properties in Graphite Fibers, Part III," AFML-TR-72-133,. 1975.
46. R. J. Diefendorf, D. M. Riggs, and I. W. Sorensen, "The Relationships of Structure to Properties in Graphite Fibers, Part IV," AFML-TR-72-133, 1975.
47. S. Otani et al., "On the Raw Materials of MP Carbon Fiber," *Carbon* **4**, 425 (1966).
48. M. Morishita et al., Japanese Patent 6,902,510, 1969.
49. A. Tadashi and G. Shimpei, "Production of Molten Pitch Carbon Fiber," *Appl. Polym. Symp.* No. 9, 331 (1969).
50. S. Otani, "The Fundamental Structure of MP Carbon Fiber," *Carbon* **3**, 213 (1965).
51. S. Otani et al., "Effects of Heat Treatment under Stress on MP Carbon Fiber," *Appl. Polym. Symp.* No. 9, 325 (1969).
52. H. M. Hawthorne et al., "High Strength, High Modulus Graphite Fibres from Pitch," *Nature* **227**, 946 (1970).
53. H. M. Hawthorne, "Carbon Fibers: Their Composites and Applications," in: *Proceedings of the First International Conference on Carbon Fibres, London, February 1971*, Plastics Institute, London 1971, p. 81.
54. R. Didchenko, "Graphite Fibers from Pitch, Part I," AFML-TR-73-147, 1973.
55. L. S. Singer, Belgian Patent 797,543, 1973.
56. J. D. Brooks and G. H. Taylor, "The Formation of Graphitizing Carbons from the Liquid Phase," *Carbon* **3**, 185 (1965).
57. J. B. Barr et al., "High Modulus Carbon Fibers from Pitch Precursor," *Appl. Polym. Symp.* No. 29, 161 (1976).
58. H. Fugimaki et al., *Tanso*, No. 80, 3 (1975).
59. D. M. Riggs, Doctoral Thesis, Rensselaer Polytechnic Institute, Troy, New York, 1976.
60. K. Kawamura and G. M. Jenkins, "Mechanical Properties of Glassy Carbon Fibers Derived from Phenolic Resin," *J. Mater. Sci.* **7**, 1099 (1972).
61. S. Otani, "On the Carbon Fiber from the Molten Pyrolysis Products," *Carbon* **3**, 31 (1965).
62. S. Otani, "Mechanism of the Carbonization of MP Carbon Fiber at the Low Temperature Range," *Carbon* **5**, 219 (1967).
63. W. Johnson and W. Watt, "Structure of High Modulus Carbon Fibres," *Nature* **215**, 384 (1967).
64. D. V. Badami et al., "Microstructure of High Strength, High Modulus Carbon Fibres," *Nature* **215**, 386 (1967).
65. H. M. Ezekiel and R. G. Spain, *J. Polym. Sci. Part C* **19**, 249 (1967).
66. F. Siclari, in: *Man-Made Fibers* (edited by H. F. Mark, S. M. Atlas, and E. Cernia), Vol. 1, Wiley-Interscience, 1967.
67. W. E. Fitzgerald and J. P. Craig, "The Extrusion Process in Solution Spinning," *Appl. Polym. Symp.* No. 6, 67 (1967).
68. D. J. Williams, *Polymer Science and Engineering*, Prentice-Hall, Englewood Cliffs, New Jersey, 1971.
69. P. J. Goodhew et al., *Sci. Eng.*, **17**, 3 (1975).
70. J. P. Craig et al., *Text. Res. J.* **32**, 435 (1962).
71. D. R. Paul, "Diffusion during the Coagulation Step of Wet-Spinning," *J. Appl. Polym. Sci.* **12**, 383 (1968).
72. J. P. Knudsen, *Text. Res. J.* **33**, 13 (1963).
73. J. P. Bell and J. H. Dumbleton, "Changes in the Structure of Wet Spun Acrylic Fibers during Processing," *Text. Res. J.* **41**, 196 (1971).
74. W. Bobeth and U. Muller, *Faserforsch. Textiltech.* **16**, 290 (1965).
75. D. J. Johnson and C. N. Tyson, *Brit. J. Appl. Phys.* **2**, 215 (1969).
76. R. D. Andrews and R. M. Kimmel, "Birefringence Effects in Acrylonitrile Polymers. I. Effects at Different Temperatures," *J. Appl. Phys.* **35**, 3194 (1964).
77. S. Okajima et al., "A New Transition Point of Polyacrylonitrile," *J. Polym. Sci.* Part A-1, **6**, 1925 (1968).
78. W. R. Krigbaum and N. Tokita, "Melting Point Depression Study of Polyacrylonitrile," *J. Polym. Sci.* **43**, 467 (1968).
79. A. J. Clarke and J. E. Bailey, "Oxidation of Acrylic Fibers for Carbon Fiber Formation," *Nature* **243**, 146 (1973).
80. J. E. Bailey and A. J. Clarke, "Carbon Fiber Formation—The Oxidation Treatment," *Nature* **234**, 529 (1971).
81. R. C. Howtz, *Text. Res. J.* **20**, 786 (1950).
82. A. Standege and R. Matkowski, "Thermal Oxidation of Polyacrylonitrile," *Eur. Polym. J.* **7**, 775 (1971).
83. D. J. Muller et al., in: *Proceedings of the First International Conference on Carbon Fibres, London, February 1971*, Plastics Institute, London, 1971, Paper No. 2.
84. N. Grassie and R. McGrichan, "Pyrolysis of Polyacrylontrile and Related Polymers. I," *Eur. Polym. J.* **6**, 1277–1291 (1970).
85. H. Friedlander et al., "On the Chromophore of Polyacrylonitrile. VI. Mechanism of Color Formation in Polyacrylonitrile," *Macromolecules* **1**, 79 (1968).

86. N. Grassie, "Novel Types of Chain Reactions in Polymer Degradation," *J. Polym. Sci.* **48**, 79 (1960).
87. W. Watt and W. Johnson, "The Effect of Length Changes during the Oxidation of Polyacrylonitrile Fibers on the Young's Modulus of Carbon Fibers," *Appl. Polym. Symp.* No. 9, 215 (1969).
88. W. Watt and W. Johnson, *Polym. Prepr., Am. Chem. Soc., Div. Polym. Chem.* **9**, 1245 (1968).
89. A. Fiedler et al., *J. Am. Chem. Soc.* **31**, 380 (1971).
90. W. Watt, in: *Proceedings of the Third Conference on Industrial Carbon and Graphite,* Society of Chemical Industry, London, 1970.
91. N. Grassie and R. McGrichan, "Pyrolysis of Polyacrylonitrile and Related Polymers. II," *Eur. Polym. J.* **7**, 1091 (1971).
92. W. N. Turner and F. C. Johnson, "The Pyrolysis of Acrylic Fiber in Inert Atmosphere. I. Reactions up to 400°C," *J. Appl. Polym. Sci.* **13**, 2073 (1969).
93. J. N. Hay, "Thermal Reactions of Polyacrylonitrile," *J. Polym. Sci. Part A-1* **6**, 2127 (1968).
94. J. K. Gilham and R. F. Schwenker, "Thermomechanical and Thermal Analysis of Fiber Forming Polymers," *Appl. Polym. Symp.* No. 2, 59 (1966).
95. E. V. Thompson, "The Thermal Behavior of Acrylonitrile Polymers," *J. Polym. Sci. Part B,* **4**, 361 (1966).
96. L. Reich, Macromol. Rev. **3**, 49 (1968).
97. T. Uchida, in: *Proceedings of the First International Conference on Carbon Fibres, London, February 1971,* Plastics Institute, London, 1971, Paper No. 1.
98. N. Grassie and R. McGrichan, "Pyrolysis of Polyacrylonitrile and Related Polymers. III," *Eur. Polym. J.* **7**, 1357 (1971).
99. K. Miyamichi et al., *J. Soc. Fibre Sci. Technol.* Japan **22**, 538 (1966).
100. S. P. Varma et al., "IR Studies on Pre-Oxidized Pan Fibers," *Carbon* **14**, 207 (1976).
101. R. H. Knibbs *J. Micros.* (Oxford) **94**, 273 (1971).
102. C. R. Rowe, "Evaluation of Stabilized Precursors for High-Strength Carbon Fibers," NOLTR-72-259, November 1972.
103. W. Watt, "Carbon Work at the Royal Aircraft Establishment," *Carbon* **10**, 121 (1972).
104. R. Bacon, "An Introduction to Carbon/Graphite Fibers," *Appl. Polym. Symp.* No. 9, 213 (1969).
105. W. Watt, *Nature* **236**, 10 (1972).
106. J. B. Donnet and P. Ehrburger, "Carbon Fibre in Polymer Reinforcement," 143 *Carbon* **15**, (1977).
107. W. Watt, in: *Proceedings of the First International Conference on Carbon Fibres, London, February, 1971,* Plastics Institute, London, 1971, Paper No. 4.
108. P. J. Goodhew et al., *Sci. Eng.* **17**, 3 (1975).
109. E. S. McCarnety and R. Walline, *Nature* **191**, 1361 (1961).
110. V. I. Kasatvchkin and V. A. Kargin, *Dokl. Phys. Chem.* **191**, 303 (1970).
111. A. Marchand and J. V. Zanchetta, "Proprietes Electroniques D'Un Carbone Dope A L'Azote," *Carbon* **3**, 483 (1966).
112. W. Ruland et al., *Compt. Rend. Acad. Sci.* **269C**, 1597 (1969).
113. W. Ruland et al., "The Microstructure of PAN-Base Carbon Fibres," *J. Appl. Crystallogr.* **3**, 525 (1970).
114. W. Ruland, "X-ray Studies on Preferred Orientation in Carbon Fibers," *J. Appl. Phys.* **38**, 3583 (1967).
115. W. Ruland, Polymer **9**, 1368 (1969).
116. D. J. Johnson and C. N. Tyson, *Brit. J. Appl. Phys.* **2**, 789 (1969).
117. A. Fourdeux et al., in: *Proceedings of the First International Conference on Carbon Fibers, London, February 1971,* Plastics Institute, London, 1971.
118. D. Crawford and D. Johnson, *J. Micros. (Oxford)* **94**, 51 (1971).
119. J. A. Hugo et al., "Intimate Structure of High Modulus Carbon Fiber," *Nature* **226**, 144 (1970).
120. D. Crawford, Doctoral Thesis, University of Leeds, 1972.
121. K. Kawamiria and G. Jenkins, "Structure of Glassy Carbon, *Nature* **231**, 175 (1971).
122. S. C. Bennett et al., "Structural Characterization of a High Modulus Carbon Fibre by High-Resolution Electron Microscopy and Electron Diffraction," *Carbon* **14**, 117 (1976).
123. J. W. Johnson, "Factors Affecting the Tensile Strength of Carbon-Graphite Fibers," *Appl. Polym. Symp.* No. 9, 229 (1969).
124. W. Watt and W. Johnson, "The Effect of Length Changes during the Oxidation of Polyacrylonitrile Fibers on the Young Modulus of Carbon Fibers," *Appl. Polym. Symp.* No. 9, 215 (1969).
125. W. Watt and W. Johnson, in: *Proceedings of the Third Conference on Industrial Carbon and Graphite*, Society of Chemical Industry, London, 1970.
126. G. A. Cooper and R. Mayer, "The Strength of Carbon Fibres," *J. Mater. Sci.* **6**, 60 (1971).
127. F. Barnet and M. Norr, "Factors Limiting the Tensile Strength of PAN-Based Carbon Fibers," NOLTR-74-172, November 1974.
128. D. J. Thoren, "Distribution of Internal Flaws in Acrylic Fibers," *J. Appl. Polym. Sci.* **14**, 103 (1970).
129. R. Moreton, *Fibre Sci. Technol.* **1**, 273 (1969).
130. J. Johnson and D. Thorne, "Effect of Internal Polymer Flaws on Strength of Carbon Fibers Prepared from an Acrylic Precursor," *Carbon* **7**, 659 (1969).
131. D. J. Johnson and C. N. Tyson, "Low-Angle X-ray Diffraction and Physical Properties of Carbon Fibres," *J. Phys. D.* **3**, 526 (1970).
132. V. J. Mimeault and D. W. McKee, *Nature* **224**, 147 (1969).
133. B. F. Jones and R. G. Duncan, "The Effect of Fibre Diameter on the Mechanical Properties of Graphite Fibres Manufactured from Polyacrylonitrile and Rayon," *J. Mater. Sci.* **6**, 289 (1971).
134. B. F. Jones, "Further Observations Concerning the Effect of Diameter on the Fracture Strength and Young's Modulus of Carbon and Graphite Fibres Made from Polyacrylonitrile," *J. Mater. Sci.* **6**, 1225 (1971).
135. E. Murphy and B. F. Jones, "Surface Flaws on Carbon Fibers," *Carbon* **9**, 91 (1971).

136. E. W. Tokarsky and R. J. Diefendorf, "The Modulus and Strength of Carbon Fibers," 12th Biennial Conference on Carbon, Pittsburgh, Pennsylvania, 1975, p. 301.
137. W. Watt, "Carbon Work at the Royal Aircraft Establishment," *Carbon* **10**, 121 (1972).
138. D. Robson et al., "An Electron Spin Resonance Study of Carbon Fibres Based on Polyacrylonitrile," *J. Phys. D* **4**, 1426 (1971).
139. D. Robson et al., "Some Electronic Properties of Polyacrylonitrile-Based Carbon Fibres," *J. Phys. D* **5**, 169 (1972).
140. D. Robson et al., "Determination of Carbon Fibre Structure by Electron Spin Resonance," *Nature* **221**, 51 (1969).
141. D. Robson et al., in: *Proceedings of the Third Conference on Industrial Carbon and Graphite*, Society of Chemical Industry, London, 1970.
142. A. Marchand and J. V. Zanchetta, "Proprietes Electroniques D'Un Carbone Dope A L'Azote," *Carbon* **3**, 483 (1966).
143. H. W. Helberg and B. Wortenburg, "Dieelektrische Leitfahigkeit von pyrolysiertem Polyacrylnitril im Temperaturberich 1, 7 bis 700K," *Phys. Status Solidi A* **3**, 401 (1970).
144. A. J. Hoiberg (editor), *Bituminous Materials: Asphalts, Tars and Pitches*, Wiley-Interscience, New York, 1964–1966.
145. P. Zakar, *Asphalt*, Chemical Publishing Company, New York, 1971.
146. R. B. Long, Exxon Corporate Research Laboratories, personal communication.
147. H. H. Voge, "Catalytic Cracking," in: *Catalysis VI* (edited by Paul Emmett), Romhold Publishing Corporation, New York, 1958.
148. J. Pfeiffer and R. Sadl, *J. Phys. Chem.* **44**, 139 (1940).
149. L. W. Corbett, *Am. Chem. Soc., Div. Petr. Chem.. Prepr.* **9(z)**, 1381 (1964).
150. "Thermal Reactivity of Aromatic Hydrocarbons," in: *Research and Development on Advanced Graphite Materials*, Vol. X, WADD-TR-61-72, August 1962.
151. Supplement, *Research and Development on Advanced Graphite Materials*, Vol. X, WADD-TR-61-72, August 1964.
152. "Carbonization Studies of Polynuclear Aromatic Hydrocarbons," in: *Research and Development on Advanced Graphite Materials*, Vol. XXVII, WADD-TR-61-72, AD 427129, November 1963.
153. "Research and Development for Improved Graphite Materials," in: *Improved Graphite Materials for High Temperature Aerospace Use*, Vol. I, ML-TDR-64-125, September 1964.
154. D. M. Riggs and R. J. Diefendorf, unpublished data.
155. J. D. Brooks and G. H. Taylor, "Formation of Graphitizing Carbons from the Liquid Phase," *Nature* **206**, 697 (1965).
156. J. D. Brooks and G. H. Taylor, *Chem. Phys. Carbon* **4**, 243 (1968).
157. J. Dubois et al., "The Carbonaceous Mesophase Formed in the Pyrolysis of Graphitizable Organic Materials," *Metallography* **3**, 337 (1970).
158. J. L. White et al., "Mesophase Microstructures in Carbonized Coal Tar Pitch," *Carbon* **5**, 517 (1967).
159. H. Honda et al., "Optical Mesophase Texture and X-ray Diffraction Pattern of The Early-Stage Carbonization of Pitches," *Carbon* **8**, 181 (1970).
160. Y. Sanada et al., *Fuel* **52**, 143 (1973).
161. S. Mrozowski, *Carbon* "Electronic Properties and Band Model of Carbons," **9**, 97 (1971).
162. I. Mochid et al., "Anisotropic Mesophase of Novel Features Found in the Refluxing Carbonization of Acenaphthylene," *Carbon* **15**, 191 (1977).
163. M. Makabe et al., "Mesophase Formation of Pitch under Reduced Pressure," *Carbon* **14**, 365 (1976).
164. J. L. White, "The Formation of Microstructure in Graphitizable Carbons," AF #SAMSO-TR-74-93, April 1974.
165. G. W. Gray, *Molecular Structure and the Properties of Liquid Crystals*, Academic Press, New York, 1962.
166. G. W. Gray and P. A. Winsor, *Liquid Crystals and Plastic Crystals*, Vol. 1, Halsted Press, New York, 1974.
167. G. W. Gray and P. A. Winsor, *Liquid Crystals and Plastic Crystals*, Vol. 2, Halsted Press, New York, 1975.
168. P. A. DeGennes, *The Physics of Liquid Crystals*, Clarendon Press, Oxford, 1974.
169. J. B. Barr et al., "High Modulus Carbon Fibers from Pitch Precursor," *Appl. Polym. Symp.* No. 29, 161 (1976).
170. D. A. Schulz, "Process for Producing High Mesophase Content Pitch Fibers," U.S. Patent 3,919,376, 1975.
171. L. S. Singer, "Process for Producing High Mesophase Content Pitch Fibers," U.S. Patent 3,919,387, 1975.
172. I. C. Lewis, "Process for Producing Carbon Fibers from Mesophase Pitch," U.S. Patent 3,995,014, 1976.
173. L. S. Singer, "High Modulus, High Strength Carbon Fibers Produced from Mesophase Pitch," U.S. Patent 4,005,183, 1977.
174. I. C. Lewis, "Process for Producing Mesophase Pitch," U.S. Patent 4,017,327, 1977.
175. E. R. McHenry, "Process for Producing Mesophase Pitch," U.S. Patent 4,026,788, 1977.
176. D. A. Schultz, "Process for Producing Self-Bonded Webs of Non-Woven Carbon Fibers," U.S. Patent 4,032,607, 1977.
177. D. M. Riggs, unpublished data.
178. D. M. Riggs and R. J. Diefendorf, "Elongational Characteristics of Mesophase Containing Pitches," Paper Presented at the 31st Pacific Coast Regional Meeting of the American Ceramic Society, October 1978.
179. R. Didchenko et al., "High Modulus Carbon Fibers from Mesophase Pitches. Part 1: Preparation and Properties of Pitches," Extended Abstracts, 12th

Biennial Conference on Carbon, Pittsburgh, Pennsylvania, 1975, pp. 329–330.

180. R. Didchenko et al., "High Modulus Carbon Fibers from Mesophase Pitches. Part 2: Fiber Properties and Structure," Extended Abstracts, 12th Biennial Conference on Carbon, Pittsburgh, Pennsylvania, 1975, pp. 331–332.

181. L. S. Singer, "The Mesophase and High-Modulus Carbon Fibers from Pitch," Paper Presented at the 13th Biennial Conference on Carbon, Irvine, California, 1977.

182. Union Carbide Corporation, "Graphite Fibers from Pitch, Part I," AFML-TR-73-147.

183. Union Carbide Corporation, "Graphite Fibers from Pitch, Part II," AFML-TR-73-147.

184. Union Carbide Corporation, "Graphite Fibers from Pitch, Part III," AFML-TR-73-147.

185. J. C. Bowman, Union Carbide Corporation, personal communication.

186. A. A. Bright and L. S. Singer, "Electronic and Structural Characteristics of Carbon Fibers from Mesophase Pitch," Paper Presented at the 13th Biennial Conference on Carbon, Irvine, California, 1977, pp. 100–101.

187. G. Montaud and J. Duflos, U.S. Patent 3,332,489, 1967.

188. I. Yoneshega and H. Teranishi, Japanese Patent Specification 2774/70, 1970.

189. S. L. Strong, *Am. Chem. Soc., Div. Org. Coat. Plast. Chem., Prepr.* **31**, 426 (1971).

190. W. Tang and W. Neile, "Effect of Flame Retardants on Pyrolysis and Combustion of X-Cellulose," *J. Polym. Sci. Part C* **6**, 75 (1964).

191. Aggarival, R. K., "Evaluation of Relative Wettability of Carbon Fibres," Carbon, *15* 291 (1977).

192. P. H. Hermans, *Physics and Chemistry of Cellulose Fibers*, Elsevier, Amsterdam, 1949.

193. R. F. Schwenker and E. Pascu, "Chemically Modifying Cellulose for Flame Resistance," *Ind. Eng. Chem.* **50**, 91 (1958).

194. R. W. Little, *Text. Res. J.* **21**, 901 (1951).

195. A. Shindo et al., "Carbon Fibers from Cellulose Fibers," *Appl. Polym. Symp.* No. 9, 271 (1969).

196. A. Shindo, U.S. Patent 3,529,934, 1970.

197. G. Mackay, Canadian Department of Forestry, Publication No. 1201, 1967.

198. A. Broido and F. J. Kilger, *Fire Res. Abstr. Rev.* **5**, 157 (1963).

199. R. C. Laible, *Am. Dyest. Rep.* **47**, 173 (1958).

200. W. Hoffman et al., *Chem. Ind. (London)* p. 95 (1960).

201. R. Moyer et al., U.S. Patent 3,333,926, 1967.

202. C. L. Gutzeit, U.S. Patent 3,479,150, 1969.

203. E. M. Peters, U.S. Patent 3,235,353, 1966.

204. R. Bacon et al., U.S. Patent 3,305,315, 1967.

205. S. E. Ross, *Text. Res. J.* **38**, 906 (1968).

206. A. Shindo *Polym. Prepr. Am. Chem. Soc., Div. Polym. Chem.* **9**, 1333 (1968).

207. C. L. Gutzeit, U.S. Patent 3,479,151, 1969.

208. D. R. Moore et al., U.S. Patent 3,527,564, 1970.

209. C. H. Mack, *Text. Res. J.* **37**, 1063 (1967).

210. H. Schuyten et al., *J. Am. Chem. Soc.* **70**, 1919 (1948).

211. M. J. Hunter, U.S. Patent 2,532,622, 1950.

212. W. I. Portnode, U.S. Patent 2,306,222, 1942.

213. J. V. Duffy, "Pyrolysis of Treated Rayon Fiber," *J. Appl. Polym. Sci.* **15**, 715 (1971).

214. S. L. Madorsky et al., "Thermal Degradation of Cellulosic Materials," *J. Res. Nat. Bur. Stand.* **60**, 343 (1958).

215. R. W. Little (editor), *Flameproofing Textile Fabrics*, Van Nostrand Reinhold, New York, 1947.

216. S. L. Madorsky et al., "Pyrolysis of Cellulose in a Vacuum," *J. Res. Nat. Bur. Stand.* **56**, 343 (1956).

217. W. Tang and W. Eickner, U.S. Forest Service Research Paper, FPL 84, January 1968.

218. W. A. Reeves et al., "Some Chemical and Physical Factors Influencing Flame Retardancy," Text. Res. J., **40**, 223 (1970).

219. W. Schalamon and R. Bacon, British Patent 1,167,007, 1969.

220. R. Bacon and W. Schalamon, Eighth Biennial Conference on Carbon, Buffalo, New York, June 1967.

221. W. J. Spry, British Patent 1,093,084, 1967.

222. D. W. Gibson and G. Langlors, *Polym. Prepr., Am. Chem. Soc., Div. Polym. Chem.* **9**, 1376 (1968).

223. R. Manley, "Fine Structure of Native Cellulose Microfibrils," *Nature* **204**, 1155 (1964).

224. R. Manley, "The Fine Structure of Regenerated Cellulose," *J. Polym. Sci. Part B* **3**, 691 (1965).

225. R. Jeffries et al., *Text. Res. J.* **38**, 234 (1968).

226. S. Y. Lin, "Accessibility of Cellulose: A Critical Review," *Fiber Sci. Technol.* **5**, 303 (1972).

227. W. Ruland, "Small-Angle Scattering Studies on Carbonized Cellulose Fibers," *J. Polym. Sci. Part C* **28**, 143 (1969).

228. W. Ruland, 9th Biennial Conference on Carbon, held at Chestnut Hill, Massachusetts, June 1969.

229. R. Bacon and W. Schalamon, "Physical Properties of High Modulus Graphite Fibers Made from Rayon Precursor," *Appl. Polym. Symp.* No 9, 285 (1969).

230. R. C. Novak, "Fracture in Graphite Filament Reinforced Epoxy Loaded in Shear," ASTM STP 460, 1969, p. 540.

231. R. Simon, S. P. Porsen, and J. Duffy, "Carbon Fibre Composites," *Nature* **213**, 1113 (1967).

232. J. C. Goan and S. P. Porsen, "Interfacial Bonding in Graphite Fiber–Resin Composites," ASTM STP 452, 1969, p. 3.

233. J. P. Favre and J. Perrin, "Carbon Fibre Adhesion to Organic Matrices," *J. Mater. Sci.* **7**, 1113 (1972).

234. D. W. McKee and V. J. Mimeault, "Surface Properties of Carbon Fibers," *Chem. Phys. Carbon* **8**, 151 (1973).

235. P. J. Dynes and D. H. Kaelble, "Surface Energy Analysis of Carbon Fibers and Films," *J. Adhes.* **6**, 195 (1974).

236. L. T. Drzal, "The Surface Composition and Energetics of Type A Graphite Fibers," *Carbon* **15**, 129 (1977).

237. J. V. Larsen, T. G. Smith, and P. W. Erickson,

"Carbon Fiber Surface Treatments,"NOLTR 71-165, 1971.

238. B. Rand and R. Robinson, "Surface Characteristics of Carbon Fibres from PAN," *Carbon* **15**, 257 (1977).

239. R. J. Dawksky, "Graphite Fiber Treatments which Affect Fiber Surface Morphology and Epoxy Bonding Characteristics," *J. Adhes.* **5**, 211 (1973).

240. D. H. Kaelble, P. J. Dynes, and L. Maus, "Surface Energy Analysis of Treated Graphite Fibers," *J. Adhes.* **6**, 239 (1974).

241. B. Harris, P. W. R. Beaumont, and A. Rosen, "Silane Coupling in Carbon Fibre-Reinforced Polyester Resin," *J. Mater. Sci.* **4**, 432 (1969).

242. D. J. Pinchin and R. T. Woodhams, "Pyrolytic Surface Treatment of Graphite Fibres," *J. Mater. Sci.* **9**, 300 (1974).

243. S. J. Mitchell, "Pyrocarbon Coating of Carbon Fibre to Increase the Interlaminar Strength of Carbon-Fibre-Reinforced Carbon," in: *Proceedings of the Second International Conference on Carbon Fibres, London, February 1974*, Plastics Institute, London, 1974, Paper No. 20.

244. D. Clark, N. J. Wadsworth, and W. Watt, "The Surface Treatment of Carbon Fibres for Increasing the Interlaminar Shear Strength of CFRP," in: *Proceedings of the Second International Conference on Carbon Fibres, London, February 1974*, Plastics Institute, London, 1974, Paper No. 7.

245. J. B. Donnet, E. Papirer, and H. Dauksch, "Surface Modification of Carbon Fibres and Their Adhesion to Epoxy Resins," in: *Proceedings of the Second International Conference on Carbon Fibres, London, February 1974*, Plastics Institute, London, 1974, Paper No. 9.

246. G. Riess and M. Bourdeaux, "Surface Treatment of Carbon Fibres with Alternating and Block Copolymers," in: *Proceedings of the Second International Conference on Carbon Fibres, London, February 1974*, Plastics Institute, London, 1974, Paper No. 8.

247. J. W. Herrick, P. E. Gruber, and F. T. Mansur, "Surface Treatments for Fibrous Carbon Reinforcements, Part I," AFML-TR-66-178, July 1966.

248. R. Didchenko, "Carbon and Graphite Surface Properties Relevant to Fiber Reinforced Composites," AFML-TR-68-45, 1968.

249. Scala, D. A. and Brooks, 25th Society of Plastics Industry Meeting, Washington, D.C., February 1970.

250. R. J. Bobka and L. P. Lowell, "Integrated Research on Carbon Composite Materials, Part I," AFML-TR-66-310, October 1966, pp. 145–152.

251. F. Tuinstra and J. L. Koenig, "Characterization of Graphite Fiber Surfaces with Raman Spectroscopy," *J. Compos. Mater.* **4**, 492 (1970).

252. J. S. Mattson and H. B. Mark, "Infrared Internal Reflectance Spectroscopic Determination of Surface Functional Groups on Carbon," *Colloid Interface Sci.* **31**, 131 (1969).

253. D. H. Kaelble, P. J. Dynes, and E. H. Cirlin, "Interfacial Bonding and Environmental Stability of Polymer Matrix Composites," *J. Adhes.* **6**, 23 (1974).

254. M. Barber, P. Swift, E. L. Evans, and J. M. Thomas, "High Energy Photoelectron Spectroscopic Study of Carbon Fibre Surfaces," *Nature* **227**, 1131 (1970).

255. F. Hopfgarten, "Surface Study of Carbon Fibres with ESCA and Auger Electron Spectroscopy," *Fibre Sci. Technol.* **11**, 67 (1978).

256. D. M. Brewis, J. Comyn, J. R. Fowler, D. Briggs, and V. A. Gibson, "Surface Treatments of Carbon Fibers Studied by X-ray Photoelectron Spectroscopy," *Fibre Sci. Technol.* **12**, 41 (1979).

257. N. C. W. Judd, "The Effect of Fiber Surface Treatment, Resin Type and Fabrication Process on the Interlaminar Shar Strength of Carbon Fibre Composites," *Brit. Polym. J.* **9**, 272 (1977).

258. S. C. Thompson, H. C. Kim, and F. R. Matthew, "The Effect of Processing on the Microstructure of CFRP," *Composites* **4**, 86 (1973).

259. N. C. W. Judd and W. W. Wright, "Voids and Their Effects on the Mechanical Properties of Composites—An Appraisal,"*SAMPE J.*, 10 (1978).

260. J. W. Herrick, "Surface Treatments for Fibrous Carbon Reinforcements, Part II," AFML-TR-66-178, June 1967.

261. J. H. Williams and P. N. Kousiounelos, "Thermoplastic Fibre Coatings Enhance Composite Strength and Toughness," *Fibre Sci. Technol.* **11**, 83 (1978).

262. E. L. McKague, R. E. Bullock, and J. W. Head, "Improved Mechanical Properties of Composites Reinforced with Neutron-Irradiated Carbon Fibers," *J. Compos. Mater.* **7**, 288 (1973).

263a. "Test Methods for Determining the Physical Properties of Carbon and Graphite Tows," Technical Data Sheet HD-SG-2-6001B, Hercules Inc., Magna, Utah, September 1974.

263b. "Torayca Information Quality Carbon Fibre," Technical Data Sheet No. TY-030, Toray Industries, Inc., Tokyo, Japan.

264. *Advanced Composites Design Guide,* 3rd Ed., Vol. IV, Air Force Flight Dynamics Laboratory, Wright-Patterson Air Force Base, Dayton, Ohio, January 1973.

265. D. R. Lovell, "Quality Control in Carbon Fibre Manufacture," in: *Proceedings of the First International Conference on Carbon Fibres, London, February 1971*, Plastics Institute, London, 1971, p. 359.

266. "Quality Assurance Testing," Technical Data Sheet No. TY-020, Toray Industries, Inc., Tokyo, Japan.

267. "Tensile Strength and Young's Modulus for High-Modulus Single Filament Materials," *Annual Book of ASTM Standards*, Part 36, D3379-75.

268. "Reporting Test Methods and Results on High Modulus Fibers," *Annual Book of ASTM Standards*, Part 36 D3544-76.

269. P. E. McMahon, "Graphite Fiber Tensile Property Evaluation," ASTM STP 521, 1973, p. 367.

270. E. Scala, "High Strength Filaments for Cables and Lines," ASTM 521, 1973, p. 390.

271. G. Jouquet and R. Schill, "Mechanical Properties of

Carbon Fibres and Their Composites," in: *Proceedings of the First International Conference on Carbon Fibres, London, February 1971*, Plastics Institute, London, 1971, Paper No. 16, p. 113.
272. J. T. Hoggatt, "Test Methods for High Modulus Carbon Yarn and Composites," ASTM STP 460, 1969, p. 48.
273. T. T. Chiao and R. L. Moore, "A Tensile Test Method for Fibers," *J. Compos. Mater.* **4**, 118 (1970).
274. T. T. Chiao, M. A. Hamstad, and E. S. Jessop, "Tensile Properties of an Ultra High-Strength Graphite Fiber in an Epoxy Matrix," ASTM STP 580, 1975, p. 612.
275. W. N. Reynolds, "Structure and Physical Properties of Carbon Fibers," *Chem. Phys. Carbon* **11**, 1 (1973).
276. B. F. Jones and R. G. Duncan, *J. Mater. Sci.* **6**, 289 (1971).
277. A. J. Perry, K. Phillips, and E. deLamotte, *Fibre Sci. Technol.* **3**, 317 (1971).
278. R. Moreton, *Fibre Sci. Technol.* **1**, 273 (1968).
279. "Density of Plastics by the Density-Gradient Techniques," *Annual Book of ASTM Standards*, Part 35, D1505-68.
280. "Specific Gravity and Density of Plastics by Displacement," *Annual Book of ASTM Standards*, Part 35, D792-66.
281. I. L. Kalnin, "Thermal Conductivity of High-Modulus Carbon Fibers," ASTM STP 580 1975, p. 560.
282. P. E. McMahon, "The Relationship between High Modulus Fiber and Unidirectional Composite Tensile Strength," 19th SAMPE National Symposium and Exhibition, Buena Park, California, 1974, p. 214.
283. J. W. Johnson and D. J. Thorne, "Effect of Internal Polymer Flaws on Strength of Carbon Fibers Prepared from an Acrylic Precursor," *Carbon* **7**, 659 (1969).
284. A. R. Nicoll and A. J. Perry, "Diameter Dependence of Carbon Fibre Mechanical Properties," *Fibre Sci. Technol.* **6**, 135 (1973).
285. C. R. Rowe and D. L. Lowe, "High Temperature Properties of Carbon Fibers," 13th Biennial Conference on Carbon, London, 1977, p. 170.
286. P. C. Pinoli and A. Ambrosio, "Fiber Density Analysis by Density Gradient Technique, 13th Biennial Conference on Carbon, London, 1977, p. 292.
287. "Hercules Magnamite Graphite Fibers," Technical Data Bulletin 200-269B 4-78, Hercules Inc., Salt Lake City, Utah.
288. 'Material Properties Composites," Technical Data Bulletin, Celanese Corporation, Chatham, New Jersey.
289. J. R. Vinson, "Advanced Composite Materials—Environmental Effects," ASTM 658, 1978.
290. A. C. Loos and G. S. Springer, "Effects of Thermal Spiking on Graphite–Epoxy Composites," J. *Compos. Mater.* **13**, 17 (1979).
291. E. Fitzer and M. Heyon, "Carbon Fibre Reinforced Composites for Applications at Elevated Temperatures," 13th Biennial Conference on Carbon, London, 1977, p. 128.
292. K. E. Hofer, M. Stander, and P. N. Rao, "A Comparison of the Elevated Temperature Strength Loss in High Tensile Strength Graphite/Epoxy Composite Laminates Due to Ambient and Accelerated Aging," *J. Test. Eval.* **3**, No. 6, 423 (1975).
293. J. R. Kerr, J. F. Haskins, and B. A. Stein, "Program Definition and Preliminary Results of a Long-Term Evaluation Program of Advanced Composites for Supersonic Cruise Aircraft Applications: Environmental Effects on Advanced Composite Materials," ASTM STP 602, 1976, p. 3.
294. W. G. Scheck and J. M. Stuckey, "Development and Evaluation of Graphite and Boron Polyimide Composites," Fourth National SAMPE Technical Conference, Palo Alto, California, October 17, 1972, p. 9.
295. V. F. Mazzio and R. L. Mehan, "Effects of Thermal Cycling on the Properties of Graphite–Epoxy Composites," in *Composite Materials: Testing and Design (Fourth Conference),* ASTM STP 617, American Society for Testing and Materials, (1977), p. 460.
296. L. G. Bevan and J. B. Sturgeon, "Fatigue Limits in CFRP," *Proceedings of the Second International Conference on Carbon Fibres, London, February 1974*, Plastics Institute, London, 1974, paper No. 32.
297. J. L. Camahort, F. H. Rennhack, and W. C. Coons, "Effects of Thermal Cycling Environment on Graphite/Epoxy Composites," ASTM STP 602, 1976, p.37.
298. P. E. McMahon, "Oxidative Resistance of Carbon Fibers and Their Composites," ASTM STP 658, 1978, p. 254.
299. H. H. Gibbs, R. C. Wendt, and F. C. Wilson, "Carbon Fiber Structure and Stability Studies," 33rd SPI Conference, Paper 24-F, 1978.
300. S. E. Wentworth, A. O. King, and R. J. Shuford, "The Potential for Accidental Release from Carbon/Graphite Fiber from Resin Matrix Composites as Determined by Thermogravimetric Analysis," Army Materials and Mechanics Research Center, Watertown, Massachusetts, TR 79-1, January 1979.
301. S. E. Wentworth, "Assessment of Accidental Release Potential by TGA and SEM," Paper Presented at the Second Annual Army Composite Materials Research Review, University of Massachusetts, May 1979.
302. N. C. W. Judd, "The Chemical Resistance of Carbon Fibres and a Carbon-Fibre/Polyester Composite," in: *Proceedings of the First International Conference on Carbon Fibres, London, February 1971,* Plastics Institute, London, 1971, Paper No. 32, p. 258.
303. J. Hertz, "Moisture Effects on the High-Temperature Strength of Fiber-Reinforced Resin Composites," Fourth National SAMPE Technical Conference, Palo Alto, California, October 17, 1972, p. 1.
304. C. E. Browning, G. E. Husman, and J. M. Whitney, "Moisture Effects in Epoxy Matrix Composites," in:

Composite Materials: Testing and Design (Fourth Conference), ASTM STP 617, American Society for Testing and Materials, 1977, p. 481.

305. C. D. Shirrell and J. Halpin, "Moisture Absorption and Desorption in Epoxy Composite Laminates," in: *Composite Materials: Testing and Design (Fourth Conference),* ASTM STP 617, American Society for Testing and Materials, 1977, p. 514.
306. N. C. W. Judd, "Absorption of Water into Carbon Fibre Composites," *Brit. Polym. J.* **9**, 36 (1977).
307. H. G. Carter and K. G. Kibler, "Rapid Moisture-Characterization of Composites and Possible Screening Applications," J. *Compos. Mater.* **10**, 355 (1976).
308. J. G. Belani and L. J. Broutman, "Moisture Induced Resistivity Changes in Graphite-Reinforced Plastics," *Composites*, **9**, 273 (1978).
309. C. E. Browning, "The Mechanisms of Elevated Temperature Property Losses in High Performance Structural Epoxy Resin Matrix Materials after Exposures to High Humidity Environments," AFML TR-76-153, March 1977.
310. P. E. Sandorff and Y. A. Tajima, Sr., "A Practical Method for Determining Moisture Distribution, Solubility and Diffusivity in Composite Laminates," *SAMPE Quart.* **10**: 21, January 1979.
311. C. E. Browning and J. T. Hartness, "Effects of Moisture on the Properties of High Performance Structural Resins and Composites," ASTM STP 546, 1974, p. 284.
312. D. H. Kaelble, P. J. Dynes, and L. Maus, "Hydrothermal Aging of Composite Materials," *J. Adhes.* **8**, 121 (1976).
313. D. H. Kaelble, P. J. Dynes, L. W. Crane, and L. Maus, "Interfacial Mechanisms of Moisture Degradation in Graphite–Epoxy Composites," *J. Adhes.* **7**, 25 (1975).
314. D. H. Kaelble and P. J. Dynes, "Methods for Detecting Moisture Degradation in Graphite–Epoxy Composites," *Mater. Eval.,* 103 (April 1977).
315. J. M. Augl and A. E. Burger, "Moisture Effect on Carbon Fiber Epoxy Composites," NSWC/WOL/TR 76-7, September 1976, AD-A034787.
316. J. M. Augl, "The Effect of Moisture on Carbon Fiber Reinforced Epoxy Composites. II. Mechanical Property Changes," NSWC/WOL/TR 76-149, February 1977, AD-A039903.
317. J. H. Powell and D. J. Zyrang, "The Moisture Absorption and Desorption Characteristics of Three Epoxy/Graphite Systems," *SAMPE J.* **13**, 4 (1977).
318. H. L. Young and W. L. Greever, "High Temperature Strength Degradation of Composites during Aging in the Ambient Atmosphere," *Compos. Mater. Eng. Design,* 695 (1973).
319. C. Shen and G. S. Springer, "Environmental Effects on the Elastic Moduli of Composite Materials," *J.* Composite Materials, *11*, 250 (1977).
320. Shen, C. and Springer, G. S., "Effects of Moisture and Temperature on the Tensile Strength of Composite Materials," *J. Compos. Mater.* **11**, 2 (1977).
321. D. J. Wilkins, "Environmental Sensitivity Tests of Graphite–Epoxy Bolt Bearing Properties," in: *Composite Materials: Testing and Design (Fourth Conference),* ASTM STP 617, American Society for Testing and Materials, 1977, p. 497.
322. K. E. Hofer, D. Larsen, and V. E. Humphreys, "Development of Engineering Data on the Mechanical and Physical Properties of Advanced Composites Materials," AFML-TR-24-266, February 1975.
323. C. Y., Lundemo and S. Thor, "Influence of Environmental Cycling on the Mechanical Properties of Composite Materials," *J. Compos. Mater.* **11**, 276 (1977).
324. J. Delmonte, "Moisture Penetration into Composites," 7th National SAMPE Technical Conference, Albuquerque, New Mexico, 1975, p. 306.
325. B. Harris, P. W. R. Beaumont, and E. Moncunill de Ferran, "Strength and Fracture Toughness of Carbon Fibre Polyester Composites," *J. Mater. Sci.* **6**, 238 (1971).
326. E. L. McKague, Jr., J. E. Halkias, and J. D. Reynolds, "Moisture on Diffusion," *J. Compos. Mater.* **9**, 2 (1975).
327. T. T. Serafini and R. D. Vannucci, "Tailor Making High Performance Graphite Fiber Reinforced PMR Polyimides," 30th Annual SPI Technical Conference, 1975, Section 14-E, pp. 1–5.
328. C. E. Browning, "HME Resin Matrix System", *SAMPE J.*, 23rd National SAMPE Symposium, Anaheim, California, 1978, p. 541.
329. B. Bloch and M. Ropars, "PSP Resins, New Thermosetting Binders for Advanced Composites," *SAMPE J.*, 23rd National SAMPE Symposium, Anaheim, California, 1978, p. 836.
330. R. J. Fabian, "Engineer's Guide to Polyimide Plastics," *Mater. Eng.*, 26–31 (August 1971).
331. S. L. Kaplan and C. W. Snyder, "Polyimide Molding Compounds," 29th Annual SPI Technical Conference, 1974, Section 11-B, pp. 1–7.
332. R. J. Kray, R. Seltzer, and R. A. E. Winter, "Thermally Stable Polyimides with 400° F Processability," 29th Annual SPI Technical Conference, 1974, Paper 11-C.
333. R. D. Vannucci, "Effect of Processing Parameters on Autoclaved PMR Polyimide Composites," 9th National SAMPE Technical Conference, Atlanta, Georgia, 1977, p. 177.
334. R. D. Vannucci, and W. B. Alston, "PMR Polyimides with Improved High Temperature Performance," 31st Annual SPI Technical Conference, 1976, Paper 20-A.
335. J. T. Hoggatt, and A. D. VonVolkli, "Evaluation of Reinforced Thermoplastic Composites and Adhesives," Report No. D18D-17503-3, Contract No. N00019-74-C-0226, March 1975.
336. H. H. Gibbs and J. R. Ness, "The Development of Quality Control Techniques for NR-150 Polyimide Adhesive and Binder," 23rd National SAMPE Symposium, 1978, p. 806.

338. T. L. St. Clair and R. A. Jewell, "Solventless LaRC-160 Polyimide Matrix Resin," 23rd National SAMPE Symposium, Anaheim, California, 1978, p. 520.
339. N. Bilow and A. L. Landis, "Recent Advances in Acetylene-Substituted Polyimides," 8th National SAMPE Technical Conference, Seattle, Washington, 1976, p. 94.
340. J. J. Aponyi, C. B. Delano, J. D. Dodson, R. J. Milligan, and J. M. Hurst, "Thermal 600 and Acetylene Terminated Quinoxaline Resins," 23rd National SAMPE Symposium, Anaheim, California, 1978, p. 763.
341. N. Bilow, L. B. Keller, A. L. Landis, R. H. Boschan, and A. A. Castillo, "New Developments in Acetylene Substituted Polyimides," 23rd National SAMPE Symposium, 1978, p. 791.
342. N. Sung and F. J. McGarry, "The Mechanical and Thermal Properties of Graphite Fiber Reinforced Polyphenylquinoxaline and Polyimide Composites," *Polym. Eng. Sci.* **16**, 426 (1976).
343. R. A. Pike and M. A. DeCrescente, "Elevated Temperature Characteristics of Borsic® and Graphite Fiber/Polyphenylquinoxaline Resin Composites," 29th Annual SPI Technical Conference, 1974, Paper 18-F.
344. R. A. Mayor, "Advanced Composites for High Temperature Applications," 9th National SAMPE Technical Conference, Atlanta, Georgia, 1977, p. 478.
345. M. G. Maximovich, "Development and Applications of Continuous Graphite Reinforced Thermoplastic Advanced Composites," 19th National SAMPE Symposium, 1974, p. 262.
346. C. L. Segal, "Properties and Applications of Graphite Fiber Reinforced Thermoplastic Molding Compounds", 19th SAMPE Symposium, 1974, p. 277.
347. B. F. Blumentritt, B. T. Vu, and S. L. Cooper, "Fracture in Oriented Short Fibre-Reinforced Thermoplastics," *Composites* **6**, 105 (1975).
348. J. T. Hoggatt, "Study of Graphite Fiber Reinforced Thermoplastic Composites," Report No. D180-18034-1, AD-778000, Contract No. N00019-73-C-0414, February 1974.
349. M. G. Maximovich, "Evaluation of Selected High-Temperature Thermoplastic Polymers for Advanced Composite and Adhesive Applications," in: *Composite Materials: Testing and Design (Fourth Conference),* ASTM STP 617, American Society for Testing and Materials, 1977, p. 123.
350. B. F. Blumentritt, B. T. Vu, and S. L. Cooper, "The Mechanical Properties of Oriented Discontinuous Fiber-Reinforced Thermoplastics. I. Unidirectional Fiber Orientation," *Polym. Eng. Sci.* **14**, 633 (1974).
351. R. C. Novak, "Graphite Fiber Reinforced Thermoplastic Resins," N77-27231, Contract No. NAS3-20077, NASA-CR-135197, April 1977.
352. J. E. Theberge, B. Arkles, and R. Robinson, "Carbon Fibers Add Muscle to Plastics," *Machine Design* **2**, 1 (1974).
353. J. E. Theberge, B. Arkles, and R. Robinson, "Carbon Fiber Reinforced Thermoplastics," 29th Annual SPI Technical Conference, 1974, Paper 20-D.
354. D. G. Chasin and Dr. J. Feltzin, "Properties and Applications of Polyethersulfones—A New Family of High Temperature Performance Engineering Thermoplastics,"7th National SAMPE Technical Conference, Albuquerque, New Mexico, 1975, p. 395.
355. D. R. Askins, "Development of Engineering Data on Advanced Composites," AFML-TR-77-151, September 1977.
356. L. N. Phillips and G. Wood, "Thermosetting and Thermoplastic Carbon-Fibre Composites," in: *Proceedings of the International Conference on Carbon Fibres, London, 1971,* Plastics Institute, London, 1971, p. 266.
357. H. S. Katz, *Plastics Compounding,* 18 (March–April 1979).
358. A. T. Dibenedetto and Gideon Salee, "The Fatigue Behavior of Graphite Fiber Reinforced Nylon," *Poly. Eng. Sci.* **18**, 634 (1978).
359. *Scientific American* **9**, No. 41 (June 24, 1854).
360. R. W. Lewis and P. F. Brake, "Polymer Matrix Composites for Automotive Applications," Symposium on Polymeric Materials and Their Use in Transportation, Polytechnic Institute of New York, Brooklyn, New York, April 1977.
361. M. J. Owen, "Fatigue of Carbon-Fiber-Reinforced Plastics," in: *Composite Materials. Vol. 5. Fatigue and Fracture* (edited by L. J. Broutman), Academic Press, New York, 1975, pp. 342–369.
362. M. G. Bader and M. Johnson, "Fatigue Strength and Failure Mechanisms in Uniaxial Carbon Fibre Reinforced Epoxy Resin Composite Systems," *Composites* **5**, 58 (1974).
363. A. K. Green and P. L. Pratt, "The Axial Fatigue Behavior of Unidirectional Type III S Carbon Fibre–Epoxy Resin Composites," *Composites* **5**, 63 (1974).
364. C. K. H. Dharan, "Fatigue Failure Mechanisms in Pultruded Graphite–Polyester," in: *Proceedings of the Second Symposium on Failure Modes in Composites*, American Institute of Mining, Metallurgical, and Petroleum Engineers, New York, May 1974.
365. C. K. H. Dharan, "Fatigue Failure in Graphite Fibre and Glass Fibre–Polymer Composites," *J. Mater. Sci.* **10**, 1665–1670 (1975).
366. J. J. Nevadunsky, J. J. Lucas, and M. J. Salkind, "Early Fatigue Damage Detection in Composite Materials," *J. Compos. Mater.* **9**, 394 (1975).
367. J. Awerbuch and H. T. Hahn, "Fatigue and Proof-Testing of Unidirectional Graphite/Epoxy Composite," in: *Fatigue of Filamentary Composite Materials* (edited by K. L. Reifsnider and K. N. Lauraitis), ASTM STP 636, American Society for Testing and Materials, 1977, pp. 248–266.
368. J. T. Ryder and E. K. Walker, "Effect of Compression on Fatigue Properties of a Quasi-Isotropic Graphite/Epoxy Composite," in: *Fatigue of Filamentary Composite Materials* (edited by K. L.

Reifsnider and K. N. Lauraitis), ASTM STP 636, American Society for Testing and Materials, 1977, pp. 3–26.

369. D. C. Hiler, "Carbon Fiber Composites," *Plastics World* (July 1977).

370. T. T. Chiao, C. C. Chiao, and R. J. Sherry, "Lifetimes of Fiber Composites under Sustained Tensile Loading," Proceedings of the 1977 International Conference on Fracture Mechanics and Technology, Hong Kong, March 21–25, 1977.

371. K. E. Hofer and L. C. Bennett, "Influence of Fiber and Matrix Variables on the Fatigue and Creep Characteristics of Hybrid Composites," Final Report to Naval Air Systems Command, Contract N00019-75-C-0470, IIT Research Institute, Chicago, Illinois, November 1977.

372. C. C. Chamis and R. F. Lark, "Hybrid Composites—State of the Art Review: Analysis, Design, Application and Fabrication," Report NASA TMX-73545, NASA/Lewis Research Center, Cleveland, Ohio, 1977.

373. J. Summerscales and D. Short, "Carbon Fiber and Glass Fibre Hybrid Reinforced Plastics," *Composites* **9**, 157–166 (1978).

374. C. H. Zweben, "Tensile Strength of Hybrid Composites," *J. Sci.* **12**, 1325–1337 (1977).

375. L. N. Phillips, "The Development and Uses of Glass/Carbon Hybrids," in: *Proceedings of the 1978 International Conference on Composite Materials*, Toronto, Canada, American Institute of Mining, Metallurgical, and Petroleum Engineers, New York, 1978, p. 1340.

376. E. Guerini and P. Lissae, "Contribution to the Study of Hybrid Laminates Fabricated from Graphite, Kevlar 49 and Glass," *Verre Text. Plast. Reinf.*, No. 6, pp. 5-12 (July–August 1978).

377. D. Yates, "Design and Fabrication of High Performance Composites," SAE International Automotive Engineering Congress and Exposition, Detroit, Michigan, February 27–March 3, 1977.

378. J. D. Helfinstine, "Charpy Impact of Unidirectional Graphite/Aramid/Epoxy Hybrid Composites," in: *Composite Materials: Testing and Design (Fourth Conference)*, ASTM STP 617, American Society for Testing and Materials, 1977, pp. 375–388.

379. C. H. Zweben "Hybrid Fiber Composite Materials," in: *Proceedings of the 1975 International Conference on Composite Materials*, Boston, Massachusetts Vol. 1, American Institute of Mining, Metallurgical, and Petroleum Engineers, New York, 1975, p. 345.

380. R. Dukes and D. L. Griffiths, "Marine Aspects of Carbon Fibre and Glass Fibre/Carbon Fibre Composites," in: *Proceedings of the First International Conference on Carbon Fibres London, February 1971*, Plastics Institute, London, 1971, Paper No. 28, pp. 226–231.

381. *Advanced Composites Design Guide*, 3rd Ed., Vol. I, Air Force Flight Dynamics Laboratory, Wright-Patterson Air Force Base, Dayton, Ohio, January 1973.

382. "Composite Box Beam Optimization," Fourth Quarterly Progress Report for AFML, Contract F33615-71-C-1605, Grumman Aerospace Corporation, Bethpage, New York, June 1972.

383. I. L. Kalnin, "Evaluation of Unidirectional Glass-Graphite Fiber/Epoxy Resin Composites," in: *Composite Materials: Testing and Design (Second Conference)*, ASTM STP 497, American Society for Testing and Materials, 1972, pp. 551–563.

384. K. E. Hofer, M. Stander, and L. C. Bennett, "Degradation and Enhancement of the Fatigue Behavior of Glass/Graphite/Epoxy Hybrid Composites after Accelerated Aging," Proceedings of the 32nd Annual Conference, SPI, Reinforced Plastics/Composites Institute, Washington, D.C., February 1977, p. 11-F.

385. K. E. Hofer, N. Rao, and M. Stander, "Fatigue Behavior of Graphite/Glass/Epoxy Hybrid Composites," in: *Proceedings of the International Conference on Carbon Fibres, London, February 1974*, Plastics Institute, London, 1974, Paper No. 31.

386. R. Kaiser, "Technology Assessment of Advanced Composite Materials," Final Report to National Science Foundation, Contract ERS77-19467, Argos Associates Inc., Winchester, Massachusetts, April 1978.

387. E. M. Trewin, "Graphite Fibres—The Expanding Scope of Applications in the Industrial Market," in: *Proceedings of the 1978 International Conference on Composite Materials*, American Institute of Mining, Metallurgical, and Petroleum Engineers, New York, 1978, pp. 1462–1473.

388. G. Lubin and P. Donohue, "Real Life Aging Properties of Composites," Proceedings of the 35th Annual Conference, SPI Reinforced Plastics/Composites Institute, New Orleans, February 1980, p. 17-E.

389. C. Staebler, G. Lubin and M. Stander, "Application and Testing of Metallic Coatings on Graphite Epoxy Composites," Proceedings of the 36th Annual Conference, SPI, Reinforced Plastics/Composites Institute, Washington, D.C., February 1981, p. 17-D.

12
ARAMID FIBERS AND COMPOSITES*

C. C. Chiao and T. T. Chiao**
Lawrence Livermore Laboratory, University of California
Livermore, California

12.1. INTRODUCTION

Aramid fiber is the generic name for aromatic polyamide fibers. As defined by the U. S. Federal Trade Commission, an aramid fiber is "a manufactured fiber in which the fiber-forming substance is a long chain synthetic polyamide in which at least 85% of the amide linkages are attached directly to two aromatic rings."

During the past decade, considerable work has been done on the preparation of fibers from wholly aromatic polymers to obtain high strength, high modulus, and heat resistance. To date, the only commercially available aramid fibers are DuPont's Kevlars,† first introduced in 1971. In addition to their high strength and modulus, the Kevlars are corrosion resistant as well as chemically and mechanically stable over a wide range of temperatures.

In this chapter, we summarize the available published data on the two aramid fibers, Kevlar 29 and Kevlar 49, and their composites.

12.2. ARAMID FIBERS AND FABRICS

12.2.1. Manufacturing

The aramid fiber-forming polymers, i.e., the aromatic polyamides, are believed to be made by solution-polycondensation of diamines and diacid halides at low temperatures. The polymers are spun from strong acid solutions (e.g., concentrated H_2SO_4) by a dry-jet wet spinning process. (Blades[1] has described the production of aramid fibers using various polyamides made from diamines and diacid halides.) Principally, the polymers are made by rapidly adding a diacid chloride to a cool (5–10° C; 41–50° F) amine solution, with stirring. The polymer thus formed is recovered from the crumbs or gel by pulverizing, washing, and drying. To form filaments, the clean polymer, mixed with a strong acid, is extruded from spinnerets at an elevated temperature (51–100° C; 124–212° F) through a 0.5–1.9-cm layer of air into cold water (0–4° C; 32–39° F). (See Fig. 12.1.) The fibers then are washed thoroughly in water and dried on bobbins.

Fiber properties can be altered by using solvent additives, by varying the spinning conditions, and by using post-spinning heat treatments.

12.2.2. Chemical Structure

Chemically, the Kevlar fibers are poly (*p*-phenylene terephthalamide),[2,3] which is a poly-

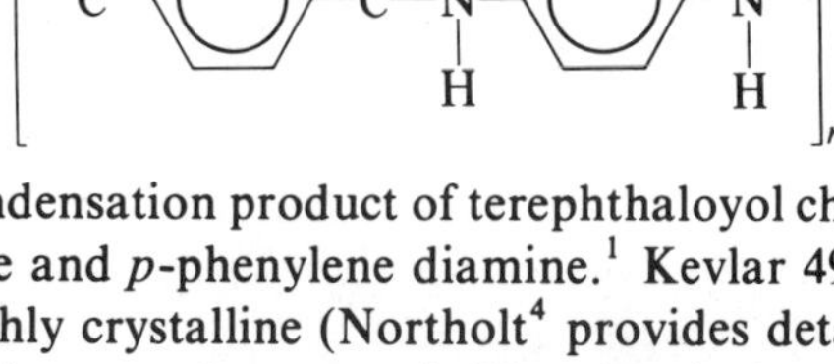

$$\left[-\overset{\displaystyle O}{\overset{\|}{C}}-\bigcirc-\overset{\displaystyle O}{\overset{\|}{C}}-\underset{\displaystyle H}{\underset{|}{N}}-\bigcirc-\underset{\displaystyle H}{\underset{|}{N}}- \right]_n$$

condensation product of terephthaloyol chloride and *p*-phenylene diamine.[1] Kevlar 49 is highly crystalline (Northolt[4] provides details on its crystal structure). The chemical struc-

*This work was performed under the auspices of the U. S. Department of Energy by the Lawrence Livermore Laboratory under contract No. W-7405-Eng-48.

**C. C. Chiao is presently with the Dow Chemical Company in Walnut Creek, California.

†Reference to a company or product name does not imply approval or recommendation of the product by the University of California or by the U. S. Department of Energy to the exclusion of other products that may be suitable.

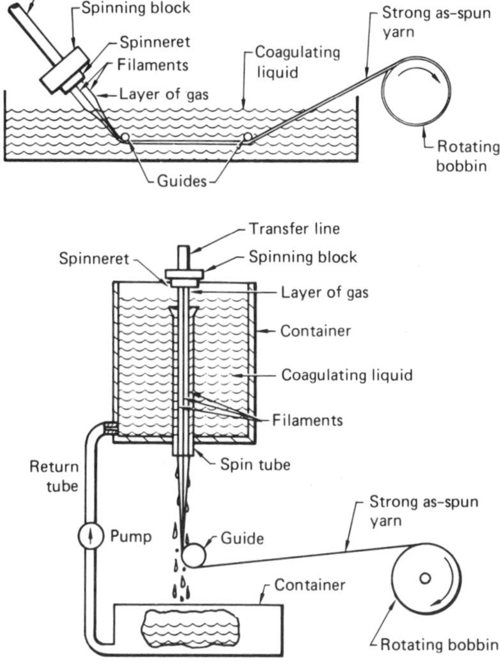

Figure 12.1. Schematic view of two apparatuses for dry-jet wet spinning.

ture of the Kevlar fibers is illustrated in Fig. 12.2.

The rigid linear molecular chains are highly oriented in the fiber-axis direction. Polymer chains are held together in the transverse direction by hydrogen bonding. The difference in bonding (strong covalent bonds in the fiber direction versus weak hydrogen bonds in the transverse direction) is responsible for the anisotropy in the mechanical properties of the Kevlar fibers; the fibers exhibit high longitudinal strength and low transverse strength.

The aromatic ring structure of the fiber induces high rigidity and forces extension of the polymer chains (instead of folding), which

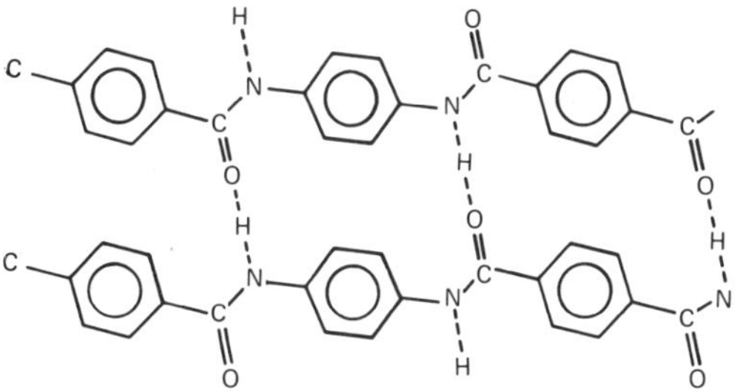

Figure 12.2. Chemical structure of Kevlar fibers.

results in a rod-like structure. This, in turn, gives the fiber a high modulus. The linearity of the polymer chain yields a high packing efficiency, enabling many polymers to be packed into a unit volume. The resulting higher density produces a fiber with higher strength.

The aromatic ring structure also contributes to the chemical stability of the fiber through the delocalization (resonance) of the electrons in the rings. In addition, the high-temperature dimensional stability of the Kevlars results from the crystalline nature of the polymer, which is due to the chain rigidity afforded by the aromatic ring structure. Thus, the fiber exhibits no drastic enthalpic changes until it decomposes at a high temperature. This enables the Kevlar fiber to resist thermoplastic behavior at relatively high temperatures and yet to be less brittle than the cross-linked polymers.

12.2.3. Fiber Morphology

Several microscopic studies on the Kevlar 49 fiber have been conducted.[5–7] Both the bare fiber and fiber composites have been examined with scanning electron microscope (SEM) and transmission electron microscope (TEM) techniques. Here, we describe the morphology of the Kevlar 49 fiber when it is subjected to various stress histories (e.g., cutting, twisting, tensile rupture, compression, and bending). More details on this subject can be found in reports by Chiao *et al.*[5] and Abbott *et al.*[6]

A SEM photomicrograph of a typical Kevlar 49 single filament is shown in Fig. 12.3. The surface of the filament appears to contain many globular particles and shallow grooves in the longitudinal direction. These grooves probably are caused by strips of material peeling off at the fiber surface, suggesting that the fiber possesses a microfibrillar structure.

In Fig. 12.4, we see that the helical groove on the surface of a twisted fiber (70 turns/cm) is similar to the longitudinal grooves on the straight fiber in Fig. 12.3; granular material also can be seen within the twisted groove. After the twisted fiber is unwound, we expect the grooves to return to the longitudinal direction. The microstructure of the Kevlar 49 filament is further revealed by the TEM photo-

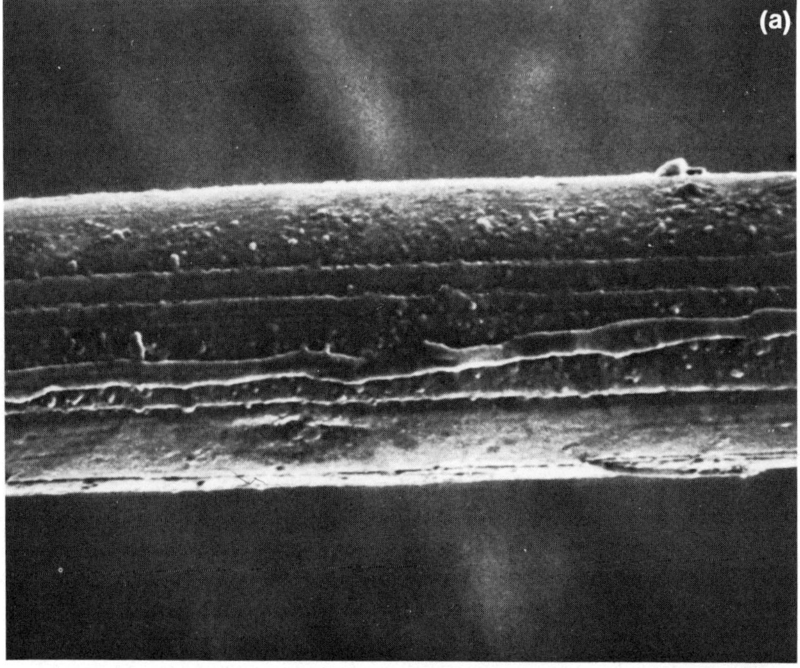

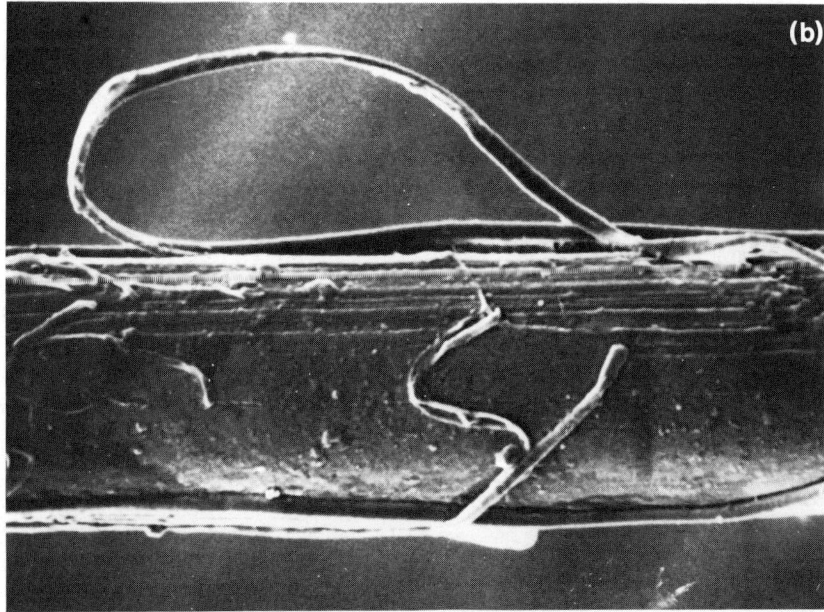

Figure 12.3. Longitudinal view of virgin Kevlar fibers (3000*X*).[6] (*Courtesy Fabric Research Laboratory.*)

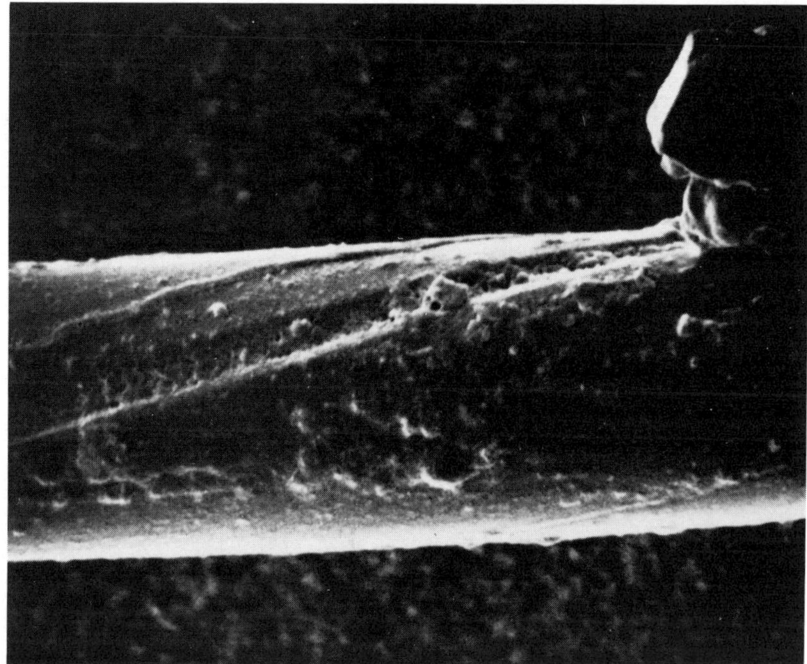

Figure 12.4. Kevlar fiber twisted 70 turns/cm (3000×).[6] (*Courtesy Fabric Research Laboratory.*)

micrographs of Fig. 12.5. Obviously, there are defects not only on the surface but also within the fiber filaments. In Fig. 12.6, a SEM photomicrograph of a fiber that has been cut off with a razor blade reveals that the cut fiber end is smeared.

The tensile fracture of the Kevlar 49 fiber is characterized by longitudinal fragmentation and splintering into subfilaments. Ends of all subfilaments tend to be pointed (see Fig. 12.7). A laterally compressed fiber remains essentially intact, but some longitudinal delamination occurs in the compressed region. (Fig. 12.8).

When a single-fiber filament of Kevlar 49 is subjected to bending, as when it is tied in a knot, the fiber buckles severely on the inside of the bend (compression). The fiber does not fracture, even at about 50% of bending strain. However, the fiber surface splinters severely when the knot is tightened (see Fig. 12.9).

12.2.4. Commercially Available Fibers and Fabrics[8]

The only known commercially available aramid fibers are Kevlar 29 and Kevlar 49 from DuPont. Both fibers are made from the same base polymer, poly (*p*-phenylene terephthalamide).

12.2.4.1. Kevlar 29

Kevlar 29 is available from DuPont as a continuous filament yarn in a range of deniers and varieties to suit many specific applications. This fiber also is available without twist and without finish. Tables 12.1 and 12.2 list the available forms of yarns for weaving, cables, and cordage.

12.2.4.2. Kevlar 49

Kevlar 49 is available from DuPont as yarn, roving, and fabric (see Tables 12.3–12.5).

12.2.5. Fiber Properties

Aramid fibers possess unique properties. Tensile strength and modulus are substantially higher and fiber elongation is significantly lower for aramid fibers than for other organic fibers. Aramid fibers are inherently resistant to flame and high temperature, as well as to organic solvents, fuels, and lubricants. They are not brittle like the glass or graphite fibers

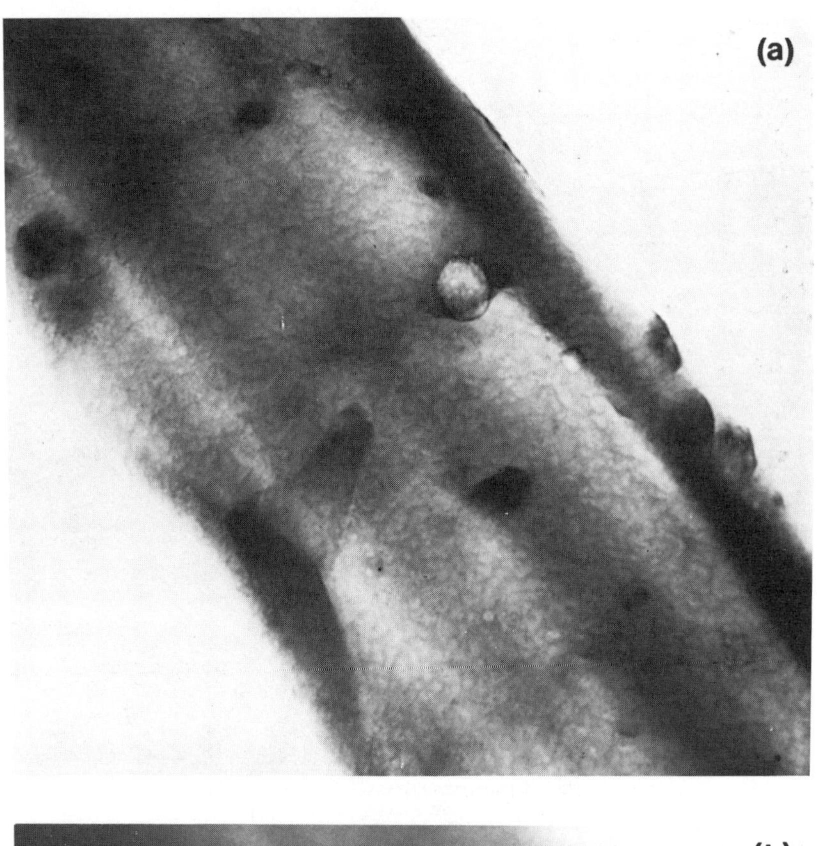

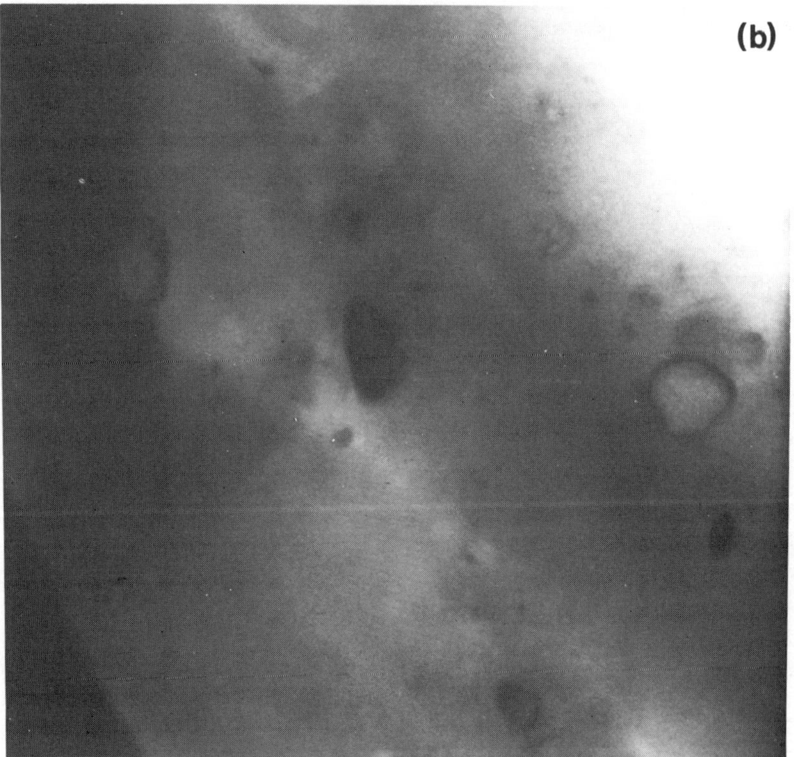

Figure 12.5. Transmission electron micrographs of bare single Kevlar filaments.[5] (*Courtesy DiGiallonardo and Hanafee, Lawrence Livermore Laboratory.*)

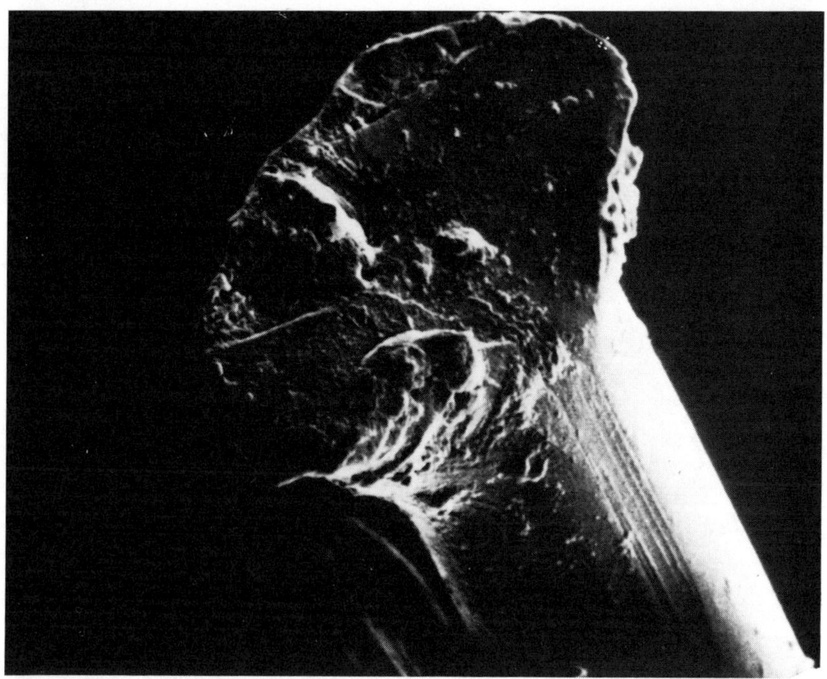

Figure 12.6. Kevlar fiber cut crosswise with a razor blade (3000×).[6] (*Courtesy Fabric Research Laboratory.*)

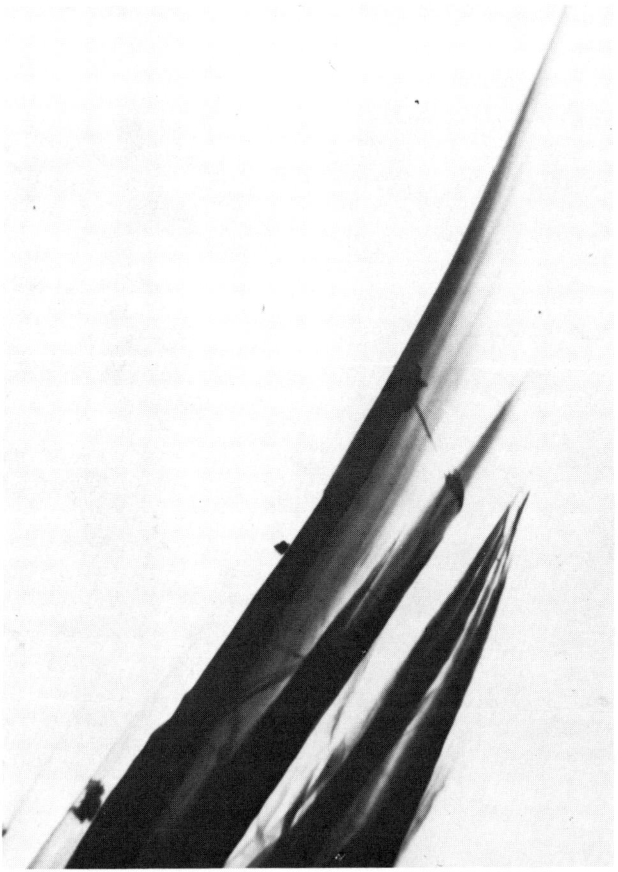

Figure 12.7. Transmission electron micrograph of fractured ends of Kevlar 49 single filaments broken in tension.[5] (*Courtesy DiGiallonardo and Hanafee, Lawrence Livermore Laboratory.*)

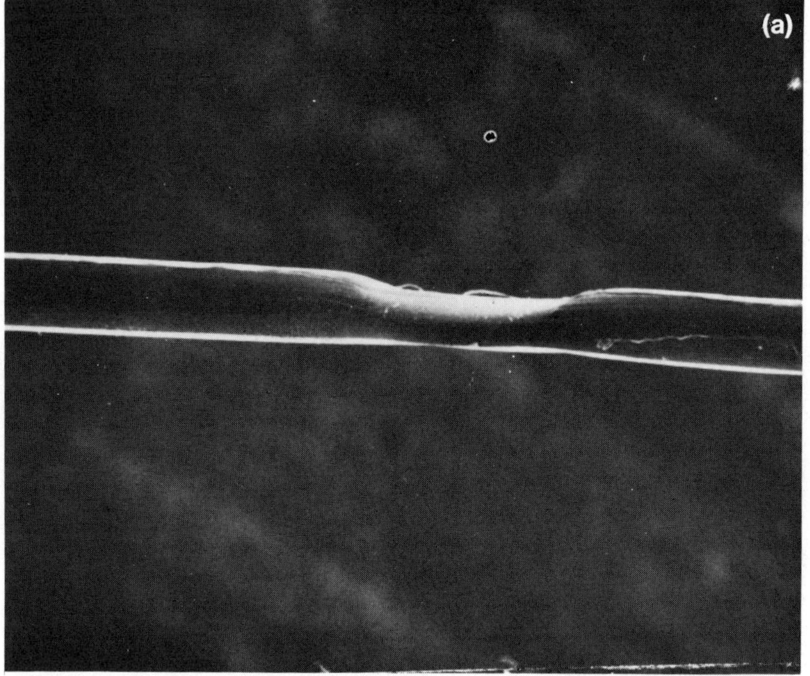

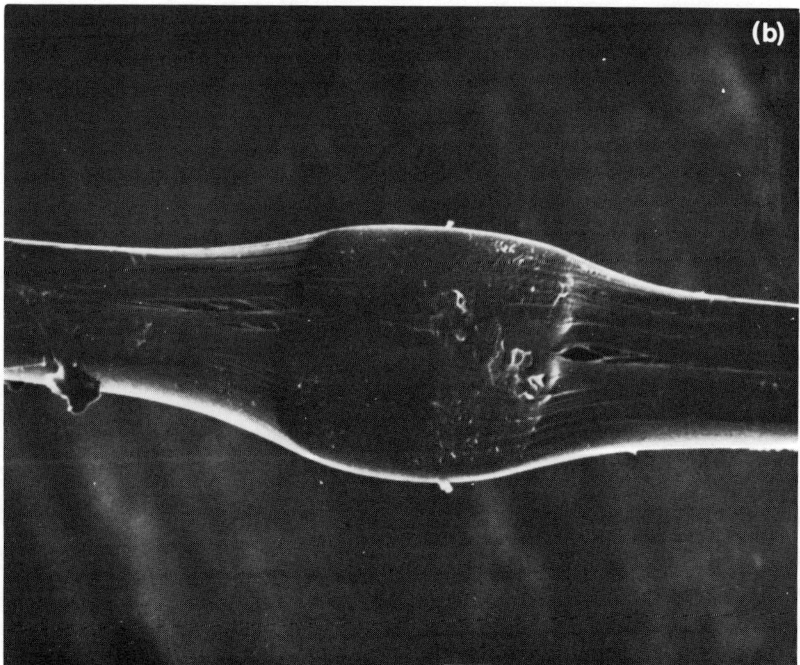

Figure 12.8. Laterally compressed Kelvar fiber (1000×). *a*) Top view. *b*) Side view.[6] (*Courtesy Fabric Research Laboratory.*)

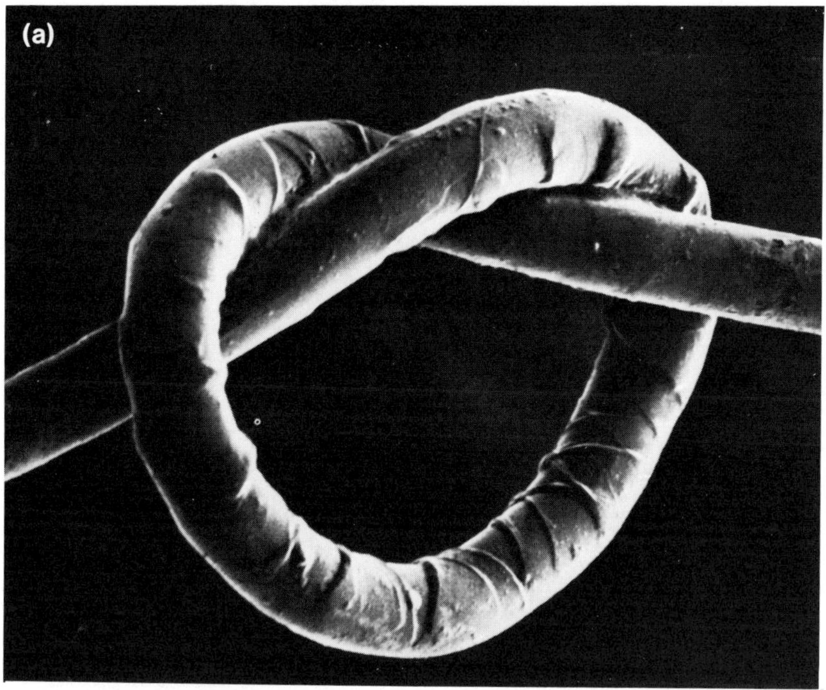

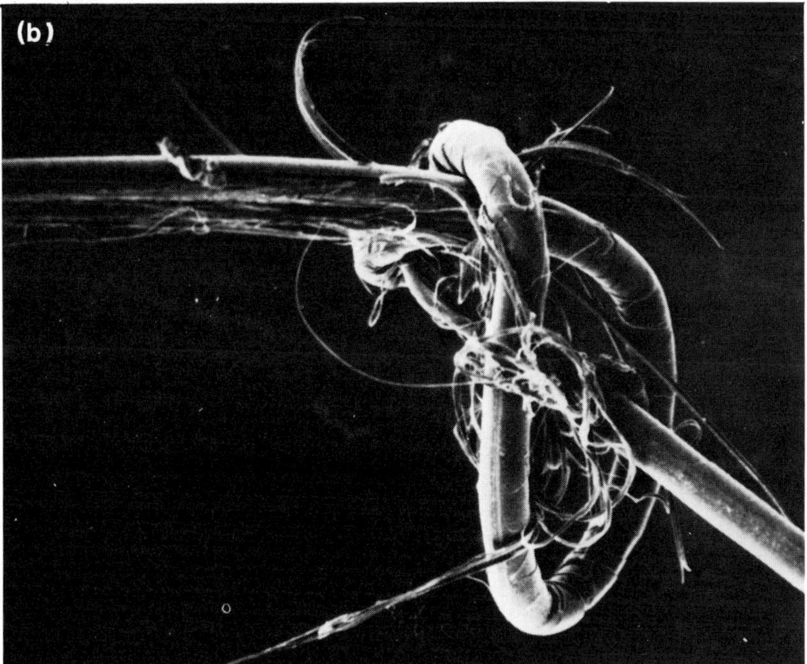

Figure 12.9. Bending strain imposed by overhand knots in single Kevlar fibers.[6] (*Courtesy Fabric Research Laboratory.*)

Table 12.1. Kevlar 29 Yarn for Weaving (Type 964)[8]

DENIER	NO. OF FILAMENTS	TWIST	FINISH
200	134	R80	Standard
400	267	R80	Standard
1000	666	R80	Standard
1500	1000	R80	Standard

Table 12.2 Kevlar 29 Yarn for Cables and Cordage (Type 960)[8]

DENIER	NO. OF FILAMENTS	TWIST	FINISH
1000	666	R80	Standard
1500	1000	R80	None Standard Cordage
15,000	10,000	0	Standard Cordage

Table 12.3. Roving of Kevlar 49[8]

DENIER	GRADE	TWIST	FINISH
4560	Aerospace	0	None
7100	Commercial	0	None

Table 12.4. Yarn of Kevlar 49[8]

DENIER	FILAMENTS	TWIST	FINISH
195	134	0	None
380	267	0	None
1420	1000	0	None

and can be woven readily on conventional fabric looms.

The commercially available aramid fibers, Kevlar 29 and Kevlar 49, are lightweight. The filaments are round in cross-section and the 1.5-denier fibers are 11.9 μm (0.00047 in.) in diameter.[8] Kevlar 49 has refractive indices of 2.0 and 1.6 when measured, respectively, parallel and perpendicular to the fiber axis. Electron radiation has no effect on the Kevlar 49 fiber. In addition, the thermal, chemical, and mechanical properties of the Kevlar fibers have been studied extensively[6, 8, 9] (see Table 12.6). The tensile properties of the Kevlar 49 filaments vary slightly, depending on the length of the test specimen and on the fiber bundle size from which the filament specimens were taken. Table 12.7 summarizes typical data from DuPont on this fiber.

12.2.5.1. Thermal Properties[8]

The thermal properties of Kevlar 49 fiber are presented in Table 12.8. Figure 12.10 shows the tensile properties of Kevlar 49 as a function of temperature in both a nitrogen and air atmosphere. Figure 12.11 shows the thermal conductivity of this fiber as a function of temperature. It is obvious that the tensile strength decreases and the thermal conductivity increases with temperature.

12.2.5.2. Chemical Properties[8, 10]

Kevlar 49 fiber is chemically quite stable; its resistance to neutral chemicals is generally

Table 12.5. Fabric of Kevlar 49[8]

STYLE	CONSTRUCTION, YARN DENIER	WEAVE	WEIGHT, g/m² (oz/yd²)	THICKNESS, μm (mill)	TENSILE STRENGTH, WARP/FILL, N/cm (lb/in.)
120	34 (195) × 34 (195)	Plain	61 (1.8)	114 (4.5)	438/438 (250/250)
143	100 (380) × 20 (195)	Crowfoot	190 (5.6)	254 (10)	2277/219 (1300/125)
181	50 (380) × 50 (380)	8-HS[a]	170 (5.0)	228 (9)	1226/1226 (700/700)
243	38 (1140) × 18 (380)	Crowfoot	227 (6.7)	330 (13)	2627/525 (1500/300)
281	17 (1140) × 17 (1140)	Plain	170 (5.0)	254 (10)	1138/1138 (650/650)
285	17 (1420) × 17 (1420)	Crowfoot	170 (5.0)	254 (10)	1138/1138 (650/650)
328	17 (1420) × 17 (1420)	Plain	231 (6.8)	330 (13)	1226/1226 (700/700)

[a] Harness satin.

Table 12.6. Properties of Kevlar Fibers[6,8,9]

PROPERTY	KEVLAR 29		KEVLAR 49	
	DATA FROM REF. 6	DATA FROM REF. 9	DATA FROM REF. 6	DATA FROM REFS. 8,9
Specific gravity	1.44	1.44	1.44	1.45
1.5-denier fiber diameter, μm	12	11.9	12	11.9
Moisture absorption, %	3.9[a]	6.0[b]	4.6[a]	3.5[b]
Rupture tenacity, MPa (gpd)	3275 (26)	2758 (21)	3034 (24)	2758 (21)
Rupture elongation, %	3.9	4.0	2.3	2.4
Initial modulus, GPa (gpd)	69.0 (550)	62.1 (500)	124.1 (990)	131.0 (1000)
Maximum modulus, GPa (gpd)	96.5 (750)	—	137.9 (1090)	—
Bending modulus, GPa (gpd)	53.1 (410)	—	105.5 (820)	—
Calculated axial compression modulus, GPa (gpd)	40.7 (320)	—	75.8 (600)	—
Dynamic modulus, GPa (gpd)	96.5 (740)	—	137.9 (1050)	—
Knot strength, % tensile strength	—	—	—	35
Flexural fatigue resistance, cycles	—	—	—	200[c]
Creep at 90% of ultimate tensile strength, cm/cm	—	—	—	0.0011 (0)[d]
Coefficient of friction	—	—	—	0.46 (0.41)[e]

[a]At 21°C (69.8°F) and 65% relative humidity.
[b]At 22°C (71.6°F) and 55% relative humidity.
[c]At 368 MPa over a 76.2-μm-diameter pin.
[d]Initial (and secondary) creep.
[e]Yarn/yarn (yarn/metal).

very high. However, the fiber is susceptible to attack by acids and bases, especially by strong acids. Table 12.9 summarizes the chemical resistance of the fiber. The ultraviolet stability of Kevlar 49 is given in Table 12.10.

12.2.5.3. Mechanical Properties

The mechanical properties of the Kevlar fibers differ from those of other organic fibers. Kevlar fibers have very high tensile strength and initial modulus, as well as low elongation. Also, their stress-strain curves are essentially linear. The tensile properties of Kevlar 49 are unique, even among the inorganic reinforcing fibers (see Fig. 12.12). Abbott *et al.*[6] have done considerable work on the mechanical properties of the Kevlar fibers, and the important points of their studies are summarized below.

Tensile Properties. The stress-strain responses of dry Kevlar 29 and Kevlar 49 yarns at various temperatures and strain rates are shown in Figs. 12.13 and 12.14. By increasing

Table 12.7. Tensile Properties of Virgin Kevlar 49 Filaments[8]

FILAMENT LENGTH, mm	STRAND, BUNDLE DENIER	NO. OF SPECIMENS	BREAKING STRESS, Mpa (CV, %)[a]	BREAKING STRAIN, % (CV, %)[a]
25.4	4560	40	3420 (13.9)	4.13 (15.5)
254	4560	39	2740 (19.0)	2.30 (19.3)
25.4	380	20	3500 (12.1)	4.00 (12.5)
254	380	20	2940 (10.8)	2.31 (8.9)

[a]Coefficient of variation.

Table 12.8. Thermal Properties of Kevlar 49 Yarn and Roving[8]

PROPERTY	VALUE
Long-term use at elevated temperature in air, °C (°F)	160 (320)
Decomposition temperature, °C (°F)	500 (932)
Tensile strength, MPa (ksi)	
At room temperature for 16 months	No strength loss
At 50°C (122°F) in air for 2 months	No strength loss
At 100°C (212°F) in air	3170 (460)
At 200°C (392°F) in air	2720 (395)
Tensile modulus, GPa (10^3 ksi)	
At room temperature for 16 months	No modulus loss
At 50°C (122°F) in air for 2 months	No modulus loss
At 100°C (212°F) in air	113.8 (16.5)
At 200°C (392°F) in air	110.3 (16.0)
Shrinkage, %/°C (%/°F)	4×10^{-4} (2.2×10^{-4})
Thermal coefficient of expansion, 10^{-6} cm/cm·°C	
Longitudinal, 0–100°C (32–212°F)	−2
Radial, 0–100°C (32–212°F)	+59
Specific heat at room temperature, J/g·°C (Btu/lb·°F)	1.42 (0.34)
Thermal conductivity at room temperature,[a] J·cm/sec·m²·°C (Btu·in./hr·ft²·°F)	
Heat flow perpendicular to fibers	4.110 (0.285)
Heat flow parallel to fibers	4.816 (0.334)
Heat of combustion, kJ/g (Btu/lb)	34.8 (15,000)

[a] Here, $\rho = 0.19$ g/cm³.

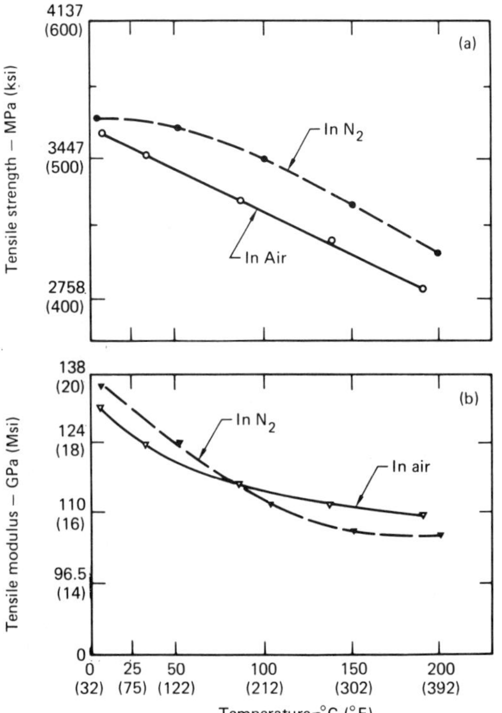

Figure 12.10. Kevlar 49 yarn strength (*a*) and modulus (*b*) versus temperature (no twist or finish on the yarn). Dashed curves give performance in a nitrogen atmosphere (30-minute soak time), and solid curves give performance in an air atmosphere (30-minute soak time).[8]

the strain rate from 0.167 to 8000% per second, the tensile strength of the dry Kevlar fibers drops an average of 14%; the rupture elongation remains virtually unchanged. The effect of temperature on the tensile properties of the dry yarn fibers is not great.

The tensile properties of wet Kevlar 29 and Kevlar 49 yarns (submerged in water for 5 minutes and then tested in water) at 21°C (69.8°F) showed little to no change. No strength loss was seen for Kevlar 29 and only

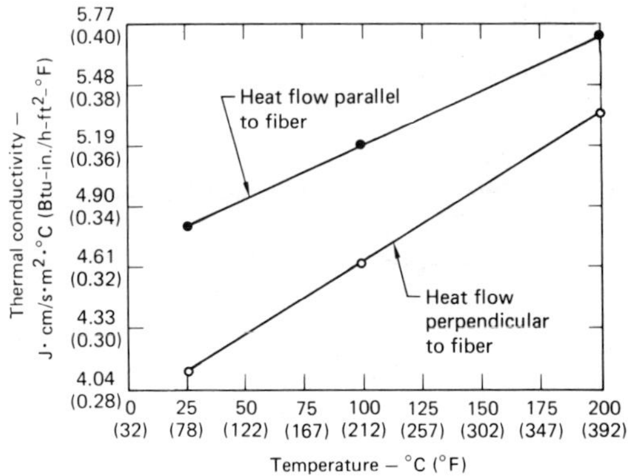

Figure 12.11. Thermal conductivity of Kevlar 49 yarns versus temperature. Open circles denote heat flow perpendicular to the fiber length; solid circles denote heat flow parallel to the fiber length.[8]

Table 12.9. Stability of Kevlar 29 and Kevlar 49 Fibers in Various Chemicals[8,10]

CHEMICAL	CONCENTRATION, %	TEMPERATURE, °C (°F)	TIME, hours	STRENGTH LOSS, %	
				KEVLAR 29	KEVLAR 49
Acetic acid	99.7	21 (69.8)	24	—	0
Benzoic acid	3	99 (210.2)	100	—	26
Formic acid	90	21 (69.8)	100	—	7
Hydrochloric acid	37	21 (69.8)	24	—	0
	37	21 (69.8)	1000	83	—
Hydrofluoric acid	5	21 (69.8)	24	—	0
	10	21 (69.8)	100	12	8
	48	21 (69.8)	24	—	10
Hydrobromic acid	10	21 (69.8)	1000	—	60
Nitric acid	1	21 (69.8)	100	18	5
	70	21 (69.8)	24	—	60
Phosphoric acid	10	21 (69.8)	100	—	1
Salicylic acid	3	99 (210.2)	1000	—	0
Sulfuric acid	1	21 (69.8)	1000	—	5
	10	21 (69.8)	100	14	—
	10	21 (69.8)	1000	—	31
	70	21 (69.8)	1000	—	59
	96	21 (69.8)	24	—	100
Sodium hydroxide	50	21 (69.8)	24	—	10
Ammonium hydroxide	28	21 (69.8)	1000	10	—
Acetone	100	21 (69.8)	24	0	0
Benzene	100	21 (69.8)	24	—	0
Carbon tetrachloride	100	21 (69.8)	24	—	0
Dimethyl formamide	100	21 (69.8)	24	—	0
Methylene chloride	100	21 (69.8)	24	—	0
Methylethyl ketone	100	21 (69.8)	24	—	0
Trichloroethylene	100	21 (69.8)	24	—	1.5
	100	88 (190.4)	387	7	—
Toluene	100	21 (69.8)	24	—	0
Benzyl alcohol	100	21 (69.8)	24	—	0
Ethyl alcohol	100	21 (69.8)	24	0	0
Methyl alcohol	100	21 (69.8)	24	—	<1
Jet fuel (JP-4)	100	21 (69.8)	300	0	4.5
	100	199 (390.2)	100	4	—
Brake fluid	100	21 (69.8)	312	2	—
	100	163 (325.4)	100	33	—
Transformer oil (Texaco No. 55)	100	60 (140)	500	4.6	0
Kerosene	100	60 (140)	500	9.9	0
Formalin	100	21 (69.8)	24	—	1.5
"Freon" 11	100	60 (140)	500	0	2.7
"Freon" 22	100	60 (140)	500	0	3.6
Tap water	100	100 (212)	100	0	2
Seawater (Ocean City, New Jersey)	100	21 (69.8)	3 mo	—	0
	100	21 (69.8)	6 mo	—	2.4
	100	—	1 y	1.5	1.5
Water at 10 ksi	100	21 (69.8)	720	0	—
Superheated water	100	138 (280.4)	40	9.3	—
Saturated steam	100	149 (300.2)	48	28	—

Table 12.10. Ultraviolet Stability of Kevlar 29 and Kevlar 49[8,10]

MATERIAL	SIZE	TWIST, TURNS/cm	TYPE OF EXPOSURE	LENGTH OF EXPOSURE	BREAK LOAD, kg		STRENGTH LOSS, %
					BEFORE	AFTER	
Kevlar 29	1500-denier yarn (no finish)	0.83	Weatherometer[a]	200 hours dry	33.1	24.3	27
Kevlar 49	1420-denier yarn (no finish)	0.71	Weatherometer	200 hours wet-dry	32.4	24.3	25
				500 hours wet-dry	32.4	20.4	37
Kevlar 49	7100-denier yarn (no finish)	0.35	Weatherometer	200 hours wet-dry	150	117	22
				500 hours wet-dry	150	100	34
Kevlar 49	7100-denier, urethane-impregnated	0	Weatherometer	200 hours wet-dry	156	103	34
				500 hours wet-dry	156	84	46
Kevlar 49	7100-denier, urethane-impregnated	0.35	Weatherometer	200 hours wet-dry	161	129	20
				500 hours wet-dry	161	107	34
Kevlar 29 + 25% dacron	12.7-mm-diameter three-strand rope	—	Florida exposure[b]	200 hours dry	5171	4808	7
				6 months dry	5171	4717	9
				12 months dry	5171	4309	17
				24 months dry	5171	4309	17
Kevlar 29	12.7-mm-diameter three-strand rope	—	Florida exposure	6 months dry	6532	5897	10
				12 months dry	6532	5262	19
				8 months dry	6532	4513	31
				24 months dry	6532	4509	31
Kevlar 49	700-denier twisted cord (530 filaments)	—	Control	—	—	17.2	—
			Fadeometer[c]	100 hours dry	—	10.1	41
				200 hours dry	—	9.1	47
			Weatherometer	100 hours dry	—	9.7	44
				100 hours wet	—	10.1	41
				200 hours dry	—	8.2	52
				200 hours wet	—	8.3	51
Kevlar 49	3.2-mm-diameter cable[d]	—	Control	—	—	600	—
			Weatherometer	100 hours dry	—	467	22
Kevlar 49	12.7-mm-diameter three-strand rope[d]	—	Control	—	—	5171	—
			Weatherometer	200 hours dry	—	4808	7
			Florida exposure	6 months dry	—	4654	10
				12 months dry	—	4191	19

[a]Weatherometer exposure, sunshine carbon arc lamp.
[b]Florida Exposure, Hialeah.
[c]Fadeometer exposure, xenon lamp.
[d]Data indicate self-screening influence of outer layers of Kevlar 49.

some loss of strength was noted for Kevlar 49. The effect of a test temperature of 88°C (190.4°F) in water is approximately twice as severe as 93°C (199.4°F) in air (see Table 12.11).

Figure 12.15 presents the stress-strain behavior of Kevlar 29 and Kevlar 49 yarns with various levels of twist. The effect of twist on the tensile properties is greater for Kevlar fibers than for conventional textile fibers. Abbott *et al.*[6] found that the high tensile strength loss in twisted Kevlar yarns could not be explained by the effect of yarn geometry alone. They believe this behavior might be due to the combined effects of yarn geometry and fiber damage occurring in the twisted yarns.

At a low strain rate of 10% per minute, Kevlar 29 fiber can withstand many cycles of tensile strain at high strain levels, if the attached surface is relatively friction-free.[6] Figure 12.16 shows the cycle lifetime of 400-denier Kevlar 29 yarns at 21 and 204°C (69.8

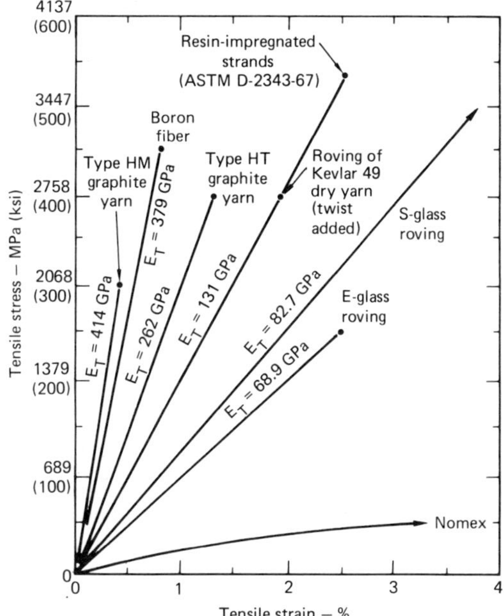

Figure 12.12. Stress-strain behavior of various reinforcing fibers (ASTM D-2343-67 resin-impregnated strand test).[8]

and 399.2° F). The fiber was attached to a machined steel surface and the resulting surface friction led to the low number of cycles to rupture.

Bending Properties. After being bent for a short time at 21° C (69.8° F) and then released, Kevlar fiber exhibits good recovery. However, the recovery decreases when the specimens are bent for longer times and/or at higher temperatures (see Figs. 12.17–12.20). Therefore, these Kevlar fibers may be heat-set. The results of a heat-setting study conducted by Abbott *et al.*[6] are given in Fig. 12.21.

Torsional-Shear Properties. Abbott *et al.*[6] studied the torsional stress-strain response of Kevlar fibers. The torsional-shear properties

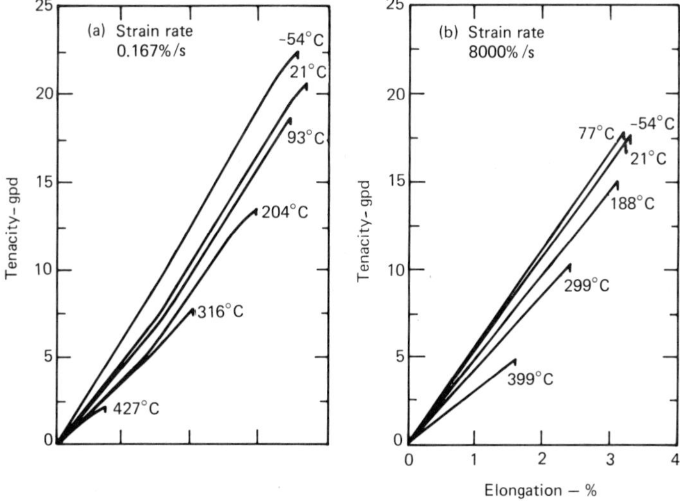

Figure 12.13. Stress-strain curves for Kevlar 29 yarn (about 0.8 turns/cm of twist), tensile-tested at various temperatures at a strain rate of *a*) 0.167% per second and *b*) 8000% per second (based on at-temperature yarn dimensions).[6]

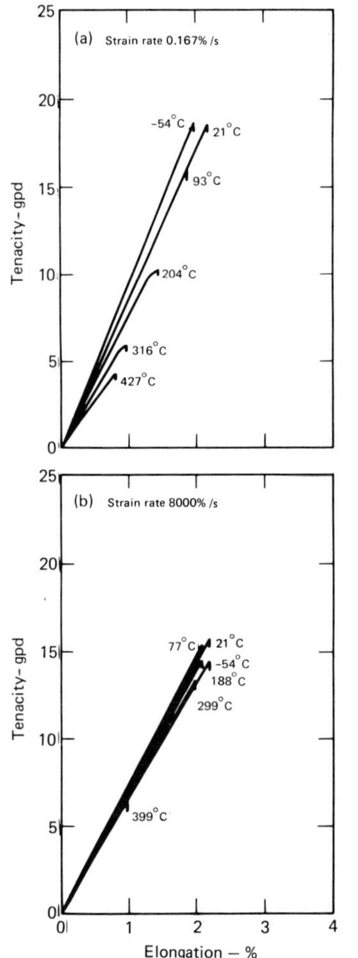

Figure 12.14. Stress-strain curves for Kevlar 49 yarn (about 0.8 turns/cm of twist), tensile-tested at various temperatures at a strain rate of *a*) 0.167% per second and *b*) 8000% per second (based on at-temperature yarn dimensions).[6]

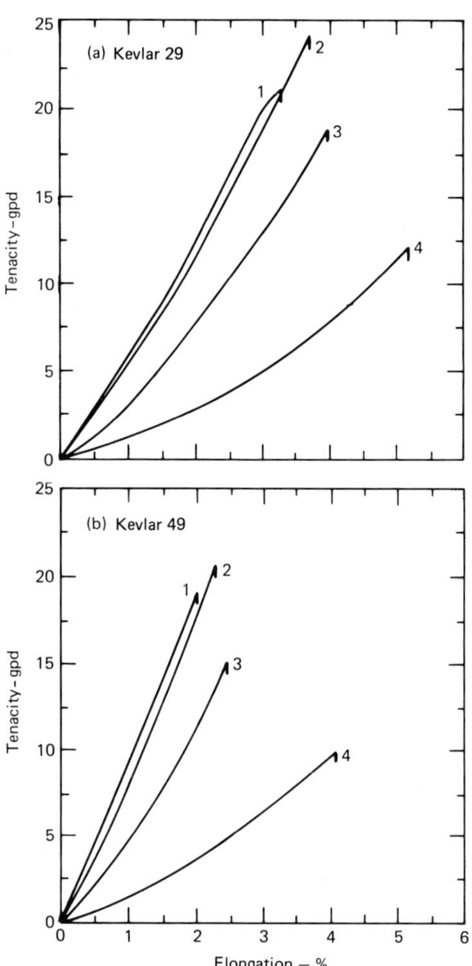

Figure 12.15. Typical stress-strain curves of *a*) Kevlar 29 and *b*) Kevlar 49 twisted yarns. Curves 1 through 4 have twists of 0, 1.4, 3.2, and 5.5 turns/cm (*a*) and twists of 0, 2.6, 5.4, and 8.8 turns/cm (*b*).[6]

Table 12.11. Tensile Properties of 1500-Denier Kevlar Yarns in Water at 21 and 88°C (69.8 and 190.4°F)[a,b]

YARN	TEST CONDITIONS	INITIAL MODULUS, gpd	CHANGE IN MODULUS, %	RUPTURE ELONGATION, %	RUPTURE TENACITY, gpd	STRENGTH RETENTION, %
Kevlar 29	Air at 21°C (69.8°F)	485	—	3.6	19.8	—
	Water at 21°C (69.8°F)	488	0	3.6	20.1	100
	Air at 93°C (199.4°F)	432	−11	3.4	18.3	92
	Water at 88°C (190.4°F)	378	−22	3.4	17.0	86
Kevlar 49	Air at 21°C (69.8°F)	904	—	2.2	20.3	—
	Water at 21°C (69.8°F)	892	−1	2.1	19.3	95
	Air at 93°C (199.4°F)	811	−10	2.0	17.6	87
	Water at 88°C (190.4°F)	726	−20	1.9	15.8	78

[a]Each value given is the average of five specimens per test.

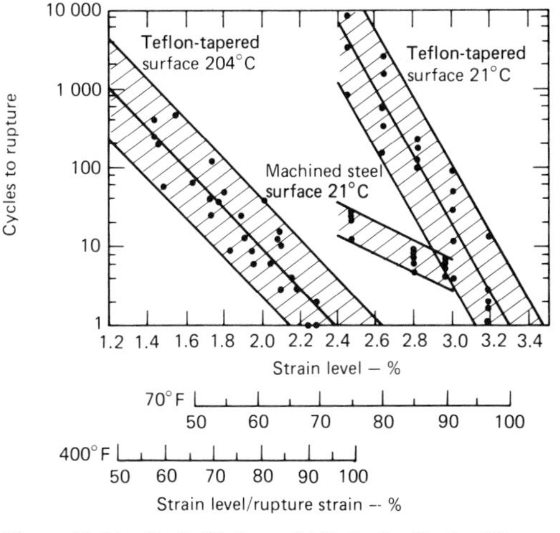

Figure 12.16. Cycle lifetime of 400-denier Kevlar 29 yarn (2 turns/cm) at 21 and 204°C (69.8 and 399.2°F).[6]

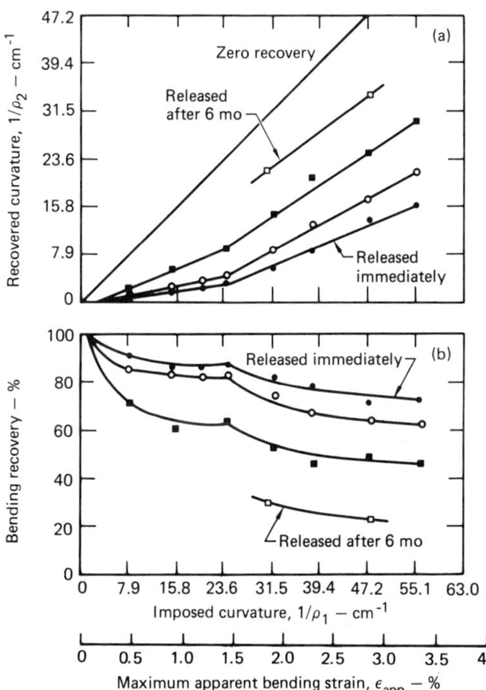

Figure 12.17. Recovered curvature (*a*) and bending recovery (*b*) of Kevlar 29 filaments as a function of imposed curvature at 21°C (69.8°F) after various times before release. Solid circles denote immediate release; open cirlces denote release after 5 minutes; solid squares denote release after 16 hours; and open squares denote release after 6 months.[6]

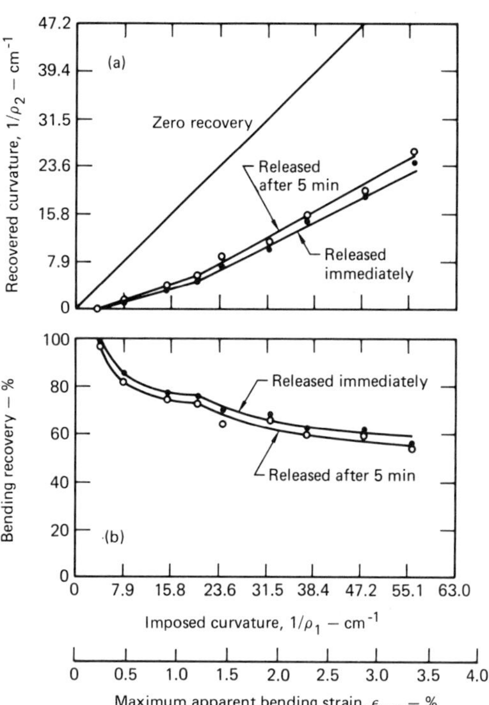

Figure 12.18. Recovered curvature (*a*) and bending recovery (*b*) of Kevlar 49 filaments as a function of imposed curvature at 21°C (69.8°F) after various times before release. Solid circles denote immediate release; open circles denote release after 5 minutes.[6]

of these fibers are presented in Table 12.12 and Fig. 12.22.

12.2.6. Fabric Properties

The tensile properties of Kevlar 49 fabrics are summarized in Table 12.5. The tensile strengths vary between 438 and 2627 N/cm (250 and 1500 lb/in.), depending on the fabric style.

12.2.7. Specifications

Two ASTM (American Society for Testing and Materials) specifications for aramid fibers are available: ASTM D-3317-74a for high-modulus organic yarn and roving and ASTM D-3318-74 for cloth woven from high-modulus organic fiber. There also are several aerospace material specifications (Society of Automotive Engineering, Inc.): AMS 3901 (July 1973) for high-modulus organic fiber, yarn, and roving for structural composites; AMS

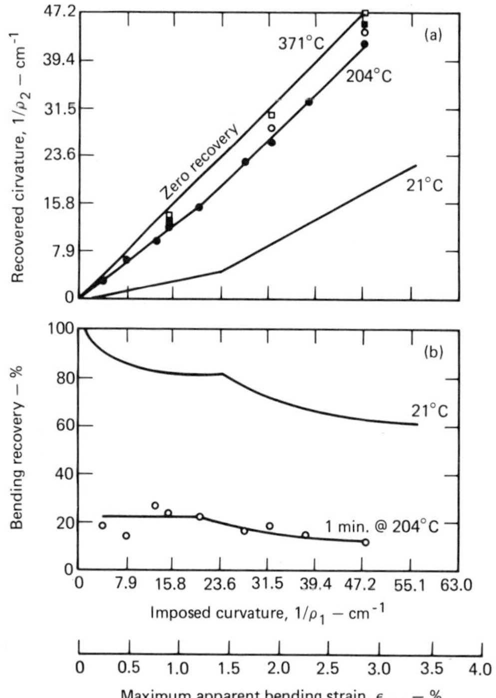

Figure 12.19. Recovered curvature (*a*) and bending recovery (*b*) of Kevlar 29 filaments after a 1-minute exposure to various temperatures as a function of imposed curvature with release after 5 minutes. In *a*, solid circles denote 204° C (399.2° F); open circles denote 260° C (500° F); solid squares denote 316° C (600.8° F); and open squares denote 371° C (699.8° F).[6]

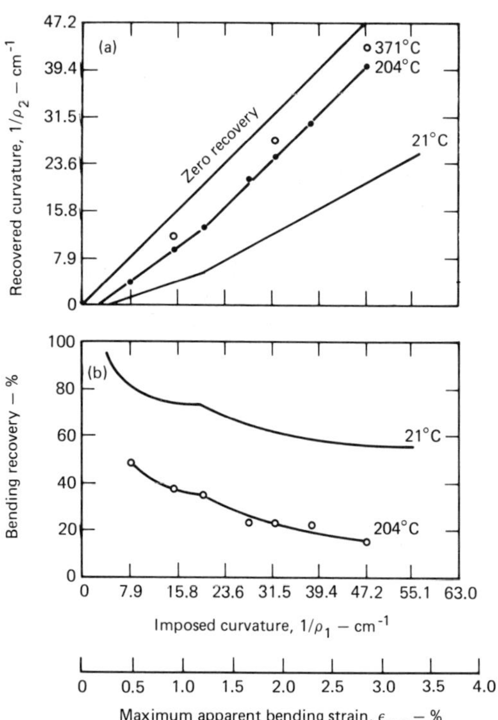

Figure 12.20. Recovered curvature (*a*) and bending recovery (*b*) of Kevlar 49 filaments after a 1-minute exposure to various temperatures as a function of imposed curvature with release after 5 minutes. In *a*, solid circles denote 204° C (399.2° F); open circles denote 371° C (699.8° F).[6]

3902 (July 1973) for high-modulus cloth and organic fiber for structural composites; and AMS 3903 (June 1975) for cloth and organic fiber impregnated with high-modulus epoxy resin. In addition, Lawrence Livermore Laboratory report UCID-17229 (August 1976) gives specifications for Kevlar 49 fiber and a sampling plan for determining fiber tensile strength.

12.3. ARAMID COMPOSITES

12.3.1. Epoxy-Impregnated Fiber Strands

Fiber strength data obtained from filament strands without matrix are of little value, in predicting fiber composite performance. The effect of the matrix on the fiber strength can be substantial. For this reason, the use of fiber/matrix composite strands has become ac-

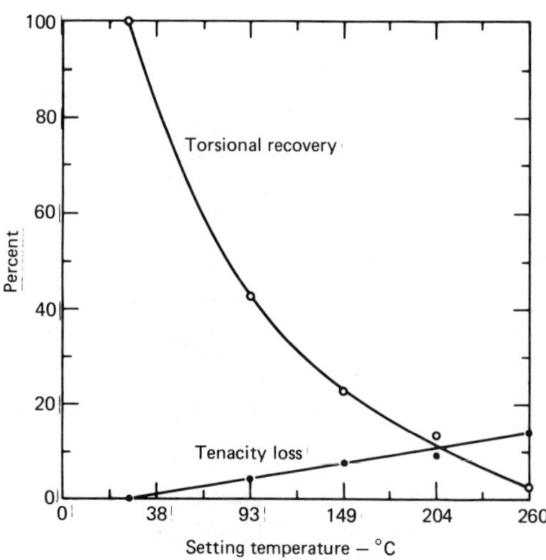

Figure 12.21. Torsional recovery and tenacity loss as a function of setting temperature for 100-denier Kevlar 29 yarn. Setting time was 1 minute and the initial twist level was 2.4 turns/cm.[6]

Table 12.12. Torsional Moduli of Kevlar Fibers[6]

PARENT YARN	NOMINAL FIBER DENIER	FIBER DIAMETER, μm	TORSIONAL SHEAR MODULUS, 10^{10} dyne/cm^2	(10^6 psi)
Kevlar 29 (1000 denier)	1.5	12	1.86	(0.27)
Kevlar 49 (400 denier)	1.5	12	1.62	(0.24)

cepted for many tests. The tensile properties of the matrix-impregnated strands, when fabricated at the optimum fiber content, can represent realistically the maximum achievable composite material properties. The epoxy-impregnated fiber strands are the basic building blocks of complex composites and can be made easily by wet winding[11] or from a prepreg roving. Figure 12.23 presents the stress-strain curves for the 380-denier Kevlar 49 fiber in an epoxy matrix (fiber content of 71.5 Volume %).[12] For a fiber bundle larger than 380 denier, the fiber strength in a composite strand is always somewhat lower (see

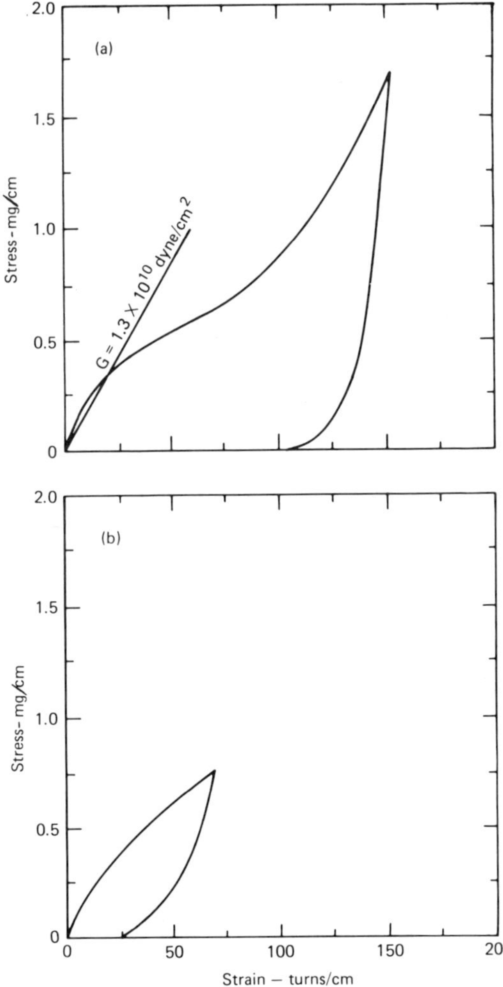

Figure 12.22. Typical torsional stress-strain curves for 1.5-denier Kevlar 29 and Kevlar 49 fibers *a* and *b*, respectively).[6]

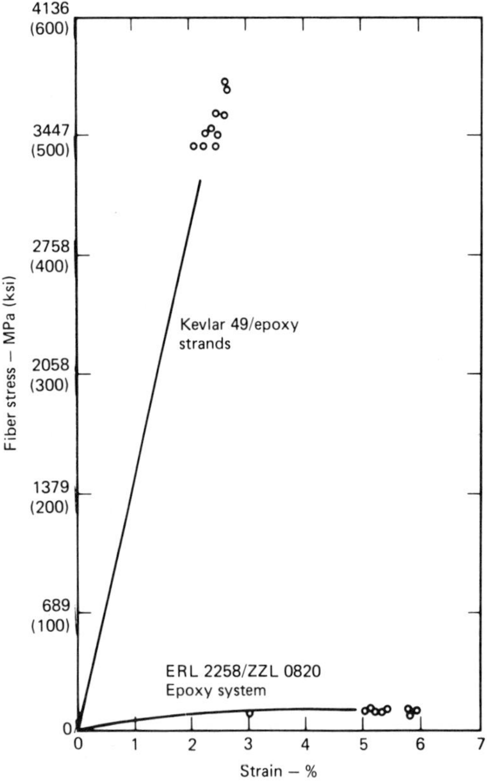

Figure 12.23. Typical stress-strain curves of Kevlar 49/epoxy strands and the epoxy resin system (Union Carbide ERL 2258/ZZL 0820), cured at 93° C (199.4° F) for 3 hours plus 163° C (325.4° F) for 2 hours. Rupture points are shown by open circles.[12]

Table 12.13). Fiber strength generally is a controversial issue because many factors (e.g., fiber bundle size, test techniques, fiber content, and matrix properties) can greatly influence the results.

12.3.1.1. Effect of Matrix Properties on Fiber Strength

Depending on its ability to transfer stresses, a matrix can greatly affect the fiber strength of a composite strand. Adhesion and modulus are the two most important properties that control the ability of the matrix to transfer stresses. The combined effect of these two properties on the composite strand under tension influences the mode of composite failure, which, in turn, results in high or low fiber strength. If the matrix modulus in a strand is very low, the filaments can act individually under tension, and the failure mode is one of progressive fiber breaking. Because there is no synergistic effect in this mode of failure, the average fiber strength is expected to be low. If, on the other hand, the matrix has adequate adhesion and modulus under tension, the filaments in a composite strand will behave as a single unit and the average fiber stress is expected to be high. In practice, however, the failure mode of a composite is usually in between these two extremes, yielding an intermediate average fiber failure stress. Published data on epoxy-impregnated Kevlar 49 strands reveal that when the epoxy properties do not vary greatly, their effect on the fiber strength is negligible (see Tables 12.14 and 12.15).[13]

Table 12.13. Fiber Strength of Simple Kevlar 49/Epoxy Composites (Union Carbide Epoxy System ERL 2258/ZZL 0820 in a 100/30 Weight Ratio)[12]

FIBER DENIER	SPECIMEN TYPE	FIBER CONTENT, VOLUME %	FIBER STRENGTH, MPa (ksi)	COEFFICIENT OF VARIATION, %	NO. OF SPECIMENS
380	Strand	71.5	3510 (510)	6.3	484
1420	Strand	69	3047 (442)	4.0	10
380	Elongated ring	64	2870 (420)	5.4	10
1420	Elongated ring	65	2690 (390)	10.8	9

Table 12.14. Summary of Epoxy Systems[13]

EPOXY SYSTEM NO.	COMPONENTS[a]	WEIGHT RATIO, g	GELLING, hours/°C (hours/°F)	CURING, hours/°C (hours/°F)
Control epoxy systems				
1	ERL 2256/Tonox 60-40	100/29.5	16/50 (16/122)	2/95 (2/203)
2	ERL 2258/ZZL 0820/ Tonox 60-40	100/31	16/60 (16/140)	3/160 (3/320)
3	EPON 826/RD 2/ Tonox 60-40	100/25/28.3	4/60 (4/140)	3/120 (3/248)
Moderate-pot-life epoxy systems				
4	XD 7818/ERL 4206/ Tonox 60-40	100/30/39.7	4/60 (4/140)	3/120 (3/248)
5	XD 7575.02/XD 7818/ ERL 4206/ Tonox 60-40	50/50/30/38.1	4.5/60 (4.5/140)	3/120 (3/248)
6	XD 7575.02/XD 7818/ XD 7114/ Tonox 60-40	50/50/45/33.7	4.5/60 (4.5/140)	3/120 (3/248)
Long-pot-life epoxy systems				
7	XD 7818/XD 7114/ Tonox LC	100/45/50.3	5/60 (5/140)	3/120 (3/248)
8	XD 7818/XD 7114/ Tonox 60-40/2,6 DAP	100/30/13.2/13.2	5/60 (5/140)	3/120 (3/248)
9	XD 7575.02/XD 7818/ XD 7114/2,6 DAP	50/50/45/14.1	51/80 (51/176)	3/120 (3/248)
10	ERE 1359/RD-2/2,6 DAP	100/12.5/23.9	12/80 (12/176)	2/100 + 2/125 + 4/150 (2/212 + 2/257 + 4/302)

[a]ERL and ZZL from Union Carbide, Tonox from UniRoyal, ERE and RD from Ciba-Geigy, EPON from Shell, and XD from Dow.

Table 12.15. Fiber Strength Data for Various Epoxy-Impregnated Kevlar 49 (380-Denier) Fiber Strands (95% Confidence Limits are Given in Parentheses)[13]

EPOXY SYSTEM NO.	NO. OF STRAND SPECIMENS[a]	FIBER CONTENT, VOLUME %	FIBER TENSILE PROPERTIES[b] RUPTURE STRESS, MPa	RUPTURE STRAIN, %	MODULUS, GPa
Control epoxy systems					
1	48	61.3	3543 (±27)	2.6 (±0.05)	137.9 (±1.6)
2	24	62.4	3578 (±100)	2.6 (±0.03)	137.9 (±2.3)
3	22	63.6	3468 (±43)	2.6 (±0.01)	135.1 (±0.2)
Moderate-pot-life epoxy systems					
4	44	63.5	3612 (±32)	2.6 (±0.01)	137.3 (±0.6)
5	68	62.5	3378 (±32)	2.4 (±0.03)	141.3 (±0.8)
6	9	61.0	3447 (±110)	2.6 (±0.10)	132.4 (±1.2)
Long-pot-life epoxy systems					
7	30	65.1	3461 (±52)	2.6 (±0.03)	134.4 (±0.6)
8	29	64.6	3247 (±56)	2.3 (±0.02)	143.4 (±1.2)
9	43	65.1	3302 (±12)	2.4 (±0.02)	139.9 (±0.8)

[a]These data represent five spools of yarn. The test strands were taken from different locations on these spools.
[b]The coefficient of variation for each resin system is less than 7% and is nominally 4.5%.

12.3.1.2. Effect of Fiber Volume on Fiber Strength

The effect of fiber volume on the fiber strength of the composite strands is not generally known. Data from the Lawrence Livermore Laboratory on 1420-denier Kevlar 49 and on 1500-denier Kevlar 29 fibers in an epoxy matrix (Dow Chemical DER 332/Jefferson Chemical Jeffamine T-403) are given in Fig. 12.24. For tensile-critical applications, it appears that a fiber content between 65 and 70 volume % is a good choice.

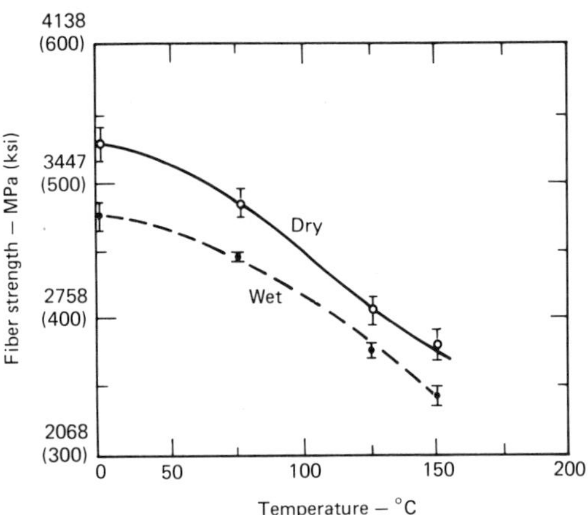

Figure 12.25. Fiber strength of epoxy-coated Kevlar 49 strands as a function of temperature (dry and water-soaked).[14] (*Courtesy J. M. Augl, U. S. Navy.*)

12.3.1.3. Effect of Moisture and Temperature on Fiber Strength

Fiber strength data for epoxy-impregnated, 4560-denier Kevlar 49 strands (both dry and soaked in water for two weeks) versus temperature are summarized in Fig. 12.25. Over the temperature range of 22–150° C (71.6–302° F) moisture causes an approximate 10% loss of tensile strength and temperature causes about 28% strength loss. However, low temperature (liquid nitrogen) or a change in strain rate (from 8.33×10^{-5} to 8.33×10^{-2} m/m/second) have little effect on the fiber strength.[14]

12.3.1.4. Stress-Rupture (Lifetime) Behavior

Design engineers often rely on static strength and strength retention data to determine a safety factor with which to estimate the stress-rupture lifetime of a material. However, Chiao *et al.*[15] found no correlation between strength retention values and the remaining life of the Kevlar/epoxy composite. To estimate long-term performance in engineering design, they recommend the use of stress-rupture data instead of strength data.

The stress-rupture behavior of Kevlar 49/epoxy composite strands (same composition as shown in Fig. 12.23) has been studied by Chiao *et al.*[14, 16] Approximately five years of data have been collected (Fig. 12.26) as part of a continuing project. Because the conven-

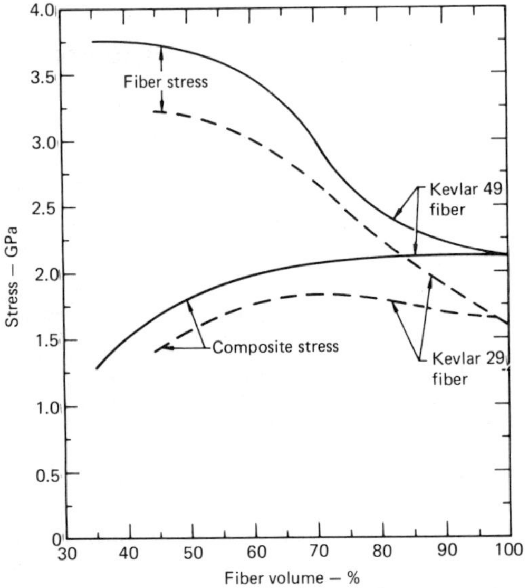

Figure 12.24. Effect of Kevlar 49 fiber volume on fiber and composite strength.

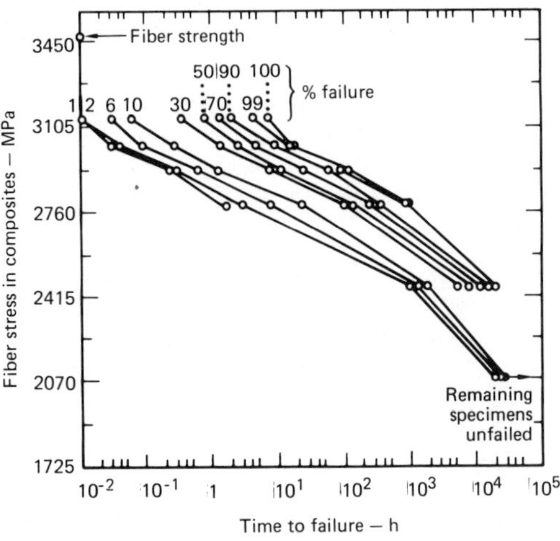

Figure 12.26. Stress-rupture failure contours of Kevlar 49/epoxy strands [resin system Union Carbide ERL 2258/ZZL 0820, 100/30, parts by weight cured at 93° C (199.4° F) for 3 hours plus 163° C (325.4° F) for 2 hours.[16]

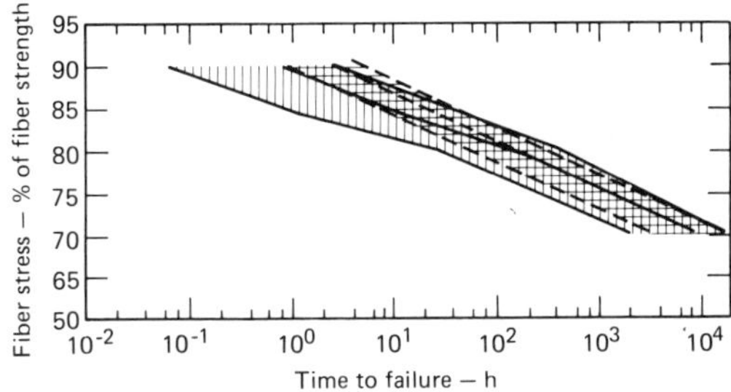

Figure 12.27. Comparison of the model-predicted stress-rupture life at 25° C (77° F) (dashed lines) with the room-temperature test data[16] (solid lines) for Kevlar 49/eposy composite strands (same resin system as in Fig. 12.16).[18] The upper, middle, and lower lines are for 90, 50 (median), and 10% failure, respectively.

tional stress-rupture techniques require an impractically long test time, Chiao[17] proposes the use of an accelerated test for predicting the lifetime of organic fiber composites. The acceleration model was tested with Kevlar 49/epoxy composite strands.[12] An expression for predicting the median stress-rupture lifetime, t_{median} (in hours), of this composite was derived[18]:

$$t_{median} = 7.9 \times 10^{-11} \exp\left[(18246 - 123\sigma)/T\right],$$

where T is the absolute temperature (° K) and σ is normalized as a percent of the fiber strength. Combined with a Weibull distribution function to describe the variability, this expression yields good predictions of the stress-rupture lifetime of this Kevlar 49/epoxy composite within a limited range of applied stresses (see Fig. 12.27).

12.3.1.5. Fatigue Behavior

For 0-deg composites, the fiber essentially controls composite behavior. Work by Bunsell[19] on the fatigue behavior of Kevlar 49 fiber is summarized below:

- For cycling at 50 Hz, there is only a narrow hysteresis curve. After the first few cycles, the fiber structure essentially stabilizes, indicating that only a low percentage of irreversible work is spent and dissipated as heat. Hence, Kevlar 49 fiber should not heat up significantly during fatigue.
- Cycling the fiber to a constant maximum load at 50 Hz, but with various oscillatory loads, can affect the lifetime significantly. A definite trend is indicated for increased fiber survival when the minimum load is raised and the maximum load is held constant.
- In general, Kevlar 49 is considered to have high resistance to fatigue damage when compared to other polymeric fibers tested under similar conditions.

In addition, limited work has been conducted on the fatigue behavior of Kevlar 49/epoxy strands (same strands as in Fig. 12.23).[20] At 10 Hz in a sinusoidal mode with applied loads varying from 1–88% of the strand failure strength, the fatigue life varied from approximately 2000–76,000 cycles. All ten specimens failed properly in the gauge length. These results clearly illustrate two points: accurate determination of fatigue life requires large numbers of specimens, and an average value for fatigue life is meaningless because there is so much scatter in the lifetime data.

12.3.2. Aramid Fiber/Epoxy Rings

The simple specimens with which to evaluate the longitudinal tensile properties of filament-wound composites are the NOL ring and the modified NOL ring (elongated racetrack shape with 25-mm straight sections between two semicircular ends). Data for Kevlar 49 fiber/epoxy rings are summarized in Table 12.13. Figure 12.28 gives the fiber and composite tensile strengths of the NOL rings as a function of fiber volume for several composite systems (the Kevlar 29 and Kevlar 49 fibers were 1500 and 1420 denier, respectively).[21]

12.3.3. Unidirectional Kevlar 49 Epoxy Laminates

Much work has been done on unidirectional composites. In this section, we summarize the available data on several properties.

12.3.3.1. Preliminary Engineering Data

Depending on the selected matrix and the fabrication variables, the composite properties can differ substantially. This is particularly true for the matrix-controlled properties. In Table 12.16 and Fig. 12.29, the properties of a unidirectional laminate made from wet

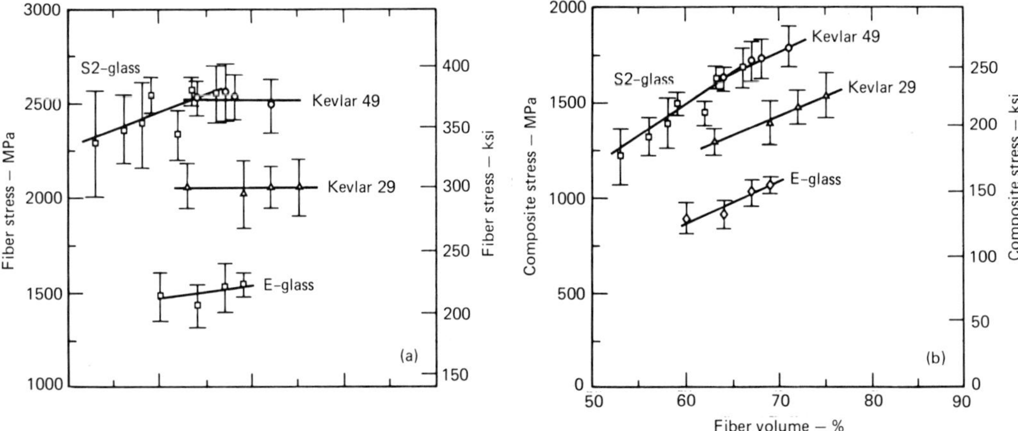

Figure 12.28. NOL-ring hydroburst test results for fiber (*a*) and composite (*b*). Stress at failure is plotted against the fiber volume fraction. Each point represents an average of from 10 to 14 tests; error bars represent ±1 standard deviation.[21]

filament-winding and press-molding are given. The fiber was 1420-denier Kevlar 49 in the epoxy, Dow Chemical DER 332/Jefferson Chemical Jeffamine T-403 (100/45 parts by weight). The system was cured for 24 hours at room temperature and for 16 hours at 85°C (185°F)[22] Data on some typical prepreg systems are given in Figs. 12.30 and 12.31.[8]

12.3.3.2. Fatigue and Creep Behavior

Limited fatigue and creep data supplied by the fiber manufacturer[8] are shown in Fig. 12.32.

Table 12.16. Properties of a Unidirectional Laminate of 1420-Denier Kevlar 49 Fiber in a DER 332/Jeffamine T-403 Epoxy[a] (100/45 Weight Ratio)[22]

Mechanical elastic constants							
Longitudinal Young's modulus, E_{11}, GPa					81.8 ± 1.5		
Transverse Young's modulus, E_{22}, GPa					5.10 ± 0.10		
Shear modulus, G_{12}, GPa					1.82 ± 0.09		
Major Poisson's ratio, ν^{12}					0.310 ± 0.035		
Minor Poisson's ratio, ν^{21}					0.0193 ± 0.0014		
Mechanical Ultimates		*Tension*			*Compression*		*Shear*
Longitudinal strength, MPa		1850 ± 50			(235 ± 3)		—
Longitudinal ultimate strain, %		2.23 ± 0.06			(0.48 ± 0.3)		—
Transverse strength, MPa		7.9 ± 1.1			(53 ± 3)		—
Transverse ultimate strain, %		0.161 ± 0.023			(1.41 ± 0.12)		—
Shear stress at 0.2% offset, MPa		—			—		24.4 ± 2.4
Shear strain at 0.2% offset, %		—			—		1.55 ± 0.16
	−50°C	−25°C	0°C	25°C	50°C	75°C	100°C
Thermal properties	(−58°F)	(−13°F)	(32°F)	(77°F)	(122°F)	(167°F)	(212°F)
Linear coefficient of thermal expansion, 10^{-6}/°C							
Longitudinal	−3.8	−3.8	−3.8	−4.0	−4.7	−6.0	—
Transverse[b]	61	66	72	79	87	150	214
Thermal conductivity, W/m·°C							
Longitudinal	2.62	2.84	3.05	3.22	3.31	3.34	—
Transverse	—	0.27	0.33	0.35	0.37	—	—
Heat capacity, J/kg·°C	840	930	1020	1120	1190	1300	—

[a]The epoxy system was cured for 24 hours at room temperature and for 16 hours at 85°C (185°F); the fiber content was nominally 60 volume %. Limits are given for 95% confidence; data without confidence limits were taken from a single specimen. Data in parentheses are estimated from results for Kevlar 49 in an XD 7818/T-403 matrix. Elastic constants are assumed to be valid for both tension and compression (1 MPa = 0.145 ksi).

[b]Transverse linear coefficient of thermal expansion at 125°C (257°F) is 214 × 10^{-6}/°C.

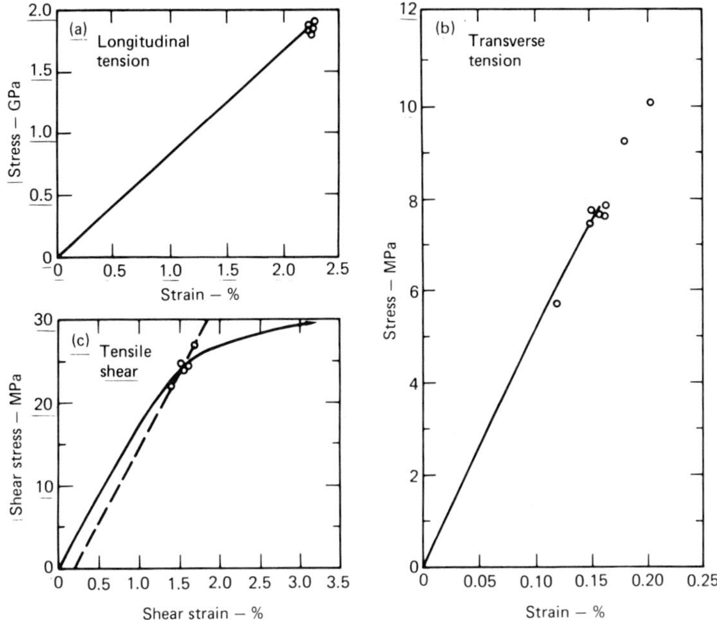

Figure 12.29. Average longitudinal tensile strength (*a*), transverse tensile strength (*b*), and shear strength at 0.2% strain offset (*c*) stress-strain curves and failure points of individual composite specimens of Kevlar 49/epoxy (Dow Chemical DER 332/Jefferson Chemical Jeffamine T-403), nominally 60 volume % fiber.

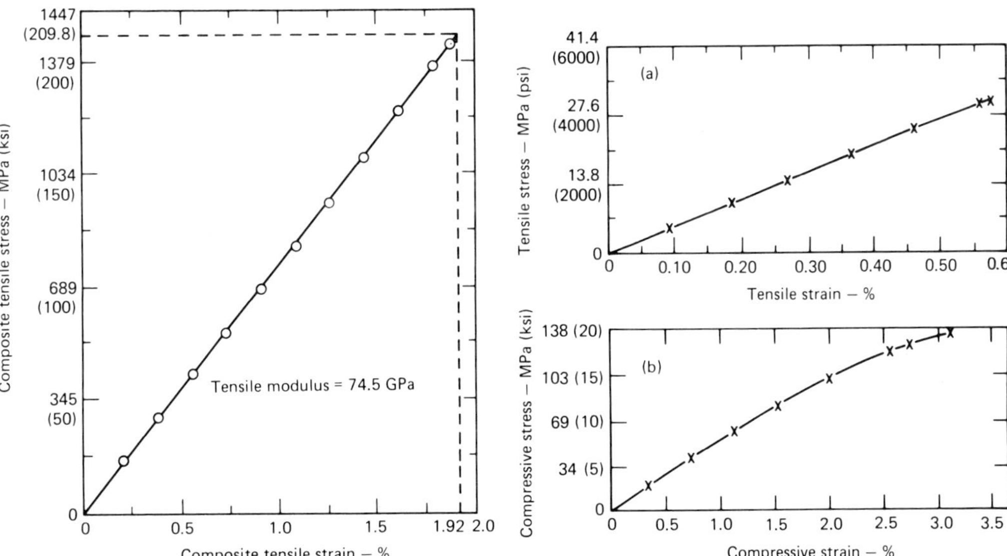

Figure 12.30. Stress-strain curves of unidirectional Kevlar 49/American Cyanamid BP-907 epoxy composites (autoclave-molded at 345 kPa, 60 volume % fiber) tested at 0 deg to fiber direction in tension.[8]

Figure 12.31. Stress-strain curves of unidirectional tape Kevlar 49/3M SP-306 epoxy composites (autoclave-molded at 1034 kPa, 60 volume % fiber) tested at 90 deg to fiber direction in tension (*a*) and in compression (*b*).[8]

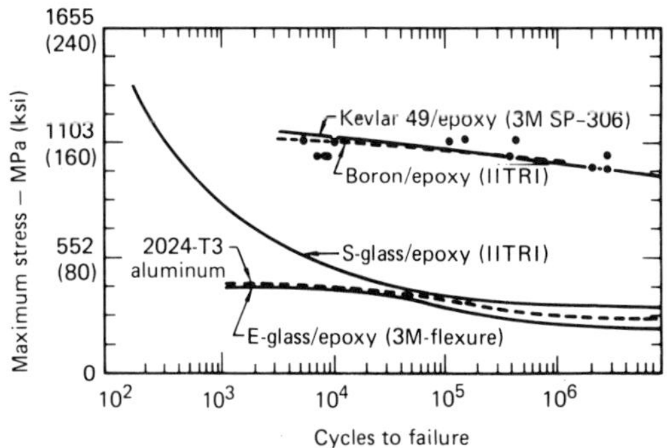

Figure 12.32. Tension-tension fatigue behavior of unidirectional composites and of aluminum at room temperature (stress ratio, $R = 0.1$).[8]

Data from several references indicate that the creep curves have a region of primary creep, followed by a nearly linear region of secondary creep.[23, 24] Work by Bunsell[19] on the creep behavior of Kevlar 49 fiber indicates that little creep occurs; after 5 minutes of loading, the fiber appears to have stabilized.

Ericksen[24] studied the creep of Kevlar 49 in a cycloaliphatic epoxy (Kevlar 49/Union Carbide ERLA 4617/MPDA); these results are summarized in Fig. 12.33. Here, creep strain equals total strain minus instantaneous load-on strain, which can be estimated from the composite stress-strain curve of Fig. 12.34. The composite tensile failure stress ranged from 1240–1380 MPa. By replotting the 0-deg composite data in log time (Fig. 12.35), Ericksen[24] draws the following conclusions.

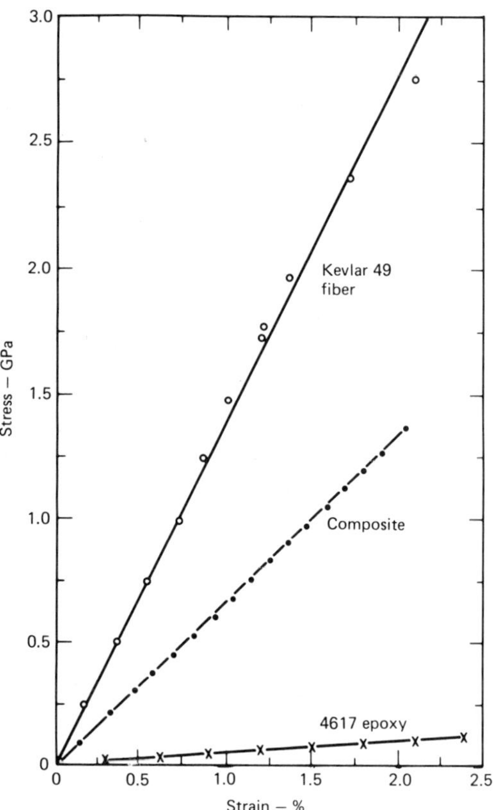

Figure 12.34. Room-temperature stress-strain curves for Kevlar 49 fibers, 4617 epoxy, and Kevlar 49/4617 epoxy composites.[24]

- Kevlar 49/epoxy composite exhibits transient creep (i.e., no steady state creep observed) at room temperature for times up to 1000 hours.
- After about 10 hours, the composite creep strain varies linearly with the log time at all stress levels tested.
- At long times, creep of the composite is

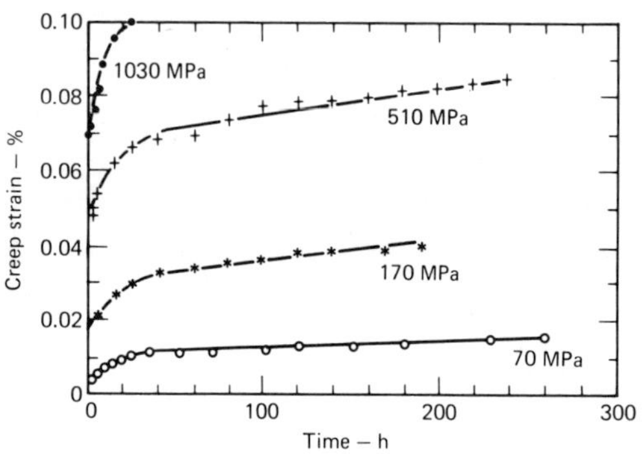

Figure 12.33. Room-temperature creep strain versus time plot of Kevlar 49/Union Carbide 4617 epoxy 0-deg composites.[24]

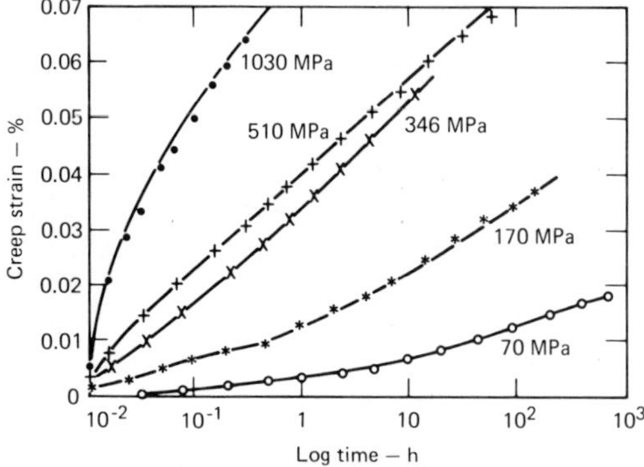

Figure 12.35. Room-temperature creep strain versus log time plot of Kevlar 49/Union Carbide 4617 epoxy 0-deg composites.[24]

controlled by the creep of the Kevlar 49 fiber.

12.3.3.3. Shear Properties

The interlaminar (short-beam) shear data of 1420-denier Kevlar 49 in a number of epoxy systems (tested with ASTM-D-2344-72) are presented in Table 12.17.[25] These short-beam shear test results do not duplicate well. However, there is as yet no good method for determining the interlaminar shear strength of fiber composites. Chiao *et al.*[26] evaluated several test methods for measuring shear properties of fiber composites: for the present, they recommend use of the ±45-deg laminate tensile shear test. Table 12.18 summarizes the shear properties of 1420-denier Kevlar 49 in several epoxy resins, determined by the ±45-deg laminate tensile shear test. Figure 12.36 shows the corresponding shear stress-strain curves of the composites.[27]

12.3.4. Filament-Wound Aramid Fiber/Epoxy Pressure Vessels

Kevlar 49 is particularly well suited for winding pressure vessels in tensile-critical applications. Published data are summarized below.

12.3.4.1. Vessel Design and Performance

Filament-wound Kevlar 49 fiber/epoxy pressure vessels (e.g., cylinders or spheres lined with rubber, polymeric film, or metal) have been studied extensively.[28-33] These vessels appear to offer many advantages, based on the performance factor, P_bV/W, where P_b is the burst pressure, V is the contained volume, and W is the composite or total vessel weight.

In a joint effort with the NASA-Lewis Research Center, researchers at the Lawrence Livermore Laboratory studied the effects of the length-to-diameter ratio (L/D), the boss-to-diameter ratio (d/D), and the shape of the closed ends of the vessel (dome contours) on

Table 12.17. Short-Beam Shear Test Results of Some Kevlar 49/Epoxy Systems (Nominal Fiber-to-Epoxy Ratio of 65/35)[25]

EPOXY SYSTEM (WEIGHT RATIO)	CURE CYCLE, HOURS/°C (HOURS/°F)	SPECIMEN TYPE[a]	INTERLAMINAR SHEAR STRENGTH,[b] MPa (psi)	VARIATION, %	NO. OF SPECIMENS
XD 7818/ERL 4206/	2.5/80 +	1	34.7 ± 0.6 (5034 ± 92)	2.6	10
Tonox 60-40	2/160	1	34.3 ± 1.1 (4974 ± 154)	5.1	13
(80/20/29.1)	(2.5/176 +	2	35.9 ± 1.1 (5216 ± 160)	4.8	12
	2/320)	2	34.6 ± 1.1 (5019 ± 142)	4.5	12
		3	38.0 ± 3.0 (5513 ± 438)	11.8	11
		3	35.9 ± 0.8 (5206 ± 119)	4.3	16
XB 2793/Tonox 60-40	2/90 +	2	32.5 ± 21. (5716 ± 308)	4.1	4
(100/25.6)	2/160	2	28.7 ± 1.7 (4156 ± 254)	3.8	4
	(2/194 +	2	28.1 ± 1.3 (4079 ± 188)	2.9	4
	2/320)	2	34.6 ± 1.8 (5022 ± 267)	5.1	6
		2	33.9 ± 1.6 (4910 ± 225)	4.1	6
		3	35.9 ± 0.9 (5203 ± 131)	3.9	12
		3	36.4 ± 1.0 (5284 ± 147)	4.4	12
XD 7818/XD 7575.02/	4.5/60 +	2	35.0 ± 1.1 (5071 ± 164)	2.0	4
Tonox 60-40/ERL	4/120	2	26.0 ± 0.7 (3771 ± 102)	1.7	4
4206 (50/50/38.1/30)	(4.5/140 +	2	29.8 ± 0.8 (4315 ± 116)	1.7	4
	4/248)	3	32.5 ± 1.1 (4720 ± 153)	5.6	14
		3	36.7 ± 1.0 (5328 ± 139)	4.1	12
XD 7818/XD 7114/	5/60 +	2	38.3 ± 1.5 (5551 ± 218)	3.7	6
Tonox 60-40	3/120	2	39.0 ± 0.8 (5661 ± 123)	1.7	5
(100/45/50.3)	(5/140 +	3	41.4 ± 1.6 (6991 ± 226)	5.9	12
	3/248)				

[a]Specimen type 1 = short-beam specimen cut from a unidirectional laminate; specimen type 2 = flat-beam specimen cut from the straight section of an elongated NOL ring; and specimen type 3 = curve-beam specimen cut from the curved section of an elongated NOL ring.

[b]The ± values indicate the 95% confidence limits.

Table 12.18. Shear Properties of Composites of Kevlar 49 Fiber in Seven Epoxy Resins[27]

RESIN NO.	EPOXY SYSTEM (WEIGHT RATIO)	CURE CYCLE, hours/°C (hours/°F)	SHEAR FAILURE STRESS, MPa (CV)[a]	SHEAR STRAIN AT FAILURE STRESS, % (CV)[a]	SECANT SHEAR MODULUS AT 0.5% SHEAR STRAIN, MPa (CV)[a]	NO. OF SPECIMENS
1	XD 7818/ERL 4206/ Tonox 60-40 (80/20/29.1)	2.5/80 + 2/160 (2.5/176 + 2/320)	21.4 (2.6)	1.35 (2.2)	1884 (3.9)	3
2	DER 332[b]/ Jeffamine T-403 (100/39)	24/60 (24/140)	29.4 (2.0)	1.73 (2.3)	1923 (4.7)	5
3	ERL 2256/Tonox 60-40 (100/29.5)	16/50 + 2/120 (16/122 + 2/248)	23.0 (8.6)	1.49 (2.2)	1775 (0.9)	3
4	Epon 826/RD 2/ Tonox 60-40 (100/25/28.3)	3/60 + 2/120 (3/140 + 2/248)	23.4 (6.3)	1.91 (6.5)	1520 (3.9)	4
5	XB 2793/Tonox 60-40 (100/25.6)	2/90 + 2/160 (2/194 + 2.320)	21.9 (0.3)	1.69 (2.9)	1600	2
6	XD 7818/XD 7575.02/XD 7114/ Tonox 60-40/DAP (50/50/45/14.1/14.1)	5/80 + 3/120 (5/176 + 3/248)	39.7 (0.9)	2.43 (2.5)	1852 (1.7)	5
7	XD 7818/XD 7114/ Tonox LC (100/45/50.3)	5/60 + 3/120 (5/140 + 3/248)	31.9 (3.4)	1.91 (4.5)	1850 (1.5)	4

[a]CV = coefficient of variation, %.
[b]Dow Chemical Company.

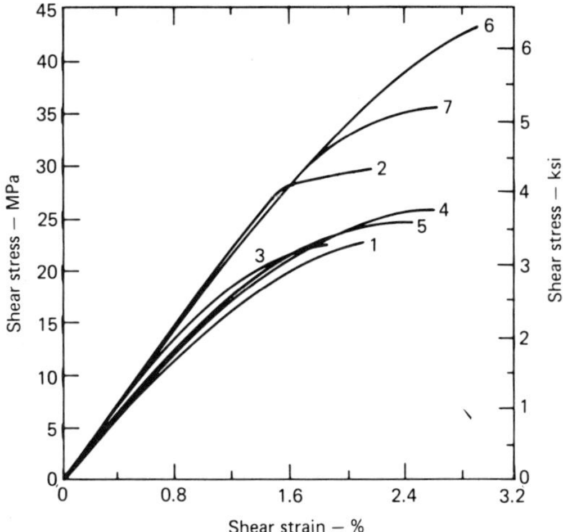

Figure 12.36. Shear stress-strain curves of Kevlar 49/Union Carbide 4617 epoxy composites.[27] Numbers correspond to the resin numbers in Table 12.18.

the performance of 102-mm-diameter Kevlar 49/epoxy pressure vessels.[5] The following types of cylindrical composite pressure vessels were filament-wound.

- In-plane-wrapped dome contours with L/D ratios of 1.37, 2.0, and 3.0.
- In-plane-wrapped dome contours with a constant L/D ratio of 1.37, and d/D ratios of 0.10-4, 0.166, and 0.234.
- In-plane helical and hemispherically shaped dome contours with a constant d/D ratio of 0.104 and a cylindrical section length of 76 mm.

All vessels were fabricated on a numerically controlled filament-winding machine. The fibers were impregnated with an epoxy as they passed through a vacuum chamber (controlled to approximately 670 Pa) and wound directly over 0.25-mm-thick elastomeric liners, supported on washout-type mandrels.

Single-cycle burst tests were conducted using hydraulic fluid pressurization; the vessels were instrumented with strain gauges to determine hoop and longitudinal strains. Test results are given in Table 12.19; columns 1–3 show the effects of varying the L/D ratio. No significant effects were observed for L/D ratios up to 3.0. The effect of the d/D ratio on vessel performance is shown in Table 12.19, columns, 1, 4, and 5. No significant differences were noted in vessel performance over a 10–23% difference in the d/D ratio. The effect of the dome contours on the structural efficiency ($P_a V/W$) is documented in columns 4, 6, and 7 of Table 12.19. The in-plane-wrapped dome contour is the most efficient, followed closely by the helical dome contour. Vessels with the hemispherically shaped dome contour performed the worst.

The Lawrence Livermore Laboratory study demonstrated that a well-designed filament-wound Kevlar 49/epoxy composite pressure vessel can exhibit hoop-fiber-stress values of about 3100 MPa (450 ksi). This value is significantly greater than that obtained for most other advanced high-modulus fibers. High fiber-stress values, in combination with the low fiber density of Kevlar 49/epoxy, yield composite pressure vessels with high structural efficiencies.

The selected winding pattern and fabrication process (e.g., winding tension, cure cycle, resin selection), in general, greatly affect the quality and consistency of highly efficient filament-wound vessels. The performance of cylindrical composite pressure vessels is optimized by use of wide-band winding (reducing composite buildup at the bosses), by in-plane-wrapped dome contours, by a d/D ratio less than 0.1, by an L/D ratio less than 3.0, by a hoop/longitudinal fiber ratio of about 1.81, and by the addition of local composite reinforcement at the knuckles of the vessels (this alone improved vessel performance by about 10%).

12.3.4.2. Effect of Rate of Pressurization and Epoxy on Vessel Performance

Researchers at Lawrence Livermore Laboratory also have done limited work on vessel performance at different pressurization rates (Table 12.20).[28] No difference in vessel performance is seen, even for a tenfold increase in the pressurization rate. Tables 12.21–12.23 show the effect of different epoxies on the performance of the Kevlar 49/epoxy vessels.[13] The vessel design used in this study (Fig. 12.37) was very similar to the above work.[28–33]

Under biaxial loading (e.g., a vessel under pressure), the epoxy formulation also significantly affects vessel performance. NASA-

Table 12.19. Test Results of 102-mm-Diameter, Rubber-Lined Kevlar 49/Epoxy Vessels[28]

PARAMETER	CONTROL[a]	VARIOUS LENGTH/DIAMETER RATIOS		VARIOUS BOSS/DIAMETER RATIOS		VARIOUS DOME CONTOURS	
	1	2	3	4	5	6	7
Design							
Winding angle, θ	13°	9°	5° 15′	7° 30′	10°	7° 30′	5° 55′
Dome contour	In-plane	In-plane	In-plane	In-plane	In-plane	Helical	Hemispherical
Boss/diameter	0.935/4	0.935/4	0.935/4	0.415/4	0.665/4	0.415/5	0.415/5
Length/diameter	5.5/4	8.0/4	12/4	5.5/4	5.5/4	5.5/4	5.5/4
Composite mass, W_c, g	44.8	66.2	97.6	45.3	46.4	45.2	50.8
Coefficient of variation, %	3.0	1.9	3.0	3.5	2.5	1.4	2.3
Number of specimens	15	10	8	10	10	10	10
Vessel volume, V, cm^3	949	1420	2200	926	928	939	1120
Coefficient of variation, %	0.3	0.3	0.2	0.2	0.5	0.2	0.2
Number of specimens	72	10	7	9	9	9	9
Fiber content, volume %	67.8	66.7	69.1	68.0	66.7	65.9	66.3
Coefficient of variation, %	5.0	3.6	3.2	3.4	2.4	1.7	3.0
Number of specimens	15	10	8	10	10	10	10
Density at 22° C (71.6° F), g/cm^3	1.375	1.377	1.382	1.382	1.386	1.372	1.382
Vessel thickness, mm	0.83	0.83	0.81	0.81	0.81	0.81	0.82
Axial winding, mm	0.35	0.35	0.31	0.33	0.32	0.30	0.34
Hoop winding, mm	0.49	0.48	0.49	0.49	0.52	0.51	0.47
Mean burst pressure, P, kPa (psi)[b]	18,600 (2700)	17,700 (2570)	17,200 (2490)	18,500 (2690)	19,100 (2770)	17,100 (2490)	6,070 (880)
95% Confidence limit, kPa (psi)	330 (48)	860 (124)	1850 (269)	340 (49)	410 (60)	710 (103)	330 (48)
Coefficient of variation, %	3.9	6.3	8.7	2.4	2.6	5.4	7.1
Number of specimens	21	9	5	9	8	9	9
Failure mode[c]	$17H + 4F$	$8H + 1K$	$3H + 2K$	$9H$	$7H + 1K$	$5H + 4K$	H_e equator
Composite modulus, GPa (10^6 psi)							
Axial	33 (4.8)	34 (4.9)	36 (5.2)	35 (5.0)	35 (5.0)	35 (5.0)	35 (5.0)
Hoop	57 (8.2)	54 (7.8)	59 (8.6)	58 (8.4)	58 (8.4)	57 (8.2)	52 (7.5)
Rupture strain, %							
Axial	1.68	1.60	.145	1.66	1.68	1.52	0.54
Hoop	1.98	2.00	1.84	1.96	2.02	1.88	0.72
Mean vessel performance,							

PV/W_c, kPa · m^3/kg	390	380	382	380	382	358	162
95% Confidence limit, kPa · m^3/kg	12	20	35	10	15	15	20
Coefficient of variation, %	4.8	7.3	7.4	3.3	4.8	5.0	15.9
Number of specimens	13	9	5	9	8	9	9
Calculated mean hoop fiber failure stress, kPa (psi)	2940 (427)	2890 (419)	2870 (416)	2990 (433)	3050 (443)	2260 (400)	903 (131)
95% Confidence limit, kPa (psi)	50 (7.0)	70 (10.0)	300 (40)	50 (7.4)	60 (9.0)	110 (15.5)	40 (5.2)
Coefficient of variation, %	3.8	3.0	8.7	2.4	2.6	5.4	5.8
Number of specimens	21	8	5	9	8	9	9

[a]Standard 102-mm-diameter model vessels.
[b]Tested at ambient temperature at a pressurization rate of 0.12 kPa/sec.
[c]Failure locations: H = hoop, K = knuckle, and F = fitting (e.g., $2H + 1K$ means two vessles failed in hoop and one at knuckle).

Table 12.20. Effect of Pressurization Rate on the Performance of Kevlar 49/ Epoxy Vessels (Six Vessels were Tested at Each Pressurization Rate)[28]

PARAMETER	PRESSURIZATION RATE	
	287 kPa · sec	2870 kPa · sec
Mean burst pressure,[a] kPa (psi)	16,400 (2380)	16,700 (2420)
95% Confidence limit, kPa (psi)	930 (140)	930 (140)
Coefficient of variation, %	5.6	5.6
Composite mass, W_c, g (lb)	43.3 (0.0953)	43.1 (0.0949)
Mean vessel performance, PV/W_c, kPa · m^3/kg	368	360
95% Confidence limit	22	22
Coefficient of variation, %	6.1	6.0
Fiber content, volume %	68.4	68.4
Coefficient of variation, %	4.0	2.1
Mean hoop fiber failure stress,[b] MPa (ksi)	2720 (395)	2680 (389)
95% Confidence limit	160 (23)	160 (23)
Coefficient of variation, %	5.5	5.7

[a]All specimens were standard 10.2-mm model vessels with a volume of 9.49×10^{-4} m^3, a hoop/axial fiber ratio of 1.5. All vessels were fabricated identically with vacuum winding, all vessels were tested at 21°C (69.8°F) and all failed in hoop.
[b]Estimated, based on a netting analysis.

sponsored work performed at Hercules, Inc. (Cumberland, Maryland) confirms this resin effect on vessel performance.[30]

12.3.4.3. Fatigue Behavior

Hamstad *et al.*[29] have studied the fatigue of rubber-lined, 102-mm-diameter Kevlar 49/ epoxy vessels. They investigated several variables by testing 25 vessels to determine the burst pressure variation, 25 vessels under sinusoidal cycling at 1 Hz between 4 and 91% of the mean burst pressure, 25 vessels under rectangular pressure pulses at 0.33 Hz between 4 and 91% of the mean burst pressure, and 25 vessels under sustained loading, also at 91% of the mean burst pressure. Their data are summarized in Table 12.24. The burst pressure of the vessels varied 4.5% with respect to the mean; the cyclic fatigue life extended over five to six decades. More important, the fatigue life of the Kevlar 49/epoxy vessels was found to depend on both the number of stress cycles and the total time at peak load. Under a constant sustained load at 91% of the mean burst pressure, vessel lifetime was much longer than the total time at peak load under rectangular wave fatigue tests. Under the same load amplitude (4–91% of mean burst pressure), the fatigue life of vessel under sinusoidal wave cycles was about six times longer than that of a vessel under rectangular wave cycles.

12.3.4.4. Rubber- and Aluminum-Lined Cylindrical and Spherical Vessels

Researchers at Lawrence Livermore Laboratory also studied the performance of the 203-mm-diameter cylindrical and spherical vessels at room and low temperatures (liquid nitrogen and hydrogen). Both butyl rubber and aluminum liners were used. The data are given in Tables 12.25–12.28.[33] The principal conclusions obtained from this study follow.

- The best vessel performance factor based only on composite weight is 441 kPa ·m^3/kg (1.8×10^6 in.) for a sphere with a 0.76-mm-thick aluminum liner. When liner weight is included, the performance factor of the total vessel decreases to 70% of the above value.
- The vessel performance factor is approximately the same for both spherical and cylindrical vessels. The choice between the two vessel shapes should be based on dimensional requirements, as well as on the availability and type of winding machine.
- There is no obvious size effect between the 102- and 203-mm-diameter vessels. In

Table 12.21. Performance of Filament-Wound Pressure Vessels of Kevlar 49/Epoxy from the Control Epoxy Systems (see Table 12.14)[13]

PROPERTY	EPOXY SYSTEM NO.							
	1	1		2		2		3[a]
Components	ERI 2256/ Tonox 60–40	ERL 2256/ Tonox 60–40		ERL 2258 ZZL 0820		ERL 2258 ZZL 0820		Epon 826/RD 2/ Tonox 60–40
Gel cycle, hours/°C (hours/°F)	2/50 (2/122)	16/50 (16/122)		16/50 (16/122)		2/90 (2/194)		4/60 (4/140)
Cure cycle, hours/°C (hours/°F)	3/150 (3/302)	2/94 + 3/150 (2/201.2 + 3/302)		3/163 (3/375.4)		3/150 (3/302)		3/120 (3/248)
Composite vessel data[b]								
Number of specimens[c]	4	4	4	4	4	4	4	4
Fiber mass, W_f, g	30.4	31.8	31.7	32.0	30.5	32.0	31.7	30.5
Composite mass, W_c, g	43.0	46.0	45.6	44.3	42.7	45.4	45.9	42.7
Fiber content, V, volume %	66.5	65.1	65.5	68.1	67.3	66.5	64.9	66.1
Burst pressure, P, MPa (ksi)	16.2 (2.3)	19.4 (2.8)	18.1 (2.6)	15.6 (2.3)	15.9 (2.3)	17.7 (2.6)	17.8 (2.6)	16.5 (2.4)
Coefficient of variation, %	1.1	2.6	2.9	6.7	4.4	1.5	1.1	1.9
Vessel performance factors[d]								
Fiber performance, PV/W_f, kPa · m^3/kg (0^6 in)	522 ± 9 (2.10)	575 ± 22 (2.31)		492 ± 26 (1.98)		543 ± 6 (2.18)		534 ± 9 (2.15)
Coefficient of variation, %	1.1	4.5		6.4		1.4		2.0
Composite performance, PV/W_c, kPa · m^3/kg (10^6 in)	368 ± 9 (1.48)	399 ± 13 (1.60)		353 ± 18 (1.42)		379 ± 3 (1.52)		380 ± 5 (1.53)
Coefficient of variation, %	1.6	3.9		5.9		1.0		1.7
Average calculated hoop fiber stress, MPa (ksi)	1974 (286)	2286 (331)		1920 (278)		2164 (314)		2091 (303)

[a]Pot life of this epoxy system is relatively short.
[b]Volume of each vessel is constant, 9.75×10^{-4} m^3.
[c]Two groups of four vessels each were fabricated from different fiber spools at different times.
[d]The ± values are the 95% confidence limits.

Table 12.22. Performance of Filament-Wound Pressure Vessels of Kevlar 49/Epoxy from the Moderate-Pot-Life Epoxy Systems (see Table 12.14)[13]

PROPERTY	EPOXY SYSTEM NO. 4		5		6	
Components	XD7818/ERL4206/ Tonox 60–40		XD 7575.02/ XD7818/ERL4206/ Tonox 60–40		XD 7575.02/ XD 7818/XD 7114/ Tonox 60–40	
Gel cycle, hours/°C (hours/°F)	40/60 (40/140)		4.5/60 (4.5/140)		16/60 (16/140)	
Cure cycle, hours/°C (hours/°F)	3/120 (3/248)		3/120 (3/248)		4/120 (4/248)	
Composite vessel data[a]						
Number of specimens[b]	4	4	4	4	4	4
Fiber mass, W_f, g	32.0	30.2	31.7	30.4	32.0	30.4
Composite mass, W_c, g	44.2	42.5	43.8	42.5	43.8	42.3
Fiber content, V, volume %	68.3	67.1	68.7	67.8	69.2	68.0
Burst pressure, P, MPa (ksi)	16.7 (2.4)	15.3 (2.2)	17.1 (2.5)	15.9 (2.3)	16.4 (2.4)	16.1 (2.3)
Coefficient of variation, %	5.7	5.5	1.2	4.6	0.7	2.6
Vessel performance data[c]						
Fiber performance, PV/W_f, kPa · m^3/kg (10^6 in)	503 ± 23 (2.02)		517 ± 15 (2.08)		508 ± 10 (2.04)	
Coefficient of variation, %	5.4		3.4		2.3	
Composite performance, PV/W_c, kPa · m^3/kg (10^6 in)	316 ± 16 (1.27)		373 ± 11 (1.50)		368 ± 6 (1.48)	
Coefficient of variation, %	5.4		3.7		1.9	
Average calculated hoop fiber stress, MPa (ksi)	1950 (283)		2011 (291)		1981 (287)	

[a]Volume of each vessel is constant, 9.75×10^{-4} m^3.
[b]Two groups of four vessels each were fabricated from different fiber spools at different times.
[c]The ± values are the 95% confidence limits.

addition, both cryogenic temperature and strain rate have little effect on the fiber failure stress or on the vessel performance factor.

12.3.4.5. Thick-Walled Kevlar 49/Epoxy Vessels

Theoretical analysis by Gerstle and Moss[31] on metal-lined, thick-walled spherical composite vessels indicates that, at a failure pressure up to 425 MPa, an aluminum-lined Kevlar 49/epoxy composite sphere is the most efficient combination. Stress gradients in anisotropic composite vessels are more pronounced than in isotropic metal vessels of the same wall thickness. The loss in performance due to stress gradients induced by wall thickness and anisotropy can be minimized by over-wrapping the Kevlar 49/epoxy composite with a stiffer, somewhat weaker material (e.g., graphite composite).

12.3.4.5. Metal-Lined, Load-Sharing Kevlar 49/Epoxy Vessels

Titanium-, 6A1-4V, and Iconcel 718-lined Kevlar 49/epoxy spheres (318- and 635-mm diameters) are being evaluated to contain high-pressure gases for the space shuttle orbiter propulsion and life-support systems by NASA.[32] This program has demonstrated several key advantages of the load-sharing, metal-lined vessel, as outlined below.

- The vessel fails by leakage at operating pressure, rather than by catastrophic rupture, as in an all-metal vessel.
- A weight saving of approximately 25% is obtained over the all-metal counterparts.
- All vessels closely meet the design goal of 1600 failure cycles: the pressure amplitude for the titanium-lined vessels ranged from 0–31 MPa (4500 psi) with 46 MPa (6600 psi) as the designated burst; the pressure

Table 12.23. Performance of Filament-Wound Pressure Vessels of Kevlar 49/Epoxy from the Long-Pot-Life Epoxy Systems (see Table 12.14)[13]

PROPERTY	EPOXY SYSTEM NO.							
	7		8		9		10	
Components	XD 7818/ XD 7114/ Tonox LC		XD 7818/ XD 7114/ Tonox 60–40 2,6 DAP		XD 7575.02/ XD 7118/ XD 7114/ Tonox 60–40		ERE 1359/ RD 2/ 2,6 DAP	
Gel cycle, hours/°C (hours/°F)	4.5/60 (4.5/140)		5/60 (5/140)		5/60 (5/140)		12/80 (12/176)	
Cure cycle, hours/°C (hours/°F)	3/120 (3/248)		4/120 + 4/155 (4/248 + 4/311)		4/120 + 4/155 (4/248 + 4/311)		2/100 + 2/125 + 4/150 (2/212 + 2/257 + 4/302)	
Composite vessel data[a]								
Number of specimens[b]	4	4	4	4	4	4	4	4
Fiber mass, W_f, g	32.1	30.6	30.4	31.8	31.8	31.9	31.8	32.7
Composite mass, W_c, g	43.9	42.1	43.0	44.9	45.2	45.2	45.8	46.0
Fiber content, V, volume %	69.3	68.8	66.6	66.7	66.2	66.5	65.3	67.4
Burst pressure, P, MPa (ksi)	15.0 (2.2)	14.6 (2.1)	13.9 (2.0)	13.7 (2.0)	14.3 (2.1)	13.4 (1.9)	14.1 (2.0)	14.1 (2.0)
Coefficient of variation, %	2.7	3.3	1.7	2.4	1.0	2.8	0.8	2.2
Vessel performance data[c]								
Fiber performance, PV/W_f, kPa · m^3/kg (10^6 in)	461 ± 11 (1.85)		432 ± 15 (1.74)		424 ± 14 (1.71)		426 ± 9 (1.71)	
Coefficient of variation, %	2.9		4.0		4.0		2.4	
Composite performance, PV/W_c, kPa · m^3/kg (10^6 in)	336 ± 9 (1.35)		305 ± 9 (1.23)		298 ± 9 (1.20)		299 ± 4 (1.20)	
Coefficient of variation, %	3.2		3.7		3.7		1.5	
Average calculated hoop fiber failure stress, MPa (ksi)	1804 (261)		1682 (244)		1688 (245)		719 (104)	

[a]Volume of each vessel is constant, 9.75×10^{-4} m^3.
[b]Two groups of four vessels each were fabricated from different fiber spools at different times.
[c]The ± values are the 95% confidence limits.

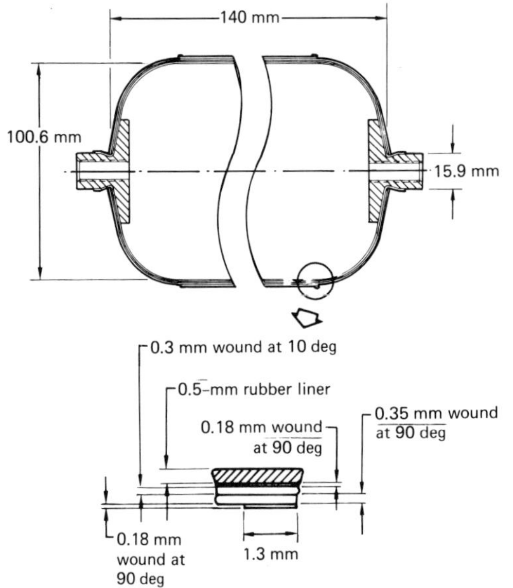

Figure 12.37. Design of the Kevlar 49/epoxy test vessel.

amplitude for the Inconel-lined vessel ranged from 0–23 MPa (3300 psi) with 34 MPa (4950 psi) as the designated burst pressure.

The design-allowable Kevlar 49 fiber strength used in this program is 2200 MPa (320 ksi) for composites with an approximately 60 volume % fiber content. It is significant that most of the difficulties encountered and all of the redesignings required are attributed to problems with the metal liners.

12.3.5. Fabric Composites

Available data on fabric composites are based mainly on Kevlar 49 fiber in epoxies or polyesters.[8] Table 12.29 lists the mechanical properties of several fabric/matrix composites;

Table 12.24. Results of Burst, Fatigue, and Sustained-Load Tests of Kevlar 49/Epoxy Vessels (102 mm in Diameter)[29]

	SINUSOIDAL PRESSURE CYCLE	RECTANGULAR PRESSURE CYCLE		SUSTAINED LOAD AT 91% OF MEAN BURST PRESSURE
BURST PRESSURE,[a] MPa (psi)	CYCLIC LIFE, CYCLES	CYCLIC LIFE, CYCLES	TOTAL TIME AT PEAK STRESS, HOURS	LIFETIME, HOURS
13.69 (1985)	1	1	0.0000	0
13.93 (2020)	9	2	0.0047	28
14.14 (2090)	635	31	0.0120	151
14.44 (2095)	1030	50	0.0190	154
14.55 (2110)	3440	83	0.0320	500+[b]
14.75 (2140)	4230	88	0.0340	500+
14.79 (2145)	11,800	242	0.0940	500+
14.89 (2160)	12,100	393	0.1500	500+
14.96 (2170)	12,600	2740	1.1000	500+
15.00 (2175)	14,300	3680	1.4000	500+
15.10 (2190)	18,800	4030	1.6000	500+
15.24 (2210)	22,300	4690	1.8000	500+
15.27 (2215)	24,500	7060	2.8000	500+
15.34 (2225)	27,100	7140	2.8000	500+
15.38 (2230)	35,600	7210	2.8000	500+
15.38 (2230)	88,700	7640	3.0000	500+
15.38 (2230)	95,200	12,000	4.7000	500+
15.51 (2250)	104,000	14,500	5.6000	500+
15.62 (2265)	129,000	14,700	5.7000	500+
15.72 (2280)	132,000	19,400	7.6000	500+
15.72 (2280)	144,000	22,200	8.6000	500+
16.00 (2320)	176,000	26,400	10.0000	500+
16.10 (2335)	193,000	45,000	18.0000	500+
16.34 (2370)	200,000	45,100	18.0000	500+
16.38 (2375)	430,000	54,300	21.0000	500+

[a]All are hoop failures [one other vessel failed on the dome at 12.27 MPa (1978 psig)].
[b]Plus (+) indicates test terminated after 500 hours.

Table 12.25. Performance of 203-mm-Diameter, Rubber-Lined Cylindrical Kevlar 49/Epoxy Vessels (0.5-mm-Thick Rubber Liner, 21°C (69.8°F— Test Temperature)[33]

PROPERTY	VESSEL NO.										
	P-142[a]	P.144[a]	P-148[b]	P-151[c]	P-152[c,d]	P-153[c,d]	P-156[c,d]	P-157[c,d]	P-158[c,d]	P-159[c,d]	P-160[c,d]
Volume, V, 10^{-3} m^3	9.28	9.28	9.28	9.28	9.28	9.28	9.28	9.28	9.28	9.28	9.28
Burst pressure, MPa	14.2	14.2	15.4	14.2	13.0	12.9	15.6	14.9	14.1	13.1	14.7
(ksi)	(2.06)	(2.06)	(2.24)	(2.06)	(1.89)	(1.88)	(2.27)	(2.17)	(2.04)	(1.90)	(2.13)
Failure mode[e]	*H*	*F*	*F*	*F*	*K + F*	*F*	*F*	*F*	*F*	*F*	*F*
Composite performance,											
PV/W_c, kPa · m^3/kg	358	368	363	381	324	316	398	383	341	319	376
(10^6in.)	(1.44)	(1.48)	(1.46)	(1.53)	(1.30)	(1.27)	(1.60)	(1.54)	(1.37)	(1.28)	(1.51)
Composite wall											
thickness, mm	1.40	1.45	1.40	1.37	1.35	1.40	1.35	1.37	1.32	1.37	1.35
Longitudinal, mm	0.56	0.56	0.53	0.51	0.51	0.53	0.51	0.51	0.48	0.51	0.51
Hoop, mm	0.84	0.89	0.86	0.84	0.86	0.84	0.86	0.86	0.84	0.86	0.84
Composite mass, W_c, g	368.0	355.0	395.0	347.0	375.0	380.0	364.0	362.0	382.0	381.0	362.0
Fiber content, V, volume %	69.4	72.4	65.8	75.0	67.3	70.1	71.6	72.4	67.8	69.2	73.8
Fiber stress,[f] GPa:											
Hoop	2.46	2.43	2.66	2.45	2.25	2.23	2.70	2.57	2.43	2.26	2.53
Longitudinal	2.21	2.19	2.40	2.21	2.02	2.01	2.44	2.32	2.19	2.04	2.28

[a]Resin system (weight ratio) DER 332/T-403 (100/36), gelled for 16 hours at room temperature and cured for 3 hours at 73.9°C (165°F).
[b]Resin system (weighr ratio) ERL 2258/ZZL 0820 (100/30), gelled for 90 hours at room temperature and cured for 2 hours at 73.9°C (165°F).
[c]Resin system (weight ratio) DER 332/T-403 (100/36), gelled for 16 hours at room temperature and cured for 3 hours at 73.9°C (165°F).
[d]Test rate of 69 kPa/sec (600 psig/min) was used.
[e]*H* = hoop failure, *F* = fitting failure, and *K* = knuckle failure.
[f]Derived (calculated) data.

Table 12.26. Performance of 203-mm-Diameter, Aluminum-Lined Cylindrical Kevlar 49/Epoxy Vessels—(0.76-mm-Thick Aluminum Liner, Test Temperature of 21°C (69.8°F)[a]

PROPERTY	P-143[a]	P-145[a]	P-147[a]	P-149[a]	P-177[a,b]	P-180[a,b]
Volume, V, 10^{-3} m^3	9.28	9.28	9.28	9.28	9.28	9.28
Burst pressure, MPa	14.3	14.7	14.0	17.1	15.2	14.2
(ksi)	(2.08)	(2.13)	(2.04)	(2.48)	(2.20)	(2.06)
Failure mode[c]	*H*	*H*	*F*	*H*	*F*	*H*
Composite performance,						
PV/W_c, kPa · m^3/kg	353	376	353	441	411	391
(10^6 in.)	(1.42)	(1.51)	(1.42)	(1.77)	(1.65)	(1.57)
Composite wall						
thickness, mm	1.47	1.42	1.45	1.37	1.40	1.37
Longitudinal, mm	0.61	0.53	0.51	0.51	0.51	0.51
Hoop, mm	0.86	0.89	0.94	0.86	0.89	0.86
Composite mass, W_c, g	377	362	369	359	343	337
Fiber content, V, volume %	67.5	69.9	68.6	70.3	71.3	69.7

[a]Resin system (weight ratio) DER 332/T-403 (100/36), gelled for 16 hours at room temperature and cured for 3 hours at 73.9° C (165° F).
[b]Test rate of 69 kPa/sec (600 psig/min) was used.
[c]H = hoop failure, F = fitting failure.

electrical and thermal properties are given in Table 12.30.

12.3.6. Hybrid Composites

Kevlar composites have poor compressive properties (i.e., low compressive yield and strength). These properties can be improved substantially by the addition of another fiber (e.g., graphite) to form a hybrid composite. Kevlar and graphite fibers are particularly well suited for hybridization because of their well-matched thermal coefficients of expansion. A Kevlar/graphite composite overcomes

Table 12.27. Performance of 203-mm-Diameter Spherical Kevlar 49/Epoxy Vessels: Resin System DER 332/T-403 (100/36 Parts by Weight), cured for About 8 Weeks at 21°C (69.8°F)[33]

PROPERTY	LINER TYPE: 0.76-mm-THICK ALUMINUM		2-mm-THICK ALUMINUM[a]		RUBBER[b]	
Boss type	Large double		Small single		Small single	
Number of vessels	10		6		4	
Volume, V, 10^{-3} m^3	4.1		4.1		4.1	
	Average	*Standard Deviation*	*Average*	*Standard Deviation*	*Average*	*Standard Deviation*
Burst pressure, MPa	28.7	2.5	39.0[c]	1.1	35.5	1.2
(ksi)	(4.17)	(0.36)	(5.65)	(0.15)	(5.14)	(0.18)
Composite performance,						
PV/W_c, kPa · m^3/kg	334	42	413[d]	17	406	12
(10^6 in.)	(1.34)	(0.17)	(1.66)	(0.07)	(1.63)	(0.05)
Composite wall						
thickness, mm	2.20	0.05	20.6	0.04	2.17	0.08
Composite mass, g	355	19	340	5	358	8
Fiber content, volume %	69.9	4.8	67.9	1.6	68.1	2.6

[a]Test rate of 86 kPa/sec (750 psig/min) was used.
[b]Rubber liner over perforated aluminum; test rate of 115 kPa/sec (1 ksig/min) was used.
[c]Liner effect was not considered in the calculations.
[d]Data are corrected for liner effect.

Table 12.28. Effect of Liquid Hydrogen Temperature on the Performance of Spherical and Cylindrical (203-mm-Diameter) Kevlar 49/Epoxy Vessels: Test Temperature of −253°C (487.4°F) Resin System of DER 332/T-403 (100/36 Parts by Weight)[33]

PROPERTY	CYLINDRICAL (0.760-m LINER)		SPHERICAL (0.76-mm LINER)		SPHERICAL (2-mm LINER)
Vessel number	P-154[a]	P-155[a]	0423[b]	0434[b]	0686[b]
Volume, V, 10^{-3} m^3	9.28		4.09		4.09
Burst pressure, MPa	15.5	16.5	31.9	30.0	43.3
(ksi)	(2.25)	(2.40)	(4.63)	(4.35)	(6.28)
Composite performance,					
PV/W_c, kPa · m^3/kg	398	423	391	336	513
(10^6 in.)	(1.60)	(1.70)	(1.57)	(1.35)	(2.06[c])
Composite wall					
thickness, mm	1.40	1.40	2.24	2.21	2.06
Composite mass, W_c, g	260.9	364.9	334.5	365.6	345.0
Fiber content, volume %	70.8	69.7	75.6	72.8	70.2

[a]Gelled for 3 hours at 60°C (140°F) and cured for 3 hours at 73.9°C (165°F).
[b]Gelled and cured at room temperature.
[c]No correction is applied for liner contribution.

Table 12.29. Mechanical Properties (Warp Direction) of Kevlar 49 Fabric Composites[8]

PROPERTY	COMPOSITE		
	181/EPOXY[a]	181/POLYESTER[b]	120/POLYESTER[c]
Fiber content, volume %	50	40	37
Specific gravity	1.33	1.30	—
Tensile strength, MPa	517	414	440
(ksi)	(75)	(60)	(64)
Tensile modulus, MPa	31	24	23
(Msi)	(4.5)	(3.5)	(3.4)
Failure strain, %	1.7	—	—
Compressive stress at			
0.02% strain offset, MPa	83	—	—
(ksi)	(12)	(—)	(—)
Compressive modulus, GPa	31	—	22
(Msi)	(4.5)	(—)	(3.2)
Flexural strength, MPa	345	207	220
(ksi)	(50)	(30)	(32)
Flexural modulus, GPa	27.6	20	19
(Msi)	(4.0)	(3.0)	(2.7)
Flexural stress at			
0.02% strain offset, MPa	172	97	—
(ksi)	(25)	(14)	(—)
"Short-beam" shear, MPa	55	21	23
(ksi)	(8)	(3)	(3.3)
"Rail" shear, MPa	38	—	—
(ksi)	(5.5)	(—)	(—)
"Rail" shear modulus, GPa	2	—	—
(Msi)	(0.3)	(—)	(—)

[a]America Cyanamid BP-907 epoxy, autoclave-molded.
[b]"Corezyn" polyester.
[c]Room-temperature-cured ICI "Atlac" 382-05 polyester with MEK peroxide.

Table 12.30. Thermal and Electrical Properties of Kevlar 49 Fabric (Style 120)/Epoxy Composites[7,8]

PROPERTY	VALUE
Thermal conductivity (46 volume % fiber)	
Across fabric layers, W/m · °K (Btu in./hr ft^2 °F)	0.22 (1.49)
Parallel to warp, W/m · °K (Btu in./hr ft^2 °F)	0.91 (6.30)
Thermal coefficient of expansion (20–100°C) (68–212°F) 10^{-6}/°C	0
Dielectric constant (58 volume % fiber)	
Perpendicular at 9.3×10^9 Hz	3.3
Parallel at 9.3×10^9 Hz (room temperature)	3.7
Perpendicular (48 volume % fiber) at 10^6 Hz	4.1
Loss tangent (58 volume % fiber)	
Perpendicular at 9.3×10^9 Hz	0.010
Parallel at 9.3×10^9 Hz (room temperature)	0.013
Perpendicular (48 volume % fiber) at 10^6 Hz	0.024
Dielectric strength (48 volume % fiber), V/mm (V/mil)	24.4 (960)
Volume resistivity (48 volume % fiber), Ω-cm	5×10^{15}
Surface resistivity (48 volume % fiber), Ω	5×10^{15}
Arc resistance (48 volume % fiber), seconds	125

the key drawbacks of graphite composites (i.e., high cost and catastrophic failures due to low toughness). A Kevlar/E-glass composite overcomes the major drawbacks of the glass composites (i.e., low stiffness). The possible material combinations for hybrid composites for specific applications are too numerous to summarize here.[34-37] We limit our discussion to some of the typical properties of unidirectional and fabric hybrid composites.

Table 12.31 presents the properties of unidirectional Thornel 300- Kevlar 49/epoxy hybrid composites made with Fiberite 934 epoxy resin.[34] Table 12.32 summarizes data on hybrid fabric composites.[35] Data obtained to date on Kevlar 49 hybrid composites indicate that the impact resistance of graphite composites can be substantially improved via hybridization with aramid fibers with only minor reduction of strengths and stiffness.[37] Also, hybrid composite fabrics offer better energy absorption and a significant cost savings over the graphite composites made from prepreg tapes. However, the mechanical properties of the hybrid fabrics are not as good as those made from the prepreg tapes.[35]

12.3.7. Short Fiber-Reinforced Composites

Blumentritt *et al.*[38] studied fracture in oriented short fiber-reinforced thermoplastics. They concluded that short fiber-reinforced thermoplastics fracture primarily by fibers pulling out of the matrix material, both in ductile and brittle composites. In addition, the ductile matrix materials produce tough composites when fiber concentrations are relatively low.

Table 12.31. Properties of Undirectional Thornel 300-Kevlar 49/Epoxy Hybrid Composites (60 Volume % Nominal Fiber Content)[34]

RATIO OF THORNEL 300 TO KEVLAR 49 FIBERS, %	SPECIFIC GRAVITY[a]	TENSION MODULUS, GPa	(Msi)	TENSION ULTIMATE STRESS, MPa	(ksi)	COMPRESSION STRESS AT 0.02% OFFSET, MPa	(ksi)	COMPRESSION ULTIMATE STRESS, MPa	(ksi)	FLEXURE STRESS AT 0.02% OFFSET, MPa	(ksi)	FLEXURE ULTIMATE STRESS, MPa	(ksi)	SHORT-BEAM SHEAR STRESS, MPa	(ksi)
100/0	1.60	145	(21.1)	1565	(227)	678	(98.4)	1007	(146)	1605	(223)	1606	(233)	91	(13.2)
75/25	1.56	120	(17.4)	1282	(186)	469	(68.8)	938	(136)	1248	(181)	1358	(197)	76	(11.0)
50/50	1.51	108	(15.7)	1213	(176)	413	(59.9)	688	(99.8)	827	(120)	1103	(160)	56	(8.1)
0/100	1.35	77	(11.2)	1262	(183)	182	(26.4)	286	(41.5)	339	(49.2)	634	(91.9)	49	(7.1)

[a]Data supplied by Fiberite.

Table 12.32. Properties of Thornel 300-Kevlar 49/Epoxy Hybrid Balanced Fabric Composites (Nominal 60 Volume % Fiber Content);[35] Balanced Fabric Contains Equal Number of Fiber in the (0-deg) and Fill (90-deg) Directions

RATIO OF THORNEL 300 TO KEVLAR 49 FIBERS, %	RESIN	SPECIFIC GRAVITY	MODULUS, GPa	(Msi)	TENSILE STRENGTH, MPa	(ksi)	COMPRESSION: STRESS AT 0.02% OFFSET, MPa	(ksi)	COMPRESSION: ULTIMATE STRESS, MPa	(ksi)	SHORT-BEAM SHEAR STRENGTH, MPa	(ksi)
0/100	Fiberite 934	1.40	35.9	(5.2)	545	(79)	76	(11)	152	(22)	26	(3.8)
50/50	Fiberite 934	1.49	48.3	(7.0)	400	(58)	159	(23)	228	(33)	29	(4.2)
75/25	Fiberite 934	1.57	57.2	(8.3)	434	(63)	221	(32)	317	(46)	32	(4.7)
100/0	Fiberite 934	1.60	60.0	(8.7)	434	(63)	324	(47)	558	(81)	40	(5.8)
50/50	American Cyanamid BP-907	1.44	46.0	(6.7)	414	(60)	165	(24)	290	(42)	48	(7.0)

Table 12.33. Properties of matrix resins[38]

RESIN	SPECIFIC GRAVITY	TENSILE YIELD STRENGTH,[a] MPa	TENSILE MODULUS, GPa	ELONGATION TO YIELDS, %	TENSILE STRESS/STRAIN CURVE AREA, m · N/m^3
Ionomer	0.95	12.5	0.162	17	1.58
Polyethylene	0.95	22.4	1.07	13	2.08
Nylon 12	1.01	42.7	1.25	12	4.10
Polycarbonate	1.20	61.5	1.93	6.2	2.34
Polymethyl methacrylate	1.19	73.1	2.63	4.6	2.08

[a]The yield point is defined as the first maximum in the engineering stress/strain curve.

Table 12.34. Tensile Properties of Kevlar 49 Short Fiber-Reinforced Composite Material (20 Volume %)[38]

RESIN	ORIENTATION	ULTIMATE TENSILE STRENGTH, MPa	TENSILE MODULUS, GPa	ULTIMATE ELONGATION, %	STRESS-STRAIN CURVE AREA, m·MN/m^3
Ionomer	Random in-plane	119.3	5.033	2.9	1.99
	Unidirectional	157.9	8.894	3.1	2.96
Polyethylene	Random in-plane	73.77	5.16	2.3	1.06
	Unidirectional	131.7	10.6	2.3	1.71
Nylon 12	Random in-plane	106.9	4.17	5.9	3.43
	Unidirectional	151.7	8.55	2.8	3.32
Polycarbonate	Random in-plane	115.8	5.06	3.0	1.94
	Unidirectional	163.4	9.79	2.1	1.83
Polymethyl methacrylate	Random in-plane	182.0	7.79	3.0	3.01
	Unidirectional	207.5	11.10	2.0	2.08

Table 12.35. Shear Strength of Unidirectional Discontinuous Kevlar 49 Fiber Composites (20 Volume %)[38]

RESIN	SHEAR STRENGTH, MPa
Ionomer	27.6
Polyethylene	27.6
Nylon 12[a]	37.9
Polycarbonate	39.9
Polymethyl methacrylate	42.1

[a]Clean shear fractures were not obtained.

However, as the fiber concentration increases, composite toughness becomes more dependent on fiber toughness. The presence of 20 volume % Kevlar 49 short fibers in the matrix materials remarkably improves the mechanical properties of the composite. Table 12.33 lists the properties of a number of matrix materials; Tables 12.34 and 12.35 summarize the tensile and shear properties of the same matrix resins reinforced with Kevlar 49 short fibers.

12.4. APPLICATIONS

Kevlar 49 aramid fiber is designed primarily to be used as a reinforcement in epoxies, polyesters, and other resins for high-performance applications, including aerospace and military designs, sporting goods, boat hulls, flywheels, cables, and tension members. Kevlar 29 aramid fiber is used mainly for ropes and cables and for protective clothing. Another grade of fiber, designated simply as Kevlar, is used only for rubber reinforcements, such as automobile tires, V-belts, and rubber hoses. Kevlar is actually a Kevlar 29 fiber coated with a special rubber-compatible finish.

12.4.1. Aerospace and Military Structures

Because of its high strength-to-weight ratio, Kevlar 49/epoxy composite is used in commercial airplanes and helicopters for floorings, doors, fairings, and radomes (see Fig. 12.38). Filament-wound pressure vessels are used in aircraft and space vehicles as containers for life-support gases and pressurants (Fig. 12.39). Military applications include rocket engine cases (Fig. 12.40) and composite helmets (Fig. 12.41).

12.4.2. Sporting Goods

In many sporting goods, Kevlar 49 composites offer significant weight savings, as well as good impact resistance, vibration damping, and stiffness (see Figs. 12.42–12.44).

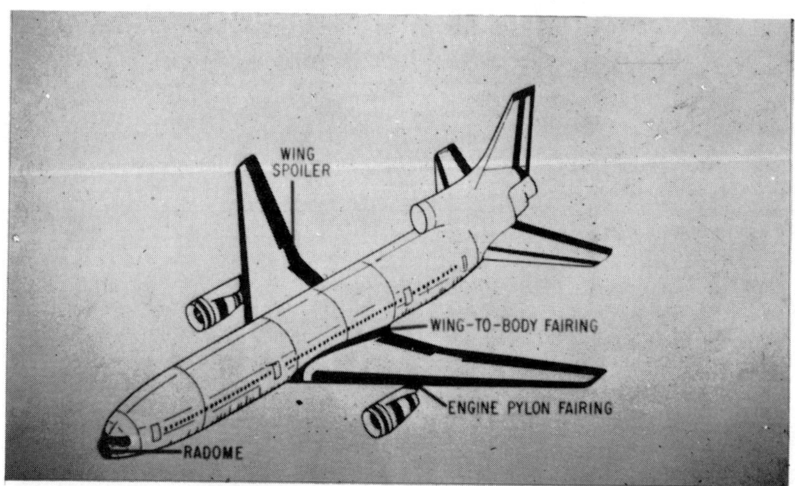

Figure 12.38. Fiber reinforcement areas of the L-1011 exterior. (*Courtesy E. I. DuPont de Nemours.*)

Figure 12.39. Pressure-tested, aluminum-lined Kevlar 49/epoxy vessel (intended for liquid-gas containment).

12.4.3. Ropes and Tension Members

In deep-sea mooring lines, mining and drilling rigs, antenna guys, oceanographic equipment, and yacht rigging, Kevlar 29 and Kevlar 49 fibers and flexible resin-coated fibers are quickly replacing steel. With the highest strength-to-density ratios of any materials known, Kevlar cables offer increased payloads and permit easier handling with smaller, lighter, and more economical equipment. In addition, the corrosion resistance and the nonconducting characteristics of Kevlar fibers are very attractive.

12.4.4. Energy Storage Flywheels

Experimental composite flywheels (Fig. 12.45) are able to store many times more energy per unit weight than metal flywheels. In the U.S., the Kevlar/epoxy composite flywheel, used in conjunction with a battery, is being evaluated as a potential power system for commuter vehicles.

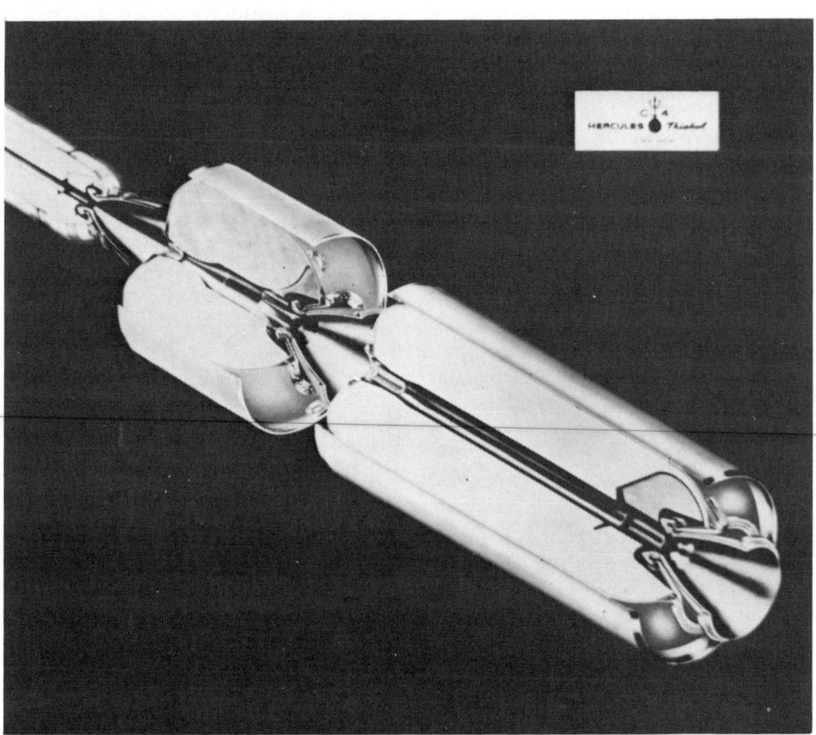

Figure 12.40. Kevlar 49/epoxy filament-wound cases for rocket engines. (*Courtesy E. I. DuPont de Nemours.*)

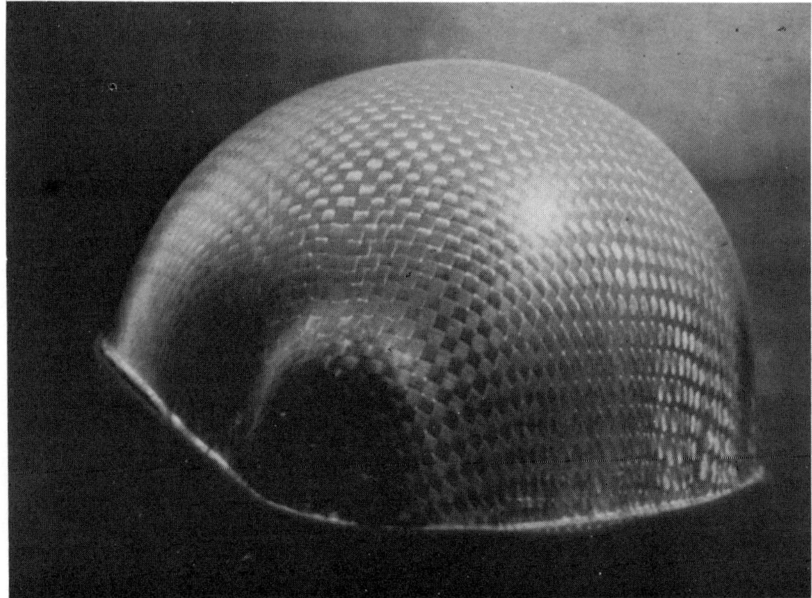

Figure 12.41. Kevlar/epoxy composite helmet. (*Courtesy E. I. DuPont de Nemours.*)

12.4.5. Vehicle Tires

Kevlar cord is being used to replace steel belts in radial automobile tires. For durability, comfort of ride, and hazard resistance, Kevlar is claimed by DuPont to out-perform steel at a significant weight savings. Truck and off-the-road vehicle tires using Kevlar cord also are being actively developed. The Kevlar cord offers higher strength and thermal stability than rayon and polyester cords; better fatigue life and impact resistance than glass cords; and better fatigue life and corrosion resistance, lower weight, and greater flexibility than steel wire.

Figure 12.42. Kayak made with Kevlar 49 fabric/epoxy composite. (*Courtesy E. I. DuPont de Nemours.*)

12.4.6. Soft Body Protection and Composite Armors

According to DuPont, properly designed fabrics of Kevlar 29 and Kevlar 49 provide twice the protection of ballistic nylon at an equal weight. For composite armor, Kevlar fabrics offer protection equivalent to that of a glass/polyester composite at a 30% weight savings. Applications in protective clothing and headgear are well established. Kevlar 29,

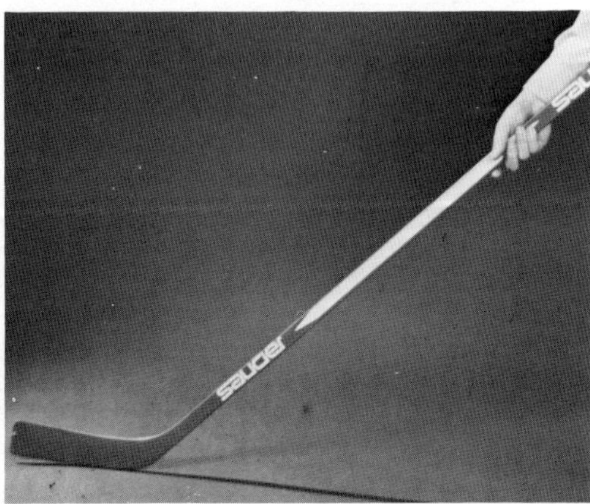

Figure 12.43. Hockey stick made with Kevlar 49 fabric. The fabric can either be laminated directly to the wood or combined with glass to make an all-composite stick. (*Courtesy E. I. DuPont de Nemours.*)

Figure 12.44. Kevlar 49/epoxy composite surfboard. (*Courtesy E. I. Dupont de Nemours.*)

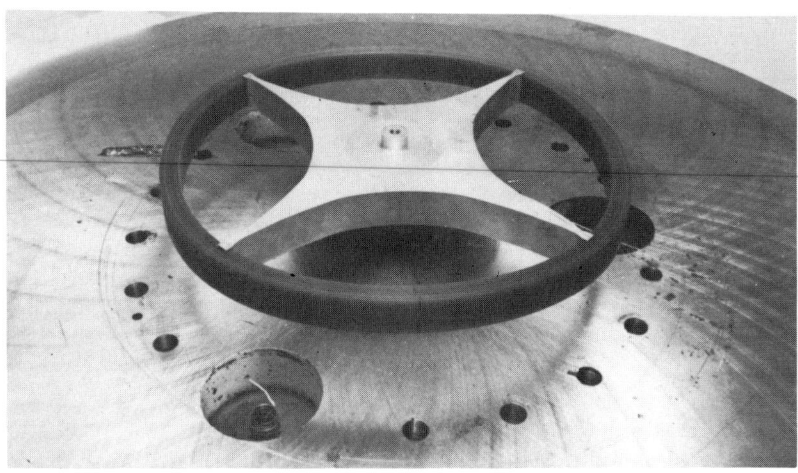

Figure 12.45. Multirim Kevlar 49/epoxy flywheel mounted on an aluminum hub. (*Courtesy Garrett AiResearch.*)

especially, combines impact resistance and cut-through resistance with low weight.

12.4.7. Other Industrial Applications

Other applications include belts, hoses, and conveyor belting; rubber- or PVC-coated architectural fabrics for space enclosures; electrical circuit boards, and high-fidelity speaker cones. Many more potential uses for aramid fiber/epoxy composites are being developed.

ACKNOWLEDGMENT

We sincerely thank Patricia Lien for her long hours of editing of this chapter.

REFERENCES

1. H. Blades, "Dry-Jet Wet Spinning Process," U. S. Patent 3,767,756, October 23, 1973.
2. L. Penn, H. A. Newey, and T. T. Chiao, "Chemical Characterization of a High-Performance Organic Fiber," *J. Mat. Sci.* **11**, 190 (1976).
3. L. Penn, Lawrence Livermore Laboratory, private communication (1976).
4. M. G. Northolt, "X-Ray Diffraction Study of Poly(*p*-Phenylene Terephthalamide) Fibers," *European Polym. J.* **10**, 799 (1974).
5. T. T. Chiao, M. A. Hamstad, M. A. Marcon, and J. E. Hanafee, *Filament-Wound Kelvar 49/Epoxy Pressure Vessels,* Lawrence Livermore Laboratory Report UCRL-51466 (1973). See also National Aeronautics and Space Administration Report NASACR-134506 (1973).
6. N. J. Abbott, J. G. Donovan, M. M. Schoppee, and J. Skelton, "Some Mechanical Properties of Kevlar and Other Heat Resistant, Nonflammable Fibers, Yarns, and Fabrics," Air Force Materials Laboratory, Technical Report AFML-TR-74-65, Part III (1975).
7. J. H. Greenwood and R. G. Rose, "Compressive Behavior of Kevlar 49 Fibers and Composites," *J. Mat. Sci.* **9**, 1809 (1974).
8. *Kevlar 49 Data Manual*, DuPont de Nemours Chemical Co., Wilmington, Delaware, 1974.
9. D. L. G. Sturgeon and T. K. Venkatachalam, "Potential Contribution of High Strength, High Modulus Aramid Fibers to the Commercial Feasibility of Lighter-Than-Air-Craft," paper presented at *Lighter-Than-Air Workshop,* sponsored by the Flight Transportation Laboratory of MIT, the U. S. Navy, and the National Aeronautics and Space Administration, Monterey, California, September 9–13, 1974.
10. P. G. Riewald and T. K. Venkatachalam, "Kevlar Aramid Fiber for Rope and Cable Applications," paper presented at the *Marine Kevlar Cable Workshop,* sponsored by the Marine Technology Society Cable and Connector Committee at the Offshore Technology Conference, Houston, Texas, May 6, 1975.
11. T. T. Chiao and R. L. Moore, "Strength of S-Glass Fiber," *SAMPE Quart.* **3** (*3*), 28 (1972).
12. T. T. Chiao, J. E. Wells, R. L. Moore, and M. A. Hamstad, "Stress-Rupture Behavior of Strands of an Organic Fiber/Epoxy Matrix," *American Society for Testing and Materials, Standard Testing Procedure,* ASTM STP-546, 209 (1974).
13. T. T. Chiao, E. S. Jessop, and M. A. Hamstad, "Performance of Filament-Wound Vessels from an Organic Fiber in Several Epoxy Matricies," in: *Proceedings of the 7th National SAMPE Technical Conference, Albuquerque, New Mexico, October 14–16, 1975,* p. 202.
14. T. T. Chiao and R. L. Moore, "Tensile Properties of PRD-49 Fiber in Epoxy Matrix," *J. Composite Mat.* **7**, 547 (1972).
15. C. C. Chiao, R. J. Sherry, and T. T. Chiao, "Strength Retention and Life of Fiber Composite Materials," *Composites* **7**, 107 (1976).
16. C. C. Chiao, "Long-Term Performance of Fiber Composites," in: *Proceedings of the 1975 Flywheel Technical Symposium*, p. 160, sponsored by the *U. S. Energy Resources and Development Administration and the Lawrence Livermore Laboratory*, Berkeley, California, November 10–12, 1975.
17. C. C. Chiao, "An Accelerated Test for Predicting the Lifetime of Organic Fiber Composites," in: *Proceedings of the 3rd Biennial AIME Symposium: Failure Modes in Composites, Las Vegas, Nevada, February 22–26, 1976*, p. 157.
18. C. C. Chiao, R. J. Sherry, and N. W. Hetherington, "Experimental Verification of an Accelerated Test for Predicting the Lifetime of Organic Fiber Composites," *J. Composite Mat.* **11**, 79 (1977).
19. A. R. Bunsell, "The Tensile and Fatigue Behavior of Kevlar-49 (PRD-49) Fibre," *J. Mat. Sci.* **10**, 1300 (1975).
20. M. A. Hamstad and T. T. Chiao, "Acoustic Emission from Stress-Rupture and Fatigue of an Organic Fiber Composite," *American Society for Testing and Materials, Standard Testing Procedure,* ASTM STP-580, 201 (1975).
21. L. S. Penn and E. S. Jessop, "Fiber-Composite Systems for Energy-Storage Flywheels," in: *Proceedings of the 22nd National SAMPLE Symposium, San Diego, California, April 26–28, 1977*, p. 442.
22. L. L. Clements and R. L. Moore, "Composite Properties for an Aramid Fiber in a Room-Temperature-Curable Epoxy," Lawrence Livermore Laboratory Report UCRL-79549 (1977). See also *SAMPE Quart.* **9** (*1*), 6 (1977).
23. M. P. Hanson, "Effect of Temperature on the Tensile and Creep Characteristics of PRD-49 Fiber/Epoxy Composites," National Aeronautics and Space Administration, Technical Memorandum X68053 (1972).
24. R. H. Ericksen, "Room Temperature Creep of Kevlar 49/Epoxy Composites," *Composites* **7** (*3*), 189 (1976).

25. C. C. Chiao and R. L. Moore, "Evaluation of Interlaminar Shear Test for Fiber Composites," Lawrence Livermore Laboratory Report, UCRL-51766 (1975).
26. C. C. Chiao, R. L. Moore, and T. T. Chiao, "Measurement of Shear Properties of Fiber Composites—I. Evaluation of Test Methods," *Composites*, 161 (July 1977).
27. C. C. Chiao, R. L. Moore, and T. T. Chiao, "Measurement of Shear Properties of Fiber Composites—II. Shear Properties of an Aramid Fiber in Several Epoxy Resins," *Composites*, 171 (July 1977).
28. T. T. Chiao and M. A. Marcon, "Filament-Wound Vessel from an Organic Fiber-Epoxy System," 28th Annual Conference, Reinforced Plastics/Composites Institute, SPI, Washington, D. C., February 6–9, 1973, Section 9-B, p. 1.
29. M. A. Hamstad, T. T. Chiao, and R. G. Patterson, "Fatigue Life of Organic Fiber/Epoxy Pressure Vessels," in: *Proceedings of the 7th National SAMPE Conference, Albuquerque, New Mexico, October 14–16, 1975*, p. 217.
30. R. F. Lark, "Recent Advances in Lightweight Filament-Wound Composite Pressure Vessel Technology," *Composites in Pressure Vessels and Piping*, ASME, Publication No. PVP-PB-021, p. 17 (1977).
31. F. P. Gerstle, Jr. and M. Moss, "Thick-Walled Spherical Composite Pressure Vessels," *Composites in Pressure Vessels and Piping*, ASME Publication No. PVP-PB-021, p. 69 (1977).
32. G. M. Ecord, "Composite Pressure Vessels for Space Shuttle Orbiter," *Composites in Pressure Vessels and Piping*, ASME Publication No. PVP-PB-021, p. 129 (1977).
33. T. T. Chiao, and M. A. Hamstad, "High-Performance Vessels from an Aromatic Polyamide Fiber/Epoxy Composite," in: *Proceedings of the 1975 International Conference on Composite Materials, Geneva, Switzerland and Boston, Massachussetts, April 7–11 and 14–18, 1975*, Vol. 2, p. 365.
34. C. H. Zweben, "Hybrid Fiber Composite Materials," in: *Proceedings of the 1975 International Conference on Composite Materials, Geneva, Switzerland and Boston, Massachussetts, April 7–11 and 14–18, 1975*, Vol. 1, p. 345.
35. C. H. Zweben and J. C. Norman, "Kevlar 49/Thornel 300 Hybrid Fabric Composites for Aerospace Applications," in: *Proceedings of the 21st National SAMPE Symposium Exhibit, Los Angeles, California, April 6–8, 1976.*
36. P. G. Riewald and C. H. Zweben, "Hybrid Composites for Commercial and Aerospace applications," 30th Annual Conference Reinforced Plastics/Composites Institute, SPI, Washington, D.C., February 6, 1975, Section 14-B, p. 1.
37. J. C. Norman, "Damage Resistance of High Modulus Aramid Fiber Composites in Aircraft Applications," paper presented at the Society of Automotive Engineers Business Aircraft Meeting, Wichita, Kansas, April 8–11, 1095.
38. B. F. Blumentritt, B. T. Vu, and S. L. Cooper, "Fracture in Oriented Short Fiber-Reinforced Thermoplastics," *Composites*, 107 (June 1975).

Section II
Processing Methods

13
HAND LAY-UP TECHNIQUES

Charles Wittman
Lawrence Wittman & Company Inc.
Copiague, New York
Gerald D. Shook
Consultant in RP and C
Huntington Station, New York

13.1. INTRODUCTION

Reinforced plastics composites are a combination of a resin matrix, fibers, and fillers, which, when cured, produce a solid structure. The characteristics of the desired end product (such as size, shape, and quantity) determine the method by which the basic building blocks are combined and molded.

In the fabrication of wood or metal products, flat sheet stock is joined together to form a structural part upon which the external paint finish is applied. The reverse of this procedure occurs in room temperature FRP lay-up fabrication, where we have formless materials that harden and take the shape of the container into which they are applied.

Starting with a mold (usually female), the negative of the final desired product, the molder applies a pigmented polyester coating called gel coat. The structure of the part is built up on this coating and consists, usually, of glass reinforcements and catalyzed polyester resin. When the cured item is removed from the mold, the gel coat will be the finished exterior side. Therefore, as opposed to metal or wood fabrication, the product is "painted" before it is made.

Contact or open mold procedures are divided into two main types. In hand lay up, after gel coating the mold, fiberglass chopped strand mat, cloth, or woven roving is laid into the mold; saturated with resin; and brushed or rolled to compact the material and remove entrapped air.

Spray-up molding differs from this in that the glass reinforcement consists of continuous roving which is chopped into short strands and sprayed onto the mold along with catalyzed resin.

With either process, the makeup of the resulting laminate can be considered as an elementary engineering material. By varying the resin/glass ratio, the type and direction of reinforcement, the type of resin, and the type and amount of fillers, we can significantly alter (control) the physical properties. Hence, it is said that the material is alloyed as it is made.

The variability of product composition, and the latitude in size, shape, and design possibilities, enables a lay-up or spray-up part to be tailored to economically fit its desired function.

Although there are some drawbacks, such as the high labor cost, the low production rate, the requirement for trained operators, and the fact that there is only one finished surface (the reverse surface is usually rough), there are several proven applications (shown in Table 13.1) for hand lay-up and spray-up FRP. (Table 13.2 lists the advantages and disadvantages of both contact molding processes, and Table 13.3 lists comparative economic and design factors for FRP laminates using all the available processes.)

Table 13.1. Applications of FRP

- *Agricultural*—fertilizer hoppers, hog slats, silos, pens, dog houses, and sheds.
- *Automotive*—car and truck bodies, fenders, roof caps, and trailers.
- *Bonding*—joining, coating, and repairing of existing structures.
- *Commercial*—signs, displays, caskets, store fixtures, and hospital furniture (see Fig. 13.1).
- *Construction*—tubs, showers, gutters, leaders, roof panels, curtain walls, facades, doors, sinks, concrete forms, modular housing, and roof systems.
- *Corrosion resistance*—plating, equipment, cisterns, tanks, storage and mixing materials, reactors, scrubbers, ducts, pipes, shelters, coatings, and liners.
- *Electrical*—transformer housings and switch gear enclosures.
- *Equipment housings*—machine guards, computer enclosures, equipment cabinets, business machine housings (see Fig. 13.2).
- *Furniture*—tables, chairs, and outdoor furniture.
- *Marine*—canoes, boats, floats,docks, buoys, minesweepers, and fishing, sail, and power boats.
- *Recreation*—recreation vehicles, snowmobiles, playground equipment, and amusement park equipment.
- *Swimming pools*—solar heating panels, pool covers, pool walls, steps, slides, diving boards, filter tanks, and starting blocks (see Fig. 13.3).
- *Water pollution*—septic tanks, coatings, weirs, and chlorination equipment.

Table 13.2. Advantages and Disadvantages of Hand Lay-Up and Spray-Up

ADVANTAGES

- Design flexibility
- Large and complex items can be produced
- Production rate requirements are low
- Minimum equipment investment is necessary
- Tooling cost is low
- Any material that will hold its shape can be used as a mold form
- The start-up lead time and cost are minimal
- Design changes are easily effected
- Molded-in inserts and structural reinforcements are possible
- Sandwich constructions are possible
- Prototyping and pre-production method for high-volume molding processes.
- Semi-skilled workers are needed, and are easily trained

DISADVANTAGES

- The process is labor-intensive
- Only one good (molded) surface is obtained
- Quality is related to the skill of the operator
- It is a low-volume process
- Longer cure times required
- Product uniformity is difficult to maintain within a single part or from one part to another
- The waste factor is high

See Figs. 13.1, 13.2, and 13.3 for applications of FRP.

13.2. PROPERTIES

The choice of resin system, reinforcement, fillers, sandwich, and laminate schedule directly affects the properties of the laminate being made.

Typical hand lay-up/spray-up properties, value ranges, and controlling factors are shown in Table 13-4.

13.3. DESIGN

The fundamental information needed for any design includes the stresses applied under storage and use and the strength of the material used. (Processing is a function of how the product is to be made.)

Assume that the size, shape, quantity, and rate of production have dictated the use of open mold techniques. Then the final thickness, orientation, and quantity of reinforcing fibers is dependent upon the stresses that must be resisted, how often, and for how long.

It is essential that the designer finds out what strength can be built into the laminate. This sets the FRP fabricator apart from the thermoplastic fabricator, since the FRP fabricator "makes the material when he makes the product"; the percentages and orientation of the reinforcement, the types of fillers, and the types of resins determine the properties of the final laminate.

The FRP fabricator commences with ingredients—resin, filler, fibers, etc.—and, using an appropriate formulation, mixes the components and pours, pulls, squeezes, or sprays this mix into molds where the resin polymerizes and this mix evolves into the part. Therefore, design of FRP products is more than determining size, shape, and so on; it also involves careful attention to what happens to the mix in or on the mold—especially what

Table 13.3. Economic and Design Factors for FRP Laminates*

MOLDING PROCESS	EQUIPMENT COST	RATE OF PRODUCTION	MOLDED PART STRENGTH	IMPORTANCE OF OPERATOR'S SKILL	PART COMPLEXITY POSSIBLE	PART REPRO-DUCIBILITY
Hand Lay-up	1	3	3	10	9	1
Vacuum bag	2	2	4	10	9	3
Pressure bag	3	1	6	6	7	4
Spray-up	4	4	3	10	8	1
Filament winding	6	6	10	2	4	9
Pultrusion	7	9	9	2	2	10
SMC	10	8	7	4	9	10
Centrifugal	9	7	8	3	3	6
Continuous laminate	10	10	5	2	1	10
Resin injection	3	2	3	7	7	8
Injection	10	10	6	2	10	10
Shell coating	9	4	3	5	7	9
Pre-mix (BMC)	9	8	7	4	8	10

*10 equals highest; 1 equals lowest.

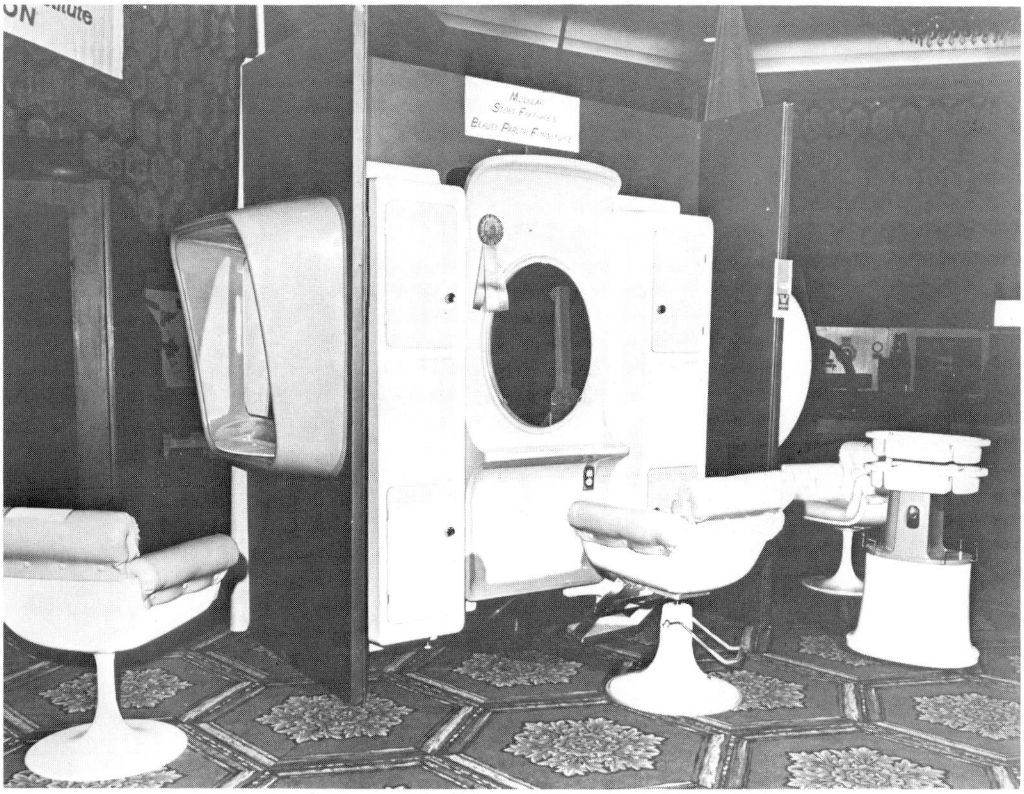

Figure 13.1. Beauty parlor equipment and furnishings. (*Courtesy Lawrence Wittman & Co. Inc., Copiague, N.Y.*)

Figure 13.2 Museum of Broadcasting console.(*Courtesy Elliot Fine, Brooklyn, New York.*)

Figure 13.3. Swimming pool steps. (*Courtesy Lawrence Wittman & Co., Inc., Copiague, N.Y.*)

happens to the fibers in the mix—and what properties can be obtained.

13.3.1. Typical Examples*

13.3.1.1. Box-Like Machine Cover (Formerly Made of Steel)

It was determined that an FRP cover would be lighter than a steel cover and would not rust. Quantity and size indicated an open mold process. The critical property here was material stiffness.

Procedure. Determine basic thickness of the part in FRP, using the empirical thickness formula:

$$t_{FRP} = t_s \sqrt[3]{\frac{E_s}{E_{FRP}}}$$

where

t_{FRP} = thickness of FRP
t_s = thickness of steel unit
E_s = modulus of elasticity of steel
E_{FRP} = modulus of elasticity of FRP.

At this point, the value of E_{FRP} must be known, and the designer should find out what values can be developed with a specific process (spray-up, fabric hand lay-up, etc. or he can select a value of E_{FRP} and formulate an actual mix from which the value that is chosen can be attained. Assume he chooses a value of $E_{FRP} = 2 \times 10^6$ psi (14 GPa).

With a 0.10-in. (2.5-mm) thickness FRP and an E_{FRP} of 2×10^6 psi (14 GPa), the designer will have to specify a fabric type laminate because continuous fibers are required to obtain such a modulus (see Fig. 13.7).

By looking at fabric specifications, a construction can be chosen to give 0.10 in. (2.5 mm) thickness with the required strength and at the least cost. For instance: use one-ply, 24 oz (814 gr/m^2) woven roving, or 0.04 in. (1.0 mm), plus two plies of 1½ oz (400 gr/m^2) mat,

*The formulae and the methods for stress and size calculations discussed here are suitable primarily for low load, contact layup structures. For highly loaded composites, see Chap. 20.

Table 13.4. Physical Properties

	HAND LAY-UP				
	MAT	FABRIC	SPRAY-UP	ALUMINUM	STEEL
Specific gravity	1.4–1.8	1.6–2.0	1.4–1.6	2.75	7.85
Tensile strength psi × 10^3(MPa)	10–20 (70–140)	20–50 (135–340)	9–18 (60–125)	6–27 (40–185)	29–33 (200–230)
Tensile modulus, psi × 10^6 (GPa)	0.8–1.8 (5.5–12.5)	1.5–4.5 (10–30)	0.8–1.8 (5.5–12.5)	10 (70)	30 (205)
Compressive strength, psi × 10^3 (MPa)	10–25 (70–175)	30–55 (140–375)	10–25 (70–175)	9 (62)	28 (195)
Flexural strength, psi × 10^3 (MPa)	20–40 (135–275)	35–65 (240–445)	16–22 (110–150)	20 (135)	30 (205)
Flexural modulus, psi × 10^6 (GPa)	0.8–1 (5.5–7)	1.2–2 (8–14)	0.8–1 (5.5–7)	10.3 (70)	28–30 (190–205)
Impact strength, ft-lb/in. (J/cm)	5–25 (2.5–13.5)	20–30 (10.5–16)	5–25 (2.5–13.5)	25 (13.5)	45 (24)
Hardness, Barcol (depends on resin)	40–55	45–65	40–55		
Moisture absorption, %	0.05–1.0	0.05–1.0	0.05–1.0		
Burning rate		Slow to self-extinguishing			
Heat resistance, continuous °F (°C)	150–350 (50–160)	150–350 (50–160)	150–350 (50–160)		
Effects of acids and alkalies		Fair to excellent			
Effect of solvents		Poor to fair			
Machining qualities	Fair to good	Use carbide or diamond tools	Fair		
Dielectric constant, IMEG/9360 MEG		4–4.6/ 4.1–4.5			
Loss tangent, IMEG/9360 MEG		0.01–0.06/ 0.01–0.03			
Glass content, %	30–40	45–60	25–40 (10–25 filled)		
Dielectric strength, volts/mil		900			
Thermal insulation	1.3–1.8	1.3–1.8	1.3–1.8	810–1620	260–460
Acoustics		Good to excellent		Poor	Poor
Weathering		Good to excellent		Corrodes	Rusts
Optical properties		Translucent to opaque		Opaque	Opaque
Coefficient of expansion, in/in/°F × 10^{-6} (m/m/°C)	10–18 (18–32.5)	10–18 (18–32.5)	10–18 (18–32.5)	12–13 (21.5–23.5)	6–10 (11–18)
Ultimate tensile elongation, %	1–2	1–2	1–2	30–40	38–40

or 0.06 in. (1.5 mm), to give about 0.10 in (2.5 mm) total thickness. Sixty percent (maximum) resin and a 0.02 in. (0.5 mm) gel coat should be specified for a total laminate of 0.110–0.120 in. (2.75–300 mm) thickness with built-in color and a weight of approximately 0.110/0.125 × 1 lb/ft^2 = 0.88 lb/ft^2 (33.6 kg/m^2).

The steel equivalent was made from a 0.040 in. (1 mm) sheet of approximately 0.040 × 144 = 0.4 lb/in.^{2} = 1.38 lb/ft^2 (52.7 kg/m^2). The FRP provides the advantages of sound damping and corrosion resistance. In addition, it will not be electrically or thermally conductive.

Alternatively, the designer could have used boat cloth, an open weave-balanced fabric about 0.010 in (0.25 mm) thick, and a lay-up plan of six plies of boat cloth and two plies of 1 oz mat to yield approximately 0.10 in (2.5 mm) thickness with a 0.020 in. (0.5 mm) gel coat to provide color.

This laminate will also supply a modulus of 2 × 10^6 (14 GPa), but it requires eight laminating operations, as opposed to the three-ply laminate in the first example. Furthermore, boat cloth costs more per pound or per yard than woven roving because it is more difficult to manufacture. Hence, cost effectiveness should always be evaluated when choosing a laminate reinforcement system.

13.3.1.2. Pressure Tank

Classical stress analysis proves that hoop stress is twice that of longitudinal stress (stress trying to push out the ends of the tank).

To build a tank of conventional materials (steel, aluminum, etc.) requires the designer to use sufficient materials to resist the hoop stresses, which results in unused strength in the longitudinal direction. In FRP, however, the designer specifies a laminate which has twice as many fibers in the hoop direction as in the longitudinal direction.

Consider a pressure tank 3 ft in diameter and 6 ft long (0.91 × 0.15 m) with semi-spherical ends. Such a tank's stress calculations (excluding the weight of both the product contained in, and the support for, the tank) are represented by the formulas $s = pd/2t$, for the hoop stress, and $s = pd/4t$, for the end stresses and longitudinal stresses, where s = stress, p = pressure, d = diameter, and t = thickness (tensile stresses are critical in tank design). The designer assumes the pressure in this application will not exceed 100 psi (700 Pa) and selects a safety factor of 5. The stress must be known so that the thickness can be determined. The stress or the strength of the final laminate is derived from the makeup and proportions of the ingredients, including the resin, the mat, and the continuous fibers. Representative panels must be made and tested, with the developed tensile stress values then used in the formula. Thus, the calculated tank thickness and method of lay up or construction can be determined.

$$t_h = \frac{pd}{2s_h} \qquad t_l = \frac{pd}{4s_l}$$

$$t_h = \frac{100 \times 3 \times 12}{2 \times \dfrac{20 \times 10^3}{5}} = 0.450 \text{ in.}$$

$t_l = \frac{1}{2} t_h = 0.225$ in. (or, the same thickness with half the load or stress)

where

t_h = hoop thickness
t_l = longitudinal thickness
s_h = hoop stress
s_l = longitudinal stress

$s_h = 20 \times 10^3$ psi (140 MPa)

safety factor = 5
p = 100 psi max (700 Pa)
d = 3 ft (1.0 m).

If the stress values had been developed from a laminate of alternating plies of woven roving and mat, the lay-up plan (or schedule) would include sufficient plies to make 0.40 in. (11 mm) or about four plies of woven roving and three plies of 1½ oz (460 gr/m^2) mat. However, the laminate would be too strong axially. To achieve a laminate with 2 to 1 hoop to axial strength, one would have to carefully specify the fibers in those two right angle directions, or filament-wind the tank so that the vector sum of the helical wraps would give a value of

2 (hoop) and 1 axial, or a wrap of approximately 54° from the axial.

Another alternative would be to select a special fabric whose weave is 2 to 1, warp to fill, and circumferentially wrap the cylindrical sections to the proper thickness, thus getting the required hoop and axial strengths with no extra, unnecessary strength in the axial (longitudinal) direction, as would inevitably be the case with a homogeneous metal tank.

As can be seen from the above, the design of FRP products, while essentially similar to conventional design, does have the added dimension that the materials are combined when the product is made. The FRP designer must consider how the load bearing fibers are placed and ensure that they stay in the proper position during the act of fabrication.

13.3.2. Resin

The choice of resin system to be used in a given product is a major design factor.

Performance characteristics, such as chemical resistance, fire retardance, flexibility, weathering, impact, warpage, electrical properties, and strength, are partially or completely determined by the type of resin used.

Processing and handling characteristics (gel time, knife trim time, peak exotherm, warpage, shrinkage, opacity, viscosity, and thixotropy) are also considerations in the determination or resin to be specified.

Resin manufacturer's specification sheets should be consulted to help choose a resin with proper characteristics to suit a particular application.

13.3.3. Filled Resins

An additional design aspect of resin selection is the possibility of incorporating various fillers in the system. Depending on the type and percent loading, fillers can reduce product cost by replacing the resin and/or the glass. Fillers can also improve flame retardance, suppress toxic smoke generation, increase stiffness, lower the peak exotherm, reduce shrinkage (resuting in less glass pattern print transfer through the gel coat), improve electrical properties, and reduce weight.

While fillers can be used in hand lay-up, their greatest usage is in spray-up, where batch mixing and spray application can ensure uniformity and minimize handling problems. Higher filler loading can be achieved by starting with a low viscosity resin.

Table 13.5 lists some of the most widely used spray-up and lay-up fillers, their range of loading, and their typical uses.

13.3.4. Fire Retardance

As noted above, resin selection plays a critical role in determining the degree to which a laminate will retard combustion or be self-extinguishing. The most fire retardant resins are the halogenated varieties, the vinyl esters, and those resins with "HET" acid bases.

In addition to the resin, the presence of fillers, the percentage of filler loading, certain aspects of construction (such as balsa sandwich), and the use of intumescent coatings have an effect on flammability rating and the cost of achieving that rating.

Table 13.6 is a compendium of the types of test procedures and methods of rating the fire retardance of FRP laminates.

13.3.5. Design Details

13.3.5.1. Parts With Severe Contour and Thickness Variations

It may seem easy to incorporate variations in contour and thickness into the design of a new product. However, in open mold products, such designs must be made with full awareness of what is involved.

The molding operation requires laying the material on the mold to follow its contour. If the angles are sharp (90° without radius), the lay-up will not follow the mold surface and will develop air bubbles behind the gel coat, in the vicinity of the angle. For instance, in inside right angle corners without radius, the laminate will not pack into the corner:

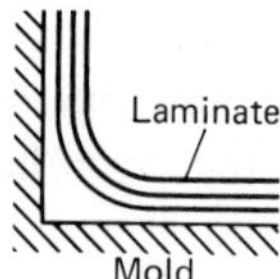

Table 13.5. Fillers for Hand Lay-Up and Spray-Up: Types of Fillers, Ranges of Loading, and Typical Uses

LOADING A = 1–10% B = 10–30% C = 30–50%	RESIN EXTENDER	FIRE RETARDANT	SMOKE SUPRESSANT	INCREASE STIFFNESS	LOWER EXOTHERM	REDUCE SHRINKAGE	ELECTRICAL PROPERTIES	REDUCE WEIGHT	SYNTACTIC FOAM	THIXOTROPE	IMPROVED MACHINABILITY	PASTES AND PUTTIES	CHEMICAL RESISTANT COATINGS	POTTING	HAND LAY-UP	SPRAY-UP
Aluminum trihydrate (ATH)	C	C	C	C	C	C	C								C	C
"Q-Cell"	B,C			C	C	C		C	C		B	B,C	B	B,C	C	C
3M Spheres	B,C			C	C	C		C	C		B	B,C	B	B,C	C	C
Potter's Beads	B,C			C	B,C	B,C							B	B,C	B,C	B,C
Suzorite Mica	C			C	C		C						B,C			C
"Micromix"	C			C	C								B,C			C
"Cabosil Aerosil"	A									A		A		A	A	A
Milled Fibers					A							A		A	A	
Sand	A				A	A								A	A	
Glass Flakes				B		A	A						A,B		A	A
Antimony Trioxide		A	A												A	A
Soapstone											A	A				
Clay	A										A	A				

Table 13.6. Test Methods—Fire Retardance* of FRP Laminates

METHOD	DESCRIPTION	RATING	COMMENTS
BENCH TOP TESTS			
ASTM D-635	Horizontal specimen ignited by gas flame	Ratings eliminated; now report burning length and time	Least severe test; used for screening
UL 94 horizontal	Same as above	94HB	Same as above
UL 94 vertical, used for UL 478 data processing equipment and UL 484 room air conditioners, etc.	Vertical specimen ignited by gas flame	(94 V-0) 5 specimens × 2 ignitions, 5 seconds average burn (94-V-1), 5 specimens × 2 ignitions, 25 seconds average burn (94-5V) specimens × 5 ignitions, 60 seconds maximum burn	Generally used for filled systems; required for UL approval; specimen thickness critical
HLT-15	Vertical specimen ignited by gas flame	0–100 (100 best rating)	Moderately severe test, suitable for medium to high halogen content resins
ASTM D-757 "Globar Test"	Electrically heated rod in contact with specimen	Burning rate for 3 minutes in./minute	Suitable only for medium to high halogen content resins
Federal Test Standard 406, Method 2023	Bureau of Ships electrically heated coil test of ½ × ½ in. (12 × 12 mm) laminate	Ignition time and burning time reported	Severe test for high heat applications, not open flame
ASTM E-162	Gas heated radiant panel	Flame spread index calculated from ignition properties and heat liberated	Severe test for surface flammability; used for filled systems
ASTM D-2863 (limiting oxygen index)	Specimen burned in atmosphere of controlled oxygen concentration	Oxygen index—minimum oxygen concentration that supports combination; index of 28 considered fire retardant	Excellent reproducibility; filled systems do not necessarily relate to high halogen

FULL-SCALE BURN TESTS			
ASTM E-84, UL 723, NPFA 255 ("Steiner Tunnel Test")	25-ft-long tunnel with test panels on top; ignition by gas flame	Compare to asbestos as 0 and red oak as 100 by flame spread and smoke generation	Standard for codes in construction and industrial uses
Factory Mutual "Corner Wall" Test and UL "Corner Room Test"	Ignite full-scale fire in a structure built of test panels		Tests being developed

*The numerical flame spread or other test results developed from the above tests are not intended to define the hazards presented under actual fire conditions.

When there are sharp outside corners, the laminate will not wrap tightly over the corner:

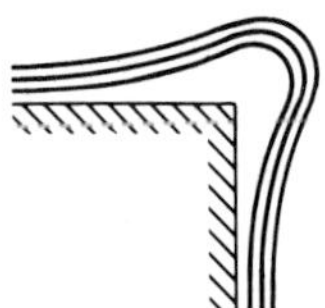

The solution to such problems is to design with a generous radius (preferably 3/16 to 1/2 in.) (4.75 to 12.75 mm) inside and out. The laminate will then follow the contour.

Abrupt changes in direction are high stress areas and tend to delaminate and crack. They should be avoided and moderate self-reinforcing curvatures used.

13.3.5.2. Changes in Thickness

To change thickness in open mold construction is to add or remove plies of material. An abrupt change means that the plies must be

carefully laid up in a precise pattern, which increases the labor cost. An abrupt change in thickness results in a stress concentration or high stress point and should be avoided as delamination is sure to occur at such a point. The solution to this problem is not to have abrupt changes but to gradually change by stepping back or "shingling" the lay-up.

13.3.5.3. Openings

The best opening is a round hole; the worst, an opening with sharp, non-rounded corners. The solutions to stresses in an opening is to use large radii in the corners, to build up thickness gradually at sharp corners, or design a molded-in flange around the opening.

13.3.5.4. Joints and Bonding*

FRP products often are made by the assembly of several separate parts. A sail boat, for example, consists of a hull, an inner liner, a cockpit, a cabin top, and a deck, and all these parts are bonded together, each bonded area representing a type of joint. We classify joints in order of their strength, as follows (high and low):

1. Lap shear
2. Butt
3. Cleavage
4. Peel.

The *lap shear joint* is perhaps the easiest to make and is quite common. Its shape and mode of loading utilize most adhesives where they have the greatest strength. You will find lap shear joints where bulkheads are joined to the hull, where decks are bonded to cabin tops, and so on.

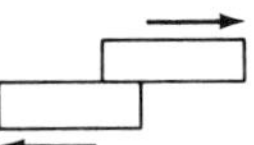

It is easy to calculate the strength of a lap shear joint. Merely figure the lap area and multiply by the nominal lap shear values published by the adhesive manufacturer (or by the use values you have discovered on your own).

A failure in a lap shear joint occurs when it degrades into a peel joint as the loads increase:

Increased loads cause the joint to rotate so as to line up the forces. This rotation causes bending in the FRP and induces peel conditions at the ends of the lap. If loads increase, then the peel stresses exceed the adhesive's strength and the joint quickly fails. However, if the edges of the lap are tapered, the stiffness is reduced: When the joint rotates to accommodate the tension force, the tapered edges of the lap, being more flexible, do not overstress the adhesive. Thus, this simple modification can increase the strength at the failure without increasing the shear area. Furthermore, by proper preparations, even higher strengths can be obtained for the same shear area by making a *scarf joint*:

A *scarf joint* is a lap shear joint that introduces *no* rotation when under load. The adhesive is in shear and there are little or no peel effects. Such joints require more labor to

*See also Chapter 22.

prepare and require good jigging or tooling to locate and secure adherends during cure of the adhesive. But they represent the best joints available and often can develop strengths nearly equal to that of the adherends.

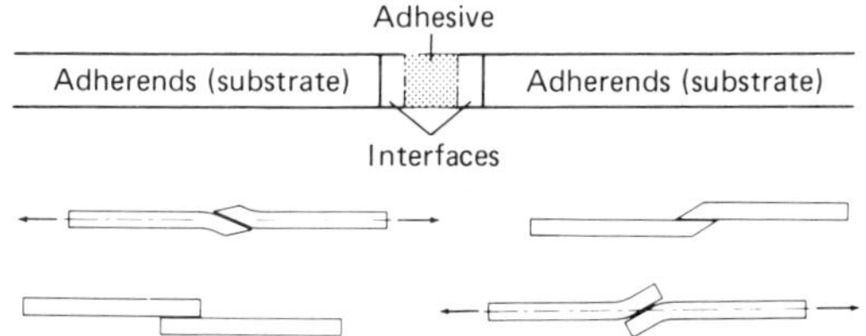

The *butt joint* is a pure tension joint between rigid adherends. It is usually low to moderate in strength and is easy to calculate, but it represents a type of joint that occurs very seldom in actual construction. Also, a butt joint must be carefully prepared—and it must be loaded so that the stresses in the joint, under load, are balanced. Often, television commercials for a new adhesive show a drop of the adhesive between two discs, with the disc assembly in the cable between a support and a large load (such as a car), demonstrating that the adhesive between the discs holds fast while the load is lifted. This represents an ideal butt joint: no side loads; only pure tension and remarkable strength:

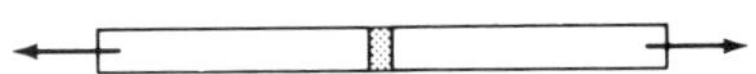

A *peel joint* represents a construction wherein the stresses are concentrated in a line where one adherend bends away from the other and puts unbalanced tension stress in the adherend:
In such a joint, only the adhesive at the point

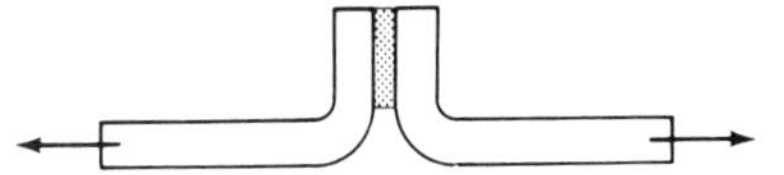

of peel is working; the rest of the adhesive is not loaded until it, too, is stressed at the point of peel (as the joint begins to fail).

13.3.6. Design Hints

- Combine and consolidate parts into single moldings wherever possible.
- Non-skid and decorative textured surfaces can be molded in. (Use flexible rubber molds for extreme texture).
- Drill-hole locations can be molded in or scribed on the mold.
- For a self-reinforcing design, use shape (such as crowning or ribbing) to provide stiffness, eliminating or reducing the need for subsequent addition of stiffening members.
- Use shape to meet the functional needs of the part—combining aesthetics, form, and function.
- Provide the maximum radius and draft possible, consistent with aesthetic and functional requirements.
- Design for producibility: avoid deep, tight, zero-draft sections.

13.3.7. Designing Considerations

- Minimum inside radius: 0.1875–0.25 in. (4.8–6.4 mm). (Tighter radii are possible, but not desirable.)
- Minimum draft recommended: 2° (0° with split molds).
- Undercuts: should be avoided but can be made by using split or rubber molds.
- Trimmed in mold: (knife trim to size).
- Molded in holes: large diameter only
- Minimum practical thickness (in.): 0.030 in. (0.8 mm), hand lay-up; 0.06 in. (1.5 mm), spray-up.
- Maximum practical thicknesss (in.): unlimited total, 0.25 in. (6 mm) per cure.
- Normal thickness variation: +0.030/−0.015 in. (+0.8/−0.4 mm), hand lay-up; +0.025/−0.025 (+0.64/−0.64 mm), spray-up.
- Maximum thickness build-up: as desired.
- Built in cores (sandwiches).
- Metal inserts.
- Metal (or other) edge stiffeners.
- Bosses: must be tapered.
- Fins: special handling required.
- Corrugated (or crowned) surfaces.
- Limiting size factor: none (mold size and handling consideration).
- Maximum size part to date: 4000 ft^2 ($372\ m^2$).

- Shape limitations: none.
- Finished surfaces: one (two with special tooling).
- Gel coat surface: Only one smooth surface, reverse side can be gel coated after molding.
- Surface mat.
- Translucency: Not transparency.
- Molded in labels: Graphics, logos, part numbers.
- Raised or recessed numbers and graphics.
- Strength orientation: random or directional (spray-up—random only).
- Typical glass loading (% by weight): 30–55%, hand lay-up; 25–45%, spray-up, 10–20%, filled resin spray-up.

13.3.8. Simplified Design Calculations

Some of the principles and formulas for the design of lightly loaded fiberglass parts of the type suitable for contact molding are given below. Additional information on the design and analysis of composite structures is included in Chapter 16, on filament winding, and Chapter 20, on Design of Composites, and Chapter 21, on sandwich structures.

The symbols used in the formulas below are as follows.

F_B = flexural (bending) stress, psi (Pa)
F_t = tensile stress
p = load, lb (kg) (for tanks, pressure of water)
l = length between supports, in. (m)
b = width of panel, in. (m)
t = thickness of panel or wall, in. (m)
t_{FRP} = thickness of fiberglass, in. (m)
t_M = thickness of metal, in. (m)
E_{FRP} = modulus of elasticity of fiberglass, psi (Pa)
E_M = modulus of elasticity of metal, psi (Pa)
d = diameter of tank, in. (m)
h = height of tank, in. (m)
F_n = hoop stress, psi (Pa)

Tables 13.7, 13.8, 13.9, and 13.10 list the properties and design requirements for various commercial fiberglass structures, such as pipes, tanks, and joints.

13.3.8.1. Flexural Strength*

$$F_s = \frac{3}{2}\,\frac{pl}{bd^2}$$

Let

$p = 150$ lb
$l = 3$ in.
$b = 1$ in.
$d = 1/8$ in.

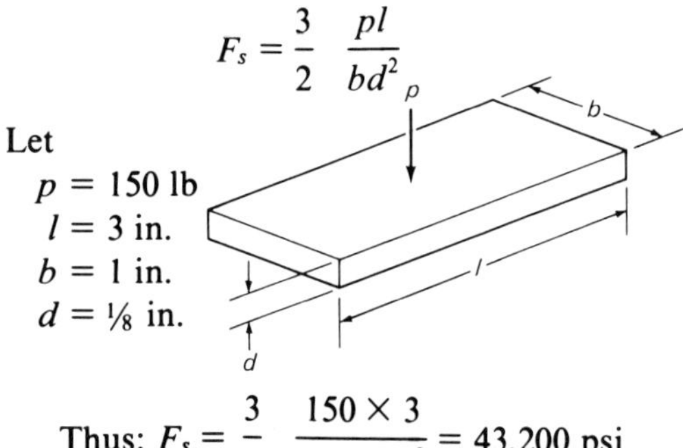

$$\text{Thus: } F_s = \frac{3}{2}\,\frac{150 \times 3}{1 \times (1/8)^2} = 43{,}200 \text{ psi.}$$

13.3.8.2. Thickness Conversion, Metal to Plastic

$$t_{FRP} = t_M \sqrt[3]{\frac{E_M}{E_{FRP}}}$$

Use:

$E_M = 30 \times 10^6$ psi (steel)
$E_{FRP} = 8 \times 10^5$ psi (spray -up laminate)
$t_M = 0.030$ in. (approximately 22 gauge).

$$\text{Thus: } t_{FRP} = 0.030 \sqrt[3]{\frac{30 \times 10^6}{8 \times 10^5}} = 0.100 \text{ in.}$$

13.3.8.3. FRP Tank Design

$$t = \frac{pd}{2F_t}$$

p = water pressure = approximately ½ psi/ft of depth

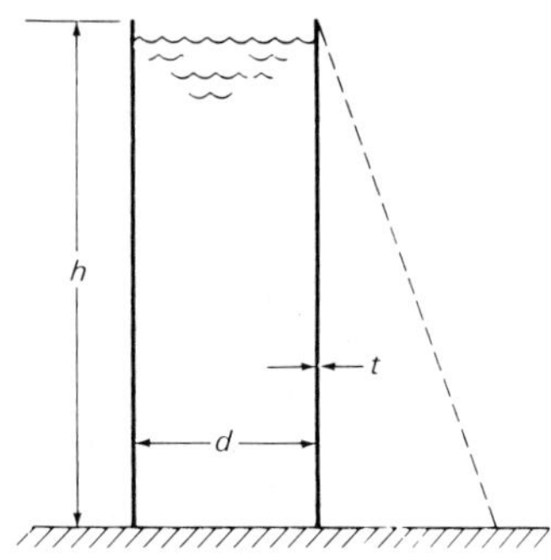

*To simplify the text, only the "English" system is used in the examples.

Let
$h = 50$ ft $\qquad p = 25$ psi
$d = 10$ ft $= 120$ in.
F_t = tensile stress equivalent of laminate (based on tests) = 15,000 (for spray-up).

$$t = \frac{25 \times 120}{2 \times 15{,}000}$$

$t = 0.100$ in. with no safety factor
Demands 10 to 1 safety factor; thus: $t = 1$ in. (at bottom).

13.3.8.4. Hoop Stresses

Tension is based on pressure and diameter.

$$\text{Hoop stress} = F_h = \frac{pd}{2t}$$

1. Choose stress and obtain t. Design composite to match .
2. Use safety factor based on use, chemical, temperature, life, etc.
3. Example: If stress is 15,000 for a spray-up and the safety factor is 4, use $\frac{15{,}000}{4}$ as E_h and solve for t.

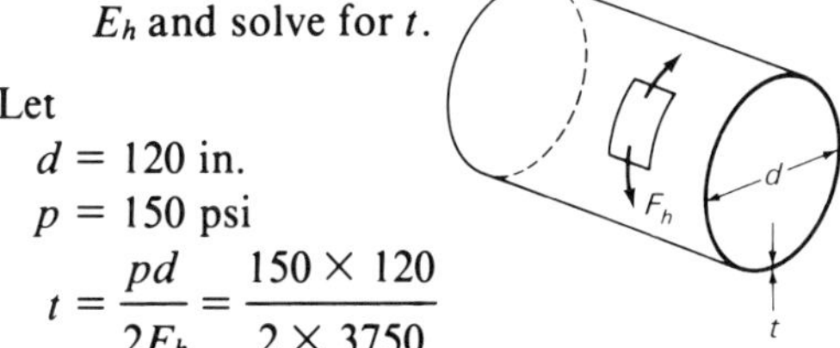

Let
$d = 120$ in.
$p = 150$ psi

$$t = \frac{pd}{2F_h} = \frac{150 \times 120}{2 \times 3750}$$

$t = 2.4$ in. (this is unrealistic; try filament winding: $F_h = 75{,}000/4 = 18{,}750$ psi)

$$t = \frac{150 \times 120}{2 \times 18{,}740}$$

$t = 0.48$ in.
(A more reasonable value).

13.4. TOOLING

13.4.1. Introduction

Tooling in most industries represents materials, equipment, or forms onto which (or into which) the product is made, assembled, or cast. Contact mold tooling differs little from the above, but has certain special conditions or limitations which are dependent upon the nature of the product's ingredients prior to consolidation—namely, liquid resins, powdered fillers and reinforcing fibers and fabrics. Tooling for contact molded laminates can range from using an actual part as a pattern, from which a mold is laid up or cast, to a precisely designed, engineered, and fabricated polished chrome-plated steel tool, which could be every bit as costly as a press molding die. The choice of tooling is based upon several interacting imputs such as:

1. Quantity of parts to be made.
2. Value of the part to be made.
3. Size of the part.
4. Tolerances of the part.
5. Complexity and appearance.

If it is assumed that a product to be made is described by an engineering drawing and specification, the normal sequence of events would be as follows.

1. A pattern in wood, plaster, plywood, etc. would be made to the blueprint dimensions and fitted with trim flanges. This pattern would be finished to the same surface finish as the desired end product and prepared for the next step by application of a suitable parting film or release system.

2. Upon this pattern would be laid up, or cast, a shell, known as the mold, the surface of which would replicate the surface of the pattern. This mold surface is generally reinforced with FRP and finished off with additional rigidizing structure made from metal or wood, to permit handling during its use without distortions and damage.

3. This mold is then carefully removed (parted) from the pattern (often the pattern is destroyed during this operation). The mold surface is inspected and repaired (if required and polished to enhance appearance and to ease removal of subsequent molded parts.

4. The mold is then prepared for use by "breaking it in" by proper application of release agents (waxes and/or films) to ensure that parts will readily separate from the mold (see pp. 343–344).

Table 13.7. Physical Properties of Laminates Made of Polyester Resin and Various Glass Compositions*

LAMINATE	TYPE OF CONSTRUCTION PLIES	MATERIAL	THICKNESS, IN. (mm)	GLASS CONTENT, %	TENSILE STRENGTH, psi (MPa)	FLEXURAL STRENGTH, psi (MPa)	MODULUS $\times 10^6$, psi (GPa)	IMPACT, ft-lb/in. OF NOTCH (J/m)	WEIGHT OF GLASS, oz/ft^2 (kg/m^2)	WEIGHT OF RESIN, oz/ft^2 (kg/m^2)	WEIGHT OF LAMINATE, OZ/FT^2 (kg/m^2)
1	2	2-oz mat	0.125	28.2	14,050	23,400	0.92	9.0	4.0	10.2	14.2
			(3.2)		(96.8)	(161.0)	(6.33)	(480)	(1.2)	(3.1)	(2.3)
2	1	1000 cloth	0.100	28.4	10,500	18,550	0.89	8.2	3.1	7.8	10.9
	1	2-oz mat	(2.5)		(72.3)	(128.0)	(6.13)	(438)	(0.9)	(2.4)	(3.3)
3	1	1000 cloth	0.110	32.0	13,850	22,100	1.10	9.8	4.1	8.7	12.8
	2	1.5-oz mat	(2.8)		(95.4)	(152.0)	(7.58)	(523)	(1.2)	(2.7)	(3.9)
4	1	1000 cloth	0.130	30.1	14,950	21,400	0.95	9.8	5.1	11.8	16.9
	2	2-oz mat	(3.3)		(103.0)	(147.5)	(6.54)	(523)	(1.5)	(3.6)	(5.1)
5	1	1000 cloth	0.180	32.0	13,200	18,750	0.92	12.0	7.1	15.1	22.2
	3	2-oz mat	(4.6)		(90.9)	(129)	(6.33)	(640)	(2.2)	(4.6)	(6.8)
6	1	1000 cloth	0.120	24.9	9090	37,600	1.47	8.2	3.7	11.2	14.9
	1	1.5-oz mat	(3.0)		(62.6)	(259)	(10.13)	(438)	(1.1)	(3.4)	(4.5)
	1	1000 cloth									
7	1	1000 cloth	0.180	22.6	10,500	28,400	1.20	9.7	5.2	17.8	23.0
	2	1.5-oz mat	(4.6)		(72.3)	(196)	(8.27)	(518)	(1.6)	(5.4)	(7.0)
	1	1000 cloth									
8	1	1.5-oz. mat	0.254	39.2	30,300	42,800	1.59	31.2	14.01	21.99	36.0
	3	2415									
		Fabmat	(6.5)		(208.7)	(295)	(10.95)	(1665)	(4.3)	(6.7)	(11.0)
9	1	1000 cloth	0.092	42.5	18,500	28,500	0.78		5.0	6.7	11.7
	1	24-oz W.R.	(2.3)		(127.5)	(196)	(5.37)		(1.5)	(210)	(3.5)
	1	1000 cloth									
10	1	1000 cloth	0.125	38.3	17,050	22,850	1.25	31.5	5.4	8.6	14.0
	1	1.5-oz mat	(3.2)		(117.5)	(157.5)	(8.61)	(1680)	(1.6)	(2.6)	(4.2)
	1	24-oz W.R.									
11	2	24-oz W.R.	0.080	52.7	38,950	44,900	1.85	36.3	5.0	4.5	9.5
			(2.0)		(268.4)	(309.5)	(12.75)	(1940)	(1.5)	(1.4)	(2.9)
12	1	24-oz W.R	0.100	53.2	29,000	45,900	2.20	45.8	7.0	6.2	13.2
	1	1.5-oz mat	(2.5)		(199.8)	(316)	(15.15)	(2445)	(2.1)	(1.9)	(4.0)
	1	24-oz W.R.									
13	1	2-oz mat	0.125	36.0	11,500	23,000	0.73		5.9	10.5	16.4
	1	24-oz W.R.	(3.2)		(79.2)	(158.5)	(5.03)		(1.8)	(3.2)	(5.0)
	1	1000 cloth									

14	1	1.5-oz mat	0.100	47.0	22,200	41,800	1.90	42.2	7.0	7.9	14.9
	2	24-oz W.R.	(2.5)		(153.0)	(288)	(13.1)	(2255)	(2.1)	(2.4)	(4.5)
15	1	1.5-oz mat	0.100	47.9	24,900	31,400	1.11	25.8	6.8	7.5	14.3
	1	24-oz W.R.	(2.5)		(171.6)	(216.5)	(7.65)	(1380)	(2.0)	(2.3)	(4.3)
	1	1.5-oz mat									
	1	1000 cloth									

NOTE: Mat—weights based on oz/ft (2-oz mat refers to 2 oz/ft^2). Cloth and woven roving—weights based on $oz/yd.^2$ (24-oz W.R. refers to 24 ox/yd^2; 1000 cloth refers to 10 oz/yd^2). Examples do not include gel coat. 2145 Fabmat is a registered trademark of Fiberglas Industries, Inc. Weights are based on 4.2 oz/ft^2.

2 oz mat = 610 g/m^2.

1.5 oz-mat = 458 g/m^2.

24 oz woven roving = 814 g/m^2.

Table 13.8. Typical Requirements for Properties of Mat Reinforced Polyester Laminates*

	PROPERTIES AT 73.4° F (23° C)			
Thickness, in. (mm)	1/8–3/16 (3–4.75)	1/4 (6)	5/16 (8)	3/8 and up (10 and up)
Ultimate Tensile Strength, Minimum, psi (MPa)	9000 (60)	12,000 (80)	13,500 (90)	15,000 (100)
Flexural Strength Minimum, psi (MPa)	16,000 (110)	19,000 (130)	20,000 (140)	22,000 (150)
Flexural Modulus of Elasticity (Tangent), Minimum, psi (GPa)	700,000 (4.8)	800,000 (5.5)	900,000 (6.2)	1×10^6 (6.9)

*From *NBS Voluntary Product Standard P. S. 15–69.*

Table 13.9. Minimum Wall and Bottom Thickness of Vertical Tanks Relative to Diameter and Distance from Top*†

DISTANCE FROM TOP (ft)	MINIMUM WALL AND BOTTOM THICKNESS (ft) FOR TANKS OF DIAMETER:														
	2 ft	2½ ft	3 ft	3½ ft	4 ft	4½ ft	5 ft	5½ ft	6 ft	7 ft	8 ft	9 ft	10 ft	11 ft	12 ft
2	3/16	3/16	3/16	3/16	3/16	3/16	3/16	3/16	3/16	3/16	3/16	3/16	3/16	3/16	3/16
4	3/16	3/16	3/16	3/16	3/16	3/16	3/16	3/16	3/16	3/16	3/16	3/16	3/16	3/16	3/16
6	3/16	3/16	3/16	3/16	3/16	3/16	3/16	3/16	3/16	3/16	3/16	3/16	1/4	1/4	1/4
8	3/16	3/16	3/16	3/16	3/16	3/16	3/16	3/16	3/16	1/4	1/4	1/4	1/4	1/4	5/16
10	3/16	3/16	3/6	3/16	3/16	3/16	3/16	1/4	1/4	1/4	1/4	1/4	5/16	5/16	5/16
12	3/16	3/16	3/16	3/16	3/16	3/16	1/4	1/4	1/4	1/4	1/4	5/16	5/16	5/16	3/8
14	3/16	3/16	3/16	3/16	1/4	1/4	1/4	1/4	1/4	5/16	5/16	5/16	5/16	3/8	3/8
16	3/16	3/16	3/16	1/4	1/4	1/4	1/4	1/4	1/4	5/16	5/16	3/8	3/8	3/8	7/16
18	3/16	3/16	3/16	1/4	1/4	1/4	1/4	5/16	5/16	5/16	3/8	3/8	3/8	7/16	1/2
20	3/16	3/16	1/4	1/4	1/4	1/4	5/16	5/16	5/16	3/8	3/8	3/8	7/16	1/2	1/2
22	3/16	1/4	1/4	1/4	1/4	5/16	5/16	5/16	5/16	3/8	3/8	7/16	1/2	1/2	9/16
24	3/16	1/4	1/4	1/4	1/4	5/16	5/16	5/16	3/8	3/8	7/16	1/2	1/2	9/16	5/8

Sample Calculation for Thickness.

$$S = \frac{\text{ultimate tensile strength}}{\text{safety factor}}$$

$$= \frac{15{,}000 \text{ psi}}{10}$$

p = 0.5 psi per ft of depth
d = diameter, in.
t = thickness, in.

Given:
distance from top = 24 ft
diameter of tank = 10 ft

$$t = \frac{pd}{2S}$$

$$t = \frac{(24 \times 0.5) \times (10 \times 12)}{2\left(\frac{15{,}000}{10}\right)}$$

= 0.48 in. (or approximately ½ in.)

*Based on a safety factor of 10 to 1 using mechanical property data and a liquid specific gravity of 1.2. For tanks intended for service above 180° F (82.2° C) consideration in design should be given to the physical properties of the material at the operating temperature. Tanks with physical loadings, such as agitation, should be given special design consideration.
†From *NBS Voluntary Product Standard P. S. 15–69.*
**For metric values see Fig. 13.10.

Table 13.10. Minimum Wall and Bottom Thickness of Vertical Tanks Relative to Diameter and Distance from Top

DISTANCE FROM TOP (m)	MINIMUM WALL AND BOTTOM THICKNESS (mm) FOR TANKS OF DIAMETER:														
	0.6	0.75	0.9	1.1	1.2	1.35	1.5	1.7	1.8	2.1	2.4	2.75	3	3.4	3.6
0.6	4.8	4.8	4.8	4.8	4.8	4.8	4.8	4.8	4.8	4.8	4.8	4.8	4.8	4.8	4.8
1.2	4.8	4.8	4.8	4.8	4.8	4.8	4.8	4.8	4.8	4.8	4.8	4.8	4.8	4.8	4.8
1.8	4.8	4.8	4.8	4.8	4.8	4.8	4.8	4.8	4.8	4.8	4.8	4.8	6.4	6.4	6.4
2.4	4.8	4.8	4.8	4.8	4.8	4.8	4.8	4.8	4.8	6.4	6.4	6.4	6.4	6.4	8
3.0	4.8	4.8	4.8	4.8	4.8	4.8	4.8	6.4	6.4	6.4	6.4	6.4	8	8	8
3.6	4.8	4.8	4.8	4.8	4.8	4.8	6.4	6.4	6.4	6.4	6.4	8	8	8	9.5
4.2	4.8	4.8	4.8	4.8	6.4	6.4	6.4	6.4	6.4	8	8	8	8	9.5	9.5
4.8	4.8	4.8	4.8	6.4	6.4	6.4	6.4	6.4	6.4	8	8	8	9.5	9.5	11
5.4	4.8	4.8	4.8	6.4	6.4	6.4	6.4	8	8	8	9.5	9.5	9.5	11	13
6.0	4.8	4.8	6.4	6.4	6.4	6.4	8	8	8	9.5	9.5	9.5	11	13	13
6.6	4.8	6.4	6.4	6.4	6.4	8	8	8	8	9.5	9.5	11	13	13	14.3
7.2	4.8	6.4	6.4	6.4	6.4	8	8	8	9.5	9.5	11	13	13	14.3	16

The product made from the tooling described above would have only one finished side, as is typical of the open mold process, and would be expected to produce a minimum of 100 parts. This type of tooling would also be suitable (with slight modifications) for vacuum bag or autoclave molding processes.

13.4.2. "One-Off"

If it is assumed that only one large FRP part is required (such as a 40-ft yacht hull), the cost of making a pattern and a mold in order to make the hull is not justified. In fact, the construction of a pattern is nearly equal to the actual hull and could be a hull. Furthermore, the construction of the mold from the pattern represents effort and cost equal to that of making the hull. Therefore the cost of one 40-ft hull made from pattern and mold will be two to three times that of the hull alone.

To avoid this cost and still get an FRP hull, "one-off" tooling for the 40-ft hull would be made, as follows.

1. A simple temporary frame to the inside contour of the hull is constructed from stations and stringers and covered with strips or sheets of wood, foam, or fiberglass splining material.
2. This surface is then sprayed or coated with plied layers of resins/mat/cloth to the thickness designated and allowed to cure.
3. The outside of the hull is now essentially complete but has a rough surface, unsuitable for use, and must be finished. The procedure is essentially one used today in an auto-body shop and consists of filling and grinding the surface in two or three stages and culminates with a final coat of a good grade of epoxy or urethane marine paint (this paint step is often postponed until the boat is completely assembled and ready for fitting out).
4. The hull is then inverted and placed in a fitted saddle or shoring and the temporary frame removed. Additions of an inner skin for sandwich construction, bulkheads, flooring, and interior complete the work.

(The deck and cockpit are usually made in the same manner, but are often made from wood in order to save labor costs.)

13.4.3. No Mold (No Pattern or Plug)

It is often desired to copy an existing item, such as a carved wooden piece of furniture. In this case, neither pattern nor plug is made, as the actual item serves as the pattern.

The sample part must be well coated with release waxes and parting films so that the mold materials will not attack or bond to it. These mold materials, usually polyester resins, urethane resins, latex, and silicone resins, are cast against the item; allowed to cure; backed up by a rigidizing structure; and eventually parted from the item. The sample can then be cleaned of release materials and returned to its original condition. The mold thus constructed is used to make duplicate items.

13.4.4. Patterns (Plugs)

A pattern is a temporary form, the exact shape, contour, and finish of that which is to be molded. (If the outer shape of the item is desired, the inside contour is used.)

Patterns are made of many materials—namely, wood, plaster, plaster/metal, and other combinations. Wood and plaster patterns are made in generally the same way that patterns have been made traditionally for the foundry. Pattern makers for steel, aluminum, and brass foundries can easily adjust to plastics. Almost any material can be considered as a pattern material if it holds its shape. The pattern maker must avoid shellac or other pattern coatings that would react with parting agents or the styrene component of polyester resins. Acceptable pattern coatings are such materials as sanding surfacers or sealers and are usually polyester or epoxy formulations.

13.4.5. Preparation/Mold Release

Preparation of a finished pattern for mold making entails the application of coating materials that prevent the mold from sticking to the pattern, while being thin enough to permit transfer of surface details from pattern to mold. After sanding the pattern with successively finer sandpaper and buffing it with a machine polishing compound, it is ready to be waxed and/or coated with polyvinyl alcohol (PVA) or some other release agent. The manufacturer's recommendations should be carefully followed. Poor or improper waxing could cause the mold to permanently adhere to the pattern.

13.4.6. Parting Boards/Split Molds

If the product to be made has a configuration that would make it impossible to remove it from the mold, it is then necessary to design a mold that will split into two or more pieces.

Provisions for splitting the mold are constructed on the pattern and are called parting boards. The location and number of parting boards is determined by configuration and aesthetics. Flash will occur at a split line, so thought must be given to locate the split where it will least be visible.

Conditions requiring split molds include the size of the product (ease of handling molds), contour (extremely deep draw sections), undercuts (return flanges), and lack of adequate draft (or no draft at all).

13.4.7. Molds (Temporary and Permanent)

Temporary molds are considered molds of sufficient strength to withstand the rigors of making one to five parts without too much damage.

Permanent molds are molds built to last for several years under normal use, or molds very seldom used but expected to maintain quality and contour even though they may be stored for extended periods between uses. Such molds will have greater thickness, sandwich construction, and usually steel or wooden supplementary reinforcement.

13.4.8. Mold Designs

Mold design is a function of cost and projected life and/or use. A production mold is made carefully from the best materials and by the most skilled workmen. Such a mold will be designed by an experienced designer who will incorporate the necessary thicknesses, materials, and structural reinforcement and hardware required for the intended use. The mold for a high production item such as a tub-shower combination will be vastly different from a mold for a prototype. For instance, a production mold made from FRP will have:

1. Tooling gel coat—slow cure.
2. High-temperature tooling resin.
3. Laminate—carefully made and with high glass content (usually over 50%). (Sandwich construction such as balsa core is often used.)
4. Back-up structure—steel tubing tied to FRP, with good support at the edges of

Table 13.11. Minimum Total Widths of Overlays for Reinforced-Polyester Tank Shell Joints*

Tank Wall Thickness, in. (mm)	3/16 (4.8)	1/4 (6)	5/16 (8)	3/8 (9.5)	7/16 (11)	1/2 (13)	9/16 (14.3)	5/8 (16)	11/16 (17.5)	3/4 (19)
Minimum of Outside Overlay Width, in. (mm)	4 (100)	4 (100)	5 (125)	6 (150)	7 (175)	8 (200)	9 (225)	10 (250)	11 (275)	12 (300)
Minimum of Inside Overlay Width, in. (mm)	4 (100)	4 (100)	5 (125)	5 (125)	6 (150)	6 (150)	6 (150)	6 (150)	6 (150)	6 (150)

*From *NBS Voluntary Product Standard P. S. 15–69.*

the mold and with provision for support during part construction (including hooks for conveyor support and axles for rotation).

5. Provisions for part removal, such as jacking devices and lugs, air ports for blow-off, hinged, pinned, or removable sections (if required), and a proper knife trim edge.

A prototype made from FRP would have:

1. General purpose gel coat.
2. Minimum lay-up—reasonable effort in the construction and glass content less than 40%.
3. Minimum or no back-up structure—wood or plywood.
4. Few, if any, provisions for handling the mold would be added.

13.4.9. Mold Design Parameters

Mold design parameters are similar to part design parameters, with the following additional points.

- Part removal considerations:
  1. Blow-out holes—locate at deep draw areas or tight sections. (Note that blow-out holes, while very necessary for part removal, will cause a blemish, so care must be taken to locate them in the least noticeable places.)
  2. Jack outs—screw jacks or wedges.
  3. Separable molds for undercut, deep draw, or no-draft situations.
- Edge configuration:
  1. Knife trim edge for parts, molded to size.
  2. Extra height to allow for secondary saw trim.
- Reinforcement:
  1. Sandwich construction (balsa, plywood, foam, etc.).
  2. Ribs, egg crate, tubes, (wood, steel, or paper tubes).
- Multi-piece (separable) molds:
  1. Used where undercuts or no-draft situations occur.
  2. Caution—misalignment will cause a step to be molded in (flash will occur, which necessitates secondary finishing—vulnerable area of mold).

13.4.9.1. Design Considerations For Multi-Piece Molds

1. Adequate mating flanges for clamping.
2. Proper and positive line-up devices—bumps, pins, bolts, etc.
3. Design the mold split line into the pattern; use proper parting boards.
4. Reinforce both halves to avoid shrinkage and warping.
5. Avoid multi-piece molds if possible.

13.5. MOLDING METHODS

13.5.1. Introduction

Hand lay-up and spray-up are more than the simple combination of fiberglass, resin, and catalyst. Numerous reinforcements, additives, and fillers can be combined with the resin to customize the mixture to meet specific performance requirements.

Since good surface appearance is usually desired, the first step in these open-mold processes is the application of a specially formulated resin layer called gel coat. After this coating cures, layers of fiberglass are laid on or sprayed on to the mold, saturated with catalyzed resin, and compacted against the mold.

In hand lay-up, the fiberglass is applied to the form of chopped strand mat, cloth, or woven roving. The resin and catalyst are either pre-mixed in a bucket and applied with a brush and squeegee, or deposited via a wet-out gun which mixes the components as it sprays them.

The spray-up process involves chopping the fiberglass into short strands and mixing it simultaneously with the resin and catalyst in one spraying operation. Once this mixture is deposited on the mold, it is compacted with serrated metal rollers.

Since hand lay-up uses pre-formed mat and woven fabrics, a high degree of product uniformity strength, and control can be achieved.

Spray-up, a partially automated form of hand lay-up, relies heavily on operator technique for product control (Fig. 13.4). While

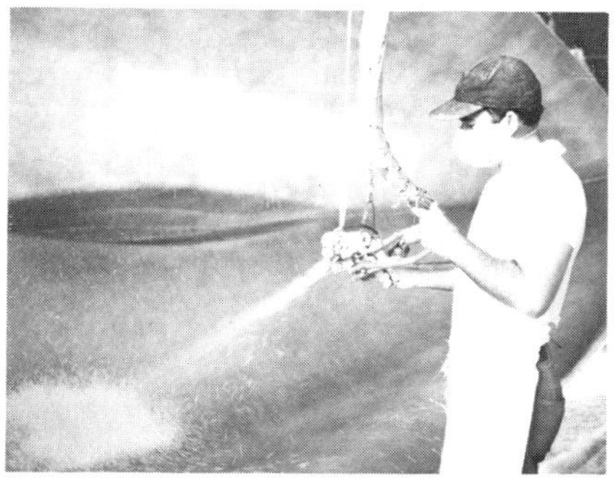

Figure 13.4. Fiberglass spray operation.

less uniform parts are made, considerable cost savings can be realized due to the lower cost of gun roving compared to mat or fabric, and a lower labor factor due to the faster rate of material deposition.

Additional cost savings are attainable with spray-up by utilizing inexpensive filler systems, which replace the more costly resin or glass components, while yielding products of comparable strength.

In practice, combinations of hand lay-up and spray-up are often used in which layers of sprayed chopped fibers are alternated with cloth or woven roving.

Cored laminates (where a layer of end-grain balsa, foam, plywood, etc. is sandwiched between layers of resin-saturated glass) are attainable with either hand lay-up or spray-up.

Except for the method of glass deposition, all other processing characteristics are generally the same in both systems.

13.5.2. Mold Preparation/Release*

The first step in making any open mold product is the application of a release material to the mold. Without such a material, the part will permanently bond to the mold surface.

Many different release systems are available (see below). The choice depends upon the type of surface to be molded, the degree of luster desired on the finished product, and whether or not painting or other secondary finishing will be required.

A production mold preparation program starts with a thorough machine buffing and polishing of a new (unused) fiberglass mold. After the desired shine has been attained, several coats (usually three or four) of paste wax are applied. Each coat should be well buffed with a clean towel and adequate time allowed between coats for the wax to "set up."

Quite often during the mold "break-in period" (first one to four parts), an additional

*Also see Chap. 23.

Mold Release Methods

TYPE	FORM	USE
Wax	Paste or liquid (carnuba)	High luster Good detail transfer Multiple runs
PVA	Liquid (usually sprayed)	Water-soluble One-time use Wash off part and mold Excellent release Provides paintable surface
Fluorocarbons silanes and silicones	Liquid or spray	Low coefficient of friction Not very high gloss More expensive
Release papers and release films	Coated paper Cellophane PVA film	One-offs Flat sheet molding
Internal Releases	Liquid (mixed into gel coat)	High luster Good detail transfer Eliminates need for mold waxing. Provides paintable surface.

coat of sprayed-on PVA (poly vinyl alcohol) is used on top of the wax to ensure part release.

Molds should always be buffed after each molding cycle. Depending on the type of wax and the molding conditions, additional wax should be applied every one to five or more cycles. Care should be taken to avoid wax build-up (evidenced by rough and/or dull parts).

Periodically (every 10 to 20 parts), the mold should be stripped down (removing all the old wax and any residual resin) using a commercial mold stripper/cleaner followed by a complete rewaxing.

Proper mold preparation is one of the most important functions in the molding cycle. If it is done well, the moldings will look good and separate from the mold easily. If it is neglected or short cuts are taken, product appearance will suffer,and if a part "hangs-up" in the mold and can't be removed, both the part and the mold might have to be scrapped.

13.5.3. Gel Coating

Gel coat is normally a polyester, mineral filled, pigmented, non-reinforced layer or coating which is applied first to the mold but which becomes the outer surface of the laminate when completed. This produces a decorative, high protective, glossy, colored surface which requires little or no subsequent finishing. (see Table 13.12.)

- Techniques of application:
  1. Paint on—uncontrolled thickness; unadvisable, non-production method.
  2. Cup gun—pre-catalyzed, air-atomized; throw away mixing cup; inexpensive, low-production method; gravity feed, pre-mixed.
  3. Pressure pot—catalyst and gel coat under pressure; mixed at nozzle; air-atomized.
  4. Airless sprayer—similar (or the same as) spray-up equipment.

Table 13.12. Troubleshooting Guide for Gel Coats

PROBLEM	CAUSE	SOLUTION
Wrinkling of gel coat during lamination (alugatoring).	Uncured or thin gel coat. Gel coat swells and separates from mold surface in confined area because of insufficient cure and action of styrene in lay-up resin.	Check with wet film gauge for minimum 15 mils thickness: Apply gel coat evenly. Allow adequate gel coat cure time.
Waviness in gel coat.	Too long a gel time in lay-up. Too long a gel time for lay-up resin.	Use more catalyst; adjust catalyst to weather conditions for 1-hour cure.
Streaks in gel coat (particularly pastel colors).	Draining of gel coat, causing color separation.	Use heavier gel coat or lay molds flat.
Hollow spaces beneath gel coat.		
Rough molded surface.	Wax build-up.	Wash off with styrene or buff with mold cleaner.
Glass pattern in mold.	Soft mold gel coat.	Use heat-resistant resin in future molds.
Star crazes in mold.	Rough handling use of mallet in removing part from mold.	1. Grind down to glass. 2. Apply mold gel coat. 3. Apply wax paper and tape. 4. Refinish.
Wrinkling of gel coat immediately after application.	Trapped acetone; water in gel coat; insufficient catalyst in gel coat.	Hold gun farther from mold; use higher atomization; use more catalyst; drain traps; check line; warm molds.
Dimples in gel coat (when using PVA film).	PVA-separating film not dried.	Allow more drying time; clean line of moisture.

Table 13.12. Continued

PROBLEM	CAUSE	SOLUTION
Cracking of gel coat.	Too heavy a coat. Back-up layer not cured; shrinks later and cracks gel coat.	Use 25-mil maximum thickness; use fast cure on first layer.
Pits in gel coat.	Foreign particles in film.	Spray film in dust-free room.
Uneven color in gel.	Air entrapment; poor hiding power; insufficient pigment.	Use styrene for good flow; consult gel coat supplier; use 10% minimum pigment.
Dull surface.	Rough mold.	Refinish mold.
Difficulty in removing part from mold.	Mold not broken in; rough mold; undercuts in mold; insufficient wax.	Use PVA; repeat mold-prep process; fill undercuts; cover all areas.
Telegraphing of glass pattern in gel coat.	Gel coat too thin; undercure.	Use 15–20 mils; wait for full gel coat cure.
Patching does not match gel coat.	Patch cured too fast.	Use thinned gel coat; use low-catalyst concentration; do not add filler.
Gel coat sticking to mold (brushed or sprayed).	Improper release agent or application.	Apply release and let cure. If wax, allow to dry thoroughly and buff. If trouble persists, use PVA-sprayed film over wax.
Hazy or non-glossy surface.	Entire part prematurely removed from mold. Contamination of release prior to application of gel coat.	Permit more complete cure of gel coat and lay-up.
Voids under gel coat.	Small or large, flat blisters caused by separation of gel coat from lay-up. Gel coat should not cure tack-free in air but should remain sticky for better bond to lay-up.	Allow first lay-up application to cure prior to adding second and third (etc.). Inspect closely for blisters after lay-up. Cut out and putty mix: 1 part resin to 3 parts $CaCO_3$.
Open bubbles, blisters, and pinholes in gel coat surface.	Trapped air, free solvent, dirt, or excessively high exotherm in gel coat or lay-up resin.	Avoid mixing air into gel coat when introducing catalyst. Let stand for short period after mixing and before spraying. Keep containers and working area clean.
Soft areas.	Uneven cure.	More thorough mixing of catalyst into gel coat.
Cratering.	Use of too high surface angle release, preventing gel coat from wetting in small spots, 1/16–1/4 in., so that lay-up shows through gel coat.	More careful selection and application of release agent.

- Quality control:
  1. Preweigh gel coat and catalyst.
  2. Mold should be a different color than the gel coat being sprayed so that uniform and complete coverage can be easily seen.
  3. Use wet film gauge (should be, at minimum, 18(±2) mils.
  4. Check for proper catalyst levels.

13.5.4. Hand Lay-up

After properly preparing the mold and gel coating it, the next step in the molding process is material preparation.

Chopped strand mat, cloth, and woven roving are all supplied in large rolls (in various widths). The desired length sheet is cut from the roll and then cut into patterns, if necessary, by using a utility razor knife, large scissors, or an electric cloth-cutting machine, like those used in the garment industry.

Pre-measured resin and catalyst are then thoroughly mixed together. The resin mixture can be applied to the glass either outside of or on the mold. To insure complete air removal and wet out, the resin is applied first with the glass placed on top. Brushes, squeegees, and serrated rollers (or mohair rollers, in some cases) are used to compact the material against the mold to remove any entrapped air.

As previously mentioned, the resin/catalyst mixture can be deposited on the glass via a spray gun, which automatically meters and combines the ingredients.

Extra care must be given to the first layer (often a skin coat of surfacing veil or a 0.75-oz mat) to ensure that no air bubbles are left between the glass and the gel coat.

Additional layers of mat and/or woven roving can then be applied until the total design thickness (and hence strength) is achieved. Layers of mat and woven roving should be alternated to ensure good interlaminar bond, to avoid entrapped air, and to get the best strength.

Figure 13.5 shows a schematic arrangement for the hand lay-up process.

In the case of an all-mat laminate, the glass content should be 25–35%. At a 30% glass to resin ratio, a useful rule of thumb is that 4.5 oz/ft^2 of mat will yield 0.125-in. (3-mm) thick laminate (see Table 13.13, showing the number of plies required for each laminate thickness).

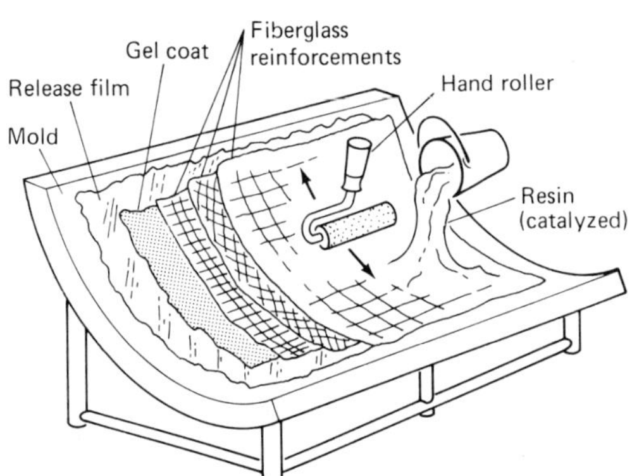

Figure 13.5. Hand lay-up process.

With a mat/woven roving laminate, glass content will be in the 35–45% range. An all-cloth molding will contain about 50% glass.

Table 13.14 is a "troubleshooting" guide for hand lay-up, and Table 13.15 lists the types of hand lay-up resins. Table 13.16 lists promoter-inhibitor systems for room temperature cure polyester resins. In Fig. 13.6, variations in gel times of a typical resin system are given.

13.5.5. Corrosion Resistance

13.5.5.1. Typical Corrosion Resistant Laminate

The laminate forming a corrosion resistant structure is composed of various layers of glass fiber mat and/or woven glass fibers saturated with carefully selected corrosion resistant resin. Experience has shown that a particular sequence of lay-up, now accepted as standard, will yield the best corrosion resistance.

The first step is to make sure there will be a resin-rich layer on the surface exposed to the corrosive environment. This is done by coating the surface with resin and then laying in and completely impregnating and encapsulating the surfacing veil, leaving no exposed fibers. This veil is usually 10–20 mils (0.25–0.5 mm) thick, depending upon the severity of the exposure, and is made of C-Glass, Dynel, or Dacron (resin content about 90%).

The resin-rich layer/surfacing veil should be followed by two layers of 1.5-oz. or 2-oz

Table 13.13. Number of Plies versus Laminate Thickness, in. (mm)

HAND LAY-UP

	1	2	3	4	5	6	7	8	9	10	11	12	13	14	15
2-oz Mat	0.058 (1.5)	0.116 (2.8)	0.175 (4.5)	0.237 (6.0)	0.299 (7.6)	0.361 (9.2)	0.423 (10.8)	0.485 (12.3)	0.546 (13.9)	0.609 (15.5)	0.671 (17.0)	0.733 (18.6)	0.795 (20.2)	0.857 (21.8)	0.919 (23.3)
24-oz Woven Roving	0.036 (0.9)	0.071 (1.8)	0.109 (2.8)	0.147 (3.7)	0.185 (4.7)	0.223 (5.7)	0.262 (6.7)	0.300 (7.6)	0.338 (8.6)	0.376 (9.6)	0.414 (10.5)	0.452 (11.5)	0.490 (12.4)	0.528 (13.4)	0.566 (14.4)
10-oz cloth	0.016 (0.40)	0.032 (0.8)	0.048 (1.2)	0.064 (1.6)	0.080 (2.0)	0.096 (2.4)	0.112 (2.8)	0.128 (3.2)	0.144 (3.6)	0.160 (4.0)	0.176 (4.5)	0.192 (4.9)	0.208 (5.3)	0.224 (5.7)	0.240 (6.1)
Fabmat 2415	0.075 (1.9)	0.150 (3.8)	0.225 (5.7)	0.300 (7.6)	0.375 (9.5)	0.450 (11.4)	0.525 (13.3)	0.600 (15.2)	0.675 (17.1)	0.750 (19.0)	0.825 (21.0)	0.900 (22.9)	0.975 (24.8)	1.050 (26.7)	1.125 (28.6)

SPRAY-UP, GLASS DEPOSISITION, oz/ft^2 (kg/m^2) VERSUS THICKNESS, in. (mm)

	2 (0.5)	4 (1.0)	6 (1.5)	8 (2.0)	10 (2.5)	12 (3.0)	14 (3.5)	16 (4.0)	18 (4.5)	20 (5.0)	22 (5.5)	24 (6.0)	26 (6.5)	28 (7.0)	30 (7.5)
Spray-up Thickness, in. (mm)	0.058 (1.2)	0.116 (2.4)	0.175 (3.6)	0.237 (4.8)	0.299 (6.0)	0.361 (7.2)	0.423 (8.4)	0.485 (9.6)	0.546 (10.8)	0.609 (12.0)	0.671 (13.2)	0.733 (14.4)	0.795 (15.6)	0.857 (16.8)	0.919 (18.0)

NOTE: Thickness of other weight mats will be proportional.
2 oz mat = 610 g/m^2.
10 oz mat = 339 g/m^2.
24 oz mat = 814 g/m^2.

Table 13.14. Troubleshooting Guide for Hand Lay-Up

PROBLEM	CAUSE	SOLUTION
Cure in thickened rods or strings.	Pregelation.	Keep mixing containers clean and free of previously catalyzed gel coat. Use throw-away mixing containers.
Cracking and fissuring.	Larger cracks caused by too thick areas of gel coat or excessive exotherm or thin point in lay-up. Fissuring because of front or reverse impact blow.	More uniform application of gel coat and better mixing with catalyst. Prevent accidental or injuring blows.
Fiber pattern: random fibers from mat or cross-hatch from woven roving weave.	High exotherm; coarse weave material; too close to gel coat.	Cure laminate in steps; use lower exotherm resin. Put more mat in front of woven roving. Best solution is application of an intermediate layer of more rigid resin-containing Vitro-Strand fibers.
Lay-up draining on vertical surfaces.	Resin too low in viscosity; resin with insufficient thixotropic agent; mold or room too warm.	Most probable correction is to increase thixotropic agent content of resin.
Bubbles.	Air entrained in reinforcement after combination with resin.	Add 0.2% green pigment to lay-up resin to see voids. Work lay-up more freely with brushes, squeegees, or serrated rollers. If possible, apply a liberal quantity of resin onto work before applying reinforcement, so that the resin forces air out from the bottom.

Table 13.14. Continued

PROBLEM	CAUSE	SOLUTION
Bridging over small radius curves such as lap-strakes, etc.	Reinforcement too stiff; curves below design-allowables.	Select more highly wettable or soluble mat or woven roving. Use loose-mixed putty to caulk small radii curvatures prior to lay-up. Redesign mold.
Thin areas.	Gaps in lapping reinforcement caused by improper placement or short-cutting, etc.	Correct placement and cutting errors. Lay in patches to correct thin spots prior to removal from mold. Try pre-wetting of reinforcement by resin prior to placement in mold.
Fibers protruding from inner lay-up surface.	Usually unavoidable if mat is sole reinforcement.	For finish layer, apply woven fabric, woven roving, or veil mat on inside. After cure, sand and apply splatter paint.
Cracked or resin-rich areas usually at bottom or wellpoint.	Drainage of resin in large lay-up to a low point and, because of high exotherm, results in cracking; possibly too high a resin to glass ratio.	Introduce more thixotropic agent into resin. Continue to squeegee excess resin out of collection points until gelation occurs. Add additional reinforcement.
Warpage of part.	Unbalanced laminate; flat surface.	Use symmetric lay-up; design slight radius in surface.
Distortion of part.	Undercured in mold.	Allow full cure in mold.
Cracking next to stiffening members.	Hard spot.	Use fillet in corner where stiffener meets laminate.
Low impact strength.	Insufficient glass; too much flexing.	Use bag molding, more woven roving, roving; use stiffener or sandwich construction.
Slow curing laminating resin.	Weather changes.	Adjust catalyst to weather changes.
Roller picks up fibers when working on mat.	Too close to gel time; styrene evaporation; rolling too fast.	Adjust gel time; adjust fans; dip roller in styrene or fresh resin; more deliberate rolling.

E-Glass mat and wet out thoroughly with the corrosion resistant resin yielding a 70% glass/30% resin ratio. The two layers of mat are followed by successive layers of mats, cloth, or woven rovings, up to the specified design thickness. Each layer is resin-saturated and rolled to ensure complete wetting out of the fibers and elimination of air bubbles.

Every effort should be made to maintain constant laminate thickness with little variation in glass/resin ratio. Care should be taken to keep poorly wetted out (dry) areas at a minimum and, conversely, to avoid excessively resin-rich sections. The laminate should contain a minimum number of air bubbles, with no evidence of foreign contamination. The surfaces must not be cracked or crazed and should be uniformly smooth, and the exterior should be coated with a finishing coat top layer of resin to protect all reinforcing fibers.

13.5.5.2. Inspection of Corrosion Resistant Structures

Visual examination provides one of the simplest and most effective ways to judge the soundness of a laminate. In some cases, appearance may not be indicative of serviceability. However, there are certain easily detected

Table 13.15. Types of Hand Lay-Up Resins

TYPE	PROPERTIES	USES
Orthophthalic Type	High modulus, low impact strength, high Barcol hardness.	Stationary, non-flexing structures. General purpose Inexpensive
Isophthalic Type	High impact strength, high Barcol hardness, fast get-to-cure cycle, good stability, good wet strength, improved chemical resistance.	Boats, bathrooms, building material, tanks, swimming pools.
Chemical-resistant, rigid Bisphenol-A	Usually high viscosity and slow wetting, high cost (consult suppliers).	Tanks, ducts, pipes, fume hoods, vats, silos, etc.
Chemical-resistant, flexible Vinyl Ester	Permits normal viscosity, thix, and cure time	Tank bottoms (vibrated) and corrosion resistant gel coats.
Fire retardant Het Acid Halogenated	Self-extinguishing—will not support combustion, high cost (consult suppliers)	Where required or specified. Govt. products Building materials Lifeboats
Epoxy	Slow wet-out, exotherm prior to gelation, good dimensional stability, dark color.	Tooling, high strength. Aircraft, Aerospace.

defects that do affect performance. Visual identification of such defects can eliminate some of the more elaborate inspection procedures, such as tensile and compression tests and glass fiber content burn-out tests, and may avert a failure in service.

By placing a light behind the wall or section of an unpigmented structure, the inspector can judge the uniformity of the laminate and detect air bubbles, dry spots, internal cracking, and other conditions indicating possible weaknesses in the structure.

Surface condition is another indication of laminate construction. A smooth, true surface is generally a sign of a well made, uniformly cured laminate. If exposed fibers exist on the corrosion surface, there is a rupture in the protective resin layer vital to corrosion resistance. Exposed fiber bundles also indicate the likelihood that a surfacing mat is absent.

Industrial specifications often call for cutting a section (test coupon) of the fabricated piece (from trim-offs, a manhole, etc.) for testing. The resulting cross-section will reveal the depth of the surfce layer and the shape and conformity of the laminate. Stressed areas often will be seen immediately in the form of internal cracks. The cross-section will also show the bond between layers and the degree of wet-out of the fibers as well as the position of the surfacing mat and the thickness of the protective resin-rich layer.

Another test often used is Barcol hardness. This test is useful only for measuring the relative hardness of pieces fabricated from a single type of resin. Results are meaningless when comparing different resins because each polyester will have its own inherent hardness. If the Barcol hardness of a laminate falls short of the average ratings found in well-cured structures of the same resin and construction, the inspector should immediately suspect an incomplete cure and resultingly poor corrosion resistance.

Table 13.16. Catalyst Promoter Inhibitor Systems for Room-Temperature Cure Polyester Resins

APPLICATION OR END USE	SYSTEM, %	GEL TIME STARTING AT ROOM TEMPERATURE, minutes	APPROXIMATE TIME, hours, AT 70–75° F FOR DEVELOPMENT OF BARCOL HARDNESS = 35
Gel coats	MEK peroxide—1.5[a] Cobalt naphthenate—0.4[b] (Accessory promoters usually omitted because of tendency to discolor)	30 (High filler content)	6–8 (can proceed with lay-up over gel coat in 30–45 minutes)
For normal lay-up resins	MEK peroxide—1.0 Cobalt naphthenate—0.4	32	6–8
For fast-cure resins	MEK peroxide—1.0 Cobalt naphthenate—0.4 Dimethylaniline—0.1	16	2–2.5
For fast-cure resins	MEK peroxide—1.0 Cobalt naphthenate—0.4 Quaternary ammonium salt—0.1	15	2–2.5
Alternate room-temperature cure	Cyclohexanone peroxide[c]—1.0 Cobalt naphthenate—0.4	30	Approximately 6–8
Alternate room-temperature cure	Bis-l-hydroxy cyclohexyl peroxide[c]—1.0	30	Approximately 6–8
Alternate room-temperature cure	Benzoyl peroxide—1.0 Dimethyl aniline—0.1	20	2
Effect of inhibitor	MEK peroxide—1.0 Cobalt naphthenate—0.4 Hydroquinone—0.1	∞	∞

[a]Percentages based on 100 parts polyester resin.
[b]Concentration cobalt metal 6%.
[c]Peroxides costlier than MEK peroxide.

13.5.6. Spray-Up

Chopped glass fibers and resin are simultaneously deposited in (or on) an open mold. The fiberglass rovings are fed through a chopper and blown into a resin stream which is directed at the mold by either of two spray systems (external or internal mix).

1. A spray nozzle ejects resin pre-mixed with catalyst, or catalyst alone, while another nozzle ejects resin pre-mixed with accelerator (see Fig. 13.7).
2. Resin and catalyst are fed into a single-gun mixing chamber ahead of the spray nozzle.

By either method, the resin mix pre-coats the strands of glass and the merged spray is shot at the mold in an even pattern by the operator.

After the resin and glass mix is deposited, it is rolled by hand methods to remove air, to compact the fibers, and to smooth the surface. Cure and knife trim characteristics are similar to those in hand lay-up.

13.5.6.1. Control Techniques

Quality assurance for hand lay-up and spray-up molding is obtained by careful incoming inspection and checking of component materials, by "building it in" through in-process controls and operator supervision, and by examination and testing of the final products.

Since spray-up is really an automated extension of hand lay-up, the same considerations regarding proper and complete roll-out of

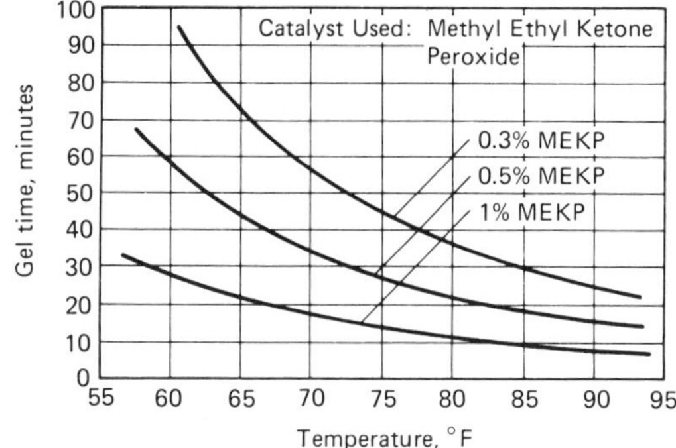

Figure 13.6. Typical variation in gel times of rapid-cure, double-promoted polyester resins with changes in catalyst and temperature, both catalyzed and uncatalyzed.

Room-temperature gel time:

1% MEK peroxide, resin at room temperature. 10–20 minutes

Catalyzed stability:

1% BPO, resin at room temperature. 1–14 days

Intermediate temperature stability:

1% BPO, resin held at 150° F. 5–30 minutes

High-temperature (SPI) gel time:

1% BPO, resin raised to 180° F. 1.5–5 minutes

Uncatalyzed stability:

Resin is stored in the dark at 70° F and periodically tested by SPI gel time to observe gel time drift.

Initial reaction:

Aging of resin induces as 25 to 50% lengthening or upward drift in gel time, indicating that the inhibitor initially incorporated is chemically converted to a more active, intermediate, inhibiting compound during storage. Time span for upward gel time drift may be from 24 hr. to 4 mo. following manufacture.

Final reaction:

Gel time will eventually show shortening of downward drift approaching zero gel time until the entire container of resin gels at room temperature. This indicates that the inhibitor in the resin has ultimately become exhausted. Time span for downward gel time drift may be from 4 to 18 months following manufacture. (Data courtesy of Allied Chemical Corporation.)

materials, catalyst concentrations, knife trim, and cure time exist.

Because the quality of the final product is so intimately dependent upon operator technique, additional control methods must be used.

Laminate composition and thickness is a function of the glass to resin ratio. Hence, proper control over any of the inputs can control the resultant laminate.

- *Gun calibration—time control.* (If the gun is set with a known resin to glass ratio, spraying for a specified time will control the total gun output.)

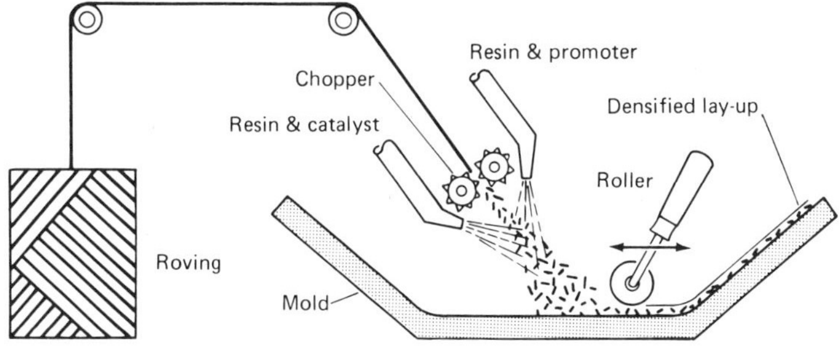

Figure 13.7. Two-pot system—airless.

1. Spray resin (catalyst off) for 15 seconds into plastic bag.
2. Spray glass for 15 seconds into separate bag.
3. Weigh results; set timing clock with bell.

- *Control weight of glass used.* (Thickness is a function of weight/ft^2 of glass.)
  1. Place the roving package on a scale.
  2. Weigh the amount of glass required.
- *Direct measurement methods:*
  1. Spray and check with thickness gauge.
  2. Weigh the amount of glass required.
  3. Burn out to determine composition
- *Uniformity:*
  1. Red tracer is available to assist the operator in obtaining a visually even and uniform coating.
  2. Pattern (the operator plans his coverage of the surface to be sprayed).
  3. Distance position, and speed of the gun with respect to the mold surface should stay constant.
  4. Gun setting—the relationship of the flow of chopped glass to the resin spray fan.

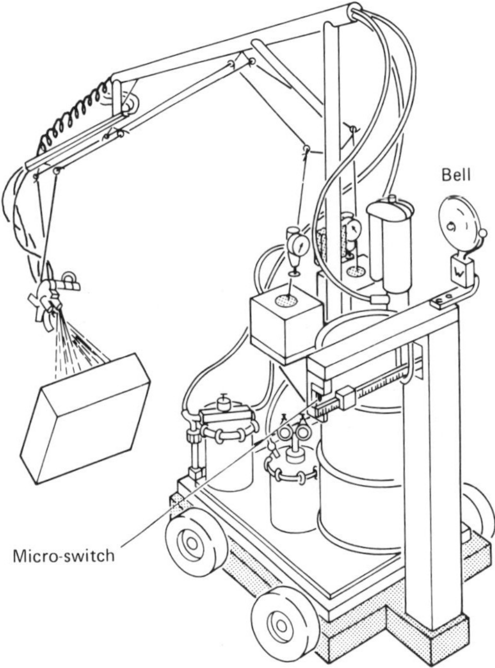

Figure 13.8. Roving and/or resin scales.

13.5.7. Equipment—Spray-Up

13.5.7.1. Definition

A spray up system consisting of a means to transport resin and catalyst through hoses to a gun which disperses the resin and catalyst along with chopped fibers of reinforcement (usually glass) onto a preformed surface, thus generating a "spray-up" laminate (see Fig. 13.8). Equipment consists of the following.

- Pumps for resin and/or catalyst.
- Pressure pots for catalyst.
- Pressure pots for flushing solvents.
- Chopper for the reinforcement.
- Various air regulators and gauges for control.
- Boom and dolly assembly or chassis.
- Various hoses for transporting materials.
- Gun assembly for dispersing materials.

13.5.7.2. Pumps

All the pumps used are positive displacement, air operated pumps, and thus can be adjusted for rate of delivery and pressure through control of the air regulator. Most pumps evolved from automotive grease pump design, modified with Teflon and Viton seals, glands, and gaskets to handle all thermosetting fluid resins, all solvents, and, with special steel, all common catalysts. Pump capacities vary with gun nozzle sizes. Most resin pumps are designed to fit through the bung hole in a standard 55-gallon drum. But recent designs have incorporated an intake hose extension or "snorkel," which is merely inserted into the resin drums, allowing the pumps to be firmly affixed to the dolly and boom assembly.

Catalyst pumps are of two types—piston and peristaltic—and are designed to dispense a catalyst ration from ¾% to 8%. The piston pumps are of two styles—the Venus type and the Binks type.

The Venus type is a true piston pump which is connected directly to the shaft of the resin pump, through a pivot beam. Each stroke of the resin pump causes a stroke of the catalyst pump and thus this system provides the *only* accurate ratio of catalyst to resin.

The Binks pump is a bellows type (acting

like a piston, but without the leakage problems of the piston type) and is driven by an air motor. The rate of catalyst is read via a flow meter and is controlled by adjusting the air pressure to the motor. The motor runs in response to the trigger on the gun.

The peristaltic pump (Glascraft) is a loop of tubing kneaded by rollers which pinch the tube, thus drawing up catalyst between the rollers and providing a relatively smooth flow. This pump is driven by an air pump, and the rest of the system is similar to the Binks system.

Both Binks and Glascraft systems pump catalyst at less than 70 psi (0.5 MPa) and draw directly from the supply jug, as does the Venus system. In addition to reducing handling time, pumping catalyst directly from the manufacturer's shipping container minimizes the risk of dangerous contamination and spillage.

13.5.7.3. Pressure Pots (Catalysts)

The majority of spray-up systems use pressure pots for catalyst dispersing, as these pots constitute a cheap and effective way to get the catalyst to the gun nozzle. However, such a method has two drawbacks:

1. A catalyst under pressure can be dangerous. Spillage occuring during the filling of pressure pots increases the risks of skin and eye injuries, and also explosions caused by contamination. Manufacturer's safety instructions *must* be stringently followed.
2. Accurate control of catalyst percentage is difficult to achieve, and many problems of under-and-over cure can be traced to the catalyst percentage straying from what is desired. There is no direct connection between resin pump output and pressure pot output.

13.5.7.4. Pressure Pots (Solvents)

Such pots are quite simple and are usually controlled through a pressure regulator and feed to the gun through a hose. Flushing systems only apply to internal mix guns—Venus, Glascraft, Polycraft, etc.

13.5.7.5. Choppers

These devices are designed to accept glass rovings and break these rovings into lengths from 0.5–2 in. (12–50 mm) long. They are usually driven by a small air motor whose speed can be controlled via valves and whose exhaust is usually ducted into the chopper chamber to help thrust the chopped roving toward the resin and catalyst fan exiting from the nozzles. The chopper is comprised of two rollers, one made of rubber or urethane (called the cott wheel), and the other of aluminum fitted with slots to hold short lengths of Schick type razor blades (called the cutter wheel). The air motor drives either wheel, and the glass fibers passing between the two wheels are broken and then blown into the resin fan to finally impinge upon the mold surface.

13.5.7.6. Regulators and Gauges

Since all of the spray-up systems are air operated, regulators and gauges constitute the only way to control their output.

Most regulators are simple servo systems and are based upon air bleeding past a needle or poppet valve which is controlled by a spring-loaded diaphragm. As long as the parts are in good condition, the regulation of air pressure (and performance) is good. The gauges monitor the input and output air pressure and should give little trouble—but should be checked yearly against a standard.

13.5.7.7. Boom and Dolly (Chassis)

Most systems are portable and have a dolly with four wheels, two of which are steerable. The dolly carries a drum of resin and a mast to which the pumps, gauges, regulators, and racks for the catalyst and the roving supply are affixed. Each manufacturer's boom is of different design, but the common function is to support the gun to minimize operator fatigue and yet provide a maximum freedom of motion. Booms vary in total length and complexity.

13.5.7.8. Hoses

Hoses are very important and must be resistant to high pressure and aggressive chemicals, which include thermosetting resins, solvents, and catalysts. The hoses must be able to withstand a moderate amount of kinking and crushing and still carry resins at 500–1000 psi (3.4–6.9 Mpa).

Most hoses are lined with Teflon, Viton, or PVF and are protected with a high-strength braid overwrap and an abrasion resistant jacket.

13.5.7.9. Gun Assemblies

All spray-up systems are essentially similar in every aspect but the gun itself. There are many recognized varieties of guns, the most common of which are:

- External mix
- Airless external mix
- Air internal mix
- Airless internal mix
- Two-pot systems.

An external mix gun (Fig. 13.9a) has four nozzles—two for the resin and two for the air-atomized catalyst—arranged in a square pattern around a central hole through which the chopped glass is blown. The chopped glass is enveloped with the resin and catalyst, and thus very little glass overspray is generated. The chopper assembly is part of the body of the gun, and there is no adjustment for glass pattern except for more or less air exhaust into the glass passage which exits at the front of the gun. The gun does not have to be solvent-flushed, but the front of the gun body should be washed clean of resin and catalyst spray at the end of any protracted spraying session. Such washing is especially important when the gun has been spraying at close quarters or into recesses where some blow-back can be expected.

There are several makes of gun using the principle of *airless external mix* (Fig. 13.9b and c). The catalyst nozzle is fed from a pressure pot and flow meter under enough pressure to atomize the catalyst without the need for air. Various arrangements of catalyst nozzle with relation to the resin nozzle exist. One type centers the catalyst nozzle between two resin nozzles which are angled so that the catalyst is caught in the cross-flow about 6 in. (15 cm) from the gun. In all such guns, with one or two resin nozzles, the chopper is mounted on top of the gun and the glass is blown into the convergence of resin and catalyst with varying success, dependent upon proper adjustment of the chopper.

Airless external mix systems require no flushing with solvent, but the nozzles should be washed clean at the end of a session and before closing down at the end of the day.

Air internal mix (Fig. 13.10a) guns (Glascraft) are the most prevalent today and mix air, catalyst and resin internally before ejecting the mix onto the mold. The chopped glass is

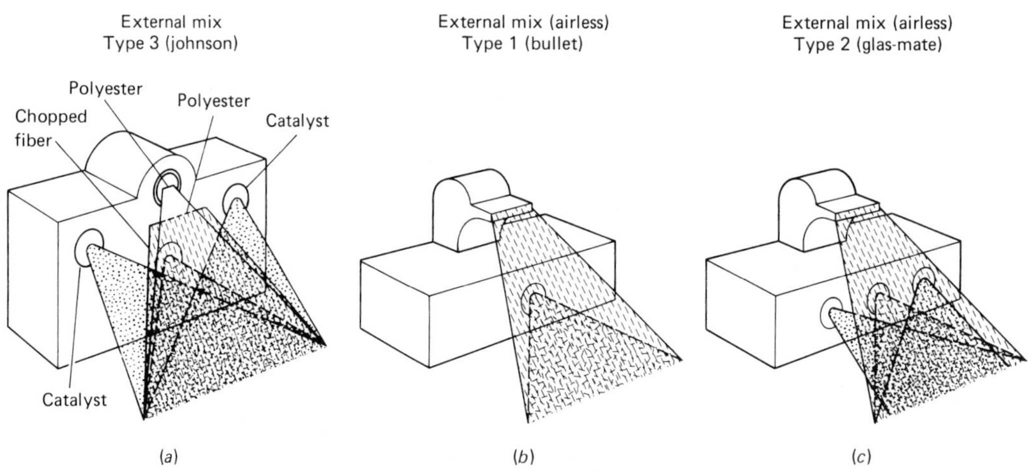

Figure 13.9.

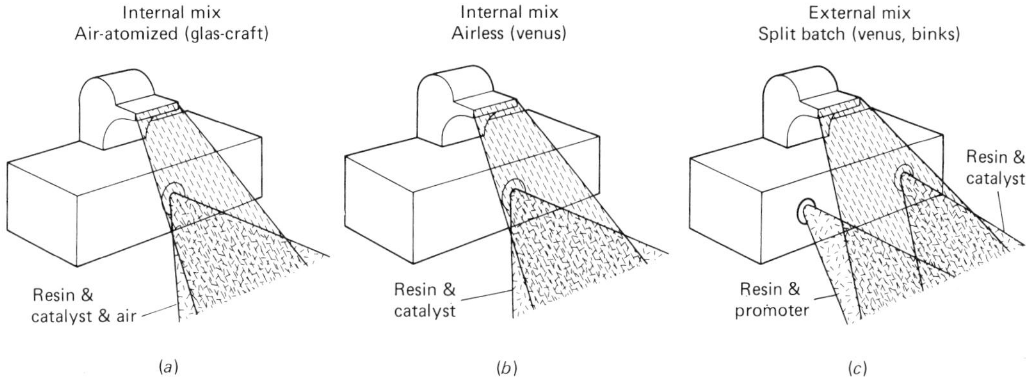

Figure 13.10. Comparison of various spray-up equipment.

blown onto the top of the mix fan and a certain amount of fiber overspray is inevitable. The air mixed with catalyst and resin tends to produce porosity in the laminate and such guns have a difficult time producing a good spray-up into deep recesses or pockets on a mold due to blow-back and "stalling out" of the chopped fibers. Furthermore, early guns were fed with a mixture of air and catalyst from a carburetor at the catalyst pressure pot. When the gun would be re-started after a short rest period, precipitated catalyst in low parts of the hose would be blown through the gun in slugs, and such high concentrations of catalyst could cause hot spots and rejected parts. As a result, such guns must be cleared of catalyst slugs before spraying on a part, and this procedure not only wastes a lot of expensive catalyst; it is also dangerous. Later designs of these guns have placed the carburetor at the gun and this has solved the problem. This type of gun requires a solvent flushing valve to wash out the resin and catalyst in the internal mixing chamber. Despite this, it is recommended that the nozzle and internal mixing element be removed at the end of each day and the chamber be flushed with solvent.

The *airless internal mix* gun (Venus) (Fig. 13.10b) introduces pressurized catalyst and resin into an internal mixing chamber and hence through a nozzle at high (at least 800 psi) pressure.

A solvent flushing valve is provided to clear the mixing chamber after each spraying session. It is recommended that the internal mix element and nozzle be removed and the chamber re-flushed at the end of the day.

The uniqueness of this airless internal mix gun among the available guns is that it can be fitted with a static (Kenics) mixer to pump mixed material into a closed mold under pressure (resin transfer molding). It also can handle 8% acid catalyst for furan resins as well as liquid BPO catalyst for chemical resistant bisphenol A polyester resins. This gun can be fitted with air or oil operated valves and is available in larger sizes for high-volume production users. The chopper is mounted on top of the gun and blows chopped glass down onto the resin catalyst fan with some fiber overspray. The mixing head is a stainless steel block and thus can be salvaged via a burn-out should a malfunction result in a plugging-up of cured resin within the head. Other internal mix guns are aluminum and cannot withstand such extreme salvage procedures with any degree of success.

Two-pot systems (airless) (Fig. 13.10c) have two nozzles designed to impinge about 6 in. (15 cm) ahead of the nozzles. The chopper is top mounted and is adjusted to blow the chopped fibers into the convergence with minimal overspray. Because the chopped fibers are trapped between the two streams of material, such guns operate at 400–1100 psi (2.7–7.5 MPa) and are the simplest of all. They require no flushing, as the blending is done in midair. Such guns can be merely turned off at the end of the day and are ready to go the next day—a feature not found in any of the other

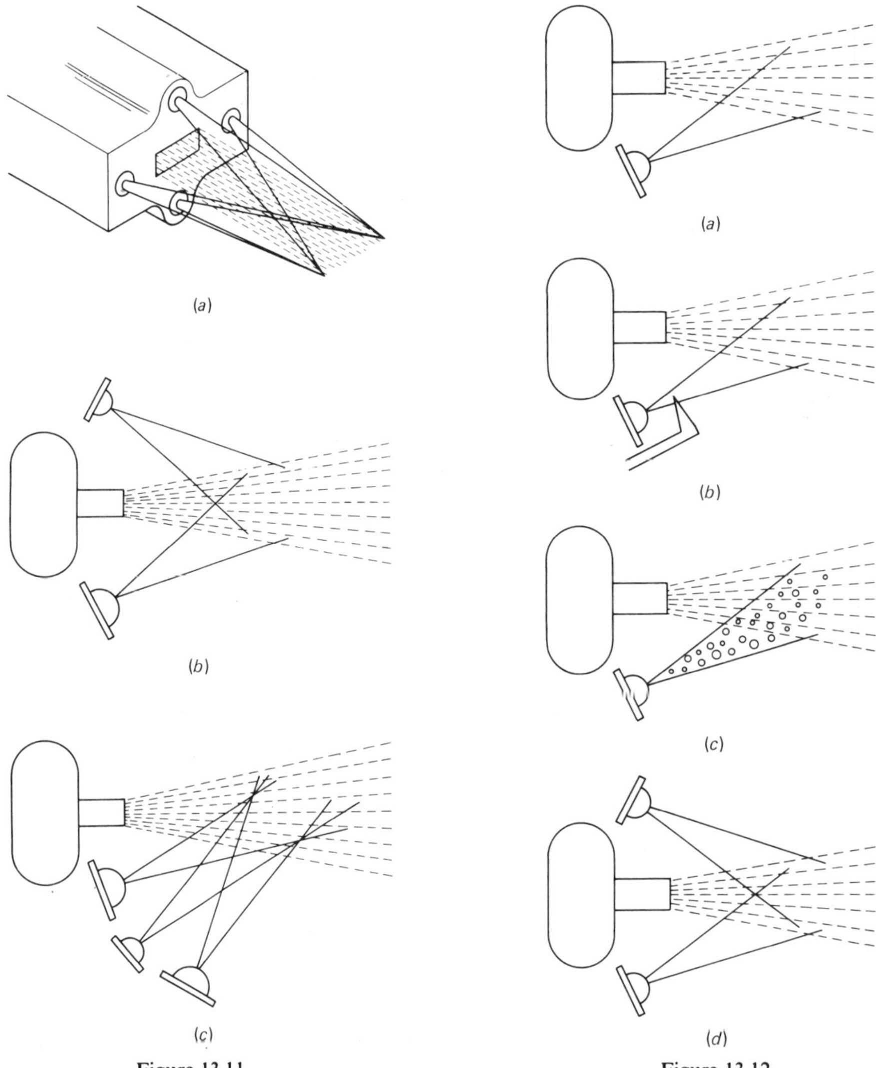

Figure 13.11.

Figure 13.12.

types of guns available. (See Figs. 13.11 and 13.12).

13.5.8. Trimming, Fabricating, and Assembly

13.5.8.1. Trimming

Depending on the required quality and accuracy of the finished edge configuration, there are several methods for trimming open mold lay-up fiberglass parts.

Most basic of these methods is known as "knife trimming." During the curing cycle of a properly catalyzed molding at the end of gellation, there exists a short span of time when the laminate exhibits a firm, non-sticky, leatherlike consistency. It is during this critically short period that the edges of the molded part can be trimmed with a sharp razor or utility knife.

If trimming the part is attempted too soon, the fibers will pull through the soft resin and will bunch up. When knife trim is attempted too late, the part will already be cured solid and the effort will be futile (and will necessitate a time-consuming and dusty grinding or sawing operation).

If in-mold trimming is contemplated, certain provisions should be considered in mold design, such as peripheral mold flange, perpendicular to the molded surface at the exact finish height of the part.

Post-molding trimming techniques of lay-up/spray-up parts are similar to that of other FRP molding methods. However, since only one side is a molded (good) surface, all alignment, banking, and measuring must be done from that surface.

Carbide, carborundum, or diamond cutting blades should be used at high cutting speeds. Toothed cutters should be avoided. If the part is gel coated, care must be taken to avoid chipping (use finer blades).

Operators must always wear proper eye protection. Dust masks should be worn if direct dust exhaust is not available.

The most common trim tools are a high-speed air router, an electric table router, a shaper, an overhead router, a band saw, an electric circular hand saw, a reciprocating or saber saw, a table saw, a disc sander, and a 7-in. electric or air sander.

Trimming is done to jigs and routing fixtures, to scribe lines on the part, to free hand, to molded-in edges, and to saw set-ups.

13.5.8.2. Drilling

Drilling requires no special techniques, although carbide bits and air drills are often used.

For large, accurate holes, pilot drilling and counter boring are suggested. Very large holes may be made by a fly-cutter or standard hole saw.

Holes with loose tolerance can be located by scribe lines or small bumps (use drive screws) on the mold. For tight tolerance holes use drill fixtures with bushings, drill press set-ups, or back drilling from other mating parts.

13.5.8.3. Assembly

Fastening can be accomplished mechanically by means of bolting, sheet metal type screws, blind fasteners (such as "pop" rivets), or traditional riveting.

Numerous adhesive systems can be used, either structurally (such as epoxy or acrylic), as a sealant (such as polysulfide or silicone), or as a temporary bond prior to laminating the mating parts with glass/resin lay-up.

Prior to any painting, bonding, or laminating operations, mating surfaces should be prepared (roughed up) and cleaned to remove any residual molding release agents, dirt, or grease.

If parts are to be painted, a water-soluble, or internal release agent may be used.

13.5.9. Repair

The basic concept of making repairs to hand lay-up/spray-up products is that the strength and quality of the repair is dependent upon the preparation and cleanliness of the interface between the original material and the fresh repair material.

In repairing thermosets such as polyester, epoxy, and furan, reliance must be placed on the adhesive properties of the resin to glue the new laminate to the original laminate. Hence, the need to be very careful in joint interface preparation. Since tensile strength is $\frac{P}{A}$, it is essential to plan the repair so the bonding area is as large as possible, thus reducing the unit stress on the resin adhesive, the actual laminating resin. For example, to repair a puncture in a boat hull, if the hull is 0.25 in. (6 mm) thick, we have found that the optimum grinding around the hole should be 10–12 times the hull thickness, or 2–3 in. (60–70 mm) and this grinding should represent a taper from a feather edge at the hole to full thickness—2–3 in. (60–70 mm) away from the hole (see Fig. 13.13).

Furthermore, this grinding should be carefully done so that the taper is uniform throughout the periphery of the area. In a multi-ply hull laminate, the trace of roving plies is often used as a form of contour line (as in map reading), each ply representing a certain level and the

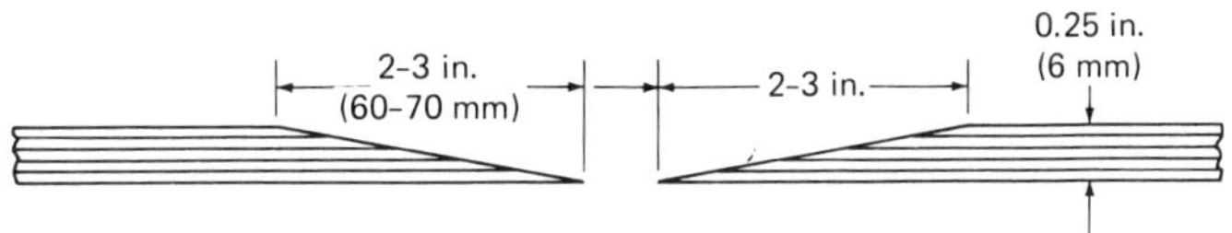

Figure 13.13.

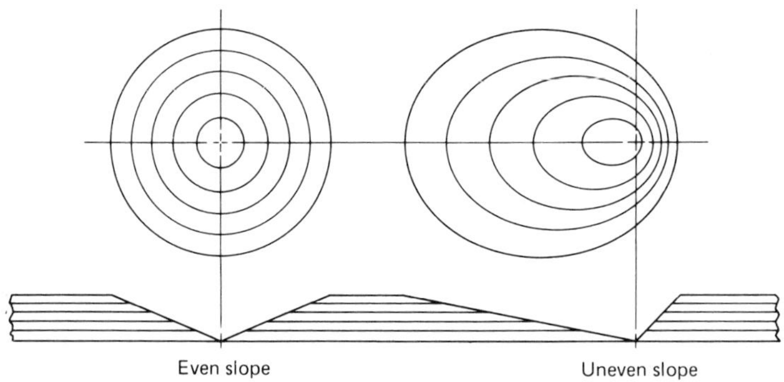

Figure 13.14.

spacing of the contour lines representing the slope of the cut.

If the lines are parallel and evenly spaced, the cut is uniform and will make an efficient repair. If the lines are as shown in Fig. 13.14, the repair will tend to fail because the stresses will not be uniform and the stress will be highest where the lines are close together—thus failing from that point.

Cleanliness is of vital importance in making repairs. All laminations are contaminated. Even laminations one day old can be contaminated because of air pollution and the presence of small airborne particles of a staggering variety of materials. One sure way to avoid such contamination is to make the joint preparation (grinding), then make the repair within minutes of preparation.

If the repair cannot be made right after part preparation, which includes grinding or sand blasting, simply cover the area with paper, being sure to avoid touching the freshly ground area.

There are two basic types of repair: cosmetic and structural. *Cosmetic* repair follows the relatively simple steps of replacing gel coats and then restoring the contour and polishing by techniques similar to those used in the automotive paint shop. *Structural* repairs are unique, as each repair is a different problem, requiring a different sequence and approach. But regardless of these differences, the new repair is glued to the old original (parent) laminate and the rules of preparation hold.

A good structural repair can be made by replacing the plies of fabric or roving, as shown in Fig. 13.15. Admittedly, this repair seems logical: replace material that was removed and put it back in the same orientation as before. Such a repair is described in almost all existing literature, but the restored strength is actually considerably lower than the original strength. A better way to make a structural repair is to plan the repair, remembering that the new material must be bonded to the old in a manner which will ensure the maximum bond strength. In other words, apply the new material parallel to the surface, as shown in Fig. 13.16. The problems of uneven resin concentrations and discontinuities at the interface are partially avoided by this method. When a flush surface is not critical, the best method (strengthwise) is to apply an overlapping patch over the repair, tapering the added material to reduce the protrusion. A maximum strength repair may be required in certain highly stressed or high vibration areas that require more than just the adhesion of the resin. Such vibration resistant repairs can be accomplished by adding in "tie screws" as shown in Fig. 13.7.

The repair is accomplished by again, first grinding to a slope of 10 or 12 to 1. Several pilot holes for sheet metal screws are then drilled in the sloped area of the parent laminate as shown in the sketch. The surface is then wetted with resin and a few layers of glass mat and fabric, also impregnated with resin are

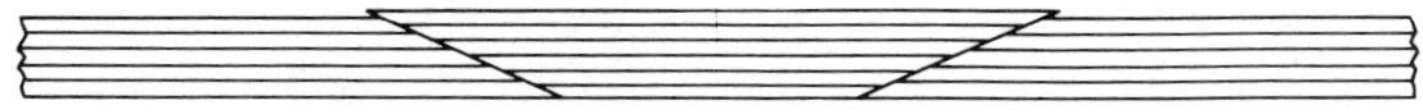

Figure 13.15.

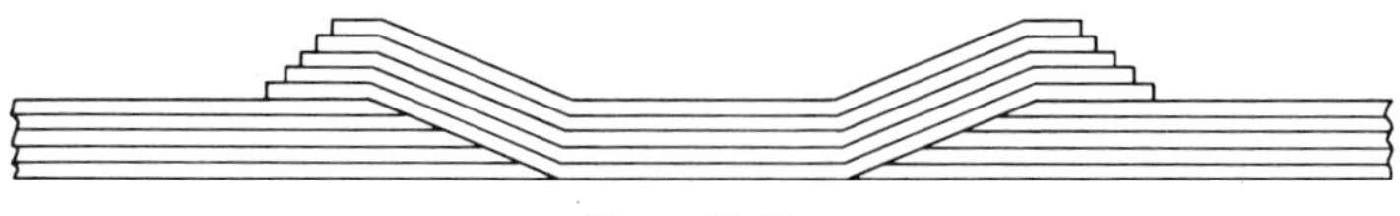

Figure 13.16.

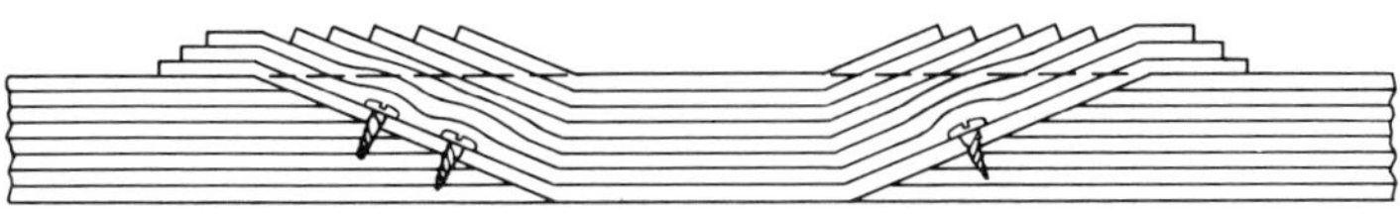

Figure 13.17.

placed over the area. The screws are then inserted but not tightened, and the remainder of the laminate is applied to fill the cavity. After cure, the repair is ground flush with the parent laminate surface and sealed with a coat of catalyzed resin and one ply of mat extending approximately 2–3 in. (5–8 cm) past the edge of repair. In cases where this type of repair is located on the get coat side, the finishing is accomplished using auto body repair techniques.

13.5.10. Finishing Cosmetic Repair and Painting

Finishing of hand lay-up and spray-up parts utilizes auto body and fender repair techniques. There is one difference between auto body painting and final gel coating or painting, and that is basically that auto body paints "dry" through evaporation of the solvent, while FRP gel coats or finishes "cure" by cross-linking of the polymer chains (polymerization).

The quality of "curing" type coatings has improved enormously in recent times, and at this writing many FRP hand lay-up and/or spray-up products are made and primed and then finally spray painted with a polyester, urethane, or epoxy-100% solid curing system. One well known company is press molding bathtubs and painting them with urethane paints. Some boat builders have found it makes more sense to mold without gel coats and then prepare the hull and paint as a final step. This avoids the frustrating and costly gel coat problems that are noticeable only after the part is removed from the mold (and after a lot of work has already been done that may have to be scrapped).

13.5.11. Final Finish Techniques

Once the final gel coat is cured, the sanding operation is carried out, using successively finer grits, 150 to 400 and even 600 grit. At the 200 grit level, wet sanding is used and the area is flushed with enough water to prevent any scratches. The next stage is compounding, usually using a rouge stick applied to a lamb wool buff with a gentle pressure on the sanded surface. Then, when compounding has brought up the shine, the surface is washed, dried, and waxed.

13.6. MATERIAL AND EQUIPMENT SOURCES

The FRP industry is one of the most rapidly growing industries in America. Many companies have responded to the challenge of supplying materials and equipment to meet the needs of this expanding market. The following is a listing of materials used in hand lay-up/spray-up processing and several typical suppliers of each item (space limitations preclude the listing of all suppliers in the FRP industry).

Accelerators

Apogee Product Division
M & T Chemicals, Inc.
Richmond, Calif. 94804

Ram Chemicals Mfg. Co., Inc.
Gardena, Calif. 90247

Witco Chemical Co., Inc.
New York, N. Y. 10017

Balsa Wood
Baltek Corp.
Northvale, N. J. 07647
("Contour Kore" balsa)

Catalysts
Lucidol Division
Penwalt Corp.
Buffalo, N. Y. 14240

Noury Chemicals
Burt, N. Y. 14028

U. S. Peroxygen
Richmond, Calif.

Fillers
Cabot Corp.
Boston, Mass.
("Cab-O-sil")

Calcium Carbonate Co.
Quincy, Ill. 62301
(calcium carbonate)

Freeport Kaolin Co.
New York, N. Y. 10017
(clay)

Philadelphia Quartz Co.
Valley Forge, Pa. 19482
(glass beads, microsheres—Q-cell)

Potters Industries
Hasbrouck Heights, N. J. 07604
(glass beads)

3M Company
St. Paul, Minn. 55101
(Glass bubbles)

Marietta Resources International, Ltd.
Bethesda, Md. 20034
(suzorite mica)

Micro Materials Corp.
Sugar Grove, Ill.
("Micro-Mix")

Mold preparation materials
Mold cleaners
Costa Chemicals
Laguna, Calif. 92651
("Formula #5" liquid cleaner)

Mirror Bright Polish Co.
Pasadena, Calif. 91109
("Mirror Glaze" machine cleaner)

Mold release agents—paste wax
Mirror Brite Polish Co.
Pasadena, Calif.
("Mirror Glaze" auto wax)

Specialty Products Co.
Jersey City, N. J. 07302
("Honeywax")

Mold release agents—liquid
Contour Chemical Co.
Woburn, Mass. 01801
(PVA)

Fiberglass Chemical Co.
Long Beach, Calif. 90813
(Oscar's 500 liquid wax)

Frekote, Inc.
Boca Ratan, Fla. 33432
(Silane, non-silicone
types FRP and 44)

Specialty Products Co.
Jersey City, N.J. 07302
(liquid "Honey Wax", PVA,
Internal Releases)

M & H Laboratories, Inc.
Chicago, Ill.

Polyester gel coats
Cook Paint & Varnish Co.
Kansas City, Mo. 64141

Color Division
Ferro Corp.
Cleveland, Oh. 44105

Advance Coatings
Westminster, Mass.

Coatings & Resin Division
PPG Industries
Pittsburgh, Pa. 15222
(sanitary ware gel coats only)

Ram Chemicals Mfg. Co., Inc.
Gardena, Calif. 90247
(sanitary ware gel coats only)

*Polyester resins**
Plastics Division

*GP = general purpose; I = isophthalic; FR = fire retardant; CR = chemical resistant; V = vinyl ester; and F = furan.

American Cyanamid Co.
Wallingford, Conn.

Chemical Products Division
Ashland Chemical Co.
Columbus, Ohio 43216
(FR, I, CR, V, F)

Cook Paint & Varnish Co.
Kansas City, Mo. 64141
(GP, I)

Freeman Chemical Corp.
Port Washington, Wisc. 53074
(GP, I)

Tar and Chemical Division
Koppers Company, Inc.
Pittsburgh, Pa. 15219
(GP, I, FR)

Owens-Corning Fiberglas Corp.
Toledo, Ohio 53659
(GP, I, FR)

Coatings & Resins Division
PPG Industries
Pittsburgh, Pa. 15222
(GP, I)

Reichhold Chemicals, Inc.
White Plains, N. Y. 10022
(GP, I, FR)

U. S. S. Chemicals
Linden, N. J. 70736
(GP, I)

I.C.I. Americas, Inc.
Willmington, Del. 19897
(CR, V, F)

Air compressors
Binks Mfg. Co.
Chicago, Ill. 60612

De Vilbiss Company
Toledo, Ohio 43601

Trim tools
Aro Corp.
Bryan, Ohio 43506

Dotco, Inc.
Hicksville, Ohio

Skil Corp.
Chicago, Ill. 60630

Tunco
(saw blades)

Remington
(band saw blades)

Air mixer
Binks Mfg. Co.
Chicago, Ill. 60612

DeVilbiss Co.
Toledo, Ohio 43601

Mixing Equipment Co., Inc.
Rochester, N. Y. 14603

Foam equipment
Binks Mfg. Co.
Chicago, Ill. 60612

Glas Craft Division
Ransburg Electro-Coating Corp.
Sun Valley, Calif. 91352

Poly Craft (Div. Binks Mfg. Co.)
Sun Valley, Calif. 91352

Martin Sweets Co.
Louisville, Ky. 40212

Glass Choppers & Systems
Plastics Division
Binks Mfg. Co.
Chicago, Ill. 60612

DeVilbiss Co.
Toledo, Ohio 43601

Glas Craft Division
Ransburg Electro-Coating Corp.
Sun Valley, Calif. 91352

Ransburg Electro-Coating Corp.
Ft. Lauderdale, Fla. 33309
("Glas-Mate")

Poly Craft (Div. Binks Mfg. Co.)
Sun Valley, Calif. 91352

Plural Components
Santa Ana, Calif.

Venus Products
Kent, Wash. 98055

Johnson & Sons
Glendale, Calif. 91208

Finn & Fram
Sun Valley, Calif. 91352

Ovens
Binks Mfg. Co.
Chicago, Ill. 60612

Blue M Electric Co.
Blue Island, Ill. 60406

Resin pumps
Plastic Division
Binks Mfg. Co.
Chicago, Ill. 60612

DeVilbiss Co.
Toledo, Ohio 43601

Glas Craft Division
Ransburg Electro-Coating Corp.
Sun Valley, Calif. 91352

Gray Co., Inc.
Minneapolis, Minn. 55413

Venus Products
Kent, Wash. 98055

Safety containers
Eagle Mfg. Co.
Wellsburg, W. Va. 26070

Protectoseal Co
Chicago, Ill. 60623

Scales
Hobart Mfg.
Troy, Ohio 45373

Toledo Scale Co.
Toledo, Oh. 43612

Detecto
Brooklyn, N.Y.
(milk scales)

O'Haus Scale Co.

Serrated rollers
Binks Mfg. Co.
Chicago, Ill. 60612

Glas Craft Division
Ransburg Electro-Coating Corp.
Sun Valley, Calif. 91352

Venus Products
Kent, Wash. 98055

Miscellaneous rollers and supplies
United Industrial Sales
Cleveland, Oh. 44113
(rollers and all other supplies)

Boatex Fiberglass
Boston, Mass.

Hastings Plastics
Santa Monica, Calif. 90404

White Mfg. Co.
St. Petersburg, Fla. 33714
(cup guns)

Spray booths
Binks Mfg. Co.
Chicago, Ill. 60612

DeVilbiss Co.
Toledo, Ohio 43601

Hess Industries
Bronx, N. Y. 10460

Spray-up equipment and accessories
Plastics Division
Binks Mfg. Co.
Chicago, Ill. 60612
(any type needed)

Glas Craft Division
Ransburg Electro-Coating Corp.
Indianapolis, Ind. 46208

DeVilbiss Co.
Toledo, Ohio 43601
(any type needed)

Glas-Craft Division
Ransburg Electro-Coating Corp.
Sun Valley, Calif. 91352
(single nozzle, catalysts
injected, air atomized,
internal mix gun)

Johnson & Sons
Glendale, Calif. 91208
(airless, external mix catalyst)

Ransburg Electro-Coating Corp.
Ft. Lauderdale, Fla. 33309
("Glas-Mate" system,
triple nozzle, catalysts
injected or two-pot system,
airless or air atomized
external mix; uses a
Glas-Mate chopper)

Venus Products
Kent, Wash. 98055
(Spray-up, Wet Out, RTM)

Plural Component Systems
Santa Ana, Calif. 92704

Poly Craft Systems (Div. Binks Mfg. Co.)
Sun Valley, Calif. 91352

Wet film gauge
Nordson Corp.
Amherst, Oh. 44001
(Gel coat); wet film thickness 0.0025–0.020 in. reading; this guage measures gel coal thickness while wet to determine whether proper amount of material has been applied).

K-D Co.
Palo Alto, Calif. 94302
(wet film gauge)

Wet thickness gauge
Owens-Corning Fiberglas Corp.
Toledo, Oh. 43659
(Measures wet laminate thickness)

Viscosity tester
Brookfield Engineering Laboratories
Stoughton, Mass. 02072

Gardner Laboratories, Inc.
Bethesda, Md.

General Electric Co.
Pittsfield, Mass.
(Zahn Viscometer)

Gel timer
Shoydu Instruments
Brooklyn, N.Y. 11231

Muffle Oven
Shoydu Instruments
Brooklyn, N.Y. 11231

Fiberglass
Certain Teed Corp.
Valley Forge, Pa. 19482
(mat, roving, woven roving)

Owens-Corning Fiberglas Corp.
Toledo, Ohio 43659
(mat, roving)

PPG Industries
Pittsburgh, Pa.
(mat, roving, woven roving)

Fiberglass Industries
Amsterdam, N. Y. 12010
(cloth, mat, woven roving)

Foam
Cook Paint & Varnish Co.
Kansas City, Mo. 64141
(polyurethane foam and foaming systems)

Reichhold Chemicals, Inc.
White Plains, N. Y. 10603
(polyurethane foam and foaming systems)

The Upjohn Co.
Texas
(polyurethane foam-sheet, slab and board)

Pigment concentrates
Cook Paint & Varnish Co.
Kansas City, Mo. 64141

Color Division
Ferro Corp.
Cleveland, Oh. 44105

Pigment Dispersions, Inc.
Iselin, N. J. 08830

Gel coats—tooling
Cook Paint & Varnish Co.
Kansas City, Mo. 64141

Ram Chemicals Mfg. Co.
Gardena, Calif. 90247

Color Division
Ferro Corp.
Cleveland, Oh. 44114

The sources mentioned above are intended for use as a starting place in a molder's search for material vendors and imply no endorsements.

For additional listings of suppliers, the reader is directed to *The Modern Plastics Encyclopedia, The SPI Directory and Buyers Guide of Members, and The Thomas Register.*

13.7. AUTOMATED OPEN MOLD SYSTEMS

As the industry expands, many companies find that their orders have risen from 25–50 to 500 or 5000 parts, and though this expansion may still not be great enough to warrant matched dies or complete automation, some automation is warranted to increase productivity and minimize the labor costs.

The earliest approach to semi-automation was to make many molds and conveyorize them to reduce handling and to increase the rate of production. This did not reduce the labor skill required, but utilized the skill where needed. Mechanized open mold systems, such as spray-up, were the next step, with the machine doing some or all of the actual application of materials. More advanced systems have been automated to virtually carry the process from basic ingredients to final product, ready to ship with a minimum of handling.

In automation, the easiest products were tackled first—anything that was simple and could utilize simple molds, and was sold in large quantities, would meet the requirements. Hence, flat and corrugated panels were the initial products. The basic machine used chopped strand mat on a conveyor belt, added resin and catalyst and cured the mixture in an oven over the belt. Later machines actually developed the chopped strand mat from rovings, transferred the mat onto a polyethylene or cellophane web, added pigmented resin, placed a second film on top of this, compressed the resin glass mixture between the two films, and then shaped the composite by drawing through shaped templates mounted in an oven. As the composite exited the oven (fully cured), panels were cut from the composite with traveling, water lubricated saws that were controlled via punch cards entered into the saw controller. The scrap or trimmings fell away into dumpster containers and the panels advanced through a stripping machine which removed the films and water and ejected the dry panels onto an inspection table.

In the early 1960's, a flat sheet machine was developed which made large panels—16 × 40 ft—using a new device called a "fiber epoxy gun," which had been developed in Europe and imported here. The gun was mounted on a bridge and reciprocated over a movable molding table depositing a band of chopped fiber and resin in a fairly uniform manner. Downstream of this application, a series of mounted, grooved rollers worked out the air and densified the laminate. The panel thus formed was allowed to cure and was removed for further processing. Several molding tables were constructed, mounted on dollies, and pulled by a chain and capstan system at the proper speed. They were linked in such a way, through a shuttle track, so as to have continuous processing tables.

The manufacture of cylinders was the next most likely candidate for automated hand lay-up. Filament winding is well known and became quite sophisticated during the 1960's, spurred on by the space program. Filament winding is an open mold process, but was precise by necessity and so special by design that filament winding is treated as a separate entity. However, in the commercial (not government) field, products such as stacks, pipes, and tanks are relatively simple, have much lower tolerances, and are used in much greater quantities than space hardware, Fig. 13.18.

Machines to build large, underground gasoline storage tanks were developed and put into use in the mid-1960's and made tanks which were combinations of hand lay-up, spray up, and filament winding. The end bells of such tanks were made via a programmed "chopper gun" (a fiber spray gun) spraying out material on a rotating mandrel which was densified with rollers in the hands of an experienced operator. Later versions of this machine eliminated the operator.

Today, there are machines to make flat panels, large panels, pipes, tanks, and stacks,

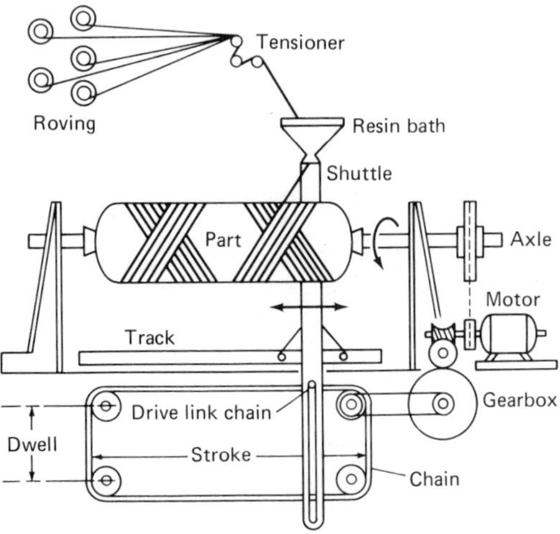

Figure 13.18. Filament winding process.

and all follow the basic principles of applying the materials and densifying them on the mold in some manner. One notable machine for making continuous cylindrical items utilizes what is called a "disappearing mandrel." The mandrel is constructed with a rotating spider on a one-piece band of stainless steel approximately 5 in. wide (12.5 cm) wrapped onto the spider with a pitch of 5 in. (12.5 cm) per revolution. When the wrapped band gets to the end of the spider, it is drawn, via several flanged rollers, through the tubular axis (thus disappearing) to be fed through several flanged rollers back onto the rotating spider to go through the cycle again. This mechanism solves the mandrel problems, as well as the problem of applying materials onto a mandrel.

In conventional filament winding, Fig. 13.18 fibers and resins are supplied from a stationary source and wound onto a rotating mandrel of finite length. There is always lost material on each end of the mandrel due to reverses, and long pipes require long mandrels and machines, and the mandrels must be extracted.

In continuous mandrel filament winders, the mandrel does not turn but moves axially through rotating fiber supply reels which have to be constantly stopped and replenished—an operation which limits the amount and kind of fibers that can be applied. After the cure, the mandrels must also be extracted.

In the "disappearing mandrel machine," (Drostholm Process) there is no limit to the length of product that can be made—as there is no mandrel. Also, since the mandrel rotates, the supply of material is a stationary source and can be programmed so that the machine need never be stopped for lack of material. Furthermore, many kinds of materials and resins can be wrapped onto this mandrel, including an actual sandwich construction, or combinations of sandwich and solid skin.

High production hand lay-up or spray-up products, such as small boats, sinks, bathtubs, lily ponds, and garden and park furniture, are being built via robots. Such multi-axial, servo-controlled and tape-directed robots are common in the auto industry, and are being adapted to the needs of the industry. However, a trained operator is still required to densify the lay-up, Fig. 13.19.

An advantage of the robot is that it will enable manufacturers to meet strict OSHA

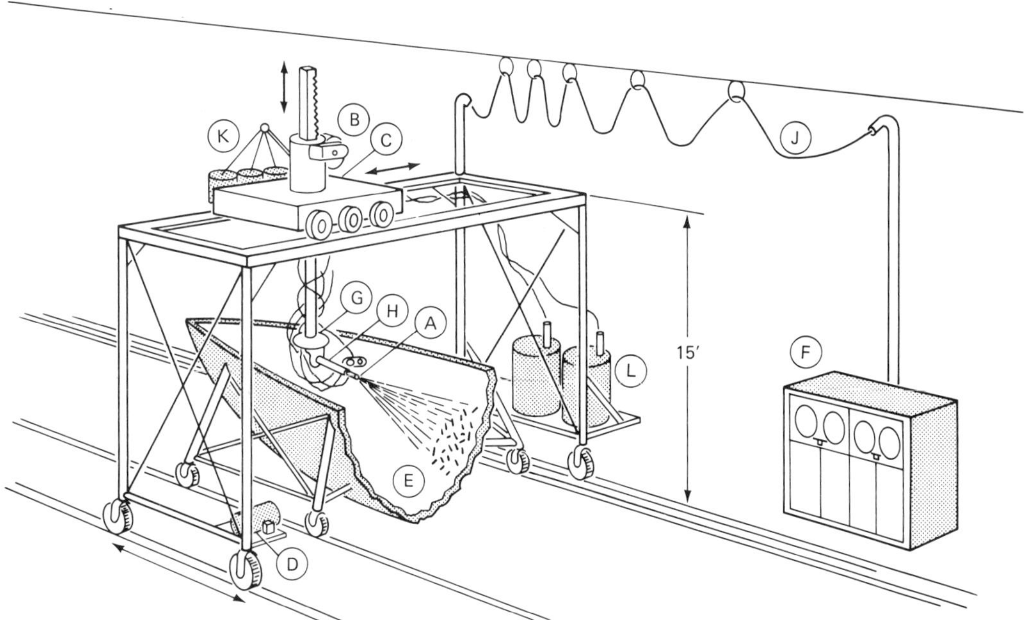

Figure 13.19. Proposed three-axis automatic spray-up machine. A = spray-up head; B = vertical actuator; C = transverse actuator; D = longitudinal actuator; E = partial mold; F = control tapes; G = azimuth head; H = elevation axis; J = power cables; K = glass supply; L = resin supply.

requirements in the spray areas. Not needing to breathe and not subject to allergies, the robot can work in an enclosed room with little or no attention except entry and removal of molds and supplies of materials.

The application of automation has extended to some vacuum bagging operations, especially in the manufacturing of structural sandwiches. The sandwich skins are laid up by reciprocating spray chopper guns and the cores are installed by automatic placing devices. The whole assembly is then fitted with a vacuum bag, sealed to the mold and air withdrawn to densify the composite and ensure tight fit of the skins to the core. The presence of the bag nearly eliminates the styrene emission during the manufacturing sequence.

13.7.1. Rigidized Formed Thermoplastics (Modified Acrylic)

The rigidized formed thermoplastic method is a process wherein thermoplastic sheets are heated until they are soft, and then are vacuum formed, pressure formed, or draped into a suitable mold and allowed to cool and stiffen. These shaped sheets are then transferred to a holding fixture and chopped glass and resin are sprayed on and densified to the back side of these sheets. The specially formulated resin used adheres to the acrylic sheet, and after cure, the sheet and back-up of fiber-reinforced resin acts as a single sheet. A strong composite evolves wherein the thermoplastic sheet, or skin, becomes the surface of the part and takes the place of gel coats and / or paints, Fig. 13.20.

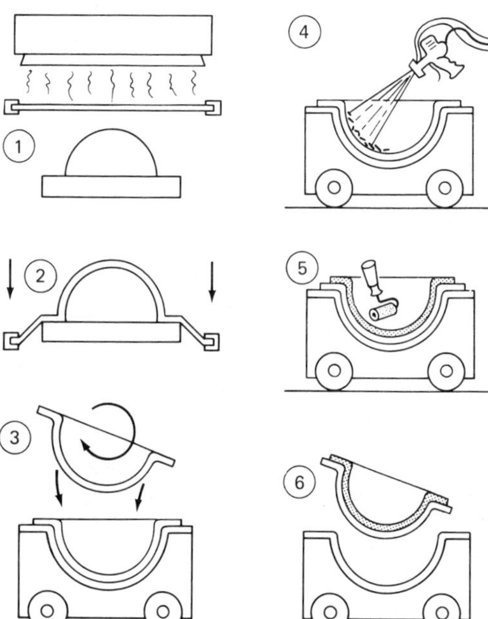

Figure 13.20. Shell coat process. 1) Heat acrylic sheet. 2) Vacuum form. 3) Remove, invert, and place in molding fixture. 4) Spray-up against acrylic. 5) Roll down and cure. 6) Remove and trim.

Rigidized formed thermoplastic is a very popular manufacturing method and is used to a great extent in making bathtub-shower combinations, sinks, commodes, small boats, car top carriers, van conversions, outdoor signs, and many products requiring resistance to water and sunlight. The most popular thermoplastic is methyl-methacrylate sheet (Plexiglas, Perspex, Swedcast, etc.) and some special grades of material which will adhere to the styrene component of polyester resins such as PVC and polycarbonate (Lexan).

The reason manufacturers go to the extra effort of using thermoformed skins on their FRP product is the difficulty of controlling a spray-applied polyester gel coat. Unless operators are well trained, gel coat problems can increase costs out of proportion to the material and the time needed to apply it. For instance, a tub-shower combination, using gel coat, can be completely built and cured, only for the operator to find out, as the part is removed from the mold, that extensive reworking must be done to repair a spotting, wrinkling, and crazed gel coat. On the other hand, an identical tub-shower combination in rigidized thermoplastic will not enter into production until the inspector agrees that the skin is perfect. Only *then* is the rigidizing FRP applied.

This process is used in a modified form to make chemical-resistant tanks for hydrofluoric acid and other corrosive chemicals. PVC, polypropylene and some fluorocarbon materials, such as polyvinyl fluoride (PVF) and polyvinylidene fluoride (Tedlar) in sheet form, are heat-bonded to a glass fabric substrate or carrier. The opposite side of the fabric is free to bond to a sprayed-up or laid-up polyester or epoxy-glass reinforcement. In production,

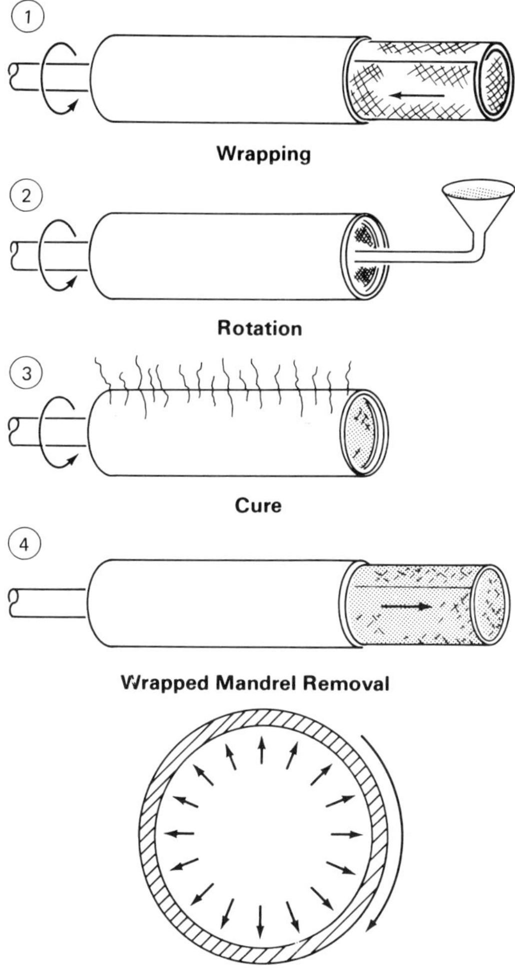

Figure 13.21. Centrifugal molding.

the chemical-resistant liner is assembled on a mandrel and the edges are welded; the rigidizing is then applied as required, and after removal from the mandrel, the joints between the panels comprising the liner are welded on the inside and the inspection is accomplished via a high-voltage spark test over each seam (a wire was attached over each weld seam during initial fabrication of the liner). Any voids will cause a spark discharge requiring rewelding and retesting.

13.7.2. Centrifugal Castings

This is a process wherein cylindrical, conical, or parabolic shaped laminates can be densified by the action of centrifugal forces applied by rotation about the longitudinal axis, Fig. 13.21.

14
BAG MOLDING PROCESSES

A. Slobodzinsky
PLASTEC, Picatinni Arsenal
Dover, New Jersey

14.1. INTRODUCTION

Bag molding fiber-reinforced organic matrix composites and bonded structures comply with different standards of quality and different criteria for reproducibility. The uses of composites range from adornments and decorative architectural panels to complex, efficient load-carrying structures. The bag molded composites that perform adequately and are competitive with alternate types of constructions most often result from optimizations of process adaptations and design refinements.

Composite designs and bag molding process adaptations are interrelated by production goals and controls of manufacturing costs, as well as by service or mission requirements. Responsibilities shared by fabricators and designers include the following.

- Reproducibility, availability, and processing characteristics of candidate materials.
- Requirements for facilities and tooling.
- Provisions for assembly, inspection, and quality control.

Innovative adaptations of bag molding processes are often required to structure the fiber-reinforced organic matrix materials into a unified, single construction.

14.1.1. Bag Molding Methods

Molding methods include vacuum bag (Fig. 14.1), pressure bag (Figs. 14.2 and 14.3), and autoclave molding (Fig. 14.4). Bags, the thin and flexible membranes or silicone rubber shapes, separate the laid-up constructions from the pressurizing gases during the composite cures. The bagged lay-ups in autoclaves are usually vented to pressures lower than those applied to the bag. Consolidations and densifications of the lay-ups are achieved by the resulting pressure differentials across the bag contents. Consolidations are achieved when the separate plies of prepreg in the lay-ups and other adherends, when present, are bonded together. Densifications result in diminutions of voids and removal of excess resin. Other results desired of bag molding methods during the cures include prevention of blistering in the composites, increased controls of pressure and heat application, and control of the fiber/resin ratio.

Consolidations and densifications of vacuum bag moldings can be achieved by atmospheric pressure alone as the bagged lay-ups are evacuated throughout the cure cycles. The pressure-bagged and autoclave-cured composites are pressurized by hot gases. Vents to the atmosphere, or vacuum, provide for the escape of the volatilized reaction by-products and the entrapped air from the curing composites. Of the three methods, vacuum bag molding is least limited as to the size of constructions which can be processed. On a few occasions, vacuum bag molded composites are room-temperature cured. Most are thermally cured to produce improved properties. Thermal cures are best attained in air-circulating ovens, but can be achieved in infrared heated or passive type convection ovens as well.

Alternate curing methods include induction, dielectric, microwave, xenon flash, ultraviolet, electron beam, and gamma radiation.

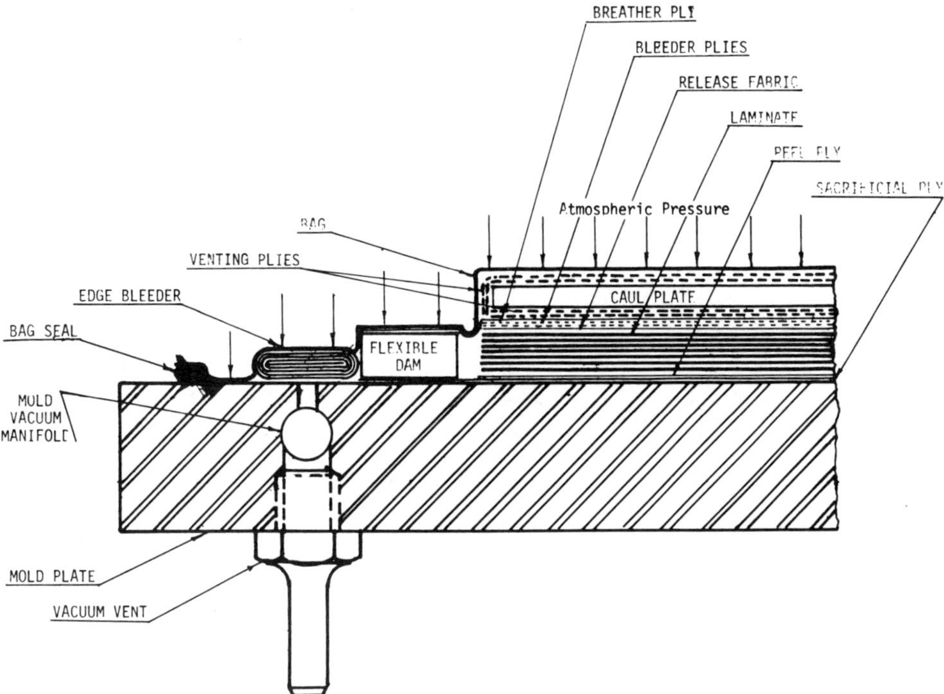

Figure 14.1. Vacuum bag molding method with vertical bleeder.

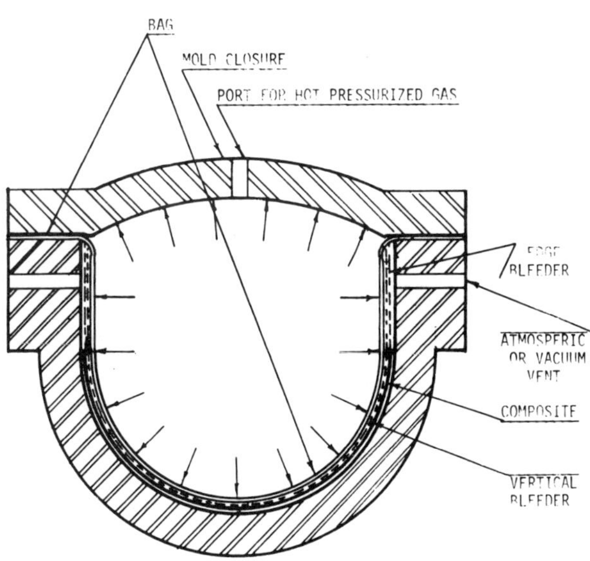

NOTE: CLAMPS ARE NOT SHOWN

Figure 14.2. Pressure bag mold.

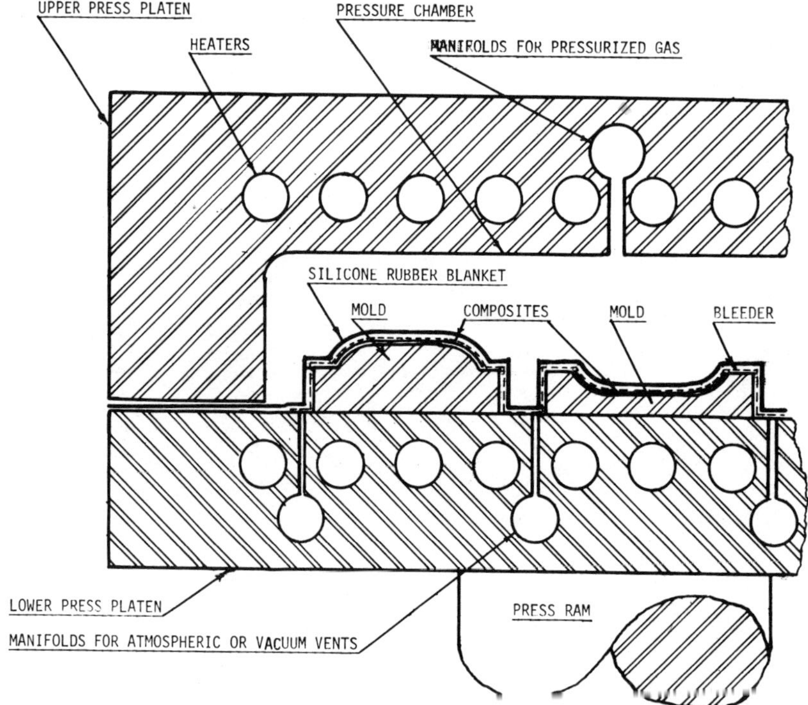

Figure 14.3. Press modification for pressure bag molding.

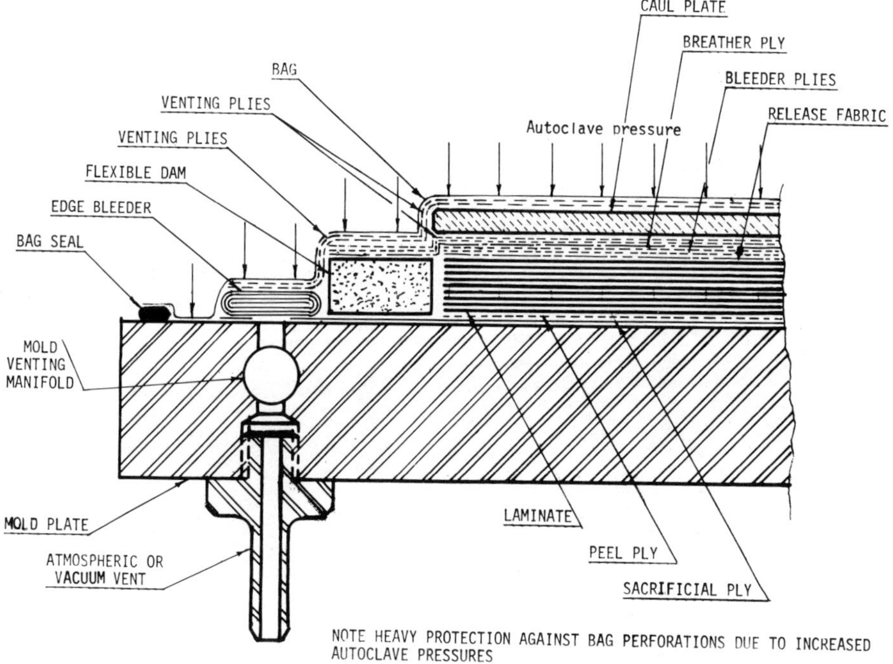

Figure 14.4. Autoclave molding method with vertical bleeder.

Of the light curing systems, the ultraviolet and xenon flash radiation are the most commercially developed. Thin, unsymmetrically laid up, and unpigmented composites,which deform excessively when thermally cured, are the more successful candidates for the light curing composite systems. Dielectric and microwave curing techniques continue to be demonstrated. Electron beam cures have been shown to be effective for thin films. Accelerated particle cures are still laboratory developments.

Tooling is less expensive for the vacuum bag and autoclave molding than for matched die molding methods. Molds and molding plates are required to withstand curing conditions without distorting or degrading, and to tolerate handling during the fabrication processes. They are not subjected to unbalanced pressures.

The higher costs of autoclave facilities can be amortized by taking advantage of their improved capabilities to mold complex constructions. Composites that may ordinarily require secondary bonding are often more economically co-cured in autoclaves. Depending on the size of the pressure vessels and their associated instrumentation or controls, increased costs for autoclaves may be reduced by curing many parts at one time.

Pressure bag molding methods are efficient for both producing deeply contoured structures and shallow composites. Sonar domes, radomes, and antenna housings are examples of deeply contoured composites. Architectural panels, door panels, and aircraft fairings are examples of shallow composites.

Heavy molds are built to reproduce the deeply contoured structures. Each specialized mold is constructed to withstand the elevated temperatures and increased pressures required for the cures. Sometimes the large molds are suitable for use as autoclaves for smaller bag molded composites.

Shallow items may often be bag molded in modified compression presses. The lower press platens contain vents and vacuum lines. The upper press platens are made hollow to enclose the mold plates together with the laid up assemblies. Frequently, the upper platens are also fitted with a pressurized heated gas line. Silicone rubber blankets or conventional pressure bags are placed over the mold plates on the lower platens. When the presses are closed, the sealed chambers are pressurized and heated to attain molding conditions similar to those of an autoclave. Unlike the specialized pressure bag molds, the modified presses are used to cure many different composite constructions.

Autoclave and pressure bag molding conditions to 350° F (177° C) and 200 psi (1379 kPa) are routinely attained. Newer, customized autoclaves attain cure conditions that exceed 500° F (260° C) and 500 psi (3447 kPa). Most of the special pressure vessels are size limited and are not readily accessible for routine production. The higher temperature and pressure curing facilities are currently being expanded to accommodate the polyimide resin composites used for higher temperature resistant structures.

Fire hazards are greatly increased at the more elevated temperatures and pressures. Pressure vessel fires are minimized by uses of fire retardant processing materials and inert pressurizing gas. Fire prevention measures include uses of silicone rubber, nylon, or Tedlar bags. Before cure cycles are initiated, the pressure vessels are purged of all enclosed air. After the thermal cure is completed, the pressure vessels and their contents are cooled to 150° F (68° C) before the pressure is relieved and the autoclave or pressure bag mold is opened.

14.1.2. Commercially Available Matrix Materials

Polymeric matrices for composites include both thermoplastics and thermosetting resin compositions. Current usage of thermoplastics is relatively minor compared to thermosetting resins. Polysulfone is the most common thermoplastic for prepregs. Varnish type thermosetting resins are used exclusively for "wet" lay-ups. Both varnish types and "B staged" solid thermosetting resins are used for prepregs. Some polyimide prepregs contain monomeric reactants which combine during "B staging" and cure.

Polysulfone prepregs are dry and stiff. The forming temperatures are above 380° F (193° C). They must be thermoformed to comply to the molding surfaces. Sometimes, they are manually formed with the help of a heat gun. Often, the plies are individually bagged and thermoformed in an autoclave. After the formed plies are stacked onto the molding surface, they are bagged and fused into integral composites at about 600° F (204° C) and 200 psi (1379 kPa). The advantages afforded by thermoplastic prepregs include the following.

- Indefinite storage life at ambient room conditions.
- Fusion within a matter of minutes at equilibrium "cure" conditions.
- Elevated temperature properties that are more unaffected by moisture.

The question remains whether special rubber-modified epoxies are to be preferred to the thermoplastic resins for uses in high temperatures after exposure to high humidities. The modified epoxies currently being developed show excellent promise for improved high temperature properties after being subjected to saturation humidity conditions. In addition, the new epoxies attain superior processing characteristics.

Varnish type epoxies, polyesters, and phenolics, without inert diluents, are preferred for the "wet" lay-ups and filament wound constructions. Solvents are undesirable because they volatilize during the cure cycle and increase the potential void formations. Solvent-diluted resin impregnating systems are acceptable since most of the unreactive solvents are volatilized when the commercial prepregs are "B staged" to attain the desired handling and processing characteristics. When no solvent content in the prepregs can be tolerated by the composite molding process, "B staged" resins are available in hot melt impregnations of reinforcements. The PMR polyimide resins, containing unpolymerized monomeric reactants, are fluid for impregnating reinforcements. They are then partially reacted on the reinforcements to attain the desired prepreg characteristics.

14.1.3. Commercial Fiber Reinforcements

Glass, aramid, graphite, and boron fiber reinforcements are described in other chapters of this handbook and are catalogued in the literature. However, it is to be emphasized that most fiber reinforcements are surface-treated to enhance coupling to specific types of resins. Fabricators' or prepreggers' requirements for the fiber-finish systems should be clearly defined in the company procurement specifications.

14.2. MOLD PREPARATIONS

Coefficients of thermal expansion for conventional tooling materials and composites are listed in Table 14.1. Of the metals, the coefficient for steel compares most closely with the coefficients for the composites. Other characteristics for steel include superior durability, ample tolerance for elevated temperature service, and good thermal conductivity.

Ceramics produce favorable characteristics for molds. They have the lowest coefficient of thermal expansion and their thermal conductivity approaches that of some hardened tool steels. However, ceramics are brittle at ambient temperatures. They must be protected from processing handling hazards. One way to afford protection is to enclose ceramic inserts in a steel case.

Both steel molds and steel molds with ceramic inserts are candidate tools for reproducing high quality composites. Due to the low thermal expansions, the ceramic inserts provide capabilities for positioning components for bondments most precisely. Such molds are prime candidates for fabricating large numbers of co-cured constructions in which adhesive bonds are cured simultaneously with the FRP materials. However, the tools are costly and production quantities are often not sufficient to amortize tooling costs

Table 14.1. Linear Thermal Expansion of Continuous FRP Composites and Tooling Materials

MATERIAL SYSTEM	EXPANSION FROM 75–400° F (24–204° C), %
FRP	
Aramid or graphite-reinforced	0.016–0.033
Boron fiber-reinforced	0.081–0.162
Glass fiber-reinforced	0.12
Slip cast and fired ceramic	0.015
Tool steel	0.20
Electroformed iron	0.21
Electroformed nickel	0.23
Semi-steel	0.24
Thermally cycled cast plaster	0.25
High-temperature epoxy	0.35
Aluminum	0.42

at competitive prices for the production items. On those occasions, less costly alternatives are desirable.

Aluminum molds are among the less costly of the wrought or cast metal tools. Although thermal conduction is better for aluminum than for steel, the tools are less durable and the thermal expansion is excessive. Shallow or flat mold plates are usually limited to cures below 350° F (177° C). Other metal tools include semi-steel (Meehanite) molds and sprayed or electroformed molds reinforced with cast backings.

Electroformed nickel tools which have been used for over 15 years have lately been optimizied to provide dense, non-porous structure, highly polished tooling surface and smooth back face. The costs for these molds have been reduced and intricate contours can now be produced. Figure 14.5a shows a typical electroformed nickel mold for an aircraft part. The coefficient of thermal expansion for nickel is also compatible with fiberglass composites.

Figure 14.5*a*. Electraformed mold. (Courtesy Grumman Aerospace Corp.)

Almost uniform wall thickness is required for successful semi-steel molds. When molds with abruptly changing profiles and cross-sections are thermally cycled, the cast metal may crack or warp. The thermal conductivity of semi-steel is low for metals. Where changes in mold wall thicknesses occur, mold face temperatures may vary considerably. As a result, controls of the composite cures are difficult to attain. When the added weight of the tool can be tolerated, and when heating and cooling through the bagged surface are adequate, the cast tools should be sufficiently massive to help attain a uniform temperature range.

Low melting point alloys, whose phase changes occur above the composite cure temperatures, are most often cast into the prepared sprayed or electroformed mold shells. Other cast backings include thermally conductive plastics and other thermal strain compatible materials in which heating elements and cooling ducts can be consolidated. Alternate types of tooling include composite molds, usually based on high temperature resistant cast or laminated epoxy resins.

Master forms for laying up composite tools can also be fabricated, using any of the materials described; a mock-up model of the item may be used; or plaster masters can be prepared from models. The quality of the plaster masters depend on the strain compatibility between the plaster and the reinforcements the masters contain and on the condition of the hardened surface.

When properly cast, set, and matured, a mix of Hydrocal and water in a weight ratio of 100:50 will develop a crushing strength of 3500–3800 psi (24.1–26.4 kPa). The setting plaster expands 0.5% and when thermally cycled to 400° F (204° C), the plaster reversibly expands another 0.25%. The plaster molding surfaces are hardened with penetrating varnishes and cured above the intended use temperatures. Prime candidate resins for the solvent varnishes are the epoxy novolacs that will cure at 400° F (204° C).

Composite tools can be laid up using fiber orientations that most closely match the expansion of the items to be produced. Fiberglass and graphite fibers are the principal reinforcements. Woven fiber reinforcements are the most economical to use.

Mold maintenance is best relegated to specialized personnel, while preparations for the bag molding processes are assigned to production personnel. A successful practice is to provide production personnel with soft tools and solvents that do not degrade the molding surfaces. MEK (methyl ethyl ketone) is the solvent most often used for cleaning molding surfaces. If the production tools and solvents are inadequate for removing debris and cleaning, the molds are taken out of service for maintenance, repair, or replacement. After the molds are returned to service, they are solvent-wiped-clean and mold release agents are applied.

14.2.1. Release Agents

Release agents for bag molded composites include carnauba paste wax; aerosol dispensed compositions that contain carnauba, fluorocarbon resins, or silicone resins; plastic films; and metal foils. On most occasions, the wax or resin mold releases do not contaminate the composite surfaces excessively or prevent subsequent secondary bonding or coating operations. Sometimes, the composite surfaces are solvent-wiped-clean and lightly sanded to remove gloss. Other times, peel plies are used to protect clean surfaces for primary adhesive bonds. Plastic films, metal foils, and sprayed metal coatings also serve as release agents when they are integrally laminated to the cocured composites.

Both the polished wax surfaces and the sprayed wax coatings are excellent mold releases for composites cured below 250° F (121° C). However, the wax degrades and discolors the composites at higher molding temperatures. Commercial fluorocarbon mold releases are used for higher cure temperatures.

The FEP (fluoroethylene propylene) mold releases form a continuous film on mold surfaces. Although the condition of the release film is easy to maintain below 350° F (177° C), the coating degrades at increased temperatures. Fluorine, which is noxious, corrosive, and highly toxic, disassociates from the polymer above 350° F (177° C).

PTFE (polytetrafluoroethylene) is stable and is often contained in mold releases for service in excess of 500° F (260° C). The mold releases contain suspensions of micropulverized Teflon in a volatile dispersant. Depositions on mold surfaces do not form continuous films, but the Teflon particles provide excellent dry lubrication for the release of the cured composite. Furthermore, the residual particles on the cured composite surfaces are easily removed with a solvent-wipe. Since a variety of commercial mold releases that contain fluorocarbon are on the market, it is essential that manufacturers' recommendations on uses and limitations be scrupulously followed.

Frekote, Inc. produces silane (non silicone) resin mold release said to be stable at 900° F (482° C). Some silicone mold releases can be used only to 400° F (204° C). These resin systems are adequate, provided that the manufacturers' recommendations are not exceeded. Some of these formulations provide surfaces which can be painted or bonded without any further sanding.

Cellophane, polyvinyl alcohol (PVA), polyethylene, Mylar, and nylon film are often used to release bag molded composites. Uses of strippable film releases for production moldings are usually limited to surfaces of single curvature or flat shapes.

Tedlar, a fluorocarbon plastic film, and metal foils are often laminated to the co-cured laminate to provide protection from hostile environments or harmful ultraviolet degradation (see Chapter 19 on environmental effects). Metal foils also provide protection from many forms of radiation and lightning strikes to aircraft parts, and often are used as decorative or reflective surfaces, for electromagnetic shielding, or for electrical conducting paths. When plastic films and metal foils are laminated while the composite is being cured, the films or foils are positioned on the mold first. The molds are often provided with vaccuum ports to anchor the thin materials so they will not move during the lay-up. Release agents are applied to the molds merely to facilitate maintenence and cleaning.

Sprayed aluminum coatings are alternatives to metal foils. The molds are first coated with a PVA spray release agent and the aluminum is sprayed to a 0.005-in. (0.13-mm) thickness. The composite is laid up directly on the sprayed metal. Sometimes, the laminated coating is polished to match the appearance of adjacent parts.

Uses of silicone oils and greases are to be avoided since they are the most persistent contaminants of molded composite surfaces. They release secondary bonds and coatings from composite surfaces as effectively as they release the cured composites from the molds. Silicone oils and greases migrate and defeat most attempts to remove them. They contaminate the solvent wetted cloths and sandpapers so that, instead of removing the silicones, the vehicles spread them. Contaminated surfaces may be salvaged for painting by sand blasting with virgin grit.

14.3. PEEL PLIES AND RELEASE FILMS OR FABRICS

Uncontaminated faying surfaces, the surfaces that are adhesively bonded together, are required to attain reproducibly strong bonded joints. Peel plies for that purpose provide maximum protection to the faying surfaces during subsequent operations prior to the application of adhesives. Peel plies are ordinarily removed just prior to the bonding or secondary coating operations.

Release films and fabrics serve many purposes. Some are used as separators between successive layers of prepreg. Some serve as backings for carrying pre-cut plies of prepreg to the mold. Others are used to provide cleavage between the composite construction and the bleeder plies used to absorb the excess resin from the lay-up during the bag molding cure.

Even though they leave no residual contaminants, some release films and fabrics are unsuitable for peel plies because they cannot adhere to the composite surface. On the other hand, greige fiberglass fabrics are often used as peel plies even though residual oil-starch size remains on the composite surface after the fabric is removed. However, before the secondary or bonding operations, the residual size is removed by sanding or by solvent-wiping. Adequacies of such surfaces for adhesive bonded structures should be experimentally verified.

Peel plies for some architectural adornments and constructions for interiors are fiberglass fabric greige goods. After the peel plies are removed, the exposed surface textures complement the patterns of the reinforcements or the decorations.

The most popular candidates for peel plies are the commercially available heat cleaned and scoured nylon, heat cleaned lightweight fiberglass, and the suitable polyester release fabrics. User preferences for target surface textures vary. Some prefer stronger fabrics and accept coarser weaves. The secondary sanding operations for refining textures are less costly than removal of finely woven peel plies. Other companies use the finer weaves that leave textures requiring no additional refinements. Most commercial peel plies are square weave fabrics not ordinarily conformable to dished or contoured shapes. Military Specification MIL Y 1140 lists more conformable satin fabrics of heat cleaned fiberglass. Styles 120 and 2120 are the preferred lighter weaves.

Table 14.2. Expendable Processing Materials

Peel Plies
- Miltex, a heat scrubbed and set nylon fabric. Miltex 3921 is a finely woven style.
- Dacron release fabrics that are available from prepreg suppliers.
- Fiberglass fabrics. Finer woven fabrics include styles 2112, 2116, and 2120. Coarser fabrics include woven rovings for decorative architectural panels.

Release Fabrics
- Teflon coated 104 fiberglass scrim.
- Dacron release fabric.

Bleeder Plies
- Mochburg fabrics.
- Fiberglass fabrics, usually styles 2120 or 7781.
- Fiberglass mat, ¼-oz increments to 3 oz/ft^2.
- Pellon, cellulose, or polyester mats.

Breather Plies
- Perforated Tedlar, 0.020-in. (0.05-mm) diameter perforations; 0.5–1 in. (1.3–2.5 cm) on centers.
- Perforated nylon, 0.020–0.030 in. (0.05–0.08 mm) hand-made perforations 0.5–1 in. (1.3–2.5 cm) on centers.
- Teflon coated 104 fiberglass scrim.

Manifold Plies
- Mochburg fabric.
- Fiberglass fabric, style 7781.

Bag Sealer
- Zinc chromate sealer tape.

Bag
- Kapton, to 600° F (316° C).
- Silicone rubber blanket, 0.125 in. (3.2 mm) thick.
- Fiberglass-reinforced silicone rubber bags.
- Nylon, to 350° F (179° C).
- PVA film, to 250° F (121° C).

Table 14.2 lists some of the preferred expendable processing materials used for bag molding.

14.4. BAGGING TECHNIQUES

For vacuum bag molding, the bags are used to evacuate the air from the laminate and to generate the atmospheric pressure required for compaction over the mold. For autoclave molding, the bags serve to contain the compacting gases during the cure. If the pressures within the bag are not reduced from those applied to the bag, the membrane remains inert and there is no compaction. Bleed-out systems are devised to maintain reduced pressures within the bag's contents. As long as the pressure within the bag remains reduced, the compacting gases bear on the bag as it presses against the lay-up. The requirements for the pressure applications include the following.

- Consolidations of the successive plies.
- Completion of the fiber impregnation with the resin.
- Elimination of the void nucleating volatiles, reaction by-products, and entrained air.
- Reduction of the excess resin in the lay-up.

The bagged lay-up includes the bleed-out system designed for the composite part.

Bagged lay-ups can be vertically or edge bled. The classical differences between the two can be seen by comparing Fig. 14.5 with Figs. 14.1 and 14.4. Fig. 14.5b shows the lay-up assembly for an edge bleed-out system. The lay-up stacking sequence is as follows.

1. The surface of the mold is prepared with the release agent.
2. A sacrificial ply is laid up on the prepared surface. Usually, it is a ply of 120 fiberglass fabric with a resin compatible with the resin used in the main lay-up.
3. The peel ply is anchored to the sacrificial ply.
4. The composite plies are oriented and laid up in the directions specified in the design, and rubbed out on top of the peel ply (see Fig. 14.6).
5. The edge bleeder is positioned at least ½ in. from the periphery of the lay-up

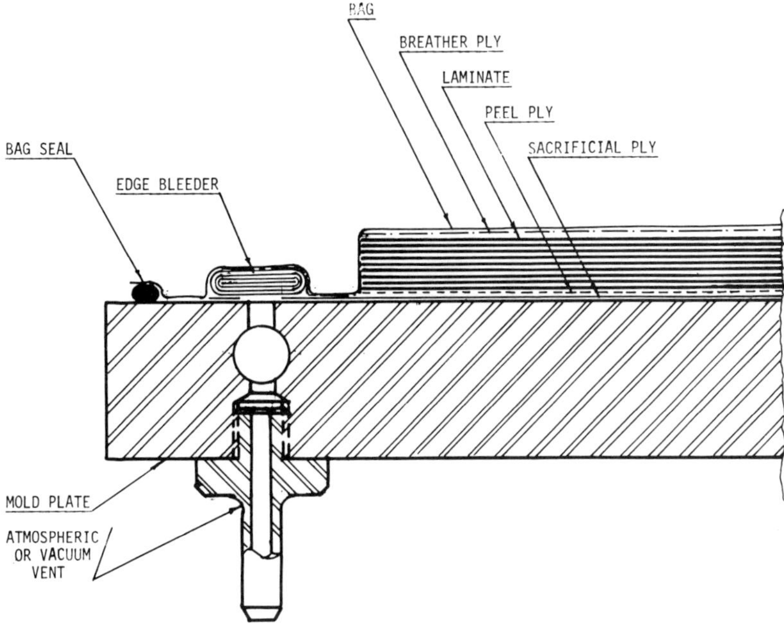

Figure 14.5*b*. Edge bleed-out system.

and is joined to the ports of the venting system.

6. Porous release fabric is laid up over the composite plies and is extended over the edge bleeder. The release fabric is made to comply to the ½-in. (1.3-cm) trough between the lay-up and the edge bleeder. The release fabric is trimmed so that it does not extend beyond the edge bleeder.
7. If a self-sealing silicone rubber bag is not used, a bag sealer is applied ½–1 in.

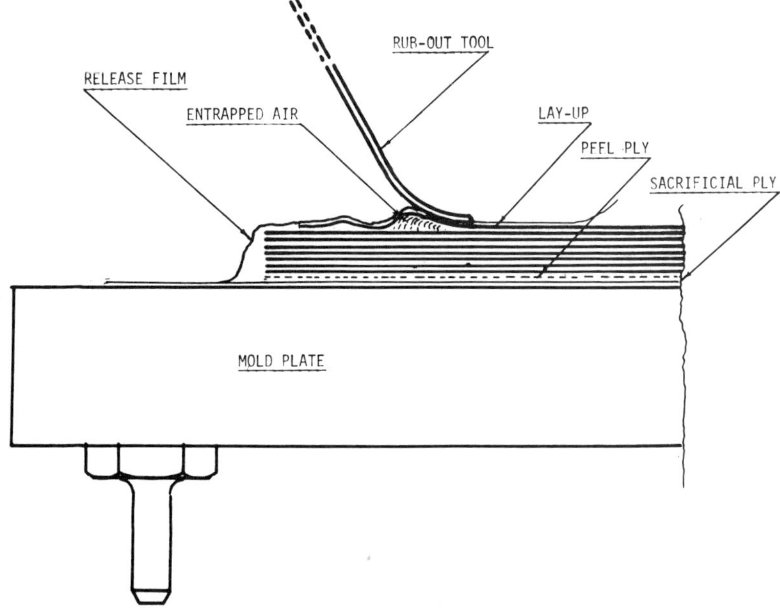

Figure 14.6. Prepreg lay-up operation.

(1.3–2.5 cm) from the periphery of the edge bleeder.

8. Special care is taken to seal the bag covering the assembly so that it will not leak.
9. The bag contents are evacuated and the system is checked for leaks. All leaks are sealed.
10. The bagged lay-up is ready to be cured.

This system does not provide for significant resin bleed-out and provides only a partial escape of volatiles. Composites cured using this system will usually have tapered edges to a distance of about 2 in. (5 cm) from the edge of the lay-up. These are normally removed from the cured composite and sometimes are used for quality control tests.

Figures 14.1 and 14.4 show lay-ups for vertical bleed-out systems. In these figures, the bleed-outs are controlled to maintain predetermined thicknesses.

1. The surface of the mold is prepared with the release agent.
2. The sacrificial ply is laid up and is extended to anchor a flexible dam.
3. The peel ply is carefully positioned and anchored so as not to interfere with anchoring the flexible dam.
4. The composite plies are indexed, laid-up, and rubbed out on top of the peel ply.
5. The flexible dam is anchored to the sacrificial ply approximately 0.125 in. (3.2-mm) from the edge of the composite lay-up.
6. The release fabric is laidup over the dam and the lay-up.
7. A predetermined number of bleeder plies are laid up over the release fabric and extended only to the perimeter of the lay-up (sometimes they are stitched to the release fabric).
8. A perforated Tedlar breather ply is laid up over the bleeders and extended to the flexible dam. [The Tedlar contains 0.030-in. (7.5-mm) perforations 1 in. (2.5 cm) on centers.]
9. An edge bleeder is connected to the venting ports.
10. Two layers of 181 style dry glass fabric are placed over the lay-up and extended to the edge bleeders.
11. Cauls or insulating layers are located over the composite lay-up. A 0.125 in. (3.2 mm) steel, aluminum, or transite plate provides adequate protection against sharp temperature increases. The intent is to maintain the same heating and cooling rates with the mold and to prevent local purging of resin during pressurization. Cauls are also used to ensure a smooth, non-wavy surface.
12. Two or more plies of dry fiberglass fabric are laid over the cauls to protect the bag against perforations.
13. The sealer is applied around the perimeter of the edge bleeder.
14. The bag is positioned and sealed.
15. The bag contents are evacuated and smoothed, and the bag is checked and sealed against leaks.
16. The bagged lay-up is ready to be cured.

In this bleed-out system, the amount of bleeder plies are calculated to reduce resin contents by predetermined amounts. Figure 14.4 contains every element that might be included in any modified system. At this point, the distinctions between edge and vertical bleed-out systems was made specifically for the purposes of explanation. The distinctions between the different modifications were not made. The chief purposes for the modifications are how best to consolidate the composite parts and how to make the most efficient use of the materials. Reference is again made to Table 14.2 for the lists and applications of expendable materials used for processing.

14.5. REUSABLE VENTING BLANKETS AND SILICONE RUBBER BAGS

There are costly disadvantages to the use of expendable venting plies and bags of plastic films for fabricating production composites. The disadvantages include:

- The excessive costs for recurring manual operations and wasteful uses of processing materials.
- The increased potentials for rejected composites because of defects caused by wrinkling of venting plies or leaks and blowouts of the bags fabricated of thin plastic films.
- The increased potential for damage to the mold surface due to additional cleaning and handling.
- The incurred costs detract from competitive advantages the composites might otherwise attain over alternate constructions.

Each of the expendable venting plies are separate layers of dry woven fabric which are draped and tailored to conform to the lay-up. Besides the material, extensive labor of trained operators is required. After the bag molded assemblies are cured, the venting plies are more costly to salvage than to replace.

The recurring costs for labor and expended materials may be reduced by combining the tailored venting plies with a suitable breather ply and sewing them together into a reusable venting blanket. Operations for laying-up venting and breather plies separately can be replaced by one operation to lay up the blanket. The sewn blankets are equally useful for venting the adhesively bonded assemblies during their cure and for venting the FRP composites. The superficial differences between the two uses for the venting blanket are that bleeder plies are usually used between the laid up composite (see Figs. 14.2 and 14.4) and that bleeder plies are seldom used for curing adhesive bonds or co-curing one-step sandwich constructions.

Expendable bags, laid up of plastic films and the associated sealants, also incur needless recurring costs. The expendable bags can be laid up only once because of degradation during handling and the thermal cures. Even on normal handling during production, the plastic films can be easily perforated. In addition, the seals for the bag may not be effective, so that the assembly may leak. The seal is obtained by pressing the sealant between the plastic film and the mold surface. Before the bag molding consolidating pressures can be attained, all leaks must be detected and repaired. Even during the thermal cure, the risk is present that weakened plastic films will blow out during pressurization. Often, the result is that a costly composite part is not properly molded because the lay-up was not consolidated or densified. Uses of thick silicone rubber blankets and self-sealing reusable bags reduce fabrication costs and defective composite parts because of the resulting work simplification and the more positive control of the bag molding cure conditions.

Uses of mechanically clamped fiberglass-reinforced and plain silicone rubber bags or blankets are illustrated in Figs. 14.2 and 14.3. The blankets are usually sealed by clamping the silicone rubber between two flat and reasonably smooth metal surfaces. No additional sealer is required.

Another, less costly edge seal provides a vacuum type closure which is maintained under separate vacuum from the bag contents.[1] In use, the bag would be positioned over the lay-up on the mold. Then, the connections to the vacuum closure would be made and the bag would be sealed to the mold by the drawn vacuum. After that, the bag contents may be evacuated or just vented to the atmosphere, as the molding conditions would require.

The vacuum line connections may be cast into the silicone rubber bags. Figures 14.7*a* and 14.7*b* show quick disconnect attachments for vacuum lines cast into fiberglass-reinforced silicone rubber bags. Figure 14.7*c* shows a bottom view of the bag shown in Fig. 14.7*b*, which illustrates how auxiliary hardware can also be cast into the silicone rubber bags. On this occasion, a perforated tubular manifold for evacuating the bag contents is shown. Similarly, rigidizing frames for large bags, integral heating elements, and thermocouples for temperature controls may be cast. Two designs of more easily assembled closures are shown in Figs. 14.8*a* and 14.8*b*. Figure 14.8*a* shows a fiberglass-reinforced silicone rubber closure cast of RTV 92-048. Figure 14.8*b* shows a modified version which incorporates a low durometer hardness precast silicone rubber insert. The seal with the precast insert is more compliant to mold surfaces. However,

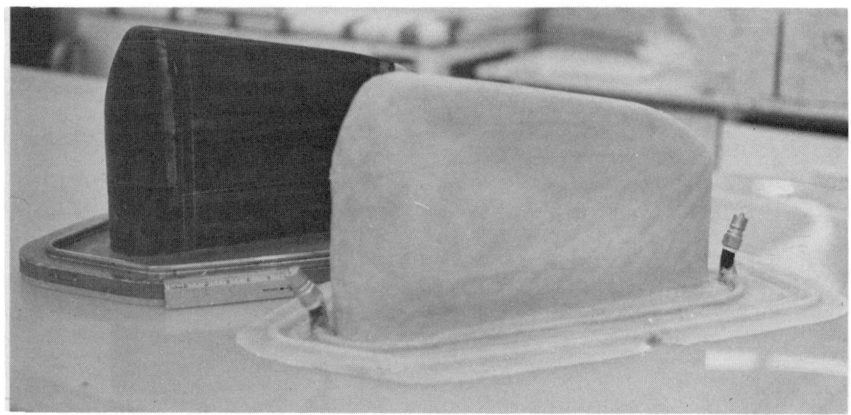

Figure 14.7*a*. Rubber cup mold and bag.

Figure 14.7*b*. Silicone bag with cast-in vacuum line, top.

Figure 14.7*c*. Silicone bag with cast-in vacuum line, bottom.

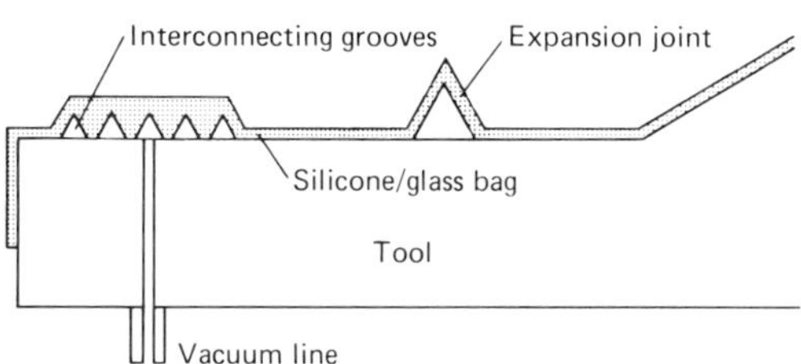

Figure 14.8*a*. Multi-groove seal.

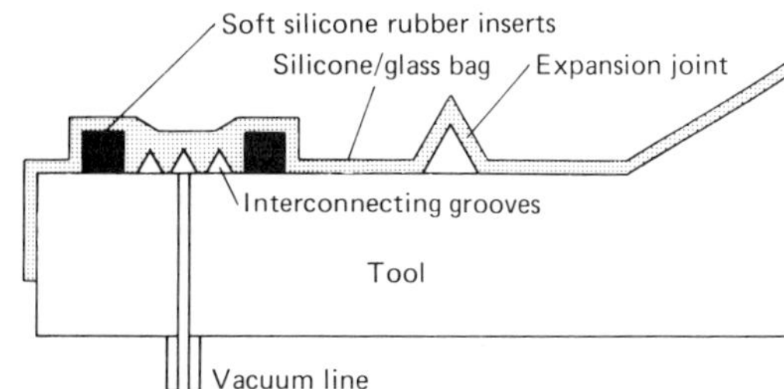

Figure 14.8*b*. Modified multi-groove seal.

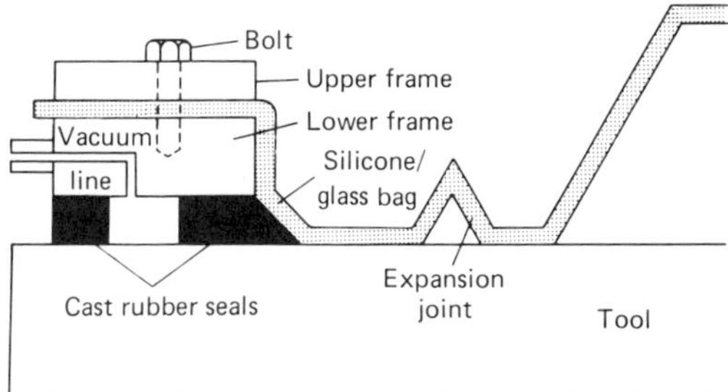

Figure 14.9*a*. External frame seal.

all surfaces under the seal must be clean and free of scratches and dents. Figure 14.9*a* shows a more durable metal frame closure with rubber seals. Figure 14.9*b* shows the actual production bag used for the F-14 stabilizer skin bonding operation.

Reusable venting blankets and silicone rubber bags have withstood up to 1600 bag molding cycles. There are occasions where expendable material costs for vertically bled

Figure 14.9*b*. F-14 stabilizer skin bonding bag.

systems were reduced over 90% and the associated labor costs were reduced over 80%.

The material requirements for the cast silicone rubber bags include the following.

- Sufficient flow to conform to complex shapes.
- Capability to wet out the fiberglass reinforcements.
- Attainment of adequate room-temperature cure.
- Sufficient softness to comply to sealing surfaces.
- Durability to withstand repeatedly long exposures to 350° F (179° C) without loss of strength or resilience.
- Dimensional stability
- Repairability.

Materials recommended by the developers of reusable bags for the Grumman Aerospace Corporation include:

- Dow Corning, 92-048 RTV silicone rubber latex.
- Dow Corning, DC 92-009 RTV dispersion, diluted 1:1 with naptha to wet out fiberglass reinforcements.
- General Electric, RTV 630.
- Style 1000 fiberglass fabric with a neutral pH heat cleaned finish.
- A United Merchants knitted glass fabric for deeply convoluted contours.
- Mosites and "D" Aircraft Products—rubber, rubber sheets, cured and uncured. cured.

The bag is usually fabricated over a bag molded lay-up with the bleeders and venting blankets in place on the mold. A release film is placed over the assembly and the silicone bag is built in place, with a vacuum drawing the assembly down onto the mold as compactly as possible. The vacuum is maintained throughout the silicone rubber bag fabrication process. Wax strips are placed around the perimeter of the mold to form the serations for the vacuum activated seals. When they are used, the soft rubber extruded strips are also located around the mold perimeter.

14.6. MATERIAL SYSTEMS

"Wet" lay-up techniques are sometimes combined with bag molding processes to enhance the properties of otherwise low stressed constructions. Because it is difficult to wet out dry fiber reinforcements with less resin, initial volumetric fraction ratios of resin to reinforcements are seldom less than 2:1. On a weight basis for fiberglass-reinforced systems, the ratio is approximately 1:1. When contact molded, candidate structures are required to tolerate increased composite thicknesses to compensate for the high resin contents. Candidates include architectural panels and enclosures, bathroom fixtures and enclosures, and composite constructions in trucks and motor vehicles. There are sometimes space limitations for critical supports. Adaptations of bleed-out techniques can be used to reduce resin contents and to enhance structural properties.

Depending on the sizes and contours for the lay-ups, some techniques will reduce resin volume by 5%. Larger and more deeply convoluted composites may tolerate only 1% resin bleed-out. Excessively bled composites result in fiber wash-outs, ply wrinking, gross thickness variations, and incomplete cures because of excessive losses of reactive volatiles. Such defects will degrade properties. When consolidations are properly conducted, laminates with void contents less than 2% by volume are produced.

Techniques for laying up dry reinforcements with catalyzed resins are similar to the methods for contact molded hand lay-ups. The resins for room-temperature cure lay-ups are usually formulated to attain longer pot lives. Catalyzed resins are applied by rollers, brushes, or other appropriate means, and to ensure the wet-out of reinforcements and elimination of entrapped air, the resins are hand worked into the fibers.

When the volume fraction of resin is less than 60%, (less than 43% by weight resin in a fiberglass composite), it is difficult to wet out the fiber reinforcements. It is often necessary to force the resin into the fibers by rolling over a thick film placed over the lay-up. Sometimes, excessive force is required which disrupts and tears up the orientation of the fibers. On those occasions, decisions are made whether to increase the resin content, tolerate the disruptions to the reinforcements, or accept increased void contents.

Uniformity of distribution of resins through reinforcements is dependent on operator skills, resin viscosity, adequate resin pot life, and orientation of the reinforcements. Difficulties encountered during fabrication of deeply contoured shapes include pooling of resin in depressions and sagging of laid up reinforcements from sloping walls. The main difference from contact molding arises from adaptations of bleed-out and bagging techniques. Low cost fiberglass reinforcements are almost exclusively used in wet laid-up composites. Generally, saved weights or increased strengths or stiffnesses are too nebulous to justify the uses of more costly reinforcements. However, attainment of higher stiffness and fatigue resistance properties may justify the uses of graphite fibers.

14.6.1. Resin System Modifiers

Liquid resins for "wet" lay-ups and filament wound prepregs contain all the constituents to attain the cured state. They are formulated to attain sufficiently long pot lives to complete the bag molding processes. Additives are usually fire retardants, ultraviolet barriers, and thickening agents. Most often, inert diluents are avoided. On occasions when diluents are used to reduce resin viscosities, attempts are made to volatilize them prior to initiating the resin cures.

The advantages of using thickening agents in liquid resins include the following.

1. Enhancement of structural integrity of the cured laminates and all co-cured bonded joints.
2. Retention of reinforcements on sloping mold surfaces during the fabrication operations.
3. Retention of the resin by the reinforcements during the fabrication operations.
4. Control of constituent contents and cured composite dimensions.

Usually, less than 10% by weight of thickeners will achieve the desired results.

14.6.2. Prepreg Winding Solvent Resin Systems

Sometimes, inert solvents must maintain resins in a sufficiently fluid state to wet the reinforcements during the filament winding of prepregs. When such systems are used, additional equipment is required. Besides devices that meter the amount of resin to remain on the fibers, other equipment is required to volatilize the solvents and stage the prepregs. Adaptations of drying tunnels and staging towers are used. Multi-layered prepregs can then be wound without appreciable solvent content to nucleate voids.

14.6.3. Filament Wound Prepregs

Filament winding processes are described in Chapter 16. On occasions, it is economical or structurally desirable to wind multi-layered prepregs. These are windings taken from the mandrel for subsequent bag molding.

Mandrels are first overwrapped with compliant plastic films which later serve as backings for the prepregs. The filament wound prepreg must contain the completed sequence of angle plies required for the lay-up. When the fibers are collimated, fiber volume content fractions of 60–65% are attained.

The number of layers the prepregs can contain are limited to the prepreg thickness that can be laid up on a molding surface without disrupting fiber orientations or wrinkling any layers. Optimum prepreg thicknesses are affected by mandrel diameters, the match with molding surface contours, and the areas of molding surface to be covered by the prepreg.

After the winding is staged, the excess of built-up windings adjacent to the poles is trimmed away. The backing, together with the wound prepreg, is cut parallel to the mandrel axis. It is then lifted off the mandrel and placed on the molding surface. It is seldom desirable to lay up the prepreg on a reversed curved surface. If a male form is used, the backing should be against the mold surface. If a female form is used, the resinous surface should be laid into the cavity. If the prepreg is adequately advanced, the backing may be removed after the lay-up is completed in a cavity or an auxiliary backing is positioned for laying up the prepreg on a male form. Otherwise, the backing must remain in place until the composite is cured.

Staged prepregs approach qualities of commercial aerospace quality grades. They are laid up as completed sequences of plies, are consolidated on the mandrel, require little if no bleed-out, and can be continuously monitored during the winding operation.

14.7. COMMERCIAL PREPREGS

Commercial prepregs are reinforcements impregnated with predetermined amounts of uniformly distributed resin and are processed to attain optimum handling characteristics and reproducible cured laminate properties. The impregnating resins include epoxy, polyester, phenolic, silicone, polyimide, and thermoplastics such as polysulfone. Resin systems are applied as liquid, hot melt, and solvent diluted compositions, as well as unpolymerized monomeric reactants.

Reinforcements, in addition to fiberglass, include boron, graphite, and high modulus aramid fibers. These reinforcements may be in the form of woven fabrics, tapes, or broadgoods. In comparison to wet lay-ups, these prepregs provide superior structural properties and are principally used to meet the more stringent requirements of the aerospace industry.

14.7.1. Quality Control Standards

The exact compositions of most prepregs are proprietary. Prepreggers certify physical and processing properties and compliance to the composite property standards delineated in company procurement specifications. Military and industry specifications establish general standards for quality controls of prepregs. Company procurement and processing specifications amend the standards to suit specific applications or particular constructions. Standards for the quality controls of compositions and processing characteristics include the following.

- Resin or reinforcement, and volatile gravimetric content fractions:
  1. Resin matrix or reinforcement contents, measured by solvent extraction, pyrolysis, or chemical digestion methods.
  2. Volatile contents, measured by partial pyrolysis at standard test conditions.
- Processing characteristics:
  1. Tack, a measure of adhesion qualities.
  2. Flow, a measure of resin that can be squeezed from a standard specimen as it cures between press platens at standard temperatures and pressures.
  3. Gel time, a measure of the time a standard specimen remains between heated platens until the resin no longer can adhere to a probe.

To verify certified prepreg data, suppliers' recommended cures for quality control test panels are required. Tensile strengths, flexural strengths, and interlaminar shear strengths relative to the principal directions of the prepreg are usually tested to prove compliance to specifications.

14.7.2. Constituent Content Controls

For structural applications, optimum properties are attained most reproducibly in structural composites that contain 2% by volume or less of voids and average reinforcement volumetric fractions (V_{f_0}) as follows.

REINFORCEMENT	V_{f_0}
0.004-in. (0.10-mm) diameter boron fibers on style 104 glass scrim	0.50
Bidirectionally woven fabrics (aerospace quality)	0.55
Directionally woven fabrics (aerospace quality), nested	0.60–0.65
Directionally woven fabrics (aerospace quality), cross-plied	0.55–0.60
Collimated fibers with no scrim	0.65

Unless resin bleed-out is limited below 2%, difficulties such as ply wrinkling, fiber washouts, and non-uniform thicknesses are encountered. The higher performance structural composites are much more critical than are the wet laid-up systems. An estimate of minimum gravimetric fiber content fraction (W_f) can be determined for prepregs. For high performance structures, the weight fraction of fibers in prepregs is determined from the following relationship.

$$W_f = \frac{(V_{f0} - 0.02)\ G_f}{(V_{f0} - 0.02)\ G_f + (1.02 - V_{f0})\ G_r}$$

Where

G_f = the fiber specific gravity

G_r = the resin specific gravity.

Excepting thermoplastics, higher reinforcement contents often result in poor resin bleed-out and increased laminate void content. Increased resin contents can result in poor reproducibility.

14.7.3. Volatile Content Controls

Quality control test results do not distinguish inert from reactive volatiles, but are used nevertheless to establish volatile content values. They provide qualitative evaluations of prepreg advancements, degrees of volatilizations of solvents, and degrees of degradation due to aging. Volatile contents measured prior to lay-up are compared to values when the prepreg was first received. Such comparisons provide indications of excessive aging.

14.7.4. Tack Controls

Tack is the adhesion characteristic which is controlled to facilitate lay-up operations. It is affected by resin and inert volatiles contents, prepreg advancement, and the lay-up room temperature and humidity. Sometimes, tack is increased by increasing resin and volatile contents, retarded prepreg advancement, or a slight increase in lay-up room temperature. At other times, resin formulations are reapportioned or new additives are blended into the resin. Alterations to the formulations should not be accepted without prior requalifications by the fabricator. Often, properties are affected so that for a cured laminate, performance and durability must be verified.

Prepregs with excessively heavy tack generally can not be handled without grossly disrupting resin distribution and fiber orientation or causing a roping (fiber bundling) of the reinforcements. Constituent contents are not reproducible since undetermined amounts of resin are always removed when the release film or backing is separated from the prepreg. In general, all the disadvantages of wet lay-up systems are inherent in the overly tacky prepregs.

Prepregs with no tack are either excessively advanced or have exceeded their normal storage life. Such materials can not attain adequate cured properties and should be discarded. Exceptions are silicones and some polyimides, which can only be prepared with no tack. Lay-ups with these materials are limited to those occasions where lower mechanical properties can be tolerated in exchange for improved heat resistance or electrical properties. No tack in thermoplastic prepregs does not interfere with their consolidations.

Most prepregs are staged to attain more manageable tack. The tack qualities should be adequate for adhering the prepreg to prepared molding surfaces or preceding plies for a lay-up and still light enough to part from the backing film without loss of resin. Tack qualities can be specified which require the prepreg to remain adhered to the backing until a predetermined force is applied to peel it off.

Provided that cured laminates will not be adversely affected, tack requirements can be modified to suit plant fabrication conditions. Local temperature and humidity sensitivities are minimized by air conditioning, or heavy tack made manageable by reducing the temperature. Drier tack can be improved with judicious use of hot air guns.

14.7.5. Flow Controls

Flow measurements indicate resin capabilities to fuse successive plies in a laminate, and to bleed out the void producing gas reaction by-products. Flow is also an indicator of prepreg age or advancement. It is often desirable to optimize resin content and tack to attain adequate flows. In some cases, flow is controlled by thickening additives in a resin.

14.7.6. Gel Times

Gel time is an indicator for the degree of prepreg advancement. The useful life of prepregs is limited by the amount of staging or advancement. Most prepregs are formulated to attain a useful life of eight days or more at standard atmospheric conditions. The life can be prolonged by storage at −40°F (−40°C), but each time the prepreg is brought to equilibrium at use conditions at lay-up room temperatures, the useful life is shortened. Gel time measurements are used as quality control assurance verifications. Criteria based on those results determine whether to initiate more costly property testing or to dispose of an overage prepreg.

14.8. PREPREG LAY-UP TECHNIQUES

Techniques for locating and orienting the resin impregnated reinforcements onto a molding surface in accordance with the composite design are adapted to the tack and drape characteristics for the prepreg. Most prepregs are advanced or staged to attain acceptable tack and drape.

14.8.1. Non-Tacky Prepreg Lay-Up Techniques (Silicones and Polyimides)

Techniques for laying up dry and non-tacky prepregs are limited and not suitable for primary structures. With the few exceptions that are reinforced with graphite fibers for aerospace vehicles, most are reinforced with fiberglass. The woven fabrics make possible uses of sewn stitches, staples, or clamps. Usually, the lay-ups are enlarged to provide allowances for trimming off the fasteners after the laminate is cured. Sometimes, the prepregs are draped over male forms with weighted edges to draw the lay-up snugly onto the molding surface. The poor draping qualities of the prepregs restrict contours to shallow dished shapes and single curvatures. Most often, successful lay-ups depend on operators' skills to innovate.

14.8.2. No Bleed Process

The curing of advanced composites without a bleeder system (no-bleed process) has been introduced to reduce costs and tooling complexity. Prepreg materials with a lower than normal resin content are used to increase the fiber volume fraction while maintaining ease of handling. For graphite/epoxy tape, the no-bleed prepreg has a nominal 35% resin content, compared to the 40% normally used. The laid up plies are sealed off with a plastic film (Tedlar) or non-porous Teflon coated glass fabric. The entrapped air is removed from the laminate by diffusion through the plastic film or glass yarns connecting the lay-up and a perimeter bleeder system. Typically, laminates made by the no-bleed process have reduced strengths due to porosity and higher-than-normal resin content, but an improvement in properties is anticipated due to current development studies.

14.8.3. Thermoplastic Matrix Prepreg Lay-Ups

Although they are stiff and dry at room temperature, thermoplastic prepregs can be thermoformed above their glass transition temperatures and fused after only a short dwell at their melting temperatures. They may be stored indefinitely at room temperature without the need for special precautions. They are not subject to moisture degradations of the same magnitudes as are the thermosetting resin matrices.

Among the candidate prepregs proposed for composites required for 350°F (179°C) service are the polysulfone prepreg systems. Prepregs containing either collimated fiber reinforcements or woven fabrics are being developed. The fabrication technology is not yet fully developed.

Currently, the prepregs contain no allowances for resin bleed-outs. Plies are separately thermoformed at about 400°F (204°C), cooled to room temperature, and stacked on the mold to form the lay-up. The lay-up is then vacuum bag or autoclave molded at about 600°F (315°C). The dwell at fusion temperature depends on the lay-up thickness. A dwell less than 30 minutes is required to fuse an eight-ply graphite fiber-reinforced laminate.

Impediments to successful fusion are degraded prepreg surfaces, contaminants, and non-uniform fusion conditions. Surface contaminations result in delaminations and increased voids. Collimated fiber-reinforced prepregs are best used for flat shapes or single-curvature surfaces. Woven fabric-reinforced prepregs comply to curved shapes more readily.

Other candidate thermoplastic matrices include polyarylsulfone, polyethersulfone, and acrylic. Poveromo et al.[2] describe a very promising potential application for acrylic/graphite tape in a space satellite solar energy generating station. Preliminary tests show very high mechanical strength and durability for this material. Figure 14.10 shows a typical graphite/acrylic Beam Cap assembly. This part was actually produced by continuous thermoforming process, but the same material can also be bag formed.

Figure 14.10. Graphite/acrylic Beam Cap assembly.

14.8.4. Production Prepreg Lay-Up Techniques

It is essential that prepregs for structural applications be staged to desirable tack and drape qualities. Tack should be adequate to adhere the prepreg to the prepared molding surface or to preceding plies for the lay-up with the application of light pressure. Tack should also be sufficiently light to allow the prepreg to part from the backing without loss of resin. Drape is sufficiently soft to permit the prepreg to conform to the contour of the molding surface. The desirable combination of manageable tack and drape is best attained by woven satin fabric-reinforced prepregs. Non-woven collimated fiber reinforcements have low strength transverse to the fibers, and boron fiber prepregs are stiff in the fiber direction. Sometimes, multi-plied or cross-plied prepregs are used to provide the transverse strengths for lay-ups of broadgoods.

To ensure that the cured laminates will not be deficient in durability and required engineering properties, prepregs are included whose tack and drape properties are modified to suit mechanized equipment or local fabricating conditions. Ideally, temperature and humidity sensitivities are minimized in air conditioned and pressurized "clean rooms." The pressure for clean rooms is maintained by filtered air kept at positive gauge pressure by blowers. The pressure is just high enough to prevent airborne contaminants in the surrounding atmosphere from entering the clean room. Usually, the ideal conditions are not achieved and it is often necessary to require seasonal adjustments to the handling characteristics for the prepreg.

14.8.5. Lay-Ups of Woven Fabric Prepregs

The drape characteristics of prepregs containing long shaft satin reinforcements are the most compliant. Prepregs with crowfoot satin weave attain less drape, while prepregs containing square weave or basket weave fabrics attain the least drape.

Woven fabric prepregs may be darted to comply to convoluted shapes of low stressed items. Darting is the practice of slitting the prepregs at locations where folds would normally occur in a lay-up. The excess material at those locations may be removed completely and the remaining edges butted together; or, the prepreg may be slit where a crease in the wrinkle would normally form, and the excess material may overlap, provided that it is wrinkle-free. When the former method is used, an additional ply is required to compensate for the weak butt joints in the lay-up. Darting is not recommended for the highly stressed lightweight constructions. On those occasions, prepregs should be cut to predetermined patterns in which joints do not coincide in any of the successive plies. The overlapping joints must be deliberately placed and joint widths must be controlled. Usually, patterns for pre-cutting the prepregs allow for 0.5-in. (1.3-cm) overlaps on the lay-up.

When woven fabric reinforcements are laid up on convoluted shapes, weave patterns become distorted and the fibers change directions. Orientations, on the order of (0°, ±60°) or (0°, ±45°, 90°) are used to compensate for undetermined deficiencies. These plying sequences provide reinforcements for laminate plane quasi-isotropic properties. However, ply alignments of heavily draped lay-ups of fiberglass-reinforced prepregs are difficult to control. Colored tracer fibers, woven into the fabrics, simplify the lay-ups and inspections of the composites.

The Burlington Industrial Products Division supplies fiberglass fabric reinforcements with brown warp tracer fibers spaced at 5-in. (12.8-cm) intervals. The J. P. Stevens Company supplies similar reinforcements with blue tracer fibers at 3-in. (7.6-cm) intervals. The tracers show through unpigmented epoxies and polyesters. The adequacies of the orientations and the ply count in a lay-up can be optically verified. Sometimes, fibers that are opaque to X-ray transmissions are included among the warp fibers of custom woven fabrics.

Structural composites to resist loads must be designed to obtain reproducible properties. The shapes should permit the plies to be oriented in predetermined directions. The principles for the lay-up techniques are similar, whether the lay-up is produced manually or is automated. When the structural shapes permit, the most reproducible properties are developed by laying up plies that are cut to size and then applied to the transfer films. These transfer films or Mylar templates are indexed with respect to specified ply locations and orientations with respect to the mold. Plies that are laid up on templates are transferred to the molds without additional distortions, and the templates are removed after the plies are laid-up and transferred onto the mold. Anisotropies of fabric-reinforced prepreg in one ply are corrected with equal but opposite anisotropy in adjacent plies. The corrections to achieve symmetry are important to avoid distortions to the cured laminates. Other corrections are sometimes made by cross-plying compensating misalignments to attain orthotropy.

14.8.6. Lay-Ups of Non-Woven and Collimated Fiber Prepregs

Collimated fiber prepregs are wound or laid up in any of the three systems of sequences of arrays. One set is the (0°, ±45°, 90°) or $\pi/4$ sequence. Another is the (0°, $\pm\theta$, 90°) set of arrays. The third is the $\pm\theta$ angle plied lay-up, that is used in filament wound constructions. The (0°, ±60°) or $\pi/3$ sequence is also used. When thickness tolerances permit, the plying sequences are made symmetrical about the laminate mid-plane to prevent bending coupling effects.

To optimize the properties of a $\pi/4$ sequence for minimum weight designs, the plies in the preferred directions are reinforced with additional plies, while plies in the overweight directions are reduced. To maintain symmetry about the principal directions, the ±45° directions are added or reduced in pairs. Often, the angles of the ±45° plies are changed to minimize ply additions. However, the lay-ups of the (0°, $\pm\theta$, 90°) arrays are more difficult to reproduce and inspect, and the weight savings are most often negligible. Optimum $\pm\theta$ wound plies are combined with hoop wraps to develop the desired directional properties for the laminate.

To expedite lay-up operations, and to make more efficient use of molds and tools, extensive use is made of auxiliary devices and equipment. Indexed Mylar templates are used to reduce the lay-up times on molds. The presized plies are first laid up and oriented on the templates. When the mold is available for the lay-up, the plies are positioned exactly onto the molds and are transferred. The exact positioning is achieved by the references used for indexing. In most instances, reference posts are located on the mold. Corresponding holes in the templates fit exactly over the posts. In some cases, the templates are shaped so that they fit only one way in the mold. The plies are rubbed out from the templates onto the mold in the proper order, the bleed-out systems are laid up, and the assemblies are bagged and cured.

Collimated fiber-reinforced prepreg tapes are supplied in 3-in. (7.6-cm) multiples of width. Broadgoods from 24 in. (60 cm) are widely used. It is customary to pre-size the laid-up ply before it is applied to the mold. Usually, an auxiliary backing is fixed in position to the lay-up tool. Sometimes, the lay-up tool is equipped with vacuum ports to anchor the auxiliary backings in position. Plies are oriented to within ±1° with the use of tape laying heads, or manually with the help of ruled lines on the table, straight edges, or drafting machine dividing heads. If either broadgoods or tapes are used, the plies are laid up in position to approximate size. Then the plies are cut to exact size, sometimes with lasers or water jets. While still in position on the lay-up table, the indexed template is laid over the ply and the ply is adhered to the template. Only after the ply is transferred to the template is the auxiliary backing removed. One or more plies can be transferred to the template in the same way.

There are lay-up machines that can lay up 3 in. (7.6 cm) and wider tapes and broadgoods directly onto the molds, but their use is restricted to simpler shapes.

14.8.7. Use of Wide or Narrow Prepreg

This is strictly a matter of economics: the use of narrow prepreg results in a minimal material loss. Normal waste can be as low as 7–10%. Narrow—3-in. (7.6-cm)—prepreg is ideally suited for very expensive material, such as boron/epoxy tape at about $15 per running foot. However, using narrow tape greatly increases labor costs and these costs must be balanced against material costs.

The labor costs for lay-ups fall into three categories. The labor costs using an automatic lay-up machine are the lowest, with a minimum of trim waste. A proper program of machine cutting of preforms is probably the most economic method of operation and can operate on wide tape as easily as on narrow ribbon. The limitation is the use of only one layer at a time. The pre-cut patterns can be used either on one mold or on multiple molds or can be stored to be used at some future time. Both tapes and broadgoods can be used.

The second method is the use of a pattern-cutting table, where a minimum of eight plies of material are laid up. The various templates are located on top of the lay-up, and the most economical arrangement determined by matching templates. The patterns are then cut and stored till required. This method is often used in modern composites shops. It is most suited for broadgoods and wide tapes.

The third method involves the unrolling of an excessive length of tape or prepreg, draping it over the tool, and trimming to size on the tool. The scrap can run in excess of 50%, depending on the size and complexity of the part. Usually, no precise control is exercised in cutting the overall length of the material. This can only be tolerated on a small number of extremely complex parts on which the use of preforms would not be practical.

14.8.8. Bag Sealing

There are four basic types of bag cleaners used for bag molding: High Temperature Sealants, Mechanical Sealing, Vacuum Sealing, and O-ring/Groove Sealing. High temperature sealants are rubber formulations with high adhesive properties and very good heat and flame resistance. They are usually supplied in tape form, approximately 1 in. (8.5 cm) wide, ready for use in sealing. These materials can be obtained from:

Fiber Resin Corp.
P.O. Box 4187
Burbank, Calif. 91503

Airtech International Inc.
P.O. Box 6207
Carlso, Calif. 90749

Mechanical, vacuum and O-ring/groove sealing are described elsewhere in the text.

REFERENCES

1. Lubin, G., Dastin, S., Mahon, J., and Tanis, C., "The Self-Sealing, Reusable Silicone Rubber Blankets for Composite Molding and Assembly Bonding Operations," 29th Annual Technical Conference, Reinforced Plastics/Composites Institute, SPI, Washington, D.C., 1974
2. Poveromo, L. M., Muench, W. K., Marx, W. and Lubin, G., "Composite Beam Builder", *SAMPE Journal*, January/February 1981, p. 7.

15
THERMOSET MATCHED DIE MOLDING

P. Robert Young
P. Robert Young Associates
Brooklyn, New York

Products made by the thermoset matched die molding process constitute half of all reinforced plastics manufactured at the present time. This process, in one or another of its many forms, is employed when production rate requirements are large and/or when precision and reproducibility are factors. It is likely to offer the best all-around quality at the lowest cost. Also, the quality of the part is only slightly influenced by operator skill.

Even when large quantities are not required, as for aerospace components and other high performance items, the necessity for precision and reproducibility sometimes dictates the use of matched die molding methods.

Thermoset matched die molding can be broadly defined as a molding process in which the loading and closing of the mold cause the molding material to conform to the designed configuration, and in which the cure takes place while the material is contained in the mold. It encompasses so many different sub-processes and/or materials that it is more common to find it identified by one of its sub-headings. In this chapter, mat molding and preform molding will be discussed under the subject heading "mat and preform"; the reinforced molding compounds premix, BMC, SMC, TMC, and XMC and the processes of compression molding, transfer molding, and thermoset injection will be covered under the topic "reinforced molding compounds"; cold press molding and comoforming will be described under "cold press molding"; and discussions of resin injection molding and foam reservoir molding can be found under their respective headings. Each of the commonly used processes is covered in some detail. However, due to their increasing dominance of the industry, the reinforced molding compounds (premix, BMC, SMC, and, most recently, TMC and XMC) will be our primary concern.

In every instance, thermoset matched die molding employs a "mold" or "matched dies." The mold, or mold set, is usually in two main parts, referred to as male and female, with one part fitting inside the other with a controlled space between them. Complex items may require intricate molds of several major elements, and there is no simple nomenclature to differentiate one of these parts from another.

Properties of common materials that are used in molding operations are covered in great detail in other chapters and are therefore not included here unless pertinent to a particular operation.

15.1. REINFORCED MOLDING COMPOUNDS

This section will be concerned with reinforced molding compounds that are now commonly identified in the plastics industry as "bulk molding compound" (BMC), also known as "premix" or "dough molding compound (DMC), and the closely related, but slightly different, product "sheet molding compound"

(SMC), also referred to in modified versions as TMC, HMC, and XMC (terms that will be explained later).

Premix has been defined as: "A fiber reinforced thermoset molding compound not requiring advancement of cure, drying of volatiles, or other processing after manufacture to make it ready for use at the molding press."[1] To this definition, I have added, "and which can be molded without reaction by-products under only sufficient pressure to flow and compact the material."*

While this description fits both BMC and SMC, there are differences, principally in the compounding methods, the forms in which they reach the molder, and mechanical properties. Each is, or can be, formulated from the same basic ingredients.

BMC is formulated by combining all the ingredients in an intensive mixing process. It emerges from the mixer in a fibrous putty form, and it can be used directly, the operator merely weighing out charges. Some formulations can be compacted and extruded into bars or "logs" of simple cross-section to facilitate handling. Sheet molding compound, as the name implies, reaches the molder in the form of a thin, semi-tacky sheet which is cut and plied to suit the article being molded. It is made in a machine that combines previously blended resin, filler, and other ingredients with reinforcing fibers in such a way that the fibers are wetted without vigorous stirring, thus bringing a high degree of their original strength to the molding compound.

Most of the parts are molded from glass fiber-reinforced polyester materials; however, BMC can be made with other solventless resin systems, such as epoxies and vinyl esters, and with other reinforcing fibers, including sisal, asbestos, carbon, aramid, chopped nylon rag, and even wood. On the other hand, SMC must be made with a resin system that will thicken to proper molding viscosity after the wetting or impregnation of the reinforcement. SMC with high strength exotic fibers appears possible, but is not yet commercially developed.

*In contrast to phenolic, melamine, or urea compounds that have water as a reaction product and require high pressure to prevent steam formation and consequent voids in the molding.

15.1.1. History

15.1.1.1. Bulk Molding Compound (BMC)

BMC first became a practical possibility when glass fiber rovings came on the market in 1949. Prior to this, some reinforced polyester molding compound was made by chopping fiberglass fabric prepreg into small pieces. In spite of many disadvantages—high cost, residual solvents, no internal release agent, no fillers, etc.—some useful products offering properties available in no other known material were successfully molded.

It is difficult to track down the first BMC of the new breed. The earliest were probably made about 1950, employing a process of impregnating roving strands with the resin, filler, etc. blend, and chopping them to length in the wet stage. Wetting glass fibers with a resin containing much filler is a difficult and slow operation, and consequently these compounds had a high glass content.

Sometime in the early 1950's, the idea of adding pre-chopped glass fibers to a resin/filler mixture was conceived, and, by the mid-1950's, premix molding was a going business. It was, however, delayed in development since neither the compounders nor the molders were experienced in the kind of business in which they found themselves. No compounders of conventional thermoset compounds were initially interested. As a result, such simple things as internal mold release were a long time coming. And no compression molders would contaminate their premises with the sticky, smelly stuff, so their extensive knowledge of mold construction, ejector mechanisms, heating devices, and so on was denied the fledgling industry.

The first move to high volume occurred with the development of a BMC based on sisal fibers and the molding of automobile heater housings, the largest (in area) items ever made in volume production from a molding compound up until that time. While this type of compound was not outstanding in its physical properties, the unique character of the sisal fiber permitted intensive mixing without fiber damage and promoted homogeneity throughout a large molding, which was not possible with the glass fibers then available.

While the easy molding, low cost sisal compound continued to dominate the high volume automobile market, the development of a resin coating for glass fiber strands that preserved their integrity (i.e., kept them intact as bundles of fibers through the mixing process) made possible large area molding with strength, chemical resistance, electrical insulation values, and other desirable properties. Consequently, heavy, large area electrical and chemical items, as well as a few commercial products (such as internal parts of appliances, where surface quality was a secondary consideration), became commonplace. Surface waviness and coloring problems limited product applications to items and/or parts not normally exposed to the consumer's view.

To obtain water and stain resistance not offered by sisal and glass fiber-reinforced compounds, the use of chopped nylon rag was introduced in the early 1960's. This material has brought about high volume manufacture of such items as shower bases with dimensions up to 3 × 6 ft and weighing as much as 75 lb each.

As mentioned earlier, the applications of glass fiber-reinforced compounds were seriously limited by irregular waviness in the molded surface. Developments in Germany and England, in about 1960, in the use of chemical thickening agents and thermoplastic additives, markedly reduced cure shrinkage, which improved the quality of the surface and limited other distortions. These improvements, unfortunately, required some compromises, and efforts to produce distortion-free products without sacrifice of other properties continue up to the present.

The use of hydrated alumina as a filler to provide both fire resistance and non-tracking qualities has made BMC the standard material for almost any electrical application.

Most recently, the refinement of BMC injection molding, which, combined with low shrink, low distortion compounds, provides precision molding, has changed the design concepts of many items, particularly small hand power tools. And some of the benefits of injection molding—good, blemish-free surfaces, insensitivity to labor skill, and high throughput—have attracted automotive applications that were formerly the exclusive province of compression molded SMC.

15.1.1.2. Sheet Molding Compound (SMC)

Most large area parts have been produced by the preform and mat die molding process. In this process, the mat or preform and the resin/filler are brought together at the press and the resin is distributed with varying degrees of uniformity throughout the reinforcement by the force exerted as the mold closes. While many techniques have been developed to overcome the inherent disadvantages, this process is messy and difficult to mechanize, and it requires substantial operator skill.

The obvious simplification of pre-combining the resin and mat, so successful with prepreg woven fabrics, defied solution. For most convenient use, the sheet of material must be dry in the uncured state. Mats with sufficient insoluble binder to permit impregnation by the conventional resin/solvent system and subsequent passes through a drying tower do not make compounds that flow properly in the molding process. The development of the chemical thickening process mentioned earlier provided a solventless, low viscosity resin system that readily impregnates the reinforcing fiber. The thickening to a moldable viscosity takes place in a period of several hours to several days without further processing other than rolling up in a polyethylene film, wrapping in a vapor barrier, and storing in a reasonably controlled warm atmosphere.

While SMC has not replaced mat and preform completely, its ability to be molded in complex shapes with ribs, bosses, cut-outs, has made it the material of choice in any but the most structurally demanding applications. Its wide use in front and rear parts of automobiles has made the auto industry the single largest user of reinforced plastics.

15.1.2. Properties

BMC and SMC offer an extraordinary range of desirable properties. The nearly infinite variability of resin, filler, and reinforcement types and contents would seem to permit "tailoring" a compound for any design or performance requirement. Within limitations, this is true, and it is probably the single most outstanding quality of these materials.

It also makes definition of detailed properties difficult. However, the important attainable characteristics, if the compound is properly formulated, are the following.

1. Excellent electrical performance—especially resistant to tracking.
2. High strength—particularly impact strength.
3. Heat and flame resistance.
4. Dimensional stability.
5. Chemical resistance.
6. Rapid cure.
7. Low molding shrinkage.
8. Low molding pressure.
9. Thick, thin, and variable thicknesses are moldable.
10. Low cost.

15.1.2.1. Electrical

An important electrical property is the ability to resist arcing. This, in combination with strength, heat resistance, dimensional stability, moldability, and reasonable cost, has resulted in widespread use of these plastics in high voltage applications. The arc suppression qualities come largely from the inclusion of such inert inorganic filler materials as hydrated alumina, silica, and china clay. Addition of small amounts of fine, powdered polyethylene (of the order of 5% by weight) and the use of nylon fiber are reliably reported to improve arc resistance. For maximum arc resistance, it is necessary to minimize both the resin and the glass fiber content, with a resultant reduction in mechanical properties.

Table 15.1 shows some relationships among resin, glass fiber, and filler contents with respect to arc resistance. If the arc resistance is satisfactory, it is likely that other properties, such as dielectric strength, dielectric constant, dissipation factor, volume resistivity, and arc quenching, will be satisfactory for most electrical purposes.

15.1.2.2. Strength

The strength and stiffness that distinguish reinforced molding compounds are due to the proportion, type, and length of the reinforcing fiber used, which is not to say that the other ingredients have no influence. Their purpose is to keep the fiber in place and to protect it from damage. The proportion and the length of fiber are the important considerations. BMC fiber lengths in excess of ¼ in (6.35 mm) produce little, if any, practical improvement in mechanical properties, and lengths over ½ in. (12.7 mm) create mixing and molding problems. A typical relationship is shown in Fig. 15.1, relating impact resistance to fiber length. Other properties vary similarly. Injection molded parts may show even less advantage with increased fiber length, probably because the longer fibers are more vulnerable to damage while flowing through sprues and runners.

The strength of BMC is much more responsive to changes in fiber content than any other factor. While the mechanical properties generally increase with greater fiber content, as shown in Table 15.2, a point of diminishing returns is reached at about 35%, when mixing, moldability, and surface finish are taken into account. Because of the different way the

Table 15.1. Arc Resistance as Affected by Formulation[6]

GLASS FIBER, %	FILLER, %	RESIN, %	FILLER	ARC RESISTANCE, SEC, ASTM D-495
15	64	21	Hydrated alumina	202
9	73	18	Hydrated alumina	240
7.5	75	17.5	Hydrated alumina	264
0	85	15	Hydrated alumina	300

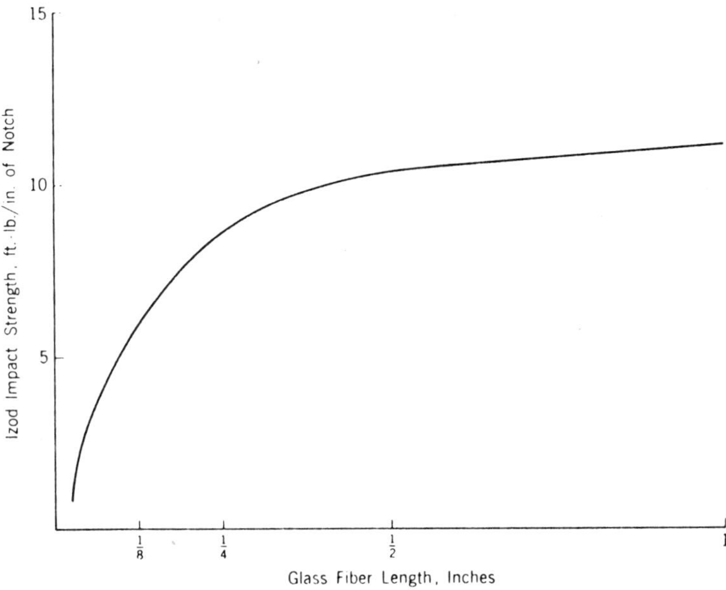

Figure 15.1. Impact resistance versus fiber length.[7]

fibers are incorporated into the compound, SMC can effectively employ a wider range of fiber lengths and fiber content than can BMC. At this time, however, almost all SMC is made with 1-in. (25.4 mm) fibers at a concentration of 25–30%. With shorter fibers—¼–½ in. (6.35–12.7 mm)—the properties are essentially the same as those of BMC at a slightly higher compounding cost. Longer fibers, 2 or 3 in. (50.8 or 76.2 mm) that can be handled on a conventional SMC machine, provide a small increase in strength with a disproportionate reduction in moldability (especially in filling of bosses, ribs, etc.) and give a poorer surface finish.

Very long or continuous fibers arranged in a deliberate pattern, parallel to each other, or at a controlled angle, provide very high strengths in the direction of the fibers and very low strength transverse to them. Arranging the mold charge so that the fibers end up in the direction of the likely loads on the finished part results in very high effective strength SMC. To compensate for the transverse strength deficiencies, these parallel fiber SMC's can be plied with conventional random fiber types (see Table 15.3, which lists some properties of a selection of random and continuous fiber SMC's and combinations thereof).

Disadvantages are poor moldability and surface finish, but where surface finish is not a problem, replacement of steel or aluminum items is a distinct possibility.

In SMC, as in BMC, the amount of fiber is all important in determining the strength.

In the past, for the most part, engineering design of SMC products has been empirical—"cut and try." If the product broke in testing or use, a rib was added or the thickness was increased locally. One problem with this ap-

Table 15.2. Mechanical Properties Versus Fiber Content

GLASS FIBER CONTENT, % 1/4-in. FIBERS	IZOD IMPACT, ft-lb/in. NOTCH (j/m)	FLEXURAL STRENGTH, psi (MPa)	FLEXURAL MODULUS, psi × 10^6 (GPa)	TENSILE STRENGTH, psi (MPa)	APPROXIMATE COST (1978) \$/lb (\$/Kg)
10	3.66 (195)	9800 (68)	1.53 (10.6)	3550 (24)	0.30 (0.66)
20	6.13 (327)	16,000 (110)	1.64 (11.3)	6340 (44)	0.38 (0.84)
30	7.37 (393)	19,600 (135)	1.66 (11.5)	7920 (55)	0.45 (0.98)
40	10.65 (568)	22,300 (154)	1.72 (11.9)	9530 (66)	0.53 (1.17)
50	11.03 (588)	23,200 (160)	1.59 (11.0)	6890 (48)	0.60 (1.32)

Table 15.3. Selected Mechanical Properties[1] of Various SMC Constructions

DESIGNATION	FIBER ARRANGEMENT	FIBER CONTENT, %	FLEXURAL STRENGTH, psi × 10^3 (MPa)	FLEXURAL MODULUS, psi × 10^6 (GPa)	TENSILE STRENGTH, psi × 10^3 (MPa)	NOTCHED IZOD IMPACT, ft-lb/in. (j-m)
TMC-25[c]	Random (1/2-in. fiber)[b]	25	17 (117)	1.0 (6.9)	7 (48)	10 (533)
TMC-28[c]	Random	28	27 (186)	1.4 (9.7)	14 (97)	10 (533)
SMC-R15[d]	Random	15	15 (103)	1.4 (9.7)	8 (55)	8 (426)
SMC-R30[d]	Random	30	23 (159)	1.9 (13.1)	11 (76)	
SMC-R40[d]	Random	40	30 (207)	2.0 (13.8)	17 (117)	
SMC-R50[d]	Random	50	37 (255)	2.2 (15.2)	23 (159)	
SMC-R65[d]	Random	65	48 (331)	2.3 (15.9)	30 (207)	
HMC-65[e]	Random	65	59 (407)	2.2 (15.2)	30.5 (210)	20.5 (1092)
SMC-C30 R20[d]	Continuous random	30/20	85 (586)	3.3 (22.8)	55 (379)	
XMC-3[e]	Continuous × 85° random	50/25	125 (862)	5.5 (38.0)	75 (517)	
SMC-C60[e]	Continuous	60	130 (897)	5.4 (37.3)	81 (559)	
XMC-2[e]	Continuous × 85° random	75	155 (1069)	5.5 (38.0)	90 (621)	

[a]In fiber direction where continuous fibers are used.
[b]All other random fiber is 1 in. (25.4 mm).
[c]USS Chemicals.
[d]Owens-Corning Fiberglas.
[e]PPG Industries.

proach was that there is no way to know where overstrength and/or overweight exist. Since SMC can, obviously, provide weight savings over competitive materials such as steel or aluminum, refinement of design was not critical. However, with energy crises demanding more efficient use of materials, the empirical design approach is no longer acceptable. Advances in structural design techniques, particularly the use of finite element analysis, can result in highly efficient designs, but they do require more extensive mechanical properties characterization than has generally been available for SMC. The mechanical properties of a 50% random fiber SMC have now been fairly thoroughly determined.[2] The serious need for this kind of information will no doubt engender more such work by material suppliers and users of SMC.

15.1.2.3. Heat Resistance—Flammability

Heat resistance, short term hot strength, and flammability tend to be considered together, but they are not necessarily common properties of a given molding compound. "Heat resistance" relates to the long term resistance to thermal degradation below the flammability temperature. "Short term hot strength" or simply "hot strength" pertains to the thermoplasticity qualities of the resin. Heat resistance and hot strength are properties derived largely from the resin, although there is evidence that certain fillers do enhance heat resistance, even though the mechanism is not explained. Flammability is, of course, the measure of overt burning. In spite of the fact that all organic materials will burn (if heat above the ignition temperature continues to be applied), compounds are variously called "non-burning," "self-extinguishing," and "fire retardant," depending on their ease of ignition, their ability to extinguish when the heat source is removed, and their rate of burning. In addition, since molding compounds are now used to make items that may form all or part of a structure occupied by people, the factors of smoke generation and flame spread must be considered. Flame resistance in polyester resins is achieved with halogenated resins (plus antimony trioxide for maximum effect), by adding

Table 15.4. Flame Resistance of Aluminum Hydrate Versus Halogen-Antimony Trioxide Systems in BMC Formulations[7]

	MIX 1	MIX 2	MIX 3	MIX 4	MIX 5
Formulations					
"Plaskon" 9520 resin	30	30	None	None	30
"Halogenated" resin (28% Cl)	None	None	34	34	None
Benzoyl peroxide paste	0.6	0.6	0.6	0.6	0.6
Aluminum hydrate (Hydral-710)	53	38	None	15	None
Clay—ASP-400	None	15	44	29	53
Antimony trioxide	None	None	5	5	None
1/4-in. fiberglass	15	15	15	15	15
Zinc stearate	1.4	1.4	1.4	1.4	1.4
Flame resistance					
By ASTM-D-635	N.B.[a]	N.B.[a]	N.B.[a]	N.B.[a]	S.E.[b]
By Federal spec meth 2023					
Ignition time, seconds	154	135	95	150	93
Burning time, seconds	31	60	56	27	234

[a]N.B. = Non-burning.
[b]S.E. = Self-extinguishing.

halogen- and phosphorous-containing compounds to conventional resins, and by using hydrated alumina as the principal or sole filler in a compound. Table 15.4 provides some data on flame resistance of various polyester BMC formulations.

Phenolic resins are inherently flame resistant. Halogen and phosphorous compounds, as well as the use of chlorendic anhydride (50% or more) as a hardener, will provide a measure of flame resistance to epoxies.

15.1.2.4. Stability

The dimensional stability of BMC and SMC formulated with polyester resin, glass fiber reinforcement and china clay, silica, alumina, and some other fillers is unexcelled. Post-molding shrinkage is minimal. Water absorption is very low, resulting in slight changes in weight and not easily detected dimensional changes. The thermal expansion coefficient is very near that of aluminum, and sometimes lower. Changes with continuous exposure to elevated temperature can be minimum. Figure 15.2 shows some stability characteristics of a BMC formulated with a general purpose polyester resin.

15.1.2.5. Corrosion Resistance

The chemical or corrosion resistance of these compounds can generally be rated excellent. By selection of resin and filler, a compound to

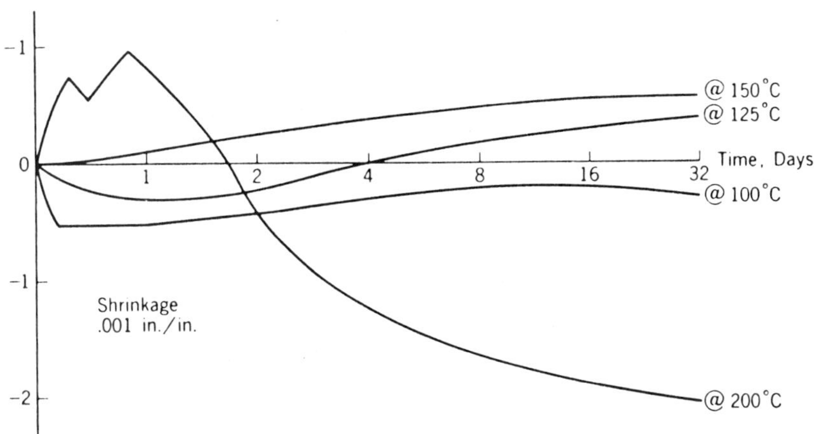

Figure 15.2. Dimensional change versus time and temperature.[8]

resist a specific exposure can usually be formulated. The properties of the resin chosen are all important.

No filler or reinforcement will compensate for the failings of the resin. Acid and alkali resistant polyesters, vinyl esters, furans, and acid resistant epoxies can be used in BMC. There are limitations on resin selection for SMC since the resin must also respond to the chemical thickening mechanism. New thickening mechanisms that function somewhat independently of the chemistry of resin may broaden the kinds of resins that may be used. Of the common fillers, china clay and silica are good all around, and calcium carbonate has good alkali resistance but is poor in acids. Better performance might be expected from so-called corrosion resistant glass fibers, but no evidence supports this.

15.1.2.6. Cure Rate

Rapid to very rapid cures at temperatures below that for other themosets are possible with polyester-based compounds. The exothermic nature of polyester curing results in an imbalance of cure rate to thickness such that thick sections cure in shorter times than might be expected. Also, many parts that are nominally thin but have thick ribs or bosses can be removed from the mold at an early stage and the cure will continue to completion in the interior of the thick sections. Standard polyester BMC and SMC require 45–60 seconds at 280–300° F (138–149° C) for a 0.1-in. (2.5-mm) thickness. Compounds made for prompt use (within two or three days) can be formulated to cure rapidly and effectively at temperatures as low as 225° F (107° C).

15.1.2.7. Shrinkage

Glass reinforced compounds can be expected to have very low molding shrinkage. A maximum of 0.004-in./in. (0.1 mm/mm) is typical, with many compounds down nearer zero. The glass fibers and inorganic fillers that have little thermal shrinkage contribute most to the low overall mold shrinkage. However, the combination of low shrink, high strength fibers with the high cure and thermal shrinkage of conventional resin systems develops considerable stress in the resin matrix between the fibers. The secondary effects of the stress are surface waviness, warping, surface cracks, and internal voids. The foregoing is true of compounds not employing thickeners or low shrink additives. The chemical thickening process used in making SMC and some BMC reduces both cure shrinkage and orientation and dissociation of the fibers as the charge flows to fill the mold, with consequent reduction of the above-mentioned faults, without impact on the mechanical properties. However, to attain a very high quality surface and almost complete freedom from distortion, it is necessary to take an additional step, the incorporation of one or another of a number of thermoplastic materials. These include polyethylene, polystyrene, polyvinyl acetate, methyl methacrylate, and others.

The mechanism by which acrylic (methyl methacrylate) monomer acts to limit shrinkage is explained as follows: The monomer does not cross-link with the polyester but homopolymerizes during the exotherm of the polyester reaction, leaving foamlike occlusions which apparently exert some pressure in their formation to resist the polyester polymerization shrinkage (see Figs. 15.3, 15.4, and 15.5). The exact mechanism by which the other thermoplastic additives function has not been fully determined.

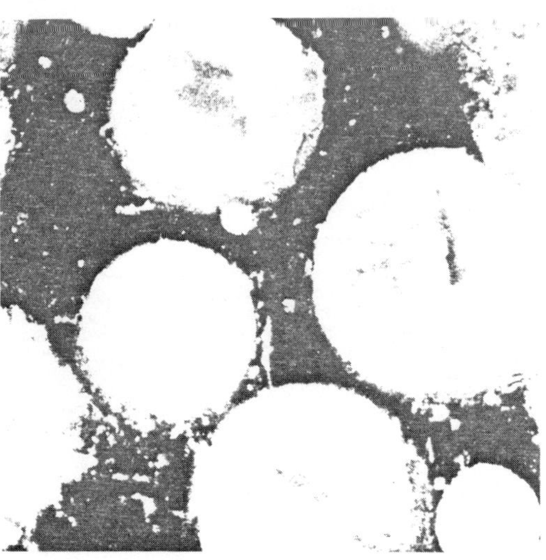

Figure 15.3. Section of cured P19A casting showing foam-like occlusions.[9]

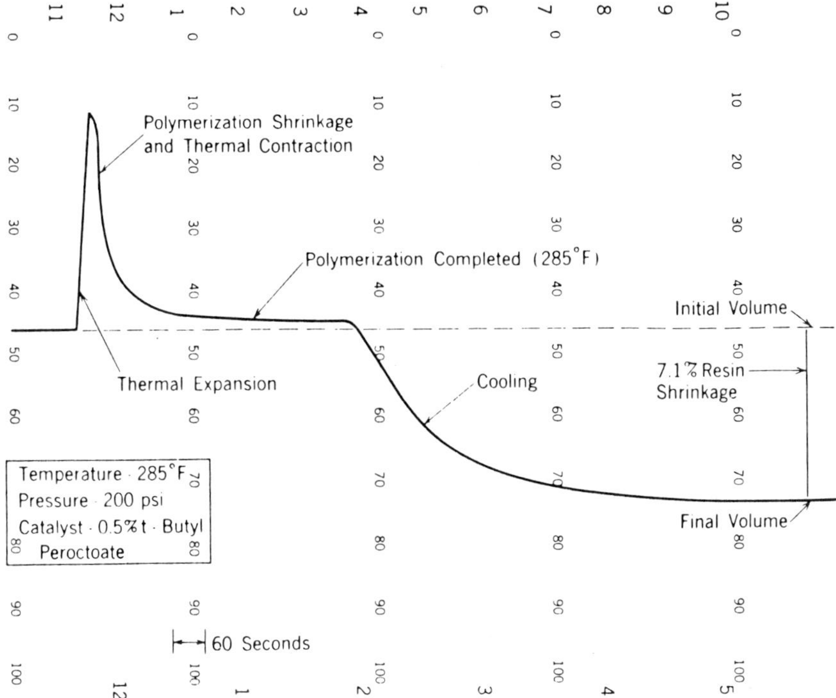

Figure 15.4. Volume change during polymerization of a conventional polyester resin.[9]

REINFORCED MOLDING COMPOUNDS

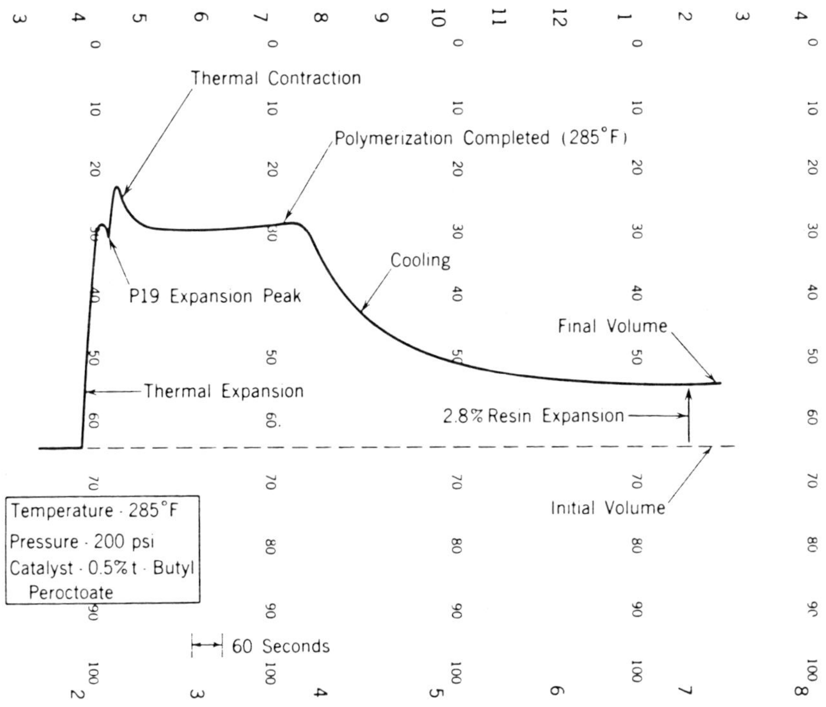

Figure 15.5. Volume change during polymerization of Paraplex P19A.[9]

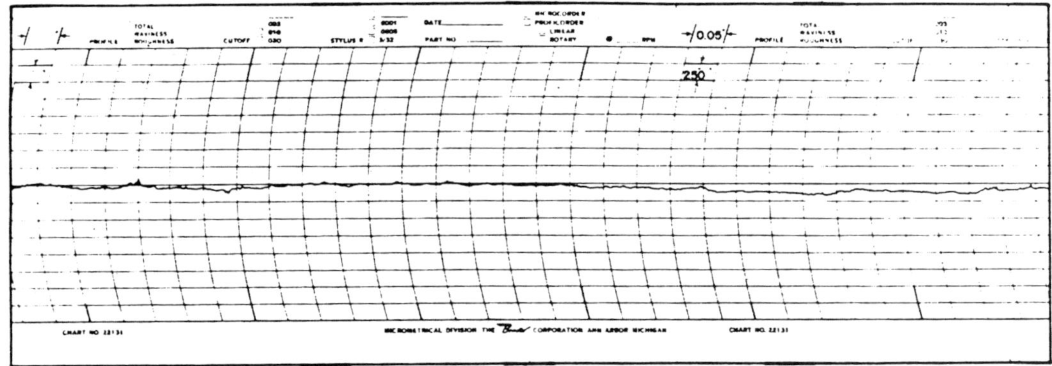

BASED ON P-19A

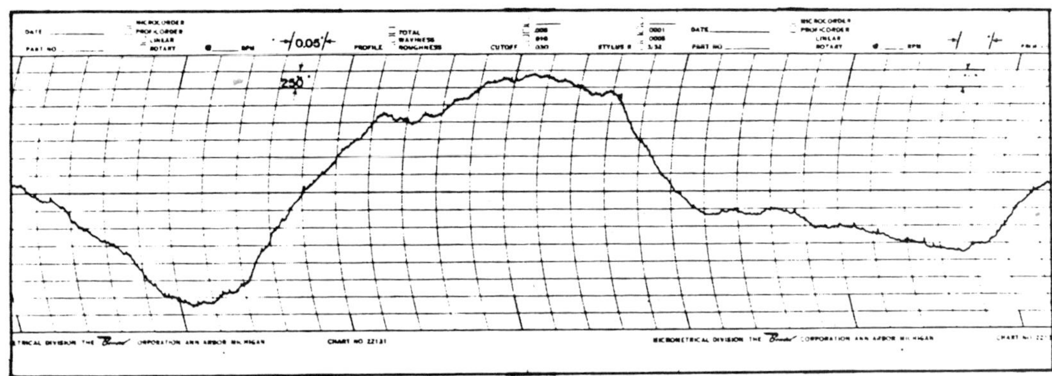

BASED ON CONVENTIONAL RESIN

Figure 15.6. Surface profiles of BMC moldings based on Paraplex P19A versus conventional polyester resin.[9]

The most obvious advantages of the low shrinkage resin systems appear in the surface smoothness. This is apparent in Fig. 15.6, which compares by Microcorder* readings the surface waviness in BMC compounds based on low shrink and conventional resins, and in Fig. 15.7, which compares sample moldings. The other common faults are similarly influenced: Fig. 15.8, for example, shows the reduced tendencies toward warpage.

*Micrometrical Division, Bendix Corp.

While there are claims of equal or superior

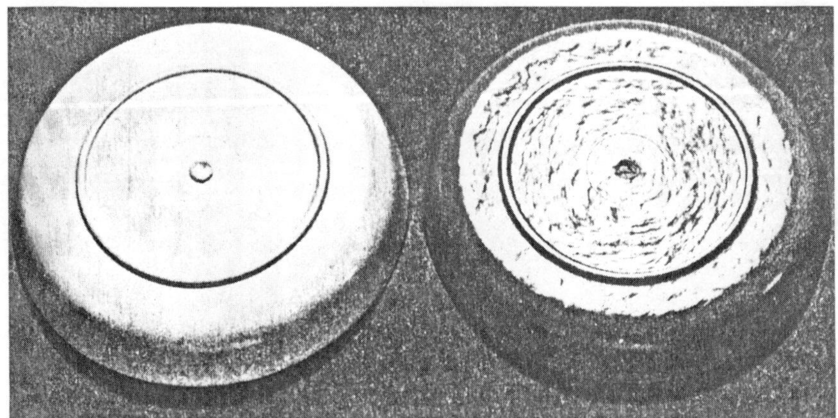

Figure 15.7. Surface smoothness of BMC moldings: Paraplex P19A versus conventional polyester.[9]

Figure 15.8. Warp resistance of BMC moldings: Paraplex P19A versus conventional resin.[9]

properties for the low shrink systems, there are so many different conditions for which a reinforced compound may be specially formulated that time has not permitted all-encompassing evaluations. In general, the mechanical properties are poorer, yet it is true that mechanical, electrical, and chemical properties adequate for a wide variety of applications can be attained with low shrink compounds, and that products made with these materials can be painted with a minimum of surface preparation and/or pigmented for satisfactory molded-in color.

15.1.2.8. Molding Pressures

Low molding pressure is a major advantage of BMC and SMC. The highest pressure required is at the low end of the pressure needed for typical phenolic, melamine, and urea compounds. The requirements are no more than those necessary to flow the material to the extremities of the mold. None is needed to keep reaction by-products such as water or residual solvents from gassing and forming blisters or voids. BMC can be formulated to mold simple shapes at 100 psi (690 kPa). Chemically thickened compounds require higher pressures in the range of 500–1500 psi (3.45–10 MPa). The higher the viscosity of the thickened compounds, the better the quality of the molding, in general, but also the higher the molding pressure. To be practical, SMC must be non-tacky so that it can be handled at the press. Some compounds can be formulated to be non-tacky at 2–12 million centipoises, but most go to 30 million and up. Figure 15.9 shows a typical relationship of SMC viscosity to molding pressure.

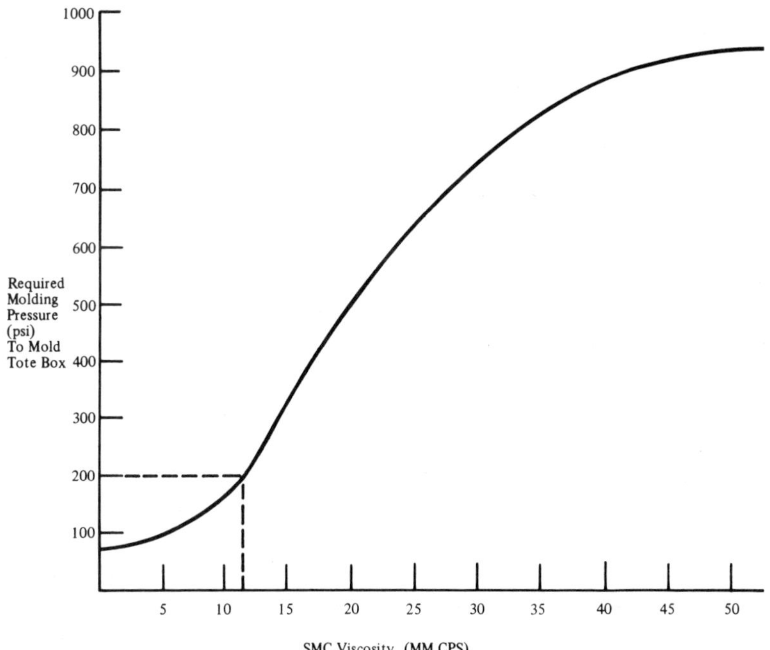

Figure 15.9. Molding pressure versus SMC viscosity. (*Courtesy Owens-Corning Fiberglas Corp., Publication 8-ccr-F129, January 1976.*)

15.1.2.9. Thickness

Wide variations in thickness are possible within the same molding without major molding problems, if proper care is taken in design of the transition from one thickness to another. With conventional compounds, sinks or depressions (often with small cracks) will appear opposite stiffening ribs in thin, flat sections unless precautions are taken. The maximum rib thickness should be no more than that of the surface it is stiffening. Radii at the juncture are desirable from a structural standpoint but contribute to the sink problem. Two small ribs are better than one thick one. Unsupported edges are better stiffened by turning than by increasing the thickness, except if a thicker edge is needed to resist handling damage.

Other faults that occur with thick and thin sections have much to do with fiber orientation. The fibers tend to align themselves with the direction of flow in thin sections, and crosswise from thin into thick sections. The results, if not appearance defects, are often wide and unexpected variations in strength. Chemical thickening and the use of flow controlling fillers help to keep the fibers in their original heterogeneous relationship.

15.1.2.10. Cost and Other Properties

Even though glass fibers and some of the other ingredients of reinforced molding compounds are relatively high in cost, the capacity to employ large quantities of low cost fillers and still maintain excellent properties puts these materials at a favorable spot on the cost-performance scale. Table 15.5 lists material costs for SMC of various glass fiber contents as of late 1978 for SMC of various glass fiber contents. Obviously, this raw material cost cannot be directly compared to that of a piece of steel ready for stamping. Taking account of waste, ancillary supplies, and processing costs, 5–30% must be added, depending on the volume and the rate of conversion.

Table 15.6 lists the properties of a selection of reinforced molding compounds. Included are commercially available reinforced compounds based on phenolics, melamines, and silicones that do not come under the strict definition of the materials considered in this chapter but are not covered elsewhere in this book. For the most part, the data are extracted from manufacturers' literature. Other data on BMC and SMC are taken from technical reports. In some instances, properties indicated are from the author's files or averages of published data.

Included as well, for comparison, are a general purpose (non-reinforced) phenolic and some typical properties of preform and mat die moldings. No plastics material is better known or more widely accepted than phenolic. Except for the pressure needed, it is the most moldable of all plastics, "moldable" meaning the ability to fill out intricate contours, blind ribs, etc. completely; no cracks and weld lines; smooth surface; easy flash removal; suitability for mechanized handling, preforming, or mold charging; rapid cure; injection or transfer molding capability; and uniformity in various parts of the molding, as well as among different moldings of the same item. While preform and mat moldings are not as familiar as phenolics, they have an established reputation as quality molded plastics. A considerable degree of success will be attained when a reinforced molding compound, whether BMC or SMC, effectively combines the properties of both the moldability of phenolic and the structural qualities of preform molding.

15.1.3. Applications

The great variability in attainable properties of reinforced molding compounds has resulted in a wide range of applications in which generally the only common characteristic is that they are structural products; i.e., they replaced or were used in place of wood, metal, ceramics, or laminated reinforced plastics. In addition to the basic structural quality, they can be specially formulated to attain low cost, corrosion resistance, heat resistance, good electrical properties, flame resistance, and so on. Reinforced molding compounds can be, and have been, applied successfully in place of such diverse materials as concrete, papier mâche, wood, sheet steel, other plastics, aluminum, zinc, and iron and bronze castings, in

Table 15.5. SMC Raw Material Costs[10]

		RESIN	FILLER	REINFORCEMENT	CATALYST	MOLD RELEASE	THICKENER		
GLASS CONTENT		POLYESTER	CALCIUM CARBONATE	FIBERGLAS ROVING	T-BUTYL PERBENZOATE	ZINC STEARATE OXIDE	MAGNESIUM OXIDE	TOTAL	
	$/lb	0.49	0.05	0.54	$2.05	0.77	0.55		$/lb
30%	Parts, lb	35	35	30	0.3	0.6	0.53	101.4	
	Cost, $	17.15	1.75	16.20	0.62	0.46	0.29	36.47	0.360
40%	Parts, lb	35	25	40	0.3	0.6	0.53	101.4	
	Cost, $	17.15	1.25	21.60	0.62	0.46	0.29	41.37	0.408
50%	Parts, lb	35	15	50	0.3	0.6	0.53	101.4	
	Cost, $	17.15	0.75	27.00	0.62	0.46	0.29	46.27	0.456
60%	Parts, lb	35	5	60	0.3	0.6	0.53	101.4	
	Cost, $	17.15	0.25	32.40	0.62	0.46	0.29	51.17	0.505
65%	Parts, lb	35	0	65	0.3	0.6	0.53	101.4	
	Cost, $	17.15	0.0	35.10	0.52	0.46	0.29	53.52	0.528

Table 15.6. Properties of Reinforced Molding Compounds

	RESIN	REINFORCEMENT TYPE, % BY WEIGHT	FORM[b]	SPECIFIC GRAVITY MOLDED	Mil-M-14F DESIGNATION	IZOD IMPACT, ft-lb/in. NOTCH ASTM D-256	(J m)	FLEXURAL STRENGTH, psi ASTM D-790	(MPa)	FLEXURAL MODULUS, 10^6 psi ASTM D-790	(GPa)	TENSILE STRENGTH, psi ASTM D-638	(MPa)
1 Wood flour filled phenolic[a]	Phenolic	None	G	1.4	CFG	0.31	(16.5)	11,000	(75.9)			7000	(48.3)
2 High impact phenolic[a]	Phenolic	GL-45	S	1.85		15.0	(799.5)	20,000	(138)			7000	(48.3)
3 Asbestos phenolic[a]	Phenolic	ASB-45	FW	1.85	MFA 30	3.5	(186.6)	14,000	(96.6)	2	(13.8)	7000	(48.3)
4 Impact melamine	Melamine	GL-45	FW	1.9	MMI 30	5.0	(266.5)	20,000	(138)			6000	(41.4)
5 Glass reinforced silicone	Silicone	GL-45	FP	1.9	MSI 30	10.0	(533)	14,000	(96.6)	2.1	(14.5)	4500	(31)
6 Mineral-glass epoxy	Epoxy	GL-15	G	2.0		0.4	(21.3)	12,500	(86.3)	1.5	(10.4)	7000	(48.3)
7 High impact alkyd	PE	GL-30	FP	2.05	MAI 60	10.0	(533)	14,500	(100)	2.3	(15.9)	7000	(48.3)
8 Medium impact alkyd	PE	GL-15	ER	2.15		2.5	(133.3)	22,500	(155.3)	2.0	(13.8)	6000	(41.4)
9 Orlon filled "Dapon"	DAP	OR-30	G	1.8	SD15	1.2	(64)	10,000	(69)	0.7	(4.8)	6000	(41.4)
10 Asbestos filled "Dapon"	DAP	ASB 45	GR	2.0	MDG	0.5	(26.6)	10,000	(69)	1.2	(8.3)	5500	(38)
11 Long glass filled "Dapon"	DAP	GL-15	FW	1.9	GDI 30	6.0	(319.8)	18,000	(124.2)	1.25	(8.6)	7000	(48.3)
12 Non-tracking flame retardant premix	PE	GL-15	ER	2.2		2.0	(106.6)	15,000	(103.5)			6000	(41.4)
13 Non-tracking SE premix	PE	GL-15	ER	1.98		3.5	(186.6)	14,000	(96.6)	1.4	(9.7)	4500	(31)
14 Low cost general purpose premix	PE	GL-15	ER	2.14		2.0	(106.6)	15,000	(103.5)			4000	(27.6)
15 Low cost general purpose premix	PE	GL-15	ER	2.03		2.0	(106.6)	14,000	(96.6)	1.36	(9.4)	5000	(34.5)
16 Flame retardant medium impact premix	PE	GL-20	FP	2.2		6.0	(319.8)	15,000	(103.5)			7000	(48.3)
17 Self-extinguishing medium impact premix	PE	GL-20	FP	1.99		6.0	(319.8)	19,000	(131.1)	1.6	(11)	6000	(41.4)
18 High strength premix	PE	GL-45	FP	1.8		15.0	(799.5)	25,000	(172.5)			8000	(55.2)
19 High strengh premix	PE	GL-30	FP	1.92		10.0	(533)	27,000	(186.3)	2.2	(15.2)	10,000	(69)
20 Low cost general purpose premix	PE	GL-10	FP	2.0		2.0	(106.6)	12,000	(82.8)			6000	(41.4)
21 Self-extinguishing corrosion resist premix	PE	GL-25	FP	1.9		6.0	(319.8)	15,000	(103.5)			7000	(48.3)
22 Non-tracking SE[c] premix	PE	GL-30	FP	1.85		12.5	(666.3)	24,500	(159)			8000	(55.2)
23 Low shrink premix	PE	GL-20	FP	1.69		8.5	(453)	14,000	(96.6)	1.25	(8.6)	8500	(58.6)
24 Low shrink premix	PE	GL-15	FP			5.0	(266.5)	15,000	(103.5)	1.2	(8.3)	6000	(41.5)
25 Acrylic monomer low strength premix	PE	GL-20	FP	1.8		2.6	(138.6)	12,100	(83.5)	1.73	(11.9)	5200	(35.9)
26 Low shrink premix	PE	GL-30	FP	1.87		7.3	(392.8)	19,600	(135.2)			7920	(54.6)
27 Alum. hydrate filled premix	PE	GL-15	FP	2.0		5.0	(266.5)						
28 Halogenated resin premix	PE	GL-15	FP	2.0		5.0	(266.5)						
29 Arc resistant premix	PE	GL-9	FP	1.96		2.6	(138.6)	10,100	(69.7)				
30 SMC	PE	GL-30	S			10.0	(533)	30,000	(207)	1.6	(11)	13,000	(89.7)
31 Flomat SMC	PE	GL-35	S	1.85		10.0	(533)	31,000	(213.9)	1.5	(10.5)	15,000	(103.5)
32 Non-tracking SMC	PE	GL-30	S	1.75		10.0	(533)	25,000	(172.5)			14,000	(96.6)

TENSILE MODULUS, 10^6 psi ASTM D-638 (GPa)	COMPRESSIVE STRENGTH, psi ASTM D-695 (MPa)	WATER ABSORPTION ASTM D-570	DEFLECTION TEMPERATURE °F at 264 psi ASTM D-648 (°C)	FLAMMABILITY, ASTM D-685	LP-406-2023-2 IGNITION BURNING TIME, SECONDS	DIELECTRIC STRENGTH SHORT TIME/STEP BY STEP VOLTS/MIL ASTM D-149	DISSIPATION FACTOR @ 60/10^6 CYCLES ASTM D-150	TRACK RESISTANCE, HOURS, ASTM D2303-64T	INSULATION RESISTANCE, MEGOHMS ASTM D-257	ARC RESISTANCE, SECONDS, ASTM D-495	BULK FACTOR
1.2 (8.3)	30,000 (207)	0.6	320 (160)		60/270	350/275	0.06/0.035				2.5
3.0 (20.7)	16,500 (113.9)	0.8	600 (316)			400/300	0.05/0.02				4.6
2.5 (17.3)	22,000 (151.8)	0.9	400 (204)			90/					6.0
2.4 (16.6)	23,000 (158.7)	0.15	400 (204)	SE	600/0	150/125	0.18/0.014		1×10^4	185	6.0
	10,000 (69)	0.12	900 (482)	SE	90/120	275/	0.006/0.004		4×10^7	240	6.0
	25,000 (172.5)	0.20	400 (204)	SE	190/55	330/350	/0.011		1×10^8	180+	2.5
	26,000 (179.4)	0.08	400 (204)	NB	90/90	375/325	0.025/0.018			180+	4.5
	31,000 (213.9)	0.20	400 (204)	SE		325/300	0.02/0.02			180+	1+
	30,000 (207)	0.20	265 (129)	SE		375/325	0.026/0.018		1×10^7	100	2
	25,000 (172.5)	0.20	325 (163)	SE		385/325	0.06/0.05		1×10^7	150	2
	26,000 (179.4)	0.08	400 (204)	NB	90/90	350/340	0.03/0.016		1×10^7	105	4.5
	18,000 (124.2)	0.10	300 (149)	SE	95/60	300/—		>400		180	1
	16,000 (110.4)	0.15	550 (288)	SE	127/90	425/—	0.03/0.013	>1375		190	1+
	20,000 (138)	0.2	450 (232)			300/—	—/0.025			180	1.1
	21,000 (144.9)	0.16	515 (321)			400/—	0.046/0.034			180	1+
	20,000 (138)	0.1	300 (149)	SE	85/85	350/—	0.03/—			130	3
	20,000 (138)	0.18	430 (221)	SE	125/55	375/—	0.038/0.013			180	3
	25,000 (172.5)	0.16	300 (149)			300/—				130	3
	22,000 (151.8)	0.1	560 (293)			400/—	0.03/0.015			130	3
	17,000 (117.3)	0.1	300 (149)			300/—				130	2
	18,000 (124.2)	0.1	400 (204)	SE	120/30	350/—				180	3
	25,000 (172.5)	0.01	400 (204)	SE	80/40	300/—		300		180	3
1.25 (8.6)		0.16	400 (204)			300/—	—/0.016			150	1.5
1.83 (12.6)		0.17	392 (201)								
		0.19		NB	154/31						
				NB	150/27						
	33,700 (232.5)	0.08		NB		345/—					
											1.1
1.3 (9)	30,000 (207)	<0.25				275/—	—/0.024		1×10^6		1.1
	40,000 (276)	0.1		SE	80/70	400/—	0.011/0.016	600		180	1.1

Table 15.6. Continued

	RESIN	REINFORCEMENT TYPE, % BY WEIGHT	FORM[b]	SPECIFIC GRAVITY MOLDED	Mil-M-14F DESIGNATION	IZOD IMPACT, ft-lb/in. NOTCH ASTM D-256 (J m)	FLEXURAL STRENGTH, psi ASTM D-790 (MPa)	FLEXURAL MODULUS, 10^6 psi ASTM D-790 (GPa)	TENSILE STRENGTH, psi ASTM D-638 (MPa)
33 Flame retardant SMC	PE	GL-30	S	1.74		6.0 (319.8)	25,500 (176)	1.4 (9.7)	14,800 (102)
34 Solid monomer SMC	PE	GL-35	S	2.0		15.4 (820.8)	36,000 (248.4)	2.18 (15)	15,800 (109)
35 Continuous strand mat laminate	PE	GL-40		1.65		20+ (1066)	35,400 (244.3)	1.37 (9.5)	21,500 (148.4)
36 Preform molding[a]	PE	GL-25		1.61		10 (533)	25,000 (172.5)	1.1 (7.6)	15,000 (103.5)

Table 15.6. Continued

	PREFORMING	MOLDING PRESSURE psi (MPa)	SHELF LIFE, DAYS	MOLDING SHRINKAGE in./in., ASTM D-955	SURFACE WAVINESS, MICROIN./in.	HANDLEABILITY	MOLDABILITY	HARDNESS, BARCOL, ROCKWELL M	COLOR	PRICE/lb[c]
1 Wood flour filled phenoic[a]	Machine	2000/4000 (13.8/27.6)	270	0.006		E	E	—/113	Black Brown	0.25
2 High impact phenolic[a]	Hand	1500/3000 (10.3/20.7)	>270	0.0009 Exp.		F	P	—/87	Natural	0.52
3 Asbestos phenolic[a]	Hand	1500/3000 (10.3/20.7)	>270	0.001		G	G		Natural	
4 Impact melamine	Hand	2000/4000 (13.8/27.6)	>180	0.003		G	G		Natural	
5 Glass reinforced silicone	Hand	1000/5000 (6.9/34.5)	>90	0.0005		G	G	—/88	Red	
6 Mineral-glass epoxy	Hand	1000/2000 (6.9/13.8)	90	0.003		G	G	65/—	Varies	1.45
7 High impact alkyd	Hand	1500/2000 (10.3/13/8)	180	0.003		G	G	70/—	Grey	0.75
8 Medium impact alkyd	Cut-off	600/1000 (4.1/6.9)	180	0.004		G	G	75/—	Grey	0.50
9 Orlon filled "Dapon"	Machine	500/3000 (3.5/20.7)	>180	0.009		G	G		All	1.00
10 Asbestos filled "Dapon"	Machine	500/3000 (3.5/20.7)	>180	0.006		G	G		All	1.00
11 Long glass filled "Dapon"	Hand	500/2000 (3.5/13.8)	>180	0.003		G	G	65/—	All	1.00
12 Non-tracking flame retardant premix	Cut-off	500/1000 (3.5/6.9)	>60	0.002	>2000	G	G		Black	0.45
13 Non-tracking SE premix	Cut-off	500/1000 (3.5/6.9)	>60	0.003	>2000	G	G	60/90	All	0.51
14 Low cost general purpose premix	Cut-off	500/1000 (3.5/6.9)	>90	0.003	>2000	G	G		Varies	0.39
15 Low cost general purpose premix	Cut-off	500/1000 (3.5/6.9)	>60	0.002	>2000	G	G	65/90	Varies	0.38
16 Flame retardant medium impact premix	Hand	500/1000 (3.5/6.9)	30	0.002	>2000	G	G		Red	0.53
17 Self-extinguishing medium impact premix	Hand	500/1000 (3.5/6.9)	60	0.001	>2000	G	G	60/85	Varies	0.37
18 High strength premix	Hand	500/1000 (3.5/6.9)	60	0.001	>2000	G	G		Natural	0.49
19 High strength premix	Hand	500/1000 (3.5/6.9)	60	0.001	>2000	G	G	70/85	Varies	0.46

TENSILE MODULUS, 10^6 psi ASTM D-638	(GPa)	COMPRESSIVE STRENGTH, PSI ASTM D-695	(MPa)	WATER ABSORPTION ASTM D-570	DEFLECTION TEMPERATURE °F at 264 psi ASTM D-648	(°C)	FLAMMABILITY, ASTM D-685	LP-406-2023-2 IGNITION BURNING TIME, SECONDS	DIELECTRIC STRENGTH SHORT TIME/STEP BY STEP VOLTS/MIL ASTM D-149	DISSIPATION FACTOR @ 60/10^6 CYCLES ASTM D-150	TRACK RESISTANCE, HOURS, ASTM D2303-64T	INSULATION RESISTANCE, MEGOHMS ASTM D-257	ARC RESISTANCE, SECONDS, ASTM D-495	BULK FACTOR
		26,000	(179.4)	0.16	400	(204)	SE						183	1.1
				0.1	375	(190)	SE		290/260	0.021/0.016			136	1.1
1.4	(9.7)	25,000	(172.5)	0.2	400	(204)								
1.1	(7.6)	20,000	(138)	0.2	400+	(204)								

	PREFORMING	MOLDING PRESSURE psi (MPa)	SHELF LIFE, DAYS	MOLDING SHRINKAGE in./in., ASTM D-955	SURFACE WAVINESS, MICROIN./in.	HANDLEABILITY	MOLDABILITY	HARDNESS, BARCOL, ROCKWELL M	COLOR	PRICE/lb[c]
20 Low cost general purpose premix	Hand	500/1000 (3.5/6.9)	60	0.001	>2000	G	G		Varies	0.28
21 Self-extinguishing corrosion resist premix	Hand	500/1000 (3.5/6.9)	60	0.001	>2000	G	G		Varies	0.33
22 Non-tracking SE[e] premix	Hand	500/1000 (3.5/6.9)	60	0.001	>2000	G	G		Varies	0.62
23 Low shrink premix	Hand	250/500 (1.7/3.5)	90	0015	200	G	G		Varies	0.42
24 Low shrink premix	Hand	500/1000 (3.5/6.9)	60	.001	150	G	G		Varies	0.14[d]
25 Acrylic monomer low strength premix	Hand	500/1000 (3.5/6.9)	60	.0001	290	G	G	66/—	White	0.15[d]
26 Low shrink premix	Hand	300/1000 (21./6.9)	60	.001	400	G	G		Varies	0.20[d]
27 Alum. hydrate filled premix	Hand	500/1000 (3.5/6.9)	60	.001	>2000	G	G	63/—	Varies	0.20[d]
28 Halogenated resin premix	Hand	500/1000 (3.5/6.9)	60	.001	>2000	G	G		Varies	0.25[d]
29 Arc resistant premix	Hand	500/1000 (3.5/6.9)	60	.001	>2000	G	G		Varies	0.20[d]
30 SMC	Cut	1000/2000 (6.9/13.8)	60	.001	325	G	G		Varies	0.35[d]
31 Flomat SMC	Cut	500/3000 (3.5/20.7)	60	.001	350	G	G		Varies	0.51
32 Non-tracking SMC	Cut	500/2000 (3.5/13.8)	60	.001	350	G	G	40/—	Varies	0.60
33 Flame retardant SMC	Cut	200/2000 (1.4/13.8)	60	.0015	500	G	G	45/—	Varies	0.57
34 Solid monomer SMC	Cut	400/1000 (2.8/6.9)	180	.001	>1000	G	G	50/—	Varies	0.55
35 Continuous strand mat laminate		75/500 (0.5/3.5)		.001	1500	F	G	55/97	Varies	0.35[d]
36 Preform molding[a]		75/500 (0.5/3.5)		.001	1000	F	G	50/94	Varies	0.25[d]

[a]General purpose phenolic and preform data included for reference.
[b]Granular, G; extruded rope, ER; sheets, S; fibrous putty, FP; sticks, ST; fibrous wads, FW.
[c]Advertised selling price of commercial compounds, except as noted.
[d]Estimated mixed cost—by author.
[e]Self extinguishing.

almost every instance offering lower cost as well as some important improvement in the performance of the product.

With the development of low shrink, low profile formulations, reinforced molding compounds are no longer limited to industrial-technical products not normally exposed to the user's view. There are now many applications where the product is decorative as well as functional in an engineering or structural sense, and the designer is free to consider these materials as either or both.

Some representative successful applications are described below to illustrate what reinforced molding compounds are good for and where they should be used. Where possible, the alternate material will be indicated with the justification and/or advantages of the reinforced molding compound.

Figure 15.10. Automobile heater housing—sisal reinforced. (*Courtesy Reichhold Chemicals, Inc.*)

15.1.3.1. Automotive

The sheer volume of material consumed makes the automotive applications of prime interest. Components of the heating and ventilating systems; front end panels that incorporate fender extensions, mounts for headlamps, and grilles; hoods and other major cab components of trucks; and many other parts have made automotive applications use the largest percentage of total reinforced plastics production.

Automobile heater housings and related ducting molded of sisal reinforced polyester BMC are used in nearly every make of car (see Fig. 15.10). They replace an assembly of many pieces of sheet metal and fiberboard with a product offering simpler installation, better performance (less noisy, smoother interior surfaces with better air flow), no necessity for painting, and low cost. Some competition is experienced from thermoplastics, but the low cost ones fall short on performance and the high performance types fail on the cost criterion.

Front ends, or as designated by the trade, "grille opening panels," in SMC are the standard of the industry. Molded-in provisions for mounting headlamps, grille trim, and other accessory components result in a one-shot molding that replaces 15 or more metal stampings and/or die castings. These parts, painted to match the exterior of the car, call for a material as smooth as sheet metal that can withstand the temperatures of the paint baking process. The low shrink, low profile resin systems offer surfaces nearly equal to sheet metal, better heat resistance than cost competitive thermoplastics, and lower cost than stamped metal assemblies, zinc die castings, or heat resistant thermoplastics.

Truck hoods and cab components have been made of reinforced plastics for 20 or more years, by hand lay-up/spray-up techniques for low volume, and by preform matched die molding methods for the higher volumes. In spite of somewhat lower and less consistent mechanical properties of SMC, there has been a shift from preform to SMC, especially in those parts where advantage can be taken of complex molding ability with consequent reduction in number of pieces and assembly costs. SMC bumpers, while not presently on American cars, are successfully used on several European cars with definite savings in weight (see Fig. 15.11). Indications are that the first American use of SMC in the bumper system will be for non-visible support brackets, where a high strength SMC molding can replace several pieces of metal.

Figure 15.11. SMC automobile bumpers. (*Courtresy Owens-Corning Fiberglas Corp.*)

Exterior body parts, such as hood and deck lids for limited production models, have been made of SMC, the principal advantage being lower tooling cost and shorter tool up time. However, station wagon tail gates, because of their great complexities due to both vertical and horizontal hinging and related multiple latching, offer great opportunities for SMC.

15.1.3.2. Electrical

Electrical and electromechanical products are somewhat "behind the scenes" applications where reinforced compounds have made a genuine technical and economic contribution. Electrical "switchgear" housings, formerly fabricated from flat plastics laminates in a metal framework, are now molded of BMC or SMC, with the advantages of lower cost, better electrical performance, and considerably decreased dimensions. The elimination of metal structure and the excellent arc resistance of the reinforced molding reduces the clearances needed to prevent arcing under high voltage conditions. Even though one of the earliest and continuing uses of SMC (in Germany) is for low voltage electrical distribution station housings, unpainted outdoor applications have been few, since weathering characteristics of reinforced molding compounds are not outstanding. However, polyurethane coatings have proven very effective and resulted in the use of SMC for such things as medium voltage insulator brackets for above-ground power distribution.

Other electrical applications include housings for hand power tools (electrical drills, sanders, etc.). Precision molding with low shrink resin systems has changed the concept of tool construction. The stability and strength of the reinforced molding compound permits its use as both the structural housing and the electrical insulation.

15.1.3.3. Appliances

Reinforced molding compounds have found many applications in appliances; for instance, in air conditioners, where advantage is taken of corrosion resistance, electrical insulation, mechanical strength, and the ability to mold complex shapes. Air conditioner housings can have molded-in blower scrolls, air ducts, mounting brackets for controls, blower motor, switches, etc. A substantial measure of thermal and sound insulation is achieved, and no painting is required (Fig. 15.12). Other appliance uses are in food disposers, refrigeration, humidifiers, dishwashers, and laundry equipment.

15.1.3.4. Miscellaneous Applications

Figure 15.13 shows a shower floor molded of polyester BMC using synthetic fiber reinforcement. Even though the molding weighs over 60 lb, it is lightweight compared to the easily broken cast concrete terrazzo product it replaces. It offers water and stain resistance, much easier installation, and lower shipping costs; it is non-porous and easy to clean; and it

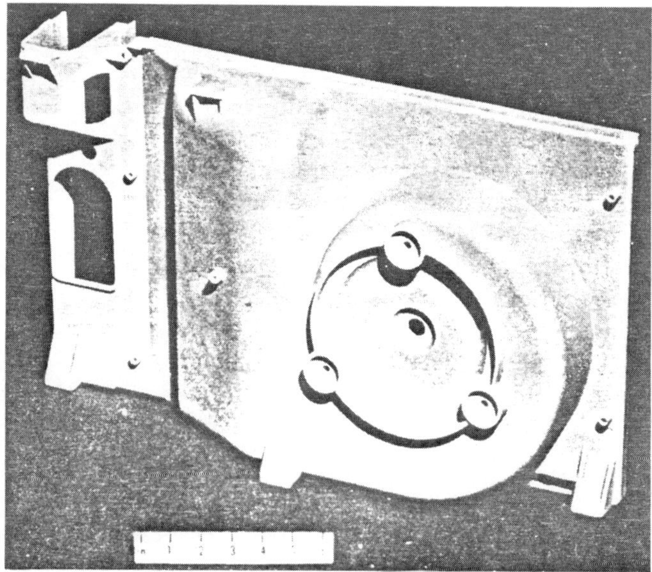

Figure 15.12. Air conditioner partition, BMC, (*Courtesy Premix, Inc.*)

has a nonslip surface, many colors, and a very competitive price.

15.2. MATERIALS

Since the basic materials are rather completely described in other chapters, the discussion in this chapter will be limited to material requirements and qualities that are specific to BMC and SMC. A typical compound contains resin, reinforcement, filler, release agent, colorant, curing agent, thickener, and, sometimes, a low profile additive.

15.2.1. Resin

The resin should ideally have a low viscosity to permit easy mixing, but high enough not to liquefy and separate from the other ingredients as the compound flows in the mold. It should cure rapidly, have high hot strength to permit removal of the part from the mold without damage, and be sufficiently resilient to permit some deformation of the part without cracking.

Most polyester resins for reinforced molding compounds are in the 25 poise viscosity

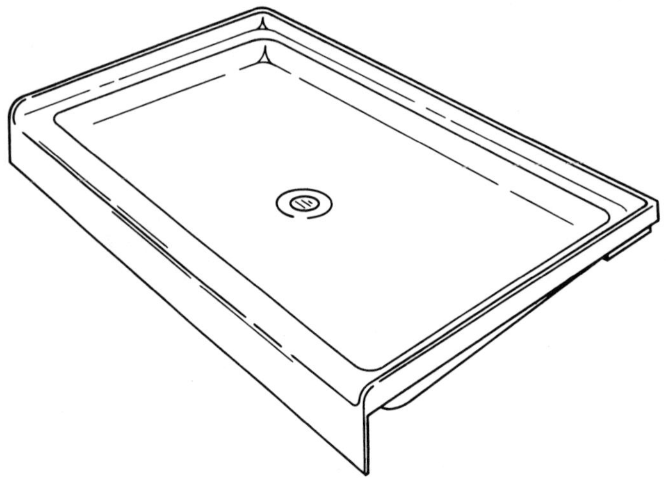

Figure 15.13. Shower floor, BMC. (*Courtesy Powers Fiat, Division of Mark Controls Inc.*)

Table 15.7. Molding Compound Resins—Properties of Unfilled Castings

POLYMER MONOMER	ORTHOPHTHALIC STYRENE	ISO- STYRENE	ISO- VINYL TOLUENE	ORTHO- STYRENE	ORTHO- DAP	VINYL ESTER STYRENE	HET ACID STYRENE
Viscosity, poises	26	27	26	24	400	5	5
Specific gravity	1.22	1.2	1.09	1.17	1.25	1.04	1.04
Heat distortion, °F (°C)	162 (72)	176 (80)	214 (101)	165 (74)	392 (200)	190 (88)	212 (100)
Flexural strength, psi (MPa)	17,500 (121)	17,600 (121.4)	18,500 (127.6)	23,000 (158.7)	12,900 (89)	20,000 (138)	16,000 (110.4)
Flexural modulus, psi (GPa)	0.56×10^6 (3.9)	0.42×10^6 (2.9)	0.55×10^6 (3.8)	0.50×10^6 (3.1)	0.45×10^6 (3.1)	0.50×10^6 (3.4)	0.5×10^6 (3.45)
Tensile strength, psi (MPa)	8900 (61.4)	9500 (65.6)	7730 (53.3)	13,000 (89.7)	8,000 (55.2)	11,000 (75.9)	12,000 (82.8)
Tensile elongation, %	1.8	6.5	2.9	4.0	1.5	5.2	4.0

range, although resins from 10–2500 poises are being used. Up to 600 poises can be mixed in conventional equipment without viscosity reducing solvents that must be evaporated. Polyester resins are often classified for molding compound purposes by: 1) their basic polymer ingredient (for instance, orthophthalic, isophthalic, bisphenol, or "Het" anhydride) or 2) their cross-linking monomer (styrene, vinyl toluene, DAP, etc.). Generally, the lowest cost is an orthophthalic based polymer with styrene monomer. Isophthalic, bisphenol, and "Het" anhydrides offer better mechanical properties, corrosion resistance, and reduced flammability, respectively. Vinyl toluene is less volatile than styrene, and compounds made with it do not suffer from monomer evaporation on exposure to air. DAP is even less volatile and provides better electrical properties as well.

Epoxy resins offer substantial advantages in strength and chemical resistance, yet their much higher price, limitations on compound selections, cure rate, etc. have restricted their applications in reinforced molding compounds. However, vinyl ester, a close relative built on an epoxy backbone, but cross-linked with styrene and peroxide-cured, offers toughness, chemical resistance,and flexibility in compounding that, in spite of its higher price, makes it a viable product for high performance applications.

In addition, there are resins especially formulated for chemical thickening purposes and other low shrink devices, as mentioned in the earlier section on properties. Some typical properties of some resin/monomer systems are shown in Table 15.7.

15.2.2. Reinforcements

Reinforcements for BMC include glass, asbestos, sisal, and various other organic fibers. The glass fibers used are of three types: chopped strand (bundles of continuous filaments), chopped spun roving (short fibers made into threads), and a resin-coated chopped strand generally identified as "high strand integrity" fiber (or HSI). They are available in various lengths with 1/4 and 1/2 in. (6.35 and 12.7-mm) commonly used. The HSI fiber resists degradation in the mixing process (i.e., the strand does not break up into its individual filaments) and provides the best overall mechanical properties, especially in parts requiring long flow. Other glass fiber forms have better color characteristics (they are less noticeable in the mix) and produce a better surface. While glass fibers generally give the maximum mechanical properties, sisal fibers make an easy flowing compound, excellent for large, moderately complex parts not requiring a high degree of water resistance, and can be used in combination with glass fibers. Asbestos fibers are used where special chemical resistance is sought. A very low cost molding with excellent water and stain resistance, plus good electrical properties, can be made with diced nylon tricot fabric. The mold shrinkage is quite high, but since there is no differential between the nylon and the resin, the surfaces are remarkably smooth. The color and weave of the nylon fabric are difficult to mask and the mechanical properties of such moldings are low. The lower molded density permits thicker sections, which can sometimes compensate for the lower strength. Some compounds have been made with carbon and aramid (Kevlar 49) fibers. They increase some mechanical properties, but not in proportion to their greater basic fiber strength.

Sheet molding compounds are commonly reinforced with fiberglass roving chopped in 1-in. (25.4-mm) lengths, although 1/2–3-in. (12.7–76.2-mm) are used. Roving is usually identified as being hard or soft, with reference to the finish on the glass. The hard types chop easily, wet out poorly, but mold well. The soft types are harder to chop, do not mold as easily, and give a poorer surface, but they wet out easily and provide higher mechanical properties. Originally, SMC was made from soluble (soft) binder chopped strand mat, 2-in. (50.8-mm) fiber length. Some SMC is still made this way (mainly in Europe) and recently, in efforts to obtain higher strengths, continuous filaments of glass, carbon, and aramid fibers are being incorporated into SMC.

15.2.3. Fillers

The fillers in common use are in four chemical groups: silicas and silicates, carbonates, sulfates, and oxides. However, almost any material that can be reduced to a particle size in the range of 1/2–50 microns can be used. The first group includes asbestos, talc, china clay (kaolin), silica (sand), diatomaceous earth, and volcanic ash; the second is comprised entirely of various calcium carbonates; the third includes barium sulfate (barytes) and calcium sulfate; and the fourth, hydrated alumina. In these groups are the so-called "natural" materials that are brought to a useful state by grinding (wet and/or dry), and those produced by chemical precipitation. The latter can be of the smallest available particle size as well as the most uniform (almost constant), which may be of questionable advantage, as will be pointed out later. The specific gravity ranges from 2, for diatomaceous earth, to 4.45, for barytes, with the predominately used materials (china clays and calcium carbonates) in the 2.6–2.7 range. China clay and calcium carbonate at 5¢/lb (11¢/kg) or less are at the low end of the price scale. Specialty fillers, like hydrated alumina, are about 15¢/lb (33¢/kg).

The bulking fillers are principally clays and calcium carbonates. Calcium carbonates have the lowest oil absorption and, consequently, more can be put into a mix; however, their characteristics are such as to give poor flow.

Clay-filled compounds have better flow, and molded properties are better in many respects, but not in color. Frequently, a combination of clays and calcium carbonates provides a high filler loading with good flow. Addition of smaller quantities of high oil absorption talc to predominately calcium carbonate-filled compounds will also improve the flow with less effect on the good color.

The following list of desirable filler properties can be used to narrow down the tremendous number of minerals offered for use in reinforced plastics:

- Low specific gravity
- Low oil absorption
- Non-porous
- Non-abrasive
- Low cost
- Ready dispersibility without agglomeration
- Chemical purity and whiteness
- Wide particle size range—about 1–15 microns, with a mean diameter of 5 microns.

There are also additives for special use that may come under the heading of fillers. These are materials for fire resistance, such as antimony trioxide, used in conjunction with halogen containing resins or additives; for chemical thickening, which include magnesium oxide and calcium hydroxide; and for lowering shrinkage effects, such as fine, powdered polyethylene.

Various other detailed characteristics and properties of fillers will be discussed below.

15.2.3.1. Particle Size

These materials are commonly classified as to particle size in terms of the fineness of sieve through which a given percentage will pass (for example, 99.8% through 325 mesh screen) or by the particle size expressed in microns. For our purposes, the figure in the example is a reasonable low limit. The space between the wires of a 325 mesh is 44 microns, and any substantial quantity of larger particles is not satisfactory.

15.2.3.2. Particle Size Distribution

The particle size classification above provides information as to the maximum particle size, but nothing as to the smallest, or to the quantities of various sizes in between. Fortunately, this information is available for most of the fillers in the form of a particle size distribution graph (see Fig. 15.14). It is reasoned and has been shown by experience that a filler with a fairly wide particle size range such that the smaller particles fit between the medium, the medium between the large (etc.) will provide a compact arrangement requiring a minimum of resin to fill the space between the particles. The greatest economy, as well as the best mechanical properties, results when there is sufficient resin to fill the spaces and not so much that it separates the particles. Other necessary characteristics of a molding resin

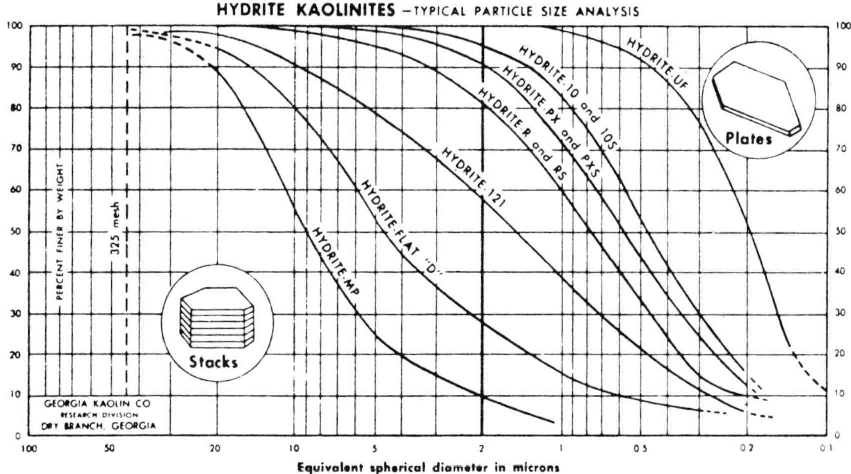

Figure 15.14. Particle size distribution graph. (*Courtesy Georgia Kaolin Co.*)

complex may limit attainment of this optimum condition.

15.2.3.3. Oil Absorption

This property, which is stated as the percentage of linseed oil required to wet out a given weight of filler, provides, when selecting fillers, an approximation of the relative amounts that could be used to attain the same viscosity. The oil absorption is a function of the specific surface of the particles. Porous particles have higher values than non-porous ones of the same size. The lowest oil absorption permits the highest proportion of filler, and in most cases the principal filler in a resin/filler system can be one of low oil absorption value.

15.2.3.4. Thixotropy

Thixotropy is a phenomenon in which the nominal viscosity of a material is markedly reduced when the material is disturbed, returning to the original state when the disturbance ceases. Some high oil absorption fillers, while providing high viscosity, also have a substantial thixotropic effect. Mold closing forces are usually sufficiently disturbing that viscous thixotropic resin/filler systems can exhibit the flow properties of lower viscosity non-thixotropic systems.

15.2.4. Release Agents

Internal mold release is used in all compounds. Zinc stearate is most common. Also used are calcium and aluminum stearate and stearic acid for lower temperature molding, and an alkyl phosphate product of Dupont, "Zelec UN." The latter is a liquid, is easy to measure, and is easy to mix. The stearates are dusty powders.

15.2.5. Colorants

Dispersions of pigments in compatible resins are widely used, but economies can sometimes be effected with the use of lower cost, dry color pigments in BMC. The intensive mixing required to make a BMC is often sufficient to blend in many dry pigments properly. Some pigments have an accelerating or inhibiting effect on resins and some consideration should be given to the effects on storage stability and cure time when choosing pigment types.

15.2.6. Curing Agents and Inhibitors

Benzoyl peroxide is a good, economical curing agent for BMC that is to be used shortly after mixing, but is not satisfactory where long-term storage is required. Tertiary butyl perbenzoate (TBP) requires higher molding temperatures but is very stable and permits higher temperature mixing. It is the standard of the BMC-SMC industry. Some recent developments include peroxyesters and peroxyketals, which give shelf life equal to TBP with slightly faster cure times. Combinations of TBP with more reactive peroxides, such as

Table 15.8. Press Molding Peroxides: Reaction Data[a,b]

PEROXIDE	CONCENTRATION, % BY WEIGHT	GEL TIME, SECONDS	EXOTHERM TIME, SECONDS	PEAK TEMPERATURE, °F (°C)	PLATEN GEL TIME AT 295°F (146°C), SECONDS
Tertiary butyl perbenzoate	1.0	152	179	329 (165)	27
Tertiary butyl perbenzoate	1.4	142	168	330 (166)	25
Peroxyester	1.0	124	152	313 (156)	22
Peroxy ketal	1.0	106	133	305 (152)	18
Tertiary butyl perbenzoate	0.80	104	133	307 (153)	19
Tertiary butyl peroctoate	0.20				
Tertiary butyl perbenzoate	0.70	99	127	310 (154)	15
Tertiary butyl peroctoate	0.30	99	127	310 (154)	15
Tertiary butyl perbenzoate	0.50	86	112	315 (157)	13
Tertiary butyl peroctoate	0.50	86	112	315 (157)	13

[a]Resin: Typical isophthalate SMC polyester.
Apparatus: modified hot block tests at 270°F (132°C).
[b]Source: U. S. Peroxygen Division, Witco Chemical Corp.

tertiary butyl peroctoate, can significantly reduce cure times, again if long term shelf life is not a factor. Table 15.8 shows the reactions of some typical peroxides and combinations in a standard isophthalic polyester SMC. The platen gel time at 295° F (146° C) is a reasonable indication of the relative mold flow time which can be expected.

Accelerators or promoters for some room-temperature cured systems cannot be used in molding compounds because of practically zero shelf life. However, an organometallic complex, trade name PEP,* provides reduction in cure times without detrimental effect on shelf life or molded properties.

Inhibitors beyond those normally present to stabilize the resin can be used to further promote storage stability to the compound, to prevent gelation during mixing, and, rarely, to control the cure rate. Hydroquinone, benzoquinone, tertiary butyl catechol, and a long list of other quinones stabilize with little effect on the cure rate. Very small quantities are used, in the range of 0.005–0.02%.

15.2.7. Thickeners

Thickeners may be defined as materials or systems that increase the viscosity of the compound without cure. Most of the thickeners used are known chemically as Group IIA metal oxides, and include the oxides and hydroxides of magnesium and calcium, MgO, $Mg(OH)_2$, CaO, and $Ca(OH)_2$. Relatively newer, and appearing to have many advantages, is a thickening system based on the in situ formation of a three-dimensional polyurethane rubber network distributed throughout the polyester matrix. This system is called the interpenetrating thickening process (ITP) and is a development of ICI America, Inc.

The function of the thickener is twofold: to make a compound that is in a form that can be conveniently handled—dry, non-tacky, and easy to cut and form; and to maintain a matrix viscosity sufficient to retain the homogeneity of the reinforcement/filler/pigment/resin mix as the compound flows to fill the mold. The optimum thickener would be one that does not start its thickening reaction until the resin has thoroughly wet out the balance of the ingredients, and then rapidly thickens to the desired viscosity, and remains there until molded.

Unfortunately, the metal oxide/hydroxide systems start to thicken immediately on mixing into the resin, and never stop thickening, even though this process slows down and there is a period of time during which compounds will fall in a satisfactory molding range.

If a short thickening or "maturation" time is desired, say 24 hours, the compound will have a short moldable life, as little as 3 or 4 days. If a longer moldable life is necessary, or if compounding does not permit any initial vis-

*Air Products & Chemicals, Inc.

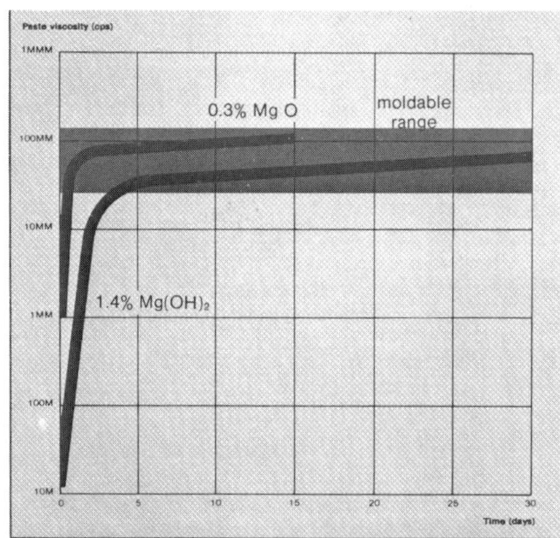

Figure 15.15. Viscosity increase versus time. Thickening agents in low profile resin. (*Courtesy Owens-Fiberglas Corp., Publication 5-TM-6991-A.*)

cosity build-up, a longer maturation time, perhaps 3–5 days, will be required (see Fig. 15.15). An additional problem with the continuous viscosity increase of the metal oxide/hydroxide system is that the molding conditions may have to be altered in some proportion, especially as regards molding pressure (see Fig. 15.9).

The ITP or polyurethane rubber system involves a finite chemical reaction. When the ingredients—an isocyanate, catalysts, and a polyol to react with the isocyanate—are converted, the thickening process is complete, and the viscosity will remain constant. Maturation time can be very short, but practically, just as for the metal oxide/hydroxide systems, time must be allowed for wet-out of the other ingredients. A practical maturation time is 16–24 hours.[3]

15.2.8. Low Profile Additives

The use of low profile additives has been a major factor in the increased replacement of metal, and other plastics as well, by BMC and SMC. At this time, all low profile additives are thermoplastics, which, when added to a thermosetting compound, impart a smooth surface and dimensional stability to the molded part. The part can often be painted with little or no abrasive surface treatment. With some of these additives, molded-in color is excellent. Warpage, a serious problem with conventional resin systems, is practically eliminated. Moldings can now be made with tolerances equal to, or better than, precision cast metals, and are in high volume production.

The use of many common thermoplastics has been reported, all with some degree of success, either alone or in combinations. These include:

Polyethylene
Acrylic
Polystyrene
Cellulose acetate butyrate (CAB)
Polyvinyl chloride (PVC)
Polyvinyl acetate
Polycaprolactone.

Finely powdered polyethylene, initially used to reduce small voids, increase weather resistance, and improve electrical properties, was an unpublicized low shrink agent. Acrylic homopolymers, their use patented in 1968, continue to be standards of performance comparison for all later developments.

Low profile additives usually reduce all the mechanical properties except impact strength, which is often increased. Some of the low profile additives apparently have a substantially lower elastic modulus than the basic resins and tend to act similarly to the rubber-like tougheners, increasing the ability of a molding to distort without breaking.

Some have an effect on the thickening rate, usually increasing it, if so. And some, for example acid modified polyvinyl acetate, thicken independently of the resin.

15.2.9. Tougheners

A number of synthetic rubber compounds have been found to increase the impact resistance and elongation before breakage of molding compounds, usually with a reduction of flexural strength and stiffness. Some of the thermoplastic low profile additives act in a similar manner. One of the major faults of thermoset-reinforced plastics is brittleness. These additives, while only providing a small improvement, are worthwhile in many instances.

15.3. FORMULATION

Formulation is the design of a compound to bring about the desired properties in the finished molding, within the limitations of the molding conditions. Compounds may be formulated to provide strength, stiffness, toughness, electrical insulation, corrosion resistance, fire resistance, and so on, and often combinations of two or more of these qualities, but first and foremost they must be moldable.

Moldability has been generally defined earlier in this chapter. In addition, a reinforced compound must, to the maximum degree, remain homogeneous as it flows through and to the extremities of the mold. If the resin/filler/reinforcement separate, serious variations in properties will occur throughout the molding, and much of the usefulness of the reinforcement will be lost.

The ideal compound should also flow easily and fill extremities and details of the mold. These two molding characteristics are seldom easy to combine with the other necessary requirements, and usually the best formulation is a compromise.

The flow properties are, in large part, a function of the degree to which the resin is absorbed by, or adsorbed on, the filler and the reinforcement. Each of the dry ingredients employed in compounds has its own particular resin absorption, or drying effect on the resin. For example, in the case of two very common fillers, china clay has more than twice the absorptive power of calcium carbonate. Longer glass fibers are less drying than short, and coated fibers (HSI types) less than conventional. The dryer the compound, the lower the plasticity or flow. Unfortunately for the formulator, the resin absorption qualities of all the various ingredients are not fully determined or published. Some guidance on fillers is available from the linseed oil absorption data published for the paint industry. Table 15.9 lists oil absorption properties of some commonly used fillers.

Not only do the fillers vary in resin absorption, but resins vary in their ability to wet the fillers, depending on their viscosity, basic

Table 15.9. Oil Absorption of Some Fillers Used in Reinforced Molding Compounds

FILLER	PARTICLE SIZE, MICRONS	SPECIFIC GRAVITY	OIL ABSORPTION[a]	COST, ¢/lb (¢/kg)
Antimony trioxide	44	5.70	11	150 (330)
Asbestos	50	2.56	38	6 (13.2)
Barytes	7.5	4.40	11	5.5 (12.1)
Calcium carbonate	2.5	2.71	14–16	4.1 (9)
Calcium carbonate	5	2.71	9–10	3.25 (7.15)
Calcium carbonate	7.5	2.71	5.5–6.5	2.2 (4.8)
Calcium carbonate	14	2.71	5–6	1.25 (2.75)
Georgia kaolin (China clay)	1	2.58	60	5.5 (12.1)
Georgia kaolin (China clay)	5	2.58	32	5.5 (12.1)
Diatomaceous earth	7	2.05	88	7 (15.4)
Talc	5	2.71	55–59	2.3 (5.1)
Aluminum trihydrate	12	2.42	30	8 (17.6)
Aluminum trihydrate	1	2.42	60	14 (30.8)
Feldspar	14	2.60	30	2 (4.4)
Feldspar	9	2.60	35	2.7 (5.9)

[a]Grams linseed oil per 100cc filler.

chemical construction, type and quantity of monomer, and so on.

The answer to the problem of formulating a compound that will have satisfactory flow and maintain reasonable homogeneity lies mostly in the selection of the resin and filler combinations. A high viscosity resin will carry reinforcement and filler well, but high viscosity makes mixing more difficult. Combinations of smaller amounts of high absorption fillers, such as china clay and asbestos, with low absorption ones, such as calcium carbonate or silica, work fairly well to solve both the flow and homogeneity problems.

The chemical thickening process used to make SMC can also be used as a flow control device in making BMC. A resin with a low initial viscosity readily permits addition of large quantities of filler or reinforcement. The thickening effect subsequent to mixing simulates the effect of a high viscosity resin, or the use of high absorption fillers.

15.3.1. INGREDIENTS

15.3.1.1. Resins

Resin content may range from approximately 18–50% by weight, with 30% a good starting point for almost any compound requirements. Where very low absorption fillers, such as calcium carbonates, can be used, a resin content at the low level will make a moldable compound. Where the special properties of a particular high absorption filler such as asbestos are needed, the resin requirement will be on the high side of the range.

15.3.1.2. Fillers

The filler content of a given compound is inversely proportional to the amount of reinforcement needed to satisfy the mechanical property requirements. However, in formulating for electrical or fire retardance, the opposite is true. Sufficient filler, such as aluminum hydrate, must be included to achieve the desired properties, then the maximum amount of reinforcement possible.

Usually fillers can replace the reinforcement with minimal effect on the general moldability of the compound.

15.3.1.3. Reinforcements

Glass fiber contents in BMC may vary from as little as 5% to about 50%. Less than 5% provides no detectable structural advantages, and more than 35% creates molding problems. Up to 20% provides compounds that can be extruded or compacted for easy handling. Higher glass fiber content makes fluffy, springy compounds that do not hang together well; conversely, compounds high in resin content are very wet and sticky. Sisal fiber can practicably be used up to about 20%, with as little as 5% offering worthwhile reinforcing effect. Up to 10% handles well. Nylon tricot fabric chopped or diced can be used up to approximately 15% by weight. It makes a light, fluffy mix that is difficult to compact and is most easily handled by using a deep tray.

Commonly, SMC formulations contain about 30% glass fiber usually chopped in 1-in. (2.5-cm) lengths. SMC can be, and is being, made with reinforcement content as little as 18%. On the higher side, compounds are made with up to 65% with short fibers and to 75% with continuous fibers or combinations of both. Where continuous fibers are used these may be other than glass; i.e., carbon or aramid (Kevlar 49).

15.3.1.4. Pigments

Concentrations vary widely, depending on the depth of color required, and it is difficult to indicate a realistic maximum or the point at which some pigments start becoming fillers.

15.3.1.5. Internal Mold Release

The stearates—aluminum, calcium, and zinc—are used at the 1–3% level, stearic acid about the same. Zelec* is effective at 0.5%. Some reduction in properties may be expected with excessive use of internal mold release, so its quantity should be minimized.

15.3.1.6. Curing Agents

The catalyst concentration is in proportion to the amount of resin and is usually specified as a percentage of the resin content. Average

*E. I. Du Pont de Neumours & Co.

concentrations for common polyester curing agents are shown in Table 15.8. Smaller or larger concentrations can be used, depending on shelf life requirements, curing temperature, and acceptable molding cycles. Combinations of peroxides are often used with one kind, reacting at a lower temperature, this latter often referred to as the "kickoff" catalyst.

15.3.1.7. Thickeners

The amount and type of thickener used are functions of a wide number of variables, particularly the properties of the resin system employed, the thickening rate desired and practicable (long enough to permit proper wet-out of fillers, etc., but no longer), the time required to reach a handleable, tack-free state, and the period the compound will remain in a moldable viscosity range. The most common thickeners, MgO and $Mg(OH)_2$, are used in the range of 1–1½% and 3–5% of the resin component of the compound, respectively. The MgO thickens faster and the $Mg(OH)_2$ increases the viscosity more slowly, but the compound remains in a moldable range for a longer time (see Fig. 15.15). The urethane thickening process, ITP,* employs a polyisocyanate in the range of 3% of the resin component.

15.3.1.8. Low Profile Additives

Most low profile additives (acrylics, polystyrene, polyvinyl acetate, for example) are pre-dissolved in styrene at a level of 30–40% solids. This solution is used in the compound at a level of about 50% of the basic resin component of the compound. Low molecular weight, finely powdered polyethylene is usually added directly in concentrations of 3–5% of the resin.

Some typical formulations for different purposes are listed in Table 15.10.

15.4. COMPOUNDING

While, in principle, there are many similarities in the compounding of BMC and SMC, there are differences. BMC, even when made in large quantities, tends to be made in batches, while SMC is usually made in a continuous process. In either case, there are always two distinct phases—one where the resin, colorant, catalyst, release agent, and part or all of the filler are combined, and a second where these materials are combined with the reinforcement.

In the first phase, the mixing can hardly be too intense or too long (except that excessive heat may develop). In the second phase, the mixing (or combining) should only be sufficient to ensure wetting and uniform distribution of the reinforcing fibers. Intense mixing of glass fibers can drastically reduce their effectiveness as a reinforcement by breaking or degrading them.

With compounds using fast-acting thickeners, such as MgO, there can be an intermediate stage where the thickener is added to the pre-batch immediately prior to combining with the reinforcement. Therefore, there is not time for the viscosity to increase sufficiently to prevent proper wet-out. Below, the methods of compounding BMC, the compounding of the SMC resin/filler paste, and, finally, the combining of the resin/filler paste with the fiber reinforcement to make the SMC as used by the molder will be examined.

15.4.1. BMC—Batch Process

15.4.1.1. Mixers

The basic equipment consists of two mixers—one for a pre-batch of resin, color, catalyst, release agent, and, occasionally, part of the filler; the other for blending the final mix of pre-batch with additional filler and reinforcement (glass, sisal, or other fibers).

The first mixer may be of any of several kinds, from the simple propeller type, such as the "Lightnin"—a 3 hp (2.24 kW) gear reduction model suitable for slowly mixing 25–30 gal (98–114 ℓ) in a 55-gal (208 ℓ) drum—to the dissolvers or dispersers in common use by the paint industry. A 10 hp (7.47 kW) Cowles "Dissolver" will rapidly mix a 400–500 lb (182–227 Kg) batch. For convenience of use, elevating and swinging aside of the mixer are important, but prime concerns are the mixing

*ITP is a trademark of ICI America, Inc.

Table 15.10. Typical Compound Formulations

	GENERAL PURPOSE GLASS REINFORCED PREMIX	ARC RESISTANT PREMIX	FAST CURE GLASS REINFORCED PREMIX	FIRE RETARDANT ALUMINUM HYDRATE FILLED PREMIX	FIRE RETARDANT HALOGENATED RESIN PREMIX	CHEMICALLY THICKENED PREMIX	SHEET MOLDING COMPOUND
Resin	Ortho-styrene, 28%	Iso vinyl toluene, 18%	"Acpol 42-2671," 30%	"Plaskon 9520," 30%	Halogenated resin, 34%	Ortho styrene, 32.9%	"Derakane QX3923," 33%
Catalyst	Benzoyl peroxide, 0.3%	Benzoyl peroxide, 0.2%	"Luperox 118," 0.3%	Benzoyl peroxide paste (50%), 0.6%	Benzoyl peroxide paste (50%), 0.6%	"Dicup 40C," 0.8%	Benzoyl peroxide, 0.3%
Release	Zinc stearate, 1%	Zinc stearate, 1%	Zinc stearate, 0.7%	Zinc stearate, 1.4%	Zinc stearate, 1.4%	Zinc stearate, 0.8%	Zinc stearate, 1%
Colorant	TiO_2, 5%						
Filler	China clay, 15%	Hydrated alumina, 72%	China clay, 26%	Hydrated alumina, 53%	China clay, 44%	$CaCO_3$, 49.3%	China clay 33%
Filler	$CaCO_3$, 35%		$CaCO_3$, 20%			MgO, 1.7%	MgO, 1%
Filler			Asbestos floats, 3%				
Reinforcement	HSI glass fiber, 15%	HSI glass fiber, 9%	HSI glass fiber, 20%	¼-in. HSI glass fiber, 15%	¼-in. HSI glass fiber, 15%	¼-in. HSI glassfiber, 14.5%	Soluble binder mat, 32%

effectiveness and speed. The more elaborate and faster models have advantages easily justifying their higher cost.

The final mixing is done in what is variously called a dough mixer, a kneader, a sigma blade, or a double arm mixer (see Fig. 15.16). For BMC, this must be the heaviest duty of this type of mixer available. Table 15.11 shows some typical capacity horsepower ratings.

While many existing mixers have so-called sigma blades, a kind of zigzag design, the single curve helix type with a substantial clearance—1/4–3/8 in. (6.35–9.5mm)—between the blades and the trough seems to be an adequate compromise between thorough mixing and a minimum degradation of the reinforcing fibers. The blade speeds should be low and different, with the higher speed one about 30–40 rpm, maximum. The mixer trough should be jacketed to permit control of the temperature with hot water, as considerable advantage can be realized with some formulations mixed at elevated temperatures. Other types of mixers can also be used. Especially effective is a strengthened version of the double planetary blade type (see Fig. 15.17). Advantages include easy cleaning and lower degradation of fiber reinforcement. With multi-speed drives, this type of mixer can be used at high speed for the early mixing phase and at low speed for mixing in heavy fillers and reinforcements.

Figure 15.16. Dumping finished BMC into hopper cart.

15.4.1.2. Mixing

A typical mixing schedule for a general purpose batch process BMC would be as follows.

1. In a Cowles Dissolver (or similar mixer), blend: resin, catalyst, release agent, and colorant, and mix 10 minutes.
2. Add 50% filler and mix 5 minutes.
3. Transfer to double arm mixer.
4. Add balance of filler and mix 10 minutes.

Table 5.10. Continued

LOW PROFILE ACRYLIC MONOMER MODIFIED PREMIX	LOW PROFILE THERMOPLASTIC ADDITIVE PREMIX	LOW PROFILE CHEMICALLY THICKENED THERMOPLASTIC ADDITIVE PREMIX	LOW COST GENERAL PURPOSE CHOPPED NYLON RAG-GLASS FIBER PREMIX	CHOPPED NYLON RAG REINFORCED PREMIX	SHEET MOLDING COMPOUND
"Paraplex P-19A," 35%	"Atlac 382-13," 43.5%	"Stypol 40-2366," 25%	"Laminac PDL-7-991," 25%	"Laminac PDL-7-991," 32%	"Stypol 40-2732," 40%
Tert butyl peroctoate, 0.3%	"Luperox 118," 0.8%	"Dicup 40C," 0.44% "Luperox 118," 0.06%	Benzoyl peroxide paste (50%), 0.4%	Benzoyl peroxide paste (50%), 0.4%	Tertiary butyl perbenzoate, 0.5%
"Zelec UN," 0.2%	Zinc stearate, 2%	Zinc stearate, 0.5%	Zinc stearate, 1.8%	Zince stearate, 1.8%	"Zelec UN," 0.3%
			Pigment, 0.8%	Pigment, 0.8%	
China clay, 34.5%	$CaCO_3$ "Surfex MM," 17%	$CaCO_3$, 49.6%	China clay, 17%	China clay, 18%	China clay ASP 400, 25.7%
	China clay ASP-100, 17%	$Ca(OH)_2$, 0.7%	$CaCO_3$, 34%	$CaCO_3$, 36%	
	Microthene FN 510 polyethylene, 5%	Microthene FN510 Polyethylene, 3.8%			
¼-in. HSI glass fiber, 25%	¼-in. HSI glass fiber, 22%	¼-in. HSI glass fiber, 20%	Chopped nylon rag, 8% ¼-in. HSI glass fiber, 8%	Chopped nylon rag, 12%	Soluble binder mat, 33.5%

5. Add reinforcement and mix 3 minutes.
6. Transfer to sealed container (to inhibit loss of monomer).
7. Age at 77°F (25°C) for 4 hours.

If a thickened formulation is called for, the finished mix must be processed further as it comes from the mixer by either extrusion or calendaring; otherwise it will be impossible to process for molding when it reaches its molding viscosity.

15.4.1.3. Extruding

Compounds of lower fiber content (less than 20% glass fiber reinforcement) and some higher content thickened compounds can be effectively extruded into bars or logs. Extruders are usually equipped with guillotine type cutters to cut the extrusion into mold charges as it leaves the extruder nozzle (see Fig. 15.18). Commercially available extruders are of the screw type and cause some degradation of the fibers. Some compounders and

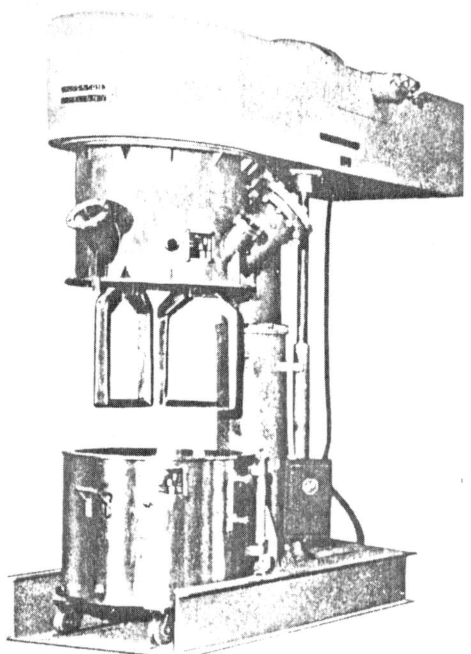

Figure 15.17. Ross double blade planetary mixer.

Table 15.11. Capacity Versus Horsepower of Double Arm Mixers

WORKING CAPACITY, (*l*)	CAPACITY, lb (Kg)[a]	POWER, hp (kW)
20 (75.7)	150 (68)	20 (14.9)
35 (132.5)	300 (136)	30 (22.4)
50 (189.3)	400 (182)	30 (22.4)
100 (378.5)	800 (364)	50 (37.3)

[a]Average compound.

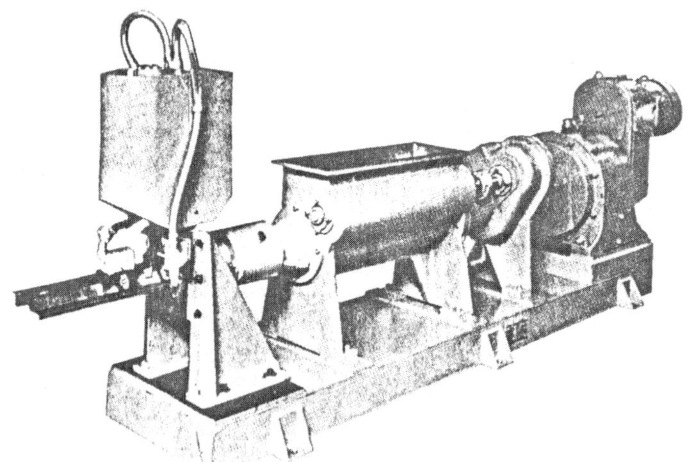

Figure 15.18. Bonnot extruder with guillotine slug cutter.

molders have custom built piston or ram extruders that operate at a low rate but with low fiber degradation.

15.4.1.4. Other Considerations

Handling material in and out of the various mixers, while seemingly elementary, can turn out to be a major problem if proper precautions are not taken. Even a modest size dough mixer is quite tall, and putting it in a pit would only complicate the problem of getting the finished mix out. Drum handling and hoisting trucks take up space and are quite expensive if powered. Most satisfactory seems to be arranging all the equipment—scales, mixers, etc.—in a line with a single beam trolley and hoist overhead. Simple drum hoists with a chain operating tilting mechanism permit pouring from resin drums into mixers or mixing containers.

15.4.1.5. Preparation for Molding

Between mixing and molding, some time elapses during which the material must be protected from various kinds of degradation, the most common of which is loss of monomer by evaporation. Keeping the mix well enclosed in a bag or blanket of plastic film (cellophane or polyethylene) will usually suffice for short periods. If the mix must be kept for long periods, sealed containers and refrigerated storage areas are necessary, and, of course, formulation plays an important role in this problem.

15.4.2. BMC—Continuous Mixing

Continuous mixing involves the use of moving belt weigh feeders for the dry ingredients, speed controlled choppers for the fibrous reinforcement, and metering pumps for the liquids, all feeding to a flow-through mixer. It is obvious that such a system requires sophisticated controls for accurately interrelating a variety of feed rates, and is not amenable to frequent rate or proportion changes. In practice, there are no truly continuous systems starting with bulk forms of all the ingredients. Small percentage ingredients, such as catalysts, release agents, thickeners, and colorants, are usually batch mixed into one or the other of the bulk items. Figure 15.19 is a schematic diagram of a process for continuous compounding and molding of BMC.

15.4.3. SMC—Resin/Filler Paste Batch Process

The first phase of SMC compounding by a batch process is essentially the same as that for BMC, the main difference being that SMC always has a thickener. The fast-reacting thickeners require that the batch size must be fairly small, usually less than one hour's flow through the SMC machine, lest the viscosity build too high for proper wet-out of the fiber reinforcement. This is not a small amount. Even a 2-ft width SMC machine can use as much as 2000 lb of resin paste per hour. A typical mixing schedule for the resin/filler

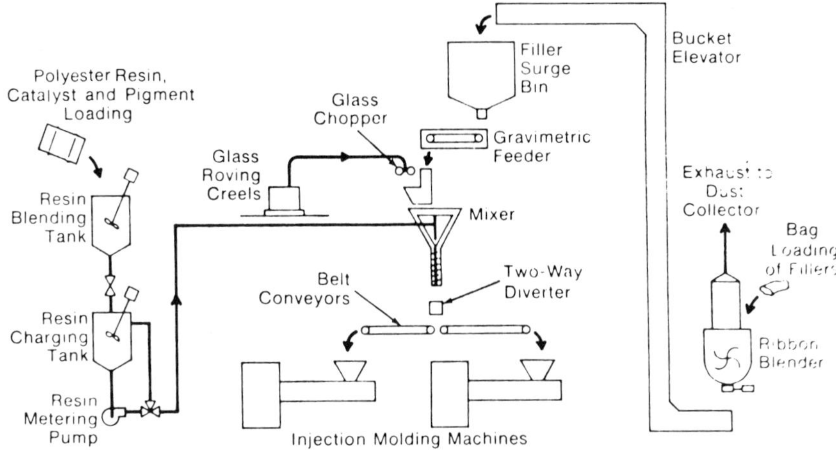

Figure 15.19. Schematic of continuous compounding system. (*Courtesy Farrel Machinery Group USM Corp.*)

paste pre-batch for an SMC compound might be as follows:

1. In a Cowles Dissolver (or similar mixer), blend: resin, catalyst, release agent, colorant, and low profile or low shrink additive, and mix as required (at least 10 minutes).
2. Add fillers and mix 15–20 minutes.

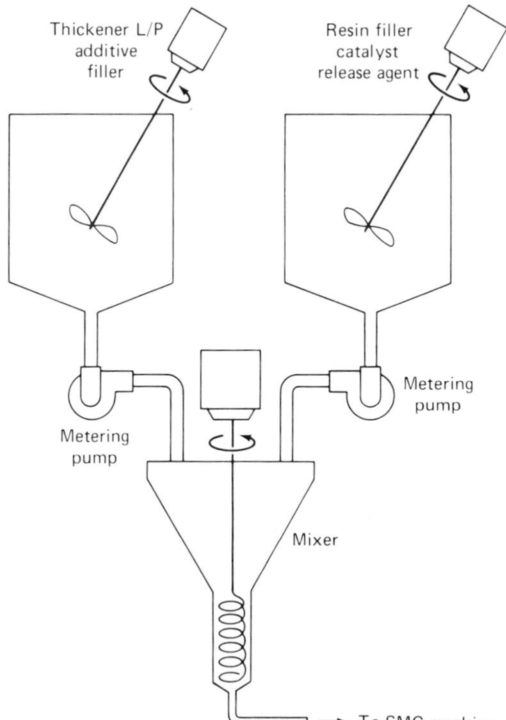

Figure 15.20. Schematic—SMC batch/continuous mixing system.

3. Add thickener and mix 5 minutes.
4. Transfer to paste metering section of SMC machine.

15.4.4. SMC—Resin/Filler Paste, Batch/Continuous System

A modification of the above system, sometimes referred to as the batch/continuous system, reduces the problem of variable thickening of the paste over the batch consumption time. It employs two mixing tanks, into one of which is charged the thickener in a non-thickenable carrier, such as the low profile additive solution. The other tank is charged with the basic resin and the balance of the ingredients. After mixing, the material is metered from the two tanks through a static or dynamic mixer to the paste metering section of the SMC machine (see Fig. 15.20).

For continuous operation of an SMC machine, two such sets of tanks are required.

15.4.5. SMC—Resin/Filler Paste, Continuous Mixing

Continuous mixing systems are similar to that described above and in Fig. 15.19 for BMC, using weigh feeders for the dry ingredients and metering pumps for the liquids. As was pointed out, practical systems are something less than fully continuous from bulk sources of all ingredients. Items used in small percentages, such as catalysts, thickeners, and release agents, are pre-batched into one or more of the higher bulk ingredients.

15.4.6. SMC Phase Two: Combining the Resin-Paste and Fiber

As was mentioned earlier, the compounding of both BMC and SMC is divided into two phases. It is in the second phase that the real differences arise.

15.4.6.1. SMC Machines and their Operating Principles

There are, at this time, three basically different kinds of SMC machines. These can broadly be divided into:

1. Standard
2. TMC
3. XMC

The so-called standard machine has a number of different versions, but the basic operation is the same in each. Two plastic film tapes or belts are continuously coated with a resin paste, and fiber reinforcement is continuously chopped and deposited on one of the paste-coated films. The films are brought together through a series of rollers in such a way as to trap the chopped fiber between them and to knead the resin paste into the fiber reinforcement. Finally, the whole is wound on a take-up roller (see Fig. 15.21). Other versions of this type of machine have chain belts instead of kneading rollers for working the resin paste into the fibers. Also, there can be provisions for incorporating continuous strands of fibers to provide particular mechanical properties to the products molded from the compound.

TMC (thick molding compound) is made in a machine that bears a superficial resemblance to a simplified, conventional SMC machine but is really more akin to a two-roll mill (see Fig. 15.22). The resin paste is contained between two contrarotating rolls. The fibers are chopped into the paste and mixed in by the action of the rolls. The mixed compound passes between the rolls and is deposited on and between plastic films which are further carried between a belt and roller conveyor that controls the thickness by a combination of roll position and belt speed. Both a thicker sheet—up to 2 in. (51 mm)—and a higher throughput are possible, in contrast to conventional SMC. High fiber content and directional fiber inclusion for high strength compounds, on the other hand, are not feasible. In addition to making a form of SMC, the TMC machine can also be adapted to make high strength BMC on a continuous basis.

An XMC machine, as such, does not really exist. XMC can be made on almost any filament winding machine. XMC is a form of SMC with continuous fibers arranged in an X pattern, providing very high strength in the dominant fiber direction but low strength in the transverse direction. To balance the

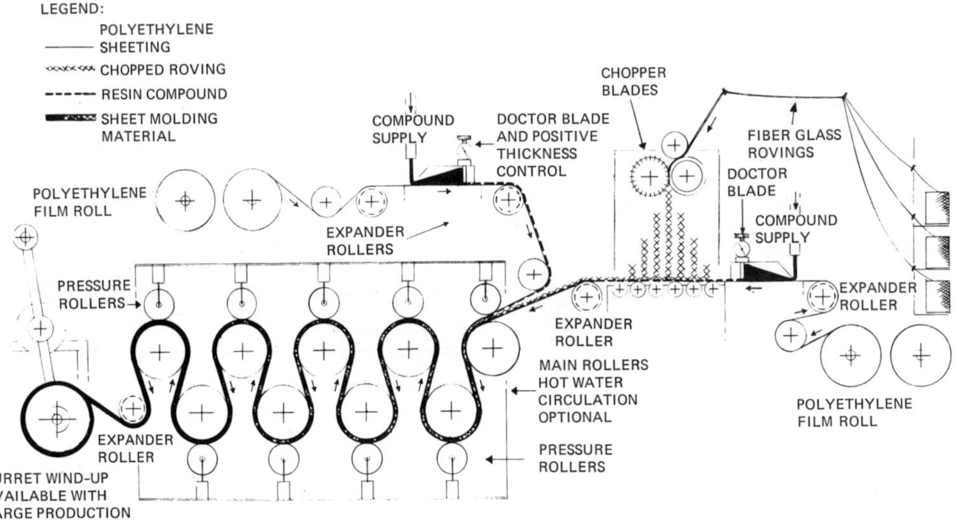

Figure 15.21. Schematic—E. B. Blue SMC Machine. (*Courtesy E. B. Blue Co.*)

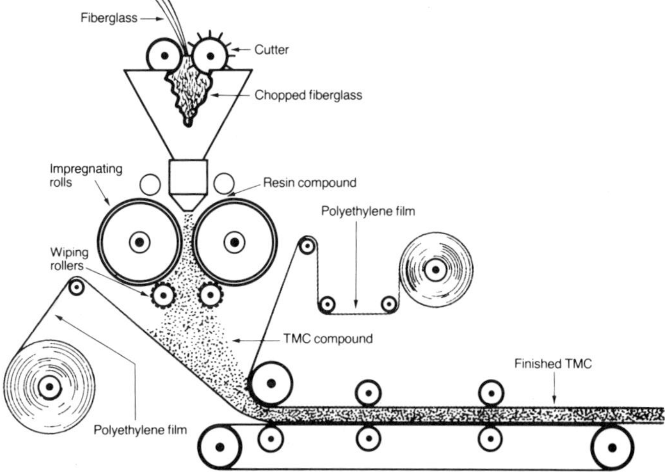

Figure 15.22. Schematic—TMC process. (*Courtesy Polyester Unit, USS Chemicals.*)

strength, some short random fibers can be added.

In its simplest form, XMC is made by drawing reinforcing fibers through a resin bath and winding them on a mandrel in some pattern so as to cover the mandrel (see Fig. 15.23). When the requisite thickness has been achieved—maximum about 3/8 in. (9.6-mm) on a 30-in. (762-mm) diameter mandrel—it is wrapped in a protective film, cut to remove it from the mandrel, and flattened for storage and maturation. Various winding angles can be used to control the directional properties, but most common is a relatively high angle (85°), with the addition of short chopped fibers to enhance the transverse strength. It is practical to add up to 60% of the total fiber content in short chopped fibers. XMC can be made with higher fiber content, lower resin content, and higher strength than any other reinforced molding compound. However, its production rate is low, 750 lb (341 kg) per hour versus 4000 lb (1818 kg) per hour for a conventional SMC machine, and there are substantial limitations on its general moldability.

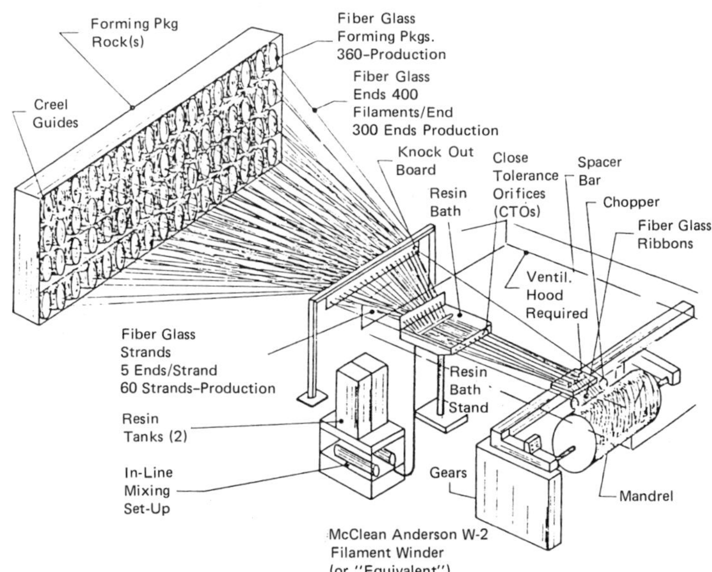

Figure 15.23. Schematic—XMC process. (*Courtesy PPG Industries.*)

15.4.6.2. Maturation

While not strictly a part of the compounding process, maturation is a key step between the raw material state and molding. It may be defined as the time between compounding and readiness for molding, perhaps more specifically as the time it takes for the compound to attain a non-tacky state to permit easy handling—removal of protective film, patterning without sticking to tools, etc.—and to attain a satisfactory molding viscosity. It usually takes too long and there is some uncertainty as to when the reaction is complete.

To minimize the variability, the maturation is carried out at a fixed temperature. To speed it up, it is usually at an elevated temperature. The compound is protected from loss of monomer, usually by wrapping each roll in aluminum foil, although sealed containers are sometimes used. Maturation time depends on the type and quantity of thickener used, which is further influenced by the compounding situation and the period of time over which the compound must stay in the moldable range. It can be as little as 24 hours to as much as seven days.

The foregoing applies mostly to compounds employing so-called group II metal oxides and hydroxides as thickeners. The ITP* polyurethane thickening process is not quite as sensitive, although it still responds better to controlled conditions.

15.4.6.3. Other Designations

In addition to SMC, TMC, and XMC, there are many sub-designations or alternate designations for what are basically the same kinds of products and are made on the same devices. These include:

- SMC-II, a low pressure molding SMC.
- SMC-R, usually with a numeral at the end, as SMC-R50, the R50 indicating 50% random chopped strand content.
- HMC, a designation for a high glass content SMC, usually without fillers.
- SMC-C, an SMC made with continuous fibers on a modified standard machine, sometimes designated SMC-C, R with numerals (SMC-C30, R20) indicating 30% continuous fibers and 20% random chopped fibers. Also made on a modified standard machine.
- SMC-D, the D indicating finite lengths of fiber arranged in one direction.
- XMC-3, an SMC with short chopped fiber as well as continuous strands.

*ITP is a trademark of ICI America Inc.

15.4.6.4. Compounding Facilities

This part of a molding plant is of the greatest concern to agencies that deal with environmental protection and employee safety. Insurance underwriters and building code authorities consider it to be a hazardous area that must be effectively separated from other parts of the plant—with fireproof walls, self-closing doors, etc. Explosion venting, sprinklers, and drains or dams at doors to prevent flow out of burning liquids are common requirements. A balance must be struck between exposure of the workers to noxious and toxic fumes and contamination of the outside atmosphere. While the entire space should be well ventilated and, ideally, the temperature and humidity controlled, each fume emitting unit should have its own specific ventilating outlet as close to the source of the undesirable emission as possible. Most of the likely emissions are best handled with small, high velocity ducts with an air velocity of 2500–3000 ft/minute (13 to 16 m/second).

15.5. MOLDING

BMC and SMC can be molded by compression, transfer, and thermoset injection methods. The largest volume is processed by compression molding.

The advantages of transfer or thermoset injection that apply to non-reinforced compounds apply only in part to reinforced compounds. However, the advantages of improved surface finish and precise dimensions offered by low profile, low shrink compounds can be more fully realized with injection and transfer molding techniques. Parts are free of flash and have closely reproducible dimensions (the mold is closed before it is charged; it cannot be held open by an overcharge). Mechanical properties are lower. The selection of the length and the quantity of reinforcing fibers is limited to the shorter and lesser. And, in most cases, a further reduction of mechanical properties can be expected due to the orientation of fibers as the material flows through sprues, runners, and gates. There is little to be gained in cure time by preheating in the transfer chamber as is experienced with, say, general purpose phenolics. In some instances, mold temperatures must be lowered and cure cycles lengthened because the transfer time is longer than compression mold closing time would be. There is obvious advantage in material handling and mold loading when lots of small items are to be made in a multi-cavity mold. For large parts, the greatest advantage seems to be in surface finish, where injection molding offers a distinct improvement. For most products, the substantial penalty of increased capital cost of equivalent capacity injection-transfer equipment over compression equipment does not seem warranted.

A possible alternative to gain the surface finish advantages of injection molding is the so-called "in-mold" coating process wherein the molds are partially opened at one point in the mold cycle and a paint-like material injected between the molded part and the mold surface.

Compression molding of reinforced compounds is not greatly different from molding conventional thermoset materials in principle. The main difference lies in the nature of the compound. Instead of free-flowing powders or neat, dry preforms, the compound is either a sticky, fibrous mass, a slug that must be cut, or a sheet from which the polyethylene must be stripped—each of which must be accurately weighed if the mold is to be filled completely or not be held open by an overcharge.

The rapid cure rate of polyester-based compounds is certainly a big advantage, yet it precipitates some molding problems. The time from placement of the charge to the final closing of the mold must be as short as possible to prevent premature gelation of compound in contact with the hot mold. Similarly, compound flowing in thin sections of a mold will tend to gel if the flow is not rapid. On the other hand, the flow rate can reach a very high velocity in thin sections, with consequent disturbance of the reinforcement. The result of some of these factors is wide variation in the strength of the finished molding.

The variability has been found to be of several kinds, which occur within:

1. Groups of identical pieces of the same or different batches.
2. Different parts of individual pieces.
3. Different kinds of pieces employing the same compound.

There is a tendency among molders and users of reinforced molding compounds to consider them as homogeneous materials, which they definitely are not.

In compounding BMC, a compromise must always be made between mixing long enough to distribute the fibers uniformly among the resin/filler ingredients and not mixing so much that it destroys the integrity of the fibers. The results of mixing are always imperfect. The fiber concentration, degree of interlocking, and fiber degradation all vary throughout the mix. Sheet molding compounds are subject to less variability and minimum damage to fibers in the compounding process, but they are not completely uniform either, and nothing that occurs in the molding operation does anything but compound existing non-uniformity. The first step, weighing the charge, is a likely cause of variability. The charge is usually brought to the correct weight by adding or taking away small wads of material. If wads are added, there is no interlocking of the fibers between these wads and the large mass, and planes of weakness result. If wads are taken away to bring the charge to correct weight they are usually thrown back into the "pot" to contribute their damage to later pieces in the run.

Shaping of the charge and loading the mold can have their influences, mostly on fiber orientation and knit or weld lines. The fibers tend to align themselves with the direction of flow in thin sections and at right angles to the direction of flow on going from thin to thick sections. It can readily be seen that a different location of the charge in the mold could change the fiber orientation.

Table 15.12. Troubleshooting Guide

PROBLEM: Mold is not filled.
DESCRIPTION: The mold is not filled at the edges.

POSSIBLE CAUSE	REMEDY
Charge weight is too low.	Increase the charge weight until the material appears at the telescoping edge.
Temperature is too high; compound gels before mold is filled.	Lower the temperature.
Closing time for the press is too long; compound gels before mold is filled.	Shorten the closing time.
Pressure is too low.	Increase the pressure.
Area of the charge is too small.	Choose a charge with a larger area.
Toolshift or deflection.	Move the charge toward the non-fill area.

PROBLEM: Mold is not filled.
DESCRIPTION: The mold remains unfilled at the edges in only a few spots.

POSSIBLE CAUSE	REMEDY
Charge weight is too low.	Increase the charge weight until the material appears at the telescoping edge.
The charge escapes before the mold is closed.	Place the charge more carefully.
Clearance of the telescoping edge is too large or the length of the telescope is too short and allows the compound to escape before the mold is filled.	Make the clearance smaller or the telescope deeper. (If the fault is minor, a higher temperature or excess material may help.)

PROBLEM: Mold is not filled.
DESCRIPTION: Mold is not filled in some spots, although the entire edge is filled.

POSSIBLE CAUSE	REMEDY
Charge weight is too low.	Increase the charge weight until the material exudes from the telescoping edge.
Air cannot escape from the mold.	Arrange the charge in such a manner that air cannot be trapped and so that the mat pushes the air ahead during the flow.
Blind holes or pockets make it impossible for air to escape.	De-aeration of the enclosures by a three-part construction of the mold or by bleeding air past ejector pins. (If the fault is small, increase of pressure may help.)

PROBLEM: Burning.
DESCRIPTION: Dark brown or sooty surface in places where the part is not completely filled.

POSSIBLE CAUSE	REMEDY
By compressing trapped air and styrene vapors, the temperature is raised to the ignition point.	Choose a charge which will not trap air, but pushes the air with it as it flows. Slow the closing speed. (If these brown spots appear on blind holes or pockets, they have to be de-aerated by a three-part mold construction or by venting around ejector pins.)

PROBLEM: Blisters.
DESCRIPTION: Round elevations on the surface of the cured part.

POSSIBLE CAUSE	REMEDY
Air entrapped between the layers of the resin mat.	Remove trapped air from the charge by prior compression. Decrease the area of the charge so that air can escape better. Increase the molding viscosity.
Too high a mold temperature (monomer vapors).	Lower the mold temperature.

Table 15.12. Continued

Curing time is too short (monomer vapors).	Increase the curing time.
Unwetted glass or air is in compcund.	Increase SMC compaction.

PROBLEM: Blisters.
DESCRIPTION: Round elevations on the surface of the cured part of the heavy section.

POSSIBLE CAUSE	REMEDY
Only with heavy wall thickness. Internal stress tears the laminate between the individual layers.	Decrease the area of the cut piece so that the glass fibers of the various layers mesh better. Lower the mold temperature.
Weak spot along a knit line.	Change of the catalyst system.
Decrease in strength in one direction in spots with extremely long flow paths (glass fiber orientation).	Shape the charge in such a way that no knit line can form. Shorten the flow path by increasing the area of the charge.
Damage during removal from the mold caused by:	
a. Undercuts (unintentional).	a. Remove undercuts.
b. Ejection pins with too small an ejection area.	b. Increase the ejection area.
c. An insufficient number of ejection pins.	c. Increase the number of ejection pins.
d. Sticking to the mold.	d. (See "sticking.")
e. Incomplete curing.	e. Increase the curing time or temperature.

PROBLEM: Internal cracks.

POSSIBLE CAUSE	REMEDY
Only with heavy wall thickness. The laminate cracks because of strong shrink stress between the individual layers.	Decrease the area of the charge so that the glass fibers of the various layers mesh better. Lower mold temperature. Change the catalyst.

PROBLEM: Sticking.
DESCRIPTION: It is hard to remove the finished part from the mold. In some spots, the material sticks to the mold.

POSSIBLE CAUSE	REMEDY
Mold temperature is too low.	Increase the mold temperature.
Curing time is too short.	Increase the curing time.
SMC was unpacked too long. With rolls of SMC, open only the outer layers.	Keep rolls sealed in Barrier film until used.
Mold is not broken in. The mold is new or has not been used for a long time, or has been used to mold a different material.	Use a mold release on the first few moldings.
Mold surface is too rough.	Polish the surface.
Non-chrome surface.	Increase the release level in the compound.

PROBLEM: Sticking.
DESCRIPTION: The cured part is hard to remove. In spots, material sticks to the mold. At the same time, pores and scars show at the surface.

POSSIBLE CAUSE	REMEDY
Area of the charge is too large. Air on the surface cannot escape due to short flow path. Trapped air delays cure.	Decrease the area of the charge. Add a small charge on top of the larger charge.

PROBLEM: Hanging.
DESCRIPTION: Cured parts are hard to remove. No apparent sticking to mold.

POSSIBLE CAUSE	REMEDY
Insufficient shrinkage.	Change the resin formula.
Sticking in flash area.	Increase the internal mold release. Clean and wax the flash area.

Table 15.12. Continued

PROBLEM: Surface porosity.
DESCRIPTION: If these pores are numerous, the part is difficult to remove.

POSSIBLE CAUSE	REMEDY
Area of the charge is too large. Air on the surface cannot escape because the flow path is too short.	Decrease the area of the charge. Add a small charge on top of the larger charge.
Pre-gel.	Decrease the temperature or close more quickly.
Unwetted glass or air is in compound.	Increase compaction.
Low viscosity compound.	Increase maturation viscosity.

PROBLEM: Mold abrasion.
DESCRIPTION: Dark to black spots on the surface of the cured part.

POSSIBLE CAUSE	REMEDY
Abrasion from the mold.	Chrome-plate the mold. (Nickel plating will not help.) Change to softer filler and/or pigments. NOTE: Titanium dioxide is a major contributor to this problem. Replacement by zinc sulfide (SC33-33) is recommended.

PROBLEM: Warpage.
DESCRIPTION: The part is slightly warped.

POSSIBLE CAUSE	REMEDY
Warpage is due to shrinkage during hardening and cooling.	Cool the part in a jig. Employ low-shrink or zero-shrink resins in compounding.
One mold is much hotter than the other mold.	Reduce the temperature differential of molds.
Unbalanced construction	Put extra reinforcing mat in the center rather than on the surface.

PROBLEM: Warpage.
DESCRIPTION: The part is badly warped.

POSSIBLE CAUSE	REMEDY
Warpage is due to glass fiber orientation caused by particularly long flow path.	Shorten the flow path by increasing the area of the charge. Employ low-shrink or zero-shrink resins in compounding.

PROBLEM: Wavy surface.
DESCRIPTION: Waves are found on long, vertical, thin walls at a right angle to the direction of flow. Also, with other adverse flow conditions (large differences in wall thickness), an irregularly wavy surface may occur.

POSSIBLE CAUSE	REMEDY
Complex design interrupts uniform flow.	In most cases this cannot be eliminated completely. Improvement can be obtained by: 1. Increasing pressure. 2. Changing design of mold. 3. Altering position of charge. Employ low-shrink or zero-shrink resins in compounding.

PROBLEM: Sink marks.

POSSIBLE CAUSE	REMEDY
Non-uniform shrinkage during molding.	Employ low-shrink or zero-shrink resins in compounding. Increase the temperature of half of the mold. A difference of 10°F is usually sufficient. Shorten the length of the chopped fibers. Change the mold design. Alter the position of the charge. Narrow the clearance of the telescoping edge.

Table 15.12. Continued

PROBLEM: Erosion of cut-off on mold.
DESCRIPTION: Metal breaks off in the direction of applied thrust.

POSSIBLE CAUSE	REMEDY
Inaccurate or weak guide pins.	Provide accurate mold guidance (heel blocks). Strengthen guide pins. Place the charge to minimize side thrust.
Improperly set mold.	Adjust positioning.

PROBLEM: Laking.
DESCRIPTION: Areas of low gloss on cured part.

POSSIBLE CAUSE	REMEDY
Lack of follow-through pressure.	Increase pressure; mold off stops both internal and external.
Mold contamination.	Clean and condition.

PROBLEM: Dull surface.
DESCRIPTION: Surface is not shiny enough.

POSSIBLE CAUSE	REMEDY
Pressure is too low.	Increase the pressure.
Mold temperature is too low.	Increase the mold temperature.
Unsatisfactory mold surface.	Repolish and chrome-plate mold.
Undercure.	Increase the cure time.

PROBLEM: Flow lines.
DESCRIPTION: Local waviness on surface.

POSSIBLE CAUSE	REMEDY
Mold closure is improperly designed or is damaged.	Follow recommendations on tool design.
Mold temperature is too low.	Increase the temperature.
Glass fiber orientation in places with extremely long or adverse flow paths.	Shorten flow paths by increasing the area of the charge.
Mold shifting causing excessive pressure drop at one edge.	Improve mold guidance.

While molding pressures for reinforced molding compounds are low when considered in the light of all molding materials, they do range up to 3000 psi (20.7 MPa), even though some very soft-flow BMC's can be molded at 100 psi (0.69 MPa). The minimum pressure indicated by vendors of compounds is for very shallow parts of little complexity. Deep, complex items can easily require 1000 psi (6.9 MPa) with soft-flow compounds.

There are many problems encountered in molding reinforced compounds. In Table 15.12 are listed most of the difficulties and/or defects that may be encountered in molding reinforced compounds, with possible solutions. If the reader comes across problems that are not listed, or that do not respond to the suggested solutions, we hope he or she will find the answers and make a contribution to the state of the art.

15.6. MOLD CONSTRUCTION

While it is frequently stated that reinforced molding compounds can be molded in equipment and molds designed for other compression molding materials, the claim really applies only to switching from another material to BMC or SMC in existing equipment. When starting from scratch on a new product, initial mold costs and molding problems can be reduced by designing for reinforced molding compounds. Reinforced molding compound materials vary widely in their molding properties, but in general, their molding pressures are above those of preform or mat-reinforced plastics and below general purpose phenolics, in the range of 500–1000 psi (3.45–6.9 MPa) and occasionally up to 2500 psi (17.3 MPa)—with many items (having no extra-special requirements) falling in under the 500-psi (3.45-MPa) category.

15.6.1. Mold Material Selection

Molds for large parts are machined from a selection of steels, including AISI 1045 and 4140, as well as a special mold steel, P20. Some molds, where slight surface porosity may not be of serious concern, can be cast of Meehanite and other casting steels, with considerable cost saving over machining from steel billets. Cast-in steam cavities can provide better and quicker heat transfer than do drilled steam passages. For medium to smaller molds, pre-hardened steels (32–35 Rockwell "C" scale hardness) are a good choice. They can be freely machined and polished to a high finish. For small, high production molds, air hardening tool steels, easy to machine in the annealed state, are readily heat treated to 50–55 Rockwell "C" with a minimum of distortion. The resin seal area should be flame hardened to 50–52 Rockwell "C" (see Fig. 15.24).

15.6.2. Mold Finishing

The very high polish used on some molds for thermoplastics does not seem warranted for reinforced compounds. The filler and fiber appear to set a limit on gloss regardless of the polish of the mold in which they are made. A so-called 600 finish will do for most molds. The 600 refers to the grit fineness of the abrasive used in the last step in mold finishing between machining and buffing. Successively finer grits of abrasive materials are used in approximately 100-grit intervals, starting with a sufficiently coarse grit to just remove the machining tool marks. Hard chrome plating from 0.0003–0.001 in. (0.00001–0.00004 mm) thickness protects the mold surface from corrosion and minor abrasive damage and promotes release of the part. A high gloss which accentuates surface irregularities can be masked by a uniformly dull finish imparted to molds by the liquid honing or vapor blast processes used to remove heat treat scale.

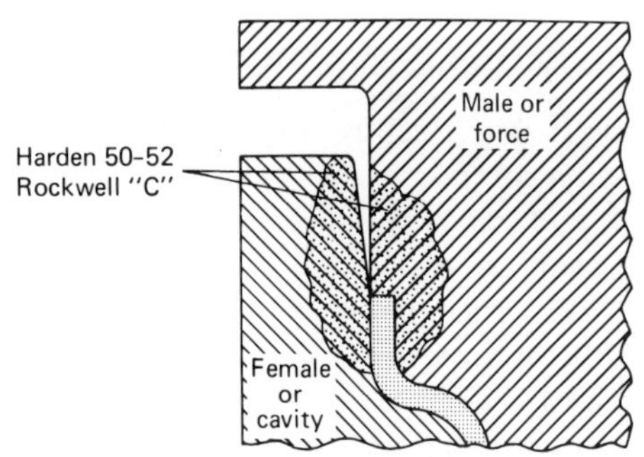

Figure 15.24. Resin seal detail.

15.6.3. Mold Construction

General construction is the same as that for phenolic compression and transfer molds, except for taking into account that lower molding pressures permit lighter sections and use of aluminum alloys for pillars, ejection bar spacers, and other supporting parts for better heat transfer when molds are heated from platens.

15.6.4. Mold Seals

Most compression molded parts are made in one or the other of two types of molds that are called positive or semi-positive. In the positive mold, the seal or juncture of the male and female mold sections at the edge of the molded piece (see Fig. 15.25*a*) telescopes with only sufficient clearance to permit the escape of air and not the molding material. This type of mold, while very critical as to charge weight if pieces of uniform size are to be made, compresses the material to its maximum density and has, in some instances, been found to be the only satisfactory way to mold parts meeting critical electrical or mechanical requirements. The positive mold is most commonly used with an external land, which, when combined with a slight (perhaps 1%) overcharging permits molding dense parts with a minimum size variation. In a semi-positive

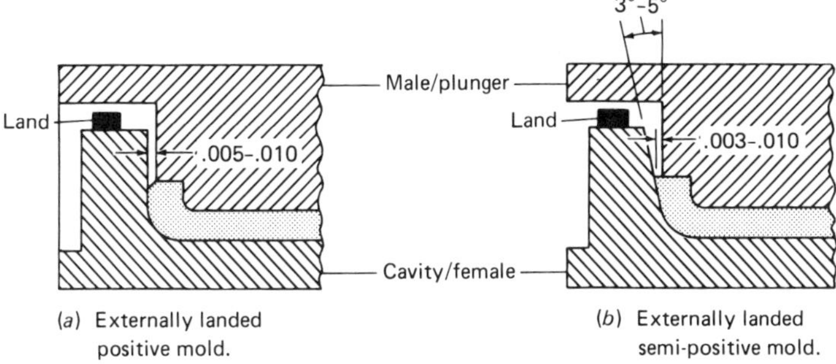

(a) Externally landed positive mold.

(b) Externally landed semi-positive mold.

Figure 15.25. Details of mold seals.

mold, the female or cavity section is relieved with a taper (see Fig. 15.26*b*) such that the excess material can escape so the molds close, yet is increasingly restricted until the closure is positive over a final short length of travel. This type of seal is not sensitive to overcharging in the range of 2–5%, yet provides parts of uniform size, weight, and density. There are, of course, numerous variations of these two types of mold seals to obtain horizontal flash or accomplish other purposes.

Injection and transfer molds generally do not telescope. The male and the female come together in a horizontal plane (see Fig. 15.26). Since the mold halves go together before the mold is charged, there is no need to provide relief for excess material. Since it is practically impossible for the two surfaces to meet exactly, there is usually enough space to vent air that is trapped in the compound. For particularly difficult venting problems, local areas may be ground away a few thousandths of an inch.

Venting of blind pockets has been successfully accomplished by the use of:

1. Three-part molds; and
2. Special vent pins.

Deep draw parts with blind pockets at the bottom can be vented by using three-part molds which provide a second parting line at the bottom of the mold. Although extra mold expense is incurrèd, the sections so vented will be void-free. Vent pins can be operated by the ejector plate (and will also act as ejector pins). The stroke must be long enough to expose the reduced diameter of the pin so that it can be completely cleaned (see Fig. 15.27).

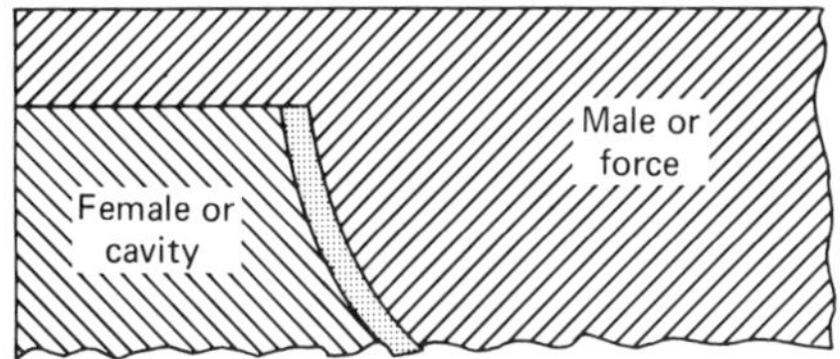

Figure 15.26. Detail of seal—flash type mold.

15.6.5. Nonferrous Mold Materials

Molds can be constructed of materials other than those mentioned above, but it is the author's opinion that none but aluminum are useful except for the most limited prototype work. Molds machined from low copper aluminum alloys and subsequently anodized and impregnated with a fluorocarbon can be very satisfactory for limited production. This process, commercially know as "Tufram,"* provides excellent release and a non-marring surface with a Rockwell hardness up to C-70.

Prototype molds can also be made of plastic materials and of a zinc alloy, "Kirksite." The

*General Magnaplate Corp., Linden, N. J.

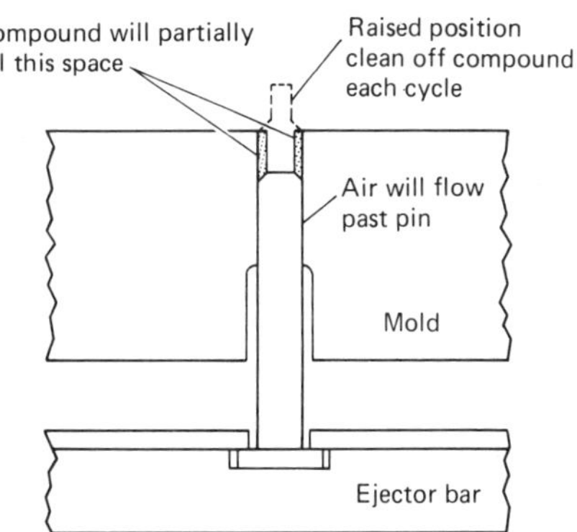

Figure 15.27. Vent pin detail.

cost is usually at least 50% of that of a steel mold. Since the technology of reinforced plastics is sufficiently advanced that the product performance can be reasonably predicted, there seems no justification for such expensive prototyping. Test pieces for marketing analysis or other such purposes can be made by hand lay-up or a similar low tooling-cost process.

15.6.6. Release and Ejection of Parts from the Mold

The first step in effective mold release occurs during the part design. Every item, regardless of its material of construction, must be designed with respect to its manufacturing process, and removal of parts from the mold is a very critical facet of the molding process.

Step two is deciding in which half of the mold the part should remain as the mold opens. Obviously, the part should remain with the mold half that permits the more ready handling of the part by the operator(s).

Step three is choosing the method for keeping the part in the desired half of the mold. One or the other of the following can usually be used.

1. Incorporate areas in the part without draft.
2. Provide shallow undercuts, preferably in large flat areas. They will keep the part in place as the mold opens but will spring out when the part is ejected.
3. If all the part surfaces are critical, undercuts in the flash can sometimes keep the part in the desired place. The flash may have to be thickened for this method to be effective.
4. Provide follower pins (similar to ejector pins), actuated for a short distance as the mold opens to force the part to stay on the desired mold half.

Step four is devising the means of getting the part out of the mold once it is effectively retained. If it is a simple shape and is retained in the cavity, the cure shrinkage plus the thermal shrinkage as the part cools will often release it so that it can be picked out by hand (with suction cup assist if there are sufficiently large, flat areas). Some parts can be blown out with an air blast around the periphery. Air can also be used effectively, if in a somewhat more complex way, by incorporating in the mold spring-loaded poppet valves connected to air passages. Air pressure will lift the valve slightly and allow the air to flow in and around the part to eject it.

Most common and effective are ejector pins of the same type as those used in conventional compression molding. The greater strength of the reinforced plastics usually permits far fewer pins, but its generally lower hot strength dictates larger diameter pins. Straight pins with close clearance work well in BMC and SMC, but the hole should be relieved from a point 1½–2 diameters below the mold surface to provide space for material that may flow past the pin.

15.7. MOLDING PRESSES

There are several kinds of presses made for diverse purposes which have been adapted to reinforced plastics molding with varying degrees of success. Most common and satisfactory is a four-column, upward or downward moving platen, direct acting hydraulic press (see Fig. 15.28).

Because of increasing demands for rigidity and parallelism, welded frame presses are becoming more popular (see Fig. 15.29). These have four rectangular parallel ways on which the moving platen rides. Adjustable wear plates (gibs) on the moving platen permit very close control of its parallelism. Disadvantages compared to round column presses are that, for the same useful die space, they are larger, and access for mold changes, press loading, and so on is inhibited.

It is only in recent years that the major press manufacturers have listed stock designs for reinforced plastic molding. New entrants into press molding frequently try the second-hand market, where there is little available specifically constructed for reinforced plastics.

Some of the more important factors in selecting a press are discussed below.

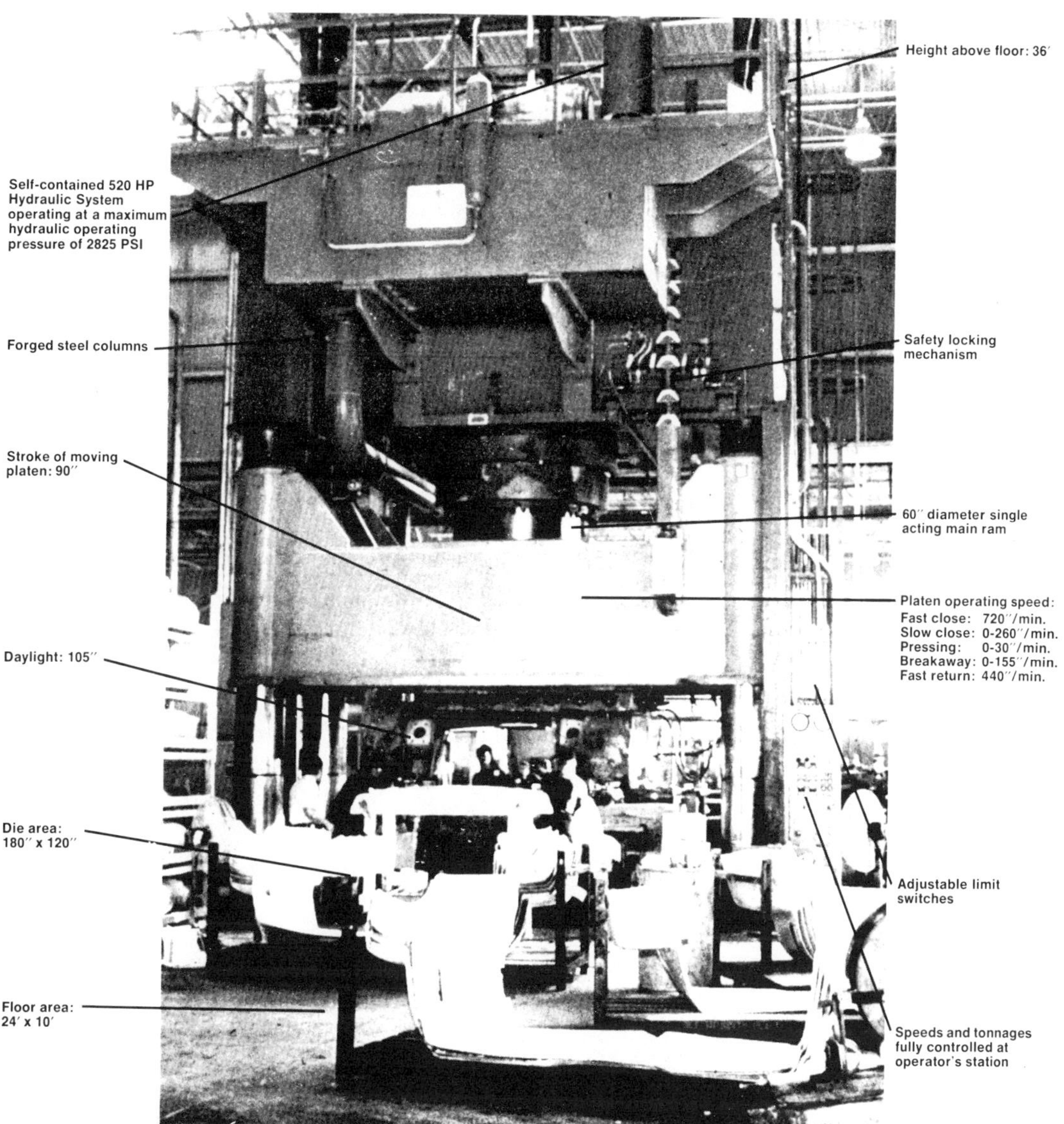

Figure 15.28. Column type press. (*Courtesy Erie Press systems.*)

Figure 15.29. Welded frame press. (*Courtesy Erie Press Systems.*)

15.7.1. Pressing Capacity

A press purchased for BMC and SMC should ideally be able to apply 4000 psi (27.6 MPa) on the part. This would require a 200-ton press for a 10 × 10 in. (2.5 × 2.5 cm) molding. Since practical molding pressures range downward to 500 psi (3.45 MPa), the platen area should be sufficient for a 25 × 25 in. (64 × 64 cm) molding.

15.7.2. Breakaway or Return Force

This requirement is sometimes overlooked. Single acting gravity return presses satisfactory for some kinds of molding cannot be used. (20–25% of the pressing capacity is common; this may be attained with the use of auxiliary cylinders or with a double acting main cylinder).

15.7.3. Stroke and Daylight

These two features in combination determine the depth of the part that can be molded. The daylight must be three times the depth of the largest part, plus allowance for mold thickness, knockout mechanisms, heating platens, and so on. The stroke must be at least twice the depth of the largest part. While it can be longer to permit easier handling of shallow parts, it is more economical to absorb the extra daylight with bolsters (spacers) under the smaller molds, or with a ram extender (a spacer that fits between the moving platen and the ram). Some presses have movable strain heads to vary the daylight (see Fig. 15.30).

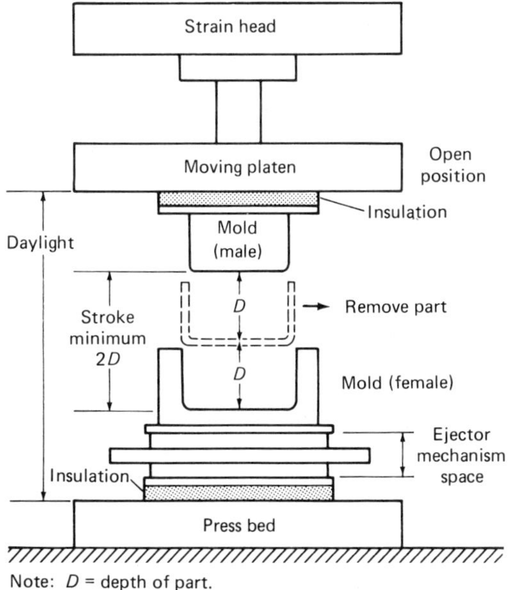

Figure 15.30. Mold space requirements—stroke and daylight.

15.7.4. Mold Space

The mold (or die) space is often given as the space between strain rods one way (say, left to right) and the dimension of the platen in the other direction (front to back). This is misleading, since the most common way of securing molds to platens is with bolts and "dogs," which use up some of the area. Most molds can be secured adequately from two opposite sides so a minimum mold space should be about 12 in. (30.5 cm) greater than the largest mold in one direction (either front to back or left to right).

15.7.5. Operating Speeds

That the speed should be as great as possible consistent with cost and technical requirements is axiomatic. In that part of the cycle where the molds are not yet engaged (closing or opening), the minimum time is desirable. In other phases, the nature of the molding usually limits the speed.

15.7.6. Closing

Two speeds are required and three are preferred—a rapid advance until the molds are engaged, an intermediate, and a slow final pressing speed. Typical speeds of some late model presses are:

Rapid advance—400 in./minute (10 m/minute)
Intermediate—100 in./minute (2.54 m/minute), adjustable.
Final press—0–15 in./minute (0–0.38 m/minute), adjustable.

15.7.7. Opening

Three speeds are required: a slow breakaway, a rapid return, and a slowdown before the ejector mechanism operates. The first and third are usually about the same as the final pressing, and the second corresponds to the rapid advance speed.

15.7.8. Controls

Most presses are semi-automatic. The operator starts the cycle with dual pushbuttons after loading the mold. As the platen moves through its stroke adjustable cams contact limit switches to effect the changes in speed from rapid advance to intermediate to final pressing. When the pressing tonnage is reached, a pressure switch starts the cure cycle timer. When the timer runs out, the return phase of the cycle is automatically started with a slow breakaway, a rapid return, a slowdown near the end of the return stroke, and a return to the initial position of the platen.

For die set-up and other purposes, a selector switch to manual control permitting upward and downward movement at the same rate as the final pressing speed should be provided. Presses are available with manual control at a saving of 10–15%, but in the author's opinion, it is an ill-advised economy even on laboratory or experimental equipment.

15.7.9. Upward or Downward Acting

Upward acting presses of even moderate size require pits if the platen is to be at a convenient height for the worker. Access must be provided in the way of removable floor panels. The hydraulic pump must be mounted separately, using up additional floor space. Downward acting presses can be mounted directly on the floor. It is usually possible to mount all the hydraulic apparatus on the strain head. A minor disadvantage is a hole in the roof of a low ceilinged plant, with a kind of penthouse around the top of the press. This usually costs no more than a pit, and an unanticipated advantage is the ease of lowering even a large press through a hole in the roof with a crane, versus the difficulty of erecting a press inside the building.

15.8. DESIGN

General principles of design for any kind of molding apply to reinforced molding compounds. Almost any shape can be molded if the cost and complexity of the mold are not factors. As mentioned earlier, thick and thin sections can be incorporated in the same molding and the examples in the applications section show great complexity in many of the details. However, generally abrupt changes from thick to thin sections should be avoided.

Square corners can be made, but generous radii are more desirable. Extremely thin sections, less than 1/16 in. (1.6 mm) can be a problem, depending on the molding direction. Molded holes in the direction of molding are easy to accomplish but some loss in strength from knit lines may occur as the compound flows around the pins making the holes. Draft requirements are small compared to metal castings, and occasionally the minimum draft can be obtained from tolerance on overall dimensions. However, good practice is to get all the draft the application will permit. Threaded metal or other inserts can be molded in with some of the same problems as molding holes. Generally, inserts that are molded in resist torque and axial pull-out better than those inserted after molding.

Some special design consideration must be given to reinforced compounds because of their low elongation or extensibility. In terms of metal characteristics, their yield and ultimate strengths are identical. This means that a comparatively small distortion (even though at a high stress level) will result in breakage of a part. To cite an example: a 1/8-in. (3.18-mm) thick × 1 in. (25.4-mm) wide flexural test specimen with a 20,000-psi (138-MPa) flexural strength will break at an applied load of 25 lb (172 KPa) a 1/16-in. (1.58-mm) thick specimen at about 6 lb (41 KPa), even though the full unit stress of 20,000 psi (138 MPa) will be developed. It can thus be seen that comparatively slight forces applied in handling such items with exposed thin sections during removal from the mold, or in subsequent assembly operations, etc., might result in damage even though the part were not subject to such forces in its ultimate application.

The general answer to the problem is to make all sections of any BMC or SMC part thick enough or contoured in such a way that the likely applied local forces will not result in stresses above the ultimate. Large, flat areas should be divided up with convolutions or ribs. Edges of parts should be flanged or increased in thickness. If a part has a mounting flange or extending feet with bolt holes, generous gussets should be provided extending from the adjacent wall to beyond the bolt holes.

It might be pointed out that increases in thickness cause very little more than an increased material cost, and sometimes the material added in one place can be removed from another less critical section of the part. While increased thickness does have its effect on the molding cycle, it is slight as compared to increasing the thickness of a thermoplastic part, for instance, which not only will result in a disproportionately longer cycle, but may even require the use of a larger molding machine.

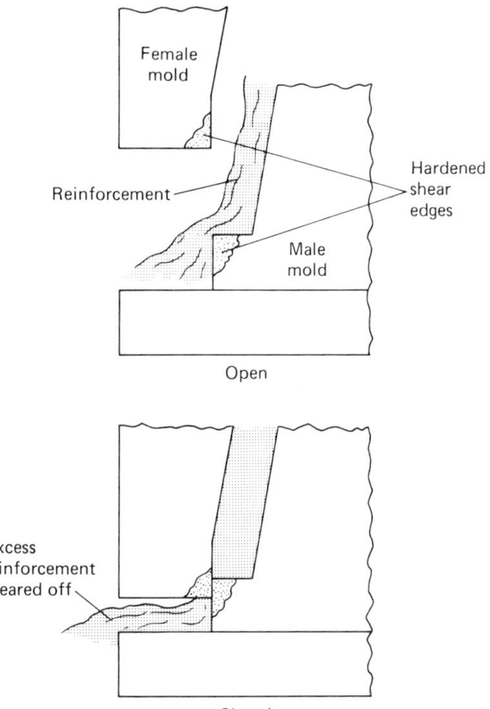

Figure 15.31. Preform mold—shear action.

15.9. MAT AND PREFORM MOLDING

Mat and preform molding are essentially the same insofar as the molding techniques are concerned, differing only in the pre-molding preparation and in the intricacy of products attainable.

The reinforcement, preform, or mat, is made or patterned to cover a bit more area than the entire mold. The resin (with fillers, pigments, release agents, and catalyst already mixed in) is applied to the reinforcement, usually after it is placed in or on the mold. Closing the mold does two things:

1. Forces the resin mix to flow and fill and surround the fibers of the reinforcement.
2. Shears off the excess fiber at the point of mold engagement (the shear edge or seal), at the same time trapping the resin mix until it cures under heat and pressure (see Fig. 15.31).

Preform molding, developed about 1949 to solve the problem of molding a complex-shaped washing machine basket, enabled reinforced plastics to compete with steel and aluminum in the medium volume production level for the first time. In not too many years, sports cars (Chevrolet Corvette), truck front ends, tractor hoods, tote boxes, food trays, furniture, boat hulls, and a host of other items were in regular production. In some instances, preform moldings offered properties not attainable in other materials, usually corrosion resistance, light weight, low tooling cost, low capital equipment cost, and pleasing appearance. Substantial deficiencies existed especially with surface quality and paintability. Preform molding was in the forefront of the expansion of the use of reinforced plastics until the advent of SMC in the early 1960's. SMC has displaced preform in many instances, especially in complex parts, but preform molding is still a useful process, even though it has limitations with respect to complexity of detail. Bosses, ribs, abrupt variations in thickness, inserts, and so on are all difficult. When a product can be designed around the limitations, preform matched die molded items are at the high end of the strength-to-cost ratio list. Uniformity of mechanical properties throughout is excellent since the reinforcing fibers are evenly distributed over the whole surface of the molding. Where preform is a viable design alternative to SMC, it is possible for it to cost 10% less.

15.9.1. Properties

As with reinforced molding compounds, the mechanical properties of mat and preform moldings are largely influenced by the amount and kind of reinforcement. Where continuous strand glass mat can be used, it provides the highest strength, but its use is limited to parts with simple shapes and shallow draws.

The properties of chopped strand mat and preform reinforced moldings are essentially the same and about 20% less than those reinforced with continuous strand mat. The practical range of glass fiber content in mat and preform molding is 25–50%, with most falling in the 25–35% range. Table 15.6 includes some properties of preform and continuous strand mat reinforced moldings.

While mechanical properties of mat and preform moldings are high compared to SMC and BMC, properties influenced by the filler, such as electrical insulation and fire resistance, are usually inferior. Since the resin/filler mix must flow through and wet the reinforcement as the mold closes, its maximum viscosity, and therefore its filler content, is limited. The different methods of wetting the reinforcement in BMC and SMC processing readily permit as much as 50% filler versus a maximum of 35% with preform or mat, each with 25% fiber reinforcement. It is also practicable to make BMC and SMC with lower fiber contents (less than 25%) and higher filler content.

15.9.2. Applications

Potential and existing applications include any item of simple or complex shape that requires corrosion resistance, electrical and heat insulation characteristics, high strength-to-weight ratio, and toughness. Products include automobile, truck, tractor, and other modestly high volume transportation products, underground electrical junction and transformer boxes, forms for casting concrete, furniture, bus and subway seating, high production small boat hulls, and many other products.

15.9.3. Preform Process

This is a method for collecting chopped fibers in the shape of the item to be molded and retaining them in that shape until they have been effectively impregnated with the molding resin. The process involves collecting chopped fibers on a screen that is made in the shape of the part. A high volume of air flowing through the screen draws chopped fibers onto it, with a relatively even distribution. A resin binder, usually in an aqueous solution, is sprayed on the fibers to hold them to the shape. The emulsion is dried or cured and the preform is then removed from the screen and placed on the mold. Approximately 5% of the solid binder based on the preform weight is usually sufficient to give the required bond, but this figure varies with the shape and size of the preform. The glass is purchased in the form of continuous roving wound on a spool. These rovings are sent through a roving cutter to be chopped into lengths varying from 1/2–3 in. (12.7–76.2 mm) depending on the machine used and the final application. Combinations of various fiber lengths may be used for better contour control.

For deep draw, relatively straight sided moldings, the preforms must be very compact; otherwise, they will be damaged by the shear edge of the cavity when the mold closes. High air velocity and consequent high horsepower

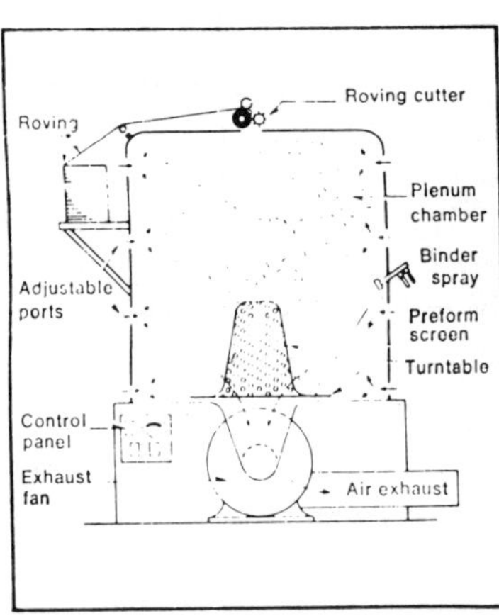

Figure 15.32. Plenum chamber type preform machine.

is required to make compact preforms. The possible thickness of parts made with preforms is limited by available suction of the preform machine. In most instances, 1/4 in. (6.5 mm) is the maximum. This will require about 3000 CFM (85 m^3/minute) and about 6 hp/ft^2(4.5 kW/m^2). For greater thickness, it is possible to make two preforms and stack them. This does mean two different sizes of preform screens. If the greater thickness is required only in a localized part of the molding, the preform can be augmented with fiberglass mat.

There are two basic types of preform machines: the plenum chamber type and the directed fiber type. The plenum chamber machine is shown in Fig. 15.32. The rovings are fed into a cutter on top of the plenum chamber. Chopped strands are directed onto a spinning fiber distributor to separate the strands and distribute them uniformly in the plenum chamber. Falling strands are drawn onto the preform screen by suction and the resinous binder is sprayed on. The screen is usually mounted on a rotating turntable for better distribution of deposited glass. After sufficient glass has been deposited, the preform is placed with the screen into an oven, where it is cured and dried. The preform is then removed from the screen and the screen is returned to the plenum chamber. The process can be mechanized so that the screen is moved directly into the oven while another screen moves into the plenum chamber.

In the directed fiber process, air carries a stream of chopped glass directed by the operator onto a perforated preform screen (see Fig. 15.33). The air inside the screen is exhausted by a powerful fan, creating a suction which draws the chopped glass to the screen and holds it there. The binding resin is sprayed on simultaneously with the glass from a separate resin spray gun. The operator manipulates the stream of glass to suit the contours of the preform screen, building up thicker or thinner deposits as required. The preform screen usually rotates to ensure that each surface is brought into the range of the blower. The roving cutter can be preset to turn off when the proper amount of glass has been deposited. The rate of deposition depends on the chopper but is usually 1 lb/minute (0.4 kg/minute). After the stream of glass is turned off, the operator continues to spray the binder to ensure thorough wetting of the glass by the resin. The operation may be interrupted to add additional mat layers where required for build-up, and then more chopped glass may be sprayed on. The preform with the screen is then placed in an oven to remove the water from the emulsion and cure the resin.

The preform can also be dried by rotating the whole turntable set-up into an oven as shown in Fig. 15.34. Here a three-step opera-

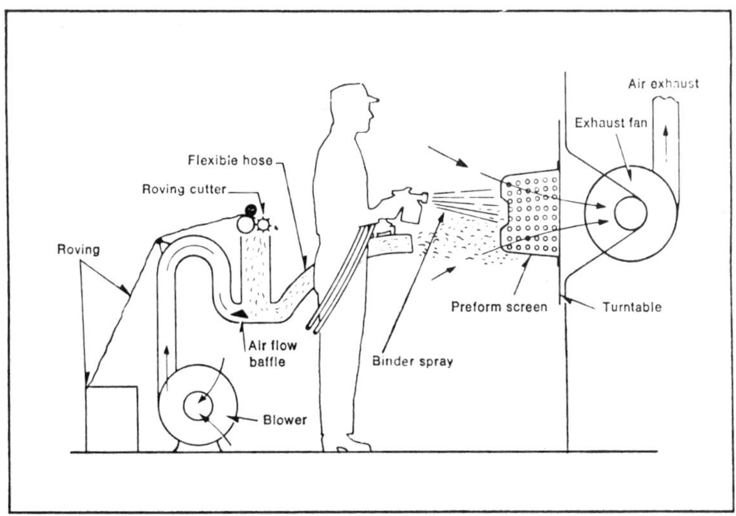

Figure 15.33. Directed fiber preform machine.

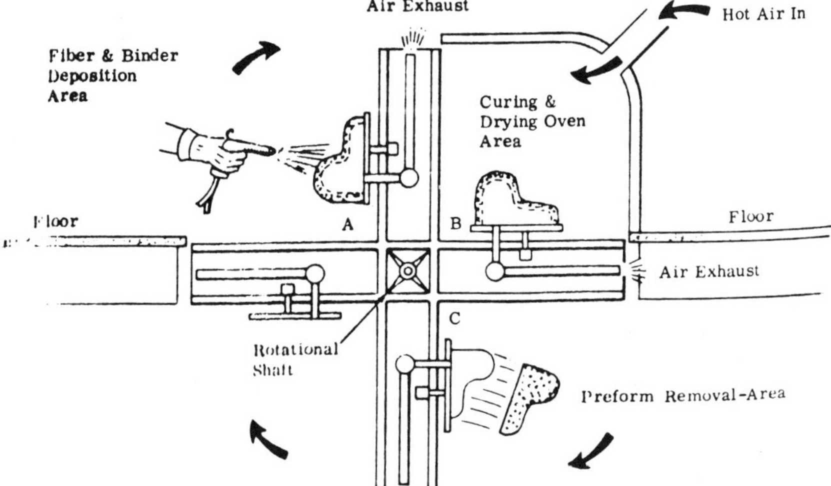

Figure 15.34. Mechanized directed fiber preform machine.

tion is accomplished—step A is the deposition of glass, step B is the cure, and step C involves the removal of the dried preform and cleaning of the screen. A parting agent is usually applied to the screen to prevent sticking.

15.9.4. Preform Screens

Preform screens are fabricated from either a strong wire mesh or perforated metal. In the latter case, 16–18-gauge mild steel sheet is usually used, with approximately 40% of the area perforated. The holes may be 1/8 in. (3.18 mm) in diameter on approximately 3/16-in. (4.8-mm) centers. Other hole sizes and spacings may be used if desired. The shape of the screen will determine the actual hole spacing. If the part is round or ogive shaped, no problem will be encountered in hole spacing. If the part is rectangular or complex, larger holes will be required at the corners and edges to ensure even deposition of the fibers. Frequently, the perforations are not sufficient and internal baffling may be required. In most cases, the preform screens reproduce the shape of the male die.

TFE emulsion can be baked on, to form a permanent non-stick surface that requires no other release material. Otherwise, the screens should be coated with a release material, such as silicone wax, or polyethylene spray. Screens should be cleaned periodically to remove excess release material and caked on resin.

15.9.5. Preform Binders

Preform binders are an important part of the preform process. The binder must hold the preform together through the handling steps to the mold, yet it must not inhibit the wetting of, or the bond of the molding resin to, the fiber. It must hold the fibers in place against displacement as the molding resin flows through the preform at a high velocity. Most commonly used are water emulsions of a high reactivity polyester resin with a catalyst and possibly a wetting agent. Because the water must be evaporated, the oven curing time is longer than that required to cure the resin alone. As a consequence, some molders substitute a solvent carrier for the water, which adds to the ventilation problem. Dry powder polyesters that melt in the curing oven heat, as well as acrylic and polystyrene emulsions, have also been used. Whichever binder is used, the amount falls in the range of 5–10% of the preform weight.

15.9.6. Molding

When chopped strand mat is used, it is usually patterned in some way, or even made up into a preform in the shape of the part and held in shape by stitching or by a small amount of adhesive. Continuous strand mat has more ability to conform to the mold shape. Unless the part is very complex, continuous strand mat is usually used without patterning other

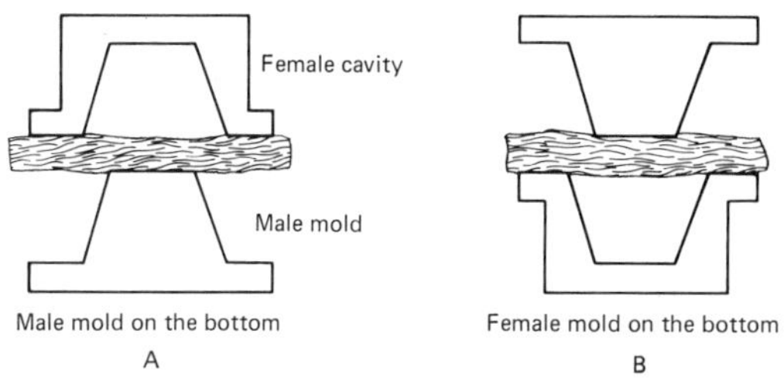

Figure 15.35. Matched molds using mat.

than cutting to the peripheral shape required. In most instances, the mat materials are molded with the cavity (female) down (see Fig. 15.35). The bottom part of the male mold pushes the reinforcement into place with a minimum of disturbance or tearing.

Most preforms are loaded onto the male mold, as shown in Fig. 15.36. This usually minimizes the disturbance of the preform in most configurations. However, there will be some cases (parts with lots of draft) where there is an advantage in loading the preform into the cavity.

In small parts,the resin/filler mix can usually be applied in one spot on top of the preform, never on the mold surface. In large parts, it is necessary to spread it over a fairly large area. A charge pattern usually must be worked out for each part by trial and error to arrive at one that will assure proper fill-out of the preform. Careful control of the final pressing speed is necessary so that the resin flow is not so fast that it disturbs the reinforcement, yet fast enough to prevent pre-cure of the resin in contact with the hot mold surface.

Molding conditions for preform and mat are similar to those for reinforced molding compounds except that the unit pressure needed is substantially less. A maximum molding pressure of 500 psi (3.45 MPa) may be required, and 200 psi (1.38 MPa) is usually sufficient. Often, the pressure needed is dictated by the shearing force required to cut the reinforcement, in thick parts—1/4 in. (6.35 mm)--as much as several hundred lb/in.

Molding temperature and cure time are controlled principally by the time required to charge the resin and to fill out the mold. To prevent pre-cure during this part of the cycle, it is often necessary to compromise and use a longer cure time.

Small parts may have an overall molding cycle of as little as 3 minutes, but very large parts can require as long as 20 minutes. Mold temperatures are in the range of 220–300°F (105–149°C), with larger parts tending to the lower temperatures. Usually, internal mold release is now used in the formulation, but mold releases in the form of waxes and low

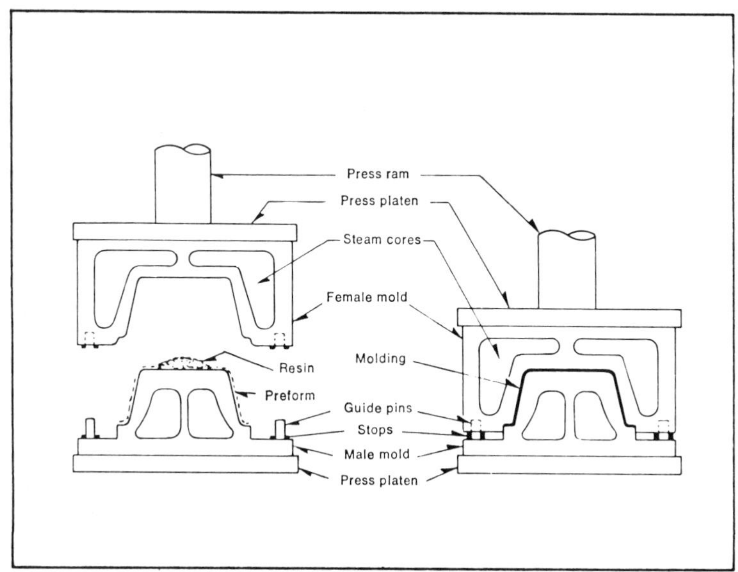

Figure 15.36. Matched molds using preform.

molecular weight polyethylene are applied to the mold surfaces.

Removal of parts from the mold is usually accomplished with the aid of air blasts, suction cup lifters, brass chisels, and so on. Ejector mechanisms used with molding compounds are not as practicable or as necessary with mat and preform. After removal, most parts must be placed on a cooling fixture to prevent distortion. Since small, molded holes are generally not feasible in preform and mat molding, provisions for punching, drilling, routing, and other machining operations may be included in the cooling fixture.

15.9.7. Formulation

While most kinds of polyester resins—isophthallic, ortho-phthallic, bisphenol, het anhydride, vinyl ester, etc.—can be used, they must be formulated to a relatively low viscosity, say about 800–1200 centipoises, or must be able to be reduced in viscosity by additional styrene monomer. This is because of the need for easy flow through the reinforcement at the time of molding. This need for relatively low viscosity prevents the use of low shrink thickeners. Although low profile additives are used, their effect on surface quality improvement is diminished by the absence of the thickener.

The amount of filler that can be used is also limited by the viscosity requirement, and the dominant filler must be one that does not separate from the resin as the resin/filler system flows in the mold. China clay is commonly used, along, or with, some low oil absorption filler such as calcium carbonate.

Aluminum hydrate, effective in BMC and SMC for improved fire resistance and electrical properties, cannot be used in sufficient quantity to have a similar effect in preform and mat formulations. However, when aluminum hydrate is used with a halogenated resin, some reduction in smoke generation and flame spread can be realized.

Some typical formulations are given in Table 15.13.

15.9.8. Mold Construction

Molds for mat and preform differ from molds previously described under reinforced molding compounds in two major particulars:

1. The need for a shearing action to cut off the excess reinforcement at the edge of the part (see Fig. 15.31).
2. A requirement of close control of the die space dimensions.

The so-called "shear edges" on both the male and the female molds must be hardened to cut highly abrasive glass fibers effectively. The fit between the molds at this shear edge must be quite close—0.002–0.004 in. (0.05–

Table 15.13. Typical Formulations for Preform and Mat Molding

INGREDIENTS	PERCENT OF RESIN MIX 1	2	3	4	PERCENT OF TOTAL
Resin	65	49	69	33	60–80
Low profile additive				16	
Catalyst	1/2–1	1/2–1	1/2–1	1/2–1	
Release agent	1	1	1	1	
Pigment	5	5	5	5	
Fillers					
China clay	29	20	25	20	
Calcium carbonate		25		25	
Preform or mat reinforcement					40–20[a]

[a]At higher reinforcement content some of the dry ingredients (fillers) mav have to be reduced for moldability.

0.1 mm)—to ensure a shearing action. This degree of fit is difficult to attain and retain. Even small temperature differences in large molds can change the clearance from too much to interference (which usually results in mold damage). Guide pins must be substantial and close fitting to prevent displacement and damage to the shear edge. Since the reinforcement will not flow to accommodate variations in die space, both the die space and the reinforcement thickness must be closely controlled to prevent resin-rich and resin-starved areas in the molded part. Variations in excess of 0.010 in. (0.25 mm) in a 0.100-in. (2.5-mm) thick part can cause problems. When variations occur, the only solution to get a good appearing part is to reduce the amount of reinforcement to accommodate the minimum die space. This, of course, reduces the strength of the part.

The two factors of very close fitting shear edges and close tolerance on the die space tend to make mat and preform molds more costly at the machining stage of their manufacture than BMC and SMC molds. The latter always appear, finally, to cost more, but they usually include ejector mechanisms, pins for holes and inserts, ribs, and other details not commonly incorporated in a mat or preform mold.

15.9.9. Presses

Press specifications, except for lower tonnage requirements, are the same as for BMC and SMC. This low tonnage requirement permits use of existing low tonnage, large platen presses for mat and preform molding; however, if a new press is purchased, it seems wise that it should have the capacity to accommodate BMC and SMC as well.

15.9.10. Design

In general, mat and preform molding is best used in products of relatively constant thickness with generous radii at any change in direction. The practical thickness range is from about 1/16 in. (1.6 mm) to about 1/4 in. (6.35 mm). Transitions from one thickness to another should be made over a distance of several times the thickness difference but every effort should be made to use a constant thickness. Radii should be about four times the thickness. Molded holes require shear edges in the mold around their periphery with a consequent large increase in mold cost. Bosses and ribs can be molded, but usually require insertion of bits of BMC or extra pieces of mat at the time of molding. Preferable treatment for bosses and ribs is shown in Fig. 15.37. Stiffening of unsupported edges is best accomplished by turning (flanging) rather than thickening (see Fig. 15.38). And, finally, but most important, the draft angle should be the maximum allowed by the use of the product, never less than 1° and preferably 2°. Sometimes one side of a part can be straight if the other has generous draft and it is so designed that the mold can be tilted to provide molding draft.

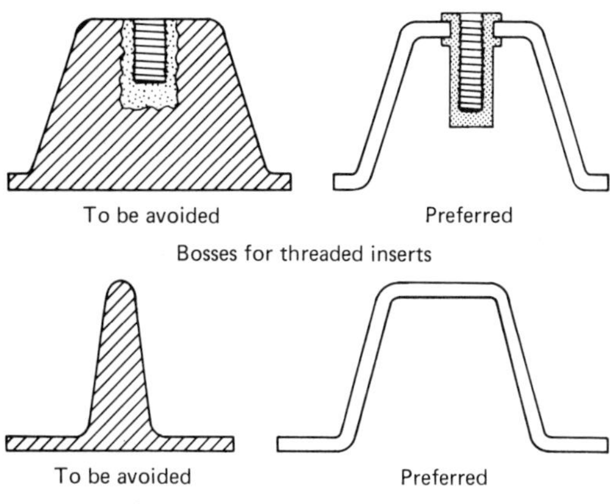

Figure 15.37. Boss and rib design details.

15.10. COLD PRESS MOLDING

Cold press molding is a matched die molding process using low pressure and unheated

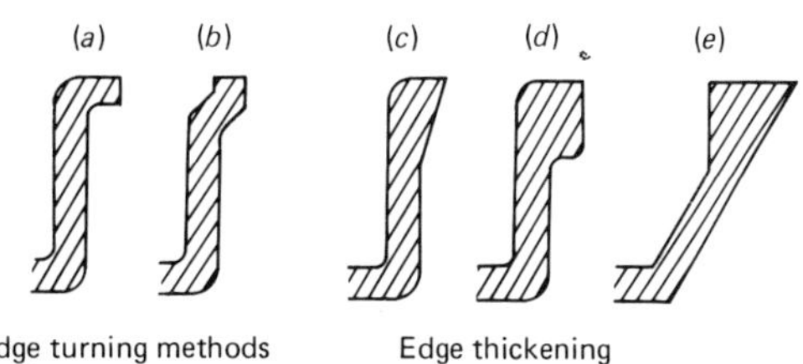

Five ways to stiffen a part edge. Edge turning, or plane change (a), (b), is preferable to edge thickening (c), (d), (e).

Figure 15.38. Edge stiffening design details.

molds, usually constructed of reinforced plastics.

The first phase of the process is essentially the same as making a hand lay-up, spray-up, or rigidized thermoplastic product, except that it is done on one of a set of matched molds which are closed before the cure takes place.

Principal advantages over the open mold processes are:

1. Modestly smooth and dimensionally controlled back surface.
2. Relatively low tooling cost (compared to heated metal molds).
3. Low press cost. Not much pressure is needed—about 50 psi (345 kPa).
4. Rapid tool-up time—only slightly longer than for open mold processes.
5. Gel coats or thermoformed thermoplastic surface is possible on one side, at least.

Disadvantages include:

1. Long molding cycle—not too different from open mold processes.
2. Short mold life, due to high exotherm during resin cure with no way for heat to dissipate from poorly conductive plastic molds. (Molds can be made more elaborate, and costly, with spray metal faces and cast in cooling coils.)
3. Parts must be machine-trimmed after molding.
4. Surfaces not gel coated (or otherwise treated) need substantial post-mold finishing if a quality finish is required on the part.
5. Ribs, bosses, inserts, and complex shapes are not practical.

Materials include those normally suitable for open mold processes:

1. Low viscosity, high reactivity polyester.
2. Chopped strand, continuous strand mats.
3. Spray-up roving chopped in place.
4. Open mold gel coats.
5. Room temperature cure catalyst-accelerator systems.

Applications include any product of relatively simple shape that could be made by an open mold process but needs more precision in its thickness or relatively good finish on both sides, and which will be made in sufficient volume to justify the mold cost.[4,5]

15.11. RESIN INJECTION MOLDING

This process, also known as "resin transfer molding," is another process for obtaining a product with two relatively good appearing surfaces with better dimensional control than open mold processes offer.

In this process, the reinforcement, without the resin, is loaded into the mold, the mold is closed, and the resin is injected or transferred into the mold, in such a manner as to completely impregnate the enclosed reinforcement. The process has most of the advantages and disadvantages of the cold press molding process. However, a molding press is not required. With simple parts, injected at low pressure, trunk latches can be strong enough to clamp the molds shut. However, when more complex parts are attempted, the mold structure and clamping devices can become quite massive. While a press is not required to apply pressure for molding, one may be required to handle the mold. At the least, some type of lifting device, such as a chain hoist, will be needed. When the mold is closed, catalyzed resin is injected at a low point. As it rises in the mold, it pushes the entrapped air out, at the edge of the part, or from vents. Where the two halves of the mold join, the reinforcement is usually partially compressed so that air can pass but flow of the viscous resin is slightly restricted (see Fig. 15.39). As with cold press molding, the exotherm of the cure has a tendency to degrade the plastic molds, and again more elaborate and costly molds can be employed.

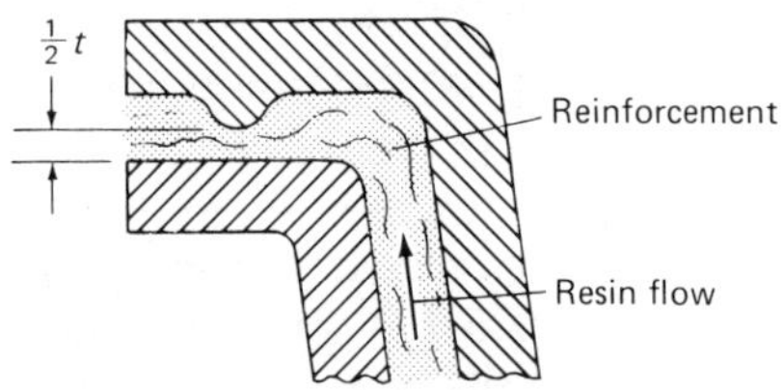

Figure 15.39. Seal detail resin injection mold.

Most of the resin injection products are made with room-temperature cure polyesters in reinforced plastic molds, but for many years some very complex items for aircraft applications have been made in aluminum molds using high-temperature cured epoxies and polyesters. These products generally have used satin weave cloth reinforcement, in contrast to mats. To impregnate this material effectively, it is necessary to draw a vacuum to evacuate the air and to pump the resin in under modest pressure. After charging the mold, it is brought to curing temperature by use of built-in heating devices or by placing in an oven.[11]

Resin injection molding requires the same kind of two-component mixing and injection device similar to spray-up. The catalyst and resin must be brought together and mixed just before injecting into the mold. Usually, it is a two-pump system feeding a static mixer at the point of resin injection.

More complex parts can be made by resin injection (compared to cold press molding), but this process suffers many of the same problems. Parts must be machined, trimmed, etc., after molding.

The applications are similar—any modest volume production item that needs two relatively precise surfaces and is required in sufficient quantity to write off relatively costly tooling.

15.12. FOAM RESERVOIR MOLDING

This process, also known as "elastic reservoir molding," consists of making a sandwich of resin impregnated open-celled flexible polyurethane foam between face layers of fibrous reinforcement. When the product is placed in a mold and squeezed, the foam is compressed, forcing the resin out and into the face layers. The elastic foam exerts sufficient pressure to force the face layers into contact with the mold surfaces.

When the components are properly proportioned, it is possible to end up with a molding having the advantages of sandwich construction; that is, a low density core with strong faces providing a favorable stiffness-to-weight ratio.

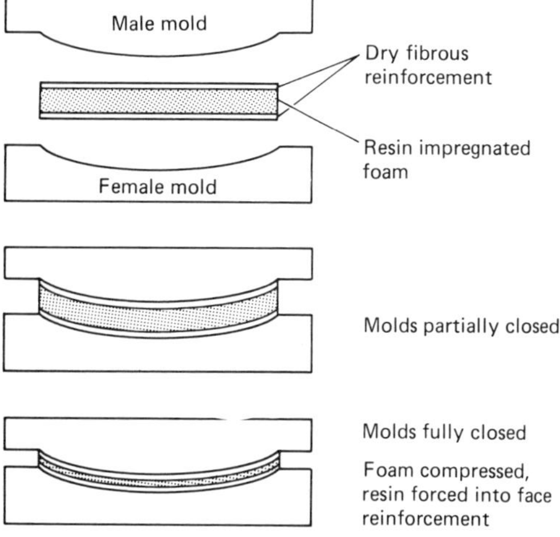

Figure 15.40. Schematic—foam reservoir molding.

Figure 15.40 schematically shows the process. The molding pressures are low, and less than 100 psi (690 kPa) may be used. Applications include relatively large, flat shapes, such as decks, hoods, and roofs of vehicles.

The advantages include:

1. Low density parts.
2. High impact strength.
3. Good bending strength.
4. Fast molding cycle.

REFERENCES

1. White, Roger B., *Premix Molding*, Reinhold Book Corp., New York, 1964.
2. Denton, Douglas L., "Mechanical Properties Characterization of an SMC-R50 Composite," Paper 11-F, 34th Annual Conference, Reinforced Plastics/Composites Institute, SPI, Washington, D.C., February 1979.
3. Ferrarini, J. *et al.*, "Development of a Unique Method for the Preparation of High Quality SMC," Paper 9D, 33rd Annual Conference, Reinforced Plastics/Composites Institute, SPI, February 1978.
4. Owens-Corning Fiberglas Corp., Publication 5-PL-5361-A, *Cold Press Molding Manual.*
5. Rohm & Haas, Inc., *Rigidizing of Plexiglass.*
6. Pratt, B. D., "Factors Affecting the Arc Resistance of Premix," Annual Conference, Reinforced Plastics/Composites Institute, SPI, February 1965.
7. Connolly, W. J. and Thornton, A. M., "A New Polyester Resin and Filler System Producing Excellent Flame Resistance and Heat Aging Properties," Annual Conference, Reinforced Plastics/Composites Institute, SPI, February 1965.

8. Sutcliffe, M. R., "Degree of Cure in Polyester Dough Molding Compounds," Third International Reinforced Plastics Conference of the British Plastics Federation, Reinforced Plastics Group, London, November, 1962.
9. Kroekel, Charles H. and Bartkus, Edward L., "Low Shrink Polyester Resins: Performance and Application," Annual Conference, Reinforced Plastics/Composites Institute, SPI, February 1968.
10. Owens-Corning Fiberglas Corp., Publication 5-TM-8364, October 1978.
11. *Handbook of Fiberglass and Aavanced Plastics Composites* edited by G. Lubin. Van Nostrand Reinhold Publishing Co., New York, 1969, pp. 349–355.

16
FILAMENT WINDING

A. M. Shibley
Consultant, Formerly at
Plastics Technical Evaluation Center
Picatinny Arsenal, Dover, New Jersey

16.1. INTRODUCTION

Filament winding is a comparatively simple operation in which continuous reinforcements in the form of rovings or monofilaments are wound over a rotating mandrel. Specially designed machines, traversing at speeds synchronized with the mandrel rotation, control the winding angles and the placement of the reinforcements. These may be wrapped in adjacent bands or in repeating patterns which ultimately cover the mandrel surface. Successive layers are added at the same or different winding angles until the finished thickness is reached. The wrap angle may vary from low angle "longitudinals" to high angle "hoops" approaching 90° relative to the mandrel axis, and any "helical" angle may be wound between these limits. A thermoset resin serves as a binder for the reinforcements. In "wet winding," the resin is applied during the winding stage. The alternate "dry winding" method utilizes pre-impregnated B-staged rovings. Normal curing is conducted at an elevated temperature without pressurization, and mandrel removal completes the process. Finishing operations, such as machining or grinding, may be performed if required.

The basic process is subject to numerous variations offering a broad spectrum of structural types, design features, material combinations, and equipment. Structures are necessarily wound as surfaces of revolution, although within limitations other shapes can be formed by internal pressurization of an uncured winding in a closed mold. Structures may be plain cylinders, pipe, or tubing, varying from a few inches to several feet in diameter. Spherical, conical, and geodesic shapes are within winding capability. End closures can be incorporated into the winding to produce pressure vessels and storage tanks. The structures can be designed for specific loading conditions, such as internal pressure, external pressure, and torsional or compressive members. Combination winding is another possibility, in which hoop overwraps are added as reinforcement to thermoplastic pipe or metal pressure vessels. Parts can be designed and fabricated with a high degree of precision. On the other hand, the winding may be geared to less critical fast rate production.

Potentially almost any continuous reinforcement is suitable for winding purposes. In practice, filament winding is essentially a processing of fiberglass. Graphite fibers and the aramid fiber, Kevlar 49, have been adapted to the more exacting space and aircraft applications, where high specific strength and modulus are major considerations.

The principal matrix materials are epoxy, polyester, and vinyl ester resins. Polyimides, phenolics, and silicones, which yield condensation products on curing, are processed with greater difficulty. Currently, interest is being directed to the utilization of certain thermoplastics for specific applications.

Winding machines vary from lathe types or chain driven machines to more complicated computerized equipment with three or four axes of motion. Machines have also been built for the production of continuous pipe. In another variation, portable equipment has been designed for winding large storage tanks on location. These machines normally lay down only hoop windings, while the longitudinal reinforcement is provided by either chopped strands or broadgoods.

The major aspects of filament winding will be discussed in this chapter. Emphasis will be placed on the basic reinforcing and matrix materials, winding procedures, process controls, and the properties of the cured composite.

16.2. BASIC RAW MATERIALS

16.2.1. Reinforcements

Nearly all filament winding is conducted with continuous E-glass roving as the reinforcement. A somewhat stronger, but higher priced, S-glass roving is used less frequently, principally in the aerospace industry. Both types are characterized by high tensile strengths well suited for filament winding. Specific strengths range from 3-6 million in. (0.75–1.50 MPa·m^3/kg). In regard to unidirectional composites, tensile strengths as high as 350,000 psi (2.4 GPa) have been reported. A limiting weakness of the roving is its low modulus of elasticity compared to other structural materials. Values for specific moduli are 110–140 million in. (27.4–34.9 MPa·m^3/kg), depending on the glass type. The maximum moduli to be achieved in unidirectional composites are in the order of 6–8 million psi (41.4–55.2 GPa). Most commercial pipe, tanks, and pressure vessels are not subjected to excessive bending or buckling loads, so that the low modulus glass can be successfully adapted to these structures. When the loading modes are more conducive to buckling, wall thicknesses can be increased or localized stiffeners incorporated into the design without incurring excessive penalties in weight or cost.

The alternate method for imparting greater rigidity is to utilize higher modulus fibers in the winding. Of the high modulus materials available for this purpose, graphite filaments and the DuPont aramid fiber, Kevlar 49, have

Table 16.1. Comparative Filament Properties

FILAMENT	DENSITY lb/in.^{3} (kg/m^3)	TENSILE, STRENGTH, ksi (GPa)	SPECIFIC STRENGTH, in. × 10^6 (MPa · m^3/kg)	MODULUS, psi × 10^6 (GPa)	SPECIFIC MODULUS, in. × 10^7 (MPa · m^3/kg)	ELONGATION, %
E-glass						
Monofilament	0.092	500 (3.45)	5.4 (1.35)	10.5 (72.4)	11.4 (28.4)	4.8
Strand	(2547)	485 (3.34)	5.3 (1.31)			
Roving[a]		280 (1.93)	3.0 (0.76)			
Roving[b]		200 (1.38)	2.2 (0.54)			
S-glass						
Monofilament	0.090	665 (4.59)	7.4 (1.84)	12.6 (96.9)	14.0 (38.9)	5.3
Strand	(2491)	550 (3.79)	6.1 (1.52)			
Roving[c]		500 (3.45)	5.6 (1.38)			
Roving[d]		400 (2.76)	4.4 (1.11)			
S-1014 MF		656 (4.52)	7.3 (1.81)			5.2
Graphite						
Thornel 300[e]	0.064 (1772)	325 (2.24)	5.1 (1.26)	34.0 (234)	53.0 (132)	1.0
Celion 3000[f]		360 (2.48)	5.6 (1.40)			1.1
Aramid						
Kevlar 49[g]	0.052	400 (2.76)	7.7 (1.92)	19.0 (131)	36.0 (91)	2.0

[a]G-dia, per Mil-R-60346.
[b]Other than G-dia, Mil-R-60346.
[c]Mil-R-60346.
[d]General Purpose, Mil-R-60346.
[e]Union Carbide.
[f]Celanese.
[g]DuPont.

received the most attention. These reinforcements have specific moduli from three to four times greater than glass roving. Another advantage is their low density, yielding specific strengths in the composite which compare favorably with fiberglass (see Table 16.1 for a comparison of fiber properties). The price of graphite and aramid fibers, though greatly reduced in recent years, still remains prohibitive for widespread commercial acceptance. Currently, only the aerospace manufacturers have been able to exploit the graphite and Kevlar. "Hybrids," combinations of fiberglass with higher modulus fibers, have been suggested as a compromise for structures which are optimized as to weight and rigidity versus cost. A typical example is a pressure vessel wound with graphite to give longitudinal stiffness, overwound with fiberglass to resist the forces in the hoop direction.

Numerous other filamentary materials have been tested as the reinforcing media in filament wound constructions. The list includes steel wire, boron, beryllium, polyamides, polyesters, and asbestos. None, however, has attained commercial significance.

16.2.1.1. Strength of Fiberglass

The monofilament strengths of E-glass and S-glass are 500,000 and 665,000 psi (3.4 and 4.5 GPa), respectively. The standard deviation is approximately ±10%. These values are the result of a large number of individual readings. Strength distribution of these measurements generally follows the histogram shown in Fig. 16.1. Values will range from close to zero on the low side and approaching the theoretical limit—1.5–2.0 million psi (10.3–13.8 GPa)—on the high side. This variability results from existing flaws in the fiber and the interaction of environmental factors on these flaws.[1] Humidity is the major factor. Atmospheric moisture attacks the fiber at the flaw sites, particularly during the stressed condition, leading to flaw growth and ultimate failure. This stress corrosion mechanism is controlling in static fatigue as well as tension. The flaw structure consists of severe surface defects incident to the fiber drawing process or subsequent handling to form the roving package, and less severe surface pitting, which may have been present in the filament as drawn, or may have resulted from corrosion with or without stress. Additionally, the glass may contain internal flaws.

When a strand or bundle of fibers is tested in tension, the results are approximately 20% below the monofilament average. As individual filaments within the bundle fail, the remaining fibers are forced to carry a greater portion of the load. The overall strength is thereby decreased. In actuality, the bundle strength can be calculated from the strength distribution of the monofilament with a high degree of accuracy. Catenary (unequal tension of filaments within a strand or roving) produces a similar progressive failure effect.

According to the fiberglass manufacturers, multiple end rovings, normally not exceeding 60 ends, have approximately the same unit strength as the single end. Such a conclusion assumes that no excessive catenary or friction

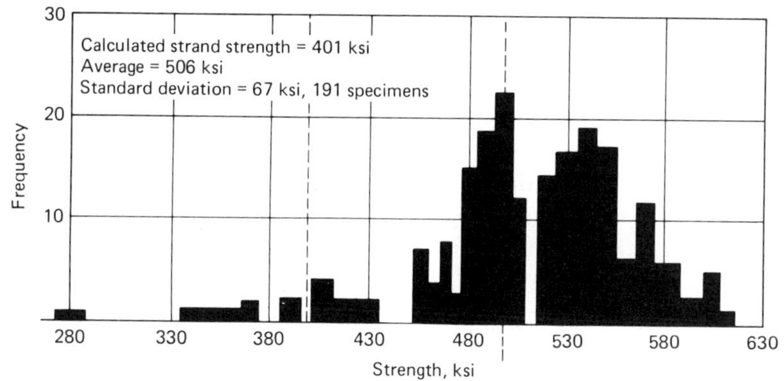

Figure 16.1. Tensile strength distribution, E-glass virgin fiber. (*Courtesy Owens-Corning Fiberglass Corp.*)

damage is introduced while the individual ends are gathered into the roving.

The filament diameter is another variable influencing tensile strength. Under closely controlled conditions, it has been demonstrated that monofilament strength does not decrease as the diameter is increased up to the maximum range of commercial fibers. For practical purposes, however, it is clearly evident that the large diameter filaments are weaker than the finer fibers. The difference may be gauged from the requirements of Military Specification R-60346 for filament winding roving. The minimum value for E-glass roving made from G-fibers—0.35-0.40 mils (0.009—0.010 mm)—is specified as 280,000 psi (1.93 GPa). For fibers other than G, that is, larger diameters up to T-filament—0.90–0.95 mils (0.023-0.024 mm)—the minimum allowable tensile strength is 200,000 psi (1.38 GPa).

Filament strengths are also determined by testing cured composites. When fiber alignment and uniform loading are maintained, unidirectional composites attain values equal or superior to bundle strengths. Filament strength in an NOL ring can be as high as 450,000 psi (3.1 GPa). In a thin wall pressure vessel, it ranges from 400,000–450,000 psi (2.76 to 3.1 GPa). A larger and thicker winding, on the other hand, will have a maximum strength of approximately 300,000 psi (2.07 GPa). Gross estimates from the larger structures are lower for several reasons: 1) fiber damage during winding; 2) misalignment or poor collimation; 3) non-uniform tension in the winding layers; 4) variation in stress from the inner to the outer layer; and 5) undetermined local stress conditions.

As a general conclusion, estimates of fiber strength for design purposes are more accurate when based on composite tests. Comparison with data taken from strand tests will indicate the degree of efficiency in the fabrication process. A detailed stress analysis is required to determine the actual fiber stress at failure.

16.2.1.3. Roving Construction

During the early period of filament winding, all strands (single ends) consisted of 204 filaments of G-diameter and had a nominal yield of 13,500–15,000 yards/lb (27,200–30,250 m/kg). These strands were combined to form multiple end roving. Current practice is to increase fiber diameter as well as the number of filaments. Strands of 408 or 816 filaments are common. A further innovation is to draw several thousand filaments into a single end, which also serves as the roving. Fiber diameters and typical yields of E-glass filament winding grades now available are listed in Tables 16.2 and 16.3. S-glass is still drawn as G-diameter, 204 filaments. Specific finishes are applied for epoxy and polyester resins. A single finish is sometimes used which is said to be compatible with epoxy, polyester, or vinyl ester resins.

Table 16.2. Filament Diameters

CODE	DIAMETER, in. × 10^{-5} (micrometers)
G	35-40 (9-10)
J	45-50 (11-13)
K	50-55 (13-14)
M	60-65 (15-17)
T	90-95 (23-24)

16.2.2. Resin Systems

The major matrix systems for filament winding are based on epoxy, polyester, or vinyl ester resins. Phenolics, silicones, and polyimides find infrequent use in specific high temperature or electrical applications. These latter three thermosets are difficult to process under the usual winding conditions and require pressurization during cure to remove reaction products and residual solvents. Currently, thermoplastics are being investigated as binders. A promising candidate is polysulfone resin, which has comparatively high strength properties and heat resistance at elevated temperatures. The obvious and intriguing advantages of a thermoplastic are that no curing cycle is necessary and that pot-life and shelf-life problems are obviated. Efficient processing techniques for winding a thermoplastic, however, have not yet been demonstrated. Procedures for coating the fiber with resin and for consolidating the

Table 16.3. Typical Roving Yields for Filament Winding Grades,[2,3] Nominal Yield, yards/lb

NUMBER OF ENDS									
SINGLE ENDS		MULTIPLE ENDS							
TYPE 30[a]	HYBON 2079[b]	8	12	15	16	20	30	60	INPUT STRAND
			1125			675	450	225	13,500
			1167			700	467	233	14,000
			1231			738		246	15,000
		462		247			123	61.5	3700
					225				3600
1800		225							1800
1200									1200
900									900
675	675								675
450									450
225									225
	250								250

[a]Owens-Corning.
[b]PPG Industries.

composite on the mandrel must be established before thermoplastic winding becomes a reality.

16.2.2.1. Epoxy Resins

Traditionally, epoxies have been used in aircraft, space, and military applications, which tend to be more strength and weight critical than commercial windings. Selection of this resin rather than the cheaper polyesters has been justified by its superior mechanical properties, fatigue behavior, heat resistance, stronger bonding to the reinforcement, and lower shrinkage during curing. A long history of satisfactory performance and attendant reliability have been decisive factors favoring continued epoxy usage by aerospace fabricators.

Epoxies for filament winding are essentially the same as the laminating resins. Minor system modifications are made to meet specific processing conditions. The diglycidyl ether of bisphenol-A (DGEBA) is the most important polymer type. Epoxy novolacs and cycloaliphatics are utilized to a lesser extent. Other available systems are based on brominated epoxy for improved ignition resistance, resorcinol diglycidyl ether for extended processibility, and flexible epoxies for impact resistance and greater elongation.

A variety of curing agents and catalysts are used with the epoxies. Nadic methyl anhydride (NMA) and metaphenylene diamine (MPDA) are the most popular curing agents. Commercial curing agents are frequently modified by the addition of accelerators to increase the cure rate, by partial reaction with a small amount of resin to slow the cure rate, and by incorporating various compounds to improve solubility in the resin and to prevent crystallization of the curing agent. The principal curing agents and a number of commercial adaptations are listed in Table 16.4.

The cast resin properties of several systems are presented in Table 16.5. The system based on ERLA 2256 has been widely accepted for high strength windings. (This resin is no longer being manufactured; its properties are included for comparative purposes.)

The selection of a resin system for a specific application is prompted by its processing characteristics, its curing temperature, and its effect on composite properties. The viscosity and pot life of the catalyzed system, which is to say, the initial viscosity and the change in viscosity with time, are major processing considerations. Gel time and resin flow under winding tension and during cure are other rheological factors. A low viscosity is essential for complete wetting of the reinforcement and

Table 16.4. Curing Agents for Epoxy Resins

AGENT	COMPOSITION	COMMENTS
MDA	Methylene dianiline	Melts at 185° F (85° C), pot life 4–6 hours at room temperature.
MPDA	Metaphenylene diamine	Melts at 140° F (60° C), pot life 4–6 hours at room temperature.
DDS	Diamino, diphenyl sulfone	Melts at 347° F (60° C), BF_3MEA used as accelerator.
NMA	Nadic methyl anhydride	Liquid, long pot life, BDMA and others used as accelerators.
HHPA	Hexahydrophthalic anhydride	Melts at 95° F (35° C), soluble in liquid resin at room temperature.
BDMA	Benzyl dimethyl amine	Used as accelerator and to control B-staging.
BF_3MEA	Borontrifluoro-mono-ethyl amine	Used as accelerator and to control B-staging. (Moisture sensitive.)
DICY	Dicyandiamide	Used as accelerator and to control B-staging.
DAP	2,6-diaminopyridene	Cured properties similar to MDA, slower reacting.
DMP-30	Tridimethylaminomethyl phenol	Used as accelerator.
Tonox 6040	60% MDA, 40% MPDA	Viscous liquid.
Jeffamine T-403	Polyether triamine	Low viscosity, adds flexibility and toughness.
Ciba 906	Modified hydrophthalic anhydride	Somewhat like NMA, usually adds accelerator.
TETA	Triethylene triamine	General purpose room temperature, high exotherm.
DETA	Diethylene triamine	General purpose room temperature, high exotherm.

for removal of entrapped air and volatile solvents. Suitable working viscosities fall in the range of from 350–1500 centipoise, measured at 77° F (25° C). If the system is too fluid, problems in the control and uniformity of resin content will be encountered. Some fibers, such as the graphites, will not pick up enough resin. In other instances, resin will migrate to the outer layers of the winding, leaving dry inner layers—a source for premature composite failure. Too viscous a resin, on the other hand, causes fiber fuzzing in the resin bath and feed eye, uneven fiber coating, and excessive tension, in addition to air entrapment. With high viscosity and solid resins, reactive diluents are required to bring the system into a working range. Typical diluents for this purpose are butyl glycidyl ether (BGE), phenyl glycidyl ether (PGE), the diglycidyl ether of neopentyl glycol (Dow XD7114), and Ciba-Geigy's diglycidyl ether of 1,4-butanediol (RD-2). Non-reactive diluents are generally to be avoided. It may also be necessary with certain optimized systems to apply heat to reduce the resin viscosity.

A pot life of at least several hours is a normal requirement to prevent gellation before the winding is completed. Winding over a gelled or partially gelled layer is not advisable. Such a practice again results in

Table 16.5. Properties of Cured Epoxy Resins

Type	DGEBA	DGEBA	Novolac	DGEBA	DGEBA	DGEBA	DGEBA
Resin	ERLA 2256	Epon 826	DEN 438	Ciba 6005	DER 332	DER 383	DER 331
Curing agent	Tonox 6040	Tonox 6040	BF_3MEA[b]	NMA	NMA	MDA	MDA
Concentration, phr	29.5	28.3	3.0	90	87.5	27	27
Diluent (phr)	ERLA 0400[a]	RD-2 (25)					
Initial viscosity, CPS	1590	1200					
Cure, hours/°C	16/49, 2/93, 3/167	3/60, 2/121	4/100, 16/150	2/100, 2–4/150–200	2/90, 4/165, 16/200	2/80, 2/160	2/80, 2/160
Tensile strength, ksi (MPa)	14.8–15.6 (102–108)	10.9–13.0 (75–90)	6.1 (42)	9.0 (62)	6.3 (43)	10.3 (71)	10.2 (70)
Tensile modulus, psi × 10^5 (GPa)	4.9–5.0 (3.4–3.5)	3.7–3.9 (2.5–2.7)	4.8 (3.3)	3.0–4.5 (2.1–3.1)		3.6 (2.5)	3.2 (2.2)
Maximum strain, %	7.1–7.3	7.6–7.8			1.6		
Water absorbtion, %	1.45	0.93		0.4–1.0			
Heat deflection temperature, °F (°C)	272 (133)	250 (121)	338	338 (170)	275 (135)	325 (163)	320 (160)

[a]bis (2,3-epoxycyclopentyl) ether, 37 phr.
[b]Subject to moisture degradation.

non-uniform resin distribution and weakened structures. In this respect, continuous metering of the resin/curing agent mixture or a controlled batch mixing will reduce the problems associated with premature gellation.

Time/temperature/viscosity relationships for various systems are available from the resin manufacturers. For more critical control, these relations are determined experimentally, using resin volumes close to the anticipated operating range. The characteristics of several systems are shown in Fig. 16.2.

Flow properties are of greater significance in dry winding. Here, sufficient tack and flow of the prepreg are needed to ensure bonding of the successive winding layers. The flow properties are functions of the degree of B-staging during fabrication of the prepreg and, to some extent, of the residual volatiles. Stored under refrigeration, commercial prepregs will maintain initial flow up to six months. Improved flow and tack are obtained by heating the strand during winding.

Closely controlled curing cycles are of primary importance if optimum composite properties are to be attained. The curing characteristics of the system, such as the time at specific temperatures and the rate of heating, are functions of the curing agents and the catalysts in the system. The anhydride curing agents tend toward higher curing temperatures and longer curing times, but yield composites with higher heat deflection temperatures. The aromatic amines are cured at an intermediate temperature and are faster curing. Some systems based on aliphatic amines have been formulated with an extended pot life and can be cured at low temperatures. The normal practice with all systems is to gel the resin at a reduced temperature, continue curing at a higher temperature, and follow with a post-cure at the maximum prescribed temperature. This stage curing prevents excessive resin flow, minimizes the effects of high exotherm, and ensures a high degree of cure in the product. As a rough approximation, general purpose systems are cured in a temperature range of 250–275°F (121–135°C). Heat resistant systems are cured at 350–375°F (177-191°C). Post-cures may be as high as 400°F (205°C). Higher temperature curing, it is noted, may increase resin shrinkage and may require more costly mandrels.

A high strength, modulus, and elongation are considered desirable attributes of the resin. These properties determine the transverse and shear strength of the composite and, in turn, will influence burst strength under internal or external pressure. The exact

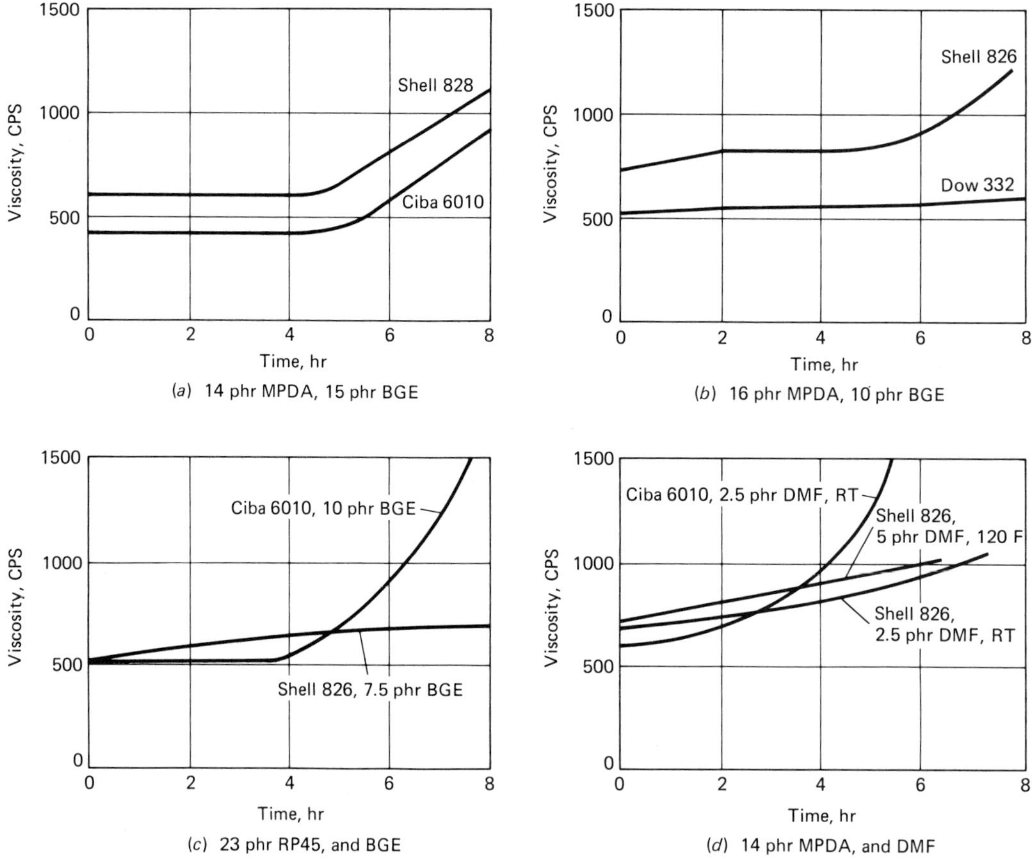

Figure 16.2. Time versus viscosity for various epoxy systems.[4]

magnitude of such effects, however, is not established with the available data.

The shrinkage of epoxies during cure is lower than that of the other filament winding resins. Shrinkage varies with the resin type, curing agent, rate of heating, and cure temperature. Test results with DGEBA, for example, will range from 2–8%, depending on the catalyst and the cure. A low shrinkage is an advantage in reducing internal composite stresses and facilitates the removal of a winding from its mandrel.[5,6,7,8]

16.2.2.2. Polyesters and Vinyl Esters

Due to their lower cost and a balance of physical and chemical properties, these resins find extensive use in commercial practice. Their handling characteristics are readily adapted to filament winding. Processing viscosity is comparatively easy to control. As with the epoxies, no fundamental distinction can be made between filament winding and laminating systems.

The unsaturated polyesters for reinforced plastics are esterification products of an unsaturated dibasic acid or anhydride with a glycol, and usually contain a saturated dibasic acid or anhydride as a third reactant. The vinyl esters are reacted from epoxy resins with acrylic, methacrylic, or similar acids. In both cases, the reaction product or prepolymer is dissolved in a monomer, normally styrene, to form the liquid resin. During cure, the monomer combines with the prepolymer. Numerous acids, anhydrides, glycols, and monomers can enter into these reactions so that a multiplicity of products with a broad range of properties is available. Many of these find applications in filament winding. The orthophthalic polyesters with styrene monomer are the least expensive, but are prone to chemical attack. Isophthalic resins have better corrosion and chemical resistance to justify their higher prices. Several of the vinyl esters have become popular in the corrosion field, particularly where service temperatures are higher. Resins based on chlorendic, tetrabromophthalic, and other saturated dibasics impart improved flame resistance, but at the expense of lowered mechanical properties at elevated temperatures.

The normal method of polymerizing the base resin with the monomer is by the action of organic peroxide catalysts, of which, again, there are many. The catalyst selected determines the temperature at which curing takes place. Ambient temperature cure, which is of minor interest to filament winders, is initiated by systems consisting of methyl ethyl ketone peroxide (MEKP) plus cobalt napthanate (CoN) or dimethyl aniline (DMA) with benzoyl peroxide (BPO). Tertiary-butyl perbenzoate alone, or in combination with BPO, is frequently used for cures in temperature ranges of 200–300°F (93–149°C). Promoters or accelerators are added when lower cure temperatures are desired. Retarders serve to reduce peak exotherm, while inhibitors prevent premature gelation. Cure rates of the various resins are functions of the constituent acids and monomers. The glycols have a lesser effect. For a specific system, cure rate and gel time depend on catalyst and promoter concentration. The general procedure is to establish concentrations which minimize exotherm.

Curing is also induced by exposure to such energy sources as ultraviolet light or high intensity xenon lamps. The polyester is precatalyzed and reacts under the light. These systems, it is anticipated, will be more fully exploited in future commercial developments; particularly with small diameter pipe.

In contrast to the epoxies, mechanical properties show only minor differences as a result of variation in the catalyst type. There are indications, however, that corrosion resistance may be affected. Of greater significance is the degree of cure. More complete cures and improved properties are obtained at lower cure temperatures with longer cure times. Heat deflection tests, flexural or tensile tests, and Barcol hardness are useful tools in determining optimum cures.

Processing viscosity is controlled by the amount of monomer present. For styrene, the normal content is 40–45%. It may be judicious, in some cases, to increase the viscosity by the addition of thixotropic agents. The effect of one such agent is shown in Table 16.6. Impregnating viscosities of 250–1000 CPS are typical. This range, somewhat lower than is the case with epoxies, reflects the higher winding speeds with commercial polyesters.

The mechanical properties of several resins are listed in Table 16.7. Maximum service temperatures for the polyesters and vinyl esters in a non-corrosive environment is normally between 200 and 225°F (93 and 107°C). Most resin suppliers will furnish the maximum recommended service temperatures for specific chemicals. An ultimate

Table 16.6. Viscosity of Shell Epocryl 480 Vinyl Ester[9]

			BROOKFIELD VISCOSITY AT 77°F, CPS (25°C)	
	ADDITIVE	SPINDLE[a]	SPEED	RPM
ADDITIVE	CONTENT, phr[b]	1	5	50
		500	500	500
Styrene	10[c]	250	250	250
Styrene	20[d]	100	100	100
Cab-o-sil[e]	1[f]	5,000	2600	600
Cab-o-sil[e]	2[f]	45,000	15,000	2500

[a]Brookfield Model RVT, #4 spindle.
[b]Parts by weight added to 100 parts of EPOCRYL Resin 480.
[c]This results in a final styrene content of 45% weight.
[d]This results in a final styrene content of 50% weight.
[e]Cabot Corp.
[f]Mixed for 3 minutes in a high shear blender.

Table 16.7. Properties of Polyester and Vinyl Ester Resins

Resin	Derakane 411	Derakane 510-A	Epocryl 480	Epoxy Novolac	6100-10
Manufacturer	Dow	Dow	Shell		Koppers
Type	VE	VE	VE	VE	Iso-PE
Tensile					
Ultimate, ksi (MPa)	11.5 (79)	10.5 (72)	12.2 (84)	11.3 (78)	8.0 (55)
Modulus, psi × 10^5 (GPa)	4.9 (3.4)	5.0 (3.4)	4.7 (3.2)	4.7 (3.2)	
Elongation, %	4.7	6.5	5.3	5.3	12–15
Flexural,					
Ultimate, ksi (MPa)	17.0 (117)	17.0 (117)	17.0 (117)	16.3 (105)	13.8 (94)
Modulus, psi × 10^5 (GPa)	4.5 (3.1)	5.2 (3.6)	4.7 (3.2)	4.6 (3.2)	3.9 (2.7)
Heat deflection temperature, °F (°C)	220 (104)	230 (110)	250 (121)	250 (121)	

elongation of 3–7% is a desirable property. Cure shrinkages are estimated to be in the order of 7–10%. Shrinkage increases at higher styrene contents.

16.3. THE WINDING PROCESS

16.3.1. Winding Methods and Patterns

The two basic processes are identified as polar and helical winding. Each method produces a distinctive filament pattern. In polar (also called planar) winding, the mandrel remains stationary while a fiber feed arm rotates about the longitudinal axis, inclined at the prescribed angle of wind. The mandrel is indexed to advance one fiber bandwidth for each rotation of the feed arm. This pattern is described as a single circuit polar wrap (Fig. 16.3). The fiber bands lie adjacent to each other; a completed layer consists of two plies oriented at plus and minus the winding angle.

In the helical mode, the mandrel rotates continuously while the fiber feed carriage traverses back and forth. The carriage speed and mandrel rotation are regulated to generate the desired winding angle. The normal pattern is multi-circuit helical. After the first traverse, the fiber bands are not adjacent. Several circuits are required before the pattern repeats. (A typical ten-circuit pattern is represented in Fig. 16.4.) The filament path advances one-tenth of the circumference plus one-tenth of the bandwidth per circuit; the eleventh path then falls alongside the first. A layer again contains two plies. This configuration characteristically results in fiber intersections (fiber crossovers) at certain sections. Crossovers may occur at more than one section, depending on the wind angle.

Winding patterns are determined by trial and error machine adjustments or are calculated from the geometry. The following simplified example illustrates a method of establishing the circuits per pattern and the total number of circuits for complete coverage of the mandrel.

Example: To wind a 10 in. (25.4 cm) diameter by 40 in (101.6 cm) long cylinder at a 45°

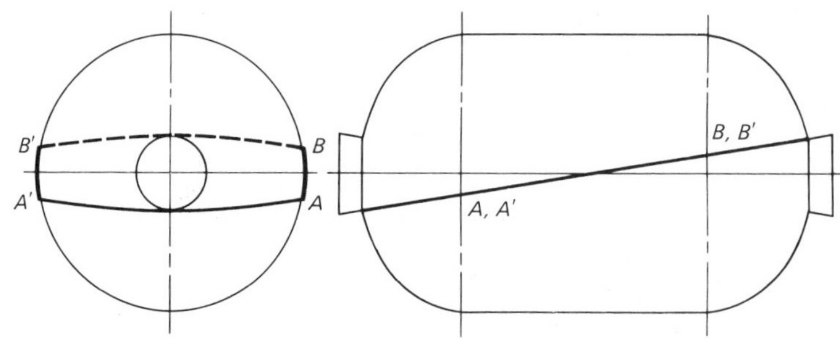

Figure 16.3. Single circuit planar path.

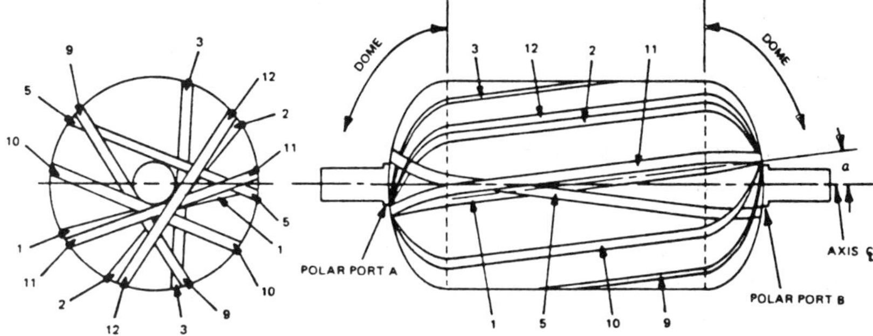

Figure 16.4. Ten-circuit helical pattern.[24]

wrap angle. The fiber bandwidth is assumed to be 0.250 in. (0.64 cm) and the dwell angle 180°. (The dwell angle is a measure of mandrel rotation at each end before the travel direction of the feed is reversed.)

In one circuit, the feed moves 80 in. (203.2 cm), while the mandrel rotates $80/10\pi$ revolutions plus one revolution of dwell, or a total of 1276.8°. The fiber path is thus advanced 196.8° (1276.8° − 3 × 360°). This angle will not give a closed pattern. The simplest adjustment is to increase the dwell at each end by 1.6°, so that the advance will be 200°. At this angle, the fiber paths will coincide after nine circuits. The total number of circuits per layer is:

$$C_L = \pi D / S_C$$

where D is the mean composite diameter, S_C is the circumferential component of bandwidth, and $S_C = 0.354$ for a 45° wind.

$$C_L = 10\pi / 0.354 = 88.8.$$

The bandwidth is reduced slightly so that C_L is equal to 90, which allows for ten repeating patterns. The ratio of mandrel speed to traverse speed is finally adjusted so that the fiber path advances an additional 0.354 in. (0.9 cm) in the nine circuits.

The degree of precision in controlling machine speeds is apparent from the example. It is seen that bandwidth and dwell angle are varied to obtain complete coverage. In practice, the cylindrical length and wind angle are also considered as variable in realizing a satisfactory pattern.

Other winding types include the following.

Hoop Winding. Hoop or circumferential layers are wound close to 90°; the feed advances one bandwidth per revolution. The layer is considered to be a single ply. (Hoop layers may also be applied as doublers or localized stiffeners at strategic points along the cylinder.)

Longitudinal Winding. This term applies to low angle wraps which are either planar or helical. For closed pressure vessels, the minimum angle is determined by the polar openings at each end.

Combination Winding. Longitudinals are reinforced with hoop layers; the customary practice with pressure vessels is to place the bulk of the hoop wraps in the outer layer. A balance of hoop and longitudinal reinforcement can also be achieved by winding at two or more helical angles.

Miscellaneous. Included here are multicircuit planar wraps analogous to multicircuit helicals, a single circuit helical wrap similar to a planar wind. Both patterns differ from their counterparts only in regard to the fiber path over the end closures.

16.3.2. Winding Machines

These machines are designed for either polar or helical winding. Variations are made in each type to accommodate hoop windings and to add versatility. Polar winders usually operate with the mandrel in a vertical position which eliminates deflections due to weight and permits a simpler construction of the rotating feed arm. A major advantage of the polar machine is that simpler control of the machine motions is possible. The rotation of the feed arm is continuous and at a uniform speed, so that there are no inertial effects which can occur when speeds are varied or directions reversed. On the other hand, operation is limited to prepreg feedstock in most cases, since a wet winding system is difficult to install.

The basic movements of the helical winder are mandrel rotation and feed traverse. To these can be added a cross-slide perpendicular to the mandrel axis and a fourth axis of motion, rotation of the feed eye. These latter two permit more accurate fiber placement over the ends. Controls may be either mechanical or numerical. Mechanical control usually implies a single drive system in which rotation and traverse are fixed by gear trains, link chains, or drive screws. The motions of the numerical winder are controlled by punched tape operation of hydraulic servo drives, each axis having its own hydraulic motor. A recent innovation by one manufacturer utilizes a microcomputer to control servo motors. A silicon chip performs the logic, memory, and calculations for machine function.

Other machines are designed for specific structures. Figure 16.5 is a schematic representation of a spherical winder.[10] A 12-bit minicomputer furnishes control of the machine, which can wind spheres 3–15 in. (7.6–38 cm) in diameter.

Companies which manufacture filament winding machines and auxiliary equipment are listed below:

Brenner, I. G. Co., Newark, Ohio
Engineering Technology, Inc., Salt Lake City, Utah.
Gavlick Machinery Corp., Bristol, Connecticut.

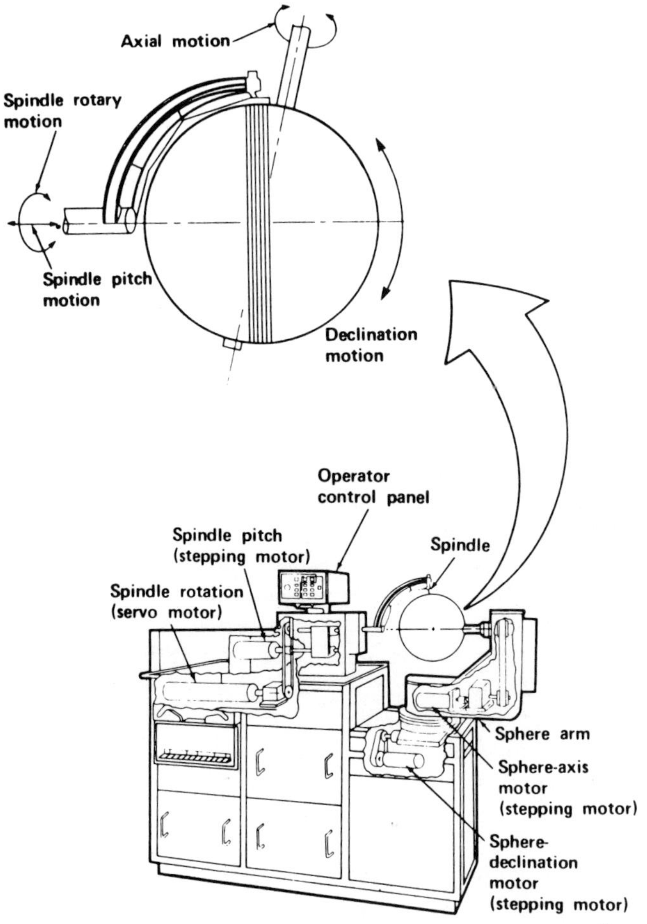

Figure 16.5. Lawrence Livermore Laboratory spherical winder.[10]

Goldsworty Engineering, Inc., Torrance, California.

McClean-Anderson, Inc., Milwaukee, Wisconsin.

Thaden Engineering Co., High Point, North Carolina.

Venus Products, Inc., Kent, Washington.

Vermont Instrument Co., Inc., Burlington, Vermont.

16.3.3. Mandrels

Mandrel design is comparatively simple for open end structures such as cylinders or conical shapes. Either cored or solid steel or aluminum serves satisfactorily. When end closures are integrally wound, as in pressure vessels, careful consideration must be given to mandrel design and selection of a suitable material. A proper design will minimize fiber damage during part removal as well as dimensional intolerances and excessive residual stresses. The mandrel must resist sagging due to its weight and applied winding tension. It must retain sufficient strength during cure at elevated temperatures and be easy to remove after cure. Construction concepts include the following.

1. *Segmented collapsible metal.* These are costly and are not warranted for less than 25 parts. The suggested diameter range is 3–5 feet (0.91–1.52 m). Removal can be complicated with small polar openings.

2. *Low melting alloys.* These are high in density and tend to creep under moderate winding tension. They are limited to small vessels in the order of 1 ft (0.3 m) in diameter by 1 ft (0.3 m) in length.

3. *Eutectic salts.* These are better suited than the alloys and are applicable up to 2 ft (0.6 m) in diameter. With care, they can be slush molded, and they are easy to remove.

4. *Soluble plasters.* These have a long plastic stage and can be wiped to contour. They are easily washed out.

5. *Frangible or break-out plasters.* These are used best with large diameters. Internal support is required, and break-out is difficult and can cause damage. Chains can be imbedded to facilitate removal.

6. *Sand—PVA.* This material is an excellent choice for diameters up to 5 ft (1.5 m) and for limited quantities. It dissolves readily in hot water, but requires careful molding control. Low compressive strength is a limitation.

7. *Inflatables.* These are not suitable where it is necessary to resist torque loads. One technique for improving the torque resistance is to fill the mandrel with a material such as sand and to apply a vacuum. Another use for inflatables is to transfer the uncured winding to a closed mold and to cure with pressurization through the mandrel.

The properties of several mandrel materials are given in Table 16.8. Figure 16.6 is a schematic of the support structure for a wash-

Table 16.8. Properties of Mandrel Materials

Material	Kerr DMM[a]	Hydrocal B-11[b]	Brak-Away[c]	Paraplast 36[c]	Sand/PVA
Type	Wash-out plaster	Frangible plaster	Frangible plaster	Soluble salt	Wash-out
Compressive strength, psi (MPa)	700 (4.8)	3800 (26.2)	375 (2.6)	14,000 (96.5)	500 (3.5)
Specific gravity	1.25	1.36	0.8	2.08	
Moisture pick-up					
% at 75% RH	0.07	0.45			
% at 70% RH			86.5		
Maximum use temperature, °F (°C)	400 (204)	400+ (204+)	400+ (204+)	350 (177)	350 (177)
Set time, minutes	20–25	45–55	10–15		
Set expansion, in./in.	0.043	0.0004	0.0015		

[a]Kerr Mfg Co
[b]US Gypsum
[c]Rezolin, Inc

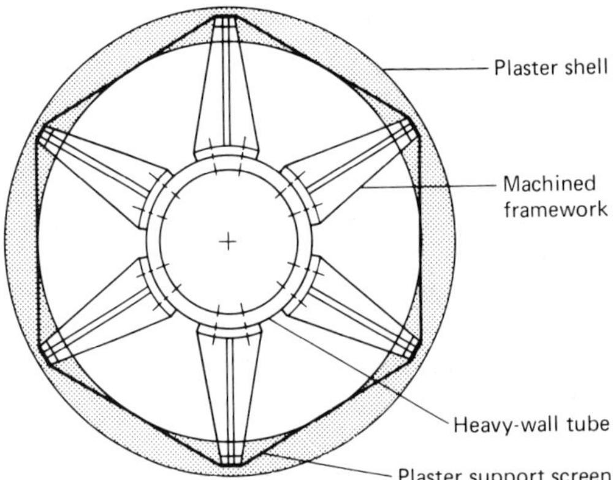

Figure 16.6. Framework for plaster mandrel.[11]

out plaster mandrel. Figure 16.7 plots mandrel thickness versus diameter for three materials. In this latter case, the vessel is designed for a pre-stress equal to 20% of operating pressure.

16.3.4. End Closures

End closures for pressure vessels are either mechanically fastened to the cylindrical portion or are integrally wound. The former method has been found practical in some commercial applications. For high performance vessels, typified by rocket motor cases, integral heads are essential. Head contours deviate from a spherical shape, which is less efficient. The fiber path yields a balance of meridional and circumferential forces and is consistent with winding conditions so that no slippage occurs. The head contours and related polar bosses are critical in vessel design. Below, the contours associated with helical and polar winding are briefly summarized.[12,16]

16.3.4.1. Geodesic Isotensoid

This contour is normally adapted to helical winding. The fiber path is taken as tangential to the polar boss (Fig. 16.8). Each point on the path is defined by its meridional and circumferential radii, r_1 and r_2, respectively. These radii are related to the X, Y coordinates, which establish the contour, in the following manner.

$$r_1 = \frac{-[1 + (Y')^2]^{3/2}}{Y''} \qquad 16\text{-}1$$

$$r_2 = \frac{-X[1 + (Y')^2]^{1/2}}{Y'} \qquad 16\text{-}2$$

where Y' and Y'' are the first and second derivatives of Y with respect to X.

The principal membrane forces due to an internal pressure, P, are:

$$\text{Meridional: } N_\phi = \frac{Pr_2}{2} \qquad 16\text{-}3$$

$$\text{Circumferential: } N_\theta = \frac{Pr_2}{2}[2 - r_2/r_1] \qquad 16\text{-}4$$

For a balanced stress condition, the fila-

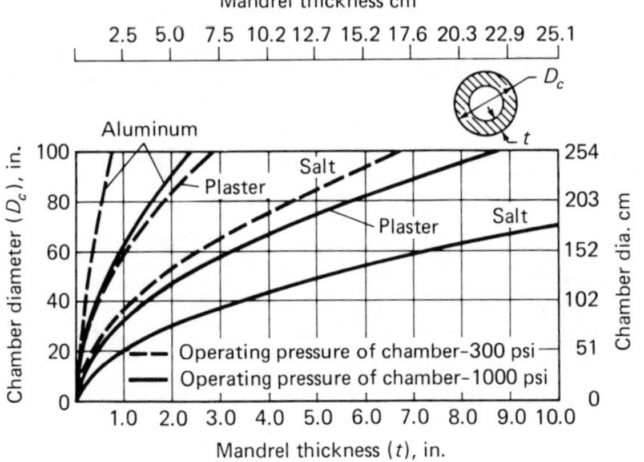

Assumptions: Radial pressure from filament winding – 20% of operating pressure of chamber. Allowable radial deflection of mandrel under winding tension – 0.020 in.

Figure 16.7. Mandrel thickness versus cylinder diameter.[12]

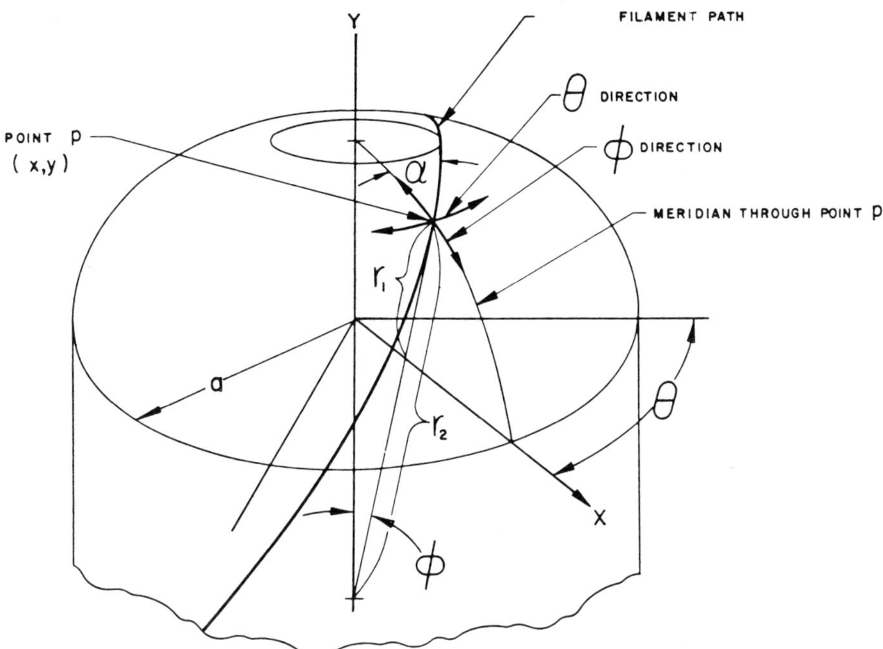

Figure 16.8. Geometry for geodesic path.[15]

ment strengths in the principal directions are equated to these forces, and the following equation results.

$$\frac{r_2}{r_1} = \frac{XY''}{Y'[1(Y')^2]} = 2 - \tan^2 \alpha \qquad 16\text{-}5$$

where α = the winding angle. For the path to be geodesic, Eq. 16-6 holds true:

$$X \sin \alpha = \text{constant} \qquad 16\text{-}6$$

At the tangency point, $\alpha = 90°$ and:

$$\text{Sin } \alpha = X_0 / X \qquad 16\text{-}7$$

where X_0 is the boss diameter.

Solution of Eqs. 16-5 and 16-7 yield the coordinates of the contour. The solution is generally obtained by a computerized step integration. Graphical solutions are also possible.[14,16] When $\tan^2 \alpha$ equals 2, Eq. 16-5 is no longer usable. The simplest solution beyond this inflection point is to continue a spherical radius to a junction with the polar boss.

16.3.4.2. Planar or Balanced in Plane

The fiber path for this contour is described as the intersection of a plane with the end closure (Fig. 16.9). Here, the principal stresses are balanced by adjusting the instantaneous radii of curvature at each point. The equation of the wrap plane is:

$$X \cos \theta = B + AY \qquad 16\text{-}8$$

The filament angle at any point is given by Eq. 16-9.

$$\text{Tan } \alpha = \frac{A \sin \phi + \cos \phi \cos \theta}{\sin \theta} \qquad 16\text{-}9$$

$$\text{Where: } \tan \phi = -Y' \qquad 16\text{-}10$$

Equation 16-5, in conjunction with Eqs. 16-8, 16-9, and 16-10 establish the contour and the filament path.

Table 16.9 lists the coordinates for a geodesic-isotensoid contour as determined by a graphical solution. Figure 16.10 is the contour for a typical planar wrap. The calculations for both types of contours, it should be noted, are based on a netting analysis, and all the assumptions of this method are implied.

16.3.5. Material Handling and Process Controls

The equipment for transporting the strands from the roving package to the mandrel forms

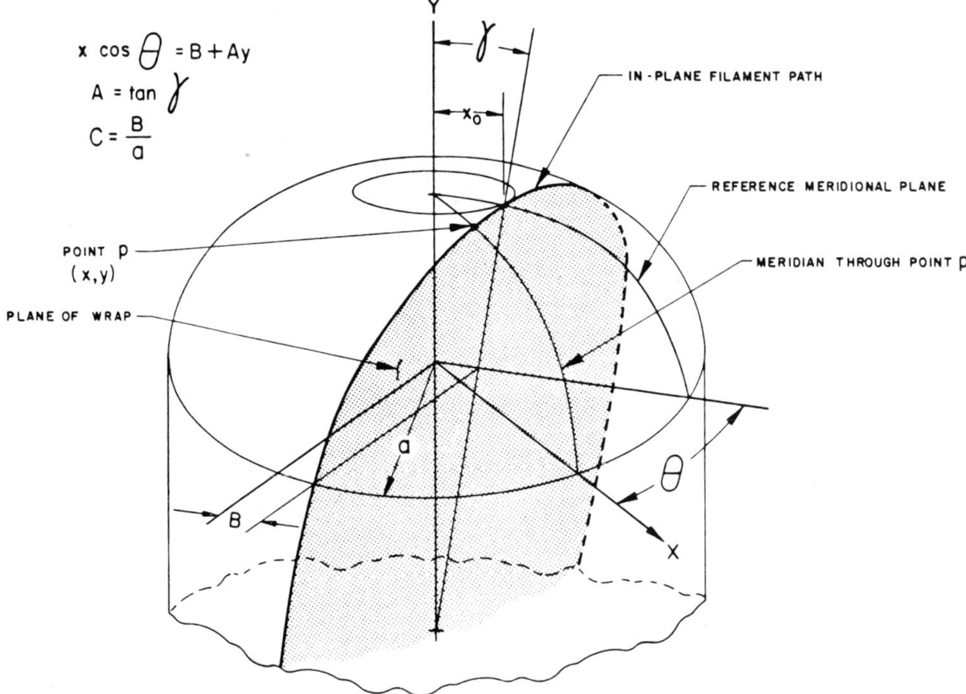

Figure 16.9. Geometry of a planar path.[15]

Table 16.9. Head Contour for a Helical Wrap[16]

ANGLE DEGREES	MINUTES	RADIAL X	AXIAL Y
17	6	1.00[a]	0.000
18		0.95	0.222
19	6	0.90	0.309
20	15	0.85	0.368
21	38	0.80	0.417
23	5	0.75	0.457
24	50	0.70	0.489
26	48	0.65	0.516
29	12	0.60	0.539
32	16	0.55	0.559
36		0.50	0.575
40	46	0.45	0.590
47	18	0.40	0.600
57	2	0.35[b]	0.613
78	32	0.30	0.622
90		0.294[a]	0.625

[a]Equator.
[b]Inflection point.
[c]Polar boss.

an essential feature of the winding. Such equipment provides the means for controlling tension, resin content, bandwidth, and layer thickness. Careful attention to details at this point prevents fiber damage and leads to optimum mechanical properties and product uniformity.

Tension is provided by guide eyes, drum type brakes, scissor bars, and the drag through the resin tank. Typically, tension ranges from 0.25–1.0 lb (1–4.4 N) per end. Normal procedure is to keep the tension on dry fibers at a minimum to prevent excessive abrasion and snarling. The tension is then increased as the strand passes through the resin tank and feed bar. With the tubeless inside pull roving package, sufficient initial tension can be applied by guide eyes alone. When the roving is tube-wound for an outside pull, it may be necessary to install some form of braking on each roll. After initial tensioning, a minimum number of guides is employed to direct the strands to the resin tank. The ceramic guides used for handling textile fibers are considered superior to other materials, such as stainless

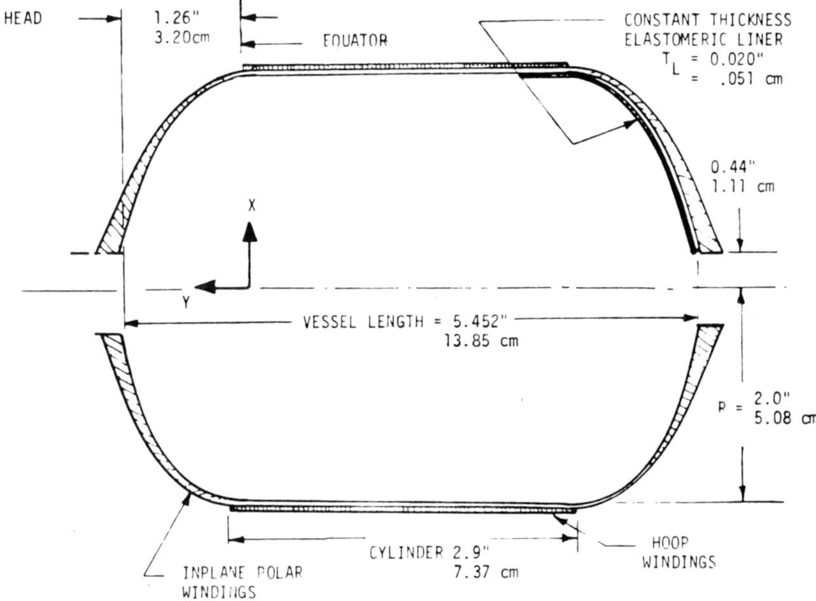

DOME CONTOUR				
	Inches		cm	
	Y	X	Y	X
EQUATOR	0	2.	0	5.08
	0.319	1.950	.810	4.95
	0.439	1.900	1.115	4.826
	0.526	1.850	1.336	4.699
	0.612	1.800	1.554	4.572
	0.840	1.59	2.134	4.039
	0.979	1.380	2.487	3.505
	1.060	1.210	2.642	3.073
	1.131	1.000	2.872	2.54
	1.179	0.800	2.995	2.032
	1.209	0.597	3.071	1.516
BOSS	1.222	0.490	3.104	1.245

Figure 16.10. Schematic for planar wind.[17]

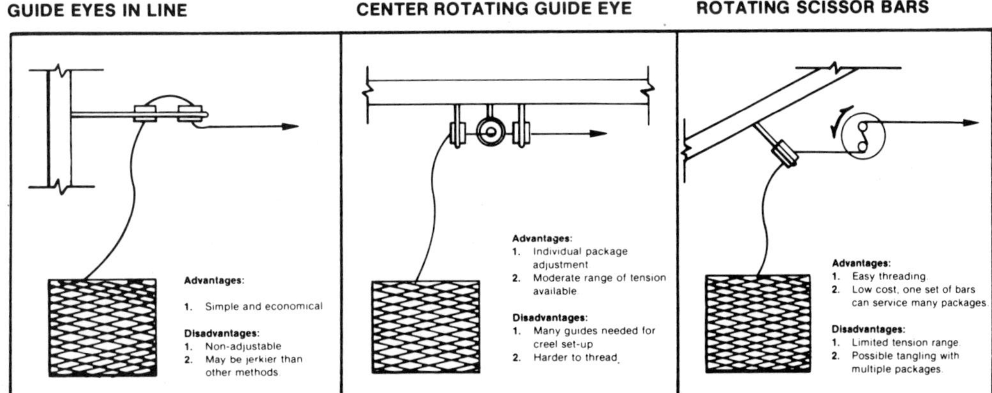

Figure 16.11. Roving guide arrangements.[2]

steel and Teflon. Several guide arrangements are shown in Fig. 16.11.

The design of the impregnation tank depends on the number of strands being processed, the process speed, the resin viscosity, the resin gel time, and whether or not the resin is to be heated. The two basic methods for strand impregnation are illustrated in Fig. 16.12. In one case, the strands pass under rollers located in the resin. The alternate method is to pass the strands over a metering roll. A more elaborate design, with a heated bath and rollers, is shown in Fig. 16.13. This tank handles a 15-end roving band with a tension of 4-6 lb (17.8-26.7 N). In most cases, a wiping device or doctor blade is required to remove excess resin from the strands as they leave the tank.

Several methods for forming a flat roving band are depicted in Fig. 16.14. In these illustrations, the feed bar or eye is allowed to swivel freely as the feed direction is reversed. For a more precise band placement on the mandrel, the feed head is machine controlled in a path tangent to the mandrel. A uniform flat band will result in improved strength as well as a more uniform thickness.

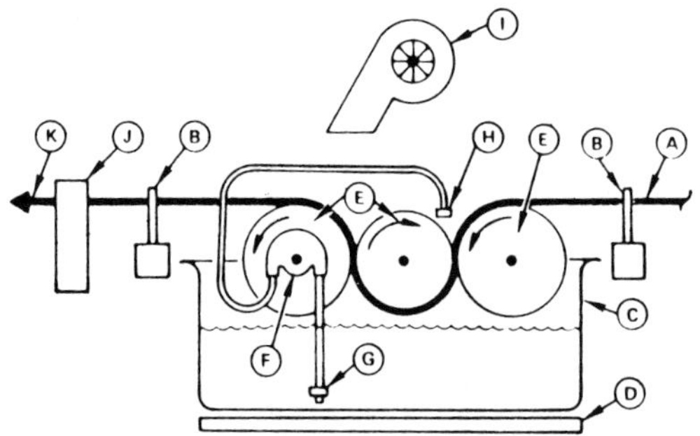

Figure 16.13. Resin impregnator.[18]

IMPREGNATION TANK WITH FEED FROM DOUBLE LEVEL

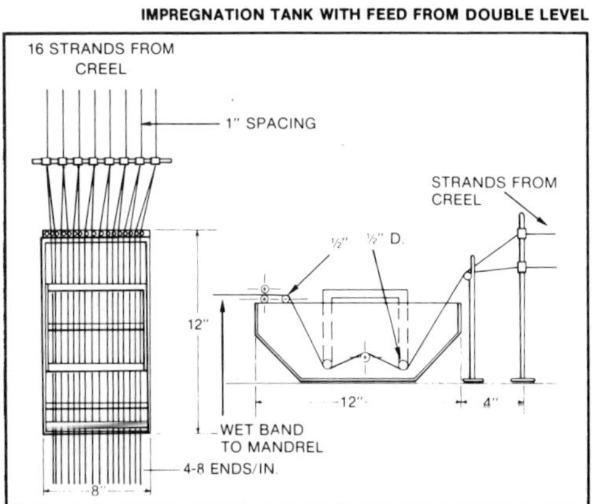

ROLLER IMPREGNATION WITH FEED FROM SINGLE LEVEL

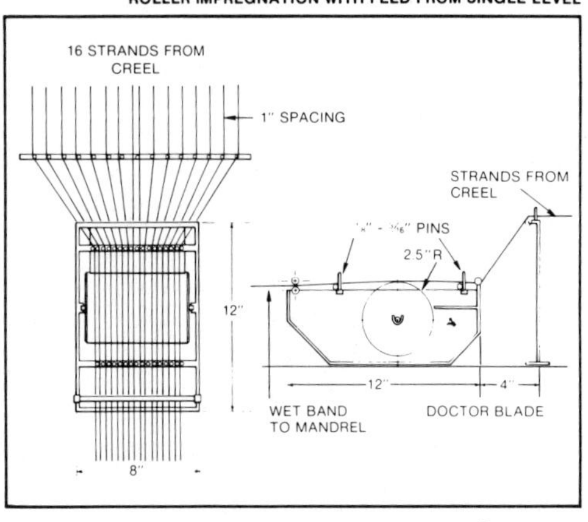

Figure 16.12. Roving impregnation.[2]

The thickness of a single layer can be calculated for a specific band density (ends/in. or cm) and glass content. The results of such calculations for a Type 30 roving are graphically illustrated in Fig. 16.15. Similar charts can be prepared for other rovings. Experimentally, an angle-ply of G-fiber strands will have a normal thickness range of 10–15 mils (0.25–0.38 mm). A single ply hoop layer will be 4–7 mils (0.10–0.18 mm). These figures, it should be noted, are rough estimates. The average layer thickness will depend on the accuracy with which the glass/resin ratio is maintained, the amount of voids, and the degree of compaction on the mandrel.

Winding speeds vary over a broad range. Some newer winding machines are said to have capabilities up to 450 linear ft/minute (137 m/minute). This speed would appear to be excessive. An upper limit of 300–350 linear ft/minute (91.4–106.7 m/minute) seems more practical. For more precise winding, particularly with graphite and aramid fibers, speeds of 50–100 lfpm (15.2–30.5 m/minute) are typical.

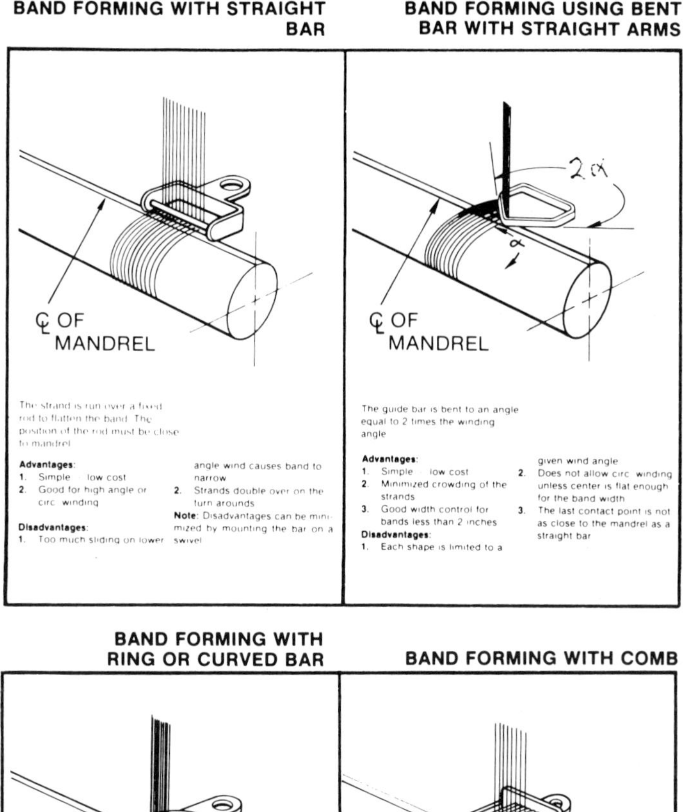

Figure 16.14. Band formation.[2]

16.4. MECHANICAL PROPERTIES

16.4.1. General

A summation of the properties of filament wound composites, as with other fiber reinforced materials, is complicated by a number of factors.

- A wide selection of reinforcement types and resin systems is currently in use, each of which requires separate evaluation, and many of which have not been fully tested.
- Process variables, such as uniformity of fiber/resin ratio, air entrapment, degree of compaction, and curing, strongly influence properties; variation from one fabricator to another is great.
- Design features, such as wind angle, layer

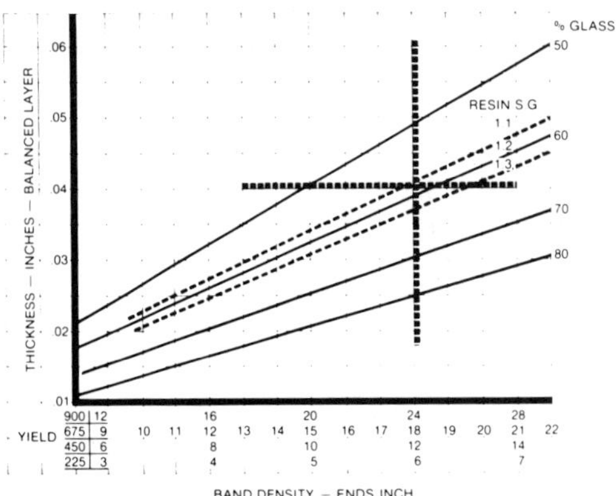

Figure 16.15. Layer thickness versus band density.[2]

sequence, and end closures, produce differing results.

- The testing of FRP, in general, tends to show wide scatter.
- Evaluation approaches exemplified by aerospace and commercial fabricators lead to disparate results.
- Structures designed for specific mechanical, chemical, or electrical properties may be weakened in some properties as a result of the optimization.

Consequently, only certain mechanical properties are reviewed here. Property values of typical pressure vessels and general purpose pipes, together with some theoretical results, are included. The chemical, electrical, and environmental properties are the major concern in numerous winding applications. These properties are not within the scope of this chapter. The chemical and electrical response of the filament wound composites is essentially a function of the resin system. Recourse to data on specific resins will therefore provide indications of the composite behavior.

16.4.2. Analytical Methods

The behavior of filament wound composites is analogous to that of other angle-plied laminates so that the analytical methods developed for laminates can be applied to filament wound structures. In the macromechanics approach, the composite is analyzed by assuming each lamina to be an anisotropic, homogeneous monolayer. The monolayer consists of filaments oriented at plus and minus an angle, alpha, or else it is unidirectional. The properties of the monolayer are usually determined experimentally, and the analysis proceeds on a layer by layer basis. Micromechanics, on the other hand, is a study of the response characteristics of the composite constituents; that is, the stress and strain distribution within the fibers and matrix. The properties of the reinforcement and resin, as well as the geometry of the composite, are taken into account in determining point by point stresses and strains. This microstress analysis establishes what loads the composite can withstand before yielding at some point or reaching ultimate failure. Micromechanics is also used to calculate the properties of the composite from the constituent properties and to estimate the variable effects of these properties on the composite.

Both procedures are discussed in other sections of this handbook. Additional treatments dealing specifically with filament wound cylinders are found in the literature.[14,19,20,21]

The netting analysis is a simplified procedure used mainly to estimate fiber stresses in a cylindrical vessel subjected to internal pressure. This method is based on the assumptions that only the reinforcing fibers have a load carrying capability and that all fibers are uniformly stressed in tension. Figure 16.16 represents a two-layered system of parallel fibers from which the netting equations can be derived.

The stress in each fiber is σ_f acting on the cross-sectional area of the fiber. These forces are resolved into components in the X and Y directions, corresponding to the hoop and longitudinal directions, respectively. A summation of the forces in both directions leads to the following equations.

$$\sigma_x = \sigma_f \cos^2 \alpha \qquad 16\text{-}11$$

$$\sigma_y = \sigma_f \sin^2 \alpha \qquad 16\text{-}12$$

and

$$\sigma_y / \sigma_x = \tan^2 \alpha \qquad 16\text{-}13$$

where α is the angle of wind.

When the winding consists of two layers at different angles, the following equations can be derived.

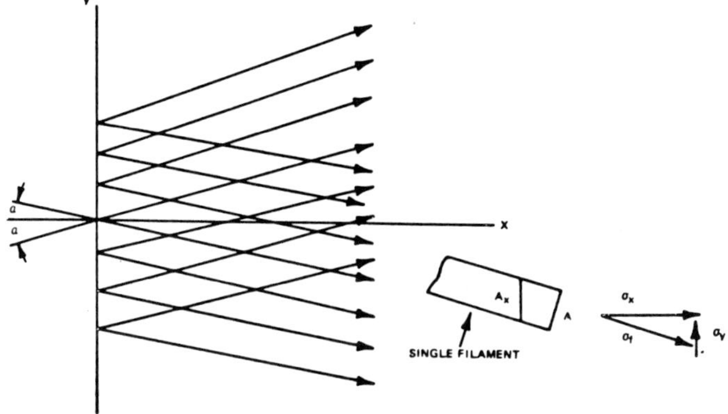

Figure 16.16. Fiber orientation for netting analysis.

Longitudinal:

$$\sigma_L(t_1 + t_2) = \sigma_1 t_1 \cos^2 \alpha_1 + \sigma_2 t_2 \cos^2 \alpha_2 \quad 16\text{-}14$$

Hoop:

$$\sigma_H(t_1 + t_2) = \sigma_1 t_1 \sin^2 \alpha_1 + \sigma_2 t_2 \sin^2 \alpha_2 \quad 16\text{-}15$$

When the outer layer is composed of hoop wraps only, Eqs. 16-14 and 16-15 are reduced to:

$$\sigma_L t_1 = \sigma_1 t_1 \cos^2 \alpha_1 \quad 16\text{-}16$$

$$\sigma_H(t_1 + t_2) = \sigma_1 t_1 \sin^2 \alpha_1 + \sigma_2 t_2 \quad 16\text{-}17$$

Where σ_L = longitudinal stress
σ_H = hoop stress
σ_1 = fiber stress in layer 1
σ_2 = fiber stress in layer 2
t_1 and t_2 = thickness of layer 1 and 2
α_1 and α_2 = wind angles in layer 1 and 2

Normally, $\sigma_1 = \sigma_2 = \sigma_f$ and the ratio of hoop to helical thickness in a closed pressure vessel can be expressed as Eq. 16-18.

$$\frac{t_{hoop}}{t_{helix}} = 2\cos^2\alpha_1 - \sin^2\alpha_1 = 3\cos^2\alpha_1 - 1 \quad 16\text{-}18$$

The tensile modulus in the fiber direction and Poisson's ratio are frequently estimated through the "rule of mixtures."

Longitudinal modulus:

$$E_x = E_f V_f + E_m V_m \quad 16\text{-}19$$

Poisson's ratio:

$$\mu_{xy} = \mu_f V_f + \mu_m V_m \quad 16\text{-}20$$

where V is the volume fraction, and the subscripts f and m refer to the fiber and matrix.

16.4.3. Analytical Solutions

The theoretical approximations of composite principal properties serve a useful purpose in preliminary design and in understanding material behavior. It is also possible to establish a correlation between theoretical and empirical values. Table 16.10 is a typical computer printout showing the composite properties versus the wind angle. The four reinforcements listed are each at a 50% fiber volume.

Some earlier results are presented in Figs. 16.17–16.20, in which test values are compared with theory for the elastic constants and Poisson's ratio.

Most analyses assume a linear material response even though it is conceded that such is not the actual case. An analysis[21] takes into account the observable non-linearity. A comparison is made in Table 16.11 between linear and non-linear behavior of a six-ply glass/epoxy cylinder under internal pressure of 4000 lb (27.58 MPa). The four inner plies are wound at an angle of 54°, while the two outer layers are at 83°. As indicated, the non-linearity reduces the stress in the inner wraps while increasing it in the outer plies. Fig. 16.21 shows typical effective modulus vs. wind angle relationship.

16.4.4. Empirical Results

Burst strengths of pressure vessels of various diameters are listed in Table 16.12. These

Table 16.10. Monolayer Properties at Various Wind Angles[22]

	Msi					Ksi						GPa			MPa				
ALPHA	EX	EY	GXY	UXY	UYX	FXTU	FYTU	FXCU	FYCU	FXY	ALPHA	EX	EY	GXY	FXTU	FYTU	FXCU	FYCU	FXY
E-Glass/Epoxy, 50% Fiber Volume, 0.0666 Density (1843 Kg, m^3)																			
0	5.49	1.77	0.26	0.285	0.092	137.5	0.0	92.5	29.4	5.5	0	37.9	12.2	1.8	948.1	000.0	637.8	202.7	37.9
5	5.39	1.75	0.30	0.309	0.101	132.2	0.2	91.1	29.1	7.0	5	37.2	12.1	2.1	911.6	1.4	628.1	200.6	48.3
10	5.10	1.69	0.42	0.380	0.126	118.3	0.6	86.9	28.1	9.7	10	35.2	11.7	2.9	815.7	4.1	599.2	193.7	66.9
15	4.62	1.60	0.60	0.491	0.169	100.1	1.9	79.5	26.6	13.2	15	31.9	11.0	4.1	690.2	13.1	548.1	183.4	91.0
20	3.96	1.46	0.82	0.627	0.232	81.3	3.4	69.0	24.6	17.3	20	27.3	10.1	5.7	560.5	23.4	475.7	169.6	119.3
25	3.17	1.31	1.05	0.762	0.314	64.5	5.5	56.0	22.1	21.5	25	21.9	9.0	7.2	444.7	37.9	386.1	152.4	148.2
30	2.36	1.14	1.27	0.862	0.415	50.2	8.3	42.1	19.4	25.5	30	16.3	7.9	8.8	346.1	57.2	290.3	133.8	175.8
35	1.66	0.98	1.45	0.901	0.533	38.7	11.8	29.7	17.0	28.8	35	11.0	6.8	10.0	266.8	81.4	204.8	117.2	198.6
40	1.18	0.90	1.56	0.868	0.660	29.4	16.4	21.0	15.6	31.0	40	8.1	6.2	10.8	202.7	113.1	144.8	107.6	213.7
45	0.94	0.94	1.60	0.780	0.780	22.1	22.1	16.6	16.6	31.7	45	6.5	6.5	11.0	152.4	152.4	114.5	114.5	218.6
S-Glass Epoxy, 50% Fiber Volume, 0.0656 Density (1816 Kg/m^3)																			
0	6.54	1.82	0.26	0.285	0.080	162.5	0.0	107.5	29.5	5.4	0	45.1	12.5	1.8	1120.4	000.0	741.2	203.4	37.2
5	6.42	1.80	0.31	0.313	0.088	155.1	0.2	105.8	29.2	7.2	5	44.3	12.4	2.1	1069.4	1.4	729.5	201.3	49.6
10	6.07	1.74	0.45	0.395	0.113	136.2	0.8	100.7	28.2	10.1	10	41.8	12.0	3.1	939.1	5.5	694.3	194.4	69.6
15	5.47	1.64	0.67	0.523	0.157	112.4	1.9	91.8	26.7	14.2	15	37.7	11.3	4.6	775.0	13.1	632.9	184.1	97.9
20	4.65	1.50	0.93	0.678	0.219	89.2	3.4	79.0	24.6	19.0	20	32.1	10.3	6.4	615.0	23.4	544.7	169.6	131.0
25	3.66	1.34	1.21	0.828	0.302	69.1	5.5	63.0	22.0	24.2	25	25.2	9.2	8.3	476.4	37.9	434.4	151.7	166.9
30	2.65	1.16	1.47	0.933	0.407	52.9	8.3	46.1	19.2	29.1	30	18.3	8.0	10.1	364.7	57.2	317.8	132.4	200.6
35	1.80	0.99	1.68	0.962	0.531	40.2	11.9	31.4	16.7	33.3	35	12.4	6.8	11.6	277.2	82.0	216.5	115.1	229.6
40	1.23	0.90	1.82	0.911	0.668	30.2	16.5	21.3	15.3	36.1	40	8.5	6.2	12.6	208.2	113.8	146.9	105.5	248.9
45	0.95	0.95	1.87	0.803	0.803	22.5	22.5	16.3	16.3	37.1	45	6.6	6.6	12.9	155.1	155.1	112.4	112.4	255.8
Aramid/Epoxy, 50% Fiber Volume, 0.0468 Density (1295 Kg/m^3)																			
0	9.74	0.76	0.23	0.285	0.022	162.5	0.0	35.0	2.7	1.4	0	67.2	5.2	1.6	1120.4	000.0	241.3	18.6	9.7
5	9.55	0.76	0.30	0.376	0.030	150.3	0.1	34.4	2.7	1.9	5	65.8	5.2	2.1	1036.3	0.7	237.2	18.6	13.1
10	8.94	0.75	0.50	0.633	0.053	122.1	0.5	32.4	2.7	2.9	10	61.6	5.2	3.4	841.9	3.4	223.4	18.6	20.0
15	7.83	0.73	0.80	0.996	0.092	92.1	1.1	28.6	2.6	4.5	15	54.0	5.0	5.5	635.0	7.6	197.2	17.9	31.0
20	6.22	0.70	1.18	1.340	0.150	67.5	2.0	22.9	2.5	6.8	20	42.9	4.8	8.1	465.4	13.8	157.9	17.2	46.9
25	4.44	0.67	1.58	1.525	0.230	49.0	3.3	16.4	2.5	9.5	25	30.6	4.6	10.9	337.8	22.8	113.1	17.2	65.5
30	2.90	0.65	1.96	1.499	0.335	35.8	4.9	10.7	2.3	12.1	30	20.0	4.5	13.5	246.8	33.8	73.8	15.9	83.4

35	1.82	0.65	2.27	1.321	0.470	26.1	7.2	6.7	2.3	14.0	35	12.5	4.5	15.7	180.0	49.6	46.2	15.9	96.5
40	1.18	0.70	2.46	1.083	0.640	19.1	10.1	4.3	2.5	15.2	40	8.1	4.8	17.0	131.7	69.6	29.6	17.2	104.8
45	0.85	0.85	2.53	0.847	0.847	14.0	14.0	3.1	3.1	15.6	45	5.9	5.9	17.4	96.5	96.5	21.4	21.4	107.6
Graphite 300/Epoxy, 50% Fiber Volume, 0.0524 Density (1450 Kg/m^3)																			
0	17.23	0.78	0.51	0.285	0.015	162.5	0.0	107.5	5.4	4.0	0	118.8	6.0	3.5	1120.4	000.0	741.2	37.2	27.6
5	16.91	0.88	0.63	0.416	0.022	152.6	0.1	105.7	5.5	5.4	5	116.6	6.1	4.3	1052.4	0.7	728.8	37.9	37.2
10	15.81	0.89	0.97	0.771	0.043	128.7	0.6	99.3	5.5	7.6	10	109.0	6.1	6.7	887.4	4.1	684.7	37.9	52.4
15	13.72	0.90	1.49	1.225	0.081	101.1	1.4	86.8	5.6	10.9	15	94.6	6.2	10.3	697.1	9.7	598.5	38.6	75.2
20	11.83	0.93	2.13	1.574	0.136	76.6	2.5	68.8	5.8	15.3	20	81.6	6.4	14.7	528.1	17.2	474.4	40.0	105.5
25	7.83	0.98	2.81	1.676	0.211	57.2	4.1	49.7	6.1	21.0	25	54.0	6.8	19.4	394.4	28.3	342.7	42.1	144.8
30	5.33	1.07	3.44	1.551	0.310	42.6	6.2	33.8	6.6	26.9	30	36.7	7.4	23.7	293.7	42.7	233.0	45.4	185.5
35	3.60	1.20	3.96	1.309	0.439	31.6	8.9	22.8	7.5	32.3	35	24.8	8.3	27.3	217.9	61.4	157.2	51.7	222.7
40	2.50	1.44	4.30	1.044	0.601	23.4	12.5	15.8	9.0	35.9	40	17.2	9.9	29.6	151.1	85.2	108.9	62.1	247.5
45	1.84	1.84	4.42	0.803	0.803	17.2	17.2	11.5	11.5	37.1	45	12.7	12.7	30.5	118.6	118.6	79.3	79.3	255.8

EX, EY, GXY—MODULUS; UXY, UYX—POISSON'S RATIO
FXTU, FYTU—TENSILE; FXCY, FYCU—COMPRESSION
FXY—SHEAR

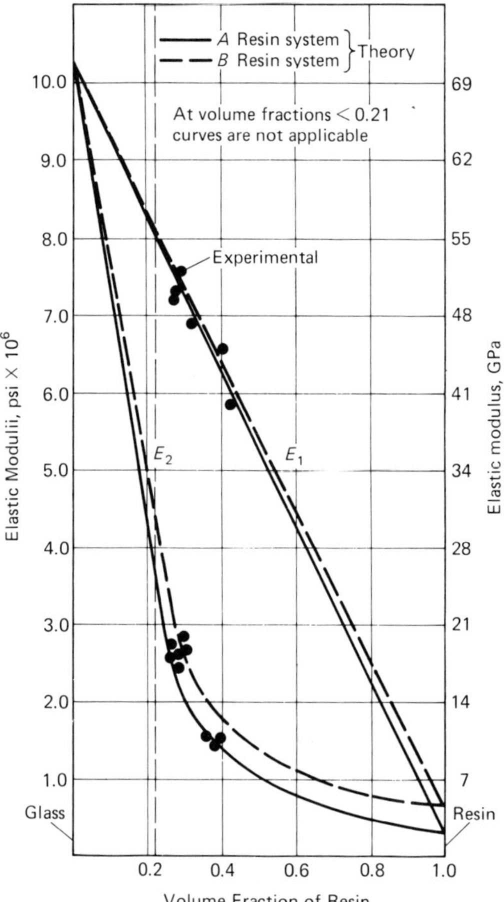

Resin Properties

A: $E_R = 4.29 \times 10^5$; $G_R = 1.67 \times 10^5$; $\mu = 0.284$.

B: $E_R = 5.41 \times 10^5$; $G_R = 1.96 \times 10^5$; $\mu = 0.380$.

Glass Same For Both

$E_G = 10.5 \times 10^6$; $G_G = 4.3 \times 10^6$; $\mu_G = 0.22$.

Figure 16.17. Modulus versus resin volume, unidirectional laminate.[23]

vessels are wound from S-glass/epoxy, with fiber volumes ranging from 55–65%. Burst pressures are converted to fiber and composite stresses by netting analysis, which yields an average stress. Burst pressure, it must be noted, varies considerably with the rate of pressurization and the time held at any pressure. The previously mentioned decrease in strength with increasing diameter is apparent. Design factors have been used to counteract this effect by modifying the allowable filament stress. Table 16.13 summarizes the factors determined by one manufacturer. Here, the results are experimentally verified up to a 54-in. (1.37-m) diameter and are extrapolated to 300 in. (7.62 m). Similar factors may be applied for variations in thickness and L/D ratios.

Aramid and graphite have proven superior to S-glass in the high pressure service associated with liquid gas and fluid containment at ambient and cryogenic temperatures. A summary of NASA programs in this area is listed in the references,[25] and is briefly described here. Three types of liners have been developed for these vessels; namely, elastomeric, thin metal, and load sharing metal. The relatively low modulus of S-glass limits its serviceability to elastomeric liners. Such vessels, in turn, are applicable only to moderate pressures and temperatures. Aramid/epoxy with a thin aluminum liner has a performance factor in the order of 1.2 million in. (3 million cm). The performance factor is defined as burst pressure times volume divided by total weight. There is no distinction in performance between spherical or cylindrical vessels. Planar winding appears superior to helical winding in the range of test vessel diameters. Graphite/epoxy with a titanium load sharing liner has the lowest weight and a longer cyclic life, 3000 cycles at 50% of average burst strength. The aramid vessels are only slightly heavier, have a medium cyclic life, and are less costly than the graphite version. Tables 16.14, 16.15, and 16.16 show representative results obtained with experimental vessels.

Burst strength under pressurization varies with the method of testing. Table 16.17 summarizes the results of tests under two conditions: 1) closed ends, in which the hoop stress is twice the axial; and 2) ends free to slide, so that the axial stress is zero. The pipe specimens for these tests were wound of fiberglass/polyester at an angle of 54°44′ and an average fiber volume of 53%. The values shown are the average of five tests. With the closed ends, resin cracking occurred relatively early in the pressure cycle and lead to a non-linear response. As pressure was increased, the pipe behaved as though complete resin debonding had taken place. Weeping also began early, as small droplets in a line parallel to the fibers.

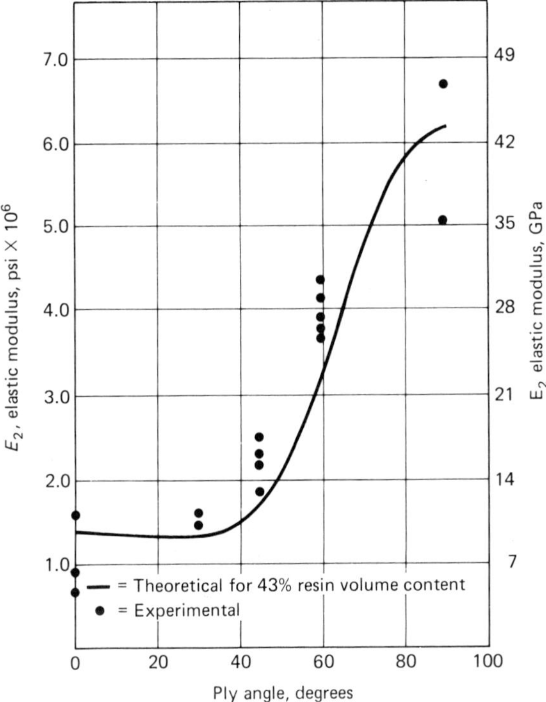

Figure 16.18. Elastic modulus versus ply angle, hoop loading.[23]

Failure involved large scale fracture of the fibers. With free ends, resin shear appeared to dominate the failure mode. Weeping occurred more suddenly and at a pressure level closer to the ultimate.[28]

Tables 16.18 and 16.19 list the test results for wound pipe of various fiber orientations in uniaxial compression and tension. Also included are shear properties in torsion and bending. Properties of commercial pipe made by the "Drostholm" process are shown in Table 16.20. This method utilizes continuous

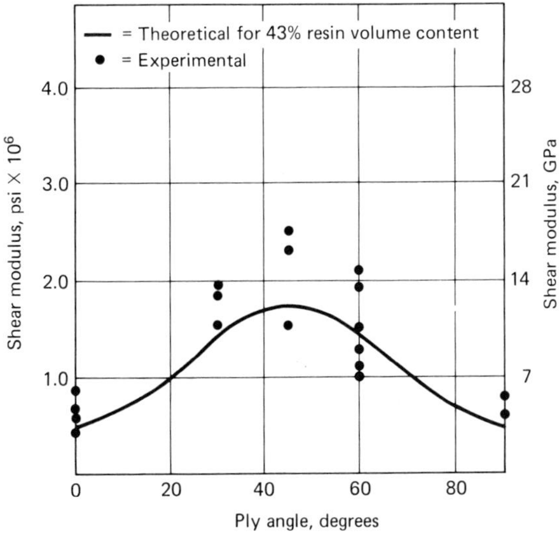

Figure 16.19. Shear modulus versus ply angle, torsion.[23]

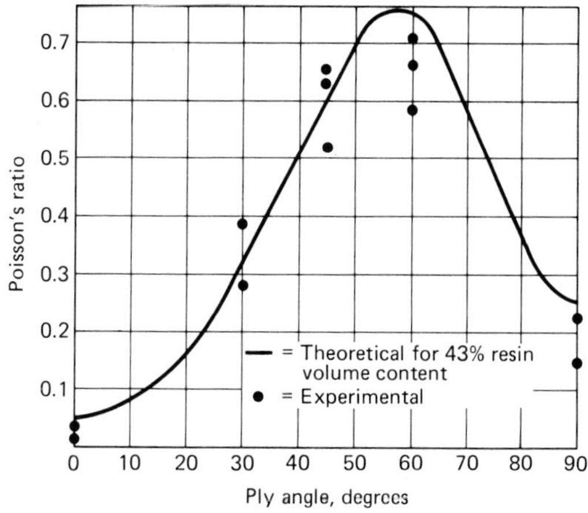

Figure 16.20. Poisson's ratio versus ply angle, hoop tension.[23]

Table 16.11. Comparison of Fiber Stress[21]

PLY NUMBER	LINEAR ANALYSIS, ksi (MPa)	NON-LINEAR ANALYSIS, ksi (MPa)
1	58.7 (407.2)	45.6 (316.6)
2	58.4 (404.1)	47.6 (329.4)
3	58.4 (404.3)	49.3 (340.4)
4	58.0 (401.5)	48.8 (338.0)
5	115.0 (792.8)	139.0 (967.3)
6	113.0 (784.6)	138.0 (949.7)

Table 16.12. Burst Strength of Pressure Vessels

DIAMETER, in. (cm)	FIBER STRESS, ksi (GPa)	COMPOSITE STRESS, ksi (GPa)
4 (10.16)	400–480 (2.76–3.31)	180–210 (1.24–1.45)
6 (15.24)	380–420 (2.62–2.90)	180–210 (1.24–1.45)
12 (30.48)	360–400 (2.48–2.76)	160–180 (1.10–1.24)
18 (45.72)	340–400 (2.34–2.76)	160–180 (1.10–1.24)
44 (111.76)	300–370 (2.07–2.55)	130–160 (0.90–1.10)
54 (137.16)	290–350 (2.00–2.41)	125–150 (0.86–1.03)
260[a] (660.4)	260–290 (1.79–2.00)	125–175 (0.86–1.21)

[a]Design values.

Table 16.13. Correction Factor for Allowable Stress[12]

DIAMETER, in. (cm)	HOOP FILAMENT	LONGITUDINAL FILAMENT
4 (10)	0.99	0.95
18 (46)	0.96	0.89
36 (91)	0.94	0.87
44 (112)	0.93	0.87
54 (137)	0.92	0.86
300[a] (762)	0.87	0.83

[a]Extrapolated.

Table 16.14. Rubber Lined Kevlar/Epoxy Vessels[25]

PROPERTY	RANGE
TYPE OF WINDING	PLANAR
Diameter	4 in. (10.2 cm)
L/D	1.3-3.0
Boss diameter % diameter	10-23
Burst pressure, psi (MPa)	2450-2770 (16.9-19.1)
Composite modulus at burst, longitudinal, psi × 10^6 (GPa)	3.2-5.0 (22.1-34.5)
Hoop modulus, psi × 10^6 (GPa)	7.2-9.4 (49.6-62.7)
Maximum strain, %	
Axial	1.5-1.7
Hoop	1.5-2.0
P_BV/W, composite, in. × 10^6 (KPa · m^3/kg)	1.49-1.57 (371-391)
Maximum fiber stress, hoop, ksi (GPa)	419-443 (2.89-3.05)

Table 16.15. Aluminum Lined Kevlar/Epoxy Vessels[25]

PROPERTY	RANGE
DIAMETER	6 in. (15.24 cm)
P_BV/W, composite, in. × 10^6 (KPa · m^3/kg)	1.42-1.76 (354-438)
Maximum fiber stress, hoop, ksi (GPa)	352-428 (2.43-2.95)
Maximum fiber stress, axial, ksi (GPa)	321-428 (2.21-2.98)

Table 16.16. Titanium Lined Graphite/Epoxy Vessels[25]

PROPERTY	RANGE	MEAN
DIAMETER		4 in. (10.2 cm)
P_BV/W, composite, in. × 10^6 (KPa · m^3/kg)	1.44-1.76 (359-438)	1.63 (406)
P_BV/W, total vessel, in. × 10^6 (KPa · m^3/kg)	0.86-1.04 (214-259)	0.97 (241)

Table 16.17. Glass/Polyester Pipe[28]

	CLOSED ENDS	FREE ENDS
Hoop stress, ksi (MPa)		
At burst	67.1 (462)	41.3 (285)
At weep	14.5 (101)	37.6 (259)
At non-linearity	6.4 (44)	7.0 (48)
Percent strain		
At burst		
Hoop	1.75	2.4
Axial	1.62	−1.6
At weep		
Hoop	0.33	2.26
Axial	0.22	−1.53
At non-linearity		
Hoop	0.13	0.22
Axial	0.04	−0.13

Table 16.18. Uniaxial Compression, Torsion, and Bending Strengths[26]

ORIENTATION				
ANGLE	% HOOP	COMPRESSION, ksi (MPa)	TORSION, ksi (MPa)	BENDING, ksi (MPa)
15°	0	40–61 (276–421)	25–30 (172–207)	26–31 (179–214)
15°	9		40–52 (276–290)	
15°	18		39–43 (269–296)	
15°	27		36–38 (248–262)	
30°	0	31–43 (214–296)	50–69 (345–476)	27–30 (186–207)
45°	0	31–33 (214–228)	53–72 (365–496)	33–40 (228–276)
45°	9	43–45 (296–310)	62–72 (427–496)	43–49 (296–338)
45°	18	39–49 (269–338)	62–68 (427–469)	49–54 (338–372)
45°	27	45–50 (310–345)	52–62 (359–427)	47–50 (324–345)

Table 16.19. Filament Wound Pipe[29]

ORIENTATION	0° LONG, 90° HOOP
Tension	59,500 psi (410 MPa)
Tensile modulus	2.4 million psi (16.5 GPa)
Compression	31,500 psi (217 MPa)
Compression modulus	2.6 million psi (17.9 GPa)
Shear modulus	700,000 psi (4.8 GPa)
Poisson's ratio	0.132
Specific gravity	2.0

Table 16.20. Combination Pipe Made on Drostholm Machine[27]

COMPOSITION AND PROPERTIES	TYPE OF PIPE		
	TANK	SEWER	PRESSURE
% hoop	2	5	40–50
% chopped fiber	24	15	15–20
% sand		40	
% polyester	74	40	30–45
Tensile, ksi (MPa)			
Hoop	12.6–14.1 (87–97)	14–15.5 (97–107)	70–85 (483–586)
Axial	11.3 (78)	7.0 (48.3)	7.0 (48.3)
Flexural, ksi (MPa)	19.6 (135)	22.5 (155)	70–85 (483–586)
Flexural modulus, psi × 10^6 (GPa)	0.85 (5.86)	1.25 (8.6)	3.5 (24.1)
Compression, ksi (MPa)	22.6 (156)	25.0 (172)	56–70 (386–483)
Specific gravity	1.40	1.75	1.80

glass strands in the hoop direction only. Chopped glass furnishes the longitudinal reinforcement. The resin mix is applied after the glass has been deposited. The overall composition is formulated for specific applications, such as sewer pipe, storage tanks, or pressure pipe, with resulting variations in properties.

16.4.5. Test Methods

The following ASTM test methods are directly applicable to filament wound composites.

ASTM NO.	TITLE
D 2344	Apparent Horizontal Shear Strength of Reinforced Plastics by Short Beam Method, Test for.
D 2290	Apparent Tensile Strength of Ring or Tubular Plastics by Split Disk Method, Test for.
D 2291	Fabrication of Ring Test Specimens for Reinforced Plastics, Recommended Practice for.
D 2586	Hydrostatic Compression Strength of Glass-Reinforced Plastic Cylinders, Test for.
D 2585	Preparation and Tension Testing of Filament-Wound Pressure Vessels.
D 2343	Tensile Properties of Glass Fiber Strands, Yarns, and Rovings Used in Reinforced Plastics, Test for.
D 3039	Tensile Properties of Oriented Fiber Composites, Test for.
D 2355	Fiber Content of Unidirectional Fiber/Polymer Composites, Test for.
D 3299	Filament-Wound Glass Fiber-Reinforced Polyester Chemical Resistant Tanks, Specifications for.
D 2924	External Pressure Resistance of Reinforced Thermosetting Plastic Pipe, Test for.

16.5. STATUS OF FILAMENT WINDING

Filament winding is now well established as a versatile method for the processing of reinforced plastics. Storage tanks and pipes for the chemical and other industries furnish the principal commercial outlets. Polyester resin and fiberglass are the major constituent materials and will continue to be so in the foreseeable future. An increased use of graphite and aramid reinforcements is indicated, particularly for the more critical pressure

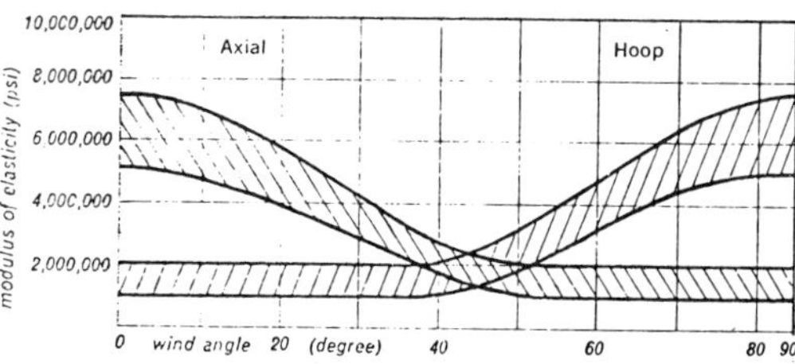

Figure 16.21. Effective modulus versus wind angle, glass/polyester. (*Courtesy Owens-Corning Fiberglass Corp.*)

vessel applications. Here the matrix material is likely to be an epoxy. New developments may be expected in on-site winding and combination winding, such as a fiberglass overwrap of PVC pipe. The compression molding of filament wound lay-ups is another potentiality to be investigated. Such moldings may provide unique solutions for the fabrication of weight-sensitive structural components.

REFERENCES

1. Schmitz, G. K. and Metcalfe, A. G., "Characterization of Flaws on Fiberglass," 20th Annual Technical Conference, Reinforced Plastics/Composites, SPI, Washington, D.C., 1965.
2. Owens-Corning Fiberglas, Publication No. 5-CR-6516, 1974.
3. PPG Industries, Bulletins F-131A, F-132B, Hybon 2000 Series.
4. Goldsworthy Engineering, USAAMRDL-TR-76-5, 1976.
5. Whisenhunt, F. S., Jr., NRL 6161, 1964.
6. Chiao, T. T., Jessup, E. E., and Penn, L., "Screening of Epoxy Systems for High Performance Filament Wound Applications," 7th Natl SAMPE Conference, October 1975.
7. Duffy, J. V., NSWC/WOL/TR 75-48, 1975.
8. Rinde, J. A. and Newey, H. A., UCID-17219 (Lawrence Livermore Laboratories), 1976.
9. Shell Chemical Co., Technical Bulletin SC-122-76.
10. Mahler, J. P. and Bradley, R. T., UCRL-51934 (Lawrence Livermore Laboratories), 1975.
11. Grumman Aircraft, ADR 08-18-67.2, 1967.
12. Aerojet-General, ML-TDR-64-43 Volumes I & II, 1964.
13. Darms, F. and Litvak, S., "Optimum Design for Filament-Wound Rocket Motor Cases," 19th Annual Technical Conference, Reinforced Plastics/Composites, SPI, Washington, D.C., 6-D, 1964.
14. Hofeditz, J. T., "Structural Design Considerations for Fiberglass Pressure Vessels," 19th Annual Technical Conference, Reinforced Plastics/Composites, SPI, Washington, D.C., 7-C, 1964.
15. Aerojet-General, AFML-TR-68-126, 1968.
16. Picatinny Arsenal, Technical Report 3257, 1965.
17. Hoggatt, J. T., NASA-CR-120835.

18. Hughes Aircraft, USAAMRDL-TR-77-19A, 1977.
19. *Advanced Composites Design Manual*, Volume IV, Rockwell International, 1973.
20. Reuter, R. C., Jr., *J. Composite Materials* **6** (January 1972).
21. Craddak, J. N. and Zak, A. R., *J. Composite Materials* **11** (April 1977).
22. Fiber Science, Inc., USARTL-TR-77-53, February 1978.
23. B. F. Goodrich, ABL Subcontract 89, 1964.
24. Young, E. C., 18th Petroleum Mechanical Engineering Conference, 1963.
25. Lark, R. F., NASA-TM-73699, 1977.
26. Vogt, C. W., Haniuk, E. S., and Trice, J. M., Jr., AFML-TR-66-274, 1966.
27. Gilbu, A., "New Developments in Connection with Production of Pipes and Tanks on Dristholm Machine." 31st Annual Technical Conference, Reinforced Plastics/Composites, SPI, Washington, D.C., 9-B, February 1976.
28. Hull, D., Legg, M. J., and Spencer, B., *Composites* (January 1978).
29. Chiao, T. T., SPE Regional Conference, Cleveland, 1965.

17
CONTINUOUS MANUFACTURING PROCESSES

W. B. Goldsworthy
Goldsworthy Engineering, Inc.
Torrance, California

The rapidly expanding usage of composite components in automotive and other mass production industries has focused attention on continuous production techniques as they apply to these engineering materials. It is recognized that where product volume warrants, a continuous processing system from raw material to finished product approaches optimum manufacturing efficiency. In composites, where properties are dependent almost solely on fiber orientation, continuous processing has the additional advantage of providing dependable control of filament orientation and tensioning. The combination of these and other more subtle processing advantages results in product economies not achievable by other processing techniques.

With the accelerated evolution of composite processing, new continuous processing methods are evolving rapidly. To provide some insight into this area, we will review some of the continuous machines currently in production.

17.1. PULTRUSION

This technique for the manufacture of structural profiles from composites on a continuous basis is the exact analogy of an aluminum extruder or a thermoplastic extrusion machine. All three produce constant cross-sectional profiles from their respective materials (see Fig. 17.1).

Pultrusion was originally viewed as a method of making simple solid sections reinforced with unidirectional fiber. It has evolved into a system capable of producing an almost infinite variety of solid and hollow profiles. Concurrently has come the ability to tailor the properties of these profiles to fit a wide scope of engineering and structural requirements.

Chemical resistance, temperature resistance, impact strength, and fatigue properties may be enhanced by proper resin selection. Throughput rates continue to climb as resin and catalyst systems designed specifically for this method become commercially available. Where running rates of 2–3 ft/minute (61–91 cm/minute) have been the norm, these rates are now climbing into the 15–20 ft/minute (457–610 cm/minute) range.

These dramatic production increases are due to the synergistic interplay between resin modification and advanced curing techniques. The use of radio frequency heating in conjunction with traditional heating methods not only increases running speeds but permits the manufacture of massive profiles and those that have widely varying masses within the profile (see Figs. 17.2 and 17.3).

Now that the production of complex profiles has become commonplace, the development area currently receiving the most attention is that of providing exact reinforcing fiber orientation to allow property optimization for specific profile requirements. To illustrate the flexibility available with modern approaches, consider that tubing is being produced that has an inner ply of polyester veil for surface lubricity, a ply of continuous strand mat for omni directional properties, three ply of longitudinal roving for bending strength, a ply of circumferential roving for burst strength, a

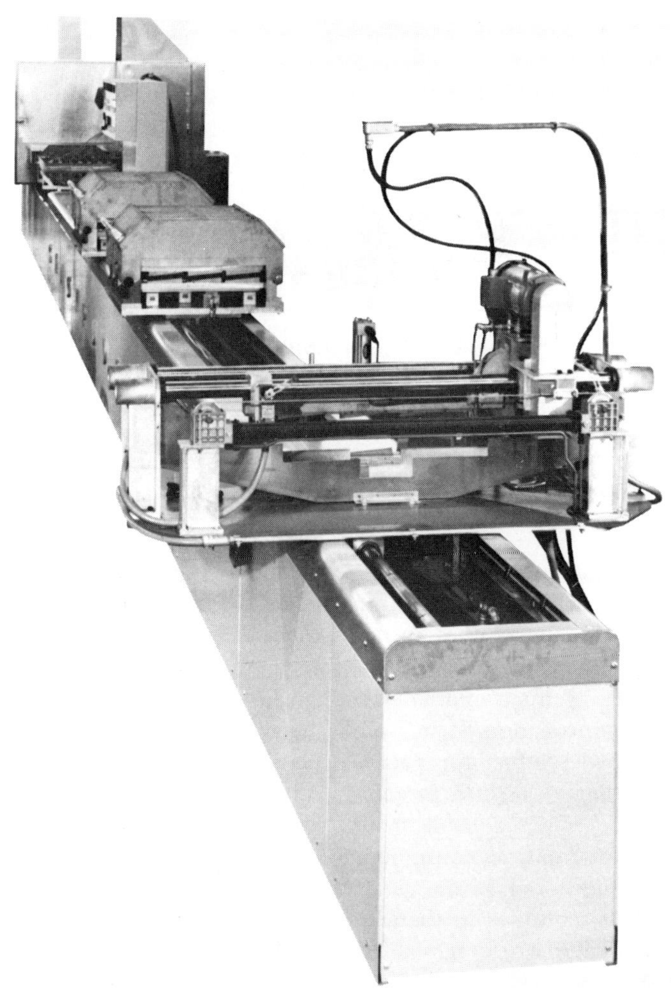

Figure 17.1. Glastruder® pultrusion machine. (*Courtesy Goldsworthy Engineering, Inc.*)

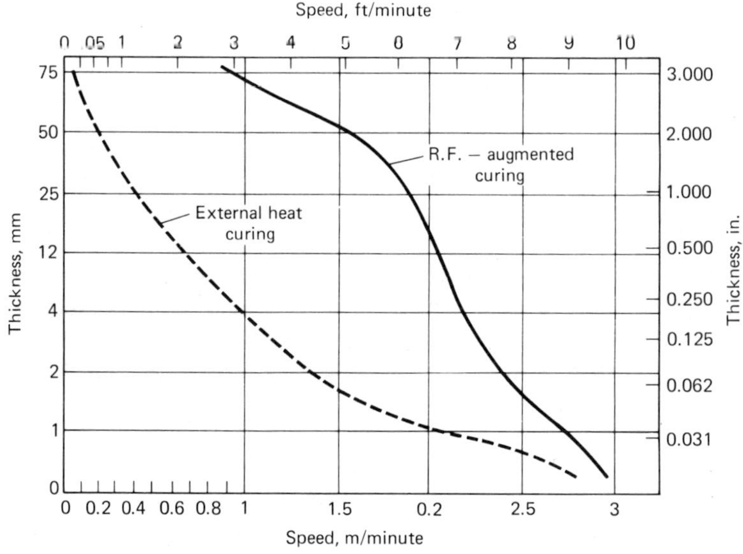

Figure 17.2. Thickness versus speed. (*Data from Goldsworthy Engineering, Inc.*)

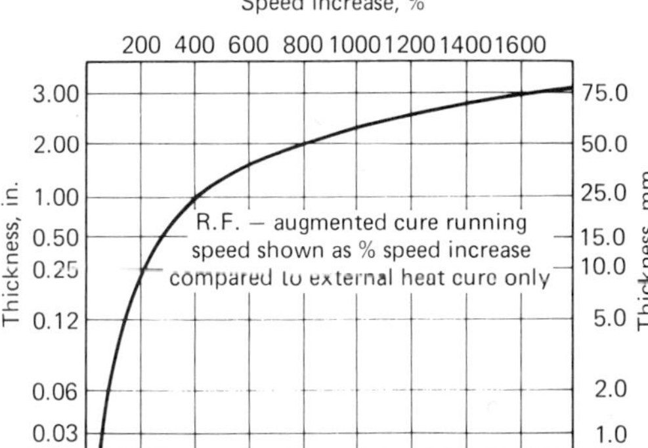

Figure 17.3. Thickness versus percent speed increase. (*Data from Goldsworthy Engineering, Inc.*)

+45° and a −45° roving ply for torsional loading, another continuous strand mat ply for impact enhancement, and a final outer ply of chemical glass veil for corrosion resistance! This is an excellent example of the property versatility available in these structural elements (see Fig. 17.4).

Typically, the oriented fiber package is consolidated dry and then impregnated by pumping the resin system through the dry package. This approach tends to eliminate entrained air. Excess resin drains back into the sump and is recirculated (see Fig. 17.5).

In most instances, the dry oriented fiber package is continuously formed on a cylindrical mandrel and transitioned into the desired final shape (see Fig. 17.6). The increase in availability and continuing cost reductions have encouraged the use of high strength fibers in the pultrusion process. Graphite, aramid, and S-glass fibers are being used to produce structural shapes with properties far superior to those of the traditional materials. Hybridizing of these fibers to satisfy both engineering and economic parameters has become a commercial reality.

Although properties of pultruded shapes reinforced with various forms of E-glass fiber have been well documented (see Table 17.1), very little has been published on the properties of profiles using higher strength fibers. The values shown in Tables 17.2–17.10 are the result of a government sponsored test program, and although they are far from comprehensive, they provide some reasonable baselines and, in some cases, allow a comparison between pultrusion and the most commonly used advanced composite molding technique, vacuum bag molding.

It is apparent that properties of the products obtained by pultrusion methods compare favorably with those resulting from more traditional fabricating processes. The trend toward somewhat higher values can be attributed to several advantages inherent in the process—consistent controlled fiber tensioning and orientation, lower void content, and uniform fiber fraction retention. Even the property most likely to be suspect, inter-

Figure 17.4. Fiber orientation tooling on Glastruder.® (*Courtesy Goldsworthy Engineering, Inc.*)

Figure 17.5. Square tubing production showing impregnation, R.F. zone, and forming die. (Courtesy Goldsworthy Engineering, Inc.)

Figure 17.6. Square tubing tooling showing perforated mandrel impregnation zone, R.F. electrodes, and transition area. (*Courtesy Goldsworthy Engineering, Inc.*)

Table 17.1. Typical Properties—E-Glass Pultrusions[a]

	UNIDIRECTIONALLY REINFORCED	MAT/ROVING MIXTURE REINFORCED
Flexural strength, KSI(MPa)	100–180 (690–1240)	24–38 (165–262)
Flexural modulus, MSI(GPa)	4–6 (27.6–41.4)	1.4–1.8 (9.65–12.4)
Tensile strength, KSI(MPa)	60–180 (414–1241)	25–45 (172–310)
Tensile modulus, MSI(GPa)	4–6 (27.6–41.4)	2–4 (13.8–27.6)
Compressive strength, KSI(MPa)	30–70 (207–483)	28–40 (193–276)
Compressive modulus, MSI(GPa)		2–3 (13.8–20.7)

[a]All properties taken along longitudinal axis.

Table 17.2. Tension Test Result Averages

TEST NUMBER	FIBER ORIENTATION TO LOAD	RESIN	FIBER	FIBER CONTENT (V/O)	STRENGTH, psi × 10^3 (10^3KN/M^2)	YOUNG'S MODULUS, psi × 10^6 (10^6KN/M^2)	ULTIMATE STRAIN, 10^{-6} in./in. (10^{-6} mm/mm)	POISSON'S RATIO
Pultruded								
1	$(0^\circ)_c$	826	Hercules AS	59	175.0 (1206.6)	20.0 (137.9)	8500	0.20
5	$(0^\circ)_c$	5208	Hercules AS	59	154.0 (1061.8)	19.4 (133.8)	7500	0.18
7	$(0^\circ)_c$	826	S-Glass	68	228.0 (1572.0)	7.23 (49.9)	32,800	0.34
11	$(0^\circ)_c$	826	Kevlar	68	160.0 (1103.2)			
2	$(90^\circ)_c$	826	Hercules AS	59	5.4 (37.2)	1.66 (11.4)	3700	0.023
6	$(90^\circ)_c$	5208	Hercules AS	59	2.9 (20.0)	2.06 (14.2)	1740	0.029
8	$(90^\circ)_c$	826	S-Glass	68	7.0 (48.3)	1.81 (12.5)	2850	0.094
12	$(90^\circ)_c$	826	Kevlar	68	1.6 (11.0)			
3	$(\pm 45^\circ)_c$	826	Hercules AS	59	59.0 (406.8)	8.01 (55.2)	8470	0.86
4	$(0^\circ \pm 45^\circ)_s$	826	Hercules AS	59	93.0 (641.2)	12.5 (86.2)	7400	0.98
9	$(\pm 20^\circ)_c$	826	S-Glass	68	92.0 (634.3)	5.89 (40.6)	17,900	0.49
10	$(0^\circ \pm 20^\circ)_s$	826	S-Glass	68	118.0 (313.6)	5.96 (41.1)	21,600	0.34
Vacuum bagged								
1	$(0^\circ)_c$	826	Hercules AS	60	124.0 (855.0)	17.9 (123.4)	7800	0.32
5	$(0^\circ)_c$	5208	Hercules AS	60	134.0 (923.9)	17.3 (119.3)	8300	0.25
3	$(0^\circ)_c$	826	S-Glass	67	152.0 (1048.0)	7.97 (55.0)	25,000	0.20
7	$(0^\circ)_c$	5208	S-Glass	62	213.0 (1468.6)	8.19 (56.5)	32,300	0.28
2	$(90^\circ)_c$	826	Hercules AS	60	2.5 (17.2)	1.16 (13.5)	2200	0.097
6	$(90^\circ)_c$	5208	Hercules AS	60	1.4 (09.7)	2.13 (14.7)	850	
4	$(90^\circ)_c$	826	S-Glass	67	5.2 (35.9)	1.38 (9.5)	2500	0.072
8	$(90^\circ)_c$	5208	S-Glass	62	4.7 (32.4)	1.90 (13.1)	1500	
8A	$(0_2^\circ \pm 45^\circ, 0_2^\circ)_c$	826	Hercules AS	60	78.4 (540.6)	11.0 (75.8)	7900	0.36
8B	$(0_2^\circ \pm 45^\circ, 0_2^\circ)_c$	5208	Hercules AS	60	103.0 (710.2)	12.0 (82.7)	9100	0.48
8C	$(0_2^\circ \pm 45^\circ, 0_2^\circ)_c$	826	S-Glass	67	32.4 (223.4)	4.41 (30.4)	4930	0.70
8D	$(0_2^\circ \pm 45^\circ, 0_2^\circ)_c$	5208	S-Glass	62	160.0 (1103.2)	5.45 (37.6)	33,600	0.47

Table 17.3. Compression Test Result Averages

TEST NUMBER	FIBER ORIENTATION TO LOAD	RESIN	FIBER	FIBER CONTENT (V/O)	STRENGTH, psi × 10^3 (10^3KN/M^2)	YOUNG'S MODULUS, psi × 10^6 (10^6KN/M^2)	ULTIMATE STRAIN, 10^{-6} in./in. (10^{-6} mm/mm)	POISSON'S RATIO
Pultruded								
13	$(0^\circ)_c$	826	Hercules AS	59	96.0 (66.19)	20.5 (141.3)	4700	(0.15)
17	$(0^\circ)_c$	5208	Hercules AS	59	133.0 (91.70)	18.9 (130.3)	6800	(0.17)
19	$(0^\circ)_c$	826	S-Glass	68	82.0 (56.54)	9.3 (64.1)	9000	(0.28)
23	$(0^\circ)_c$	826	Kevlar	68	39.0 (26.89)			
14	$(90^\circ)_c$	826	Hercules AS	59	6.4 (4.41)	1.4 (9.7)	7400	(0.023)
18	$(90^\circ)_c$	5208	Hercules AS	59	13.6 (9.38)	1.93 (13.3)	4400	(0.046)
20	$(90^\circ)_c$	826	S-Glass	68	24.0 (16.55)	1.64 (11.3)	7600	(0.037)
24	$(90^\circ)_c$	826	Kevlar	68	5.1 (3.52)			
16	$(\pm 45^\circ)_c$	826	Hercules AS	59	25.0 (17.24)	8.56 (59.0)	5700	(0.58)
15	$(0^\circ \pm 45^\circ)_s$	826	Hercules AS	59	51.0 (35.16)	11.2 (77.2)	4500	(0.36)
22	$(\pm 20^\circ)_c$	826	S-Glass	68	58.0 (39.99)	7.01 (48.3)	8500	(0.51)
21	$(0^\circ \pm 20^\circ)_s$	826	S-Glass	68	66.0 (45.50)	7.35 (50.7)	9400	(0.49)
Vacuum Bagged								
9	$(0^\circ)_c$	826	Hercules AS	60	70 (48.26)	17.2 (118.6)	3900	(0.48)
13	$(0^\circ)_c$	5208	Hercules AS	60	57 (39.30)	13.1 (90.3)	4700	(0.46)
11	$(0^\circ)_c$	826	S-Glass	67	72 (49.64)	13.6 (97.8)	5600	(0.25)
15	$(0^\circ)_c$	5208	S-Glass	62	98 (67.57)	7.91 (54.5)	11,900	(0.35)
10	$(90^\circ)_c$	826	Hercules AS	60	14 (9.65)	1.17 (8.1)	13,700	(0.011)
14	$(90^\circ)_c$	5208	Hercules AS	60	12 (8.27)	1.44 (9.9)	8200	(0.0087)
12	$(90^\circ)_c$	826	S-Glass	67	22 (15.17)	1.19 (8.2)	32,300	(0.34)
16	$(90^\circ)_c$	5208	S-Glass	62	17 (11.72)	2.43 (16.8)	9100	(0.060)

Table 17.4. In-Plane Shear Test Result Averages

TEST NUMBER	FIBER ORIENTATION TO LOAD	RESIN	FIBER	FIBER CONTENT (V/O)	STRENGTH, ksi(MPa)	FLEXURAL MODULUS, Msi(GPa)
Pultruded						
25	$(0°)_c$	826	Hercules AS	59	6.7 (46.2)	1.49 (10.3)
28	$(0°)_c$	826	S-Glass	68	3.0 (20.7)	0.80 (5.5)
27	$(\pm 45°)_c$	826	Hercules AS	59	13.0 (89.6)	2.78 (19.2)
26	$(0° \pm 45°)_c$	826	Hercules AS	59	9.5 (65.5)	1.98 (13.7)
30	$(\pm 20°)_c$	826	S-Glass	68	12.0 (82.7)	2.29 (15.8)
29	$(0° \pm 20°)_s$	826	S-Glass	68	10.0 (68.9)	2.05 (14.1)

Table 17.5. Four-Point Flexure Test Result Averages

TEST NUMBER	FIBER ORIENTATION TO LOAD	RESIN	FIBER	FIBER CONTENT (V/O)	STRENGTH, ksi(MPa)	FLEXURAL MODULUS, Msi(GPa)
Pultruded						
31	(0°)	826	Hercules AS	59	180 (1241.1)	24.2 (166.9)
32	(0°)	826	Celanese GT30	60	115 (792.9)	17.2 (118.6)
33	(0°)	826	Celanese GY50	60	207 (1427.3)	16.7 (115.2)
34	(0°)	826	Celanese GY70	60	93 (641.2)	33.3 (229.6)
35	(0°)	826	S-Glass	67	184 (1268.7)	7.2 (49.6)
36	(0°)	826	Kevlar	68	72 (496.4)	9.8 (67.6)
37	(0°)	826	Thornell 300	59	181 (1248.0)	17.0 (117.2)
37A	(0°)	5208	Hercules AS	59	210 (1448.0)	18.4 (126.9)

Table 17.6. Short Beam Interlaminar Shear Test Result Averages

TEST NUMBER	FIBER ORIENTATION TO LOAD	RESIN	FIBER	FIBER CONTENT (V/O)	STRENGTH, ksi(MPa)
Pultruded					
38	(0°)	826	Hercules AS	60	14.0 (96.5)
39	(0°)	826	Thornell 300	60	12.0 (82.7)
40	(0°)	826	Celanese GT-30	60	8.3 (57.2)
41	(0°)	826	Celanese GY-50	60	9.5 (65.5)
42	(0°)	826	Celanese GY-70	60	8.1 (55.9)
43	(0°)	826	E-Glass	68	9.6 (66.2)
44	(0°)	826	Kevlar	68	6.5 (44.8)
44A	(0°)	5208	Hercules AS	60	11.4 (78.6)
44B	(0°)	826	S-Glass	68	11.9 (82.1)
Vacuum Bagged					
17	(0°)	826	Hercules AS	60	8.9 (61.4)
18	(0°)	5208	Hercules AS	60	6.3 (43.4)
19	(0°)	826	S-Glass	67	12.0 (82.7)
20	(0°)	5208	S-Glass	62	6.7 (46.2)

Table 17.7. Tensile Fatigue Test Results with Hercules AS/826 Resin, Fiber Oriented 0° to Load—Tensile Strength Maximum = 175,800 psi, $K = 1$, $R = 0.05$

SPECIMEN NUMBER	PERCENT OF STATIC	TEST STRESS, ksi(MPa)	CYCLES TO FAILURE
45-1	90	156.9 (1081.8)	1
45-2	68	119.6 (824.6)	1
45-3	90	156.9 (1081.8)	4
45-4	65	114.3 (788.1)	3.614×10^4
45-5	80	140.6 (969.4)	7.774×10^4
45-6	85	149.4 (1030.1)	1.123×10^4
45-7	85	149.4 (1030.1)	15
45-8	85	149.4 (1030.1)	440
45-9	85	149.4 (1030.1)	310
45-10	80	140.6 (969.4)	2.331×10^4
45-11	80	140.6 (969.4)	490
45-12	75	131.8 (908.8)	130
45-13	75	131.8 (908.8)	140
45-14	70	123.1 (848.8)	4
45-15	70	123.1 (848.8)	12
45-16	65	114.3 (788.1)	1.076×10^4
45-17	65	114.3 (788.1)	No failure in 1.889×10^6

laminar shear, shows a marked improvement (see Fig. 17.7).

Some interesting and commercially valuable offshoots of the basic pultrusion process have occurred that take advantage of its high production rates, excellent dimensional tolerances, and good physical, chemical, electrical, and thermal properties. One of these is the manufacture, by pultrusion methods, of continuous plate and sheet stock. This material is produced in standard 48-in. (1.22-m) widths in thicknesses up to 1½ in. (3.81 cm) (see Fig. 17.8).

The mobility of this type of processing is underscored by its adaption to on-site, large diameter tank construction. The heart of this

Table 17.8. Tensile Fatigue Test Results with Owns-Corning HTS-904 Fiberglass/826 Resin, Oriented 0° to Load—Tensile Strength Maximum $K = 1$, $R = 0.05$

SPECIMEN NUMBER	PERCENT OF STATIC	TEST STRESS, ksi(MPa)	CYCLES TO FAILURE
48-1	65	133.6 (921.2)	1
48-2	65	133.6 (921.2)	390
48-3	65	133.6 (921.2)	140
48-4	65	133.6 (921.2)	270
48-5	55	111.1 (779.8)	530
48-6	45	92.5 (637.8)	1480
48-7	45	92.5 (637.8)	700
48-8	45	92.5 (637.8)	1.25×10^4
48-9	35	72.0 (496.4)	1.605×10^4
48-10	35	72.0 (496.4)	1.566×10^4
48-11	35	72.0 (496.4)	7.772×10^4
48-12	25	51.4 (354.4)	3.9712×10^5
48-13	25	51.4 (354.4)	No failure in 1×10^6

Table 17.9. Torsional Fatigue of Hercules AS/826 Unidirectional Composite—Static Interlaminar Shear Stress = 14 Ksi (97 MPa) $K = 1.0, R = 0.10$

SPECIMEN NUMBER	TEST STRESS ksi(MPa)	CYCLES TO FAILURE
51-1	6.74 (46.5)	2150
51-2	7.31 (50.4)	870
51-3	6.43 (44.3)	150
51-4	5.35 (36.9)	1650
51-5	4.59 (31.6)	4000
51-6	4.20 (29.0)	4800
51-7	5.35 (36.9)	2800
51-8	3.82 (26.3)	8000
51-9	10.30 (71.0)	10
51-10	8.20 (56.5)	18
51-11	7.50 (51.7)	96
51-12	6.10 (42.1)	450

Table 17.10. Torsional Fatigue of S-Glass/826 Unidirectional Composite—Static Interlaminar Shear Stress = 11.9 Ksi (82 MPa), $K = 1.0$, $R = 0.10$

SPECIMEN NUMBER	TEST STRESS ksi(MPa)	CYCLES TO FAILURE
52-1	12.66 (87.3)	45
52-2	13.14 (90.6)	30
52-3	7.90 (54.5)	1500
52-4	11.99 (82.7)	50
52-5	8.25 (56.9)	1200
52-6	7.58 (52.3)	1200
52-7	7.14 (49.2)	2700
52-8	6.37 (43.9)	5800
52-9	6.18 (42.6)	6500
52-10	11.85 (81.7)	96
52-11	10.20 (70.3)	330
52-12	14.80 (102.1)	11
52-13	5.10 (35.2)	10,000[a]

[a]Run out.

tankage erection system is a pultrusion machine operating in a modified 40-ft (12.2-m) trailer van (see Fig. 17.9). This unit is driven to the tank foundation site, where the total length of a specifically designed profile is pultruded into a storage cannister. The exact length required is determined by the tank diameter, the profile width, and the tank wall height. When sufficient stock has been loaded

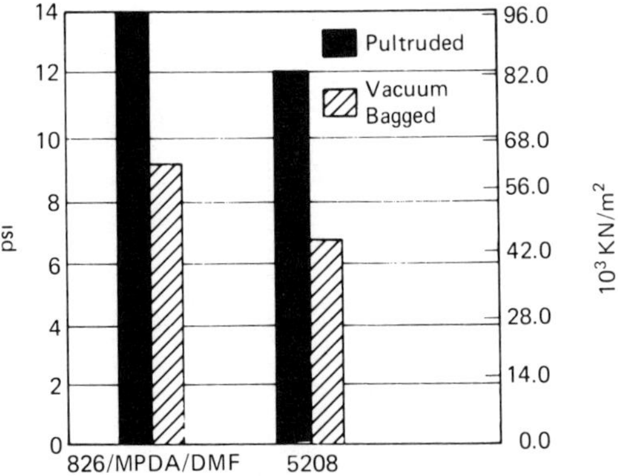

Figure 17.7. Short beam interlaminar shear strength comparison.

into the cannister, the pultrusion trailer proceeds to the next site.

The pultrusion, which normally has some form of groove on one edge and tongue on the other, is back-fed from the cannister into an erection machine, which locks the first two courses together at the desired tank diameter and proceeds to interlock each course to the prior one in helical fashion until the finished tank height is achieved (see Fig. 17.10). As this occurs, a bonding resin is being metered into the groove while the tongue is passed through the hardening agent. Mating the two continuously bonds the structure together. The ability to produce these large structures on site with unskilled labor makes them economically attractive.

The ready mobility typified by the tank erection unit has led to some other developments, such as its shipboard use for laying subaqueous pipe and its possible intergration into the space shuttle for manufacturing large structures in space.

17.2. CONTINUOUS PIPE MACHINES

Continuous production of pipe in all sizes, ranging from 1 in. (2.54 cm) through 120 in. (3.05 m) is now a well proven and highly developed composite product (see Fig. 17.11).

Pipe manufactured in this manner competes well economically with that produced from traditional materials (concrete, steel, asbestos, cement, clay, etc.) and therefore is making a transition from the specialized mar-

Figure 17.8. Continuous pultruded plate and sheet machine. (*Courtesy Morrison Molded Fiberglass.*)

kets in which it was introduced to the broad field of pipe applications. As developments in the areas of fitting manufacture, bedding, and trenching technology and mechanical joining techniques reach fruition, composite pipe should capture a substantial share of the overall market.

17.3. MISCELLANEOUS CONTINUOUS SYSTEMS

Much in the same fashion that chopped strand SMC (sheet molding compound) has filled a manufacturing need, in many industries there is high volume potential for unidirectionally reinforced, continuous strand, formable stock. This need is being filled by machines which continuously impregnate and package bar stock for subsequent processing (see Fig. 17.12).

The sporting goods industry (golf shafts, fishing rods, ski poles, etc.), as well as other industrial and commercial industries, has created a high volume demand for tapered tubing made from a variety of composite systems. Additionally, the wide spectrum of

Figure 17.9. On-site tank erection based on mobile pultrusion. (*Courtesy Goldsworthy Engineering, Inc. A. U.S. Army Photograph.*)

Figure 17.10 Finished pultruded tank with erector unit being moved to next site (*Courtesy Goldsworthy Engineering, Inc. A U.S. Army Photograph.*)

Figure 17.11. Continuous manufacture of large diameter pipe. (*Courtesy Owens-Corning Fiberglas Corp.*)

Figure 17.12. Machine for continuous production of formable composite bar stock. (*Courtesy Goldsworthy Engineering, Inc.*)

end product requirements dictates a machine with versatility in producing a variety of lengths, taper angles, wall thickness taper, and ply angle programming for longitudinal and torsional stiffness variation. These, coupled with high throughput requirements, indicate continuous processing as the most feasible approach (see Fig. 17.13).

The production of corrugated and flat translucent sheet represents the oldest form of continuous composite processing in the industry. However, microprocessor controlled

Figure 17.13. Continuous production of tapered composite tubing. (*Courtesy Goldsworthy Engineering, Inc.*)

machines for the continuous composite facing of plywood and other core materials, structural and metal clad sheet, three-dimensionally reinforced insulating panel, and straight and curved pulforming of varying cross-section and varying volume shapes are all entering production or are on the drawing boards as continuous composite production technology keeps pace with industry needs.

REFERENCES

1. Goldsworthy Engineering, Inc., "MM&T—Pultruded Composite Structural Elements," December 1976.
2. "Evaluation of Curing Rates of Flat Sheet Materials," Report No. 1, GEI R-13, August 13, 1971.
3. Goldsworthy Engineering, Inc., "Large Diameter Reinforced Plastic Pipe International Achievements," 1974 Reinforced Plastics Congress, November 12, 1974.
4. Goldsworthy Engineering, Inc., "An Approach To The Logistics, Manufacturing Techniques and Construction of the Trans Canada Pipeline Utilizing Composite Materials," North Slope Pipeline—Steel or Composites Workshop, February 15, 1971.
5. Goldsworthy Engineering, Inc., "Manufacturing Methods and Cost Factors to Produce a Full Line of Pipe Products, 2-inch through 108-inch, from Glass Reinforced Plastics," October 27, 1971.
6. Goldsworthy Engineering, Inc., "Continuously Filament Wound Pipe."
7. Goldsworthy Engineering, Inc., "Underwater Piping Systems," April 1967.
8. Jones, Brian H., Ph.D., "Design and Manufacturing Technology for Pultruded Composite Structural Elements," Society of the Plastics Industry.
9. "Fight Corrosion with Fiber Glass Reinforced Plastics," Society of the Plastics Industry, 1978 Technical Conference, November 7–10, 1978.
10. Jones, Brian H., Ph.D., Goldsworthy Engineering, Inc., "The Design and Production of Economical FRP Energy Absorbing Systems for Transportation Applications."
11. Jones, Brian H., Ph.D., "Pultruding Filamentary Composites—An Experimental and Analytical Determination of Process Parameters," Society of the Plastics Industry, 1974 Technical Conference.

18
FABRICATION OF ADVANCED COMPOSITES*

L. E. "Roy" Meade
Lockheed Aircraft
Marietta, Georgia

18.1. INTRODUCTION

The basic operations involved in fabricating composite parts are 1) lay-up of the raw materials-reinforcements and matrices, 2) application of heat and pressure, and cure and consolidation into the final desired form, 3) assembly, and 4) joining. To achieve these operations, however, a number of possible avenues are open.

Methods for placing continuous and short filaments in proper orientations and consolidation within the plies of a composite structure are described and analyzed for cost-effectiveness and process reliability. Techniques for resin and metal matrix composites are presented.

The fabrication process is one of the most critical factors in the successful use of advanced composites in various components. Fabrication is so particularly important because of:

1. The necessity to fabricate the basic structural macro-material (prepreg tape or monoplayer sheets) from precursor unidirectional or woven laminae (filament and matrix) prior to, or concurrent with, the fabrication of actual parts (normally accomplished by suppliers of materials).
2. Complexities caused by the increase in variables inherent in processing heterogeneous combinations of material.
3. The intractable nature of some of the constituents (e.g., the brittleness, inflexibility, and extreme hardness of boron).
4. The critical importance of process controls in achieving product integrity and consistency.
5. The complex requirements of joint and attachment areas.

This chapter provides the engineer with sufficient insight into manufacturing processes to assure the generation of practicable, maintainable, repairable, and cost-effective/competitive designs. It is intended to provide sufficient information for making valid decisions between competitive designs based on other than technical merit alone, by enabling the engineer to include in the decision process such relative factors as ease and cost of fabrication, feasibility of effective inspection, and practicality of economical maintenance.

18.1.1. Vacuum Bag Molding

Vacuum bagging techniques have been developed for fabricating a variety of components and structures. Complex shapes, including double contours, and relatively large parts are readily handled. The process is principally suited for those cases where higher pressure molding cannot be used.

The essential steps in the process are the lay-up, preparation of a bleeder system, and

*This chapter was accumulated from information acquired during the generation of the DD/NASA-sponsored "Structural Composite Fabrication Guide." Acknowledgment is hereby made to virtually every aerospace company in the U.S. involved in the development and fabrication of composite hardware which contributed information to the "FAB Guide" effort.

the bagging operation. The required number of plies for a layup are pre-cut to size and positioned in the mold (or mold form) one ply at a time. Each ply is separately worked to remove trapped air and wrinkles to ensure intimate contact with the previously laid-up ply. The complete lay-up is covered with a porous, non-adhering material which provides easy release and vacuum access to the lay-up. Various perforated films or coated fabrics are available for this purpose.

A controlled-capacity bleeder system is normally required to absorb excess resin and permit the escape of volatiles while maintaining the specified fiber-volume ratio of a particular laminate. Bleed-out may be through the edges (edge bleed-out), through the top surface (vertical bleed-out), or a combination of both. An edge bleed-out is constructed by placing a narrow width of bleeder cloth (about 1 in. (2.54 cm) of uncoated fiberglass, burlap, or a similar material) around the periphery of the lay-up. Vertical bleeder plies of the same materials are placed directly over the perforated film. Edge-bled parts usually require trimming to remove resin-rich or resin-starved edges. When parts cannot be trimmed, "flexible dams" are used in conjunction with vertical bleeders to minimize edge bleed-out. Bleed-out of unidirectional tapes or wide goods normally requires the addition of a thin, flexible caul plate over the bleeder plies to prevent excessive resin wash-out and provide a smooth bag-side surface. The lay-up may contain peel plies when subsequent bonding operations are to be performed. The peel ply is placed in direct contact with the lay-up, below the release film.

Vacuum connections are placed over the vertical bleeders or outside the edge bleeders. A sufficient number of ports is provided to ensure a uniform flow of resin and volatiles. The bagging film, tailored to fit the part, is placed over the lay-up, bleeder system, and vacuum connectors and is sealed to the mold plate. A partial vacuum is usually drawn to smooth the bag surface prior to applying full vacuum and heat. The bagged mold is transferred to an oven for curing with full vacuum applied. In most cases, vacuum is maintained during the entire heating and cooling cycle.

18.1.2. Autoclave Molding

Autoclave molding is similar to the vacuum bag process, except that the lay-up is subjected to greater pressures and denser parts are produced.[4] The bagged lay-up is cured in an autoclave by the simultaneous application of heat and pressure. Most autoclave processes also use vacuum to assist in the removal of trapped air or other volatiles. The vacuum and autoclave pressure cycles are adjusted to permit maximum removal of air without incurring an excessive resin flow. Vacuum is usually applied only in the initial stages of the curing cycle, while autoclave pressure is maintained during the entire heating and cooling cycles. Curing pressures are normally in a range of 50–100 psi (377.5–755 kg/m^2) Compared with vacuum bag molding, this process yields laminates with closer control of thickness and lower void content. (See Chap. 14).

18.1.3. Matched Die Consolidation of Boron/Aluminum, Boron/Epoxy, or Graphite/Epoxy Composites

The complexity and relatively small size of turbine engine parts, such as blades and vanes, lend themselves to matched die molding or bonding of such components. End attachments, such as shrouds or roots, will normally be incorporated in the die molding process. Processing parameters will range from about 500–5000 psi (3775–37,750 kg/m^2), 350–1050° F (348–566° C), depending on the materials used; and the environment may be air, inert gas, or vacuum—again, depending upon the materials.

18.1.4. Matched Die Molding of Chopped Fiber Prepregs

Some structures have complex shapes and require high stiffness-to-density ratios, but only moderate strength. Matched die molding of such structures, either from short fiber preforms or from combinations of short fiber with continuous fiber inlays, can offer an economical means of producing complex parts or assemblies "net"; they require little or no finish machining. Since complex dies are themselves expensive, especially when machined from steel, the economic advantage lies in large production runs of identical parts from the same die set.

Processing procedures leading to the matched die molded structure have three basic steps:[5]

1. *Prepreg mold charge preparation.* The required amount of preimpregnated tape is cut into 1-in. (2.54-cm) lengths for the randomly oriented mold charge.
2. *Preform staging and shaping.* The continuous fiber plies are positioned in a sub-assembly preform mold, vacuum bagged, and staged in the autoclave for 2 hours at 150° F (65.5° C) under 50 psi (377.5 kg/m^2) pressure. These shaped and staged preforms are then positioned in an assembly preform mold, and the chopped material—1-in. (2.54-cm) lengths—is distributed in random pattern and required volume. The assembly preform is then bagged and staged in the autoclave for 2 hours at 150° F (65.5° C) using 100 psi (755 kg/m^2) autoclave pressure.
3. *High pressure cure in die set.* The steel matched die set is installed in the press, mold release is applied to the surfaces, and the mold temperature is increased to 350° F (348° C). The preformed assembly is installed in the female half of the die set, the press is closed, and the assembly is cured for 30 to 90 min. under 2000 psi (15,000 kg/m^2) pressure.

18.1.5. Thermal Expansion Molding

One method of molding integrally stiffened structures having complex shapes is to wrap prepreg layers over blocks of rubber and restrain the layup and rubber blocks in a metal cavity. As temperature is increased, the rubber blocks or mandrels expand more than the metal restraining tool thereby generating curing pressure eliminating the need for external pressure application, such as with an autoclave.

Very high pressures can be generated—up to 800 psi (6040 kg/m^2)—if the mass of rubber versus the cavity isn't controlled. Tooling is simple, inexpensive, and easily changed for prototype hardware changes. A designer has considerable design latitude to mold complex assemblies in a single cycle. Thus, overall manufacturing costs can be reduced by decreasing the number of parts which have to be handled, by eliminating joints and decreasing assembly and fastening operations.

18.1.6. Braiding

Braiding is a mechanized textile process which dates from the early 1800's. In the braiding operation, a mandrel is fed through the center of the machine at a uniform rate, and fibers from moving carriers on the machine braid about the mandrel at a controlled angle. The machine operates like a maypole with the carriers working in pairs to accomplish the over/under braiding sequence. This process is applicable to channel sections and webs.

18.1.7. Filament Winding

Filament winding is a process wherein fibers or tape are wound onto a rotating mandrel from a relatively stationary position. The fibers are dispensed from a translating head at controlled angles normal to the rotating mandrel axis. Lay-up compaction can be applied by the degree of tension maintained on the fibers during the application. (See Chap. 15).

18.1.8. Sandwich Fabrication

Two methods of bonding advanced composite facings to core materials are used in fabricating sandwich structures. The first involves a two-step operation, in which the facings are laid up and cured independently before bonding to the core material; the second is a one-step or "co-curing" method, in which the facings are laid up and applied to the core in the "B" staged condition. In this case, the sandwich assembly is then cured and bonded in one autoclave operation. As new manufacturing methods and techniques were investigated, it was found that quality sandwich panels could be fabricated by co-curing, while achieving the same structural integrity and considerable savings in labor and tooling costs.[6]

In special cases of complex core machining

and forming, it may be desirable in either of the above methods to bond only one face sheet at a time. For example, to machine one side of the core and bond the facing to core in a formed mold, and then to machine the other side of the core in the restrained position and co-cure the second face sheet.

Panels are manufactured by the vacuum/autoclave process, lay-up, and bagging techniques described in Sections 18.1.1 and 18.1.2. In the co-curing method, facings may be laminated from *adhesive* type prepregs made from a modified resin system used for adhesive bonding, in which resin content and flow of the prepreg material are such that a uniform filleted bond between facings and core is obtained. Strength properties of the laminate, however, are somewhat lower when the modified adhesive resin systems are used. It is suggested that higher strength epoxy resin prepreg be used, with an adhesive film applied between the core and laminated facing, as is done in the two-step pre-cured face sheet-to-core bonding. To prevent damage to the core and facing in the co-cure method, the total molding pressure is normally not in excess of 50 psi (377.5 kg/m^2). Cure and post-cure temperatures are also limited to avoid degradation of the core. Fiber-glass facings on graphite skins are recommended to prevent corrosion of aluminum honeycomb core.

18.1.9. Pultrusions

Pultrusions, the composite equivalent of metallic extrusions, can produce low-cost, constant cross-sectional structural shapes. Preimpregnated uniaxial tapes or sewn cross-plied strips are pulled through a ceramic die to form the shape. The resin is concurrently cured by RF or microwave energy. Channel sections, "Z" sections, hat sections, and flat bars to 40 ft (12.19 m) long have been pultruded using several resin systems. (See Chap. 17).

18.2. RESIN-MATRIX COMPOSITES

Resin-matrix composites are defined as reinforcing filaments consolidated by means of any resinous material matrix (see Fig. 18.1). Aerospace users of composites do not generally make their own composite materials from the raw filament and matrix materials. The "raw material" of the aerospace user is usually preassembled for him, by a supplier, from the proper proportions of the proper constituents in the proper array, and partially processed to a stage where it can be conveniently shipped and handled. These embryo composite materials are referred to as "prepregs" (from filaments preimpregnated with matrix material). Fabrication of high-quality composite structures is highly dependent on prepreg quality and uniformity in such particulars as uniformity of fiber spacing, number and distribution of broken filaments, and resin tack. To ensure standards of quality, visual inspection and acceptance strength testing are re-

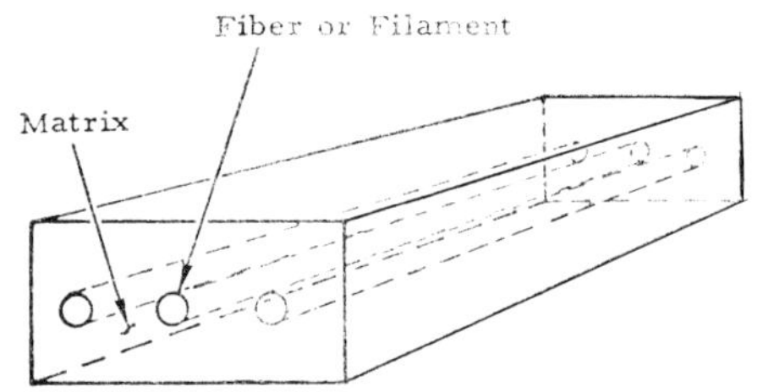

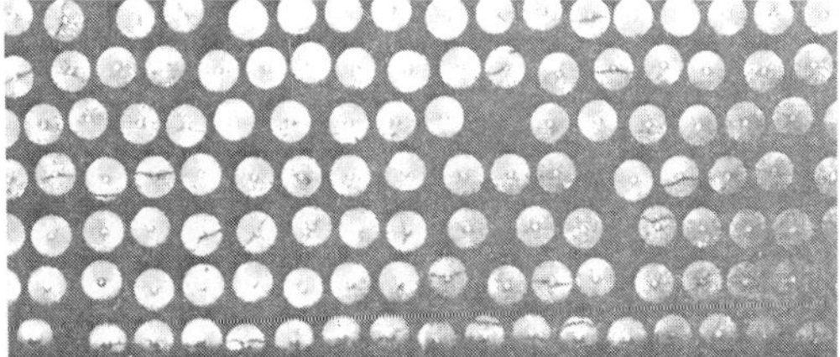

Figure 18.1. Reinforcing filaments in a matrix.

Figure 18.2. Three-in. (0.076-m) wide boron/epoxy tape.

Figure 18.3. Twelve-in. (0.305-m) wide cross-stitched graphite/epoxy tape.

quired. Properties which must be demonstrated through testing are normally specified in procurement specifications.

Boron and graphite filaments are manufactured and fabricated in tape form, normally up to 3 in. (0.076 m) wide for boron and 12 in. (0.304 m) for graphite (see Fig. 18.2). Some graphite has been fabricated into cross-stitched tapes 12 in. (0.304 m) wide, as shown in Fig. 18.3, and in some cases is being made commercially available with cross-stitching up to 60 in. (1.524 m) wide. These tapes are impregnated with resin by wet impregnation (solution coated), or by warm pressing of the fibers into a film of "B" staged resin.

18.2.1. Manual Lay-up

Fabrication of advanced composite structure, as with any filamentary reinforced composite, requires the placement of the reinforcing fibers within a matrix in an orientation that will provide the properties required by engineering. Manual lay-up means the placement of layer after layer of tape in a prescribed orientation on separate ply templates with subsequent stacking of the plies, or by directly laying up of ply on ply without separate templates. For prototype parts, manual lay-up is often the most economical method, unless the parts generally lend themselves, by similarity, to other parts being made (Fig. 18.4 shows tape being manually laid up on mylar templates). These separate plies are subsequently stacked with the orientation controlled as required by the engineering drawing, as shown in Fig. 18.5.

In laying up unidirectional tape with all plies in one direction, the use of tape of the exact size required saves the need for side trimming. The tape merely needs to be laid up, ply-on-ply, with an end trim, as shown in Fig. 18.6.

Composite laminates are assembled by proper placement of wide-goods or prepreg tapes ply by ply, following the required ply orientation and stacking order, as shown in Fig. 18.7. When hand lay-up procedures are used, it is common practice to pre-assemble

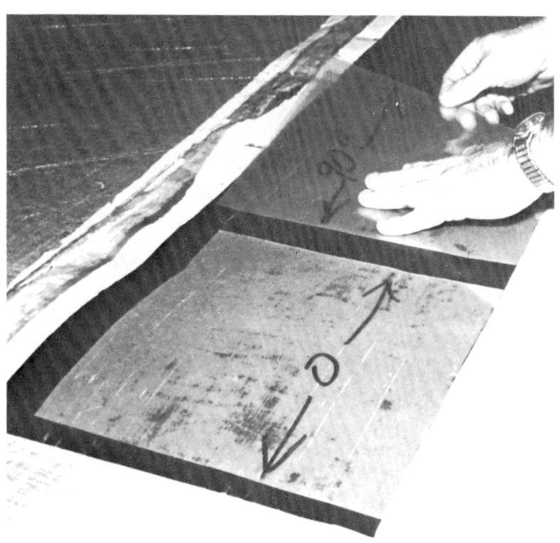

Figure 18.4. Mylar being applied to 12-in. (0.305-m) wide graphite/epoxy tape.

Figure 18.5. Stacking plies.

Figure 18.6. Net width boron/epoxy tape being laid up ply-on-ply.

Figure 18.7. Adding Mylar covered "90°" ply to lay-up.

plies on special plastic templates. This procedure is performed in a controlled environment area typical of those used in fiberglass and adhesive-bonding lay-up operations. The details are essentially as described below.

18.2.1.1. Preparation of Individual Plies

- Allow tape raw material to warm to room temperature before opening the sealed package, to prevent moisture from condensing on tape. Place the tape on a tape dispenser.
- Record the spool or sheet and batch numbers of each ply of material at the time it is used.
- Wipe exposed templates with clean cheesecloth moistened with solvent; then air dry.
- Unroll sufficient prepreg for a strip of lay-up, and cut with sharp scissors, knife, or other suitable cutting tool.
- Place prepreg tape in the area indicated on the templates. Place adjacent tapes so that no gap in excess of 0.030 in. (0.7 mm) occurs. Overlapped filaments are not permissible for boron lay-ups. For boron parts, specify glass-fabric scrim cloth in design as either up or down with respect to the template; lay up all subsequent plies the same way.
- Work the prepreg lay-up against the template, using a wiping motion to ensure intimate contact. Remove the separator strip, and place it in a waste container.
- Inspect each ply lay-up for filament (or film) breaks, cross-overs, gaps between tapes, and foreign matter contamination.
- If ply lay-up storage is required, seal the ply in a clear, clean moisture-proof protective film, and store it flat until needed.

18.2.1.2. Laminate Lay-up (Fig. 18.8)

Prepare the lay-up tool as follows.

- Clean to remove all foreign material.
- Polish to develop a smooth surface.
- Wipe with suitable solvent, and air dry.
- Apply a suitable release or parting agent.

Cover the surface of the tool with a nonporous, Teflon-coated fabric if the part does

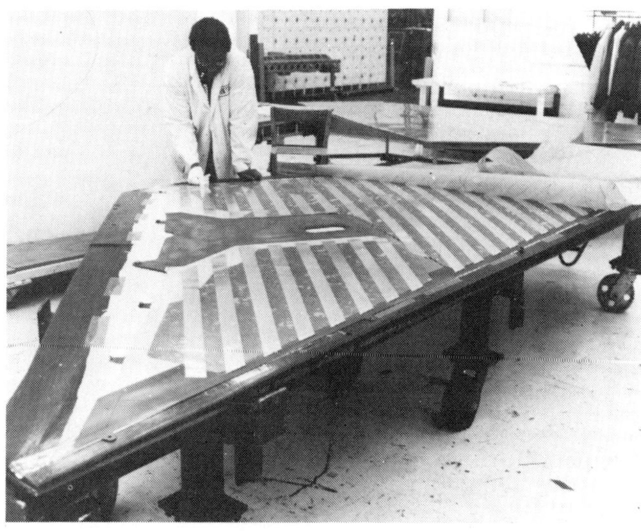

Figure 18.8. View of structural lay-up and tooling.

not get cured directly against the tool surface. Then prepare a peel ply (only if surfaces of the cured part are to be adhesively bonded or painted). Position one ply of set and heat-scoured nylon peel ply with ½-in. (12.7-mm) excess on all sides. Rub out with a squeegee to remove all air pockets and wrinkles. All splices must be of the butt type. Remove the remaining separator film.

Tool-pin holes or locators are provided to align the template precisely with respect to the tool. Occasionally, there is more than one set of locators; therefore, take care to match the right set of locators with the correct ply orientation. Next, using the roller or squeegee method, remove any air entrapped between plies.

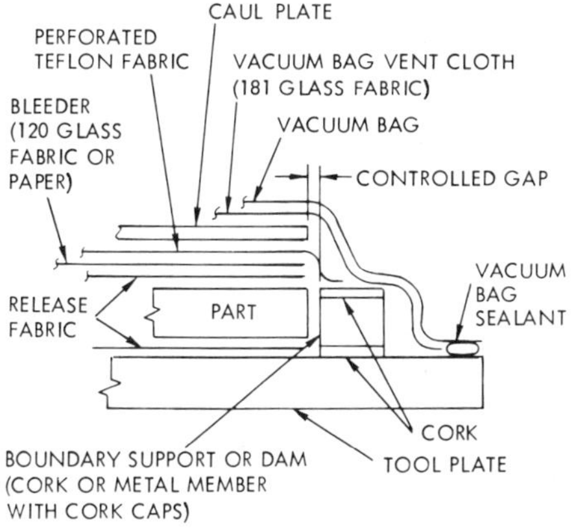

Figure 18.9. Controlled-flow bleed system.

If the prepreg tack is relatively dry, the Mylar templates can be removed easily by a peeling action. If prepreg is more tacky, remove the template by a jerking, peeling action. Inspect each ply after the template has been removed.

Repeat the steps outlined above, beginning with preparation of the peel ply, until all required plies are laid up. Each ply lay-up should be inspected and accepted on the proper shop order paperwork before the next ply is started. Use a balance of 104 glass scrim, impregnated with the same resin, as the last ply of a boron laminate.

18.2.1.3. Preparation for Cure

Cover the part with a release material perforated for resin bleed. Although the material should cover the entire part, it should not extend beyond the periphery. Boundary supports or dams are required to prevent crowning by the bag and to restrict side bleeding. Normally, these supports are cork, rubber, or metal sheet and are carefully sealed before use. Resin can flow through unsealed separations and mismatches between segments of the dam or between the dam and tool. Boundary supports for a top-bleed system should be within 0.060 in. (1.52 mm) of the part edge to prevent filament washing during the resin flow stage. Figure 18.9 shows a typical boundary support system.

During the cure, each ply debulks (reduces from its original thickness) by voiding itself of entrapped air and excess resin. Therefore, each advanced composite material and part requires a different boundary support (dam). Dams can be fabricated by using one of the following techniques.

1. All cork or rubber—for use when the part is less than 40 plies and is to be trimmed.
2. All metal—for use when the part is less than 100 plies and is fabricated net.
3. Combination metal and cork or rubber—for use when the part is over 40 plies.
4. Air dam—a polymeric moldable tape.

The height of the dam should always be less than the width to prevent lateral movement. The ratio of height to width should be no greater than 2:3. For laminates of less than 100 plies, metal dams should be the same height as the desired cured laminate thickness, while cork dams should be 10:7 greater than this thickness to allow for cork debulking. (Note that, occasionally, a cork or other non-metallic dam conforms and seals to the tool surface, preventing adequate air escape from laminate edge and thereby trapping voids in the laminate.)

For laminates over 100 plies, and when a combination cork or rubber and metal dam material is used, the metal portion of the dam should be of the same height as three-fourths of the desired cured laminate thickness, while the remainder of the dam should be cork or Corprene rubber of a height equal to the remaining one-fourth of the laminate thickness times a factor of 10:7. Aluminum pressure-sensitive tape may be placed under the dam to act as a seal to prevent excess resin loss.

One ply of dry 120 glass fabric should be applied for each five plies of boron prepreg. The 120 glass should be the same size as the part and should not extend beyond the periphery of the part. This 1-to-5 relationship between dry 120 glass and the prepreg provides a controlled bleed of the resin. This same control can be accomplished with a combination of 116 glass and felt. The relationship is shown in the table below.

PLIES OF BORON EPOXY	PLIES OF 116 GLASS	THICKNESS OF FELT, IN. (mm)
10	2	0
15	4	0
20	2	1/16 (1.588)
30	2	1/8 (3.175)
40	2	3/16 (4.762)
50	2	1/4 (6.35)
60	2	5/16 (7.938)
70	2	3/8 (9.525)
80	2	7/16 (11.112)
90	2	1/2 (12.7)
100	2	9/16 (14.288)

For laminates in excess of 100 plies, the felt should be increased 1/16 in. (1.588 mm) for each 10 additional plies of prepreg.

A layer of perforated plastic film such as Teflon should be placed over the entire lay-up so that the edges extend well beyond the periphery of the part. The holes should be on approximately 2-in. (50.88-mm) centers. Cover the entire lay-up with two plies of 181 glass fabric for vacuum bag venting.

The entire lay-up should be held securely against the tool during all subsequent handling until the part is in the autoclave for cure. The controlled-flow bleeder system is shown in Fig. 18.9.

In the laminate bleed system shown, a portion of the resin flow occurs through the controlled gap between the laminate edge and the dam or boundary supports. The volume of resin to be bled from the composite prepreg is calculated and equated to the gap voume. A simple trial-and-error solution is used to determine the gap dimension. Additional tailoring of the gap dimension may be accomplished if the edge bleeding is limited to laminate lateral edges. Laminates up to 10 by 65 in. (25.4 by 165.1 cm) have been fabricated with this controlled-flow bleed system at the desired ply thickness for the cured composite. It is necessary that the boundary supports be principally metal and that their position be properly maintained, since the gap void must be maintained under the cure temperature and pressure conditions preceding resin flow. Approximately 1/16-in. (1.588 mm) aluminum plates, sized to the laminate lay-up dimensions, may be used as a precaution against cured laminate bag surface irregularities and excessive resin loss from the top composite plies into the vacuum bag vent cloth. (Note that each organization should conduct a controlled resin-flow edge-bleed process investigation for the raw material type, laminate thickness, and related cure cycle for its own application.)

18.2.2. Semi-Automatic Lay-up

Semi-automatic methods of lay-up are applicable to almost any degree of automation short of a fully automatic short tape laying machine. One form involves semi-automation to make broadgoods by winding tape onto a large mylar tube; plies are then cut from broadgoods, as shown in Figs. 18.10, 18.11, and 18.12, using Mylar or similar templates.

Another form of semi-automation is shown in Figs. 18.13, 18.14, and 18.15, where the alignment of side by side lay-up of tape is controlled by a manually operated alignment payoff machine. In a manner similar to a "drafting machine," the payoff spool of tape is manually indexed but mechanically controlled for parallel lay-up. This mechanical method of semi-automation is commonly referred to as "a la flintstone" at several companies. Cutting of the tape at the trim line is still a manual portion of all the existing semi-automatic operations.

Making ply-on-ply lay-ups is difficult to control unless each ply is trimmed to the same edge trim line. One method being used for superimposing trim lines of irregularly shaped plies in a ply-on-ply lay-up is accomplished by visual projection. The ply outline is projected

Figure 18.10. Broadgoods being laid up on Mylar tube.

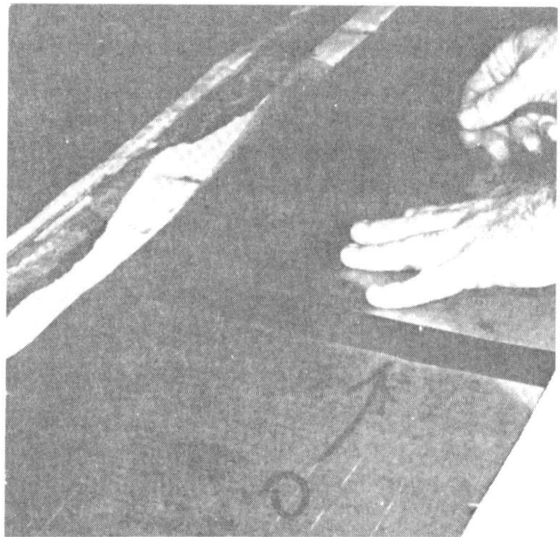

Figure 18.11 Templates being used to trim plies from broadgoods.

onto the part lay-up, using carefully coordinated, sized, and aligned photo masters and projection equipment. The material is then trimmed to the photographically projected image. Caution must be exercised in trimming so as not to damage underlying plies.

Semi-automated techniques for laying up composite materials can be extremely beneficial in cost reductions and lay-up uniformities. In most cases, the method used must be tailored to the part or shape being made to realize maximum benefits. Multi-layer broadgoods can easily be made from which simple shapes may be performed, rather than laying up each ply to the final shape.

The objectives of machine lay-up are to eliminate the human error, improve the laminate quality, and speed up the fabrication process (lower the cost). Some of the problems encountered with developing tape lay-up machines have been tape tack, tape alignment, indexing of tape orientation through numerical control, and lay-up of curved surfaces. A major hindrance to the use of these machines in sub-structure fabrication has been the inability to wrap tapered components while maintaining constant laminate orientation.

Tape laying machines, as they exist today, range from hand-operated dispensing heads on a track to fully automated machines. The simplest of these is shown in Fig. 18.13. As shown in this example, one person manually

Figure 18.12. Ply cut from 3-in. (0.076-m) wide tape lay-up on Mylar template.

operates the lay-down head and manually cuts the tape. Many variations can be applied to this basic concept, such as a cutting device on the head, a table base that rotates for different tape orientations, a continuous belt system for indexing the Mylar template, and others that can be adapted to suit a particular application.

18.2.3. Automatic Lay-up

The ultimate goal is to make lay-ups, either flat or contoured, by numerically controlled machines with tape laying heads capable of uniformly and consistently laying up ply after ply as programmed without manual assistance. In fact, it is envisioned that several machines might be operated by one console with a single operator. This would provide high-quality, uniformly correct parts with a high degree of repeatability and reliability, and at the lowest possible labor cost.

Several independent efforts have been halted because of inconsistencies in the available tape forms. Gap/overlap control still remains the primary problem. Several low-keyed independent developments are in progress with varying degrees of success. Two major Air Force sponsored development efforts have made great progress. One deals extensively with the mechanics and quality assurance devices of the tape-laying head using existing 3-in. (7.6-cm) wide tape. Aspects of typical automatic machines are shown in Figs. 18.16 and 18.17.

Another program uses a state-of-the-art tape laying head to identify and develop tape quality requirements that will result in composite structure of acceptable quality. The problem of fiber gap/overlap is accentuated with the stiffer, larger diameter—0.004-in. (0.0102-mm) and 0.0056-in. (0.0142-mm)—boron fibers. The extremely small fiber diameter—0.0004 in. (0.0102 mm)—and bend-

Figure 18.13. Manually operated tape laying equipment.

Figure 18.14. "Flintstone" semi-automatic tape laying (Grumman).

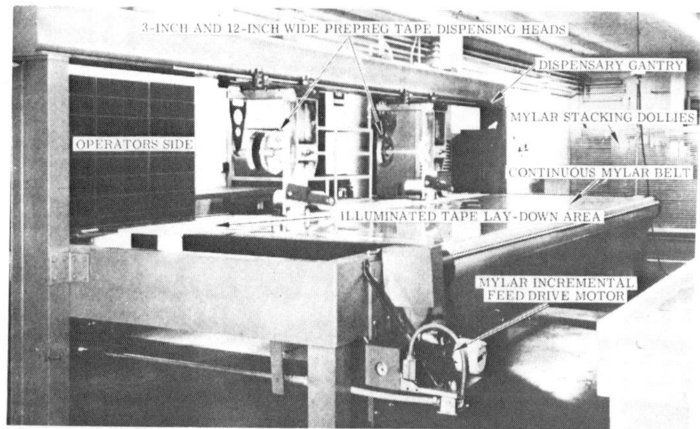

Figure 18.15. Three-in. (0.076-m) and 12-in. (0.305-m) tape semi-automatic laying machine (Northrop).

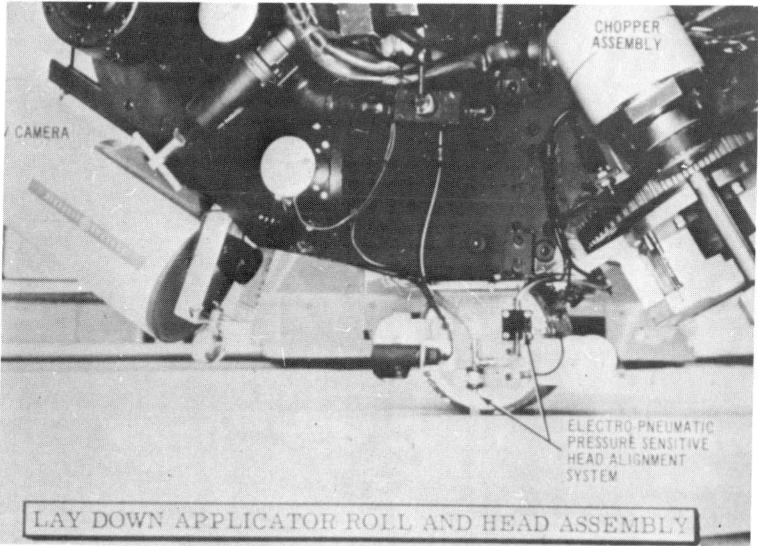

Figure 18.16. Automatic tape laying head close-up.

Figure 18.17. Grumman automatic tape laying machine.

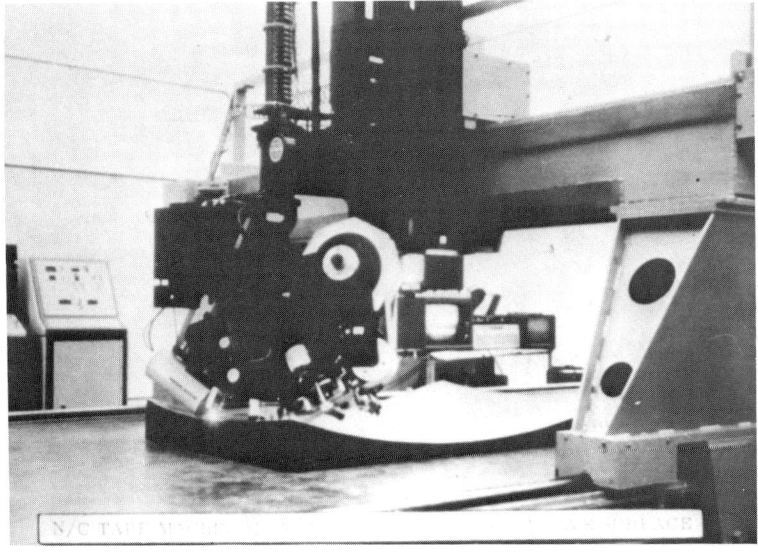

Figure 18.18. Five axis tape laying machine.

ability of graphite will tolerate some fiber overlap without sacrificing composite property. Figures 18.18–18.21 show the highlights of this program.

Two ongoing production programs are using fully automatic lay-up for the production of composite parts: the Grumman F14 Horizontal Stabilizer B/Ep Skins and the GD/Ft. Worth F111 Wing Pivot Fitting Gr/Ep Fairings.

A typical first-generation, automated, three-axis tape laying machine was constructed to

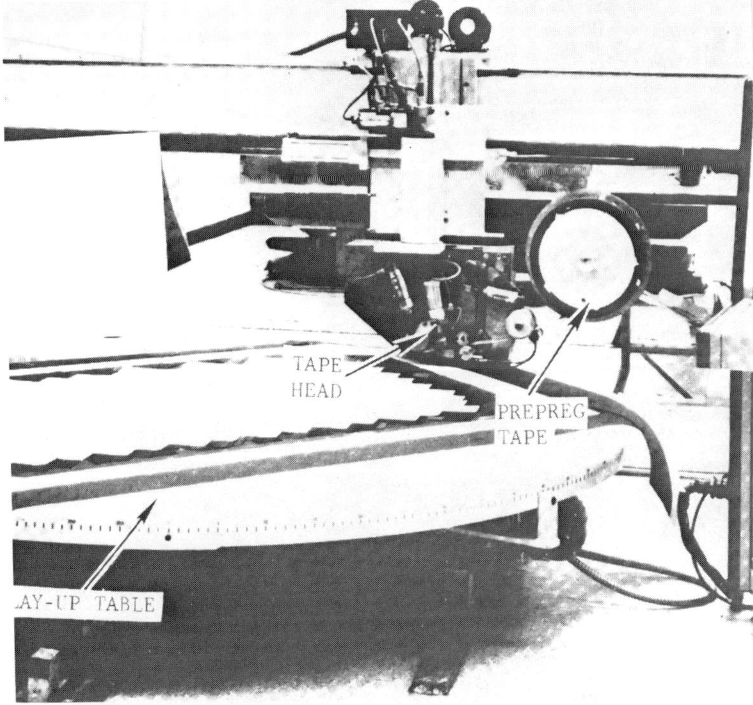

Figure 18.19. Automated 3-in. (0.076-m) wide prepreg tape laying machine and control console.

Figure 18.20. Close-up of Grumman's tape laying head.

Figure 18.21. Flat automatic lay-up.

have a laying area of 72 by 80 in. (1.83 by 2.032 m), a three-axis control system, a vertical head motion of 10 in. (0.0102 mm), by pneumatic actuator, a maximum tape lay of 2 lb (0.907 kg) of 4.0-mil (0.0102-mm) boron or 2.5 lb (1.159 kg) of 5.6-mil (0.0142-mm) boron per hour, and an average production rate of 1.25 lb/hour (0.567 kg/hour). The backing paper on the boron tape has edge perforations so that the tape can be fed in a manner similar to motion picture film.

A second-generation five-axis machine which represents the current state of automated technology has the following features.

- A laying area of 72 by 360 in. (1.829 by 9.144 m).
- Manual, semi-automatic, and automatic five-axis control system:
  - *X*-axis—longitudinal: 720 in. (18.288 m)/min. max.
  - *Y*-axis—transverse: 720 in. (18.288 m)/min. max.
  - *C*-axis-head rotation: ±180°
  - *D*-axis—shear rotation: ±60°
  - *P*-axis—tape transport: 40 in. (1.016 m)/min.
- Vertical head motion—7 in. (0.1778 m) pneumatic actuator.
- Air bearing work table which will support 125 lb/ft^2 (410 kg/m^2) tool loads.
- Shear which will operate at all tape laying speeds.
- Maximum tape lay of 36 lb/hour (16.326 kg/hour) of 4.0-mil (0.0102-mm) boron or 45 lb/hour (20.408 kg/hour) of 5.6-mil (0.0142-mm) boron.
- An average production rate of 13 lb/hour (5.896 kg/hour) of 4.0-mil (0.0102-mm) boron or 16 lb/hour (7.256 kg/hour) of 5.6-mil (0.0142-mm) boron.

Improved automatic tape lay-up machines are being developed. Table 18.1 shows the approximate lengths of 3-in. (7.6-cm) wide tape in 1 lb (0.4535 kg) of material.

Those persons using machine lay-up techniques now find that the ply template previously discussed may be eliminated, and the machine can be programmed to place the tape automatically in the desired orientation for the required number of plies. The preparation and the cure of a machine laid-up composite are the same as described for a hand laid-up composite.

Table 18.1 Three-in. Tape Length Versus Weight

MATERIAL	LENGTH, ft/lb (m/kg)
Graphite	64–80 (43–53.8)
Boron (4.0 mil)	63 (42.3)
Boron (5.6 mil)	50 (33.5)
Organic fibers	91 (61.2)

18.2.3.1. Two-Dimensional Lay-ups

Flat lay-ups can be made rather easily within the gap control provided by the present tape uniformity. Several commercially available tape laying heads, can accomplish this. The primary machine controls involve parallel tracking, side-by-side indexing of adjacent layers of tape, and end-trim cut-off of the tape to predetermined trim lines. Unidirectional glass tape as a low-cost trial material has been used in machine development. The transition from glass to advanced composite materials has not been a major problem. An example of an automatic head making a flat lay-up is shown in Fig. 18.21.

18.2.3.2. Three-Dimensional Lay-ups

Simple and complex contoured, automated lay-ups present the machine design with an entirely new set of problems. These are in addition to the already existing difficulty of tape run-out or convergence due to complex contour (three-dimensional shape/two dimensional non-stretch tape). The latest automatic machines have provisions to accomodate moderate contours by either using flexible membranes or self-adjusting mechanical pressure devices.

18.2.3.3. No Bleed Resin Molding

The commonly used form of resin matrix prepreg has a resin content in the range of 41 ± 3% and requires a significant amount of resin bleedout during cure to achieve a cured laminate resin content of 28–32%. More emphasis is now being placed on lower resin content prepregs with their attendant low or no resin bleed processes. Economic advantages exist in two areas: (1) user buys less resin, and (2) less breather/bleeder material is used. However, when no-bleed processes are used, removal of entrapped air becomes a more critical aspect of process control. More attention to debulking/preconsolidation of the laminate is necessary in a no-bleed process to ensure removal of entrapped air.

This process is now being used successfully for wing components for Boeing 757 and 767 aircraft. In the use of this method, it has been found necessary to end up with a higher resin content in the 34–38% range.

The process is used to fabricate complete units with honeycomb core and co-cured skins. Usually, the uncured skins are placed under tension during the cure cycle to prevent core crushing and dimpling.

Another advantage of this process is greater control of the resin content in the cured laminate. In the case of the Boeing 757 and 767 applications, both co-cured assemblies, there is the weight advantage of using an adhesive resin system as a prepreg.

There is still the inability to satisfy the smoothness requirements of the aerodynamic surface due to surface porosity. This is presently being corrected by applying a layer of adhesive to the tool surface. It is expected that this problem will be resolved with greater experience on the part of the prepregers as well as the composites product fabricators.

18.2.4. Filament Winding

This technique has been used primarily with glass fiber, but some advanced composite designs using graphite/epoxy have been fabricated. Examples of the use of this method with advanced composites are in landing gear or power transmission tubes. The prime area of application appears to be in the latter example, where tubes, radomes, or cylindrical components are required.

For cylindrical pressure vessels, the ratio of hoop to longitudinal stress is 2:1. The designer of filament wound pressure vessels is able to take advantage of filament orientation to produce an orthotropic material which is nearly twice as strong in the hoop direction as in the longitudinal direction. In an idealized filament wound pressure vessel, the filaments can be arranged in a cross-ply configuration so that they are oriented ±54.75° with respect to the longitudinal axis. This angle, whose tangent is 2, provides a symmetrical filament lay-up. Within approximate assumptions connected with netting analysis, the strength in the hoop direction is twice the strength in the longitudinal direction.

In practice, closing the ends of the cylindrical filament wound vessel imposes a compromise which is different from the ideal ±54.75° cross-ply construction. To bring fibers around to form dome end-closures, it is necessary to use other winding angles. The filaments that are wound near the edges of the polar boss (or the opening usually left in the end domes) must, for stability reasons, pass from the cylinder to the dome regions at very low winding angles. Those filaments are continued at the same low winding angle along the cylinder to cover the dome at the other end of the pressure vessel. Since the low-angle windings, which are frequently called "longitudinal wraps," are applied at less than the optimum ±54.75° angle, the usual practice is to compensate for the longitudinal wraps with a set of circumferential windings applied at some high winding angle, such as 85°. A mixture of high and low windings, or "covers," is designed to yield a so-called "balanced" winding pattern.

This process, developed primarily by Whittaker Research and Development, has been used to produce fishing rods and arrow shafts initially. Later, it was adapted to produce air foil leading edges, 29-in. (0.74-m) diameter sub-scale missile interstage, molded cylindrical shapes, aircraft fuel tanks, and concepts for modular housing units. (See Chap. 15).

18.2.5. Plywrapping

Plywrapping is a low-cost fabrication method wherein plies of composite material, fed onto

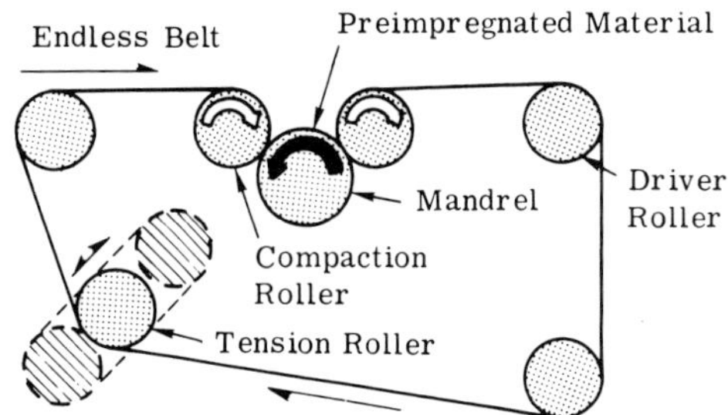

Figure 18.22. Schematic of plywrap machine.

a continuous belt, are wrapped and compacted on a mandrel. The principle of this lay-up method is schematically shown in Fig. 18.22. The mandrel, trapped in the endless belt "pocket" formed by the adjusted position of the compaction roller, is rotated by the belt movement. Plies are placed on the belt ahead of the mandrel and subsequently are wrapped and compacted as seen in Fig. 18.23. Once the required number of plies have been wrapped onto the mandrel, heat and pressure are applied to cure the part. An example of a fuel tank mold is shown in Fig. 18.24 and a cost comparison of this 300-gal (1135.5-1) fuel tank is shown in Fig. 18.25.[1]

The balanced winding pattern is intended to yield the same orthotropic set of mechanical properties as the idealized ±54.75° lay-up. A compromise has been made in the area of strain compatibility, because greater interlaminar shear force develops between layers of low and high angle covers than would occur in the (±54.75°) pattern. When a low modulus resin is used as a matrix for glass-reinforcing filaments, a moderate degree of shear distortion can be accommodated. However, with

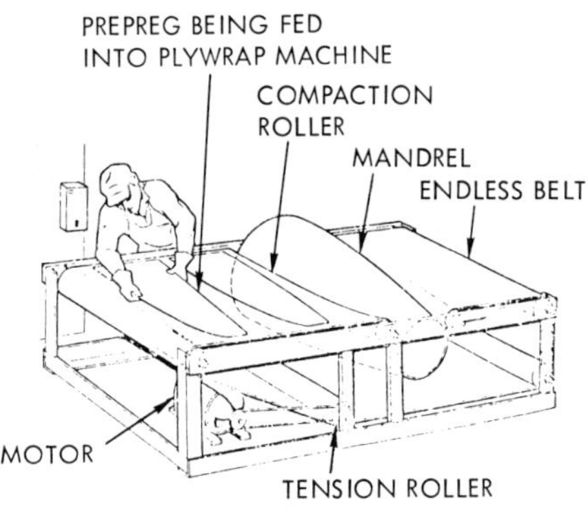

Figure 18.23. Plywrap operation.

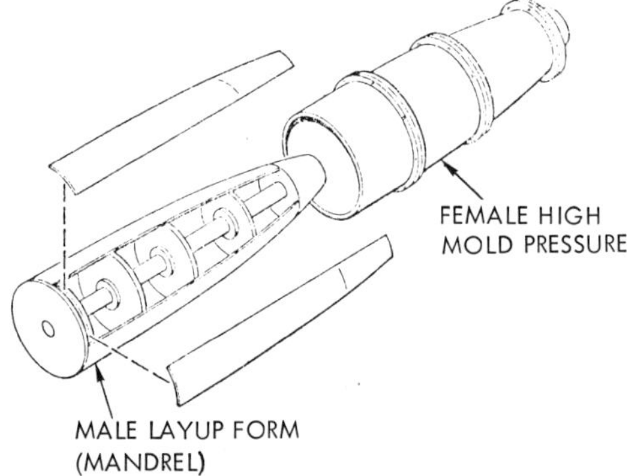

Figure 18.24. Plywrapped fuel tank tool.

the trend toward high temperature performance requirements, the resins that must be used are less forgiving by nature, and structural failure by crazing in the matrix resin or delamination between low and high angle covers may develop. A further advantage is using high modulus, filament wound prepreg tapes is that good strain compatibility is provided with metal liners that are often required in filament wound pressure vessels to ensure against leakage of the contents.

18.2.6. Secondary Fabrication Operation

Two methods of attachment are discussed here: adhesive bonding and mechanical fastening. The bonding technique is emphasized because of its importance in saving structural weight and cost in composite components.

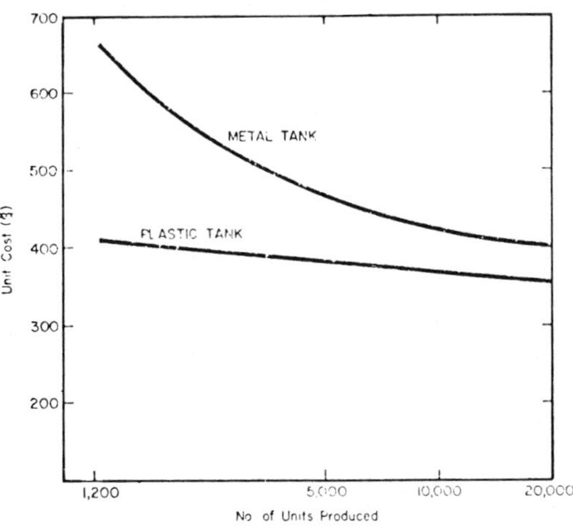

Figure 18.25. Cost comparison—300-gal. (113.5-l) fuel tank.

18.2.6.1. Adhesive Bonding

There are two forms of composite bonding: co-curing and secondary bonding. Investigations have shown that co-curing requires fewer tools and, hence, less labor hours; however, when co-curing on honeycomb core, some properties are degraded because of dimpling effects. "Embossing" (the imprint of honeycomb cells through the co-cured facings) also degrades the skin properties under compressive loading. The extent of property loss as result of co-curing honeycomb panels is dependent upon cell size, skin thickness (number of plies), ply orientation, and cure pressure. Co-curing has been developed as an acceptable manufacturing process to reduce the cost of advanced composite components.

Basically, the overall bonding technique is a three-step process: selection of adhesive, surface preparation, and bond cycle. These steps will be discussed in succeeding paragraphs.

Adhesive Selection. One of the most important steps in bonding is to choose the correct adhesive system. The following guidelines apply for choosing an adhesive.

1. It must meet the strength requirements of a particular design in the stated environment.
2. It must be compatible with the laminate resin system.
3. It must have adequate tack for the particular number of bond cycles required.
4. It must have a low enough modulus relative to the adherends to permit a fairly uniform stress distribution without excessive edge concentration.

Surface Preparation. Perhaps the most important step in bonding concerns proper surface preparation. Below, surface preparation methods for the various types of materials used in component fabrication are described. Aluminum sheet will not be discussed, since there is a vast data bank on this material.

Honeycomb Core. Aluminum and fiberglass cores are being used in sandwich structures which are both bonded and co-cured. The method of cleaning cores depends on the condition of the core after previous processing. In the case where core has been machined in the before-expansion state or held in position by double-backed tape, the core is cleaned by the standard vapor-degreaser method. When core is "set" by polyethylene glycol ether (polyglycol), the core (after machining) should be washed thoroughly with hot water—over 160° F (71.1° C)—alkaline-cleaned, rinsed with deionized water, and quickly dried. Aluminum core should be carefully examined for stain or possible corrosion after the cleaning operation.

Titanium. Research efforts to develop a surface preparation method for titanium have produced inconsistent results. The one technique that has proved to be the most consistent is the vapor honing method. This has worked consistently and has produced cohesive failure within the adhesive. The procedure is detailed in the following steps.

1. Vapor hone to a uniform satin finish using 400-grit aluminum oxide at 60–90 psi (4530–6039 kg/m^2). Quickly rinse in tap water, and observe uniformity and water break. (*Caution:* The surface finish must be uniform and satin. Repeat, if necessary.) Dry grit blasting may be substituted for vapor honing, provided the proper safety conditions have been incorporated.
2. Wipe the grit-blasted surfaces quickly with methyl ethyl ketone (MEK) until dry. There should be no change from a uniform surface. (Stains are not acceptable.)
3. Alkaline-clean faying surfaces by standard methods; rinse, and dry.
4. Treat faying surfaces with Pasa-Jell 107; rinse, and dry.
5. Prime surfaces or apply adhesive within 4 hours after the drying operation of step 4.

Other techniques for preparing adhesive or resin bonded joints have been used by various organizations involved in the use of titanium structures. A review of the literature on titanium bonding reveals that bond strengths may be highly variable and, under some

Table 18.2. Surface Preparation of Titanium (Phosphate-Fluoride Treatment)

STEP	OPERATION	MATEEIALS	TIME (MINUTES)	TEMPERATURE, °F (°C)	REMARKS
1	Clean	Alkaline cleaner	10–20	170–210 (76.6–98.9)	Parts shall be "water-break" free.
2	Acid pickle	63–70% volume 42° Bé nitric acid; 4.2–11.4 oz/gal. (31.45–85.4 kg/m^3) ammonium bifluoride	1–2	55–90 (12.8–32.2)	Ammonium bifluoride may be added to maximum indicated to maintain bath activity.
3	Rinse	Deionized water	1–2	60–150 (15.5–65.6)	Constant overflow.
4	Phosphate-fluoride treatment	7.3 ±0.2 oz/gal (54.7 ±1.5 kg/m^3) $Na_3PO_4 \cdot 12\ H_2O$; 1.5 ±0.1 oz/gal (11.2 ± 0.75 kg/m^3) NaF; 7.3–7.4% by volume HF (60% solution)	½–1	80–90 (26.7–32.2)	Dissolve chemicals in warm water: then cool to working temperature. During the coating operation, rapid bubbling is usually observed for the first few seconds.
5	Rinse-immersion	Deionized water	15–20		Constant overflow or spray.
6	Spray or immersion rinse	Deionized water			Air deionized water blast may be used as required
7	Force dry	Hot circulating air	Until dry	140–250 (60–121.1)	If non-uniform coating is observed before priming, parts may be re-worked.

circumstances, quite unpredictable. The experience of previous investigators is not suprising in light of current knowledge that bond strength is highly sensitive to the chemical condition of the metal surface.

A second method is the immersion fluoride treatment, which is outlined in Table 18.2. This procedure has produced the best results thus far. Scanning electron microphotographs of prepared surfaces have revealed that slight changes in processing and solution concentration can have an impact on the bondability of the metal. Studies also show that surface activity continues during the first 24 hours after the drying operation, and that the best results are obtained on surfaces primed and bonded immediately after the surface preparation.

Steel. Fortunately, steel can be more readily prepared for bonding than titanium. The following method has been generally accepted by the industry.

1. Remove oil and grease by MEK wiping or vapor-degreasing.
2. Grit-blast with 180-grit aluminum oxide, using 60–90 psi (4350–6039 kg/m^2)
3. Wipe with a clean cloth dampened with MEK.
4. Alkaline-clean in accordance with industry standards.
5. Rinse with tap water.
6. Immerse in 20–55% HNO_3 at a temperature of 65° F (18.3° C) minimum for 90 (±30) seconds.
7. Rinse in deionized water, and dry.
8. Prime surfaces or apply adhesives within 4 hours after grit blast operation described above.

Composites. Three methods have proved successful in composite surface preparation. The first is the peel ply method, where heat sealed and scoured nylon peel ply material is applied to the "B" stage laminate in the area to be bonded. This material does not wrinkle easily and produces a suitable surface. The ultrasonic NDT methods are applicable with the peel ply intact. After NDT, the peel ply can be removed, providing a surface ready for bonding. The second method is grit-blasting with a fine aluminum oxide. The third method utilizes a hand-sanding operation followed by a solvent wipe and an air dry step. All methods have produced approximately the same shear ultimate strength of bonding using Shell 951 adhesive and graphite/epoxy laminates.

Bond Cycle. To achieve proper bonds in a structure, correct heat-up rates for the adhesive must be maintained. This is usually ascertained through sub-component tests conducted during the design stage.

18.2.6.2 Mechanical Fastening

Mechanical fastening is normally used to attach advanced composite structures when high peel stresses are present, when fail-safe criteria are required, or when periodic disassembly of the joint is mandatory. For example, internal fuel pressure gives rise to unacceptable peel (or flatwise tension) stresses within bonded joints in a wing integral fuel tank; hence, mechanical attachment is required. Use of mechanical fasteners, however, gives rise to serious stress concentrations. Two ways to alleviate these concentrations involve replacing 0° laminae adjacent to the holes with 1) metallic interleaf strips and 2) softening strips of fiberglass or ±45° graphite/epoxy.

Bucked rivets are not recommended for advanced composite joints, but squeezed or pull rivets have been used. When bolts are used, the choice between standard heads (exterior or counter-bored) or counter-sunk heads is largely a question of design constraints and the reduced net section produced. When aluminum fasteners are used with graphite/epoxy, it is suggested the fasteners have a protective treatment to prevent corrosion problems.

18.2.6.3 Machining Epoxy Laminates

Boron/epoxy composites present a challenge in the area of machining. The epoxy matrix is relatively easy to machine, but its comparative weakness in some cases does not provide adequate support to the boron filaments to prevent fiber break-out. The boron filament itself presents other difficulties because its hardness approaches 9.5 on the Mohs scale. Conventional hard cutting materials, such as tungsten carbide, aluminum oxide, silicon

Table 18.3. Machining Boron/Epoxy Composites

OPERATION	DIAMOND TOOL DESCRIPTION	CUTTING SPEED	FEED RATE	COMMENTS
Drilling				
¼–¾-in. (6.35–19.05-mm) diameter	Core drill; sintered metal bond, 60/80 grit; or plated, 60/90 grit	50–300 SFM (15.24–91.4 SmM) or 3000 RPM	Hand oscillating feed, light pressure	Flood coolant through drill.
Less than ¼-in. (6.35 m) diameter	HSS twist drill	350 RPM	0.007 in. (0.178 mm)/ revolution, automatic	Flood coolant.
Reaming ¼–½-in. (6.35–12.7-mm) diameter	Diamond reamer, 120 grit	50–125 SFM (15.24–38.1 SmM)	Light pressure	Core drills were used for drilling reaming holes.
Counter-sinking (rough)	Sintered, 100° angle, 60 grit or Plated, 100° angle, 60 grit	120 SFM (36.58 SmM)	Hand oscillating feed, light pressure	Flood or mist coolant
Counter-sinking (finish)	Sintered, 100° angle, 100–200 grit	120 SFM (36.58 SmM)	Hand oscillating feed, light pressure	Flood or mist coolant.
	or Plated, 100° angle, 100–200 grit	120 SFM (36.58 SmM)	Hand oscillating feed, light pressure	Flood or mist coolant.
Routing	Plated, slotted 60/80 grit	450 SFM (137.16 SmM)	Depth of cut controls feed rate	Mist coolant.
Bandsawing	Coated-edge blade, 40/60 grit	1600 SFM (487.7 SmM)	Hand-feed	Mist coolant.
Radial slitting or cutting	Sintered metal bond, 6-in. (15.2-mm,) diameter × 0.050-in. (0.127-mm) thick, 100 grit	2350–6000 SFM (716.3–1828.8 SmM)	1½–22 in. (38.1–558 mm)/minute	Back-up material required to reduce edge delamination.
	Plated, 6-in. (15.2-mm) and 8-in. (20.3-mm) slotted wheel 0.060 in. (0.152-mm) thick, 60/80 grit	2350–7100 SFM (716.3–2164.1 SmM)	3–20 in. (76.2–508 mm)/minute	Produces rougher surface cut.
Grinding	Diamond, sintered metal, bond cup wheel, 3½-in. (8.89-mm) diameter, 100 grit	1400 SFM (326.7 SmM)	½–3 in. (12.7–76.2 mm)/minute	Edge grinding performed on milling machine. Mist coolant.

Table 18.4. Machining Boron/Epoxy 17-7PH Steel Interleaf Composite—Interleaf 0.005-0.010-in. Thickness

OPERATION	TOOL DESCRIPTION	CUTTING SPEED	FEED RATE	COMMENTS
Drilling ¼–½-in. (6.35–12.7-mm) diameter	a. Diamond core drill, sintered bond, 100 grit	5–15 minutes per hole	Hand oscillating feed, light pressure	Flood coolant through drill.
	b. Rough drill—HSS Drill—then ream	350 RPM	0.007 in. (0.178 mm)/ revolution constant-feed drill motor	Flood coolant.
	c. 60–80 grit	1300 RPM 6 minutes	⅛ in. (3.175 mm)/ minute	Special flood coolant.
Reaming	Diamond-plated reamer		Hand drill motion	Mist coolant
Counter-sinking	a. Diamond, sintered metal bond, ¾-in. (19.05-mm) diameter	400 RPM	Hand oscillating feed	Mist or flood coolant.
	b. Rough with 36 grit diamond-plated tool	400 RPM	Hand feed	Mist coolant.
	c. Finish with 100 grit diamond-plated tool	400 RPM	Hand feed	Mist coolant.
Routing				Routing on edge of panel unsuccessful when cut touches metal.
Bandsawing				Same as for straight baron/ epoxy.
Radial slitting or cutting	a. Silicon carbide wheel 16-in. (406-mm) diameter × $^{1}/_{16}$-in. (1.588-mm) thickness	4200 SFM (1280 SmM)	½–1 in. (12.7–25.4 mm)/ minute	Mist coolant.
	b. Diamond-impregnated saw	6000 RPM	2 in. (50.8 mm)/ minute	Mist cool with backing.
Grinding				Do not recommend.

Tabvle 18.5 Matching Boron/Epoxy Titanium 6-4 Interleaf Composite—Interleaf 0.005–0.010-in. Thickness

OPERATION	TOOL DESCRIPTION	CUTTING SPEED	FEED RATE	COMMENTS
Drilling ¼–½-in. (6.35–12.7-mm) diameter				Same as for 17–7PH steel interleaf.
Reaming ¼–½-in. (6.35–12.7-mm) diameter				Core drills were used for both drilling and sizing holes.
Counter-sinking				Same as for 17–7PH steel interleaf.
Routing				Routing on edge of panel unsuccessful when cut touches metal.
Bandsawing				Same as for straight borox/boron/epoxy
Radial slotting or cutting	Diamond, sintered metal bond, 6-in. (152.4-mm) diameter, 0.050-in. (1.27-mm) thick, 100 grit	2400 SFM (731.5 SmM) or 1265 RPM	½–1¼ in. (12.7–31.75 mm)/minute 2 in. (50.8 mm)/minute	Mist coolant. Mist coolant.
Grinding	Silicon carbide wheel 12-in. (304-mm) diameter × 2-in. (50.8-mm) width	5600 SFM (1706.9 SmM)	100 in. (2.54 m)/minute	Grinding was done 90° to lay of interleaf, edge grinding with metal backup.

carbide, and tool steel, are softer than boron. While diamond, with a hardness of 10 on the Mohs scale, has been shown to be effective for machining boron/epoxy composites, considerable problems have been experienced when composite and interleaved metallic laminae must be machined together. For certain edge attachment designs, for example, drills have been changed at each interface between metal and composite. When thin foil metallic interleaves are used, this method is obviously impractical, and a compromise machining system is necessary.

A number of less conventional machining approaches have been investigated, including laser cutting, electromagnetic machining, and various combinations of tools with ultrasonic energy. Problems have included resin burning and poor tolerance control. In addition to diamond-faced tools, some manufacturers have used a variety of drills for metallic composite combinations. High-speed steel drills, piloted-core cobalt twist drills, and carbide space drills have been tried frequently followed by a diamond-surfaced reamer for final sizing and finish.

Table 18.3 lists typical machining methods which have been found to be successful with boron/epoxy composites. Table 18.4 gives similar information for composite stainless steel combinations. For all of these operations, adequate backup and clamping support is required during machining to prevent breakout of delamination. Table 18.5 gives machining information for composite-titanium combinations. Again, in this case, adequate support is required during machining.

In the case of graphite and other organic fibers, standard steel twist drills, countersinks, and routers have been used extensively and have not shown the need for specialized techniques or equipment. However, the drilling of certain of the organic fibers (e.g., Kevlar 49) may produce excessive tool wear because of their high toughness. Delamination can occur when drilling through composite. To prevent push-out when the drill breaks through the back side, masking tape should be placed over the area and the hole drilled slowly.

Water jet cutting has been found to be highly effective for graphite and aramid composites. An ultra-high speed jet of water is used and high cutting speeds and clean edge surfaces are usually obtained.

REFERENCES

1. "Panel Reports of Composites Recast, An Air Force/NASA Long Range Planning Study," February 22, 1972.
2. Hercules Product Data Sheets 815-3 and 831.
3. "Plastics for Aerospace Vehicles, Part I, Reinforced Plastics," *Military Handbook*, MIL-HDBK-17A, January 1971.
4. "Advanced Composite Wing Structures—Vol. I, Engineering," AFML-TR-70-231, Vol. I, Contract F33615-68-C-1301, Grumman Aerospace Corp., December 1970.
5. "Structural Design Guide for Advanced Composite Applications," Contract F33615-69-C-1368, Rockwell International, Los Angeles Division, Third Edition, January 1973.
6. "Manufacturing Methods for Cocuring Advanced Composites Materials," Contract F33615-71-C-1824, Northrop Corp.
7. "Manufacturing Methods for Thermal Expansion Molding of Advanced Composites," AFML-TR-75-10, Lockheed-Georgia Company, August 1975.

19
ENVIRONMENTAL EFFECTS ON PROPERTIES OF COMPOSITES

R. Staunton
Electronic Systems Division
Hanscom Air Force Base, Massachusetts

19.1. INTRODUCTION

Environmental effects on composite materials must be considered in the earliest stages of a system design. Failure to do so can lead to expensive design iterations during the later design stages or to a great increase in the cost of ownership as a result of failures caused by the environment. The design engineer must consider both natural and induced environments. Most important, when considering environmental effects, is to address the combined effects of the total environmental factors to which the system will be exposed.

19.2. CORROSION

Fiber reinforced plastics (FRP) exhibit inherent corrosion resistance and may be used to effect corrosion prevention and control by substitution for commonly corroded metal structure or by blending the composite into the metal design to take advantage of the positive synergistic effects of the metal and reinforced plastic component. While composites should not be thought of as cure-alls for corrosion prevention, they can, when used selectively, in conjunction with other materials, all but eliminate corrosive effects.

FRP composites have been successfully employed in various applications, with one of the primary objectives being corrosion prevention.[1] FRP shipboard structures, such as masts and spars, deckhouses, tanks, radar equipment, floats, and buoys; chemical plant equipment; and aerospace structures have all been successfully designed and used in highly corrosive environments for many years. Chemical corrosion can be prevented if the proper resins, reinforcements, and additives are chosen with care. The basic resin is of greatest importance, as is the selection of the chemical additive in the basic resin, which could leach out of the composite and cause corrosive attack.

Galvanic corrosion is one of the more common forms of corrosion. Galvanic corrosion can occur when metallic fasteners are used juxtaposed to graphite/epoxy composites. The same is true for many other combinations of metal structure in contact or in close proximity with the more noble graphite/epoxy composite. Given the proper conditions, with two or more dissimilar materials in the same structure, the more anodic material of the two will sacrificially corrode away, leaving the cathodic material behind. The severity of this corrosion is controlled by the galvanic cell strength, which, in turn, is controlled by the separation of the materials in the galvanic series, the extent of polarization, and the magnitude of the resultant current. In the proper electrolyte, these factors can drive the corrosive attack between two dissimilar materials.

Figure 19.1[2] indicates the highly noble nature of graphite/epoxy composites in comparison to various metals. Graphite/epoxy composites can be used in conjunction with less noble metals if the tactic of isolation is employed. Suitable coatings are applied to the

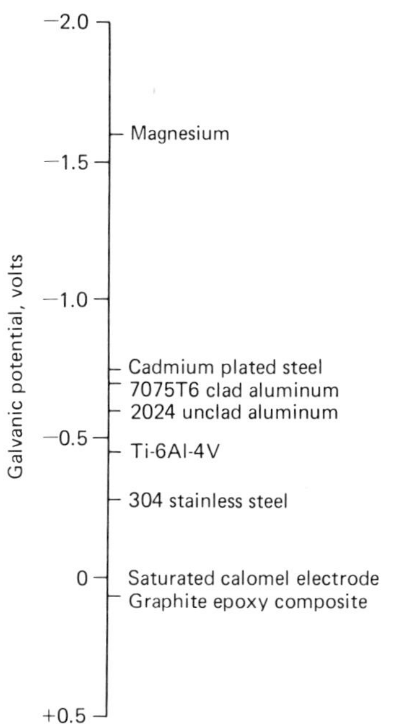

Figure 19.1. Glavanic potential in 3.5% NaCl solution (aerated and stirred) (25° C°).

faying surfaces which interrupt the galvanic cell.

Corrosion prevention and control is one of the most important factors in any design, and in many cases prevents corrosion from becoming the driving factor in ownership costs. Corrosion control can be affected by insulation, reduction in the cathode area, the use of corrosion inhibitors, and the elimination of moisture from the immediate environment.

19.3. ELECTROMAGNETIC EFFECTS

Electromagnetic pulse (EMP) effects or electromagnetic interference (EMI) can be caused by various sources, including nuclear detonations, lightning, and p-static or corona discharge. Its effects can be annoying or catastrophic. EM effects are not considered for the damage they can cause to composites; of interest is the amount of protection that can be provided by the composites for the underlying electronics. EM shielding has been evaluated over the various frequency ranges to include the threats listed above. Non-conductive composites provide little shielding effectiveness, while graphite/epoxy and boron/epoxy composites provide varying degrees of protection. Graphite/epoxy is more conductive than boron/epoxy, but neither can provide (on an equivalent thickness basis) the level of protection that can be provided with metals. The level of shielding provided by composites is significant, and each system design must be considered in terms of the threat to determine if added protection is required. Skouby[3] studied the shielding effectiveness of advanced composites over the frequency range from 50 kHz to 18 GHz. He concluded that graphite/epoxy composites 8 plies thick could provide shielding to plane wave threats as high as approximately 70 dB, and that similar boron panels could provide 30-dB shielding. The shielding effectiveness of fiber-reinforced composites can be greatly improved by metal coatings. (The various coatings are discussed in Section 19.7.)

19.4. FATIGUE

Fatigue implies repeated loading at very low loads to effect localized plastic deformations which can eventually cause damage to the structural capacity of the material. The applied loads are not high enough to cause immediate damage, yet, over a prolonged period of time, the cumulative effects can cause a catastrophic failure. Fatigue includes acoustic vibrations or mechanical loading. In general, advanced composites are highly resistant to the general effects of fatigue and can be used as substitute materials for metals in high fatigue environments. An all-mechanically fastened hybrid composite structure, representative of the B-1 horizontal stabilizer torque box, was exposed for 181 hours of acoustic fatigue that varied from 152–167 db without any composite failures.[4] Full-scale composite structures have been exposed to the B-1 service environment, at a two-times-life fatigue environment, with no degradation of the composite structure. Subelement tests, under the B-1 horizontal stabilizer program, were conducted with built-in production flaws. Elements were subjected to moisture, mechanical fatigue, and temperature excursions from 10–260° F (−12.2–126.7° C) representative of the B-1 service environment with no detectable flaw growth.[5]

19.5. FIRE

FRP can be classified in all ranges of flammability, from highly flammable to non-burning. The relative flammability of these materials can be altered significantly by the addition of fire retardants, which either impede the burning rate, render the plastic self-extinguishing, or provide for non-burning of the plastic. An excellent summary paper has been published[6] that discusses flame retardant additives and their effects on the properties of plastics. The amount of fire retardant required to effectively decrease the relative flammability is also discussed in this paper. The reinforcements used in FRP are inherently fire resistant, with the exception of the organic fibers. Dependent on the matrix, coupled with the reinforcement, the fiber can have either a negative or a positive effect on the flammability of the composite. If the matrix would normally tend to have the molten matrix drip away from the main source, this can, in some instances, extinguish the flames. The reinforcements can alter this procedure by trapping the flame base in place, thus allowing the flame to propagate. The reinforcements can also act as a flame barrier by minimizing the matrix availability to the fire. In general, one should expect the addition of flame retardants to lower some of the more desirable properties of the composite, such as strength and stiffness. Depending whether the additive is a plasticizer or a non-plasticizer, the impact resistance can be improved or degraded. As in all formula alterations, the effect on all of the properties of the resultant product must be considered. Epoxy resins used in advanced composites are rated as self-extinguishing without the addition of flame retardants.

19.6. IMPACT

Impact infers damage due to foreign objects striking the surface of the composite, causing possible localized damage or gross delaminations. It includes balistic damage, damage from sand, dust, and stones, and physical abuse. The impact resistance of composite materials can be controlled by the choice of reinforcement and matrix. The matrix can be altered by addition of plasticizers, which increase the strain to failure. This characteristic is also temperature dependent. Thermoplastic matrix materials tend to get softer with an increase in temperature, until such time that the resin will flow. Thermoset materials also become less brittle with a temperature increase, with a significant change at the glass transition temperature. Brittle reinforcements, such as boron and graphite, have a low strain to failure (less than 1%). The substitution of a less brittle fiber—for example, glass—or high strength organic fibers can significantly increase the impact resistance of composites. The impact resistance of many combinations of materials in composites has been investigated by numerous authors.[8,9] It is important, in tailoring the impact resistance of composites by addition of fillers or more ductile fibers, that close attention be given to the strength and stiffness of the final product. In general, the stiffness will go down as the impact strength goes up.

19.7. LIGHTNING

In general, FRP composites are considered to be electrically insulative, due to the non-conductive nature of the resins and the majority of the reinforcements used in composites. High or low conductivity in any structure can be an advantage or a disadvantage when the structure is exposed to lightning. Conductive structures can be used to carry off the electrical charge caused by a strike, while non-conductive structures are less attractive to strike initiation. Advanced composites, such as boron/epoxy and graphite/epoxy, are more conductive than the more common fiberglass composites, especially in the plane of the fiber axis. If a lightning strike attaches to a graphite/epoxy composite, the strike may actually be carried along the surface of the structure in the direction of the fiber axis in the surface plys. Boron/epoxy composites require special attention because of the conductive nature of the inner tungsten core and the electrically insulative outer cylinder of boron, which constitutes the most commonly available boron fibers. Boron on carbon substrate will have a similar relationship, only somewhat decreased by the less conductive nature of carbon in comparison to tungsten. Regardless of the makeup of the constituents involved, if lightning strikes are considered a threat to the system utilizing composites, then a lightning protection scheme must be incorporated into the system being designed. A number of protection schemes have been developed for use in conjunction with composites. The most common technique is to provide a conductive path on the surface of the composite to allow for a non-destructive current path for charge dissipation. The most common coatings are wire mesh, flame sprayed aluminum, and aluminum foil. Lightning diverter strips or bars have also been devised to accept direct attachment of a 200,000-ampere strike and then transport the strike harmlessly into the conductive portion of the vehicle. The less conductive composite materials can be used to isolate the interior of a vehicle to ensure that the strike remains on the exterior surface. Various designs have been proposed, and, depending on the particular application, most of them are effective. For example, it is not necessary to cover the entire exterior surface with a conductive material to successfully ward off lightning. If the area considered is susceptible to direct attachment of a strike, then total coverage is recommended. However, if the area is unlikely to see direct strike, then a smaller percentage of the total area need be addressed. This becomes extremely important in terms of cost and weight savings. One example where 50% coverage was employed in a secondary zone of an aircraft structure is in the design of the B-1 bomber horizontal tail made from graphite epoxy (see Chapter 28). In this particular design, flame sprayed aluminum was used as the conductive path. Lightning protection schemes of this nature have seen many hours of flight time, as evidenced by the highly successful application to the F-14 stabilizer, which uses 50% coverage consisting of aluminum foil strips bonded to the outer surface. There are several recent documents that convey the different types of lightning protection schemes available which incorporated manufacturing techniques, cost and weight tradeoffs, and numerous other evaluations.[10, 11, 12, 13]

Fiber-reinforced composites can also be employed in the absence of a lightning strike protection scheme when the design concept involves a deliberate attempt to prevent strikes. In general, FRP materials are less susceptible to attracting a strike when compared to metals. These materials can be factored into a design to electrically insulate more susceptible components. Depending on the manufacturing technique, and on the design of the protection scheme, any one of the above techniques can be shown effective in terms of cost and weight savings. Review of the above referenced documents provides considerable insight into the available techniques. One factor that has been common to all of the studies to date is that when a coating technique is used to provide protection, the minimum allowable thickness of the coating should be 5 mils (0.127 mm) or greater.

19.8. MOISTURE

Moisture is constantly present in the operational environment in which a composite is manufactured and throughout its useful life. The effects of moisture on composite materials has been studied in detail and it is known that moisture has a potentially degrading effect on matrix materials. Moisture is present in many forms, and eventually penetrates all organic materials by a diffusion controlled process until the moisture equilibrium concentration is achieved. Applied heat can cause matrix alterations that allow the matrix water holding capacity to increase. In general, moisture causes a lowering of the matrix mechanical properties. In epoxies, removal of the moisture by slowly drying out the matrix restores the matrix material to its original mechanical property level. The moisture effectively plays the role of a resin plasticizer which softens the matrix and lowers the glass transition temperature. Recently, a great deal of attention has been given to the effects of moisture in epoxy resins because of their use in aerospace structures. It is now believed that the moisture disrupts the secondary bonding between polymer chains and that it does not disrupt primary bonding. The process is reversible, without permanent damage to the matrix. The end result of the effects of mois-

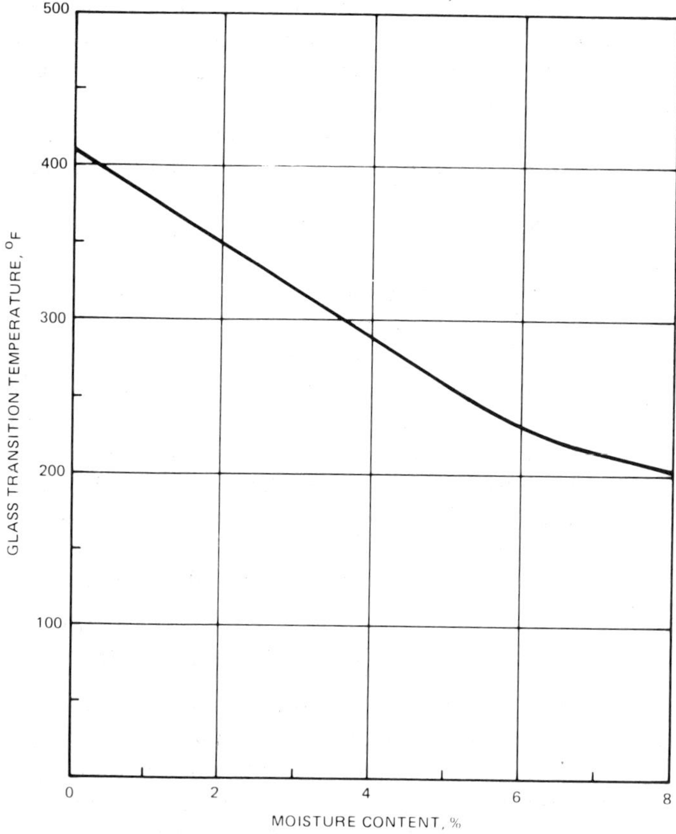

Figure 19.2. The effects of moisture on the glass transition of temperature of epoxy resins.

ture on composites is that the designer must consider moisture effects in combination with other environmental factors. Moisture effectively lowers the glass transition temperature of the matrix, as shown in Fig. 19.2. As the amount of moisture increases, the glass transition temperature decreases.

The effect of moisture on the mechanical properties of glass/epoxy and graphite/epoxy has been measured in recent studies.[14] AS/3501-6 graphite/epoxy was fabricated into 18-ply (0.45, 90:8,8,2) panels and exposed to conditions of humidity—140° F (60° C) and 98% R.H.—and humidity-thermal spiking—3 days at 140° F (60° C) and 98% R.H., followed by 2 hours at 260° F (126.6° C). The increase in moisture content resulting from 90-day humidity exposure and from 40 thermal spiking cycles is shown in Figs. 19.3 and 19.4.

A reduction in the mechanical properties of the composite is the result of this moisture absorption (see Table 19.1). The reduction in flexural stress at 260° F (126.6° C) was 44% after 90-day humidity exposure and 49% after 40 thermal cycles (see Fig. 19.5). The corresponding reduction in horizontal shear strength at 260° F (126.6° C) was 51% and 56% (see Fig. 19.6). These measurements are of importance in the supersonic aircraft environment, where temperatures of 260° F (126.6° C) can be reached in relatively short time intervals. Utilization of advanced composite structures to their design limits necessitates the protection of these structures against the strength-degrading effects of moisture.

Two coatings have been developed which are capable of protecting laminates from moisture penetration.[14] Both are aluminum foil coatings, one using solid foil applied in a secondary bonding operation, the other using co-cured foil applied by co-curing with the laminate. Moisture pick-up by the laminate was reduced by up to 65% after humidity exposure and after thermal cycling (Figs. 19.3

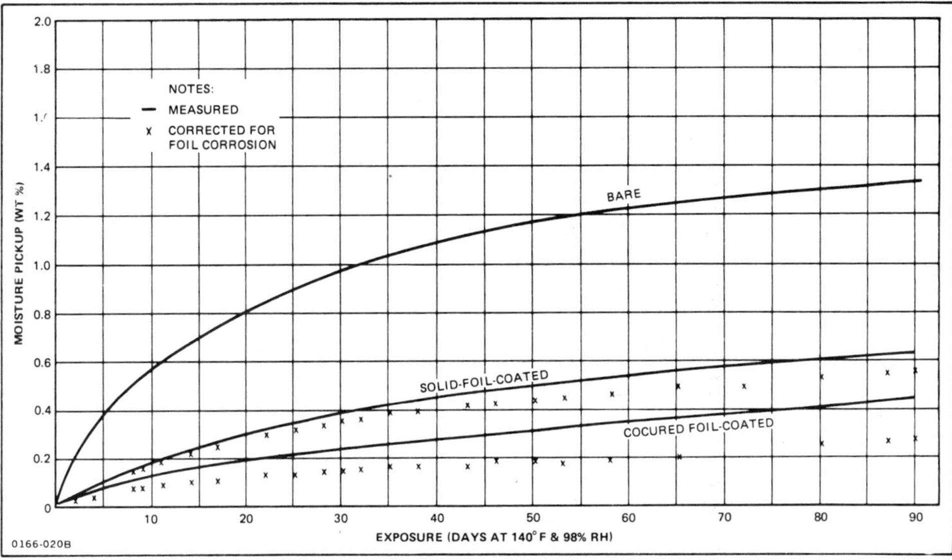

Figure 19.3. Moisture pickup of bare and foil coated graphite/epoxy (AS/3501-6) under humidity exposure.

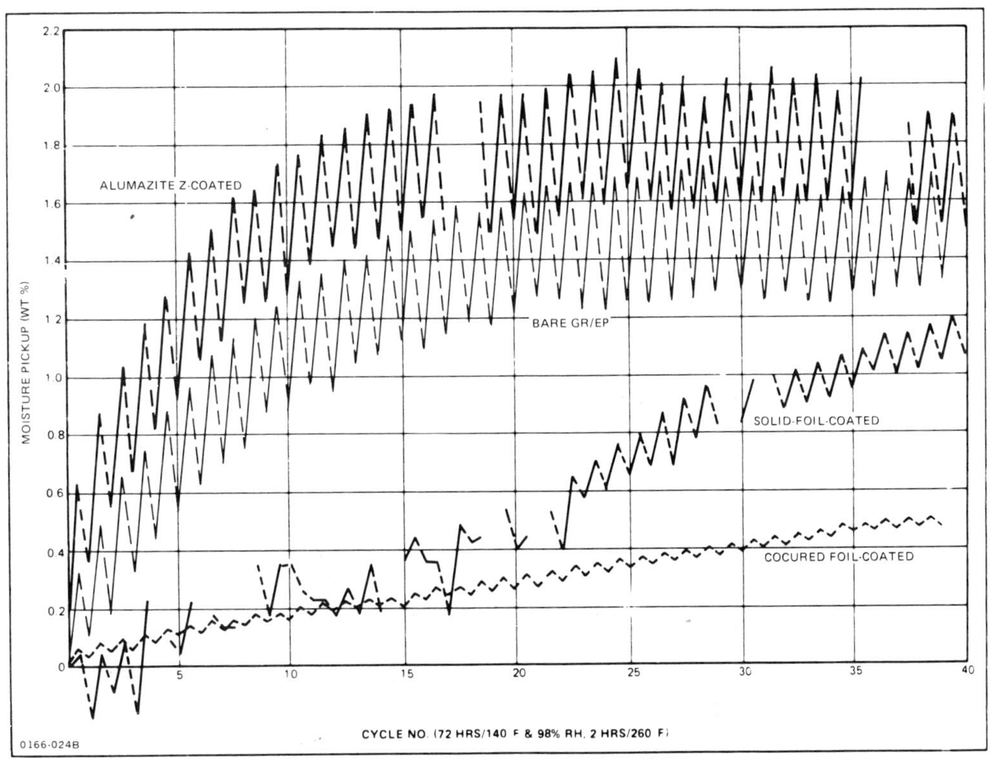

Figure 19.4. Relative coating effectiveness in moisture penetration under thermal spiking exposure (6 × 6 in. panels).

Table 19.1. Effectiveness of Moisture Protection Coatings on Strength Retention of Graphite/Epoxy

PARAMETER	BARE GRAPHITE/EPOXY			SOLID ALUMINUM FOIL COATING			CO-CURED ALUMINIUM FOIL COATING		
	CONTROL	HUMIDITY	SPIKED	CONTROL	HUMIDITY	SPIKED	CONTROL	HUMIDITY	SPIKED
Length of exposure		91 days	40 cycles	91 days	91 days	40 cycles	90 days	90 days	39 cycles
Moisture level, %		1.4	1.4	0.2	0.7	1.1	0	0.5	0.5
Flexural stress, 260° F (126.6° C), ksi	153.4	85.6	78.8	148.6	150.5	64.1	175.3	173.1	130.1
Flexural modulus, 260° F (126.6° C), msi	7.5	6.6	6.3	6.9	7.2	5.6	7.1	8.0	17.2
Horizontal shear strength, 260° F (126.6° C), ksi	8.2	4.0	3.6	8.4	7.0	5.6	9.2	8.6	7.6

NOTES: 1. Exposure conditions: control—desiccated; humidity—140° F (60° C), 98% relative humidity; spiked—3 days at 140° F (60° C), 98% relative humidity, followed by 2 hours at 260° F (126.6° C), each cycle.
2. AS/3501-6 GR/EP, 18 ply, $(\pm 45/O_2/90/O_2/\pm 45)$ fiber orientation.
3. Coatings: solid aluminum foil—2024T3 aluminum foil, 0.002 in. (0.0508 mm), secondary, bonded to graphite/epoxy (both sides); co-cured aluminum foil—5036 perforated (top) foil, 0.003 in. (0.0762 mm), and 2024T3, a solid, aluminum (bottom) foil, 0.002 in. (0.0508 mm); co-cured with graphite/epoxy.

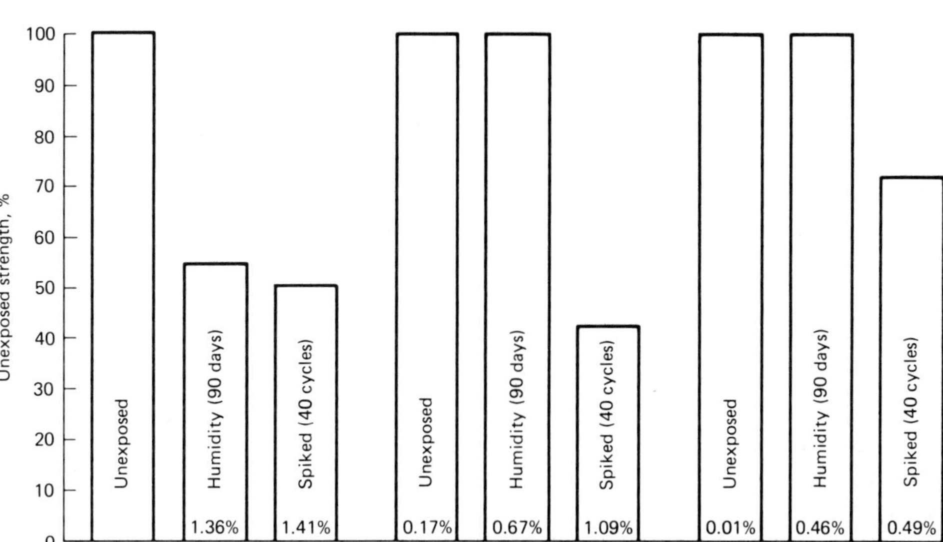

Figure 19.5. Flexural stress (260° F/127° C) as percent of unexposed strength.

and 19.4) using these coatings, and additional studies indicate that even greater reductions are possible when the foil is painted. The reduction in moisture pick-up results in a corresponding improvement in strength retention of the composite laminate (Table 19.1). The improvement in flexural strength retention after humidity exposure and after thermal cycling can be seen in Fig. 19.5, and the improvement in shear strength retention is shown in Fig. 19.6. Further studies (in progress) indicate that even greater improvements will be obtained with painted foil coatings. Previous studies at Grumman and elsewhere have shown that organic coatings do not provide the moisture barrier needed to protect graphite/epoxy components from critical strength-degrading moisture absorption.

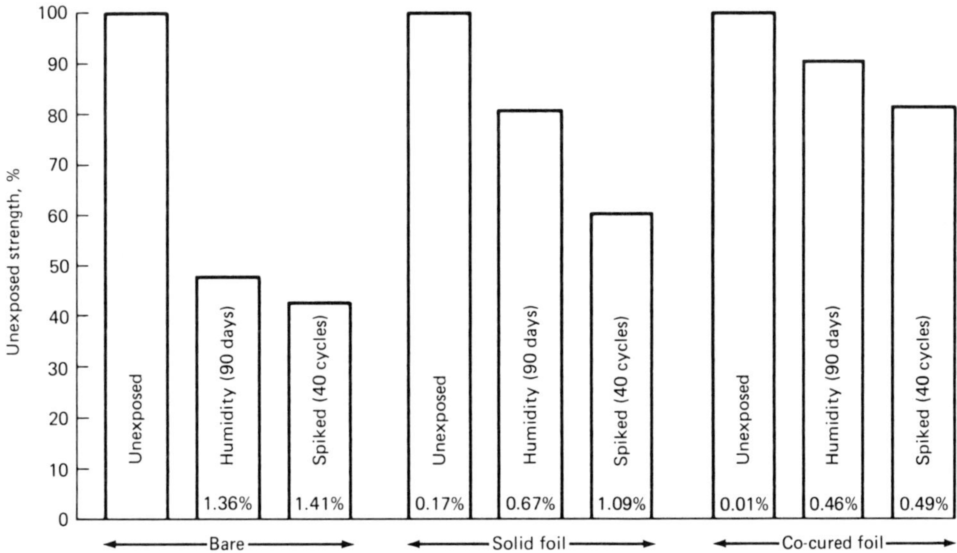

Figure 19.6. Horizontal shear strength (260° F/127° C) as percent of unexposed strength.

19.9. NUCLEAR

In a nuclear environment, various aspects of the nuclear threat can affect the performance of composite materials. The designer must consider thermal, overpressure, and radiation effects. As in all thermal exposures of materials, the combined stresses applied to the system as would be seen in service must be applied to the structure in order to evaluate the material's performance. The thermal nuclear threat results in a high energy, short term blast with the combined effects of overpressure and radiation, accompanied by the mechanical stresses due to normal service loads to the system under evaluation. The spectral content of a nuclear source is such that composites, unprotected, absorb large amounts of the emitted energy. Depending on the thickness of the composite, the thermal resistance of the resin, the reinforcement used in the composite, and the amount and rate of the absorbed energy, the composite will receive varying degrees of damage. Tests conducted using various coatings to reflect the incident energy yielded varying degrees of damage.[15] In general, when composites are used in a nuclear thermal environment, an appropriate reflective coating should be used to ward off the thermal pulse to prevent delaminations.

Composites have been exposed to nuclear radiation of varying degrees. Exposures to thermal neutron, fast neutron, and gamma rays had little or no effect on the physical properties of the composites.[16]

19.10. OZONE

Ozone is an allotropic form of oxygen and is suspect because of its potential deleterious effects on rubber and metals at high altitudes, where it can exist in harmful quantities. Ozone can degrade materials by oxidation, which can cause loss of mechanical properties and lead to more rapid attack from other environmental factors. A literature survey conducted to determine the effects of ozone on the matrices used in advanced composites yielded no known cases where composites were degraded by ozone.

19.11. P-STATIC

Tribo-electrification of dielectric surfaces is caused by friction charging due to contact with particles in flight, rubbing together various materials, or parting two materials from each other. Precipitation static charge can be both annoying and harmful. Fuel tanks, armaments, and electrical equipment require isolation from static discharge. Lightning protection schemes can sometimes serve the dual purpose of providing lightning protection and p-static protection. The objective is to bleed off the static charge prior to any significant build-up that could cause a fire or explosion, or precipitate electromagnetic interference with the on-board electrical equipment. If a separate protection concept is to be employed, then one of the more common approaches is to coat the substrate with an electrically conductive coating, such as a carbon filled or metal oxide filled coating. Application of a p-static discharge system is specific to the system design. One such concept was evolved for the B-1 aircraft by the Lightning & Transients Research Institute under contract to Rockwell International.[17] In order to incorporate the effective design to ward off p-static build-up, a continuous conductive path must be provided, suitable dischargers used, and conductive coatings applied.

19.12. TEMPERATURE

FRP composites are exposed to temperature extremes as cold as −65° F (−53.9° C) and elevated temperatures up to 250° F (121.1° C) for military applications, and much higher temperatures where other than natural heating sources are encountered. Strength and stiffness properties are generally unimpaired by cold temperatures, and, in some instances, actually increase. Cold temperatures tend to make polymers less flexible, and thus susceptible to damage due to fatigue induced by mechanical leading. All resins have a useful operating temperature range, and most failures are caused by the improper selection of a matrix material for a given thermal environment. Thermal fatigue, or excursions from hot to cold repeatedly, can cause localized mechanical stress due to thermal expansions and contractions. This can be a prime source of trouble if the resin system is incompatible with the reinforcements.

High temperatures can cause resins to chemically degrade, and almost all chemical reactions are accelerated by an increase in temperature. Reinforced plastics can be employed in high thermal environments—once again, by the selection of the proper resin system. For temperature extremes above 250° F (121.1° C), most room-temperature-processed matrix materials should not be used. High temperature-curing commercial resins available today can be used in thermal environments in excess of 600° F (315.6° C). One of the most readily available, and, in most cases, reliable, pieces of data that can be obtained from the product literature is the useful temperature range of the materials. Operating temperature extremes of some resins are given by Rasato.[18] In selecting a resin for a particular environment, the resin can be tailored to that environment to excell in one aspect, but usually at the expense of some other property.

As the temperature of the matrix is elevated, the properties of the composite remain unchanged until the point at which the matrix begins to soften. With sufficient application of heat, the solid matrix will reach the glass transition temperature, at which the polymer passes from a glass-like state to a rubbery state. At this point, substantial losses in mechanical properties of the matrix occur. For most applications, matrices should not be used above the glass transition temperature.

19.13. RAIN EROSION

Many materials have relatively poor rain erosion resistance when they encounter rain, snow, or ice in flight. The speed of the drops, their angle of impact and their frequency, and their mass determine the rate of erosion for any given composite. The composition of the composite can be altered internally to increase its toughness or resistance to impact, but more commonly the composite is coated with a rain erosion resistant finish, which is capable of dissipating frequent and discrete amounts of energy without significant damage to the substrate. Mainly, the concern is aircraft structures, such as radomes and antenna covers experiencing high speed flight, or the leading edges of rapidly rotating blades, such as on helicopters. Empirical studies can be done in controlled test facilities to demonstrate the relative resistance of different coatings,[19] or the system can be modeled mathematically and then proven by empirical test.[20] Studies are also available in the literature which demonstrate the effect of varying the constituents that make up the composite system.[21]

Several years ago, graphite/epoxy composites were thought to be too brittle to be incorporated into the leading edges of aircraft structures; however, with the advent of improved developments in polyurethane rain erosion resistant coatings, and the incorporation of the newly developed graphite fibers, graphite foils have been shown to provide the necessary resistance in the above applications.

19.14. WEATHERING

Weathering refers to the combined effects of sunlight, moisture, heat, cold, wind, and contaminants. When studying these combined effects, it is necessary to apply the service loads. The only true test of this phenomenon is to use the composite materials and periodically evaluate those components that have been in service. The effect of weathering on composites is to degrade the mechanical performance by leaching out chemical constituents, hydrolyzing the resin, and eroding away the protective matrix from the fibers. Sunlight can cause further cross-linking of the polymer, causing it to become embrittled, or it can break chemical bonds, causing the outer layers to degrade, followed by delaminations and further damage.

The deleterious effects of weathering on composites can all but be eliminated by the use of gel coats or proper coating techniques. Inhibitors can also be added to the base resin to deter the effects of ultraviolet energy. Depending on the application, another common technique is to incorporate sacrificial coatings, such as braided sleeves around the load carrying substrate. The braid degrades and may even eventually fall off, leaving the composite virtually unaffected.

19.15. REAL TIME WEATHERING

Most of the data on weathering of composites have been, of necessity, obtained by using accelerated weathering techniques, where the exposure conditions were deliberately made more severe than those encountered in service. With these exposures, the damage to composite structures could be obtained in a relatively short time. However, it was always difficult to correlate the results of accelerated tests with real life conditions. An attempt was made at Grumman to locate old in-service fiberglass parts that had seen severe conditions during use. These parts were tested, and the test results were compared with the test data obtained during the original design investigations. These parts included a large, 24-ft, (8-m) fiberglass sandwich construction rotodome from an E-2A aircraft, serial number 1, which had been exposed for 19 years, several A-6A nose radomes, which have been in service 11-15 years, and a section of E 2A tail, which has been flying for 12 years.

In the case of the rotodome, this part had been scrapped and was found in a storge area with some of the paint peeled off. Samples were selected both in areas where there was no paint and in areas where the paint was intact. Also, samples were removed from inner and outer and top and bottom skins—wherever there was sufficient area. The leading edge cap, which was a solid laminate of eight plies and was badly eroded, was also sampled. Flexural and tensile coupons were obtained wherever possible and compared with original

Table 19.2. Flexural Tests on E-2A Rotodome, Serial Number 1, at 77°F (25°C)

	NORMALIZED 0.011 in. (0.279 mm)/PLY		ORIGINAL (POLAR PLOTS)	
TEST AREA	STRENGTH, ksi (MPa)	MODULUS, msi (GPa)	STRENGTH, ksi (MPa)	MODULUS, msi (GPa)
Upper skin, painted	53.8 (371) 56.6 (390) 57.0 (393)		61.5 (424) 63.5 (438) 63.6 (438)	
Average	55.8 (384)		62.8 (433)	
Percent retention	89%			
Bottom skin, painted	76.5 (526) 65.9 (454) 76.4 (526)	2.34 (16.1) 2.37 (16.3) 2.37 (16.3)	64.0 (441) 57.5 (396) 62.0 (427)	2.75 (18.9) 2.30 (15.8) 2.50 (17.2)
Average	72.9 (502)	2.36 (16.2)	61.2 (411)	2.52 (17.4)
Percent retention	+100	93.7		
Upper skin, no paint	57.2 (394) 59.6 (411) 59.9 (413)	1.88 (13.0) 1.88 (13.0) 1.90 (13.1)	63.0 (434) 66.0 (455) 67.3 (464)	1.9 (13.1) 2.0 (13.8) 2.1 (14.5)
Average	58.9 (406)	1.89 (13.0)	65.9 (454)	2.0 (13.8)
Percent retention	89%	95		
Inner skin, no paint (moisture content: 1.1%)	53.5 (369) 56.8 (391) 57.3 (395)	1.36 (9.4) 1.45 (10.0) 1.63 (11.2)	61.5 (424) 59.0 (407) 60.0 (413)	1.8 (12.4) 1.9 (12.4) 1.7 (11.7)
Average	55.9 (395)	1.48 (10.2)	60.2 (415)	1.8 (12.4)
Percent retention	93%	92%		
Cap, no paint, eroded (moisture content: 0.84%)	43.9 (302)		65.0 (488)	
Percent retention	68%			

Table 19.3. Tensile Tests on E-2A Rotodome, Serial Number 1, at 77°F (25°C)

	NORMALIZED, 0.011 in. (0.279 mm) PLY		ORIGINAL DATA (POLAR PLOTS)	
TEST AREA	STRENGTH, ksi (MPa)	MODULUS, msi (GPa)	STRENGTH, ksi (MPa)	MODULUS, msi (GPa)
Upper skin, no paint	35.0 (241)	1.77 (12.2)	37.0 (255)	2.5 (17.2)
(moisture content:	34.0 (234)	1.88 (13.0)	40.0 (276)	2.8 (19.3)
1.00%)	38.0 (265)	1.94 (13.4)	39.0 (269)	2.5 (17.2)
	32.3 (223)	1.74 (12.0)	40.0 (276)	2.4 (17.2)
	38.5 (265)	1.98 (13.6)	40.0 (276)	2.7 (18.6)
Average	35.7 (246)	1.86 (12.8)	39.0 (269)	2.6 (17.9)
Percent retention	91.5	71.5		
Bottom skin, painted	37.2 (256)	2.07 (14.3)	32.0 (220)	2.1 (14.4)
(moisture content:	25.3 (174)	1.98 (13.6)	32.0 (220)	1.7 (11.7)
1.32%)	34.4 (237)	1.94 (13.4)	31.5 (215)	2.2 (15.2)
	35.1 (242)	1.89 (13.0)	36.5 (251)	2.3 (15.8)
	37.8 (260)	1.88 (13.0)	36.5 (251)	2.3 (15.8)
Average	34.0 (234)	1.95 (13.4)	33.7 (232)	2.1 (14.4)
Percent retention	+100	92.9		
Upper skin, painted	39.5 (272)	1.83 (12.6)	36.3 (250)	2.3 (15.8)
(moisture content:	42.8 (296)	2.09 (14.4)	39.4 (271)	2.5 (17.2)
1.46%)	43.5 (300)	2.27 (15.6)	38.4 (265)	2.6 (17.9)
Average	42.0 (209)	2.06 (14.2)	38.0 (262)	2.5 (17.2)
Percent retention	+100	82.4		

data. Strength and modulus of elasticity were determined. The test results were compared with original data corrected from fabric orientation using polar plots.[1,4,5] The results are shown on Tables 19.2 and 19.3. The following conclusions were obtained.[22]

- The leading edge cap, which was badly eroded, retained only 68% of its original strength and showed the worst degradation.
- The upper skin areas where the paint had peeled showed 89% flexural strength retention and 91% tensile strength retention. The respective moduli retentions were 95% for flexure and 72% in tension.
- The upper skin area where the paint was intact showed 89% flexural strength retention and over 100% for tension. The flexural modulus was 89% in flexure and 93% of the original in tension.
- The bottom skin, where the paint was intact, showed 100% retention in tensile strength and 93% in tensile moduls. The flexural retention was +100% and the modulus 94%.
- The inner skin, which was not painted and not directly exposed to weathering, showed 93% flexural strength retention and 82% modulus retention.

The construction here was fiberglass fabric, style 181, with Shell Epon 828 resin with CL hardener. It can be concluded that where the paint remained intact, the strength retention was in the +90% range, while the modulus figures were slightly lower, in the 82–94% range. These results, after 19 years of exposure, show that under normally ambient exposure conditions, and tested at ambient temperature, the fiberglass structures do not degrade below their design value. It is to be noted that painted areas retained more moisture than areas where the paint had degraded.

The nose radome on A-6A aircraft was of filament wound construction. The serial number 38 unit was made of Shell Epon 828 resin and BF_3-400 curing agent. This composition was found to be hygroscopic. Serial number 50 was changed to 828/MNA/BDMA formula, which was used for the remainder of several hundred units. The outer surface of these radomes was painted with a rain erosion coating. It can be seen from the test results that BF_3-400 formulation degraded badly, while the MNA/BDMA formulation showed

Table 19.4. Tensile and Flexural Tests on A-6 Nose Radome at 77°F (25°C)

TYPE	SERIAL NUMBER	DATE	RESIN CONTENT %	MOISTURE CONTENT %	TENSILE STRENGTH, ksi (MPa)		FLEXURAL STRENGTH, ksi (MPa)		FLEXURE, msi (GPa)	
					AGED	ORIGINAL	AGED	ORIGINAL	AGED	ORIGINAL
828/$BF_3$400	38	5-5-63	16.6	0.50	75.8 (522)	80.5 (555)	38.9 (268)	68.9 (475)	3.13 (21.6)	2.25 (1.15)
828/MNA/BDMA	195	9-2-65	17.1	0.20	94.7 (652)	93.8 (646)	87.1 (600)	88.9 (613)	4.00 (27.6)	3.14 (21.6)
828/MNA/BDMA	266	4-29-66	17.7	0.20	88.0 (606)	88.1 (607)	82.5 (568)	86.3 (595)	3.05 (21.0)	2.64 (18.2)
828/MNA/BDMA	299	10-19-66	16.7	0.18	89.5 (617)	87.4 (602)	86.0 (593)	91.3 (629)	3.13 (21.6)	3.16 (21.8)
828/MNA/BDMA	369	4-12-67	16.6	0.25	89.5 (617)	91.0 (627)	96.1 (662)	80.2 (553)		3.00 (20.7)
Min Spec Regmt					50.0 (345)		50.0		2.0	

Table 19.5. Effects of Real Time Aging—Painted Graphite/Epoxy[a,b]

		TEST SAMPLES PAINTED WITH AIRCRAFT PAINT							
		50% RELATIVE HUMIDITY EXPOSURE				EXPOSED ON STAND AT FIRE ISLAND NEW YORK			
	EXPOSURE	MOISTURE	FLEX STR.	FLEX. MOD	HOR SHEAR	MOISTURE	FLEX STR.	FLEX MOD.	HOR SHEAR
TEST TEMPERATURE[c]	CONDITION	PICKUP %	ksi MPa	msi GPa	ksi MPa	PICKUP %	ksi MPa	msi GPa	ksi MPa
77° F (25° C)	Control		153.7 1059	8.24 56.7	7.55 52.0		153.7 1059	8.24 56.7	7.55 52.0
	3 months exposure		168.5 1161	8.75 60.3	7.50 51.7		146.3 1008	8.13 56.0	8.54 58.8
	% retention		100+	100+	99.3		95.2	98.7	100+
	12 months exposure	1.06	167.3 1153	8.46 58.3	9.38 64.6	1.34	162.9 1122	8.38 57.7	8.80 60.6
	% retention		100+	100+	100+		100+	100+	100+
	24 months exposure	0.96	147.8 1018	8.50 58.6	8.60 59.3	1.21	159.2 1097	8.22 56.6	9.74 67.1
	% retention		96.2	100+	100+		100+	100+	100+
260° F(127° C)	Control		156.7 1080	8.36 57.6	-		156.7 1080	8.36 57.6	-
	12 months exposure		104.2 718	7.17 49.4	-		113.7 783	7.47 51.5	-
	% retention		66.5	85.8	-		72.6	89.4	-
	24 months exposure		110.1 759	7.65 52.7	5.27 36.3		113.9 784	7.75 53.4	5.47 37.7
	% retention		70.3	91.5			72.7	92.7	
	Dried to Const wt		151.1 1041	8.05 55.5	6.45 44.4		153.6 1058	7.93 54.6	6.78 46.7
350° F(177° C)	Control		131.5 906	6.90 47.5	4.69 32.3		131.5 906	6.90 47.5	4.69 32.3
	3 months exposure		110.1 759	7.95 54.8	4.73 32.6		111.5 768	7.44 51.3	5.66 39.0
	% retention		83.7	100+	100+		84.8	100+	100+
	12 months exposure		82.4 568	7.16 49.3	4.49 30.9		75.9 523	6.66 45.9	3.71 25.6
	% retention		62.7	100+	95.7		57.7	96.5	79.1
	24 months exposure		62.2 428	6.30 43.4	2.95 20.3		50.3 347	6.13 42.2	2.80 19.3
	% Retention		47.3	91.3	62.9		38.3	88.8	59.7
	24 months exposure								
	Dried to const. wt		118.9 819	7.76 53.5	4.97 34.2		108.6 748	7.86 54.2	5.34 36.8
	% retention		90.4	100+	100+		82.6	100+	100+

[a]Comparisons are based on scope of original study.
[b]Flexural properties are normalized to 0.00525 in. (0.133 mm)/ply.
[c]30-minute soak before loading.

Table 19.6. Effects of Real Time Aging—Bare Graphite/Epoxy[a,b]

TEST TEMPERATURE[c]	EXPOSURE CONDITION	50% RELATIVE HUMIDITY EXPOSURE							EXPOSED ON STAND AT FIRE ISLAND NEW YORK						
		MOISTURE	FLEX STR.		FLEX. MOD		HOR SHEAR		MOISTURE	FLEX STR.		FLEX MOD.		HOR SHEAR	
		PICKUP %	ksi	MPa	msi	GPa	ksi	MPa	PICKUP %	ksi	MPa	msi	GPa	ksi	MPa
77° F (25° C)(2)	Zero time, Dry		148.7	1025	8.12	55.9	9.78	67.4		148.7	1025	8.12	55.9	9.78	67.4
(2)	48 Hrs @ 50% RH		156.1	1076	8.63	59.5	8.69	59.9		156.1	1076	8.63	59.5	8.69	59.9
(Control)	2–4 wks. in shop		152.7	1047	8.05	55.5	9.79	67.5		152.7	1047	8.05	55.5	9.79	67.5
	3 months exposure		157.7	1087	8.00	51.1	10.15	69.9		151.6	1045	7.76	53.5	8.63	59.5
	% retention		100+		99.4		100+			99.3		96.4		88.1	
	12 months exposure	0.89	157.9	1088	7.95	54.8	9.38	64.6	0.96	154.2	1062	7.95	54.8	8.83	60.8
	% retention		100+		98.4		95.8			100+		98.8		90.2	
	24 months exposure	0.88	151.9	1047	7.84	54.0	9.29	64.0	0.94	143.3	987	7.89	54.5	8.57	59.0
	% retention		99.5		97.4		94.9			93.8		98.0		87.5	
	Dried to const. Wt		146.8	1011	8.08	55.7	8.99	61.9		140.0	965	7.91	54.5	6.53	45.0
260° F(127° C)(2)	Zero time, dry		156.2	1076	8.09	55.7	7.74	53.5		156.2	1076	8.09	55.7	7.74	53.3
(2)	48 hrs @ 50% RH		152.0	1047	8.11	55.9	6.88	47.4		152.0	1047	8.11	55.9	6.88	47.4
(Control)	2–4 wks. in shop		-		-		-			-		-		-	
	12 months exposure		109.7	756	6.76	46.6	-			107.0	737	7.10	48.9	-	
	% retention		72.2		83.4		-			70.4		87.5		-	
	24 months exposure		118.8	818	7.44	51.3	5.33	36.7		112.4	774	6.02	41.5	5.17	35.6
	% retention		78.2		91.7		77.5			73.9		74.2		75.1	
	Dried to const. wt		152.7	1052	7.60	52.4	7.59	52.3		143.7	990	7.36	50.7	7.33	50.5
350° F(177° C)(2)	Zero time, dry		124.4	857	8.04	55.4	5.43	37.4		124.4	857	8.04	55.4	5.43	37.4
(2)	48 hrs @ 50% RH		126.9	874	7.85	54.1	5.72	39.4		126.9	874	7.85	54.1	5.72	39.4
(Control)	2–4 wks in shop		116.0	799	7.40	51.0	5.62	38.7		116.0	799	7.40	51.0	5.62	38.7
	3 months exposure		104.5	721	7.46	51.4	6.56	45.2		105.4	726	7.37	50.8	5.31	36.6
	% retention		90.2		100+		100+			90.9		99.6		94.4	
	12 months exposure		77.5	534	6.21	42.8	3.61	24.9		73.6	507	6.42	44.2	3.48	24.0
	% retention		66.8		83.9		64.2			63.4		86.8		61.9	
	24 months exposure		53.5	369	5.75	39.6	3.05	21.0		66.4	457	6.41	44.2	3.30	22.7
	% retention		46.1		77.7		54.3			57.2		86.6		58.7	
	Dried to const. wt		111.4	768	7.18	49.5	5.46	37.6		105.6	728	7.23	49.8	5.77	39.8
	% retention		96.0		97.0		97.2			91.0		97.7		100+	

[a]Comparisons are raised on scope of original study.
[b]Flexural properties are normalized to 0.00525 in. (0.133 mm)/ply.
[c]30-minute soak before loading.

no drop-off in tensile strength, flexural strength, or modulus (see Table 19.4).

The test results on a painted E-2A tail fin, showed 84 to 100+% strength retention on the skins and 80 to 100+% retention in modulus. The study of the residual adhesive strength in the honeycomb area shows 80–94% retention in flatwise tension, 88% in compression, and 96% in beam flexure. (All after 12–14 years of "in service" exposure.)[22]

Tests on real time aging of graphite/epoxy specimens, both painted and unpainted, exposed both outdoors at Fire Island at ambient conditions and under constant laboratory 50% humidity, show that, at room temperature, there is no degradation at up to two years for painted specimens. At 260° F (126.6° C), there is a minimum amount of degradation, but at 350° F (177° C), the degradation is so severe that the strength drops off to 50%. Here, the strength drops off more than the modulus. It can be noted, however, that this degradation is completely reversible. If exposed specimens are slowly dried and retested at 350° F (177° C), all of the original strength is recovered. The conclusions are that the standard epoxy resins with graphite reinforcement should not be exposed above 260° F (126.6° C). For higher temperatures, polyimide resins must be used. The results are shown on Tables 19.5, 19.6, and 19.7.

Table 19.7. Strength of Real Time Aged Graphite/Epoxy Bolted Specimens

EXPOSURE TIME, MONTHS	BOLTED JOINT STRENGTH, lb			
	CONTROL, 50% HUMIDITY		FIRE ISLAND EXPOSURE	
	BARE	PAINTED	BARE	PAINTED
1	4550	4510		
3	4210	4584	4445	4215
12	4350	4440	4505	4410
24	4525	4405	4710	4890
36	4585	4530	4590	4705

Conditions:
1. AS/3501-5 graphite/epoxy bolted to Ti-6AJ-4V with Ti6AJ-4V fasteners.
2. All failures by bearing or net tension of composite.

Acknowledgement—The author is indebted to Mr. C. Stabler and Mrs. B. Simpers of Grumman Aerospace Corp. for the text and data on moisture and metallic coatings for composites, and to G. Lubin for the real life aging data.

REFERENCES

1. Lubin, George, *Handbook of Fiberglass and Advanced Plastics Composites*, Van Nostrand Reinhold, New York, 1969.
2. Prince, Daniel E., "Corrosion Behavior of Metal Fasteners in Graphite/Epoxy Composites," AFML-TR-75-53, July 1975.
3. Skouby, C. D., "Electromagnetic Effects of Advanced Composites," McDonnel Aircraft, January 1975.
4. Ludwig, Walter, *et al.*, "B-1 Composite Horizontal Stabilizer Development," in *Proceedings of 21st National SAMPE Symposium*, April 1976.
5. Erbacher, H. E., "Advanced Development of Conceptual Hardware Horizontal Stabilizer," *14th Quarterly Progress Reports*, Contract F33615-73-C-5173, January 15, 1977.
6. Howarth, J. T., "What Designers and Processors Should Know about Flame Retardant Additives," *Plastics World* (March 1973).
7. Novak, Richard, "Materials Variables Affecting The Impact Resistance of Graphite and Boron Composites," AFML-TR-74-196, September 1974; and "Part II," June 1975.
8. Husman, George *et al.*, "Residual Strength Characterization of Laminated Composites Subjected to Impact Loading," AFML-TR-73-309, February 1974.
9. Beaumont, P. W. R. *et al.*, "Methods for Improving the Impact Resistance of Composite Materials," *ASTM Symposium on Foreign Object Impact Behavior of Composites*, September 1973.
10. Lubin, George, "Effect of Lightning Strikes on Boron-Epoxy Single Skin and Honeycomb Sandwich Panels," Report No. ADR 02-06-70.3, December 1970.
11. Penton, A. P. *et al.*, "The Effects of High Intensity Electrical Currents on Advanced Composite Materials," N00019-71-C-0063, March 21, 1972; and N00019-72-C-0205, March 1973.
12. Quinlivan, J. T., "Coatings For Lightning Protection of Structural Reinforced Plastics," AFML-TR-70-303-PT-I, March 1971; and "Part II," January 1972.
13. Erbacher, H. E., "Advanced Development of Conceptual Hardware Horizontal Stabilizer," *Quarterly Progress Report*, Contract F33615-73-C-5173.
14. Staebler, C. J. and Simpers, B. F., "Metallic Coatings for Graphite/Epoxy Composites," *Final Report*, January 1979.
15. Weaver, J. H. *et al.*, "Thermal Flux Protection for Aircraft Systems," AFML-TR-75-167, March 1976.

16. General Dynamics, "Radiation Effects on Boron Filaments and Composites," ERR-FW-716, December 1967.
17. Robb, J. D., "Precipitation—Static Control for the B-1 Aircraft," L & T Report No. 536, February 1972.
18. Rosato, D. V., "Heat Resistant Resins," *Plastics World* March 26–32, 1968.
19. Schmitt, G. F. *et al.*, "Joint Air Force Navy Supersonic Rain Erosion Evaluations of Materials," AFML-TR-67-164, December 1964.
20. Springer, G. S. *et al.*, "Analysis of Rain Erosion of Coated and Uncoated Fiber Reinforced Composite Materials," AFML-TR-74-180, August 1974.
21. Kimmel, B. G., "Development of Composites Constructions With Improved Rain Erosion Resistance," Report No. P74237, July 1974.
22. Lubin, G. and Donohue, P. "Real Life Aging Properties of Composites," 35th Annual Technical Conference, 1980, Reinforced Plastics/Composites Institute, SPI.

Section III Design

20
DESIGN AND ANALYSIS OF ADVANCED COMPOSITE STRUCTURES

R. N. Hadcock
Grumman Aerospace Corporation

20.1. INTRODUCTION

During the past fifteen years, advanced composite structures have developed from laboratory curiosities into mature production primary and secondary aircraft and spacecraft structures.

Structural design is an iterative process which starts with trade-offs of various conceptual designs utilizing different materials. Some of the major factors which must be considered in the trade-off process are structural weight; costs associated with development, production, certification, and operations; availability of special facilities; experience; and confidence. The importance of these factors varies with the application. Weight savings may be the most important consideration if a structure is being designed for a new spacecraft system. Procurement and operational cost reductions would be the major consideration if a structure were being designed for replacement use in a current commercial aircraft. Weight savings can be transformed into reduced aircraft size in the design process of a new aircraft system, and this becomes as much or even more, of a consideration as the component part cost in the trade-off.

The design process evolves from these initial trade-offs of structural concepts. Iterations are performed in greater and greater depth on fewer and fewer candidate designs until a single configuration and material distribution is chosen. The final design process can now take place. However, this is more complex with composite materials than it is with metals, since the material itself is designed for optimum performance throughout the structure.

The detail design and analysis methodology associated with composite materials is performed at one level lower than is usual for metal structures. Since the material is built up from a number of individual layers, each oriented in a given direction, each particular element of the material within the structure can theoretically be designed to be optimum for a number of design conditions. In practice, a number of constraints are imposed by detail design, fabrication, inspection, and maintenance considerations. Even so, the lightest and, generally, the least expensive design will be tailored in such a way that gross material properties will vary throughout the part.

Design and analysis are complicated by this new dimension, and the iterative final design process can only be readily accomplished by utilizing a computer. However, the process can be greatly simplified for preliminary design.

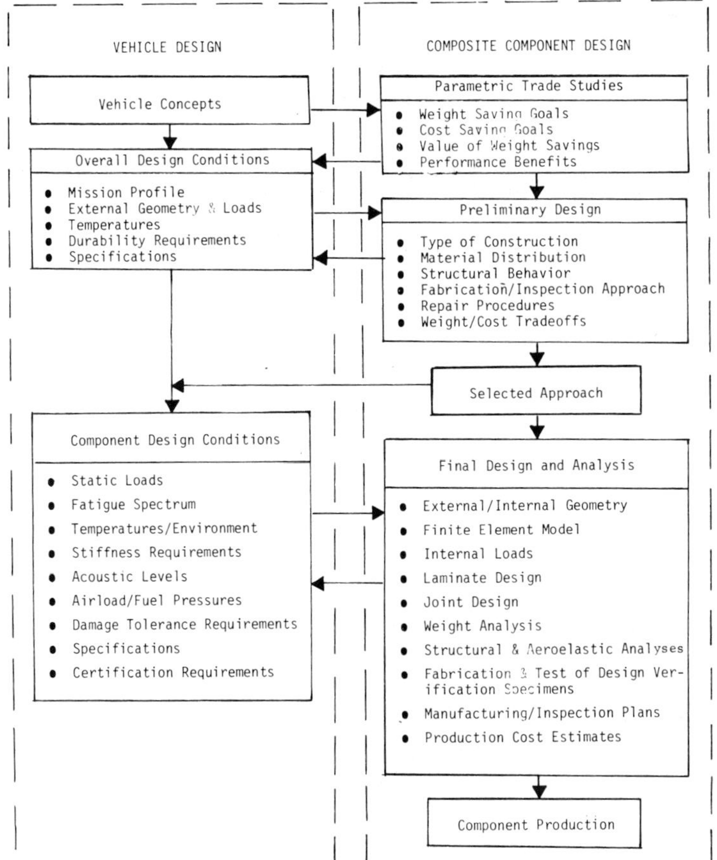

Figure 20.1. Composite structure design cycle.

This simplified design approach, which is quite adequate for preliminary design, is the subject of this chapter and is shown diagrammatically in Fig. 20.1.

20.2. PARAMETRIC TRADE STUDIES

The preliminary design of composite structures begins with trade-offs of the impact of using different types of construction and materials on the weight and cost of either a part or on the total structure associated with an aircraft or space vehicle. Lately, the starting point of this process has been somewhat simplified by the availability of data from weighed composite components or from extensive design studies which can be utilized to obtain an initial assessment of potential weight savings. These data, which are in terms of percent weight savings over an all metal baseline component versus the percentage of composite materials in the composite design, were generated for the Advanced Design Composite Aircraft Study.[1, 2, 3] The components which provide the weight savings data base are listed in Table 20.1.

Figure 20.2 shows the weight savings achieved utilizing advanced composite materials for horizontal and vertical stabilizers as well as wing torque boxes. Nine of the points shown are weighed data; twelve points are from results of reasonably extensive design studies. As can be seen from the figure, there are two rather distinct bands of data. The upper band represents first and second generation composite structures, where achievement of weight savings was the major objective, often at some increase in cost. The lower band

Table 20.1. Components Providing Composites Weight Savings Data Base

WINGS AND STABILIZER (Fig. 20.29)

WING BOXES		HORIZONTAL STABILIZERS		VERTICAL STABILIZERS
F-5	ACWS	F-5	B-1	F-5
F-15	CBBO	F-111		F-15
F-16	YC-15	F-14		B-1
AV-8B	A-7	F-15		B-737
B-1	ADCA	A-5		L-1011

FUSELAGES (Fig. 20.3)

F-5	CH-53
F-111	AH-16
YF-16	
ADCA	

RUDDERS, FLAPS, ELEVATORS, DOORS, ETC. (Fig. 20.4)

F-4 rudder	S-3 spoiler	B-737 elevator
A-9 rudder	B-737 spoiler	A-37 side brake
DC-10 rudder	F-15 speed brake	B-1 longerons
A-4 flap	A-7 speed brake	C-4 cone
F-5 TE flap	F-5 landing door	Booster thrust structure
F-5 LE flap	F-14 landing door	Metering truss
C-5A slat	F-14 overwing frg.	

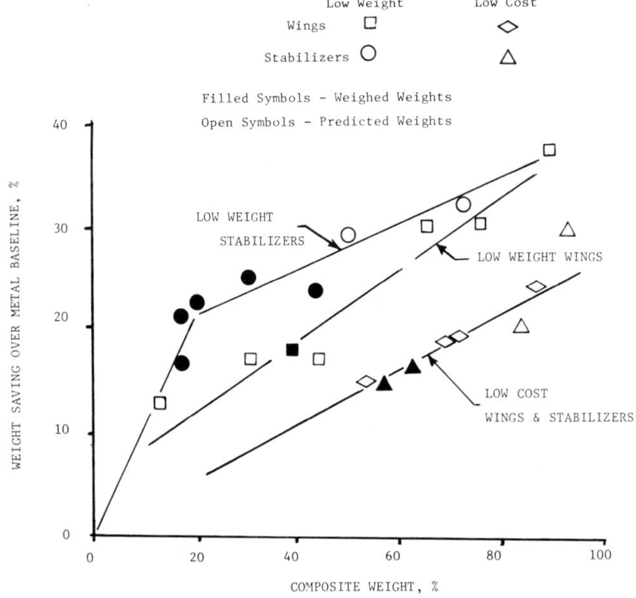

Figure 20.2. Stabilizer and wing torque box weight savings.

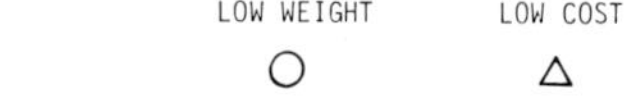

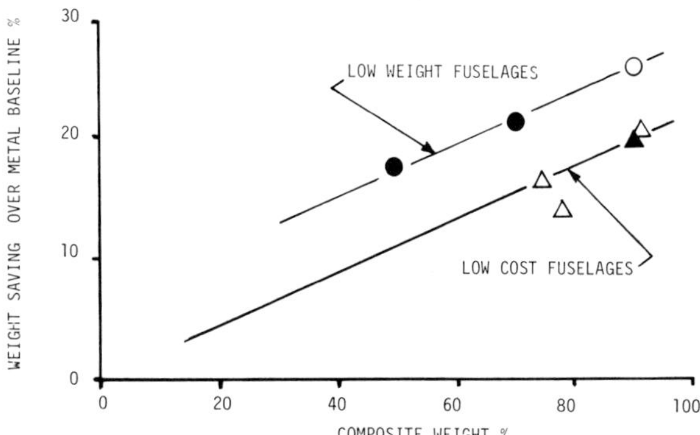

Figure 20.3. Fuselage weight savings.

represents third generation structures, which were designed to achieve weight savings in conjunction with cost savings.

A smaller data base exists for fuselages (seven points), as shown in Fig. 20.3, but there is a relatively large data base for such components as flaps, slats, spoilers, and fairings, as shown in Fig. 20.4. The trends are similar to those seen in the wing and empennage structures: there is an upper band, generally associated with the earlier minimum weight designs, and a lower band, associated with later parts which were designed to achieve both weight and cost savings.

Preliminary estimates of weight savings which can be achieved by using advanced

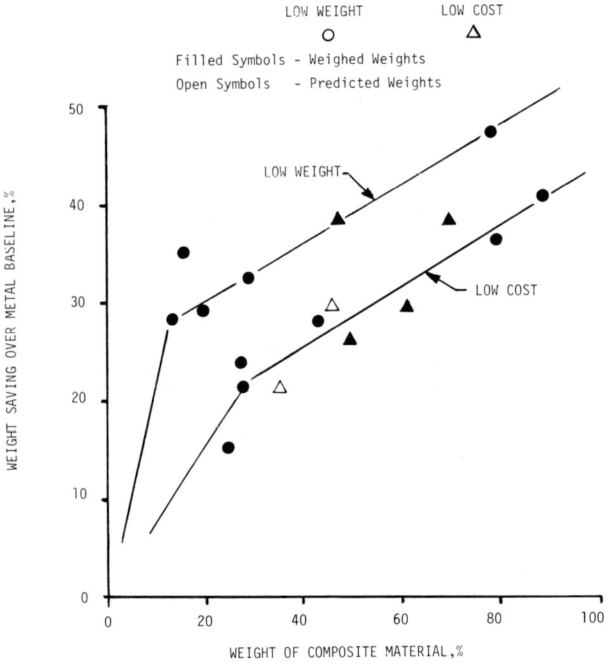

Figure 20.4. Weight savings for rudders, elevators, flaps, etc.

Table 20.2. Weight Reductions Over Metal Airframes (as Used for Sizing Studies)

COMPONENT	WEIGHT SAVINGS, %	COMPOSITE GOAL, %
Fixed wing		
Conventional	28	87
Double delta	23.5	87
Swing wing	20	65
Tails and canards		
Slab	23	79
Fixed	30	79
Body	20	72
Air induction		
Fixed inlet	22	80
Var inlet	20	80
Landing gear	16	40

composites to replace metals on existing parts should be estimated utilizing the lower curve in the appropriate figure, since this represents a compromise design which saves weight and, on a unit for unit comparison, also saves cost.

It is recommended that the values shown in Table 20.2 be used to predict the weight savings using composites in a new design vehicle. These weight savings are representative of components designed to save maximum weight consistent with some cost savings. The associated percentages of composite materials are shown in the table. These data were generated under the Advanced Design Composite Aircraft Study and were used for the vehicle preliminary sizing studies. The resulting preferred vehicle structure was defined sufficiently well to confirm the validity of the initial weight savings data.

The effect of utilizing 75% composites on the aircraft which was designed to perform a given supersonic penetration interdiction mission is shown in Fig. 20.5, and the aircraft configuration is shown in Fig. 20.6. The overall benefits of the application of these materials to a new design are clearly shown in Fig. 20.5. The composite aircraft is significantly smaller and lighter than its all metal counterpart. The 35% reduction in structural weight and 26% reduction in takeoff gross

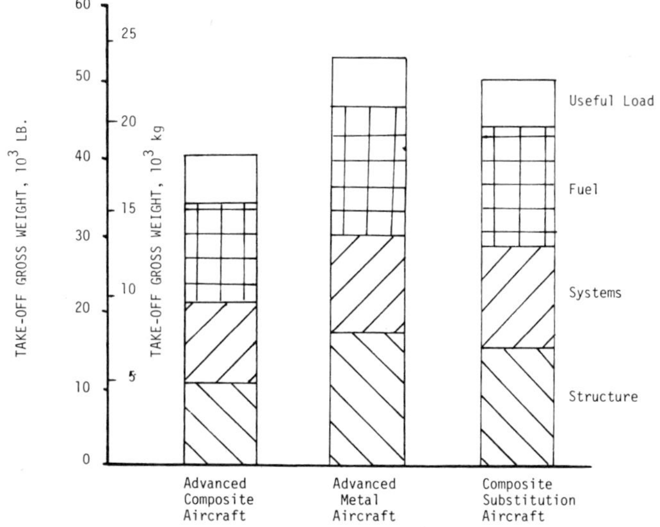

Figure 20.5. Composite versus metal aircraft weight comparison.

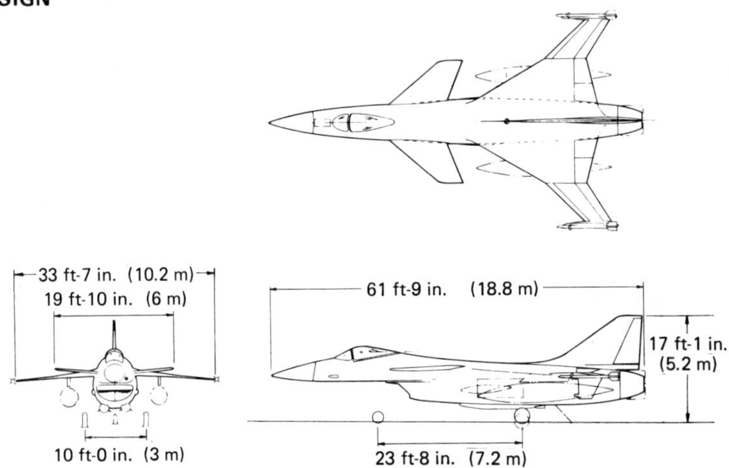

Figure 20.6. Advanced design composite aircraft configuration.

weight are benefits which can only be gained by designing with composites from the outset. A preliminary cost analysis indicated a 21% savings in production costs, with a projected fuel savings of 30%.

Fig. 20.7 shows the results of one of the trade studies. The vehicle was sized for various wing thickness/chord (t/c) ratios to determine the optimum ratio and determine the effect of vehicle wave drag.

This particular study shows that a t/c ratio of 0.035 is optimum for this particular wing configuration, and this study is typical of the many trade studies which must be performed during a preliminary design phase.

Cost data are needed to perform parametric trade studies, but these data cannot be easily

WEIGHT (LB/SHIPSET)		
COMPONENT	METAL	COMPOSITE
COVERS (INCLUDING SURFACE PROTECTION AND FASTENERS)	1687.6	1267.7
FRONT AND REAR SPARS	192.0	131.9
INTERMEDIATE SPARS	244.4	217.1
RIBS	153.1	124.5
MISC (SHIM AND HOIST FITTING)	.....	46.3
TOTAL BOX	2277.1	1787.5
LEADING EDGE, TRAILING EDGE TIP	575.5	575.5
SEAL (INBOARD)	46.3	15.4
FINISH	40.3	40.3
BEARING SUPPORT FITTING	376.0	395.0
TOTAL	3315.2	2813.7*
GOAL		2866
SAVING OVER METAL DESIGN		501.5 (15%)
* ACTUAL WEIGHT OF STATIC (1400 LB) AND FATIGUE (1413.7 LB) ARTICLES		

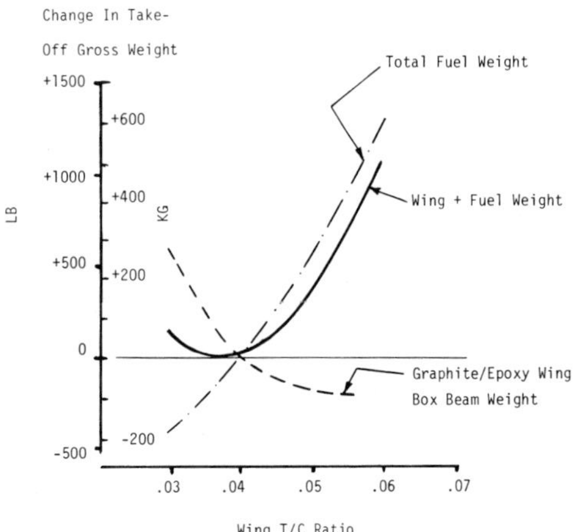

Figure 20.7. ADCA weight versus wing thickness/chord ratio.

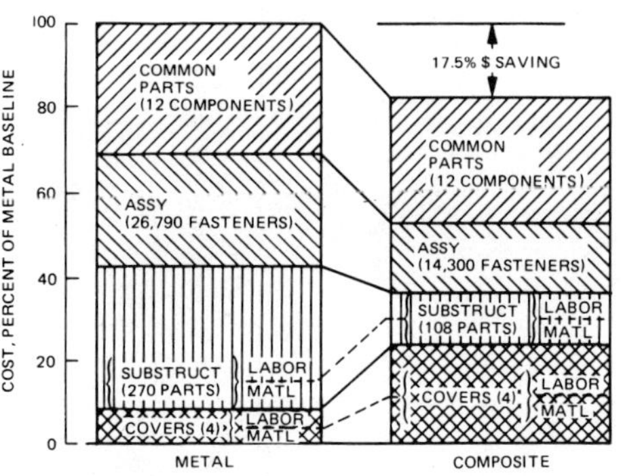

Figure 20.8. B-1 horizontal stabilizer production weight and cost comparisons.

generalized since costs are very sensitive to material prices, part count, and complexity, as well as production quantity and rate. Some indication of potential cost savings can be obtained from the results of a study of the production costs of the B-1 horizontal stabilizer.[2,4] The results, shown in Fig. 20.8, project an overall cost savings of 17.5%, in conjunction with a total weight savings of 15%. The cost and weight savings associated with the redesigned composite torque box are both 22%.

A reasonable guideline based on graphite/epoxy prices forecast for the mid-1980 time period is that percentage cost savings should be at least one-half the percentage weight savings of the composite parts. These cost savings will be much greater if the composite material is used to achieve a major reduction in part count as well as weight savings.

20.3. MULTIDIRECTIONAL LAMINATE STRUCTURAL BEHAVIOR

The in-plane structural behavior of a multidirectional laminate is the basic starting point in the analysis of Advanced Composite structures.

Each of the individual layers has fibers in only one direction—defined as the longitudinal direction of that layer—with the matrix providing transverse, normal, and in-plane loading capability and stiffness to the layer.

Each of the layers will generally be under a state of combined stress and strain, even if the laminate is loaded uniaxially.

The layer stress-strain relations were developed in terms of functions of angles of rotation of the layers by Tsai and Pagano.[5]

$$\begin{bmatrix} f_x \\ f_y \\ f_{xy} \end{bmatrix} = \begin{bmatrix} Q_{11} & Q_{12} & Q_{16} \\ Q_{21} & Q_{22} & Q_{26} \\ Q_{61} & Q_{62} & Q_{66} \end{bmatrix} \begin{bmatrix} \epsilon_x \\ \epsilon_y \\ \gamma_{xy} \end{bmatrix} \quad (20\text{-}1)$$

where

f_x, f_y, and f_{xy} = stresses applied to the layer
ϵ_x, ϵ_y, and γ_{xy} = layer strains

and

$$Q_{11} = 3U_1 + U_2 + U_3 \cos 2\theta + U_4 \cos 4\theta$$
$$Q_{22} = 3U_1 + U_2 + U_3 \cos 2\theta + U_4 \cos 4\theta$$
$$Q_{21} = Q_{12} = U_1 - U_2 - U_4 \cos 4\theta$$
$$Q_{66} = U_1 + U_2 + U_4 \cos 4\theta$$
$$Q_{61} = Q_{16} = \tfrac{1}{2}U_3 \sin 2\theta + U_4 \sin 4\theta$$
$$Q_{62} = Q_{26} = \tfrac{1}{2}U_3 \sin 2\theta + U_4 \sin 4\theta$$

θ = angle between layer axis and laminate axis.

Also:

$$U_1 = \frac{1}{8\psi}\left[E_{11} + E_{22} + \nu_{21}E_{11} + \nu_{12}E_{22}\right]$$
$$U_2 = \frac{1}{2\psi}\left[\psi G_{12} - \frac{1}{4}(\nu_{21}E_{11} + \nu_{12}E_{22})\right]$$
$$U_3 = \frac{1}{2\psi}\left[E_{11} - E_{22}\right]$$
$$U_4 = \frac{1}{8\psi}\left[E_{11} + E_{22} - (\nu_{21}E_{11} + \nu_{12}E_{22}) - 4\,\psi G_{12}\right]$$

where

E_{11}, E_{22}, and G_{12} = layer longitudinal, transverse, and shear moduli
ν_{12} = major Poisson's ratio
ν_{21} = minor Poisson's ratio
$\psi = 1 - \nu_{12}\nu_{21}$.

By Maxwell's reciprocal theorem,

$$\nu_{12}E_{22} = \nu_{21}E_{11}.$$

Sign conventions for stress and ply orientation are shown on Fig. 20.9.

Most types of structures have combined loading conditions which vary from location to location. Since the fiber orientation in any given layer cannot be changed without cutting the fibers, laminates with layers oriented in four directions (0°, 45°, 90°, and −45°) have

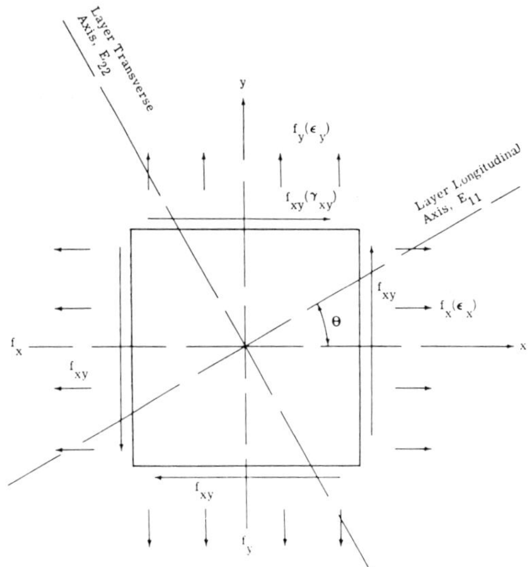

Figure 20.9. Layer stress conventions.

become the most generally used family. Laminates within this family can be designed to be near optimum at each location of the structure. The number of layers in each of the four directions can be varied throughout the structure. It has, however, been general practice to maintain symmetry about the laminate centerline and to pair the +45° and −45° layers. This reduces the possibility of warping, which can be very pronounced with unsymmetrical laminates.

In order to minimize the stresses in the matrix and to provide fiber control of the in-plane properties of the laminate, layers must be oriented in a minimum of three directions (e.g., 0°, +45°, and −45°). Many designs have layers oriented in all four directions to minimize the contribution of the matrix and to obtain more consistent structural behavior.

Therefore, this chapter will cover design and analysis of structures utilizing the family of symmetric laminates which have fibers oriented in the four directions (0°, 45°, 90°, and −45°) and where the +45° and −45° layers are applied in pairs.

Considering now the [0°/90°/±45°] laminates with equal numbers of layers in the +45° and −45° directions, the elastic stress-strain coefficients for the various layers are summarized in Table 20.3.

Using L, M, and N to denote the proportion of layers in the 0°, 90°, and ±45° directions, assuming plain strain through the laminate and using the relations of equilibrium, then:

$$Lf_x^L + Mf_x^M + Nf_x^N = f_x$$

$$Lf_y^L + Mf_y^M + Nf_y^N = f_y \qquad (20\text{-}2)$$

$$Lf_{xy}^L + Mf_{xy}^M + Nf_{xy}^N = f_{xy}$$

where f_x, f_y, and f_{xy} are, respectively, the longitudinal, transverse, and shear stresses applied to the laminate; f_{ij}^L and f_{ij}^M are the 0° and 90° layer stresses, and f_{ij}^N is the stress in the ±45° pair of layers.

Substituting Eq. 20-1 into 20-2 results in Eq. 20-3.

$$\begin{aligned} f_x &= (\Sigma RQ_{11}^R)\,\epsilon_x + (\Sigma RQ_{12}^R)\,\epsilon_y = A_{11}\,\epsilon_x + A_{12}\,\epsilon_y \\ f_y &= (\Sigma RQ_{12}^R)\,\epsilon_x + (\Sigma RQ_{22}^R)\,\epsilon_y = A_{12}\,\epsilon_x + A_{22}\,\epsilon_y \\ f_{xy} &= (\Sigma RQ_{66}^R)\,\gamma_{xy} = A_{66}\,\gamma_{xy} \end{aligned} \qquad (20\text{-}3)$$

Table 20.3. Elastic Stress-Strain Coefficient Equations

COEFFICIENT	0° LAYERS	90° LAYERS	±45° LAYERS (Pairs)
Q_{11}	E_{11}/ψ	E_{22}/ψ	$[1/4(E_{11}+E_{22}) + 1/2\nu_{12}E_{22} + \psi G_{12}]/\psi$
Q_{22}	E_{22}/ψ	E_{11}/ψ	$[1/4(E_{11}+E_{22}) + 1/2\nu_{12}E_{22} + \psi G_{12}]/\psi$
Q_{12}	$\nu_{12}E_{22}/\psi$	$\nu_{12}E_{22}/\psi$	$[1/4(E_{11}+E_{22}) + 1/2\nu_{12}E_{22} - \psi G_{12}]/\psi$
Q_{66}	G_{12}	G_{12}	$[1/4(E_{11}+E_{22}) - 1/2\nu_{12}E_{22}]/\psi$*
$Q_{16} = Q_{26}$	0	0	0

*For ±45° layers, Q_{66} is obtained using average tension and compression values of E_{11} and E_{22}; i.e., $E_{11av} = 1/2(E_{11}^t\ E_{11}^c)$; $E_{22av} = 1/2(E_{22}^t + E_{22}^c)$.

where ΣRQ_{ij}^{R} represents the summation of the proportion of each layer thickness multiplied by its stress-strain coefficient; that is:

$$\Sigma RQ_{11}^{R} = LQ_{11}^{L} + MQ_{11}^{M} + NQ_{11}^{N}.$$

And A_{ij} are the stiffness coefficients of the laminate.

Inverting Eq. 20-3, the strain-stress relations become:

$$\epsilon_x = \frac{[\Sigma RQ_{22}^{R}]f_x - [\Sigma RQ_{12}^{R}]f_y}{[\Sigma RQ_{11}^{R}]\,[\Sigma RQ_{22}^{R}] - [\Sigma RQ_{12}^{R}]^2}$$

$$\epsilon_y = \frac{-[\Sigma RQ_{12}^{R}]f_x + [\Sigma RQ_{11}^{R}]f_y}{[\Sigma RQ_{11}^{R}]\,[\Sigma RQ_{22}^{R}] - [\Sigma RQ_{12}^{R}]^2}$$

$$\gamma_{xy} = \frac{f_{xy}}{\Sigma RQ_{66}^{R}} \qquad (20\text{-}4)$$

The laminate overall elastic properties in the elastic range are given by Eq. 20-5.

$$A_{11} = A_x = \frac{E_x}{1 - \nu_{xy}\nu_{yx}} = \Sigma RQ_{11}^{R}$$

$$A_{22} = A_y = \frac{E_y}{1 - \nu_{xy}\nu_{yx}} = \Sigma RQ_{22}^{R}$$

$$A_{12} = \frac{\nu_{yx}E_x}{1 - \nu_{xy}\nu_{yx}} = \frac{\nu_{xy}E_y}{1 - \nu_{xy}\nu_{yx}} = \Sigma RQ_{12}^{R}$$

$$A_{66} = A_{xy} = G_{xy} = \Sigma RQ_{66}^{R} \qquad (20\text{-}5)$$

$$\nu_{xy} = \frac{\Sigma RQ_{12}^{R}}{\Sigma RQ_{22}^{R}}$$

$$\nu_{yx} = \frac{\Sigma RQ_{12}^{R}}{\Sigma RQ_{11}^{R}}$$

The stress applied to each layer is obtained by substituting Eq. 20-4 into Eq. 20-1, which results in Eq. 20-6.

A major complication is caused by the relatively large data base of properties needed for design and analysis when a part is subjected to various conditions of applied loads over a range of temperatures. In all, seven properties are needed, and each varies with temperature. The values of E_{11}, F_1^{tu}, F^{cu}, and ν_{12}, the layer longitudinal modulus of elasticity, tension strength, compression strength, and major Poisson's ratio, respectively, do not vary significantly with temperature. However, the values of E_{22}, G_{12}, G_{44}, and G_{55}, the transverse modulus of elasticity, the in-plane shear modulus, and the longitudinal and transverse interlaminar shear moduli respectively, do vary significantly, since they are all matrix controlled.

Simplification is needed for preliminary design. This can be accomplished by accepting maximum errors of 10% for boron/epoxy (B/Ep) and 5% for graphite/epoxy (Gr/Ep). The result is the simplified version of Table 20.3, which is shown in Table 20.4, where $\overline{E}_{11}$ is substituted for $(E_{11} - E_{22})$.

The major Poisson's ratio is a constant 0.025 over the −67–350° F (−55–177° C) temperature range for unidirectional B/Ep and Gr/Ep. The value of $\overline{E}_{11}$ is predominant and varies by a maximum of 2% with B/Ep (AVCO 5505) over the −67–350° F (−55–177° C) temperature range. The comparative variationin the value for Gr/Ep (3501/AS) is 4%. This is considered quite acceptable for preliminary design and analysis. The values of E_{22} do vary significantly with temperature, but the values of $K(G_{12}/E_{22})$ are assured constant at $\frac{1}{4}$ for B/Ep and 2/5 for Gr/Ep. Thus, the simplification reduces to the stiffness coefficients needed for analysis, and only one, E_{22}, has variation with temperature. The error using E

$$f_x^{R} = \frac{[Q_{11}^{R}\Sigma RQ_{22}^{R} - Q_{12}^{R}\Sigma RQ_{12}^{R}]f_x - [Q_{11}^{R}\Sigma RQ_{12}^{R} - Q_{12}^{R}\Sigma RQ_{11}^{R}]f_y}{[\Sigma RQ_{11}^{R}][\Sigma RQ_{22}^{R}] - [\Sigma RQ_{12}^{R}]^2}$$

$$f_y^{R} = \frac{-[Q_{22}^{R}\Sigma RQ_{12}^{R} - Q_{12}^{R}\Sigma RQ_{22}^{R}]f_x + [Q_{22}^{R}\Sigma RQ_{11}^{R} - Q_{12}^{R}\Sigma RQ_{12}^{R}]f_y}{[\Sigma RQ_{11}^{R}][\Sigma RQ_{22}^{R}] - [\Sigma RQ_{12}^{R}]^2} \qquad (20\text{-}6)$$

$$f_{xy}^{R} = \frac{Q_{66}^{R}f_{xy}}{\Sigma RQ_{66}^{R}}$$

Table 20.4. Simplified Elastic Stress-Strain Coefficient Equations

COEFFICIENT	0° LAYERS	90° LAYERS	±45° LAYERS (Pairs)
Q_{11}	$\bar{E}_{11} + E_{22}$	E_{22}	$\frac{1}{4}\bar{E}_{11} + E_{22}$
Q_{22}	E_{22}	$\bar{E}_{11} + E_{22}$	$\frac{1}{4}\bar{E}_{11} + E_{22}$
Q_{12}	$\frac{1}{4}E_{22}$	$\frac{1}{4}E_{22}$	$\frac{1}{4}\bar{E}_{11} + E_{22}$
Q_{66}	KE_{22}	KE_{22}	$\frac{1}{4}\bar{E}_{11} + KE_{22}$
$Q_{16} = Q_{26}$	0	0	0

instead of E/ψ is less than 1%. Figure 20.10 shows the variation of α and K with temperature for B/Ep and Gr/Ep.

The simplification has one further advantage. The laminate can now be modeled as individual layers of fibers embedded in an isotropic matrix. Substituting $\bar{E}_{11}$ for $(E_{11} - E_{22})$, the overall elastic properties of the laminate in the elastic range are now given by Eq. 20-7.

$$E_x = \Sigma RQ_{22}{}^R - \frac{(\Sigma RQ_{12}{}^R)^2}{\Sigma RQ_{22}{}^R} \qquad (20\text{-}9)$$

$$G_{xy} = \Sigma RQ_{66}{}^R$$

and the major and minor Poisson's ratios are:

$$\nu_{xy} = \frac{\Sigma RQ_{12}{}^R}{\Sigma RQ_{22}{}^R} : \nu_{xy} = \frac{\Sigma RQ_{12}{}^R}{\Sigma RQ_{11}{}^R}. \qquad (20\text{-}10)$$

$$A_{11} = A_x = \frac{E_x}{1 - \nu_{xy}\nu_{yx}} \simeq \bar{E}_{11}(L + N/4) + E_{22}$$

$$A_{22} = A_y = \frac{E_y}{1 - \nu_{xy}\nu_{yx}} \simeq \bar{E}_{11}(M + N/4) + E_{22} \qquad (20\text{-}7)$$

$$A_{12} = A_{21} = \frac{\nu_{xy}E_x}{1 - \nu_{xy}\nu_{yx}} = \frac{\nu_{xy}E_x}{1 - \nu_{xy}\nu_{yx}} \simeq \frac{1}{4}(N\bar{E}_{11} + E_{22})$$

$$A_{66} = \frac{N}{4}\bar{E}_{11} + KE_{22}$$

$$\nu_{xy} = \frac{\frac{1}{4}E_{22}}{\bar{E}_{11}(M + N/4) + E_{22}}; \quad \nu_{yx} = \frac{\frac{1}{4}E_{22}}{\bar{E}_{11}(L + N/4) + E_{11}}$$

and the basic equation governing the response of the orthotropic laminate is, from Eq. 20-1:

$$\begin{bmatrix} f_x \\ f_y \\ f_{xy} \end{bmatrix} = \begin{bmatrix} A_{11} & A_{12} & 0 \\ A_{21} & A_{22} & 0 \\ 0 & 0 & A_{66} \end{bmatrix} \begin{bmatrix} \epsilon_x \\ \epsilon_y \\ \gamma_{xy} \end{bmatrix} \qquad (20\text{-}8)$$

The longitudinal and transverse elastic moduli, E_x and E_y, and the in-plane shear modulus, G_{xy}, are obtained from Eq. 20-7.

$$E_x = \Sigma RQ_{11}{}^R - \frac{(\Sigma RQ_{12}{}^R)^2}{\Sigma RQ_{22}{}^R}$$

Using the simplified elastic stress-strain coefficients from Table 20.4, substituting α for $E_{22}/\bar{E}_{11}$, and neglecting secondary terms in α, since α is small (<0.12):

$$E_x \simeq \bar{E}_{11}\left(L + \frac{MN}{4M + N} + \alpha\right)$$

$$E_y \simeq \bar{E}_{11}\left(M + \frac{LN}{4L + N} + \alpha\right) \qquad (20\text{-}11)$$

$$G_{xy} = \bar{E}_{11}\left(\frac{N}{4} + K\alpha\right)$$

and

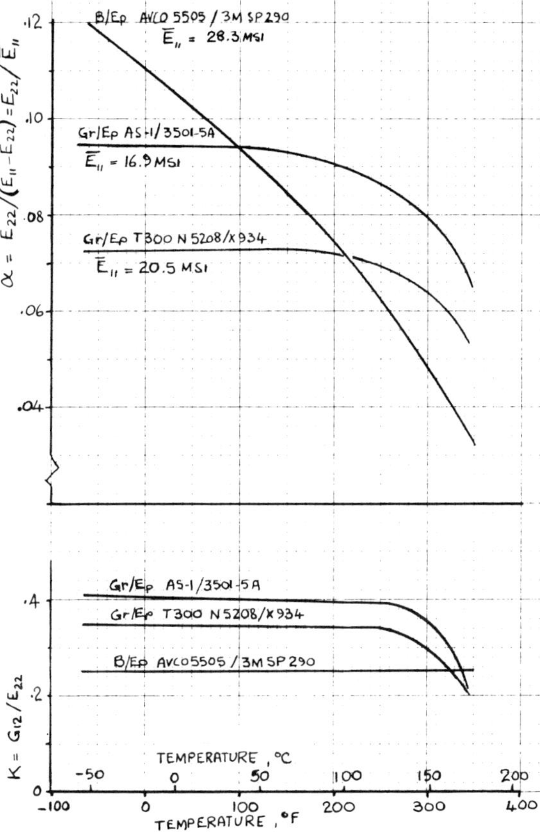

Figure 20.10. Transverse modulus and shear modulus ratios.

$$\nu_{xy} = \frac{N + \alpha}{(4M + N + 4\alpha)} \qquad (20\text{-}12)$$

$$\nu_{xy} = \frac{N + \alpha}{(4L + N + 4\alpha)}$$

Values of the ratios of laminate to layer moduli are shown in Figs. 20.11 and 20.12. These are used in conjunction with the properties shown Fig. 20.10 and Table 20.5 to obtain the laminate moduli.

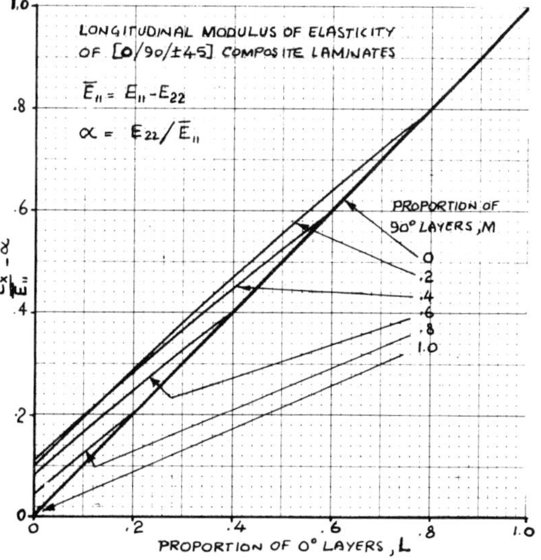

Figure 20.11. Longitudinal modulus of elasticity of [0°/90°/±45°] composite laminates.

20.3.1. In-Plane Strength of Laminates

Various failure hypotheses have been used to define the strength of multidirectional laminates. The two hypotheses which have resulted in good agreement with test data are the maximum layer longitudinal layer stress approach and the maximum layer strain approach.

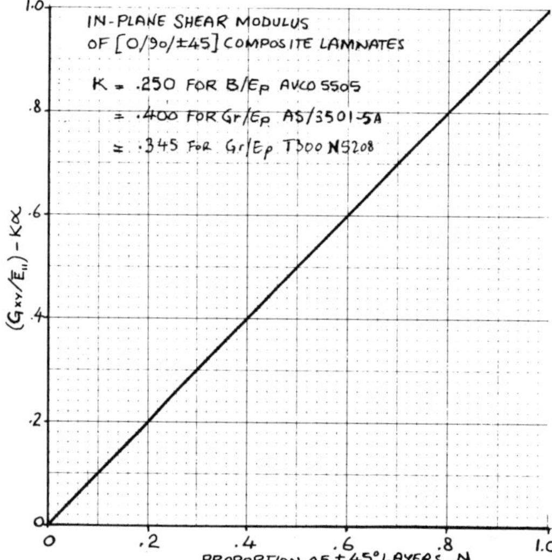

Figure 20.12. In-plane shear modulus of elasticity of [0° / 90° / ±45°] laminates.

The first hypothesis equates the fiber controlled longitudinal stresses in each layer to an allowable value based on tests of unidirectional laminates. The matrix contribution to in-plane strength is effectively neglected, and transverse or shear stresses in the matrix are not considered to be the cause of overall failure.

The second hypothesis equates the longitudinal, transverse, or shear strains in each layer to allowable strains which are again based on tests or unidirectional laminates. Strains induced in the matrix by the large differences in longitudinal and transverse coefficients of expansion in each layer during cool-down in the cure cycle are neglected, even though these can be quite high.

20.4. LAMINATE STRENGTH UNDER UNIAXIAL LOADING

The strength of the [0° / 90° / ±45°] family of laminates is determined by equating the longitudinal stresses (or strains) in each of the layers to their allowable stress (or strain):

$$\frac{F_x}{E_X} = \frac{F^{LU}}{E_{11}{}^L} = \epsilon^{LU} \qquad (20\text{-}13)$$

where F_x and E_x are the ultimate longitudinal strength and the elastic modulus of the laminate; $F_1{}^{LU}$ and $E_{11}{}^L$ are the longitudinal strength and the elastic modulus of the 0° layers; and ϵ^{LU} is the ultimate longitudinal strain of the 0° layers.

The ultimate longitudinal strength is now obtained using Eq. 20-13 with Eq. 20-11.

$$F_x = \frac{F_1{}^{LU}}{(1+\alpha)}\left[L + \frac{MN}{4M+N} + \alpha\right]$$

$$= F^{LU}\left[1 - \frac{(2M+N)^2}{(1+\alpha)\,(4M+N)}\right] \qquad (20\text{-}14)$$

This equation is, again, in the form of the basic terms $L + \dfrac{MN}{4M+N}$, which are not affected by temperature, plus terms in α, which *is* temperature dependent. The variation of the ratio F_x/F^{LU} with lay-up and α is shown in Fig. 20.13.

The in-plane shear strength of the laminate is obtained in a similar manner and is defined by Eq. 20-15.

$$F_{xy} = \pm\tfrac{1}{2}F^{NU}\,\frac{[N + 4K\alpha]}{[1+\alpha]} \qquad (20\text{-}15)$$

where F^{NU} is the tension or, if lower, the compression strength of the ±45° layers.

The variation of in-plane shear strength with lay-up is shown for Gr/Ep material ($\alpha = 0.1$; $K = 0.375$) in Fig. 20.14.

It should be noted that Eqs. 20-14 and 20-15 relate to filament fracture failure modes, and these may be greater than local instability modes. Local instability limits design compression and shear stresses to approximately 0.833 G_Z and 0.25 G_Z, respectively.

Under tension conditions, the matrix can be loaded in shear and normal tension at the boundary of a laminate or a hole. Matrix stresses can be sufficiently high to cause interlaminar failure and reduce the overall load capability of the laminate.[2] The matrix is also loaded in shear microscopically, since tension failure of the fibers is progressive. As the weaker fibers break, the matrix has some capability of transferring load from a broken fiber to surrounding unbroken fibers.

Table 20.5. Design Properties of Unidirectional Non-Woven Composite Materials (Room Temperature)

PROPERTY	SYMBOL	GRAPHITE/EPOXY AS-1/3501-5A (Hercules)	GRAPHITE/EPOXY T300/N5208 (Narmco)	GRAPHITE/EPOXY GY70/HYE 1534 (Celanese)	BORON/EPOXY AVCO 5505* (Avco)	ARAMID/EPOXY KEVLAR 49/CE3305 (DuPont)
STRENGTHS, ksi (MPa) — NO HOLES (Gross Area Stress W/D = 6)						
Tension						
Longitudinal	F_1^{tu}	169 (1165)	169 (1165)	90 (621)	165 (1138)	168 (1158)
Transverse	F_2^{tu}	6 (41)	4 (30)	2 (13.8)	10 (73)	1.6 (11.0)
Compression**						
Longitudinal	F_1^{cu}	162 (1116)	141 (972)	90 (621)	480 (3309)	40 (276)
Transverse	F_2^{cu}	25 (172)	20 (138)	28 (193)	40 (276)	12.1 (83)
Interlaminar shear	F_{ILS}	7 (49)	14 (98)	4 (27.6)	7 (48)	10 (69)
STRENGTHS, ksi (MPa) — WITH HOLES						
Tension†						
Longitudinal	F_1^{tu}	76 (524)	76 (524)	21 (145)	80 (551)	95 (655)
Transverse	F_2^{tu}	3 (22)	2 (14)	1 (6.9)	5 (34)	0.8 (55)
Compression**†						
Longitudinal	F_1^{cu}	76 (524)	76 (524)	21 (145)	185 (1275)	30 (207)
Transverse	F_2^{cu}	12 (86)	10 (69)	14 (96.5)	12 (83)	6.1 (42)
Bearing, $D/t > 2.0$	F_{bru}	66 (455)	66 (455)	66 (455)	140 (965)	22 (152)
MODULI, MSI (GPa)						
Longitudinal	E_{11}	18.50 (127.6)	22.0 (151.7)	42 (289.6)	30.3 (208.9)	11.9 (82)
Transverse	E_{22}	1.60 (11.0)	1.50 (10.3)	1 (6.9)	2.8 (19.3)	0.6 (4)
In-plane shear	G_{12}	0.65 (4.5)	0.52 (3.6)	0.7 (4.8)	0.72 (5.0)	0.4 (2.8)
Interlaminar shear†	G_z	0.104 (0.72)	0.14 (1.0)	>0.08 (0.55)	0.25 (1.72)	>0.04 (0.28)
$(E_{11} - E_{22})$	$\overline{E}_{11}$	16.90 (116.6)	20.50 (141.4)	41 (282)	27.5 (189.6)	11.3 (78)
Major Poisson's ratio	ν_{12}	0.25	0.25	0.25	0.25	0.25
Coefficient of thermal expansion, in./in./°F (m/m/°C)						
Longitudinal	α_{11}	0.25 (0.45)	0.30 (0.54)	−0.58 (1.04)	2.5 (4.5)	−2.0 (−3.6)
Transverse	α_{22}	15.20 (27.4)	11.0 (19.8)	16.5 (29.7)	13.0 (23.4)	32.0 (57.6)
Average cured layer thickness, in. (mm)	t'	0.00525 (0.133)	0.00525 (0.133)	0.0055 (0.140)	0.00525 (0.133)	0.00716 (0.181)
Density, lb/in.^{3} (g/cm^3)	ρ	0.055 (1.52)	0.055 (1.52)	0.061 (1.69)	0.075 (2.07)	0.048 (1.32)

*In hybrids.
**Laminate compression strength cannot exceed G_Z.
†5/16-in. (7.9-mm) diameter holes. Strength/diameter correction is $(0.6 + 5\,(D_o/D) - 0.1\,(D_o/D)^2)$, where D_o = 5/16 in. (7.9 mm) and D = actual hole diameter $(0.625 \leqslant \frac{D_o}{D} \leqslant 2.50)$.

Gross area stress, $\frac{W}{D} = 6$.

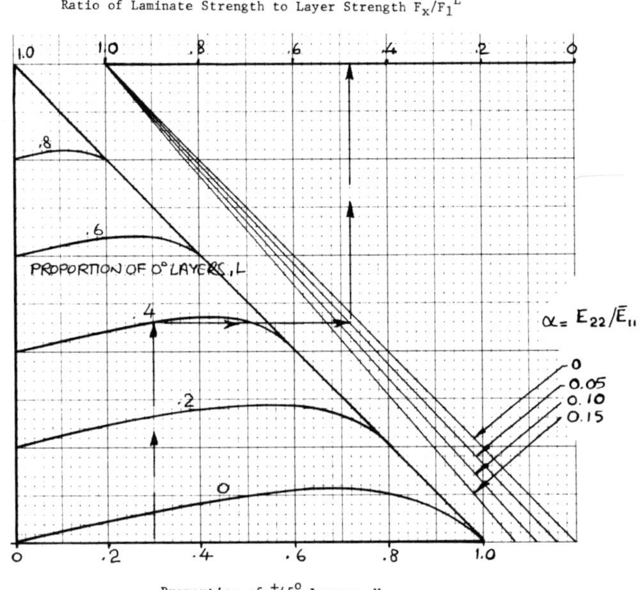

Figure 20.13. In-plane strength of multidirectional laminates. (Worked example: L =.4, M = .3, N = .3, = .05 F_x/F_1^L = .48)

In the case of B/Ep material, this matrix capability can increase the room temperature strength by 21.5% when 0° layers are stacked in groups rather than individually dispersed between the 90° and ±45° layers. This improvement, first noted by Tsai, Adams, and Doner, who compared the strength of dry bundles of fibers with similar bundles of fibers embedded in a matrix, is defined as the matrix effectiveness factor, β. The variation of the B/Ep matrix effectiveness factor with temperature is shown in Fig. 20.15, and the de-

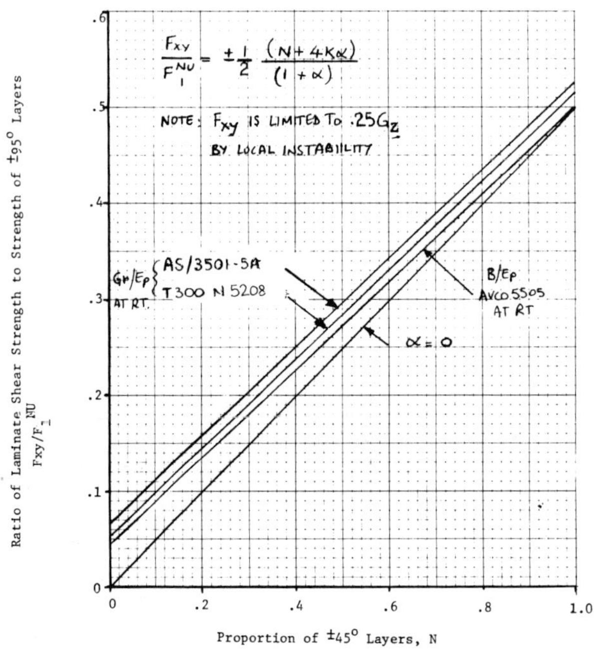

Figure 20.14. In-plane shear strength of multidirectional laiminates.

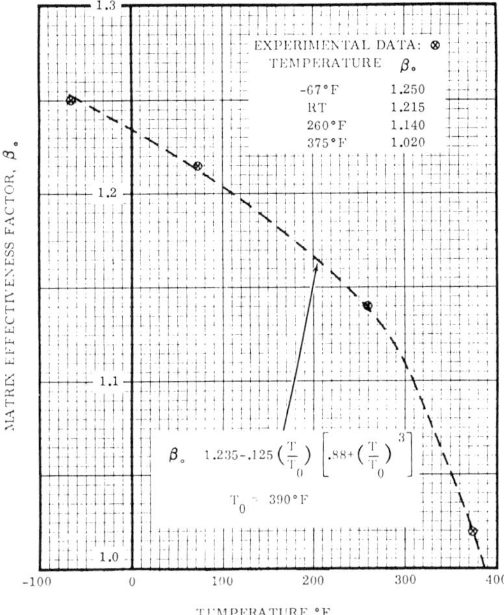

Figure 20.15. Variation of matrix effectiveness factor with temperature.

pendence of the factor on lay-up is shown in Fig. 20.16.

This phenomenon does not appear to be nearly as pronounced with G/Ep. A possible explanation is that the large number of surrounding fibers within a layer are capable of carrying the small load transferred from a single fractured small diameter graphite fiber. In the case of B/Ep, the much higher load from a fractured large diameter fiber is transferred directly by the matrix only to the few adjacent fibers within that layer, should that layer be isolated from other layers which are also carrying the load. Thus, for B/Ep material only:

$$F^{LU} = \frac{F_1{}^{tu}}{\beta} \tag{20-16}$$

20.4.1. Laminate Instability

Under compression loading, both the fibers and the layers are stabilized by the matrix. The relatively low stiffness of the matrix can have significant influence on the compression strength of the laminate. Failure can occur at fiber stresses which are considerably lower than those needed to cause filament fracture. There are a number of different local instability modes of failure which have been noted on tests. Suarez *et al.*[6] have developed an approach of predicting these modes which gives good agreement with experimental data. This approach models both the fibers and the layers as beams, with or without initial curvature, supported by an elastic foundation.

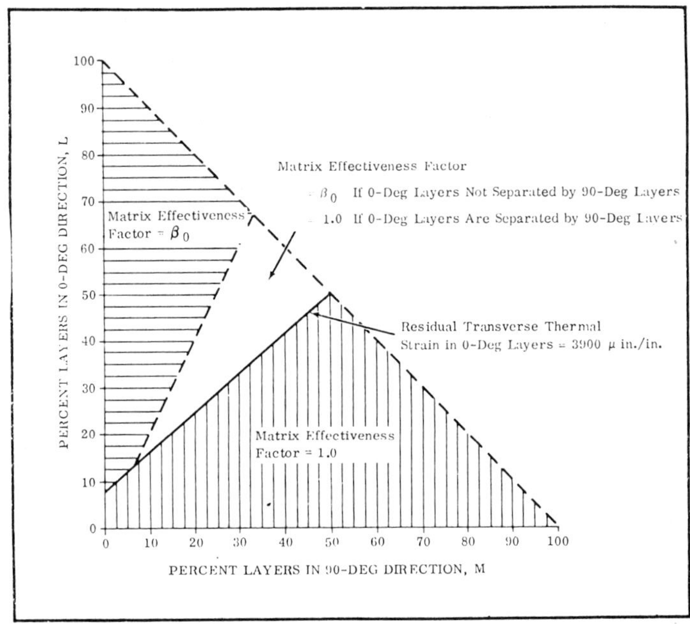

Figure 20.16. Variation of matrix effectiveness factor with lay-up.

Equating the strain energies of bending and shear of the beam and the strain energy of the elastic foundation to the work done by the compressive force results in equations for the various possible local instability modes.

20.4.2. Layer Instability

This mode of failure results from instability of the outermost layer oriented parallel to the load. The expression used to determine the layer instability stress is derived by considering the remainder of the laminate as the elastic foundation.

$$\frac{F_{cr}}{G_z} = \xi(2 - \xi), \xi \leqslant 1 \qquad (20\text{-}17)$$
$$= 1, \xi \geqslant 1$$

where

F_{cr} = buckling stress
G_z = interlaminar shear modulus of the composite (Fig. 20.17)
ξ = stiffness parameter = $\dfrac{\sqrt{DB}}{S}$.

For filamentary composites with stiff, regularly spaced circular filaments, the stiffness terms can be written as:

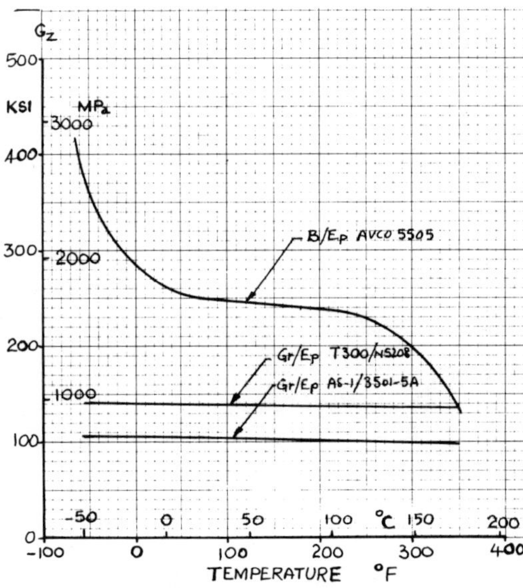

Figure 20.17. Design interlaminar shear stiffness for stability analysis (design $G_z = G_z$).

$$D = \frac{\pi d_f^4 E_f b}{64 w_f}$$

$$B = \frac{Lb\ E_z}{t'} \qquad (20\text{-}18)$$

$$S = \frac{bt'G_z}{L}.$$

where D is the flexural stiffness of the fibers and B and S are the normal and shear stiffnesses of the laminate.

Hence:

$$\xi = \frac{d_f^2 L^{3/2}}{8t' G_z}\left[\frac{\pi E_f E_z}{t' w_f}\right]^{1/2}$$

where

d_f = filament diameter
w_f = in-plane filament spacing
E_f = filament elastic modulus
E_z = normal tensile modulus of laminate
L = proportion of layers in load direction
t' = average layer thickness.

20.4.3. Face Wrinkling

20.4.3.1. Laminate Instability

This type of failure occurs in composite sandwich construction. Here, the model for analysis consists of the laminate treated as a beam, with a sandwich core acting as the elastic foundation. The critical facing stress is determined by using the general expression for instability, Eq. 20-17, and the stiffness parameter, ξ, given by

$$\xi = \frac{\sqrt{DB}}{S}$$

where

D = longitudinal flexural stiffness of the laminate
B = foundation stiffness
S = interlaminar shear stiffness of the composite.

Also:

$$\frac{B}{b} = \frac{E_c}{t} + \frac{1.1007}{t_f}\left(\frac{E_c^2 G_c^2}{E_f}\right)^{1/3} \qquad (20\text{-}19)$$

$$t = 0.9805\, t_f \left(\frac{E_f E_c}{G_c^2}\right)^{1/3}$$

where

E_c = core compression modulus of elasticity
G_c = core shear modulus in load direction
E_f = facing compression modulus of elasticity
t_f = facing thickness
b = width of laminate.

For sandwich panels with thin cores ($t_c \leqslant 2t$), the foundation stiffness is taken as

$$\frac{B}{b} = \frac{2E_c}{t_c} \tag{20-20}$$

where t_c = core thickness.

20.4.3.2. Separation from Core

This mode of failure results from an initial waviness, δ_0, in the compression face of the honeycomb sandwich panel. The facing stress at which this failure mode occurs is given as Eq. 20-21.

$$F^{cu} = \frac{F_{cr}}{1 + \dfrac{B\delta_0}{F_z b}} \tag{20-21}$$

where F_z is the core to facing bond tensile strength, or the core tensile strength.

20.4.3.3. Core Crushing

The facing stress that will cause this type of failure in honeycomb sandwich columns and panels is obtained by using Eq. 20-21 and substituting the core compressive strength for F_z.

For honeycomb sandwich beams, in addition to the above stress, compressive stresses are induced in the core due to beam curvature. The effect of these should be considered in addition to the stresses included by face wrinkling. The total stress applied to the core is equated to the compressive strength of the core, yielding the following expression for the facing stress at failure.

$$F_z = F_c^2 \frac{K_0}{E_c} + K_1 E_c \left(\frac{F_c / F_{cr}}{1 - \dfrac{F_c}{F_{cr}}}\right) \tag{20-22}$$

where

$$K_0 = \frac{t_f E_c}{d E_{1c}} \left(1 + \frac{E_{1c} t_f}{E_{1t} t_t}\right)$$

$K_1 = \dfrac{\delta_0}{t}$, t is obtained from the second equation of Eq. 20-19
E_{1c} = Young's modulus of compression face
E_{1t} = Young's modulus of tension face
t_f = thickness of compression face
t_t = thickness of tension face
d = distance between facing centroids
E_c = core compressive modulus of elasticity
F_c = stress on compression face
t_c = core thickness
δ_0 = initial waviness
F_z = core compressive strength.

20.4.3.4. Core Shear Failure

Shear failure of the foundation may result from initial waviness. The facing stress for this failure mode is given by Eq. 20-23.

$$F = \frac{F_{cr}}{1 + \dfrac{1.9065\delta_0 G_c}{F_{sc} t_f}\left(\dfrac{E_c G_c}{E_f^2}\right)^{1/6}} \tag{20-23}$$

where F_{sc} = core shear strength. If the facing core bond shear strength is less than F_{sc}, the lower value should be used.

20.4.4. Face Dimpling

This mode of failure, intercell buckling, is found in honeycomb sandwich constructions where instability occurs in unsupported areas between the cell walls. The facing stress at which dimpling of the sandwich facing occurs is given by the empirical formula given as Eq. 20-24.

$$F^{cu} = 2\frac{E'}{\psi}\left(\frac{t_f}{s}\right)^2 \tag{20-24}$$

where

$E' = \sqrt{E_1 E_2}$
t_f = facing thickness
E_1, E_2 = moduli of elasticity of composite facing along axes of orthotropy

ψ = 1 minus product of major and minor Poisson's ratios

s = core cell size (diameter of inscribed circle).

20.4.5. Overall Stability Analysis

Plain and honeycomb sandwich panels may buckle as columns. The effect of normal or interlaminar shear stiffness on the Euler buckling load, P_{eu}, has been derived by Suarez *et al.*[6] and is given by Eq. 20-25.

$$P_{cr} = \frac{P_{eu}}{1 + \dfrac{P_{eu}}{W_{xz}b}} \tag{20-25}$$

where

W_{xz} = the normal shear stiffness.

An exact expression for the normal shear stiffness, W_{xz}, of honeycomb sandwich panels with laminated facings is derived in Appendix II of the Suarez *et al.* paper.[6] However, for preliminary design purposes, the following equations can be used.

$W_{xz} = 5/6\, G_z t$ for symmetrically laminated plain panels (20-26)

$W_{xz} = G_c t_c$ for honeycomb sandwich panels.

20.4.6. Composite Plate Buckling Analysis

Compression and shear buckling data for composite plates are obtained using Eqs. 20-27–20-35, Tables 20.6 and 20.7. (These equations are derived in *Theory of Elastic Stability*.[8]) The equations apply to plates which have the following constraints.

- The plates are rectangular, orthotropic, homogeneous, and of constant thickness.
- Ends and sides of the plates are either simply supported or fully fixed.
- The laminate is essentially elastic at buckling stress.
- Effects of normal shear stiffness, which can reduce the buckling stress significantly, are not included.

The normal shear stiffness of composite plates is very low in relation to the in-plane flexural and shear stiffnesses. This can have considerable effect on the compression buckling stress of the plate, since the buckling stress cannot exceed 2/3 the normal shear modulus of the plate. This phenomenon is especially significant for compression buckling of composite plates at high temperatures, since the normal shear stiffness is matrix dependent and the matrix stiffness is significantly reduced at high temperature.

To obtain the actual compression buckling stress, the effects of the normal shear stiffness, W_{xz}, of the plate must be considered. From Eq. 20-26:

$$W_{xz} = \tfrac{5}{6}\, G_z t.$$

Assuming that the effect of normal shear deformation on a plate is similar to that on a column, the critical buckling stress is expressed as Eq. 20-36.

$$F_{xcr} = \frac{C_c F'_{xcr}}{1 + C_c\, \dfrac{1.2\, F'_{xcr}}{G_z}} \tag{20-36}$$

where F'_{xcr} = buckling stress of plate with infinite normal shear stiffness. The reduction factor, C_c, may be evaluated as a function of F'_{xcr}/G_z and the lay-up of the plate using the exact compression buckling equation given in the Department of Defense paper, "Structural Sandwich Composites."[7] Shear buckling of a flat plate is also given by the expression from this paper:

$$F_{xycr} = F'_{xycr}\left[\frac{1}{1 + 4\left(\dfrac{K_{mo}}{K_m} - 1\right)}\right]. \tag{20-37}$$

F'_{xycr} is obtained using the appropriate expression from Table 20.6 or 20.7, and K_m is the buckling stress coefficient including normal shear stiffness effects from the Department of Defense paper.[7] K_{mo} is the buckling stress coefficient for a plate with infinite normal shear stiffness, obtained using the same expression.

When $K_{mo}/K_m = 1$, Eq. 20-28 gives the critical shear buckling stress of a thin plate,

Table 20.6. Equations for Composite Plate Buckling Stresses (Neglecting Normal Shear Stiffness Effects)

LOADING	EDGE CONDITIONS	a/b	BUCKLING STRESS	EQUATION NO.
COMPRESSION	Simply supported ends and sides	Gen	$F'_{xcr} = \frac{1}{12}\left(\frac{\pi t}{b}\right)^2 \left\{ A^{11}\left(\frac{mb}{a}\right)^2 + 2\,(A_{12} + 2A_{66}) + A_{22}\left(\frac{a}{mb}\right)^2 \right\}$	20-27
	Simply supported ends and sides	∞	$F'_{xcr} = \frac{1}{6}\left(\frac{\pi t}{b}\right)^2 \{(A_{11}A_{22})^{\frac{1}{2}} + A_{12} + 2A_{66}\}$	20-28
	Fixed ends and sides	Gen	$F'_{xcr} = \frac{1}{12}\left(\frac{\pi t}{b}\right)^2 \left\{ A_{11}\left(\frac{mb}{a}\right)^2 + 2.67A_{12} + 5.33\left(A_{22}\left(\frac{a}{mb}\right)^2 + A_{66}\right) \right\}$	20-29
	Fixed ends and sides	∞	$F'_{xcr} = \frac{1}{12}\left(\frac{\pi t}{b}\right)^2 \{4.6(A_{11}A_{22})^{\frac{1}{2}} + 2.67A_{12} + 5.33A_{66}\}$	20-30
SHEAR	Simply supported ends and sides	∞	$F'_{xycr} = \frac{1}{3}\left(\frac{t}{b}\right)^2 \{A_{11}A_{22}^3\}^{\frac{1}{4}} \left\{ 8.125 + \frac{5.05}{\theta} \right\};\ \theta = \frac{(A_{11}A_{22})^{\frac{1}{2}}}{(A_{12} + 2A_{66})} > 1$	20-31
			$F'_{xycr} = \frac{1}{3}\left(\frac{t}{b}\right)^2 \{A_{22}(A_{12} + 2A_{66})\}^{\frac{1}{2}}\{11.7 + 0.532\theta + 0.938\theta^2\} : \theta < 1$	
	Fixed ends and sides	∞	$F'_{xycr} = \frac{1}{3}\left(\frac{t}{b}\right)^2 \{A_{11}A_{22}^3\}^{\frac{1}{4}} \left\{15.1 + \frac{7.0}{\theta}\right\} : \theta > 1$	20-32
			$F'_{xycr} = \frac{1}{3}\left(\frac{t}{b}\right)^2 \{A_{22}(A_{12} + A_{66})\}^{\frac{1}{2}} \{18.6 + 1.65\,\theta + 1.90\,\theta^2\} : \theta < 1$	

Table 20.7. Equations for Composite Plate Buckling Stresses (Continued) (Neglecting Normal Shear Stiffness Effects)

LOADING	EDGE CONDITIONS	a/b	BUCKLING STRESS	EQUATION NO.
SHEAR	Simply supported ends and sides	1.0	$F'_{xycr} = \frac{k}{3}\left(\frac{t}{b}\right)^2 \{A_{11}A_{22}{}^3\}^{\frac{1}{4}} : \theta > 1$	20-33
			$k \simeq 8.20 + \frac{5}{\theta} + 10^{(A/\beta + B/\beta)}$	
			$A = 0.185/\theta - 0.270:$ $B = 0.82 + 0.46/\theta - 0.2/\theta^2 : \beta = \left(\frac{A_{11}}{A_{22}}\right)^{\frac{1}{4}}$	
	Fixed ends and sides	1.0	$F'_{xycr} = \frac{j}{3}\left(\frac{t}{b}\right)^2 \{A_{11}A_{22}{}^3\}^{\frac{1}{4}} : \theta > 1$	20-34
			$j \simeq j_o + 10^{(A/\beta + B\beta)} : \beta = \left\{\frac{A_{11}}{A_{22}}\right\}^{\frac{1}{4}}$	
			$\left.\begin{array}{l} j_o = 15 + 7.5/\theta \\ A = -0.100 \\ B = 1.30 + 0.2/\theta \end{array}\right\} \beta \leqslant .6$ $\left.\begin{array}{l} j_o = 19.3 + 8.45/\theta \\ A = 0.040 + 0.035/\theta \\ B = 1.99 + 0.38/\theta \end{array}\right\} \beta \geqslant .6$	
COMBINED COMPRESSION AND SHEAR	Simply supported or fixed ends		$\frac{f_x}{F_{xcr}} + \left\{\frac{f_{xy}}{F_{xycr}}\right\}^2 = 1$	20-35

neglecting normal shear deformation, F_{xycr} can be written as

$$F_{xycr} = F_{xy'cr}.$$

By writing the term that includes the normal shear stiffness of the plate as

$$V = \frac{\pi^2}{12}\sqrt{A_{11}A_{22}} \qquad (20\text{-}38)$$

$$\tau = (t/b)^2 \Big/ \frac{5}{6} G_z \qquad (20\text{-}39)$$

then, for long, flat panels with simply supported or clamped edges:

$$4\left(\frac{K_{mo}}{K_m} - 1\right) = \frac{\pi^2}{3} A_{22}\tau. \qquad (20\text{-}40)$$

A_{22} is the transverse stiffness of the plate. The critical shear buckling stress, including normal shear deformation for long, flat plates, is given as

$$F_{xycr} = F_{xy'cr}/(1 + C_1\tau). \qquad (20\text{-}41)$$

$C_1 = \frac{\pi^2}{3} A_{22}$ for a simply-supported or clamped plate.

An equation for the shear buckling stress coefficient of an orthotropic plate with an aspect ratio a/b is given by Plantema[9] [p. 143, 5.5(3)]. This equation is similar to an equation for isotropic plates in Timoshenko and Gere,[8] p. 383. The equation, though approximate, may be used for preliminary design, since it yields a slightly conservative estimate of buckling stress. The equation is:

$$F_{xycr} = F_{xycr}^{\infty} + (b/a)^2(F_{xy'cr} - F_{xycr}^{\infty}) \quad (20\text{-}42)$$

where

F_{xycr} = shear buckling stress of plate with aspect ratio a/b

F_{xycr}^{∞} = shear buckling stress for an infinitely long plate

$F_{xy'r}$ = buckling stress coefficient for a square plate.

For preliminary design, an equation similar

to Eq. 20-42 can be used to obtain the compression buckling stress for a plate with aspect ratio a/b. This expression is given as Eq. 20-43.

$$F_{xcr} = F_{xcr}^{\infty} + (b/a)^2(F_{xc'r} - F_{xcr}) \quad (20\text{-}43)$$

where

F_{xcr} = compression buckling stress of plate with aspect ratio a/b

F_{xcr}^{∞} = compression buckling stress for an infinitely long plate

$F_{xc'r}$ = buckling stress coefficient for a square plate.

20.5. PRELIMINARY ANALYSIS AND DESIGN OF COMPOSITE PANELS

Preliminary analysis of long, simply supported $(0°/90°/\pm45°)$ composite panels can be accomplished using the curves shown in Fig. 20.18. These curves were derived by assuming that the panels are homogeneously orthotropic, and using Eq. 20-28 for axial compression and Eq. 20-31 for in-plane shear, in conjunction with the values of A_{ij} from Eq. 20.7. The curves are drawn using the values of α $(= E_{22}/\bar{E}_{11})$ of 0.1 and K $(= G_{12}/E_{22})$ of 0.375—values which are reasonably typical of

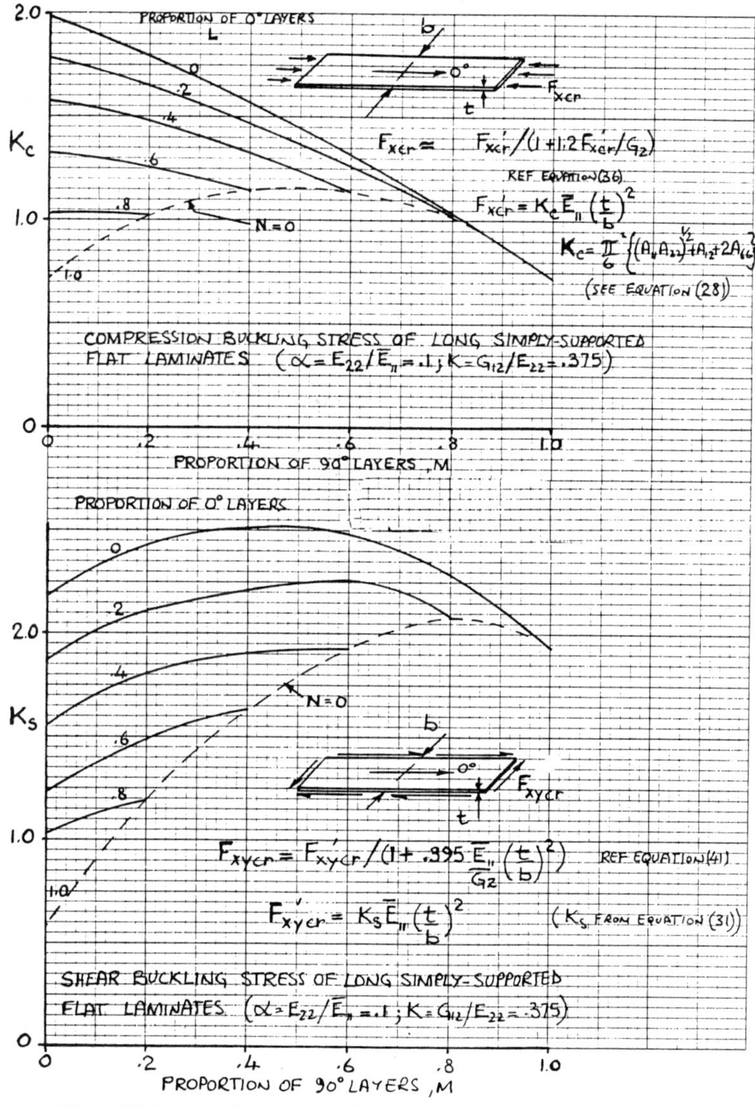

Figure 20.18. Buckling stresses of long, simply supported flat laminates.

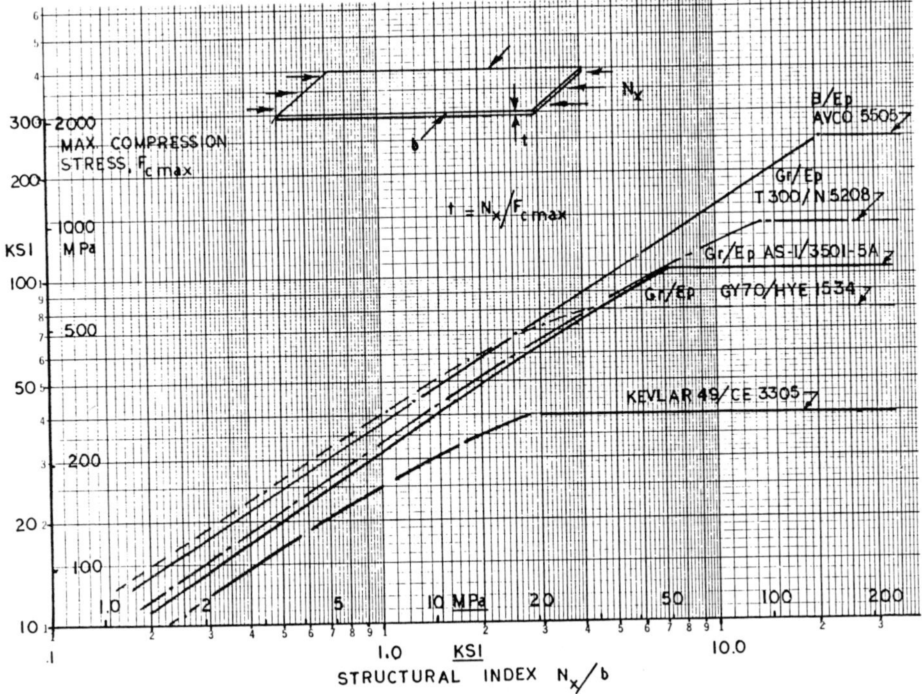

Figure 20.19. Optimum compression stress for long, simply supported flat laminates.

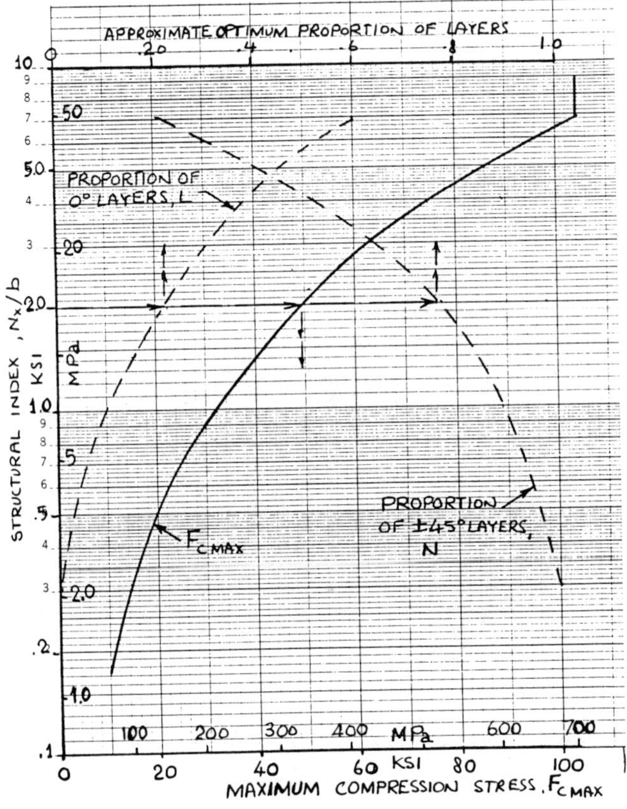

Figure 20.20. Optimum lay-up for type AS-1/3501-5A graphite/epoxy compression panels.

B/Ep and Gr/Ep properties at room temperature (refer to Fig. 20.10).

Preliminary design of compression panels can be accomplished using the curves in Figs. 20.19 and 20.20. Figure 20.19 shows the variation of optimum compression stresses with structural index. These compression stress curves are plotted for laminates where the proportion of 90° layers, M, is zero. However, the optimum stress is not significantly different provided M is less than 0.20. Figure 20.20 shows the lay-up associated with the maximum stresses for Type AS-1/3501-5A Gr/Ep.

The curves in Figs. 20.19 and 20.20 were derived using the relationships:

$$F_{xcr} \simeq K_c \bar{E}_{11} \left(\frac{t}{b}\right)^2$$

and

$$F_x = \frac{N_x}{t}$$

(neglecting the effects of interlaminar shear stiffness) where K_c is the compression buckling stress coefficient, N_x is the axial compression loading, b is the panel width, and t is the panel thickness. For designs where buckling constitutes failure, the maximum (or optimum) stress. F_{cMAX}, is obtained by equating F_{xcr} to F_x. Then:

$$F^3_{cMAX} = F_{xcr} F_x^2 = \left(\frac{N_x}{b}\right)^2 K_c \bar{E}_{11}$$

or

$$F_{cMAX} = (K_c \bar{E}_{11})^{1/3} \left(\frac{N_x}{b}\right)^{2/3}. \qquad (20\text{-}44)$$

The values of K_c from Fig. 20.18 are used in conjunction with the appropriate values of F_x from Eq. 20-14, and the material properties from Table 20.5 to generate the curves shown in Figs. 20.19 and 20.20.

The same approach was used to derive the curve shown in Fig. 20.21 for the optimum shear stress of long, flat panels. For this load-

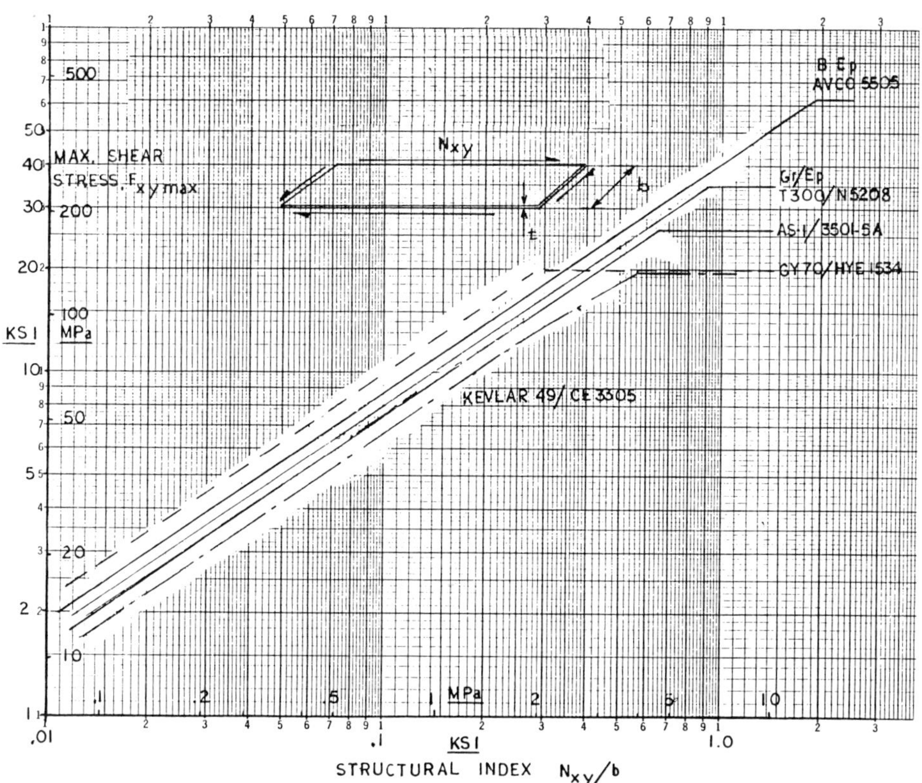

Figure 20.21. Optimum shear stress for long, simply supported flat laminates.

ing condition, the equation becomes

$$F_{xyMAX} = (K_s\bar{E}_{11})^{1/3}\left(\frac{N_{xy}}{b}\right)^{2/3}. \qquad (20\text{-}45)$$

where N_{xy} is the applied shear loading. A shear stress cut-off of 0.25 G_z has been used, since this represents a practical maximum stress level for shear panels. The optimum lay-up for long, flat panels is 50–90%; 50% ±45°, since this results in the maximum buckling stress coefficient (see Fig. 20.18).

REFERENCES

1. Bannink, E., Hadcock, R., and Forsch, H., "Advanced Design Composite Aircraft Study," Paper No. 77-393, presented at the 18th AIAA/ASME Structures, Structural Dynamics and Materials Conference, San Diego, California, March 1977.
2. Hadcock, R., "The Application of Advanced Composites to Military Aircraft," ICAS Paper No. 76-09, presented at the 10th Congress of the International Council of the Aeronautical Sciences, Ottawa, Canada, October 1976.
3. Schwartz, P., "Advanced Design Composite Aircraft, the Next Step for Composites," SAWE Paper No. 1105, Index Category No. 27., presented at the 35th Annual Conference of the Society of Allied Weight Engineers, Philadelphia, Pennsylvania, May 1976.
4. Ludwig, W., Erbacher, H., and Lubin, G., "Composite Horizontal Stabilizer for the B-1: Design, Fabrication and Test," 32nd Annual Technical Conference, Reinforced Plastics/Composites Institute, SPI, Section 15B, Washington, D.C., February 1977.
5. Tsai, S.W. and Pagano, N.J., "Invariant Properties of Composite Materials," Composite Materials Workshop, Technomic, January 1968.
6. Suarez, J.A., Whiteside, J.B., and Hadcock, R.N., "The Influence of Local Failure Modes on the Compressive Strength of Boron/Epoxy Composites," ASTM Special Technical Publication 497, 1972.
7. Department of Defense, "Structural Sandwich Composites", MIL-HDBK-23A, Washington, D.C., June 1968.
8. Timoshenko, S.P. and Gere, J.M., *Theory of Elastic Stability*, Second Edition, McGraw-Hill, New York, 1961.
9. Plantema, F.J., *Sandwich Construction*, John Wiley Sons, New York, 1966.

21
SANDWICH CONSTRUCTION

A. Marshall
Consultant
Dublin, California

21.1. INTRODUCTION

This chapter covers a unique form of composites known as "structural sandwich construction."

A structural sandwich consists of three elements, as shown in Fig. 21.1: 1) a pair of thin, strong facings; 2) a thick, lightweight core to separate the facings and carry loads from one facing to the other; and 3) an attachment which is capable of transmitting shear and axial loads to and from the core.

The following pages will provide a general background and a brief summary of the various materials in common use; the design steps used to calculate loads; some design details for solving load point, edging, and attachment problems; and a number of tables, charts, and graphs containing useful information for the designer. An attempt is also made throughout the chapter to provide suggestions and perspectives to help a new users of sandwich structures technology to avoid some of the errors of his predecessors.

Structural sandwich construction is one of the first forms of composite structures to have attained broad acceptance and usage. Virtually all commercial airliners and helicopters, and nearly all military air and space vehicles, make extensive usage of sandwich construction. The effectiveness of sandwich construction is shown in Fig. 21.2.

In addition to air and space vehicles, the system is commonly used in the manufacture of cargo containers, movable shelters and airfield surfacing, navy ship interiors, small boats and yachts, die models and production parts in the automobile and recreational vehicle industry, snow skis, display cases, residential construction materials, interior partitions, doors, cabinets, and a great many other everyday items.

Although the employment of a sandwich design to produce lightweight or special purpose load-carrying members is thought to have originated as early as 1820, routine commercial use of the idea did not occur until about 110 years later. What started this sudden acceptance was the successful commercial production of structural adhesives, starting in both England and the United States in the 1930's. This early production resulted from the development of the rubber-phenolics and the vinyl-phenolics. Such materials as "cycleweld," "plycosite," and "Redux" adhered well to both wood and metals, possessed rather high and predictable strength, and began a revolution in bonding technology. Many further developments followed in only a few years. They included improved cleaning methods for metal skins; low weight, high strength/stiffness honeycomb core materials; "B" staged tape adhesives; glass fabrics and collimated tapes preimpregnated with accu-

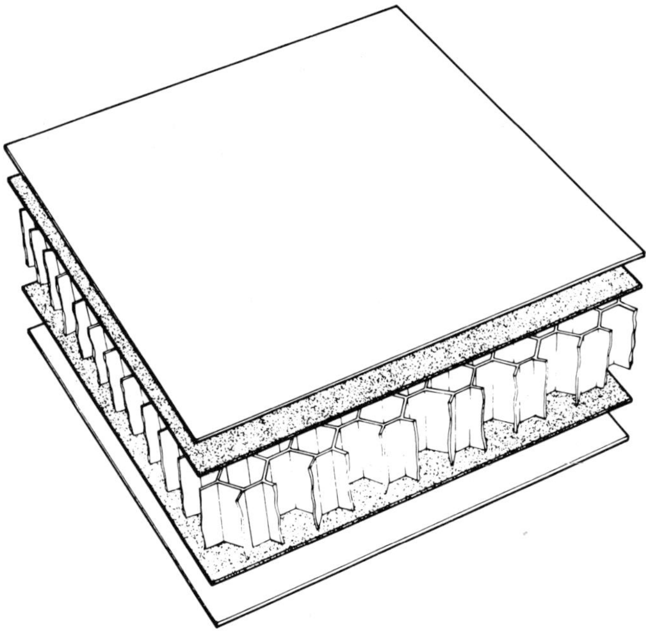

Figure 21.1. The elements of a sandwich structure are as follows: 1) two rigid, thin, high strength facings; 2) one thick, low density core; and 3) an adhesive attachment which forces the core and facings to act as a continous structure. The facings of a sandwich panel act similarly to the flanges of an I-beam by taking the bending loads—one facing in compression and the other in tension. The honeycomb core, like the web of the I-beam, resists the shear loads, and increases the bending stiffness of the structure by spreading the facings apart, but unlike the I-beam's web, it gives continuous support to the flanges or facings.

rately measured amounts of "B" staged resins; high strength resins; tough, high peel adhesives; and lower cure temperature and pressures; as well as the discovery of the resistance of sandwich to sonic fatigue.

21.2. FACING MATERIAL

The primary function of the face sheets is to provide the required bending and in-plane shear stiffness and to carry the axial, bending, and in-plane shear loading. In the aerospace field, facings most commonly chosen are resin impregnated fiberglass cloth (usually, "prepreg"), graphite prepreg (either unidirectional tape or woven fabric), 2024 or 7075 aluminum alloy, titanium, or stainless steel. Even the most economical of these products represents a substantial cost, and customary practice is to choose among them very carefully on a

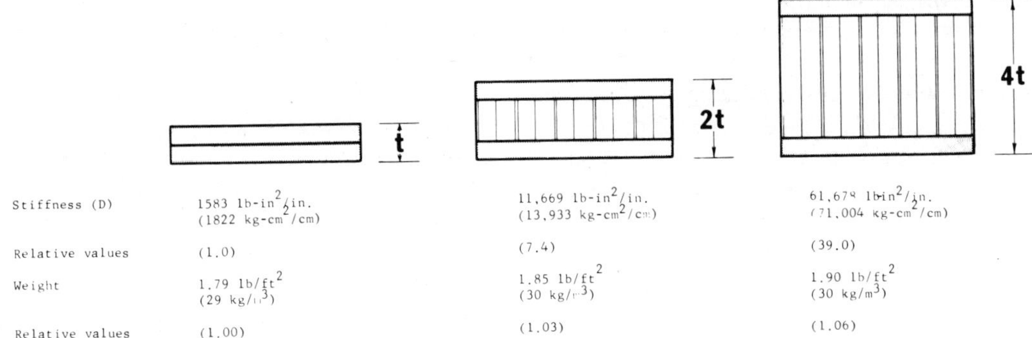

Stiffness (D)	1583 lb-in^2/in. (1822 kg-cm^2/cm)	11,669 lb-in^2/in. (13,933 kg-cm^2/cm)	61,678 lb-in^2/in. (71,004 kg-cm^2/cm)
Relative values	(1.0)	(7.4)	(39.0)
Weight	1.79 lb/ft^2 (29 kg/m^3)	1.85 lb/ft^2 (30 kg/m^3)	1.90 lb/ft^2 (30 kg/m^3)
Relative values	(1.00)	(1.03)	(1.06)

Figure 21.2. A striking example of how conversion to sandwich stiffens a structure without materially increasing its weight. This example uses 0.063-in. (2-mm) thick aluminum facings and 1/4-5052-2.3, lb/ft^3 (37 kg/m^3)

value engineering, or lowest lifetime cost, basis.

21.2.1. Suitability of Materials

When choosing facing materials (as well as the core, adhesive, or other materials) for an application, it is wise to examine the less obvious properties of the material, such as toughness or brittleness, mode of fracture, durability and weatherability, compatibility with rivets and bolts, and other such attributes which may directly affect the usability or success of the end product, even though not directly involved in stress analysis or weight savings. An understanding of these requirements has resulted in a switch from aluminum to fiberglass skins and from fiberglass to aramid cores for most aircraft cabin interior panels. This understanding is also the reason for the use of hybrid skin composed of graphite fabric with a surface layer of aramid fabric on much of the sandwich structure of several Boeing airliners scheduled for initial service in the 1980's. Table 21.1 lists the mechanical properties of several commonly used facing materials.

21.3. CORE MATERIALS

The primary function of a core in sandwich structures is that of stabilizing the facings and carrying most of the shear loads through the thickness. In order to perform this job efficiently, the core must be as rigid and as light as possible, and must deliver uniformly predictable properties in the environment (i.e., high humidity) in which the finished part is to perform.

21.3.1. Types of Core Materials

21.3.1.1. Wood

Several different materials are used extensively as sandwich cores. The oldest of these is wood, which is used in many applications as a core for doors, partitions, and many other "builder's supply" items. It is also used in the majority of snow skis, either flat-grain or end-grain, although the trend is toward honeycomb, foam, or reinforced plastic cores. End-grain balsa has broad acceptance in boat hulls up to lengths of 50 ft (15.2 m) or more and is still used for replacement flooring for many older and a few new aircraft. The traditional advantage of the low cost of wood has been progressively eroded in the 1970's, and many users report difficulty in supply, even at prices higher than foam, and sometimes approaching that of honeycomb. Even so, the ease of use and excellent durability of the end product has led to sharply increased usage, particularly of the carefully selected grades of end-grain balsa, in applications such as boat hulls, large tanks, and airborne pallets and containers. This broadening usage is also prompted by its excellent compressive strength and modulus properties when compared to all but the aramid paper honeycombs, which are much more expensive. The mechanical properties of three commonly used densities of balsa wood are listed in Table 21.2.

21.3.1.2. Foam

The use of foam as a structural core has been, and is now, extensive. Recent developments in the technology of foam injection have sharply increased its use. The most novel of these is use of a cold-cavity die, in which the foam is injection molded in a single production step. A careful adjustment of the curing reaction of the foam and the heat-sink effect of the mold results in a part with facings which are simply a higher density form of the foam which constitutes the core. The high productivity and modest cost of this scheme have resulted in many applications in the automotive and industrial fields. Another fast-growing form of the material is in cores for fiberglass snow skis and tennis rackets, in which an assembly of facings and close-out details is placed in a closed cavity and foam injected to form both the core and the adhesive attachment to the pre-cured glass fiber skins. The resultant saving in labor has resulted in rapid acceptance of the process and the recent construction of many new factories in both Europe and the United States. Foams also provide special properties such as insulation or radar transparency.

Table 21.1a. Mechanical Properties of Typical Sandwich Facing Materials

FACING MATERIAL	YIELD STRENGTH,[a] psi	MODULUS OF ELASTICITY, psi	λ_δ $(1 - \mu^2)$	WEIGHT OF 1-MIL THICKNESS 1 ft²	COMMENTS
Aluminum:					
2024-T3	42,000	10×10^6	0.89	0.014	Good strength, moderate cost
5052-H34	24,000	10×10^6	0.89	0.014	Moderate cost, corrosion resistant
6061-T6	35,000	10×10^6	0.89	0.014	Workable
7075-T6	66,000	10×10^6	0.89	0.014	High strength
Mild carbon steel	50,000	30×10^6	0.91	0.040	Low cost, high weight
Stainless steel:					
316	60,000	30×10^6	0.94	0.040	Corrosion resistant
17-7	200,000	30×10^6	0.94	0.040	High strength
Titanium:					
Annealed Ti-75A	80,000	15×10^6	0.94	0.0235	Corrosion resistant
Heat treated 6A1-4V	143,000	16.8×10^6	0.94	0.0230	High cost
Graphite woven	80,000	10×10^6	0.99	0.008	High cost, high strength
Graphite unidirectional	220,000	20×10^6	0.99	0.008	High modulus
Fiberglass mat/PE	14,000	0.92×10^6	0.98	0.007	
Woven roving/PE	38,000	1.85×10^6	0.98	0.007	
Fiberglass Prepreg:[b]					
Epoxy	63,000	3.5×10^6	0.98	0.0095	Low temperature cure
Epoxy	62,000	3.5×10^6	0.98	0.0088	Heat resistant
Phenolic	48,000	3.5×10^6	0.98	0.0094	High
Polyester	48,000	3.5×10^6	0.98	0.010	Low
Polyimide	60,000	3.5×10^6	0.98	0.0095	High temperature resistant
Kevlar[c] epoxy prepreg	60,000/28,000	4.4×10^6	0.99	0.007	Very tough
Plywood, 36 lb/ft³					
Douglas fir, exterior grade	2650	1.80×10^6	0.99	0.003	
Southern Pine	2650	1.80×10^6	0.99	0.003	
Lauan	2250	1.80×10^6	0.99	0.002	
Tempered hardboard, 70 lb/ft³	3600	0.65×10^6	0.99	0.005	
Gypsum board, 50 lb/ft³	120	0.30×10^6	0.98	0.004	

NOTE: When working with plywood facing, "effective" thickness should be used except for locating centroid.

Actual Thickness	"Effective Thickness"	Weight
0.250 in.	0.14 in.	0.8 psi
0.375 in.	0.14 in.	1.1 psi
0.500 in.	0.26 in.	1.5 psi

[a]Typical yield strength is either tensile or compressive, whichever is the lower value.
[b]Values shown are with 181 fabric style reinforcement autoclave cured.
[c]Yield is both tensile and compressive. Kevlar is a registered trademark of DuPont.

The very low cost polystyrene foams are used primarily in non-sandwich applications, their role in structural parts for refrigerated vehicles and buildings having been largely taken over by the urethanes. The polyvinyl chloride (PVC) foams, which made an impact on the aircraft industry as flooring cores, have also been largely replaced by the more effi- ient high density aramid honeycombs.

The foam-in-place system of producing sandwich structures has been used for more than 25 years because of its simple concept.

Table 21.1b. Mechanical Properties of Typical Sandwich Facing Materials (Metric)

FACING MATERIAL	YIELD STRENGTH, MPa	MODULUS OF ELASTICITY, MPa	λ_δ $(1-\mu^2)$	WEIGHT OF 1-mm THICKNESS 1 m²
Aluminum:				
2024-T3	800	160×10^3	0.89	2.69
5052-H34	416	160×10^3	0.89	2.69
6061-T6	560	160×10^3	0.89	2.69
7075-T6	1169	160×10^3	0.89	2.69
Mild carbon steel	800	480×10^3	0.91	7.68
Stainless steel:				
316	961	480×10^3	0.94	7.68
17-7	3200	480×10^3	0.94	7.68
Titanium:				
Annealed Tl-75A	1280	240×10^3	0.94	4.52
Heat treated 6A1-4V	2290	269×10^3	0.94	4.42
Graphite woven	1280	160×10^3	0.99	1.54
Graphite unidirectional	3500	320×10^3	0.99	1.54
Fiberglass mat/PE	224	14.7×10^3	0.98	1.35
Woven roving/PE	608	29.6×10^3	0.98	1.35
Fiberglass Prepreg:				
Epoxy	1009	56×10^3	0.98	1.83
Epoxy	990	56×10^3	0.98	1.69
Phenolic	769	56×10^3	0.98	1.81
Polyester	769	56×10^3	0.98	1.92
Polyimide	961	56×10^3	0.98	1.83
Kevlar epoxy prepreg	961/448	70×10^3	0.99	1.35
Plywood				
Douglas fir, exterior grade	42	29×10^3	0.99	0.58
Southern Pine	42	29×10^3	0.99	0.58
Lauan	36	29×10^3	0.99	0.38
Tempered hardboard	57	10×10^3	0.99	1.12
Gypsum board	1.9	4.8×10^3	0.98	.80

NOTE: When working with plywood facing, "Effective" thickness should be used except for locating centroid.

Actual Thickness	"Effective Thickness"	Weight Kg/M²
6 mm	4 mm	3.90
10 mm	4 mm	5.37
13 mm	7 mm	2.32

However, users of this system have always been plagued by a continuing problem in producing uniform properties from one mix to the next and in achieving uniformly high core and bond strengths to the metal or pre-cured glass fiber skins. The use of systematic incoming inspection, automatic mixing and dispensing equipment, and, in the case of critical airframe parts, test coupons, produced integrally with the basic part, have all helped to keep the problems under control. It will be noted that Table 21.3 does not list the shear strength of many of the various foams, even though this value is needed for sandwich panel design. This property, even where listed, must be determined for the actual materials and

Table 21.2. Mechanical Properties of Balsa Wood When Used as a Sandwich Core

			6 lb/ft^3 (96 kg/m^3)		11 lb/ft^3 (176 kg/m^3)		15.5 lb/ft^3 (248 kg/m^3)	
	DENSITY		psi	MPa	psi	MPa	psi	MPa
SHEAR VALUES	End grain —strength	Typically high	500	3.45	1450	10.0	2310	15.9
		Typically Low	750	5.17	1910	13.2	2950	20.3
	End grain —modulus	Typical	330,000	2275	768,000	5295	1,169,000	8025
	Flat grain —strength	Typically high	84	0.58	144	0.993	198	1.36
		Typically low	50	0.34	100	0.689	145	1.0
	Flat grain —modulus	Typically high	16,000	110	37,000	255	55,000	379
		Typically low	5,100	35.1	13,000	89.6	19,900	137
TENSILE VALUES	End grain —strength	Typically high	1375	9.48	3050	21.0	4,525	31.2
	Flat grain —strength	Typically high	112	0.77	170	1.17	223	1.54
		Typically low	72	0.49	118	.814	156	1.07
COMPRESSIVE VALUES	Strength	Typically high	180	1.24	360	2.48	522	3.59
		Typically low	158	1.09	298	2.03	425	2.93
	Modulus	Typical	16,000	110	37,000	255	55,000	379

conditions of use in order to be considered reliable. When a value for shear strength is not available, it may be estimated to be about 0.7 times the compressive strength shown.

21.3.2. Honeycomb

21.3.2.1. General Honeycomb Facts

Honeycomb in common usage includes products made from uncoated and resin-impregnated kraft paper, various aluminum alloys, aramid paper, and glass-reinforced plastic in a number of cloth weaves and resin systems. Titanium, stainless steel, and many others are used in lesser quantities. Most honeycomb cores are constructed by adhesively bonding strips of thin material together, as shown in Fig. 21.3.

In the case of aramid paper honeycomb, the inherent toughness and abuse resistance of the material makes cores of 1–3 lb/ft^3 (16–48 kg/m^3) an excellent choice for cabin interior walls and ceilings, even with glass fabric-reinforced skins as low as 0.010-in. (0.254 mm) thickness.

Physical and mechanical properties of the honeycomb core materials are strongly influenced by the properties of the materials from which they are manufactured. Some of these differences are obvious in the thermal conductivity information shown in Fig. 21.4, and in the thickness correction factor shown in Fig. 21.5. However, several significant properties of honeycomb cores are peculiar to the honeycomb geometry rather than the basic materials, and should be separately noted. Some of these are listed below.

- *Density*. All mechanical properties increase with higher density, as shown in Fig. 21.6.

Table 21.3. Properties of Several Foam Materials Used as Cores*

TYPE	DENSITY		TENSILE STRENGTH (ASTM D1623)		COMPRESSIVE STRENGTH AT 10% DEFLECTION (ASTM D1621)		MAXIMUM SERVICE TEMPERATURE		THERMAL CONDUCTIVITY		SHEAR STRENGTH		SHEAR MODULUS	
	lb/ft³	kg/M³	psi	MPa	psi	MPa	°F	°C	BTU-in./hr-ft²-°F	W/m-°K	psi	MPa	psi	MPa
ABS (acrylonitrile-butadiene-styrene)														
Injection molding type pellets	40–56	641–897	2000–4000	13.8–27.6	2300–3700	15.8–25.5	176–180	80–82	0.58–2.1	0.08–0.30				
Cellulois acetate														
Boards and rods (rigid, closed cell foam)	6.0–8.0	96–128	170	1.2	125	0.86	350	177	0.31	0.04				
Epoxies														
Rigid closed cell,	5.0	80	51	0.35	90	0.62	350	177	0.26	0.04				
precast blocks,	10.0	160	180	1.2	260	1.8	350	177	0.28	0.04				
slabs, sheet	20.0	320	650	4.5	1080	7.4	350	177	0.32	0.05				
Phenolics														
Foam-in-place liquid resin	½–1½	5–24	3–17	0.021–0.12	2–15	0.014–0.10		0.21–0.28	0.03–0.04					
	2–5	32–80	20–54	0.138–0.372	22–85	0.15–0.58	Continuous service at 300		0.20–0.22	0.03–0.04				
	7–10	112–160	80–130	0.552–0.896	158–300	1.09–2.07		145	0.24–0.28	0.03–0.04				
Polycarbonate														
Pellets	50	801	5500	37.9	7500	51.7	270	132	1.05	0.15				
Polypropylene														
High density foam, molded parts and shapes with solid integral skin	35.0	561	1600	11.03	2100	14.4			4.2	0.61				
Polyurethane														
Rigid (closed cell)	1.3–3.0	21–48	15–96	0.10–0.65	15–60	0.10–0.41	180–250	82–121	0.11–0.21	0.2–0.4	20	0.14	226	1.56
molded parts;	4–8	64–128	90–290	0.62–1.99	70–275	0.48–1.90	200–250	93–131	0.15–0.29	0.02–0.04	90	0.62	1500	10.3
boards, blocks,	9–12	144–192	230–450	1.58–3.10	290–550	1.99–3.79	250–275	121–135	0.19–0.35	0.03–0.05	180	1.24	4500	31.0
slabs; pipe covering;	13–18	208–288	475–700	3.28–4.83	650–1100	4.48–7.58	250–300	121–149	0.26–0.40	0.04–0.06				
one-shot, two-	19–25	304–400	775–1300	5.34–8.96	1200–2000	8.27–13.8	250–300	121–149	0.34–0.52	0.05–0.07	450	3.1	15000	103.5
and three-package systems for foam-														

Table 21.3 Continued

TYPE	DENSITY lb/ft^3	kg/M^3	TENSILE STRENGTH (ASTM D1623) psi	MPa	COMPRESSIVE STRENGTH AT 10% DEFLECTION (ASTM D1621) psi	MPa	MAXIMUM SERVICE TEMPERATURE °F	°C	THERMAL CONDUCTIVITY BTU-in./hr-ft^2-°F	W/m-°K	SHEAR STRENGTH psi	MPa	SHEAR MODULUS psi	MPa
in-place; for spray, pour, or froth-pour techniques														
Skinned molded (rigid)														
Skin	25–65	400–1041	100–2700	0.68–18.6	40–3000	0.28–20.7	150–250	66–121	0.12–0.80	0.02–0.12				
Core	3–30	48–481	15–1500		15–1500		150–250	66–121	0.21–0.55		20–500		225–15000	
Polyvinyl chloride														
Rigid closed cell boards and billets	3	48	1000 and up	6.90 and up	95	0.65			2.0 at 70		65	.45	1200	8.3
	6	96			200	1.38					120	.83	2200	15.2

*Where shear strength and modulus properties are not shown, use a figure of 0.7 times the compressive strength shown as a first approximation for design feasibility consideration. Always test actual material used for true value of shear strength and modulus.

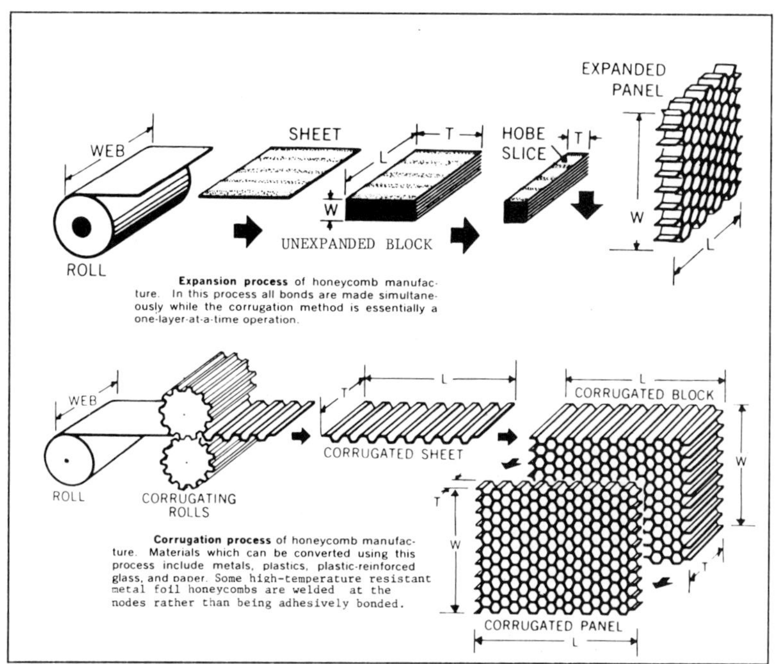

Figure 21.3. Most honeycomb is produced by the expansion process. Actual cell shape produced by either method may vary greatly.

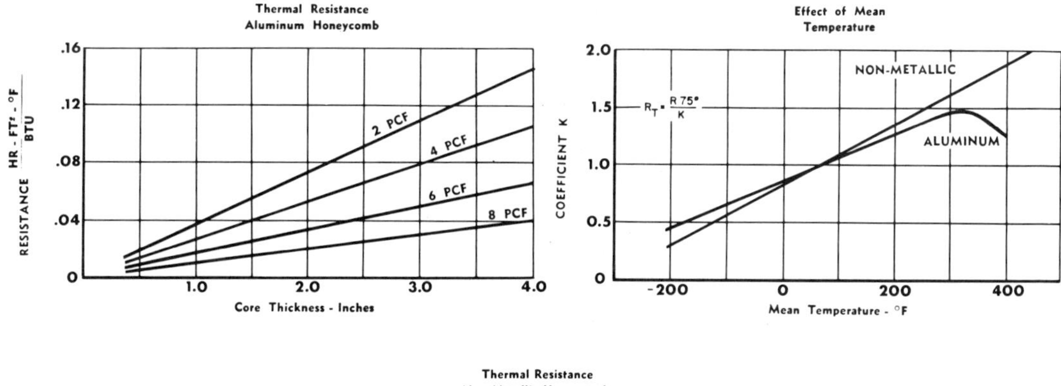

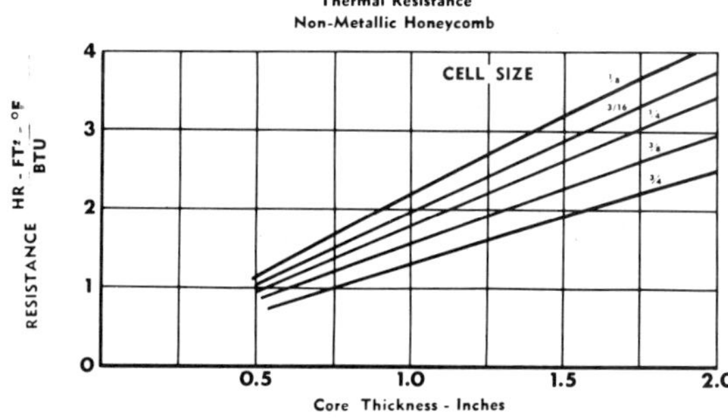

Figure 21.4. Thermal conductivity through sandwich panels can be isolated into the contribution of each component: facings, core, and adhesive. The resistances (or reciprocal of conductivity) can simply be added—including the effect of boundary layer conditions. The thermal properties of typical facing materials may be found in handbooks. Thermal resistance values for typical core to facing adhesives are 0.03 for film adhesives with a scrim cloth support, and 0.01 for unsupported adhesives. These graphs give the resistance for aluminum and non-metallic honeycomb at a mean temperature of 75° F (23.9° C). Note that for non-metallic honeycomb, it has been found that the cell size is more critical than core density. The reverse is true with aluminum honeycomb. To correct for mean temperature, divide the resistance at 75° F (23.9° C) by coefficient K as shown above.

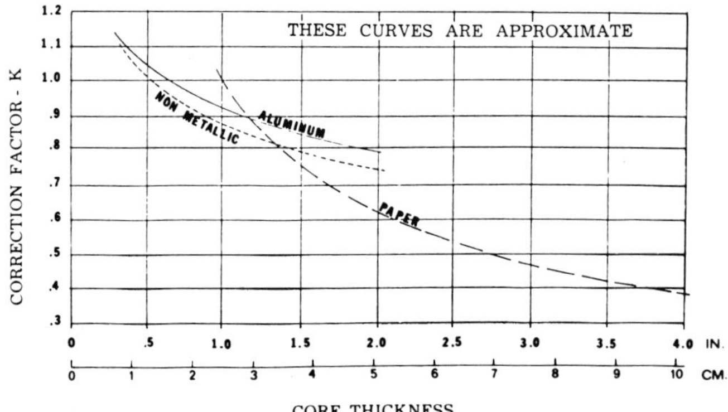

Figure 21.5. Measured core shear strength will vary, depending upon the test method, core thickness, skin thickness, and many other factors. The above curves may be used only for preliminary correction factors. Physical tests of the final design must be used to confirm actual values obtained (the curves are approximate).

- *Cell shape.* All honeycombs are anisotropic, and the resulting directional properties should be adapted to the loads anticipated. Figure 21.7 shows typical differences in shear strength for the *L* and *W* directions. In addition, some cell shapes allow easy forming or curving at a small loss in strength/weight ratio. This attribute can be of great importance in manufacturing curved parts of appreciable thickness. Cell shape variations may be either furnished to specification by the core manufacturer, or, in certain materials such as aluminum, shapes may be intentionally or inadvertently altered by the core user. It should be noted that the under- or over-expansion of the core changes its cell shape *and density*. The over-expanded version of Fig. 21.8*c* changes directional properties such that the *L* direction becomes slightly the weaker of the two major axes. Since the drop in the *L* direction strength can amount to 30%, such changes in cell shape must not be allowed to occur by error.
- *Cell size.* Although the cell size tends to be a secondary variable for most mechanical properties of core materials, it is primary in fixing the strength level of the core-to-face attachment (or, conversely, in fixing the lower limit on core-to-panel adhesive weight), and in determining stress levels at which intracell buckling or face dimpling of facings occurs.
- *Thickness.* The shear and compressive properties noted for a specific core type can only be realized when test methods are carefully specified and controlled and the correct thickness of core tested. Failure to allow for the effect of thickness can affect observed values by a factor of 4 or more, as noted in Fig. 21.5. It should be emphasized that the correction factor shown may be considerably different, depending on skin material and thickness.
- *Specimen geometry and test method.* Like thickness, these must be specified and carefully controlled in order to obtain comparability with values obtained by others. Shear strength values obtained using plate shear test methods of Fig. 21.9 are quite normally up to 25% below those obtained when using the flexure method shown in Fig. 21.10. Both methods are accepted and used, and any lack of understanding of the differences can lead to monumental, if nonsensical, problems. It will be noted that the tables of mechanical properties for various honeycombs, Tables 21.4–21.12, specify the shear test method used in producing the data shown.

21.3.2.2. Paper Honeycomb

Paper honeycomb is the first predecessor of all the types of honeycomb, having been produced for some 2000 years. Early forms were not used as structural cores, but were

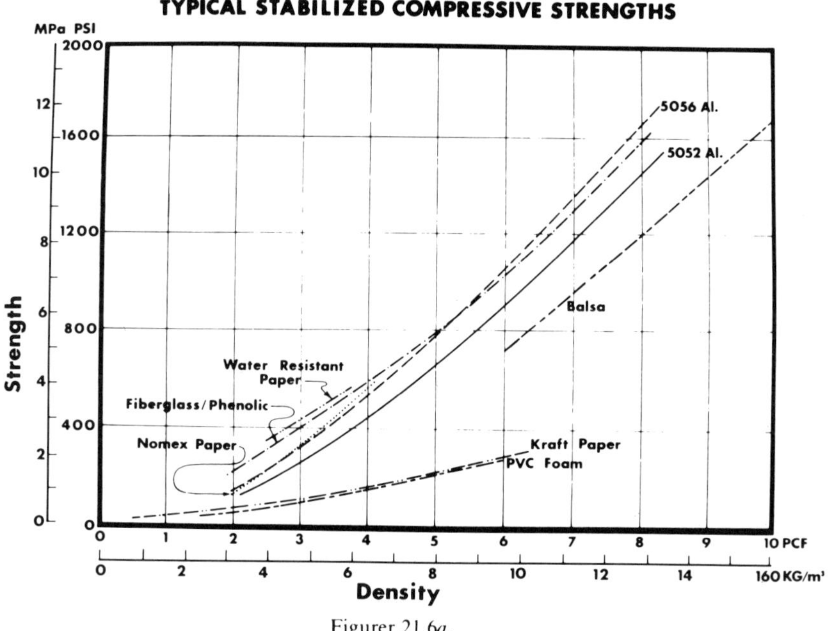

Figurer 21.6*a*.

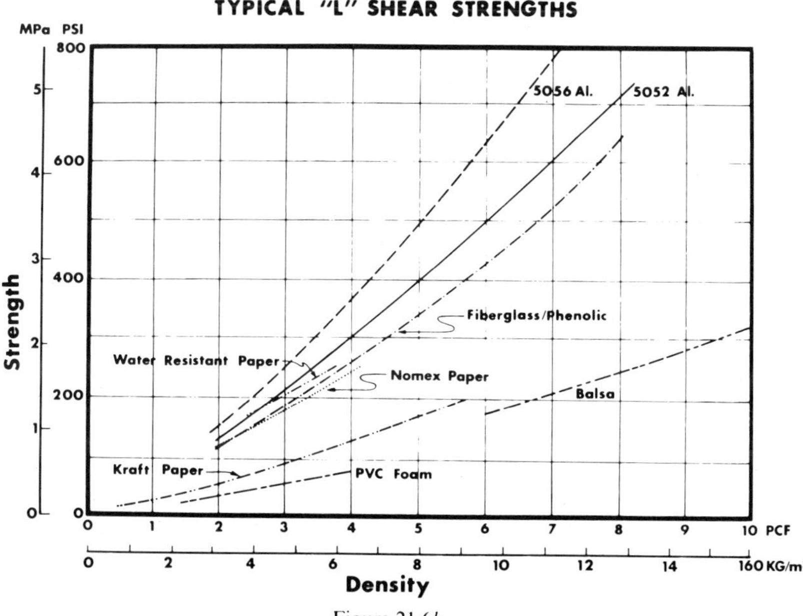

Figure 21.6*b*

employed as decoration—and are still frequently seen today as seasonal decorations in department stores in the form of expanded bells, spheres, and so forth. Current materials used as sandwich cores are different, in that much stronger kraft paper is employed, and 11–35% phenolic resin is frequently used to improve mechanical properties, moisture, and fungus resistance. Many variations are available in cell sizes of 3/8, 1/2, and 3/4 in. (10, 13, and 19 mm) or even larger sizes. The higher strength versions are only produced in the smaller cell size, with the 3/8-in. (10-mm) cell available as a water-migration resistant grade meeting military specification MIL-H-2104Q. Most applications are found in non-aircraft uses, where cost saving is the one primary objective. Usage is growing rapidly in

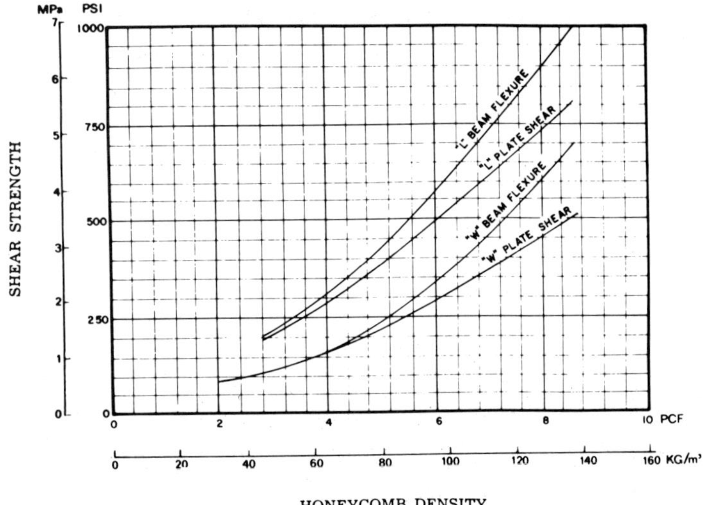

Figure 21.7. Plate shear versus flexures, typical values, 5052 aluminum honeycomb.

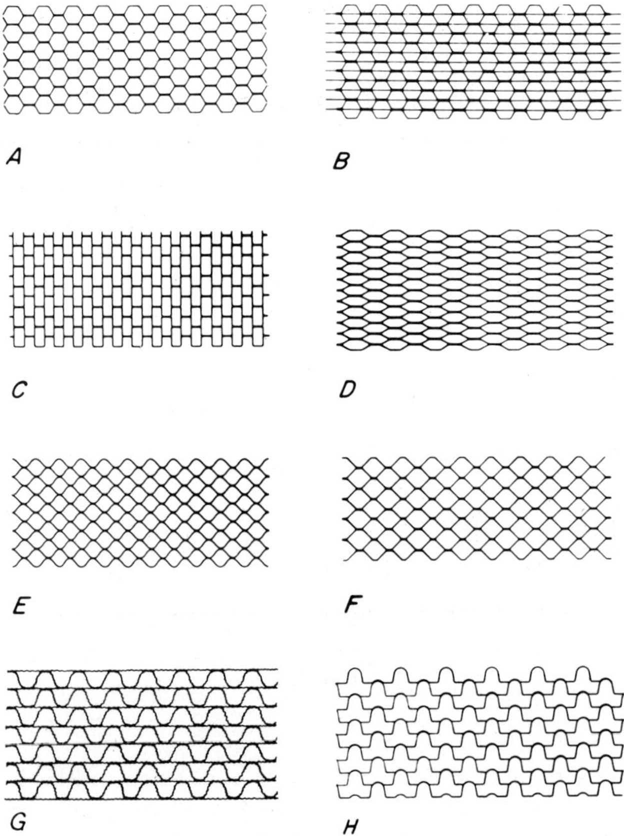

Figure 21.8. Various cell configurations in common usage. B and C are only produced by the corrugating method. F is a cell configuration nearly always used in the manufacture of welded metal honeycomb. C is flexible in one axis, while G and H are flexible in both axes. A, C, and D are expanded from identical unexpanded slices, A being normal expansion, C fully over-expanded, and D 50% expanded. B is reinforced corrugated core, with an extra layer of uncorrugated material placed between each layer of corrugated material. Reinforcing layers may be added in discrete locations or patterns and may be of the same or different web material or thickness.

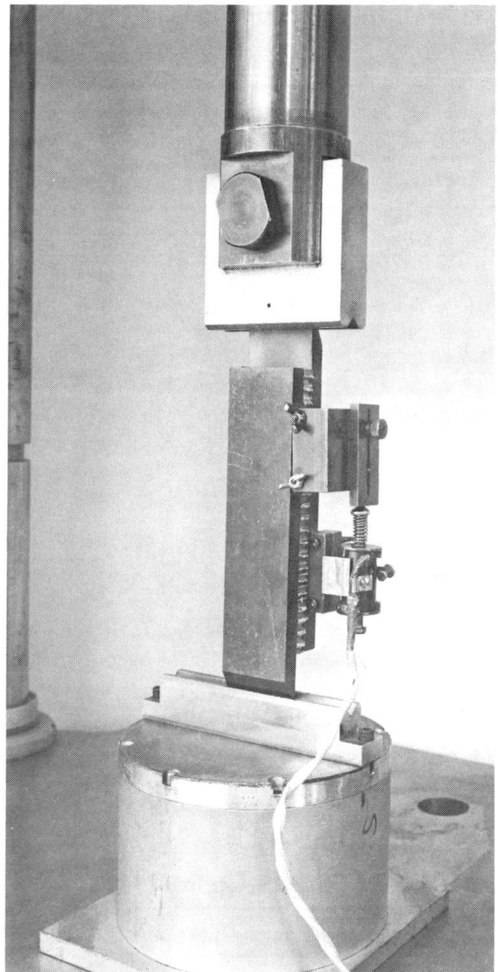

Figure 21.9. Plate shear test for honeycomb shear strength and modulus [0.50-in. (1.27-cm) thick steel plates are oven-cleaned and reused many times].

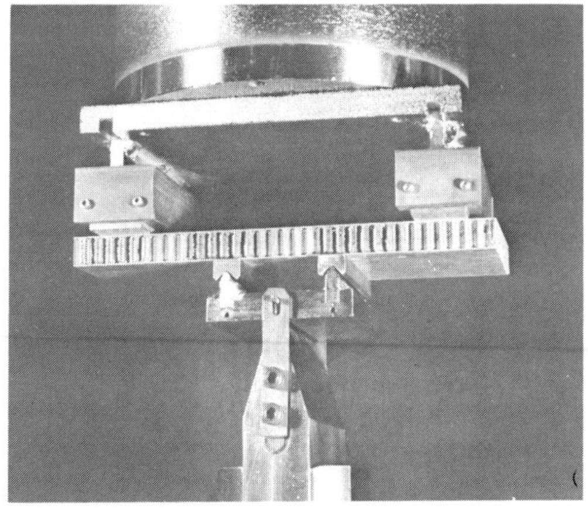

Figure 21.10. Short beam shear test for honeycomb core. Note the bearing area provided to preclude core crushing prior to core shear failure.

recreational vehicles; for doors, walls, and partitions; for factory produced kitchen cabinets; in packaged patio room additions for homes; in curtain wall panels; and in bearing walls for commercial building. Mechanical properties of several commercially available paper honeycomb cores are listed in Table 21.4 (*a* and *b*).

21.3.2.3. Aluminum Honeycomb

This family of materials has been in production and growing since about 1950. Aluminum honeycomb now includes four alloys, five cell shapes, and many foil gauges to provide a range of densities. The alloys available are as follows.

- 3003-H19, the lowest strength of the group, usually used for non-aircraft applications.
- 5052-H39, the most often used aircraft grade, available with a corrosion resistant surface treatment. Mechanical properties are listed in Table 21.5 (*a* and *b*).
- 5056-H39, the strongest of the regular aircraft grades, available with a corrosion resistant surface treatment. Mechanical properties are listed in Table 21.6 (*a* and *b*).
- 2024-T3 or -T81, the most heat-resistant alloy, and slightly stronger in some properties than 5056-H39. Available with a corrosion resistant surface treatment. Mechanical properties are listed in Table 21.6 (*a* and *b*).

Some of the above alloys are also available as corrugated, corrugated and reinforced, over-expanded, and flexible cell configurations. Some have also been produced in a specially tailored geometry to make all the axes lie on a true radius of a cylinder, a sphere, or other unique configurations. These same alloy foils can also be wound as a corrugated spiral to form a cylinder or tube for energy absorption applications.

The aluminum honeycomb cores remain the most used, as well as the most versatile of the various core materials obtainable and are very often found to possess the most favorable performance/cost ratio available. The range

Table 21.4*a*. Properties of Several Kraft Paper Honeycombs*

DESIGNATION	NOMINAL DENSITY, lb/ft³	COMPRESSIVE STABILIZED STRENGTH, psi	COMPRESSIVE STABILIZED MODULUS, ksi	PLATE SHEAR "L" DIRECTION STRENGTH, psi	PLATE SHEAR "L" DIRECTION MODULUS, ksi	PLATE SHEAR "W" DIRECTION STRENGTH, psi	PLATE SHEAR "W" DIRECTION MODULUS, ksi
		typical	typical	typical	typical	typical	typical
Resin impregnated (%)							
KP 1/4 - 80(11)	5.0	400	66.9	192	30.2	86	6.5
KP 3/8 - 60(25)	2.3	200	45.0	110	115.5	60	5.8
KP 1/2 - 60(25)	1.9	130	35.0	90	12.0	46	5.0
KP 1/2 - 80(11)	1.9	125	29.0	70	10.5	36	3.8
KP 1/2 - 80(18)	2.2	140	34.0	79	11.9	41	4.4
KP 3/4 - 80(11)	1.3	72	16.0	44	6.4	24	2.3
KP 3/4 - 80(18)	1.5	90	20.0	45	6.5	28	3.0
KP 3/4 - 99(11)	1.5	67	9.0	52	6.4	28	3.7
KP 3/4 - 99(18)	1.5	67	9.5	52	5.2	31	3.5
Unimpregnated							
KP 3/8 - 80(0)	2.1	72	18.5	46	7.4	31	3.7
KP 1/2 - 80(0)	1.9	64	16.5	42	6.9	28	3.4
KP 3/4 - 80(0)	1.2	39	9.5	26	4.8	18	2.3
KP 1 - 80(0)	1.0	29	7.0	20	3.8	13	1.8
KP 1½ - 80(0)	0.6	14	2.5	10	2.2	7	1.0

*Test data obtained at 1.00 in. thickness. Actual values may vary.

Table 21.4*b*. Properties of Several Kraft Paper Honeycombs*

HEXCEL HONEYCOMB DESIGNATION	NOMINAL DENSITY, kg/m³	COMPRESSIVE STABILIZED STRENGTH, KPa	COMPRESSIVE STABILIZED MODULUS, MPa	PLATE SHEAR "L" DIRECTION STRENGTH, KPa	PLATE SHEAR "L" DIRECTION MODULUS, MPa	PLATE SHEAR "W" DIRECTION STRENGTH, KPa	PLATE SHEAR "W" DIRECTION MODULUS, MPa
		typical	typical	typical	typical	typical	typical
Resin impregnated (%)							
KP 1/4 - 80(11)	80	2758	461	1324	208	593	45
KP 3/8 - 60(25)	37	1379	310	758	107	414	40
KP 1/2 - 60(25)	30	896	241	621	83	317	34
KP 1/2 - 80(11)	30	862	200	483	72	248	26
KP 1/2 - 80(18)	35	965	234	545	82	283	30
KP 3/4 - 80(11)	21	496	110	303	44	165	16
KP 3/4 - 80(18)	24	621	138	310	45	193	21
KP 3/4 - 99(11)	24	462	62	359	44	193	26
KP 3/4 - 99(18)	24	462	66	359	36	214	24
Unimpregnated							
KP 3/8 - 80(0)	34	496	128	317	51	214	26
KP 1/2 - 80(0)	30	441	114	290	48	193	23
KP 3/4 - 80(0)	19	269	66	179	33	124	16
KP 1 - 80(0)	16	200	48	138	26	90	12
KP 1½-80(0)	10	97	17	69	15	48	7

*Test data obtained at 2.54-cm thickness. Actual values may vary.

Table 21.5*a*. Properties of 5052 Alloy Hexagonal Aluminum Honeycomb*

HONEYCOMB DESIGNATION CELL - MATERIAL - GAUGE	NOMINAL DENSITY lb/ft	COMPRESSIVE					CRUSH STRENGTH,** psi	PLATE SHEAR					
		BARE		STABILIZED				"L" DIRECTION			"W" DIRECTION		
		STRENGTH, psi		STRENGTH, psi		MODULUS, ksi		STRENGTH, psi		MODULUS, ksi	STRENGTH, psi		MODULUS, ksi
		typical	minimum	typical	minimum	typical	typical	typical	minimum	typical	typical	minimum	typical
1/16 - 5052 - 0.0007	6.3	870p		910p		275p		510p		90p	320p		40p
1/16 - 5052 - 0.001	9.0	1480x		1500x		420x		775x		105x	520x		53x
1/8 - 5052 - 0.0007†	3.1	270	200	290	215	75	130	210	155	45.0	130	90	22.0
1/8 - 5052 - 0.001†	4.5	520	375	545	405	150	260	340	285	70.0	220	168	31.0
1/8 - 5052 - 0.0015†	6.1	870	650	910	680	240	450	505	455	98.0	320	272	41.0
1/8 - 5052 - 0.002†	8.1	1400	1000	1470	1100	350	750	725	670	135	455	400	54.0
1/8 - 5052 - 0.003	12.0	2200p		2325p		900p		1100p			625p		
5/32 - 5052 - 0.0007	2.6	200	150	215	160	55	90	165	120	37.0	100	70	19.0
5/32 - 5052 - 0.001	3.8	395	285	410	300	110	185	270	215	56.0	175	125	26.4
5/32 - 5052 - 0.0015	5.3	690	490	720	535	195	340	420	370	84.0	270	215	36.0
5/32 - 5052 - 0.002	6.9	1080	770	1130	800	285	575	590	540	114	375	328	46.4
5/32 - 5052 - 0.0025††	8.4	1530	1070	1600	1180	370	800	760	690	140	475	420	56.0
3/16 - 5052 - 0.0007	2.0	130	90	135	100	34	60	120	80	27.0	70	46	14.3
3/16 - 5052 - 0.001†	3.1	270	200	290	215	75	130	210	155	45.0	130	90	22.0
3/16 - 5052 - 0.0015†	4.4	500	360	525	385	145	250	330	280	68.0	215	160	30.0
3/16 - 5052 - 0.002†	5.7	770	560	810	600	220	390	460	410	90.0	300	244	38.5
3/16 - 5052 - 0.0025	6.9	1080	770	1130	800	285	575	590	540	114	375	328	46.4
3/16 - 5052 - 0.003	8.1	1400	1000	1470	1100	350	750	725	670	135	455	400	54.0
1/4 - 5052 - 0.0007	1.6	85	60	95	70	20	40	85	60	21.0	50	32	11.0
1/4 - 5052 - 0.001†	2.3	165	120	175	130	45	75	140	100	32.0	85	57	16.2
1/4 - 5052 - 0.0015	3.4	320	240	340	250	90	150	235	180	50.0	150	105	24.0
1/4 - 5052 - 0.002	4.3	480	350	505	370	140	230	320	265	66.0	210	155	29.8
1/4 - 5052 - 0.0025	5.2	670	500	690	510	190	335	410	360	82.0	265	200	35.4
1/4 - 5052 - 0.003	6.0	850	630	880	660	235	430	495	445	96.0	315	265	40.5
1/4 - 5052 - 0.004†	7.9	1360	970	1420	1050	340	725	700	650	130	440	390	52.8
3/8 - 5052 - 0.0007	1.0	30	20	45	20	10	25	45	32	12.0	30	20	7.0
3/8 - 5052 - 0.001	1.6	85	60	95	70	20	40	85	60	21.0	50	32	11.0
3/8 - 5052 - 0.0015	2.3	165	120	175	130	45	75	140	100	32.0	85	57	16.2
3/8 - 5052 - 0.002	3.0	260	190	270	200	70	120	200	145	43.0	125	85	21.2
3/8 - 5052 - 0.0025††	3.7	370	270	390	285	105	180	260	200	55.0	170	115	26.0
3/8 - 5052 - 0.003	4.2	460	335	485	355	135	220	310	255	65.0	200	150	29.0
3/8 - 5052 - 0.004	5.4	720	500	745	535	200	360	430	380	86.0	280	228	36.8
3/8 - 5052 - 0.005††	6.5	970	700	1020	750	265	505	545	500	105	350	300	43.5

*Corrugated 5052 and 5056 aluminum honeycomb is available in higher densities with crush strengths up to 6000 psi. Test data obtained at 0.625-in. thickness. p = preliminary properties; x = predicted values.

**Crush strength values shown are average or typical; actual values may vary because of density tolerances, etc.

†Core types most readily available.

††Minimum order may be required.

Table 21.5b. Properties of 5052 Alloy Hexagonal Aluminum Honeycomb (Metric)*

HEXCEL HONEYCOMB DESIGNATION CELL - MATERIAL - GAUGE	NOMINAL DENSITY, kg/m³	COMPRESSIVE					CRUSH STRENGTH,** KPa	PLATE SHEAR					
		BARE		STABILIZED				"L" DIRECTION			"W" DIRECTION		
		STRENGTH, KPa		STRENGTH, KPa		MODULUS, MPa		STRENGTH, KPa		MODULUS, MPa	STRENGTH, KPa		MODULUS, MPa
		typical	minimum	typical	minimum	typical	typical	typical	minimum	typical	typical	minimum	typical
1/16 - 5052 - 0.0007	101	5998p		6274p		1896p		3516p		621p	2206p		276p
1/16 - 5052 - 0.001	144	10204x		10342x		2896x		5343x		724x	3585x		365x
1/8 - 5052 - 0.0007†	50	1862	1379	1999	1482	517	896	1448	1069	310	896	621	152
1/8 - 5052 - 0.001†	72	3585	2586	3758	2792	1034	1793	2344	1965	483	1517	1158	214
1/8 - 5052 - 0.0015†	98	5998	4482	6274	4688	1655	3103	3482	3137	676	2206	1875	283
1/8 - 5052 - 0.002†	130	9653	6895	10135	7584	2413	5171	4999	4619	931	3137	2758	372
1/8 - 5052 - 0.003	192	15168p		16030p		6205p		7584p			4309p		
5/32 - 5052 - 0.0007	42	1379	1034	1482	1103	379	621	1138	827	255	689	483	131
5/32 - 5052 - 0.001	61	2723	1965	2827	2068	758	1276	1862	1482	386	1207	862	182
5/32 - 5052 - 0.0015	85	4757	3378	4964	3689	1344	2344	2896	2551	579	1862	1482	248
5/32 - 5052 - 0.002	111	7446	5309	7791	5516	1965	3964	4068	3723	786	2586	2261	320
5/32 - 5052 - 0.0025††	135	10549	7377	11032	8136	2551	5516	5240	4757	965	3275	2896	386
3/16 - 5052 - 0.0007	32	896	621	931	689	234	414	827	552	186	483	317	99
3/16 - 5052 - 0.001†	50	1862	1379	1999	1482	517	896	1448	1069	310	896	621	152
3/16 - 5052 - 0.0015†	70	3447	2482	3620	2654	1000	1724	2275	1931	469	1482	1103	207
3/16 - 5052 - 0.002†	91	5309	3861	5585	4137	1517	2689	3172	2827	621	2068	1682	265
3/16 - 5052 - 0.0025	111	7446	5309	7791	5516	1965	3964	4068	3723	786	2586	2261	320
3/16 - 5052 - 0.003	130	9653	6895	10135	7584	2413	5171	4999	4619	931	3137	2758	372
1/4 - 5052 - 0.0007	26	586	414	655	483	138	276	586	414	145	345	221	76
1/4 - 5052 - 0.001	37	1138	827	1207	896	310	517	965	689	221	586	393	112
1/4 - 5052 - 0.0015	54	2206	1655	2344	1724	621	1034	1620	1241	345	1034	724	165
1/4 - 5052 - 0.002	69	3309	2413	3482	2551	965	1586	2206	1827	455	1448	1069	205
1/4 - 5052 - 0.0025	83	4619	3447	4757	3516	1310	2310	2827	2482	565	1827	1379	244
1/4 - 5052 - 0.003	96	5861	4344	6067	4551	1620	2965	3413	3068	662	2172	1827	279
1/4 - 5052 - 0.004†	127	9377	6688	9791	7239	2344	4999	4826	4482	896	3034	2689	364
3/8 - 5052 - 0.0007	16	207	138	310	138	69	172	310	221	83	207	138	48
3/8 - 5052 - 0.001	26	586	414	655	483	138	276	586	414	145	345	221	76
3/8 - 5052 - 0.0015	37	1138	827	1207	896	310	517	965	689	221	586	393	112
3/8 - 5052 - 0.002	48	1793	1310	1862	1379	483	827	1379	1000	296	862	586	146
3/8 - 5052 - 0.0025††	59	2551	1862	2689	1965	724	1241	1793	1379	379	1172	793	179
3/8 - 5052 - 0.003	67	3172	2310	3344	2448	931	1517	2137	1758	448	1379	1034	200
3/8 - 5052 - 0.004	86	4964	3447	5137	3689	1379	2482	2965	2620	593	1931	1572	254
3/8 - 5052 - 0.005††	104	6688	4826	7033	5171	1827	3482	3758	3447	724	2413	2068	300

*Corrugated 5052 and 5056 aluminum honeycomb are available in higher densities with crush strengths up to 41000 KPa. Test data obtained at 16-mm thickness. p = preliminary properties; x = predicted values.

**Crush strength values shown are average or typical; actual values may vary because of density tolerances, etc.

†Core types shown most readily available.

††Minimum order may be required.

Table 21.6a. Properties of 5056 and 2024 Hexagonal Aluminum Honeycomb*

5056 Hexagonal Aluminum Honeycomb

HONEYCOMB DESIGNATION CELL - MATERIAL - GAUGE	NOMINAL DENSITY, lb/ft³	COMPRESSIVE: BARE STRENGTH, psi		COMPRESSIVE: STABILIZED STRENGTH, psi		COMPRESSIVE: STABILIZED MODULUS, ksi	CRUSH STRENGTH,** psi	PLATE SHEAR: "L" DIRECTION STRENGTH, psi		PLATE SHEAR: "L" DIRECTION MODULUS, ksi	PLATE SHEAR: "W" DIRECTION STRENGTH, psi		PLATE SHEAR: "W" DIRECTION MODULUS, ksi
		typical	minimum	typical	minimum	typical	typical	typical	minimum	typical	typical	minimum	typical
1/16 - 5056 - 0.0007	6.3	1000x		1100x		330x		645x		95x	370x		38x
1/16 - 5056 - 0.001	9.0	1700p		1800p		500p		980p		110p	600p		50p
1/8 - 5056 - 0.0007†	3.1	340	250	360	260	97	170	250	200	45.0	155	110	20.0
1/8 - 5056 - 0.001†	4.5	630	475	670	500	185	320	425	350	70.0	255	205	38.0
1/8 - 5056 - 0.0015	6.1	1000	760	1100	825	295	535	640	525	102	370	305	38.0
1/8 - 5056 - 0.002	8.1	1520	1200	1700	1300	435	810	900	740	143	520	440	51.0
5/32 - 5-56 - 0.0007	2.6	255	180	260	185	70	120	200	152	36.0	120	80	17.0
5/32 - 5056 - 0.001	3.8	475	360	500	375	140	235	335	272	57.0	205	155	24.0
5/32 - 5056 - 0.0015	5.3	820	615	865	650	240	420	530	435	85.0	310	250	33.0
5/32 - 5056 - 0.002	6.9	1220	920	1340	1000	350	650	760	610	118	430	360	43.0
3/16 - 5056 - 0.0007	2.0	155	110	160	120	45	75	140	105	27.0	85	50	13.0
3/16 - 5056 - 0.001†	3.1	340	250	360	260	97	170	255	200	45.0	155	110	20.0
3/16 - 5056 - 0.0015	4.4	600	460	650	490	180	310	410	340	68.0	245	198	27.5
3/16 - 5056 - 0.002	5.7	910	685	980	735	270	480	585	480	94.0	340	280	36.0
1/4 - 5056 - 0.0007	1.6	100	75	110	80	30	50	90	78	20.0	60	38	12.0
1/4 - 5056 - 0.001†	2.3	205	145	210	155	58	100	170	130	32.0	105	62	15.0
1/4 - 5056 - 0.0015	3.4	395	300	420	315	115	200	290	230	50.0	175	130	22.0
1/4 - 5056 - 0.002	4.3	580	440	620	465	172	300	400	325	67.0	240	190	27.0
1/4 - 5056 - 0.0025	5.2	790	600	820	645	230	410	500	425	84.0	300	245	32.0
3/8 - 5056 - 0.0007	1.0	35	25	50	35	15	35	60	45	15.0	35	25	9.0
3/8 - 5056 - 0.001	1.6	100	75	110	80	30	50	90	78	20.0	60	38	12.0
3/8 - 5056 - 0.0015	2.3	205	155	210	155	58	100	170	130	32.0	105	62	15.0
3/8 - 5056 - 0.002	3.0	320	240	340	260	92	160	245	190	43.0	145	100	19.0

2024 Hexagonal Aluminum Honeycomb

HONEYCOMB DESIGNATION CELL - MATERIAL - GAUGE	NOMINAL DENSITY, lb/ft³	COMPRESSIVE: BARE STRENGTH, psi		COMPRESSIVE: STABILIZED STRENGTH, psi		COMPRESSIVE: STABILIZED MODULUS, ksi	CRUSH STRENGTH,** psi	PLATE SHEAR: "L" DIRECTION STRENGTH, psi		PLATE SHEAR: "L" DIRECTION MODULUS, ksi	PLATE SHEAR: "W" DIRECTION STRENGTH, psi		PLATE SHEAR: "W" DIRECTION MODULUS, ksi
		typical	minimum	typical	minimum	typical	typical	typical	minimum	typical	typical	minimum	typical
1/8 - 2024 - 0.0015	5.0	700	525	780	620	200	425	500	400	82.0	315	250	33.0
1/8 - 2024 - 0.002	6.7	1100	825	1225	980	300	640	760	600	118	470	375	45.0
1/8 - 2024 - 0.0025	8.0	1480	1100	1650	1320	380	840	960	770	148	590	470	54.0
1/8 - 2024 - 0.003	9.5	1970	1475	2300	1725	480	1120	1150	950	170	650	585	64.0
3/16 - 2024 - 0.0015	3.5	330	250	370	290	86	200	290	230	55.0	180	143	23.0
1/4 - 2024 - 0.0015	2.8	220	165	250	175	40	110	200	140	42.0	120	88	19.0

*Corrugated 5052 and 5056 aluminum honeycomb are available in higher densities with crush strengths up to 6000 psi. Test data obtained at 0.625-in. thickness. p = preliminary properties; x = predicted values.

**Crush strength values shown are average or typical; actual values may vary because of density tolerances, etc.

†Core types print most readily available.

Table 21.6*b*. Properties of 5056 and 2024 Hexagonal Aluminum Honeycomb (Metric)*

5056 Hexagonal Aluminum Honeycomb

HEXCEL HONEYCOMB DESIGNATION CELL - MATERIAL - GAUGE	NOMINAL DENSITY, kg/m^3	COMPRESSIVE: BARE STRENGTH, KPa		COMPRESSIVE: STABILIZED STRENGTH, KPa		STABILIZED MODULUS, MPa	CRUSH STRENGTH,** KPa	PLATE SHEAR "L" DIRECTION STRENGTH, KPa		"L" DIRECTION MODULUS, MPa	PLATE SHEAR "W" DIRECTION STRENGTH, KPa		"W" DIRECTION MODULUS, MPa
		typical	minimum	typical	minimum	typical	typical	typical	minimum	typical	typical	minimum	typical
1/16 - 5056 - 0.0007	101	6894x		7584x		2275x		4447x		655x	2551x		262x
1/16 - 5056 - 0.001	144	11721p		12410p		3447p		6756p		758p	4136p		344p
1/8 - 5056 - 0.0007†	50	2344	1723	2482	1792	668	1172	1723	1378	310	1068	758	137
1/8 - 5056 - 0.001†	72	4343	3275	4519	3447	1275	2206	2930	2413	482	1758	1413	262
1/8 - 5056 - 0.0015	98	6895	5240	7584	5688	2034	3689	4413	3620	703	2551	2103	262
1/8 - 5056 - 0.002	130	10480	8274	11721	8963	2999	5585	6205	5102	986	3585	3034	352
5/32 - 5056 - 0.0007	42	1758	1241	1793	1276	483	827	1379	1048	248	827	552	117
5/32 - 5056 - 0.001	61	3275	2482	3447	2586	965	1620	2310	1875	393	1413	1069	165
5/32 - 5056 - 0.0015	85	5654	4240	5964	4482	1655	2896	3654	2999	586	2137	1724	228
5/32 - 5056 - 0.002	111	8412	6343	9239	6895	2413	4482	5240	4206	814	2965	2482	296
3/16 - 5056 - 0.0007	32	1069	758	1103	827	310	517	965	724	186	586	345	90
3/16 - 5056 - 0.001†	50	2344	1724	2482	1793	669	1172	1758	1379	310	1069	758	138
3/16 - 5056 - 0.0015	70	4137	3172	4482	3378	1241	2137	2827	2344	469	1689	1365	190
3/16 - 5056 - 0.002	91	6274	4723	6757	5068	1862	3309	4033	3309	648	2344	1931	248
1/4 - 5056 - 0.0007	26	689	517	758	552	207	345	621	538	138	414	262	83
1/4 - 5056 - 0.001†	37	1413	1000	1448	1069	400	689	1172	896	221	724	427	103
1/4 - 5056 - 0.0015	54	2723	2068	2896	2172	793	1379	1999	1586	345	1207	896	152
1/4 - 5056 - 0.002	69	3999	3034	4275	3206	1186	2068	2758	2241	462	1655	1310	186
1/4 - 5056 - 0.0025	83	5447	4137	5654	4447	1586	2827	3447	2930	579	2068	1689	221
3/8 - 5056 - 0.0007	16	241	172	345	241	103	241	414	310	103	241	172	62
3/8 - 5056 - 0.001	26	689	517	758	552	207	345	621	538	138	414	262	83
3/8 - 5056 - 0.0015	37	1413	1069	1448	1069	400	689	1172	896	221	724	427	103
3/8 - 5056 - 0.002	48	2206	1655	2344	1793	634	1103	1689	1310	296	1000	689	131

2024 Hexagonal Aluminum Honeycomb

HEXCEL HONEYCOMB DESIGNATION CELL - MATERIAL - GAUGE	NOMINAL DENSITY, kg/m^3	BARE STRENGTH, KPa typical	BARE STRENGTH, KPa minimum	STABILIZED STRENGTH, KPa typical	STABILIZED STRENGTH, KPa minimum	STABILIZED MODULUS, MPa typical	CRUSH STRENGTH,** KPa typical	"L" STRENGTH, KPa typical	"L" STRENGTH, KPa minimum	"L" MODULUS, MPa typical	"W" STRENGTH, KPa typical	"W" STRENGTH, KPa minimum	"W" MODULUS, MPa typical
1/8 - 2024 - 0.0015	80	4826	3620	5378	4275	1379	2930	3447	2758	565	2172	1724	228
1/8 - 2024 - 0.002	107	7584	5688	8446	6757	2068	4413	5240	4137	814	3241	2586	310
1/8 - 2024 - 0.0025	128	10204	7584	11376	9101	2620	5792	6619	5309	1020	4068	3241	372
1/8 - 2024 - 0.003	152	13583	10170	15858	11893	3309	7722	7929	6550	1172	4482	4033	441
3/16 - 2024 - 0.0015	56	2275	1724	2551	1999	593	1379	1999	1586	379	1241	986	159
1/4 - 2024 - 0.0015	45	1517	1138	1724	1207	276	758	1379	965	290	827	607	131

*Corrugated 5052 and 5056 aluminum honeycomb are available in higher densities with crush strengths up to 41000 KPa. Test data obtained at 16-mm thickness. p = preliminary properties; x = predicted values.

**Crush strength values shown are average or typical; actual values may vary because of density tolerances, etc.

†Core types most readily available.

of expanded aluminum cores commercially available runs from a low of about 2 lb/ft^3 (32 kg/m^3) to a high of 12.0 lb/ft^3 (192 kg/m^3). Corrugated aluminum cores, however, start at under 8 lb/ft^3 (128 kg/m^3) and can be purchased up to 55 lb/ft^3 (880 kg/m^3). At densities below 8 lb/ft^3 (128 kg/m^3) corrugated core suffers a serious penalty in shear properties.

21.3.2.4. Glass Fiber-Reinforced Plastic Honeycomb

This family of materials is most commonly used in electrically sensitive parts, such as radomes and antennae, or where a heat resistant resin and low thermal conductivity make it a natural choice. It has also seen distinguished service as a matrix for retaining nonstructural ablative materials, such as soft silicone rubbers or syntactic rigid epoxy foams, which otherwise could not have been used effectively as ablative heat shields on the Gemini and Apollo re-entry vehicles. Only polyester coated nylon-phenolic, high temperature phenolic, and polyimide cores are produced in the United States. They are available in cell sizes of 3/16, 1/4, and 3/8 in. (5, 6.3, and 10 mm) with a 1/8-in. (3-mm) cell available in a bias weave glass reinforcement. Densities range from 2–12 lb/ft^3 (32–192 kg/m^3). Mechanical properties of several commercially available glass fiber-reinforced cores are shown in Tables 21.6 through 21.11.

21.3.2.5. Aramid Paper Honeycomb

This is an especially tough and damage resistant product, based on a completely synthetic, calendered "Nomex" paper material produced by DuPont. The core is expanded very much like aluminum or glass fabric honeycomb, and then dip-coated with phenolic or some other suitable resin system. The mechanical properties of the material as a structural core are somewhat lower than aluminum, especially in modulus, but it possesses a unique ability to survive overloads in local areas without permanent damage. This translates into abuse resistance when applied to very light interior aircraft panels or flooring, and gives the material a competitive edge even at the higher cost it represents. The base material is relatively incombustible, and the small amounts of material present in typical panels result in low volumes of smoke and gases given off in fire tests. Typical applications make use of these properties very effectively, and they have grown to the second largest volume core material (behind aluminum) used in aircraft structures. Uses outside the aerospace industry are limited due to the high cost of the material, but despite this, it has seen some application in boat hulls up to 40 ft (1.02 m) in length, as well as in skis, racing shells, and several other products.

Aramid core is normally produced in cell sizes of 1/8, 3/16, 1/4 and 3/8 in. (3, 5, 6.5, and 10 mm), in densities of 1.5–9 lb/ft^3 (24–144 kg/m^3). Densities higher than 4 lb/ft^3 (64 kg/m^3) are almost entirely used for aircraft flooring. Mechanical properties of some of these core materials are shown in Table 21.12 (*a* and *b*).

21.4. ADHESIVE MATERIALS

Adhesives, as they apply to sandwich structures, constitute a somewhat different family of materials than those not required to bond an open cellular core to a stiff and continuous facing. Although these differences are less important with some of the newer modified epoxy materials, they remain basic, and must be understood by the designer and fabricator in order for the otherwise inevitable problems to be avoided. Some factors which merit attention are discussed below.

21.4.1. Products Given Off During Cure

Some adhesive types, such as phenolic, give off a vapor as a product of the curing reaction, and the presence of these secondary materials can lead to several problems.

- Internal pressure, resulting in little or no bond in some areas, or "blisters".
- Core splitting, as the gas forces its way to a lower pressure area.
- Core movement, sometimes several inches, resulting in an unusable cured part.

Table 21.7*a*. Properties of Several Commonly Used Glass-Reinforced Plastic Honeycombs*†

HONEYCOMB DESIGNATION MATERIAL - CELL - DENSITY	COMPRESSIVE					PLATE SHEAR					
	BARE		STABILIZED			"L" DIRECTION			"W" DIRECTION		
	STRENGTH, psi		STRENGTH, psi		MODULUS, ksi	STRENGTH, psi		MODULUS, ksi	STRENGTH, psi		MODULUS, ksi
	typical	minimum	typical	minimum	typical	typical	minimum	typical	typical	minimum	typical
Hexagonal											
HRP - 3/16 - 4.0	500	350	600	480	57	260	210	11.5	140	110	5.0
HRP - 3/16 - 5.5	800	600	940	750	95	425	370	19.5	220	190	8.5
HRP - 3/16 - 7.0	1150	900	1230	1000	136	500		28.0	290		12.5
HRP - 3/16 - 8.0	1400	1100	1600	1280	164	660	600	34.0	400	370	15.0
HRP - 3/16 - 12.0	2280	1600	2300	1800	260p	940p	815	55p	570	500	25p
HRP - 1/4 - 3.5	350	260	500	400	46	230	170	9.0	120	100	3.5
HRP - 1/4 - 4.5	630	450	700	560	70	300	250	14.0	170	140	6.0
HRP - 1/4 - 5.0	700	510	820	660	84	340		17.0	200		7.5
HRP - 1/4 - 6.5	1025	850	1180	900	120	450		25.0	260		11.0
HRP - 3/8 - 2.2	150	105	200	145	13	105	75	5.0	60	45	2.0
HRP - 3/8 - 3.2	320	245	440	350	38	200	160	8.0	105	85	3.0
HRP - 3/8 - 4.5	610	450	690	550	65	300	260	14.0	170	150	6.0
HRP - 3/8 - 6.0	900	750	1000	750	100	400	340	22.5	260	210	10.0
HRP - 3/8 - 8.0	1060	920	1200		150p	520		31p	320		13p
Ox-Core											
HRP/OX - 1/4 - 4.5	520	350	625p		43p	210		8.0	250		15.2
HRP/OX - 1/4 - 5.5	810	600	950p		65p	270p		10.5	330		18.0
HRP/OX - 1/4 - 7.0	1150		1230p		84p	395p		14p	450p		20p
HRP/OX - 3/8 - 3.2	340	260	425p		32p	140		4.5p	150p		9p
HRP/OX - 3/8 - 5.5	700	580	820p		60p	240		10p	300p		17p
Flex-Core											
HRP/F35 - 2.5	180		240		25	125p		12.5p	70p		7.0p
HRP/F35 - 3.5	320		400	300	37	200	140	15.0	105	75	10.0
HRP/F35 - 4.5	440		600		49	280		22.0	140		12.0
HRP/F50 - 3.5	300		425	300	37	195	140	20.0	100	75	10.0
HRP/F50 - 4.5	400		600	500	49	265	200	25.0	140	100	13.0
HRP/F50 - 5.5	600p		880p		61p	390p		31.5p	205p		16.0p

*Test data obtained at 0.500-in. thickness.
†p = preliminary properties.

Table 21.7*b*. Properties of Several Commonly Used Glass-Reinforced Plastic Honeycombs (Metric)*†

HEXCEL HONEYCOMB DESIGNATION MATERIAL - CELL - DENSITY	COMPRESSIVE					PLATE SHEAR					
	BARE		STABILIZED			"L" DIRECTION			"W" DIRECTION		
	STRENGTH, KPa		STRENGTH, KPa		MODULUS, MPa	STRENGTH, KPa		MODULUS, MPa	STRENGTH, KPa		MODULUS, MPa
Hexagonal	typical	minimum	typical	minimum	typical	typical	minimum	typical	typical	minimum	typical
HRP - 3/16 - 4.0	3447	2413	4137	3309	393	1793	1448	79	965	758	34
HRP - 3/16 - 5.5	5516	4137	6481	5171	655	2930	2551	134	1517	1310	59
HRP - 3/16 - 7.0	7929	6205	8481	6895	938	3447		193	1999		86
HRP - 3/16 - 8.0	9653	7584	11032	8825	1131	4551	4137	234	2758	2551	103
HRP - 3/16 - 12.0	15720	11032	15858	12411	1793p	6481p	5619p	379p	3930p	3447p	172p
HPR - 1/4 - 3.5	2413	1793	3447	2758	317	1586	1172	62	827	689	24
HPR - 1/4 - 4.5	4344	3103	4826	3861	483	2068	1724	97	1172	965	41
HPR - 1/4 - 5.0	4826	3516	5654	4551	579	2344		117	1379		52
HPR - 1/4 - 6.5	7067	5861	8136	6205	827	3103		172	1793		76
HPR - 3/8 - 2.2	1034	724	1379	1000	90	724	517	34	414	310	14
HPR - 3/8 - 3.2	2206	1689	3034	2413	262	1379	1103	55	724	586	21
HPR - 3/8 - 4.5	4206	3103	4757	3792	448	2068	1793	97	1172	1034	41
HPR - 3/8 - 6.0	6205	5171	6895	5171	689	2758	2344	155	1793	1448	69
HPR - 3/8 - 8.0	7308	6343	8274		1034p	3585		214p	2206		90p
Ox-Core											
HRP/OX - 1/4 - 4.5	3585	2413	4309p		296p	1448		55	1724		105
HPR/OX - 1/4 - 5.5	5585	4137	6550p		448p	1862p		72	2275		124
HRP/OX - 1/4 - 7.0	7929		8481p		579p	2723p		97p	3103p		138p
HRP/OX - 3/8 - 3.2	2344	1793	2930p		221p	965		31p	1034p		62p
HRP/OX - 3/8 - 5.5	4826	3999	5654p		414p	1655		69p	2068p		117p
Flex-Core											
HRP/F35 - 2.5	1241		1655		172	862p		86p	483p		48p
HRP/F35 - 3.5	2206		2758	2068	255	1379	965	103	724	517	69
HRP/F35 - 4.5	3034		4137		338	1931		152	965		83
HRP/F50 - 3.5	2068		2930	2068	255	1344	965	138	689	517	69
HRP/F50 - 4.5	2758		4137	3447	338	1827	1379	172	965	689	90
HRP/F50 - 5.5	4137p		6067p		421p	2689p		217p	1413p		110p

*Test data obtained at 12.70-mm thickness.
†p = preliminary properties.

Table 21.8*a*. Properties of Several Special-Purpose Glass-Reinforced Plastic Honeycombs (Note the Sharp Improvement in Shear Modulus Shown by the Bias-Weave-Reinforced Material When Compared to the Equivalent Square-Weave-Reinforced Material Shown in Table 21.7)*†

HONEYCOMB DESIGNATION MATERIAL - CELL - DENSITY	COMPRESSIVE			PLATE SHEAR					
	STABILIZED			"L" DIRECTION			"W" DIRECTION		
	STRENGTH, psi		MODULUS, ksi	STRENGTH, psi		MODULUS, ksi	STRENGTH, psi		MODULUS, ksi
	typical	minimum	typical	typical	minimum	typical	typical	minimum	typical
HRH 327 - 3/16 - 4.0	440		50	280		29	130		10
HRH 327 - 3/16 - 4.5	520	400	58	320	220	33	150	110	11
HRH 327 - 3/16 - 5.0	600		68	370		37	180		12.5
HRH 327 - 3/16 - 6.0	780	625	87	460	345	45	230	170	15
HRH 327 - 3/16 - 8.0	1300	1000	126	650	500	62	410	330	22
HRH 327 - 3/8 - 4.0	440	325	50	280	195	29	150	100	12
HRH 327 - 3/8 - 5.5	680	540	78	420	300	41	210	160	13.5
HRH 327 - 3/8 - 7.0	1000p		106p	550p		53p	310p		18.5p

Glass-Reinforced Phenolic Honeycomb (Bias Weave Reinforcement)

HONEYCOMB DESIGNATION MATERIAL - CELL - DENSITY	COMPRESSIVE			PLATE SHEAR			
	BARE	STABILIZED		"L" DIRECTION		"W" DIRECTION	
	STRENGTH, psi	STRENGTH, psi	MODULUS, ksi	STRENGTH, psi	MODULUS, ksi	STRENGTH, psi	MODULUS, ksi
	typical	typical	typical	typical	typical	typical	typical
HFT - 1/8 - 3.0	300p	350p	22p	185p	17p	95p	7p
HFT - 1/8 - 4.0	390p	575p	45p	300p	32p	150p	12p
HFT - 1/8 - 5.5	525p	960p	67p	425p	42p	225p	17p
HFT - 1/8 - 8.0	1450p	1625p	100p	575p	48p	340p	25p
HFT - 3/16 - 1.8	75p	120p	14p	105p	13p	50p	4p
HFT - 3/16 - 2.0	100p	170p	17p	115p	15p	60p	5p
HFT - 3/16 - 3.0	275p	375p	32p	200p	24p	100p	9p
HFT - 3/16 - 4.0	435p	550p	45p	275p	30p	140p	12p
HFT/OX - 3/16 - 6.0	1000p	1100p	67p	290p	13p	335p	30p

*Test data obtained at 0.500-in. thickness. Honeycomb is normally not tested for bare compressive strength.
†p = preliminary properties.

Table 21.8*b*. Properties of Several Special-Purpose Glass-Reinforced Plastic Honeycombs (Metric)*†

Glass-Reinforced Polyimide Honeycomb

HEXCEL HONEYCOMB DESIGNATION MATERIAL - CELL - DENSITY	COMPRESSIVE			PLATE SHEAR					
	STABILIZED			"L" DIRECTION			"W" DIRECTION		
	STRENGTH, KPa		MODULUS, MPa	STRENGTH, KPa		MODULUS, MPa	STRENGTH, KPa		MODULUS, MPa
	typical	minimum	typical	typical	minimum	typical	typical	minimum	typical
HRH 327 - 3/16 - 4.0	3033		344	1930		199	896		68
HRH 327 - 3/16 - 4.5	3585	2757	399	2206	1516	227	1034	758	75
HRH 327 - 3/16 - 5.0	4136		468	2551		255	1241		86
HRH 327 - 3/16 - 6.0	5377	4309	599	3171	2378	310	1585	1172	103
HRH 327 - 3/16 - 8.0	8963	6894	868	4481	3447	427	2826	2275	151
HRH 327 - 3/8 - 4.0	3033	2240	344	1930	1344	199	1034	689	82
HRH 327 - 3/8 - 5.5	4688	3723	537	2895	2068	282	1447	1103	93
HRH 327 - 3/8 - 7.0	6894p		730p	3792p		365p	2137p		127p

Glass-Reinforced Phenolic Honeycomb (Bias Weave Reinforcement)

HEXCEL HONEYCOMB DESIGNATION MATERIAL - CELL - DENSITY	COMPRESSIVE			PLATE SHEAR			
	BARE	STABILIZED		"L" DIRECTION		"W" DIRECTION	
	STRENGTH, KPa	STRENGTH, KPa	MODULUS, MPa	STRENGTH, KPa	MODULUS, MPa	STRENGTH, KPa	MODULUS, MPa
	typical	typical	typical	typical	typical	typical	typical
HFT - 1/8 - 3.0	2068p	2413p	151p	1275p	117p	655p	48p
HFT - 1/8 - 4.0	2688p	3964p	310p	2068p	220p	1034p	82p
HFT - 1/8 - 5.5	3619p	6618p	461p	2930p	289p	1551p	117p
HFT - 1/8 - 8.0	9997p	11203p	689p	3964p	331p	2344p	172p
HFT - 3/16 - 1.8	517p	827p	97p	724p	89p	344p	27p
HFT - 3/16 - 2.0	689p	1172p	117p	792p	103p	413p	34p
HFT - 3/16 - 3.0	1896p	2585p	220p	1378p	165p	689p	62p
HFT - 3/16 - 4.0	2999p	3792p	310p	1896p	206p	965p	82p
HFT/OX - 3/16 - 6.0	6894p	7584p	461p	1999p	89p	2309p	206p

*Test data obtained at 12.70-mm. thickness. Honeycomb is normally not tested for bare compressive strength.
†p = preliminary properties.

Table 21.9. HRH-327 Glass-Reinforced Polyimide Honeycomb*

HEXCEL HONEYCOMB DESIGNATION MATERIAL-CELL-DENSITY	COMPRESSIVE			PLATE SHEAR					
	STABILIZED			"L" DIRECTION			"W" DIRECTION		
	STRENGTH, psi		MODULUS, ksi	STRENGTH, psi		MODULUS, ksi	STRENGTH, psi		MODULUS, ksi
	typical	minimum	typical	typical	minimum	typical	typical	minimum	typical
HRH 327 - 3/16 - 4.0	440		50	280		29	130		10
HRH 327 - 3/16 - 4.5	520	400	58	320	220	33	150	110	11
HRH 327 - 3/16 - 5.0	600		68	370		37	180		12.5
HRH 327 - 3/16 - 6.0	780	625	87	460	345	45	230	170	15
HRH 327 - 3/16 - 8.0	1300	1000	126	650	500	62	410	330	22
HRH 327 - 3/8 - 4.0	440	325	50	280	195	29	150	100	12
HRH 327 - 3/8 - 5.5	680	540	78	420	300	41	210	160	13.5
HRH 327 - 3/8 - 7.0	1000p		106p	550p		53p	310p		18.5p

*Test data obtained at 0.500-in. thickness. Honeycomb is normally not tested for bare compressive strength. p = preliminary properties.

Table 21.10. NP Glass-Reinforced Polyester Honeycomb*

HEXCEL HONEYCOMB DESIGNATION MATERIAL-CELL-DENSITY	COMPRESSIVE					PLATE SHEAR					
	BARE		STABILIZED			"L" DIRECTION			"W" DIRECTION		
	STRENGTH, psi		STRENGTH, psi		MODULUS, ksi	STRENGTH, psi		MODULUS, ksi	STRENGTH, psi		MODULUS, ksi
Hexagonal	typical	minimum	typical	minimum	typical	typical	minimum	typical	typical	minimum	typical
NP 3/16 - 4.5	520	365	670	470	80	280	195	13.5	130	90	5.2
NP 3/16 - 6.0	880	615	1050	735	116	330	230	15.0	155	110	5.8
NP 3/16 - 9.0	1700	1200	1800	1260	180	460	320	20.0	230	160	7.5
NP 1/4 - 4.0	420	295	560	390	68	260	180	13.0	120	85	5.0
NP 1/4 - 6.0	880	615	1050	736	116	330	230	15.0	155	110	5.8
NP 1/4 - 8.0	1400	980	1540	1080	160	410	290	18.0	205	145	7.0
NP 3/8 - 2.5	200	140	280	195	34	170	120	10.0	100	70	4.0
NP 3/8 - 4.5	520	365	670	470	80	280	195	13.5	130	90	5.2
Ox-Core											
NP/OX 1/4 - 4.0	350					160		5.0	190		12.0
NP/OX 1/4 - 6.0	700					275		7.5	375		19.5
NP/OX 3/8 - 4.5	420					190		5.5	285		15.0

*Test data obtained at 0.500-in. thickness.

Table 21.11. HFT Glass-Reinforced Phenolic Honeycomb (Fibertruss Bias Weave)*

HEXCEL HONEYCOMB DESIGNATION MATERIAL-CELL-DENSITY	COMPRESSIVE			PLATE SHEAR			
	BARE	STABILIZED		"L" DIRECTION		"W" DIRECTION	
	STRENGTH, psi	STRENGTH, psi	MODULUS, ksi	STRENGTH, psi	MODULUS, ksi	STRENGTH, psi	MODULUS, ksi
	typical	typical	typical	typical	typical	typical	typical
HFT - 1/8 - 3.0	250p	360p	21p	185p	16p	96p	6.4p
HFT - 1/8 - 4.0	460p	530p	45p	310p	25p	150p	9.5p
HFT - 1/8 - 5.5	850p	950p	65p	460p	34p	240p	13.5p
HFT - 1/8 - 8.0	1600p	1750p	95p	600p	43p	340p	20.0p
HFT - 3/16 - 2.0	90p	140p	17p	118p	15p	55p	4.3p
HFT - 3/16 - 3.0	250p	320p	32p	170p	20p	90p	6.5p
HFT - 3/16 - 4.0	460p	530p	45p	310p	25p	150p	9.5p
HFT/OX - 3/16 - 6.0	1000p	1100p	67p	290p	13p	335p	30.0p

*Test data obtained at 0.500-in. thickness. p = preliminary properties.

- Subsequent corrosion of core or skins by the chemical action of the vapor or its condensate.

21.4.2. Bonding Pressure

Adhesives such as the phenolics and some others actually require more than atmospheric pressure in order to prevent excessive porosity. Certain forms may be suitable for solid cores like balsa, but cannot be used at all in open cores such as honeycomb or large cell foams. Also, most core materials will not themselves stand compressive bonding loads exceeding a few atmospheres, and consequently cannot be used with any adhesive system requiring higher pressures.

21.4.3. Fillet Forming

In order to achieve a good attachment to an open cell core, such as honeycomb, the adhesive must have a unique combination of surface wetting and controlled flow during early stages of cure. Controlled flow prevents the adhesive from flowing down the cell wall and leaving a low strength top skin attachment and an overweight bottom skin attachment.

21.4.4. Adaptability

The requirements noted above must all be met while also meeting all the requirements of a skin-to-skin or skin-to-doubler attachment. In the case of contoured parts, the adhesive must also be a good "gap-filler" without appreciable strength penalty, since tolerance control of details is much more difficult to achieve on contoured than on flat panels, and a greater degree of latitude for misfit must usually be allowed.

21.4.5. Bond Line Control

This is a need which exists because of misfitting details, and is approximately the opposite of adaptability. It is the capability of the adhesive to resist being squeezed out from between faying surfaces when excessive pressure is applied during cure to a local area of the part. Many adhesives are formulated to achieve good core filleting and are subsequently given controlled flow by adding an open weave cloth or fibrous web, cast within a thicker film of adhesive. This "scrim cloth" then prevents the faying surfaces from squeezing out all the adhesive, which would result in an area of low bond strength.

21.4.6. Toughness

The word "toughness" has many meanings in the world of adhesives. Usually, it refers to the resistance shown by the adhesive to permitting bondline cracks to grow under impact loading. In the area of sandwich core-to-facing bonds, it refers to the resistance shown by the adhesive toward loads which act to separate the facings from the core under either

Table 21.12*a*. Properties of Nomex Paper Honeycomb*†

HONEYCOMB DESIGNATION MATERIAL - CELL - DENSITY GAUGE	COMPRESSIVE					PLATE SHEAR					
	BARE		STABILIZED			"L" DIRECTION			"W" DIRECTION		
	STRENGTH, psi		STRENGTH, psi		MODULUS, ksi	STRENGTH, psi		MODULUS, ksi	STRENGTH, psi		MODULUS, ksi
Hexagonal	typical	minimum	typical	minimum	typical	typical	minimum	typical	typical	minimum	typical
HRH 10 - 1/8 - 1.8 (1.5)	110	70	130	85		90	65	3.7	50	36	2.0
HRH 10 - 1/8 - 3.0 (2)	300	180	330	270	20	180	162	7.0	95	85	3.5
HRH 10 - 1/8 - 4.0 (2)	500	330	560	470	28	245	225	9.2	140	110	4.7
HRH 10 - 1/8 - 5.0 (3)	900	600	925	660		325	235		175	120	
HRH 10 - 1/8 - 6.0 (3)	1075	800	1125	825	60	370	260	13.0	200	135	6.0
HRH 10 - 1/8 - 8.0 (3)	1575	1100	1700	1250	78	490	355	16.0	250	190	7.8
HRH 10 - 1/8 - 9.0 (3)	1700	1400	1800	1600	90	520	370	17.0	270	240	9.0
HRH 10 - 5/32 - 5.0 (4)	800p		900p			360p		11.5p	180p		5.0p
HRH 10 - 5/32 - 9.0 (4)	1775p		2050p			525p		18.0p	285p		9.5p
HRH 10 - 3/16 - 2.0 (2)	150	90	170	105	11	110	72	4.2	55	40	2.2
HRH 10 - 3/16 - 3.0 (2)	300	180	330	270	20	150	130	5.0	95	67	3.5
HRH 10 - 3/16 - 4.0 (3)	500	320	560	470	28	245	215	7.8	140	110	4.7
HRH 10 - 3/16 - 4.5 (5)	425	320	475	400		290	225	9.5	145	110	4.0
HRH 10 - 3/16 - 6.0 (5)	650	580	700	650		390	330	14.5	185	150	6.0
HRH 10 - 1/4 - 1.5 (2)	90	45	95	55	6	75	45	3.0	35	23	1.5
HRH 10 - 1/4 - 2.0 (2)	150	80	170	105	11	110	72	4.2	55	36	2.8
HRH 10 - 1/4 - 3.1 (5)	275	180	285	240		170	135	7.0	85	60	3.0
HRH 10 - 1/4 - 4.0 (5)	370	310	400	360		240	200	7.5	125	95	3.5
HRH 10 - 3/8 - 1.5 (2)	90	45	95	55	6	75	45	3.0	35	23	1.5
HRH 10 - 3/8 - 2.0 (2)	150	80	170	105	11	110	72	4.2	55	36	2.2
HRH 10 - 3/8 - 3.0 (5)	285p		300p		17p	170		5.6p	95p		3.0p
OX-CORE											
HRH 10/OX - 3/16 - 1.8 (2)	110	70	130			60	45	2.0	60	35	3.0
HRH 10/OX - 3/16 - 3.0 (2)	365	250	400	270	17	115	95	3.0	125	95	6.0
HRH 10/OX - 1/4 - 3.0 (2)	350	210	385	250	17	110	90	3.0	115	90	6.0
FLEX-CORE											
HRH 10/F35 - 2.5 (3)	150	105	170	119	12p	70	49	4.0p	40	28	1.9p
HRH 10/F35 - 3.5 (5)	300p		350p		24p	150p		5.7p	80p		2.8p
HRH 10/F35 - 4.5 (5)	450p		490p		33p	270p		7.3p	150p		3.7p
HRH 10/F50 - 3.5 (3)	300	189	350	217	24	150	105	5.7p	80	56	2.8p
HRH 10/F50 - 4.5 (5)	450p		490p		33p	270p		7.3p	150p		3.7p
HRH 10/F50 - 5.0 (5)	550		625	525	37	330	300	8.0	190	160	4.1
HRH 10/F50 - 5.5 (5)	650p		700p		42p	390p		8.8p	235p		4.6p

*Test data obtained at 0.500-in. thickness. Nomex is a registered trademark of DuPont.
†p = preliminary properties

Table 21.12*b*. Properties of Nomex Paper Honeycomb*†

HEXCEL HONEYCOMB DESIGNATION MATERIAL - CELL - DENSITY GAUGE	COMPRESSIVE					PLATE SHEAR					
	BARE		STABILIZED			"L" DIRECTION			"W" DIRECTION		
	STRENGTH, KPa		STRENGTH, KPa		MODULUS, MPa	STRENGTH, KPa		MODULUS, MPa	STRENGTH, KPa		MODULUS, MPa
Hexagonal	typical	minimum	typical	minimum	typical	typical	minimum	typical	typical	minimum	typical
HRH 10 - 1/8 - 1.8 (1.5)	758	482	896	586		620	448	25	344	248	13
HRH 10 - 1/8 - 3.0 (2)	2068	1241	2275	1861	137	1241	1116	48	655	586	24
HRH 10 - 1/8 - 4.0 (2)	34473	2275	3861	3240	193	1689	1551	63	965	758	32
HRH 10 - 1/8 - 5.0 (3)	6205	4136	6377	4550		2240	1620		1206	827	
HRH 10 - 1/8 - 6.0 (3)	7411	5515	7756	5688	413	2551	1792	89	1378	930	41
HRH 10 - 1/8 - 8.0 (3)	10859	7584	11721	8618	537	3378	2447	110	1723	1310	53
HRH 10 - 1/8 - 9.0 (3)	11721	9652	12410	11031	620	3585	2551	117	1861	1654	62
HRH 10 - 5/32 - 5.0 (4)	5515p		6205p			2482p		72	1241p		34
HRH 10 - 5/32 - 9.0 (4)	12238p		14134p			3619p		124	1965p		65
HRH 10 - 3/16 - 2.0 (2)	1034	620	1172	723	75	758	496	28	379	275	15
HRH 10 - 3/16 - 3.0 (2)	2068	1241	2275	1861	137	1034	896	34	655	461	24
HRH 10 - 3/16 - 4.0 (2)	3447	2206	3861	3240	193	1689	1482	53	965	758	32
HRH 10 - 3/16 - 4.5 (5)	2930	2206	3275	2757		1999	1551	65	999	758	27
HRH 10 - 3/16 - 6.0 (5)	4481	3998	4826	4481		2688	2275	99	1275	1034	41
HRH 10 - 1/4 - 1.5 (2)	620	310	655	379	41	517	310	20	241	158	10
HRH 10 - 1/4 - 2.0 (2)	1034	551	1172	723	75	758	496	28	379	248	19
HRH 10 - 1/4 - 3.1 (5)	1896	1241	1965	1654		1172	930	48	586	413	20
HRH 10 - 1/4 - 4.0 (5)	2551	2137	2757	2482		1654	1378	51	861	655	24
HRH 10 - 3/8 - 1.5 (2)	620	310	655	379	41	517	310	20	241	158	10
HRH 10 - 3/8 - 2.0 (2)	1034	551	1172	723	75	758	496	28	379	248	15
HRH 10 - 3/8 - 3.0 (5)	1965p		2068p		117p	1172p		38p	655p		20p
OX-Core											
HRH 10/OX - 3/16 - 1.8 (2)	758	482	896			413	310	13	413	241	20
HRH 10/OX - 3/16 - 3.0 (2)	2516	1723	2757	1861	11	792	655	20	861	655	41
HRH 10/OX - 1/4 - 3.0 (2)	2413	1447	2654	1723	117	758	620	20	792	620	41
Flex-Core											
HRH 10/F35 - 2.5 (3)	1034	723	1172	820	82p	482	337	27p	275	193	13p
HRH 10/F35 - 3.5 (5)	2068p		2413p		16p	1034p		39p	551p		19p
HRH 10/F35 - 4.5 (5)	3102p		3378p		227p	1861p		59p	1034p		25p
HRH 10/F50 - 3.5 (3)	2068	1303	2413p	1496	16	1034	6929	39p	551	386	19p
HRH 10/F50 - 4.5 (5)	3102p		3378p		227p	1861p		50p	1034p		25p
HRH 10/F50 - 5.0 (5)	3792		4309	3619	255	2275	2068	55	1310	1103	28
HRH 10/F50 - 5.5 (5)	4481p		4826p		289p	2688		60p	1620		31p

*Test data obtained at 12.70-mm thickness. Nomex is a registered trademark of DuPont.
†p = preliminary properties

static or dynamic conditions. It has been found from experience that greater toughness in the bondline usually equates to greater durability, and thus to longer service life.

Many types of tests have been devised to measure toughness, but the most common one used for sandwich structures is the climbing drum peel test (Fig. 21.11). This test has the virtue of being easily duplicated, as well as possessing an obvious relationship to the toughness whose value is sought. Values of peel strength will vary considerably, depending upon:

Figure 21.11. Climbing drum peel test for adequacy of skin adhesion. The difference in diameter of the cylinders to which the straps are attached and the cylinder to which the skin is attached causes the drum to rotate clockwise when tension is applied by the universal testing machine This arrangement allows for easier duplication of test results from one shop to another.

- Toughness of the adhesive
- Amount of adhesive used
- Density of the core
- Cell size of the core
- Direction of the peel (with or across the ribbon direction)
- Adequacy of the surface preparation
- Degradation of the adherend surface subsequent to bonding.

Because these variables can lead to widely differing peel strengths for the very same adhesive, all of them must be properly understood and controlled if the peel test is to be used and its value compared to other test results.

The peel test is used to control quality throughout the sandwich industry. Values obtained, provided the adhesive weight and core material are in balance, will give indications of tooling or cure problems, and of adherend surface preparation problems. It is particularly useful for this when an environmental exposure involving both elevated temperature and high humidity is interposed between manufacture and test. It is also adaptable to use with nearly any skin material, except that it becomes impractical with very thick or very stiff skins.

It can be readily seen that a number of points of difference separate the sandwich adhesives from other structural adhesives. Fortunately for the sandwich user, many adhesives are available which satisfactorily meet both sets of requirements. The types available, along with some salient features, are as follows.

21.4.7. Phenolics Blended with Vinyls, Rubbers, or Epoxy

All of these families of adhesives give off at least some water during cure, and are therefore used only where their high strength, durability, or high temperature mechanical properties are essential. Since the out-gassing cure products usually require venting or perforating the core material, and a number of non-out-gassing, high temperature adhesives have become available, their use as sandwich adhesives has declined in recent years.

21.4.8. Epoxies Modified with Nylon or Other Polyamide Polymers

These adhesives were the first to have excellent filleting and controlled flow along with both high strength and high toughness, although they are somewhat moisture sensitive. Some versions are provided as one side of a two-sided tape adhesive, in which the other side is a rubber or vinyl-phenolic, to provide both excellent peel and durability at the skin side with excellent peel at the core side.

21.4.9. Nitrile Modified Epoxies

These make up a broad group of more recent materials which provide much of the flow and toughness shown by the nylon-epoxies, along with the durability and weather resistance of the vinyl-phenolics. They are the most common of the "toughened" thermosetting adhesives and are usually limited to about 300°F (149°C) service temperature. Some of these materials routinely achieve shear strengths of 5000 psi (34,500 kPa), and most can be cured over a wide range of temperatures and pressures.

21.4.10. Urethanes

Urethane based adhesives are used in many commercial structures. Both moisture-cured and two-part systems are available.

21.4.11. Other Polyimides, Thermoplastics, and Highly Specialized Adhesives

These are used in a number of applications ranging up to about 700°F (371°C) service temperature, but do not represent a very large group of materials. In addition to categorizing the available adhesives by chemical type, they can be grouped by the form in which they are available. Generally these are as follows.

- *Light liquids, heavy liquids, pastes, putties, or syntactic foams.* Only a few are used as a core-to-face bond, but many such materials are used in sandwich construction to splice pieces of core to each other; to provide high strength edges, areas, or surfaces; to carry shear loads from fittings, inserts, or end ribs; and for other such uses. Most of the materials so used are epoxies, modified epoxies, or epoxy polyimides. Curing temperatures vary from as low as 40°F (4.4°C) for some two-part systems up to 420°F (216°C) for some of the materials intended for service at elevated temperatures.
- *Supported films having a carrier of light glass fiber, cotton, nylon, or polyester cloth, or spunbonded synthetic fiber.* These tapes are provided either dry or with slight to moderate "tack" or stickiness, so that the parts of the assembly stay in place as they are being assembled.
- *Unsupported films, containing only the adhesive, with no carrier.* The very low weight films are nearly always furnished without a carrier, as the weight of the carrier itself becomes quite appreciable in very light sandwich structures. They are often hard to handle and sometimes have bondline control problems.
- *Reticulating films.* These are intended for use at very low weights, with the adhesive being melted after placing on the core, so that it draws back to the cell edge and leaves material to form the largest possible fillet without wasting any on the inside facing surface in the middle of the cell.
- *Cell-edge adhesive.* This is a material placed on the cell edge by the honeycomb manufacturer to provide the same results as those produced with reticulating films.
- *Self-adhesive skins.* These skins are structural fabrics of glass, graphite, quartz, or aluminum coated glass fibers preimpregnated with a resin, which is then cured so that the fiber-filled resin becomes both the face structure and the attaching material.

Table 21.13. Common Adhesives in Current Use*

ADHESIVE TYPE	TRADE DESIGNATION	MANUFACTURER
Nitrile phenolic	FM 238	American Cyanamid
	AF 30, AF 31, AF 32	3M Company
	Metlbond 402	Narmco
	Plastilock 655, 638, 620, 650	BF Goodrich
Vinyl phenolic	FM 47	American Cyanamid
Epoxy phenolic, 350° F (176.6° C)	HT 424	American Cyanamid
Modified epoxy liquid and pastes	M-6803, M-6860	Narmco
	R471, R380	Ciba-Geigy
	EA 9320, Ea 9309, Ea 934	Hysol
	MA 1015, Ma 2021	McCann Adhesives
	Crest 471, 3135	Crest Products
	EC 2216	3M Company
	A1396B, A1444B, A1446B	BF Goodrich
Modified epoxy, 250° F (121° C) cure and below	Crest 3181	Crest Products
	EA 9601, EA 9602.3, EA 9628	Hysol
	MA 229, MA 429, MA 456	McCann Adhesives
	FM 24, FM 53, FM 73, FM 123, FM 137	American Cyanamid
	Metlbond 117, M1113, M1133, M1137, M1204	Narmco
	Plastilock 717B	BF Goodrich
	Reliabond R382, R393, R7114	Ciba-Geigy
	AF 126, AF 126-2	3M Company
	HP 347	Hexcel Corporation
Modified Epoxy, 350° F (176.6° C) cure	Metlbond 328, 329, 329-7, 1515	Narmco
	AF 130, AF 143, AF 147	3M Company
	FM 61, FM 96, FM 150-2, FM 300, FM 400	American Cyanamid
	Plastilock 729-3	BF Goodrich
	Reliabond 398	Ciba-Geigy
	MA 529	McCann Adhesives
	EA 9649	Hysol
Epoxy/polyamide	A1177B, A1273B	BF Goodrich
	EA 951	Hysol
	FM 1000	American Cyanamid
Polyimide	FM34B-18	American Cyanamid
	HP955, HP956	Hexcel Corporation
Modified urethanes	Crest 7410, 7450, 7395	Crest Products
	Metlbond 6872, 6875, 6876	Narmco
	EC 3549	3M Company
	Plastilock 550	BF Goodrich
Core splicing adhesives	Crest 3181, 3158	Crest Products
	Metlbond 6602, 6607, 6601, 6603	Narmco
	FM 37, FM 39, FM 40, FM 41, FM 47-type O, FM 404	American Cyanamid
	Reliabond 370B, 370C	Ciba-Geigy
	MA 550, MA 2150	McCann Adhesives
	EA 9815, Thermofoam 3050, Thermofoam 3056	Hysol
	AF 3002, AF 3015	3M Company
	HP905, HP906	Hexcel Corporation
	Plastilock 654, 658	BF Goodrich

*The use of a corrosion inhibiting primer on metal facings is strongly advised. This step improves resistance to bondline corrosion and greatly increases durability. It also allows for lengthy storage of ready-to-bond details, or making them into riveted subassemblies, without the need for re-cleaning immediately prior to bonding. Only one primer, BR 127 (American Cyanamid), seems to have gained acceptance for use with a number of adhesive systems. However, conditions of use must be very carefully controlled and be consistent with recommendations of *both* the adhesive supplier and the primer supplier, if optimum results are to be achieved.

All the above forms of adhesive are in current use at substantial volume, and most are available from many sources, as listed in Table 21.13.

21.5. DESIGNING A SANDWICH

The usual objective of a sandwich design is to save weight or increase stiffness or to use less of an expensive skin material, or both. Sometimes other objectives, such as reducing tooling or other costs, achieving aerodynamic smoothness, reducing reflected noise, or increasing durability under exposure to acoustic energy, are also involved. The designer's problems sift down to relatively few, such as getting the loads in, getting the loads out, and attaching small or large load-carrying members, under constraints of deflection, contour, weight, and cost. Below are a few suggestions.

1. *Understand the fabrication sequence and methods.* The cost of a sandwich structure is fundamentally fixed at the design stage, and a considerable difference in cost results from alternate solutions to the design problem. Both of the edge close-out details shown in Fig. 21.12 perform essentially the same job at the same weight. Placing the legs of the channel facing outward instead of inward saves the cost of two relief cuts into the core and the very difficult step of sliding the edge of the core and adhesive into the channel. Another alternative at even lower cost for either fixed or simply supported edges is shown in Fig. 21.13. Additional edge treatments, joints, and corners are shown in Figs. 21.14–21.16.
2. *Use the right core.* Several densities of core can be used in a single panel, each appropriate to the load carried in the area and adhesively bonded to its neighbor, as shown in Fig. 21.17. In many cases, however, the weight saved in lower density areas of core is added back in the form of core splice adhesive weight. Core splices, such as those shown in Fig. 21.18 *B* or *C*, have been used to produce ablative matrix structures for large re-entry heat shields, but become prohibitively expensive to produce for splices more than a few inches long.
3. *Do not hesitate to use several joining methods in the same part.* Fittings to be included in a bonded sandwich may be produced from weldments, forgings, or riveted assemblies, or may themselves be bonded assemblies. Available adhesives permit secondary bonding to be performed at temperatures from 60° F (16° C) up to 350° F (177° C) without degrading the integrity of the bonded sub-assemblies.
4. *Use bolts and rivets for carrying loads (not soothing fears).* Where space is not available for progressive doublers or wide-area bonded overlaps to carry high loads, the addition of rivets or bolts is sometimes the only solution. Their use, however, often results in lower fatigue life of the structure, in addition to increased weight. The use of "chicken rivets" for the sole purpose of appearance is to be particularly avoided, since they often defeat much of the advantage resulting from use of the bonded structure.
5. *Use doublers where needed, instead of a heavier facing over the entire part.* The use of doublers, although adding labor cost in assembly, often improves the part quality. Where skins are formed of glass or graphite prepreg, the problem is even simpler, since extra plies can be added to

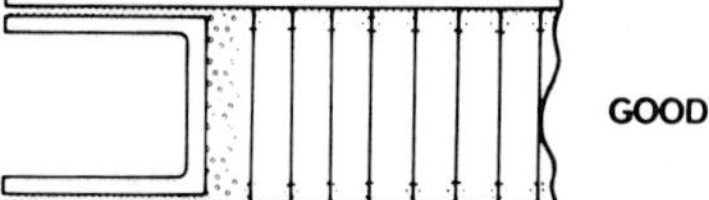

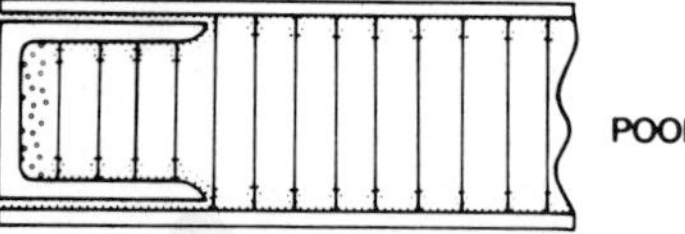

Figure 21.12. The square edge close-out shown here using a channel may result in a neat, clean edge, but requires machining top and bottom of the core and squeezing adhesive and core into the channel during assembly. The alternative shown would be much better.

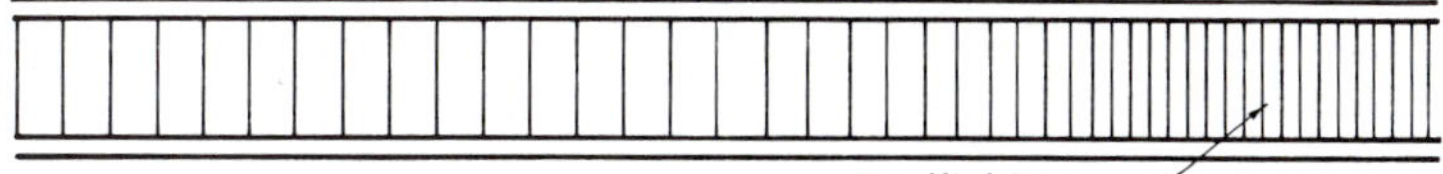

Figure 21.13. Densified core edge treatment.

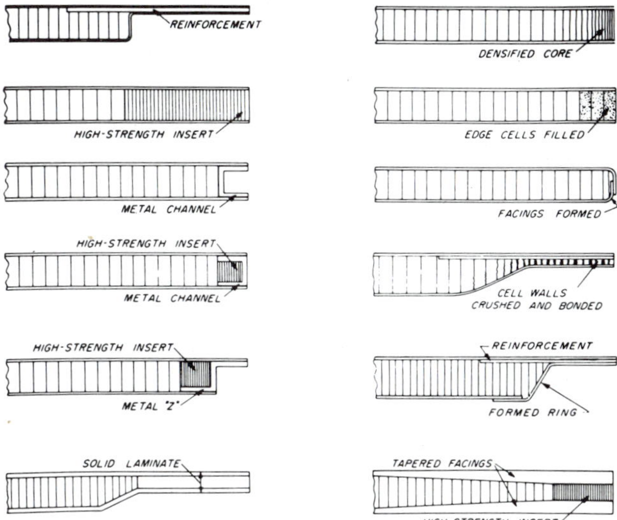

Figure 21.14. Edge treatments.

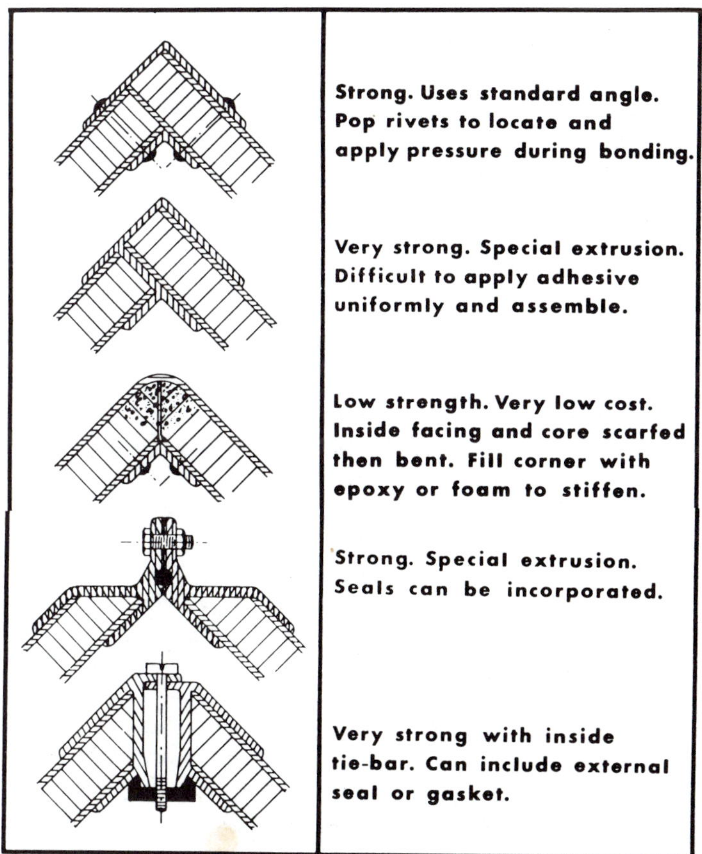

Figure 21.15. Several suggestions for corner designs, edge close-outs, and splices.

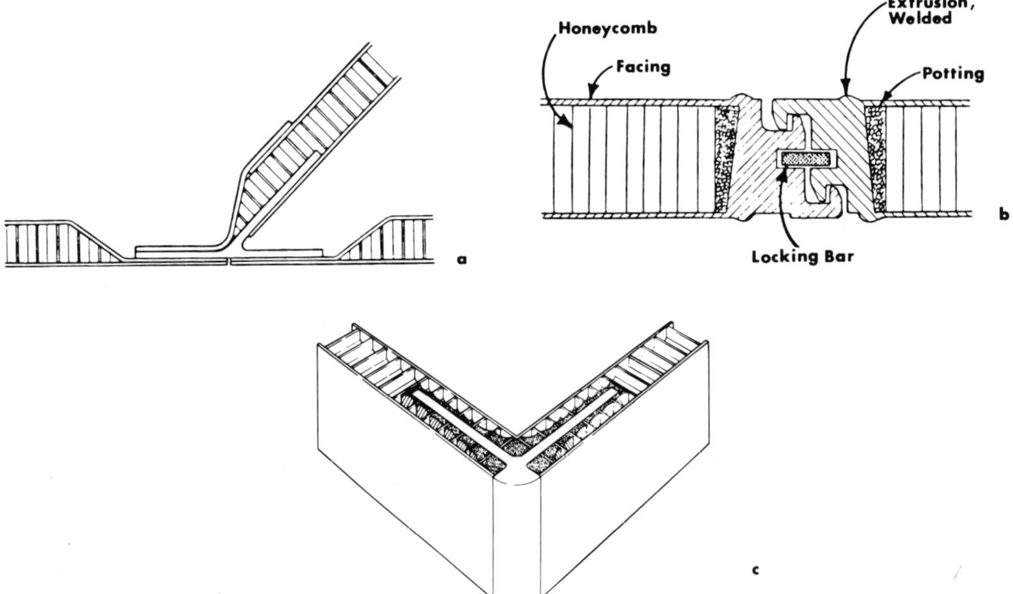

Figure 21.16. Additional joints and corner treatments.

carry extra loads exactly where and as needed.

6. *Use external doublers rather than internal doublers wherever possible.* The use of internal doublers usually means that a relief cut must be made in the core to prevent bridging and a consequent unbonded area where the doubler ends. Figure 21.19 shows a panel where the loads which can be carried are the same at each end of the panel. The design detail on the left end can cost substantially more to manufacture than that on the right end. Figure 21.20 shows the same panel with both ends produced at low cost, while still achieving an unbroken outer skin line on one side. In the case of some skin materials, such as 0.010-in. (0.25-mm) aluminum, or most weights of preimpregnated glass or

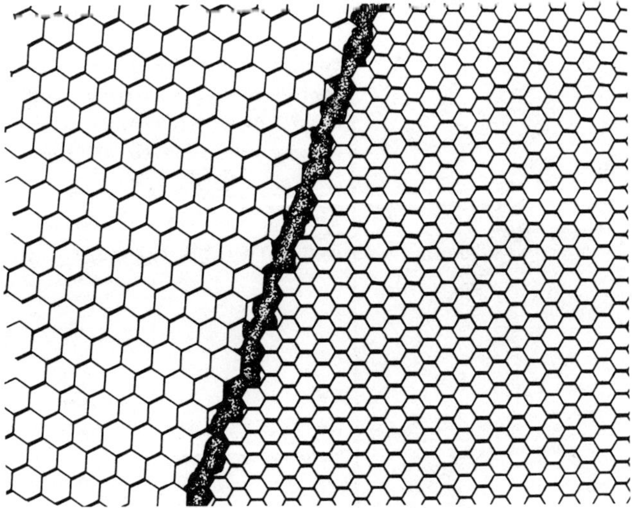

Figure 21.17. Typical core splice using a foam-tape adhesive. Foaming of tape allows less-than-perfect fit of core details, but requires that the core be fixed in position during the cure.

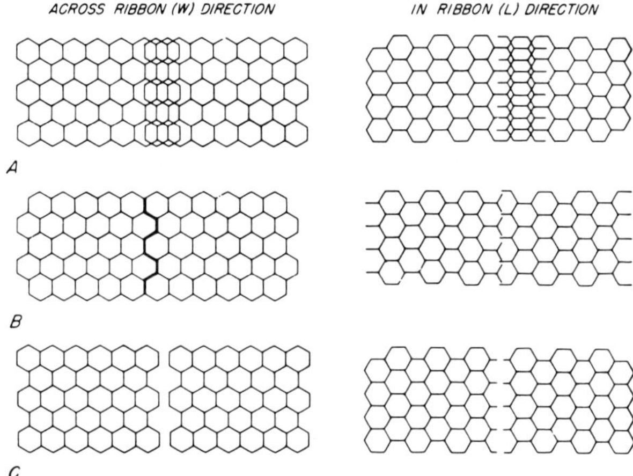

Figure 21.18. Joint A may be formed by simply crushing one piece of glass fabric honeycomb into the adjoining section. This method will work to some extent with some aluminum honeycombs, but not with most other core materials. Joints B and C require a perfect match of cell shape and cell pitch, and become nearly impossible to produce on a realistic and cost-effective basis.

graphite cloth, it is feasible to use thin doublers without a relief cut in the core, so that the gap caused by bridging is small enough to be within the capacity of the adhesive to fill. Sometimes an extra layer of adhesive film is added to help. An example of a double skin splice using this method is shown in Fig. 21.21.

21.6. STRUCTURAL ANALYSIS FOR SPECIFIC CASES

The following notations are utilized in sandwich design formulas.*

D = flexural stiffness; $D = \dfrac{E_f t_f h^2}{2\lambda_f}$

E_c = modulus of elasticity of the honeycomb, psi (Pa)

E_f = modulus of elasticity of facing material, psi (Pa)

G_c = shear modulus of rigidity of the honeycomb psi (Pa)

K_b = bending deflection constant

*This chapter's formulas are only for honeycomb beams and columns which have the same facings on each side of the core.

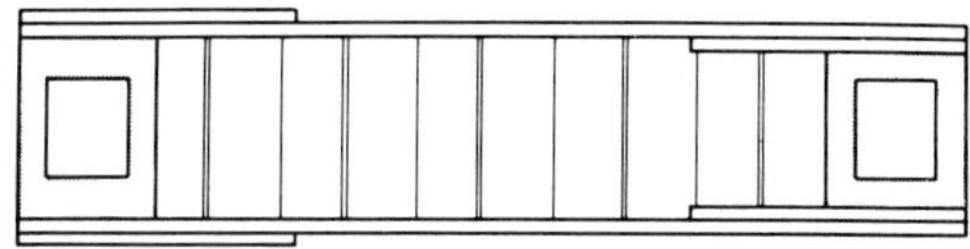

Figure 21.19. Internal and external doubler treatment.

K_s = shear deflection constant

L = beam span length or column height, in. (mm)

b = beam width, in. (mm)

M = maximum moment, in.-lb/in.width (m = kg/m width)

P = load, lb (kg)

P_{cr} = column critical load, lb/in. (kg/m)

P_y = column facing yield load, lb/in. (kg/m)

V = maximum shear force, lb/in. width (kg/m width)

d = sandwich total thickness, in. (mm)

h = distance between facing centroids, in. (mm); $h = t_c + t_f$

s = core cell size, in. (mm)

t_c = core thickness, in. (mm)

t_f = facing thickness, in. (mm)

w = uniform beam load, psi (Pa)

Δ = maximum beam deflection, in. (mm)

λ_δ = 1 − Poisson's ratio of the facing material squared $= 1 - \mu^2$

μ = facing material's Poisson's ratio.

σ_f = maximum facing stress, psi (Pa)

σ_y = yield stress of facing material, psi (Pa)

$\tau_{c\text{-}c}$ = maximum core compressive stress, psi (Pa)

τ_{cs} = maximum core shear stress, psi (Pa)

The need to be able to accurately calculate the exact performance for many forms of sand-

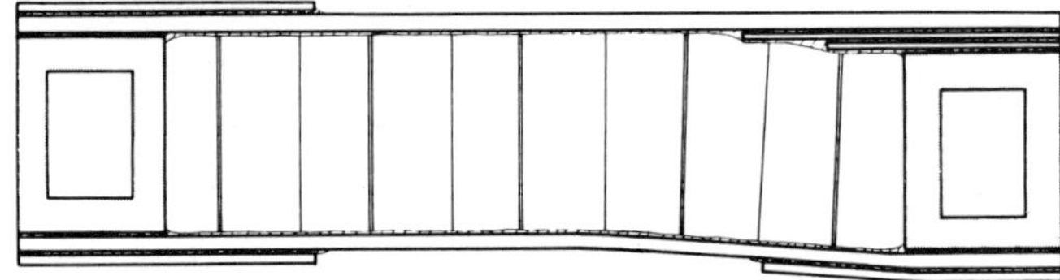

Figure 21.20. Low cost doubler treatment.

wich structures had led to the development of a substantial body of literature on the subject. This chapter will cover in detail only the commonly used analyses, and will provide reference sources for a number of others.

21.6.1. Design Requirements

Sandwich structures should be designed to meet the basic structural criteria listed below, when these criteria pertain to the type of loading under consideration.

1. The facings should be thick enough to withstand the tensile, compressive, and shear stresses induced by the design load.
2. The core should have sufficient strength to withstand the shear stresses induced by the design loads.
3. The core should be thick enough and have sufficient shear modulus to prevent overall buckling of the sandwich under load.
4. Compressive modulus of the core and the compressive strength of the facings should be sufficient to prevent wrinkling of the faces under the design load.
5. The core cells should be small enough to prevent intracell dimpling of the facings under design load.
6. The core should have sufficient compressive strength to resist crushing by design loads acting normal to the panel facings or by compressive stresses induced through flexure.

21.6.2. Modes of Failure

Typical modes of failure are shown in Fig. 21.22.

21.6.3. Design Steps

1. *Define loads.* For multi-point loadings, use the formulas in Roark's *Formulas for Stress and Strain.*[16]
2. *Define beam type.* The values of Fig. 21.23 provide the simple starting point for these calculations. Some care in using the fixed end type of support is needed, as in actual practice total fixity is not realized, and the resulting deflection is greater than that calculated.
3. *Determine deflection limitations.* For most applications, the allowable deflection of the structure is usually limited to L/360. In some cases, greater deflections may be used, or, as in the case of snow skis, very much greater deflections may be a normal part of the function of the structure.
4. *Select skin material.* Table 21.1 gives data for common sandwich skins. Skin considerations include the weight target, possible abuse, and local (denting) loads, corrosion or decorative constraints, and costs. Select standard thicknesses and make the initial calculation as outlined below. The facing thickness directly affects both the skin stress and the deflection.

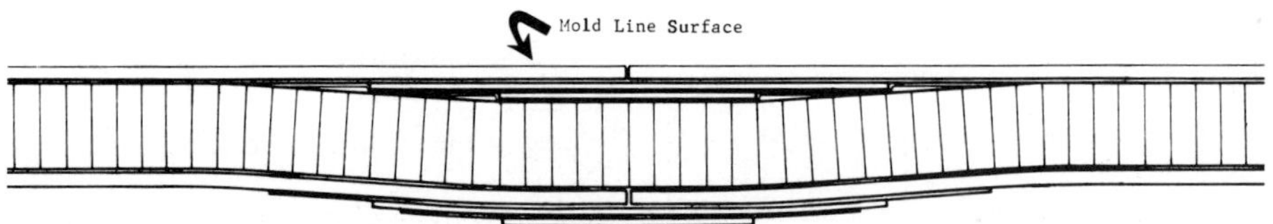

Figure 21.21. Double skin splice.

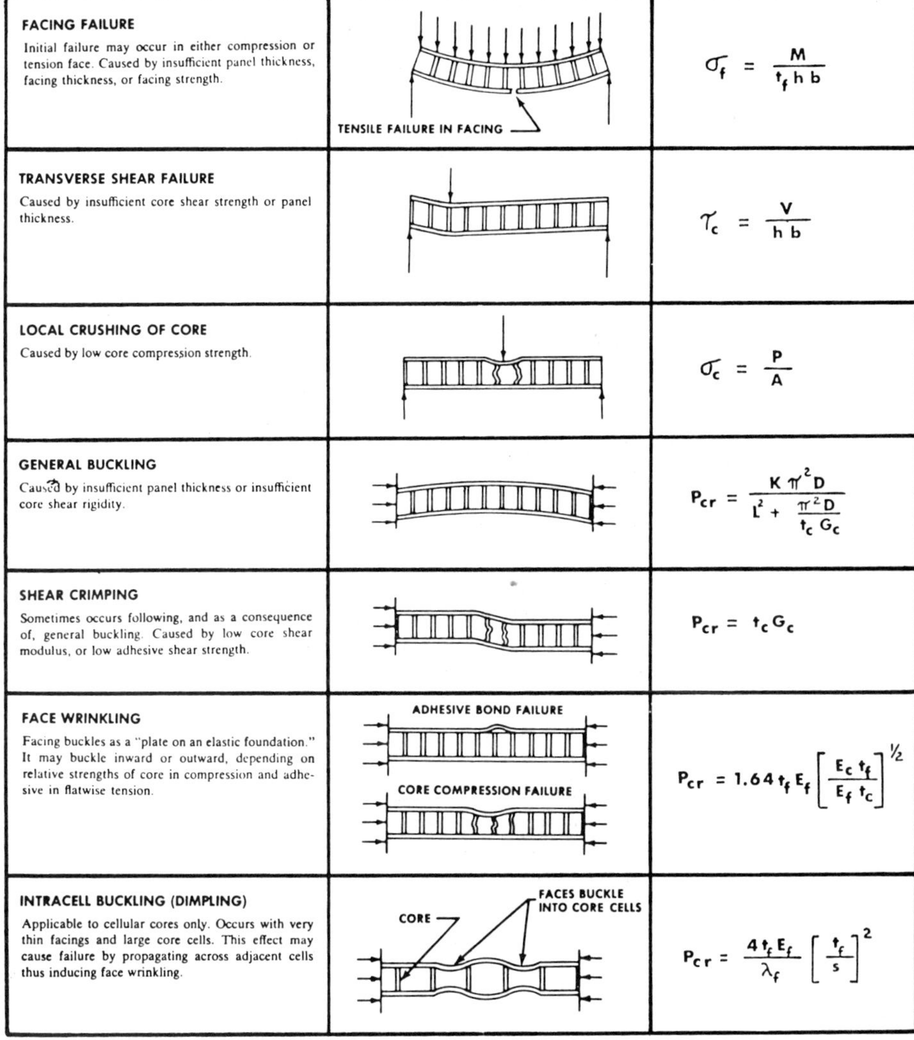

Figure 21.22. Modes of failure in sandwich structures. Sandwich structures must be designed to resist these modes of failure. Failures may occur which combine certain of the modes shown.

5. *Calculate first approximation.* After the first sandwich thickness, h, is determined, another selection of t_f or E_f may be made to arrive at more desirable or practical values of h. Most sandwich structures in ordinary usage are in the thickness range of 0.06–6.00 in. (1.5–150 mm).
6. *Select skin thickness.* Keep in mind that materials such as fiberglass cloth and aluminum are available in specific, standard thicknesses. After the skin thickness for deflection is selected, it should be checked for stress. The formula for δ_f is used, and a factor of safety determined.
7. *Select core.* Calculate the core shear stress, $\tau_{c\text{-}c}$. Make a preliminary selection from the materials listed in Tables 21.1–21.8. Note that the core strength is not the same in the L and W directions. Refine the selection, including considerations of material compatibility, cell sizes, and types. Determine the corrections needed to account for the effects of thickness on strength, as shown in Fig. 21.5. Check the factor of safety using the calculated stress and

BEAM TYPE	MAXIMUM SHEAR FORCE V	MAXIMUM BENDING MOMENT M	BENDING DEFLECTION CONSTANT K_B	SHEAR DEFLECTION CONSTANT K_S
SIMPLE SUPPORT, UNIFORM LOAD, P = ql	$\frac{P}{2}$	$\frac{PL}{8}$	$\frac{5}{384}$	$\frac{1}{8}$
BOTH ENDS FIXED, UNIFORM LOAD, P = ql	$\frac{P}{2}$	$\frac{PL}{12}$	$\frac{1}{384}$	$\frac{1}{8}$
SIMPLE SUPPORT, CENTER LOAD, P	$\frac{P}{2}$	$\frac{PL}{4}$	$\frac{4}{192}$	$\frac{1}{4}$
BOTH ENDS FIXED, CENTER LOAD, P	$\frac{P}{2}$	$\frac{PL}{8}$	$\frac{1}{192}$	$\frac{1}{4}$
CANTILEVER, UNIFORM LOAD, P = ql	P	$\frac{PL}{2}$	$\frac{1}{8}$	$\frac{1}{2}$
CANTILEVER, END LOAD, P	P	PL	$\frac{1}{3}$	1
CANTILEVER, TRIANGULAR LOAD, P = l/2 ql	P	$\frac{PL}{3}$	$\frac{1}{15}$	$\frac{1}{3}$
ONE END SIMPLY SUPPORTED, ONE END FIXED, UNIFORM LOAD, P = ql	$\frac{5P}{8}$	$\frac{PL}{8}$	$\frac{1}{185}$	$\frac{1}{14.2}$

Figure 21.23. Loaded beam chart P = total load (per unit width), L = span, σ_f = facing stress, t_f = skin thickness, h = centroid distance, τcs = core stress, shear, τcc = core stress, compressive, $\lambda = (1-\mu)^2$ facing property, E_f = modulus of elasticity of facings, G_c = modulus of elasticity of core in shear, I_{sw} = moment of inertia, sandwich, s = cell size, (cm), E_c = modulus of elasticity of core in compression, FS = factor of safety, T = total sandwich thickness (note that P must be determined for a beam unit width). If deflections are critical, actual deflections should be verified by tests.

the corrected allowable stress. Other considerations include crushing and compression strengths, modulus in shear, weight, and costs. For rolling wheel loadings, the crushing strength and the skin thickness are often the most important considerations.

8. *Check deflection.* For many applications, the calculation of the expected deflection may omit the shear deflection portion. With a very small deflection limitation, with a very thick sandwich, or with a very short span, the shear component should be calculated and the core selection may be influenced by the shear modulus needed.
9. *Face wrinkling and intracell dimpling.* With thin skins, a local failure of the skin in buckling may be encountered. A check on the $\sigma_{f_{crit}}$ will determine whether this may be a design consideration.
10. *Other considerations.* Often, honeycomb panels are supported on more than two sides. If the ratio of length to width is greater than 3:1, the calculations using the shorter span and designing as a unit beam are quite adequate. The formulas in Roark[16] are useful where the shear deflection may be ignored, using the following formulas.

$$I_{\text{sandwich}} = \frac{t_f h^2 b}{2}; \qquad I_{\text{solid}} = \frac{bt^3}{12}.$$

So, for plate calculations:

$$I_{\text{solid}} = 6 t_f h^2$$

Use of these formulas for deflections may

give lower values than actually experienced, since the shear deflection may be important. Table 88 of Roark[16] gives some approximate multipliers to use for plates when supported as noted.

21.6.4. Simple Formulas Which May Be Used

Bending stress in facings:

$$\sigma_f = \frac{M}{t_f h}$$

(where M is determined by Fig. 21-23).

Core shear stress:

$$\tau_{cs} = \frac{V}{h} \text{ (where } V \text{ is from Fig. 21-23).}$$

Deflection:

$$\Delta = \frac{K_B P L^3 2}{E_f t_f h^2} + \frac{K_s P L}{h G_c}$$

(K_B and K_s from Fig. 21.23).

For most beams, the second term is relatively small, but should be checked if deflection is critical or span is short.

Moment of inertia:

$$I_{sw} = \frac{t_f h^2 b}{2}.$$

Face dimpling:

$$\sigma_{f\text{crit}} = \frac{2E_f}{\lambda}\left[\frac{t_f}{s}\right]^2.$$

Face wrinkling:

$$\sigma_{f\text{crit}}\ 0.82 = E_f\left[\frac{Ect_f}{E_f tc}\right]^{1/2}.$$

Factor of safety:

$$FS = \frac{\text{Allowable or typical stress}}{\text{calculated stress}}.$$

21.6.5. Sample Problem: Analysis of Flat Rectangular Sandwich Beams

Design a flat roof panel for a bus stop. Use a snow load of 120 pounds per square foot. Use a simple panel with a simply supported span of 8 ft (0.203 m). Deflection is to be limited to L/270 and the factor of safety is to be greater than 2.0. Skin material is to be woven roving, polyester, and core to be KP-3/8-60(25).

Load, P': $P' = 120/144 = 0.833$ psi.

Span, L: $L = 8 \times 12 = 96$ in.

K_B, K_S, M, V from Fig. 21.23:

$K_B = 0.013$, $K_S = 0.125$, $M = 8$, $V = 2$.

Skin, t_f: Try $t_f = 0.090$ in.

λ, E_f from Table 21.1, fiberglass:

$\lambda = 0.98$, $E_f = 1.85 \times 10^6$.

Calculate h:

$$h = \frac{K_B P' L^4 2\lambda}{t_f\, \Delta\, E_f}^{1/2}$$

$$= \frac{0.013 \times 0.833 \times 96^4 \times 2 \times 0.98}{0.090 \times 96/270 \times 1.85 \times 10^6}^{1/2}$$

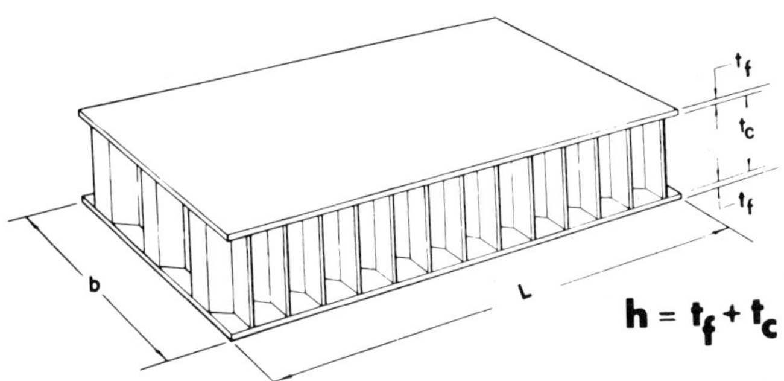

Figure 21.24. Schematic diagram of a flat sandwich panel.

$h = 5.518$ (round out to 5.5 = panel total thickness, $h = 5.41$)

Try thicker skin, $t_f = 0.150$.

$$h = \left[\frac{0.013 \times 0.833 \times 96^4 \times 2 \times 0.98}{0.150 \times 96/270 \times 1.85 \times 10^6}\right]^{1/2}$$

$$h = 4.274.$$

Use 4.00 overall thickness, $h = 3.850$. Since either construction is practical, check out the skin and core stresses:

$$\sigma_f = \frac{P'L^2}{t_f h M_c} = \frac{0.833 \times 96^2}{0.090 \times 5.41 \times 8} = 1971 \text{ psi;}$$

or,
$$\sigma_f = \frac{0.833 \times 96^2}{0.150 \times 3.85 \times 8} = 1662 \text{ psi.}$$

Note that the skin stresses are quite close; therefore, the factors of safety would be similar.

$$FS = \frac{38{,}000}{1971} = 19; \; FS = \frac{38{,}000}{1662} = 23.$$

For core:

$$\tau_{cs} = \frac{P'L}{hV} = \frac{0.833 \times 96}{5.41 \times 2} = 7.39 \text{ psi;}$$

or,
$$\frac{.833 \times 96}{3.85 \times 2} = 10.38 \text{ psi.}$$

Note that the core stresses are quite low, and there is not much difference in the stresses for the two thicknesses chosen. Select a core from Table 21.4. For KP-3/8-60(25), *W* shear strength = 60 psi. From Fig. 21.5: thickness factor = 0.42, *W shear modulus* = 5800.

$$W\text{, shear, corrected} = 60 \times 0.42 = 25 \text{ psi}$$

$$FS = 25/10.4 = 2.4.$$

The use of KP-3/8-60(25) with a factor of safety of 2.4 could be marginal, as Table 21.4 lists typical values, which may vary from lot to lot of material. The other properties, compression strength, and density are acceptable. Note that if the core is oriented to utilize the *L* shear properties, KP-1/2-80(11), with $\tau_{cs} = 70 \times 0.42 = 29.4$ might be satisfactory.

Calculate deflection:

For 5.50 T,
$$\Delta = \frac{K_B P' L^4 2}{t_f h^2 E_f} + \frac{K_s P' L^2}{G_c h}$$

$$= \frac{0.013 \times 0.883 \times 96^4 \times 2}{0.090 \times 5.41^2 \times 1.85 \times 10^6} + \frac{0.125 \times 0.833 \times 96^2}{5800 \times 5.41}$$

$$= 0.377 + 0.032 = 0.0409 \text{ in.}$$

Note that the added shear deformation is only 9% of the total deflection.

For 4.00 T,

$$\Delta = \frac{0.013 \times 0.833 \times 96^4 \times 2}{0.150 \times 3.85^2 \times 1.85 \times 10^6} + \frac{0.125 \times 0.833 \times 96^2}{5800 \times 3.85}$$

$$= 0.447 + 0.042 = 0.489 \text{ in.}$$

21.6.6. Analysis of Flat Rectangular Sandwich Columns: Column Design Example

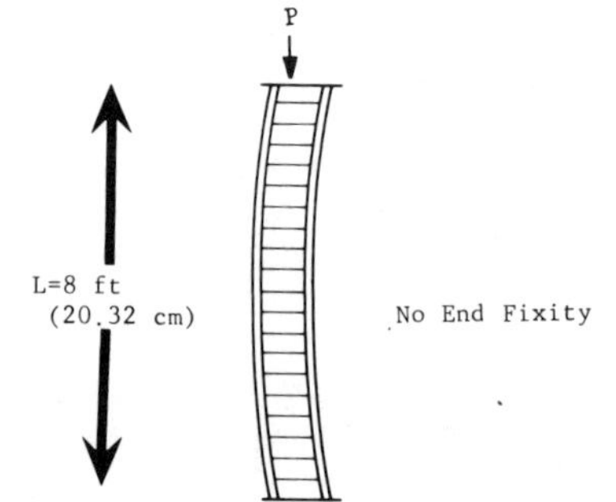

Facings:

Tempered hardboard, $\sigma_f = 3600$ psi.

From Table 21.1:

$$\mu = 0.99$$

$$E_f = 0.65 \times 10^6 \text{ psi}$$

Core:

Urethane foam, 6 lb/ft^3, $t_c = 3$ in.

From Table 21.3:

$$\lambda_{cs} = 90 \text{ psi}$$

$$\lambda_{c\text{-}c} = 170 \text{ psi}$$

$$G_c = 1500 \text{ psi.}$$

Check facing yielding:

$$P_y = 2t_f\sigma_y = 2(0.25)(3600) = 1800 \text{ lb/in.}$$

Check general buckling:

$$P_{cr} = \frac{\pi^2 D}{L^2 + \dfrac{\pi^2 D}{t_c G_c}} \qquad \left(\text{where } D = \frac{E_f t_f h^2}{2\lambda}\right)$$

$$D = \frac{0.65 \times 10^6(0.25)(3.25)^2}{2(0.99)}$$

$$= 866{,}872 \text{ lb-in./in. of width}$$

$$P_{cr} = \frac{\pi^2(866872)}{(96)^2 + \dfrac{\pi^2(866872)}{3.0(1500)}}$$

$$= 112 \text{ lb/in. or } 1352 \text{ lb/ft.}$$

Check shear crimping:

$$P_{cr} = t_c G_c = (3.00)(1500)$$

$$= 3214 \text{ lb-in./in. of width.}$$

Check dimpling and wrinkling: Since facings are relatively thick and continuously supported by foam core, dimpling or wrinkling will probably not occur.

21.6.7. Design Conditions

In-depth treatments for the design conditions listed below can be found in MIL-HDBK-23, available from the U. S. Government Printing Office.*

SUBJECT	MIL-HDBK-23 CHAPTER
Wrinkling of sandwich facings under edgewise load	3
Dimpling of sandwich facings under edgewise load	4
Design of flat, rectangular sandwich panels under edgewise compression load	5
Design of flat, rectangular sandwich panels under edgewise shear load	6
Design of flat, rectangular sandwich panels under edgewise bending moment	7
Design of flat, rectangular sandwich panels under combined loads	8
Design of flat sandwich panels under uniformly distributed normal load	9
Design of sandwich cylinders under external radial pressure	10
Design of sandwich cylinders under torsion	11
Design of sandwich cylinders under axial compression or bending	12
Design of sandwich cylinders under combined loads	13
Design of sandwich strips under torsion load	19
Design of flat circular sandwich panels loaded at an insert	20

*MIL-HDBK-23 is revised from time to time, with new chapters sometimes added and older material updated. A check with the Plastics Technical Evaluation Center, U. S. Army Armament Research and Development Command, Dover, New Jersey, 07801, can verify if you are in possession of the most recent revision.

21.7. MANUFACTURING SANDWICH STRUCTURES

The manufacture of sandwich structures requires three conditions to be met: the application of pressure; the application of temperature (both pressure and temperature in the precise amounts required for adhesive cure; and the provision for tooling and fixtures to hold the assembly in the desired shape and keep all the details in their proper positions during cure. Many different ways of providing these conditions are currently used, from vacuum bags or simple presses to autoclaves and unit tools, where volume and complexity can justify them. Most of the equipment is similar to equipment used in producing bonded structures or reinforced plastic parts where no sandwich structure is involved. However, bonding of sandwich structures is nearly always performed at lower pressures than is the bonding of structures which do not have a low density core, and tooling is sometimes lower in cost as a result. Aside from the need for lower maximum pressure, there is little noticeable difference between a sandwich bonding facility and one which only handles non-sandwich bonding.

A few suggestions can be offered to aid in living with the problems of sandwich bonding.

1. Make sure the core is properly sized to fit the space it is intended to occupy. If it has been stretched a little, to make the distance from one edge member to the opposite one, it will probably shrink back as the cure cycles starts, leaving mysterious voids next to an edge member. If it is undersize in thickness at an edge, the adjoining edge member or fitting will hold the facing away from the core and result in an unbonded area.
2. If a honeycomb core is being used, remember that the adhesive between the core and the faces will end up much thinner than the same adhesive between the edges or solid inserts and the facings. For this reason, it is common to require the core to be as much as 0.010 in. (0.25 mm) thicker than adjoining solid parts in the same assembly.
3. The elevated temperatures which most core-to-facing adhesives require for curing are often inaccurately measured. A good point to remember is that only the adhesive being cured can give you the cure temperature you are trying to measure. Some shops insert thermocouples directly into the bond line to determine temperature, and leave the thermocouple permanently in the part after cure.
4. Most adhesives flow at an early point in the cure cycle. At this time, the bond lines will change in thickness by substantial amounts. The tooling employed to establish the shape of the part and hold details in place must also allow the details to move into the final cured position. Simple examples are a hot platen press, in which the platens close on the sandwich as the bondlines grow thinner, or an autoclave, in which a flexible bag follows the details as the adhesive flows, continuously transmitting the autoclave pressure to the shrinking assembly. Most adhesives are very weak and crack-prone as they go through the gel point.
5. Inserts or heavy members being cured as a part of a very light assembly will heat up much more slowly, giving rise to warpage problems on cool-down. Warpage on very light parts can also be caused by one side cooling down too fast as a result of being removed from one side of still-hot tooling while the other side continues to stay at the temperature of the tool, or by one side heating faster or to a higher temperature than the opposite face. Slower heat-up rates or better heat distribution in the tool design will prevent these problems.
6. Be sure to provide a route for the escape of trapped air and gases from a totally enclosed part while it is being cured. This is particularly important in parts which are vacuum bagged to a female tool and cured in an autoclave. A coarse cloth bleeder should be enclosed inside the bag to prevent the bag from sealing off portions of the assembly as pressure is being applied. Critical or expensive assemblies should have several vacuum lines attached at different points of the bag, with each monitored separately by a pressure recorder.
7. Caul plates should be carefully matched to the job they are expected to perform. These tooling aids are often used to cover the top of an assembly containing several different pieces of core, inserts, edges, etc., so that a thin skin will not push each detail to the minimum bondline thickness and result in an uneven outer surface. When the caul plate is moderately stiffer than the top skin, the bonding pressure is transmitted more to the thicker inserts and less to the undersized inserts, allowing all of the details to "float" in the adhesive before cure, resulting in optimum relative placement of all the internal details in the sandwich. If the caul plate is extremely stiff—or thick—

this effect is changed to one of simply bridging over the most oversized details, and the danger of producing voids or unbonded areas over the thinner details is substantially increased. Generally, the caul plate should not be more than two or three times the thickness of the sandwich facing material. Where thicker caul plates are used, the dimensional control over the size of detail parts in the assembly must be correspondingly better. The advantage of using such a thick caul plate derives from the ability to make both sides of a sandwich part have the smooth appearance usually associated only with the "tool side."

8. Make sure that core, pre-cured or rigid edges, inserts, skins, and other relatively unyielding details assembled in the layup have close enough dimensional control to allow adhesives or resins to achieve the target strengths. In simple bonded assemblies, a figure of ±0.005 in. (±0.1 mm) is necessary, while assemblies having multiple layers of prepreg or many layers of thin metal doublers can sometimes be successfully produced with much less demanding dimensional control.

21.7.1. Core Shaping

When core materials must be cut, trimmed, carved, or shaped, many special purpose tools are available. Sawing is the most common machining method, using either conventional blade tooth patterns, or, for some trimming operations, a special "honeycomb band," in which the blade appears to be running backward, with the teeth sharpened on the back side so that each tooth acts as a slicing knife blade. A different type of saw is also used as a mandrel-mounted router bit. Such tools, shown in Fig. 21.25, are very common where sculpturing of honeycomb or foam is to be accomplished. Router speeds vary from 1200–30,000 rpm for blade diameters of 3/4–4 in. (1.8–10 cm). Roll forming can be accomplished on metal cores, as shown in Fig. 21.26, while non-metal cores must usually be heat formed. In either case, forming can be much easier if an inherently formable cell configuration is used.

Figure 21.25. Honeycomb carving bits employing a slitting saw 0.010 in. (0.254 mm) thick × 32 teeth per in., 2 in. (50.8 mm) in diameter at the cutting edge. Turning at 12–30000 rpm, these tools leave a smooth, burr-free surface on nearly any core material. The coarse teeth on the closer tool are for the purpose of breaking up and removing the excessive amounts of core removed in cut depths of 0.2–2.0 in. (5.08–50.8 mm).

Figure 21.26. Metal honeycomb may be roll-formed using ordinary slip rolls. The surface sometimes must be protected during the operation by a loose sheet of thin sheet metal.

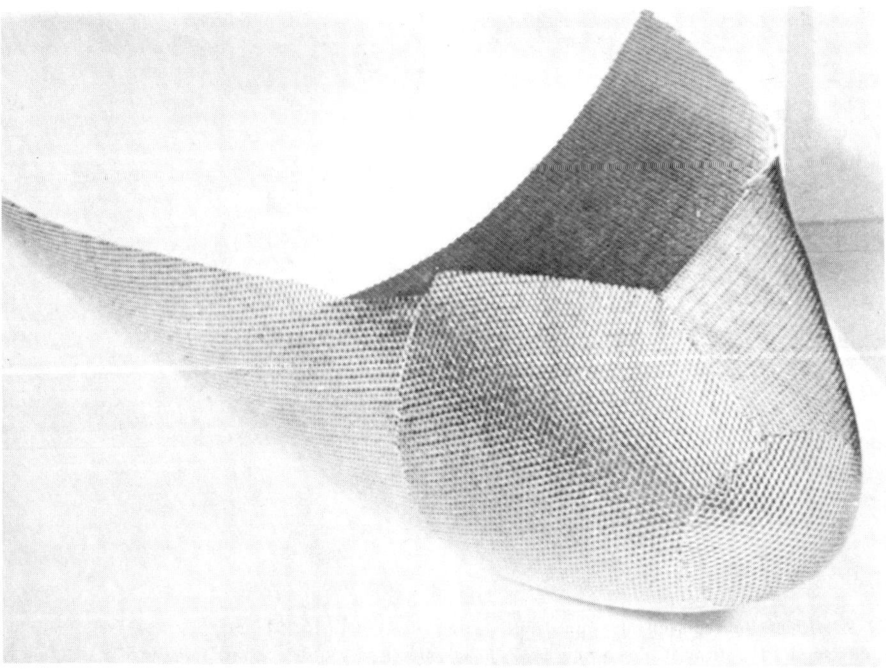

Figure 21.27. Nose radome core assembly assembled by edge-bonding together and post-formed sections of glass fabric-phenolic honeycomb.

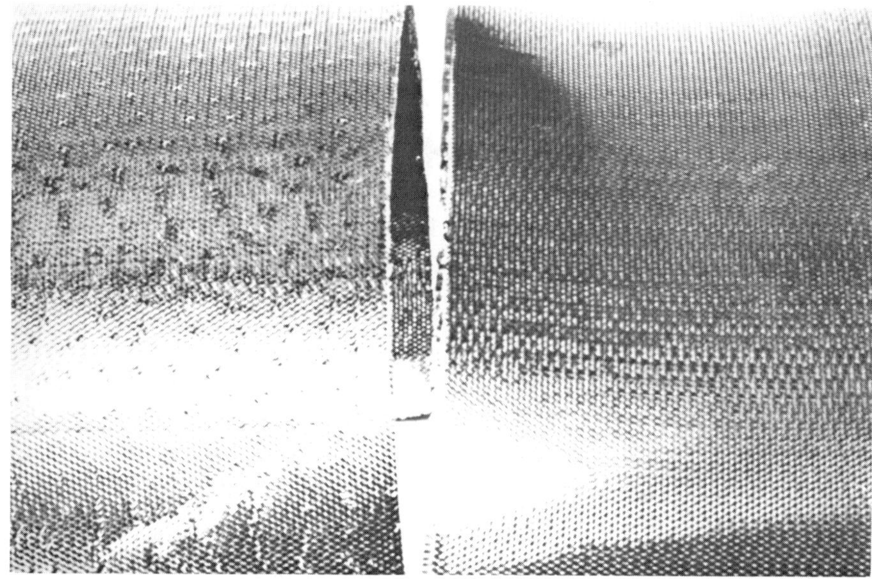

Figure 21.28. Effect of roll-forming on aluminum honeycomb. The core on the left has been roll-formed in sheet metal, forming rolls, while the piece on the right has not been pre-formed at all. Note the anticlastic, or "saddle shape," of the unformed piece when forced to a cylindrical form.

REFERENCES

1. MIL-HDBK-23, U. S. Government Printing Office, Washington, D.C.
2. MIL-HDBK-17, U. S. Government Printing Office, Washington, D.C.
3. MIL-HDBK-5, U. S. Government Printing Office, Washington, D.C.
4. MMM-A-132, U. S. Government Printing Office, Washington, D.C.
5. MIL-A-25463, U. S. Government Printing Office, Washington, D.C.
6. MIL-STD-401, U. S. Government Printing Office, Washington, D.C.
7. Adhesive Bonded Aerospace Structures Standard Repair Handbook, U. S. Government Printing Office, Washington, D.C.
8. Hexcel Corporation, TSB-120.
9. Hexcel Corporation, TSB-123.
10. Hexcel Corporation, TSB-124.
11. American Cyanamid, TR-112.
12. American Cyanamid, *Handbook of Adhesives*.
13. Baltek Corporation, *Baltek Catalogue*.
14. Plantema, Fredric J., *Sandwich Construction*, John Wiley & Sons, New York, N.Y. 1966
15. American Plywood Association, *Plywood Design Specification*.
16. Roark, R. J., *Formulas for Stress and Strain*, McGraw-Hill, New York, N.Y., 5th Edition, 1975
17. Timoshenko, S., Woinowsky-Krieger, S., *Theory of Plates & Shells*, McGraw-Hill, New York, N.Y. 2nd Edition, 1959.

*Publications of the U. S. Government may be updated and revised from time to time. Be sure you have the most recent edition. This can be checked by contacting the Plastics Technical Center, (201) 328-3189.

22
JOINING AND MACHINING TECHNIQUES

S. J. Dastin
Grumman Aerospace Corporation
Bethpage, New York

22.1. INTRODUCTION

Composites, like other structural materials, must be joined and machined to create useful structures. The manner in which these processes are performed is a determining factor in the efficiency and suitability of such components.

Joint and machining design guidelines for composites must be understood to avoid premature failures or excess weight and cost. Industry and government research agencies have investigated machining and joining of composites, and data have been published in MIL-HDBK-17 and the *Advanced Composite Design Guide.*

22.2. MECHANICAL FASTENING OF COMPOSITES

22.2.1. General

Mechanical fastening of laminated composites is an effective joining technique when consideration is given to tensile and bending stresses, as well as to strength and flexibility of the fastener, loss of tensile stress capability in the adherend due to the drilling, shear distribution in the joint, friction between parts, residual stresses, allowable bearing stresses, types of fasteners, and fatigue consideration.

The selection of the mechanical joining method for composites is as broad as it is with metals; namely, riveting, bolting, pinning, and fastening. The joining selection is based on consideration of component required strength (static and fatigue), reliability, ease of fabrication, cost, and special joint functions (removable, replaceable, etc.).

Composites are mechanically fastened in a manner similar to the fastening of metals (i.e., adherends are drilled and countersunk and joined with rivets, bolts, screws, or pins).

22.2.2. Joint Design—Mechanically Fastened Composites

Composite materials are orthotropic, possessing three mutually perpendicular axes of elastic symmetry which correspond to the two directions in the plane of the sheet and the other in the direction of the thickness. In woven fiberglass-reinforced plastics (FRP), where the reinforcement of each ply has two mutually perpendicular strength directions, the bearing strength of the composite appears to be unaffected by fiber orientation, but other strength properties are markedly different, with different reinforcement orientation.

Although many composites are possible, the glass fiber-, boron-, aramid (Kevlar)-, and graphite-reinforced thermosetting resin matrix types are the most utilized for structures. The composite is a laminate, composed of multiple layers of reinforcement, one layer over the other. The reinforcement provides resistance to tensile and compressive loads, while the matrix transmits the loads from fiber to fiber and prevents fiber buckling. Generally, the service temperature of the composite is related to the type of matrix selected.

The stress-strain diagrams for composites are such that they resemble cast iron and other "brittle" materials rather than steel and other "ductile" materials. Most composites have no yield point but do show a secondary linear stress to strain ratio. Most composites are low elongation materials with rupture occurring at 1–2% strain. The primary considerations of composite joint design are bearing, shear tear-out, tensile/compressive strengths in the composite, and shear/bending stresses in the fastener. Additional factors of concern are thermal stresses, fatigue requirements, and the operational environment.

22.2.3. Joint Geometry

Several types of mechanical joints have been successfully used in composites: laps, butt-plate (single and double shear), tapered butt-plate, offset lap, and others, shown in Fig. 22.1. In the double shear butt joints, the bending stresses common to the other joints are avoided, while the tapered butt-plate joint minimizes excessive loads at the joint edges.

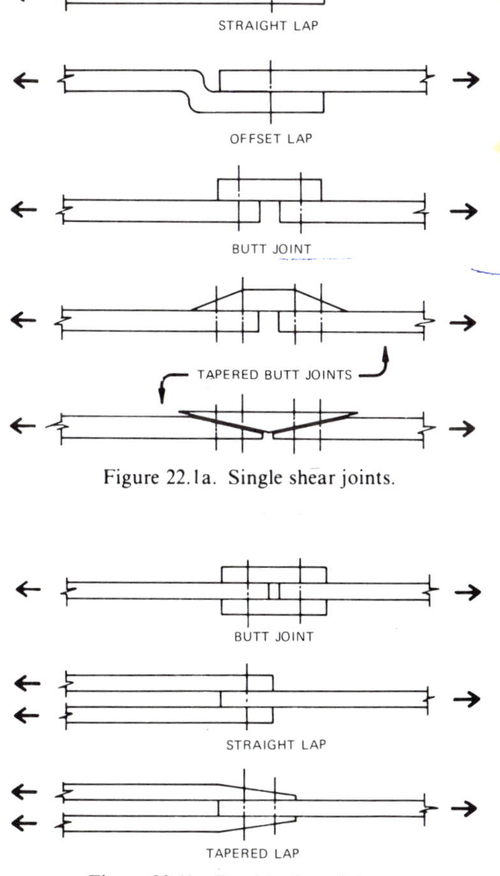

Figure 22.1a. Single shear joints.

Figure 22.1b. Double shear joints.

Typical stress concentration in the composite adherend in the immediate vicinity of a fastener hole is shown in Fig. 22.2*a*.

22.2.4. Design Criteria

Generalized design criteria are difficult to present unless the type of structure, the purpose of the product, classification of the joint, and many other variables are established. However, the following guidelines have been used in structurally sound components for positive margins on ultimate loading, considering bearing and in-plane strengths with stress concentrations due to the presence of holes. Joint design should be verified by subscale element static and fatigue tests, including the effects of the environment. Cut-off bearing strength of composites is the stress that produces 4% elongation of the diameter of the hole carrying the fastener. Typical 4% hole elongation bearing strengths are given in Table 22.1. The bearing strength of composites generally decreases with increasing D/t ratios (hole diameter divided by composite thickness), as shown in Fig. 22.2*b*. Conservative bearing load designs are those that utilize a D/t ratio of 1 for fiberglass-reinforced composites greater in thickness than 0.090 in. (2.29 mm), while using somewhat larger D/t ratios along with the reduced allowable bearing strength of large D/t ratios for thinner laminates. Extremely thin laminates, 0.030 in. (0.76 mm), should be reinforced locally at the attachment area to provide a greater thickness and thus to avoid the extremely reduced bearing allowables that result at D/t ratios greater than 4. For boron- and graphite-reinforced composites, D/t ratios of up to 3 are generally satisfactory for most thicknesses. Composites can be countersunk for flush fasteners, but the adherend thickness should be selected so that at least 0.030 in. (0.76 mm) of the thickness is maintained beyond the depth of the countersink. Recommended thicknesses of fiberglass-reinforced composites for various size flush fasteners are given in Table 22.2, using the optimum 100° countersink. Design practice for fiberglass-reinforced

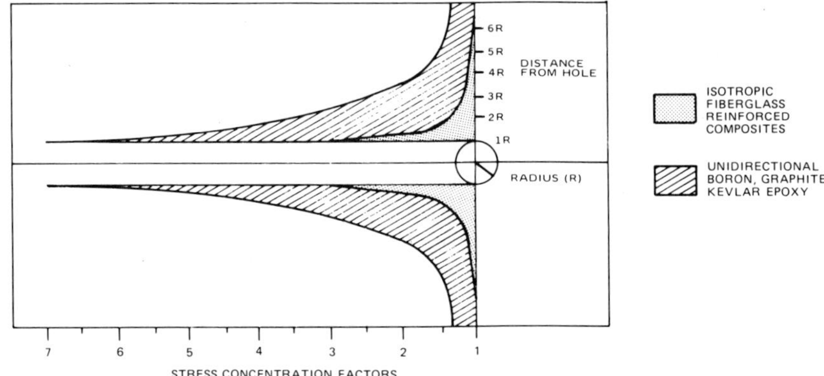

Figure 22.2a. Stress distribution is most concentrated in the immediate vicinity of a hole in composites (modified from Ref. 1).

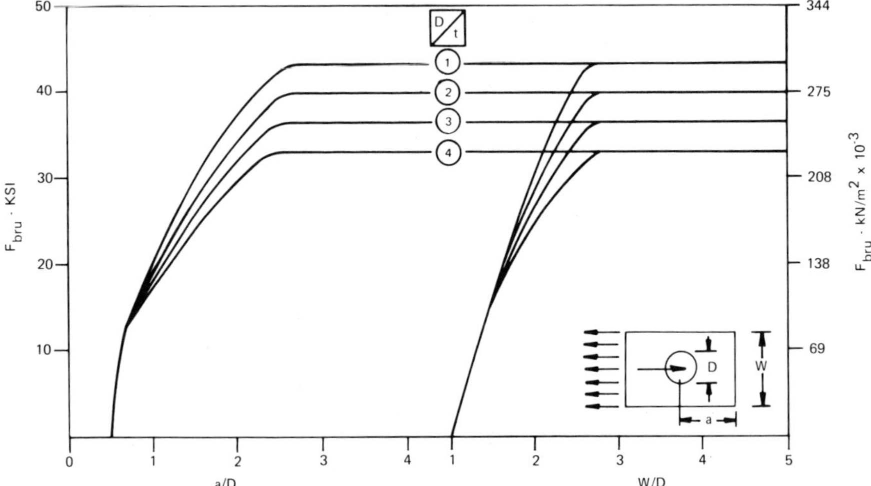

Figure 22.2b. Bearing strength parallel lay-up (style 181 fabric-polyester resin) loaded in warp direction (adapted from Ref. 2).

composite bearing areas for flush fasteners is based on the uncountersunk thickness and diameter of the hole. For joints requiring minimum weights and low margins of safety, the bearing area is based on the uncountersunk thickness plus one-half of the countersunk depth and the hole diameter. The bearing area for boron- and graphite- reinforced composite flush fasteners utilizes the complete countersunk depth, since bolt fit-up tolerances are tightly held. Conservative design of the net tensile area across a row of countersunk holes is the width of the joint minus the effective hole diameters and the adherend thickness. Effective hole diameters (equivalent hole diameter) for 100° countersinks in fiberglass-reinforced composites are given in Figure 22.3. For boron and graphite composites, net tensile allowables are predetermined for specific orientations. Fastener type and diameter are based on consideration of D/t ratio, minimum countersink depth, fastener strengths, service temperature, joint loading, available fastener area, stress concentration factors, durability, and environmental stability.

Small fasteners, 1/16-in. (1.59-mm) diameter, are seldom used in aerospace structures since they are difficult to handle and are prone to locally damaging the composite during installation. Conversely, large diameter fasteners above 5/16-in. (7.94-mm) diameter, must be used with care when joining aramid (Kevlar)- or fiberglass-reinforced composites,

Table 22.1. Bearing Strengths for a D/t Ratio of One

	ULTIMATE STRENGTH, psi (MPa)	DESIGN ALLOWABLE (4% HOLE ELONGATION), psi (MPa)
Fiberglass-reinforced composite		
Woven 181—polyester	43,200 (297.6)	35,000 (241.1)
Mat—polyester	30,000 (206.7)	20,000 (137.8)
Woven 181 glass fabric—epoxy	46,500 (319.5)	37,000 (254.9)
Boron-reinforced composite		
0°/90° epoxy	200,000 (1,378.0)	150,000 (1033.5)
0°/90°/± 45° epoxy	150,000 (1,033.5)	120,000 (826.8)
Aramid (Kevlar)-reinforced composite		
Woven 181 glass fabric—epoxy	55,000 (378.9)	45,000 (310.0)
Graphite-reinforced composite		
0°/90° epoxy	65,000 (447.8)	55,000 (378.9)
0°/90°/± 45° epoxy	50,000 (344.5)	45,000 (310.0)

since high installation loading is required. For boron- and graphite-reinforced composites, this is of lesser importance. Generally, à fastener is selected that is the smallest in diameter consistent with the strength required for the design. Wherever possible, when design requires composites less than 0.060 in. (1.52 mm) in thickness, metallic washers are used.

Edge and side distance considerations are of prime importance in mechanical fastener design. Edge distance is defined as the distance from the end of the joint to the center of the closest hole, and side distance is the distance from the sides of the joint to the center of the nearest holes (see Fig. 22.4*a*).

Table 22.3 gives the recommended edge and side distances for composites of various thicknesses.

Design of joints of multiple fasteners in composites should be based on controlled pitch. Pitch is defined as fastener spacing, the distance between the centers of a row of fasteners (Fig. 22.4*a*). Generally, a pitch of $5D$ (5 times the fastener diameter) is acceptable for joints of large safety factors, while $4D$ is a minimum value. In designs of multiple rows of fasteners, pitch on outer rows should be greater than that on inner rows, since large tensile loading can thus be tolerated in the joint.

Designs must consider eccentricity of the

Table 22.2. Recommended Thicknesses for Countersinks

FASTENER SIZE, in. (mm)	MINIMUM THICKNESS FOR 100° COUNTERSINK, in. (mm)
3/32 (2.38)	0.07 (1.78)
1/8 (3.18) (#4)	0.08 (2.03)
5/32 (3.97) (#6)	0.09 (2.29)
3/16 (4.76)	0.10 (2.54)
7/32 (5.56) (#8)	0.11 (2.79)
1/4 (6.35)	0.125 (3.18)
9/32 (7.14) (#10)	0.140 (3.56)
5/16 (7.94)	0.190 (4.83)
3/8 (9.52)	0.220 (5.59)

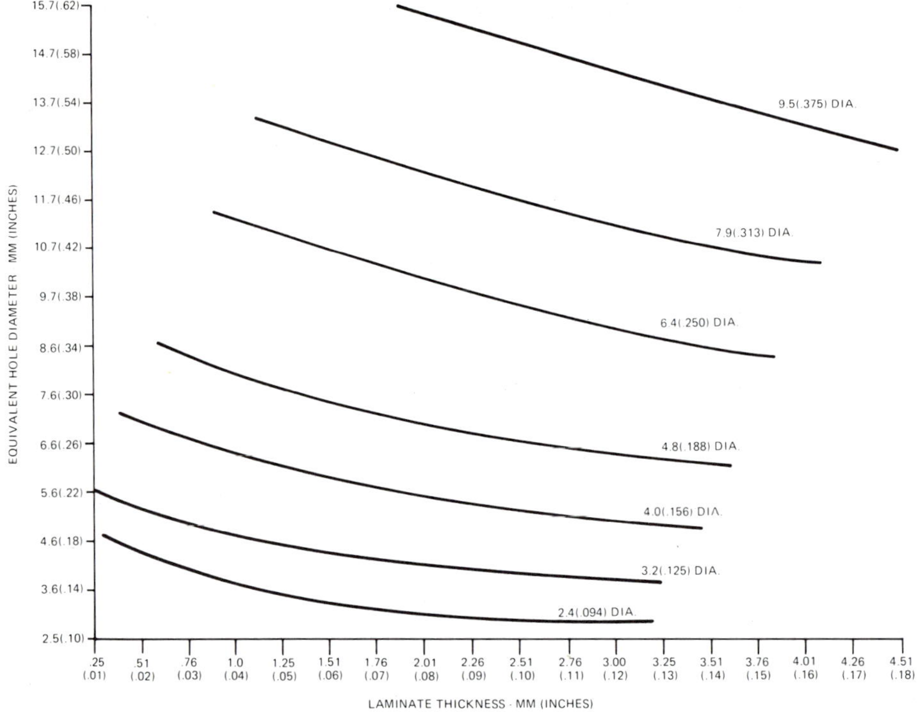

Figure 22.3. Equivalent diameters for holes countersunk for 100° flush rivets (adapted from Ref. 1).

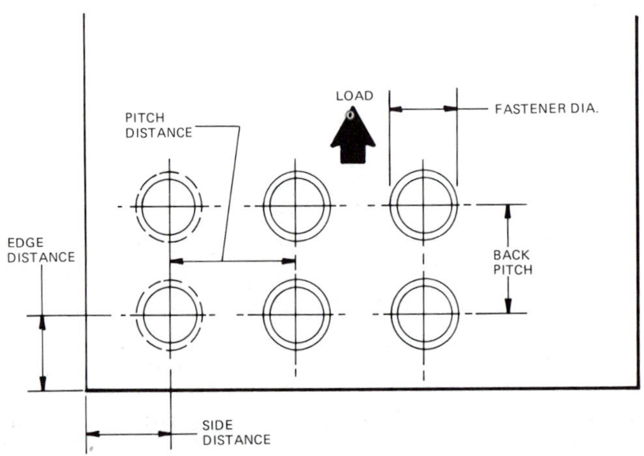

Figure 22.4a. Spacing of holes.

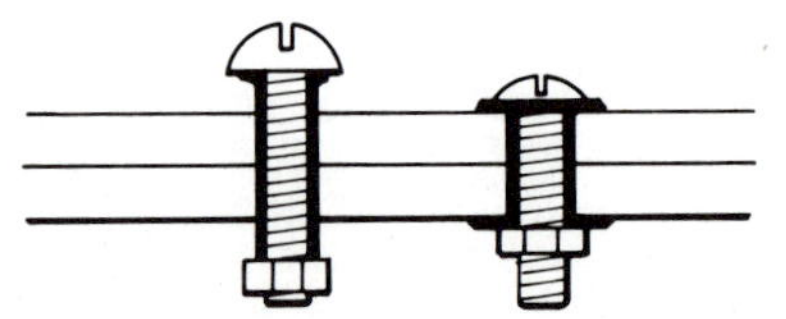

Figure 22.4b. Spacers for bolted connection, in solid laminates (adapted from Ref. 3).

Table 22.3. Recommended Fastener Distances for Various Composite Thicknesses (Based on Fiberglass/Epoxy)

THICKNESS OF LAMINATE, in. (mm)	EDGE DISTANCE RATIO (EDGE DISTANCE/HOLE DIAMETER)	SIDE DISTANCE RATIO (SIDE DISTANCE/HOLE DIAMETER)
1/8 (3.18) or less	3.0	2.0
1/8–3/16 (3.18–4.76)	2.5	1.50
3/16 (4.76) and greater	2.0	1.25

joint, which is inherent in all unsymmetrical type fastening. Eccentricity produces a loading couple which is reacted as a bending stress in the joint. Multiple rows of fasteners minimize damage due to bending but also minimize axial load carrying capacity. In such a case, a single row of fasteners displaced every second fastener could be an improvement for this condition. In eccentric joints, fastener back pitch—distance between adjacent rows (see Fig. 22.4*a*)—should be no less than $4D$ (four times fastener diameter). The high back pitch (symmetrical joint back pitch is $2D$) provides the displacement needed to minimize stresses caused by joint eccentricity.

Designs of composite joints can require local built-in added thickness in joint areas to provide ease of countersinks and to maximize bearing loading. The added reinforcement should be molded in a stepwise fashion (tapering) to avoid sharp changes in part stiffness and to minimize stress concentrations.

22.2.5. Joint Efficiency

The joint efficiency of metallic assemblies is different than for composite assemblies in that the former is based on net strength while the latter is based on load carrying ability comparison. Joint efficiency of either basis is the ratio of the strength of the joint members compared to the strength of an unjoined continuous member of equal size. Complete analysis of "effective" joint efficiency would require a weight factor as well as a strength factor. The double standard of efficiency between materials of construction stems from the fact that metallic materials are ductile, while composites are "brittle" in the sense that they do not exhibit yielding. Joint efficiency for metallic components is as follows.

$$\text{Efficiency} = \frac{(w - nd)}{w}$$

where

w = width of member, in. (mm)
n = number of fasteners in a row
d = fastener diameter, in. (mm).

Joint efficiency for composites is as follows.

$$\text{Efficiency} = \frac{L_f}{L_c}$$

where

L_f = load producing failure, lb (N), in joint
L_c = load producing failure, lb (N), in continuous member.

22.2.6. Material Design Allowables for Joints

Composite design allowables are of necessity lower than average strengths reported to provide for the following considerations.

- High probability (90–95%) at high confidence level (95%) of reproducibility of strengths.
- Variations in materials and processes.
- Inspection limitations.
- Size and geometry effects.

Design allowables for composite joints, in addition to the bearing strength allowable given in Table 22.1, working allowables for the various type composites in edgewise shear, shear tear-out, and interlaminar shear loading, must be defined. Table 22.4 lists recommended shear values.

It is to be noted that the values shown in Table 22.4 are based on the assumption that stress concentrations are not severe. Addi-

Table 22.4. Shear Properties of Composites at Room Temperature

PROPERTY	GRAPHITE/EPOXY, CROSS-PLY	FIBERGLASS-EPOXY, 181 WEAVE	BORON/EPOXY, CROSS-PLY	ARAMID (KEVLAR)/EPOXY, 181 WEAVE
Panel shear (edgewise) strength, psi (MPa)	20,000(137.8)	16,000 (110.2)	30,000 (206.7)	18,000 (124.0)
Modulus, msi (GPa)	1.3 (8.96)	0.8 (5.51)	1.5 (10.34)	1.0 (6.89)
Shear tear-out, psi (MPa),	10,000 (68.9)	8000 (55.1)	15,000 (103.3)	8000 (55.1)
Interlaminar shear, psi (MPa)	8000 (55.1)	4000 (27.6)	10,000 (68.9)	6000 (41.3)

tional reductions should be considered, especially for tear-out and interlaminar shear loading, if designs are such that high stress concentrations are unavoidable.

22.2.7. Safety Factor

Like metallic joints, composite joints are designed with safety factors. Factor of safety (*FS*) is the ratio of the design allowable stress divided by the maximum stress level encountered in service. Aircraft designs for metallic components are geneally based on *FS* of 1.5 on ultimate allowables, or 1.15 on yield allowables. In composite structures, which display no true yield point, *FS* values are generally based on ultimate allowable strengths. A conservative *FS* for fiberglass-reinforced structures is 3, while designs that are weight critical utilize an *FS* of 2 and require greater quality assurance to avoid premature failures. For boron-aramid (Kevlar)-, and graphite-reinforced composites, the *FS* is 1.5, along with a practical margin of safety of 15%.

22.2.8. Joint Analysis

Generally, a joint is suitable for the intended application if the load required to produce failure is equal to the design load. Design load is defined as the product of the maximum service load anticipated and the safety factor selected for the design. The design requires modification if the joint analysis shows the joint failing load is either less than or excessively greater than the design load. The analysis of composite joints is comprised of the following.

- Establishment of maximum service loading and *FS* required and comparison to the ascertained material design allowable.
- Determination of controlling conditions (i.e., static versus fatigue, thermal environments, sonic conditions, service life, exposures, and corresponding margins of safety).
- Determination of effects of combined loading on joint allowables.
- Determination of the joint mode of failure and loading conditions inducing failures. Classic failure modes for composite joints are shear in the fastener, and tensile or bearing failure in the base material. Bearing failures in joints are not catastrophic, while matrix tensile failures are rapid and with little warning. In some cases, mechanically fastened composite joints fail by shear tear-out if small edge distances are used, or by fastener pull-through if the joint has excessive normal loading or high edge moments (peel loading).

22.2.9. Fasteners for Composites—Types and Design Allowables

Selection of mechanical fasteners for composite joints is based not only on the imposed service loads but on the installation loads as well. The requisite service load on the fastener can be calculated, but determination of loading during assembly is difficult to assess. Installation loads, especially on small fasteners, of as low as 10% of the total axial load of the fastener used can cause undetected damage to the composite, which can lead to a

Table 22.5. Holding Forces of Fasteners in Fiberglass-Reinforced Composite (Courtesy of SPI)

	POLYESTER 181 CLOTH, TYPE OF FASTENER-MACHINE SCREW							
	AXIAL HOLDING FORCE				LATERAL HOLDING FORCE			
	MINIMUM		MAXIMUM		MINIMUM		MAXIMUM	
FASTENER SIZE-THDS	DoP*, 0.625 in. (1.58 mm)	FORCE, lb (N)	DoP*, 0.625 in. (1.58 mm)	FORCE, lb (N)	DoP*, 0.625 in. (1.58 mm)	FORCE, lb (N)	DoP*, 0.625 in. (1.58 mm)	FORCE, lb (N)
4–40	2 (3.17)	130 (579)	5 (7.94)	500 (2225)	1 (1.58)	210 (934)	2 (3.17)	350 (1557)
6–32	2 (3.17)	170 (757)	6 (9.53)	820 (3649)	1 (1.58)	250 (1112)	3 (4.76)	700 (3115)
8–32	3 (4.76)	320 (1424)	7 (11.11)	1140 (5073)	2 (3.17)	440 (1958)	3 (4.76)	810 (3604)
10–32	3 (4.76)	420 (1869)	8 (12.7)	1500 (6675)	2 (3.17)	850 (3782)	4 (6.35)	1200 (5340)
1/4–20	4 (6.35)	700 (3115)	10 (15.9)	2800 (12,460)	2 (3.17)	1200 (5340)	5 (7.94)	1900 (8455)
5/16–18	4 (6.35)	820 (3649)	12 (19.1)	4200 (18,690)	3 (4.76)	2100 (9348)	6 (9.53)	2800 (12,460)
3/8–16	5 (7.94)	1160 (5162)	14 (22.2)	6000 (26,700)	4 (4.76)	3300 (14,685)	7 (11.11)	3900 (17,355)
7/16–14	5 (7.94)	1300 (5785)	16 (25.4)	8000 (35,600)	4 (4.76)	3800 (16,910)	8 (12.7)	5200 (23,140)
1/2–13	6 (9.53)	1660 (7387)	17 (26.9)	9800 (43,610)	5 (7.94)	5300 (23,585)	10 (15.9)	7500 (33,375)
9/16–12	6 (9.53)	1820 (8099)	18 (28.6)	2000 (53,400)	5 (7.94)	5800 (25,810)	11 (17.5)	9000 (40,050)
5/8–11	7 (11.11)	2200 (9790)	20 (31.7)	15,300 (68,085)	6 (9.53)	7300 (32,485)	13 (20.6)	11,900 (52,955)
3/4–10	7 (11.11)	2550 (11,347)	22 (34.9)	20,600 (91,672)	7 (11.11)	8900 (39,605)	15 (23.8)	15,000 (66,750)

*DoP = depth of penetration.

Laminate mechanical properties:

Tensile strength—35,000–45,000 psi (241.1–310.0 MPa).

Edge compressive strength—25,000–37,000 psi (172.2–254.9 MPa).

Shear strength (double shear)—12,000–17,000 psi (82.7–117.1 MPa).

premature tensile or compressive failure. Threaded fasteners cause such conditions; therefore, whenever possible, washers should be used to avoid installation damage.

General procedures and tolerances for countersinking, counterboring, drilling, and tapping are given in Section 22.4.

Types of threaded fasteners for aramid (Kevlar)- and fiberglass-reinforced composites, typical fastener pull-out strengths, both axially and laterally, for several screw sizes are given in Table 22.5 for fiberglass woven Style 181 laminates. Whenever possible, especially for applications of multiple assembly/disassembly, bonded threaded inserts should be used. In applications where high installation loads cannot be avoided, metallic spacers should be used (threaded or unthreaded) to preclude compression or bearing premature failures. Figure 22.4*b* shows typical spacers that can be used.

Threaded fasteners used for metallic structures are applicable to composite structures. Standard threaded fasteners and related hardware are usually described in the applicable government specifications of the *Military Standards Manual, Volume 1, Part 1* of either *AN* (*Air Force-Navy Aeronautical Standard*) or *MS* (*Military Standard*) for primary structural joints or various industry specifications. For graphite-reinforced composites, noble fasteners (A286 stainless steel or titanium) should be used to minimize potential galvanic corrosion.

Riveted joints should be designed to avoid loading the rivet in tension. Unless ultra-high fastener strength is required, soft rivets should be used. Since these rivets are not considered structural fasteners, guaranteed minimum strengths at known reproducibility are not published. The design allowables for such rivets are generally taken at 3/4 of the published typical strengths of such rivets. The selection of rivet length should be based on the thickness of the components, as shown in Fig. 22.5*a*. Whenever possible, trusshead rivets should be employed to take advantage of the 30% larger area under the fastener head, as compared with button-head rivets (see Fig. 22.5*b*). Rivet holes should be large enough to permit rapid assembly and yet not so large as

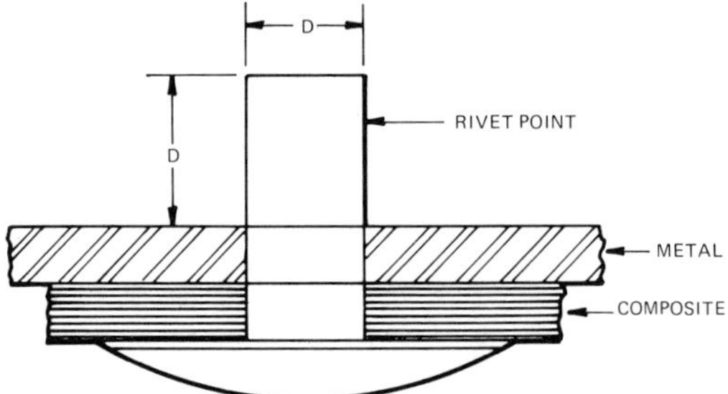

Figure 22.5a. Rivet length.

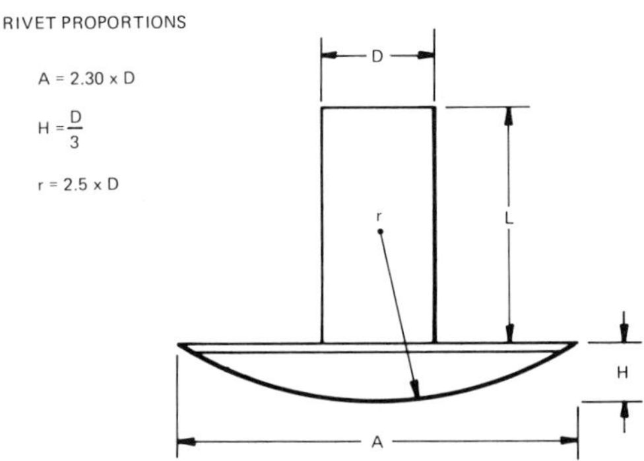

Figure 22.5b. Proportions of truss head rivet (adapted from Ref. 3).

to cause bending of the rivet shank. Generally, a hole 1/64 in. (0.39 mm) larger than the rivet/shank is satisfactory.

For joint loading requiring structural fasteners, Hi Loks, Hi Shear, Huck bolts, or Riv nuts should be used, and for the blind side, pull-up fasteners such as Jo bolts should be considered (see Fig. 22.6).

Control of aluminum rivets for aramid (Kevlar)- or fiberglass-reinforced composite joints used for aircraft applications may be in accordance with the applicable government specifications of the *Military Standards Manual* (*MS*) and *National Aerospace Standards* (*NAS*) or industry specifications.

22.3. ADHESIVE BONDING OF COMPOSITES

22.3.1. General

Structural adhesive bonding of composites to composites, as well as composites to metallic components, has developed rapidly and at present is utilized in many military as well as commercial aircraft. Adhesive bonding materials and processes are clearly defined by

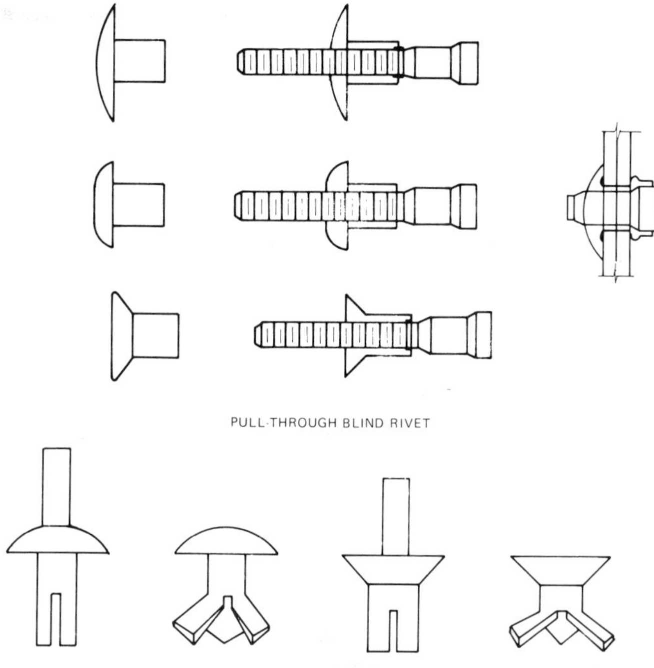

Figure 22.6. Blind fasteners (adapted from Ref. 3).

government and industry specifications both for built-up structures and for sandwich components. Structural adhesive bonding for aircraft became a significant joining method in 1945, and with the passage of time has seen increased usage. Adhesive suppliers have continually improved their products so that, with presently existing materials, the range of service temperature is −423–500°F (−253–260°C). Comparison of bonded versus mechanically fastened joints shows the following generalities.

1. After first loading, bonded joints show less permanent set than equivalent mechanical joints.
2. Bonded joints exhibit lower stress concentrations than mechanical joints where penetration of the member to be joined is needed, and thus provides increased static strength.
3. Bonded joints possess the potential of 25% weight savings in secondary structures, and between 5 and 10% for many primary structures, as compared to mechanical joints (Friendship F-27 aircraft).
4. Bonded joints enable the design of smooth external surfaces, and integrally sealed joints with minimum sensitivity to crack propagation.
5. On large area joints, bonded assemblies are generally less costly than the mechanical joint counterparts.
6. Bonded joints allow for the assembly of dissimilar materials with respect to galvanic corrosion, but thermal stresses due to the dissimilar adherends must be considered.

The disadvantages of adhesive joining are:

1. Elevated temperature creep resistance is poor for some types of adhesives.
2. Bonded joints require designs that minimize peel loading.
3. Bonded joints are difficult to completely inspect non-destructively.
4. A cured bonded joint is permanent and thus cannot be considered for removable assemblies.
5. Efficient structural bonded joints require accurate mating of adherends (similar to welding).

6. Durability and life of bonded joints must be validated by costly laboratory accelerated testing.

22.3.2. Joint Design

22.3.2.1. Basic Considerations

External loads to primary aircraft structures seldom produce uniform fields of stress, even in homogeneous metallic structures. In composite structures, the field of stress is almost always anisotropic, and thus the adhesive joint contains areas of asymmetric stresses. The adhesive joint is further burdened by stress variations due to the elevated temperature setting processes for the adhesive. The designed adhesive joint geometry as well as the adhesive selected determines the magnitude of stress differences, stress-rupture time, and dimensional stability of the joint.

In general, the "non-elastic" type adhesives, adhesives with non-linear relationships between stress and strain (high shear strength, and low shear modulus), tend to sharply reduce stress concentrations in bonded joints. On the other hand, this type of behavior in the composite adherends tends to increase the magnitude of joint stress concentration.

It should be realized that many of the existing analyses of stresses in composite adhesive joints are based on the assumptions that the adherends are elastic. Therefore, consideration of anisotropy, material time-dependent properties, and inelasticity are accounted for by generous safety factors.

22.3.3. Joint Geometry

The classic types of adhesive bonded joints of structural members are scarfed joints, tapering lap joints, simple lap joints, double lap joints, and modified lap joints. Figure 22.7 shows these joints in schematic form. These various joints react differently under the many possible types and directions of load. Figure 22.8 shows a typical aircraft panel loaded in

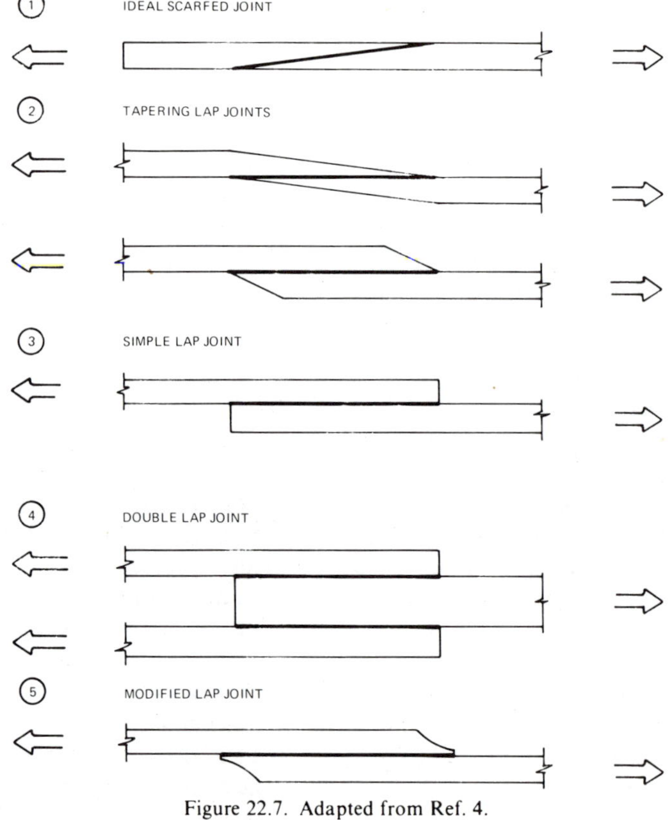

Figure 22.7. Adapted from Ref. 4.

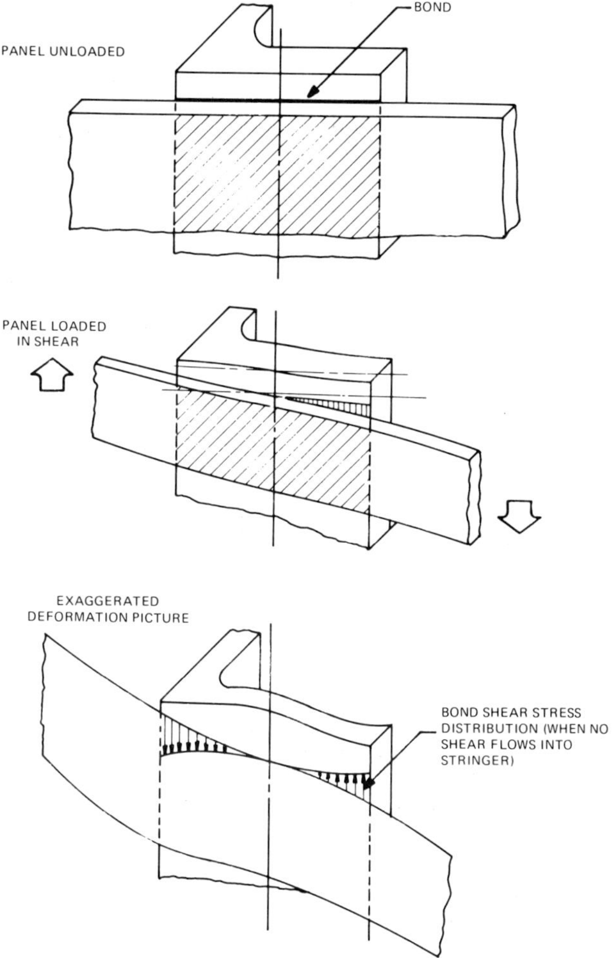

Figure 22.8. Adapted from Ref. 4.

shear. A joint which may be ideal for compression loading may be poor in tension or bending. Figures 22.9–22.12 represent tendencies of various joints to develop stress concentrations under load. These tendencies are presented merely as a guide, and each specific design should be verified by mechanical tests.

The typical stress distribution of bonded composite structures is shown in Fig. 22.13. As shown by Fig. 22.14, for simple overlaps, when the length of overlap increases, the load carrying ability increases somewhat—lb/in (kg/m) of width—but the allowable adhesive shear stress, psi (kN/m^2) remains constant or decreases. Testing of various adhesives have shown that those materials that have low moduli of rigidity (flexible adhesives) have the greatest ability to maintain shear stress resistance with increasing length of overlap.

Structural composite bonded joints are primarily of the overlap type (single or double overlap). Scarf joints are structurally efficient but are difficult or costly to manufacture; they are generally employed for repairs. Butt joints, although simple to process, are not considered for primary structural joints.

22.3.4. Linear Differential Joint Stresses (Thermal Stresses)

The elevated temperature bonding of composite structures generally requires joining of composites to metallics, and thus thermal stresses arise due to adherends of dissimilar coefficient of linear thermal expansion. In

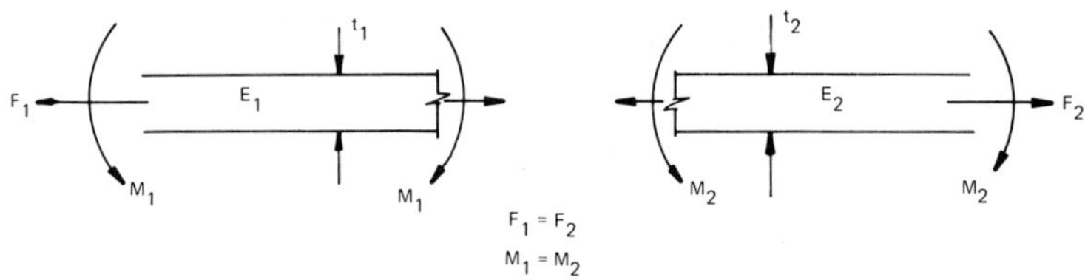

	ADHERENDS	STRESS CONCENTRATIONS UNDER:		
		F	-F	M
SCARF	$E_1 = E_2$	MINOR	MINOR	MAJOR
	$E_1 >> E_2$	MAJOR	MAJOR	MAJOR
OFFSET LAP	$E_1 = E_2$	MAJOR	MAJOR	MAJOR
	$E_1 >> E_2$	MAJOR	MAJOR	MAJOR
DOUBLE LAP	$E_1 = E_2$ $E_1 >> E_2$	MAJOR	MODERATE	MAJOR
	$E_1t_1 = E_2t_2$	MAJOR	MINOR	MODERATF
DOUBLE BUTT LAP	$E_1 = E_2$	MODERATE	MINOR	MAJOR
	$E_1 >> E_2$	MAJOR	MINOR	MAJOR

Figure 22.9. Behavior of collinear flat joints under load (adapted from Ref. 5).

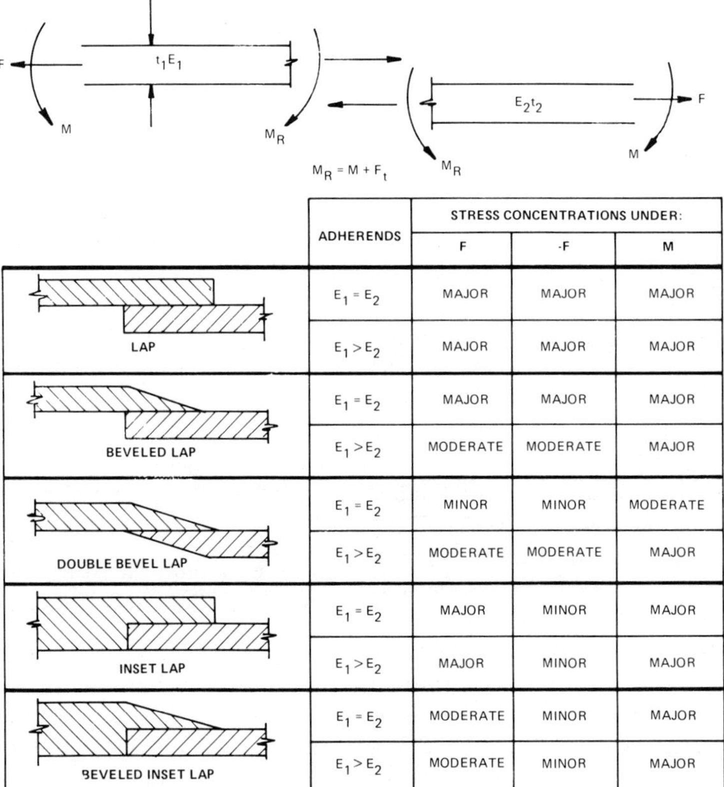

	ADHERENDS	STRESS CONCENTRATIONS UNDER:		
		F	-F	M
LAP	$E_1 = E_2$	MAJOR	MAJOR	MAJOR
	$E_1 > E_2$	MAJOR	MAJOR	MAJOR
BEVELED LAP	$E_1 = E_2$	MAJOR	MAJOR	MAJOR
	$E_1 > E_2$	MODERATE	MODERATE	MAJOR
DOUBLE BEVEL LAP	$E_1 = E_2$	MINOR	MINOR	MODERATE
	$E_1 > E_2$	MODERATE	MODERATE	MAJOR
INSET LAP	$E_1 = E_2$	MAJOR	MINOR	MAJOR
	$E_1 > E_2$	MAJOR	MINOR	MAJOR
BEVELED INSET LAP	$E_1 = E_2$	MODERATE	MINOR	MAJOR
	$E_1 > E_2$	MODERATE	MINOR	MAJOR

Figure 22.10. Behavior of offset lap joints under load (adapted from Ref. 5).

RIGIDITY OF ADHERENDS	STRESS CONCENTARTIONS				
	F	-F	$P_1 > P_2$	$P_2 < P_1$	M
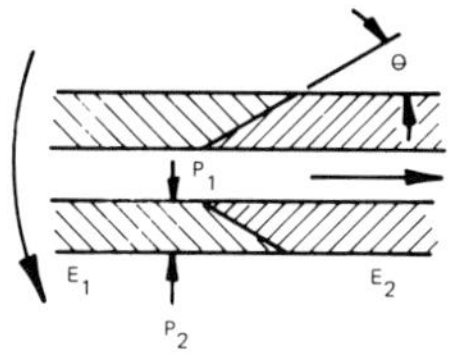 $E_1 = E_2$	MINOR	MINOR	MINOR	MINOR	MINOR
$E_1 > E_2$	MAJOR	MAJOR	MINOR	MAJOR	MAJOR
$E_1 < E_2$	MAJOR	MAJOR	MAJOR	MINOR	MAJOR
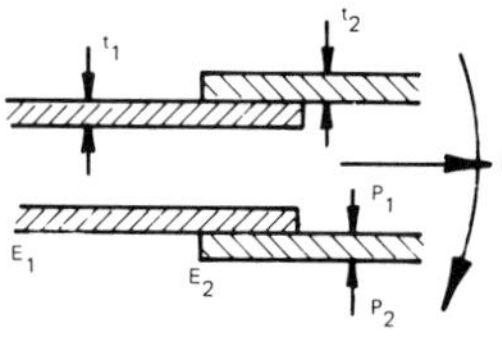 $\frac{E_1 t_1}{d_1} = \frac{E_2 t_2}{d_2}$	MAJOR	MAJOR	MINOR	MINOR	MAJOR
$\frac{E_1 t_1}{d_1} > \frac{E_2 t_2}{d_2}$	MAJOR	MAJOR	MAJOR	MINOR	MAJOR
$\frac{E_1 t_1}{d_1} < \frac{E_2 t_2}{d_2}$	MAJOR	MAJOR	MINOR	MAJOR	MAJOR
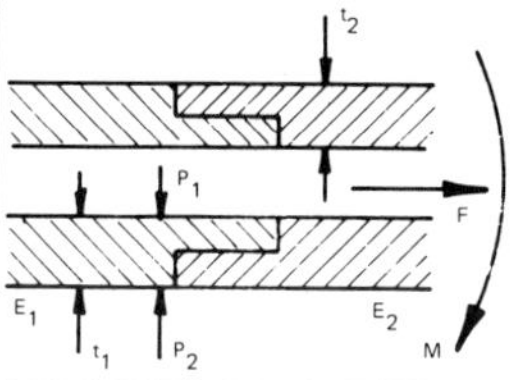 $\frac{E_1 t_1}{d_1} = \frac{E_2 t_2}{d_2}$	MAJOR	MAJOR	MINOR	MINOR	MINOR
$\frac{E_1 t_1}{d} > \frac{E_2 t_2}{d}$	MAJOR	MINOR	MAJOR	MINOR	MAJOR
$\frac{E_1 t_1}{d_1} < \frac{E_2 t_2}{d_2}$	MAJOR	MINOR	MINOR	MAJOR	MAJOR
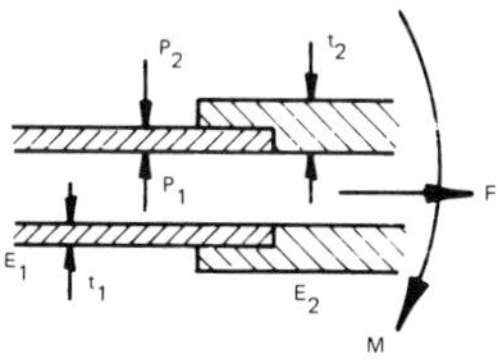 $\frac{E_1 t_1}{d_1} = \frac{E_2 t_2}{d_2}$	MAJOR	MINOR	MINOR	MINOR	MAJOR
$\frac{E_1 t_1}{d} > \frac{E_2 t_2}{d}$	MAJOR	MINOR	MAJOR	MINOR	MAJOR
$\frac{E_1 t_1}{d_1} < \frac{E_2 t_2}{d_2}$	MAJOR	MINOR	MINOR	MINOR	MAJOR

Figure 22.11. Behavior of tubular joints under load (adapted from Ref. 5).

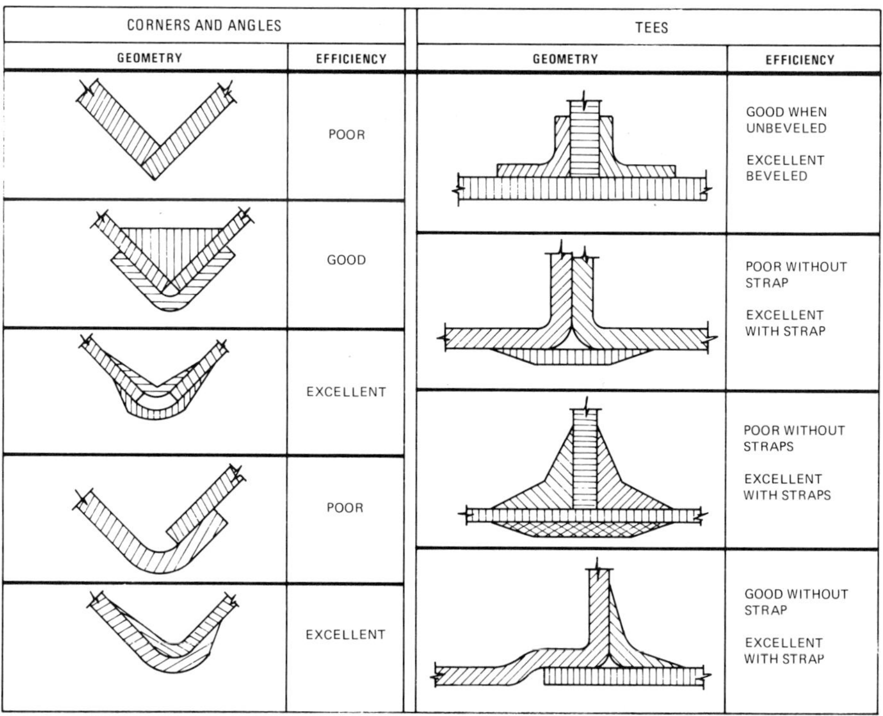

Figure 22.12. Behavior of ells and tees under load (adapted from Ref. 5).

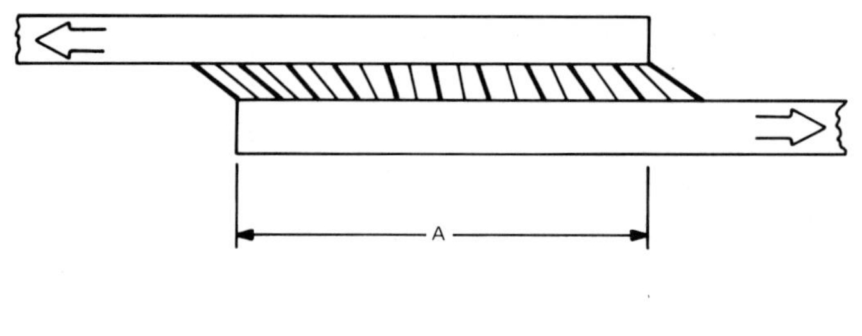

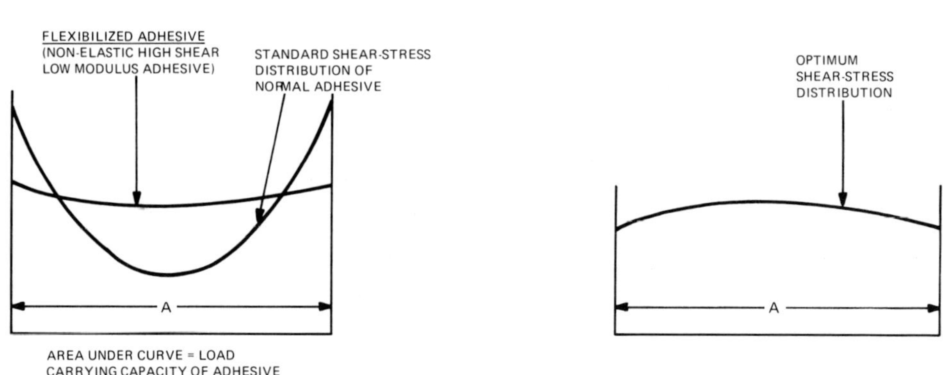

Figure 22.13. Shear-stress pattern of bonded lap joints.

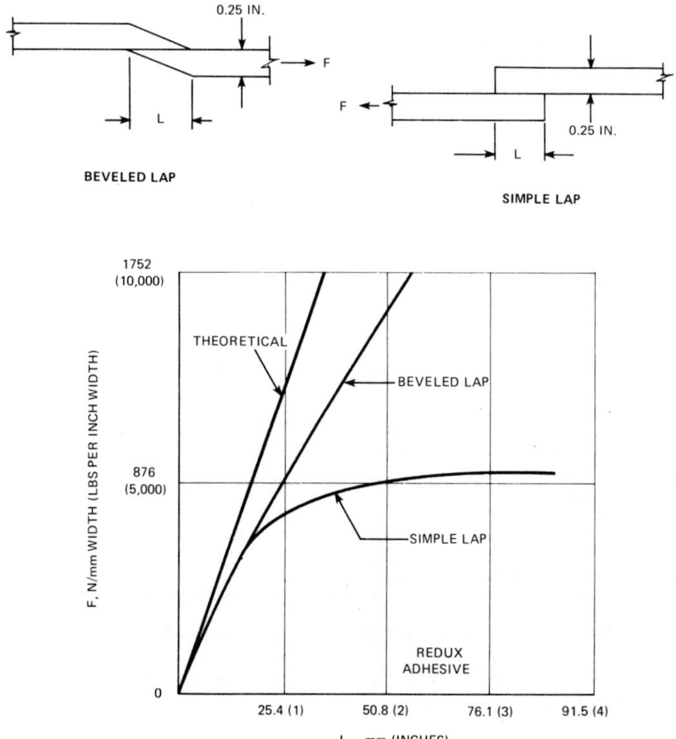

Figure 22.14. Influence of bonded lengths in flat lap joints (adapted from Ref. 6).

addition, the joint is further stressed since the linear coefficient of the adhesive is markedly different than either the composite or metallic adherend. Finally, since most structural components operate in a broad range of temperatures, transient thermal stresses must be considered. Typical stress gradients due to these parameters are shown in Fig. 22.15, along with the mathematical expression generally used to estimate the thermal stress.

22.3.5. Creep Characteristics

Creep deflection and creep rupture of structural joints require consideration. In the bonded joint, creep is a function of temperature and time, joint configuration, stress level, type of adhesive, and orientation of the composite adherend.

Creep, the phenomenon of movement, strain, or deformation in excess of the normal movement that results from the elastic qualities of the joint, passes through three stages,

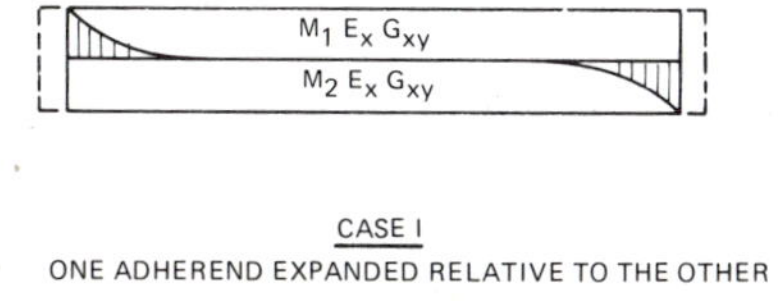

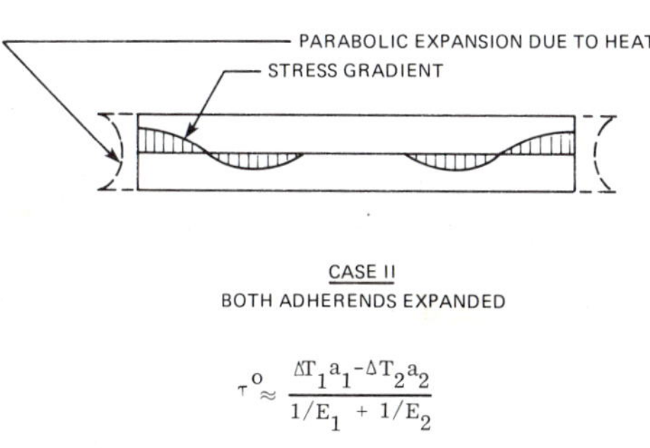

$$\tau^{o} \approx \frac{\Delta T_1 a_1 - \Delta T_2 a_2}{1/E_1 + 1/E_2}$$

WHERE:

τ^{o} = AVERAGE SHEAR STRESS ON ADHESIVE DUE TO DIFFERING RATES OF THERMAL EXPANSION IN ADHERENDS

E_1, E_2 = YOUNG'S MODULI OF ADHERENDS

a_1, a_2 = COEFFICIENTS OF THERMAL EXPANSION OF ADHERENDS

$\Delta T_1, \Delta T_2$ = DIFFERENCES IN TEMPERATURES OF ADHERENDS FROM CURING TEMPERATURE

Figure 22.15. Stresses due to differential expansion of adherends (adapted from Ref. 6).

as shown in Fig. 22.16. The first and second stages of the creep curve, the initial and constant rate deflection, should be accounted for in proper joint design.

The generalized characteristics of creep in adhesives joints are as follows.

1. Brittle (high shear modulus) adhesives exhibit less creep than ductile adhesives.
2. A reduction of 30% in either stress or operating temperature level increases the creep durability by more than 500 times.
3. Creep rates of adhesives follow exponential laws similar to those of stress-rupture analysis (generally, the creep rate is equal to $\underline{Ae^{-Q/RT}}$, where A and Q are constants based on stress level and material properties, R is the gas constant, and T the temperature in the joint).

22.3.6. Fatigue Characteristics

Fatigue endurance limits of bonded composite joints are generally limited to the endurance of the adhesive. Symmetrical adhesive joints have shown that they have fatigue endurance limits higher than unsymmetrical joints (simple overlap type). The general characteristics of fatigue resistance of adhesive joints are given below.

1. An increase in depth of lap creates increased fatigue life.
2. Adhesive fatigue life decreases slightly with increased thickness of adherends.
3. The fatigue endurance limit of adhesive joints tested at various temperatures shows that some adhesives provide greater resistance as the temperature of the joint is increased, while other adhesives have lower endurance as the temperature is raised.

Safe trends of fatigue endurance of bonded composites can be concluded from Figs. 22.17 and 22.18, which show the fatigue behavior of 2024-T3 aluminum adherends. It can be seen from these figures that the position of the fiberglass/epoxy composite simple adhesive

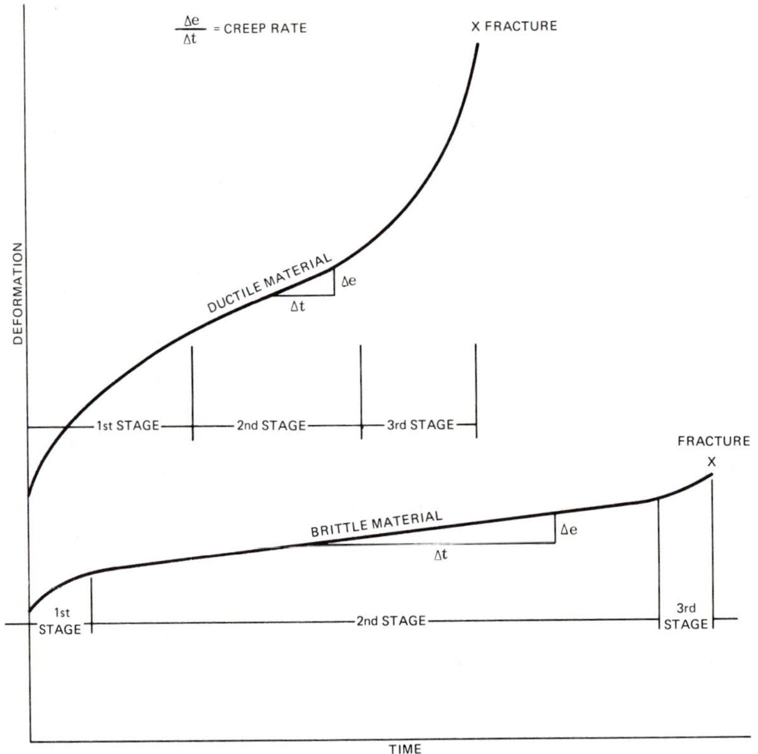

Figure 22.16. Typical creep-rupture time curves.

lap joint is between the fatigue curve of the metal itself and the metal with a hole, as would be used in a riveted structure. Further, the fatigue curve for boron/epoxy or graphite/epoxy adherend supported adhesive joints is significantly superior to metallic adherends. Figure 22.19 summarizes the fatigue curves for bonded versus riveted specimens on both 2024 and 7075 aluminum adherends.

Fatigue tests on specimens of stringers bonded to sheet (at low stress levels) versus riveted specimens (Fig. 22.20) showed the following.

- Tensile loading indicated equal fatigue resistance, but none of the failures were in the adhesive.
- Compressive loading showed that the bonded specimens provided 75% greater life.

An additional advantage provided by bonded composite structures is increased sonic fatigue endurance. The typical improvement to be expected is shown in Fig. 22.21.

22.3.7. Adhesive Types

The materials commonly employed in structural adhesive bonding of composite structures are thermosetting resins. Once cured, these resins are cross-linked into three-dimensional molecular networks that are insoluble and non-meltable. The thermosetting resins are subdivided into five basic chemical classes: polyesters, epoxies, phenolics, polyimides, and silicones. The structural adhesives used for aircraft joining of composites are primarily of the epoxy type, although various blends of the above classes are often compounded to provide a compromise between the valuable properties of each. In addition, to adjust the shear moduli of the thermoset adhesives, thermoplastic resin additives have been effective. Among the thermoplasts presently employed in the structural adhesives are the synthetic rubber derivatives (nitriles), polyamides (nylon), and various vinyl resins. These adhesive blends are materials that are readily miscible and co-react during the curing process to form inert solids. Table 22.6 is a

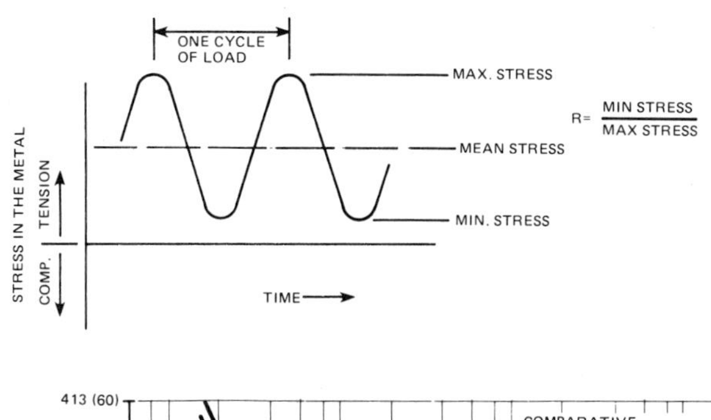

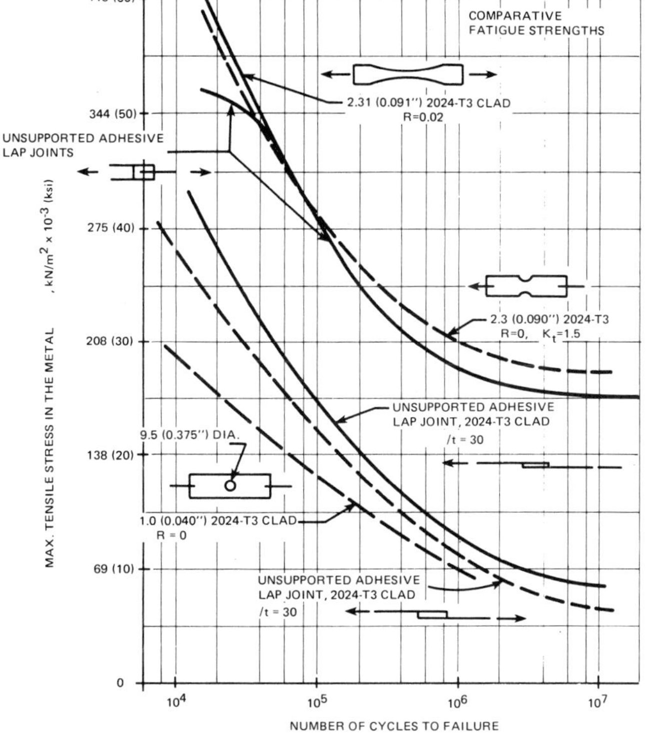

Figure 22.17. Typical fatigue strength of bonded joints vs. parent metal (adapted from Ref. 6).

typical listing of the various types of adhesives commercially available.

22.3.8. Characteristics of Thermoset Adhesives

The general characteristics of the thermoset adhesives and some of the adhesive blends are given in Table 22.7.

The structural adhesives used in the aircraft industry are in one of two basic forms. In general, the adhesives providing the greatest load carrying ability are in the form of a semi-solid ("B" stage) broadgoods (prepreg). In this form, the adhesive is precatalyzed and cast or extruded on carrier fabrics or on separator films and aged to produce an artificial solid. The "B" staged artificial solid is stored (in roll form) at low temperatures, 40° F (4.4° C), and when ready for use is activated by heat and pressure. The adhesive then liquefies for a brief period and cures to its insoluble infuse-able state.

The second form is the thixotropic paste. These viscous liquids are either one or two component systems. This type of adhesive has the ability to join vertical members, since adhesive drainage is minimal. Such adhesives find widespread usage in secondary bonding, where the application of pressure on the bond

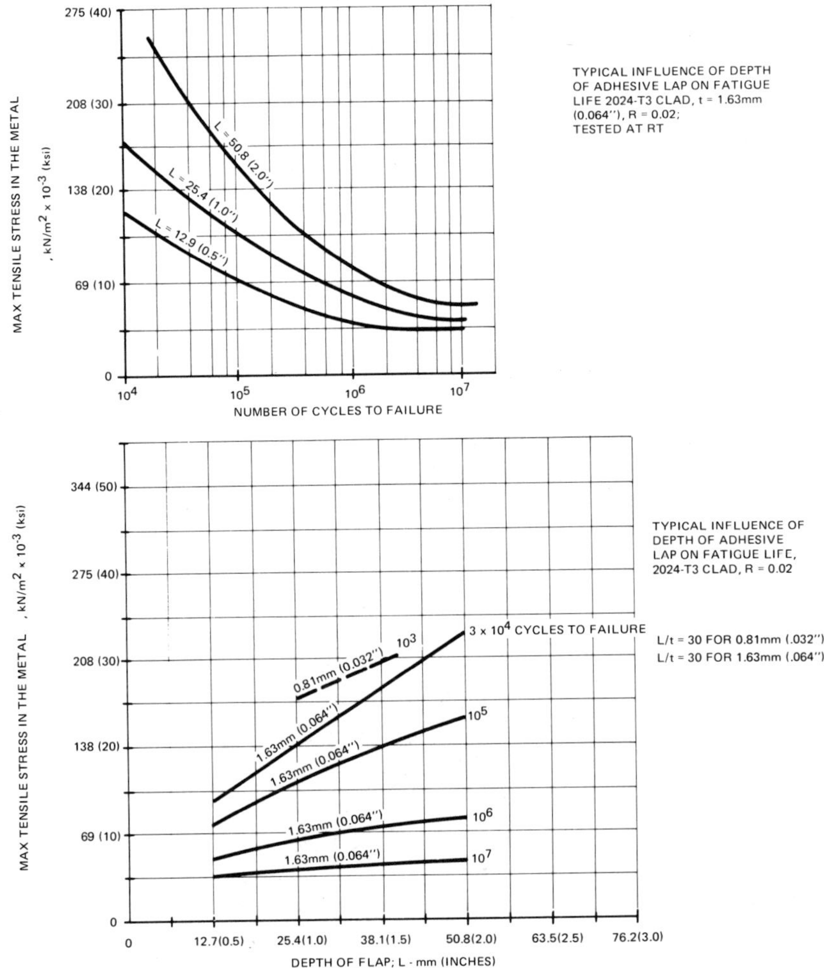

Figure 22.18. Typical influence of depth of adhesive lap on fatigue properties. (Courtesy of American Cyanamid-FM47 Adhesive) (adapted from Ref. 6).

is unattainable, or the mating of joining parts is uncontrolled. The one-component systems require elevated temperatures to achieve cure, while the two-component systems can be cured at room temperature.

22.3.9. Epoxy Adhesives

The epoxy and modified epoxy adhesives are most frequently employed in composite bonded structures. These adhesives are the most versatile and provide the best balance of properties required for joining composites to itself or to a metallic structure. The epoxies are easy to process, exhibit low shrinkage on cure, possess excellent wetability on many types of adherends, and provide high load carrying capability. The general recommended glue-line thickness is 0.004–0.008 in. (0.1–0.2 mm) for maximum joint efficiency, although adhesives up to 1/8 in. thick (3.17 mm) thick have performed satisfactorily. The epoxy adhesives perform well at low temperatures but are somewhat degraded in strength at temperatures above 350° F (176° C). Maximum long time service temperature of existing epoxy-phenolic adhesives is 400° F (204° C) or at temperatures above 260° F (126.6° C) when saturated with moisture (approximately 2% by weight).

To overcome the temperature stability or low peel resistance of the epoxies, modified epoxy adhesives are employed. These adhesives are available in both liquid and prepreg forms and are classified as follows:

CLASSIFICATION	MAXIMUM OPERATING TEMPERATURE, °F (°C), SHORT TIME, DRY	TYPICAL TENSILE SHEAR ALLOWABLE, 0.5 in.-(12.7-mm) OVERLAP AT ROOM TEMPERATURE, psi (MPa) (METAL TO METAL BONDS)	TYPICAL PEEL STRENGTH AT ROOM TEMPERATURE, in.-lb/in. (mm-N/mm) WIDTH (METAL TO METAL)
Epoxy	300 (149)	3500 (24.1)	3 (13.32)
Epoxy-nylon	200 (93)	6500 (44.8)	150 (666.0)
Epoxy-nitrile	250 (121)	4000 (27.6)	30 (133.2)
Epoxy-novolacs	350 (176)	3800 (26.2)	20 (88.8)
Epoxy-phenolic	400 (204)	3000 (20.7)	10 (44.4)

22.3.10. Adhesive Allowables

Allowable adhesive joint mechanical properties for composite to composite or to metallic adherends are difficult to define on an absolute scale, since changes in composite type, orientation, operating conditions, and/or processing result in large changes in the working allowable. Therefore, the strength performance of the composite joints presented in Table 22.8 is a guide only. Exact strength allowables for a particular primary structural design should be based on a controlled testing program simulating the actual conditions. Table 22.9 shows the effect of overlap length on the tensile shear strength of a typical liquid epoxy adhesive.

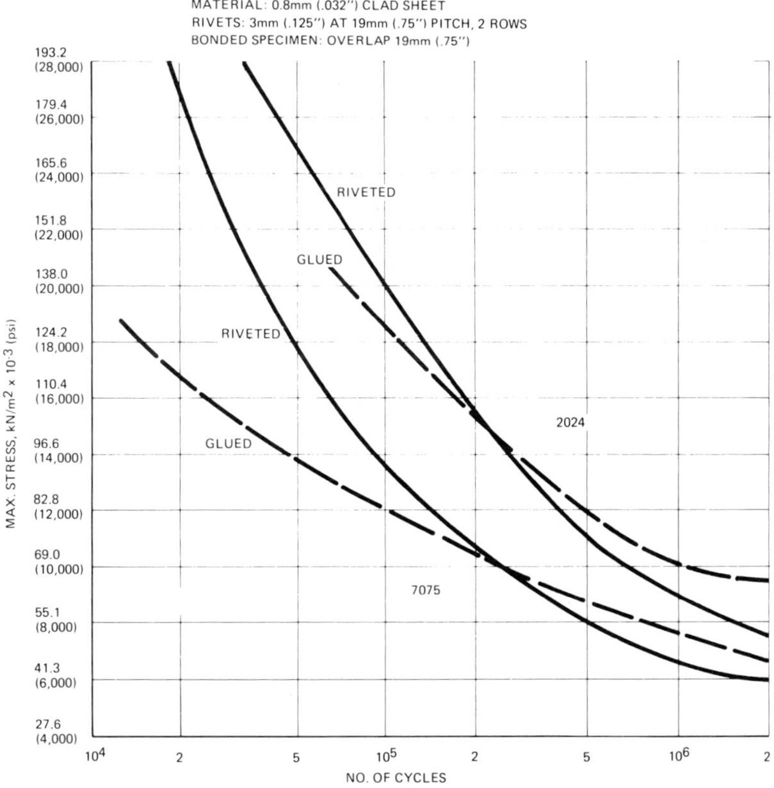

Figure 22.19. Fatigue curves for bonded and riveted lap joints of 2024 and 7075 alloys (adapted from Ref. 6).

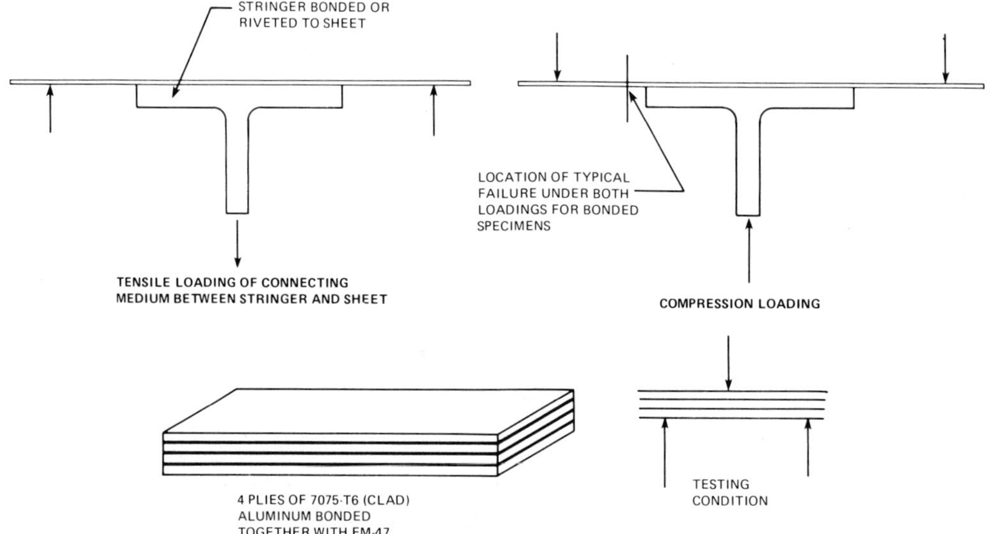

Figure 22.20. Specimens used for fatigue studies (adapted from Ref. 6).

22.4. MACHINING TECHNIQUES FOR COMPOSITES

(This section was prepared by Warren Marx, Grumman Aerospace Corp.)

22.4.1. General

The machining processes available for composites can be divided into four basic types of operations:

- Cutting uncured preimpregnated material
- Drilling cured composites
- Trimming or profiling cured composites
- Finishing cured composites.

Specific methods and techniques for machining composites are related to the type of composite; i.e. thermoplastic, thermoset, or metal matrix, and the reinforcement type, such as continuous or non-continuous and inorganic, organic, or metaloid reinforcement. The three present major material categories are thermosets, thermoplastics, and high modulus composites—boron/epoxy,

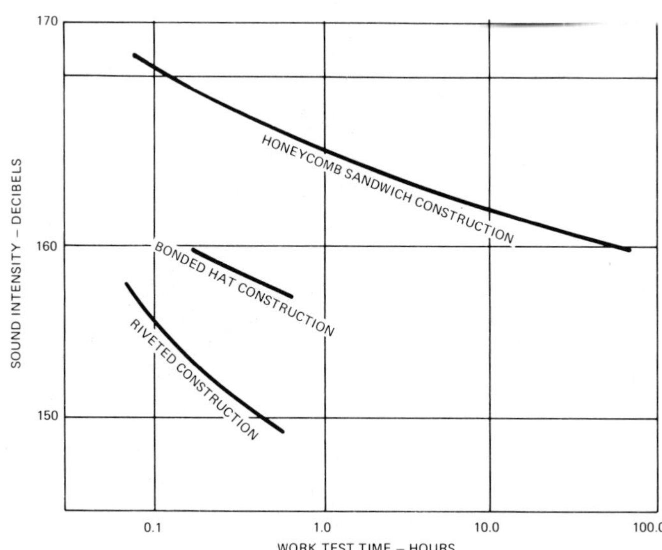

Figure 22.21. The sonic fatigue life of typical trailing-edge structures (adapted from Ref. 7).

Table 22.6. Types of Commercially Available Adhesives

Thermosetting adhesives	*Thermoplastic adhesives*
Epoxy resins	Acrylic resins
Epichlorohydrin-bisphenol A	Methylmethacrylate
Cycloaliphatics	Cellulosics
Epoxy-novolacs	Cellulose acetate
Epoxy-nitriles	Acetate-butyrate
Epoxy-phenolics	Cellulose-nitrate
Epoxy-polyamides	Ethyl cellulose
Epoxy-polysulfides	Sulfones
Polyester resins	Polysulfone
Polyester-DAP	Polyethersulfone
Polyester-TAC	Polyarylsulfone
Phenolic resins	Vinyl resins
Phenol-formaldehyde	Acetal
Vinyl-phenolics	Acetate
Nitrile-phenolics	Alcohol
Polyamide-phenolics	Chloride-acetate
Resorcinol-formaldehyde	*Rubber-base adhesives*
Melamine-formaldehyde	Rubber resins
Silicone resins	Natural
Dimethyldichlorosilanes	Reclaim
Phenyl-silicones	Butyl
Silicone-alkyds	Butadiene-nitrile
Silicone-epoxies	Butadiene-styrene
Elastomerica silicones (RTV)	Polysulfides
Polyimides	*Miscellaneous adhesives*
Dianhydride-diamine (PI)	Inorganic resins
Polybenzimidazole (PBI)	Sodium silicate
Amide-imides	Magnesium oxychloride
Chain terminated imides	Cement
	Plaster
	Natural resins
	Asphalts
	Ester gum
	Rosin/casein

aramid (Kevlar)/epoxy, and graphite/epoxy. For all "chip making" machining processes, particulate removal equipment should be employed.

22.4.2. Cutting Uncured Preimpregnated Composites

The time-honored techniques for cutting uncured composite materials have been manual cutting techniques with either a carbide disc cutter, scissors, or a power shear. However, alternate and/or advanced technology processes exist which are more amenable to rate production. Such processes include water-jet cutting, laser cutting, reciprocating mechanical cutting, and steel rule die blanking. Each of these processes poses different advantages or limitations and must be considered in terms of the application requirements and constraints.

22.4.3. Water-Jet Cutting

Water-jet cutting provides a fast—up to 300 in./minute (7.6 m/minute) at 60 ksi—omnidirectional approach to composite cutting for both single and multiple ply thicknesses. Its main advantages are a point cutting source, finished cuts, and low cutting forces, as well as its being a clean process and adaptable to automation. Process limitations are that access is required to both sides of the workpiece, that the capital cost is high, and that much noise is generated. The process can successfully sever most uncured materials

Table 22.7. Characteristics of Thermoset Adhesives

RESIN BASE	DISADVANTAGE	ADVANTAGE	MAJOR USES
Polyesters	Considerable shrinkage, brittle on impact, poor hot strength.	Fair strength, low viscosity, low-temperature cure, good electrically.	Repairs, compatible with explosives and radomes.
Epoxy	Generally rigid, poor hot strength, somewhat toxic.	High strength, low shrinkage, low-temperature cure, fair electrically.	FRP to metal joints, radomes, and aircraft structural parts.
Phenolics	Requires solvents and high-temperature cures, bad electrically, may be corrosive.	Good hot strength, nontoxic, inexpensive.	High-temperature ceramic to FRP joints.
Rubber phenolics	Require solvents and high-temperature cures, bad electrically.	Moderate strength and peel resistance.	Metallic joints, shock absorbing parts.
Epoxy phenolics	Rigid, requires heat cures, bad electrically.	High heat resistance, fair strength, good cryogenic strength.	Aircraft structural parts where high temperature or extremely low temperature is required.
Silicones	Low strength, requires solvents.	Extremely high heat resistance, good arc resistance.	High temperature, silicone adherends, and in co-polymer adhesives.
Polyimides	Rigid, requires heat cure, may be corrosive.	Extremely high heat resistance, good electrically.	High temperature, long age metal-to-metal aircraft parts.

reliably except boron/epoxy and is particularly effective for cutting Aramids.

22.4.4. Laser Cutting

Laser systems provide the most universal capability to cutting uncured materials, since they are not affected by the hardness, stiffness, or abrasiveness of the workpiece. Cutting systems are usually in the 250–500-watt power range, with single ply cutting capabilities of 300–600 in./minute (7.6–15.2 m/minute). Other features of laser systems are omnidirectional capabilities, adaptability to automation, and a requirement for access from only one side. Process limitations include heat damage, capital cost, and decreased cutting speed with increased ply count.

22.4.5. Reciprocating Mechanical Cutting

This technology evolved from the garment industry and is particularly applicable to broadgoods cutting. The reciprocating cutter requires access from only one side and does not impose any heat damage to the edge of the composite. The mechanical cutter is sensitive to abrasiveness and, therefore, not applicable to boron/epoxy; a wide selection of other composites can be cut at speeds of 600–900 in./minute (15.2–22.9 m/minute). System limitations include capital cost and use of a cover material for protection from the compacting foot which rides on the surface of the workpiece.

22.4.6. Steel Rule-Die Blanking

Steel rule-die blanking is a fast method for trimming an entire part periphery in single or multi-ply thicknesses. Generally, this method uses the steel rule-die positioned above a flat, mild steel plate mounted in a press so as to permit blanking on the down stroke. The hardened steel strip which is used for cutting utilizes a one-side-leveled configuration. The limitations of the processes are related to die requirements; namely, handling, storage, and flexibility to change.

22.4.7. Machining of Thermosets

The machining techniques for thermosets, such as glass-reinforced laminates, have been thoroughly evaluated and are well established. Standard metal and woodworking machining equipment can be used with modification to increase spindle speeds along with reduced feeds. Since standard cutting tools are suitable only for short production runs, presently used tools are carbide- or diamond-tipped. Tools used on such composites are required to be sharp, not only for clean cuts but to minimize possibilities of delamination. Generalized machining hints are as follows.

- Control heat-up of work.
- Support of work minimizes delaminations.
- Grind cutters similar to cutters used on brass.
- Use drills designed for laminates (slow twist, polished flutes, and thin webs).
- Feed work slowly.
- Use light drill pressure.
- Control usage of drill (replace when dull).

22.4.8. Drilling and Routing

Generally, drilling and routing are done dry, although finer finish results with free flowing lubricant or air blasting. Since most laminated thermosets tend to shrink somewhat after these operations, slightly oversized drilling/routing should be considered for highly accurate work. Drill and routing fixtures should be employed whenever practical and should be designed to avoid breakout at bottoms and lifting at the top of the work. For accurate holes on thick parts, the drill should be removed several times to avoid drill runoff. Drill and routing parallel to lamination should be avoided, since delamination often results. If required, however, the drill point included angle should be increased and the workpiece completely clamped. Regular twist drills and standard routers are in use, although drills with high helix angles and wide, polished flutes are recommended. Drills and routers that are carbide- or diamond-tipped are required for production runs. Drills should always be sharp, and techniques for rapid chip removal and minimum workpiece heat-up should be employed. Drill speed is generally between 200 and 400 ft/minute (61 and 122 m/minute) under light pressure. Laminates are effectively routed and drilled dry, although dust and heat removal, along with superior finish, is obtained with coolants, lubricants, and compressed air.

22.4.9. Tapping and Threading

Thermoset laminates can be threaded using standard metal working taps. In general, the hole to be tapped is chamfered to the maximum thread diameter; the tap used is of high speed steel and is chrome plated. Air cooling while tapping has proved advantageous, and most often a coarse rather than fine thread is employed. As in brass working, threads of laminates can be installed on a lathe, as well as with taps and dies. Lathe threading is usually done at high speed with oil coolant and fine cutting techniques. For either machine or hand tapping, it has been found that a 200-ft/minute (61-m/minute) speed is optimum. Taps used for high production runs should be carbide if long tool life is desired.

22.4.10. Milling

Laminates can be milled on standard machines employing generalized techniques used for brass. Conventional holding devices (fixtures) are applicable to laminates, although special care is needed to prevent delamination during cutting. High speed steel, carbide, or diamond cutters should be used. Cutter clearance angles should be ground from 7–12, with sharp cutting edges. Cutting speeds of 600–1000 ft/minute (183–305 m/minute), along with feeds of 0.002–0.005 in. (0.05–0.13 mm) per revolution are utilized for most grades of laminates. The lower range of the speed and feed are for high speed steel cutters, while the upper range is for carbide or diamond cutters. Air or vapor mist coolants are preferred, and cuts should not be greater than 0.010 in. (0.25 mm) per pass.

22.4.11. Turning

Laminates can be readily turned, bored, faced, and cut off on conventional engine or bench

Table 22.8. Mechanical and Physical Properties of Bonded Fiberglass-Reinforced Epoxy Laminate Joints

ADHESIVE	TYPE	MILITARY SPECIFICATION	VENDOR (ADHESIVE SUPPLIER)	COLOR AFTER CURE	CHEMICAL BASE	CURE SCHEDULE TEMPERATURE, °F (°C)/PRESSURE, psi (kPa)	SERVICE TEMPERATURE RANGE, °F (°C)
BR 92/A (former designation, E-92/A)	Thixotropic paste	*MIL-A-5090*	American Cyanamid	light yellow	Epoxy	200 (93)/Contact	−73/93 (−100/200)
Epon 934	Thixotropic paste	*MIL-A-5090*	Dexter Manufacturing Hysol Division	grey	Epoxy	200 (93)/Contact	−195/176 (−320/350)
EC 1933	Thixotropic paste	*MIL-A-5090*	Minnesota 3M Mining & Manufacturing	grey	Epoxy	150 (65)/10 (69)	−65/93 (−85/200)
EC 2186	Thixotropic paste	*MIL-A-5090*	3M	light cream	Epoxy	350 (176)/10 (69)	−54/93 (−67/200)
FM 97	Supported film	*MIL-A-5090*	American Cyanamid	cream	Epoxy	200–350 (93–176)/ 50 (345)	−74/149 (−100/300)
AF 110	Supported film	*MIL-A-5090*	3M	white	Epoxy	350 (176)/10 (69)	−54/93 (−67/200)
AF 111	Supported film	*MIL-A-5090*	3M	off-white	Epoxy-nitrile	250 (12)/10 (69)	−54/121 (−67/250)
FM 1000	Unsupported film	*MIL-A-5090*	American Cyanamid	white	Epoxy-nylon	350 (176)/16 (172)	−253/93 (−423/200)
HT 424	Supported film	*MIL-A-5090*	American Cyanamid	grey	Epoxy-phenolic	350 (176)/40 (276)	−253/260 (−423/500)
FM 123 Metlbond 225 AF 126-2	Supported film	*MMM-A-132*	Cyanamid Celenese- Narmco 3M	purple amber blue	Epoxy-nitrile	250 (121)/40 (276)	−54/121 (−67/250)
AF 40	Unsupported film	*MIL-A-5090*	3M	reddish brown	Epoxy-polyamide	350 (176)/50 (345)	−54/93 (−67/200)
Metlbond 329	Supported film	*MIL-A-5090*	Celenese-Narmco	grey	Epoxy-novolac	350 (176)/25 (172)	−54/176 (−67/350)

NOTES

1. Fatigue allowable for joints shown above are 30–35% of room temperature static value for 10^6 cycles of fatigue at $R = 0.1$ (minimum stress/maximum stress).
2. Adherends were Style 181—epoxy laminates 0.06 in. (1.5 mm) thick. Interlaminar tensile shear strength of the adherend is 4200 psi (28,938 kN/m^2) supported and 2800 psi (19,292 kN/m^2) unsupported.
3. Elevated temperature properties shown are based on tests at the indicated temperature after a 10-minute soak at the temperature as per *MIL-A-5090D*.
4. Tensile overlap shear tests were conducted at approximately 1200 psi/minute (8268 kN/m^2/minute) loading rate.
5. Shear moduli values are estimated based on tensile moduli tests of bulk adhesive and estimates of Poisson's ratio and not values obtained on actual joints. These data are to be used as a guide for adhesive shear deformation.
6. Values given for fiberglass/epoxy-to-fiberglass/epoxy joints are applicable to fiberglass/epoxy-to-metal joints at room temperature. At other temperatures, reduce values by 25% to allow for thermal stresses.

Table 22.8 Continued

ENVIRONMENTAL RESISTANCE	0.5-in. (12.7-mm) OVERLAP, MINIMUM TENSILE SHEAR STRENGTH, psi (MPa) ROOM TEMPERATURE	−67°F (−54°C)	160°F (71°C)	250°F (121°C)	SHEAR MODULUS, psi (MPa) AT ROOM TEMPERATURE	PEEL STRENGTH, in.-lb/in. (mn-N/mm) (METAL-TO-METAL) JOINTS	CHARACTERISTICS AND USES
Salt spray, humidity, hydraulics	1300 (8.95)	1200 (8.27)	1100 (7.58)	580 (4.0)	250,000 (1722)	3.0 (13.3)	Non-metallic filled adhesive effective for foam sandwich construction and as a repair adhesive.
Salt spray, humidity, hydraulics	1600 (11.0)	1200 (8.27)	1200 (8.27)	850 (5.86)	250,000 (1722)	3.0 (13.3)	Metal filled adhesive with good retention of strength at elevated temperature. Good creep and nuclear radiation resistance.
Salt spray, humidity, hydraulics	1500 (10.3)	1100 (7.58)	1350 (9.3)	540 (3.72)	225,000 (1550)	3.0 (13.3)	For monolithic joints where low dielectric constant paste adhesive is needed.
Salt spray, humidity, hydraulics	1200 (8.27)	1200 (8.27)	1200 (8.27)	650 (4.48)	200,000 (1378)	35.0 (155)	For monolithic joints where fit-up can't be controlled—high impact and bend strength.
Salt spray, humidity, hydraulics	1800 (12.4)	1400 (9.65)	1800 (12.4)	1200 (8.27)	350,000 (2411)	20.0 (89)	For monolithic and sandwich joints used on W2F rotodome (Grumman).
Salt spray, humidity, hydraulics	1450 (9.99)	1350 (9.3)	1450 (9.99)	350 (2.41)	250,000 (1722)	25.0 (111)	For monolithic and sandwich used on EA/6A pods.
Salt spray, humidity, hydraulics	1600 (11.0)	1300 (8.96)	1600 (11.0)	600 (4.13)	200,000 (1378)	30.0 (133)	For monolithic and sandwich used on E-2A vertical fins.
JP-4, Skydrol hydraulics	2000 (13.78)	1700 (11.71)	1400 (9.65)	400 (2.76)	50,000 (345)	150.0 (666)	For monolithic and sandwich joints, high impact, peel, and fatigue strengths.
Salt spray, humidity, hydraulics	1600 (11.0)	1600 (11.0)	1600 (11.0)	1200 (8.27)	500,000 (3450)	10.0 (44)	For monolithic and sandwich joints, outstanding temperature resistance, not used for electronic applications.
Salt spray, humidity, hydraulics	2000 (13.78)	1500 (10.34)	2000 (13.78)	850 (5.86)	175,000 (1206)	50.0 (222)	For monolithic and sandwich joints, high creep resistance.
Salt spray, JP-4, Skydrol	2000 (13.78)	1500 (10.34)	1000 (6.89)	400 (2.76)	50,000 (345)	75.0 (333)	For monolithic joints, high impact and peel strengths.
Salt spray, JP-4, hydraulics	1600 (11.0)	1200 (8.27)	1600 (11.0)	1400 (9.65)	350,000 (2411)	20.0 (89)	For monolithic and sandwich joints, good creep resistance.

Table 22.9. Typical Effect on Length of Overlap Versus Tensile Shear Strength for Fiberglass/Epoxy Laminates to Fiberglass/Epoxy Joints

	TENSILE SHEAR AT STRESS, psi (MPa), ULTIMATE, AT ROOM TEMPERATURE				
Type of joint (overlap length), in. (mm)	1/2 (12.7)	1 (25.4)	2 (50.8)	3 (76.2)	4 (101.6)
Single overlap	1500 (10.4)	1000 (6.9)	900 (6.2)	650 (4.5)	550 (3.8)
Double overlap	2000 (13.8)	1600 (11.0)	600 (4.1)	450 (3.1)	

lathes. Specialized holding fixtures are required only when the shape of the workpiece is non-standard. High speed steel-, carbide-, or diamond-tipped tool bits are used. Cutting edges of the tool should have small radius and be well honed. Rounded nose lathe tools with little clearance are utilized when smooth polish-like finishes are needed. For general cutting, side and front clearances should be similar or slightly greater than for metal cutting. Surface speeds can vary from 600–1000 ft/minute (183–305 m/minute), but cuts should be shallow and feed rates low. Coolants can be used, especially if coarse cuts or high feed rates are required.

22.4.12. Sanding and Grinding

Laminates can be finished to close tolerances by sanding and/or grinding. Belt and drum type sanding machines have been employed successfully with grit sizes of 30–240. Silicon carbide abrasives bonded with a synthetic resin are recommended, although most commercially available abrasives are satisfactory. Sanding speed is usually on the order of 4000 ft/minute (1220 m/minute), dry for the coarse grit (up to 80), and wet for finer grit, to prevent clogging and burning of the abrasive binder. Grinding procedures are similar to that of sanding, with centerless grinding being used extensively for tubes and rod forms of laminates. A coarse, open-grit grinding wheel is preferred, along with coolants or lubricants, to prevent loading and glazing of the wheel.

22.4.13. Shearing/Punching

Laminates that are thin, less than 1/8 in. (3.17 mm), and are used for semi- or non-structural parts, can be sheared on conventional sheet metal shears. To inhibit potential edge delamination, the shearing can be done at elevated temperatures.

Laminates can be punched using the same technique as that used in the metal industry, although this method is not recommended for primary structural parts. Generally, glass-reinforced laminates, 1/8 in. (3.17 mm) or less in thickness, can be cold punched in dies that are designed with little clearance between punch and die. The punch and die block is usually constructed of carbide for long life; however, carbon steels, chrome steels, and tool steel have been used.

22.4.14. Sawing

Laminated fiberglass-reinforced thermosets are cut by bandsaws and circular saws. Special precautions to overcome the material's low thermal conductivity, along with delamination caused by vibration, are required. Furthermore, the material is abrasive, and therefore, maintaining cutter sharpness is difficult. Satisfactory techniques that have overcome these potential problem areas are sawing jigs and fixtures, circular saw rim speeds of 40.5 8000–13,000 ft/minute (2440–3965 m/minute), bandsaw speeds of 2000–8000 ft/minute (610–2440 m/minute), chrome-plated steel or carbide-tipped blades, wet cutting, staggered teeth on bandsaws, and disc-type circular saws designed especially for laminates.

22.4.15. Specialty Machining

Specialty machining, such as automatic screw machining, gear cutting, shaving, reaming, blanking, and piercing, have been successfully conducted with laminated thermosets. Because these processes vary with the specific type of equipment used, standardized techniques are not applicable.

22.4.16. Finishing and Polishing

Laminates are finished by sand blasting, honing, lapping buffing, and polishing. Sand blasting employs both sand and alumina abrasives with a wide range of grit size (980–240 grit) and a range of air pressure. Sand blasting is used either to obtain a desired nongloss finish or as a step in an assembly process. Honing can be conducted dry or wet, and several types of honing machines are available. Coarse and fine grit alumina abrasives are used in water and oil vehicles; the technique utilized is a direct function of the specific honing machine. Laminates have been honed to a 30-micro-inch (7600-micrometer) finish without difficulty. Finer finishes are possible for laminates by lapping, by hand or on machines. Buffing and polishing are generally the finishing methods for obtaining a mirror-smooth surface. The buffing wheel consists of muslin discs, and the polish is a greaseless composition (silica powder). The wheel is operated at slow speeds with light workpiece pressures.

22.4.17. Machining of Thermoplastics

Although the machining of unreinforced thermoplastics is a well known art, it is complicated by the wide property differences among a large number of available materials and reinforced thermoplastics—(fiberglass, aramid (Kevlar), and graphite. The generalized machining hints given are for the largest majority of thermoplastics, although it must be recognized that anomalies do exist.

- Coolants should be used to avoid overheating and subsequent melting of the workpiece.
- Operate machinery at high speeds.
- Provide liberal clearances on cutting tools.
- Use light cuts and slow feed of workpiece.
- Turning tools should be ground to provide rake angles that minimize tool cutting and thrust forces.
- Slow spiral drills designed for thermoplastics should be used.
- Tools should be carbide-tipped or utilize the special high speed steel tools designed for plastics.
- The workpiece must be properly supported to avoid distortion under cutting pressures.
- Allowances for plastic memory and shop room temperature should be made to insure accurate machining.
- Tool bits and cutters should be sharp, since dull tools increase forces on the workpiece.

22.4.18. Sawing

As the primary consideration for sawing thermoplastics is the removal of generated heat, coolants are highly recommended. Conventional equipment, such as bandsaws and circular saws, are used. Carbide-tipped blades are recommended, along with a high degree of tool sharpness. Circular saws and bandsaws are in common usage at speeds of 2000 to 6000 rpm.

22.4.19. Drilling

Drilling of thermoplastics requires the avoidance of chip packing and overheating. Therefore, drills with wide, polished flutes, specific helix angles, and point angle related to the specific material being drilled should be used. Generally, helix angles are on the order of 10–50°. Clearances are 9–20° and point angles are 60–120°, with the drills ground sharp. Air blasts or liquid coolants are needed to maintain hole accuracy, as well as to prevent overheating. Drill speeds vary not only with the material but with the size and depth of the holes. Speeds in common usage are 900 rpm. for 1/2-in. (12.7-mm) diameter holes. High speed steel drills can be used, but carbide-tipped tools are more responsive to operation at higher speeds and generally produce smoother holes.

22.4.20. Tapping and Threading

As is the case for thermosets, thermoplastics should be chamfered on the tap entry side of the hole to prevent stripping the first few threads. Threads with rounded roots are preferred to sharp, "V" type threads, to minimize stress concentrations. Threads can be prepared in thermoplastics, using lathes or taps and dies. High speed steel, straight fluted taps are generally used and are slightly oversized to allow for elastic recovery. A tap rake angle between 10 and 15°, with a heel rake angle approximately 5°, is in common usage. Air blasting or coolants have been used to minimize tap clogging and to permit high tapping rates.

22.4.21. Milling and Threading

Standard metalworking lathes and milling machines are used for thermoplastics. High speed steel tools can be used effectively if maintained sharp, although carbide- or diamond-tipped tools are preferred. Although cutting speed and feeds are functions of the workpiece material and type of cut, they are generally between 300 and 1000 ft/minute (9.15 and 305 m/minute), with 5–10-in./minute (12.7–25.4-cm/minute) feed. Whenever possible, fixturing should be employed to prevent workpiece deflection, vibration, and chatter. Experience has shown that box tools are best for turning operations; slight nose radii on cutting edges are also preferred for both light and heavy cuts.

22.4.22. Sanding and Grinding

Thermoplastics can be ground and sanded on conventional machines using coolants to prevent clogging and glazing of the abrasive. Sanding is done both with wet belt sanders and dry and wet on abrasive discs rotating at 3000 rpm. Grinding, using silicon carbide or alumina grit wheels, requires free flowing coolants to prevent workpiece melting. Chip clearance is generally assured by using coarse grinding wheels (30–80 grit).

22.4.23. Other Machining Methods

Thermoplastics can be blanked, pierced, punched, sheared, thermally severed, reamed, tumbled-debarred, and honed and polished. Blanking is generally accomplished using steel ruled dies and "clicker presses." Piercing, punching, and shearing are accomplished on standard metalworking equipment with the workpiece either hot or cold. The thermal cutting techniques for non-reinforced thermoplastics include heated wires, and flames to melt the workpiece at the cut-off line. The feed rate for these processes is adjusted to match the rate of melting. The only precaution for honing and polishing is to ensure that the workpiece is not overheated. The techniques and equipment used on thermosets are applicable to polishing thermoplastics.

22.4.24. Machining of High Modulus Composites

Machining techniques for high modulus composites have been developed over the past few years, primarily in the aerospace industry. As such, some of the methodology is still being refined. Most of the current work is devoted to boron/epoxy, aramid (Kevlar)/epoxy, and graphite/epoxy. Each of these reinforcements have characteristics peculiar to themselves and, in turn, may impose unique machining requirements. In general, all of the conventional detail part operations (drilling, turning, and finishing) can be applied to the high modulus composites, along with advanced technology processes such as water-jet cutting and ultrasonic machining.

22.4.25. Conventional Drilling

Drilling of graphite/epoxy can be accomplished with standard carbide drills. Drills should be kept sharp to minimize breakout, and the use of particulate removal equipment is required since most operations are performed dry. As most laminated thermosets tend to shrink slightly after drilling, oversized cutting tools should be utilized to increase accuracy and maximize tool life. Drilling parallel to the lamination should be avoided, since delamination often results, but, if required, the drill point included angle should be increased and the workpiece completely clamped. Drill fixtures should be employed whenever practical and should be designed to avoid breakout at bottoms and lifting at the top of the work. Generally, a drill speed of 400 ft/minute (122 m/minute) is adequate except for graphite/epoxy, where 900–1050 ft/minute (275–320 m/minute) is desirable.

Drilling of aramid (Kevlar)/epoxy requires that alternate drill configuration must be used. One effective method is to use a Jancy type counter-bore. This drill, operating at 300–400 sfm, can produce good quality holes (minimal fuzzing) even with HSS tool material. However, the drill should be cleaned of epoxy and fiber residue after every five holes to maximize cutting ability.

Boron-reinforced laminates require diamond-impregnated tools due to the abrasive reinforcement characteristics. Diamond tools are utilized in a metal matrix, but are sensitive to heat and thus require forced water coolant. Typical speeds are 3000–5000 rpm at a 1-in./minute (2.54-cm/minute) feed.

22.4.26. Ultrasonic Drilling of Thermoset Laminates

Ultrasonics have been shown to effectively increase diamond cutting tool life in boron-reinforced laminates. A typical ultrasonic unit can supply up to 600 watts to the drill spindle resonator at a frequency of 20 kHz. The drills can be either plated or sintered diamond tools at 80–100 grit size. These cutting tools must be water cooled during operation. The advantage of ultrasonics is that drill life can be extended by 100% as compared to conventional approaches. Typical operating conditions are 2250–4000 rpm at 1-in./minute (2.54-cm/minute) feed in drill sizes up to 0.5 in. (1.27 cm).

22.4.27. Trimming Operations

A host of processes are available for cut-off and/or trimming of high modulus composites. The basic factors which enter into the process selection are straight versus curvilinear cutting; part size and configuration; rate and quantity requirements; and equipment availability or cost. In addition to correct process selection, appropriate cutting tools must be utilized to achieve efficient fabrication. These cutting tools may range from conventional high speed tool materials to the more exotic diamond-impregnated materials. The following discussion will address both equipment and cutting tool considerations.

22.4.28. Sawing

Standard metal and woodworking equipment can be used for routine radial sawing and bandsawing operations. Radial sawing provides a fast and accurate approach to composite cutting which can be achieved with either stationary or portable equipment. The single biggest drawback is its straight line cutting limitation.

Both graphite/epoxy and boron/epoxy laminates can be readily cut to a finished edge using 60-grit diamond-impregnated tools operating at 7000 sfm. Diamond-impregnated tools will quickly clog, however, if used on aramid (Kevlar)/epoxy. For this application, a hollow ground, high speed steel saw blade with 16 teeth/in. (6.3 teeth/cm) is very effective when run backwards.

Bandsawing offers the advantage of producing contoured cuts, but edge quality is generally poor and must receive a post-process finishing operation. The types of blades used for bandsawing graphite/epoxy, boron/epoxy, and aramid (Kevlar)/epoxy are, respectively, medium grit carbide, 60-grit diamond plated, and HSS with a lapped raker set configuration. Operating speeds are generally 2000–5000 ft/minute (610–1525 m/minute), with the higher speeds used on aramid (Kevlar)/epoxy.

22.4.29. Water-Jet Cutting

The water-jet cutting process shows some potential for trimming high modulus composites at pressures up to 60,000 psi (1.45×10^6 MPa). At this pressure, aramid (Kevlar)/epoxy can be cleanly trimmed, while thin graphite/epoxy laminates may demonstrate minor delaminations.

22.4.30. Routing

Routing is an edge trimming operation which may be performed with manual or machine controlled equipment. The basic difference in routing of the various high modulus composites is the type of cutting tool used. Diamond-cut carbides are effective for graphite/epoxy laminates. Diamond-coated (40–50-grit) router bits are required for the abrasive boron/epoxy, and carbide, opposed helix router bits are effective for aramid (Kevlar)/epoxy composites. Cutting speeds for these operations are generally obtained with 13,000–21,000-rpm routers.

22.4.31. Grinding

Grinding and abrasive cut-off are effective methods for machining the high modulus composites. Diamond wheels appear best, although silicon carbide and alumina wheels can be used. Coolants are required to prevent matrix thermal degradation, and speeds have been in the range of 3000–8000 ft/minute (915–2440 m/minute).

REFERENCES

1. Strauss, E. L., "How to Design Mechanical Joints," *Materials in Design Engineering* (February 1963).
2. Lockheed Aircraft Corporation, Memorandum 107a, October, 1957.
3. Society of the Plastics Industry, *Handbook of Reinforced Plastics*, Reinhold, New York, 1964.
4. Jungstrom, O. L., "Design Aspects of Bonded Structures," *Aircraft Bulletin #4 Bonded Structures*, LTD, Duxford, England (May, 1959).
5. Perry, H. A., *Adhesive Bonding of Reinforced Plastics*, McGraw-Hill, New York, 1959.
6. Bloomingdale Rubber Division of American Cyanamid Co., *Handbook for Adhesives*, 1957.
7. Catchpole, E. J., "Bonding and Sandwich Construction," *Aeroplane and Astronautics* (June 1, 1961).

23 RELEASE AGENTS

Sheldon L. Clark
Frekote, Incorporated
Indianapolis, Indiana

23.1. INTRODUCTION

The term "release agent" or "parting agent" or "lubricant" is used to describe a wide variety of chemicals which provide a barrier between a mold and the surface of a part being molded. (See Table 23.1). There are two basic types of release agents, internal and external. An internal release agent is an additive which goes directly into the resin formulation. An external release agent is applied to the surface of the mold.

Release agents are important because, while two solid surfaces generally do not adhere to each other, the use of a release agent becomes important when a solid and a liquid or a solid and a paste or dough form an interface and adhere to one another. Among some of the factors that influence the adhesion of two materials to each other are penetration, chemical reaction, surface tension, surface configuration, and polarity differences. In addition, a vacuum is formed many times which can be prevented by the use of a release agent. Suffice it to say, that release agents have become such an integral part of the manufacturing operation that today there are at least 70 suppliers. Of these, only a few manufacture their own basic ingredients. These are the suppliers who have proprietary resins offering properties that are unique among release agents. Many suppliers merely repackage and offer standard chemicals in the forms of aerosols or bulk shipment.

Many industries could not exist today without the use of release agents and, with the increasing industrial use of polymeric materials, the commercial use of release agents has become an integral part of the overall operation. Among some of the industries in which release agents are critical are: aircraft and aerospace, automotive, metal processing, rubber processing, the manufacturer and shaping of composites, polymer and plastic processing, food handling, glass and paper coating, just to mention a few. As an example, pressure sensitive tapes could not exist if the tape backing did not have a release agent.

23.2. PROPERTIES OF RELEASE AGENTS

Since the release agent can affect the properties of the part itself as well as the quality of release, proper selection and application are vital to the successful use of release agents. The optimum release agent will prevent damage to the part, provide many releases per application, not build up on the mold or transfer to the part, and will preserve mold detail and design. It will make the production line faster, more economical and more profitable. It will keep the mold in production longer by minimizing mold build-up. While there is still considerable discussion and argument among molders concerning the relative merits of internal versus external release agents, the following factors are those which should be considered in selecting the proper release agent, which ever type is used:

1. The particular polymer composition with which the release agent is to be used: Will it release and, if so, how easily?

2. The process and process conditions to

Table 23.1. Molding Operations Requiring A Release Agent

PLASTICS	AEROSPACE/COMPOSITES	RUBBER
Compression	Graphite/Boron Composites	Injection
Transfer	Fiberglass Laminates	Compression
Injection	Adhesive Bonding	Transfer
Laminating	Hand Lay-Up	Tire Manufacturer
Reinforced Plastic (FRP)		
Polyester Resin Injection		
Rotational		
	OTHER	
	Electronic and Electrical	
	Foundry	

which the composition is to be subjected: Will this result in excessive build-up on the mold? Will the release agent be able to tolerate operating conditions? Will the mold stay clean? Will the down time be minimal? If the mold must be cleaned frequently, will it have any deleterious effect on the mold? Will the release agent, be it internal or external, be compatible with the subsequent steps in the operation, such as painting and bonding? If an external agent, is there sufficient time on the production line to apply properly the release agent?

3. The properties desired in the final product: If an internal release agent, will it have any undesirable effect on the overall properties? If an external release agent, what type finish is desired? Are there cosmetic affects to be considered?

4. Safety: If an external release agent, what solvents will be used? Is there sufficient ventilation? Are the solvents acceptable? Is there any possibility of skin dermatitis? If an internal agent, will there be constant blooming of the agent resulting in undesirable surface effects?

5. Economics: What affect will the price of the release agent have on the unit cost of the part? How does it affect subsequent steps in the operation? Have all associated costs been considered? It is most important when considering the purchase of a release agent, to look at the cost per use rather than the basic cost of the item since, in the long run, this is what really counts.

As stated above, it is sometimes possible to choose between an external and an internal release agent. In many cases, however, such a choice is not available. As an example, hand lay-up and spray-up operations require an external release agent. Glass reinforced polyester composites need external release agents. On the other hand, SMC, BMC, and premixes which are compression molded in metal dies usually depend on an internal release agent. Many injection molding operations make use of internal release agents and many molders will design molds in such a manner as to avoid the use of either type of release agent. In the early days of the composite industry, the common method of assuring release was to apply waxes to external molding surfaces. With the increasing sophistication of high speed production operations, however, a faster method was necessary. This has resulted in the development of proprietary multiple release agents such as those manufactured by FreKote, Inc., Contour Chemical Co., and Axel Plastics Research Labs., and the development of specific internal release agents to eliminate the time consuming application of the agents to the die. While silicones also are widely used, they must be handled with care in certain operations because silicones can easily become airborne and, as a result, surface contamination of parts in the immediate vicinity is likely. This surface contamination can affect subsequent steps in the operation such as bonding and painting. These steps are quite critical in the aerospace industry, for example,

Table 23.2. Some Commercially Available Release Agents[2,3]

COMPANY	TRADE NAMES	ADDRESS
Frekote, Inc.	Frekote 31, 33, 33C, 33H, 34, 34H, 44, RRM, HMT, FRP (Proprietary)	Boca Raton, Florida
Dow Corning	Dow Corning 200, 1101, HV-490, 347, 24, 203, 290, 7, 233A, 20 (Silicones)	Midland, Michigan
Union Carbide	Union Carbide L-45, -7001, -7002; LE-42, -45, -46, -420, -460, -467HS; LS-46 (Silicones)	New York, New York
General Electric	SM-2140, -2154, -2162; SF-96, -1080, G-662 SF-18, SF-1221, SF-1066, SM-2068 (Silicones)	Waterford, New York
Axel Plastics Research Lab.	MOLD-WIZ (more than 100 formulations, both internal and external)	Woodside, New York
Chem-Trend	CT-31, -45, -51, -61 (assorted)	Howell, Michigan
Contour Chemical Co.	KRAXO, EXITT, LIFFT, NONSTICKENSTOFFE, RIMLEASE, 1711 (Silicones and assorted)	Woburn, Massachusetts
DuPont	VYDAX (Tetrafluoroethylene telomer)	Wilmington, Delaware

See text references.

and some aerospace companies will not permit silicones on their property. Table 23.2 shows some commercially available release agents.

23.3. EXTERNAL RELEASE AGENT

An external release agent does not destroy the characteristics of the polymer. It is often less expensive than an internal release agent. While there are some who believe that the best external release agent is one that leaves the mold on the surface of the part so as not to contaminate the mold, this is not true. In fact, it is not even desirable because the release agent, by leaving the mold on the surface of the part, is thus contaminating the part. This then requires a subsequent cleaning step and, in many cases, the solvents that would be necessary to remove the release agent may not be compatible with the part. Consequently, in numerous cases, a non-transferring release agent such as supplied by FreKote, Inc., is the most desirable type to be used where painting and adhesion are subsequent steps in the manufacturing operation.

Most external release agents can be applied by spraying, brushing or dipping. Since they are being applied to the surface of the mold, the condition of that mold surface becomes important. The preparation of the mold surface can be a major step in securing a properly coated mold particularly when a change in a release agent is contemplated. Silicone oils are generally removed with toluene or mineral spirits. Waxes, on the other hand, must be removed with chlorinated solvents such as methylene chloride, trichloroethylene or perchloroethylene. Mold surfaces should be cleaned of both the polymer and excess release agent and the cleaning method chosen depends upon the material of construction. For example, aluminum molds can be subjected to an acid wash with a material such as formic acid. Steel molds are easily cleaned with an alkaline detergent. Copper molds can be cleaned with an acid-based product such as Copperbrite and nickel molds can be cleaned with a commercially available product such as Spic N' Span. These methods, of course, are in addition to the usual abrasive techniques of using glass beads, sandblasting, lime blasting or walnut shells. In any case, all traces of oils, waxes or any other foreign matter must be removed prior to application of the release agent. Once the release agent has been applied, if it is of the proprietary nature, it may require a curing cycle. The optimum film durability and releasing capability are obtained if the manufacturers directions are followed explicitly.

The number of releases to be obtained will, of course, vary with the configuration of the mold and the abrasive nature of the polymer. Reinforced polymers, naturally, tend to be more abrasive. It is not necessary, however, to recoat the entire mold when slight sticking is experienced. The mold surface, even when hot, can be touched-up with fresh release agent in the areas where excessive wear has been noted. If the mold surface is hot, one should use a release agent especially designed to be applied at an elevated temperature. There will, of course, come a time when the mold must be thoroughly cleaned and it is most prudent to follow the directions of the release agent manufacturer. This will ensure that the cleaned mold surface will be receptive to the application of fresh release agent.

In general, a noncontaminating release agent is preferred for most composite applications and this is especially true for aerospace use where laminating and hand lay-up are predominant. As more efforts are made to raise the operating temperature of composites, greater demand is placed on the release agent. This is especially true, not just for achieving extreme temperature use but, because in most composite applications, especially in aerospace, the mold is usually a plastic reinforced material such as epoxy. If the mold release fails to do its job, the part not only does not release, it usually results in extensive damage to the mold surface as well since the part then must be pried loose and during this operation much chipping and gouging of the mold surface will result. Here, too, because of the size and shape of the parts, it is not practical to clean the surface of any release agent which has transferred to the part. Consequently, non-transferability is a paramount requirement of any release agent being used in the aerospace industry.

With the exception of the perfluorinated products such as Vydax sold by DuPont, and one or two silicone release agents advertised as paintable, the non-contaminating release agents usually are those of a proprietary nature. As previously stated, silicones in general are not desirable in aerospace applications and in many cases are not allowed.[1] The systems offered by FreKote, for example, are stable solvent solutions of highly reactive materials which polymerize and bond securely to a clean surface when applied properly. When cured, they form a hard, tough, dry, nontransferable microthin film on the mold. The unusual characteristics of such a release agent are that the polymerization reaction is completed after the formulation is applied to the surface of the mold. Another unique property offered by FreKote is that of thermal stability. These products can be used up to temperatures of 900° F, the highest reported temperature of any available external mold release. Such a high temperature range makes these types of products ideal for rotomolding and, indeed, these products are the release agent of choice. When reotomolding high density cross-linked polyethylene, the use of these products is practically mandatory.

In addition to the proprietary release agents, there are a great number of others. Among these may be found the following:

1. Waxes: Here both natural and synthetic waxes find application as release agents. Parafin and microcrystalline waxes, waxes of vegetable origin and waxes of animal origin are all used. The synthetic waxes have attained considerable importance; practically all materials from C_{10} and up have found use as release agents.
2. Metal Salts: In this category the fatty acid having the widest use is stearic acid. It has a sharply defined melting point and does have good wetting properties. The main derivatives of stearic acid such as the calcium, zinc and lead salts also serve as release agents. Which metallic stearate is chosen for a specific application will depend primarily on the polymers and the other surfaces. Calcium and lead stearate are the dominant ones. Zinc stearate is substituted for lead where non-toxicity is specified, but is not nearly as stable as lead stearate. In polyvinyl chloride, calcium stearate is probably the most effective. In rubber processing

both aluminum and magnesium salts are preferred.

3. Polyvinyl alcohol: This material is generally applied as a coating from a water solution in the form of a cast or extruded film.
4. Polyamines: These find application only in the form of extruded films since they are mostly insoluble in the commonly used solvents.
5. Polyethylene: Polyethylene is used as a film in the processing and shipping of uncured rubber and many times as a paper laminator in packaging.
6. Silicones: The commercially available silicones are all polymeric in order to obtain high boiling points and low volatilities as well as heat resistance and resistance to oxidation. Silicones are applied in the forms of fluids, resins, and greases.
7. Fluorocarbons: Fluorocarbon polymers are available in the forms of sheets and as dispersions. One special application is as a coating for metal frying pans in home use. In this case, the coating is applied as a dispersion which is then dried and fused, at approximately 250° C. The most widely used such polymer is that of tetrafluoroethylene. Care must be exercised with its use and it should not be sprayed on tobacco nor should smoking be permitted while spraying.

   Tedlar®, polyvinyl fluoride, a Du Pont film, usually 0.002 in. (0.051 mm) thick material is used extensively as a release agent in autoclave molding operations. Since curing agent BF_3 complex, which is present in some prepregs tends to decompose Tedlar, it should be sprayed with a non-transferable spray release agent prior to use with these prepregs.
8. Inorganic compounds: These are probably the oldest release agents known. Because of their insolubility they are used strictly in the form of powders which exert their release properties because of their flake-like crystal structure. The most important members of this class are talcum and mica. They are usually applied as a fine powder, sprayed or dusted onto a surface. In many cases they can be blended with metal stearates to improve the release action.

23.4. INTERNAL RELEASE AGENTS

Among the many advantages of internal release agents are the elimination of application, the elimination of a wipe down and scrubbing of the mold, and elimination of vapors. In some cases, internal agents actually improve the impact strength of rigid polymers; therefore, where appropriate, internal release agents can be very economical. As stated earlier, however, care must be taken with the use of internal release agents to guarantee that they have no adverse affect on physical properties or any other specifications. In particular, internal release agents are used in pultrusion processes where pulling of the part through the mold would tend to wear off an external release.

Lecithin was an early internal release agent but, because it was not readily available, and its greasy consistency made it difficult to meter, it is not used very much these days. Currently, metallic stearates are among the most popular internal release agents. They have the advantage that they can be varied in composition and physical characteristics to produce the desired properties in some polyester molding compounds. The most commonly used are calcium and zinc stearates although calcium produces a better surface gloss and pigmentation in some formulations. While the zinc is more expensive, it does provide a better mix viscosity. Because the choice of an internal release agent is so critical, it is advisable to test a number of formulations for their affect on the ultimate end product before making a decision.

While molders may mix internal release agents in with the resin by tumbling or batch mixing, the resin suppliers will compound them directly into thermoplastic pellets. In thermosetting resins, internal release agents are mixed in at a selected weight prior to the addition of the catalyst or any other additives. In addition to the stearates, organophosphates, soaps, silicone oils, waxes, and proprietary resins have found use as internal release agents. As stated, since the internal release agents must function in the presence of various chemicals such as catalysts, promoters, and other additives, they must be carefully chosen. Also, as previously stated, in a number of cases they provide added improvement such as antistatic characteristics or may improve impact and tensil strength as well as even the hardness of the finished product.

In summary, regardless of whether one chooses an internal or an external release agent, certain key factors should be considered in making the choice. Certainly an external agent is called for if one is working with a polymer having very narrowly defined specifications. While ease of release is certainly the first major factor to be considered, all the other factors mentioned earlier must be taken into consideration. Cost, while sometimes considered to be a factor, is really not all that important. If the release agent works, gives you efficiency of production and has minimal affect on other related costs and subsequent processing operations, then the cost of the release is usually insignificant with respect to the overall cost of the product. In general, external release agents are more readily available, more widely applicable and certainly easier to use and, for composite applications, are probably the release agent of choice.

References

1. A. Kingsbury, "Poly(dimethylsiloxanes) as Release Agents," Pentacal, **74**, 12, 52–53 (1974).
2. G. Kovach, "Release Agents," Encyclopedia of Polymer Science and Technology, **12**, 57–65, John Wiley & Sons, New York, 1970.
3. C. Kirkland, "The Mold Release Quandry: No Easy Answers for Molders," Plastics Technology, **26**, 65–70 (1980).

24
TESTING OF REINFORCED PLASTICS

George Epstein
The Aerospace Corporation
El Segundo, California

24.1. INTRODUCTION

Tests are conducted to determine the suitability of the materials, process(es), and/or design for the intended application (see Table 24.1). The five major causes of product failure are 1) misapplication of the product, 2) poor design, 3) inadequate control of materials, 4) poorly controlled manufacturing techniques, and 5) misapplication of materials. Adequate testing is required as means for preventing such failures.

Testing of reinforced plastics is especially important because the properties and concomitant performance are subject to significant variations associated with the materials, processing, and design parameters.

Standardized and/or special tests are necessary to aid in materials selection, process development, design, and quality control (including verification of product performance). Tests may be destructive or nondestructive, depending on whether or not the sample is destroyed or degraded during the test.

In testing for materials selection, a series of candidate materials (generally chosen based upon prior experience) is subjected to one or more tests to determine which of the materials is most appropriate for the intended application. The specific tests employed are designed to evaluate those properties considered important for the application.

Table 24.1. Purposes of Testing for Reinforced Plastics Products[1]

- Assess quality of raw materials.
- Evaluate and optimize materials.
- Evaluate and optimize manufacturing process variables.
- Determine effects of equipment and tool design.
- Establish engineering design information.
- Measure quality and reproducibility of end item.

As an expedient, a series of screening tests may be devised so as to select or approve materials by a process of elimination. When a material fails one test in the series, the material is then eliminated from further consideration; thus, the extent of testing can be considerably reduced.

The fabrication process influences the properties of reinforced plastics products. The curing cycle, the equipment and tooling, and variables associated therewith determine to a large measure the chemical, physical and mechanical properties. Storage conditions and handling of materials can also influence properties. Resin-curative/catalyst systems must be measured out, mixed, and applied to reinforcements prior to cure; optimum procedures and tolerances must be established. Tests, therefore, frequently are necessary to evaluate the parameters associated with processing and handling. The objectives of such tests are to ascertain the optimum (or, at least, acceptable) conditions and tolerances for processing and the requirements for handling of the material(s) prior to and during fabrication of the end item.

In order to properly design the product, certain properties (allowables) must be available to the designer (see Table 24.2). Depending on the operational requirements, various properties may be significant. Thus, if the product is to be exposed to severe environ-

Table 24.2. Properties of Interest for Reinforced Plastics

PHYSICAL PROPERTIES

MECHANICAL PROPERTIES

1. *Tensile properties*
   - Tensile strength
   - Modulus of elasticity
   - Elastic limit
   - Yield strength
2. *Compressive properties*
   - Strength
   - Modulus
3. *Flexural properties*
   - Strength
   - Modulus
4. *Shear properties*
5. *Impact strength*
6. *Properties at high rates of loading (dynamic properties)*
7. *Shear properties*
   - Strength
   - Modulus of rigidity
8. *Bearing strength*
9. *Surface hardness*
   - Indentation hardness
   - Scratch resistance
   - Abrasion resistance
   - Mar resistance
10. *Creep properties (creep-rupture and stress-relaxation)*
11. *Fatigue (cyclic properties)*
12. *Poisson's ratio*
13. *Notch sensitivity*
14. *Shatterproofness*
15. *Shockproofness*
16. *Tear resistance*

OPTICAL PROPERTIES

1. *Spectral transmission (haze)*
2. *Index of refraction (refractive index)*
3. *Light diffusion*
4. *Crazing resistance*
5. *Internal stress; transparent plastics*
6. *Optical uniformity and distortion*
7. *Surface stability, optical*

THERMAL PROPERTIES

1. *Thermal expansion*
2. *Thermal shrinkage*
3. *Thermal conductivity*
4. *Specific heat*
5. *Heat distortion temperature (deflection temperature under load)*
6. *Flammability (flame resistance)*
7. *Ignition properties*
8. *Maximum safe operating temperature*
9. *Flow temperature*
10. *Brittleness temperature*

ELECTRICAL PROPERTIES

1. *Electrical resistance (insulation resistance—volume and surface*
2. *Dielectric strength and dielectric breakdown voltage*
3. *Dielectric constant and power factor (loss factor)*
4. *Arc resistance*

MISCELLANEOUS PHYSICAL PROPERTIES

1. *Specific gravity (density)*
2. *Porosity*
3. *Machinability*
4. *Punching quality*

CHEMICAL AND PERMANENCE PROPERTIES

1. *Resistance to chemical reagents*
   - Acids
   - Bases
   - Solvents
   - Fuels
   - Bacteria and fungi
   - Salt spray
2. *Water absorption*
3. *Water vapor permeability (diffusion)—gas transmission rate*
4. *Accelerated service (temperature and humidity)*
5. *Sunlight and weather exposure (aging)*
6. *Effects of radiation*
7. *Impact sensitivity (LOX)*
8. *Toxicity*
9. *Volatile loss (outgassing)*
10. *Stress-crazing*

ments and/or loading conditions, the effects thereof on critical physical and mechanical properties must be established and considered in the design of the product.

Quality control plays an especially important role in the production of reinforced plastic components. Reproducibility and uniformity are necessary to ensure that all units will perform as anticipated. Toward this end, it is desirable to control—by testing—the quality of all constituent materials to the extent possible or practical, to control the quality of the product while it is in process, and to evaluate the quality of the end item.

Batch-to-batch variations and changes during storage are common for resins, preimpregnated fabrics, and molding materials. To preclude difficulties attendant thereto, material qualification and batch acceptance tests are frequently required. Qualification generally requires a very extensive series of tests to ensure compliance for specified properties. Acceptance may involve a few tests, most likely selected from the qualification test series, that are considered adequate to ensure essentially equivalent performance. To make sure properties have not been degraded during storage, such tests are conducted at intervals during extended storage periods. The frequency of the tests depends on the susceptibility of the material to storage.

In many cases, the end item may be subjected to extensive tests, destructive and nondestructive, as a requirement for qualification and/or acceptance. Frequently, such tests are necessary during development of the product, to determine how the product will respond to anticipated loads and environments. Depending on the results, the design may be modified accordingly.

Qualification tests often are required to ensure that the end product manufactured with the selected materials, and in accordance with specified manufacturing procedures, will provide the desired response and the ability to withstand required operational conditions. Having qualified the product, subsequent units are subject to acceptance testing for consistent quality and reproducibility. Whereas the qualification tests may be quite extensive in scope (depending on anticipated operational requirements), acceptance testing is generally limited to one or a few tests selected so as to evaluate quality and performance, consistent with costs and schedule. In some cases, a limited number of units from each lot may be tested to destruction. Frequently, test coupons are prepared along with the end item. Such coupons (specimens) may either be machined from the unit (e.g., "tab ends") or be prepared from the same batch of materials and cured simultaneously with the end item.

24.2. DOCUMENTS ON TEST METHODS FOR REINFORCED PLASTICS

Standard test methods and the necessary equipment and procedures for the tests are described in various sources. Those sources listed below are of particular importance for reinforced plastics.

- ASTM (American Society for Testing and Materials).
- IPC (Institute of Printed Circuits).
- NEMA (National Electrical Manufacturers Association).
- SAE (Society of Automotive Engineers).
- AMS (Aeronautical Materials Specifications).
- Federal Test Method Standard No. 406, "Plastics: Methods of Testing" (supercedes Federal Specification L-P-406).
- *Military Handbook MIL-HDBK-17,* "Plastics for Military Vehicles" Part II—Reinforced Plastics.
- *Military Handbook MIL-HDBK-23,* "Composite Construction for Flight Vehicles; Part I—Fabrication, Inspection, Durability, and Repair" (supercedes ANC-23 Bulletin).

The last document deals with sandwich constructions. An entire chapter is devoted to inspection and test methods, including raw materials (facings, cores, and adhesives) used in the fabrication, and completed parts; both destructive and nondestructive tests are discussed.

Numerous test method recommendations have been issued by the International Organization for Standardization (ISO), some of which are applicable to reinforced plastics.*

Table 24.3 is an alphabetical index of test methods contained in the Federal Test Method Standard (FTMS) No. 406. Corresponding ASTM methods (which generally differ somewhat) are also indicated. The types of tests are numerically included as follows.

TYPE TESTS	FTMS NUMBERS
Mechanical	1011-1131
Thermal	2011-2051
Optical	3011-3051
Electrical	4011-4052
Miscellaneous physical	5011-5041
Permanence	6011-6091
Chemical	7011-7081

*Available from the American National Standards Institute, 1430 Broadway, New York, N.Y. 10018.

Table 24.3. Alphabetical Index of Test Methods*

METHOD	FTMS 406, METHOD NUMBER	ASTM NUMBER
Abrasion Wear	1091	D1242-56
Accelerated Service Tests (Temperature and Humidity Extremes)	6011	D756-56
Accelerated Weathering Test; Carbon Arc Without Filters (Alternate Navy Test)	6022	E42-65
Accelerated Weathering Test; Soaking, Freezing, Drying, Ultraviolet Cycle (Alternate Navy Test)	6023	
Acetone Extraction Test For Degree of Cure of Phenolics	7021	D494-46
Arc Resistance	4011	D495-73
Bearing Strength	1051	D953-75
Blocking	1131	D1893-67
Bonding Strength	1111	D229-77, Paragraph 40-4
Brittleness Temperature of Plastics by Impact	2051	D746-73
Calibration of Durometers Type "A" and Type "D"	1084	
Colorfastness to Light	6031	
Compatibility of Plastic-Explosive Mixtures	7081	
Compressive Properties of Rigid Plastics	1021	D695-77
Constant-Strain Flexural Fatigue Strength	1061	
Constant-Stress Flexural Fatigue Strength	1062	
Crazing Resistance Under Stress	6053	
Deflection Temperature Under Load	2011	D648-72
Deformation Under Load	1101	D621-64, Method A
Determining the Corrosivity Index (Water Extract Conductance) of Plastics and Fillers	7071	
Dielectric Breakdown Voltage and Dielectric Strength	4031	D149-75
Diffuse Luminous Transmittance Factor of Reinforced Plastic Panels	3032	D1494-60
Dissipation Factor and Dielectric Constant	4021	D150-75
Drying Test (for Weight Loss)	7041	
Effect of Hot Hydrocarbons on Surface Stability	6062	
Electrical Insulation Resistance of Plastic Films and Sheets	4052	
Electrical Resistance (Insulation, Volume, Surface)	4041	D257-76
Falling Ball Impact	1074	
Fatigue		D671-71
Flame Resistance	2023	
Flammability of Plastics 0.050 in. (0.127 cm) and Under in Thickness	2022	D568-77
Flammability of Plastics over 0.050 in. (0.127 cm) in Thickness	2021	D635-77
Flexural Properties of Plastics	1031	D790-71
Flow Temperature Test for Thermoplastic Molding Materials	2041	D569-59, Method A
Gloss	3051	D523-67
Hot Oil Bath	6061	
Indentation Hardness of Nonrigid Plastics by Means of a Durometer	1082	D2240-75
Indentation Hardness of Rigid Plastics by Means of a Durometer	1083	D2240-75
Index of Refraction	3011	D542-50, Method A
Interlaminar and Secondary Bond Shear Strength of Structural Plastic Laminates	1042	
Internal Stress in Plastic Sheets	6052	
Izod Impact Strength	1071	D256-73, Method A
Light Diffusion	3031	D1494-60
Linear Thermal Expansion (Fused-Quartz Tube Method)	2031	D696-70
Luminous Transmittance and Haze of Transparent Plastics	3022	D1003-61
Machinability	5041	
Mar Resistance	1093	D673-70

Table 24.3. Continued

METHOD	FTMS 406, METHOD NUMBER	ASTM NUMBER
Mildew Resistance of Plastics, Mixed Culture Method, Agar Medium	6091	
Optical Uniformity and Distortion	3041	D637-50
Porosity	5021	
Punching Quality of Phenolic Laminated Sheets	5031	
Resin in Inorganic-Filled Plastics	7061	
Resistance of Plastics to Artificial Weathering Using Fluorescent Sunlamp and Fog Chamber	6024	D1501-71
Resistance of Plastics to Chemical Reagents	7011	D543-67
Rockwell Indentation Hardness Test	1081	D785-65
Salt Spray Test	6071	
Shatterproofness	1073	
Shatterproofness (Gauge Windows)	1075	
Shear Strength	1041	
Shockproofness	1072	
Short-Time Stability at Elevated Temperatures of Plastics Containing Chlorine	7051	D793-49
Specific Gravity by Displacement of Water	5011	D792-66
Specific Gravity from Weight and Volume Measurements	5012	
Surface Abrasion	1092	D1044-76
Tear Resistance of Film and Sheeting	1121	D1004-66
Tensile Properties of Plastics	1011	D638-77; D2290-76
Tensile Properties of Thin Plastic Sheets and Films	1013	D882-75
Tensile Strength of Molded Electrical Insulating Materials	1012	
Tensile Time-Fracture and Creep	1063	
Thermal Expansion Test (Strip Method)	2032	
Transverse Load of Corrugated Reinforced Plastic Panels	1032	D1502-60
Volatile Loss	6081	D1203-67
Volume Resistivity of Casting Resins	4042	
Warpage	6051	
Warpage of Sheet Plastics	6054	D1181-56
Water Absorption of Plastics	7031	D570-77
Water Vapor Permeability	7032	E96-66

*There may be a difference between the ASTM method and the method specified here, although in some cases the methods are technically identical.

It should be noted that these tests relate to all types of plastics—thermoplastic as well as thermosetting, both non-reinforced and reinforced.

A bibliography of various sources of information dealing with testing of reinforced plastics is found at the end of this chapter.

24.2.1. Filament Wound Plastics

The rapid advancement and utilization of filament wound plastics has led to the issuance of a number of specialized test specifications/standards. The following ASTM standards are of particular interest.

- ASTM D2290-76, "Apparent Tensile Strength of Ring or Tubular Plastics by Split Disk Method."
- ASTM D2291-76, "Fabrication of Ring Test Specimens for Reinforced Plastics."
- ASTM D2343-67, "Tensile Properties of Glass Fiber Strands, Yarns, and Rovings Used in Reinforced Plastics."
- ASTM D2344-76, "Apparent Horizontal Shear Strength of Reinforced Plastics by Short-Beam Method."
- ASTM D2585-68, "Preparation and Tension Testing of Filament Wound Pressure Vessels."
- ASTM D2586-68, "Hydrostatic Compres-

sive Strength of Glass Reinforced Plastic Cylinders."

- ASTM D2587-68, "Acetone Extraction and Ignition of Strands, Yarns, and Roving for Reinforced Plastics."
- ASTM D2996-71, "Filament Wound Reinforced Thermosetting Resin Pipe."
- ASTM D3299-74, "Specification for Filament Wound Glass Fiber Reinforced Chemical-Resistant Tanks."

24.2.2. Reinforced Thermoplastics

Increasing use of fiber-reinforced thermoplastics has necessitated the development of standards and test procedures for this particular family of reinforced plastics. The following ASTM standards have been issued to satisfy this need.

- ASTM D2848, "Reinforced Polycarbonate Injection Molding and Extrusion Materials."
- ASTM D2853, "Reinforced Olefin Polymers for Injection Molding or Extrusion."
- ASTM D2897, "Reinforced and Filled Nylon Injection Molding and Extrusion Materials."
- ASTM D2948, "Glass-Reinforced Acetal Plastics for Molding and Extrusion."
- ASTM D2990, "Tensile Creep and Creep Rupture of Plastics."
- ASTM D3011, "Reinforced and Filled Polystyrene, Styrene-Acrylonitrile, and Acrylonitrile Butadiene-Styrene for Molding and Extrusion."
- ASTM D3220, "Reinforced Polyterephthalate Plastic Molding and Extrusion Materials."

24.2.3. Reinforced Plastic Pipe

A large market has developed for reinforced plastic pipe, tubing, and fittings. Pertinent test standards and specifications include the following:

- ASTM D1598-76, "Time-to-Failure of Plastic Pipe Under Constant Internal Pressure."
- ASTM D1599-74, "Short-Time Rupture Strength of Plastic Pipe, Tubing, and Fittings."
- ASTM D1694-67, "Threads for Reinforced Thermosetting Plastic Pipe."
- ASTM D2105-67, "Longitudinal Tensile Properties of Reinforced Thermosetting Plastic Pipe and Tube."
- ASTM D2143-69, "Cyclic Pressure Strength of Reinforced Thermosetting Plastic Pipe."
- ASTM D2310-71, "Reinforced Epoxy Resin Gas Pressure Pipe and Fittings" (describes requirements and test procedures).
- ASTM D2517-73, "Machine Made Reinforced Thermosetting Resin Pipe."
- ASTM D2925-70, "Beam Deflection of Reinforced Thermosetting Plastic Pipe Under Full Bore Flow, Measuring."
- ASTM D2992-71, "Obtaining Hydrostatic Design Basis for Reinforced Thermosetting Resin Pipe and Fittings."
- ASTM D2996-71, "Filament Wound Reinforced Thermosetting Resin Pipe."
- ASTM D2997-71, "Centrifugally Cast Reinforced Thermosetting Resin Pipe."
- ASTM D3615-77, "Chemical Resistance of Thermoset Molding Compounds Used in the Manufacture of Molded Fittings."

24.2.4. Corrugated Panels

Several test methods have been developed specifically for corrugated reinforced plastic building panels. ASTM D1494-60, "Diffuse Light Transmission Factor of Reinforced Plastics Panels," is applicable to both flat and corrugated panels. ASTM D1502-60, "Transverse Load of Corrugated Reinforced Plastic Panels," provides a method for determining a useful strength property without involving the corrugation shape factor, and is useful for quality control and procurement purposes. However, because of differences in the mode of attachment, the test cannot be used to determine actual load-carrying capacity for panels installed in buildings. ASTM D1602-60, "Bearing Load of Corrugated Reinforced Plastics Panels," is intended to apply where rivets, bolts, or other mechanical fasteners are used in joining panels. Specimens are tested both from the crowns and from the valleys of the corrugated panels.

24.2.5. Metal Matrix Composites

Developments in fiber-reinforced metal matrix composites have necessitated the preparation of special test methods. ASTM D3552-77 describes a standard test method for determining the tensile properties of metal matrix composites reinforced by continuous and discontinuous high modulus fibers. The method applies to tests parallel or perpendicular to the direction of the reinforcement. ASTM D3553-76, "Fiber Content by Digestion of Reinforced Metal Matrix Composites," involves the chemical (acid or base, as appropriate, depending on the matrix) digestion of the metal matrix of a weighed composite sample. The residue (fibers) is then filtered, washed, dried, and weighed to determine the fiber content for the sample.

24.3. GENERAL CONSIDERATIONS

24.3.1. Specimen Configuration

The configuration of the test specimen is critical in order to achieve reproducible results. Specimen configuration is generally specified for each test procedure. In some cases, it may be necessary to modify the configuration in order to obtain acceptable values (to ensure failure in the gauge area and not within the area in contact with the test grips).

24.3.2. Conditioning for Testing and Exposures

The humidity and temperature environment to which a test specimen is exposed can significantly influence its properties. Accordingly, standardized temperature and humidity conditions are usually specified for conditioning of specimens prior to actual testing. Table 24.4 presents the conditioning requirements of ASTM and FTMS No. 406, respectively.

Often, a wider range of atmospheric conditions is permitted for testing. For example, Military Specification MIL-R-9299, "Resin, Phenolic, Low Pressure Laminating," states that "atmospheric conditions surrounding tests shall be 70 to 80° F (21 to 27° C) and 30 to 60 percent relative humidity." The standard conditions of 73.4 ± 3.6° F (23 ± 2° C) and 50 ± 4% relative humidity are required, in this case, only if test results appear to be affected by the wider range of test conditions.

On the other hand, in cases where there may be some disagreement for tests conducted under the standard laboratory atmosphere, the tolerances may be reduced to ±1.8° F (±1° C) and ±2% relative humidity.

Special tests are often utilized to assess the affects of environmental conditions on performance of composite materials. Investigations of moisture exposure, for example, indicate that resin content, fiber orientation, specimen geometry, relative humidity, and temperature will influence the degradation. Dimensional effects are also noted due to absorption and subsequent desorption of moisture. Relative

Table 24.4. Standardized Specimen Conditioning Requirements (Prior to Testing)

	ASTM D618-61	FTMS No. 406
Temperature	73.4 ± 3.6° F (23 ± 2° C)	73.5 ± 2° F (23 ± 1.1° C)
Relative humidity*	50 ± 5%	50 ± 4%
Duration (minimum) relative to specimen dimensions**	40 hours for specimens ⩽ 0.25 in. (6.35 mm) thick; 88 hours for specimens > 0.25 in. (6.35 mm) thick.	48 hours for specimens ⩽ 0.125 in. (3.175 mm) thick; 96 hours for specimens > 0.125 in. (3.175 mm) thick.

*See, for example, ASTME 104-51 for laboratory methods for obtaining constant relative humidities.

**ASTM D618-61 also describes other standard procedures for conditioning of plastics. The procedure shown is generally recommended unless other methods are specified.

humidity can affect the stiffness properties of composites subjected to cyclic bending loads.[2]

Interest in reinforced plastics for combatting corrosion in various industrial processing plants has spurred considerable testing of the effects of corrosive environments. Both laboratory and field service tests have been conducted by government and private industry, often resulting in the selection of particular reinforced plastic materials for specific chemical processing systems.[3]

When specimens and testing facilities are in controlled-atmosphere buildings (rooms), strict adherence to these conditioning environments is less critical. In any case, it is recommended that the conditioning environment be specified and recorded.

In many cases, it is necessary to determine properties after and/or during various exposure conditions (e.g., elevated temperature, cryogenic temperatures, humidity, vacuum, and radiation). Sufficient time of temperature exposure ("soak period") must be allowed for the specimen to reach equilibrium—generally 10–30 minutes. Combined environments are often required to assess possible synergistic effects on properties. In such cases, the exposure conditions and tolerances should be specified, and recorded with the test results.

ASTM D759, "Conducting Physical Property Tests of Plastics at Subnormal Temperatures," is the recommended practice for use in determining physical properties by means of appropriate ASTM test methods, at temperatures of −452 to +1022° F (−269 to +550° C), excluding the standard laboratory temperature, 73.4° F (23° C).

"Determining Permanent Effect of Heat on Plastics," ASTM D794, defines the conditions for testing the resistance of plastics to changes in properties due to exposure at elevated temperatures. The procedure includes both continuous heat tests, with exposures from minutes to weeks, and cycling heat tests.

Effects of weathering exposure may be determined in accordance with ASTM D1435, "Outdoor Weathering of Plastics."

ASTM C581 describes a relatively rapid test to evaluate the chemical resistance of thermosetting resins used in reinforced plastics of glass fiber-reinforced plastic structures under anticipated service conditions. The method provides for the determination of various property changes due to exposure in a wide variety of test reagents (acids, bases, solvents).

24.3.3. Directional Properties

Since reinforced plastics are anisotropic (i.e., properties vary with direction), it is often necessary to conduct tests in various directions. The direction(s) of a test (e.g., load application or heat conductivity) should be specified and recorded with the test results. For example, fabric-reinforced plastic laminates will have different properties perpendicular to the laminations (the "thickness direction") and parallel to the laminations (in the plane of the laminate). Depending on the weave of the fabric and the lay-up pattern, properties will also vary with direction in the plane of the laminate. Loading/testing direction may be specified with respect to the fabric warp direction, or with respect to the part configuration (e.g., axial or circumferential).

24.3.4. Loading Rate

Mechanical properties are often sensitive to the rate of load application during testing.* In accordance with ASTM D638-77, "Tensile Properties of Plastics," the standard speed of testing is selected from a table, from a specification for the material to be tested, or by mutual agreement among those concerned. When not specifically stated otherwise, the testing speed should be the lowest speed shown in the ASTM D638 table for the specimen geometry being used, which produces failure within ½ to 5 minutes testing time. For rigid and semi-rigid plastics, speed of testing ranges from 0.2 in./minute (5 mm/minute) up to 20 in./minute (500 mm/minute) for most

*The reader should also refer to the following ASTM standards relative to mechanical testing: ASTM E4-72, "Verification of Testing Machines," which describes procedures for verifying testing equipment, through the use of various standard calibrating devices, and ASTM E6-76, "Methods of Mechanical Testing," which provides a convenient source for standardized terminology and definitions relating to mechanical testing.

materials and specimen configurations. When only a limited amount of test material is available with a thickness of 0.28 in. (7 mm) or less, or if a large number of specimens are to be exposed in a limited space (e.g., thermal and environmental stability tests), a smaller specimen configuration is specified, and speed of testing then ranges from 0.05 in./minute (1 mm/minute) to 5 in./minute (100 mm/minute).

ASTM D2289, "Tensile Properties of Plastics at High Speeds," covers the determination of the tensile properties over a wide range of testing speeds, extending from the conventional speeds specified in ASTM D638 to those at which stress wave propagation effects may become important. When the speed of testing (defined as the relative rate of motion of the grips during the test) is not specified (by the material specification or otherwise), tests are often conducted at three speeds: 100 in./minute (2.5 m/minute), 1000 in./minute (25 m/minute), and 10,000 in./minute (250 m/minute).

24.3.5. Sampling, Calculations, and Reporting

The number of samples required generally is specified for each test in order to obtain a reasonably reliable test value for the material under evaluation. Variation of test values (sample-to-sample) frequently is employed in establishing design allowables based on desired confidence limits. (See also ASTM D2188, "Statistical Design in Interlaboratory Testing of Plastics.")

Sampling procedures and practices are described in the documents referenced in the bibliography at the end of Chapter 25. ASTM D1898 is of particular interest, presenting details of sampling procedures.

In establishing design properties (i.e., "design allowables") for a material, the test data must be analyzed statistically. The data are treated on a probability basis so that the end items can be expected to function with a high reliability. In so doing, it is necessary to consider the uncertainty in the estimates of the population mean for each property and the corresponding standard deviation by applying a confidence level to the estimate, either of the population statistics or of the minimum value.[4] Thus, for example, a design allowable might be defined as that value for a given property above which at least 99% of the population of values is expected to fall, with a confidence of 95%.

Standardized methods for calculating and reporting test results generally are included in each specification covering test procedures. Frequently, it is required that a complete test report be prepared. This report should include material identification, the specimen fabrication procedure, pertinent compositional information, the specimen configuration and dimensions, the conditioning procedure, atmospheric conditions in the test area, the test procedure, the date of the test, and the test results. Reporting of test results generally consists of individual specimen results, average value, and standard deviation. The arithmetic mean is given as the average value of a test series. The estimated standard deviation, s, (usually reported to two significant figures) is calculated:

$$s = \sqrt{\frac{\Sigma X^2 - n\bar{X}^2}{n - 1}}$$

where X = the value of single observation, n = the number of observations (test specimens in the series), and $\bar{X}$ = the arithmetic mean of the set of observations.

24.4. TESTING OF RAW MATERIALS FOR REINFORCED PLASTICS

The component, or the raw materials, to be used in a reinforced plastic construction is frequently tested—especially for quality control purposes. Productive raw materials of interest are the reinforcement, the resin and catalyst/curative, and (if applicable) preimpregnated/molding materials. Ancillary materials (e.g., release agents, vacuum bags, and other processing accessory materials) are rarely subject to testing unless difficulty has been experienced or anticipated.

24.4.1. Resins, Gel Coats, and Catalysts/Curatives

The following properties frequently are determined for resins and gel coat materials. Typical specifications are indicated for each test.

24.4.1.1. Viscosity, As Received, and After Catalysis

Variations in viscosity for a given resin system may result from such causes as deviations in the molecular proportions of the reactants used in the synthesis, changes in the synthesis process, the presence of impurities, and chemical changes—possibly due to storage conditions.

Various types of equipment are available for measuring the fluidity of liquid resins or resin solutions. The Brookfield Viscometer* utilizes a rotating spindle; measurement is made of the shear force to rotate the spindle through the test resin/solution. ASTM D2393-68 describes the use of the Brookfield Model RVF Viscometer (or equivalent) for measuring the viscosity of epoxy resins, modifiers, and diluents used in formulating epoxy systems, liquid curing agents, and epoxy resin-curing agent systems or mixtures. The MacMichael Viscosimeter determines viscosity by measuring the torque on a plunger suspended within the resin, which is contained in a cup revolving at constant speed. Other types of viscosity measurement equipment also are available, especially for low viscosity fluids—e.g., devices which utilize flow through orifices or capillaries and rate of fall of a sphere within the test resin.

For extremely viscous and solid resins, viscosity determinations are made after dissolving the resin in a specified quantity of a suitable solvent. ASTM D2857, "Dilute Solution Viscosity of Polymers," describes a method for determining the dilute solution viscosity of polymers that dissolve without chemical reaction or degradation to form stable solutions.

When employed with catalyzed resin systems, viscosity measurement serves as an indicator of working life.

*Available from the Brookfield Engineering Laboratories, Inc., Stoughton, Massachusetts.

Viscosity limits are of particular interest in achieving uniform impregnation and wetting of fiber/fabric reinforcements, and flow during molding and curing operations.

24.4.1.2. Visual Inspection

Resins and solutions thereof should be examined visually both for batch acceptance and production control purposes. Impurities and color changes often are apparent to the naked eye. Color may be determined by comparison with known standards (e.g., the Gardner Color Scale, ASTM D1544). Refractive index (FTMS 406 Method No. 3011 or ASTM D542-50) is often determined for resins to be used in translucent glass-reinforced laminates.

24.4.1.3. Storage Life and Working Life—Gel Time

The life of the uncatalyzed resin during storage may be affected by storage conditions (especially temperature and humidity), light and air exposure, and impurities. As indicated above, viscosity is often a good measure of storage life.

After catalysis (or addition of curatives), the processor is concerned with the "working" or "pot" life of the resin system. Viscosity change with time or the gel time of the resin-catalyst mixture can serve to determine the working life characteristics.

The Society of the Plastics Industry (SPI) and other organizations have established standardized procedures for measuring gel time. A quantity of catalyzed resin is placed in a test tube of specified dimensions and heated in a constant-temperature bath, e.g., 180° F (82.2° C), such that the resin level is below that of the bath. The temperature of the resin is continuously monitored; the gel time is defined as the time for the resin temperature to rise from a given value—e.g., 150° F (65.6° C) if a 180° F (82.2° C) bath is used—to 10° F (5.5° C) above the bath temperature.[5]

During the gel time measurement, the maximum temperature reached by the resin, or the peak exotherm, may be determined.[6] Resins producing high exothermic heat are more susceptible to cure shrinkage and craze-cracking; thus, peak exothermic temperatures usually are specified. It has been noted that laminates which experience excessive heat build-up (exotherms) during cure tend to contain weak interlaminar bonds and hence relatively poor physical properties.[7]

Changes in gel time and peak exotherm during resin and/or catalyst (curative) storage may be indicative of chemical changes therein. To insure against storage changes, as well as batch variations, the fabricator will frequently measure the viscosity and gel characteristics shortly before use.

ASTM D2471-71, "Gel Time and Peak Exothermic Temperature of Reacting Thermosetting Plastic Compositions," is useful for application evaluation, quality control, and material characterization. In addition to describing the test procedure, test sample volume and working volume are suggested for various typical applications.

ASTM D3532-76, "Gel Time of Carbon Fiber-Epoxy Prepreg," covers the determination of the gel time of carbon (graphite) fiber-reinforced epoxy preimpregnated tapes and sheet. The method is suitable for resin systems having either high or low viscosity. Samples cut from the prepreg are heated on a hot plate to a specified test temperature; and, observation is made for a bead of resin squeezed from the prepreg to form strings when probed. The gel time is reported as the point when strings cannot be formed with the probe (a toothpick or fine glass rod).

24.4.1.4. Specific Gravity (Density)

The specific gravities (or densities) of resins and curatives are usually specified and may be employed for evaluation of batch quality and for process control in some cases. Methods are available for both liquid and solid materials. Measurements may be made before or after cure.

For liquids, a method of test often employed is ASTM D1475, "Density of Paint, Varnish, Lacquer, and Related Products."

As noted in military specifications for thermosetting resins (e.g., MIL-R-7575, "Resin, Polyester, Low-Pressure Laminating," and MIL-R-9300, "Resin, Epoxy, Low-Pressure Laminating"), the specific gravities of the unactivated (uncatalyzed) and the cured unfilled resin are determined simply by weighing a specific volume of the resin in lb (g) and dividing by the volume in in.^{3} (cm^3). The Westphal Balance is convenient for this purpose, providing the relative bouyancy for a known weight in the test liquid (resin) and in water.

For solid materials, the following test standards are available.

- FTMS 406 Method 5011, "Specific Gravity by Displacement of Water."
- FTMS 406 Method 5012, "Specific Gravity from Weight and Volume Measurements."
- ASTM D792, "Specific Gravity and Density of Plastics by Displacement."
- ASTM D1505, "Density of Plastics by the Density-Gradient Technique."
- ASTM D1895, "Apparent Density, Bulk Factor, and Pourability of Plastic Materials."

Generally, these latter methods are limited to solid plastics; however, they may also be applied to plastic materials in the form of molding powders, flakes, or pellets.

24.4.1.5. Hardness

The extent of the state of cure of a resin is generally determined by a measurement of its hardness. For quality/process control, some processors employ a minimum time-to-cure test, i.e., a test of the minimum time for the resin-curative (catalyst) system to reach a specified hardness value when maintained at a given temperature.

FTMS Method No. 1081 (ASTM D785) describes the method for determining the indentation hardness of plastics by means of the Rockwell hardness tester. A small steel ball indenter is forced into the material, using a load depending on the desired Rockwell scale. The hardness value is a measure of the resistance to penetration offered by the test sample. Rockwell hardness tester. A small steel ball indenter is forced into the material, using a load, and dial scale used. ASTM D785 Procedure A measures the indentation of the specimen remaining 15 seconds after a specified major load is released to a standard 22-lb (10-kg) minor load for 15 seconds. Procedure B measures the indentation into the specimen after a 15-second application of the specified major load, while the load is still applied.

The Barcol hardness tester (manufactured by Barber-Colman Co., Rockford, Illinois) is widely used with plastics. This apparatus employs a sharp-pointed indenter which is spring-loaded against the surface of the test specimen. ASTM D2583-75, "Indentation Hardness of Plastics by Means of a Barcol Impressor," describes the method for determining hardness of both reinforced and non-reinforced rigid plastics using the Model No. 934-1 Barcol Impressor. Because the unit is small and portable, it is suitable for testing the hardness of fabricated parts and individual test specimens for production control purposes. It is necessary that the indenter be applied perpendicular to the test surface. Also, the specimens should be supported by a hard, firm surface if they might otherwise bend or deform under the applied load from the indenter.

For flexible resins, the (Shore) Durometer is commonly used to determine indentation hardness. A needle-like steel penetrant is spring-loaded against the specimen surface, and the depth of penetration is read on a calibrated dial gauge. The requirements for the use of such a device are presented in FTMS 406 Method Nos. 1082, 1083, and 1084, and ASTM Standard D2240-75. The Type A durometer is used for testing non-rigid and some semi-rigid materials (especially plastics and elastomers); the Type D durometer is used for rigid and some semi-rigid materials.

A variety of other testing devices are available. Those described above are most frequently employed in the reinforced plastics industry.

24.4.1.6. Chemical Composition

In order to ensure high quality, it is often required that certain features of the chemical composition be determined. This serves to verify the reactivity of the resin system and to make sure certain contaminants are not present in excessive amounts. The presence of contaminants can impair physical, chemical, and electrical properties. The following discussion indicates the types of tests conducted for these purposes.

For epoxy resins, tests of epoxide equivalent and hydroxyl content provide measures of the reactivity of the resin. Chlorine and moisture content are measures of purity; excessive values of such impurities can influence reactivity as well as the electrical and other properties of the end item. Where appropriate, the melting (softening) point may also be required.

By way of illustration, ASTM D1763-76 is a specification for reactive epoxy resins supplied as liquids or solids, for use in castings, coatings, tooling, potting, adhesives, or reinforced plastics. Hardeners (curatives) are not included as part of this specification. Six types of resins are described based on their chemical nature: Type I—epichlorohydrin and bisphenol A; Type II—a reaction product of phenol and formaldehyde (novolac resin) and epichlorohydrin; Type III—cycloaliphatic and peracid epoxies; Type IV—glycidyl esters; Type V—a reaction product of *p*-aminophenol and epichlorohydrin; and Type VI—a reaction product of glyoxal tetraphenol and epichlorohydrin. For each epoxy resin type, there are two grades, depending on whether a reactive diluent has been added.

Tests specified include viscosity (ASTM D1544), weight per epoxy equivalent (ASTM D1652), color (ASTM D1544), and Durrans' softening point (for solid resins). For this latter test, a 0.004-lb (2-g) sample is placed into a test tube and then heated in an oil bath to melt; then, after cooling, with a thermometer immersed in the melt, 0.11 lb (50 g) of mercury are added to the test tube. The sample is then immersed in the oil bath and heated at a rate of 3.6° F/minute (2° C/minute). The softening point is the temperature at which the molten resin becomes visible on top of the mercury surface.

For uncatalyzed polyester resins, the determination of the acid number is of special interest. This is the quantity of potassium hydroxide required to neutralize 0.0022 lb (1 g) of the resin (see, for example, ASTM D2849).

This serves to ensure batch uniformity with respect to reactivity. Since monomeric styrene, vinyl toluene, diallyl phthalate or some other cross-linking species is generally added, many polyester resin material specifications require that the polyester concentration be within established acceptable levels.

For resin solutions and resins containing volatile species, especially solvents, determination of the non-volatile (or solids) content is often required. ASTM D1259 describes the determination of non-volatile content of resin solutions, and ASTM D1644-75 covers the measurement of non-volatile content for varnishes. Various methods may be employed, depending on the time and temperature of exposure of the test sample. In ASTM D1644, Method A requires exposure at 221 ± 3.6° F (105 ± 2° C) for 3 hours (in a ventilated oven); Method B requires exposure at 300 ± 6° F (149 ± 3.3° C) for 10 minutes (on a hot plate). Non-volatile matter is determined in each case by weighing the sample before and after the exposure.

Catalysts and curatives are also subjected to tests to ensure purity and composition. For example, a typical organic peroxide catalyst used with unsaturated polyester resins may be required to have a value of active oxygen above a specified minimum allowable level, an iron content not greater than a specified value, and a specified peroxide assay. These values are determined by standard analytical chemical techniques. Similarly, amines and anhydrides employed in curing epoxy resins are often tested to determine chemical assay, acid content for anhydrides (both free acid and total acid), amine content, and water content (as an impurity).

Infrared spectrographic analysis can serve for purposes of chemical identification of resins, curatives, and modifying ingredients, and as a means for determining purity thereof.

24.4.1.7. Physical, Mechanical, and Electrical Properties of Cured Unfilled Resins

In addition to the tests of the resins and other material system components described above, it is frequently required that selected properties be determined for the cured unfilled (non-reinforced) resins. This is especially appropriate for 100% solids resin systems which do not produce volatile by-products during cure (e.g., epoxy and unsaturated polyester resins).

Specific gravity and Barcol hardness are often determined to verify the quality of the reactive materials, proper mixing, and curing procedures. Nominal values of the dielectric constant and loss tangent for cured unfilled resins may also be required, especially for resins to be employed in electrical applications.

Occasionally, selected mechanical properties (for unfilled, non-reinforced castings) may also be required, especially for resin systems to be employed in primary load-bearing structures. Strength, modulus, and ultimate elongation values may be determined (usually in tension) to ensure that the materials and the mixing, curing, and other procedures will provide the desired characteristics; minimum acceptable values may be specified toward this end.

Methods for performing electrical and mechanical tests are those typical of all other plastics materials (see Table 24.3).

24.4.2. Reinforcement Materials

The following standards and specifications describe methods for testing various commonly used types and forms of fibrous reinforcements.

24.4.2.1. Standards/Specifications of General Interest

- ASTM D76-77, "Tensile Testing Machines for Textile Materials."
- ASTM D123-77, "Definition of Terms Relating to Textile Materials."
- ASTM D1117-74, "Testing Nonwoven Fabrics."
- ASTM D1175-71, "Abrasion Resistance of Textile Fabrics."
- ASTM D1682-64, "Breaking Load and Elongation of Textile Fabrics."
- ASTM D1776-74, "Conditioning Textiles and Textile Products for Testing."
- ASTM D1777-64, "Measuring Thickness of Textile Materials."

- ASTM D2654-76, "Test for Amount of Moisture Content and Moisture Regain in Textile Materials."

24.4.2.2. ASTM Standards for Fibrous Glass

- ASTM D578-61, "Specification and Testing of Glass Yarns."
- ASTM D579-66, "Specification for and Testing Glass Fabrics."
- ASTM D580-49, "Testing and Tolerances for Woven Glass Tapes."
- ASTM D2150-70, "Woven Roving Glass Fabric for Polyester-Glass Laminates."
- ASTM D2343-67, "Test for Tensile Properties of Glass Fiber Strands, Yarns, and Rovings Used in Reinforced Plastics."
- ASTM D2408-67, "Woven Glass Fabric Cleaned and After-Finished with Amino-Silane Type Finishes, for Plastic Laminates."
- ASTM D2409-67, "Woven Glass Fabric, Cleaned and After-Finished with Vinyl-Silane Type Finishes, for Plastic Laminates."
- ASTM D2410-67, "Woven Glass Fabric, Cleaned and After-Finished with Chrome Complexes, for Plastic Laminates."
- ASTM D2587-68, "Test for Acetone Extraction and Ignition of Strands, Yarns, and Roving for Reinforced Plastics."
- ASTM D2660-70, "Woven Glass Fabric, Cleaned and After-Finished with Acrylic-Silane Finishes, for Plastic Laminates."
- ASTM D3098-72, "Woven Glass Fabric, Cleaned and After-Finished with Epoxy-Functional Silane Type Finishes for Plastic Laminates."

24.4.2.3. Military Specifications

- MIL-Y-1140, "Yarn, Cord, Sleeving, Cloth and Tape—Glass."
- MIL-C-9084, "Cloth, Glass, Finished, for Polyester Laminates."
- MIL-F-9118, "Finish, for Glass Cloth."
- MIL-F-12298, "Fabric, Glass Woven."
- MIL-M-15617, "Mats, Fibrous Glass, for Reinforcing Plastics."
- MIL-C-19663, "Cloth, Glass, Woven Roving, for Plastic Laminates."
- MIL-R-60346, "Roving, Glass, Fibrous (for Filament Winding Applications)."

In addition to the documents listed above, Federal Specification CCC-T-191 describes various methods for testing textiles. Another source of information on testing of textiles is the *Wellington Sears Handbook of Industrial Textiles* by E.R. Kaswell (Wellington Sears Company, Inc., New York, 1963).

While glass fiber reinforcements are the most widely employed in reinforced plastics, many others are also used, including asbestos, carbon and graphite, quartz, polyaramid (especially Kevlar, manufactured by E. I. DuPont de Nemours & Co., Inc.), nylon, rayon, cotton, paper, sisal, and, undoubtedly, other natural and synthetic fibers. With the advent of high performance/advanced composites, filaments (or wires) of boron, beryllium, silicon carbide, and boron nitride have become of interest. In addition to the latter, significant advances have been made in metallic and metallized glass fibers (especially for applications requiring improved electrical or thermal conductivity), mineral fibers from basalt (from lava deposits), and hybrid fabrics woven from two or more different fibers.[8] The tests described above are generally applicable; often, however, special tests must be developed to evaluate and control the quality of these newer reinforcement materials.

Recent developments in high modulus fibers have led to the issuance of the following ASTM test methods.

- ASTM D3317-74a, "High Modulus Organic Yarn and Roving."
- ASTM D3318-76, "Woven Cloth from High Modulus Organic Fiber."
- ASTM D3544-76, "Guide for Reporting Test Results on High Modulus Fibers."

With these developments and the application of new and improved reinforcements, rather extensive studies have been conducted of pertinent test methods. AFML-TR-67-159 describes techniques for determining the following properties of fibers and yarns.[9]

- Tensile properties
  1. Tensile strength

2. Breaking elongation
3. Work or energy to break
4. Tensile recovery (strain and energy), both immediate and delayed
5. Unrecovered strain, immediate and permanent
6. Initial modulus

- Shear modulus
- Bending modulus and flexural strength
- Flexural (bending) cycle life
- Density (or specific gravity)
- Linear density
- Diameter and cross-sectional area

Tensile properties are determined most often using a tensile testing machine. The optical strain analyzer[10] is used when very accurate measurements are required of elongation or modulus, and for fibers that are extremely brittle. The Dual Head Tensile Tester may also be employed for fibers at least 4 in. (10.16 cm) in length and less than approximately 0.002 in. (0.05 mm) in diameter; this instrument is capable of testing at elevated temperatures up to 2000° F (1093° C). Other types of testing machines can also be equipped with special heating chambers (ovens) for elevated temperature testing.

Fiber modulus is most often determined from the slope of the tangent to the initial linear portion of the load elongation curve. This requires that the elongation be measured with a high degree of accuracy, and slippage of the test specimen within the jaws (grips) of the testing machine, or deflection of the equipment, cannot be permitted. The sonic modulus is obtained from the wavelength of standing waves in the filament under constant tension and driven at one end by a suitable audio oscillator. The sonic modulus is then calculated from the oscillator frequency, wavelength, and fiber density (Method FMT-13[9]). The dynamic modulus of a fiber can be determined from the resonant frequency of vibration; the fiber is excited at one end by an audio oscillator which is tuned to produce the maximum amplitude of vibration of the fiber. (This method is often termed the vibrating reed technique.) Methods for determining the shear modulus (using a torsion pendulum; from the period of torsional oscillation) and the bending modulus (using two pendulums set into torsional oscillation such that the fiber is continually bent in two opposite directions) have been described.[9]

For very stiff fibers and yarns, bending modulus and flexural strength can be determined using a flexural jig. The specimen is supported horizontally near the ends and loaded at the center. Modulus and strength are calculated from the applied loads and deflection.

Flexural cycle life of a yarn can be determined by holding the yarn under a constant tension and bending it repeatedly through a fixed angle over a rod until failure occurs.

The density of fibers can be determined by direct weighing, provided they are of known and uniform cross-section. A microbalance is required to provide adequate accuracy. For fibers of unknown or non-uniform cross-section, the density gradient method is preferred. In this method, short lengths of fiber are immersed in a density gradient column (as described in ASTM D1505) prepared using liquids of specific gravity covering the range of interest (1–3.5). The Westphal Balance (ASTM D1505) is also suitable. Weighing in air and in water (or other liquid of density less than that of the sample) may also be employed (ASTM D792, "Specific Gravity and Density of Plastics by Displacement").

The linear density of a fiber (usually expressed in denier units—the weight, in g, of 9000 m of the material) can be determined by three methods. A known length can be weighed on a microbalance, and linear density is then calculated directly; this method may be used provided the specimen weight is greater than 100 μg. The denier balance is a torsion balance calibrated to read directly in denier when a 90-cm length of fiber is weighed. The vibrascope method utilizes the measurement of the resonant frequency of the fiber when vibrated under a known tensile load. The vibrascope is useful for fine fibers of linear density less than 50 denier, and is especially suitable for flexible materials. ASTM D1577 describes this method in considerable detail.

Fiber diameter can be determined by various methods. Fibers having circular cross-sections can be measured directly with a microscope using a calibrated eyepiece. Fibers of irregular cross-section are best measured using a planimeter to measure the area of the projected image of the cross-section.

Fiber diameter can also be calculated from linear density and specific gravity.

Of the various military specifications covering fiber reinforcements, MIL-C-9084 is most widely employed. This specification describes the requirements for woven glass fabrics that have been cleaned and treated with a finish for subsequent fabrication into glass fabric-reinforced polyester-resin laminates. The fabrics must conform to specified construction and physical properties. In addition, for acceptance testing, samples of the fabric are required to be fabricated into plastic laminates and tested for flexural strength under both dry and wet conditions.

Prior to acceptance on delivery, visual examination is conducted to observe defects: biased or bowed filling yarns; baggy, ridgy, or wavy cloth; holes, cuts, or tears; spots, streaks, or stains; tender or weak spots; smashed areas; broken or missing ends or picks; floats and skips; coarse or light areas; selvage defects; creases; uneven finish; brittle or fused areas; width exceeding specified tolerance; overall uncleanliness; objectionable odor; color not characteristic of the applied finish; or uneven weaving throughout.

24.4.3. Testing of Preimpregnated Reinforcements and Molding Materials

Resin-impregnated reinforcements are widely used in laminating and molding of reinforced plastic parts. Properties of the end item depend on both material and processing parameters; hence, a variety of tests are conducted for these materials. For example, the resin content, volatile content, and degree of advancement influence the handling and laminating/molding characteristics.

24.4.3.1. Volatiles Content

Volatiles content measurements are made to determine the quantity of volatile products produced from the material when heated at an elevated temperature—usually in the range of 212–347° F (100–175° C). Military Specification MIL-R-7575 ("Resin, Polyester, Low-Pressure Laminating") specifies that the preimpregnated fabric—one ply, cut into a 4-in.^{2} (10.16-cm^2) test specimen, taken at 1 in. (2.54 cm) from the edge of the fabric and bias cut to minimize raveling—is tested by weighing before and after drying by suspending the specimen for 15 ± 1 minutes in a circulating air oven at 325 ± 5° F (163 ± 3° C). The specimen is cooled in a desiccator to prevent moisture condensation. Volatiles content is calculated:

$$\text{Volatiles content, \%} = \frac{\text{Initial weight} - \text{dry weight}}{\text{Initial weight}} \times 100.$$

Usually, three or more specimens are tested from representative areas of the material.

ASTM D3530-76, "Volatiles Content for Carbon Fiber-Epoxy Prepreg," determines the percent change in weight when carbon (graphite) fiber tapes and sheet, preimpregnated with epoxy resin, are exposed to either of two standard test temperatures, 250° F (121° C) or 350° F (177° C), approximating the maximum temperature in many cure cycles. This method does not provide an accurate measure of the volatiles content of other resins that will release volatiles at higher temperatures, such as polyimides. Heating is for 10 ± 0.5 minutes.

MIL-G-83410 (USAF), "Graphite Fiber Resin Impregnated Tape and Sheet, for Hand Layup," exposes the sample—approximately 0.0022 lb (1 g)—in a preheated oven at a temperature specified depending on the resin. The sample is heated for 60 ± 5 minutes.

24.4.3.2. Resin Content

For those cases where volatiles content has been determined (above), the dried specimens are then used for determining "dry resin content." The specimen is placed in a muffle furnace, usually specified at 1050 ± 50° F (566 ± 28° C)* for 30 minutes or until all evidence of resin is absent (i.e., reinforcement appears to be bare) and the specimen reaches a con-

*Some specifications designate exposure at 1050 ± 25° F (565.5 ± 14° C).

stant weight. The ignition residue is cooled in a desiccator and weighed, and the content is calculated:

$$\text{Dry resin content, \%} = \frac{\text{Dry weight} - \text{ignited weight}}{\text{Dry weight}} \times 100.$$

During ignition, precautions must be taken to prevent loss (e.g., by blowing) of the residue.

For graphite or other carbonaceous fiber-reinforced plastics, this method is not suitable because of oxidation of the fiber. MIL-G-83410 describes two methods for determining resin (non-fiber) content: solvent extraction and acid digestion (for resin systems that cannot be solvent extracted). Solvent extraction involves immersing the specimen, approximately 0.0022 lb (1 g), in a selected solvent brought to a boil for at least 2 minutes. The solvent-extracted specimens are then dried in a circulating air oven preheated to 325 ± 10° F (163 ± 5.5° C) and weighed. Resin content is then calculated as above.

Acid digestion may use nitric acid, sulfuric acid, or sulfuric acid/hydrogen peroxide. (Digestion temperature should not exceed 140° F (60° C) when nitric acid is used.) Two samples, each approximately 0.0022 lb (1 g) in weight, are first weighed and then immersed in the acid to dissolve and digest the resin. The fibers are then filtered, washed, dried, and weighed.

ASTM D3529-76, "Resin Solids Content of Carbon Fiber-Epoxy Prepreg," covers a method for determining the resin solids content of carbon (graphite) fiber-epoxy tape and sheet. Specimens are cut adjacent to those used for determining the volatiles content (ASTM D3530). After weighing, the specimens are subjected to boiling solvent until the resin is completely dissolved; the fiber residue is then dried and weighed. This method is similar to that described in MIL-G-83410. Suggested solvents are methyl ethyl ketone, acetone, and dimethylformamide. (Suitable precautions are necessary to avoid hazards in using these solvents.)

ASTM C613-67 describes a standard test method for "Resin Content of Carbon and Graphite Prepregs by Solvent Extraction." In this case, the sample is placed in a Soxhlet extraction assembly using a suitable solvent (e.g., ethyl alcohol or dimethylformamide). The sample is cut into approximately 1/2-in. (12.7-mm) squares and weighed. The sample is then placed in a pre-dried thimble and inserted into the Soxhlet extractor. The solvent is refluxed for a minimum of 4 hours to extract the resin from the sample in the thimble. The sample is then dried and weighed. To obtain the true dry resin content (volatiles excluded), it is necessary to also determine the volatiles content and use this value in calculating the dry resin content (DRC) of the prepreg:

$$\text{DRC, mass \%} = 1 - \left(\frac{C}{A - VA/100}\right) \times 100$$

where A = the original net weight of the prepreg sample, lb (g); C = the net weight of the carrier remaining after extraction, lb (g); and V = volatiles content, weight %.

24.4.3.3. Flow Testing

The flow test is conducted to determine how the resin will flow during subsequent laminating/molding operations. (It is desirable that flow be sufficient to produce uniform wetting of the reinforcement material and not be so great that resin-starved parts may result.)

Flow test specimens usually are prepared from four 4-in. (10.16-cm) squares (one ply each) cut (on a bias for fabrics) from the resin-impregnated material. The specimens are weighed, loosely wrapped with cellophane or Teflon film (as a parting agent), and then placed (four piles stacked) between the platens of a press and cured according to a prescribed set of conditions. After curing, the specimen is cooled (usually in a desiccator), the parting film and flash are carefully removed, and the specimen is reweighed.

$$\text{Flow, \%} = \frac{\text{Initial weight} - \text{final weight}}{\text{initial weight}} \times 100.$$

Curing conditions for this test depend on the resin system. MIL-R-7575 ("Resin, Polyester, Low-Pressure Laminating") specifies 325° F (162.8° C) and 15 psi (105 kPa) for 5 minutes. MIL-R-9300 ("Resin, Epoxy, Low-Pressure Laminating") calls for a temperature as specified in the instruction sheet and a pressure of 30 psi (210 kPa) for 5 minutes. MIL-R-25506 ("Resin, Silicone, Low-Pressure Laminating") requires that samples first be precured (resin advancement) in a circulating air oven at 230° F (110° C) for 5 minutes, and then laminated (three plies used in this case) at 347° F (175° C) under 10 psi (70 kPa) for 5 minutes. MIL-R-25042 ("Resin, Polyester, High Temperature Resistant, Low-Pressure Laminating") requires that the four plies be laminated at 325° F (162.8° C) under 15 psi (105 kPa) for 5 minutes. On the other hand, MIL-R-9299 ("Resin, Phenolic Low-Pressure Laminating") takes into account inherent variations in resins from different sources; curing conditions are those specified by the manufacturer.

ASTM D3531-76, "Resin Flow for Carbon Fiber-Epoxy Prepreg," covers the determination of resin flow that will result from prepreg tape or sheet material under specified conditions of temperature and pressure. A weighed specimen consisting of two plies, approximately 2 in. (50 mm) square, cross-plied, is sandwiched between bleeder material and release film. This assembly is then heated in a platen press at a pressure of 100 psi (690 kPa) for 15 minutes or until the resin has gelled. Platen temperature is either 250° F (121° C) or 350° F (177° C), depending on the resin system. After cooling the assembly, the resin flowed to the edges is removed, and the specimen is reweighed. The change in weight as a percent of the original weight is reported as the percent flow.

Resin flow for molding compounds is often determined according to ASTM D731-67, "Measuring the Molding Index of Thermosetting Molding Powder." This method is particularly suited to flow determination of stiff-flow materials, especially those used in compression molding at moderate to high pressures. A cup-type mold is employed, designed with stops so that the resulting flash or fin thickness cannot be less than 0.0055 in. (0.14 mm). The mold is preheated, usually to 320 ± 1.8° F (160 ± 1° C), the molding material is introduced, and pressure is applied by closing the mold. Sufficient force (pressure) is applied to completely close the mold and reach a specified fin thickness. The time of flow (in seconds) is measured from the instant that the hydraulic gauge (on the molding press) shows an applied pressure to the instant that the fin has reached 0.008-in. (0.203-mm) thickness for materials with an Izod impact strength less than 0.50 ft-lb/in. (26.7 J/m) of notch, or 0.026 in. (0.66 mm) for materials with an Izod impact strength above 0.50 ft-lb/in. (26.7 J/m) of notch. This method is recommended for batch-control purposes (to determine uniformity and reproducibility), but it should not be used arbitrarily for selection of molding compounds or the design of molds.

24.4.3.4. Properties of Cured Material

Mechanical, physical, electrical, and/or chemical properties are often determined to establish the suitability of the resin-impregnated material and the specified processing operations. These tests are typical of those for cured reinforced plastics (see below).

24.5. TESTING OF MOLDINGS AND LAMINATES

Various physical, mechanical, electrical, and chemical tests are conducted for reinforced plastics and related materials.

24.5.1. Moldings

It is of interest to consider the tests typical of those required for molded plastic parts. Military Specification MIL-M-14, "Molding Plastics and Molded Plastic Parts, Thermosetting," covers the basic properties of molding compounds and methods for determining these properties, as well as parts molded using such compounds. Molding compounds covered by this specification encompass materials using phenolic, melamine, polyester, diallyl phthalate, and silicone resins; and fibrous reinforcements of glass, asbestos, mineral filler, cellulose filler, acrylic polymer fiber, and polyethylene terephthalate fiber. Standard test specimens are employed in conformance with the applicable method of Federal Test Method Standard No. 406 (see Table 24.3). The required tests and test methods are shown in Table 24.5.

Table 24.5. MIL-M-14 Test Requirements for Molding Compounds

TEST PROPERTY	TEST METHOD*	COMMENTS
Tensile strength	1012	0.25-in. (6.35-mm) specimen
Compressive strength	1021	
Flexural strength	1031	Span/depth is 16/1
Heat resistance	1031	Flexural strength at elevated temperatures, 302 and 392° F (150 and 200° C)
Impact strength	1071	Both plug and chase surfaces
Water absorption	7031	With modifications
Heat distortion temperature	2011	Chase surface; with modifications
Arc resistance	4011	
Dielectric constant	4021	At 1 KHz and 1 MHz
Dissipation factor	4021	At 1 KHz and 1 MHz
Dielectric strength	4031	Short-time and step-by-step; with modifications
Dielectric breakdown	4031	Short-time and step-by-step; with modifications
Volume and surface Resistance (electrical)	See MIL-M-14 for details	Three-terminal method; measure in humidity/heating chamber
Track resistance	ASTM D2303	
Flame resistance	ASTM D229	With modifications
Dimensional stability	See below	
Toxicity when heated	Report No. AD297-457, Defense Documentation Center	

*FTMS 406 unless otherwise indicated.

Dimensional stability tests for MIL-M-14 consist of exposure of test specimens to 10 thermal cycles wherein the temperature is varied from room temperature 73.4 ± 2° F (23 ± 1.1° C), to an elevated temperature, 257 ± 9° F (125 ± 5° C), in a circulating air oven with a relative humidity of 50 ± 2% at room temperature. Test specimens are first conditioned for 96 hours at 73.4 ± 2° F (23 ± 1.1° C) and 50 ± 2% relative humidity. Specimen length is then measured to the nearest 0.001 in. (0.025 mm). During thermal cycling, the upper temperature, 257 ± 9° F (125 ± 5° C), is held for 48 hours, and the lower temperature, 73.4 ± 2° F (23 ± 1.1° C), is then held for 24 hours. After the 10 cycles, the final length of the specimen is measured, and dimensional change (percent) is calculated.

The flame resistance test for MIL-M-14 involves minor modifications to ASTM D229 ("Testing Rigid Sheet and Plate Materials Used for Electrical Insulation"), Method II, Flame Resistance. Specimens are molded 0.5 × 0.5 × 5 in. (1.27 × 1.27 × 12.7 cm). A heater coil surrounding the test specimen is energized simultaneously with arc electrodes (two sets) located horizontally on both sides of the test specimen. Ignition time is the elapsed time (in seconds) to ignition of the specimen (not the gases released by heating). After ignition, the heater and spark gaps are de-energized, and the burning time is determined as the time that the specimen continues to burn until all flame has disappeared. The weight of the specimen before and after burning is determined to calculate weight loss. A modification specified by MIL-M-14 requires that five values of burning time (five specimens tested) be arranged in increasing order: T_1, T_2, . . . T_5. The ratios $(T_2 - T_1)/(T_5 - T_1)$ and $(T_5 - T_4)/(T_5 - T_1)$ are computed; if either of these ratios exceeds 0.642, then T_1 or T_5 is judged to be abnormal and is eliminated. The reported burning time, then, is the arithmetic mean of the remaining four values.

It should be noted that the flame resistance tests described above are conducted in air. A ventilating blower is turned on during the test. The manned space programs have necessitated somewhat similar tests in a pure oxygen environment, often at pressures greater than 1 atmosphere.[11, 12]

To satisfy the "toxicity-when-heated" requirements of MIL-M-14, the following toxic gases must be measured and shown to be present at less than specified maximum concentrations: carbon dioxide, carbon monoxide, ammonia, aldehydes (as H·CHO), cyanides (as HCN), oxides of nitrogen (as NO_2), and hydrogen chloride.

Quality conformance testing of the end item per MIL-M-14 is conducted to determine that the specified requirements are satisfied for the following properties (and any others that may be specified as appropriate for the intended application).

- Arc resistance—per FTMS 406, Method 4011.
- Dielectric constant and dissipation factor—per FTMS 406, Method 4021.
- Dielectric strength, step-by-step—per FTMS 406, Method 4031, modified.
- Flexural strength—per FTMS 406, Method 1031, modified.
- Impact strength—per FTMS 406, Method 1071.
- Water absorption—per FTMS 406, Method 7031.

24.5.2. Laminates

The test requirements for reinforced plastic laminates are illustrated by Military Specification MIL-P-25421, "Plastic Materials, Glass Fiber Base–Epoxy Resin, Low-Pressure Laminated." This specification covers the requirements for such materials to be used for airframe and non-airframe structural parts. The test methods are those described in the Federal Test Method Standard No. 406, "Plastics: Methods of Testing" (see Table 24.3).

Physical properties determined are specific gravity, resin content, and Barcol hardness. (Required values are as specified in the approved contractor's process specification.)

Mechanical properties required to be measured are tensile, edgewise compressive, and flexural strengths, and flexural modulus of elasticity. Values are determined under: 1) standard conditions of 73.4 ± 1.8°F (23 ± 1°C) and 50 ± 4% relative humidity (after four days of exposure),* and 2) wet conditions—2 hours immersion in boiling distilled water. In this latter case, the specimens are cooled in water, and are tested wet immediately after removal from the water. If there is any question as to the validity of the test results tested wet (as described above), then wet conditions are required to be 30 days soaking in distilled water at room temperature.

24.5.2.1. ASTM Tests for Reinforced-Plastic Laminates

- ASTM D494, "Acetone Extraction of Phenolic Molded or Laminated Products."
- ASTM D1867, "Copper-Clad Thermosetting Laminates for Printed Wiring."
- ASTM D1823, "Etching and Cleaning Copper-Clad Electrical Insulating Materials and Thermosetting Laminates for Electrical Testing."
- ASTM D2861, "Flexible Composites of Copper Foil with Dielectric Film or Treated Fabrics."
- ASTM D709, "Laminated Thermosetting Materials."
- ASTM D1532, "Polyester Glass-Mat Sheet Laminate."
- ASTM C582, "Reinforced Plastic Laminates for Self-Supporting Structures for Use in a Chemical Environment."
- ASTM D3039, "Tensile Properties of Oriented Fiber Composites."
- ASTM D2408, "Woven Glass Fabric, Cleaned and After-Finished with Amino-Silane Type Finishes, for Plastic Laminates."
- ASTM D2410, "Woven Glass Fabric, Cleaned and After-Finished with Chrome Complexes, for Plastic Laminates."
- ASTM D2660, "Woven Glass Fabric, Cleaned and After-Finished with Acrylic-

*Specimens may be tested without this conditioning except in cases of doubt of the materials' meeting the requirement where this condition is specified.

Silane Type Finishes, for Plastic Laminates."

- ASTM D3098, "Woven Glass Fabric, Cleaned and After-Finished with Epoxy-Functional Silane Type Finishes for Plastic Laminates."
- ASTM D2150, "Woven Roving Glass Fabric for Polyester-Glass Laminates."

24.5.2.2. ASTM Tests for Physical Properties

- Specific gravity or density, ASTM D792.
- Specific volume, ASTM D792.
- Tensile strength and modulus of elasticity, ASTM D638.
- Compressive strength and modulus, ASTM D695.
- Flexural strength and modulus, ASTM D790.
- Shear strength and modulus, ASTM D732.
- Bearing strength, ASTM D953.
- Impact strength (Izod), ASTM D256.
- Bond or cohesive strength, ASTM D952.
- Hardness, Rockwell, ASTM D785.
- Hardness, Barcol, ASTM D2583.
- Water absorption, ASTM D570.

24.5.2.3. ASTM Tests for Thermal Properties

- Thermal conductivity, ASTM C177.
- Coefficient of linear thermal expansion, ASTM D696.
- Specific heat, ASTM C351 and ASTM D2766.

24.5.2.4. ASTM Tests for Electrical Properties

- Insulation resistance and volume resistivity, ASTM D257.
- Dielectric strength, ASTM D149.
- Dielectric constant and dissipation factor, ASTM D150.
- Arc resistance, ASTM D495.

24.5.2.5. ASTM Tests for Chemical Properties

- Resistance to chemical reagents (acids, alkalies, solvents), ASTM D543.

When testing laminates for flexural strength, specimens tested with the mold-side up (assuming vacuum-bag on opposite side) produced higher strength than similar specimens tested with the mold-side down.[13] Also, it has been found that for satin weave fabric laminates, the mechanical strengths at various angles to the warp direction are not symmetrical.[13] Special lay-up procedures are needed for balanced properties.

The tensile strength specimens for composites include the standard, so called "dog bone", ASTM D 638 specimen, the elongated "bow tie" specimen used in Military and AMS specifications, Fig. 24.1, and the straight specimen with bonded on doublers, Fig. 24.2 used for high performance composites. For woven fabric laminates, the use of the "bow tie" specimen results in higher apparent tensile strengths and breaks in the gage section. The straight specimen with doublers is the only type that gives consistent results for unidirec-

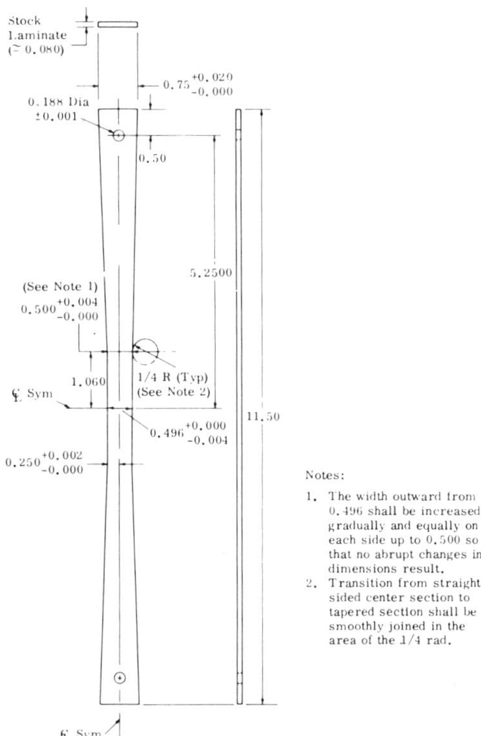

Figure 24.1. Standardized, elongated bow tie tension specimen dimensions.

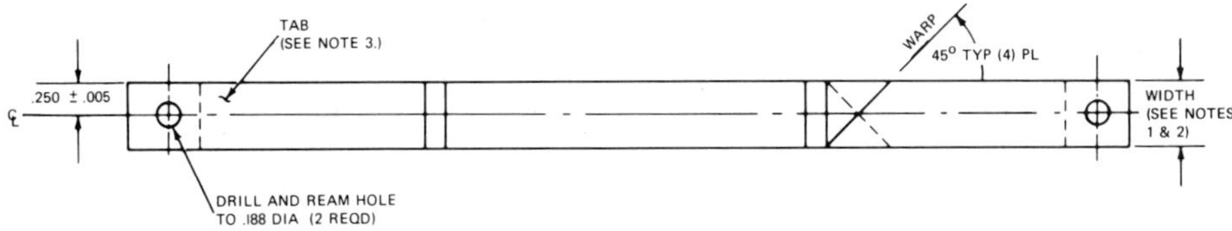

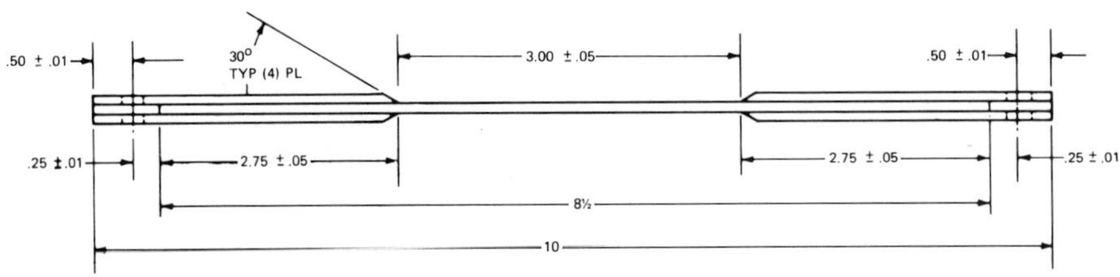

NOTES: 1. NOMINAL SPECIMEN WIDTH —: 496-.500
2. EDGES PARALLEL WITHIN —.002: SQUARE TO FACES WITHIN ±3°; FREE OF DEFECTS
3. TABS: MAT'L — SCOTCHPLY 1002 (8 CROSS PLIED LAYERS)
ADH. — METLBOND 329
ORIENTATION — 45 · 135° TO MEAN DIRECTION OF SPECIMEN AXIS

Figure 24.2. Straight-sided tension specimen.

tional, high performance fiber composites.[17]

The tensile modulus, like other moduli for glass-reinforced plastic laminates, usually is given as primary and secondary because of the shape of the stress-strain curve, which consists of two distinct straight lines with different slopes.

There are several types of compressive tests, most of which result in either a shear failure near the end of the specimen or brooming of the ends in the area of load application. Unrestrained or partially restrained compressive specimens fail at stresses lower than the tensile strength. Restrained or modified compressive specimens may have strengths of more than double the tensile strength.

Flexural tests are the easiest mechanical tests to perform; results are primarily useful for quality control and material grading. The span-depth ratio will greatly affect the results. The failure mode may be in tension, compression, or shear, depending on the nature of the specimen.

There are several types of shear tests for laminates. The following is a partial list of common shear tests and typical values for a 181 style fabric laminate.

SHEAR TEST	AVERAGE STRENGTH, psi (MPa)
Interlaminar shear (notched)	2000 (290,276)
Bending shear (short beam)	8000–12,000 (1,106,110–1,741,066)
Panel shear (picture frame)	1000–8000 (145,138–1,106,110)
Rail shear	16,000 (2,212,220)
Johnson shear	8000 (1,106,110)
Torsion shear	Varies

An excellent description of modern shear testing techniques is given by Pagano.[14]

24.6. MISCELLANEOUS TESTS

Tests are also required at elevated temperatures, depending on the type of laminate (i.e., general purpose or heat resistant). General purpose laminates are tested for flexural strength at 160°F (71.1°C) after 1/2-hour exposure at 160°F (71.1°C), as well as for flexural strength and modulus at room temperatures. Heat resistant laminates are tested for flexural strength at 160°F (71.1°C) after 1/2-hour exposure at 160°F (71.1°C); tensile and compressive strengths and the flexural modulus of elasticity (initial) are tested at 500°F (260°C) after 1/2-hour exposure at 550°F (287.8°C); and flexural strength and modulus at 500°F (260°C) after 192 hours at 500°F (260°C).

Electrical properties are required for laminates to be used at radio and radar frequencies. Dielectric constant and loss tangent are determined under both standard and wet conditions.* For laminates to be employed in radio frequency applications, these properties are determined at 1 MHz in accordance with FTMS 406, Method No. 4021. For laminates to be employed at radar frequencies, properties are determined at 8500–10,000 MHz using the shorted wave-guide method.

Requirements for testing of other laminates may differ somewhat; i.e., other tests and exposure conditions may be specified. For example, shear strength (e.g., per FTMS 406, Method No. 1041 or 1042), bearing strength (e.g., per FTMS 406, Method No. 1051), impact strength (e.g., per FTMS 406, Method No. 1071), water absorption (e.g., per FTMS 406, Method No. 7031), and/or flammability (e.g., per FTMS 406, Method No. 2021) may also be required to be determined.

24.6.1. Flammability Tests

The increased use of plastics in building construction has led to considerable interest in the evaluation of flammability properties.[15] The Underwriters' Laboratories* have issued a complete set of flammability tests (Bulletin 94). This includes a standard burning test applied to test specimens placed horizontally and vertically to provide a more or less standard flammability rating for comparing various materials, but not considered suitable for design purposes. Other tests measure susceptibility to ignition from electrical energy sources (hot wire and electrical arc ignition tests).

ASTM D635 describes a test for measuring the rate and/or extent and time of burning of self-supporting plastics in a horizontal position. ASTM E84 describes a method for determining the surface flammability of materials in an 8-ft (2.44-m) tunnel furnace.** Measurement of the "oxygen index" is covered in ASTM D2863, and is designed to overcome the limitations of "Bunsen burner tests" often conducted in laboratory evaluations. The "oxygen index" is defined as the percentage of oxygen in an oxygen-nitrogen mixture that will just sustain combustion of a test specimen, vertically mounted, which has been ignited with an external gas flame applied to its upper end. ASTM D2843 describes a means for determining the density of smoke generated from the burning or thermal decomposition of plastics.

24.6.2. Outgassing/Contamination

There has been growing use of reinforced plastics and other nonmetallic polymeric materials in applications wherein critical surfaces are sensitive to contamination from volatile species that can condense on the sensitive surface. Optical components and thermal control surfaces in satellites are examples of such applications, and are particularly sensitive when operated at reduced temperatures, lower than those of adjacent components.

As a consequence, test methods have been established for measuring mass loss and volatile condensable materials in a vacuum environment. ASTM E595-77 describes a standard test method for these tests, intended to be used as a screening technique for evaluating candidate materials or the effects of processing changes.

After conditioning at 73.4°F (23°C) and 50% relative humidity for 24 hours, the test specimen is heated under vacuum to 257°F (125°C) for 24 hours. A portion of the vapors from the heated specimen is passed into a collector chamber in which some vapor condenses on a collector plate maintained at 77°F (25°C). The total mass loss (TML) and collected volatile condensable materials (CVCM) are determined by weighing the specimen and collector before and after exposure.

After the specimen has been weighed to

*For these tests, wet conditions are defined as immersion in water for 24 hours.

*Underwriters' Laboratories have offices in several U. S. cities, including an office at 1285 Walt Whitman Rd., Melville, N. Y. 11746.

**Similar to UL-723 test; often termed the "tunnel test."

determine the TML, the water vapor regained (WVR) can be determined. The specimen is stored for 24 hours at 73.4° F (23° C) and 50% relative humidity and then weighed to determine the weight of moisture absorbed by and/or adsorbed on the specimen.

24.6.3. Dimensional Stability

The development of graphite and polyaramid fiber-reinforced plastic composites, characterized by very low values of coefficient of thermal expansion (CTE), has led to applications requiring dimensional stability over a wide range of thermal and other environmental conditions. Several methods are being used for precise dimensional measurement and determination of dimensional changes.[16]

Electrical displacement techniques using bonded wire or foil resistance strain gauges measure the average strain over a small area in one, two, or three directions simultaneously. Linear variable-differential transducers (LVDT's) can measure length changes with resolutions less than 4×10^{-7} in. (10^{-8}m). Many dilatometers employ LVDT's as the sensing element; however, these are of marginal value for low-expansion materials because of errors due to contacts, sideways motions of the core and windings in the LVDT, friction, alignment, and heating effects within the LVDT.

Electro-optical techniques include autocollimators which use collimated light beams to detect small angular displacements of reflecting surfaces. Optical levers involve methods for expanding reflected light beams from surfaces of test specimens. Fiber optics bundles also show promise for non-contact measurements.

Interferometers of various types, especially using lasers, are used for measurement of length changes during environmental exposures. The interferometer recombines two or more parts of a split beam into a single beam, the intensity of which varies with differences in the path length, relative to the reference beam.

REFERENCES

1. Epstein, G., "Testing Considerations for High-Performance Filament-Wound Structures," Paper No. 440, *ASTME Collected Papers* **62**, *Book 2*, 1962.
2. Kriener, J. H. and Almon, M., "A Study of Environmental Effects on Aerospace Grade Composites," in *Advanced Composites Technology*, Technology Conferences Associates, Box 842, El Segundo, California, March 1978.
3. National Association of Corrosion Engineers, *Managing Corrosion Problems with Plastics*, Houston, Texas, 1975.
4. Moon, D. P., Shinn, D. A., and Tyler, W. S., "Use of Statistical Considerations in Establishing Design Allowables for Military Handbooks," 5th Reliability and Maintainability Conference, July 18, 1966.
5. "Resin, Polyester, Low-Pressure Laminating," Military Specification MIL-R-7575C, June 29, 1966.
6. "SPI Procedures for Running Exotherm Curves—Polyester Resins," September 1960.
7. Calderwood, R. H., "Some Factors which Determine Glass Reinforced Polyester Laminate Quality," 16th Annual Technical and Management Conference, Reinforced Plastics Division, SPI, Chicago, Illinois, February 1961.
8. *Advanced Composites Technology*, Technology Conferences Associates, Box 842, El Segundo, California, March 1978.
9. "Evaluation Techniques for Fibers and Yarn Used by the Fibrous Materials Branch, Nonmetallic Materials Division, Air Force Materials Laboratory," AFML-TR-67-159, September 1967.
10. AFML-TR-65-366, December 1965.
11. "Apollo Spacecraft Nonmetallic Materials Requirements," NASA Document MSC-PA-D-67-13, NASA Manned Spacecraft Center, February 9, 1968.
12. ASTM D2863-77, "Measuring the Minimum Oxygen Concentration to Support Candlelike Combustion of Plastics (Oxygen Index)."
13. Lubin, G. and Tappe, W. C., "Factors Causing Scatter and Unreliable Design Data for Fiberglass Fabric Laminates for Aerospace Applications," 22nd Annual Technical Conference, Reinforced Plastics Division, SPI, Washington, D. C., January 31–February 3, 1967.
14. Pagano, N. J., AFML Technical Memorandum MAN 67-16, September 1967.
15. Reymers, H., "New Flammability Indexes and What They Mean," *Modern Plastics* (October 1970).
16. Wolff, E. G., Measurement Techniques for Low Expansion Materials," in *9th National SAMPE Technical Conference* **9**, October 4–6, 1977.
17. Dastin, S., Lubin, G., Mungak, J., and Slobodzinski, A., "Mechanical Properties and Test Techniques for Reinforced Plastic Laminates." Technical Publication #460, ASTM, 1970.

BIBLIOGRAPHY

(A Listing of Publications Discussing Tests for Reinforced Plastics)

General

Advanced Composites Technology, Technology Conferences Associates, Box 842, El Segundo, California, 1978.

Beach, N. E., "Defense Specifications and Standards for and Relating to Reinforced Plastics," PLASTEC Note 3 (AD402-225), Picatinny Arsenal, Dover, New Jersey, March 1963.

Beach, N. E., "Government Specifications and Standards for Plastics, Covering Defense Engineering Materials and Applications," PLASTEC Note 6 (AD410-401), Picatinny Arsenal, Dover, New Jersey, May 1963.

Beach, N. E., "Guide to Test Methods for Plastics and Related Materials," PLASTEC Note 17, Picatinny Arsenal, Dover, New Jersey, August 1967.

Beach, N. E., *Plastic Laminate Materials—Their Properties and Usage*, Foster Publishing Co., Long Beach, California, 1967.

Boller, K. H., "Strength Properties of Reinforced Plastic Laminates at Elevated Temperatures," ASD TR61-482, 1962.

"Comparison of United States and British Methods for Testing Plastic Materials," PLASTEC Note N32 (AD A034-734), Picatinny Arsenal, Dover, New Jersey, September 1976.

Composite Materials: Testing and Design, STP 617 (ASTM), 1977.

Environmental Effects on Advanced Composite Materials, STP 602 (ASTM), 1976.

Epstein, G., "Testing Considerations for High-Performance Filament-Wound Structures," American Society of Tool and Manufacturing Engineers, Paper No. 440, 1962.

"Filament-Wound Reinforced Plastics: State of the Art," Special Report No. 174, *Materials in Design Engineering* (August 1960).

"A Glossary of Physical Properties and Tests," Manual No. 159, *Materials in Design Engineering* (now *Materials Engineering*) (June 1959).

"International Symposium on Plastics Testing and Standardization," ASTM Special Technical Publication No. 247, 1959.

Kobler, Ruth S. and McNally, Cecilia U., "Guide to Specifications for Rigid Laminated Plastics," PLASTEC Report 7 (AD276-142), Picatinny Arsenal, Dover, New Jersey, March 1962.

Lever, A. E. and Rhys, J., *The Properties and Testing of Plastics Materials*, Chemical Publishing Co., New York, 1962.

Litvak, S., "Conference on Structural Plastics, Adhesives and Filament Wound Composites," ASD-TDR-63-396, December 1962.

"Mechanical Properties and Tests," Manual No. 106, *Materials and Methods* (now *Materials Engineering*) (July 1954).

Oleesky, S. S. and Mohr, J. G., *Handbook of Reinforced Plastics*. Van Nostrand Reinhold Co., New York, 1963.

Frados, J., ed., *Plastics Engineering Handbook*, Society of the Plastics Industry, Van Nostrand Reinhold Co., New York, 1976.

Simulated Service Testing in the Plastics Industry, STP 375 (ASTM), June 1964.

Thirty-Third Annual Technical Conference, Reinforced Plastics/Composites Institute, the Society of the Plastics Industry, 1978.

Nondestructive Testing/Inspection

Anzalone, A. M., "The Application of Nondestructive Testing to Plastics," PLASTEC

Note 1 (AD 261-550), Picatinny Arsenal, Dover, New Jersey, July 1961.

ASTM Fourth Pacific Area National Meeting, October 1962.

Baldanza, N. T., "A Review of Nondestructive Testing for Plastics: Methods and Applications," PLASTEC Report 22, Picatinny Arsenal, Dover, New Jersey, August 1965.

"Nondestructive Testing: Trends and Techniques," NASA SP-5082, October 1966.

Sampling

"Sampling Procedures and Tables for Inspection by Attributes," Military Standard MIL-STD-105.

"Sampling Procedures and Tables for Inspection by Variables for Per Cent Defective," Military Standard MIL-STD-414.

"Tentative Recommended Practice for Sampling of Plastics," ASTM D1898-68.

25
NONDESTRUCTIVE TEST METHODS FOR REINFORCED PLASTICS

George Epstein
The Aerospace Corporation
El Segundo, California

25.1. INTRODUCTION

It is the purpose of nondestructive testing (NDT) to identify and measure abnormal conditions, especially defects, within the sample, and to do so without degrading or impairing the utility of the sample in any way.* Acceptability of the sample depends on engineering judgment and/or correlation of the "observed" defects with performance (ideally the latter).

NDT methods are playing a vital role in helping to assure the integrity, reliability, and safety of industrial and defense materials.[1]

Plastec Report 22[2] presents an extensive review of the state-of-the-art of NDT for plastics.[3,4]

25.2. VISUAL NDT

Careful visual examination is the most convenient and widely used NDT inspection method. Defects that may be observed include discoloration (possibly due to overheating), foreign matter, crazing cracks, scratches, dents, blisters, orange peeling, pitting, air bubbles, porosity, resin-rich and resin-starved areas, wrinkles, and, to some extent, voids and delaminations. Aids to the eye (e.g., intense light and magnifying glass) can be helpful. Reflected light is used for observing surface irregularities and other defects; transmitted light (assuming both surfaces are accessible and the material is translucent) helps to reveal defects within the specimen. Inherently, visual examination is limited in that only relatively large defects can be observed even by a trained operator. ASTM D2563-70, "Classifying Visual Defects in Glass-Reinforced Laminates and Parts Made Therefrom," provides further details.

In addition to the NDT methods described above for reinforced plastics in general, the use of reinforced plastics in sandwich constructions introduces further needs for inspection. Visual inspection immediately after cure, upon removal of adhesive-bonded honeycomb sandwich parts from the heat source (while still hot/warm) often can reveal blisters due to unbonded or delaminated areas. These blisters may disappear when the part is cooled, reducing the internal pressure. When void-free laminate facings are employed, visual examination, preferably aided by special lighting arrangements, can help in observing defects within the structure. Generally, only gross defects can be thus observed.

Having located various observable defects, the task remains to assess the acceptability of the part. Only limited efforts have been made to correlate performance with the various types and sizes of defects in reinforced plastic composites.[5] Thus, engineering judgment and previous experience usually are the only means for establishing acceptance criteria.

*The importance of NDT is stressed by the following statement by Lt. Gen. H. M. Estes, Jr., USAF, in his keynote address at the 10th National Symposium on Reliability and Quality Control: "We cannot always afford to consume the product to prove its quality and we must therefore adapt nondestructive testing know-how to the reliability tasks."

25.3. ULTRASONICS

Several methods of NDT are available utilizing sound energy at frequencies above 20 kHz (i.e., ultrasonics). Frequencies used are generally 100 kHz to 25 mHz. Lower frequency, audible sound waves have wavelengths that are large in comparison with the size of a defect, and the sound travels around the defect. The availability of reliable methods for generating and detecting ultrasonic waves permits the location of small defects.[6]

Figure 25.1 shows a C-scan (plan view) recording of a glass filament wound test specimen, demonstrating the ability to resolve unbonded or delaminated areas/voids of approximately 1/4 in. (0.635 cm) and larger.[7] The darker areas correspond to the defects.

When an ultrasonic wave reaches an interface or a discontinuity, a portion of the energy

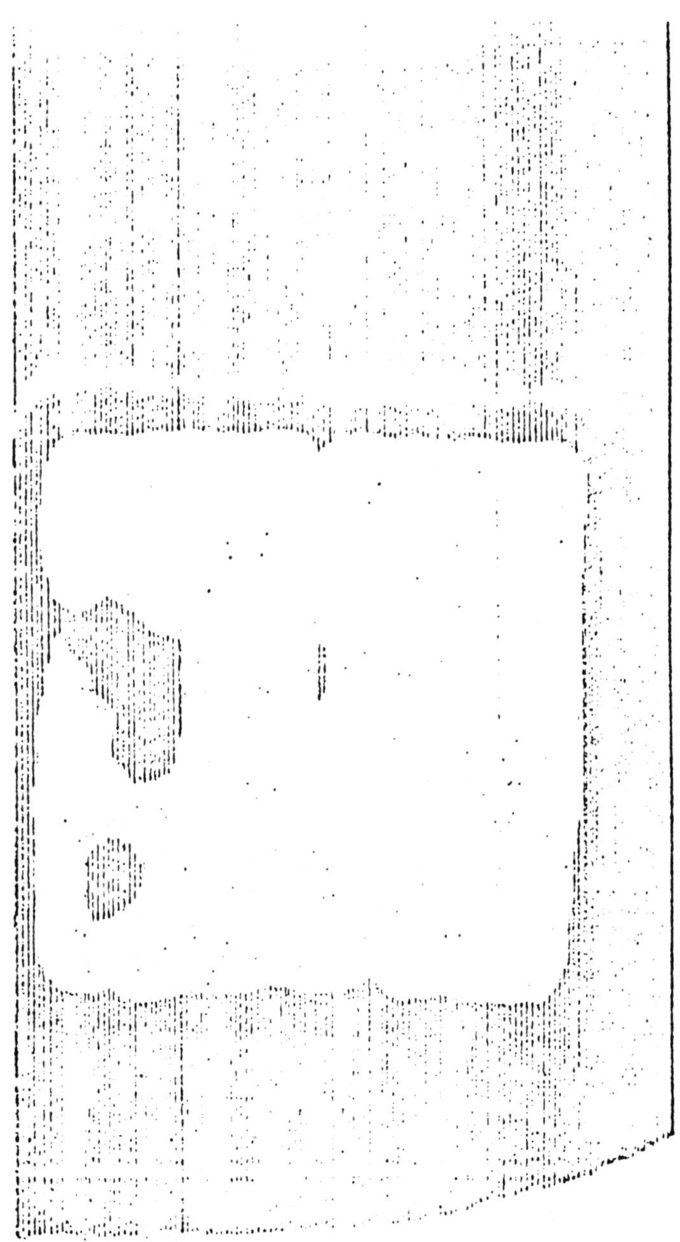

Figure 25.1. Ultrasonic NDT Inspection (C-scan) Of Filament Wound Composite Plate.

is reflected. The amount of reflected energy depends on the specific acoustic impedance of each media, Z, which is the product of the velocity, V, of the ultrasonic wave in the medium and the density of the medium, ρ. The energy transmitted through the material is reduced due to energy reflections and attenuation within the sample. Thus, the variations in reflected and/or transmitted energy can serve as means for locating defects (discontinuities) in the path of the ultrasonic wave.

In practice, a pulse generator transmits an electrical pulse through a coaxial cable to a transducer, generally a piezoelectric device which converts the electrical pulse to mechanical energy. The mechanical energy is then fed into the test specimen using a coupling medium (couplant)—water, oil, glycerine, or water-gel (other means for coupling include strippable rubber coatings, a localized stream or well of water, and water-dampened rubber wheels). Sound energy reflected by or transmitted through the specimen is picked up by a transducer, converting sound to electrical energy, which is then analyzed and displayed (on a cathode ray tube, a CRT, or a scanning-type recorder, for instance).

The pulse-echo method of ultrasonic NDT utilizes the ultrasonic energy reflected from a discontinuity for locating defects. Only one surface of the structure needs to be accessible. A defect is determined to be present if energy is reflected prior to reaching the back surface. This can be observed as a peak on the CRT between those pulses corresponding to the initial pulse and the back-surface reflection. The location of the defect can be estimated based on the relative position of the resulting peak with respect to the initial pulse and back-surface reflection peaks.

In through-transmission, the ultrasonic energy is measured after it passes through the material. Both surfaces must be accessible to the transducers. Reduction of energy due to a defect results in reduced amplitude of the peak on the CRT. Through-transmission, if practical, is generally preferred for thicker parts.

Both techniques have been found satisfactory for locating delaminations that are large in size and located perpendicular to the direction of ultrasonic transmission. Small delaminations or voids, resin-starved areas, and porosity are more difficult to detect. Through-transmission is preferred in such cases, utilizing the attenuation of the ultrasonic wave due to the presence of such defects. Pulse-echo may be used provided the part is not too thin, so that the initial pulse and back-surface reflection interfere, nor too thick, so that excessive attenuation results even for defect-free parts.

A shear-wave ultrasonic technique has been used for testing complex shapes which do not lend themselves to pulse-echo or through-transmission testing. The signal is sent into the part at an angle of 17.5–30° from the normal. If no defects are present, the sound wave is reflected between the surfaces of the part until attenuated, and there is no return signal. Any defect present will reflect the signal and an echo will be received by the transducer.[8]

In addition to the location of defects, ultrasonics have been employed in measuring the thickness of reinforced plastics. Both pulse-echo and through-transmission have been found suitable for thickness gauging of glass-reinforced laminates, with an error of approximately 2% for laminates 1/8–1/2-in. (0.3175–1.27-cm) thickness.[9] For laminates less than 1/8 in. (0.3175 cm) thick, pulse-echo is too inaccurate because of interference between the initial pulse and the back reflection; through-transmission is generally preferred for thickness measurement.

Resonance techniques of ultrasonic NDT utilize the change in resonant frequency or the energy decrease at the resonant frequency due to the presence of a defect in the reinforced plastic part. The resonant frequency is first determined, and depends on the material composition and part thickness. Inherent variations in composition and/or thickness in the part detract from the sensitivity and resolution in locating defects. This technique can also serve to determine part thickness, t, based on the relationship:

$$t = \frac{n\lambda}{2}$$

where λ is the wavelength at resonance and n is an integer corresponding to the harmonic of oscillation (for the fundamental harmonic, $n = 1$ at resonance).

Changes in the velocity of the ultrasonic wave through a known thickness can be used to indicate variability in density and elastic modulus,[10] based on the relationship:

$$V_L = \frac{E}{\rho}\left(\frac{(1-\sigma)}{(1+\sigma)(1-2\sigma)}\right)^{1/2}$$

where V_L is the longitudinal wave velocity, E is the bulk modulus of elasticity, ρ is the bulk density, and σ is Poisson's ratio. Based on measurements of laminates of Type 181 "E" glass fabric with epoxy, phenolic, polyester, silicone, and polybenzimidazole (PBI) resins, AVCO reported the relationship:

$$V_L^2\,\rho = 0.0124E$$

where the longitudinal wave velocity was measured parallel to the fabric plies.

Using through-transmission, AVCO has demonstrated that V_L measurements give reasonably good correlation with ultimate tensile strength for chopped-fibrous glass-reinforced epoxy cylinders[11] (likewise, V_L measurements across the filament direction for boron filament-reinforced epoxy parallel-lay-up laminates, correlated with interlaminar shear strength[12]). In both cases, higher strengths were characterized by the higher longitudinal wave velocities (indicating that elastic modulus is the prime variable rather than density).

An extensive investigation of ultrasonic NDT for filament wound reinforced plastics was conducted by Douglas Aircraft Company under Air Force contract.[13] Through-transmission techniques were employed for samples up to 2 in. (5.08 cm) thick, using water immersion for transducer-signal coupling. Defects included delaminations, porosity, and variations in resin content. Discontinuities in filament wound pressure vessels were reliably detected.

To improve on the ability to locate flaws in depth and to identify and characterize the type of flaw detected, Bar-Cohen and co-workers[14] have developed a shock-wave pulse-echo method with RF display. The method involves the use of pulses of very short duration; i.e., shock waves. A reference pattern is established on a CRT with a defect-free sample panel. On subsequent measurements, any defect in the area under examination manifests itself as a change from this reference pattern. Such changes may be observed as additional reflections, a velocity change, a variation in attenuation, or a reversal of the reflection phase. The method was evaluated with composite sandwich panels consisting of graphite-epoxy laminate skins, bonded to aluminum honeycomb core. It was found capable of locating defects at various depths, including delaminations between plies; skin-to-core debonds, and gaps between fiber tows within the laminate skins.

The following ASTM standards deal with ultrasonic nondestructive testing.

- ASTM E113-67, "Ultrasonic Testing by the Resonance Method."
- ASTM E114-75, "Ultrasonic Pulse-Echo Short-Beam Testing by the Contact Method."
- ASTM E214-68, "Immersed Ultrasonic Testing by the Reflection Method, Using Pulsed Longitudinal Waves."
- ASTM E317-68, "Evaluating Performance Characteristics of Pulse-Echo Ultrasonic Testing Systems."
- ASTM E494-75, "Measuring Ultrasonic Velocity in Materials."

The U. S. Navy Applied Science Laboratory has published a series of reports dealing with the use of ultrasonic techniques for NDT inspection of various types of reinforced plastics.[15] The U. S. Army Materials Research Agency in Watertown, Massachusetts has issued "A Report Guide to Ultrasonic Testing Literature, Volume I" (March 1966).

The Fokker Bond Tester (developed by the Royal Netherlands Aircraft Factories Fokker) is an ultrasonic resonance technique wherein a piezoelectric transducer vibrates at a natural frequency. (Other commercially available devices include the Coinda-Scope, STUB-Meter, and Sonizon.) When placed in contact with the sample being inspected, the transducer vibration is influenced. A voltmeter (B-scale) indicates the amount of damping of the transducer vibration; a cathode ray tube (A-scale) indicates the shift in resonant frequency.

The Fokker Bond Tester is widely used for nondestructive inspection of adhesive-bonded constructions. It has been most effective for use with metal-to-metal adhesive bonds. Some limited success has also been noted for bonded aluminum honeycomb sandwich structures; unbonded (void) areas can be detected consistently.[16]

To utilize such an instrument, the unit is first calibrated using bonded specimens of known quality. Tensile shear strength is generally used for metal-to-metal bonds, and flatwise tensile strength for honeycomb-sandwich bonds. The quality of the bond is then determined by the magnitude of the resonance frequency shift (A-scale) and/or the damping of the peak amplitude (B-scale).

25.4. SONICS

Sonic testing utilizes frequencies in the audible region, from about 10 Hz to 20 kHz. "Coin tapping" is a common technique for locating gross cracks or delaminations. A clear, sharp, ringing sound is indicative of a well-bonded solid structure. A dull sound, or thud, indicates a delamination or relatively large void area (since small voids generally cannot be detected by sonic methods).

In addition to manual tapping, automated apparatus can be utilized to facilitate the operation and improve effectiveness. The Flawmeter, for example, is an electronically controlled apparatus to control tapping force and dwell time, with a microphone pickup of the sound which is fed into an oscilloscope for visual display.[17] Used with bonded helicopter blade components, unbonded (delaminated) areas are characterized by a decreasing frequency as the wave dies away, whereas a well-bonded structure produces a relatively constant frequency.

The Sonic Resonator developed by North American Rockwell has been employed to locate unbonded areas and other gross defects in laminated and honeycomb-sandwich structures. A vibrating crystal is used to excite the structure acoustically. Resonance occurs when the frequency of the applied force corresponds to the natural frequency of the structure/specimen under inspection. Glycerine, water, or other media must be used to couple the sample and the crystal probe. Defects result in variations in the localized elastic properties and concomitant vibrational characteristics of the structure; these result in changes in the loading of the vibrating crystal. The resulting changes are processed electronically to obtain a visual indication or record. Only one surface of the structure need be accessible for this method.

The Eddy-Sonic[18] technique utilizes electromagnetic means to energize the structure with eddy currents. The resulting acoustical response is monitored to locate defects. Contact between the structure and the transducer is not required; liquid or oil couplants are not used. However, in order to produce the eddy currents, an electrically conductive material must be present in the structure to be inspected.

The sonic resonance technique was evaluated for NDT inspection of defects in reinforced plastic honeycomb sandwich ablative nozzle extensions.[19] It was determined that delaminations could be located in the inner and outer phenolic-glass laminates by measurements from one side of the structure, with a minimum detection area of about 1 in.^{2} (2.54 cm^2). Unbonded areas in the adhesive bonds of the core-to-laminate facings could also be detected. Difficulties in coupling were overcome by sealing the probed surface of the structure with a pressure-sensitive Mylar film. (Attempts to employ a strippable coating for coupling presented a problem due to the presence of a residue on the laminate surface after removal of the coating.) Fluctuations in ambient temperature tend to cause unsteady readings, and so must be avoided.

Sound (noise) generated while proof loading of a reinforced plastic structure has been used as an indication of the formation of crazing or cracks. A change in sound level accompanies the formation or extension of cracks within the structure. By the use of suitable equipment, suspect areas can be located. This technique is not entirely nondestructive in nature, since it requires the application of a proof load to the structure, and such loadings frequently degrade the strength of the part. It was found that, during hydrotesting, noises produced at lower stress levels could be correlated with ultimate strength of the structure (e.g., pressure vessels and rocket-motor cases). Green *et al.*[20] used the acoustic emission technique to verify the integrity of Polaris A3 glass filament wound rocket motor chambers.

Acoustic emission has also been used with boron fiber-reinforced plastics. It was demonstrated that this method could identify fiber fracture, matrix cracking, and interface debonding as failure mechanisms during load application.[21]

In application, acoustic transducers are bonded to the surface of the structure under test. Specialized equipment is available for this purpose. A thorough description of the instrumentation and data analysis procedures is given by Bailey *et al.*[22]

The following ASTM standards deal with acoustic emission tests.

- ASTM E569-76, "Acoustic Emission Monitoring of Structures During Controlled Stimulation."
- ASTM E610-77, "Definition of Terms Relating to Acoustic Emission."

25.5. PENETRATING RADIATION—RADIOGRAPHY

In radiography, the internal structure of a solid material is visualized by exposing the sample to a source of penetrating radiation and recording the shadow image on a photographic film or plate placed on the opposite side of the sample. An internal void or gap decreases the amount of solid material through which the radiation passes, and thus increases the intensity of the radiation reaching the film (plate) at a position corresponding to the void location. The resulting darker area of the film indicates the outline of the defect and, depending on the degree of darkening of the film, provides an estimate of the thickness (depth) of the void/gap.

X-ray inspection is the most widely used penetrating radiation method of NDT. The location of defects depends upon variations in thickness and/or density in the path of the radiation. Areas of lower density within the specimen will absorb less radiation so that the detector (film or plate) will show a stronger reaction in the corresponding area. Because plastics are of lower density than most metals, the presence of a metallic inclusion (as a foreign contaminant, for example) can also be observed.

Radiography has been employed for locating large voids, delaminations and cracks in reinforced plastic parts. Such defects can be observed provided the defect is sufficiently large with respect to the direction of the radiation path. Defects oriented normal to the radiation often are difficult to detect.

The X-ray detection of surface-connected defects can be enhanced by the application of tetrabromoethane (TBE).[23] This technique depends on the ability of the TBE to penetrate into the damaged area, and the fact that TBE is opaque to X-rays.

In applying this method to assess impact damage in graphite-epoxy composites, Bailey *et al.*[24] observed that TBE-enhanced X-ray radiographic techniques produced images that were indicative of both the size of the damage area and the extent of fiber fracture, provided that the flaws were surface-connected.

Military Specification MIL-STD-271 describes equipment and methods for conducting radiographic inspections.

ASTM standards dealing with radiographic NDT include the following.

- ASTM E94-68, "Radiographic Testing."
- ASTM E142-77, "Controlling Quality of Radiographic Testing."
- ASTM E586-76, "Definition of Terms Relating to Gamma and X-Radiography."

Betatron inspection utilizes high energy X-rays (produced by impacting a heavy metal target with high speed electrons) which have considerably greater penetrating ability than the more conventional X-rays. Such radiation is less scattered and can be focused over a very small area. Linear accelerators provide a further refinement over betatrons.

The Ohio State University, under contract with the U. S. Army Materials Research Agency, has developed equipment and techniques for TV image display of X-ray inspection. This equipment has been used to detect a variety of defects within glass fabric-reinforced laminates.[25]

In fluoroscopy, the X-ray image is projected onto a fluorescent screen. Tests have shown this technique to be somewhat less sensitive than the better radiographic techniques.[26] With an image intensifier, fluoroscopy can detect voids and cracks corresponding to a thickness change of approximately 2–3%. Large variations in resin content and nonuniform fiber orientations can also be noted.

Collimated scanning radiography has been used to improve X-ray inspection for defects such as yarn alignment, paired yarns, missing yarns, and matrix/yarn inteface quality.[27] The technique involves the use of two stationary lead apertures, aligned with a standard X-ray source, while the subject and film are moved between the apertures. The resulting image reveals axial and lateral yarns over the entire dimensions of the sample.

Other radiographic inspection techniques also are available which depend on the principle of variation in the absorption of the radiation due to defects within the material. These include Xeroradiography, tracer radiography,[28] and isotope radiography.

Beta radiation (high speed electrons) has been used to determine glass/resin ratio with a reported accuracy of approximately ±2%. The technique is based upon beta ray backscattering (reflection), which is greater for the higher density (higher atomic number) glass reinforcement compared to the resin matrix.[29] However, sensitivity is limited to relatively thin structures—permitting penetration to about 0.02 in. (0.508 mm).

A beta gauge technique was investigated for continuous measurement (monitoring) of the weight per unit length and resin content of preimpregnated glass filament roving tapes.[30] The major difficulty was the necessity for maintaining precise positioning of the roving in the beta ray field. The size of the equipment and its cost present practical problems in incorporating such a method in manufacturing processes. The accuracy of the measurements was somewhat poorer than that desired.

Absorption of thermal neutrons by the boron nuclei present in "E" glass has been employed as a nondestructive method for determining laminate thickness (number of plies) and glass content.[31]

Gamma radioisotopes are available from sources such as cobalt-60, cadmium-109, and cesium-137, producing a narrow range of frequencies. A through-transmission gamma radiation technique can be used to indicate variations in material density based on the relationship:[4]

$$I_n = I_o e^{-\mu \rho d}$$

where I_n = energy (gamma photons) transmitted through the specimen per unit time; I_o = energy source (incident gamma radiation) in counts per unit time; ρ = material bulk density; μ = linear absorption coefficient; and d = thickness of the specimen.

25.6. ELECTRICAL PROPERTIES FOR NDT

The dielectric constant and dissipation factor (or loss tangent) can be used as nondestructive methods for determining variability in a reinforced plastic structure. For a given thickness and composition, changes in degree of cure can be assessed by dielectric constant and dissipation factor measurements, the values for these parameters decreasing as the resin is cured. Similarly, moisture content can be determined with an accuracy estimated to be within ±1%.

A dielectric probe operating in the frequency range of 1 kHz to 1 mHz has been employed by AVCO.[32]

The Delsen D/K Analyzer was used to study thermal degradation of epoxy-glass laminates and the effect of cure.[33,34] An increase in dissipation factor indicated the onset of thermal degradation (loss in flexural strength) at temperatures of 300–500° F (148.9–260° C). Dielectric constant measurements were not as sensitive as dissipation factor measurements under these conditions. The changes in dielectric constant and dissipation factor during cure can serve to determine the optimum curing condition (temperature and time) and whether the part has been fully cured. Capacitance measurements have also been used to determine moisture content of laminates and sandwich construction.

Volume electrical resistivity has been used to determine optimum curing cycles, composition effects, and the presence of moisture.[35,36,37] Continuous resistivity measurement has been considered for monitoring the cure of critical reinforced plastic structures. Because electrical resistance of plastics is highly sensitive to temperature (resistivity decreases exponentially as temperature is increased), the temperature during the test must be carefully controlled. Volume resistance measurements require that the probe (parallel plate resistance cell) have access to both surfaces (sides) of the part.

Corona discharge has been employed for determining the presence of voids within plastic laminates. When subjected to a high electrical potential, ionization occurs of the gas (air, moisture, and possible gaseous contaminants from the resin) within a void, causing acceleration of electrons to the wall of the void. Detection and evaluation can be made by measuring the resulting voltage change, the light produced, or the noise emitted. The high voltage (to produce the corona discharge) must be of short duration so that the part will not be damaged.

25.7. MICROWAVE TECHNIQUES

Microwaves are electromagnetic radiations of extremely high frequencies, with a range of about 0.5–1000 GHz (with the higher frequencies widely used in radar).

Microwave techniques can be used for locating defects, and for measuring thickness, moisture content, and dielectric properties for nonmetallic materials.

Defects include voids, delaminations, porosity, foreign inclusions, resin-rich and resin-starved areas, and variations in the degree of cure and the moisture content.[38,39,40] Because of the high frequencies used, the microwave radiation beam can be focused on a small area to achieve high resolution in locating defects.

A microwave source (horn) directs the radiation at the test specimen. Energy reflected or transmitted by the specimen can then be used for the evaluation. A crystal detector converts the resultant wave into an electrical response. Negligible attenuation occurs when microwaves pass through plastic materials.

Voids have been detected at the core-to-skin interface of honeycomb ablative re-entry materials adhesive-bonded to a metallic structure.[41] Scattering of the microwaves by the voids causes a reduction in signal amplitude, which can be observed on a meter or CRT.

Using microwave signals at 30–40 GHz, separations of about 1/16 in.^{2} (0.159 cm^2) in area have been located in complex composite reinforced plastic honeycomb-sandwich structures.

Microwaves can be used to monitor the curing process and to indicate the degree of cure. This results from the changes in dielectric constant and loss tangent as a resin cures, and the microwave reflections responding to these changes.

Since the dielectric properties of a reinforced plastic depend on chemical composition, variations in composition and impurities can be detected by microwave techniques.

Microwaves strongly interact with the water molecule, and thus serve as means for measurement of moisture in plastics. When present as free water molecules, moisture absorbs and reflects the incident microwave energy, changing both the amplitude and the phase of the transmitted and reflected microwave beam.[42]

When employed with reinforced plastic radomes, microwaves serve to nondestructively measure the parameters which determine the performance of the structure. Dielectric constant and loss tangent are directly measured. In addition, changes in thickness, resin/glass ratio, or delaminations can be detected.[42]

In addition to utilizing signal attenuation due to microwave energy loss, phase shift in the standing wave can also be used for NDT measurements. A phase shift will occur if the dielectric properties of the material are varied. Part thickness and geometry can influence the results of NDT measurements for both signal attenuation and phase shift.

25.8. INFRARED/THERMAL NDT

Infrared or thermal techniques of NDT utilize differences in heat flow due to the presence of defects within the structure being inspected.[43] The structure is first heated; as it heats or cools, the surface temperature is observed through the use of a sensitive infrared measuring device (radiometer). Commercially available radiometers can measure temperature changes to a very high degree of accuracy—reported to be less than 0.18°F (0.1°C).

When a reinforced-plastic structure is heated, heat transfer (and dissipation) occurs much more readily if there are no defects, e.g., unbonded areas. The result is a higher temperature for that part of the surface nearest the defect.[44]

Single point measurements, thermal profiles, and area measurements are temperature read-out methods used in thermal NDT. An inspection system developed for honeycomb sandwich structures uniformly applies a "paint-brush" pattern of heat across the surface of the structure. After a brief dwell time, the surface temperature is measured by an oscillating radiometer. A permanent recording of the surface temperature scan can be produced directly.[45] This technique does not require contact with the structure being inspected.

Another thermal NDT approach utilizes the temperature-color sensitivity of many liquid crystals, especially those containing cholesterol in their molecular structures. Although colorless as isotropic liquids, cholesteric substances pass through a series of bright colors as they cool through their liquid-crystal phase. By combining mixtures of cholesterol esters, researchers at the Boeing Co. have been able to vary the temperature at which the color change occurs.

Color changes (e.g., red to blue) occur over a temperature range of 33–35° F (2–4° C.).[46,47] Applied from aqueous solution, the coating can be washed off after completing the measurements. The coated part is heated uniformly. A defect is observed by the color change at the surface (coating) nearest to the defect, as the structure heats or cools. (Note: Because this method is based on thermal flow, defects can be observed during heating as well as during cooling. Areas containing, or in the vicinity of, defects appear as "warmer" colors because of the slower heat transmission in these regions.) The simplicity and low cost of equipment make the liquid crystal NDT technique very attractive.

Investigations at the Air Force Materials Laboratory have demonstrated the practical application of photochromic "paints" to provide a low-cost technique for locating defects in reinforced plastic/composite structures.[48] Coatings have been developed which can be removed by peeling (strippable photochromic coatings) or by washing with a mild solvent. The coating is "activated" by exposure to an ultraviolet light. The specimen is then inspected by slowly heating the activated coating and observing color changes associated with reduced thermal conductivity in the vicinity of voids/delaminations.

It should be noted, however, that infrared/thermal methods of NDT cannot detect weakened areas—only separations (sufficiently large voids, delaminations, or unbonded areas). Defects near the surface (coating) are more readily detected; sensitivity decreases as part thickness increases.

25.9. OTHER NDT METHODS

In addition to the methods described above, there are other approaches for NDT of reinforced plastics that have been employed or are under development.

Some of these methods are not unique to plastics and need not be discussed at length in this text. Holographic technology has been applied to nondestructive inspection of composite and other structures. The structure must be mounted on a very stable platform, and is then subjected to a stress condition, either thermal or mechanical. Holography permits rather precise observation of dimensional deviations from the reference configuration, such that any anomalies can be located. Methods for detecting surface defects include liquid penetrants (using white or "black" light for observation); strain gauges to measure dimensional changes or distortions when load is applied to the part; strain-sensitive "brittle" coatings to indicate stress distribution over the surface of the part when a load is applied; and photoelastic stress measurement. This last method uses a birefringent plastic coating/film bonded to the surface of the structure being inspected. When a load is applied, the stress distribution can be observed by noting the fringe patterns (color variations) in the photoelastic coating under polarized light. The photoelastic technique[49] has been used to detect and measure the effects of strength-reducing factors in filament wound pressure vessels.

REFERENCES

1. Ballard, D. W., "NDT Testing," *Industrial Research* (October 1965).
2. Baldanza, N. T., Plastec Report 22, "A Review of Nondestructive Testing for Plastics: Methods and Applications," August 1965.
3. McGonnagle, W. and Park, F., "Nondestructive Testing," *International Science and Technology* (July 1964).
4. Zurbrick, J. R., "The Mystery of Reinforced Plastics Variability: Nondestructive Testing Holds the Key," AVCO Space Systems Division, Annual Technical Conference of the Society of Plastics Engineers, Detroit, Michigan, May 1967.
5. Werren, F. and Heebink, B. G., "Effects of Defects on the Tensile and Compressive Properties of Glass-Fabric-Base Plastic Laminate," U. S. Forest Products Laboratory Report 1814.
6. McGonnagle, W. J., *Nondestructive Testing*, McGraw-Hill, New York, 1961.
7. Hitt, W. C., Automation Industries, Inc., Santa Clara, California.
8. Kramer, J. M., Nuzzo, A. F., and Epstein, G., "Large Plastic Moldings OK'd by Ultrasonic Inspection," *Materials in Design Engineering* (February 1961).
9. Hand, W., "Testing Reinforced Plastics with Ultrasonics," *Plastics Technology* (February 1962).
10. Zubrick, J. R., "Development of Nondestructive Methods for the Quantitative Evaluation of Glass-Reinforced Plastics," AFML Technical Report TR-66-269, June 1966.
11. Lockyer, G. E., "Evaluation of a Resin-Ceramic Heat Shield Material by Ultrasonic Techniques," *Materials Evaluation* **23,** *No. 3* (March 1965).
12. "Properties Determination and Process Control of Boron Filament Composites Using Nondestructive Test Methods," SAMPE Symposium on Advanced Fibrous Reinforced Composites, San Diego, California, 1966.
13. Adams, C. J., Radtke, N. H., and Klein, J. D., "Ultrasonic Techniques and Standards for Testing Filament-Wound Structures," AFML Rept. TDR-64-117, May 1964.
14. Bar-Cohen, Y., Arnon, U., and Meron, M., "Defect Detection and Characterization in Composite Sandwich Structure by Ultrasonics," *SAMPE Journal* **14,** *No. 1* (January/February 1978).
15. New York Naval Shipyard, Laboratory Project 6188, Project No. SR007-03-04, Identification No. 18-1010-1.
16. Miller, N. B. and Boruff, V. H., "Adhesive Bonds Tested Ultrasonically," *Adhesives Age* (June 1963).
17. Arvin, M. J. and Howell, W. D., "Development of General Specification for a Device for Inspecting Helicopter Bonding," FAA, Bureau of R&D, Test and Experiment Division, November 1969.
18. North American Aviation Sonic Test System, NA-67-491 and STS 2-3, 1967.
19. Hribar, V. F., "NDT Challenged—Proving the Structural Integrity of Ablative Extensions by Sonic Resonance," Aerospace Corp., October 1967.
20. Steele, R. K., Green, A. T., and Lockman, C. S., "Use of Acoustic Techniques for Verification of Structural Integrity of Polaris Filament-Wound Chambers," 20th Annual Technical Conference, Society of Plastics Engineers, Atlantic City, New Jersey, January 1964.
21. Mehan, R. L. and Mullin, J. V., "Analysis of Composite Failure Mechanisms Using Acoustic Emissions," *J. Composite Materials* **5** (April 1971).
22. Bailey, C. D., Hamilton, J. M., and Pless, W. M., "AE Monitoring of Rapid Crack Growth in a Production-Size Wing Fatigue Test Article," *NDT International* (December 1976).
23. Rose, J. L. and Shelton, W., "Damage Analysis in Composite Materials," *Composite Reliability*, ASTM STP 580, 1975, p. 215 f.
24. Bailey, C. D., Freeman, S. M., and Hamilton, J. M., "Detection and Evaluation of Impact Damage in Graphite/Epoxy Composites," *9th National SAMPE Conference* **9,** October 4–6, 1977.
25. Mitchell, J. P., Rhoton, M. L., and McMaster, R. C., "Nondestructive System for Inspection of Fiber Glass Reinforced Missile Cases and Other Structural Materials," Report No. WAL-TR-142.5/2-9, December 1963.
26. Hund, F. C., "Fluoroscopy of Filament Wound Fiberglass Missile Motor Case Feasibility Study," Report No. QE/CD-AGC-139, Naval Weapons Station, Concord, California, September 1961.
27. Littleton, H. E., "Collimated Scanning Radiography," *SAMPE Journal* **14,** *No. 3* (May/June 1978).
28. Perry, H. A., "Tracer-Radiography of Glass Fiber Reinforced Plastics," Naval Ordinance Laboratory, 17th Annual Technical Conference, Reinforced Plastics Division, SPI, February 1962.
29. Baldanza, N. T., PLASTEC Report 22, "A Review of Nondestructive Testing for Plastics: Methods and Applications," Picatinny Arsenal, August 1965.

30. Brown, G. and Novkov, R. L., "Development of Nondestructive Dynamic Monitoring Instrumentation for Resin Impregnated Glass Roving," 22nd Annual Technical Conference, Reinforced Plastics Division, SPI, Washington D.C., February 1967.
31. Kalinsky, J. L. and DiLauro, S., "Application of Radioisotopes to Nondestructive Testing of Fiber Glass Reinforced Plastics," 12th Annual Technical and Management Conference, Reinforced Plastics Division, SPI, Chicago, Illinois, February 1957.
32. Lynnworth, L. C., "Nondestructive Testing of Plastics for Aerospace Applications," presented at Society of Nondestructive Testing, New York, January 1962.
33. Delmonte, Julian, "Prediction of Mechanical Properties Through Determination of Electrical Properties of Plastics," SPE Regional Technical Conference on Plastics for Airborne Applications, November 1957.
34. Knudsen, J. J., "Electrical Properties during Cure," SPE Regional Technical Conference on Plastics for Airborne Applications, November 1957.
35. Warfield, R. W., "Studying the Electrical Properties of Casting Resins," *SPE Journal* (November 1958).
36. Warfield, R. W. and Petree, M. C., "A Study of the Polymerization of Epoxide Polymers by Electrical Resistivity Techniques," American Chemical Society, Division of Paint, Plastics, and Printing Ink Chemistry, Atlantic City, New Jersey, September 1959.
37. Warfield, R. W. and Petree, M. C., "New Curing Techniques for Resins," U. S. Patent 3,049,410, August 14, 1962.
38. Hendron, J. A. *et al.*, "Corona and Microwave Methods for the Detection of Voids in Glass-Epoxy Structures," *Materials Evaluation* (July 1964).
39. Lindsay, E. W. and Works, C. N., "Corona Detection Techniques as a Nondestructive Method for Locating Voids in Filament-Wound Structures," *Joint ASTM-Navy Symposium on Standards for Filament Wound Reinforced Plastics*, June 1962.
40. Hochschild, R., "Microwave Nondestructive Testing in One (Not-So-Easy) Lesson," *Materials Evaluation* **26,** *No. 1* (January 1968).
41. Rockowitz, M. and McGuire, L. J., "A Microwave Technique for the Detection of Voids in Honeycombed Ablative Materials," Society of Nondestructive Testing Convention, October 1964.
42. Botsco, R. J., "Microwave Applications in Moisture Measurement and Nondestructive Testing," presented at Instrument Society of America, San Francisco, California, December 11, 1967.
43. Martin, B. G., "Infrared Technology and Nondestructive Testing," Redstone Arsenal, February 1963 (available as AD402-888, from Office of Technical Services, U. S. Department of Commerce, Washington, D. C. 20230).
44. Gericke, O. R. and Vogel, P. E. J., "Infrared Bond Defect Detection System," *Materials Evaluation* **22** (February 1964).
45. Birks, A. S. and Apple, R., "TempTest Sheds New Light on Nondestructive Tests of Aerospace Materials," Automation Industries, Inc., Boulder, Colorado, February 1967.
46. Sabourin, L., "Nondestructive Testing of Bonded Structures with Liquid Crystals," presented at Structural Adhesives Bonding Conference, NASA-Marshall Space Flight Center, March 15–16, 1966.
47. Woodmansee, W. E., "Cholesteric Liquid Crystals and their Application to Thermal Nondestructive Testing," *Materials Evaluation* (October 1966).
48. Allinikov, S., "Application of Photochromic Coatings for Nondestructive Inspection," Air Force Materials Laboratory Technical Report, AFML-TR-70-245, December 1970.
49. Eshbaugh, R. W., "Photoelastic Stress Analysis of Filament Wound Internal Pressure Vessels," 18th Annual Technical and Management Conference, Reinforced Plastics Division, SPI, Chicago, Illinois, February 1963.

Section IV
Applications

26
COMPOSITES IN LAND TRANSPORTATION

Morgan Martin
Molded Fiber Glass Companies
Ashtabula, Ohio
John F. Dockum, Jr.
PPG Industries, Inc. Fiber Glass Division,
Pittsburgh, Pennsylvania

26.1. INTRODUCTION

The magnitude of both the quantities and varieties of materials used, or available for use, in the manufacture of automobiles, trucks, buses, vans, rapid transit vehicles, and rail cars is almost greater than is manageable. The development of materials and fabrication processes to provide parts that are at once cost-effective, energy conserving, functional, and aesthetically appealing presents a formidable challenge. New technologies and rapidly changing priorities demanded by energy availability considerations, consumer protection, product liability and warranty, and environmental protection regulations have mandated modifications in the traditional material-of-construction selection process. Any candidate material or process for use in the transportation industry must be judged in the light of these criteria, as well as by the overall (auto industry) economics involved.

The automobile industry has grown to be the giant it is because of vigorous competition and a willing market. Vast sums of money and time have been committed to the design, the selection of materials, and the tooling involved in bringing a new vehicle from the concept stage to the marketplace. New materials and innovations have been the handmaidens of this industrial growth. However, the introduction of an appropriate new material or an innovative use of existing material does not guarantee automatic acceptance by the industry. Because car companies are willing to stake their reputations as well as their continued growth on innovation, those interested in the promotion of new products or concepts must appreciate the necessity of assuring that what is supplied will perform the necessary function, that quality control will ensure acceptable performance, and that production capability is adequate to satisfy the demand. Requirements of the market also demand rigid discipline in terms of timing. These are the yardsticks—or meter sticks, if metrication prevails—against which reinforced plastics must be measured.

26.2. ECONOMICS AND MARKET VOLUME

Early applications of reinforced plastics were generally based on simple break-even arithmetic. Naturally, there had to be engineering assurance that the part would perform its intended function. Beyond that, the total cost of tooling and the projected number of parts required had to be equal to, or less than, those same costs for parts produced from traditional materials before reinforced plastics would be specified.

Because the production process and tooling could be tailored to the quantities needed, reinforced plastics quickly gained acceptance for low-volume applications. The tool cost for the most sophisticated dies available for compression molding was only a fraction of the cost of dies for sheet metal stamping. For production runs of 20,000–40,000 parts, depending upon design, the higher material costs of reinforced plastics versus sheet metal could be justified. Such low-volume capability offers the opportunity to design vehicles for more specific market segments, and to change designs quickly when a competitive advantage is recognized. In addition to the lower tooling investment, shorter lead time from design to production reduces the total risk involved. Early model Chevrolet Corvettes (1953–1970) were the most visible results of these considerations.

As reinforced plastic (RP) materials and production technologies advanced, major components of the Corvette were converted from open mold and preform molding processes to compression molding of sheet molding compounds for higher productivity. Also, in 1970, the first front-end grille opening panel of sheet molding compound was introduced on a high-volume production car model. In addition to a 50% weight saving, a significant cost saving was effected through parts consolidation. Innumerable metal stamping, machining, and fastening operations, as well as the associated dies and fixtures, were eliminated in the replacement of 16 steel and die cast parts with a single RP molding. In 1979, over 35 car models used sheet molding compound for front-end panels, which may include the housings for headlamps, parking lights, turn signals, and side marker lights.

Long before the energy crisis dictated "downsizing" of the automobile, it was recognized that weight saving in any part of the vehicle could reduce the size, and thus the cost, of many other components that function to support, restrain, accelerate, stop, or otherwise contain the vehicle. As auto weight reduction is carried to the ultimate, however, there is a trade-off whereby an average penalty of 50¢/lb ($1.10/kg) for lightweight materials can be absorbed for 1 lb (0.4536 kg) of weight saved. In the case of commercial vehicles, the penalty may be as high as $5.00/lb ($11.00/kg), since each pound (0.4536 kg) of structural weight saved represents an additional pound (0.4536 kg) of pay-load that can be transported. Using steel as the basis of comparison, RP materials offer weight savings of 40–60%, and total energy savings, from production of raw materials to fuel consumed over the life of the vehicle, of 50%.

Table 26.1 illustrates the total energy saving and increase in fuel economy obtained by designing a single-component front-end grille opening panel of sheet molding compound to replace a multi-component metal assembly on a production model automobile. The data assume a vehicle life of 9.2 years.

Estimates of plastics usage in 1978 U. S. automobiles are 180 lb (81.6 kg) per car, of which 30 lb (13.6 kg) are fiberglass-reinforced plastics (FGRP) in over 350 large and small parts. Based on 9 million autos, FGRP use was 2.80×10^8 lb (1.27×10^5 metric tons), of which 1.65×10^8 lb (7.5×10^4 metric tons) were thermoset and 1.15×10^8 lb (5.2×10^4 metric tons) were thermoplastic.

The history of reinforced plastics sales to the land transportation market as a whole is given in Table 26.2.

26.3. COMPOSITES, PROPERTIES, AND APPLICATIONS

The ease of fabrication and the versatile properties of RP materials, coupled with the various processes that can be used to produce RP parts, represent a wide latitude of possibilities in product design. Although generally less rigid than metals, reinforced plastic

Table 26.1. Energy Savings and Fuel Economy*

FRONT-END PANEL	METAL	RP
Zinc die cast and stamped steel, weight, lb (kg)	20 (9.07)	9 (4.08)
BTU/finished part (hp-hours/finished part)	400,000 (157.24)	200,000 (78.62)
Fuel economy increase, %		0.2
Fuel saving, gal (l)		13 (49.205)

*Source: Ford Motor Co.

Table 26.2. Sales of Reinforced Plastics to the Land Transportation Market*

YEAR	MILLIONS OF POUNDS		THOUSANDS OF METRIC TONS
1968	167		75.7
1969	220		99.8
1970	167		75.7
1971	195		88.4
1972	250		113.4
1973	307		139.2
1974	293		132.9
1975	265		120.2
1976	398		180.5
1977	475		215.4
1978	535		242.7
1979	550		249.5
1980	436	(Estimate)	197.8
1981	474	(Estimate)	215.0

NOTE: Figures include both thermosetting and thermoplastic resins, reinforcements, and fillers.
*Source: SPI Reinforced Plastics/Composites Institute.

parts and assemblies can readily be designed to function much like sheet steel. In other designs, castings, forgings, and extrusions can be replaced with RP parts. In each design, these parts offer weight savings and corrosion resistance, and often superior resistance to impact and fatigue. All of these properties were considered extremely important in the hoods and fenders for medium and heavy trucks in which fiber glass-reinforced polyester has replaced steel. Since the materials have performed well in these applications and have received customer approval, they are now being specified for cab roofs, skirt panels, door assemblies, and entire heavy truck cabs.

An orders-of-magnitude comparison of some key mechanical properties of RP composites and metals is presented in Table 26.3. Fatigue data on the composites are particularly generalized because of incomplete testing to date, as well as because of the many possible variations in RP compositions, the several fatigue test procedures available, and the various failure criteria that can be used.

The development of low-shrink polyester resins utilized in bulk and sheet molding compounds has led to the acceptance of reinforced plastics as logical materials for many semi-structural, exterior, appearance parts for large volume passenger cars, such as body panels on the Corvette, fender extensions, radiator grille opening panels, valance panels, window opening panels, and wheel skirts. Other applications of sheet molding compound include spoilers for sports cars, roof panels for sports vans, and full front and rear panels on motor homes. Hoods comprised of two-piece bonded assemblies have been used on limited volume sports cars for specific design advantages. The Corvette replacement hood has been converted to a one-piece design with integrally molded, structural ribs on the back side. Large, complex shapes that can be produced to close dimensional tolerances in a single-step molding operation, followed by some hand finishing and painting on an existing paint line,

Table 26.3. Mechanical Properties of Materials*

	FATIGUE PERCENT OF ULTIMATE STATIC STRENGTH (10^7 CYCLES)	FLEXURAL MODULUS, psi × 10^6 (GPa)	TENSILE STRENGTH, psi × 10^3 (MPa)
Auto grade sheet molding compound	+	1.8 (12.4)	10 (69)
High chopped glass (random) sheet molding compound	+	2.0 (13.8)	30 (207)
High continuous glass (parallel) sheet molding compound	30	5.4 (37.2)	80 (552)
Aluminum	30	10 (68.9)	60 (414)
Steel	45	30 (207)	200 (1380)
Carbon fiber/epoxy-composite	80	30 (207)	180 (1242)

*Source: Ford Motor Co.
+In separate studies by Heimbuch and Sanders, General Motors Corp., Manufacturing Development, representatives of these types of compounds exhibited flexural fatigue stress levels at 10^6 cycles of 40% and 25% of their static flexural strengths, respectively. Fatigue stress levels were higher in a tension-tension mode.

make RP the most economical material of choice. Thermoset polyester is used almost exclusively as the resin matrix in external body parts that must be finish painted to match sheet metal parts.

Thermoplastics with short fiberglass reinforcement to upgrade mechanical properties, permit thinner sections, and enhance dimensional stability (particularly at elevated temperature), while preserving electrical properties and corrosion resistance, are being injection molded into many functional parts, usually hidden. Principal types of reinforced resins are the styrenics, polypropylenes, nylons, thermoplastic polyesters, polycarbonates, and acetals. Examples of applications include, respectively: instrument panel retainers and floor consoles; fan shrouds, fender liners, and tail lamp housings; windshield washer and lift gears, latches, door handles, engine cooling fan blades, brake cylinder reservoirs, oil filler caps, and ignition housings; decorative louvers, windshield wiper blade holders, lamp sockets, connectors, computer modules, and distributor caps and rotors; headlamp housings; and electric fuel pumps. In addition, fiberglass-reinforced polypropylene sheet is being stamped into front-end retainer panels, load floor shelves, battery trays, and hinged seat shells for the Corvette.

For years, virtually all refrigerated and many dry goods trailers have been lined with glass-reinforced polyester panels. Produced as either flat or corrugated sheets, they may be utilized also on the exterior and as an important structural part of the trailer. They withstand heavy wear and abuse, except a frontal attack by a fork lift, and may be steam cleaned.

Railroad hopper cars that carry edible grain as well as other bulk commodities, including salt and corrosive fertilizers, are fitted with RP hatch covers that seal properly, and that have proven to be durable, lightweight (for handling), and economical. Other rail car applications in use or under consideration are doors, running boards, and various lading devices, and an entire hopper car.

Federal regulations of the National Highway and Transportation Safety Administration (NHTSA) have mandated a Corporate Average Fuel Economy (CAFE) of 27.5 mpg (11.7 km/l) of fuel consumed for automobiles in 1985, and 17.2 mpg (7.3 km/l) for light trucks and vans in 1981. Beyond this, the Department of Transportation is calling for a socially responsive car for the 1990's which provides the consumer safe, clean, fuel-efficient and acceptable operation. CAFE objectives of 40–50 mpg (17–21.3 km/l) by the year 2000 were realized and a National Vehicle Development Policy may be proposed to help achieve these goals.

The greatest opportunity for final success is to reduce the weight of the vehicle, and this

Table 26.4. Average Auto Weight by Model Year, lb (kg)*

	1979	1985
Lightweight materials	Weight % of Total	Weight % of Total
Plastic	190 (86) 5.6%	300 (136) 11.0%
Aluminum	115 (52) 3.4%	180 (82) 6.6%
HSLA steel	190 (86) 5.6%	465 (211) 17.1%
Sub-total	495 (224) 14.6%	945 (429) 34.7%
Standard materials		
Steel	1900 (862) 55.9%	1180 (535) 43.5%
Glass	80 (36) 2.3%	70 (32) 2.6%
Other	300 (136) 8.8%	190 (86) 7.0%
Components	625 (284) 18.4%	330 (150) 12.2%
Sub-total	2905 (1318) 85.4%	1770 (803) 65.3%
Total dead weight	3400 (1542)	2715 (1232)
Total inertia weight**	3700 (1678)	3015 (1368)

*Source: PPG Industries, Inc.

**Inertia weight (i.e., curb weight) includes fuel, oil, coolant, and two people. NHTSA's CAFE requirements are based on inertia weight.

suggests both smaller cars and lighter materials of construction. Assuming engines are downsized with weight reduction to obtain the same (engine drive train) efficiency, a 1-mpg (0.425-km/l) gain in fuel economy requires a 183-lb (83-kg) weight reduction. Weight reduction through downsizing is being used as an initial solution, while lightweight materials are cautiously investigated and implemented.

Higher strength (thus lighter weight) steels, lighter metals, and all plastic materials are being re-evaluated on a weight/performance basis. In many cases, cost is of secondary importance. It is anticipated that by 1985 the quantity of ferrous metal in the average passenger car will be 25% less than it was in the 1979 models, which have already gone through rigorous downsizing. Conversely, the use of aluminum and plastics will undoubtedly increase 60% in that same time period.

Table 26.4 illustrates material used and projected use in the average U. S. automobile.

Many of the applications of plastic must be structural, load-bearing components requiring the use of reinforced plastics. Hoods, doors, and deck lids are practically a certainty since existing technology and experience suggest little, if any, risk is involved in these applications. An example of the weight reduction potential is General Motors "B" body station wagon tailgate, saving 20 lb (9 kg). Chassis components such as transmission supports, radiator support panels, bumper components, springs, and wheels also represent very significant potential for weight reduction. These structural applications will

Table 26.5. Specific Tensile Strengths of Materials

MATERIAL	STRENGTH PER UNIT WEIGHT in. (cm) × 10^6*
SAE 1015 steel	0.74 (1.88)
Sheet molding compound	3.08 (7.82)
5052 aluminum	0.60 (1.52)
HMC composite**	4.61 (11.7)
XMC composite (directional)**	12.3 (31.2)

*Specific tensile strength obtained by dividing tensile strength by specific gravity. Compared to SAE 1015 steel and 5052 aluminum, XMC composite of the highest tensile strength offers 8.2 and 5.0 times the respective beam strengths of the metals per unit of part weight.

**HMC and XMC are registered trademarks of PPG Industries, Inc.

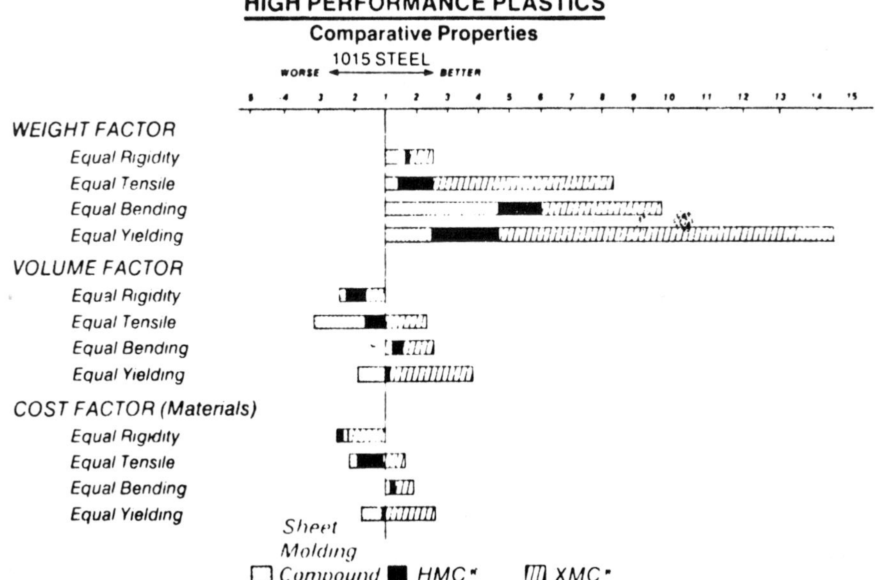

Figure 26.1. At equal rigidity, tensile load, or yield load, parts of steel would weigh 2.4, 8, and 14 times the weights, respectively, of parts molded of the highest strength XMC composite when determined in the direction of continuous fiber glass reinforcement. In other words, a part molded of XMC composite exhibits (directionally) eight times the tensile strength of a steel part of equal weight, or, conversely, XMC composite would weigh one-eighth that of steel of equal tensile.

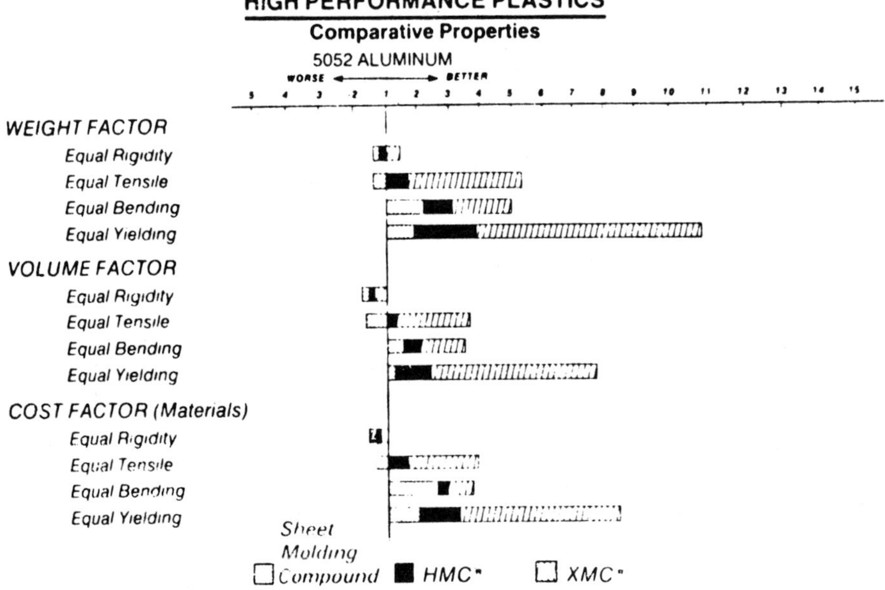

Figure 26.2. XMC composite would require slightly more volume, or thickness, to equal the rigidity of aluminum, or, conversely, aluminum exhibits somewhat more rigidity at equal thickness.

require use of the new high strength, high fiberglass (or sometimes hybrids with carbon fiber) composites whose properties relative to metals are presented in Table 26.5 and Figs. 26.1 and 26.2.

Most of these future applications were illustrated as prototypes on General Motors' "Phoenix" weight reduction concept vehicle. The hood and rear deck lid inner panels are structural FRP, as are the cat-walks; fender supports; radiator support—12 lb (5.5 kg), weight reduction replacing seven metal pieces; transmission support—save 4.5 lb (2 kg); bumper back-up beams; door beams; and wheels—save 10 lb (4.5 kg) each. Design studies, development of prototypes and fabrication techniques, and extensive testing are well along on these critical parts for ultimate use on production model vehicles. A transverse rear suspension spring of fiber glass reinforced epoxy resin is already in commercial use on the Corvette.

Some limited practical experience has been gained from not only U. S. applications but also from commercial applications on European cars. A fiberglass hard top is used on a Renault, as are bumpers on Renault and Porsche models. Several models of Simca and Lotus have fiberglass bodies. Volkswagen has a fiberglass-nylon radiator end tank, as does the Ford Escort. Fiberglass wheels were used on a special model of Citroen until production of the car was discontinued. Wheels in the U.S. are ready for marketing. Additional developments at Ford, Europe, awaiting the appropriate car models are engine compartment parts in RP that operate under stress, such as the intake manifold, and other hang-on parts requiring resistance to high temperatures. Japanese car manufacturers are looking to RP for parts consolidation and cost reduction.

With many financial institutions now granting 60-month loans for the purchase of new cars, the corrosion resistance of reinforced plastics becomes quite significant when compared to steel. A 36-month-old vehicle that is showing serious rust-through in various areas while there are still 24 payments remaining to be made is a very strong inducement to investigate RP materials.

26.4. PROCESSES AND MATERIALS

All of the RP processes are used to varying extents in the fabrication of components for the land transportation industry. Methods most commonly employed are the open mold processes (hand lay-up, spray-up, and continuous panel) and the closed mold processes, the most important of which are compression molding using composite polyester molding compounds and, less frequently, preforms, injection molding of both reinforced thermoplastics and thermoset polyester compounds, and stamping of reinforced thermoplastic sheet. Pultrusion is also used for constant cross-sectional shapes and in a form of filament winding for springs. The reinforcements of urethanes for replacement of some sheet metal body panels (e.g., fenders and doors) ushers reinforced reaction injection molding (RRIM) into the family of RP processes and materials.

Hand lay-up, spray-up, and bag modifications in open molds, although labor intensive, offer some advantages in low tool cost, short lead time, design freedom, and the capability of fabricating very large, complex parts of fiberglass/polyester. In low-volume automotive applications in the U. S., these processes compete favorably with sheet metal that is difficult to form without expensive equipment or that may require extensive assembly by welding, riveting, or soldering. In many of the less developed countries, where labor costs are low, these fabrication techniques are attractive even when relatively large quantities of parts are required. FRP hoods, fenders, doors, roofs, and some interior parts fabricated by these techniques have been used predominantly on buses, ambulances, and some heavy trucks.

A typical composition of a hand lay-up would consist of a sandable gel coat, multiple layers of glass mat or woven rovings in an unfilled polyester resin matrix built to a thickness of 1/8–3/16 in. (3.2–4.8 mm). Blocks of wood or other materials may be encapsulated at the proper location for mechanical attachment to mating parts. Stiffening ribs formed by lay-up over non-structural core materials such as corrugated cardboard or balsa wood are frequently included.

Limitations of these techniques, in addition to relatively high labor content, are the difficulty of holding close tolerance on material thickness and trim lines, the rough, unfinished appearance of the non-molded side, and finish painting problems which may require rework. Because of flexing and vibrational fatigue, the gel coat thickness must be carefully controlled and limited.

For higher volume applications of reinforced thermoset polyester, especially those with more critical dimensional tolerances, compression molding in hydraulic presses using heated, matched-steel dies is the process usually specified. The simple economic break-even point between compression molding and hand lay-up techniques will normally fall between 1000 and 5000 total parts, depending upon many factors. No gel coat is required with compression molding. Composition of finished parts will range between 15 and 40% non-woven glass, between 35 and 45% polyester resin, and the remainder (between 15 and 50%) mineral filler, for the majority of conventional applications compression molded for the transportation industry.

Large, relatively simple parts have generally been "wet" molded with continuous-strand glass mat or chopped strand preforms, with or without veil surfacing mat or low-shrink resin additives, depending on surface requirements. These parts will have consistently good mechanical properties in structural applications, as well as dimensional stability and a good finished appearance. Assembly of the parts using structural adhesives is simple and reliable, and thousands of complete automobile bodies and truck cabs have been assembled from parts compression molded in this manner. Smaller parts, such as heater housings, have long been compression molded from tacky polyester "gunk" or premix, an unsophisticated blend of fiberglass chopped strands or other fibrous reinforcement, catalyzed resin, and filler.

In recent years, the development of chemically thickened composite molding compounds has benefited the industry. Sheet molding compound and bulk molding compound have become the standard materials for many high-volume, semi-structural automotive body parts, such as the radiator grille opening/headlamp panel and fender extensions used on most passenger cars. With the use of low-shrink, low-profile polyester resins in the compounds, and relatively high molding pressures—approximately 1000 psi (6.9

MPa)—complex parts can be compression molded at a rate of 30/hour/cavity. Since stiffening ribs, bosses for attachment, and variations in wall thickness can be included in the molded part, subsequent machining, fabrication, and assembly operations are greatly reduced when compared to similar parts made from sheet steel stampings or die castings.

Although several stamping dies may be required to progressively form the same part in metal (if indeed the contours can be stamped), the total tool cost may be lower than that for RP in some high-volume applications. As many as six identical molding dies could be needed to produce the required number of RP parts to meet automotive production schedules. However, this may be turned to an advantage, in that, with minor design modifications, a different appearance may be given to similar, but lower volume, car lines that bear different names.

Low pressure—500 psi (3.45 MPa)—sheet molding compounds provide an alternative to investment in high-tonnage presses to mold more parts. Good quality parts as large as truck front ends (some having been converted from preform molding) are being compression molded in matched steel dies on presses with capacity in the range of 500–1000 tons (453,593–907,185 kg). In other words, low-pressure sheet molding compound is a way to extend the productivity of such low-tonnage presses.

Not all applications can justify the cost of matched steel dies. Under these circumstances, cold molding or resin transfer molding should be considered. Many of the limitations of hand lay-up can be circumvented at a cost lower than that required for sheet or bulk molding compound. Large, simple parts that are to be reinforced with glass mat or preform are logical candidates for these techniques. Still another alternative under investigation is the conventional autoclaving of sheet molding compounds using molds of intermediate cost.

The interest in composites for structural applications in vehicles to help achieve the goals of weight reduction and fuel efficiency has led to the development of higher strength compositions for matched die compression molding. Chopped glass loadings in the 50–65% range, with little or no filler in the resin, can be formulated into sheet molding compounds that will yield parts having relatively high and nearly isotropic (balanced) strengths; i.e., tensile strength up to 30,000 psi (207 MPa) and flexural strength up to 58,000 psi (400 MPa). If even higher directional properties are desirable, as in beam type applications (e.g., bumper back-up beam reinforcements, integral reinforcement sections of radiator supports, and door and tail gate components), continuous fiber reinforcement can be introduced and oriented longitudinally in the compounds as they are being prepared on modified SMC machines. Tensile strengths of 50×10^3–80×10^3 psi (345–550 MPa) and flexural moduli of 3×10^6 to over 5×10^6 psi (21 to over 34 GPa) can be achieved in the direction of continuous reinforcement.

Another family of unique directional compounds, known as XMC* moldable composites, consisting solely of reinforcement and resin, are prepared on standard filament winding machines. By means of programmed wind angles, the resin saturated, continuous reinforcements are precisely laid down and nested in an "X" pattern resulting in hundreds of cross-overs which act as multiple stress transfer points in the molded part. Chopped strands may be introduced as in XMC-3 composite to provide lateral strength to resist torsional and bending stresses. The foregoing, coupled with closely controlled reinforcement content, enables somewhat higher overall properties; i.e., tensile strengths of 75×10^3 to over 100×10^3 psi (517 to over 689 MPa), flexural strengths of 125×10^3–155×10^3 psi (862–1069 MPa), and flexural moduli of 5.5×10^6–7×10^6 psi (38–48 GPa) in the direction of continuous reinforcement; and property coefficients of variation are in the order of metals. The addition of carbon fiber (i.e., hybrid reinforcement) will further increase strength and modulus. Applications under development include safety related components, such as transmission supports, door intrusion beams, springs, and road wheels.

In any case, the design of the component and the molding technique must assure that

*A registered trademark of PPG Industries, Inc.

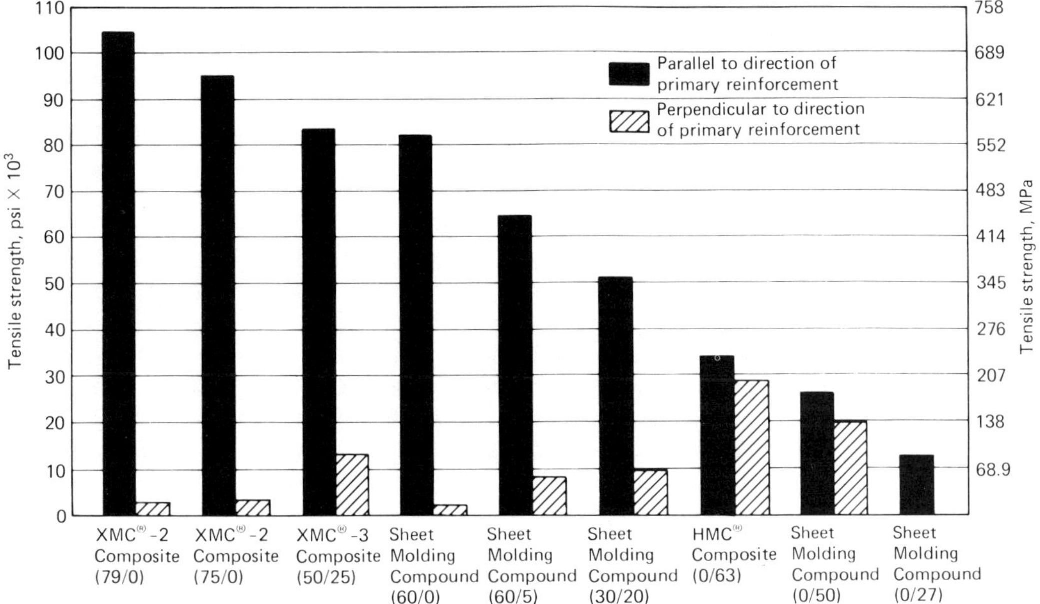

NOTE: Numbers in parentheses represent percent of glass forms, i.e., 79/0 equals 79% continuous and 0% chopped; 0/27 equals 9% continuous and 27% chopped — 1-in. (2.54-cm) — lengths.

Figure 26.3. Comparative tensile strengths (Fiberglass/polyester composites).

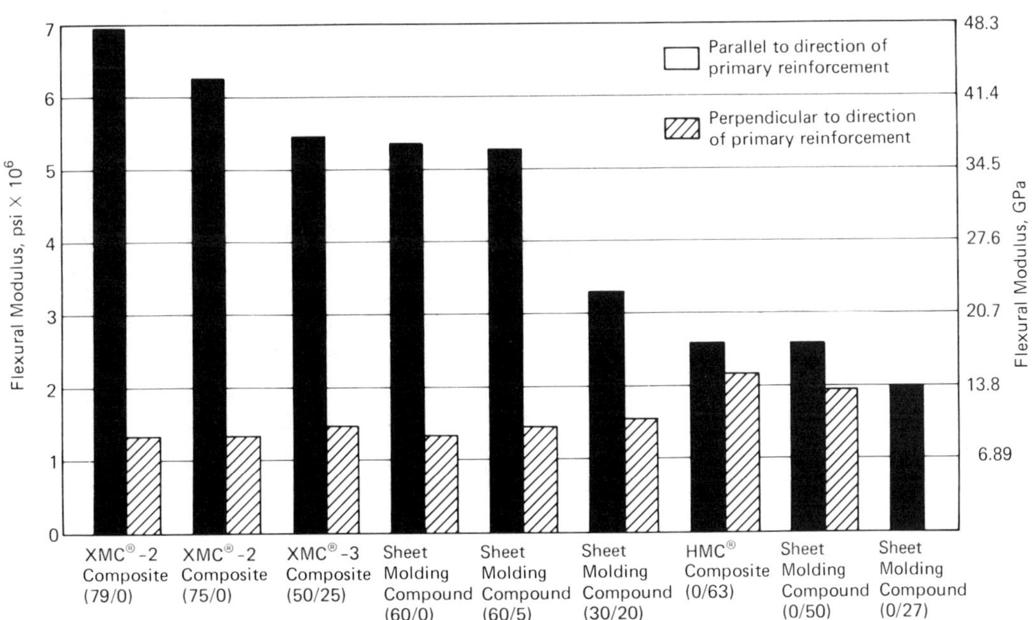

NOTE: Numbers in parentheses represent percent of glass forms, i.e., 79/0 equals 79% continuous and 0% chopped; 0/27 equals 0% continuous and 27% chopped — 1-in. (2.54-cm) — lengths.

Figure 26.4. Comparative moduli (Fiberglass/polyester composites).

the directional fibers remain comparatively straight and in the direction of the anticipated stresses. Figures 26.3 and 26.4 illustrate the relative strengths and moduli of these high-strength materials and conventional sheet molding compound.

Injection molding of both reinforced thermoplastics (RTP) and thermoset polyester compounds (and phenolics in brake pistons) is attractive for many high-volume automotive parts because of short cycles, automation (potential or realized), familiarity with design parameters (thermoplastics), and reduced post finishing and scrap. Utilizing heated plasticating cylinders and cool matched steel molds to first melt and then harden the resin, respectively, RTP injection molding is done with either physical "dry" blends of short glass fibers and resin flake or granules (i.e., direct injection molding) or with pre-compounded melt blends of fiber and resin in the form of pellets. The choice has depended upon the type of thermoplastic resin to be molded and its physical characteristics, the glass dispersion and mechanical property requirements of the composite, the anticipated volumes of any one composite system, and economic considerations of capital investment for materials blending and handling versus the purchased compound.

Thermoset polyester injection molding is a relatively new innovation, particularly in large automotive parts such as the grille opening panel and the front-end retainer. Molding is also accomplished in matched steel molds, heated to effect cure of the resin. Feedstock usually is bulk molding compound, although the TMC* (thick molding compound) process of material preparation would seem to lend itself well to injection molding. Another alternative under investigation is sheet molding compound in the form of strip for continuous feeding. The advantages of cycle time in the vicinity of one-minute, part-to-part reproducibility and cosmetic molded surfaces on the large parts have been brought about by continued technological improvements in equipment, processing, mold design, materials, and compound formulation. Improvements in the latter have also been instrumental in protecting the glass and thereby increasing the average mechanical properties (although they are flow orientation dependent) to 50% or more of those obtained in compression molding.

Glass-reinforced thermoplastic (Azdel*) and nylon (STX**) sheets that can be presoftened to the melting point with radiant heat and stamped or molded in cool dies can be used to produce parts with unique properties. Compared to traditional molded glass-polyester, strengths are similar and isotropic and moduli are somewhat lower, but impact resisttance and toughness are much higher. Specific gravity of the polypropylene sheets is especially low (1.19), despite the high (40%) continuous glass content. Finish painting is difficult with polypropylene, and satisfactory with nylon. Structural applications involving the impact of stones, abrasion, or other such abuse (e.g., bumpers and package shelves), exposure to corrosive environments (e.g., battery trays), and large areas requiring durability (e.g., seat shells) are suggested. Dies similar to those for sheet molding compound are utilized in either hydraulic presses or flywheel (stamping) presses which have been modified to stop (pause) on bottom dead center. Molding cycles of less than one minute are typical, thus permitting high production rates with a single press/mold.

Pultrusion is an economical process, both in tooling and unit cost, for fabricating high volume parts of constant cross-section. Continuous reinforcement, having been saturated with thermoset resin, is pulled through a heated steel die to form the shape. The process is versatile in that the composition and orientation of reinforcing fibers (continuous roving, mat, and woven cloth) can be tailored to the requirements of the final product. Structural and decorative parts for buses, truck bodies, and highway trailers are fabricated by pultrusion. Lading bars for use in rail cars are widely used. Pulmolding (Pultrusion) is a potential modification for greater versatility.

*A registered trademark of USS Chemicals, Division of United States Steel Corporation.

*A registered trademark of PPG Industries, Inc.

**A registered trademark of Allied Chemical Corporation.

Filament winding and pin winding with resin impregnated, continuous reinforcement are emerging as RP processes for springs and steering wheels, respectively. Eventually, rigid, geodesic frame systems may be fabricated in one step by the latter method.

Reinforced reaction injection molding (RRIM) of urethane (and, potentially, polyester) monomers is developing and could lead to impact resistant and corrosion resistant, semi-rigid parts for sheet metal replacement, especially in damage-prone areas of the vehicle. The process involves mixing of reinforcement with, usually, one of the two urethane monomers, isocyanate and polyol, followed by combining each stream in a mixhead and injecting under low pressure—50 psi (345 kPa)—into the heated, matched steel mold. Polymerization of the monomers takes place in-situ (in the mold), forming a self-skinning part with good paintable surfaces. Problems yet to be resolved are viscosity build with most reinforcements and delays in accomplishing part release and mold cleaning between shots. Because of the low investment, unlimited part size, and potentially high productivity (1 1/2-minute cycles), auto manufacturers are keenly interested in the process/material for exterior appearance applications.

26.5. PRODUCT DESIGN

For the purpose of commercial product design, the average mechanical properties given in Table 26.6 can be assumed for fiberglass-polyester composites.

The strength and modulus of composites vary almost directly with the glass (or hybrid) fiber content. However, these values for glass fiber composites represent usual automotive specifications that take cost, weight, moldability, and surface quality into consideration. Values for specific reinforced thermoplastics, may be found in Chap. 7.

In order to anticipate fatigue in transportation products, and to recognize process variables, working stresses of 20–25% of average property values are normally considered conservative. Minimum strength values for sheet and bulk molding compounds may drop as much as 80% from the average because of knit lines and undesirable orientation of the reinforcing fibers. Special care must be exercised in the part/mold design and molding operation either to avoid these problems or to ensure that they do not occur in critical stress areas.

The specific gravity of "E" grade fiberglass is 2.55; of thermoset polyester resin, 1.2; and of mineral filler (calcium carbonate), 2.70. Depending upon the ratio of these ingredients in the formulation, the density of the molded composite will vary from 0.054–0.075 lb/in.^{3} (1.49–2.08 g/cm^3). Replacement of a portion of the mineral filler with hollow glass microspheres will be done in some applications in order to reduce the density, albeit at higher cost.

The original Corvette body design was based on the assumption that fiberglass-reinforced polyester would respond to stress about the same way as steel, provided the composite was approximately three times the thickness of the steel it was to replace. Thus, 0.10-in. (2.5-mm) FGRP was used to replace 20-gauge

Table 26.6 Design Properties

COMPOSITE	TENSILE STRENGTH, psi × 10^3 (MPa)	FLEXURAL STRENGTH, psi × 10^3 (MPa)	FLEXURAL MODULUS, psi × 10^6 (GPa)
Hand lay-up/spray-up	10.0 (68.9)	17.0 (117.2)	1.0 (6.9)
Bulk molding compound	3.0 (20.7)	10.0 (68.9)	1.6 (11.0)
Sheet molding compound	10.0 (68.9)	22.0 (151.7)	1.5 (10.3)
Mat/preform—structural	15.0 (103.4)	30.0 (206.8)	1.3 (9.0)
Mat/preform—low shrink	10.5 (72.4)	24.0 (165.5)	1.1 (7.6)
High-glass sheet molding compounds*	19.5 (134.4)	36.0 (248.2)	1.9 (13.1)
XMC-3 composite (lengthwise)	75 (517)	125 (862)	5.5 (37.9)
Pultrusion (lengthwise)	30.0 (206.8)	30.0 (206.8)	2.5 (17.2)

*Minimum averages for a range of high-strength compounds.

steel—0.0375 in. (0.95 mm)—in the body panels, which resulted in a 40% weight reduction. While this is a useful "rule-of-thumb," it is obviously an oversimplification. The greater thickness of the RP compensates for the lower modulus of elasticity in bending, and the resulting panels are approximately equal in rigidity to panels made of 20-gauge steel. The strength of the composite panel in bending will be many times the strength of steel based on its elastic limit. Working stresses, therefore, are low, and fatigue is seldom a problem. Resistance to casual damage is also far greater than steel.

In automotive design, it is important to remember that the RP composite is not ductile. While a damaged (bent) metal part may still function, a broken composite part would not. RP composites have the capability of storing far more energy (as in a spring) than ductile metals, but they are not good materials for absorbing energy when stressed beyond their ultimate. This combination of properties suggests, for example, bumpers and radiator supports made of RP, but not structures that must be relied upon for initial or principal energy management in a serious collision.

It is generally frustrating, if not an exercise in futility, to attempt an efficient design of an RP part that is intended simply to replace an isolated existing part already designed for another material. The opportunity is not available to engineer total functionality, to modify adjacent parts, in order to take advantage of the unique properties and capabilities of RP in terms of weight and mass, attachment, finish, and structural contribution. Minor year-to-year modifications and model changes in an existing vehicle are a case in point.

On the other hand, a completely new body and chassis provide the design engineer the opportunity to use the new, lightweight materials of construction extensively and efficiently. In many ways, properly designed parts may be utilized to perform multiple functions. Structural parts can also be decorative and corrosion resistant (obviating additional corrosion protection) and function as insulation and damping (reducing inside noise levels, vibrations, and temperature variations). This was demonstrated by the Chevrolet XP-898, which was an all-plastic concept vehicle using monocoque construction. Large body panels and structural members were of "stressed-skin" sandwich structures providing integral stiffness and strength. Ductwork can serve as a structural member. The inclusion of pigment and/or texture may eliminate a finish painting operation. All savings in body weight result in further cost and weight savings in chassis and drive train components; thus, engines, transmissions, axles, brakes, springs, and wheels can be made smaller and lighter with no sacrifice in performance.

Although composites have proven to be extremely durable in a great many automotive semi-structural applications, there is far less experience on which to draw than there is with sheet metal parts. It is only natural that engineers are inclined to over-design to some extent, with greater wall thicknesses than may be needed. In order to properly place RP material for optimum stiffness, strength, and minimum weight, computerized finite element modeling for stress analysis under anticipated loading conditions is especially useful. It can be expected to improve confidence levels without the use of excessive safety factors.

Relatively simple stress analysis can be used to design the most obvious load-carrying portions of a structure for acceptable deflection under load. Compared with a corresponding design in steel, flanges, channels, and hat sections are necessarily larger to compensate for the lower modulus of elasticity of RP. In the calculation and evaluation, it is important that the section maintain its shape under load and, wherever possible, parts should be bonded together to form closed box sections for increased stability. In some cases, gussets or ribs can be used to prevent the buckling of sections. Proper design will permit interior components to be bonded to exterior panels without serious distortion. This provides an opportunity to form closed sections, as well as a better mechanism for load transfer to and from exterior panels.

Assemblies are normally fabricated by adhesive bonding with filled polyester, epoxy, or urethane adhesives. Tensile shear strength of the bonds of 600–800 psi (4.14–5.52 MPa) is typical. The bonding and/or mechanical joining of RP composites to metal is also feasible. Joints should be designed to be loaded in shear wherever possible. If peel or cleavage loads are anticipated, rivets are used in addition to the adhesive bond. Tapered bond flanges reduce stress concentrations and minimize the possibility of peeling stresses. If there are exterior butt joints to be filled and finished, they should be backed up with bond strips or flanges thicker or stiffer than the exterior panels to relieve the stress in the filled joint.

Self-tapping, high-low screws or studs are often used, with or without bosses, for the attachment of accessories, trim, and other components. For heavy loads (such as door hinges), weld nuts or cage nuts, which are on steel plates bonded and riveted to the opposite side of the panel, are used. For proper load transfer, the steel plate should be approximately one-third as thick as the composite to which it is bonded, so that the two materials will offer the same resistance to bending under load, and the plate edges will not tend to shear through the composite.

The removable tops used on the Chevrolet Blazer and Ford Bronco, and the Pickup Truck Box Cover offered by Ford, are examples of adhesive-bonded, compression molded assemblies. Ford, Mack, and International Harvester medium and heavy trucks also use bonded assemblies for the tilting hood-fender combination. General Motors Truck and Coach hoods and fenders are moldings (of conventional and low-pressure sheet molding compounds) assembled with bolts or studs at molded-in attachment points. One heavy truck manufacturer has introduced a new cab assembled from mat and preform moldings. It is a stressed-skin design with a minimum of local metal reinforcements at high-stress points.

The models made from engineering layouts in automotive body engineering can be used to make simple, inexpensive molds for prototype parts by hand lay-up and clamp and bake techniques. If modest differences in physical properties of the prototype parts and high pressure production parts are taken into account, the prototypes can be assembled into complete chassis, bodies, or cabs for static and dynamic testing. The decision to proceed with production tooling can be made after the prototypes have been tested. Experience indicates that a high degree of confidence in the design can be expected when prototypes have passed the required tests. Critical applications, such as the road wheel, require the molding of test parts in steel tooling in order to assess the advantages of optimum designs in XMC/HMC composite combinations over corresponding metal wheel "bogies". Typical automotive FRP parts are shown in Figs. 26.5–26.8. In addition, the Grumman Banded Tire® is being developed in association with The Goodyear Tire & Rubber Co. In a light truck size, it has already demonstrated the ability to run flat (valve core removed) under full load for 30 miles at 30 mph. A passenger car version has run more than twice as far at 50 mph. Refinements are underway to assure acceptable handling during pressurized operation (Fig. 26.9).

Basically, the Banded Tire is a radial tire of

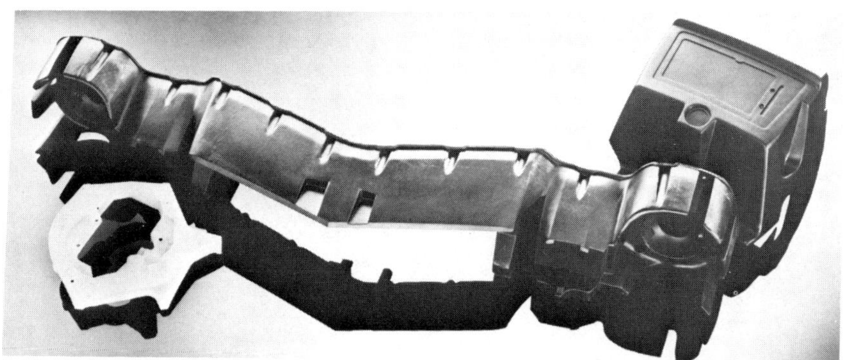

Figure 26.5. Automotive support bracket.

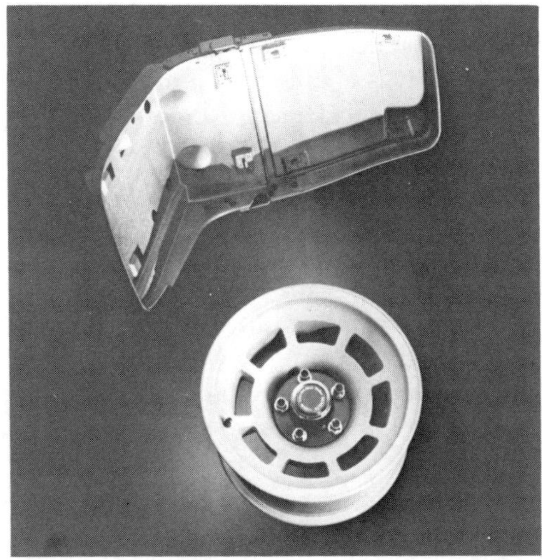

Figure 26.6. Seat back and wheel.

conventional construction, in which the belting is replaced by a thin, high-strength band of fiberglass/epoxy composite. Fig. 26.10 shows a band sliced from a filament-wound cylinder. When the tire is deflated, the band acts in conjunction with the radial sidewall reinforcing strands to create a hoop-and-tension-spoke structure which supports the load. The tire deflects approximately halfway and each sidewall bulges outward locally, but does not touch itself or the ground. The remainder of each sidewall straightens in tension. The band acts as a compressive arch, stabilized by the sidewalls. In contrast to other run-flat designs, which rely on heavy compression sidewalls or auxiliary inserts, the Grumman approach adds very little weight and does not require a special rim.

The Banded Tire offers advantages in the pressurized mode as well, e.g., lower rolling resistance and low wear rate, because of the tread-stabilizing influence of the band. Also, the Banded Tire can be deliberately deflated, partially or completely, to enhance mobility in weak soil. The low pressure footprint has a shallow entry angle (which reduces bulldozing resistance) and an even pressure distribution. Finally, the band greatly enhances resistance to puncture through the tread.

26.6. OPPORTUNITIES

The high strength-to-weight ratio of RP is extremely important in any application involving transportation because of its direct effect on energy consumption. It is expected that the American automobile buyer of the future will be willing to pay a premium price for a vehicle that offers the comfort, convenience, and status to which he has become accustomed, rather than accept a car that has been designed to meet mandated fuel economy simply by sacrificing size, performance, and comfort. This is not to say that manufacturers are being forced to use RP composites. All lightweight and higher strength materials are being scrutinized and improved, and such materials are being used for better designs, all of which leads to improved performance and more efficient use. However, total cost within the design constraints is still the controlling factor.

Basic RP materials that will receive greater attention in the transportation industry are the new breeds of thermoset polyester resins

Figure 26.7. Experimental plastic body for Chevrolet.

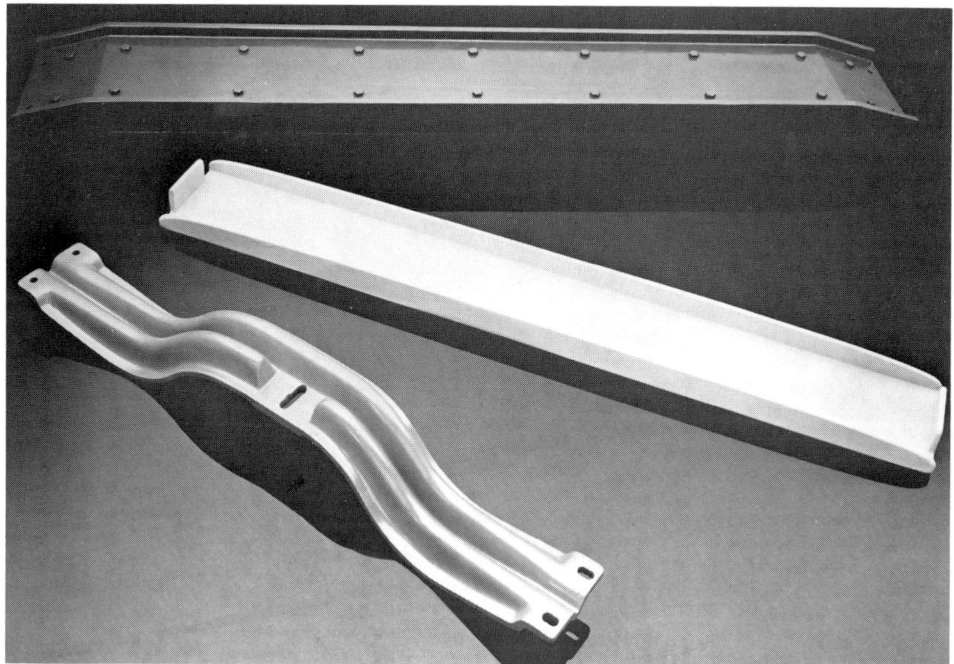

Figure 26.8. Various automotive parts.

Figure 26.9. Grumman banded tire.

Figure 26.10. Filament-wound tire reinforcing insert for Grumman tire.

(i.e., "flexible," tougher grades); high-temperature resistant grades; and the quick cure systems—less than 1 minute for 0.100-in. (2.54 mm) thickness; automotive grades of vinyl ester resins; and epoxy resins with new curing systems permitting processability more in line with automotive requirements. Carbon fiber may see a minor use, particularly in hybrid reinforcement systems with fiberglass (see Fig. 26.11), as the price decreases from a current low of $18.00/lb ($40.00/kg) to $10/lb $22/kg) in the late 1980's. Meanwhile, basic design data are being accumulated for the newer, high-strength RP composites on energy absorption and impact characteristics, creep, and fatigue to facilitate efficient design and development and to accurately predict performance.

In the beginning, low tooling cost was the most important advantage of composite materials. Subsequent improvements in both materials and processes have created opportunities to increase production quantities competitively. This, in turn, has justified better, more reliable tooling and process automation. Process controls have been added to molding equipment which will lead to improved quality. In-mold, prime coating with urethane, first used on the 1980 Corvette body panels and followed by the 1981 Olds station wagon tailgate, is an additional means of improving part surface quality, with the promise of eventual lower costs. A mold transfer system is under investigation which may reduce by 80% the number of high-pressure molding presses required. The most difficult design and production line assembly problems involve RP components that become integral parts of the body structure, such as roof panels and floor pans. Provisions must be made in part design and assembly procedures for rapid joining of dissimilar materials of construction. As these improvements, and others to come, in materials and processes are implemented,

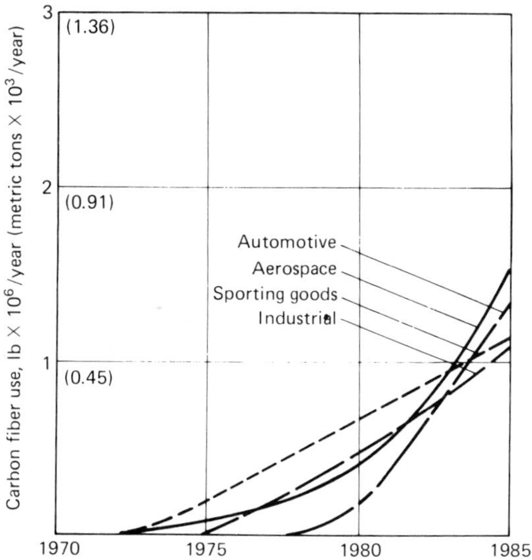

Figure. 26.11. Predicted use of carbon fibers. (*Source*: *NASH HW RW 79 212* [1].)

the production rates of RP composite parts and assemblies can be expected to rival the productivity of metals in body structures at equal or lower investment. Finally, with in-plant recyclability, a fact with thermoplastics and showing promise with thermosets, the loop will have been closed.

The foregoing factors, together with the weight reduction potential for composites and other materials, having been taken into account, Table 26.7 presents a forecast of FRP use in the U.S. land transportation market, principally autos, vans, and trucks, through 1990 (see Figs. 26.12–26.17). By that time, high-strength composites for structural applications should reach 15%; sheet and bulk molding compounds, 70%; and reinforced reaction injection molding, 15% of the thermoset portion of the total FRP. Reinforced thermoplastics will continue to consist mainly of polypropylenes, nylons, styrenics, and thermoplastic polyesters, with a significant portion in flow-molded (stamped) fiberglass-reinforced sheet. Parts will be both captive and custom molded, divided about equally between the two source groups. Auto production should be about 9.5 million units.

26.7. CONCLUSION

It is inconceivable that our society will become less mobile in the foreseeable future. As long

Table 26.7. Forecast of Fiberglass-Reinforced Plastics in the Land Transportation Market, lb × 10⁶ (metric tons × 10³)*

FRP by segment**	Estimated 1980	Projected 1985	Projected 1990	1980–1990 CGR
Auto	261 (118)	350 (159)	538 (244)	7.5%
Truck and van	87 (39)	161 (73)	228 (103)	10.1%
Other	62 (28)	89 (40)	114 (52)	6.3%
Total	410 (186)	600 (272)	880 (399)	7.9%
FRP by resin				
Thermosets	283 (128)	455 (206)	700 (318)	9.5%
Thermoplastics	127 (58)	145 (66)	180 (82)	3.5%

*Source: PPG Industries, Inc.

**Figures include both thermosetting and thermoplastic resins, reinforcements and fillers for automobiles, vans, buses, trucks, trailers, leisure and mass transit vehicles, railroads, motor homes, etc. normally included in the land transportation market.

Figure 26.12. Various plastic automotive parts.

as this mobility continues, vehicles will wear out, become obsolete, or need replacement for other reasons. In spite of the speed and economy of air travel, the convenience of personal transportation, especially for trips of less than 150 miles (240 km), will continue to command a substantial share of the family budget. The rapid evolution in design of land transportation vehicles represents a constantly growing opportunity for new and improved materials, processes, and products. The light weight, strength, corrosion resistance, and

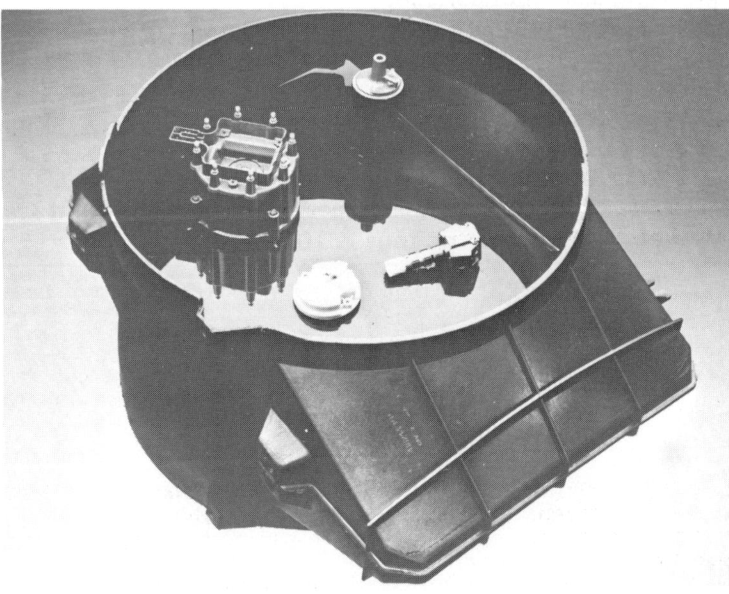

Figure 26.13. Plastic automotive engine parts.

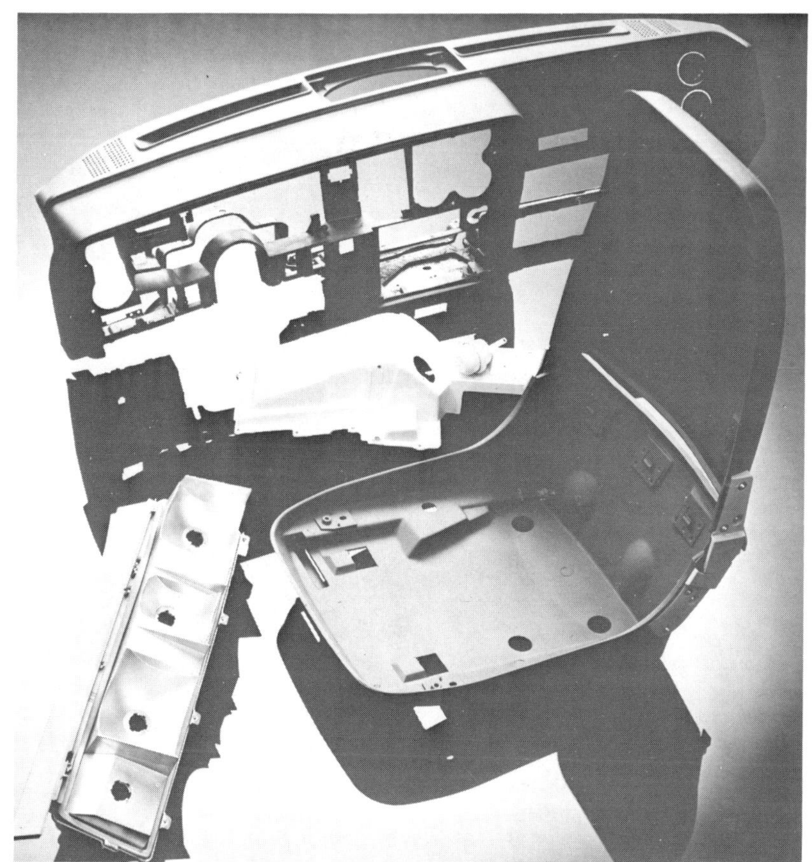

Figure 26.14. Plastic gear shift cover.

Figure. 26.15. Plastic automotive body parts (hood and headlight holder).

Figure 26.16. Plastic body front.

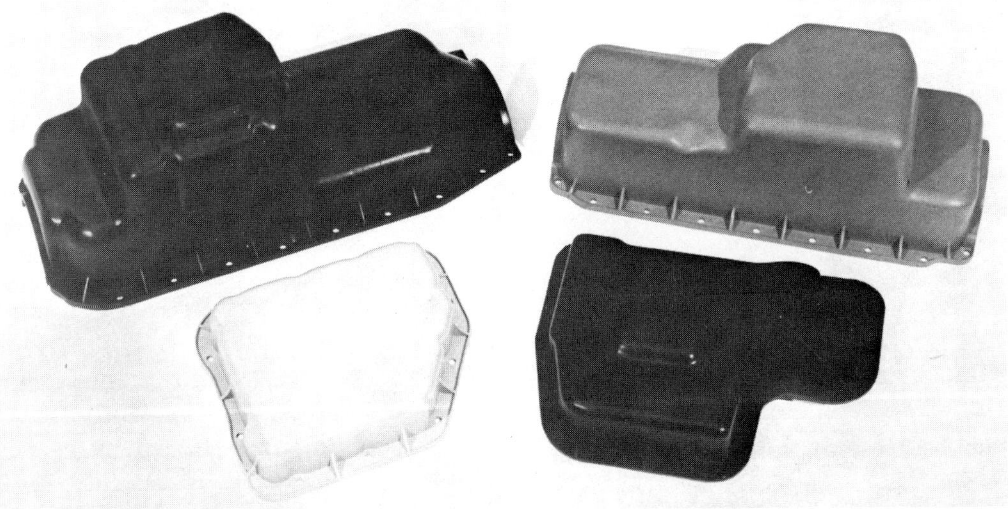

Figure 26.17. Nylon-6 composite prototype oil pans, 50% lighter than steel.

versatility of composites in design, formulation, and fabrication that permit tailoring to a specific need make them attractive for the automotive industry. Good product design, reliable tooling and process equipment, and adequate raw material supplies will provide the assurance of availability and performance necessary for confidence in the planning for, specifying of, and production with reinforced plastic composites.

27
MARINE APPLICATIONS

William R. Graner
Artech Corp.
Falls Church, Virginia
(formerly) U.S. Naval Ship Engineering Center
Hyattsville, Maryland

27.1. INTRODUCTION

Fiberglass-reinforced plastic laminates have found wide use in marine applications since their introduction as commercial materials in the late 1940's. In many applications, they are the material of choice because of their unique characteristics: their high strength-to-weight ratio; their durability and resistance to the marine environment; their ease of maintenance and repair; their toughness, particularly at very low temperatures; their non-magnetic and dielectric properties; and their low thermal conductivity, as compared to metals. In addition, they offer to the marine design engineer a considerable degree of flexibility which is not available with conventional metal construction. For example, proper selection of basic materials and manufacturing processes, as well as the orientation of fiber reinforcement to suit the specific structural requirements, can provide a structure which is lighter in weight and more efficient. The use of monolithic, seamless construction minimizes leakage and eliminates many costly secondary assembly processes (e.g., mechanical attachment such as welding or riveting).

This chapter will briefly describe the basic materials used, their properties, and their response to the marine environment, as well as the history of a few of the marine applications of glass fiber-reinforced plastic laminates. Some more recent developmental investigations will also be discussed.

27.2. BASIC MATERIALS

27.2.1. Resin Systems

Because they are generally less costly and offer greater flexibility with regard to handling and curing characteristics, the unsaturated alkyl-styrene type polyesters have been widely used in marine applications. They may be formulated for room temperature cure by the addition of an organic peroxide catalyst plus an accelerator (or promoter) such as cobalt naphthenate to initiate the reaction. "General purpose" resins are used primarily; there are also halogen-containing types available which provide a measure of fire retardance.[50] Thixotropic agents may be added to minimize resin run-off, and fillers or pigment may also be used to meet specific requirements.

Epoxy resins are used in applications requiring superior physical or mechanical properties or specific chemical resistance characteristics. They are also widely used as adhesives and, in combination with fillers, as putties or "gunks" for gap filling or fairing purposes. The epoxies may be cured at room temperature by formulating with an active hardener such as diethylene triamine. For optimum properties, however, laminates should be heat cured and pressure applied during cure (either vacuum or autoclave).

27.2.2. Reinforcements

Many types of reinforcements, such as sisal, jute, or asbestos, have been used in the past for reasons of economy or to meet special requirements. At the present time, however, borosilicate type "E" glass fibers are used almost exclusively, although there is a growing interest in higher modulus fibers, such as graphite, boron, or DuPont's aramid (Kevlar). Both graphite and boron fibers, for example, were used in the boom of the America's Cup winner, Intrepid, to provide greater stiffness with no increase in weight.[6]

Practically all of the commercial types of glass fiber reinforcement available, from chopped fiber in mat or spray-up form, through various types of woven textiles and unwoven roving, are employed to meet specific application requirements. The glass filament must be treated with an appropriate finish or sizing, specific to the type of resin used, to enhance wet-out of the glass by the resin and reduce the effect of moisture or humidity on the glass-resin bond.

27.2.3. Production Processes

Large structures, such as boat hulls, are generally produced by wet hand lay-up or spray-up methods, using polyester resins formulated for room-temperature cure. Less commonly, where a better quality, higher glass content laminate is required, vacuum bag molding is employed. For large production runs, matched metal molding has been used.[42]

The selection of the production process to

Table 27.1. Properties of Marine Structural Glass Fiber-Reinforced Plastic Laminates*

			VALUE REQUIRED FOR EACH GRADE OF LAMINATE					
PROPERTY	CONDITION	UNIT OF VALUE	GRADE 1	GRADE 2	GRADE 3	GRADE 4	GRADE 5	GRADE W
Flexural strength, flatwise	Standard	psi (minimum average) MPa	50,000 345	37,000 255	31,000 214	23,000 159	18,000 124	32,000 221
Flexural strength, flatwise	Wet**	psi (minimum average) MPa	45,000 310	33,000 228	27,000 186	20,000 138	15,000 103	29,000 200
Flexural modulus, flatwise	Standard	psi (minimum average) GPa	2,500,000 17	2,000,000 14	1,450,000 10	1,100,000 7.6	850,000 5.9	1,650,000 11.4
Flexural modulus, flatwise	Wet	psi (minimum average) GPa	2,300,000 16	1,800,000 12	1,250,000 8.6	990,000 6.8	770,000 5.3	1,500,000 10
Tensile strength	Standard	psi (minimum average) MPa	37,000 255	28,000 193	20,000 138	14,000 97	9000 62	35,000 241
Compressive strength, edgewise	Standard	psi (minimum average) MPa	33,000 228	25,000 172	21,000 145	17,000 117	16,000 110	18,000 124
Compressive strength, edgewise	Wet	psi (minimum average) MPa	28,000 193	23,000 159	19,000 131	15,000 103	14,000 97	17,000 117
Void content		% (maximum)	1.5	2	3	4	5	4
Resin content, range		% (by weight)	35–43	42–52	49–59	55–65	65–75	45–56

*Specification MIL-P-17549.
**Two-hour boil.
NOTE: Grade 1—Style 181 cloth laminate; 2—Style 1000 cloth laminate; 3—Style 1044 cloth laminate; 4—Glass mat laminate; 5—Glass mat laminate—high resin content; W—Woven roving laminate.

be used is generally dictated by cost considerations, especially where glass-reinforced plastics must be competitive with other structural materials, such as wood or aluminum. However, where economically feasible, or where a particular application may warrant added expense, marine structures may be, and have been, produced by practically every fabrication process developed for reinforced plastics, including heat curing or post-curing, autoclaving, matched die molding, and filament winding.

Wide differences have been noted in the properties of the laminates obtained from different fabricators. This makes it necessary to define minimum material requirements and to enforce inspection procedures to assure that the material meets design requirements. The U. S. Navy, for example, requires that laminate be translucent to permit visual through-inspection, and specifies acceptable resin and void contents and minimum mechanical properties.[49]

27.3. GLASS FIBER-REINFORCED LAMINATE MATERIALS

27.3.1. Mechanical Properties

As implied above, materials with a wide range of properties are available for marine construction. Table 27.1 lists the average minimum requirements for glass-reinforced polyester laminates for marine structural use, as given in a material specification for this type of material.[49] A similar compilation, intended as a general guide to laminate properties for boat construction, is shown in Table 27.2.[52] Scott[42] provides data showing the effect of type of reinforcement, orientation, and resin content on mechanical properties of laminate from various fabricators.

Where mechanical properties superior to those indicated above are required, consideration must be given to more sophisticated fabrication processes (e.g., vacuum bag or autoclave), to the use of epoxy rather than

Table 27.2. Physical Properties of Typical Marine Laminates (Average Values for Guidance Only)*

PHYSICAL PROPERTY**	CHOPPED STRAND MAT LAMINATE, LOW GLASS CONTENT	COMPOSITE LAMINATE,+ MEDIUM GLASS CONTENT	WOVEN ROVING LAMINATE, HIGH GLASS CONTENT
Percent glass by weight	25-30	30-40	40-55
Specific gravity	1.40-1.50	1.40-1.50	1.65-1.80
Flexural strength, psi $\times 10^3$ (MPa)	18-25 (124-172)	25-30 (172-207)	30-35 (207-241)
Flexural modulus, psi $\times 10^6$ (GPa)	0.8-1.2 (5.5-8.3)	1.1-1.5 (7.6-10.3)	1.5-2.2 (10.3-15.2)
Tensile strength, psi $\times 10^3$ (MPa)	11-15 (76-103)	18-25 (124-172)	28-32 (193-221)
Tensile modulus, psi $\times 10^6$ (GPa)	0.9-1.2 (6.2-8.3)	1.0-1.4 (6.9-9.7)	1.5-2.0 (10.3-13.8)
Compressive strength, psi $\times 10^3$ (MPa)	17-21 (117-145)	17-21 (117-145)	17-22 (117-152)
Compressive modulus, psi $\times 10^6$ (GPa)	0.9-1.3 (6.2-9.0)	1.0-1.6 (6.9-11.0)	1.7-2.4 (11.7-16.5)
Shear strength, perpendicular, psi $\times 10^3$ (MPa)	10-13 (69-90)	11-14 (76-97)	13-15 (90-103)
Shear strength, parallel, psi $\times 10^3$ (MPa)	10-12 (69-83)	9-12 (62-83)	8-11 (55-75)
Shear modulus, parallel, psi $\times 10^6$ (GPa)	0.4 (2.8)	0.45 (3.1)	0.5 (3.4)

* *Technical and Research Bulletin Number 2-52.*
**Properties from short term loading tests, wet condition. Composite and woven roving values for warp direction.
+Based on typical alternate plies of 2-oz/ft^2 (610-g/m^2) mat and 24 oz/yard^2 (809 g/m^2).

polyester resin, and to the use of directional reinforcements such as style #143 woven glass cloth, or unwoven, unidirectional, preimpregnated glass fiber tapes or roving.

27.3.2. Response to Marine Environment

While the favorable strength-to-weight ratio of glass-reinforced plastics may provide the basis for initial consideration for marine-structural use, the ultimate utility of these materials depends upon their stability in the marine environment and their durability under service conditions. Of primary importance is the effect of water on the properties of the materials.

27.3.2.1. Effect of Extended Water Exposure on Static Properties

Although present day glass-reinforced plastics, fabricated with reinforcements treated with improved sizings or finishes, are much less sensitive to water than earlier materials of this type, a degree of water sensitivity still exists.

No detailed discussion of the effects of water on glass-reinforced plastics will be undertaken herein in view of the extensive literature on the subject (see, Fried[12] and Rawe[36]; also see the Interface Seminars in *Proceedings of the Annual Technical Conferences, Reinforced Plastics/Composites Division, SPI*). A brief explanation of water's effects follows.

Water may be absorbed by the resin and act as a plasticizer, reducing both strength and stiffness of the laminate. Water may also migrate to the site of the glass-resin interface by diffusion through the resin, penetration through cracks, voids, and other defects, or migration along the fiber surfaces. At this site, the glass-resin bond may be weakened by the water through a number of possible mechanisms, with consequent adverse effects on the properties of the laminate.[11]

In general, the net effect of extended water exposure is to cause some degradation in static mechanical properties. Typically, there is an initial decrease in mechanical properties to a particular level (within the first one or two months), with no further deleterious effects upon extended exposure beyond this point. In the case of polyester laminates, strength reductions generally amount to 10–15%. Lesser effects may be expected in epoxy laminates, while deleterious effects may be accentuated in laminates which are made with improperly finished reinforcement, those that are under-cured, or those that contain an excessive amount of voids. Tensile properties are generally not as seriously affected as those which are controlled by the resin or interface (i.e., compression, flexure, and shear).

In recognition of the importance of water resistance in laminates designed for marine applications, specifications for these materials invariably set a limit on permissible degradation in properties due to water exposure. This is indicated in Table 27.1e, which includes a requirement for "wet" strength and stiffness as determined by a two-hour boil test. The latter procedure is in common use as an accelerated method for determining the effects of extended water exposure, being roughly equivalent, for polyester resins in many instances, to one-month immersion at normal temperature.[36] While this test can provide useful information, data thus developed should be viewed judiciously, particularly when applied to room-temperature-cured laminates, since the temperature of boiling water may produce curing effects which can mask degradative effects of the water. In such instances, it is advisable to compare properties in the "wet" condition (two-hour boil) with properties of the material in the fully cured condition, or at least after a two-hour oven cure at 212° F (100° C).

27.3.2.2. Effect of Water Under Pressure

Since glass-reinforced plastics may be employed in deep underwater applications, it is of some interest to consider the effects of exposure to water at high pressure on the properties of these materials. An indication of this is provided in Table 27.3. The data shown were obtained in an investigation in which a ⅜-in. (1-cm) thick #1000-Volan polyester laminate was exposed at a depth of 5700 ft (1737.4 m) in the Tongue of the Ocean, Bahamas, for a period of three years. It is apparent that the properties of the laminate were essentially unaffected by this deep ocean exposure. Additional data bearing on this subject, relating

Table 27.3. Effect of Extended Exposure in the Ocean at 5700 ft (1737.4 m) on Polyester-Glass Laminate

PROPERTY	INITIAL	AFTER 1045-DAY EXPOSURE AT 5700 ft (1737.4 m)
Compressive strength, psi (MPa)	33,000 (227)	33,500 (231)
Compressive modulus, psi (GPa)	2.72×10^6 (19)	2.9×10^6 (20)
Flexural strength, psi (MPa)	38,800 (267)	38,100 (263)
Flexural modulus, psi (GPa)	2.60×10^6 (18)	2.50×10^6 (17)
Interlaminar shear strength, psi (MPa)	1810 (12)	1880 (13)

specifically to filament wound materials, are presented in a later section of this chapter. Evidence to date indicates that good quality (i.e., low void content) reinforced plastic laminates are no more adversely affected by high pressure water exposure than by immersion in water at atmospheric pressure.

27.3.2.3. Recovery of Properties on Drying

Although glass-reinforced plastics show decreased mechanical properties in the wet state, following extended water immersion, it is notable that this effect is generally reversible in that properties tend to be recovered upon drying. Winans[53] has described this phenomenon for a number of reinforced plastic materials. While materials with water sensitive reinforcements were permanently degraded by extended water exposure, glass-reinforced laminates which were allowed to dry thoroughly following extended water immersion (4500 hours) showed essentially complete recovery of properties. A further indication of this effect is provided in Table 27.3, which shows little change in properties of a glass-reinforced polyester laminate after three years' water immersion followed by drying.

27.3.2.4. Effect of Water on Long Term Loading Properties

In common with other materials, glass-reinforced plastics subjected to cyclic fatigue or long term static loading (stress-rupture) will fail at stress levels below ultimate short term loading stresses; these stress-rupture and fatigue characteristics are also affected by water exposure. Information on the effect of water immersion on stress-rupture characteristics is provided in Chapter 3 of *MIL-HDBK-17*.[29] In a typical case, specimens of a style #181 polyester laminate sustained a flexural stress (in air) of approximately 68% of ultimate for 1000 hours, whereas specimens loaded under water sustained 48% of ultimate for this period of time. For an epoxy laminate of similar construction, these 1000-hour values were 70 and 60% of ultimate for "dry" and "wet" specimens, respectively.

Fatigue data developed to date on a variety of reinforced plastic laminates indicate endurance limits (or fatigue strength) of 25–30% of ultimate in the range of 10^7 cycles and beyond. Under conditions of completely reversed stress cycling (zero mean stress), limited test conducted on polyester laminates in the "wet" condition have shown an appreciable reduction in fatigue strengths in the low cycle 10^3–10^4 range but essentially no difference between dry and wet fatigue strengths in the area of 10^7 cycles and beyond.[22]

27.3.2.5. Weathering Effects

A comprehensive treatise on the effects of weathering on glass-reinforced plastics has been provided by Rugger and Titus.[41] Weathering is a complex phenomenon, involving a variety of influences, and effects vary with different climatological conditions. The consensus of reports indicates that warm, moist climates are most detrimental to glass-reinforced polyester materials. Extended weathering may be expected to erode the surface resin and result in decreases in tensile strength of up to 20%, with lesser effects on flexural and compressive properties.[41] The effects of weathering on glass-reinforced epoxy laminates are not markedly different, extended exposure resulting in decreased tensile and flexural strengths in the order of 10–20%. It is notable that there are now resin systems available which are especially designed to be weather resistant, and this type of resin may be specified where the utmost in resistance to a marine atmosphere is required. The use of pigmented gel coat or a standard marine type paint is recommended for protecting the laminate surface. The paint itself does not prevent water absorption since all existing paint formulations allow water vapor penetration. Paints, however, will protect against deleterious UV light effects and surface abrasion. The combination of UV and moisture causes the major amount of degradation in fiberglass structures.

27.3.2.6. Biological Attack—Fouling

Glass-reinforced plastic materials are not attacked by marine boring organisms. In point of fact, Rugger[41] indicates that biological attack may be ignored as a factor affecting the mechanical properties of these materials. On the other hand, marine organisms will grow upon glass-reinforced plastic surfaces, just as they will on other materials. Figure 27.1 shows fouling of glass-reinforced polyester panels immersed in shallow water at Kure Breach, North Carolina for an extended period of time. Figure 27.2 shows these panels after removal of marine growth. This was accomplished by mild scraping, indicating that although fouling occurs, removal is a relatively simple matter. The construction of glass-reinforced plastic laminates offers the intriguing possibility of incorporating biocidal compounds in the resin within the laminate, or in a gel coat, to provide lasting protection against fouling, as proposed by Graner and Stander.[19] Development work in this area is currently underway.[11,30]

27.4. MAINTENANCE AND REPAIR

Low maintenance requirements and ease of repair are fundamental advantages of fiberglass-reinforced plastic laminates.[42]

Reinforced plastic structures should be inspected periodically to determine whether there has been any deterioration in surface characteristics (such as fouling or erosion) or damage (such as scratches, gouges, or fractures, or delamination of bonds). A good coat of paint or gel coat will provide significant protection to laminated structures that are exposed to UV, moisture, or high humidity. The coating of exposed, cut edges with paint or resin is also beneficial.

It is not within the scope of this chapter to provide detailed instructions on the repair of fiberglass laminates. Such repairs are within the capabilities of unskilled workers using ordinary hand tools.[42] Reinforcements similar to those used in the original laminate and room-temperature-curing resins may be applied if material replacement is required. The environment in which the repair is to be

Figure 27.1. Marine fouling on glass-reinforced plastic test panels after 14-month immersion in the ocean.

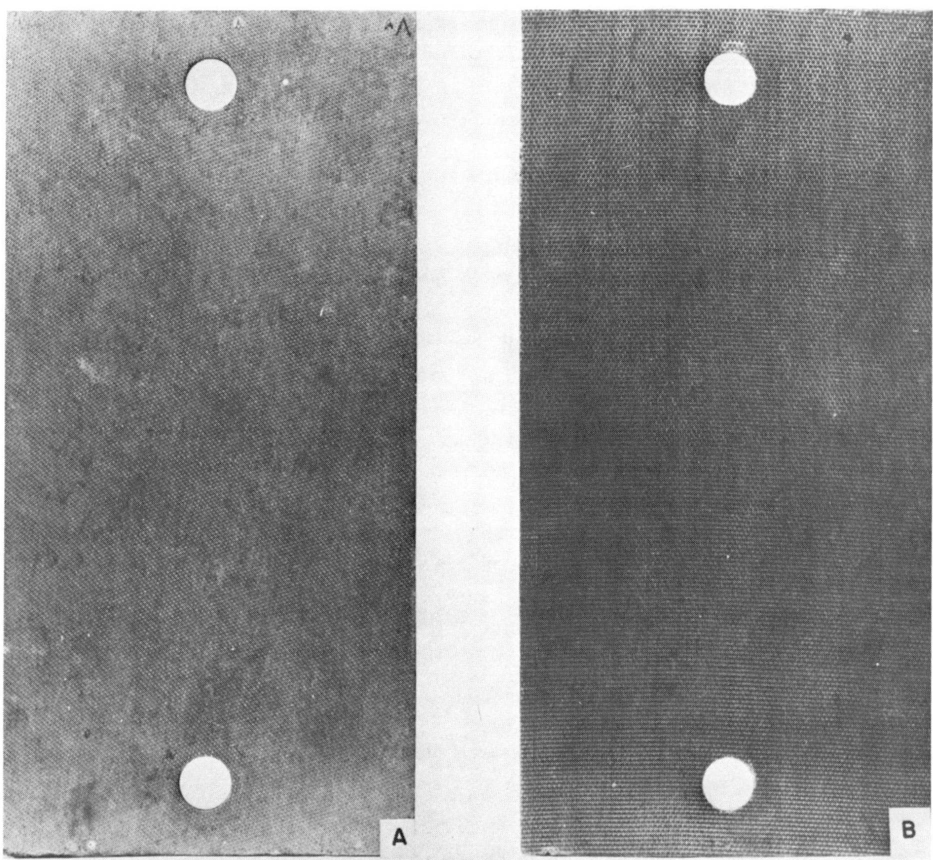

Figure 27.2. Panels in Fig. 27.1 after cleaning.

effected should be clean, dry, and maintained at a temperature of about 60–70° F (15–21° C). The resin system may not properly cure at lower temperatures and will cure much faster at elevated temperatures, thus reducing the allowable working time.

There are excellent technical brochures available from many of the basic materials manufacturers on maintenance and repair of laminates (e.g., the *Fiber Glass Repair Manual*, Ferro Corporation, and *Repair of Fiberglas Boats*, Owens/Corning Fiberglas Corp.).

27.5. DESIGN OF MARINE STRUCTURES

Structural design using glass-reinforced materials involves not only the design of the structure but also the design of the laminate itself. The designer must make some basic decisions as to the material strength requirements which are necessary to fulfill his objective. He must, therefore, have some general knowledge about the basic laminating materials, as well as fabrication and assembly techniques and the effect of variables on the laminate properties, so that he can determine that his design requirements are feasible and can be achieved within his cost limitations. His decision with regard to safety factors must take into consideration both the nature of loads (e.g., whether long term static or cyclic) and the environmental service conditions to be experienced. If the application is a new one, without the precedent of similar structures to draw upon, a prototype development program to optimize the design is in order. This is particularly necessary if the application is a critical one with regard to weight, safety, severe service abuse, or impact loads.

Fortunately, over the years, a considerable amount of information on materials and design performance has evolved which the designer of marine structures can draw upon. While design is not the subject of this chapter, some suggestions as to sources of such information specifically for marine applications will be indicated.

A general treatise on the design of glass-reinforced plastic structures has been developed by Gibbs and Cox;[15] although perhaps somewhat dated, this reference contains much general information and should prove useful. A more recent and more concise treatment is given by Scott.[42] Design data on joining of reinforced plastics and adhesive bonding are also available.[32,40] Secondary bonds require careful quality control and are highly specific to the type of joint, the adhesive, and the adherend, as well as the application conditions; the use of self-tapping screws has been found beneficial to limit peeling, particularly under cyclic or high impact loading conditions. The *Proceedings of the Annual Technical Conferences, Reinforced Plastics/Composites Division, SPI* provides an excellent source of up-to-date information on materials, design, and case histories for reinforced plastic structural applications. An example of the latter is a survey of the condition of U. S. Navy GRP boats in service for up to 15 years, which describes the performance of various design details.[17] There are also many excellent technical brochures available from the basic materials manufacturers.

The development of design data and criteria for reinforced plastics is still evolving as new materials or combinations of materials become available. Design techniques are also becoming more sophisticated through the use of computers and through better understanding of failure mechanisms.

27.6. APPLICATIONS

Reinforced plastic composites have been employed in a wide variety of structural and semistructural marine applications, and their use in such applications is constantly growing as design and service experience is acquired. From their earliest employment in small boats, 30 years ago, the use of glass-reinforced plastics has been extended to the construction of larger vessels, both commercial and military; a variety of shipboard structures, including fairings, deckhouses, masts, and tanks; and such other marine structures as floats and buoys.

A number of representative applications are discussed below to illustrate the versatility and wide applicability of glass-reinforced plastics as marine structural materials.

27.6.1. Boat Construction

By far the most important application of glass-reinforced plastics in marine structures, particularly with respect to volume consumed,[4,31,33,34] has been in boat construction, both civilian and military, and growth is continuing, as year after year glass-reinforced plastic construction increasingly dominates the small boat market. The "1977 Directory of Boats and Equipments" in *Boating* (*Boat Show Issue*) **41**, *No. 1* (January 1977) lists some 2400 boats of several types (inboard, stern drive, outboard, daysailers, and cruising sailboats) ranging in size from 16–81 ft (4.88–24.7 m). Of these, 2200 (92%) are constructed of fiberglass. The largest boats listed are a 70-ft (21.4-m) motor yacht and an 81-ft (24.7-m) cruising sailboat.

Glass-reinforced plastics are also being increasingly used in naval boat construction. The U. S. Navy pioneered in fiberglass boat construction with the production of a 28-ft (8.5-m) hull in 1947[44] and was influential in the use of hand lay-up procedures using woven roving[48] and fire retardant polyester resins[50] for boat hulls. Navy craft, which are classified as small boats, vary in size from 12-ft (3.7-m) wherries to a 57-ft (17.4-m) minesweeper hull and include 26-, 33-, and 40-ft (7.9-, 10.1-, and 12.2-m) personnel boats and utility boats 50 ft (15.3 m) in length. The British Navy launched the world's largest fiberglass ship, the 153-ft (46.7-m) H. M. S. Wilton, in 1972;[43] this ship was active in clearing the Suez Canal of mines and is the forerunner of a new class of minehunters now under construction. A consortium of the Netherlands, Belgian, West German, and

French navies is also developing a large minesweeper craft.

27.6.1.1. Fabrication Processes

Up to the early 1950's, the use of vacuum, pressure bag, or vacuum injection techniques was fairly common. The development of woven roving, thixotropic resin formulations and airless spray guns, however, facilitated hand lay-up and made this the most common boat hull fabrication technique in use today. Chopped strand, either in mat form or spray-up, is another construction method used in commercial production.

Boat construction can be broadly classified into two styles, single skin and sandwich. In the U. S., single skin predominates; it is generally cheaper and all the laminate is in the outside skin of the hull, where it can resist the local impact abuse encountered in service. Sandwich construction appears to be popular in Europe, where a polyvinyl chloride foam core, which is resilient rather than rigid, is often used and is said to offer advantages with regard to impact resistance.[38]

Single skin hulls are generally laid up in a female mold. Since the outer surface takes on the quality of the mold surface, further finishing is unnecessary except for light sanding if a paint is to be applied. Frames, foundations, and decking are installed in a secondary bonding operation; screw fasteners such as self-tapping screws are often used to apply pressure to the glue-line during bonding. They also act as crack stoppers and prevent delamination.

Sandwich skin hulls are often laid up over a male plug. Two different procedures may be followed. In the first, the so-called no-mold process, the core material is tacked to the plug, which is a lattice construction; the outer skin is laid up and cured; then the hull is taken off the plug and turned over, and the inner skin is laid up. In the second procedure, the inner skin is first laid up over the male mold; then the core material is bonded to the skin; and, finally, the outer skin is laid up on the core material. Proponents claim advantages for each method. Both, however, require considerable work to achieve a smooth outer surface. Provision must be made within the core for mounting various attachments, such as ribs, foundations, and deck fastenings.

There are many good sources of information for those interested in greater detail on boat hull construction and fabrication.[6, 7, 8, 9, 15, 17, 26, 28, 42, 43, 46, 52]

27.6.2. Naval Ship Construction—Large Surface Ships

27.6.2.1. Naval Construction

As mentioned previously, there is a trend toward construction of large fiberglass naval craft, principally for minesweepers. Fiberglass construction is particularly adaptable to minesweepers, not only because these materials are non-magnetic (a primary consideration), but also because they offer potential advantages of decreased weight and maintenance.

27.6.2.2. United Kingdom Program

The Royal Navy program on development of 150–200-ft (45.8–61-m) glass-reinforced plastic minesweepers was originally described by Henton.[26] The first design investigated consisted of a sandwich hull, decks, and bulkheads. The core was a unique, molded fiberglass, interlocking box type construction. Woven roving reinforcement and a room-temperature-curing isophthalic polyester resin were used for the skins. A full scale, 34-ft (10.4-m) long midship test section was laid up in a female mold—first the outer skin; then the box core was bonded to the outer skin; and then the inner skin. After structural and explosion tests, it was decided to go to single skin construction for the first ship, H. M. S. Wilton, as described by R. H. Dixon and C. S. Smith.[9, 43] The single skin hull is transversely framed (Fig. 27.3); the frames were secondarily laid up on the cured shell after a peel-strip of fiberglass cloth was removed. Closely spaced through-bolts were used to reinforce the frame-to-shell connection against shock loading. As a result of the experience with the Wilton, the British are now constructing a fleet of minehunters.

27.6.2.3. U. S. Program

The feasibility of using fiberglass for construction of 112-ft (34.2-m) MSI (inshore), 145-ft (44.2-m) MSC (coastal) and 189-ft (57.6-m)

Figure 27.3. A 34-ft midship section of a British minesweeper.

MSO (ocean) class minesweepers was established by the U. S. Navy in 1965[2]+1966.[45,47] As a result of these studies, a transversely stiffened, single skin hull design was decided upon. A materials test program indicated that by tightening up on quality control procedures and enforcing better workmanship, the strength properties of what has become standard Navy boat laiminate (woven roving reinforcement and fire retardant polyester resin) could be much improved. Using the British mold, a 34-ft (10.4-m) long midship test section hull was fabricated; transverse frames were secondarily laid up against the hull, and self-tapping screws were used to reinforce the connection against shock loads. The hull, outfitted with fiberglass decks, bulkheads, and machinery foundations and dummy loads to simulate engines and other equipment, was subjected to a series of tests to determine its suitability for U. S. Navy minesweepers.[28] According to Pohler,[35] the program was not continued beyond these tests due to a change in projected minesweeping operations making more extensive use of helicopters, and the subsequent lack of a requirement for new minesweepers.

27.6.2.4. Large Surface Ships

In addition to the large naval vessels previously described, there has been a continuing worldwide trend toward larger commercial and pleasure craft. Experience gained over the years in the production of glass-reinforced plastic boats, together with advances in boat building technology, processing methods, and materials, and a growing fund of service experience and design information, have provided a strong foundation for the construction of larger craft. Spaulding and Silvia[46] surveyed the design and construction of such larger vessels—60–120 ft (18.3–36.6 m) in length—and reported on activities in this direction in many countries. Among the craft described is a 60-ft (18.3-m) river launch constructed in France as early as 1958. This boat has seen continuous service in the Seine River. A 54-ft (16.5-m) fishing boat in Japan, a 79-ft (24.1-m) excursion boat in Sweden, 64-ft (19.5-m) and 67-ft (20.4-m) power boats in England, and an 80-ft (24.4-m) hydrofoil river passenger boat in the Soviet Union are other examples. A particularly interesting vessel is the 77-ft (23.5-m) pilot boat built in the Netherlands (Fig. 27.4). This boat has a double skin hull with a polyvinyl chloride core, designed to withstand the severe impact encountered in service. Two such craft have been in use since 1965 and have performed well. Two more have more recently been put into service.

Fiberglass boat construction has also been very active in the Republic of South Africa. Particularly noteworthy has been the development of fishing trawlers 74 ft (22.6 m) and 83 ft (25.3 m) in length. The 74-ft (22.6-m) trawler (Fig. 27.5) is of single skin construction, fabricated in a female mold with a combination mat-woven roving lay-up. A unique feature is the incorporation of integrally molded longitudinal stiffeners which give the hull a corrugated appearance. Excellent performance of the trawlers in service has resulted in considerable production and the extension of fiberglass construction to larger trawlers.

27.6.3. Fairings and Housings

Fairings are used in naval craft to "fair" the water flow around protrusions or discontinuities, and to minimize turbulence and provide a good hydrodynamic shape for optimum performance, much as "streamlining" accomplishes this in aerodynamics. A variety of such

Figure 27.4. A 77-ft Netherlands pilot boat. (*Courtesy Owens-Corning Fiberglas Corp.*)

Figure 27.5. A 74-ft Republic of South Africa fishing trawler. (*Courtesy Owens-Corning Fiberglas Corp.*)

fairings or housings is in use, from shrouds installed around propeller shaft couplings, to complete outer hulls for deep diving vessels (which have outboard flotation or control equipment which must be contained in an appropriate housing. Reinforced plastic materials are being used increasingly in such applications, as illustrated in the typical cases described below.

27.6.3.1. Submarine Fairwaters

Submarine fairwaters, or sails, are free flooding structures installed around the elements of the ship which protrude from the hull, to minimize turbulent flow.[18] These structures have generally been made with metallic materials, principally aluminum, to save weight. However, the use of aluminum fairwaters resulted in continuing operational difficulties due to electrolytic corrosion and maintenance problems. In order to obviate this, the Bureau of Ships, U. S. Navy (now the Naval Ship Systems Command), in 1952, turned to a consideration of an alternative fairwater construction and undertook the development of a glass-reinforced plastic structure, because of the strength, lightness, and corrosion resistance of these materials. Accordingly, the decision was made to install a reinforced plastic fairwater on the USS Halfbeak (SS-352). The construction and service history of this structure has been described by Fried and Graner.[13]

At the time of fabrication, the fairwater represented the largest structural application of fiberglass construction ever undertaken. The basic laminate was fabricated by a vacuum bag process, in nominal ¼-in. (6.35-mm) thickness, with style 181 Volan glass cloth and a general purpose polyester resin, blended with 10% of a resilient resin for increased toughness. The completed structure is shown, assembled on the Halfbeak, in Fig. 27.6.

This fairwater went through 11 years of arduous service, with excellent performance and very little maintenance, before it was removed and replaced with a structure of more modern design. At the time of its removal, the fairwater appeared to be entirely

Figure 27.6. Submarine fairwater on U. S. S. Halfbear.

serviceable. An investigation into the properties of the basic laminate showed that little change had occurred over the service period and that the material was still within specification limits (Table 27.4). Graner[18] indicates that fiberglass fairwaters of a new design have been installed on over 50 Guppy class submarines and are giving excellent performance.

27.6.3.2. Outer Hull Structures

Fiberglass outer hulls have been installed on such deep diving vehicles as Alvin,[51] among others, because of their good strength-to-weight characteristics and durability in the marine environment. A very exacting application of this type was developed for the Navy's DSRV (Deep Submergence Rescue Vehicle). This vehicle is extremely weight critical and therefore the outer hull must be as light as possible consistent with proper performance. For this reason, a high quality reinforced plastic was selected as a material of construction.

According to a recent description of fabrication details of the DSRV outer hull,[1,2] the structure was fabricated in seven sections, using glass cloth and roving preimpregnated with epoxy resin, with vacuum bag autoclave curing at 100 psi (0.69 MPa) at 325° F (163° C) for 4 hours, and post-curing in the mold at 350° F (177° C) for 10 hours. The laminate is required to be wrinkle free and to have a void content below 1.5% by volume. Thickness of the hull ranges from 0.264–0.50 in. (6.7–12.7 mm) and the overall length is 49 ft, 4 in. (15.06 m). The outer hull, which is free flooding, is required to operate at 3500 ft (1069 m) and is expected to provide a service life of 10–20 years.

27.6.3.3. Shipboard Structures

In addition to their use in boat hull construction, reinforced plastics have been applied in a variety of shipboard structures, generally to save weight or to eliminate corrosion problems inherent in the use of aluminum or other

Table 27.4. Properties of Glass-Reinforced Plastic Laminate in USS Halfbeak Fairwater

	CONDITION	ORIGINAL* DATA	DATA AFTER 11 YEARS OF SERVICE			SPECIFICATION REQUIREMENT**
			PANEL 1	PANEL 2	AVERAGE	
Flexural strength, psi (MPa)	Dry	52,400 (361)	51,900 (358)	51,900 (358)	51,900 (358)	50,000 (345)
	Wet[+]	54,300 (374)	46,400 (320)	47,300 (326)	46,900 (323)	45,000 (310)
Flexural modulus, psi $\times 10^6$ (GPa)	Dry	2.54 (17.5)	2.62 (18.1)	2.41 (16.6)	2.52 (17.3)	2.50 (17.2)
	Wet	2.49 (17.2)	2.45 (16.9)	2.28 (15.7)	2.37 (16.3)	2.30 (15.9)
Compressive strength, psi (MPa)	Dry		40,200 (277)	38,000 (262)	39,100 (269)	33,000 (227)
	Wet		35,900 (247)	35,200 (243)	35,600 (245)	28,000 (193)
Barcol hardness	Dry	55	53	50	52	
Specific gravity	Dry	1.68	1.69	1.66	1.68	
Resin content, % (by weight)	Dry	47.6	47.4	48.2	47.8	35–43

*Average of three panels.
***MIL-P-17549.*
[+]Two-hour boil.

metallic construction. Several of these applications are briefly described below by way of exemplification.

27.6.3.4. Masts and Spars

Figure 27.7 shows an 89-ft (27.1-m) reinforced plastic mast mounted aboard a naval vessel. A number of such structures of various designs were developed by the U. S. Navy.[18] The mast shown was fabricated with a woven roving reinforced polyester construction and was made in two halves and assembled with a longitudinal seam, but other fabrication processes, including mandrel wrapping and filament winding, have been used. Cobb[6] indicates that fiberglass materials have been used in commercial practice for the construction of booms, masts, and spinnaker poles for some years; however, the use of glass-reinforced plastics in these applications has not grown, due to lack of standardization in design and limitations in stiffness of the material. With proper development, there is reason to believe that fiberglass could become the preferred material for construction of masts and spars (over wood and aluminum) because of light weight, durability, and superior performance.[6]

Figure 27.7. An 89-ft (27.1-m) mast for a communications ship.

27.6.3.5. Deckhouses

A glass-reinforced polyester deckhouse is shown in Fig. 27.8. This structure was designed as an alternative to aluminum construction, and was fabricated completely in fiberglass, including internal structural elements and stiffeners. The deckhouse was designed to withstand severe shock loading, and has performed well in comparison with an aluminum counterpart. Fiberglass deckhouses of this and other types are still under development in the Navy and are also being used in commercial craft.[52] Duffin,[10] for example, describes a 188-ft (57.3-m) fiberglass superstructure installed on a mine-planter hull which was converted to a luxury yacht. This construction included a 50-ft (15.2-m) bridge housing, as well as a radio room and even a stack.

27.6.3.6. Tanks

Lightweight, corrosion resistant reinforced plastic tanks, for storage of both fuel and potable water, are used extensively in shipboard application. These tanks are generally fabricated with glass-reinforced polyester materials and may be of standardized shape or custom made to fit a particular contour. Special attention is given to the elimination of voids or pinholes to prevent leakage. For water storage, the laminate must be completely cured to assure inertness and to prevent the development of taste problems. Fiberglass materials are generally more resistant to organic fuels than to water and are therefore even more applicable to construction of fuel tanks. It is notable, however, that some polyester systems may be somewhat sensitive to certain fuels, particularly if aromatic compounds are present. It is apparent that the

Figure 27.8. Destroyer deckhouse.

resin system used in the construction of both water and fuel tanks should be selected judiciously, following the recommendations of the resin supplier. A typical fuel storage tank is shown in Fig. 27.9.

27.6.3.7. Other Shipboard Structures—Sonar Domes

Reinforced plastics have also been used in shipboard ventilation ducting and piping systems, in the fabrication of reefer boxes and hatch covers, and in stacks.[52] More extensive, perhaps, has been the use of reinforced plastics in Sonar domes[18] and radomes. The successful application of these materials to radome construction, not only for shipboard, but for airborne and land installations as well, is discussed elsewhere in this volume (Chapter 28) and will not be dwelt on here. Reinforced plastic Sonar domes, on the other hand, although in use for some time, are still undergoing development. Reinforced plastics, in addition to meeting weight requirements, offer the advantages of smooth outer surfaces and monolithic construction, with a minimum of interfering internal stiffening structure. Sonar domes are fabricated of high quality laminate having a high glass content and a low percentage of voids. A typical structure, installed aboard a submarine, is shown in Fig. 27.10.

27.6.3.8. Floats and Buoys

The use of a reinforced plastics for the construction of floats and buoys would appear, in the first instance, to be based on durability and ease of maintenance in the marine environ-

Figure 27.9. A 2850-gal (10,788-l) fuel storage tank.

Figure 27.10. Sonar dome on a submarine.

ment. However, such structures are frequently subjected to arduous service conditions where strength and toughness are important considerations. A case in point is the Navy minesweep float, one model of which is shown in Fig. 27.11. This particular float was developed in reinforced plastics primarily because the material is non-magnetic, and was designed as a replacement for stainless steel construction. Specifications included restrictive weight limitations and a severe impact resistance requirement involving a drop test on the nose from a height of 10 ft (3.05 m). The fiberglass float was originally made by fabri-

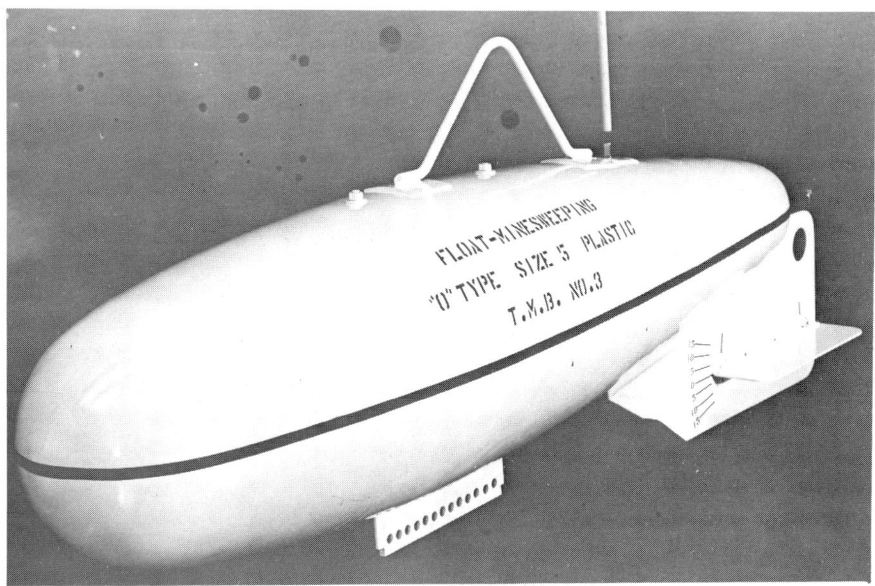

Figure 27.11. A minesweep float.

cating two half-shells and assembling with adhesive bonding and mechanical fasteners along a longitudinal seam. The shells were fabricated with style 1000 Volan glass cloth and polyester resin, in a female mold, by a vacuum bag process. Before assembly, the entire interior volume was filled with a low density unicellular plastic foam to make the float practically unsinkable even if damaged.

The fiberglass float was lighter than its stainless steel counterpart and successfully withstood the 10-ft (3.05-m) drop test on its nose, a test which severely damaged the steel float. Floats of this type have performed well in service tests and are currently being used in the Fleet.

27.6.3.9. Protective Coatings

In addition to structural and semistructural applications, reinforced plastic materials are also used as coatings to protect substrate materials from degradation in the marine environment. The process of "fiberglassing" wooden hulls is now widely used, even reaching into the do-it-yourself market. In this process, as described, for example, by Cobb,[6] the dried, prepared, outer surface of the wooden hull is covered with one or more plies of reinforced plastics. A common process employs a room-temperature-curing polyester resin formulation and style 1000 glass cloth or equivalent ("boat cloth").

Fiberglassing is commonly done on older wooden boat hulls but it is used in new construction as well.[55] Fiberglass is also used for covering wooden decks, having largely replaced canvas for this application.[6] Variations in the process include the use of alternative reinforcements and resins. In the "Cascover" process, for example, the covering consists of a nylon fabric impregnated with resorcinol resin.[3] In another interesting application, glass-reinforced plastics are being used to protect outboard propeller shafting against corrosion. This is a serious problem in oceangoing vessels, particularly where steel shafting is used with propellers made of nonferrous metals such as bronze. Previously, shafting had been protected with rubber coatings, but these coverings are expensive to install and have not always performed well in service. The alternative coatings of glass-reinforced plastics which were developed by the U. S. Navy have proven to be less expensive to install and have performed well in service. Such coatings are now being widely used on commercial vessels, with like success.

A manual for coating propeller shafting with reinforced plastics, developed by the Navy, describes the process in detail.[16] Essentially, the procedure involves the preparation of the shafting to assure removal of oil or loose rust, and the application of four plies of resin-impregnated glass fabric tape, with proper working of each ply to assure elimination of air and complete resin impregnation (Fig. 27.12). Room-temperature-curing resins are employed. For the most part, epoxy resin systems have been used, but polyester resins would appear to be applicable as well. Service tests on Navy vessels over a period of years indicate excellent performance and a minimum

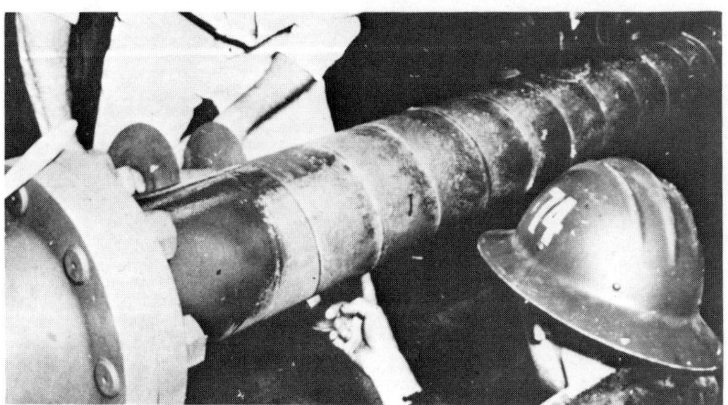

Figure 27.12. Application of a glass-reinforced plastic propeller shaft covering.

of maintenance problems, if coatings are properly applied.

27.7. CURRENT AND FUTURE DEVELOPMENT

The use of reinforcements other than fiberglass (e.g., boron, graphite, or aramid) has created a new group of reinforced plastic composite materials often referred to as advanced composites. By virtue of their light weight and superior mechanical properties, these materials have opened up new, primary structural application areas which were previously not accessible to fiberglass composites because of their relatively low modulus. These new materials are already being widely exploited in aircraft, missile, and space vehicle applications where weight is a critical consideration.[22] In marine service, however, the application of these materials has proceeded rather slowly, primarily because of cost considerations, but also because the driving force of weight savings is not as compelling as in aerospace. The U. S. Navy, however, is investigating the applicability of advanced composites to high performance craft.

27.7.1. Advanced Composites

Dr. D. A. Jewell describes many types of novel marine craft, such as the hydrofoil, surface effect ship (SES) (Figs. 27.13 and 27.14), and small waterplane area twin hull (SWATH), which are being looked at for the U. S. Navy of tomorrow.[27] All of these have a common design limitation—they are weight sensitive if their full operational capability is to be attained. Grumman Aerospace, for example, in an investigation of hydrofoil craft between 80- and 2000-ton (88- and 2205-metric ton) displacement, concluded that weight savings due to the use of advanced composites could provide significant improvement in mission payload capability (fuel and armaments); the initial cost increase in certain instances could be compensated for by lower life-cycle operating costs or negated by the vastly increased payload capacity.[25]

Figure 27.13. A U. S. Navy hydrofoil craft.

Advanced composites may be defined as plastic laminates which are reinforced with fibers having higher Young's modulus than fiberglass; they may be used alone or in combination with fiberglass (to reduce cost) or in any combination to obtain the desired strength properties. Table 27.5 shows the range of properties for such fiber materials. As can be seen, a large number of different graphite fibers are available with a wide range

Figure 27.14. A U. S. Navy surface effect ship (SES).

Table 27.5. Range of Properties of Fiber Materials

FIBER MATERIAL	VARIETY OF FIBERS AVAILABLE	RANGE OF PROPERTIES: TENSILE STRENGTH, ksi (MPa)	YOUNG'S MODULUS, msi (GPa)	DENSITY, lb/in.^{3} (g/cc)	COST,* \$/lb (\$/kg)
Glass	4	500–600	10–12	0.092	0.7–6
		(3400–4100)	(69–83)	(2.549)	(1.54–13.2)
Boron	2	400–500	55	0.092–0.098	180–225
		(2700–3400)	(379)	(2.549–2.715)	(396–495)
PRD49-111	3	400–430	12–19	0.05	20
		(2700–2900)	(83–130)	(1.385)	(45)
Graphite	26	200–470	20–75	0.054–0.071	20–200
		(1400–3200)	(138–517)	(1.496–1.967)	(45–450)

*For 100-lb (45.36-kg) quantities in 1981.

of moduli. Aramid, also known as Kevlar 49, is a polyamide fiber proprietary to DuPont; it is now being evaluated commercially for small boat construction. Boron has found application in many aerospace structures.[22] The costs indicated should be considered more relative than actual, since they are greatly affected by quantity requirements and availability. It is projected that the cost of some graphite fibers will be significantly reduced in the future because of innovations in fiber drawing and increased production.

Table 27.6 indicates some typical properties of bidirectional laminates. The laminates are most often fabricated using epoxy resin prepregs with vacuum bag or autoclave pressure. It can be seen that a wide variety of strengths, stiffnesses, and densities are available.

Greszczuk[20,21,22,23] has studied the application of advanced composites, specifically to the 100-ft (30.48-m) Patrol Craft Hydrofoil (PCH-1). Several advanced composite materials and structural details were considered, and he concluded that the application of advanced composites to the hull and decking of a hydrofoil could result in a weight savings of 16–51%, whereas the application of struts and foils (Fig. 27.15) shows potential weight

Table 27.6. Typical Properties of Bidirectional Composites

COMPOSITE	STRENGTH, ksi (MPa): TENSILE	STRENGTH, ksi (MPa): COMPRESSIVE	YOUNG'S MODULUS, msi (GPa)	DENSITY, lb/in.^{3} (g/cc)	PREPREG COST (1972) \$/lb (\$/kg)
S-glass	150	62	5.6	0.075	12
	(1030)	(427)	(38.6)	(2.08)	(26.4)
PRD49-111	94	22	5.9	0.050	75
	(64.8)	(152)	(40.7)	(1.39)	(165)
Boron	113	186	16.7	0.075	285
	(779)	(1280)	(115.1)	(2.08)	(627)
High strength graphite*	91–132	85–108	9.7–11.7	0.054–0.057	75–205
	(627–910)	(586–745)	(66.9–80.7)	(1.50–1.58)	(165–451)
High E-graphite**	62–74	52–67	15.6–17	0.054–0.058	145–205
	(427–510)	(359–462)	(107.5–117.25)	(1.50–1.61)	(319–451)
Thornel 75	109	49	22.4	0.057	275
	(752)	(338)	(154.4)	(1.579)	(605)
GY 70	46	46	21.4	0.061	75
	(317)	(317)	(147.5)	(1.690)	(165)

*Includes Thornel 300, Thornel 400, Courtaulds Hts. A, Modmor II.
**Includes Thornel 50, Courtaulds Hms, Modmor I.

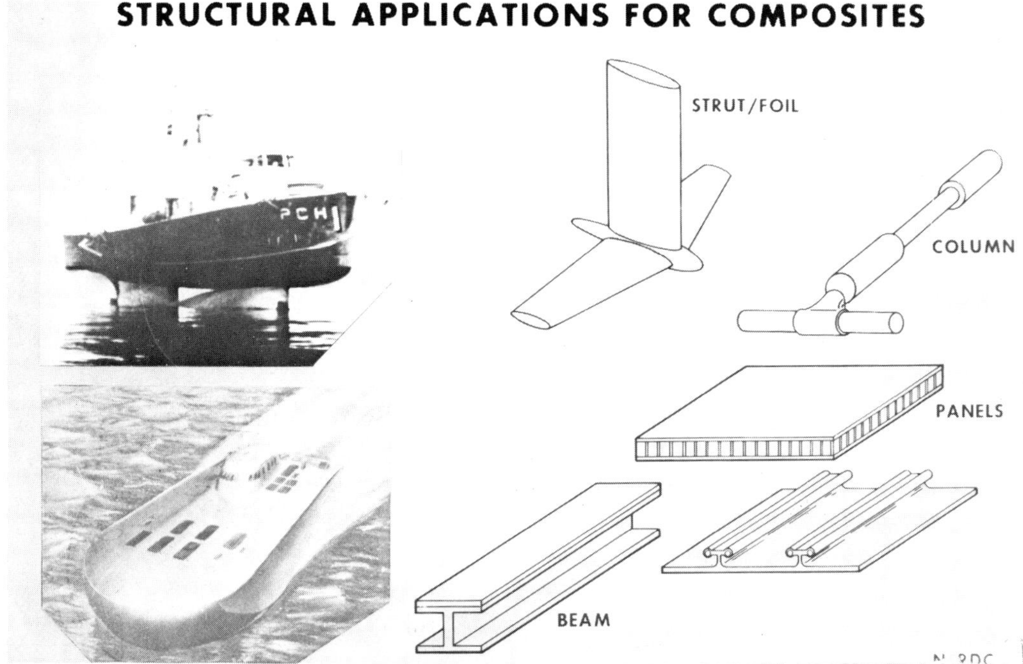

Figure 27.15. PCH-1 strut/foil and other structures for high performance craft.

savings of approximately 60%, compared to steel counterparts. Based on this study, the Navy decided to continue its evaluation of graphite composites. McDonnell Douglas Astronautics has fabricated composite box beams for test by the David Taylor Naval Ship Research and Development Center (DTNSRDC). These box beams represent a typical structural detail of the foil. They will be tested as cantilever beams in contact with seawater under cyclic loading conditions similar to those encountered in service; their performance will be compared to beams fabricated of high strength steel and titanium. Boeing is also developing and evaluating a small graphite foil control flap, which, it is anticipated, will be service tested on the PCH-1. In-house laboratory programs are being conducted at DTNSRDC and the Naval Research Laboratory (NRL) to characterize the response of various graphite composites to the severe service conditions experienced in naval applications, such as long term exposure to seawater, long term static and cyclic fatigue, impact and local abuse, and fire. Recent information [56] indicates that graphite composites are probably superior to fiberglass composites in regard to resistance to cyclic fatigue. Impact resistance[24] is of great concern, and ways of mitigating this obvious short coming of the graphite composite structure will be investigated. The effect of large scale shipboard fires on structural integrity of aluminum structures has already come under considerable scrutiny[54] and must be considered for any new structural materials. Needless to say, the results of these ongoing evaluations in the next couple of years will determine the future of advanced composites in the Navy's high performance craft.

27.8. SUMMARY

In this chapter, an attempt has been made to very briefly describe basic materials, processes, and responses of composite laminate materials, particularly related to marine application and environment. Because each of these is a very broad and complex subject, it was impossible to treat them in very great detail. A large number of references are cited from which more detailed information in specific areas of interest may be obtained.

From the materials viewpoint, there are

some obvious areas in which technological improvement would further the cause of composite materials. Some examples are improved shear strength, truly fire resistant resins, high temperature resistant resins which can be easily handled and cured, and fool-proof, reliable adhesive bonding systems suitable for shipyard use. Considering fabrication processes, greater improvement is needed in regard to automation for construction of large boat hulls and other applications which would lead to cheaper, higher quality structures with improved reliability. Related to this, improvements in quality control and inspection techniques are desirable. Better underwater repair procedures are needed to take care of damage to large boat hulls which cannot readily be drydocked. Also, integral, longer lasting antifoul coatings would reduce maintenance costs. On a more basic level, better understanding of the mechanics of reinforcement and failure could contribute to significant improvement in laminate performance.

While the use of fiberglass composites in marine structural applications has increased over the years, it has proceeded at a relatively slow pace. This has partly been because of a general lack of knowledge or familiarity on the part of the naval designer with the properties of, and design criteria for, composites. Also, there has been an understandable reluctance on the part of both the designer and the shipbuilder to become involved with a material so different in handling characteristics than conventional metal: it is not ductile; it cannot be welded; and its design must take into consideration the basic materials as well as the fabrication process, long term environmental effects, etc. However, experience has shown that composites, properly used, offer significant potential in terms of reduced cost and weight, improved performance, greater durability, reduced maintenance, and longer service life. There is today a considerable body of literature and experience on which the naval designer can draw. With the development of advanced composites, there are materials available which on a specific strength-to-stiffness basis cannot be approached by any metals now existing or any which may be available in the foreseeable future. The next 20–30 years may well develop into the Composite Age!

REFERENCES

1. *Aerospace Technology* **21:** 39 (November 6, 1967).
2. Breidenbach, L., "Advances in Structural Composites," *Science of Advanced Materials and Process Engineering Series* **12,** SAMPE Section DS-2, 1967.
3. *Brit. Plastics* **38:** 55 (April 1965).
4. *Brit. Plastics* **40:** 50 (August 1967).
5. Chance, C. *et al.*, 27th SPI Conference, Section 16-B, 1972.
6. Cobb, B., Jr., *Fiberglass Boats—Construction and Maintenance*, 2nd Edition, Yachting Publishing Corp., New York, 1967.
7. DellaRocca, R. and Scott, R., 6th SAMPE Technical Conference, October 8–10, 1974.
8. DellaRocca, R. and Scott, R., *Engineering Applications of Composites* **3,** Academic Press, New York, London (1974).
9. Dixon, R. H. *et al.*, *Proceedings of Symposium on GRP Ship Construction*, RINA, London, October 1972.
10. Duffin, D. J., *Laminated Plastics*, 2nd Edition, Reinhold Book Corp., New York, 1966.
11. Dyckman, E. J. et. al., *Naval Engineers Journal* (April 1974).
12. Fried, N., *International Conference on the Mechanics of Composite Materials*, Office of Naval Research and Missiles and Space Division, General Electric, May 1967.
13. Fried, N. and Graner, W. R., *Marine Technology* **3:** 321 (1966).
14. Fried, N. and Winans, R. R., "Symposium on Standards for Filament Wound Reinforced Plastics," *Special Technical Publication Number 27, ASTM* **83** (1963).
15. Gibbs and Cox, *Marine Design Manual for Fiberglass Reinforced Plastics*, McGraw-Hill, New York, 1960.
16. Goldfarb, P. M. and Fried, N., "Instruction Manual—Glass Reinforced Plastic Coatings for Propeller Shafting," Navships 250-634-4, Department of the Navy, January 1964.
17. Graner, W. R. and DellaRocca, R. J., "Evaluation of U. S. Navy GRP Boats for Material Durability," 26th SPI Conference, Section 7-F, 1971.
18. Graner, W. R., *Ocean Engineering* **1:** 353–372, Pergammon Press, New York, 1969.
19. Graner, W. R. and Stander, M., U. S. Patent 3,154,460.
20. Greszczuk, L. B. and Hawley, A. V., "Application of Advanced Composites to Patrol Craft Hydrofoils," Final Report, U. S. Naval Ship System Command, Contract N00024-78-C-5536, April 1973.
21. Greszczuk, L. B. and Hawley, A. V., "Application of Advanced Composites to Hydrofoil Strut," Final Report, U. S. Naval Ship System Command, Contract N00024-78-C-5536, December 1973.

22. Greszczuk, L. B. *et al.*, MDAC Paper WD 2295, AIAA/NAME Advanced Marine Vehicle Conference, San Diego, California, February 25–27, 1974.
23. Greszczuk, L. B. *et al.*, *Journal of Hydronautics* **9,** *No. 3* (July 1975).
24. Greszczuk, L. B. and Chao, H., MDAC Paper WD 2645, ASTM 4th Conference on Composite Materials: Testing and Design, May 1976.
25. Grumman Aerospace Corp., "Investigation of Composite Material for Use in Lightweight Hydrofoil Structures," U. S. Naval Ship System Command, Contract N00024-74-C-5048, September 30, 1974.
26. Henton, D., *Marine Systems* (London) **2:** 52 (May/June 1967).
27. Jewell, D. A., *Naval Research Reviews* **XXIX,** *No. 10* (October 1976).
28. Lankford, B. W. and Angerer, J. F., *Naval Engineers Journal,* **83,** *No. 5,* (October 1971).
29. *MIL-HDBK-17*, "Plastics for Flight Vehicles, Part I—Reinforced Plastics," November 5, 1959.
30. Montemarano, J. A. *et al.*, "Anti-Fouling Glass Reinforced Composite Materials," Report MAT-75-33, David W. Taylor Naval Ship Research and Development Center, January 1976.
31. Parducci, M., *Materie Plastiche ed Elastomeri* **32:** 372 (April 1966).
32. Perry, H. A., *Adhesive Bonding of Reinforced Plastics*, McGraw-Hill, New York, 1959.
33. *Plastics World* **24:** 30 (March 1966).
34. *Plastverarbeiter* **17:** 319 (May 1966).
35. Pohler *et al., Naval Engineers Journal* **87,** *No. 2* (April 1975).
36. Rawe, A. W., *Journal of the Plastics Institute* (London) *Trans. J.* **30:** 27 (1962).
37. *Reinforced Plastics* (London) **4:** 12 (September/October 1965).
38. *Reinforced Plastics* (London) **10:** 280 (May/June 1966).
39. Reitman, H. E., *Modern Plastics* **44:** 141 (December 1966).
40. Rufolo, A., "Design Manual for Joining Glass Reinforced Structural Plastics," Navships 250-634-1, 1961.
41. Rugger, G. R. and Titus, J. B., "Weathering of Glass Reinforced Plastics," Plastec Report 24, Plastics Technical Evaluation Center, Picatinny Arsenal, Dover, New Jersey, January 1966.
42. Scott, R. J., *Fiberglass Boat Design and Construction*, John De Graff, 1973.
43. Smith, C. S., "Applications of Fiber Reinforced Composites in Marine Technology," *Conference Proceedings, Composites—Standards Testing and Design*, IPC Science and Technology Press, England, April 8–9, 1974.
44. Spaulding, K. B., Jr., *Naval Engineers Journal* **78:** 333 (April 1966).
45. Spaulding, K. B., Jr. and DellaRocca, R. J., *Transactions SNAME* **73:** 415 (1965).
46. Spaulding, K. B., Jr. and Silvia, P. A., 22nd SPI Conference, Section 11-A, 1967.
47. Spaulding, K. B., Jr. and Silvia, P. A., *Materials Design Engineering* **65:** 22 (March 1967).
48. Specification MIL-C-19663, "Cloth, Glass, Woven Roving, for Plastic Laminates."
49. Specification MIL-P-17549, "Plastic Laminates Fibrous Glass Reinforced, Marine Structural."
50. Specification MIL-R-21607, "Resins, Polyester, Low Pressure Laminating, Fire Resistant."
51. Tangerman, E. J., *Product Engineering* **37:** 84 (March 14, 1966).
52. *Technical and Research Bulletin Number 2-12, Society of Naval Architects and Marine Engineers,* "Guide for the Selection of Fiberglass Reinforced Plastics for Marine Structures," March 1965.
53. Winans, R. R., Fried, N., and Hand, W., Electrical Manufacturing **56:** 106 (July 1955).
54. Winer, A. and Butler, F., *Naval Engineers Journal* (December 1975).
55. *Wood and Wood Products* **72:** 38 (March 1967).
56. Hofer, K. E., Jr., Stander, M., and Bennett, L. C., "Degradation and Enhancement of the Fatigue Behaviour of Glass/Graphite/Epoxy Hybrid Composites after Accelerated Aging," 32nd SPI Conference, Section 11-F, 1977.

The contributions of the late Nathan Fried (U.S. Naval Applied Science Laboratory) in the original draft of this chapter are acknowledged.

28
AEROSPACE APPLICATIONS OF COMPOSITES

George Lubin
Consultant
Samuel J. Dastin
Grumman Aerospace Corp.
Bethpage, New York

28.1. INTRODUCTION

While the use of composites in aerospace presently constitutes a relatively small percentage of total composite use, composite materials find some of their most sophisticated and glamorous applications in this industry. It may now be said that composites are a production reality as a substitute material, within the next 10 years, new aircraft structures will be at least 40% composites.

In many of the present and future applications of composites, the high performance or "advanced composites" provide the most spectacular applications, but fiberglass, either alone or in hybrid structures, still provides most of the volume.

In aerospace, the demands upon materials are usually greater than in other applications. The four most important requirements are light weight, high strength, high stiffness, and good fatigue resistance. Composites, particularly the high performance types, are the only existing materials efficiently meeting these requirements. The specific tensile strength of a graphite composite is about 3.6×10^6 in. (9.2×10^6 cm), as compared to aluminum, 0.8×10^6 in. (2.0×10^6 cm). The specific moduli are, respectively, 3.3×10^8 in. (8.4×10^8 cm) and 1.0×10^8 in (2.54×10^8 cm). The fatigue endurance limit of graphite is close to 80% of static strength, as compared to 35% for aluminum.

The use of composites in aircraft is increasing rapidly, especially in military aircraft, where the pay-off is the greatest. The potential applications are limited only by the present decrease in new military aircraft. In commercial aircraft, the acceptance of composites as primary structures has been slower, but is now increasing rapidly. In spacecraft, where the weight is of the greatest importance, the composites are accepted as primary materials.

The main disadvantages of present composites are the lack of extensive performance history and the temperature limitations of moisture-saturated graphite composites. It appears that most of the new advances in these materials have been in the fiber reinforcement of the composite, while the matrix (resins) component has not yet progressed to the stage at which the full potential of these materials can be utilized.

28.2. PROGRESS

Since inception, aircraft have required the use of plastics to operate. The wings of the earliest aircraft were covered with cellulose nitrate-doped fabrics, and phenolic resin woods were used for structural parts. The Wright brothers' lurch into the air at Kittyhawk in 1903 was a collection of wood, wire, and fabric representing the state-of-the-art in aeronautical science and aircraft materials development. Today, the Wright brothers' structure is considered primitive and flimsy. The structural weight of this pioneer aircraft was approximately 450 lb (200 kg). By contrast the world's largest aircraft, built in 1968, Lockheed's C-5A Galaxy, has a structural weight of some 318,000 lb (144,000 kg). Wood and fabric remained the principal building blocks of aircraft for nearly 20 years after the Wright brothers' first flight. Although there were minor advances in materials technology, it seemed as though designers were reluctant to try any new materials.

The Lockheed Vega, designed in 1923, was probably the first successful departure from fabric-covered aircraft. Its structure was all wood, using a pressure-molding process to form cross-plied cover panels. A distinct advantage of the famous monoplane was that it did not require a new fabric covering after two or three years use.

In the early 1920's, Junkers Aircraft developed an all-aluminum airplane. It took another 15 years before the all-metal airframe became common. Following the way of mankind with most progress, the controversy over all-metal construction was marked by retreats back to wood and fabric. As late as World War II, with the British Mosquito Bomber and the design of the giant Hughes flying boat, wood was a major material in many aircraft concepts. The tendency to resist change during this period is demonstrated by wood-oriented designs for early steel and aluminum airframes.

It took foresight to understand and apply the knowledge that the use of metals in aircraft construction required new approaches and design philosophies. This same barrier existed in applying the relatively new and exotic structural plastics. Once designers assessed the capabilities of metals they realized that their new concepts would have to take full advantage of higher strength materials to offset weight inherent in denser structures. As shown in Table 28.1, in 1944 a reinforced plastics (RP) sandwich aft monoque fuselage was flown, showing that RP could be used (cost competition was the principal deterrent against its growth).

Milestones such as the development of acrylic for glazing, polyethylene for radar cable insulation, glass fiber-reinforced polyester resin laminates for radomes, adhesive bonded structures, and special elastomers for tires occurred during World War II. An example of a present-day development is the use of FEP-fluorocarbon/nylon wire insulation replacing polyvinyl chloride/glass/nylon in certain areas; this saves 10% weight, requires only two-thirds the space, and permits continuous service at 250° F (120° C).

28.3. Early Aircraft Applications

Military aircraft were probably the first to use fiberglass composites in significant quantities, around 1940. The first applications were in radomes, then in secondary structures, and later in interior parts. A fiberglass-reinforced plastic (FRP) fuselage was constructed and flown at Wright Field Air Force Base in 1944. FRP wings were flown on an AT-6 in 1953. Since that time, FRP has been used extensively

Table 28.1. Primary Structural Data Analysis on First Aircraft (BT-15) to "Fly" with Reinforced Plastics*

TYPE OF STRUCTURE	WEIGHT, lb (kg)	STRUCTURAL TEST LOAD, % OF ULTIMATE	STRUCTURAL TO WEIGHT EFFICIENCY RATIO, %
Aluminum	70 (32)	108	100
Wood	86 (39)	110	82
RP Sandwich, glass fabric-polyester resin			
Balsawood core	78 (35)	180	150
RP honeycomb core	75 (34)	180	155

*RP sandwich aft monoque fuselage first flew March 24, 1944.

for radomes, rudders, flaps, and fairings.[1] However, reliability and quality control considerations and the low stiffness and compression strength of those materials have generally limited their use to secondary structural applications for military aircraft, though they have been used successfully for the complete airframe of two light civil aircraft, the Piper Cub and the Eagle I by the Windecker Research Inc.[2] (see Fig. 28.1).

It appeared that the selected high-speed production processes can reduce the initial high cost of hand fabrication methods. The fiberglass composite airframe structures promise advantages over commonly used metals including an excellent strength-to-weight ratio, fatigue resistance, superior surface finish, freedom of aerodynamic design and contouring, unusual dimensional stability, material uniformity, simplified engineering design, and fabrication simplicity; they also offer high resistance to abrasion, dents, rain erosion, and corrosion.

The rapid introduction of fiberglass reinforcements in the 1940's created a new technology—that of composite materials. The necessary design allowables and performance characteristics of such composites were generally reduced for aircraft, since the material's modulus is low compared to metals, and is thus not suitable for use in highly loaded structures. In 1966, with the introduction of boron fibers, under Air Force sponsorship, there was a quantum increase in interest in composite structures, and with funds available for study, new data and new design techniques were developed, enabling advanced composites to become practical materials for aircraft construction. This new science also spilled over to fiberglass, greatly optimizing its design. Whereas previous fiberglass components were commonly designed to 400–500% over ultimate strength, now, with the new computerized technology, the margins of safety have been drastically reduced to more practical values resulting in optimum weight savings and cost reductions.

Until recently, composites have only been substituted into products originally designed for isotropic materials. The full potential of anisotropic composites can only be exploited when their use is taken into consideration in the preliminary design stage.

Because the weight of an aircraft has a direct bearing on its performance (range, payload, fuel consumption, and many other interrelated factors all depend on weight), a value can be set against any weight saved. The limiting value for this rate—or the maximum amount the customer is prepared to spend per pound of weight saved—can usually be established at a very early stage in the design of the aircraft. This philosophy applies to both civil and military aircraft, although the basis for calculation is different in each case.

For a small, conventional civil airplane which will probably seldom fly with either a full load or over its maximum range, the value to the operator of weight saved is not likely to be more than \$25/lb (\$55/kg). However, this figure can increase by a factor of 10 or more for sophisticated aircraft such as supersonic

Figure 28.1. Windecker Eagle I—all fiberglass aircraft. Photo courtesy Composite Aircraft Corporation.

transports or vertical take-off aircraft, where the ratio of payload to aircraft weight is quite critical. It is in relation to this widely varying pay-off rate that the cost of using new materials and the benefits to be obtained must be considered. Figure 28-2 shows the value of a unit weight (lb or kg) saved for a number of aircraft and spacecraft.[3,4,5] These values are approximate and may vary considerably, depending upon the mission or the stage of design. For each new aircraft or spacecraft, there are usually very tight weight/performance goals to meet for the manufacturer at the design stage. At this stage, the cost per unit weight saved may be very high to enable the designer to meet the requirements. For a military aircraft, cost at this stage may approach \$200/lb (\$440/kg). After the design is completed, this cost may drop to \$50/lb (\$110/kg). When the aircraft is in production, any additional weight savings can only be considered on the advantages to be gained for the total life cycle of the plane—primarily higher load or reduced fuel consumption. Here the value can drop to as low as \$25–10/lb (\$55-11/kg).

In space applications, where the cost per unit weight is extremely high because of the large amount of fuel required to lift the vehicle into a space orbit, the cost savings based on the use of composites are very attractive.

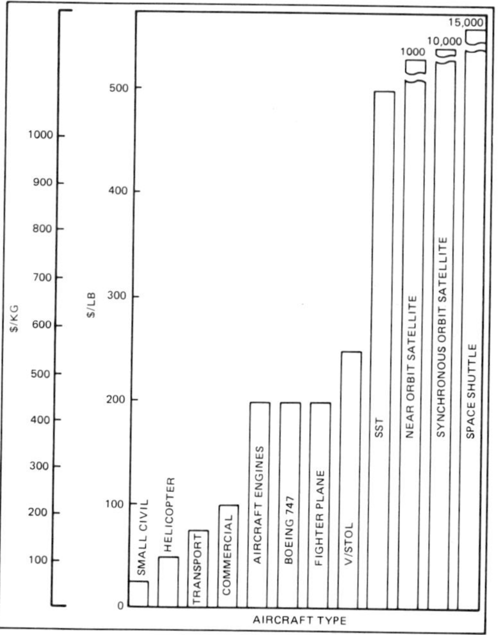

Figure 28.2. Value of weight saved in aircraft and spacecraft.

28.4. COMPOSITE PROPERTIES USED FOR AIRCRAFT DESIGN

There are numerous tables included in this book listing the various properties of composites. A summary of the mechanical properties can be found in Table 28.2 for the most commonly used aircraft materials. A list of aircraft-oriented forms of materials is shown in Table 28.3. Where the weight is of primary importance, it must be noted that glass fiber composites have a specific strength five to six times that of aluminum, the main competitive material. Where stiffness is required, note that graphite and boron/epoxy composites have about five times higher specific stiffness. Aramid fiber composites are used either alone or as hybrids with glass or graphite, and offer the greatest weight savings with an intermediate stiffness.

Additional data required for detailed design can be found in the design chapter. For further sources, the *Aircraft Design Manual* or *Military Handbook-17* are recommended. The quality of the available design allowables is reflected in the reports on the static and fatigue tests of all of the present high performance aircraft component parts designs. The average part failure is planned to occur at around 125% of design ultimate value. The failure commonly occurs within 10% of this value. For properly fabricated parts, the fatigue resistance usually exceeds two to four lifetimes. Even the higher fatigue requirements of commercial aircraft are met and exceeded.

28.5. EARLY ADVANCED COMPOSITE AIRCRAFT APPLICATIONS

Some of the first advanced composite structures include the C-141 wing tip by Lockheed, a boron/epoxy wing tip panel which became the first FAA-certified boron part.[6]

The A6-A wing fence fabricated by Grumman was an early boron/epoxy part to be

Table 28.2. Comparison of Mechanical Properties of Some Structural Materials

MATERIAL	YOUNG'S MODULUS, E, psi × 10^6 (GPa)	ULTIMATE TENSILE STRENGTH, σ, psi × 10^3 (MPa)	DENSITY, ρ, lb/in.^{3} (kg × 10^3/m^3)	SPECIFIC MODULUS, E/p, in. × 10^8 (m × 10^7)	SPECIFIC STRENGTH, σ/p, in. × 10^6(m × 10^6)
E-glass fiber*	10.5 (72.3)	460 (3170)	0.092 (2.55)	1.1 (2.8)	5.00 (1.24)
S-glass fiber*	12.0 (82.7)	600 (4130)	0.090 (2.50)	1.3 (3.3)	6.66 (1.65)
E-glass in epoxy	7.5 (51.7)	200 (1380)	0.070 (1.94)	1.1 (2.8)	2.86 (0.71)
S-glass in epoxy	7.5 (51.7)	300 (2070)	0.070 (1.94)	1.1 (2.8)	4.29 (1.07)
Aramid fiber*	20.0 (137.8)	500 (3445)	0.060 (1.69)	3.3 (8.1)	8.33 (2.04)
Aramid in epoxy	12.0 (82.7)	280 (1930)	0.055 (1.40)	5.9 (3.6)	5.09 (1.38)
HM graphite fiber*	55 (379)	300 (2070)	0.069 (1.90)	7.8 (19.8)	4.3 (1.09)
HT graphite fiber*	35 (241)	350 (2410)	0.064 (1.77)	5.6 (14.2)	5.5 (1.36)
AS or T-300 fiber*	30 (207)	400 (2760)	0.067 (1.85)	6.0 (11.2)	6.0 (1.49)
HM graphite in epoxy	30 (207)	135 (930)	0.058 (1.61)	5.2 (13.2)	2.3 (0.58)
HT graphite in epoxy	22 (152)	205 (1410)	0.054 (1.50)	4.1 (10.4)	3.8 (0.94)
AS or T-300 in epoxy	17 (117)	230 (1580)	0.056 (1.55)	4.1 (10.0)	4.1 (1.01)
Boron filaments*	60 (143)	400 (2760)	0.095 (2.63)	6.3 (16.0)	4.2 (1.05)
Boron in epoxy	31 (214)	220 (1520)	0.075 (2.08)	4.1 (10.4)	2.9 (0.73)
Maraging steel	28 (193)	300 (2070)	0.289 (8.00)	0.97 (2.5)	1.0 (0.26)
Aluminum 7075	10 (68.9)	82 (565)	0.100 (2.77)	1.00 (2.5)	0.8 (0.20)
Titanium 6A1-4V	15 (103)	155 (1070)	0.155 (4.29)	0.97 (2.5)	1.0 (0.25)
Beryllium	35 (241)	90 (620)	0.066 (1.83)	5.3 (13.5)	1.4 (0.34)

*Fibers only; does not include resin.

NOTE: Fibers and composites are unidirectional.
Metals are isotropic.

Table 28.3. Forms of Composites of Interest for Aerospace Applications

MATERIAL	APPLICATION
Chopped glass/polyester	Molded secondary components, substitution for metal castings, electrical housings, and parts.
Chopped E-glass/epoxy	Complex electrical components, leading and trailing edges, and highly loaded complex shapes.
E-glass fabric/epoxy	Primary and secondary structure for subsonic aircraft, ducts, housings, bulkheads, intake manifolds, helicopter blades, radomes, etc. (Probably the most versatile material.)
E-glass fabric/polyimide	Higher temperature resistant applications; high energy radomes, engine fairings.
E-glass—unidirectional/epoxy tapes	Higher strength application; rotor blades, wing components for smaller aircraft.
S-glass—unidirectional tape/epoxy	Stiffer and stronger than E-glass for more critical applications.
E- and S-glass, filament wound/epoxy	Radomes, high pressure tanks.
E- and S-glass, filament wound/polyimide	Same as above for higher temperature use.
Graphite/epoxy	Structural application for higher stiffness and fatigue resistance; suitable for most higher loaded structural parts.
Boron/epoxy	Same as above, but limited to shapes with simple curvature.
Graphite/polyimides	Used for high speed aircraft for high temperature resistance.
Aramid (Kevlar)/epoxy	Higher efficiency radomes, high impact resistance, lower weight.
Glass/graphite/aramid/boron/epoxy hybrids	Excellent for helicopters and ITL aircraft. Many combinations of fibers may produce a better part than individual fibers alone.

flown on an extensive basis. The examination of the surface after 200 hours of flying showed that protective coatings are recommended with these materials.

McDonnell fabricated a series of F-4 boron/epoxy rudders and installed them on a large number of service aircraft. This was the first large scale use of advanced composites and it supplied much data on their performance. Several flight-related problems were encountered, but most of them were found not to be related to the material, but instead were the result of unfamiliarity with the thickness limits and producibility criteria.

The first production flying part designed specifically as a composite part was the F-14 horizontal stabilizer.[7] This part was about 10×10 ft (2.54×2.54 m), was trapezoidal in shape (Fig. 28.3), and consisted of a full depth honeycomb core and boron/epoxy skins; the

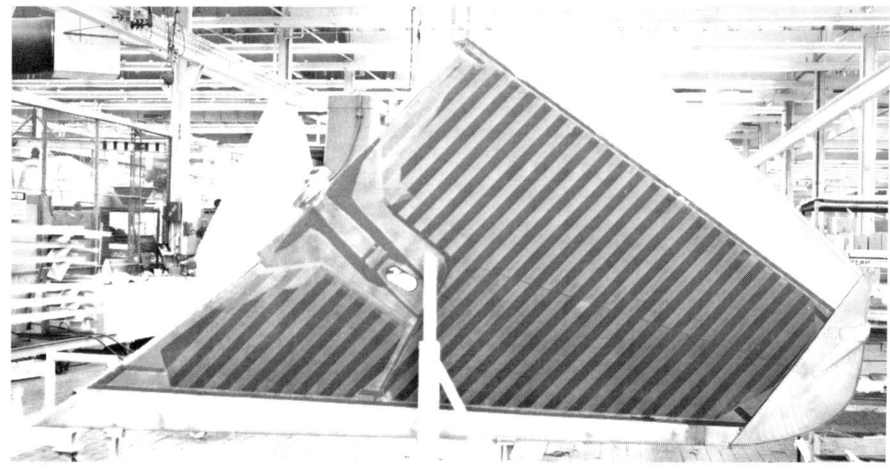

Figure 28.3. F-14A Horizontal stabilizer.

skin thickness varied from 7 plies at the tip to 56 plies at the root section; and 182 lb (82.6 kg) of weight were saved by the use of boron (this can be considered to be double the actual savings, since the weight was removed at the tail, allowing an equal decrease forward for control of the aircraft's center of gravity). These stabilizers have been flying since 1970, and no major service problems have been encountered. An interesting development was noted: the learning curve for this construction was steeper than the learning curves for metal construction, resulting in a steady decrease in the manufacturing costs.

The most comprehensive government-sponsored transport hardware program involved the C-5A slat[6] (Fig. 28.4). The purpose was to take the existing aluminum design and to redesign it completely in boron/epoxy. The final design comprised a so-called "eyebrow" panel over the upper surface, with no surface at all on the lower rib caps. The concept worked quite well and yielded a 22% weight savings. The new component had a total of 79 parts, as compared to a total of 800 in the production aluminum slat, exclusive of fasteners. Performance in the fleet has been excellent, with not a single maintenance discrepancy traceable to the composite material. Handling mishaps have caused some units to be returned to the manufacturing plant for repairs, resulting in a peripheral benefit by demonstrating that the units were quite repairable.

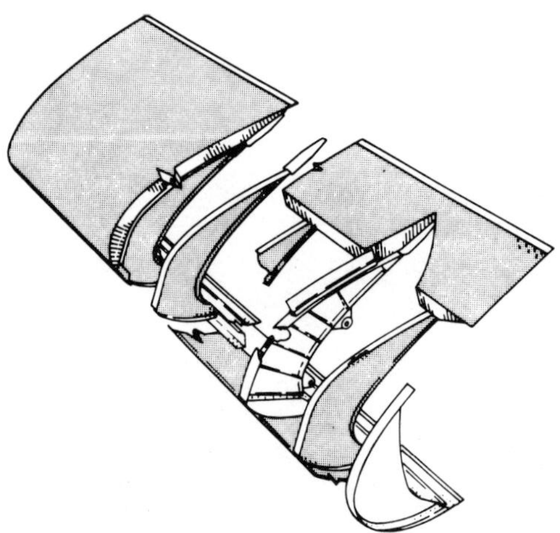

Figure 28.4. Boron slat design concept.

Typical weight savings for current production military aircraft components are 20%; for future commercial aircraft, 25%; for future military aircraft, 30–35%; and for spacecraft structures, 40%. Table 28.4 lists composite components evaluated, along with weight savings over their metal counterparts. Material and process developments have lowered the initial high costs for composites, and thus today acquisition costs for composite aircraft and spacecraft components are equal to, or less than, their metallic counterparts. Composite parts for commercial aircraft are being developed and are expected to be 10–15% less costly in production than equivalent metal parts. Initial automation and low cost composites have been developed for secondary structures and are currently being developed for primary structures.

28.6. CURRENT ADVANCED COMPOSITE AIRCRAFT APPLICATIONS

The success achieved by the introduction of advanced composites for improved aircraft structures prompted accelerated research, development, and test and evaluation of low-cost composites. Among the reinforcing fibers developed during the late 1960's and early 1970's were high strength, high stiffness graphite and a series of aramid (Kevlar) fibers. This broad range of advanced composite materials provided aircraft designers the freedom to balance structural optimization and affordable cost. Fixed wing aircraft structures utilized graphite and hybrid mixtures of graphite and boron, while helicopter structures highlighted aramid and hybrid mixtures of aramid and fiberglass, as shown in Table 28.5. Over the years, military aircraft have required increased capabilities, along with more stringent aircraft specification requirements. A historical cost trend for U. S. military aircraft with year of initial operational capability[3,8] is given in Fig. 28.5.

The costs of most of the modern aircraft more than doubles after three to four years in production. The rapid rise in the use of composites is an attempt to reduce the rapidly rising production costs. As mentioned pre-

Table 28.4. Composite Component Development

COMPONENT	MATERIAL	WEIGHT SAVINGS, %	CONTRACTOR
F-4 rudder	Boron/epoxy	35	McDonnell Aircraft Co.
F-15 stabilator	Boron/epoxy	22	McDonnell Aircraft Co.
F-111 stabilator	Boron/epoxy	25	General Dynamics Convair Aerospace, Fort Worth Operation
F-5 landing gear door	Boron/epoxy	29	Northrop Corp, Norair Division
F-14 stabilator	Boron/epoxy	20	Grumman Aircraft Engineering Corp.
A-4 flap	Boron/epoxy	32	Douglas Aircraft Co.
A-4 flap	Graphite/epoxy	47	Douglas Aircraft Co.
A-4 stabilator	Graphite/epoxy	32	Douglas Aircraft Co.
VC-10 aileron strut	Graphite/epoxy	43	Royal Aircraft Establishment
C-5A leading edge slat	Boron/epoxy	15	Lockheed-Georgia Corp.
T-39A wing box section	Boron/epoxy	37	North American Rockwell-LAD
Advanced composite wing structure	Boron/epoxy	20	Grumman
F-5 leading edge section	Graphite/epoxy	21	Northrop Corp. Norair Division
F-111 fuselage	Boron/epoxy Graphite/epoxy Boron/aluminum	18	General Dynamics Corvair Aerospace, Forth Worth Operation
Tubular struts	Boron/epoxy	30	North American Rockwell Space Division

Table 28.5. Helicopter Components—Major Material Applications

	GRAPHITE EPOXY	FIBERGLS EPOXY	KEVLAR EPOXY
BELL UH-I TRANSMISSION HOUSING	√		
BELL UH-ID MAIN ROTOR BLADE		√	√
BOEING CH-47C MAIN ROTOR BLADE		√	
HUGHES OH-6A TAIL ROTOR BLADE			√
SIKORSKY CH-54B ROTOR HUB		√	
SIKORSKY S-76 MAIN & TAIL ROTOR BLADES; EMPENNAGE			√

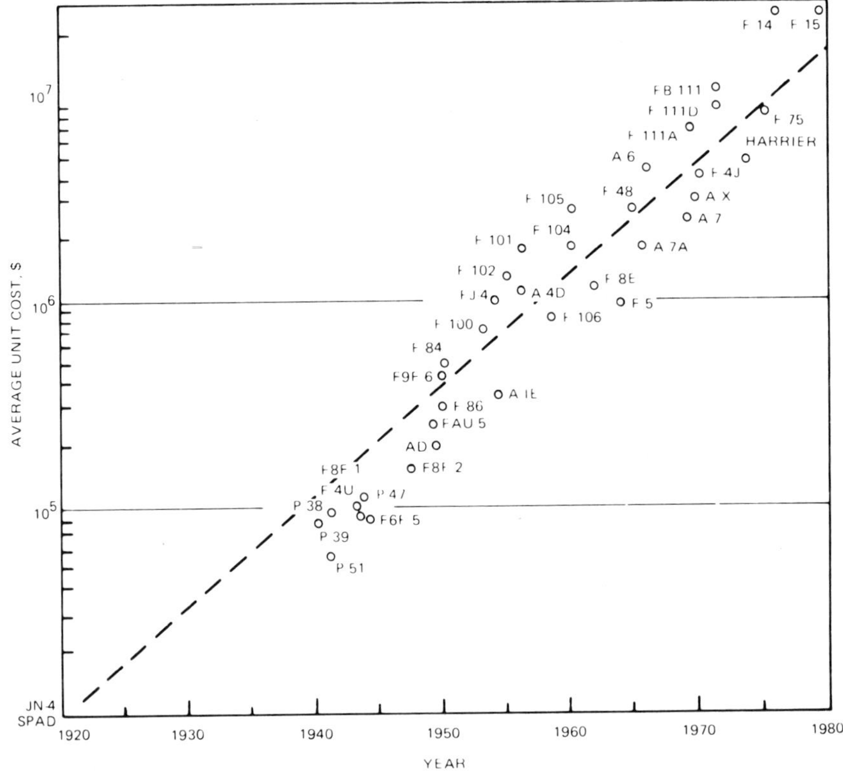

Figure 28.5. Military aircraft costs—historical trends in the United States.

viously, the greatest cost savings lie in the introduction of composites as part of the original design. Substitution for previously designed or existing metal components offers very limited pay-offs.

Composites can also be effective in improving the performance of the aircraft by reducing the weight or by providing higher heat resistance, better stiffness, and superior fatigue resistance.

Utilizing the AS or T-300 type fiber reinforcement and high temperature resistant multifunctional epoxy matrices, several graphite aircraft structures were built and successfully flown. Typical early structures were the wing leading edge of the Northrop F-5, the shoulder fuselage panels of the F-111, the main landing gear doors of several fighter aircraft, and secondary structures (flaps, spoilers, etc.) for commercial aircraft such as the B-737, the DC-10 and the L-1011.

Late in 1971, the first production application of a graphite/epoxy structure was introduced, the F-111 underwing fairing. Currently, the production applications are the F-15 speed brakes, F-16 vertical stabilizer skins, and the F-16 horizontal stabilizer. These structures, shown in Table 28.6, have clearly demonstrated structural integrity, improved performance, and up to 10-year durability to date.

Advanced composite structural weight fraction is clearly on the upswing, as shown in Fig. 28.6, and the aircraft of the future may be as much as half composites. Grumman has shown that an Advanced Design Composite Aircraft (ADCA) can accomplish a specific aircraft mission with a significantly smaller and lighter aircraft than is possible with metallics, and thus can save both weight and costs.[1]

Relative typical strengths and densities of advanced composites and metallics are given in Fig. 28.7. As can be noted, composites provide a dramatic weight reduction and are strength and stiffness competitive with current grade metals. Typical raw material costs and ratios of material purchased to material utilized (buy-to-fly ratio) are shown in Fig. 28.8. Although the composite materials are cur-

Table 28.6. Production of Military Aircraft—Composites Components

	BORON EPOXY	GRAPHITE EPOXY	1ST FLIGHT
F-14 HORIZ STABILIZER	√		1970
F-111 WING BOX DOUBLER	√		1971
F-111 UNDERWING FAIRING		√	1971
F-15 HORIZ/VERT STABILIZER	√		1971
B-1 LONGERON DOUBLERS	√		1974
F-15 SPEED BRAKES		√	1975
F-16 VERT STABILIZER SKINS		√	1976
F-16 HORIZ STABILIZER		√	1976

rently more costly than aluminum, the costs are comparable to titanium. Current trends are for low composite material costs as volume increases, along with higher metal costs due to inflation. Also, it can be seen that the buy-to-fly ratio is more favorable for composites, and, considering that composite components are at least 25% lighter than comparable metal components, material costs differences are not very significant.

One of the most ambitious composite programs was initiated by the Grumman-USAF Advanced Development Program in July 1973. This was the fabrication and testing of the B-1 composite horizontal stabilizer.[9]

The design of this unit is shown in exploded view in Fig. 28.9. The stabilizer has an area of 240 ft^2 (22.3 m^2), with a root cord of 17 ft (95.2 m) and a length of 30 ft (9.1 m), with a depth at the root of approximately 14 in. (36 cm).

The horizontal stabilizer composite torque

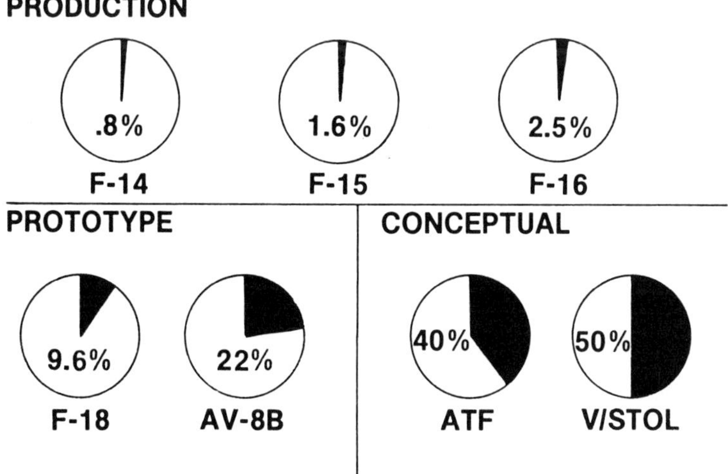

Figure 28.6. Aircraft structural weight fraction trend.

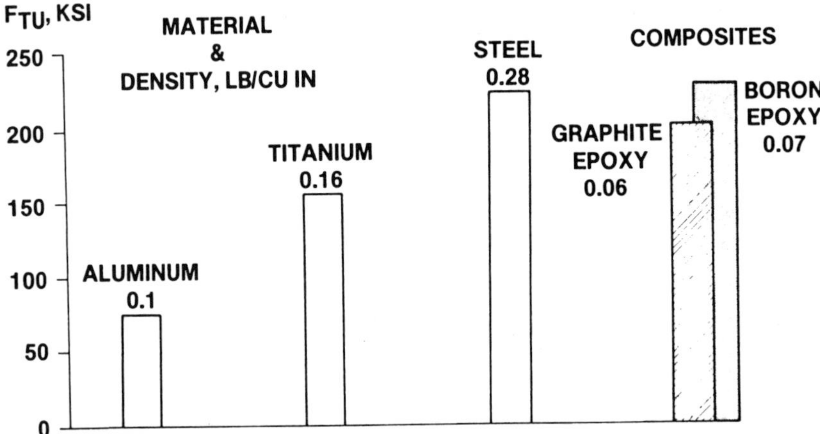

Figure 28.7. Relative strength of aircraft materials.

box was designed to minimize the number of detail parts and thus reduce assembly costs. Bonded assembly sandwich and bonded metal splice plate construction techniques were eliminated, since both entail excessive costs on a structure of this size. Instead, the torque box is assembled by drilling the composite parts and fastening with stainless steel bolts and titanium Hi-Lok fasteners.

The covers, which are the largest composite parts of the stabilizer assembly, are made using graphite/epoxy throughout the skin area, with a hybrid construction of boron/epoxy and graphite in the root and at the spot cap areas. The basic skin is composed of 106 graphite/epoxy layers. The boron/epoxy strips were incorporated into the outer skin over the intermediate spars to provide longitudinal bending stiffness. Boron/epoxy plies in the inboard skin area, next to the pivot fitting, aid in load transfer from the covers into the bearing support fitting.

The stabilizer substructure is composed of graphite/epoxy sine wave intermediate spars and ribs and flat-web front and rear spars. The sine wave spars generally have webs of only six plies, 0.03 in. (0.8 mm) thick, made in a single operation on split matched metal tools. The

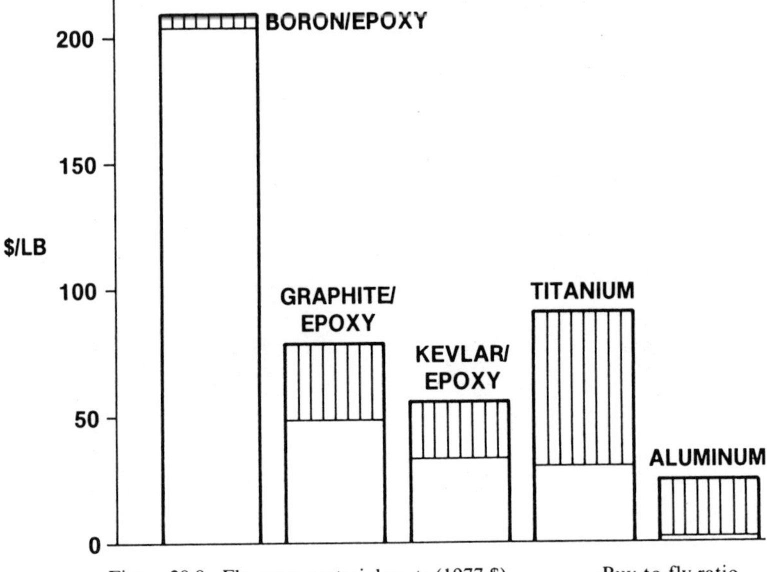

Figure 28.8. Fly-away material costs (1977 $). -Buy-to-fly ratio.

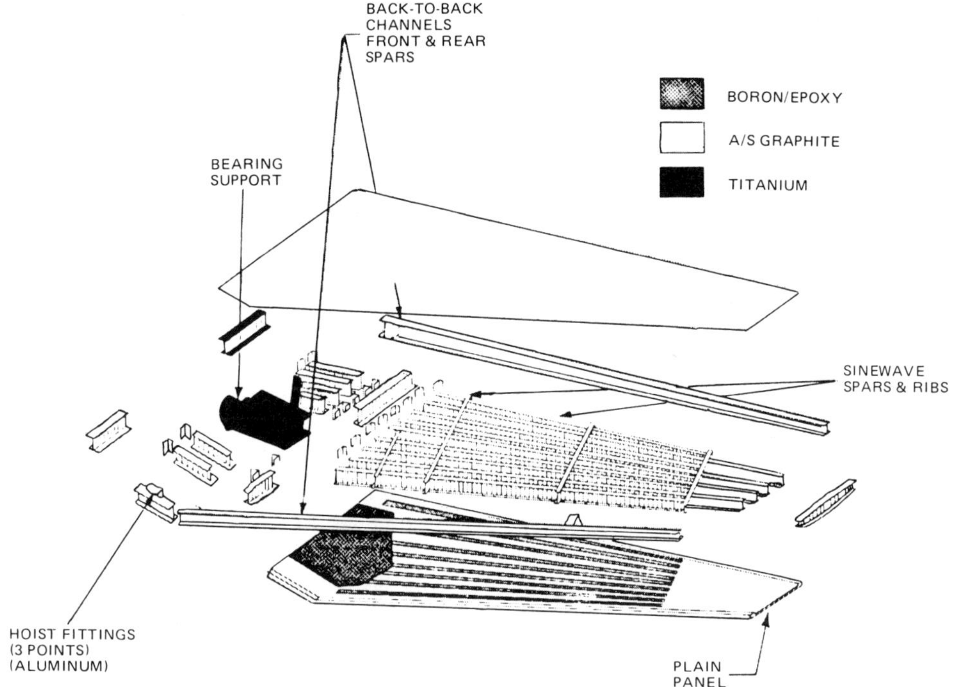

Figure 28.9. B-1 horizontal stabilizer composite torque box.

design criteria are described in the design chapter.

Integrally molded sprayed aluminum for lightning strike protection is used on 50% of the surface area of the skins.

The static test unit failed in compression at 132% of design ultimate load. The fatigue unit survived limit load after two lifetimes of spectrum fatigue.

Actual weights of the composite test stabilizer were 501 lb (227 kg) less than the equivalent metal construction. This represents an overall savings of 15% of the weight. The cost comparisons (Fig. 28.10) show that the increased costs of the composite covers are more than offset by reduction in substructure and assembly costs due to the smaller number of component parts. Overall, the production cost savings were forecasted at 17.5% over the all-metal stabilizer.

During the past several years, the composite industry in the U. S. concentrated on the inhibitors for full-scale implementation: cost, confidence, and durability. The cost issue will

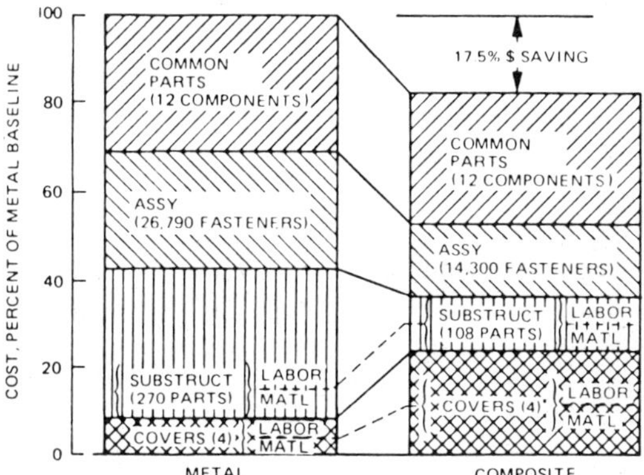

Figure 28.10. B-1 horizontal stabilizer production cost composite.

always be a concern, but progress has been made. Forecasts for future large volume for graphite and Kevlar fibers have overcome escalation, and future projections are attractive for high technology composite components. Sizable effort has been, and continues to be, made in development for low cost tooling for organic matrix composite manufacture. Automation of the process has shown that process costs for composites can be below those for many equivalent metallic structures. Grumman has developed and commissioned an "Integrated Laminating Center" (Fig. 28.11), which has reduced the production labor to one-half for the composite F-14A horizontal stabilizer.

The confidence issue is slowly but surely being overcome. A major step was the production approval of an all-composite horizontal stabilizer for the B-1 bomber previously discussed. After a five-year in-depth development program, this ultra-large composite structure satisfied all details of aircraft specifications, and both Rockwell International's weapons system manager and the U. S. Air Force accepted the component for production. Continued confidence is shown by the use of composite wing skins for the F-18 being developed for the U. S. Navy. The advanced Harrier, the AV-8B, is taking maximum advantage of composites, utilizing the material for approximately 15% of the structure's weight.

The durability of advanced composites has been under study for several years, with emphasis on moisture resistance. Extensive testing, along with in-depth analysis, has shown that graphite/epoxy laminates can absorb up to 1.5% moisture. In the moisture-saturated condition, the laminate is stable and only degrades if service temperatures are above 260° F (126.6° C). Current material specifications include mechanical tests of moisture-saturated laminates to provide assurance of durability. Evaluation of service data on the F-14A horizontal stabilizer has been undertaken. In service since 1970, with lead aircraft logging over 5000 hours and more than 300 aircraft (over 150,000 flight hours) deployed, durability requirements have been proven. Further, maintenance actions have been few, and life cycle costs have been low. Similar data have been noted for the composite empennage of the F-15.

Figure 28.11. Integrated laminating center.

Because such data are currently limited and only qualify the specific composite material and design, Grumman has established a laboratory service life-testing facility. This facility (Fig. 28.12) simultaneously simulates loading, humidity, thermal spikes, and temperatures in both accelerated and quasi-real time to provide the lead-the-fleet data for critical composite components.[10] Currently, the composites are being evaluated for damage tolerance, effects of defects and impact, and crash worthiness.

28.7. COMMERCIAL AIRCRAFT COMPOSITES

Advanced composite structures for U. S. commercial aircraft have been under study by NASA for the past several years. Initial studies were for secondary structures such as wing-to-body fairings and control surfaces, and for fatigue life extension of metallic components. Service life actual testing showed that non-sandwich components were structurally efficient, durable, and readily maintained and repaired. Corrosion, moisture ingress, and debonding were the major limitations of sandwich structures using aluminum honeycomb core. Overall, these early studies showed composites will provide significant advantages, and thus current development is for primary structures. A typical impact study on the percent composite application versus weight, cost, return on investment, and operational weight is given in Fig. 28.13.

During the past two years, NASA has funded Boeing, McDonnell, and Lockheed to design, test, and implement empennage structures for current commercial aircraft. The prime driver for these programs is energy conservation, and the weight reductions furnished by composites can provide significant strides in that direction. Suitability studies of composite fail-safe designs for wing and body are also underway, and it appears that the commercial aircraft of the 2000's may contain composites in significant quantity. Composite structures successfully flown are given in Fig. 28.14. Currently, the new Boeing series of aircraft, B767 and B757, will contain graphite/epoxy composites for the rudders, elevators, ailerons, and spoilers. In addition, aramid/graphite/epoxy composite hybrids will be used for trailing edges and fairings. Zero bleed molding process is used to mold these parts to reduce layup and material costs.

Other areas being evaluated for composites are floor beams, decks, nacelles, and engine components. The recently completed quiet engine study program has highlighted the need for composite components. It appears to be a certainty that future engines will utilize composites, especially for static components.

Figure 28.12. Life testing facility.

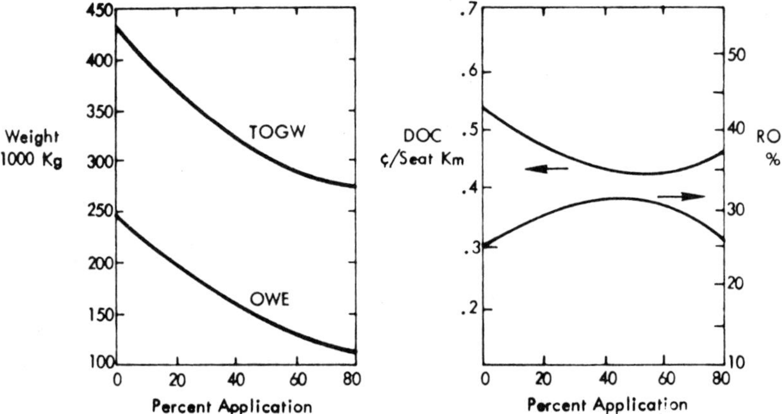

Figure 28.13. Impact of composites on large transport aircraft.[6] (TOGW = take-off gross weight; DOC = direct operating costs; ROI = return on investment; and OWE = operating weight empty.)

Engine parts already designed and tested using composites are vanes, splitters, and perforated duct skins. Presently under study are frames, augmentor ducts, and nozzle flaps. Previous study of composites for engine blades have shown a foreign object damage (FOD) limitation. The industry is still developing composites for this application utilizing hybrids to provide the resistance to bird ingestion impact. Martin-Marietta Company has successfully fabricated and tested composite thrust reversers using graphite/Kevlar/fiberglass hybrids using a highly sophisticated multi-layer structure.

28.8. SPACECRAFT COMPOSITES

Composites for use in space and on spacecraft have been developed in the U. S. by both NASA and DOD. A recent example is the payload doors of the Space Shuttle Orbiter. These components are the largest assembled composite, 12 ft wide by 60 ft long (3.66 × 18.29 m). Weight savings are at a premium for space applications, and thus composite growth in this field is rapid. Other special features of composites for spacecraft are the controllable coefficient of thermal expansion, cryogenic stability, load tailorability, and high specific stiffness. Advanced composites for space applications are available in ultra-thin thicknesses, 0.001 in. (0.0254 mm) per ply, and thus optimum structures for large area solar energy collectors will be a future reality.

Near zero coefficient of expansion composite tubular structures have been produced by Grumman for the Large Space Telescope (LST; Fig. 28.15). Composites have been successfully used for precision mounts, optical benches, and electromagnetic antennas. Structures for future space applications will be primarily advanced composites; therefore, the industry is presently developing the required data base.

Due to the high cost of sending craft into space, weight saving composite structures offer the greatest pay-offs and have been used at higher rates than they have been for aircraft. The early space capsules and rockets had nose cones, shrouds, and heat shields fabricated from high temperature resistant ablative materials. Many of the rocket nozzles also utilize ablative construction. In the original design of the Apollo command module and the SEM, many composites were used inside and outside of the cabin. After the

BOEING
- 737 SPOILER

McDONNELL-DOUGLAS
- DC-10 UPPER RUDDER
- DC-10 AFT PYLON SKIN

LOCKHEED
- L1011 WING TO BODY FAIRING
- L1011 CENTER ENGINE FAIRING
- L1011 WING TO BODY FILLET

Figure 28.14. Composite structures flown on commercial aircraft.

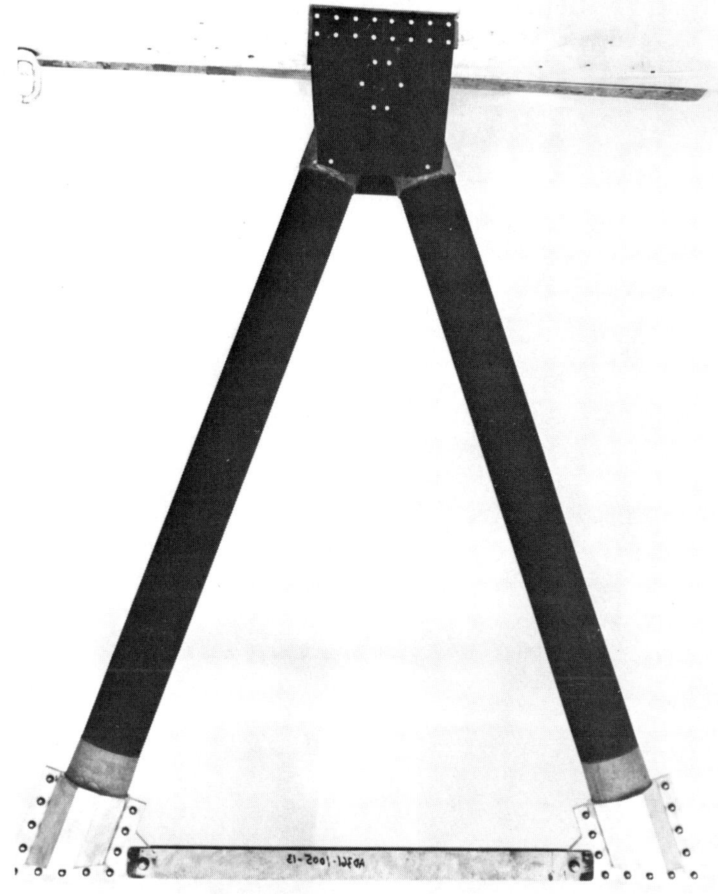

Figure 28.15. LST metering truss.

tragic Apollo fire, however, the use of composites was drastically reduced inside the cabin to non-combustible materials. For long term space exposures, resistance to outgassing and the effects of strong UV radiation was also required. Table 28.7 lists some of the composites meeting the stringent NASA requirements for space materials. Generally, polyesters were found to be not stable, while most of the epoxies and some polyimides were acceptable. One phenolic formulation was also approved. Most thermoplastics, with or without glass fibers, passed the test requirements.[11]

Boron/epoxy was used by Grumman Aerospace to fabricate a space truss for NASA. Boron tubing was fabricated by pressurizing a nylon bag inside a metal tubular female mold. This resulted in a wrinkle-free construction realizing the properties of flat material in a circular part. The end connections were metal fittings bonded to the tubing after cure. The strength and stiffness of boron/epoxy provided substantial weight savings over the original metal design.

Numerous filament wound tanks were made to contain pressurized gasses and were successfully used in all Lunar missions. For future applications, NASA has recommended the use of aramid (Kevlar 49) fibers for use in filament wound construction. These fibers are the lightest of all non-metallic fibers, and provide the highest tensile specific strength. Since these fibers are not self-abrasive, they also maintain their strength during and after processing. They also appear to be more uniform than the graphite or glass filaments.

One of the plans for future space missions

Table 28.7. List of Composites Meeting Johnson Space Center Vacuum Stability Requirements for Polymeric Materials[11]

MATERIAL	TOTAL WEIGHT LOSS, %	VOLATILE CONDENSABLE MATERIALS, %	PRECONDITIONING TIME, hours	PRECONDITIONING TEMPERATURE, °F (°C)
Thermosets				
Epoxy/glass (Hexcel F-161)	0.30	0.02	2.75	325°F (163°C)
Epoxy/glass (GE-101)	0.48	0.05		
Epoxy/glass (Ferro 2209)	0.53	0.00	As received	
Epoxy/glass (G-10)	0.10	0.01		
Epoxy/glass (G-11)	0.61	0.03		
Epoxy/glass (E-720)	0.54	0.04		
Polyimide/glass (Hexcel F-174)	0.40	0.00		
Graphite/epoxy (HY-E-1334)	0.97	0.01	1	350°F (177°C)
Epoxy/Kevlar 49 (F-164)	0.00	0.00	3	350°F (177°C)
Epoxy/graphite (Thornel-300/934)	0.62	0.00		
Phenolic/glass	0.64	0.00		
Epoxy/S-glass (Scotchply XP-251S)	0.58	0.01	0.5	284°F (140°C)
Polyimide/Kevlar (Skybond 703)	0.85	0.00		
Silicone resin/silica fibers	0.21	0.03	16	399°F (204°C)
Thermoplastics				
Teflon FEP	0.06	0.06		
Teflon TFE	0.10	0.03		
Nylon 6/6-glass (70/30)	0.81	0.04		
Acetal (Delrin)	0.48	0.07		
KEL-F	0.03	0.01		
Polycarbonate/glass laminated	0.10	0.01		
Acrylic	0.57	0.01		
Polypropylene/glass	0.13	0.04		
Polyphenyle oxide	0.04	0.03		
Polystyrene	0.26	0.01		
Polysulfone	0.33	0.00		
Polysulfone/glass (70/30)	0.24	0.01		

included the fabrication of a composite shell structure for a space transportation vehicle by McDonnell Douglas Astronautics Co.[12] A corrugated graphite/epoxy cylinderical structure, 10 ft in diameter and 10 ft long (3 × 3 m) was fabricated and tested to prove out this design concept. The final article is an open shell with external ring stiffness (hat sections). The shell was designed to sustain pure bending loads which will generate a maximum load intensity in the shell wall of 900 lb/in. (15,400 kg/m). Maximum compression loads to be sustained are 1000 lb/in. (17,100 kg/m). To reduce fabrication costs, the shell was fabricated in three 120° segments, which were molded flat and then assembled into a circular shape. The corrugated shell construction makes the skin sufficiently flexible for bending after cure. Mechanical fasteners were used for final assembly.

Another shell type structure for use in such spacecraft as Space Lab and Space Tub proposed as pay loads for the Space Shuttle System is the graphite/epoxy orthogrid panel structure developed by Martin Marietta Corp.[13]

The basic concept of a grid-stiffened composite panel is that a relatively thin skin is reinforced with a gridwork of stiffeners so that the overall panel can resist design loads without becoming structurally unstable or being overstressed. The advantage of being able to use graphite/epoxy in the design of these panels is that the skin laminate can have preferred stiffness and strength directions, and the stiffener can be designed to be very structurally efficient by using a high percentage of axially oriented fibers.

The specific graphite/epoxy orthogrid panel selected forms a part of a relatively large graphite/epoxy shell structure representative of spacecraft body structure. This shell struc-

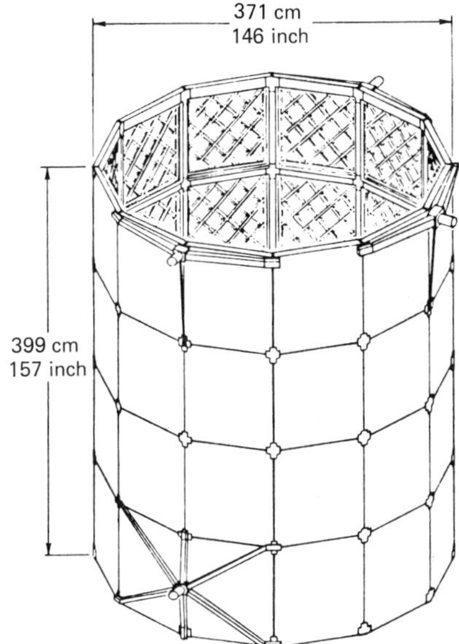

Figure 28.16. Lightweight shell structure.[13]

ture (Fig. 28.16) is 157 in. (3.99 m) high and 146 in. (3.71 m) in diameter, and was designed to resist loads of the magnitude of those expected on Space Shuttle Orbiter payloads. The detailed design for the grid structure is shown by Lager.[13] The main feature of this design is that it provides the potential for low cost structural panels when advanced to the production stage at which initial tooling costs can be amortized over a large quantity of panels. The most innovative part of the fabrication method is the foam/fiberglass stiffener web grid billet fabrication and machining to size. The fabrication of these details in large billet form results in a low number of fabrication man-hours per detail.

The skin laminates consist of four layers of T-300/934 graphite/epoxy tape oriented [+45°, −45°, +45°] with the outside edges reinforced with three layers of thin fiberglass cloth interlayer shims.

The orthogrid panel stiffener flanges are cut from a laminate consisting of eight layers of 0° oriented graphite/epoxy and three layers of style 112 fiberglass cloth.

The shear web joining the stiffener flanges to the panel skin consists of lightweight polyester foam with thin fiberglass cloth laminates bonded to the surface. After fabrication, the sample panels were placed in a shear loading frame and loaded to failure. The failure load corresponded to an in-plane edge shear of 545 ft-lb/in. (960 N/cm), well above the design load level required.

28.9. COMPOSITES FOR RE-ENTRY VEHICLES

The use of composite heat shields and nozzle liners for spacecraft is one of the most common applications of these materials. In many cases, carbon-carbon structures are used, but due to the restraints placed on the dissemination of the information on the carbon-carbon composites, they will not be discussed in this book. The phenolic tape structures are used in nonmilitary applications now and have been used since the beginning of spacecraft fabrication. Usually, the shields and nozzles are fabricated by tape winding on mandrels, followed by cure and final machining to the required shape. The tapes are usually laid up on angles which will expose the edges of the laminate to ablative forces to decrease the possibility of delamination and blistering if the composite was exposed flatwise. The heat shields and nozzles for the LEM and Apollo spacecraft were fabricated in this manner.

A new approach to the fabrication re-entry vehicle heat shields is the use of braided shingled construction, developed by McDonnell Douglas.[14]

There appears to be experimental and flight test evidence that the splices existing in the conventional wrapped heat shield result in small off-axis finlets during the ablation of the heat shield. These finlets are formed upon exposure of the splices to the ablative environment as the heat shield surface physically recesses, and appear to act as an oriented set source and contribute to unsymmetrical vehicle dynamics and roll torque.

The heat shields being developed by McDonnell are braided without splices, simulating the construction of the tape wrapped parts. The carbon fiber plies are braided directly onto the heat shield mandrel in a 20° shingled orientation with the filaments in an over-and-under bias construction. The material braided is narrow carbon tow. Excellent nesting of contiguous plies occurs and results in high filament packing at the ablative surface and minimal voids and resin-rich pockets in the cured composite.

Techniques for incorporating phenolic resin matrix into the braided reinforcement include wet impregnation of the preform after braiding and braiding of "B"-stage prepreg tows.

28.10. THERMOPLASTIC COMPOSITES FOR SPACE STRUCTURES

A materials and processes evaluation program to develop a graphite/thermoplastic system for low temperature, 170° F (77° C), space applications was conducted by several companies, including Grumman Aerospace,[15,16] Hexcel, 3M, and Fiberite. The thermoplastic matrix systems evaluated were:

- Acrylic
- Polycarbonate
- Phenoxy
- Polyester
- Polyethersulfone

Evaluation of these materials for their applicability to future space production processing, forming temperatures, and structural strength indicated that the acrylic system with woven graphite broadgoods reinforcement was the best candidate. Table 28.8 summarizes the laminating parameters and the structural test results for the resins evaluated. The tensile strength/stiffness goals of 60 ksi/7.5 msi (417 MPa/51.8 GPa) were achieved for the selected graphite/acrylic system.

The distinct advantages of this system over the others evaluated are as follows.

- Solventless prepreg.
- Liquid resin blend at room temperature for excellent fiber wetting properties and void-free laminating.
- Low laminating and forming temperature, 350° F (177° C), thus minimizing energy requirements for beam fabrication in space.
- Monomer and polymer are low in cost and widely available in quantity.
- Good strength/stiffness properties (see Table 28.8).

An acrylic monomer/polymer blend was developed with the following composition by weight:

- 77.5% methylmethacrylate monomer
- 22.0% acrylic polymer
- 0.5% benzoylperoxide

The woven graphite cloth selected for the initial beam forming tests is style 24 × 23, 8HSW, using Union Carbide's T-300 with UC 309 epoxy sizing. The per-ply thickness of this combination with approximately 65% fiber volume is 15 mils (0.762 mm). Figure 28.17 is a photomicrograph of the woven graphite/acrylic system. Table 28.9 lists flexural properties of this construction from three different sources. Table 28.10[16] shows the comparison of strength between the baseline aluminum beam and two thermoplastic composite formulations. Typical material properties are included.

28.11. CONCLUSIONS

Structures and materials technology has been, and will continue to be, an area of high interest to provide the advances needed to improve the standard of living. Since 1966, dramatic improvements have been shown by advanced composites. These improvements will continue and, by 1990, the advanced composites industry will be well established. Initial pay-offs will be in aircraft, (see Figure 28.18) followed by advances in the spacecraft industry. Composites will make significant inroads in the future energy generation fields (solar and fusion energy), as well as providing the means for increased productivity with new helicopter designs.

It appears that the current labor-intensive composites industry is heading for automation for those components requiring high-volume production, but for development of new products, as well as for low-volume parts, hand and semi-automatic methods will still be utilized. Organic matrix composites will continue to be the major material, with metal matrix composites showing growth signs in the next 8–10 years. The major high modulus fibers will be graphite and Kevlar, with boron and silicon

Table 28.8. Material Strength Properties and Process Parameters for Thermoplastic Laminates

MANUFACTURER	RESIN	CLOTH REINFORCEMENT	PROCESS PARAMETERS: TEMPERATURE, °F (°C)	PROCESS PARAMETERS: PRESSURE, psi (MPa)	PROCESS PARAMETERS: TIME, MINUTES	LONGITUDINAL TENSILE STRENGTH, ksi (MPa)	LONGITUDINAL TENSILE MODULUS, msi (GPa)	RESIN CONTENT, %	THICKNESS, in. (mm)*
3M	Polycarbonate	Graphite VC-0149	500 (260)	100 (0.69)	30	31.9 (220)	4.04 (27.9)		0.029 (0.737)
3M	Polycarbonate	Graphite 2423	500 (260)	100 (0.69)	30	28.0 (193)	3.92 (27.1)		0.026 (0.660)
3M	Polycarbonate	Graphite 2423	500 (260)	100 (0.69)	30	48.4 (334)	6.26 (43.2)		0.038 (0.914)
Hexcel	Phenoxy	Graphite 1313	350 (177)	100 (0.69)	30	48.7 (336)	4.88 (33.7)		0.029 (0.737)
Hexcel	Phenoxy	Graphite 1313	350 (177)	100 (0.69)	30	49.5 (342)	5.71 (39.4)		0.023 (0.584)
GAC	Acrylic	Graphite 1212**	300 (149)	100 (0.69)	30	68.1 (470)	9.09 (62.7)	48.3	0.031** (0.787)
GAC	Acrylic	Graphite 1212**	275 (135)	80 (0.55)	30	57.9 (400)	7.61 (52.5)		0.037** (0.940)
GAC	Acrylic	Graphite 1212**	275 (135)	80 (0.55)	30	59.6 (411)	7.50 (51.8)	38.3	0.038** (0.914)
GAC	Acrylic	Graphite 2423	300 (149)	100 (0.69)	30	64.5 (445)	8.46 (58.4)	35.4	0.031 (0.787)
GAC	Acrylic	Graphite 2423/ glass scrim	300 (149)	100 (0.69)	30	62.0 (428)	6.21 (42.9)		0.045 (1.143)
GAC	Acrylic	Graphite 2423	300 (149)	100 (0.69)	30	62.8 (433)	7.26 (50.1)		0.036 (0.914)
GAC	Acrylic	Graphite 2423	300 (149)	100 (0.69)	10	70.3 (485)	6.93 (47.8)		0.040 (1.016)
Hexcel	Polyester	Graphite 2424				36.3 (251)	4.4 (30.4)		0.055 (1.397)
Hexcel	Polyester	Graphite 2424				40.8 (282)	4.36 (30.1)		0.059 (1.499)
GAC	Acrylic	Graphite 2423	300 (149)	100 (0.69)	30	41.2 @ 170° F (284 @ 77° C)	5.24 @ 170° F (36.2 @ 77° C)		0.026 (0.660)
Hexcel	Acrylic	Graphite 2423	400 (204)	200 (1.38)	2	65.5 (452)	6.88 (47.5)		0.036 (0.914)
Hexcel	Acrylic	Graphite 2423	400 (204)	200 (1.38)	2	39.2 @ 170° F (270 @ 77° C)	5.79 @ 170° F (40.0 @ 77° C)		0.036 (0.914)
Hexcel	Polyether-sulfone	Graphite 2423	500 (260)	100 (0.69)	30	52.2 (360)	7.0 (48.2)		NA

*Laminates—two plies thick except as noted.
**Laminates—four plies thick.

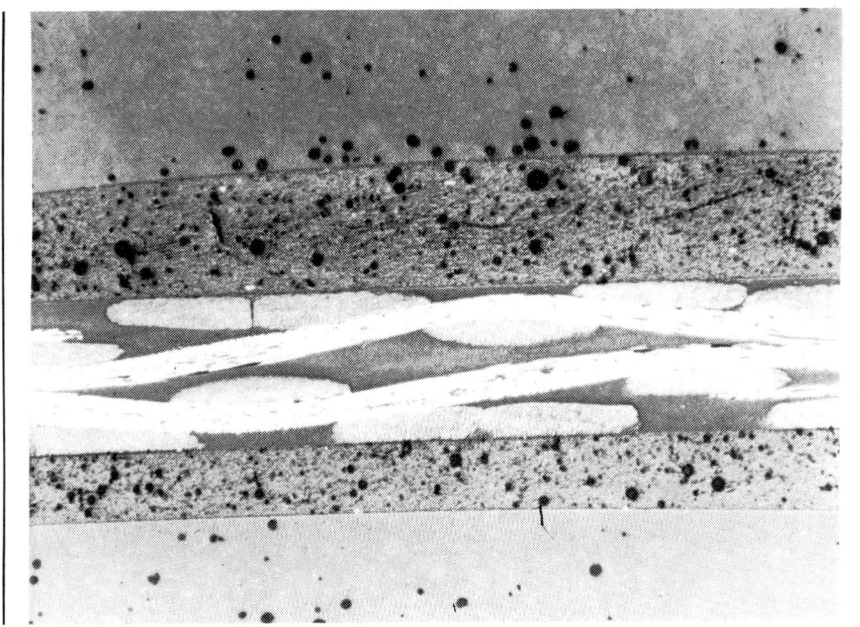

Figure 28.17. Cross-section of 0° / 90° —24 × 23 graphite / acrylic laminate (200x magnification).

Table 28.9. Room Temperature Longitudinal Flexure Properties of Woven Graphite/Acrylic

MANUFACTURER	RESIN	REINFORCEMENT	STRESS, ksi (MPa)	MODULUS, msi (GPa)	THICKNESS, in. (mm)
Hexcel	Acrylic	2423 graphite cloth	71.6 (494)	10.25 (70.7)	0.033 (0.838)**
Hexcel	Acrylic	2423 graphite cloth	44.3 (306)	7.51 (51.8)	0.036 (0.914)**
GAC	Acrylic	2423 graphite cloth	112.9 (779)	10.12 (69.8)	0.035 (0.890)**

*Laminate—four plies thick.
**Laminate—two plies thick.

Table 28.10 Comparative Data: Aluminum Beam vs Roll formed Graphite Beams

	BASELINE	THERMOPLASTIC COMPOSITE	
MATERIAL TYPE	METAL ALUMINUM 2024-T3	GRAPHITE ACRYLIC	GRAPHITE POLYETHERSULFONE
Thickness, in (cm)	0.016 (0.04)	0.030 (0.076)	0.030 (0.076)
Area, in^2 (cm^2)	0.10 (0.65)	0.18 (1.16)	0.18 (1.16)
Weight, lb/ft (Kg/m)	0.12 (0.18)	0.12 (0.18)	0.12 (0.18)
Test Length, in (m)	59.05 (1.50)	59.05 (1.50)	48.0 (1.22)
Design Ultimate, lb (Kg)	433 (197)	—	—
Failure Load, lb (Kg)	505 (229)	925 (420)	1025 (465)
Post Failure Load lb, (Kg)	—	600 (272)	800 (363)
Tensile Str. Long. Ksi (MPa)	47.0 (324)	64.5 (445)	52.2 (360)
Tensile Mod. Long. Msi (GPa)	10.5 (72.3)	8.5 (58)	7.0 (48.2)
Flex. Str. Long. Ksi (MPa)	NA	112.9 (778)	153.1 (1055)
Flex. Mod. Long. Msi (GPa)	NA	10.1 (69.6)	9.9 (68.2)
Flex. Str. Trans. Ksi (MPa)	NA	56.8 (391)	74.3 (512)
Flex. Mod. Trans. Msi (GPa)	NA	3.9 (26.9)	3.3 (22.7)

Figure 28.18. Lear Fan 2100 in flight. Photo by James Sugar.

carbide filaments used for specialty applications. The outlook for expansion of the composites industry is excellent because structural efficiency (weight savings) is becoming critical to conserve energy.

REFERENCES

1. Hadcock, R. N., "The Application of Advanced Composites to Military Aircraft," 10th ICAS Congress, Ottawa, Canada, October 3, 1978.
2. Lubin, G., *Handbook of Fiberglass and Advanced Plastics Composites*, Van Nostrand Reinhold, New York, 1969.
3. *Composite Materials 1972–1982* **I**, Center for Technological and Interdisciplinary Forecasting at Tel-Aviv University, Israel.
4. Lovelace, A. M. and Tsai, S. W., "Composites Enter the Mainstream of Aerospace Vehicle Design," *Astronautics and Aeronautics*, pp. 56–61 (July 1970).
5. Fechek, J. F., "Advanced Composite Efforts—a Status Report of Air Force Programs with Graphite Reinforced Composites," ASME Paper 71-DE-13, 1971.
6. Lassiter, L. W., "Applications and Concepts for the Incorporation of Composites in Large Military Transport Aircraft," RPG Conference—Reinforced Plastics in Aerospace Applications, Royal Aeronautical Society, April 5–6, 1973.
7. Lubin, G., and Dastin, S., "First Boron Composite Structural Production Part," SPI Reinforced Plastics/Composites Institute, 26th Annual Conference, 1971.
8. Graham, W. B., "RMV's in Aerial Warfare," *Astronautics and Aeronautics* (May 1972).
9. Ludwig, W., Erbacher, H., and Lubin, G., "Composite Horizontal Stabilizer for the B-1," SPI Reinforced Plastics/Composites Institute, 32nd Annual Conference, 1977.
10. Hedrick, I. G. and Whiteside, J. B., "Effects of Environment on Advanced Composite Structures," AIAA Paper No. 77-463, March 1977.
11. L. B. Johnson Space Center Report JSCO7572 Rev. D, 1976.
12. Penton, A. P., Johnson, R., Jr., and Freeman, V. L., "Fabrication of Composite Shell Structure for Advanced Space Transportation," 23rd National SAMPE Symposium, 1978.
13. Lager, J. R., "Graphite/Epoxy Orthogrid Panel Fabrication," 10th National SAMPE Technical Conference, 1978; and Van Hammersveld, J., "Extensive Cost Reduction Studies—Composite Empennage Component L-1011 Commercial Airliner," 23rd National SAMPE Symposium 1978.
14. Seibold, R. W. and Disser, E. F., "High Speed Braiding—An Approach for Fabrication of Reentry Vehicle Heatshields with Seamless, Shingled Construction," 10th National SAMPE Technical Conference, 1978.
15. Lubin, G., Poveromo, L. M., and Marx, W., "Reinforced Thermoplastic Composites for Space Beam Fabrication," SPI Reinforced Plastics/Composites Institute, 34th Annual Conference, 1978.
16. Poveromo, L. M., Muench, W. K., Marx, W., and Lubin, G. "Composite Beam Builder," SAMPE Journal, **17**, 1. Jan/Feb. 1981.

APPENDIX A

Typical Properties of Reinforced Composites (A)

PROPERTY	FIBERGLASS AND QUARTZ COMPOSITES											
	BMC MAT/ POLYESTER 15-25% GLASS	SMC MAT/ POLYESTER 30-40% GLASS	GLASS FABRIC POLYESTER	GLASS FABRIC EPOXY	FILAMENT WOUND EPOXY	NON-WOVEN EPOXY – UNIDIRECTIONAL	NON-WOVEN EPOXY – XPLIED	GLASS FABRIC POLYIMIDE	GLASS FABRIC PHENOLIC	GLASS FABRIC SILICONE	QUARTZ FABRIC EPOXY	S-GLASS UNIDIRECTIONAL EPOXY
Physical												
Specific gravity	1.65 - 1.78	1.7	1.7 - 1.8	1.8 - 1.9	1.9 - 2.2	1.9 - 2.1	1.9 - 2.1	2.05	1.73	1.7 - 1.9	1.73 - 1.80	1.92
Density, lb/cu. in.	.055	0.062	0.062	0.065	0.07 - 0.08	0.07 - 0.08	0.07 - 0.08	0.08	0.065	0.065	0.062	0.069
Rockwell hardness (M)	95	100 - 110	110	105 - 120	115 - 120	110 - 120	110 - 120	110 - 120	100	100		
Barcol hardness	45 - 50	50 - 60	60	55 - 60	60	60	60	75	60 - 74	60 - 70	70	
Water absorption – 24 hr, %	1.8	1.7	0.6	0.20 - 0.30	0.4 - 1.0	0.3 - 0.5	0.3 - 0.5		0.5	0.08 - 0.12		
Mechanical at room temp. at 0° to warp												
Tensile strength, KSI ult MPa ult	10.0 69	20 - 25 138 - 172	44 303	55 - 75 379 - 517	150 - 200 1030 - 1380	160 1112	75 521	50 - 82 345 - 565	50 - 60 345 - 413	30 - 40 207 - 276	79 545	215 1480
Tensile modulus, MSI GPa	1.6 11	1.7 - 1.9 12 - 13	2.4 - 3.0 17 - 21	3.4 - 3.8 23 - 26	4.5 - 7.5 31 - 52	5.3 37	3.5 24	2.8 - 4.2 20 - 29	2.5 - 3.5 17 - 24	2.7 19	2.8 - 31 19 - 22	7.1 49
Flexural strength, KSI ult MPa ult	15 103	20 - 25 138 - 172	31 214	75 - 95 517 - 654	90 - 200 620 - 1380	165 1147	120 834	75 - 100 517 - 695	60 - 70 413 - 482	33 227	99 680	220 1520
Compressive strength, KSI ult MPa ult	20 138	25 - 28 172 - 193	40 276	50 - 60 345 - 413	45 - 80 310 - 551	90 626	75 521	58 - 80 475 - 550	40 - 45 276 - 310	29 200	66.4 458	95 660
Shear strength												
Interlaminar, KSI (Short beam), MPa	2.0 13.8	2.5 - 4.0 17.2 - 27.6	3.5 24.0	4.0 27.6		3.8 26.4	3.0 20.9	3.8 26.4	2.5 17.2	0.8 5.5		9.0 - 11 63 - 76
Bearing strength, KSI MPa	25 172	30 - 36 207 - 248	33 - 43 227 - 296	38 - 46 262 - 317	27 186	30 - 40 207 - 276			50. 345	34 234		
Izod impact str, ft-lb/in. o/notch J/cm	8 - 12 4.3 - 6.4	12 - 16 6.4 - 8.5	14 - 18 7.5 - 9.6	30 16	60 32	60 32			8 - 15 4.3 - 8.0			
Poisson's ratio	0.11	0.11	0.11 - 0.12	0.12 - 0.16	0.1 - 0.3	0.1 - 0.3						
Thermal												
Thermal conductivity, Btu-in./hr/sq ft/°F W/m/°K	0.7 - 0.9 0.10 - 0.13	0.7 - 0.9 0.10 - 0.13	1.1 - 1.2 0.16 - 0.17	1.12 - 2.3 0.16 - 0.33	1.92 0.28	2.0 0.40	2.32 0.33		1.2 0.17	0.9 - 1.2 0.13 - 0.17		
Thermal expansion, in/in/°F x 10^{-6} cm/cm/°C x 10^{-6}	1.0 - 1.7 1.8 - 3.1	1.2 - 2.0 2.2 - 3.6	1.0 - 1.8 1.8 - 3.2	5.9 10.6	4.7 7.2 - 12.6	4.8	7.1		4.6 7.2 - 10.8	2.6 4.7		2.3 - 3.5 4.1 - 6.3
Heat distortion @ 264 psi, °F @ 1.8 MPa °C	300 149	350 177	350 177	400 204	100 204				500 - 550 260 - 288	550 288		
Maximum oper. temp, °F °C	250 121	250 - 300 121 - 149	250 - 300 121-149	300 149	300 - 400 149 - 204	300 149	300 149	500 - 700 288 - 371	300 - 400 149 - 204	550 - 700 288 - 371	400 204	350 177
Specific heat, Btu/lb/°F Cal/gm °C	0.27	0.28	0.28	0.25 - 0.28	0.25	0.25 0.21	0.25 0.21		0.30	0.25	0.25	0.25
Diffusivity, 10^{-3} sq ft/hr	2.6	2.6	2.6	2.5					2.5			
Emissivity, Total normal	0.82	0.82	0.82	0.79	0.79				0.82	0.83		
Mechanical properties at elev. temp												
Tensile strength @ temp, KSI MPa		15 @ 300°F 103 @ 149°C		38 @ 300°F 262 @ 149°C		50 @ 300°F 345 @ 149°C	40 @ 300°F 276 @ 149°C	64.8 @ 550°F 447 @ 288°C	40 @ 500°F 276 @ 260°C	30 @ 500°F 207 @ 260°C	654 @ 350°F 451 @ 177°C	
Flexural strength @ temp, KSI MPa		15-20 @ 300°F 103-138 @ 148°C		30 @ 300°F 207 @ 149°C		30 @ 300°F 207 @ 149°C	25 @ 300°F 172 @ 149°C	29.0 @ 600°F 200 @ 316°C	40 @ 500°F 276 @ 260°C	18 @ 500°F 124 @ 260°C	54.0 @ 350°F 372 @ 177°C	

Typical Properties of Reinforced Composites (B)

PROPERTY	BORON/EPOXY UNIDIRECTIONAL	GRAPHITE/ EPOXY INT. MODULUS	GRAPHITE/ EPOXY HIGH MODULUS	GRAPHITE/ POLYIMIDE INT. MODULUS	GRAPHITE/ EPOXY WOVEN	ARAMID/ EPOXY NON-WOVEN	ARAMID/ EPOXY WOVEN
ADVANCED OR HIGH-PERFORMANCE COMPOSITES							
Physical							
Specific gravity	2.01	1.60	1.56	1.60	1.59	1.35	1.33
Density, lb/cu. in.	0.073	0.056	0.058	0.056	0.06	0.05	0.049
Mechanical at room temp. at 0° to warp-parallel plied							
Tensile strength, KSI	200	220 - 250	113 - 208	160 - 200	85 - 90	183	75
MPa	1380	1520 - 1720	783 - 1435	1100 - 1380	586 - 620	1260	517
Tensile modulus, MSI	30	20 - 30	30 - 47	17	10.2	11.2 - 11.9	4.5
GPa	207	138 - 207	207 - 324	117	70.3	77 - 82	31
Flexural strength, KSI	260	240 - 270	90 - 230	220 - 230	122 - 150	90	50
MPa	1790	1650 - 1860	620 - 1600	1520 - 1580	841 - 1034	625	345
Compressive strength, KSI	353	213 - 230	90 - 102		100	34 - 40	12
MPa	2430	1470 - 1580	620 - 703		690	235 - 276	83
Shear strength							
Interlaminar (short beam), KSI	13	8 - 16	3.5 - 8	16	8 - 9	4.2 - 7.1	8
MPa	90	55 - 110	24 - 55	110	55 - 62	28 - 49	55
Bearing strength, KSI					100 - 143		
MPa					689 - 988		
I_3 od impact str, ft lbs/in. o/notch		28				48	
J/cm		15				26	
Poisson's ratio	0.21	0.045	0.199		0.077	0.31	
Thermal							
Thermal conductivity, Btu-in./hr/sq ft/°F	0.17 - 0.21	6 - 10	28 - 35			3.22	1.49
W/m/°K	0.02 - 0.03	0.86 - 1.44	4.03 - 5.04			0.46	0.21
Thermal expansion, in./in./°F x 10^{-6}	2.3					-0.2 - 0.3	0
m/m/°C x 10^{-6}	4.14					-0.36 - 0.54	0
Heat distortion, 264 psi - °F							
°C							
Maximum oper. temp, °F	350	350	350	700	350	350	350
°C	177	177	177	371	177	177	177
Specific heat, Btu/lb/°F		0.17					
Mechanical at elev. temp							
Tensile strength @ temp, KSI	168 @ 375 °F	200 @ 350°F	115 @ 350°F	150 @ 350°F	86 @ 350°F		
MPa	1170 @ 191°C	1380 @ 191°C	793 @ 177°C	1030 @ 177°C	593 @ 177°C		
Flexural strength @ temp, KSI	220 @ 375°F	190 @ 350°F		180 @ 350°F	82 @ 350°F	55 @ 350°F	
MPa	1520 @ 191°C	1310 @ 191°C		1240 @ 177°C	566 @ 177°C	382 @ 177°C	
Tensile modulus @ temp, MSI	26 @ 375°F	16 @ 350°F	30 @ 350°F	19 @ 350°F	10.3 @ 350°F		
GPa	179 @ 191°C	110 @ 191°C	207 @ 177°C	131 @ 177°C	71.0 @ 177°C		

APPENDIX B

Typical Properties of Graphite Fiber Composites

PROPERTY / FABRIC TYPE / WEAVE	THORNEL VCC-20[a] 5H SATIN	THORNEL VCB-45[a] 8H SATIN	THORNEL VCB-20[a] 8H SATIN	THORNEL VCC-45[a] 5H SATIN	HERCULES A-193-P[b] PLAIN	HERCULES A-370-8H[b] 8H SATIN	HERCULES A-370-5H[b] 5H SATIN	CELION W-1133[c] 8H SATIN	FIBERITE HMF-133/34[c] 8H SATIN	FIBERITE HMF-341/34[c] PLAIN	HEXCEL F3T-272 4H SATIN	HEXCEL F3T-282 PLAIN
DENSITY, lb/in.^{3}	0.058	0.061	0.058	0.061	0.065	0.065	0.065	—	0.044	0.043	0.043	0.043
Mg/m^3	1.61	1.68	1.60	1.70	1.80	1.80	1.80	—	1.58	1.54	1.54	1.53
TENSILE STRENGTH, Ksi	50	44	39	48	85	90	100	90	85	90	80	86
MPa	345	300	270	330	590	620	690	620	590	620	535	590
TENSILE MODULUS, Msi	5.7	12.8	5.6	14.0	10.0	10.0	10.5	10.5	10.5	9.7	10.0	10.5
GPa	39	88	39	96	69	69	72	72	72	67	69	69
FLEXURAL STRENGTH, Ksi	57	54	54	55	—	130	150	110	122	128	—	120
MPa	390	370	370	380	—	896	1,034	760	840	883	—	—
FLEXURAL MODULUS, Msi	5.5	12.2	5.5	13.0	—	9.5	10.0	11.0	11.4	9.5	—	9.7
GPa	38	84	38	90	—	65	69	76	78	65	—	—
COMPRESSIVE STRENGTH, Ksi	50	30	48	30	95	100	100	—	—	—	80	85
MPa	345	207	330	207	655	690	690	—	—	—	550	586
SHORT-BEAM SHEAR STRENGTH, Ksi	4.6	3.4	4.4	3.2	9.5	9.0	9.1	10.0	9.7	—	8.5	9.3
MPa	32	23	30	22	65	62	63	69	67	—	59	64
CURED PLY THICKNESS, in.	0.0105	0.017	0.0185	0.010	0.007	0.0135	0.0135	0.012	0.013	0.005	0.0072	0.0072
nm	0.27	0.43	0.47	0.25	0.18	0.34	0.34	0.30	0.33	0.125	0.18	0.18
COEFFICIENT OF THERMAL EXPANSION, in./in./°F	9×10^{-6}	9×10^{-6}	9×10^{-6}	9×10^{-6}	—	—	—	—	—	—	—	—
mm/mm/°C	5×10^{-6}	5×10^{-6}	5×10^{-6}	5×10^{-6}	—	—	—	—	—	—	—	—

[a] UNION CARBIDE DATA

[b] HERCULES DATA

[c] FIBERITE DATA

[d] HEXCEL DATA

Typical Properties of Several Types of Graphite Fabrics

PROPERTY / FABRIC TYPE	THORNEL VCC-20[a]	THORNEL VCB-45[a]	THORNEL VCB-20[a]	THORNEL VCC-45[a]	HERCULES A-193-P[b]	HERCULES A-370-8H[b]	HERCULES A-370-5H[b]	FIBERITE W-133[c]	FIBERITE W-131[c]	HEXCEL F3T-272[d]	HEXCEL F3T-282[d]
WEAVE	5H SATIN	8H SATIN	8H SATIN	5H SATIN	PLAIN	8H SATIN	5H SATIN	8H SATIN	PLAIN	4H SATIN	PLAIN
COUNT, YARNS/in.	22 x 21	20-½ x 19	20-½ x 19	22 x 21	11-½ x 11-½	21-½ x 21-½	11 x 11	24 x 23	24 x 24	12 x 12	12-½ x 12-½
YARNS/5 cm	43 x 41	40 x 37	40 x 37	43 x 41	22-½ x 22-½	42-½ x 42-½	21-½ x 21-½	47 x 45	47 x 47	23-½ x 23-½	24-½ x 24-½
WEIGHT, oz/yd^2	8.2	15	15	8.2	5.7	10.9	10.9	10.5	3.77	5.5	5.7
g/m^2	280	510	510	280	193	370	370	356	128	186	194
FIBER DENSITY, lb/in.^{3}	0.069	0.074	0.068	0.074	0.065	0.065	0.065	0.057	0.057	–	–
Mg/m^3	1.91	2.07	1.88	2.07	1.80	1.80	1.80	1.59	1.59	–	–
THICKNESS, in.	0.015	0.026	0.028	0.014	0.0105	0.021	0.021	0.017	0.007	0.010	0.010
mm	0.38	0.66	0.70	0.38	0.27	0.53	0.53	0.432	0.178	0.254	0.254
BREAK STRENGTH											
WARP, lb/in.	65	190	155	120	–	–	–	–	–	–	–
n/2.5 cm	290	845	680	550	–	–	–	–	–	–	–
FILL, lb/in.	60	100	135	70	–	–	–	–	–	–	–
n/2.5 cm	270	450	590	310	–	–	–	–	–	–	–
FILAMENT COUNT/YARN BUNDLE	1,000	2,000	2,000	1,000	3,000	3,000	6,000	3,000	1,000	3,000	3,000

[a] UNION CARBIDE DATA

[b] HERCULES DATA

[c] FIBERITE DATA

[d] HEXCEL DATA

APPENDIX C

Typical Data for Commercially Produced Glass Fiber Thermoplastic Composites

	GLASS FIBER CONTENT, WEIGHT %	SPECIFIC GRAVITY	TENSILE STRENGTH, Ksi (MPa)	TENSILE ELONGATION,* %	TENSILE MODULUS, Msi (GPa)	FLEXURAL STRENGTH, Ksi (MPa)	FLEXURAL MODULUS, Msi (GPa)	COMPRESSIVE STRENGTH, Ksi (MPa)
TEST METHOD (ASTM)		D792	D6 38	D638	D638	D790	D790	D695
ABS	10	1.10	8.5 (59)	3.0	6.7 (46)	14.8 (102)	6.5 (45)	12.0 (83)
(acrylonitrile-	20	1.22	11.0 (76)	2.0	7.4 (51)	15.5 (107)	7.1 (49)	14.0 (96)
butadiene-styrene)	30	1.28	13.0 (90)	1.4	9.1 (63)	16.3 (112)	9.3 (64)	15.0 (103)
Acetal	10	1.54	10.5 (72)	2.4	9.6 (66)	15.5 (107)	8.8 (61)	10.0 (69)
	30	1.68	12.0 (83)	2.0	11.2 (77)	16.5 (114)	10.5 (72)	11.8 (81)
Nylon	15	1.25	15.0 (103)	4.0	8.5 (59)	23.0 (158)	7.8 (54)	14.0 (96)
	30	1.37	24.0 (165)	3.0	10.5 (72)	29.0 (200)	10.0 (69)	24.0 (165)
Nylon 6/6	13	1.23	14.0 (96)	4.0	9.0 (62)	25.0 (172)	6.5 (45)	13.5 (93)
	30	1.37	25.0 (172)	3.0	13.0 (90)	34.0 (234)	13.0 (90)	27.0 (186)
Nylon 6/12	30	1.30	19.6 (135)	4.0	12.0 (83)	28.0 (193)	11.0 (76)	20.0 (138)
Polycarbonate	10	1.26	12.0 (83)	9.0	7.5 (52)	16.0 (110)	6.0 (41)	14.0 (96)
	30	1.43	17.5 (121)	2.0	12.5 (86)	20.5 (141)	10.0 (69)	17.0 (117)
Polyester (thermoplastic)	30	1.52	19.0 (131)	4.0	12.0 (83)	28.0 (193)	11.5 (79)	18.0 (124)
Polyethylene	10	1.04	5.2 (36)	4.0	3.7 (25)	6.6 (45)	3.6 (25)	5.0 (34)
	30	1.18	8.6 (59)	3.0	7.3 (50)	12.9 (89)	7.1 (49)	6.0 (41)
Polyphenylene Oxide (modified)	20	1.21	14.5 (100)	5.0	9.25 (64)	18.5 (127)	7.5 (52)	17.6 (121)
Polyphenylene Sulfide	40	1.64	22.0 (152)	3.0	20.5 (141)	37.0 (255)	19.0 (131)	21.0 (145)
Polypropylene	10	0.98	6.3 (43)	4.0	3.6 (24)	7.8 (54)	3.5 (24)	6.0 (41)
	20	1.04	6.5 (45)	3.0	5.4 (37)	8.3 (57)	5.2 (36)	6.5 (45)
	30	1.12	6.8 (47)	2.0	6.4 (44)	9.2 (63)	6.2 (43)	6.8 (47)
Polypropylene	10	0.98	7.2 to 8.6 (50–59)	4.0	5.3 (37)	10.4 to 13.6 (72–94)	5.1 (35)	6.2 to 6.4 (42–44)
(chemically	20	1.04	8.2 to 9.8 (56–67)	3.0	5.6 (39)	11.8 to 15.4 (81–106)	5.4 (37)	6.4 to 6.8 (44–47)
coupled)	30	1.12	9.8 to 12.0 (67–83)	2.0	6.7 (46)	13.1 to 19.0 (90–131)	6.6 (45)	6.5 to 7.0 (45–48)
Polystyrene	20	1.20	11.0 (76)	1.5	11.0 (76)	15.5 (107)	9.9 (68)	15.0 (103)
	30	1.29	13.5 (93)	1.2	13.1 (30)	17.5 (121)	12.4 (85)	18.0 (124)
Polystyrene High Heat Copolymer	20	1.22	13.0 (90)	1.2	12.0 (83)	19.0 (131)	11.5 (79)	16.0 (110)
Polystyrene High Heat Terpolymer	30	1.35	12.0 (83)	1.8	9.4 (65)	17.8 (123)	8.3 (57)	11.0 (76)
Polysulfone	20	1.38	14.0 (96)	2.5	8.7 (60)	20.0 (138)	8.5 (59)	18.0 (124)
	40	1.55	18.0 (124)	1.5	16.8 (116)	25.0 (172)	15.5 (107)	20.0 (138)
Polyurethane	10	1.22	4.8 (33)	48.0	0.95 (6.5)	6.2 (43)	0.90 (6.2)	5.0 (34)
PVC (polyvinyl chloride)	20	1.58	14.0 (96)	3.0	1.1 (7.6)	21.0 (145)	10.0 (69)	12.0 (83)
SAN	20	1.22	14.5 (100)	1.8	12.5 (86)	19.0 (131)	11.0 (76)	17.5 (121)
(styrene acrylonitrile)	35	1.35	16.0 (110)	1.4	16.0 (110)	22.5 (155)	13.5 (93)	21.0 (145)

HEAT DEFLECTION TEMPERATURE @ 264 psi, (1.8 MPa) °F (°C)	COEFFICIENT OF LINEAR THERMAL EXPANSION, 10^4 in./in./°F (10 m/m/°C)	IMPACT STRENGTH, IZOD NOTCHED, ft-lb/in. (J/M)	MAXIMUM TEMPERATURE CONTINUOUS USE, °F (°C)	WATER ABSORPTION, 24 HOURS, %	VOLUME RESISTIVITY, ohm-cm	DIELECTRIC STRENGTH, DRY, V/mil	MOLD SHRINKAGE, in./in. or cm/cm	COST/$/in³ ($/cm³ × 10³)
D648	D256	D696		D570	D257	D149	D965	
208 (98)	1.2 (65)	2.3 (4.1)	170 (77)	0.3	10^{15}	450	0.003	.0276 (1.68)
210 (99)	1.1 (59)	2.1 (3.8)	180 (82)	0.3	10^{16}	465	0.002	.0317 (1.93)
212 (100)	1.0 (54)	1.7 (3.1)	180 (82)	0.2	10^{14}	480	0.002	.0345 (2.10)
255 (124)	1.0 (54)	2.9 (5.2)	230 (110)	0.22	10^{14}	510	0.006	.0557 (3.39)
325 (163)	0.8 (43)	2.4 (4.3)	260 (127)	0.2	10^{14}	480	0.005	.0506 (3.08)
385 (192)	1.5 (81)	1.7 (3.1)	200 (93)	1.8	10^{13}	420	0.007	.0550 (3.35)
400 (204)	2.2 (119)	1.5 (2.7)	230 (110)	1.3	10^{13}	410	0.004	.0558 (3.40)
475 (246)	1.0 (54)	1.5 (2.7)	225 (107)	1.0	10^{14}	530	0.007	.0542 (3.30)
485 (252)	2.0 (108)	1.3 (2.3)	260 (127)	0.9	10^{14}	500	0.004	.0558 (3.40)
390 (199)	2.2 (119)	2.2 (4.0)	230 (110)	0.2	10^{13}	500	0.004	.0939 (5.73)
280 (138)	2.0 (108)	1.8 (3.2)	260 (127)	0.14	10^{13}	440	0.005	.0646 (3.94)
290 (143)	2.4 (130)	1.3 (2.3)	270 (132)	0.12	10^{13}	480	0.003	.0779 (4.75)
415 (213)	1.8 (97)	1.4 (2.5)	250 (121)	0.06	10^{15}	600	0.003	.0532 (3.24)
230 (110)	1.4 (76)	3.0 (5.4)	180 (82)	0.08	10^{14}	680	0.005	.0201 (1.23)
255 (124)	1.7 (92)	2.1 (3.8)	200 (93)	0.06	10^{14}	610	0.003	.0218 (1.33)
290 (143)	1.8 (97)	2.0 (3.6)	240 (116)	0.06	10^{13}	420	0.003	.0489 (2.98)
510 (266)	1.5 (81)	1.1 (2.0)	450 (232)					
				0.01	10^{14}	510	0.002	.1450 (8.84)
260 (127)	0.8 (43)	2.6 (4.7)	180 (82)	0.05	10^{14}	440	0.007	.0203 (1.24)
270 (132)	1.1 (59)	2.4 (4.3)	190 (88)	0.05	10^{14}	440	0.006	.0222 (1.35)
280 (138)	1.3 (70)	2.1 (3.8)	210 (99)	0.04	10^{14}	420	0.006	.0243 (1.48)
280 (138)	1.3 (70)	2.5 (4.5)	200 (93)	0.05	10^{14}	430	0.006	.0221 (1.35)
292 (144)	1.4 (76)	2.3 (4.1)	230 (110)	0.04	10^{14}	425	0.006	.0244 (1.49)
297 (147)	1.5 (81)	2.0 (3.6)	250 (121)	0.04	10^{14}	425	0.006	.0271 (1.65)
210 (99)	1.0 (54)	2.3 (4.1)	170 (77)	0.1	10^{14}	410	0.002	.0238 (1.45)
212 (100)	1.0 (54)	1.8 (3.2)	180 (82)	0.1	10^{14}	415	0.001	.0279 (1.70)
240 (116)	1.1 (59)	2.2 (4.0)	220 (106)	0.28	10^{14}	400	0.003	.0303 (1.85)
300 (149)	1.5 (81)	2.0 (3.6)	240 (116)	0.10	10^{14}	400	0.002	.0542 (3.30)
355 (180)	1.2 (65)	1.5 (2.7)	325 (163)	0.2	10^{14}	500	0.004	.1181 (7.20)
365 (185)	1.5(81)	1.3 (2.3)	340 (171)					
				0.18	10^{14}	480	0.002	.1192 (7.27)
130 (54)	14.0 (760)	3.4 (6.1)	110 (43)	0.4	10^{14}	380	0.007	.1000 (6.10)
180 (82)	1.5 (81)	2.3 (4.1)	150 (66)	0.09	10^{14}	420	0.002	.0445 (2.71)
215 (102)	1.2 (65)	2.1 (3.8)	180 (82)	0.24	10^{14}	490	0.002	.0286 (1.74)
220 (106)	1.0 (54)	1.6 (2.9)	190 (88)	0.21	10^{14}	500	0.001	.0336 (2.05)

GLOSSARY

The definitions listed here were obtained from the following sources: 1. PLASTEC NOTE 14, Glossary of Plastic Terms, A Consensus. 2. Glossary to the Science of Composites. Joint Report #HPC 66-11 by Monsanto Company/Washington University and Advanced Research Projects Agency (ARPA) of the Department of Defense. 3. Grumman Technical Manual (Unpublished). Non-Metallic materials.

A-Stage An early stage in the polymerization reaction of certain thermosetting resins (especially phenolic) in which the material, after application to the reinforcement, is still soluble in certain liquids and is fusible; sometimes referred to as resole. (*See B-Stage and C-Stage.*)

ABL Bottle An internal pressure test vessel about 18 inches in diameter and 24 inches long, used to determine the quality and properties of the filament wound material in the vessel.

Abhesive A film or coating which is applied to one solid to prevent (or greatly decrease) the adhesion to another solid with which it is to be placed in intimate contact, e.g., a "parting" of "mold-release" agent.

Ablation An orderly heat and mass transfer process in which a large amount of thermal energy is expended by sacrificial loss of surface region material. The heat input from the environment is absorbed, dissipated, blocked, and generated by numerous mechanisms. The energy adsorption processes take place automatically and simultaneously, serve to control the surface temperature, and greatly restrict the flow of heat into the substrate interior.

Ablative Plastic A material which absorbs heat (while part of it is being consumed by heat) through a decomposition process (pyrolysis) which takes place near the surface exposed to the heat.

Accelerator A material which, when mixed with a catalyzed resin, will speedup the chemical reaction between the catalyst and resin: either in polymerizing of resins or vulcanization of rubbers. Also known as "promoter."

Activation The process of inducing radioactivity in a specimen by bombardment.

Activator An additive used to promote the curing of matrix resins and reduce curing time. (*See accelerator.*)

Additive Any substance added to another substance, usually to improve properties.

Adherend A body which is held to another body by an adhesive.

Adhesion The state in which two surfaces are held together at an interface by forces or interlocking action or both.

Adhesion, Mechanical Adhesion between surfaces in which the adhesive holds the parts together by interlocking action.

Adhesive, Contact An adhesive which requires that for satisfactory bonding, the surfaces to be joined shall be no further apart than about 0.1 mm.

Adhesive Film A synthetic resin adhesive, usually of the thermosetting type, in the form of a thin dry film of resin, used under heat and pressure as an interleaf in the production of laminated materials (particularly plywood and densified wood).

Adhesiveness The property defined by the adhesion stress $A = F/S$ where F is the perpendicular force to the glue line and S its surfce. It is expressed in kilograms per square millimeter.

Addition Polymerization A chemical reaction in which simple molecules (monomers) are added to each other to form long-chain molecules (polymers) and no byproducts are formed.

After-Bake See *post-cure*)

Aggregate A hard fragmented material used with an epoxy binder as a flooring or surfacing medium, or in epoxy tools.

Aging The effect, on materials, of exposure to an environment for an interval of time; to process of exposing materials to an environment for an interval of time.

Air-Bubble Void Air entrapment within and between the plies of reinforcement; noninterconnected, spherical in shape.

Air Locks Surface depressions on a molded part, caused by trapped air between the mold surface and the plastics material.

Air Vent Small outlet, to prevent entrapment of gases.

Alkyd Plastics Plastics based on resins composed principally of polymeric esters, in which the recurring ester groups are an integral part of the main polymer chain, and in which ester groups occur in most crosslinks that may be present between chains. (See *polyester plastics.*)

Alkyd Resins Products resulting from the condensation of polybasic alcohol with a polybasic organic acid.

Allyl Plastics Plastics based on resins made by addition polymerization of monomers containing allyl groups; diallyl phthalate (DAP), diallyl isophthalate (DAIP).

Alternating Stress Amplitude A test parameter of a dynamic fatigue test. One half the algebraic difference between the maximum and minimum stress in one cycle

$$= \frac{1}{2} (\sigma_{max} - \sigma_{min}).$$

Alternative Stress A stress varying between two maximum values which are equal but with opposite signs, according to a law determined in terms of the time.

Ambient The surrounding environmental conditions such as pressure or temperature.

Amine Resin A synthetic resin derived from the reaction urea, thiourea, melamine or allied compounds with aldehydes, particularly formaldehyde.

Anistropic Exhibiting different properties when tested along axes in different directions. (Anistropy)

Anisotropic Laminate One in which the strength properties are different in different directions.

Anisotropy of Laminates The difference of the properties along the directions parallel to the length or width into the lamination planes; or parallel to the thickness into the planes perpendicular to the lamination.

Anti-Static Agents Agents which, when added to the molding material or applied on the surface of the molded object, make it less conducting (thus hindering the fixation of dust).

Arc Resistance The total time in seconds that an intermittent arc may play across a plastic surface without rendering the surface conductive.

Ash Content The solid residue remaining after a reinforcing substance has been incinerated (or strongly heated).

Aspect Ratio The ratio of length to diameter of a fiber.

Atecticity The degree of random location that the side chains exhibit off the back bone chain.

Attenuation The process of making thin and slender; as applied to the formation of fiber from molten glass.

Autoclave A closed vessel for conducting a chemical reaction or other operation under pressure and heat.

Autoclave Molding After lay-up, entire assembly is placed in steam autoclave at 50 to 100 psi. Additional pressure achieves higher reinforcement loadings and improved removal of air. (Modification of pressure bag method.)

Automatic Mold A mold for injection or compression molding that repeatedly goes through the entire cycle, including ejection, without human assistance.

Automatic Press A hydraulic press for compression molding or an injection machine which operates continuously, being controlled mechanically, electrically, hydraulically or by a combination of any of these methods.

Axial Winding In filament wound reinforced plastics, a winding with the filaments parallel to the axis (zero degree helix angle).

B-Stage An intermediate stage in the reaction of certain thermosetting resins in which the material swells when in contact with certain liquids and softens when heated, but may not entirely dissolve or fuse; sometimes referred to as resistol. The resin in an uncured prepreg or premix is usually in this stage (See *A-stage and C-stage.*)

Back Draft An area of interference in an otherwise smooth-drafted encasement; an obstruction in the taper which would interfere with the withdrawal of the model from the mold.

Back Pressure Resistance of a material, because of its viscosity, to continued flow when mold is closing.

Bag Molding A technique in which the consolidation of the material in the mold is effected by the application of fluid pressure through a flexible membrane.

Balanced Construction Plywood which has an odd number of plies, and is symmetrical on both sides of its center line.

Balanced Design In filament wound reinforced plastics, a winding pattern so designed that the stresses in all filaments are equal.

Balanced-in-Plane Contour In a filament wound part, a head contour in which the filaments are oriented within a plane and the radii of curvature are adjusted to balance the stresses along the filaments with the pressure loading.

Balanced Twist An arrangement of twist in a plied yarn or cord which will not cause twisting on itself when the yarn or cord is held in the form of an open loop.

Barcol Hardness A hardness value obtained by measuring the resistance to penetration of a sharp steel point under a spring load. The instrument, called the Barcol Impressor, gives a direct reading on a 0-100 scale. The hardness value is often used as a measure of the degree of cure of a plastic.

Bare Glass Glass (yarns, rovings, fabrics) from which the sizing or finish has been removed; also, such glass before the application of sizing or finish.

Base The reinforcing material (glass fiber, paper, cotton, asbestos, etc.) which is impregnated with resin in the forming of laminates; an insulating support for an electrical printed pattern.

Batch A measured mix of various materials. (See *lot.*)

Batt Felted fabrics: structures built by the interlocking action of fibers themselves without spinning, weaving or knitting.

Bearing Area The diameter of the hole times the thickness of the material.

Bearing Strength The bearing stress at that point on the stress-strain curve where the tangent is equal to the bearing stress divided by *n* percent of the bearing hole diameter.

Bearing Stress The applied load in pounds divided by the bearing area. (Maximum bearing stress is the maximum load in pounds sustained by the specimen during the test, divided by the original bearing area.)

Biaxial Load A loading condition in which a laminate is stressed in at least two different directions in the plane of the laminate; a loading condition of a pressure vessel under internal pressure and with unrestrained ends.

Biaxial Winding In filament winding, a type of winding in which the helical band is laid in sequence, side by side, with crossover of the fibers eliminated.

Bi-directional Laminate A reinforced plastic laminate with the fibers oriented in various directions in the plane of the laminate; a cross laminate. (See *unidirectional laminate.*)

Binder The resin or cementing constituent or a plastic compound which holds the other components together; the agent applied to glass mat or performs to bond the fibers prior to laminating or molding.

Bisphenol A A condensation product formed by reaction of two (*bi*) molecules of phenol with acetone (A). This polyhydric phenol is a standard resin intermediate along with epichlorohydrin in the production of epoxy resins.

Blanket Plies which have been laid up in a complete assemby and placed on or in the mold all at one time (flexible bag process): also the form of "bag" in which the edges are sealed against the mold.

Bleedout In filament winding, the excess liquid resin that migrates to the surface of a winding.

Blister Undesirable rounded elevation of the surface of a plastic, whose boundaries may be more or less sharply defined, resembling in shape a blister on the human skin. The blister may burst and become flattened.

Blooming Irregular spotting or clouding on plastic surfaces caused by migration to the surface of incompatible particles, or by faulty coloring or lubricating methods.

Blush See *chalking.*

Bond Strength The amount of adhesion between bonded surfaces; a measure of the stress required to separate a layer of material from the base to which it is bonded (See *peel strength.*)

Boss Protuberance on a plastic part designed to add strength, to facilitate alignment during assembly, to provide for fastenings, etc.

Bottom Plate A steel plate fixed to the lower section of a mold. It is often used to join the lower section of the mold to the platen of the press.

Branched Polymer A polymer in which side chains brance out from the main chain at irregular intervals.

Breaking Extension (ϵ_B) The elongation necessary to cause rupture of the test specimen. The tensile strain at the moment of rupture.

$$\epsilon_B = \frac{\Delta L_{max}}{L_o} \quad \text{dimensions: unity}$$

Breaking Factor Breaking load divided by the original width of the test specimen, expressed in pounds per inch.

Breathing The opening and closing of a mold to allow gases to escape early in the molding cycle (also called "degassing"): permeability to air (plastic sheeting).

Broadgoods Woven glass or synthetic fiber or combination thereof, over eighteen inches in width.

Bubble A spherical, internal void; globule of air or other gas trapped within a plastic.

Buckling Crimping of the fibers in a composite material, often occuring in glass-reinforced thermosets due to resin shrinkage during cure.

Bulk Density The density of a molding material in loose form (granular, nodular, etc.) expressed as a ratio of weight to volume.

Bulk Factor The ratio of the volume of a molding compound or powdered plastic to the volume of the solid piece produced therefrom: the ratio of the density of the solid plastic object to the apparent density of the loose molding powder.

Bulk Modulus (B) The ratio of the hydrostatic pressure P to the volume strain.

$$B = \frac{P}{\Delta V / V_o} = \frac{PV_o}{\Delta V}$$

where V_o = original volume
ΔV = volume change due to applied pressure

See also *Young's Modulus.*

Burned Showing evidence of thermal decomposition through some discoloration, distortion, or destruction of the surface of the plastic.

Burst Strength (1) Hydraulic pressure required to burst a vessel of given thickness. Commonly used in testing filament-wound composite structures. (2) Pressure required to break a fabric by expanding a flexible diaphram or pushing a smooth spherical surface against a securely held circular area of fabric. The Mullen expanding diaphram and Scott ball burst machine are examples of equipment used for this purpose.

Bushing An electrically heated alloy container incased in insulating material, used for melting and feeding glass in the forming of individual fibers or filaments; the outer ring of any type of a circular tubing or pipe die which forms the outer surface of the extruded tube or pipe.

Butt Wrap Tape wrapped around an object in an edge-to-edge condition.

C-Stage The final stage in the reaction of certain thermosetting resins in which the material is relatively insoluble and infusible; sometimes referred to as resite. The resin in a fully cured thermoset molding is in this stage. (See *A-stage and B-stage.*)

Carbonization The process of pyrolyzation in an inert atmosphere at temperatures ranging from 1000–1500°C. All non carbon elements are driven off in the process.

Caul In plywood manufacture, a sheet the size of the platens used in hot pressing. Aluminum sheet (1/16 inch thick) is generally used.

Catalyst A substance which changes the rate of a chemical reaction without itself undergoing permanent change in its composition; a substance which markedly speeds up the cure of a compound when added in minor quantity as compared to the amounts of primary reactants. (See *hardener, inhibitor, promoter,* and *curing agent.*)

Catastropic Failures Failures of a mechanical and unpredictable nature.

Catenary A measure of the difference in length of the strands in a specified length of roving as a result of unequal tension; the tendency of some strands in a taut horizontal roving to sag lower than the others.

Cavity Depression in mold; the space inside a mold wherein a resin is poured; the female portion of a mold; that portion of the mold which encloses the molded article; which forms the outer surface of the molded article (often referred to as the die): also, the space between matched molds. (Depending on number of such depressions, molds are designated as Single-Cavity or Multiple-Cavity.)

Centerless Grinding A technique for machining parts having a circular cross-section, consisting of grinding the rod which is fed without mounting it on centers. Grinding is accomplished by working the material between wheels, which rotate at different speeds, the faster wheel being the abrasive wheel which cuts the Variations of the basic principle can be used to grind internal surfaces.

Centrifugal Casting A high production technique for cylindrical composites, such as pipe, in which chopped strand mat is positioned inside a hollow mandrel designed to be heated and rotated as resin is added and cured.

Chain Length The length of the stretched linear macromolecule. (It is most often expressed by the number of identical links; that is, the "degree of polymerization.")

Chain Scission The breaking of a molecular bond causing the loss of a side group or shortening of the overall chain.

Chalking Dry, chalk-like appearance of or deposit on the surface of a plastic.

Charge The measurement or weight of material, either liquid, preformed, or powder, used to load a mold at one time or during one cycle.

Chase The main body of the mold which contains the molding cavity or cavities, or cores, the mold pins, the guide pins or the bushings, etc.: an enclosure of any shape used to shrink-fit parts of a mold cavity in place, to prevent spreading or distortion in hobbing, or to enclose an assembly of two or more parts of a split cavity block.

Chill To cool a mold by circulating water through it; to cool a molding with an air blast or by immersing it in water.

Circuit (Filament Winding) One complete traverse of the fiber feed mechanism of a winding machine; one complete traverse of a winding band from one arbitrary point along the winding path to another point on a plane through the starting point and perpendicular to the axis.

Circuit-Board A sheet of insulating material laminated to foil, which is etched to produce a circuit pattern on one or both sides; printed circuit; printed wiring.

Circumferential ("Circ") Winding In filament wound reinforced plastics, a winding with the filaments essentially perpendicular to the axis (90° or level winding).

Clamping Plate A mold plate fitted to the mold and used to fasten the mold to the machine.

Clamping Pressure In injection molding and in transfer molding, the pressure which is applied to the mold to keep it closed; in opposition to the fluid pressure of the compressed molding material.

Coefficient of Cubical Expansion See *coefficient of expansion.*

Coefficient of Elasticity The reciprocal of Young's modulus in a tension test.

Coefficient of Expansion The fractional change in dimension of a material for a unit change in temperature. Also, "coefficient of thermal expansion."

Coefficient of Friction A measure of the resistance to sliding of one surface in contact with another surface.

Coefficient of Linear Expansion The change in length per unit resulting from a one degree rise in temperature.

$$= \frac{1}{L}\left(\frac{\partial L}{\partial T}\right) \qquad \text{where: } L = \text{Length}, \; T = \text{Temperature}$$

Coefficient of Thermal Expansion (α) The change in volume per unit volume produced by a one degree rise in temperature.

$$\alpha = \frac{1}{V}\left(\frac{\partial V}{\partial T}\right) = \left(\frac{\partial \ln V}{\partial T}\right)P$$

where V = Length
T = Temperature

Cohesion The propensity of a single substance to adhere to itself; the internal attraction of molecular particles toward each other; the ability to resist partition from the mass; internal adhesion; the force holding a single substance together.

Cold Flow The distortion which takes place in materials under continuous load at temperatures within working range. (See *creep, strain relaxation.*)

Cold-Setting Adhesive A synthetic resin adhesive capable of hardening at normal room temperature in the presence of a hardener.

Compatibility The ability of two or more substances combined with each other to form a homogeneous composition of useful plastic properties; for example, the suitability of a sizing or finish for use with certain general resin types; nonreactivity or negligible reactivity between materials in contact.

Compliance Tensile compliance; the reciprocal of Young's modulus; shear compliance; the reciprocal of shear modulus.

Composite A homogeneous material created by the synthetic assembly of two or more materials (a selected filler or reinforcing elements and compatible matrix binder) to obtain specific characteristics and properties.

Composites are subdivided into classes on the basis of the form of the structural constituents; Laminar: Composed of layer or laminar constituents; Particulate: The dispersed phase consists of small particles; Fibrous: The dispersed phase consists of fibers; Flake: The dispersed phase consists of flat flales; Skeletal: Composed of a continuous skeletal matrix filled by a second material.

Compression Mold A mold which is open when the material is introduced and which shapes the material by heat and by the pressure of closing. Also "compression molding."

Compressive Modulus (E_c) Ratio of compressive stress to compressive strain below the proportional limit. Theoretically equal to Young's Modulus determined from tensile experiments.

$$E_c = \frac{\sigma_c}{\epsilon_c}$$

Compressive Strength The ability of a material to resist a force that tends to crush; the crushing load at the failure of a specimen divided by the original sectional area of the specimen.

Compressive Stress The compressive load per unit area of original cross-section carried by the specimen during the compression test.

Condensation Agent Chemical compound which, besides its catalytic action, furnishes a complement of material necessary for the achievement of the polycondensation of a resin.

Condensation Polymerization A chemical reaction in which two or more molecules combine, with the separation of water or some other simple substance. If a polymer is formed, the process is called polycondensation. (See *polymerization.*)

Condensation Resin A resin formed by polycondensation; for example, the alkyd, phenol-aldehyde, and urea formaldehyde resins.

Conductivity Reciprocal of volume resistivity; the conductance of a unit cube of any material.

Contact Adhesive An adhesive which requires that the surfaces to be joined shall be no further apart than about 0.1 mm for satisfactory bonding.

Contact Molding A process for molding reinforced plastics in which reinforcement and resin are placed on a mold, cure is either at room temperature using a catalyst-promotor system or by heat in an oven and no additional pressure is used.

Contact Pressure Resins Liquid resins which thicken or polymerize on heating and, when used for bonding laminates, require little or no pressure.

Continuous Filament An individual rod of glass of small diameter, which is flexible and of great or indefinite length.

Continuous Filament Yarn Yarn formed by twisting two or more continuous filaments into a single, continuous strand.

Cooling Fixture A fixture used to maintain the shape or dimensional accuracy of a molding or casting after it is removed from the mold and until the material is cool enough to retain its shape.

Copolymer A long-chain molecule formed by the reaction of two or more dissimilar monomers. (See *polymer.*)

Copolymerization The building up of linear or nonlinear macromolecules (copolymers) in which many monomers, possessing molecules having one or many double bonds, have located in every macromolecule of different size which constitutes the copolymerizate, following alternations which may be regular or not. (See also *polymerization.*)

Core The central member of a sandwich construction to which the faces of the sandwich are attached; the central member of a plywood assembly; a channel in a mold for circulation of heat-transfer media; part of a complex mold that forms undercut parts.

Core-Retainer Plate See *chase.*

Count (1) Fabric: Number of warp and filling yarns per inch in woven cloth. (2) Yarn: Size based on relation of length and weight. Basic unit = tex

Coupling Agent Any chemical substance designed to react with both the reinforcement and matrix phases of a composite material to form or promote a stronger bond at the interface; a bonding link.

Crack An actual separation of molded material, visible on opposite surfaces of the part, and extending through the thickness; a fracture.

Crazing Fine cracks which may extend in a network on or under the surface of a plastic material.

Creel A device for holding the required number of roving balls or supply packages in desired position for unwinding onto the next processing step.

Creep The change in dimension of a plastic under load over a period of time, not including the initial instantaneous elastic deformation. (Creep at room temperature is called "cold flow.")

Crimp The waviness of a fiber. It determines the capacity of fibers to cohere under light pressure. Measured either by (1) number of crimps or waves per unit length, or (2) the percent increase in extent of the fiber on removal of the crimp.

Critical Strain The strain at the yield point.

Critical Longitudinal Stress (Fibers) The longitudinal stress necessary to cause internal slippage and separation of a spun yarn; the stress necessary to overcome the inter-fiber friction developed as a result of twist.

Cross-Laminated Laminated so that some of the layers of material are oriented at right angles to the remaining layers with respect to the grain or strongest direction in tension. Balanced construction about the center line of the thickness of the laminate is normally assumed.

Crosswise Direction Crosswise refers to the cutting of specimens and to the application of load. For rods and tubes, crosswise is the direction perpendicular to the long axis. For other shapes or materials that are stronger in one direction than in another, crosswise is the direction that is weaker. For materials that are equally strong in both directions, crosswise is an arbitrarily designated direction at right angles to the lengthwise direction.

Crystallite A small crystal.

Cull Material remaining in a transfer chamber after the mold has been filled. (Unless there is a slight excess in the charge, the operator cannot be sure the cavity will be filled.)

Cure To change the properties of a resin by chemical reaction, which may be condensation or addition; usually accomplished by the action of either heat or catalyst, or both, and with or without pressure.

Curing Agent A catalytic or reactive agent which when added to a resin causes polymerization; synonymous with hardener.

Curing Temperature Temperature at which a cast, molded, or extruded product, a resin-impregnated reinforcement, an adhesive, etc., is subjected to curing.

Curing Time The period of time during which a part is subjected to heat or pressure, or both, to cure the resin; interval of time between the instant of cessation of relative movement between the moving parts of a mold and the instant that pressure is released. (Further cure may take place after removal of the assembly from the conditions of heat or pressure.)

Cut Number of 100 yard lengths per pound (asbestos, wool, glass).

Cut-Off The line where the two halves of a compression mold come together. (Also called "flash groove" or "pinch-off.")

Cycle The complete, repeating sequence of operations in a process or part of a process. In molding, the cycle time is the period (or elapsed time) between a certain point in one cycle and the same point in the next.

D-Glass A high boron content glass made especially for laminates requiring a precisely controlled dielectric constant.

Damping (Mechanical) Mechanical damping gives the amount of energy dissipated as heat during the deformation of a material. Perfectly elastic materials have no mechanical damping.
Damping terms may be calculated by many methods, which include use of the logarithmic decrement, the area of hysteresis loops, and others.

Deep-Draw Mold A mold having a core which is long in relation to the wall thickness.

Daylight The distance, in the open position, between the moving and fixed tables or the platens of a hydraulic press. In the case of a multidaylight press, daylight is the distance between adjacent platens.

Deflection Temperature Under Load The temperature at which a simple beam has deflected a given amount under load (formerly called heat distortion temperature).

Deformation Under Load The dimensional change of a material under load for a specific time following the instantaneous elastic deformation caused by the initial application of the load. (Also, "cold flow" or "creep.")

Degassing See *breathing*.

Delaminate: Delamination To split a laminated plastic material along the plane of its layers. (See *laminate*.) Physical separation or loss of bond between laminate plies.

Denier A yarn and filament numbering system in which the yarn number is equal numerically to the weight in grams of 9000 meters. (Used for continuous filaments.) The lower the denier, the finer the yarn.

Desizing The process of eliminating sizing, which is generally starch, from gray goods prior to applying special finishes or bleaches (for yarn such as glass or cotton).

Diamagnetic Susceptibility A relative measure of the degree to which a molecule or atom is repelled by a magnetic field.

Dielectric A nonconductor of electricity; the ability of a material to resist the flow of an electrical current.

Dielectric Constant (ϵ) The ratio of the capacity of a condenser having a dielectric material between the plates to that of the same condenser when the dielectric is replaced by a vaccum; a measure of the electrical charge stored per unit volume at unit potential.

$$\epsilon = \frac{C_p}{C_o} = \frac{Q}{C_o V}$$

Dielectric Curing The curing of a synthetic thermosetting resin by the passage of an electric charge produced from a high frequency generator through the resin.

Dielectric Heating (Electronic Heating) The plastic to be heated forms the dielectric of a condenser to which is applied a high-frequency (20-to-80 mc) voltage. Dielectric loss in the material is the basis. (See *high-frequency heating*.)

Dielectric Loss A loss of energy eventually showing through the rise in heat of a dielectric placed in an alternative electric field.

Dielectric Loss Angle (Dielectric Phase Difference The difference between ninety degrees (90°) and the dielectric phase angle.

Dielectric Phase Angle The angular difference in phase between the sinusoidal voltage applied to the dielectric and the resulting current.

Dielectric Strength The average potential per unit thickness at which failure of the dielectric material occurs.

Dimensional Stability Ability of a plastic part to retain the precise shape to which it was molded, cast, or otherwise fabricated.

Displacement Angle (Filament Winding) The advancement distance of the winding ribbon on the equator after one complete circuit.

Dissipation Factor-Electrical (D_e) The ratio of the power loss in a dielectric material to the total power transmitted through the dielectric, the imperfection of the dielectrtic. Equal to the tangent of the loss angle.

$$D_c = \frac{\epsilon''}{\epsilon'} = \tan \delta = \frac{1}{2\pi f C_p R_p}$$

where: f = frequency of applied voltage in cps
C_p = equivalent parallel capacity
R_p = equivalent parallel resistance

Doctor Roll; Doctor Bar A device for regulating the amount of liquid material on the rollers of a spreader.

Doily (Filament Winding) The planar reinforcement that is applied to a local area between windings to provide extra strength in an area where a cut-out is to be made; for example, port openings.

Dome (Filament Winding) In a cylindrical container, that portion which forms the integral ends of the container.

Doubler (Filament Winding) A local area with extra wound reinforcement, wound integrally with the part, or wound separately and fastened to the part.

Draft The taper or slope of the vertical surfaces of a mold designed to facilitate removal of molded parts.

Draft Angle The angle made by the tangent to the surface in that point and the direction of ejection.

Drape The ability of preimpregnated broadgoods to conform to an irregular shape; textile conformity.

Dry Spot (of a Laminate) Area of incomplete surface film on laminated plastics; in laminated glass, an area over which the interlayer and the glass have not become bounded. (See *resin-starved.*)

Dry Winding A term used to describe filament winding using pre-impregnated roving, as differentiated from wet winding where unimpregnated roving is pulled through a resin bath just prior to winding on a mandrel. (See *wet winding.*)

Dry Layup Construction of a laminate by the layering of preimpregnated reinforcement (partly cured resin) in a female mold or on a male mold, usually followed by bag-molding or autoclave molding.

Dwell A pause in the application of pressure to a mold, made just before the mold is completely closed, to allow the escape of gas from the molding material: (in filament winding) the time that the traverse mechanism is stationary while the mandrel continues to rotate to the appropriate point for the traverse to begin a new pass.

Dynamic Modulus The ratio of stress to strain under vibratory condtions (calculated from data obtained from either free or forced vibration tests, in shear, compression, or elongation).

E-Glass A borosilicate glass; the type most used for glass fibers for reinforced plastics; suitable for electrical laminates because of its high resistivity. (Also called "electric glass.")

E.S.C. See *environmental stress cracking.*

Edge Distance Ratio The distance from the center of the bearing hole to the edge of the specimen in the direction of the principal stress, divided by the diameter of the hole.

Edgewise Refers to the cutting of specimens and to the application of load. The load is applied edgewise when it is applied to the edge of the original sheet or specimen. For compression molded specimens of square cross section, the edge is the surface parallel to the direction of motion of the molding plunger; for injection molded specimens of square cross section, this surface is selected arbitrarily; for laminates the edge is the surface perpendicular to the laminae. (See *Flatwise.*)

Ejection The process of removing a molding from the mold impression; by mechanical means, by hand, or by the use of compressed air.

Ejection Plate A metal plate used to operate ejector pins; designed to apply a uniform pressure to them in the process of ejection.

Ejection Ram A small hydraulic ram fitted to a press for the purpose of operation ejector pins.

Elastic Deformation That part of the total strain in a stressed body which disappears upon removal of the stress.

Elasticity That property of plastic materials by virtue of which they tend to recover their original size and shape after deformation.

Elasticity, Coefficient of See *coefficient of elasticity.*

Elastic Limit The greatest stress which a material is capable of sustaining without permanent strain remaining upon the complete release of the stress. A material is said to have passed its elastic limit when the load is sufficient to initiate plastic, or nonrecoverable, deformation.

Elastic Recovery That fraction of a given deformation which behaves elastically.

$$\text{elastic recovery} = \frac{\text{elastic extension}}{\text{total extension}}$$

A perfectly elastic material has an elastic recovery = 1.

A perfectly plastic material has an elastic recovery = 0.

Electroformed Molds A mold made by electroplating metal on the reverse pattern on the cavity.

Elongation Deformation caused by stretching; the fractional increase in length of a material stressed in tension. (When expressed as percentage of the original gage length it is called percentage elongation.)

Elongation at Break Elongation recorded at the moment of rupture of the specimen, often expressed as a percentage of the original length.

Elongation Between Gages Changes in length produced between fixed gage points on the specimen by a load.

Emisivity The ratio of the total heat radiating power of a surface to that of a black body of the same area and of the same temperature.

End A strand of roving consisting of a given number of filaments gathered together. (The group of filaments is considered an "end" or strand before twisting; a "yarn" after twist has been applied); an individual warp yarn, thread, fiber, or roving.

End Count An exact number of ends supplied on a ball or roving.

Endurance Limit See *fatigue limit.*

Environmental Stress Cracking The susceptibility of the thermoplastic resin to crack or craze when in the presence of surface active agents or other environments. (E.S.C.)

Epichlorohydrin The basic epoxidizing resin intermediate in the production of epoxy resins. It contains an epoxy group and is highly reactive with polyhydric phenols such as bisphenol A.

Epoxide Equivalent The weight of a resin in grams which contains one gram equivalent of epoxy.

Epoxide A reactive group in which an oxygen atom is joined to each of two carbon atoms which are already united in some other way.

Epoxy Plastics Plastics based on resins made by the reaction of epoxides or oxiranes with other materials such as amines, alcohols, phenols, carboxylic acids, acid anhydrides and unsaturated compounds.

Equator (Filament Winding) In a pressure vessel, the line described by the junction of the cylindrical portion and the end dome.

Even Tension Describes the process whereby each end of roving is kept in the same degree of tension as the other ends making up that ball of roving. (See *catenary.*)

Exotherm The liberation or evolution of heat during the curing of a plastic product.

Extend To add fillers or lower cost materials in an economy-producing endeavor; to add inert materials to improve void filling characteristics and to reduce crazing.

Extenders Low cost materials used to dilute or extend high cost resins without much lessening of properties. (See *filler.*)

Extensibility The ability of a material to extend or elongate upon application of sufficient force. Expressed as percent of the original length.

$$\text{extensibility} = \frac{\Delta L_{max}}{L_o} \times 100 = \epsilon_B \times 100$$

Fabric A material constructed of interlaced yarns, fibers, or filaments; usually a planar structure. Nonwovens are sometimes included in this classification.

Fabricating, Fabrication The manufacture of plastic products from molded parts, rods, tubes, sheeting, extrusions, or other form by appropriate operations such as punching, cutting, drilling and tapping. Fabrication includes fastening plastic parts together or to other parts by mechanical devices, adhesives, heat sealing, or other means.

Fan In glass fiber forming, the fan-shape that is made by the filaments between the bushing and the shoe.

Fatigue The failure or decay of mechanical properties after repeated applications of stress. (Fatigue tests give information on the ability of a material to resist the development of cracks which eventually bring about failure as a result of a large number of cycles.)

Fatigue Life The number of cycles of deformation required to bring about failure of the test specimen under a given set of oscillating conditions.

Fatigue Limit The stress below which a material can be stressed cyclically for an infinite number of times without failure.

Fatigue Notch Factor The ratio of the fatigue strength of a specimen with no stress concentrator to the fatigue strength of a similar specimen with a stress concentrator.

Fatigue Ratio The ratio of fatigue strength to tensile strength. Mean stress and alternating stress must be stated.

Fatigue Strength The maximum cyclic stress a material can withstand for a given number of cycles before failure occurs; the residual strength after being subjected to fatigue.

Felt A fibrous material made up of interlocked fibers by mechanical or chemical action, moisture or heat; made from asbestos, cotton, glass, etc. (See *batt.*)

Fiber Relatively short lengths of very small cross section of various materials can be made by chopping filaments (converting). Also "filament," "thread," "bristle." (See *staple fibers.*)

Fiber Glass An individual filament made by attenuating molten glass. A continuous filament is a glass fiber of great or indefinite length; a staple fiber is a glass fiber of relatively short length (generally less than 17 inches.)

Fiber Diameter The measurement (expressed in hundred-thousandths) of the diameter of individual filaments.

Fiber Orientation Fiber alignment in a nonwoven or a mat laminate where the majority of fibers are in the same direction, resulting in a higher strength in that direction.

Fiber Pattern Visible fibers on the surface of laminates or moldings; the thread size and weave of glass cloth.

Fiber Strain in Flexure The maximum strain in the outer fiber occurring at midspan.

Fiber Stress in Flexure When a beam of homogeneous, elastic material is tested in flexure as a simple beam supported at two points and loaded at the midpoint, the maximum stress in the outer fiber occurs at mid-span.

Fibrillar A fiber-like aggregation of molecules.

Filaments Individual glass fibers of indefinite length, usually as pulled from a stream of molten glass flowing through an orifice of the bushing. In the operation, a number are gathered together to make a strand or end of roving or yarn.

Filament Weight Ratio In a composite material, the ratio of filament weight to total weight of the composite.

Filament Winding A process for fabricating a composite structure in which continuous reinforcements (filament, wire, yarn, tape, or other) either previously impregnated with a matrix material or impregnated during the winding are placed over a rotating and removable form or mandrel in a previously prescribed way to meet certain stress conditions. Generally the shape is a surface of revolution and may or may not include end closures. When the right number of layers are applied the wound form is cured and the mandrel removed.

Fill Yarn running from selvage to selvage at right angles to the warp in a woven fabric.

Filler A relatively inert material added to a plastic mixture to reduce cost, to modify mechanical properties, to serve as a base for color effects, or to improve the surface texture. (See *reinforced plastic, binder, extenders.*)

Fillet A rounded filling of the internal angle between two surfaces of a plastic molding.

Filling Yarn The transverse threads or fibers in a woven fabric; those fibers running perpendicular to the warp. (Also called "weft.")

Film Adhesive A synthetic resin adhesive usually of the thermosetting type, in the form of a thin dry film of resin with or without a paper carrier.

Finish A material applied to the surface of fibers in a fabric used to reinforce plastics and intended to improve the physical properties of such reinforced plastics over that obtained using reinforcement without finish.

Fish-Eye Small globular mass which has not blended completely into the surrounding material and is particularly evident in a transparent or translucent material.

Flame Resistance Ability of a material to extinguish flame once the source of heat is removed. (See *self-extinguishing resin.*)

Flame Retardants Certain chemicals which are used to reduce or eliminate a resin's tendency to burn. (For polyethylene and similar resins, chemicals such as antimony trioxide and chlorinated paraffins are useful.)

Flame Retarded Resin A resin which is compounded with certain chemicals to reduce or eliminate its tendency to burn.

Flame Spraying Method of applying a plastic coating in which finely powdered fragments of the plastic, together with suitable fluxes, are projected through a cone of flame onto a surface.

Flammability Measure of the extent to which a material will support combustion.

Flash That portion of the charge which flows from or is extruded from the mold cavity during the molding; extra plastic attached to a molding along the parting line. (It must be removed before the part is considered finished.)

Flash Mold A mold designed to permit the escape of excess molding material. (Such a mold relies upon back pressure to seal the mold and put the piece under pressure.

Flatwise Refers to the cutting of specimens and to the application of load. The load is applied flatwise when it is applied to the face of the original sheet or specimen. For compression molded specimens of square cross section, this is the surface perpendicular to the direction of motion of the molding plunger; for injection molded specimens of square cross section, this surface is selected arbitrarily; for laminates this is the surface parallel to the laminae.

Flexibilizer An additive that makes a finished plastic more flexible or tough.

Flexible Molds Molds made of rubber or elastomeric plastics used for casting plastics. They can be stretched to remove cured pieces with undercuts.

Flexural Modulus The ratio, within the elastic limit, of the applied stress on a test specimen in flexure to the corresponding strain in the outermost fibers of the specimen.

Flexural Rigidity (Fibers) A measure of the rigidity of individual strands or fibers; the force couple required to bend a specimen to unit radius of curvature.

Flexural Rigidity (*D*)-Plate A measure of the rigidity of a plate.

$$D = \frac{Eh^3}{12(1 - \nu)}$$

where E = Young's modulus
h = thickness of plate
ν = Poisson's ratio

units: inches pound

Flexural Strength The resistance of a material to being broken by bending stresses; the strength of a material in bending, expressed as the tensile stress of the outermost fibers of a bent test sample at the instant of failure. (With plastics, this value is usually higher than the straight tensile strength.)

Flow The movement of resin under pressure, allowing it to fill all parts of a mold; flow or creep—the gradual but continuous distortion of a material under continued load, usually at high temperatures.

Flow Line A mark on a molded piece made by the meeting of two flow fronts during molding. (Also, "striae," or "weld-mark," or "weld-line.")

Fluidized-Bed Coating A coating technique which consists of heating the object or surface to be coated well above the melting point of the plastic to be used as a coating; and then dipping the object in a fluidized bed of the pulverized or finely divided plastic. The powdered plastic melts and adheres to the object, which is then withdrawn and cooled.

Fluted Core An integrally woven reinforcement material consisting of ribs between two skins for unitized sandwich construction.

Foamed Plastics Resins in sponge form, flexible or rigid, cells closed or interconnected, density anywhere from that of the solid parent resin to 2 pounds/cubic feet. Compressive strength of rigid foams is fair, making them useful as core materials for sandwich constructions. Both types are good heat barriers. Also, a chemical cellular plastic whose structure is produced by gases generated from the chemical interaction of its constituents.

Foam-in-Place Refers to the deposition of foams which requires that the foaming machine be brought to the work which is "in place" as opposed to bringing the work to the foaming machine.

Force The male half of the mold which enters the cavity, exerting pressure on the resin and causing it to flow (also refered to as "punch"); either part of a compression mold; top force and bottom force.

Fracture Rupture of the surface without complete separation of laminate.

Free-Radical Polymerization A type of polymerization in which the propagating species is a long chain free-radical initiated by the introduction of free-radicals by thermal or photochemical decomposition.

Friction, Coefficient of See *coefficient of friction.*

FRP Fibrous-glass-reinforced plastic; a general term covering any type of plastic reinforced cloth, mat, strands, or any other form of fibrous glass.

Fungus Resistance The resistance of a material to attack by fungi in conditions promoting their growth.

Furane Plastics Plastics based on resins in which the furane ring is an integral part of the polymer chain, made by the polymerization or polycondensation of furfural, furfural alcohol, or other compounds containing a furane ring, or by the reaction of these furane compounds with not more than an equal weight of other compounds. ("Furane is also spelled "furan.")

Gage Length Length over which deformation is measured.

Gap (Filament Winding) The space between successive windings, which windings are usually intended to lay next to each other.

Gel A semisolid system consisting of a network of solid aggregates in which liquid is held; the initial jell-like solid phase that develops during the formation of a resin from a liquid.

Gelation Time The interval of time, in connection with the use of synthetic thermosetting resins, extending from the introduction of a catalyst into a liquid adhesive system until the interval of gel formation.

Gel Coat A resin applied to the surface of a mold and gelled prior to lay-up. (The gel coat becomes an integral part of the finished laminate, and is usually used to improve surface appearance, etc.)

Gel Point The stage at which a liquid begins to exhibit pseudoelastic properties. (This stage may be conveniently observed from the inflection point on a viscosity-time plot). Also, "gel time."

Geodesic The shortest distance between two points on a surface.

Geodesic Isotensoid Constant stress level in any given filament at all points in its path.

Geodesic-Isotensoid Contour In filament wound reinforced plastic pressure vessels, a dome contour in which the filaments are placed on geodesic paths so that the filaments will exhibit uniform tensions throughout their length under pressure loading.

Geodesic Ovaloid A contour for end domes, the fibers forming a geodesic line—the shortest distance between two points on a surface of revolution. The forces exerted by the filaments are proportioned to meet hoop and meridional stresses at any point.

Glass An inorganic product of fusion which has cooled to a rigid condition without crystallizing. (Glass is typically hard and relatively brittle, and has a conchoidal fracture.)

Glass Fiber A glass filament that has been cut to a measurable length. Staple fibers of relatively short length are suitable for spinning into yarn.

Glass Filament A form of glass that has been drawn to a small diameter and extreme length. (Most filaments are less than 0.005 inch in diameter.)

Glass Filament Bushing The unit through which molten glass is drawn in making glass filaments.

Glass Finish A material applied to the surface of a glass reinforcement to improve its effect upon the physical properties of the reinforced plastic. (Also "bonding agent.")

Glass Flake Very thin, irregularly shaped flakes of glass made typically by shattering a continuous thin-walled tube of glass.

Glass Former An oxide which forms a glass easily; also one which contributes to the network of silica glass when added to it.

Glass, Percent by Volume The product of the specific gravity of a laminate and the percent glass by weight, divided by the specific gravity of the glass.

Glass Stress In a filament wound part, usually a pressure vessel, the stress calculated using the load and the cross sectional area of the reinforcement only.

GPD Grams per denier. (See *tenacity.*)

Graphitization The process of pyrolyzation in an inert atmosphere at temperatures in excess of 1800° C, usually as high as 2700° C. This produces a turbostratic "graphite" crystal structure.

Gravity Closing In a down-stroke press, the closing motion is actuated by the weight of the ram and associated parts only.

Greige, Gray Any fabric before finishing, as well as any yarn or fiber before bleaching or dyeing. Also "gray goods," "greige goods."

Grex A universal yarn numbering system in which the yarn number is equal numerically to the weight in grams of 10,000 meters.

Guide Pin A pin which guides mold halves into alignment on closing.

Guide Pin Bushing The bushing through which the guide pin moves in closing of the mold.

Gusset A piece used to give added size or strength in a particular location of an object; the folded-in portion of a flattened tubular film.

Hand The softenss of a piece of fabric, as determined by the touch (individual judgment).

Hand Layup The process of placing (and working) successive plies of reinforcing material or resin-impregnated reinforcement in position on a mold by hand.

Hardener A substance or mixture added to a plastic composition to promote or control the curing action by taking part in it. Also, a substance added to control the degree of hardness of the cured film. (See *catalyst.*)

Hardness The resistance to surface indentation usually measured by the depth of penetration (or arbitrary units related to depth of penetration) of a blunt point under a given load using a particular instrument according to a prescribed procedure. (See *Barcol hardness, Rockwell hardness number.*)

Heat Build-up The temperature rise in a part resulting from the dissipation of applied strain energy as heat.

Heat-Convertible Resin A thermosetting resin convertible by heat to an infusible and insoluble mass.

Heat Distortion Point The temperature at which a standard test bar deflects under a stated load.

Heat Endurance The time of heat aging that a material can withstand before failing a specific physical test.

Heat Resistance The property or ability of plastics and elastomers to resist the deteriorating effects of elevated temperatures.

Hexa Hexamethylenetetramine: source of reactive methylene for curing novolaks.

High-Frequency Heating The heating of materials by dielectric loss in a high-frequency electrostatic field. The material is exposed between electrodes and is heated quickly and uniformly by absorption of energy from the electrical field.

High Pressure Laminates Laminates molded and cured at pressures not lower than 1000 psi, and more commonly in the range of 1200–2000 psi.

High Pressure Molding A molding process in which the pressure used is greater than 1000 psi.

High Pressure Spot An area containing very little resin; usually due to an excess of reinforcing materials.

Honeycomb Manufactured product of resin-impregnated sheet material (paper, glass fabric, etc.) or sheet metal, formed into hexagonal-shaped cells. Used as a core material in sandwich constructions (*which see*).

Hoop Stress The circumferential stress in a material of cylindrical form subjected to internal or external pressure.

Hydraulic Press A press in which the molding force is created by the pressure exerted on a fluid.

Hydromechanical Press A press in which the molding forces are created partly by a mechanical system and partly by an hydraulic system.

Hydrophilic Capable of adsorbing or absorbing water.

Hydrophobic Capable of repelling water.

Hygroscopic Capable of adsorbing and retaining atmospheric moisture.

Ignition Loss The difference in weight before and after burning; as with glass, the burning off of the binder or size.

Immediate Set The deformation found by measurement immediately after the removal of the load causing the deformation.

Impact Bar A test specimen of specified dimensions, utilized to determine the relative resistance of a plastic to fracture by shock.

Impact Shock A stress transmitted to an adhesive interface which results from the sudden jarring or vibrating of the bonded assembly.

Impact Strength The ability of a material to withstand shock loading; the work done in fracturing a test specimen in a specified manner under shock loading.

Impact Tests The impact test measures the energy necessary to fracture a standard notched bar by an impulse load. The notched test specimen is struck and fractured by a heavy pendulum, released from a known height, at the nadir of its swing. From a knowledge of the mass of the pendulum and the difference in the initial and final heights, the energy absorbed in fracture is calculated. The Izod Impact Test is typical of this technique.

A standard ASTM notch is: 45° included angle, 0.010 inch radius at the tip of the notch, and a notch depth of 0.100 inches in a 0.500 inch specimen.

Dimensions: usually reported in foot-pounds or ergs. For some plastics work, energy absorbed is given in foot-pounds per inch notch width.

Impregnate In reinforced plastics, the saturation of the reinforcement with a resin.

Impregnated Fabric A fabric impregnated with a synthetic resin (See *prepreg.*)

Inert Filler A material added to a plastic to alter the end item properties through physical rather than chemical means.

Infra-Red Part of the electromagnetic spectrum between visible light range and the radar range. Radiant heat is in this range, and infra-red heaters are much used in sheet thermoforming.

Inhibitor A substance which retards a chemical reaction; used in certain types of monomers and resins to prolong storage life.

Initial Modulus See *modulus, initial;* also *Young's modulus.*

Inorganic Designating or pertaining to the chemistry of all elements and compounds not classified as organic; matter other than animal or vegetable, such as earthy or mineral matter.

Inorganic Pigments Natural or synthetic metallic oxides, sulfides, and other salts, calcined during processing at 1200–2100 F. They impart heat- and light-stability, weathering resistance, and migration resistance to plastics.

Insert An integral part of a plastics molding consisting of metal or other material which may be molded into position or pressed into the molding after the molding is completed.

Insert, Eyelet-Type Insert having a section which protrudes from the material and is used for spinning over in assembly.

Instron An instrument utilized to determine the tensile and compressive properties of materials.

Insulation Resistance The electrical resistance between two conductors or systems of conductors separated only by insulating material; the ratio of the applied voltage to the total current between two electrodes in contact with a specified insulator.

Insulator A material of such low electrical conductivity that the flow of current through it can usually be neglected. Similarly, a material of low thermal conductivity.

Interface The junction point or surface between two different media; on glass fibers, the contact area between glass and sizing or finish; in a laminate, the contact area between the reinforcement and the laminating resin.

Interlaminar Shear Strength The maximum shear stress existing between layers of a laminated material.

Intermesh The positioning of adjacent blocks of honeycomb so that the outermost edge of one block falls within the outermost edge of the adjacent block.

Internal Stress Stress created within an adhesive layer by the movement of the adherends at differential rates or by the contraction or expansion of the adhesive layer. (See *stress*.)

Irradiation As applied to plastics, the bombardment with a variety of subatomic particles, generally alpha-, beta-, or gamma-rays. Used to initiate polymerization and copolymerization of plastics and in some cases to bring about changes in the physical properties of a plastic.

Irreversible Not capable of redissolving or remelting; chemical reactions which proceed in a single direction and are not capable of reversal (as applied to thermosetting resins.)

Isocyanate Plastics Plastics based on resins made by the condensation of organic isocyanates with other compounds. (See *urethane plastics;* generally reacted with polyols such as polyester or polyethers, and the reactants are joined through the formation of the urethane linkage).

Isomeric Composed of the same elements united in the same proportion by weight, but differing in one or more properties because of difference in structure.

Isotropic Laminate One in which the strength properties are equal in all directions.

Izod Impact Test A destructive test designed to determine the resistance of a plastic to the impact of a suddenly applied force.

Joint The location at which two adherends are held together with a layer of adhesive; the general area of contact for a bonded structure.

Joint, Butt A type of edge joint in which the edge faces of the two adherends are at right angles to the other faces of the adherents.

Joint, Edge A joint made by bonding the edge faces of two adherends.

Joint, Scarf A joint made by cutting away similar angular segments of two adherends and bonding them with the cut areas fitted together.

Joint, Lap A joint made by placing one adherend partly over another and bonding together the overlapped portions.

L/D Ratio A term used to define an extrusion screw, which denotes the ratio of the screw length to the screw diameter.

Lack of Fillout An area, occurring usually at the edge of a laminated plastic, where the reinforcement has not been wetted with resin.

Lacquer Solution of natural or synthetic resins in readily evaporating solvents, used as a protective coating.

Laminate To unite sheets of material by a bonding material usually with pressure and heat (normally used with reference to flat sheets); a product made by so bonding. (See also *bidirectional laminate, unidirectional laminate*.)

Laminated Parallel See *parallel-laminated.*

Laminated Molding A molded plastic article produced by bonding together, under heat and pressure in a mold, layers of resin-impregnated laminating reinforcement. Also, "laminated plastics."

Laminate Ply One layer of a product which is evolved by bonding together two or more layers of materials.

Land The portion of a mold which provides the separation or cut-off of the flash from the molded article; in the screw of an extruder, the bearing surface along the top of the flights; in an extrusion die, the surface parallel to the flow of material; in a semipositive or flash mold, the horizontal bearing surface; in a two-piece mold, a platform built up to the split line.

Land Area The whole of the area of contact, perpendicular to the direction of application of pressure of the seating faces of a mold; those faces which come into contact when the mold is closed.

Landed Force A force with a shoulder which seats on land in a landed positive mold. (Also "landed plunger.")

Lap In filament winding, the amount of overlay between successive windings, usually intended to minimize gapping; in textiles, a matted sheet of cotton wound on a spindle, produced by the picker. (Cotton lap is used extensively in preparing asbestos-cotton mixes in the carding machine.)

Lap Joint A joint made by placing one adherend partly over another and bonding the overlapped portions. (See *scarf joint.*)

Lay In glass fiber, the spacing of the roving bands on the roving package expressed in the number of bands per inch; in filament winding, the orientation of the ribbon with some reference, usually the axis of rotation.

Lay-Flat The property of non-warping in laminating adhesives; an adhesive material with good non-curling and non-distension characteristics.

Lay-up As used in reinforced plastics, the reinforcing material placed in position in the mold; the process of placing the reinforcing material in position in the mold; the resin-impregnated reinforcement; a description of the component materials, geometry, etc., of a laminate.

Lengthwise Direction Refers to the cutting of specimens and to the application of loads. For rods and tubes, lengthwise is the direction of the long axis. For other shapes of materials that are stronger in one direction than in the other, lengthwise is the direction that is stronger. For materials that are equally strong in both directions, lengthwise is an arbitrarily designated direction that may be with the grain, direction of flow in manufacture, longer direction, etc. (See *crosswise direction.*)

Level Winding See *circumferential winding.*

Light-Fastness Satisfactory resistance to light, particularly the colorants or other additives entering into the composition of a plastic material; light resistance.

Linear Expansion The increase of a planar dimension, measured by the linear elongation of a sample in the form of a beam which is exposed to two given temperatures. (See also *coefficient of linear expansion.*)

Line Pressure The pressure under which an air or hydraulic system operates.

Liner In a filament wound pressure vessel, the continuous, usually flexible coating on the inside surface of the vessel, used to protect the laminate from chemical attack or to prevent leakage under stress.

Liquidus Temperature The maximum temperature at which equilibrium exists between the molten glass and its primary crystalline phase.

Load-Deflection Curve A curve in which the increasing flexural loads are plotted on the ordinate axis and the deflections caused by those loads are plotted on the abcissae axis.

Loaded Pertains to roving or mat.

Longos Low angle helical or longitudinal windings.

Loop Tenacity The tenacity or strength value obtained by pulling two loops, as two links in a chain, against each other in order to demonstrate the susceptibility that a fibrous material has for cutting or crushing itself: loop strength.

Loss Angle The anti-tangent of the electrical dissipation factor. (See *dielectric loss angle.*)

Loss, Dielectric See *dielectric loss.*

Loss Factor The product of the dissipation factor and the dielectric constant of a dielectric material. See *dielectric loss factor.*

Loss Index A measure of a dielectric loss defined by the product of the power factor and the permittivity (dielectric constant).

Loss Modulus A damping term describing the dissipation of energy into heat when a material is deformed.

Loss on Ignition Weight loss, usually expressed as percent of total, after burning off an organic sizing from glass fibers, or an organic resin from a glass fiber laminate.

Loss Tangent See *dissipation factor.*

Lot A specific amount of material produced at one time and offered for sale as a unit quantity.

Low-Pressure Laminates In general, laminates molded and cured in the range of pressures from 400 psi down to and including pressure obtained by the mere contact of the plies.

Low-Pressure Molding The distribution of relatively uniform low pressure (200 psi or less) over a resin-bearing fibrous assembly of cellulose, glass, asbestos, or other material, with or without application of heat from external source, to form a structure possessing definite physical properties.

M-Glass A high beryllia content glass designed especially for high modulus of elasticity.

MVT See *moisture vapor transmission.*

Macerate To chop or shred fabric for use as a filler for a molding resin; the molding compound obtained when so filled.

Magneto Resistance The change of electrical resistance of a crystal when a magnetic field is applied.

Mandrel The core around which paper-, fabric-, or resin-impregnated glass is wound to form pipes, tubes, or vessels; in extrusion, the central finger of a pipe or tubing die.

Mass Stress (Fibers) Force per unit mass per unit length; grams per denier, etc. (Used the same way as force per unit area.)

Mat A fibrous material for reinforced plastic consisting of randomly oriented chopped filaments or swirled filaments with a binder; and available in blankets of various widths, weights, and lengths.

Mat Binder Resin applied to glass fiber and cured during the manufacture of mat, to hold the fibers in place and maintain the shape of the mat.

Matched Metal Molding A reinforced plastics manufacturing process in which matching male and female metal molds are used (similar to compression molding) to form the part — as opposed to low pressure laminating or spray-up.

Matrix See *resin.*

Matte A nonspecular surface having diffused reflective powers.

Mean Strain Analogous to mean stress.

Mean Stress (σ) A dynamic fatigue parameter. The algebraic mean of the maximum and minimum stress in one cycle.

units: psi

$$\sigma = \frac{1}{2}(\sigma_1 + \sigma_2)$$

where: σ_1 = maximum stress
σ_2 = minimum stress

Mechanical Adhesion Adhesion between surfaces in which the adhesive holds the parts together by interlocking action.

Melamine Formaldehyde A synthetic resin derived from the condensation reaction of melamine with formaldehyde or its polymers.

Melamine Plastics Plastics based on these resins.

Microminiaturization The technique of packaging a microminiature part or assembly composed of elements radically different in shape and form factor. Electronic parts are replaced by active and passive elements, through use of fabrication processes such as screening, vapor deposition, diffusion.

Micron One micron = .001 millimeter = .00003937 inch.

Mil The unit used in measuring the diameter of glass fiber strands, wire, etc. (1 mil = .001 inch).

Milled Fibers Continuous glass strands hammer-milled into small modules of filamentized glass. Useful as anticrazing reinforcing fillers for adhesives.

Modulus A number which expresses a measure of some property of a material: modulus of elasticity, shear modulus, etc.: a coefficient of numerical measurement of a property. (Note: The use of the word without modifying terms may be confusing; and such use should not be encouraged.)

Modulus, Dynamic—See *dynamic modulus*

Modulus in Compression The ratio of the compression stress to the strain in the material, over the range for which this value is constant. (See *compressive modulus.*)

Modulus in Flexure The ratio of the flexure stress to the strain in the material, over the range for which this value is constant.

Modulus, Initial Young's modulus.

Modulus in Shear The ratio of the shear stress to the strain in the material, over the range for which this value is constant.

Modulus in Tension The ratio of the tension stress to the strain in the material over the range for which this value is constant.

Modulus of Elasticity The ratio of the stress or load applied to the strain or deformation produced in a material that is elastically deformed. If a tensile strength of 2000 pounds per square inch results in an elongation of one percent the modulus of elasticity is 2000 divided by 0.01, or 200,000 per square inch (Young's modulus)

Modulus of Elasticity in Torsion The ratio of the torsion stress to the strain in the material, over the range for which this value is constant.

Modulus of Resilience The energy that can be absorbed per unit volume without creating a permanent distortion. Calculated by integrating the stress-strain curve from zero to the elastic limit and dividing by the original volume of the specimen.

Modulus of Rigidity Flexural rigidity.

Modulus of Rupture See *flexural strength.*

Mohs Hardness A measure of the scratch resistance of a material; the higher the number, the greater the scratch resistance (No. 10 being "diamond.")—

Moisture Absoprtion The pick-up of water vapor from air by a material. It relates only to vapor withdrawn from the air by a material and must be distinguished from water absorption, which is the gain in weight due to the take-up of water by immersion.

Moisture Vapor Transmission A rate at which water vapor will pass through a material at a specified temperature and relative humidity (gms-mil/24 hrs-100 in.).

Mold The cavity or matrix into or on which the plastic composition is placed and from which it takes form; to shape plastic parts or finished articles by heat and pressure; the assembly of all the parts that function collectively in the molding process.

Mold-Release Agent A liquid or powder used to prevent sticking or molded articles in the cavity. (See *parting agent.*)

Mold Seam Line on a molded or laminated piece, differing in color or appearance from the general surface, caused by the parting line of the mold.

Mold Shrinkage The immediate shrinkage which a molded part undergoes when it is removed from a mold and cooled to room temperature; the difference in dimensions, expressed in inches per inch between a molding and the mold cavity in which it was molded (at normal temperature measurement): the incremental difference between the dimensions of the molding and the mold from which it was made, expressed as a percentage of the dimensions of the mold.

Molding The shaping of a plastic composition within or on a mold, normally accomplished under heat and pressure; sometimes used to denote the finished part.

Molding Compounds Plastics in a wide range of forms to meet specific processing requirements. Granules or pellets are popular forms.

Molding Cycle The period of time occupied by the complete sequence of operations on a molding press requisite for the production of one set of moldings; the operations necessary to produce a set of moldings without reference to the time taken.

Molding Powder Plastic material in varying stages of granulation, and comprising resin, filler, pigments, plasticizers, and other ingredients, ready for use in the molding operation.

Molding Pressure The pressure applied to the ram of an injection machine or press to force the softened plastic completely to fill the mold cavities.

Molding, Pressure Bag See *pressure bag molding.*

Molding Pressure, Compression The unit pressure applied to the molding material in the mold. The area is calculated from the projected area taken at right angles to the direction of applied force and includes all areas under pressure during complete closing of the mold. The unit pressure is calculated by dividing the total force applied by this projected area, and is expressed in pounds per square inch.

Molding Pressure, Transfer The pressure applied to the cross-sectional area of the material pot or cylinder, expressed in psi.

Monofilament A single fiber or filament of indefinite length generally produced by extrusion; a continuous fiber of sufficient size to serve as yarn in normal textile operations. (Also "monofil.")

Monomer A simple molecule which is capable of reacting with like or unlike molecules to form a polymer; the smallest repeating structure of a polymer (mers); for addition polymers, this represents the original unpolymerized compound.

Multiple-Cavity Mold A mold with two or more mold impressions; that is, a mold which produces more than one molding per molding cycle.

Multi-Circuit Winding In filament winding, a winding that requires more than one circuit before the band repeats by laying adjacent to the first band.

Multifilament Yarn A multitude of fine, continuous filaments (often 5 to 100 individual filaments) usually with some twist in the yarn to facilitate handling. Sizes range from 5–10 denier up to a few hundred denier. Individual filaments in a multifilament yarn are usually about 1 to 5 denier.

Nol Ring A parallel filament wound test specimen used for measuring various mechanical strength properties of the material by testing the entire ring, or segments of it.

Nesting In reinforced plastics, the placing of plies of fabric so that the yarns of one ply lie in the valleys between the yarns of the adjacent ply (nested cloth).

Netting Analysis The analysis of filament-wound structures which assumes that the stresses induced in the structure are carried entirely by the filaments, and the strength of the resin is neglected; and also that the filaments possess no bending or shearing stiffness, and carry only the axial tensile loads.

Non—Hygroscopic Lacking the property of absorbing and retaining the appreciable quantity of moisture from the air (water vapor).

Nonpolar Having no concentration of electrical charges on a molecular scale; thus, incapable of significant dielectric loss. Examples among resins are polystyrene and polyethylene.

Nonrigid Plastic A plastic which has a stiffness or apparent modulus of elasticity of not over 10,000 psi at 23° C, when determined in accordance with ASTM D 747.

Non-Woven Fabric A planar structure produced by loosely bonding together yarns, rovings, etc. (See *fabric*.)

Notch Factor Ratio of the resilience determined on a plain specimen, to the resilience determined on a notched specimen.

Notch Sensitivity The extent to which the sensitivity of a material to fracture is increased by the presence of a surface inhomogeneity such as a notch, a sudden change in section, a crack or a scratch. Low notch sensitivity is usually associated with ductile materials and high notch sensitivity with brittle materials.

Novolak A phenolic-aldehyde resin which, unless a source of methylene groups is added, remains permanently thermoplastic; a linear thermoplastic B-staged phenolic resin. (See *thermoplastic*.)

Offset Yield Strength The stress at which the strain exceeds by a specific amount (the offset) an extension of the initial proportional portion of the stress-strain curve. It is expressed in force per unit area, usually pounds per square inch.

Oil Resistance The ability to withstand contact with an oil without deterioration of physical properties or geometric change to a degree which would impair part performance.

Open Cell Foamed A cellular plastic in which there is a predominance of interconnected cells.

Orange Peel An uneven surface somewhat resembling that of an orange peel; said of injection moldings that have unintentionally rugged surfaces.

Organic Designating or composed of matter originating in plant or animal life or composed of chemicals of hydrocarbon origin, either natural or synthetic.

Oriented Materials Materials, particularly amorphous polymers and composites, whose molecules and/or macroconstituents are aligned in a specific way. Oriented materials are anisotropic.

Orientation can generally be divided into two classes: uniaxial and biaxial. Ideal uniaxial and biaxial orientations are illustrated below.

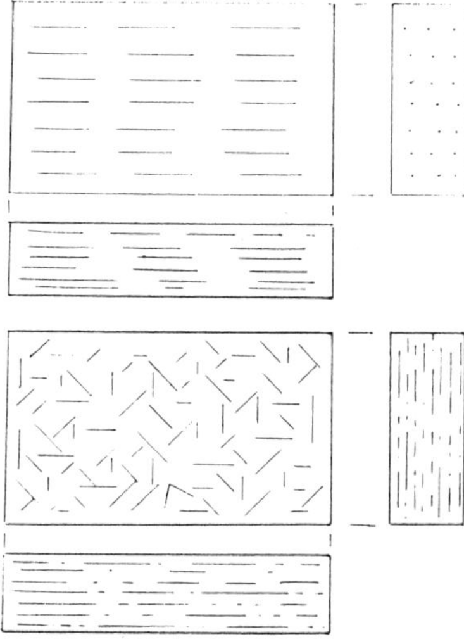

From L. E. Nielsen, "The Mechanical Properties of Polymers," New York, Reinhold Publishing Corp., 1962.

Orthotropic Having three mutually perpendicular planes of elasticity symmetry.

Oven-Dry Weight The constant weight obtained by drying at 105 ± 3C.

Overcuring The beginning of a thermal decomposition owing to too high a temperature or too long a molding time.

Overflow Groove Small groove used in molds to allow material to flow freely to prevent weld-lines and low density, and to dispose of excess material.

Overlap A simple adhesive joint, in which the surface of one adherend extends past the leading edge of another.

Overlay Sheet A nonwoven fibrous mat (in glass, synthetic fiber, etc.) used as the top layer in a cloth or mat layup, to provide a smoother finish or minimize the appearance of the fibrous pattern. (Surfacing mat.)

Package Describes the method of supply of the roving or yarn (that is, "milk bottle package," "twisted tube," etc.).

Pallet Usually refers to a number of cartons of roving placed on a wooden skid and strapped together as a shipping unit. Normally a pallet load consists of either 36 or 48 cartons of roving.

Parallel-Laminated Laminated so that all the layers of material are oriented approximately parallel with respect to the grain or strongest direction in tension (See *cross-laminated.*)

Parameter An arbitrary constant, as distinguished from a fixed or absolute constant. Any desired numerical value may be given a parameter.

Parting Agent A lubricant or an anti-adhesion (release) agent used to coat a mold cavity or a mold to prevent a molded piece from sticking to it, and thus facilitates its removal. (See *release agent.*)

Parting Line A mark on a molded piece where the sections of a mold have met in closing.

Peel Ply The outside layer of a laminate which is removed or sacrificed to achieve improved bonding of additional plies.

Peel Strength Bond strength, in pounds per inch of width, obtained by peeling the layer. (See *bond strength.*)

Permanence The property of a plastic which describes its resistance to appreciable changes in characteristics with time and environment.

Permanent Set The deformation remaining after a specimen has been stressed in tension a prescribed amount for a definite period and released for a definite period; for creep tests, permanent set is the residual unrecoverable deformation after the load causing the creep has been removed for a substantial and definite period of time.

Permeability The passage or diffusion (or rate of passage) of a gas, vapor, liquid, or solid through a barrier without physically or chemically affecting it.

Phase Angle See *dielectric phase angle.*

Phenolic, Phenolic Resin A synthetic resin produced by the condensation of an aromatic alcohol with an aldehyde, particularly of phenol with formaldehyde. (See *A-stage, B-stage, C-stage, and novolak.*)

Phenylsilane Resins Thermosetting copolymers of silicone and phenolic resins; furnished in solution form.

Photoelasticity Changes in the optical properties of isotropic, transparent dielectrics when subjected to stress.

Physical Catalyst Radiant energy capable of promoting or modifying a chemical reaction.

Pick An individual filling yarn, running the width of a woven fabric at right angles to the warp ("fill," "woof," "weft"); to experience tack; to transfer unevenly from an adhesive applicator mechanism due to high surface tack; to offset onto opposing surfaces; the relative integral strength of a cellulosic substrate relating to its ability to resist fiber distention when applied to a tacky surface and removed.

Pinch-Off In blow molding, a raised edge around the cavity in the mold which seals off the part and separates the excess material as the mold closes around the parison.

Pinhole A tiny hold in the surface of, or through, a plastic material; usually occuring in multiples.

Pin, Insert A pin which keeps an inserted part (insert) inside the mold, by screwing or by friction; removed when withdrawing the object from the mold.

Pit Small regular or irregular crater in the surface of a plastic, usually with width approximately of the same order of magnitude as its depth.

Planar Helix Winding A winding in which the filament path on each dome lies on a plane which intersects the dome, while a helical path over the cylindrical section is connected to the dome paths.

Planar Winding A winding in which the filament path lies on a plane which intersects the winding surface.

Plastic A material that contains as an essential ingredient and organic substance of large molecular weight, is solid in its finished state, and, at some stage in its manufacture or its processing into finished articles, can be shaped by flow; made of plastic.

Plastic Deformation Change in dimensions of an object under load that is not recovered when the load is removed; opposed to elastic deformation.

Plastic Flow Deformation under the action of a sustained force; flow of semi-solids in the molding of plastics.

Plastic, Rigid A plastic with a stiffness or apparent modulus of elasticity greater than 100,000 psi at 23 C, when determined in accordance with ASTM D 747.

Plastic, Semirigid A plastic which has a stiffness or apparent modulus of elasticity between 10,000 and 100,000 psi at 23 C, when determined in accordance with ASTM D 747.

Plastics Tooling Tools (mostly for the metal forming trades) constructed of plastics, generally laminates or casting materials.

Platens The mounting plates of a press, to which the entire mold assembly is bolted.

Plied Yarn A yarn formed by twisting together two or more single yarns in one operation.

Poise A unit of viscosity in which the shearing stress is expressed in dynes per square centimeter to produce a velocity gradient of one centimeter per second per centimeter. (See *viscosity.*)

Poisson's Ratio (ν) When a material is stretched, its cross-sectional area changes as well as its length. Poisson's Ratio is the constant relating these changes in dimensions, and is defined by the equation:

$$\nu = \frac{\text{change in width per unit width}}{\text{change in length per unit length}} = \frac{\Delta C/C}{\Delta L/L}$$

This is restricted to small elongations. For rubbery materials ν is close to ½, while for crystals and glasses ν is about ¼ to ⅓.

Polarity The sign of the charge on a molecule or side chain.

Polar Winding A winding in which the filament path passes tangent to the polar opening at one end of the chamber and tangent to the opposite side of the polar opening at the other end. A one circuit pattern is inherent in the system.

Polyamide A polymer in which the structural units are linked by amide or thioamide groupings. Many polyamides are fiber-forming.

Polybenzamidazole See Chapter 5, p. 117

Polyesters Thermosetting resins, produced by dissolving unsaturated, generally linear, alkyd resins in a vinyl-type active monomer such as styrene, methyl styrene, and diallyl phthalate. Cure is effected through vinyl polymerization using peroxide catalysts and promoters, or heat, to accelerate the reaction. The resins are usually furnished in solution form, but powdered solids are also available.

Polyimide A polymer produced by heating of polyamic acid. It is a highly heat resistant resin (600° F +) suitable for use as a binder or as an adhesive.

Polymer A high-molecular-weight organic compound, natural or synthetic, whose structure can be represented by a repeated small unit, the "mer"; for example, polyethylene, rubber, cellulose. Synthetic polymers are formed by addition or condensation polymerization of monomers. Some polymers are elastomers, some are plastics. When two or more monomers are involved, the product is called a copolymer.

Polymerize To unite molecules of the same kind to form a compound having the elements in the same proportion but possessing much higher molecular weight and different physical properties.

Polymerization A chemical reaction in which the molecules of a monomer are linked together to form large molecules whose molecular weight is a multiple of that of the original substance. When two or more monomers are involved, the process is called "copolymerization" or "heteropolymerization." (See *condensation polymerization.*)

Porosity The ratio of the volume of air or void contained within the boundaries of a material to the total volume (solid material plus air or void), expressed as a percentage.

Positive Mold A mold designed to apply pressure to a piece being molded with no escape of material.

Post-Cure Additional elevated temperature cure, usually without pressure, to improve final properties and/or complete the cure. In certain resins, complete cure and ultimate mechanical properties are attained only by exposure of the cured resin to higher temperatures than those of curing.

Postforming The forming, bending, or shaping of fully cured, C-staged thermoset laminates that have been heated to make them flexible. On cooling, the formed laminate retains the contours and shape of the mold over which it has been formed.

Pot Life The length of time that a catalyzed resin system retains a viscosity low enough to be used in processing. (Also "working life.")

Power Factor (P.F.) The cosine of the phase angle. Ratio of the dielectric constant ϵ' to the absolute value of the complex dielectric constant ϵ^*. Related to the dissipation factor D as follows:

$$PF = \frac{D}{\sqrt{(1 + D^2)}} = \frac{\epsilon''/\epsilon'}{\sqrt{1 + \frac{\epsilon''}{\epsilon'}^2}} = \frac{\epsilon'}{|\epsilon^*|}$$

Precure The full or partial setting of a synthetic resin or adhesive in a joint before the clamping operation is complete, or before pressure is applied.

Preform A preshaped fibrous reinforcement formed by distribution of chopped fibers by air, water flotation, or vacuum over the surface of a perforated screen to the approximate contour and thickness desired in the finished part. Also, a preshaped fibrous reinforcement of mat or cloth formed to desired shape on a mandrel or mock-up prior to being placed in a mold press. Also, a compact "pill" formed by compressing premixed material to facilitate handling and control of uniformity of charges for mold loading.

Preform Binder A resin applied to the chopped strands of a preform, usually during its formation, and cured so that the preform will retain its shape and can be handled.

Pregel An unintentional extra layer of cured resin on part of the surface of a reinforced plastic (Not relating to "gel coat.")

Preimpregnation The practice of mixing resin and reinforcement and effecting partial cure before use or shipment to the user. (See *prepreg.*)

Premix A molding compound prepared prior to and apart from the molding operations and containing all components required for molding; resin, reinforcement, fillers, catalysts, release agents, and other compounds.

Prepreg Ready-to-mold material in sheet form which may be cloth, mat, or paper impregnated with resin and stored for use. The resin is partially cured to a "B" stage and supplied to the fabricator who lays up the finished shape and completes the cure with heat and pressure.

Pressure Force measured per unit area. *Absolute pressure* is measured with respect to zero. *Gauge pressure* is measured with respect to atmospheric pressure.

Pressure Bag Molding A process for molding reinforced plastics, in which a tailored flexible bag is placed over the contact lay-up on the mold, sealed and clamped in place. Fluid pressure, usually compressed air, is placed against the bag and the part is cured.

Primer A coating applied to a surface, prior to the application of an adhesive or lacquer, enamel or the like, to improve the performance of the bond.

Promoter A chemical itself a feeble catalyst. (See *accelerator.*)

Proportional Limit The greatest stress which a material is capable of sustaining without deviation from proportionality of stress and strain (Hooke's law). It is expressed in force per unit area, usually in pounds per square inch.

Prototype A model suitable for use in complete evaluation of form, design, and performance.

Pulp Molding Process by which a resin-impregnated pulp material is preformed by application of a vacuum and subsequently oven cured or molded.

Pultrusion Reversed "estrusion" of resin-impregnated roving in the manufacture of rods, tubes and structural shapes of a permanent cross-section. The roving, after passing through the resin dip tank, is drawn through the die to form the desired cross-section.

Ram Force The total load applied by a ram, and numerically equal to the product of the line pressure and the cross-sectional area of the ram. It is normally expressed in tons.

Ram Travel The distance the injection ram moves in filling the mold, in either injection or transfer molding.

Random Pattern A winding with no fixed pattern. If a large number of circuits are required for the pattern to repeat, a random pattern is approached; a winding in which the filaments do not lie in an even pattern.

Reinforced Molding Compound Compound supplied by raw material producer in the form of ready-to-use materials; as distinguished from premix (*which see*).

Reinforced Plastic A plastic with strength properties greatly superior to those of the base resin, resulting from the presence of reinforcements imbedded in the composition.

Reinforcement A strong inert material bonded into a plastic to improve its strength, stiffness, and impact resistance. Reinforcements are usually long fibers of glass, asbestos, sisal, cotton, etc., in woven or nonwoven form. To be effective, the reinforcing material must form a strong adhesive bond with the resin. ("Reinforcement" should not be used synonymously with "filler.")

Release Agent A material which is applied in a thin film to the surface of a mold to keep the resin from bonding to the mold.

Resilience The ratio of energy returned on recovery from deformation, to the work input required to produce the deformation (usually expressed as a percentage); the ability to quickly regain an original shape after being strained or distorted.

Resin A solid, semisolid, or pseudosolid organic material which has an indefinite and often high molecular weight, exhibits a tendency to flow when subjected to stress, usually has a softening or melting range, and usually fractures conchoidally. Most resins are polymers (*which see*). In reinforced plastics, the material used to bind together the reinforcement material; the "matrix."

Resin Applicator In filament winding, the device which deposits the liquid resin onto the reinforcement band.

Resin Content The amount of resin in a laminate expressed as either a percent of total weight or total volume.

Resin, Liquid An organic polymeric liquid which, when converted to its final state for use, becomes a solid.

Resin Pocket An apparent accumulation of excess resin in a small, localized section visible on cut edges of molded surfaces.

Resin-Rich Area Space which is filled with resin and lacking reinforcing material.

Resin-Starved Areas of insufficient resin, usually identified by low gloss, dry spots or fiber show.

Resistivity The ability of a material to resist passage of electrical current either through its bulk or on a surface. The unit of volume resistivity is the ohm-em; of surface resistivity, the ohm.

Retarder See *inhibitor.*

Reverse Helical Winding As the fiber delivery arm traverses one circuit, a continuous helix is laid down, reversing direction at the polar ends. It is contrasted to biaxial, compact or sequential winding in that the fibers cross each other at definite equators, the number depending on the helix. The minimum region of crossover would be three.

Reverse Impact Test In which one side of a sheet of material is struck by a pendulum or falling object and the reverse side is inspected for damage.

Rib A reinforcing member of a fabricated or molded part.

Rigid Plastics For purposes of general classification, a plastic that has a modulus of elasticity either in flexure or in tension greater than 7000 kg per sq em (100,000 psi) at 23 and 50% relative humidity.

Rockwell Hardness Number A value derived from the increase in depth of an impression as the load on an indenter is increased from a fixed minimum value to a higher value and then returned to the minimum value. Indenters for the Rockwell test include steel balls of several specific diameters and a diamond cone penetrator having an included angle of 120° with a spherical tip having a radius of 0.2 mm. Rockwell hardness numbers are always quoted with a prefix representing the Rockwell scale corresponding to a given combination of load and indenter.

Room Temperature Curing Adhesives Adhesives that set (to handling strength) within an hour of temperatures from 68 to 86° F, and later reach full strength without heating.

Roving (filament winding) The term roving is used to designate a collection of bundles of continuous filaments either as untwisted strands or as twisted yarns. Rovings may be lightly twisted, but for filament winding they are generally wound as bands or tapes with as little twist as possible. Glass rovings are predominantly used in filament winding.

Roving Ball A term used to describe the supply package offered to the winder. It consists of a number of ends or strands wound to a given outside diameter onto a length of cardboard tube.

Roving Cloth A textile fabric, coarse in nature, woven from rovings.

S-Glass A magnesia-alumina-silicate glass, especially designed to provide very high tensile strength glass filaments.

Safety Hardener A curing agent which causes only a minimum of toxic effect on the human body, either on contact with the skin or as concentrated vapor in the air.

Sandwich Constructions Panels composed of a lightweight core material—honeycomb, foamed plastic, etc.—to which two relatively thin, dense, high-strength faces or skins are adhered.

Sandwich Heating A method of heating a thermoplastic sheet prior to forming, by heating both sides of the sheet simultaneously.

Satin A plastic finish having a satin or velvety appearance.

Scarf Joint A joint made by cutting away similar angular segments on two adherends with the cut areas fitted together. (See *lap joint.*)

Scratch Shallow mark, groove, furrow or channel normally caused by improper handling or storage.

Scrim A low cost, nonwoven open-weave reinforcing fabric made from continuous filament yarn in an open mesh construction.

Secant Modulus The ratio of total stress to corresponding strain at any specific point on the stress-strain curve. It is expressed in force per unit area, usually in pounds per square inch, and reported together with the specified stress or strain.

Self-Extinguishing Resin A resin formulation which will burn in the presence of a flame but which will extinguish itself within a specified time after the flame is removed.

Semipositive Mold A combination of the positive and flash-type molds; a mold that allows an amount of excess material to escape when it is closed. (Used where close tolerances are required.)

Sequential Winding See *biaxial winding.*

Set To convert into a fixed or hardened state by chemical or physical action, such as condensation, polymerization, oxidation, vulcanization, gelation, hydration, or evaporation of volatiles. Or, the irrecoverable deformation or creep usually measured by a prescribed test procedure and expressed as a percentage of original dimension.

Set at Break Elongation measured ten minutes after rupture on reassembled tension specimen.

Set-up To harden, as in curing.

Shear An action or stress resulting from applied forces which causes or tends to cause two contiguous parts of a body to slide relative to each other in a direction parallel to their plane of contact. Interlaminar Shear (ILS). The plane of contact is composed of resin only.

Shear Edge The cut-off edge of the mold.

Shear Modulus (G) The ratio of shearing stress τ to shearing strain γ within the proportional limit of a material.

$$G = \frac{\tau}{\gamma} = \frac{\sigma_s}{\epsilon_s} \qquad \text{units: psi}$$

When measured dynamically with a torsion pendulum, the shear modulus of a solid rectangular beam is given by:

$$G = \frac{5.588 \times 10^{-4} LI}{CD^3 \mu P^2} \text{ (psi)}$$

where: L = length of specimen between the clamps, in inches
C = width of specimen, in inches
D = thickness of specimen, in inches
I = polar moment of inertia of the oscillating system, in g cm^2
P = period of oscillations, in seconds
μ = a shape factor depending upon the ratio of the width to thickness of the specimen

Shoe A device for gathering the numerous filaments into a strand in glass fiber forming (See *chase.*)

Shore Hardness A measure of the resistance of a material to indentation by a spring loaded indenter. A higher number indicates greater resistance.

Short Beam Shear Strength The interlaminar shear strength of a parallel fiber reinforced plastic material as determined by three-point flexural loading of a short segment cut from a ring-type specimen.

Shrinkage The relative change in dimension between the length measured on the mold when it is cold and the length on the molded object 24 hours after it has been taken out of the mold.

Silicone Plastics Based on resins in which the main polymer chain consists of alternating silicon and oxygen atoms, with carbon-containing side groups; derived from silica (sand) and methyl chloride. The various forms obtainable are characterized by their resistance to heat.

Silicones Resinous materials derived from organosiloxane polymers, furnished in different molecular weights including liquids and solid resins and elastomers.

Single Circuit Winding A winding in which the filament path makes a complete traverse of the chamber, after which the following traverse lies immediately adjacent to the previous one.

Sink Mark A shallow depression or dimple on the surface of an injection molded part due to collapsing of the surface following local internal shrinkage after the gate seals; an incipient short shot.

Size Any treatment consisting of starch, gelatine, oil, wax, or other suitable ingredient which is applied to yarn or fibers at the time of formation to protect the surface and aid the process of handling and fabrication, or to control the fiber characteristics. The treatment contains ingredients which provide surface lubricity and binding action but, unlike a finish, contains no coupling agent. Before final fabrication into a composite, the size is usually removed by heat-cleaning, and a finish is applied.

Sizing Applying a material on a surface in order to fill pores and thus reduce the absorption of the subsequently applied adhesive or coating, or to otherwise modify the surface properties of the substrate to improve adhesion; the material used for this purpose (sometimes called "size").

Sizing Content The percent of the total strand weight made up by the sizing; usually determined by burning off the organic sizing ("loss on ignition").

Skein A continuous filament, strand, yarn, roving, etc., wound up to some measureable length, and usually used to measure various physical properties.

Skin The relatively dense material that may form the surface of a cellular plastic; or of a sandwich.

Skirt An extension of the cylindrical portion of a motor case from the equator, used for interstage connections, usually wound as an integral part of the case.

Slurry Preforming Method of preparing reinforced plastics preforms by wet processing techniques similar to those used in the pulp molding industry; collection of glass fibers on a screen form by evacuation of slurry-water through the screen. (See *preform.*)

S-N Curve See *stress-strain.*

Soft Flow The behavior of a material which flows freely under conventional conditions of molding and which will, under such conditions, fill all the interstices of a deep mold where a considerable distance of flow can be demanded.

Solvent Resistance The nonswelling of a material and, of course, the impossibility for it to be dissolved by the solvent in question.

Specification A detailed description of the characteristics of a product and of the criteria which must be used to determine whether the product is in conformity with the description.

Specific Gravity The ratio of the weight of any volume of a substance to the weight of an equal volume of another substance taken as standard at a constant or stated temperature. Solids and liquids are usually compared with water at 4° C.

Specific Heat The quantity of heat required to raise the temperature of a unit mass of a substance one (1) degree under specified conditions.

Specific Stress (Fibers) The load divided by the mass per unit length of the test specimen.

Specimen An individual piece or portion of a sample used to make a specific test; of specific shape and dimensions.

Spinneret A type of extrusion die; that is, a metal plate with many tiny holes, through which a plastic melt is forced to make fine fibers and filaments which are hardened by cooling in air, water, etc., or by chemical action.

Spinning Process of making fibers by forcing plastic melt through spinnerets.

Spiral In glass fiber forming, the device that is used to traverse the strand back and forth across the forming tube.

Splice The joining of two ends of glass fiber yarn or strand, usually by means of an air drying glue.

Spline To prepare a surface to its desired contour by working a paste material with a flat-edged tool. The procedure is similar to "screeding" of concrete. Also, the tool itself.

Split-Cavity Blocks Blocks which, when assembled, contain a cavity for molding articles having undercuts.

Split Mold A mold in which the cavity is formed of two or more components held together by an outer chase. The components are known as splits.

Split-Ring Mold A mold in which a split cavity block is assembled in a chase to permit the forming of undercuts in a molded piece. These parts are ejected from the mold and then separated from the piece.

Spool A term sometimes used to identify a roving ball; "roving ball" is the preferred term.

Spray A complete set of moldings from a multi-impression injection mold, together with the associated molded material.

Spray-up Techniques in which a spray gun is used as the processing tool. In reinforced plastics, for example, fibrous glass and resin can be simultaneously deposited in a mold. In essence, roving is fed through a chopper and ejected into a resin stream which is directed at the mold by either of two spray systems. In foamed plastics, very fast-reacting urethane foams or epoxy foams are fed in liquid sreams to the gun and sprayed on the surface. On contact, the liquid starts to foam.

Sprayed Metal Molds Molds made by spraying molten metal onto a master until a shell of predetermined thickness is achieved. The shell is then removed and backed up with plaster, cement, casting resin, or other suitable material. Used primarily as a mold in sheet forming process.

Spring Constant The number of pounds required to compress a specimen one inch in a prescribed test procedure.

Spun Roving A heavy, low-cost glass fiber strand consisting of filaments that are continuous but doubled back on each other.

Stabilization The process used to render the carbon fiber precursor infusible prior to carbonization.

Standard Deviation A measure of dispersion of data from the average; the root mean square of the individual deviation from the average.

Standard Laboratory Atmosphere An atmosphere having a relative humidity of 50 ± 2% at a temperature of

23 ± 1 C (73.4 ± 1.8 F). Also:

Average room conditions—40% relative humidity at a temperature of 77 F.

Dry room conditions—15% relative humidity at a temperature of 85 F.

Moist room conditions—75% relative humidity at a temperature of 77 F.

Staple Fibers Fibers of spinnable length manufactured directly or by cutting continuous filaments to short lengths. (Usually ½ to 2 inches long; 1 to 5 denier.)

Starved Area An area in a plastic part which has an insufficient amount of resin to wet out the reinforcement completely. This condition may be due to improper wetting or impregnation or excessive molding pressure.

Starved Joint An adhesive joint which has been deprived of the proper film thickness of adhesive due to insufficient adhesive spreading or due to the application of excessive pressure during the lamination process.

Static Fatigue Failure of a part under continued static load; analogous to creep-rupture failure in metals testing, but often the result of aging accelerated by stress.

Static Modulus The ratio of stress to strain under static conditions. It is calculated from static stress-strain tests, in shear, compression, or tension; expressed in psi unit strain.

Static Stress A stress in which the force is constant or slowly increasing with time; for example, test of failure without shock.

Stiffness The relationship of load and deformation; a term often used when the relationship of stress to strain does not conform to the definition of Young's modulus. (See *stress-strain.*)

Storage Life The period of time during which a liquid resin or packaged adhesive can be stored under specified temperature conditions and remain suitable for use. (Also "shelf life.")

Strain (ϵ) Strain may have several definitions, which are dependent upon the system being considered. For small deformations, engineering strain is applicable and is the most common definition of strain. The quality called true strain is sometimes used in areas of plastic deformation.

(1) Engineering strain (ϵ)
The ratio of the change in length, ΔL, of the sample to its original length, L_o.

$$\epsilon = \frac{L - L_o}{L_o} = \frac{\Delta L}{L_o}$$

dimensions: unity

(2) True strain (ϵ_t)
The integral of the ratio of the incremental change in length to the instantaneous length of a plastically deformed sample; thus, the natural logarithm of the ratio of instantaneous length to original length of such a sample.

Strain Relaxation See *creep*.

Strands A primary bundle of continuous filaments (or slivers) combined in a single compact unit without twist. These filaments (usually 51, 102 or 204) are gathered together in the forming operations.

Strand Count The number of strands in a plied yarn; the number of strands in a roving.

Strand Integrity The degree to which the individual filaments making up the strand or end are held together by the sizing applied.

Strength, Flexural The maximum stress that can be borne by the surface fibers in a beam in bending. The flexural strength is the unit resistance to the maximum load prior to failure by bending, usually expressed in pounds per square inch.

Strength in Compression The maximum load sustained by the specimen divided by the original cross-section area of the specimen.

Stress (σ) Most commonly defined as "engineering stress," the ratio of the applied load P to the original cross-sectional area A_o.

$$\sigma = \frac{P}{A_o}$$

The "true stress" or instantaneous stress is sometimes used and is defined as the applied load P per instantaneous cross-sectional area A.

$$\sigma_t = \frac{P}{A} \quad \text{actual}$$

Stress Concentration The magnification of the level of an applied stress in the region of a notch, void or inclusion.

Stress Concentration Factor The ratio of the maximum stress in the region of a stress concentrator to the stress in a similar strained area without a stress concentrator.

Stress Corrosion Preferential attack of areas under stress in a corrosive environment, where this factor alone would not have caused corrosion.

Stress-Crack External or internal cracks in a plastic caused by tensile stresses less than that of its short-time mechanical strength. The stresses which cause cracking may be present internally or externally or may be combinations of these stresses. (See *crazing*.)

Stress Relaxation The decrease in stress under sustained constant strain. (Also "stress decay.")

Stress-Strain Stiffness, expressed in pounds per square inch or kilograms per square centimeter, at a given strain.

Stress-Strain Curve Simultaneous readings of load and deformation, converted to stress and strain, are plotted as ordinates and abscissae, respectively, to obtain a stress-strain diagram.

Structural Bond A bond that joins basic load-bearing parts of an assembly. The load may be either static or dynamic.

Surfacing Mat A very thin mat, usually 7 to 20 mils thick, of highly filamentized fiber glass used primarily to produce a smooth surface on a reinforced plastic laminate.

Surface Resistance (Electrical) The surface resistance between two electrodes in contact with a material is the ratio of the voltage applied to the electrodes to that portion of the current between them which flows through the surface layers.

Surface Resistivity (Electrical) The surface resistivity of a material is the ratio of potential gradient parallel to the current along its surface, to the current per unit width of surface. Surface resistivity is numerically equal to the surface resistance between opposite sides of a square of any size when the current flow is uniform.

Surface Treatment A material applied to fibrous glass during the forming operation or in subsequent processes (that is, size or finish).

Surfacing Mat See *overlay sheet*.

Syntactic Foam A cellular plastic which is "put together" by incorporating preformed cells (hollow spheres or microballoons) in a resin matrix; as opposed to "foamed plastic" in which the cells are formed by gas bubbles released in the liquid plastic by either chemical or mechanical action.

Synthetic Resin A complex, substantially amorphous, organic semisolid or solid material (usually a mixture) built up by chemical reaction of comparatively simple compounds, approximating the natural resins in luster, fracture, comparative brittleness, insolubility in water, fusibility or plasticity, and some degree of rubberlike extensibility; but commonly deviating widely from natural resins in chemical constitution and behavior with reagents.

Tack Stickiness of an adhesive or filament reinforced resin prepreg material.

Tack Range The period of time in which an adhesive will remain in the tacky-dry condition after application to the adherend, and under specified conditions of temperature and humidity.

Tack Stage The interval of time during which a deposited adhesive film exhibits stickiness or tack, or resists removal or deformation of the cast adhesive.

Tangent of Loss Angle (Loss Factor) The dissipation factor of an electrical condenser of which the insulating material forms the dielectric when the electrodes of such a condenser are subjected to an alternating e.m.f.

Tangent Line In a filament wound bottle, any diameter at the equator.

Tangent Modulus The slope of the line at any point on a static stress-strain curve expressed in psi per unit strain. This is the tangent modulus at that point in shear, extension, or compression as the case may be.

Tenacity The term generally used in yarn manufacture and textile engineering to denote the strength of a yarn or of a filament of a given size. Numerically it is the grams of breaking force per denier unit of yarn or filament size; grams per denier, gpd. The yarn is usually pulled at the rate of 12 inches per minute. Tenacity equals breaking strength (grams) divided by denier.

Tensile Bar A compression or injection molded specimen of specified dimensions which is used to determine the tensile properties of a material.

Tensile Strength or Stress The maximum tensile load per unit area of original cross section, within the gage boundaries, sustained by the specimen during a tension test. It is expressed as psi. *Tensile load* is interpreted to mean the maximum tensile load sustained by the specimen during the test whether or not this coincides with the tensile load at the moment of rupture.

Thermal Conductivity Ability of a material to conduct heat. The physical constant for quantity of heat that passes through unit cube of a substance in time when the difference in temperature of two faces is 1°.

Thermal Decomposition Decomposition resulting from action by heat. It occurs at a temperature for which some components of the material are separating or associating together, with a modification of the macro or microstructure.

Thermal Expansion (Coefficient of) See *coefficient of thermal expansion*.

Thermo Electric Power The generation of an electric or magnetic field in a solid as a result of a temperature gradient.

Thermoplastic Capable of being repeatedly softened by increase of temperature and hardened by decrease in temperature; applicable to those materials whose change upon heating is substantially physical rather than chemical.

Thermoset A plastic which, when cured by application of heat or chemical means, changes into a substantially infusible and insoluble material.

Thixotropic, Thixotropy Concerning materials that are gel-like at rest but fluid when agitated; having high static shear strength and low dynamic shear strength, at the same time.

Thread Count The number of yarns (threads) per inch in either length-wise (warp) or crosswise (fill) direction of woven fabrics.

Toggle Action A mechanism which exerts pressure developed by the application of force on a knee joint; used as a method of closing presses and applying pressure at the same time.

Tolerance The guaranteed maximum deviation from the specified nominal value of a component characteristic at standard or stated environmental conditions.

Torsional Rigidity (Fibers) The resistance of a fiber to twisting.

Toughness The energy required to break a material. This energy is equal to the area under the stress-strain curve.

Transfer Molding Method of molding thermosetting materials, in which the plastic is first softened by heating and pressure in a transfer chamber, and then forced by high pressure through suitable sprues, runners and gates into the closed mold for final curing.

Transfer Pot A heating cylinder; transfer chamber in a transfer mold.

Transition Temperature The temperature at which the properties of a material change.

Turbostratic A type of crystalline structure where the basal planes have slipped sideways relative to eschother, causing the spacing between planes to be greater than ideal.

Turns Per Inch (TPI) A measure of the amount of twist produced in a yarn during its conversion from strand.

Twist The turns about its axis per unit of length observed in a yarn or other textile strand. Twist may be expressed as turns per inch (TPI) etc. "S" and "Z" refer to direction of twist, as to whether the twist conforms to the middle-section slope of the particular letter.

Twist, Balanced An arrangement of twists in a combination of two or more strands which does not cause kinking or twisting on themselves when the yarn produced is held in the form of an open loop.

U-V Stabilizer (Ultra Violet) Any chemical compound which, when admixed with a thermoplastic resin, selectively absorbs U-V rays.

Ultimate Elongation The elongation at rupture.

Ultimate Tensile Strength The ultimate or final stress sustained by a specimen in a tension test; the stress at moment of rupture.

Ultra-Violet Zone of invisible radiations beyond the violet end of the spectrum of visible radiations. Since UV wavelengths are shorter than the visible, their photons have more energy, enough to initiate some chemical reactions and to degrade most plastics.

Undercut Having a protuberance or identation that impedes the withdrawal from a two-piece, rigid mold; any such protuberance or indentation, depending on the design of the mold (that is, tilting a model in design of a mold for that model may eliminate an apparent "undercut").

Unidirectional Laminate A reinforced plastic laminate in which substantially all of the fiber are oriented in the same direction.

Vacuum Bag Molding A process for molding reinforced plastics in which a sheet of flexible transparent material is placed over the lay-up on the mold and sealed. A vacuum is applied between the sheet and the lay-up. The entrapped air is mechanically worked out of the lay-up and removed by the vacuum, and the part is cured. Also "bag molding."

Veil An ultrathin mat similar to a surface mat, often composed of organic fibers as well as glass fibers.

Virgin Filament An individual filament which has not been in contact with any other fiber or any other hard material.

Viscosity The property of resistance to flow exhibited within the body of a material, expressed in terms of relationship between applied shearing stress and resulting rate of strain in shear. Viscosity is usually taken to mean Newtonian viscosity, in which case the ratio of shearing stress to the rate of shearing strain is constant. In non-Newtonian behavior, which is the usual case with plastics, the ratio varies with the shearing stress. Such ratios are often called the "apparent viscosities" at the corresponding shearing stresses.

Void Content The percentage of voids in a laminate can be calculated by the use of the following formula:

$$\text{Percent voids} = 100 - X$$

$$x = \frac{ad}{c} + \frac{ae}{b}$$

where: x = total calculated volume of laminate
a = specific gravity of laminate (Method 5011 of Specification L P-406B)
b = Specific gravity of glass = 2.57
c = specific gravity of cured resin
d = resin content, expressed as a decimal (In accordance with Method 7061 of Specification L P-406B)
e = glass content, expressed as a decimal = $1 - d$

If the laminate or a molding contains a filler:

$$x = \frac{ad}{c} + \frac{ae}{b} + \frac{ar}{g}$$

where: e = glass content = $1 - d - f$
f = filler content, expressed as a decimal
g = specific gravity of filler

Voids Gaseous pockets that have been trapped and cured into a laminate; an unfilled space in a cellular plastic substantially larger than the characteristic individual cells.

Volatiles Materials in a sizing or a resin formulation which are capable of being driven off as a vapor at room or slightly elevated temperature.

Volatile Content The percent of volatiles which are driven off as a vapor from a plastic or an impregnated reinforcement.

Volatile Loss Weight loss by vaporization.

Volume Resistance The volume resistance between two electrodes that are in contact with or embedded in a specimen is the ratio of the direct voltage applied to the electrodes, to that portion of the current between them that is distributed through the volume of the specimen. Also, the electrical resistance between opposite faces of a 1-cm cube of insulating material, commonly expressed in ohm-centimeters. (Also "specific insulation resistance.")

Warp The yarn running lengthwise in a woven fabric; a group of yarns in long lengths and approximately parallel, put on beams or warp reels for further textile processing including weaving; a change in dimension of a cured laminate form its original molded shape.

Water Absorption Ratio of the weight of water absorbed by a material to the weight of the dry materials.

Weathering The exposure of plastics outdoors.

Weathering, Artificial The exposure of plastics to cyclic laboratory conditions comprising high and low temperatures, high and low relative humidities, and ultraviolet radiant energy, with or without direct water spray, in an attempt to produce changes in their properties similar to those observed on long-time continuous exposure outdoors. The laboratory exposure conditions are usually intensified beyond those encountered in actual outdoor exposure, in an attempt to achieve an accelerated effect.

Weave The particular manner in which a fabric is formed by interlacing yarns, and usually assigned a style number.

Web A textile fabric, paper, or a thin metal sheet of continuous length handled in roll form as contrasted with the same material cut into sheets; a thin sheet in process in a machine; (in extrusion coating) the molten web is that which issues from the die, and the substrate web is applied to the substrate material (which is being coated).

Weft The transverse threads or fibers in a woven fabric; those fibers running perpendicular to the warp. Also "filler," "filler yarn," "woof."

Wet Flexural Strength (WFS) The flexural strength after water immersion; usually after boiling the test specimen for two hours in water.

Wet Layup The reinforced plastic which has liquid resin applied as the reinforcement is laid up. The opposite of "dry lay-up," "prepreg."

Wet-Out The condition of an impregnated roving or yarn wherein substantially all voids between the sized strands and filaments are filled with resin.

Wet-Out Rate The time required for a plastic to fill the interstices of a reinforcement material and wet the surface of the reinforcement fibers; usually determined by optical or light transmission means.

Wet Strength The strength of paper when saturated with water, especially used in discussions of processes whereby the strength of paper is increased by the addition, in manufacture, or plastic resins; the strength of an adhesive joint determined immediately after removal from a liquid in which it has been immersed under specified conditions of time, temperature and pressure.

Wet Winding In filament winding the process of winding glass on a mandrel where the strand is impregnated with resin just prior to contact with the mandrel. (See *dry winding.*)

Whisker A very short fiber form of reinforcement, usually of crystalline material.

Winding, Biaxial See *biaxial winding.*

Winding Pattern A total number of individual circuits required for a winding path to begin repeating by laying down immediately adjacent to the initial circuit; a regularly recurring pattern of the filament path after a certain number of mandrel revolutions, leading to the eventual complete coverage of the mandrel.

Winding Tension In filament winding, the amount of tension on the reinforcement as it makes contact with the mandrel.

Working Life The period of time during which a liquid resin or adhesive, after mixing with catalyst, solvent, or other compounding ingredients, remains usable. (See *gelation time, pot life.*)

Woven Fabrics Those produced by interlacing strands at more or less right angles.

Woven Roving A heavy glass fiber fabric made by the weaving of roving.

Wrinkle A surface imperfection in laminated plastics that has the appearance of a crease or wrinkle in one or more outer sheets of the paper, fabric, or other base which has been pressed in.

Yarn An assemblage of twisted fibers or strands, either natural or manufactured, to form a continuous yarn suitable for use in weaving or otherwise interweaving into textile materials. See also *continuous filament.*

Yield Point The first stress in a material, less than the maximum attainable stress, at which an increase in strain occurs without an increase in stress. Only materials that exhibit this unique phenomenon of yielding have a "yield point."

Yield Strength The stress at which a material exhibits a specified limiting deviation from the proportionality of stress to strain; the lowest stress at which a material undergoes plastic deformation. Below this stress, the material is elastic; above it, viscous.

Young's Modulus The ratio of tensile stress to tensile strain below the proportional limit (See *modulus of elasticity.*)

Index